VII International Scientific Practical Conference on Innovative Technologies in Engineering

IOP Conference Series: Materials Science and Engineering
Volume 142

Yurga, Russia
19 - 21 May 2016

ISBN: 978-1-5108-2911-4
ISSN: 1757-8981

Printed from e-media with permission by:

Curran Associates, Inc.
57 Morehouse Lane
Red Hook, NY 12571

Some format issues inherent in the e-media version may also appear in this print version.

Printed by Curran Associates, Inc. (2016)

For permission requests, please contact the Institute of Physics at the address below.

Institute of Physics
Dirac House, Temple Back
Bristol BS1 6BE UK

Phone: 44 1 17 929 7481
Fax: 44 1 17 920 0979

techtracking@iop.org

Additional copies of this publication are available from:

Curran Associates, Inc.
57 Morehouse Lane
Red Hook, NY 12571 USA
Phone: 845-758-0400
Fax: 845-758-2633
Email: curran@proceedings.com
Web: www.proceedings.com

Table of contents

Volume 142

VII International Scientific Practical Conference "Innovative Technologies in Engineering"

19–21 May 2016, Yurga, Russian Federation

Accepted papers received: 18 July 2016
Published online: 2 September 2016

Preface

Papers

IOP Conf. Series: Materials Science and Engineering **142** (2016) 011001 doi:10.1088/1757-899X/142/1/011001

PREFACE

Innovative Technologies in Engineering Conference Proceedings report on the 7th International Scientific Practical Conference held at the Yurga Institute of Technology branch of the National Research Tomsk Polytechnic University from 19th to 21st May, 2016. The Proceedings contain selected scientific reports submitted to the Conference. Having started in 1996 the scientific conference at the Yurga Institute of Technology branch of the National Research Tomsk Polytechnic University acquired its international status in 2010. This year scholars from Russia, Kazakhstan, Belarus, Ukraine, USA, Azerbaijan, China, Germany and Poland have submitted their papers to the Conference.

The scientific reports published in these Proceedings have been revised and approved by the editorial team of the Conference. All of the reports exhibit clear, concise, and precise expositions that appeal to a broad international readership interested in mechanical engineering, welding, metallurgy, material science, as well as in computer-aided manufacturing. The reports present original ideas or results of general significance supported by clear reasoning and compelling evidence, and employ methods, theories and practices relevant to the research. The authors state clearly the questions and the significance of their research to theory and practice, describe how the research contributes to new knowledge; and provide tables and figures that meaningfully add to the narrative.

The Organizing Committee of the Conference thank all the participants of the Conference for their fruitful work and personal contribution to the development of these Conference Proceedings.

Dmitry A. Chinakhov, PhD,
Conference Convenor,
Vice-Chancellor, Yurga Institute of Technology
of National Research Tomsk Polytechnic University

Peer review statement

All papers published in this volume of *IOP Conference Series: Materials Science and Engineering* have been peer reviewed through processes administered by the proceedings Editors. Reviews were conducted by expert referees to the professional and scientific standards expected of a proceedings journal published by IOP Publishing.

Development and Study of Hard-Facing Materials on the Base of Heat-Resisting High-Hardness Steels for Plasma-Jet Hard-Facing in Shielding-Doping Nitrogen Atmosphere

N N Malushin [1], A P Kovalev [2], D V Valuev[3], E A Shats[4], I F Borovikov[5]

[1,2] 654000 Kemerovo region, Novokuznetsk, Cosmic highway 16, OJSC Evraz ZSMK,
[3,4] 652050, Kemerovo region, Yurga, Leningradskaya str.26 Yurga Technological Institute branch of Tomsk Polytechnic University
[5] The Moscow State Technical University of N.E. Bauman, Moscow, 5, 2nd Baumanskaya

E-mail: valuev@tpu.ru

Abstract: The authors develop hard-facing materials on the base of heat-resisting high-hardness steels for plasma-jet hard-facing in nitrogen atmosphere for manufacturing parts of mining and metallurgic equipment which significantly simplify the production process and effect a saving when producing bimetallic parts and tools.

Introduction

To strengthen machine parts for mining and metals industry the following chromium-tungsten heat-resisting high-speed steels are applied as hard-facing materials: R18, R6M5, R9, R2M8, 3H2V8, 3H2V4F, H10V14, 25H5FMS and other. Together with high performance properties these steels are characterized by insufficient weldability. Normally, to prevent cold cracking the traditional hard-facing technology involves mandatory application of high-temperature preheating and additional heating with the heating temperature $T_{heating}$ of 400-700 °C and retarded cooling of the item. The process involves formation of plastic products of austenite decomposition characterized by low hardness and wear resistance which in its turn requires complex heat treatment. In the process of heat treatment of a bimetallic product the properties of a heavily alloyed metal may not be taken the full advantage of to ensure its maximal hardness [1].

According to the developed methods of alloying with heat-resisting high-hardness steels the authors suggest application of kinetic plasticity effect ("superplasticity") and regulated low-temperature heating to prevent cold cracking. The hard-facing methods allow producing built-up metal which properties are close to those of the quenched metal and which does nor have cracks. The suggested methods of hard-facing can be implemented with application of reversed polarity plasma-jet hard-facing in shielding-doping nitrogen atmosphere which requires special materials [2-4].

Objective of the work is development and studying of hard-facing materials on the base of heat-resisting high-hardness steels for reverse polarity plasma-jet hard-facing in the shielding-doping nitrogen atmosphere parts of mining and metallurgic equipment.

Metal resistance to cold cracking may be changed to a certain extent regulating the residual welding stress increasing in the metal being cooled. The easiest and the most reliable method of such regulation is to choose the chemical composition of the hard-faced metal which determines the linear expansion coefficient, character and structure and bulk effect of structural changes. These factors have a considerable influence upon residual stress development in the process of hard-facing. When studying the kinetic plasticity effect of R18 steel it was noted that the doping level has a certain impact over temporary stresses relaxation. Assumptions were made on the possibility of regulation of residual stresses by shifting the curve of temporary stresses formation into the region of lower temperatures due to proper alloying of the hard-faced metal. Then, temporary stresses in the austenite area which have been accumulated during metal cooling are mainly relaxed in the interval of martensite transformation and do not develop under cooling to room temperature T_r [2].

Below we provide the results of the study of how hard-facing impacts temporary stresses formation in the hard-faced metal to find the ways to reduce residual stress and, thus, to reduce the probability of cold cracking. The experiments were carried out with samples produced from hard-faced metal of 3H2V8 type which was doped with carbon, chromium, tungsten, aluminium by changing the chemical composition of the flux-cored wire. The impact of hard-facing over temporary and residual stress formation during cooling the samples produced from heat-resisting steels was studied with thermal microscopy unit IMASH-5S-69. Dumb bell samples with the section of 9 mm^2 and body of 46 mm long were produced from heat-resisting high-hardness steels. Effect of kinematic plasticity was estimated by temporary stresses reduction during cooling of rigidly held samples within the interval of martensite transformation. The samples were held with electromechanic drive of the loading system of the unit. The samples were heated in vacuum 5×10^{-5} mm hg with the heat given out during electric current flow. The heating was completed up to 1200 C ensuring the high degree of carbide phase solution in austenite. After heating and holding the samples were rigidly fixed to keep their length unchanged during cooling. The heating and cooling conditions imitated the thermal cycle of plasma-jet hard-facing. Unchanged length determined the corresponding growth of elastic plastic deformation of the sample during the cooling process. Stresses in the samples were constantly measured with strain-gage sensors with accuracy of ± 1.5 %. The temperature was measured with platinum to platinum/rhodium thermocouple welded to the central section of the sample. Kinetics of temperature-induced temporary stresses formation in rigidly fixed samples was rerecorded with two-dimensional potentiometer of PDP-4 type.

In Figure 1 the authors present data on the changes of temporary thermal stresses in rigidly fixed samples produced from hard-faced metal of 3H2V8, 4H4V10Yu and R18Yu steels type. From the data in Figure 1 curve 1 we can see that when the samples produced from steel R18Yu are cooled starting from 1200°C kinetic plasticity effect is observed under phase transition revealing itself in reduction of temporary stresses within the range of martensite temperatures. Maximum stress on the curve of temporary stresses formation is ~ 240°C, which is rather close to the temperature under which martensite transformation of R18 steel starts under its hardening from heating temperature T_h = 1200°C, and minimum of 100°C corresponds to such temperatures lying over the temperature of the end of martensite transformation $M_{\text{conditional end of martensitic transformations}}$ ($M_{\text{conditional end of martensitic transformations}} > M_{\text{the end of the martensitic transformation}}$) when the transformation degree is large enough and the yield limit grows due to formation of strong “framing” from numerous martensite crystals and plasticity of the alloy begins to reduce. From the data in Figure 1 we can see that in the region of martensite transformation temporary stresses reduce by 1.5-2 times as compared to the stresses accumulated in austenite region. Under further cooling below $M_{\text{conditional end of martensitic transformations}}$ temperature the stresses increase up to 120-180 mPa due to some increase of martensite amount. The results of the study showed that in hardening tool steels increased plasticity effect is observed in the moment of martensite transformation. Residual stresses in the samples produced from metal of 3H2V8 type are 200 mPa as opposed to 100 mPa in the samples produced from metal of 4H4V10Yu type. It can be explained by shifting of the curve of temporary stresses formation in 4H4V10Yu alloy towards the region of lower temperatures. Such shifting results from additional doping of the hard-faced metal with carbon, chromium and aluminium. Further increase of carbon and tungsten concentration in the hard-faced metal of R18Yu type resulted in shifting $M_{\text{the beginning of the martensitic transformation}}$ – $M_{\text{conditional end of martensitic}}$

$_{\text{transformations}}$ into the region of still lower temperatures. After relaxation in the region of martensite transformation development of temporary stresses was not significant and residual stress in hard-faced metal R18 appeared to be lower than that in the metal of 4H4V10Yu type.

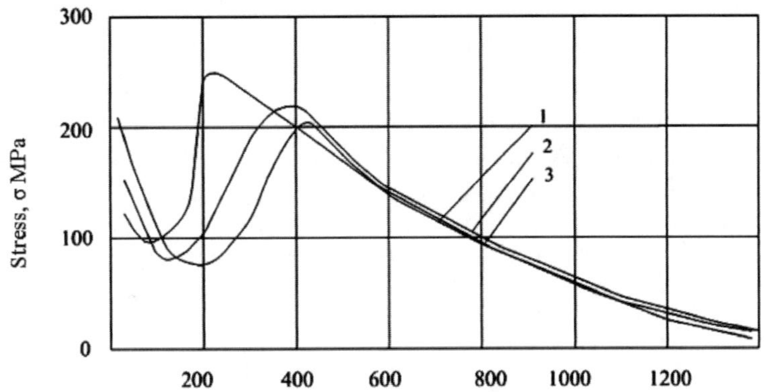

T_{Heat} = 1200° C; 1 – R18Yu; 2 - 4H4V10Yu; 3 - 3H2V8

Figure 1. Effect of complex doping on temporary stress formation in the metal

Complex doping of hard-faced metal with carbon, chromium, tungsten and aluminium allows formation of finer austenite grains, so martensite in more doped hard-faced metal of R18Yu type is more disperse than martensite in 3H2V8 steel. The answer to the question of how completely we can use the martensite transformation as a mechanism of internal peak stresses relaxation depends upon the degree of martensite dispersion. The higher degree of relaxation of temporary stresses in steel R18 can be explained by the same phenomenon. It is worth noting that, due to the small value of temperature interval $M_{\text{conditional end of martensitic transformations}} - T_k$, temporary stresses cannot develop significantly and under the room temperature the lowest level of residual stress is noted in the samples produced from hard-faced metal R18Yu [2].

According to the completed study the hard-faced metal of R18Yu steel type is hard enough. So, after solving the problem of cold cracking prevention through additional doping R18Yu steel, this grade of steel may be recommended as a base for hard-facing materials for the given methods. To prove that proper doping of hard-faced metal can be used to regulate the residual stress level and, thus, the probability of cold cracks formation, hard-faced metal of R18Yu type was additionally doped with cobalt and nickel. In Figure 2 we present the data on how doping samples produced from R18Yu metal with cobalt effects shifting the martensite transformation interval and formation of temporary stresses. As we know doping with cobalt results in shifting martensite transformation into the region of higher temperatures. So if the hard-faced metal R18Yu does not contain cobalt, point M_n determined according to maximum on the curve (see Fig. 2) equals 180 °C; if metal is doped with 1% of cobalt the same point reaches 250°C and after doping the metal with 2% of cobalt $M_{н}$ increases up to 340°C. We can also see that shifting $M_{\text{the beginning of the martensitic transformation}}$ and $M_{\text{conditional end of martensitic transformations}}$ points into the region of higher temperatures results in higher level of residual stress. Doping with non-carbide-forming nickel disimproves heat resistance to some extent due to induced lower temperature of $\gamma \rightarrow \alpha$ transformation. That is why nickel is introduced in small quantities (1-2%). As we can see from Figure 3 additional doping of R18Yu steel with nickel has led to shifting of temporary stresses formation curve into lower temperature region and significant reduction of residual stress and, thus, we can expect increased resistance of such metal to cold cracking [2-7].

The results of laboratory tests were checked by field tests of the studied grades of hard-faced metal for cold cracking resistance. Multiple-bead laser-jet hard-facing with low-temperature heating and forced cooling of 30HGSA steel workpieces was completed according to the developed method. The preheating temperature was chosen to be a bit higher than the temperature under which martensite transformation of R18Yu steel starts (T_{Heat} = 230 - 280°C) [2, 4]. When hard-facing with flux core wire doped with 2% of Co visual cracks were formed when facing the first turns of the first layer. Reducing cobalt to 1% allowed avoiding cracks only in the first layer. Three layers were faced with R18Yu wire without cracking. Doping with nickel allowed to avoid cracking under five-six-layer facing. At the same time doping with nickel significantly increases the amount and the stability of retained austenite, thus, complicating further tempering to some extent. That is why it is recommended to avoid introducing significant amounts of nickel when doping. One per cent of nickel is enough to prevent cracking of the faced metal.

Thus, it was shown that proper doping of the hard-faced metal may increase its resistance to cold cracking applying low-temperature preheating.

To complete plasma-jet hard-facing in the shielding-doping nitrogen atmosphere the scientists of Siberian State Industrial University developed flux core wires ensuring a wide range of chemical compositions and performance characteristics of the hard-faced metal.

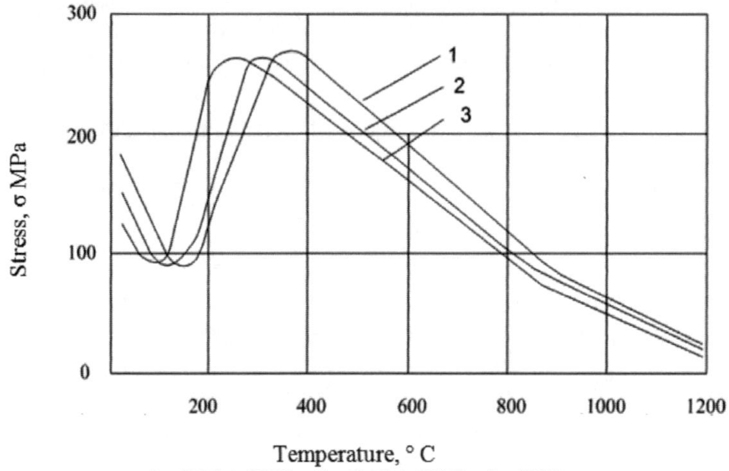

Figure 2. Effect of hard-faced metal of R18 type doping with cobalt upon the temporary stresses relaxation

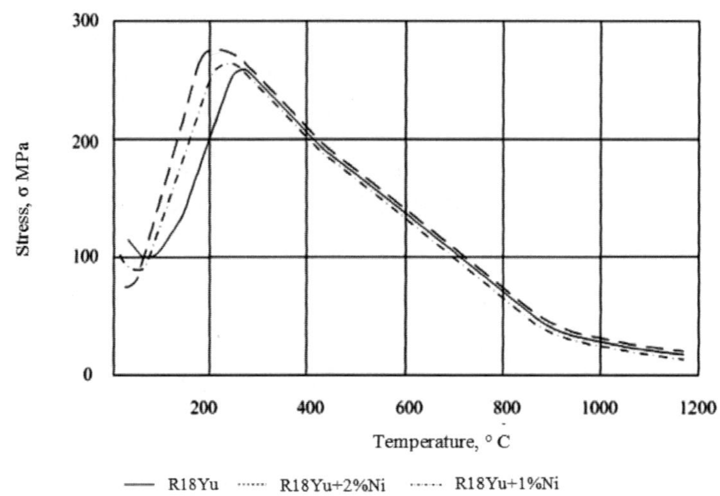

Figure 3. Effect of R18 steel doping with nickel upon the temporary stresses relaxation

Under hard-facing with constricted arc in nitrogen-containing gas atmospheres we observe intense boiling of the weld pool, as interaction between the molten metal of the weld pool and the gas phase in the process of hard-facing saturates the metal with nitrogen. When metal solidifies its ability to keep the dissolved nitrogen decreases abruptly and the excessive nitrogen escapes and, due to short duration of the hard-facing process, contributes to porosity formation. To avoid porosity aluminium was introduced into the core of the welding wire to fix the excessive nitrogen within compounds insoluble in the molten metal. Part of these compounds come to the surface of the weld pool and some stay in the hard-faced metal and can serve as the strengthening phase. On the other hand, aluminium transiting to the solid solution has a significant impact upon the process of structure formation as it increases the amount of soft structural components and, thus, contributes to destruction of the hard-faced metal.

After studying the process of structure formation of heat-resisting high-hardness tool steels hard-faced in nitrogen we established that aluminium affects metal hardness after hard-facing and tempering for secondary hardness. Aluminium participating in disperse particles formation, admittedly nitrides or compound oxides with the size of 6-12 nanometer contributes to metal strengthening but, if its amount is over 1%, it results in increased amount of δ–ferrite and softens the metal. For the purpose of obtaining maximal secondary hardness and porosity prevention optimal aluminium concentration in the hard-faced metal is between 0.8-1.0 % [2,7]. To reduce the ability for porosity formation caused by hydrogen, sodium fluorosilicate or fluoaluminic (Na_2SiF_6, Na_3AlF_6) were additionally introduced into the core of flux cored wire electrode to fix hydrogen into compound [HF}. The possibility of applying nitrogen as shielding gas instead of argon allows reducing material costs, increase hard-facing productivity, additionally strengthen the hard-faced metal by doping it with nitrogen from the gas phase in the process of hard-facing. The hard-faced metal is not liable to cold cracking and its state is close to that of tempered metal allowing application of only high-temperature tempering to increase hardness and wear resistance.

The developed flux cored wire electrodes ensure producing hard-faced metal of high-hardness steels type such as R18, R6M5, R2M9, etc. Flux core wires were designed for plasma-jet hard-facing in the shielding-doping atmosphere of nitrogen for restoration and strengthening of wearing parts of equipment and metal-working tools when especially high hardness and wear resistance of effective areas is required. Diameter of the produced flux core wire is 2.7 + 3.7 mm; metal deposit factor – 18–20 g/A·h; scattering coefficient does not exceed 3.5%. Hardness of the hard-faced metal HRC 52-60, hardness of the hard-faced metal after thermal treatment (high-temperature tempering) – HRC 62-66.

The state of the metal after hard-facing as well as after hard-facing and thermal treatment was estimated by the results of X-ray diffraction study and physicochemical phase analysis. Physicochemical analysis is completed to study the distribution of carbide-forming elements between the solid solution and the carbonitride phase. The samples were subjected to preliminary anodic dissolution on the surface to remove the layer changed after spark cutting and polishing. The physicochemical isolating of the carbonitride phase was completed in 15% water solution of sodium chlorate and 2.5% solution of tartaric acid. Current density was 0.03-0.06 A/cm^2, electrolyze time was 30-40 min. The amount of carbide-forming elements in the carbonitride phase was determined in relation to the weight of the dissolved metal. The phase composition of hard-faced high-speed steels of R18 (a, b) and R6M5 (c, d) types is shown in Fig. 4.

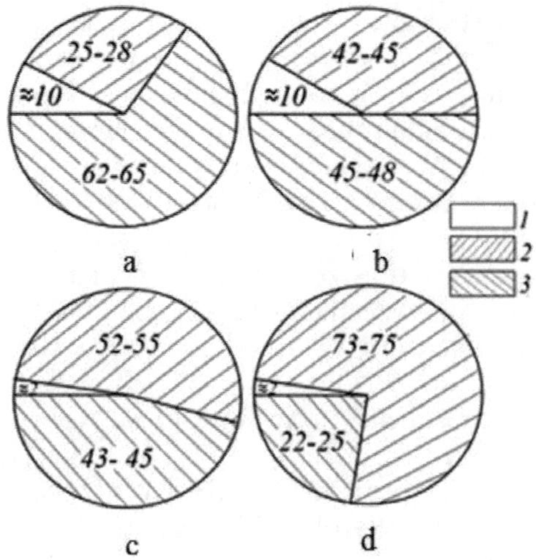

Figure 3. Phase composition of the hard-faced high-speed steels of R18 type(a, b) and R6M5 type (c, d): a, c – thermally untreated; b, d – after hard-facing and tempering; 1 – carbonitride M_6 (CN); 2 – martensite; 3 – retained austenite

In temperature-resistant steels carbides of M_6C, $M_{23}C_6$, M_2C, MC type can be formed, in steels containing nitrogen nitrides and carbonitrides of M_6 (CN) type can be also formed. X-ray diffraction study showed that the main constituent of anode deposit is the phase isomorphic to M_6C carbide. Anode deposit also contains small amount of $M_{23}C_6$ and traces of M_2C. The fact that hard-faced steels which have not been thermally treated have the form of solid solution doped with tungsten to a greater extent can be also explained by heating the doping metal in the arc as hard-facing to the temperatures significantly exceeding those of high-speed steel tempering. In this case heat resistance and secondary hardness of hard-faced high-speed steels exceed heat resistance and secondary hardness of tempered and thermally treated metal. Hard-faced high-speed steels of R18 and R6M5 types which have not been thermally treated contain the increased amount of retained austenite, their solid solution is more doped (α+γ) and their hardness is lower in comparison to that of hard-faced thermally treated steels. The increased amount of retained austenite and higher doping of the solid solution of hard-faced steels can be explained by rapid cooling of the metal from the temperatures exceeding optimal temperatures of high-speed steels tempering. Hardness after high-temperature tempering of hard-faced high-speed steels equals or exceeds hardness of thermally treated tempered steels.

Industrial tests under real production conditions proved the results of full-scale and indirect study of the properties of hard-faced heat-resistant metal. It was established that the working rollers of a cold rolling mill produced with application of plasma-jet hard-facing with the developed alloying materials on the base of heat-resistant high-hardness steels with regulated thermal cycle and low-temperature heating meet the requirements under improving their strength by 1.5…2.0 times in comparison to series rolls. Improvement of wear resistance of hard-faced rolls can be explained by fine-disperse carbides M_6C и MC and nitrides and carbonitrides M_6 (CN) in their structure.

Conclusion

1. The authors develop flux cored wires for plasma-jet hard-facing parts of mining-metallurgic equipment in nitrogen atmosphere under reverse polarity. The charge material of the wires contains carbon, chromium, tungsten, molybdenum, aluminium, nickel and sodium fluorosilicate or fluoaluminic.

2. Complex doping of the hard-faced metal with carbon, chromium, tungsten, molybdenum, aluminium, nickel, sodium fluorosilicate or fluoaluminic allows producing hard-faced metal which is not liable to crack formation and which state is close to that of the tempered metal.

3. Hardness of the metal hard-faced with the developed materials is HRC 52–60, hardness of the hard-faced metal after the thermal treatment (tempering) is HRC 62 – 66. Thermal treatment of the hard-faced parts can be limited to tempering for secondary hardness without preliminary annealing and hardening. This significantly the technological process and effects a saving when producing bimetallic parts and tools.

References

[1] Frumin I I Automated electric arc welding Metallurgizdat 1961 421
[2] Malushin N N Valuev D V Ensuring the quality of metallurgic equipment at all stages of their life cycle through plasma-jet hard-facing with heat-resistant steels Tomsk Polytechnic University Press, 2013 358
[3] Chinakhov D A Gas dynamic control of properties of welded joints from high strength alloyed steels J China Welding. Vol 23 3 (2014) 27-31
[4] Il'yaschenko D P Chinakhov D A Danilov V I Sadykov I D Pysical nature of the processes in structure forming, phase and chemical composition of pipe permanent joints when MMA welding J IOP Conference Series: Materials Science and Engineering. 125 (2016)
[5] G V Galevskiy V V Rudneva S G Galevskiy D P Il'yashchenko D S Kartsev Nanosized borides and carbides for electroplating. metal-matrix coatings: specifications, performance evaluation J IOP Conference Series: Materials Science and Engineering. 125 (2016)
[6] Chinakhov D A Grigorieva E G Mayorova E I Study of gasdynamic effet upon the weld geometry when concumable electrode welding J IOP Conference Series: Materials Science and Engineering. 127 (2016)
[7] Malushin N N Vostretsova T G Regulated thermal cycle for plasma-jet hard-facing of rollers with heat-resistant steels J Zagotovitelnye proizvodstva v machinostroyenii 13 (2014) 14–17

IOP Conf. Series: Materials Science and Engineering **142** (2016) 012002 doi:10.1088/1757-899X/142/1/012002

Procedure of Calculating the Parameters of the Inter-Pulse Period in Pulsed Arc Welding

A G Krampit [1a], Doctor of Engineering, Professor, M A Krampit [1b]

[1]Yurga Institute of Technology, TPU Affiliate, Russia, Kemerovo Region, Yurga

e-mail: [a]akrampit@mail.ru, [b]krampitm@tpu.ru

Abstract. The procedure of calculating the parameters of the inter-pulse period such as current strength, electrode supply speed and time length is presented in the paper. Their importance for the stable arc burning is demonstrated. The authors address to the inter-pulse period parameters, influencing on the temperature of the electrode in the device, where the inter-pulse current is used for heating the electrode. The data of calculations are confirmed in the process of experiments. The appropriate parameters of pulsed arc welding with preheated electrode extension are identified; the stability of the process is tested experimentally: impulse frequency, inter-pulse period, arc voltage during the inter-pulse period, speed of electrode supply, current of the inter-pulse period.

1. Introduction

Improving power efficiency of arc welding is still the urgent issue. As you know, arc welding is distinguished by considerable power consumption. Pulsed arc welding has an effect on power inputs into the welded item. However, the process of generating the modulated current requires power inputs, necessary for heating ballast resistances when decreasing welding current during the inter-pulse period.

As the results of using this process together with the robot the number of parameters is increased, their balance can hardly be detected practically; therefore, this problem requires application of mathematical modeling [1].

The researchers are currently focusing on modeling such processes as geometrical characteristics of the drop [2], modeling the origination of residual stresses [3]
when welding.

2. Description of the device

The authors have designed a device [4] (Figure 1), where a section of the electrode before the contact tip is used as a resistance, decreasing current up to 30-50A (inter-pulse period current) to improve the energy efficiency of pulsed arc welding. So it is possible to utilize the heat efficiently, which used to be emitted on the ballast resistance, causing, this way, power losses. When applying this device the amount of heat, emitted in the process of current decreasing during the inter-pulse period, is used for preheating the electrode [5] section between two contact tips. Therefore, less energy is required for melting and transfer of electrode metal drop in the pulse.

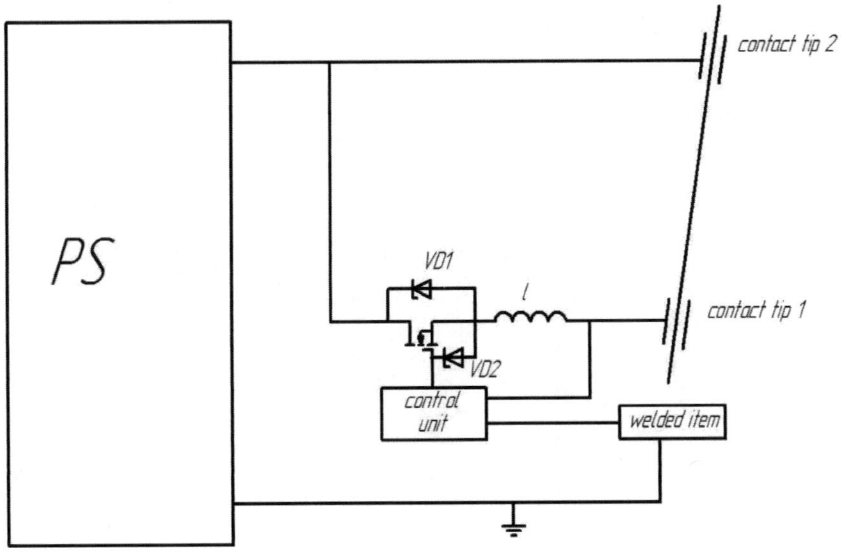

Figure 1 – Layout of the device
where: PS – power supply
VD1 – fuse diode of the overvoltage;
VD2 – protection from the input transient processes;
1 – choke.

As there are more parameters in pulsed arc welding as against arc welding, it is quite difficult to identify the interrelation between these parameters. In the proposed device we use the electric circuit "contact tip – electrode extension", which influences on the process of welding as well. Since overheating and rupture of the electrode are possible in the unstable process, it is absolutely necessary to assess importance of various parameters for the electrode section.

The stable arc burning is also an essential factor, because its behavior during the inter-pulse period (low current) depends on the arc voltage, which can't be below the critical value.

This work is aimed at developing the procedure to calculate electrode preheating and assign on its base the parameters of inter-pulse current and length.

3. Methods of Research

The inter-pulse parameters in pulsed arc welding with preheating the electrode extension by the inter-pulse current were adjusted according to the following criteria:

- temperature of pre-heating;
- stable arc burning.

The adjustment of the inter-pulse parameters in pulsed arc welding with pre-heating the electrode extension by the inter-pulse current required revealing the importance of welding mode parameters for pre-heating the electrode and stable burning of the arc. The results of experiments demonstrated the process involving pre-heating the electrode up to 680°C is the most power efficient. The rise of the temperature results in high temperature tempering [6], being the cause of electrode jam before the second contact tip. At temperatures below 500°C there is decrease in the power efficiency. Stable arc burning is to be assessed as well. Burning fails to be stable at arc voltages below 20 V [7]. Let us assume arc voltage is 22 V. The experiments to find out appropriate inter-pulse parameters are carried out in these conditions.

A model for calculations, given in Figure 2, is proposed to carry out experiments. The following assumptions are made for this model:

- electrode supply is stopped when heating the electrode and during the inter-pulse period;

- electrode is moved during the pulse time only, passing the length equal that of the complete cycle time;

electrode is heated by the passing current only, thermal conductivity is not taken into account due to the quickness of the process.

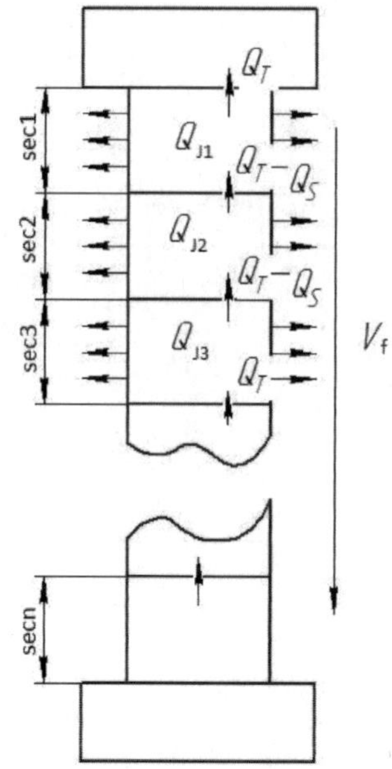

Figure 2 – A model to calculate pre-heating the electrode
Q_J – heat emitted by the passing current;
Q_T – heat consumed for thermal conductivity;
Q_S – heat consumed by surface heat emission;
vector V_f – electrode supply direction

The number of sections, the electrode between the contact tips is divided into, is dependent on the impulse frequency, electrode supply speed, and the length between two contact tips. The length between the contact tips is constant and assigned 0.4 m. Shortening the length deteriorates the effect of electrode extension pre-heating, while its elongation necessitates increasing the device dimensions. The impulse frequency is assigned 100 Hz for the purpose of calculations. Further calculations require variation of the inter-pulse period only. The electrode supply speed has a considerable effect on the temperature of pre-heating.

Physical and mechanical parameters (thermal capacity, hardness, and temperature of phase transformations) of the electrode 08Mn2Si with the diameter 1.2 mm are assigned constant for the purpose of calculations. The resistance of the electrode and its heat emission vary according to the changing temperature. All these parameters are calculated automatically in the model.

The authors do not take into consideration the gas-dynamic impact on the drop unless the model becomes more complicated [8].

The electrode is pre-heated according to Joule-Lenz law [9]. The pre-heating can be varied via changing the inter-pulse current.

Three key parameters influencing on the electrode pre-heating and its stability are ascertained.

Each parameter has a designation:

1. Inter-pulse current X
2. Electrode supply speed Y
3. Inter-pulse period Z

Table 1 provides the variability intervals with the pre-set step of parameters X, Y and Z.

IOP Conf. Series: Materials Science and Engineering **142** (2016) 012002 doi:10.1088/1757-899X/142/1/012002

Table 1. Variability intervals of the parameters

Variability interval of the parameters	The step of parametric variation
$30<X<40$	$\Delta X=5$
$400<Y<650$	$\Delta Y=50$
$5<Z<10$	$\Delta Z=1$

When planning the experiments the complete factorial experiment is used [10]. Since it is difficult to measure the temperature on the moving section of the electrode when welding, the experiments were carried out relying on the mathematical model. This model makes it possible to assess resistance vs. changing temperature, thermal conductivity, and power losses for convective and radiant heat exchange. A fragment of the model cycle is presented in Figure 3 to calculate pre-heating of the electrode section.

$$R1 \leftarrow \alpha \cdot Rv \cdot T$$
$$R \leftarrow R + R1$$
$$Q1 \leftarrow I2^2 \cdot R \cdot t$$
$$Q2 \leftarrow 1.5(T-20)\cdot t \cdot s2 \ \text{ if } \ T > 100$$
$$Q2 \leftarrow 0.5\cdot[(T-20)\cdot t \cdot s2] \ \text{ otherwise}$$
$$Q4 \leftarrow Q1 - Q2$$
$$\Delta T \leftarrow \frac{Q4}{c \cdot m}$$
$$Q3 \leftarrow k \cdot s \cdot \frac{\Delta T \cdot t}{lvileta}$$
$$Q \leftarrow Q1 - Q2 - Q3$$
$$\Delta T \leftarrow \frac{Q}{c \cdot m}$$
$$T \leftarrow T + \Delta T$$

Figure 3 – A cycle to calculate the electrode pre-heating in the software MathCAD electrode

This model makes it possible to calculate the temperature on each section (Figure 4). However, for the purpose of experiment we use the temperature in the last point (400 mm). The temperature is always maximal in this point. As one can see, this is not a linear dependence because of the changing electrical resistance and heat emission when increasing the temperature.

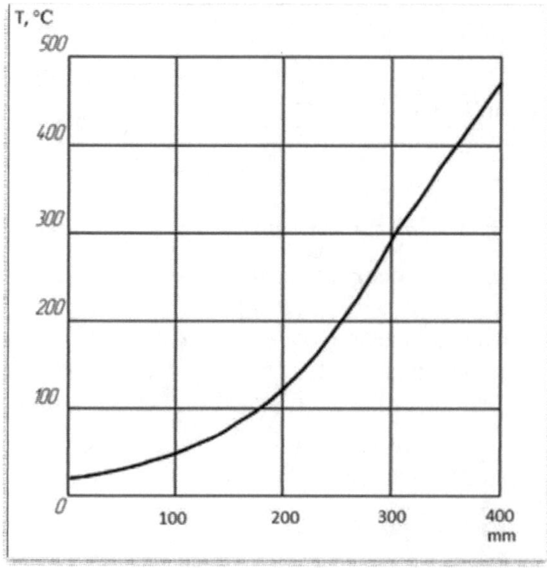

Figure 4 – The plot of temperature distribution when pre-heating the electrode wire between the contact tips 2 (point 0) and 1 (point 400 mm)

The results of numerical experiments involving the mathematical model obtained at various values of current strength, electrode supply speed and inter-pulse period are given in Figures 5, 6 and 7.

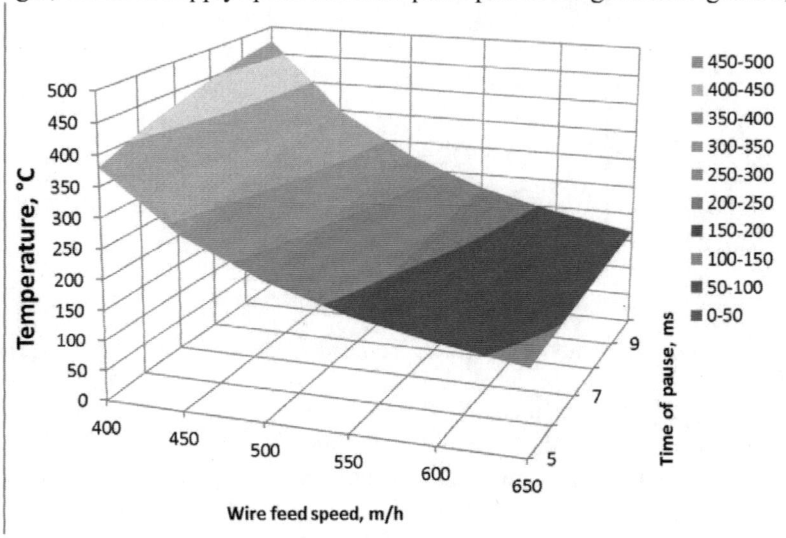

Figure 5 – The dependence of the electrode extension temperature on the supply speed and inter-pulse period, inter-pulse current 30 A

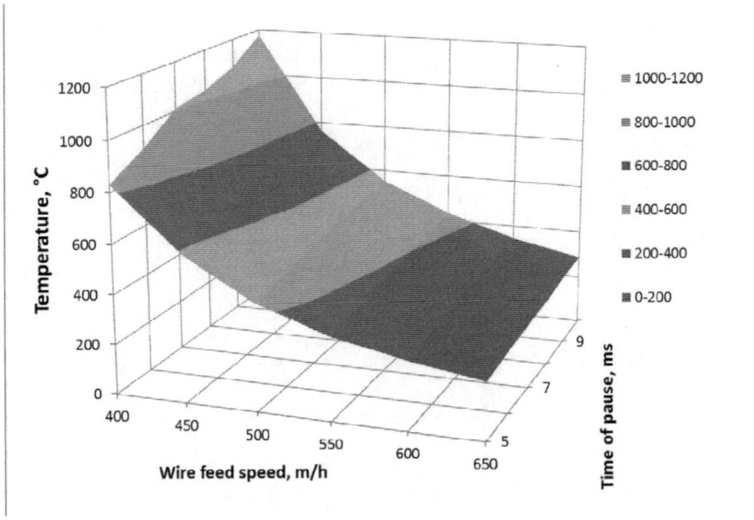

Figure 6 – The dependence of the electrode extension temperature on the supply speed and inter-pulse period, inter-pulse current 35 A

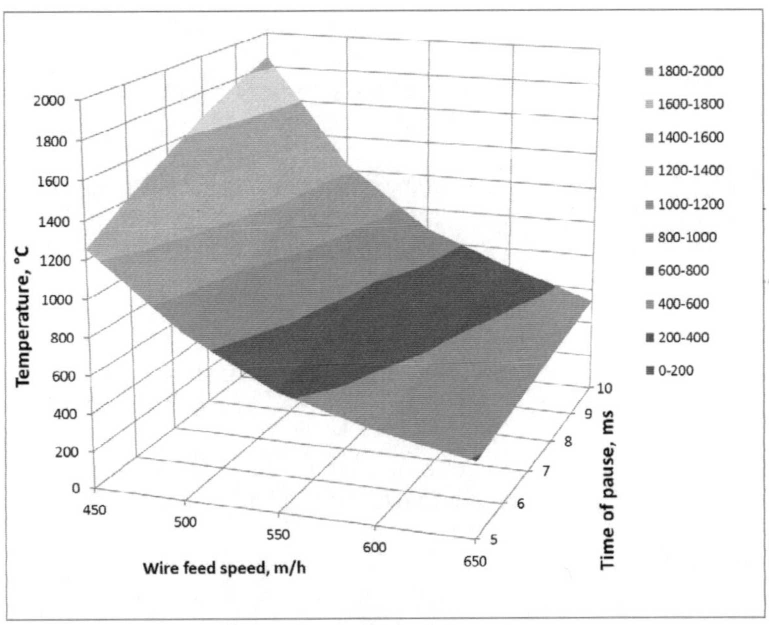

Figure 7 – The dependence of the electrode extension temperature on the supply speed and inter-pulse period, inter-pulse current 40 A

The authors have detected the influence of appropriate modes on the stable arc burning. The resistance drop is determined on the electrode extension and arc-drop voltage drop is measured.

The resistance of free running is assigned 80 V. Therefore, resistance drop on the electrode and on additional resistance can't exceed 58 V. Voltage drop is calculated as the product of current strength and resistance. The resistance of the full electrode extension between two contact tips is the sum of resistances on all its sections.

On the average, the resistance of 400 mm extension is 0.06 ohm at the temperature of 600°C before the second contact tip. The voltage drop of the electrode on the whole is 1.9 V.

The experimental research into pulsed-arc welding is carried out by the device described above. The appropriate parameters of welding are assigned: impulse frequency – 100 Hz; inter-pulse period – 7 ms; arc voltage during the inter-pulse period – 22 V; electrode supply speed – 600 m/h: inter-pulse current strength – 40 A.

Current and voltage oscillograms of pulsed-arc welding with electrode pre-heating by the inter-pulse current are given in Figure 8. Arc burning is stable over the whole range to be changed. The distinctive feature of the process is arc burning on the convex weld pool surface. Short impulses with the reduced amplitude can not cause the change in the arc length.

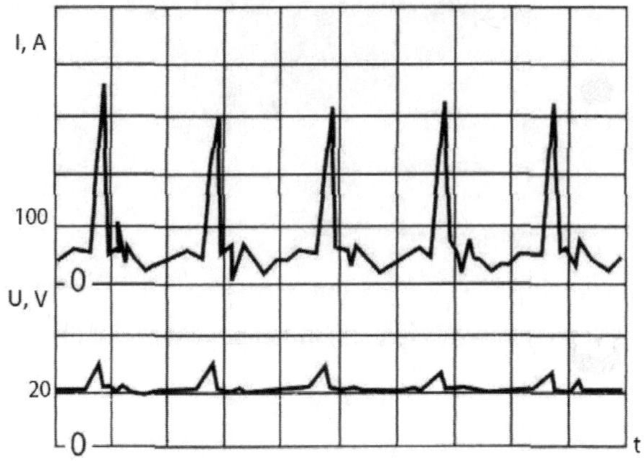

Figure 8 – Current (100 A/segment) and voltage oscillograms (20 V/segment) of pulsed-arc welding (contact tip distance l=400 mm)

Planning and carrying out the experiment on the base of the mathematical model helps to reduce its time and costs. Results assessment of the mathematical experiment provides the information for improvement of the device without causing its failure in case of wrongly adjusted parameters.

3. Conclusions

1. The appropriate modes of the inter-pulse are determined in the course of experiments and their influence on the process stability is determined. Three key parameters are identified which have an effect on electrode pre-heating and its stable behavior: speed of electrode supply, inter-pulse period and current.
2. The appropriate parameters of the inter-pulse current and time are assigned on the base of the proposed model for calculating the electrode pre-heating (impulse frequency – 100 Hz; inter-pulse period – 7 ms; arc voltage during the inter-pulse period – 22 V; electrode supply speed – 600 m/h: inter-pulse current strength – 40 A).
3. When pre-heating the electrode up to 600°C less power is required to melt the electrode and transfer the drop of electrode metal, this is made obvious by decreased amplitude and time of current pulses.

References

[1] Lin T, Chen H B, Li W H and Chen S B 2009 Intelligent methodology for sensing, modeling, and control of weld penetration in robotic welding system *Industrial Robot-an International Journal* **36** 585-93
[2] Akkas N, Karayel D, Ozkan S S, Ogur A and Topal B 2013 Modeling and Analysis of the Weld Bead Geometry in Submerged Arc Welding by Using Adaptive Neurofuzzy Inference System *Mathematical Problems in Engineering* 10
[3] Duan Y G, Vincent Y, Boitout F, Leblond J B and Bergheau J M 2007 Prediction of welding residual distortions of large structures using a local/global approach *Journal of Mechanical*

Science and Technology **21** 1700-6

[4] Krampit N.Yu., Krampit M.A. 2015 Pulsed-arc welding with preheating of electrode extension in the break period In: *Welding International.* **29** 210-212

[5] Foroutan M, Aalami-Aleagha M E and Pirmoradi S 2011 Investigation of joint groove geometry effect on the residual stresses in circumferentially butt-welded steel pipes by the FEM simulation *Advances in Materials and Processing Technologies Ii, Pts 1 and 2* **264-265** 229-34

[6] Mazaheri Y, Kermanpur A and Najafizadeh A 2015 Microstructures, Mechanical Properties, and Strain Hardening Behavior of an Ultrahigh Strength Dual Phase Steel Developed by Intercritical Annealing of Cold-Rolled Ferrite/Martensite *Metallurgical and Materials Transactions a-Physical Metallurgy and Materials Science* **46A** 3052-62

[7] Nogueira R M U, Bohorquez C E N and Zanella I G 2015 Comparison of Arc Stability and Variability of Geometry of Welds Obtained by MIG/MAG and FCAW Processes *Soldagem & Inspecao* **20** 191-204

[8] Chinakhov D A, Zuev A V and Filimonenko A G 2014 Gas-dynamic Impact of a Shielding Gas Jet on the Drop Transfer When Welding with a Consumable Electrode *High Technology: Research and Applications* **1040** 850-3

[9] Tohmyoh H and Fujimori M 2012 Microstructural and electrical characterization of Joule heat welds in ultrathin Pt wires *Physica E-Low-Dimensional Systems & Nanostructures* **46** 33-7

[10] Scinto P R 2011 Statistical Engineering Examples in the Engine Oil Additive Industry *Quality Engineering* **23** 125-33

IOP Conf. Series: Materials Science and Engineering **142** (2016) 012003 doi:10.1088/1757-899X/142/1/012003

Objectivation of the Necessity of Structural and Parametric Synthesis of the Hydraulic Drive Of Geokhod

R V Chernukhin[1], M Y Blaschuk[1], V Yu Blumenstein[2], P A Chazov[1]

[1]Yurga Institute of Technology, TPU, 26, Leningradskaya Str., 652050, Yurga, Russia
[2]T.F. Gorbachev Kuzbass State Technical University, 28 Vesennyaya str., Kemerovo, 650000, Russia

E-mail: Chernuhin@tpu.ru

Abstract. This article considers main systems of a prototype model of geokhod as well as their technical specifications. There are ways distinguished of decreasing the dimensions and the mass of elements of the hydraulic drive. As an optimization tool of structural and parametric parameters of the prototype model of geokhod, there is SimHydraulic simulation considered.

1. Introduction

The exploration of the underground space is an important task in the modern conditions of human life activities. The construction of underground facilities is aimed at the extraction of minerals, at the placement of transport communications as well as objects of urban supply, etc. Among underground facilities there are tunnels. In the publications of authors [1], there are the main tendencies marked concerning the development of tunneling, in particular, there are the following most important tendencies:

– construction of underground facilities using new technologies;
– a wider use of characteristics of rock mass;
– safety increase of tunneling excavation;
– implementation of new mechanized complexes.

As an alternative to modern shield tunneling methods, in publication [2], there is a special technology suggested which basic element is geokhod. In comparison with common tunneling machines, geokhod has several advantages which are based on an absolutely new functioning principle of the aggregate as well as on new functional and structural elements. One of these advantages is the ability to work under any slope angle of the drive working [3].

The structure of recently developed geokhod samples (Figure 1) consists of an executive element of the main forehead, executive elements of head and tail sections, transmission as well as loading system. A distinguishing feature of geokhod consists in the fact that all of the above listed systems function simultaneously, which is why there is a necessity to coordinate their functions. In publication [4], there was a mathematical model elaborated which describes the progressing of a geokhod in the rock mass. It must be mentioned that previously created mathematical models [5] did not take into account the large number of geokhod systems and their diversity. A modular principle of their structure, as well as of geokhod in general, showed the need to search for the methods of choosing preferable circuit designs depending on the exploitation conditions of geokhod [6].

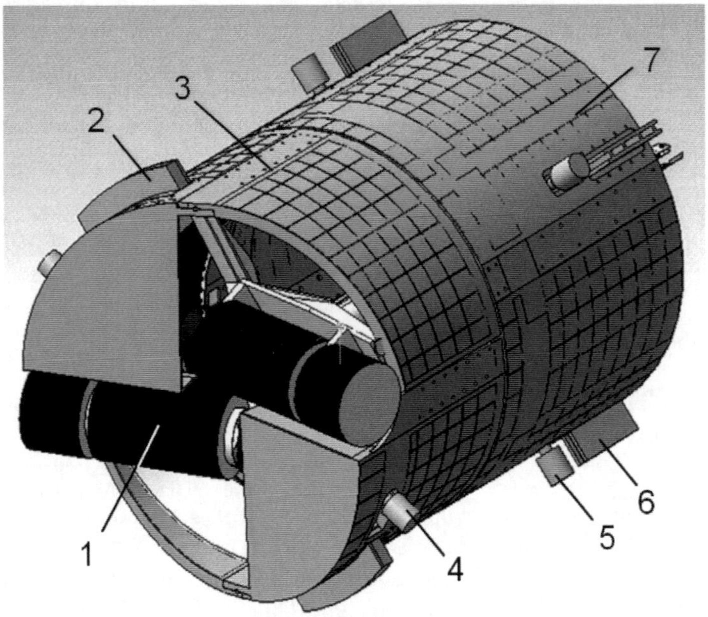

Figure 1. General view of a geokhod prototype model

1 – executive elements of the main forehead; 2 – propelling device; 3 – head section; 4 – executive element of the head section; 5 – executive element of the tail section; 6 – counter-rotating element; 7 – tail section

2. Initial data

Currently, there are not only designing works performed but also manufacturing as well as calculation of economical effectiveness of manufacture of this tunneling machine [7, 8, 9]. According to the scheme of work on the project of geokhod manufacture, researchers of Yurga institute of technology of Tomsk polytechnic university as well as of Yurga mechanical engineering plant have designed a prototype model of geokhod. Main technical specifications of the prototype model are listed in table 1.

Table 1. Main technical specifications

No.	Parameters	Measuring unit	Value
	Mining conditions		
1	Diameter of the drive working in the tunneling	m	3.2
2	Cross sectional area of the drivage	m^2	8.4
3	Angle of slope of the drive working	degrees	± 25
4	Hardness of the enclosing rock	units	up to 5
	Technical specifications		

1	Output	m^3/min	up to 0.8
2	Drilling rate	m/h	up to 6
3	Drive types of executive elements		Hydraulic
4	Number of hydraulic cylinders in the module	pieces	16
5	Nominal working pressure	MPa	20
6	Torque of the head module by nominal pressure, minimum	kN·m	1.76e03
7	Gross thrust at the tail module	kN	850
8	Lifting angle of the helical blade of the external propelling device	degrees	8°
9	Mass (without pumping stations)	kg	19,000
10	Installed capacity, minimum	kW	414
11	External diameter of geokhod modules	mm	3,200

The functioning of mining machines in general, and of the geokhod in particular, can be provided by means of power drives of various nature [10]; however, main systems of the prototype model, which perform the progressing of the geokhod down the hole, its crushing and steering along the channel, are driven hydraulically. The distinguishing features of the hydraulic drive of geokhod are high capacity as well as constant simultaneous work of all systems which is typical for recently developed geokhods [11].

The hydraulic circuit diagram of the designed sample of geokhod includes hydraulic drive of the executive element crushing the face, hydraulic drive of the motion of cylinders of the executive element, hydraulic drives of executive elements of off-circuit elements of geokhod's head and tail sections, hydraulic drive of the loading device, hydraulic drive of the transmission. In order to prevent hydraulic lines of geokhod's systems from mutual influence, they are supplied with different pumping units.

The designing of the hydraulic drive is based on the initial data listed in table 2.

Table 2. Initial data

System name	Torque at the output element M, kN·m	Rotation speed of the output element ne_{ee}, min^{-1}	Reduction ratio, u_{red}	Number of hydraulic servomotors i_{hm}, items
Executive element of the main forehead	21.4	30	66.3	2
Executive element of the head section	0.824	130	1	2
Executive element of the tail section	0.39	60	1	4
Loading system		10	15	4
Transmission	1,710	0.1	-	8

The point of departure for the determination of parameters of the hydraulic drive of the geokhod is the calculation of power characteristics of geokhod [12] as well as the necessary capacity of all geokhod systems. Total capacity of the hydraulic drive is determined by equation (1) [13].

$$N_{HS} = \sum_{i=1}^{n} N_{hm} = k_{c.tr} k_{y.tr} N_{tr} + k_{c.ou} k_{y.ou} N_{ou} + k_{c.prou.} k_{y.prou} N_{prou} + \\ + k_{c.lu} k_{y.lu} N_{lu} + \ldots + k_{ci} k_{yi} N_i, \quad (1)$$

$\sum N_{hm}$ – the total sum of power of all hydraulic motors, W.

$k_{c.tr}, k_{c.ou}, k_{c.prou.}, k_{c.lu}, k_{ci}$ – safety factor of hydraulic motor velocity, that of transmission, operating unit of geokhod, propeller operating unit, loading unit, end users i, k_c = 1,1…1,3;

$k_{y.tr}, k_{c.ou}, k_{c.prou}, k_{c.lu}, k_{yi}$ – safety factor of hydraulic motor force, that of transmission, operating unit of geokhod, propeller operating units, loading unit end users i, k_y = 1,1…1,2;

N_{tr} – power, supplied for the drive of transmission hydraulic cylinders, W;

N_{ou} – power, supplied for the drive of operating unit hydraulic motors, W;

N_{prou} – power, supplied for the drive of propeller operating units,

N_{lu} – power, supplied for loading unit drive;

N_i – power of hydraulic motor of end user i.

When designing the hydraulic drive of the prototype model of geokhod, it is necessary to judge from the principle of performance assurance of geokhod systems taking into account increased safety factors. Thus, an increased power reserve of geokhod systems is guaranteed which compensates possible designing errors and allows to carry out tests of the prototype model under critical loads.

However, such approach is unacceptable when designing production samples because the increase of capacity of hydraulic drives results in the increase of hydraulic fluid consumption which leads to the increase of volume of the hydraulic tank as well as the mass and dimensions of hydraulic units and valves. These factors make it impossible to place a pumping station inside the geokhod in case of necessity. Thus, the factors of assurance, the value of which is mostly determined by the experience of designing of hydraulic drives, in geokhod's case, demand a scientific objectivation.

3. Results and discussion

In order to achieve the assigned task, it is necessary to simulate the drive's work in the conditions imitating close to real. The library SimHydraulics of Simulink environment of the Mathlab system allows not only to receive quite exact simulation results of hydraulic systems, it also makes it possible to correct corresponding parameters based on the results of real tests. A distinguishing feature of this library consists in the fact that the applied units are represented by ready-to-use standard components of hydraulic drives. This accelerates the simulation process and makes it easier [14, 15, 16], while a wide list of units helps simulate almost any kind of hydraulic systems.

The hydraulic circuit diagram of geokhod's transmission includes two groups of hydraulic cylinders, eight pieces in each. The hydraulic fluid for a forward stroke and an idle run is delivered from different pumping units. The delivery of the hydraulic fluid is performed by means of a six-line three-position hydraulic distributor. The simulation of the transmission in the environment SimHydraulics, like any other hydraulic scheme with hydraulic cylinders, has its own peculiarities. When building a diagram from Simscape library, you should pick the proper element (in this case a double-acting hydraulic cylinder) and connect it with the distributor by means of hydraulic lines (Figure 2). In the units' parameters, you can define the technical specifications of the hydraulic cylinder and other units. Then, you must connect the hydraulic cylinder with the units of fixed bearings and simulate the load at the piston. For this, "Mass", "Translational spring", and "Translational damper" units are used. In order to measure the motions of the piston, unit "Ideal Translational Motion Sensor" is used, whereas, in order to display measured values, this unit is connected with a virtual oscillograph (unit "Scope"), and the signal transmitted is previously transformed from Simulink signal into the necessary physical quantity by means of the unit "PS S Simulink Converter".

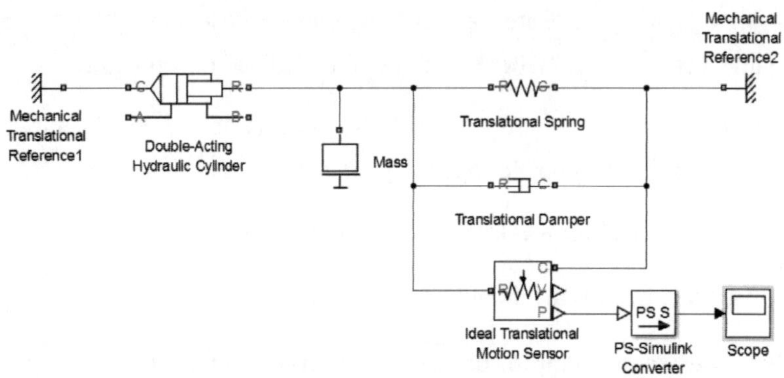

Figure 2. Simulation example of a double-acting hydraulic cylinder

As it has already been mentioned before, the transmission of the geokhod consists of two groups of hydraulic cylinders, eight items in each. This is why, in order to decrease the workload of the model, several units can be united into one subsystem. In Figure 3, such subsystems (HC1…HC16) combine hydraulic cylinder units, fixed bearings, load imitating units at piston as well as the units that allow to measure the motion of pistons.

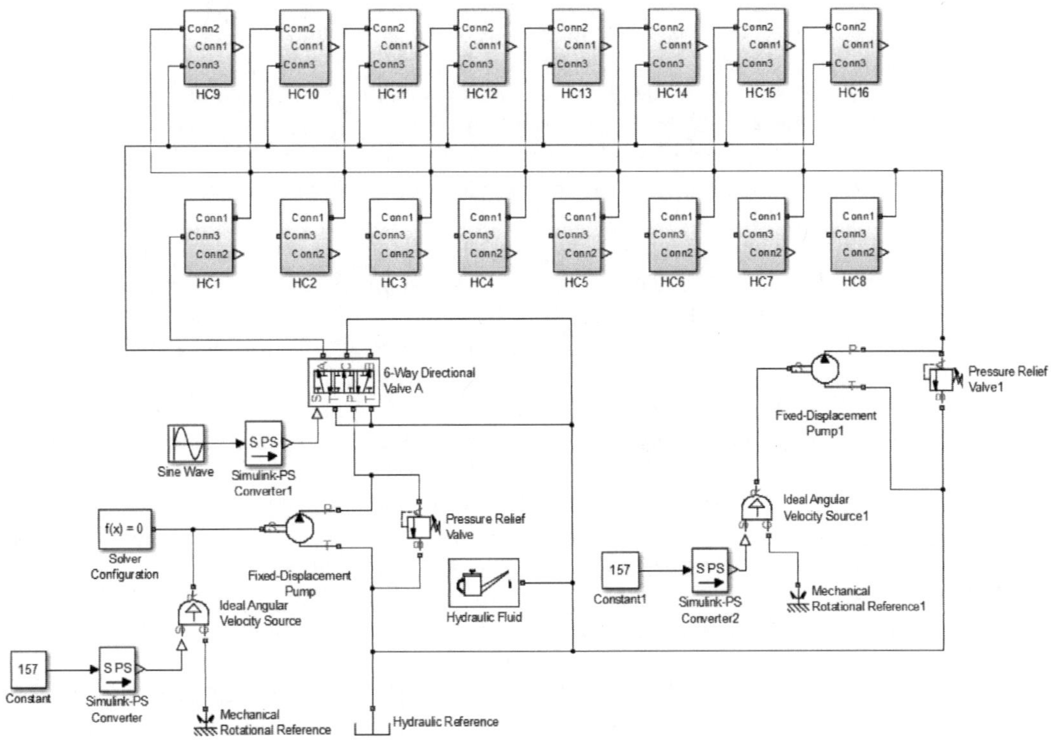

Figure 3. SimHydraulics model of geokhod transmission

Having entered the necessary parameters into the units, you can simulate the function of the transmission in various situations, e. g. to test its functioning when some of the elements fall out. It must be mentioned, that the simulation results in Simulink system can be only approximate and demand an adequacy test in experimental conditions which is, on the other hand, typical for other models as well [17].

4. Conclusions

In spite of the performed work on the search for means of decreasing the dimensions of basic elements of the pumping station, there are still open issues of determination of rational parameters of other elements of the hydraulic system of geokhod as well as structural optimization of the function of the whole hydraulic drive of geokhod, and the creation of basics of its automatic regulation system. For a comprehensive research of the hydraulic drive of geokhod, it is necessary to perform following actions: to research the sensitivity analysis of output characteristics to the changes of drive parameters, to determine extreme values of output characteristics, to conduct a statistical analysis of the drive, to determine permissible deviations of the parameters of the hydraulic drive from predefined values, and to evaluate the working capacity of the hydraulic drive in case of fall-out of its separate elements.

References

[1] S K Sengupta, P K Saxena, & S S Bakshi. Tunnel boring machine in *India-experience, prospects and challenges. In Proceedings of world tunnelling Congress (2008).* Agra, India, pp.1544-1551.

[2] A. B. Efremenkov, "Forming the subterranean space by means of a new tool (geohod)," in *6th International Forum on Strategic Technology (IFOST - 2011): Proceedings: in 2 vol.*, 2011, vol. 1, pp. 348–350.

[3] V. V. Aksenov, A. A. Khoreshok, and V. Y. Beglyakov, "Justification of creation of an external propulsor for multipurpose shield-type heading machine – GEO-WALKER," *Applied Mechanics and Materials*, vol. 379, pp. 20–23, 2013.

[4] M Y Blaschuk, A A Dronov, D A Miheev Geokhod Propel Effort Mathematical Model In *Applied Mechanics and Materials,* vol. 770, pp. 391-396, 2015.

[5] V Y Sadovets, V Y Beglyakov, V V Aksenov Development of math model of geokhod bladed working body interaction with geo-environment in *Materials Science and Engineering,* vol. 91, p.012085, 2015.

[6] R V Chernukhin, A A Dronov, M Y Blashchuk The application of the analytic hierarchy process when choosing layout schemes for a geokhod pumping station in *Materials Science and Engineering*, vol. 91, p. 012086, 2015.

[7] A V Walter, V V Aksenov Determining deviations in geometry of the geokhod shells In *Applied Mechanics and Materials,*vol. 770, pp. 439-444, 2015

[8] V V Aksenov, A V Walter, A A Gordeyev, A V Kosovets Classification of geokhod units and systems based on product cost analysis and estimation for a prototype model production in *Materials Science and Engineering*, vol. 91, p.012088, 2015.

[9] Gubaydulina R. K. , Petrushin S. I. , Galeeva A. A. Selecting an Economical Variant of the Manufacturing Method of Engineering Product Fabrication under Current Conditions in *Applied Mechanics and Materials.* vol. 682. pp. 613-616. 2014.

[10] V Y Timofeev, V V Aksenov, J I Galjamova Determination of parameters of roller of wave transmission with intermediate rolling bodies with hollow shaft for geokhod In *Applied Mechanics and Materials*, vol. 682, pp.246-250, 2014.

[11] A B Efremenkov, V. V Aksenov, & M Y Blaschuk. (2012, September). Force parameters of geohod transmission with hydraulic drive in various movement phases. In *Strategic Technology (IFOST), 2012 7th International Forum on* (pp. 1-5). IEEE.

[12] V V Aksenov, M Y Blaschuk, M V Dubrovskii Estimation of torque variation of geohod transmission with hydraulic drive In *Applied Mechanics and Materials*, vol. 379, pp.11-15, 2013.

[13] M Y Blashchuk, A A Kazantsev, R V Chernukhin Capacity Calculation of Hydraulic Motors in Geokhod Systems for Justification of Energy-Power Block Parameters In *Applied Mechanics and Materials*, vol. 682, pp 418-425, 2014.

[14] Krimbacher, N., Garstenauer, M., & Scheidl, R. (2001, September). Optimization of hydraulic systems by means of numerical simulation. In *Proceedings of the 2nd International Workshop on Computer Software for Design. Ostrava-Malenovice, Czech Republic* (pp. 76-83).

[15] Chen, Y. Q., Xu, X. H., Zhu, H. L., Ye, S. D., & Li, L. T. (2013, October). Simulation Research on the Basic Platform's Automatic Leveling System Based on SimHydraulics. In *Applied Mechanics and Materials*. vol. 345, pp. 99-103.

[16] MARIUS, P. (2010). Optimizing the Hydraulic Systems Using the SimHydraulics Programming Environment. *Journal of Engineering Studies and Research*, *16*(4), pp. 25-32.

[17] A V Walter, "Modeling of Track Overlapping Effect on Layer Integrity in the Process of Laser Sintering of Polymer Suspension," in *7th International Forum on Strategic Technology (IFOST - 2012): Proceedings: in 2 vol.*, 2012, vol. 2, pp. 125-128.

IOP Conf. Series: Materials Science and Engineering **142** (2016) 012004 doi:10.1088/1757-899X/142/1/012004

Enhancing a service life of torch components for MIG/MAG welding

A V Filonov[1], A V Kryukov[1] and M I Galimov[1]

[1]Yurga Institute of Technology National Research Tomsk Polytechnic University
26 Leningradskaya st., Yurga, Kemerovo region, 652050, Russia,
ph. +7(38451)77765

E-mail: a.filonow@gmail.com

Abstract. The paper analyzes the main vulnerable elements of torches used in mechanized gas–shielded welding. Particular attention is given to the gas nozzle designs, materials they are made of, and other welding torch elements exposed to increased electrical and thermal stresses during the welding process.

1. Introduction

For mechanized gas-shielded welding (MIG/MAG welding), torches of a GDPG type are used (Russian analogues: GS-250, A-547uM; foreign analogues: RF and MB types made by Abicor Binzel: AL and AW types by Fronius, TBi – TBI, SB and SBT by Merkle). The main advantage of these torches is their light weight and simple design. The torches of this type are designed for welding with a current of up to 300 A, and owing to their advantage they can be used for welding at higher currents after their modernization [1].

Efficiency of torches depends on their design and materials used. In welding, some parts of torches are subjected to heating from welding arcs and contact with droplets of molten metal. Nevertheless, main structural components and units in this type of torches are able to withstand increased electrical and thermal stresses without a loss of their serviceability and functionability.

2. Analysis

Nozzles of welding torches, ensuring the gas flow to a weld zone, are subject to thermal impact from the arc and are warmed to a temperature of 180-320 °C and above when continuously operating for a few minutes without insufficient cooling. In strong heating, spatters of molten metal are firmly welded to walls of metal nozzles. Building of spatter on the inside surface of nozzles causes the reduction of their inside diameter and weaker gas-shielding protection of molten metal. Arc tracking can occur between a contact tip and the nozzle. If a nozzle is made of electrically conductive material, then in cases of accidental contacts it can melt causing its failure.

According to working conditions, the following basic requirements are imposed on nozzles' designs and material they are made of: fire-resistance, mechanical strength both in a heated and cold condition, low spatter adhesive strength, a simple design and low cost of manufacture.

Heating a nozzle increases the intensity of metal spatter in it. Cooling a nozzle contributes to a sharp weakening of this process and increases the efficiency of cleaning weld spatter off the welding nozzle. Life service of nozzles when they are cooled increases approximately 3-4 times [2]. In order to increase reliable operation of nozzles and their heated parts, they are cooled with water or air, depending on welding conditions [3]. Besides, the quality of a nozzle surface coating can contribute to reducing the amount of weld spatter [2].

The nozzles cooled with air can operate at a current of 300 A. When the current exceeds 300 A, the nozzle must be cooled or structurally provided for the possibility of operating at high temperatures. The nozzles cooled with water are of limited use in industry, as in this case the weight of hoses and the nozzle are higher that results in reduced manoeuvrability, and it is not always possible to provide a working area for water. In most cases, designs of hoses and nozzles are modified in the way that to apply natural cooling.

Brass or copper are mainly used for the manufacture of nozzles. Service life of nozzles made of brass and other alloys is approximately 2 times less than of copper. The nozzle surface is often chrome plated to reduce the adhesion of spatter. The shapes of nozzles are usually cylindrical (Figure 1, b) or slightly conical (Figure 1, c) that provides a more uniform outflow of gas. Copper nozzles in cylindrical configuration are the most widely used [4].

a) b) c)

Figure 1. Copper nozzles for welding torches:
a – cylindrical; b – cylindrical with broadening; c – conical

Due to the relatively high cost of copper tubes, commonly used for the manufacture of nozzles, a lot number of attempts were made to replace copper with non-metallic materials. Non-metallic nozzles (Figure 2) are used mainly in welding with a low current.

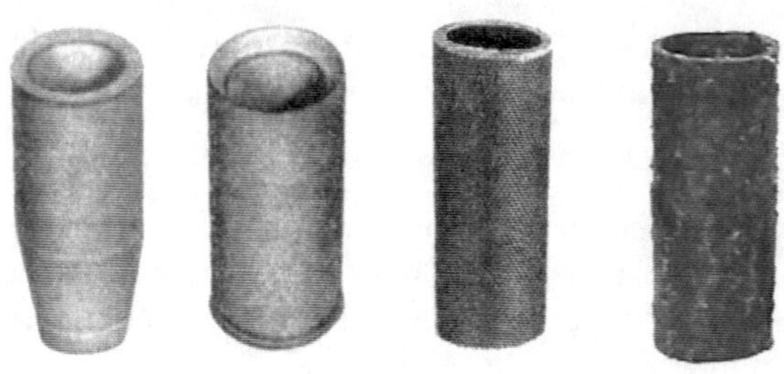

a) b) c) d)

Figure 2. Non-metallic nozzles for welding torches:
a, b – ceramic; c – fibreglass; d – paronite

One of the first attempts was made to use heat-resistant ceramics in the manufacture of nozzles in torches for argon-arc welding with non-consumable electrodes. However, the drawback of ceramic materials is their ability to break down under thermal stresses as well as low mechanical strength in cases of spatter cleaning and accidental bumps.

The manufacture of ceramic nozzles (Figure 2, a, b) by hot die casting is a more advanced technology [5]. For this purpose, dispersed slurry is prepared, where talc and clay in the dispersed phase are mixed, with paraffin as a binding agent in an amount of 8-10% of mixture's weight, which is used as a technological binder. The nozzle is 3-4 shifts in operation. Due to low mechanical strength and relative complicated manufacturing technique, the ceramic nozzles have found in limited use. There are the same limitations to the nozzles made from asbestos cement slabs 25 mm thick by cutting them out. They are not widely used as well, although their resistance is equal to 6 shifts.

Attempts were made to use combined copper and asbestos cement nozzles, i.e. asbestos cement sleeves were attached to the ends of copper nozzles. This, to some extent, excluded short circuiting and copper nozzle burning, but did not exclude mechanical damage due to shocks.

Nozzles made of fibreglass reinforced plastic AG-4B, which consists of a base (matted fibreglass) loaded with phenol-formaldehyde resin, have good performance characteristics [6]. The nozzles are made by pressing followed by heat treatment. The strength of these nozzles compared with asbestos cement nozzles is higher 2-3 times; cleaning spatter off during their use is not required.

High performance torch nozzles for wire welding from fibreglass fabrics are simple to manufacture (Figure 2, c) [7]. When FA, FB or FAU monomers are used as a binding agent, four layers of fibreglass fabrics are wound around the brass plated rods with impregnating with one of the said monomers. Then the fibreglass tube with the rod is exposed to heat treatment. After heat treatment, the tube is detached from the rod and cut to the desired lengths. The fibreglass nozzles have no drawbacks that are typical for copper, brass, ceramic and asbestos cement nozzles. Service life of fibre glass nozzles is approximately 10 shifts of continuous work with their gradual melting.

The nozzles based on fibreglass fabrics can be made using bakelite varnish, although in this case, their service life will be reduced.

Instead of monomers and varnish as a binder, in the manufacture of nozzles andesine paste can be used [8], consisting of sodium silicofluoride and andesine or diabase power, which can be replaced with powered bricks or glass mixed with liquid glass. The paste is applied between the layers of fibreglass during its winding on a rod and dried at a normal temperature. The ready tube is cut to the desired lengths. Under the influence of arc welding temperature, the nozzles become embrittled to the small extent, but their strength is sufficient. Spatter building up on the nozzles can be easily removed. Service life of the nozzle based on the andesine paste is shorter than of the nozzles on monomers. However, they do not require high-temperature treatment and therefore cheaper.

There are other variants for replacing the nozzle material (copper) with lighter, less expensive and non-conductive materials. For example, it is proposed to use a sheet paronite, folded up in a tubular lining (Figure 2, d) and fixed in the casing [9]. Such nozzles have a number of drawbacks: first, because of the gap at the spot of liner joining, a gas shield in arc and weld area can be disturbed and, second, at high temperatures paronite burns out that leads to frequent replacement of nozzles.

There is a variant of nozzles in the form of a seamless tube from heat-resistant fibreglass plastics with an epoxy organ silicon binder reinforced with fibreglass fabrics [10]. Its mass is 10 times less than copper. Spatter building-up on the inner surface of the nozzle can be easily removed. A glass-plastic nozzle is efficient for 1.5-2 shifts under heavy operating conditions of welding.

In heavy duty welding (300-450 A), copper nozzles are largely preferred for use, because of thermal and physical characteristics of copper complying with operating conditions to a larger degree.

To protect metal nozzles from molten metal spattering, special protective lubricants can be applied, which are designed for use at temperatures ranging from 100 to 250 °C. The lubricants are supplied in aerosols that allow their easy application by spraying onto nozzles. After the solvent evaporating, the surface is covered by a thin solid film that protects the surface of the nozzle against weld spatter.

As a protective cover, it is suggested using other insulation materials such as synthetic resin. Molten metal spatters appearing during the welding process do not adhere to the inner surface of the nozzle. The service life of the nozzle increases.

An effective way to increase a service life of welding torch components is applying, for example, heat-resistant coatings by ion-assisted deposition onto the working surfaces of the nozzles. One of the known coatings is a heat resistant diamond-like film [11, 12] with the use of ion-beam deposited carbon on the nozzle surface. In addition to the diamond-like coating, heat-resistant coatings based on titanium, nitride titanium and nitride zirconium have been developed. It is established [13] that the nitride titanium (TiN) coating is preferable for the nozzles used in semi-automatic welding machines. The service life of the nozzles with heat-resistant coatings increases by 30-80 % and, respectively, the demand for such nozzles decreases.

The electrically insulated nozzle has a property of accumulating heat energy from an electric arc. This heat energy causes heating of other parts. Especially dangerous is the heating of insulating sleeves. An insulating sleeve serves to isolate the current-carrying parts from the nozzle. It is a constructive element designed to hold and centre the nozzle itself on the torch (Figure 3).

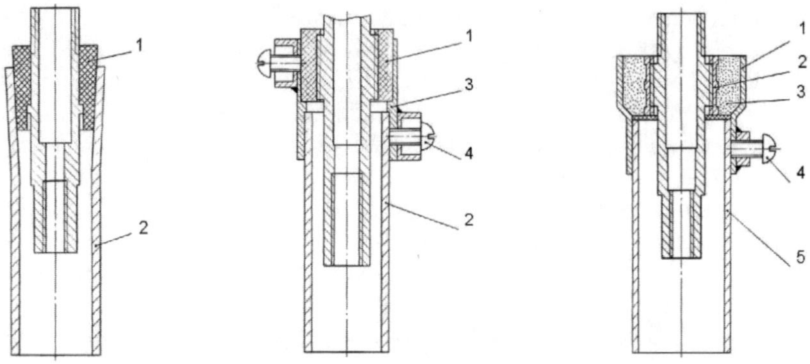

a) b) c)

Figure 3. Variants for fixing nozzles in insulating sleeves:
a – conical; b – cylindrical; c – ceramic metal

The insulating sleeves for torches used in carbon dioxide welding, operating in heavy duty environments, are exposed to repeated heating and reach high temperatures. Therefore, the materials need to meet the following requirements: high isolating properties, good thermal and mechanical properties, cheap and available for supply materials, basic manufacturing processes, and a small weight.

A range of insulating sleeves is not diverse with reference to their designs (Figure 4).

a) b) c)

Figure 4. Insulating sleeves for semiautomatic welding torches:
a - tapered; b - cylindrical; c – ceramic metal

They have a conical (Figure 4, a) or cylindrical (Figure 4, b) shape with a plain and (or) a threaded bore, depending on design features of the welding torch and contact welding tip in particular.

The insulating sleeve form determines the way of how the nozzle is attached to the welding torch. The nozzle 2 is held inside the conical sleeve 1 (Figure 3, a) by means of the tapered extension at the end of the nozzle. This attachment is not reliable because of the insulating sleeve breakdown under high temperatures that leads to spontaneous fallout of the gas nozzle. A special coupling 3 is used to hold the nozzle 2 attached to the cylindrical sleeve 1 (Figure 3, b) and screws 4 fix the attachment. This technique is much more reliable, but it also has limitations: a massive structure and inconvenience in service due to the large number of structural elements.

Electrical insulating properties are important in material used in the manufacture of sleeves. The most common are fabric-reinforced laminate and glass-fibre-base laminate. The fabric-reinforced laminate sleeves have good insulating properties but low mechanical and high temperature resistance. This type of sleeves is nondurable due to their quick embrittlement when in contact with the heated nozzle. The permissible operating temperature can reach a high of 105 °C [14] for fabric-reinforced laminate. The glass-fibre-base laminate sleeves can operate at temperatures up to 200 °C and for 1-2 hours at temperatures up to 250 °C [14]. There are also insulating sleeves from asbestos cloth, which is made by pressing and using sodium silicate for binding [15]. Operating temperatures inside the asbestos sleeves can reach as high as 450 °C.

Problems that appear while manufacturing and using the sleeves of the above materials are due to various reasons. Thus, the fabric-reinforced laminate sleeves are, first, expensive, because they require machining in manufacture, and, second, as already mentioned, easily damaged. In heavy duty operation, they burn out in 4-4.5 hours of operation on a semiautomatic welding machine [15]. As to glass-fibre-base laminate sleeves, special auxiliaries - moulds are required for their manufacture. The service life of such sleeves is seven working shifts in the average. The sleeves from asbestos cloth with sodium silicate binding have a higher resistance (112-115 hours) [15], however the manufacturing process is quite complicated.

The ceramic metal sleeves can be put to a separate category of insulating sleeves (Figure 4, c), consisting of a steel casing 1 (Figure 3, c), a steel threaded bushing 2 and a ceramic insulator 3 [1]. The nozzle 5 is fixed by means of a screw 4. Such mounting eliminates spontaneous shifting and distortion of the gas nozzle during welding. The manufacture of ceramic metal sleeves is relatively complex, and their service life is 600-800 hours.

With their remarkable resisting properties, the ceramic metal sleeves are prone to destruction as well. Due to the difference in thermal expansion coefficients of the materials, the sleeve is made of, and under the influence of mechanical actions while cleaning the nozzle from spatter, the ceramic insulator spalling takes place, the inner sleeve is loosened from the screw that is indicative of the ceramic metal sleeve losing its efficiency. At the moment, there are a number of compositions for stuff used in ceramic metal sleeves [16-19]. The main compounds in material are sodium silicate, kaolin, corundum and borax. The most widely-spread composition contains the following: sodium silicate electrode 57-62 %, corundum 28-30 %, kaolin 9-10 %, boric acid 0.5-1.5 % and borax 0.5-1.5 %.

One-piece design is known where the insulating sleeve is integral to the nozzle [20]. The insulating sleeve housing 1 (Figure 5) is rigidly connected to the heat-resistant insulation 2 and sleeve with internal thread 3.

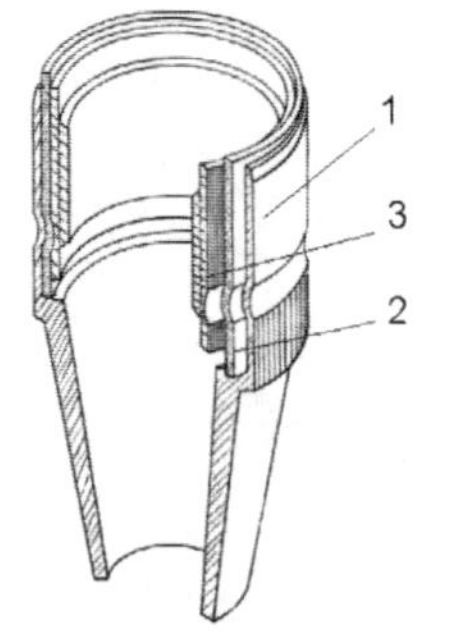

Figure 5. A Welding torch nozzle.

There is a modification wherein the insulating sleeve is not used at all. The nozzle [6], made of non-conductive heat-resistant material, is screwed directly onto the contact tube. Increased service life of the nozzle is achieved through higher thermal stability of the material and due to the possibility of replacing the contact tip upon its burning out.

3. Conclusion

Analysis of Russian and foreign findings shows that the main vulnerable elements of torches in mechanized carbon dioxide shielded welding are nozzles, insulating sleeves, contact tips and contact tubes. Their vulnerability causes their excessive consumption during the operation, and this certainly have influence on the final price of products. In addition, replacing wear parts and cleaning spatter off the nozzles lead to the loss of working time.

There are two ways to increase a service life of torch parts: the first is development of electrical and metallurgical techniques that reducing spatter; the second is development of materials and designs enabling the reduction of the amount of metal spatter that welds to and builds up on the working surface of torches.

References

[1] Fed'ko V T 1998 *Theory, technology and means for reducing spatter and content in carbon dioxide welding* (Tomsk: Publ. TPU).

[2] Dmitrik V V and Akulov A I 1991 *Reduced spattering of torch nozzles in carbon dioxide welding* (Svarochnoe Proizvodstvo (Eng: Welding production) vol 2) p 27.

[3] Belousov Yu G et al. 1968 *Torches for CO_2 welding with forced air-cooled* (Svarochnoe Proizvodstvo (Eng: Welding production) vol 11) p 46.

[4] Potapievskiy A G 2007 *Gas-shielded welding with consumable electrode. Part 1: Welding in the active gases* (Kiev: Ekotehnologija).

[5] Lobanov Yu N 1965 *Ceramic nozzles for carbon dioxide welding* (Svarochnoe Proizvodstvo (Eng: Welding production) vol 10) p 39.

[6] Naumov V I et al. 1968 *The modernization of semi-automatic machines for carbon dioxide welding* (Svarochnoe Proizvodstvo (Eng: Welding production) vol 8) p 43.

[7] Kostenko B A and Revenko V N 1967 *Glass-cloth nozzles for carbon dioxide welding* (Svarochnoe Proizvodstvo (Eng: Welding production) vol 5) p 43.

[8] Kostenko B A and Revenko V N 1968 *Nozzles based on andesite paste* (Svarochnoe Proizvodstvo (Eng: Welding production) vol 8) p 43.

[9] Ryzhkov V N, Kanel' L S et al. 1985 *Nozzles to the welding torches* A.S. 1136907 (USSR).

[10] Popov V A 1989 *Nozzles to the welding torches* (Svarochnoe Proizvodstvo (Eng: Welding production) vol 8) p 29.

[11] Dmitrik V V and Semenov A V 1991 *Heat-resistant coating* (Svarochnoe Proizvodstvo (Eng: Welding production) vol 10) p 46.

[12] Dmitrik V V et al. 1993 *On development of heat-resistant coating for nozzles and contact tubes of welding torches* (Svarochnoe Proizvodstvo (Eng: Welding production) vol 7) p 27.
[13] Danil'chenko B V et al. 1995 *Methods for reducing spatter from nozzles and contact tubes of welding torches* (Svarochnoe Proizvodstvo (Eng: Welding production) vol 9) p 24.
[14] Lysachenko I A 1971 *Electric and radio materials* (Moscow: Svjaz).
[15] Rayskiy E E and Oleynik N P 1971 *Insulating sleeves to semiautomatic torches for welding in carbon dioxide and its mixture with oxygen* (Avtomaticheskaya Svarka (Eng: Automatic welding) vol 2) p 69.
[16] Fed'ko V T, Bubenschikov Yu M et al. 1976 *The composition for the manufacture of insulating sleeves to the holders used in CO2 welding* A.S. 539853 (USSR).

[17] Fed'ko V T, Khramushin V A et al. 1981 *The composition for the manufacture of insulating sleeves for welding torches* A.S. 833764 (USSR).
[18] Fed'ko V T, Basalaev A I et al. 1984 *Insulating compounds for sleeves to the welding torches* A.S. 1099326 (USSR).
[19] Fed'ko V T, Tomas K I et al. 1999 *Composition for insulating sleeves to welding torches* Patent 2133717 (the Russian Federation).
[20] Tajbl F, Stener M 1970 *Hitzebeständige Düse für das Schweißen mit abschmelzender Elektrode* Auslegeschrift 1 440 432 (BDR).

IOP Conf. Series: Materials Science and Engineering **142** (2016) 012005 doi:10.1088/1757-899X/142/1/012005

The Influence of Shielding Gas Flow Rate on the Transfer Frequency of Electrode Metals Drops

D A Chinakhov, E G Grigorieva, E I Mayorova, D S Kartsev

Yurga Institute of Technology, National Research Tomsk Polytechnic University, 652055, Leningradskaya st., 26, Yurga, Kemerovo region, Russia

e-mail: chinakhov@tpu.ru

Abstract. Chemical composition of weld metal and properties of a joint weld are dominated by the processes occurring in the drop of molten electrode metal and in metal of the weld pool. Under certain conditions in consumable electrode welding with jet gas shielding the drop of electrode metal is influenced significantly both by main forces and the force of gas-dynamic impact of the shielding gas jet. The results of research have revealed that changing rate of shielding gas flow on the nozzle section influences the processes taking place in the welding area. There are changes in the transfer frequency of electrode metal drops and chemical composition of weld metal.

Introduction

Welding processes progress fast according to complex physical and chemical laws at high temperature. Diverse factors and phenomena dominate properties of joint welds. Formation of the structure and phase state of metal to be jointed depend on its chemical composition and conditions of thermal impact (mode and conditions of welding) [1-12]. Processes taking place in the drop of electrode metal and in metal of the weld pool are very important for properties of joint welds [1, 2, 3, 13]. The rate and character of physical and chemical processes in the drop of electrode metal influence properties of a weld and its chemical composition [12, 13].

The electrode metal transfer depends on various factors in the welding area. It is well-known [1, 3, 11] main forces like pressure force of arc plasma flows, surface tension force, reactive force of evaporating metal and emitting gas, gravitation force and electro-dynamic force effect on the drop of molten electrode metal when consumable electrode welding with traditional one-jet gas shielding. Components of these forces are dominated by the conditions of arc burning, electrode materials, welding current strength, arc voltage, traverse speed of filler wire and drop. The most forces depend on the size of electrode metal drop and its location relative to the weld pool. These forces, depending on their direction, can either block or further the transfer of electrode metal.

When consumable electrode welding with jet gas shielding in certain conditions the drop of electrode metal is influenced significantly both by main forces and the force of gas-dynamic impact of the shielding gas jet [2, 12], which depends on the mode and composition of gas shielding in the welding area, and the rate of shielding gas flow from the nozzle. Considerable influence of this force and the rate of gas flow have been revealed in conditions of two-jet gas shielding in CO_2 [12, 14].

IOP Conf. Series: Materials Science and Engineering **142** (2016) 012005 doi:10.1088/1757-899X/142/1/012005

Methodology

The force of shielding gas impact F_G depends on the rate of shielding gas flow from the nozzle; it is directed along the electrode towards the workpiece to be welded, and supports positioning of the drop along the electrode axis (Fig. 1).

The pressure on a drop is assumed to be equal to the pressure on the nozzle section in order to describe how the shielding gas jet influences the drop of electrode metal.

The gas pressure on the nozzle section was calculated by formula [15]:

$$P = \frac{bV^2}{2}, \quad (1)$$

where b – density of gas, kg/m^3, V – flow rate, m/s.

The flow rate was determined by formula:

$$V=G/S, \quad (2)$$

where G – consumption of shielding gas, m^3/s; S –area of nozzle section, m^2.

The force of shielding gas impact on the drop of electrode metal was measured by formula [12]:

$$F_G = Pc_x 2\pi r_k \sqrt{r_k^2 - r_w^2} \quad (3)$$

where P – pressure of shielding gas on a drop surface, Pa; r_k – a drop radius, m; r_w – a filler wire radius, m; c_x – aerodynamic drag coefficient of a sphere, c_x = 0.48.

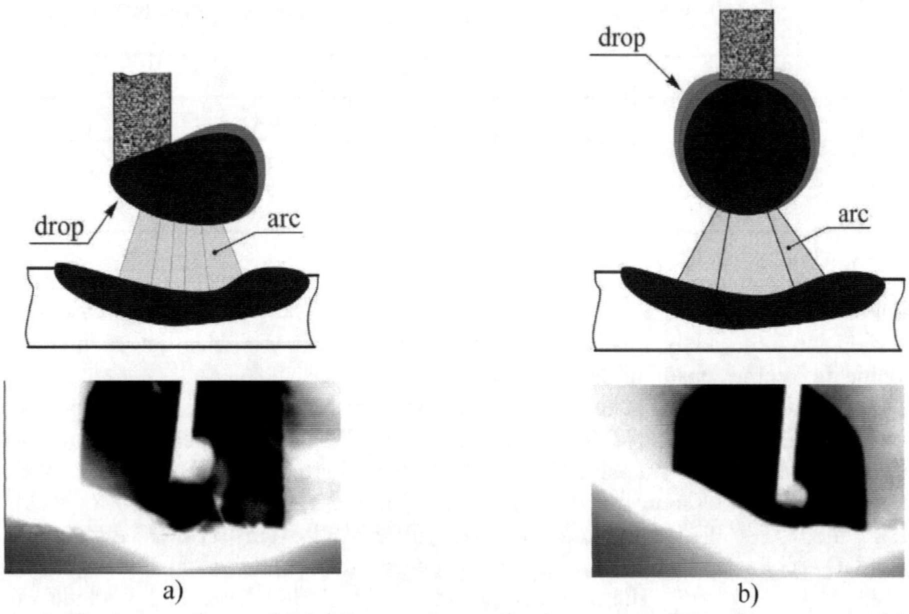

Figure. 1. The impact force of shielding gas jet on the drop: a) traditional one-jet gas shielding; b) two-jet gas shielding

The flow rate of shielding gas from the nozzle was determined, as well as the force of shielding gas impact on the drop of electrode metal was calculated. Comparative calculation was carried out for traditional one-jet (conic nozzle) and the proposed two-jet gas shielding (Tabl. 1). Shielding gas - CO_2, density of gas b =1.97 kg/m^3, consumption of gas Q = 20 l/min.

Results and Discussion

The results of calculation have stated that the rate of gas flow increases by 2.1 times when leaving the nozzle and the force of shielding gas impact augments by 4.39 times (similar consumption of gas) in conditions of double-jet shielding as against one-jet shielding (conic nozzle) [16, 17].

Table 1

The rate of gas flow on the nozzle section and the force of shielding gas jet impact

Method	Section area S, mm^2	Rate of flow V, m/s	Force of shielding gas jet F_G, $H*10^{-4}$
One-jet	173	1.93	5.02
Double -jet	82	4.05	22.04

An experiment was carried out to verify data obtained in calculations. In laboratory conditions rates of shielding gas flows were measured on the nozzle section by thermal mass flow meter Dwyer Series 471 provided that consumption of gas is different (Tabl. 2).

Table 2

Calculated and experimental rates of shielding gas

Consumption, l/min	Calculated gas rate, m/s		Experimentally obtained gas rate, m/s	
	Traditional	Double-jet	Traditional	Double-jet
10	0.97	2.02	0.94	1.96
20	1.93	4.05	1.83	3.95
30	2.89	6.07	2.7	5.82

Having compared calculation and experimental data, relative error was revealed not to exceed 10%, the developed model for calculation of shielding gas flow from the nozzle provides appropriate accuracy when solving tasks of steady-state reaction of axially symmetrical jet of shielding gas with the environment (volume consumption is different) [16, 17].

The determination of the influence of shielding gas flow rate on chemical composition of weld metal in consumable electrode welding in CO_2 required a fill-scale experiment. When experimenting, a weld bed was made on a 10 mm steel 40X plate by filler wire Sv-08G2S with diameter 1.2 mm in shielding gas CO_2. One-jet (conic nozzle) and double-jet gas shielding were used for welding a weld bed. Welding mode: I=150 A, arc voltage U = 24 V, filler wire extension L=12 mm, consumption of shielding gas Q=20 l/min. The experiments were carried out in similar conditions and modes, the speed of welding V = 3.3 mm/s. The source of power supply VS-300B, welding machine VD–1500.

In the process of experiments oscillograms of electric current and voltage were recorded by digital oscillograph Agilent Technologies DSO1012A.

The oscillograms show there is an increase in the gas-dynamic impact on the drop of electrode metal, as well as in the transfer frequency of drops from the filler wire into the weld pool due to the growing rate of gas flow from the nozzle, whereas metallurgical processes get less intensive and it takes less time for the drop to be transferred into the weld pool (Fig. 2).

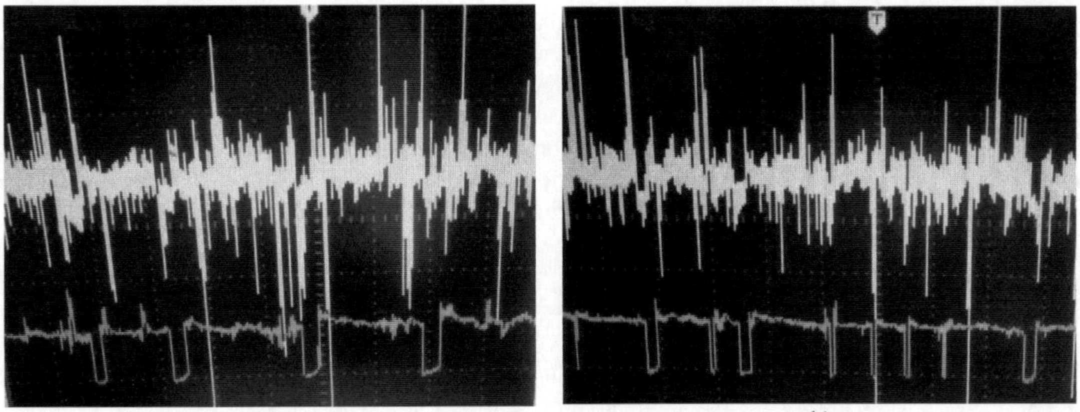

a) b)

Fig. 2. Oscillograms of electric current and voltage: a) traditional one-jet gas shielding; b) double-jet gas shielding

Tests were carried out to identify chemical composition of the deposited metal (Tabl. 3).

Table 3

Chemical tests of deposited metal

Mass concentration of elements, %					
C	Mn	Si	Fe	Cr	Ni
Traditional one-jet gas shielding					
0.12	0.92	0.23	97.89	0.42	0.10
Double-jet gas shielding					
0.12	0.85	0.20	97.98	0.41	0.10

The boiling temperature, as well as evaporation heat of manganese are the lowest ones (Tabl. 4) of all alloying elements in the chemical composition of filler wires. Therefore, in welding (filler wire Sv-08G2S) at approximate temperature 3300 K, it evaporates and oxidizes on the drop surface [13].

Table 4

Thermodynamic properties of elementary substances (normal conditions) in the chemical composition of filler wire [18]

Property	Substance					
	Mn	Si	Cr	Ni	Fe	C graphite
Density, g/cm^3	7.21	2.33	7.19	8.9	7.87	2.25
Melting temperature, K	1517	1688	2130	1726	1812	3820
Boiling temperature, K	2235	2623	2945	3005	3134	5100
Evaporation heat, kJ/mole	221	383	342	378.6	340	-

The increase in the flow rate and number of particles, reacting with the drop surface, results in high burn-out of silicon and manganese [2, 12, 13].

In welding with two-jet gas shielding the rate of gas flow is 2.1 times higher than with one-jet shielding in conditions of this experiment, so a lot of manganese is removed from the drop surface. Therefore, electrode metal with lower concentration of manganese is transferred into the weld pool. Increasing CO_2 consumption when welding leads to the decrease in manganese concentration in the drop in any conditions of gas shielding [12, 14].

IOP Conf. Series: Materials Science and Engineering **142** (2016) 012005 doi:10.1088/1757-899X/142/1/012005

A lot of researchers [1-5, 10, 13, 14 etc.] point out at more advantageous conditions for molten metal reaction with gases and slag even at the stage of a drop but not in the weld pool. It is stated in [13] specific surface of molten filler metal drops is 5-22 times more extended (relative to the size of drops) as against that of the weld pool, and the specific rate of their oxidation is approximately 39 times higher. In terms of the mentioned above a conclusion can be made it is chemical composition of weld metal that influences mainly the composition of filler metal drops. The change in chemical composition of weld metal results in modification of operational characteristics of a joint weld. Varying manganese concentration in weld metal has a significant effect on its plasticity and impact strength [12, 14, 18-19].

Double-jet shielding causes the increase in the force of active shielding gas impact on the filler metal drop and surface of the weld pool. When varying the gas-dynamic impact (consumption and rate of gas flow), transfer of filler metal drops, stability of the welding process, as well as chemical composition of weld metal, thermal and other processes in consumable electrode welding can be controlled, and required properties of joint weld can be formed [2, 14, 16, 17].

Conclusion

To sum up, in consumable electrode welding in CO_2 the rate of gas flow from the nozzle has a significant effect on the drop transfer and chemical composition of weld metal. Accurately selected gas shielding method (double-jet shielding) and consumption of gas allows of controlling the filler metal transfer and chemical composition of weld metal, making the process of welding uninterrupted, and forming necessary properties of joint welds.

The work is performed with financial support of the Russian Foundation for Basic Research. №16-38-00194 mol_a.

References

[1] Welding and welded materials: In 3 v. V. 1. Weldability of materials: reference book / edited by E.L. Makarov. Metallurgy, 1991 528 p.
[2] Chinakhov D.A., Zuev A.V., Filimonenko A.G. Gas-dynamic Impact of a Shielding Gas Jet on the Drop Transfer When Welding with a Consumable Electrode. Advanced Materials Research Vol. 1040 (2014) pp. 850-853.
[3] Lenivkin V.A., Dyurgerov N.G., Sagirov H.N. Technological properties of a welding arc in shielding gases. M.: Machinebuilding. 1989. 264 p.
[4] Shorshorov M.H. Metal science of steel and titanium alloys. «Science» Press. 1965. 337p.
[5] Livshits L.S., Khakimov A.N. Metal science of welding and thermal treatment of joint welds. – 2nd revised and enlarged edition. Machinebuilding. 1989. 339p.
[6] Brunov O.G., Solodskii S.A, Zelenkovskii A.A. Conditions of arc ignition in welding in shielding gases. Welding International. 2012. V. 26. 9. P. 710-712
[7] Krampit A.G. Calculation of the parameters of narrow gap pulsed arc welding of the root layer // Welding International. Volume 28, Issue 7, July 2014, P. 547-550.
[8] Frolov V.V. Theory of welding processes. M.: Higher school. 1988. 559p.
[9] Rykalin N.N. Calculations of thermal processes in welding. Mashgis. 1951. 296 p.
[10] Beresovsky B.M. Mathematical models of arc welding: in 7 v. V. 4. Fundamentals of thermal processes in welded workpieces. Chelyabinsk: YuUrGU Press. 2006. 547 p.
[11] Brunov O.G. , Solodskii S.A. Physico-mathematical modelling of the transfer of electrode metal droplets into the weld pool. Welding International. 2009. V. 23. 12. P. 930-933.
[12] Chinakhov D.A. Calculation of Gas-dynamic Impact of the Active Shielding Gas on the Electrode Metal Drop in Gas Jet Shielded Welding. Applied Mechanics and Materials. Vol. 379 (2013). Pp. 188-194.
[13] Novozhilov M.N. Fundamentals of metallurgy of arc welding in gases. Machinebuilding. 1979. 231p.

[14] Chinakhov D.A. Gas Dynamic Control of Properties of Welded Joints from High Strength Alloyed Steels. China Welding. Vol. 23 No.3 (2014) pp. 27-31.
[15] Novikov O.M., Rad'ko E.P., Ivanov E.N., Ivanov N.S. Development of a new technology of shielding gases arc welding on the basis of gas flows pulsations and ionization potentials. Welder-professional. 2006. 6. P. 10–13, 16.
[16] Chinakhov D.A., Chinakhova E.D., Gotovschik Y.M., Grichin S.V. Influence of welding with two-jet gas shielding on the shaping of a welding joint. IOP Conf. Series: Materials Science and Engineering 125 (2016) 012013. doi:10.1088/1757-899X/125/1/012013.
[17] Chinakhov D.A., Grigorieva E.G., Mayorova E.I. Study of gas dynamic effect upon the weld geometry when consumable electrode welding. IOP Conf. Series: Materials Science and Engineering 127 (2016) 012013. doi:10.1088/1757-899X/127/1/012013.
[18] Bigeev A.M. Steel metallurgy. Study guide, 2nd edition, Moscow, Metallurgy, 1988.
[19] Prokopenko S.A., Ludzish V.S., Kurzina I.A., Sushko A.B. Results of industry testing of multiple use rock-cutting picks// Gornyi zhurnal/ Mining Journal, 2015, №5, pp.67-71.

IOP Conf. Series: Materials Science and Engineering **142** (2016) 012006 doi:10.1088/1757-899X/142/1/012006

Effect of Some Overlay Welding Regime With Longitudinal Magnetic Field on Hardness, Phase Composition And Welded Layer Wear By Arc Method With Flux Metal Wire

D G Nosov, V V Peremitko, M H Barashkin

Dneprodzerzhinsk State Technical University
51918, Dneprodzerzhinsk, Ukraine, Str. Dneprostroevskaya 2,
tel. (+380569) 55-18-91

e-mail: ndweld@rambler.ru

Abstract. The paper defines the range of overlay welding current, frequencies and induction of a longitudinal magnetic field that enhance the wear resistance of welded layer additing the flux. The conditions of their mutual influence on the process of structure formation are stated as well as the mathematical models linking the overlay welding current, frequency and induction of a longitudinal magnetic field with hardness, wear resistance and phase composition of the welded layer, the use of which will allow to determine the welding modes to provide the necessary properties of the weld metal.

1. Introduction

The transition to automated production processes in engineering is associated with an increase in the number of interconnected machines and units operating in continuous movement, an unplanned stop of which leads to an enormous production loss [1]. During their operation the parts and components are subjected to considerable dynamic, heat, static, vibratory, and other loads, which are influenced by changing their geometrical dimensions. Therefore, the problem of increasing the wear resistance is one of the most important in engineering [2]. Restoration of the disrupted parts is carried out by means of welding. Thus the improvement of already known welding surfacing methods to provide the wear-resistance increase is very important.

One of the promising methods is weld facing with the solid cross-section under the flux layer using a longitudinal magnetic field [3,4]. This method allows to increase the weld facing productivity, to influence the geometric dimensions, the phase composition, hardness and wear resistance.

The aim is to develop mathematical models and to determine the hardness and wear resistance of the phase composition of the welded layers according to the weld facing regimes with the effect of a longitudinal magnetic field.

2. Problem setting

The object of the research is the process of arc welding surfacing adding flux with solid cross-section wire and the additional impact on the arc of an external electromagnetic field.

The subject of the research is the influence of welding surfacing parameters, frequency and induction of longitudinal magnetic field on the phase composition and properties of the welded layer.

The aim of this work is to study the joint effect of welding current (arc thermal power), induction and frequency of longitudinal magnetic field on hardness, wear resistance and phase composition of the welded layer adding flux.

3. Materials and methods

Experimental weldings were performed on carbon steel sheets St3sp GOST 380 (i.e. all-Union State Standard) adding flux AN-348A GOST 9087 with the impact of longitudinal magnetic field induction $B_z = 0 \ldots 80$ mT, frequency ω 0 to 50 Hz, wire of solid cross-section of Sv-08G2S GOST 2264 with a diameter of 3 mm. An external magnetic field was generated by a special device, consisting of a solenoid coil and its power supply unit [5]. Welding equipment: unit set of UD-209 type, power supply – universal rectifier VDU-506 (i.e. Multi-operated Arc Rectifier-506). Additional equipment: analog teslametr, EM-5511 digital multimeter, testing wear machine MI-1M, laboratory analytical balances, stationary hardness tester, microscope. The wear was evaluated by the gravimetric method. Method of hardness measuring (Brinell hardness) was performed according to GOST 9012.

When setting an experimental part of the research a non-linear planning on the matrix of non-composite plan of second order for three factors has been implemented. Experimental data processing were performed in the analytic STATISTICA system [6].

4. The results

In the research the following welding conditions have been imposed: $I_n = 460 \ldots 480$ A; $U_\partial = 26 \pm 1$B; rate of vapour deposition $v = 18$ m / h; direct current of reverse polarity. Metal thickness 10^{-2} m, diameter of the wire electrode $3 \cdot 10^{-3}$ m.

Electrical stickout l_e, and the rate of welding surfacing v weren't changed. Such independent command variables were chosen: welding current *In*, and frequency induction B_z, frequency of longitudinal magnetic field ω. Hardness *H*, wear-resistance ε and structure determined by Macrosection.

Table 1 – Experimental record

Trial №	B_z, mT	ω, Hz	*In*, A	*H*, HB	ε, units	*ferrit*,%	*perlite*, %
1	+	+	0	147	1.36	75	25
2	+	–	0	166	1.80	65	35
3	–	+	0	133	1.18	62	38
4	–	–	0	138	1.24	63	37
5	0	0	0	176	1.91	65	35
6	+	0	+	190	2.04	70	30
7	+	0	–	176	2.11	72	28
8	–	0	+	146	1.28	63	37
9	–	0	–	143	1.26	62	38
10	0	0	0	174	1.88	60	40
11	0	+	+	162	1.72	65	35
12	0	+	–	160	1.68	66	34
13	0	–	+	166	2.04	71	29
14	0	–	–	164	2.02	72	28
15	0	0	0	174	1.88	65	35

According to the data obtained during the experiment (Table 1) in the analytic STATISTICA 6.0 system, three-dimensional graphs of dependencies for hardness and wear resistance were shown.

Figure 1 shows the dependency diagrams of the hardness of the welded layer H from the frequency ω and the induction B_z of the magnetic field. The area of maximum (170…180 HB) hardness data are formed when the magnetic field B_z = 45…80 mT and frequency ω = 5…35 Hz.

Figure 2 shows a graph of the influence of the magnetic field and surfacing current on hardness of welded layers. As you can see, the dependency is complex, and the hardness of the weld beads gets maximum data (170…180 HB) under the magnetic field induction of B_z = 45…80 mT and a surfacing current of I_n = 470…480 A.

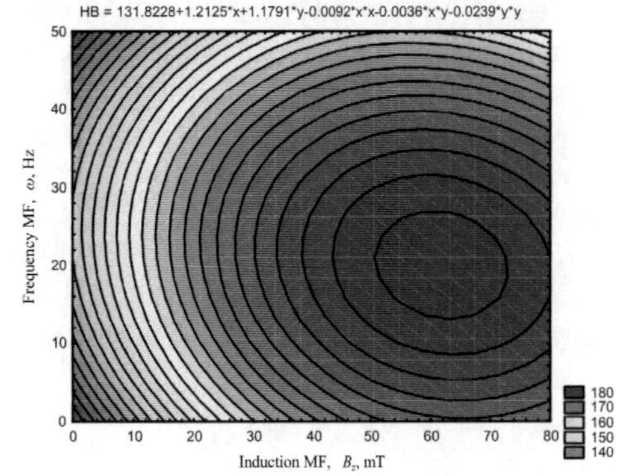

Figure 1 – Dependence of the hardness H of the build-up layer from the frequency ω and the induction B_z of the magnetic field

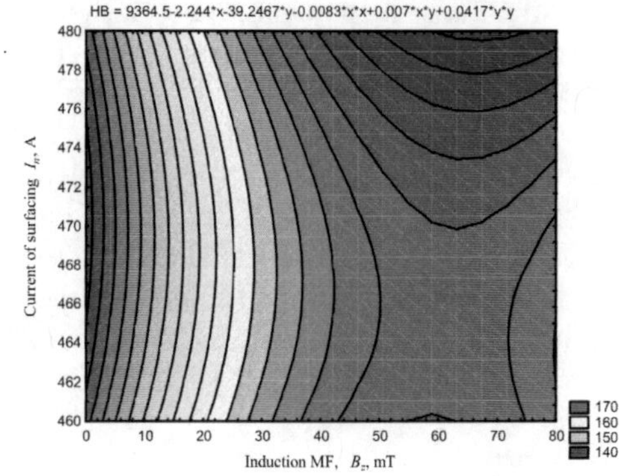

Figure 2 – Dependence of the hardness H of the welded layer from current I_n and induction B_z of the magnetic field

Figure 3 shows the effect of welding current and frequency of longitudinal magnetic field on hardness of welded layers. The dependency is nonlinear. Maximum hardness (H = 170 HB) is observed at welding surfacing with the influence of longitudinal magnetic field of ω = 15…30 Hz frequency at a current surfacing I_n = 460…465 or 470…480 A.

That is, to ensure the maximum hardness of weld metal we should use the following regimes: the magnetic field of B_z = 45…80 mT, the magnetic field frequency ω = 15…30 Hz and welding current I_n = 470…480 A.

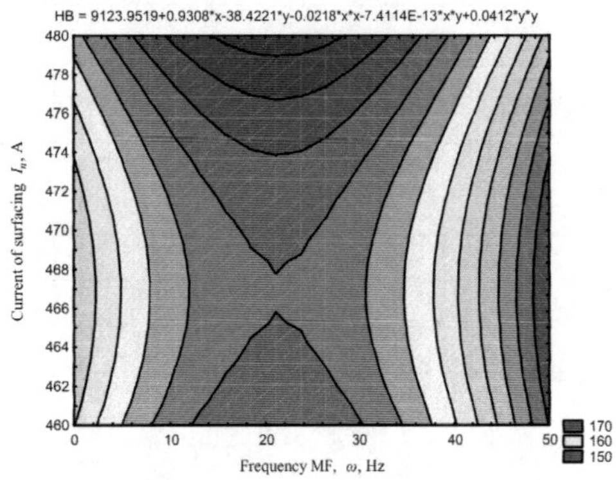

Figure 3 – Dependence of the hardness H of the welded layer from the welding current and magnetic field frequency ω

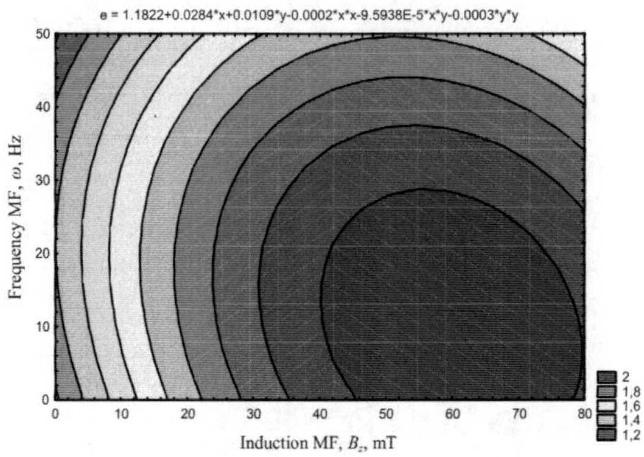

Figure 4 – Wear resistance ε dependence of the welded layer from the frequency ω and the induction B_z of the magnetic field

Figure 4 shows the effect of induction and frequency of longitudinal magnetic field on the durability of welded layers. It is evident that the increase of induction and frequency of longitudinal magnetic field results in increase of the wear-resistance data. The area of maximum wear-resistance data of the weld bead is formed at the induction of longitudinal magnetic field B_z = 45…75 mT, frequency ω = 0…25 *Hz*.

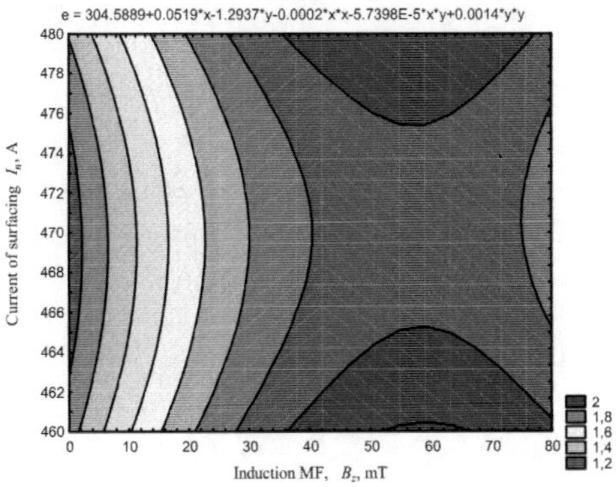

Figure 5 – Wear resistance ε dependence of welded layer from surfacing current I_n and induction B_z of the magnetic field

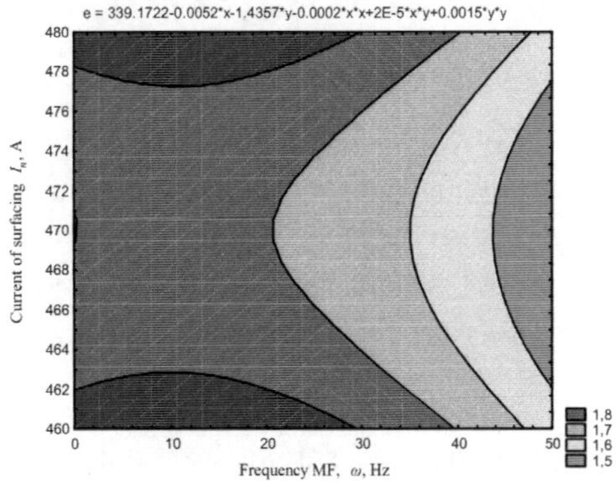

Figure 6 – Wear resistance ε dependence of welded layer from surfacing current I_n and frequency of the magnetic field ω

Figure 5 shows the wear resistance ε of welded layers from surfacing current I_n and the induction of the magnetic field B_z. As you can see, the dependency is complex and has two extreme zones. So, when surfacing with the influence of longitudinal magnetic field the durability of beads has maxima at the following modes: the magnetic field B_z = 45…75 mT, welding current I_n = 460…465 A; the magnetic field of B_z = 45…75 mT, surfacing current *In* = 475…480 A.

Figure 6 shows the effect of welding current and frequency of the magnetic field on the wear resistance. Wear resistance of welded layers obtaines the maximum figures under the following conditions: frequency of the magnetic field ω = 0…20 Hz and the current surfacing I_n = 460…465 A; frequency magnetic field ω = 0…20 Hz and welding current I_n = 475…480 A.

That is, to ensure maximum wear resistance of the welded metal figures we should use the following mode: the magnetic field of B_z = 45…75 mT, the magnetic field frequency ω = 0…20 Hz and welding current I_n = 460…465 or 475…480 A.

Thus, for maximum hardness and wear resistance figures of the welded metal, we should use the following mode: the magnetic field of B_z = 45…75 mT, the magnetic field frequency ω = 15…20 Hz and field current surfacing I_n = 475…480 A.

The regression method was used to develop mathematical models for predicting the hardness and wear resistance of the welded metal. The equation of response surface, reflecting any monitored parameter *y* can be expressed as $y=f(B_z,\omega,In)$, where v1 represents B_z, $v_2 - \omega$, $v_3 - In$, and the selected dependency is a response surface of the second order:

$$v = b_1 + b_2 \cdot v_1 + b_3 \cdot v_2 + b_4 \cdot v_3 + b_5 \cdot v_1 \cdot v_2 + b_6 \cdot v_1 \cdot v_3 + b_7 \cdot v_2 \cdot v_3 + b_8 \cdot v_1^2 + b_9 \cdot v_2^2 + b_{10} \cdot v_3^2 \quad (1)$$

The analysis of the experimental results in the calculation of the regression coefficients which were performed using STATISTICA 6.0 software package. The regression coefficients are given in Table. 2.

Table 3 – Coefficients of the regression equation (1) for the revocation of H and ε

		H	ε		H	ε
Coefficients	b_1	6858.7770	279.1011	b_6	0.007014	– 0.000057
	b_2	– 2.096000	0.054854	b_7	0	0.000020
	b_3	1.160345	0.000762	b_8	– 0.009068	– 0.000223
	b_4	– 28.611500	– 1.184810	b_9	– 0.023533	– 0.000242
	b_5	– 0.003570	– 0.000096	b_{10}	0.030417	0.001262

The obtained equations for determining the hardness and wear resistance indicators of the welded metal depending on the induction of the magnetic field and the frequency of welding current:

$$H = 6858.777 - 2.096 \cdot B_z + 1.160345 \cdot \omega - 28.6115 \cdot I_n - 3.57 \cdot 10^{-3} \cdot B_z \cdot \omega + \\ + 70.14 \cdot 10^{-4} \cdot B_z \cdot I_n + 90.68 \cdot 10^{-4} \cdot B_z^2 - 23.533 \cdot 10^{-3} \cdot \omega^2 + 30.417 \cdot 10^{-3} \cdot I_n^2 \quad (2)$$

$$\varepsilon = 279.1011 + 54.854 \cdot 10^{-3} \cdot B_z + 7.62 \cdot 10^{-3} \cdot \omega - 1.18481 \cdot I_n - 9.6 \cdot 10^{-5} \cdot B_z \cdot \omega - \\ -5.7 \cdot 10^{-5} \cdot B_z \cdot I_n + 2.0 \cdot 10^{-5} \cdot \omega \cdot I_n - 22.3 \cdot 10^{-5} \cdot B_z^2 - 24.2 \cdot 10^{-5} \cdot \omega^2 + 12.62 \cdot 10^{-4} \cdot I_n^2 \quad (3)$$

According to (Table 1) for the response function $y=f(B_z, \omega, I_n)$ phase composition (%) Ferrite (F) and pearlite (P), the following regression coefficients were obtained (1):

Table 4 - Coefficients of the regression equation (1) for the response function F and P

		F	P		F	P
Coefficients	b_1	6306.005	– 6206.01	b_6	– 0.001895	0.001895
	b_2	0.893602	– 0.893602	b_7	0	0
	b_3	– 0.315101	0.315101	b_8	0.000341	– 0.000341
	b_4	– 26.59460	26.59458	b_9	0.003733	– 0.003733
	b_5	0.002818	– 0.002818	b_{10}	0.028333	– 0.028333

The equations are calculated to determine the effect of induction, magnetic field frequency and welding current on the phase composition:

$$F = 6306.005 + 0.8936 \cdot B_z - 0.3151 \cdot \omega - 26.5946 \cdot I_n + 281.8 \cdot 10^{-5} \cdot B_z \cdot \omega - \\ - 189.5 \cdot 10^{-5} \cdot B_z \cdot I_n + 34.1 \cdot 10^{-5} \cdot B_z^2 + 373.3 \cdot 10^{-5} \cdot \omega^2 + 283.33 \cdot 10^{-4} \cdot I_n^2 \quad (4)$$

$$P = -6206.01 - 0.8936 \cdot B_z + 0.3151 \cdot \omega + 26.59458 \cdot I_n - 281.8 \cdot 10^{-5} \cdot B_z \cdot \omega + \\ + 189.5 \cdot 10^{-5} \cdot B_z \cdot I_n - 34.1 \cdot 10^{-5} \cdot B_z^2 - 373.3 \cdot 10^{-5} \cdot \omega^2 - 283.33 \cdot 10^{-4} \cdot I_n^2 \quad (5)$$

Analysis of equations (4,5) showed that with increasing frequency and induction of the magnetic field, an increase in amount of ferrite at the magnetic induction data B_z = 70…80 mT, the frequencies of the magnetic field of ω = 35 ... 50 Hz F with ferrite index up to 70…72% is also observed. With the reduction of the induction magnetic field within B_z = 0…25 mT at the welding current within I_n = 460…480 A, we observe the increase in the amount of perlite.

Using the equations 4 and 5 we can predict the phase composition of the welded metal.

One of the factors that increases the wear resistance is the reduction of the welded metal grain. To determine the effect of longitudinal magnetic field on the grain size we estimate the grain size according to GOST 5639 for microsections.

Fig. 7 shows the scale distribution of welded samples of grain (welding current I_n = 480 A) without external influence of a magnetic field (a); B_z = 40 mT, ω = 25 Hz (b); B_z = 80 mT, ω = 25 Hz (c); for the induction of B_z = 40 mT, frequency ω = 50 Hz, welding current I_n = 460 A (d).

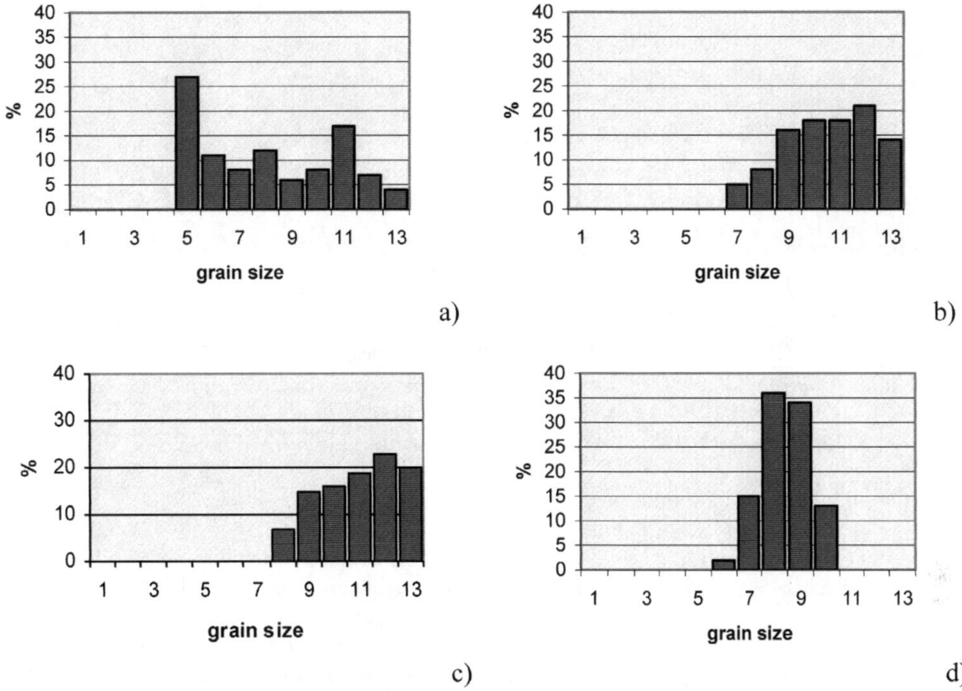

Figure 7 – The distribution scale of welded grain samples

5.Conclusions

In the analysis of experimental data we have got the nonlinear mathematical model of induction, magnetic field frequency and welding current dependence on hardness and wear resistance while surfacing that allow us to determine the effect of induction, magnetic field frequency and welding current on hardness and wear resistance of welded layers.

It was determined that for maximum wear resistance of the welded metal layer, we should use the following regimes: the magnetic field of B_z = 45…75 mT, the magnetic field frequency ω = 15…20 Hz and welding current In = 475…480 A.

References

[1] Gas-shielded welding of steels with consumable electrode. Engineering and technology of the future // Monograph / A.G. Potapyevskiy, Yu.N. Saraev, D.A. Chinakhov; Yurga Institute of Technology. – Tomsk: Tomsk polytechnic University Press, 2012. – 208 p.

[2] Nosov D.G., Razmyshlyaev A.D. effectiveness of application of combined magnetic fields in submerged arc welding / D.G.Nosov, A.D.Razmyshlyaev // The Paton Welding Jornal / № 4, 2009 (April), pp 16-20.

[3] Nosov D.G., Maltsev V.V. The influence of magnetic fields by a melting rate of wire for arc surfacing under flux. Applied Mechanics and Materials, 2013, no 379, pp. 178-182.

[4] Nosov D.G., Peremitko V.V. Influence of Frequency and Induction of Longitudinal Magnetic Field on The Electrode Metal Loss and its Spattering During MAG-Welding / D.G.Nosov, V.V.Peremitko // IOP Conf. Series: Materials Science and Engineering 91 (2015) 012011. – p 1–8.

[5] Nosov D.G. Power source of electromagnetic system for arc welding using external magnetic fields control / D.G.Nosov // Welder. -2010. -№ 4. - pp.18-19.

[6] http://www.statsoft.ru/home/textbook/default.htm

Effect Of Dynamic Characteristics of Power Supplies on Aerosol Composition While Welding With Coated Electrodes

D P Il'yaschenko[a], D A Chinakhov, I D Sadikov

Yurga Institute of Technology, TPU affiliate
Leningradskaya 26, Yurga, Russian Federation, 652055

[a]e-mail: mita8@rambler.ru

Abstract. In the context of a significant increase in production output and use of welding technologies in the manufacturing of engineering products the problem of hygienic characteristics of working conditions in arc fusion welding is becoming increasingly important. The work represents how the dynamic characteristics of a power supply affect the transfer of alloying elements from a coated electrode into a base metal, a slag phase and a solid component of welding fumes.
Short-circuit current limiting in inverters reduces overheating of electrode metal drops by 15%; welding fumes quantitative component - to 38%; manganese - to 30%; thermal radiation intensity - by 37%.

1. Introduction

While manual metal arc welding (MMA), smoke pollution of the working area atmosphere, intense optical radiation and electrode metal sprinkling cause the greatest harm to the health of workers [1]. In the arc welding course reacting of the molten metal with slag and gas occurs, thereby forming welding fumes, consisting of solid particles and gas phase. The impact of welding fumes on the organism causes occupational diseases of welders. [1,2] examine how composition of welding fumes affect the following factors: electrode coating composition; welding conditions (current and voltage); current type and polarity; composition of a base metal and an electrode; electrode coating thickness; electrode diameter.

Optical radiation is the result of high-powered heat source combustion and includes infrared and ultraviolet radiation. In the same environment the optical radiation radiates in a direct line; when interacting with the environment it can be absorbed, reflected, refracted, undergo diffraction, interference and polarization [3,4].

A problem of electrode metal scattering now takes on new significance in connection with the large-scale implementation of air purification facilities in welding areas [5]. Air-intake systems are arranged at 0.25-0.40 m distance from the welding arc to work effectively. Almost all the small drops of up to 1 mm and larger, emitted toward air scoop, can be trapped into the system of dust extraction and purification. This significantly increases the mass loading on the filter, exceeding a design smoke loading, and causes a permanent threat of filters melting or ignition.

The factors mentioned above have negative impact on the respiratory system, sight sensors, skin cover and immune system.

In recent years, [6], inverters are actively engaged in the market of power supplies for MMA. Their advantages are reduction in weight-and-size and cost of power sources as well as a significant improvement of control and quality of manufacturing methods.

2. Methods and results

The purpose is to investigate hygienic characteristics of manual arc welding with coated electrodes with the aid of different types of power supply (inverter rectifier Nebula - 315, diode rectifier VD - 306). Tests have shown [7] that using the inverter power supply changes the arc thermal effect on the drop of the electrode metal. Computations, performed by analytical method [8], show that, when using the inverter power source, an average amount of energy is Q_d = 0,13x107J per drop of the electrode metal, and when using a diode rectifier - Q_d = 0,156x107J.

Calculation of overheat temperature of the drops, during their passage through the arc gap with short circuits, is determined by the formula [9.10]:

$$\Delta T_k = \frac{1}{c}\left(\frac{q_1}{a \cdot \tau_{k.z.}^2} - \frac{1}{K}\right). \qquad (1)$$

where $\Delta T_к$ - overtemperature of the molten metal on the electrode over the melting point, k; c – average specific heat of the liquid metal, J/K (0,84); q_l – thermal capacity of the arc at the electrode end, J/c; K – coefficient characterizing the mass of metal that can be melted with an energy unit, g/J ($1{,}5 \cdot 10^{-3}$); $\tau_{k.z.}$ - time of a short circuit of the arc gap with a drop, ms: $\tau_{k.z.}$=6,37ms for the diode rectifier; $\tau_{k.z.}$ =5,37 ms for the inverter power source [9].

Overtemperature $\Delta T_к$ of electrode metal drops (average value exceeding the melting point) is $\Delta T_к$ = 410° C while welding with the inverter power supply, and 480° C with the diode power supply. Overheating of the electrode metal [1, 2, 11] leads to changes in metallurgical processes in the drop, and as a consequence, change of the welding fume chemical composition. Thus the type of power supply affects the hygienic characteristics of the air in the welding area.

To confirm this, we conducted a comprehensive laboratory testing. Two pipe samples (welded joint C17 according to GOST 5264-80) 159x6 from 09Г2С steel were welded with electrodes with use of the diode rectifier ВД-306 and the inverter power supply Nebula-315ю in the following manner: root - LB-52U (d = 2,6 mm), welding current I = 50-60 A, backfills - LB-52U (d = 3.2 mm), welding current I = 80 - 90 A.

While MMA, dust and gases gross emission was investigated in laboratory conditions (welding was performed on 09Г2С steel samples using different types of power sources and electrodes coated with LB-52U base coating). We determined the amount of dust, emitted while welding, and its chemical composition, a content of manganese dust, qualitative and quantitative composition of the dispersion medium of resulting welding fumes. Air sampling to determine the level of air pollution was taken in the welder's breathing zone (the distance from the arc was 55 cm). For the investigations the following equipment was used: an aspirator for air sampling, model 822; an aneroid БАММ -1; an aspiration psychrometer MB-4M; an analyzer «ЭЛАН-СО-50».

Table 1 shows components of welding fumes emitted while MMA (exhaust ventilation is switched off) using inverter and diode power supplies

It is obvious that, when using the inverter welding power source, fumes and manganese concentration is lower, and therefore, risks of toxic poisoning and threats of respiratory diseases is reduced for workers in the welding area. The inverter power supply limits short-circuit current, Figure 1.2 ($I_{max\ diode\ rectifier}$ = 140A;. $I_{max\ inverter}$ = 123A) resulting in less burnout of alloying elements (Table 2) Si to 0.08% and Mn to 0.2%, as confirmed by the author in [13].

Table 1 - Emissions of dust. gases and other impurities while MMA using different types of power supply

Terms of sample selection	Name of element	Unit of measurement	Power supply		Maximum permissible concentration
			Inverter	Diod rectifier	
Materials LB - 52U Ø 3.2 mm. Steel 09Г2С	carbon monoxide	mg/m^3	0.05	0.05	20.0
	nitrogen dioxide		below 0.6	below 0.6	2.0
	hydrogen fluoride		below 0.02	below 0.02	0.5
	Welding fumes		2.2±0.5	3.6±0.9	
	chromic anhydride		below 0.003	below 0.003	0.01
	chromium oxide		below 0.5	below 0.5	1.0
	Manganese		0.10±0.03	0.15±0.03	0.6

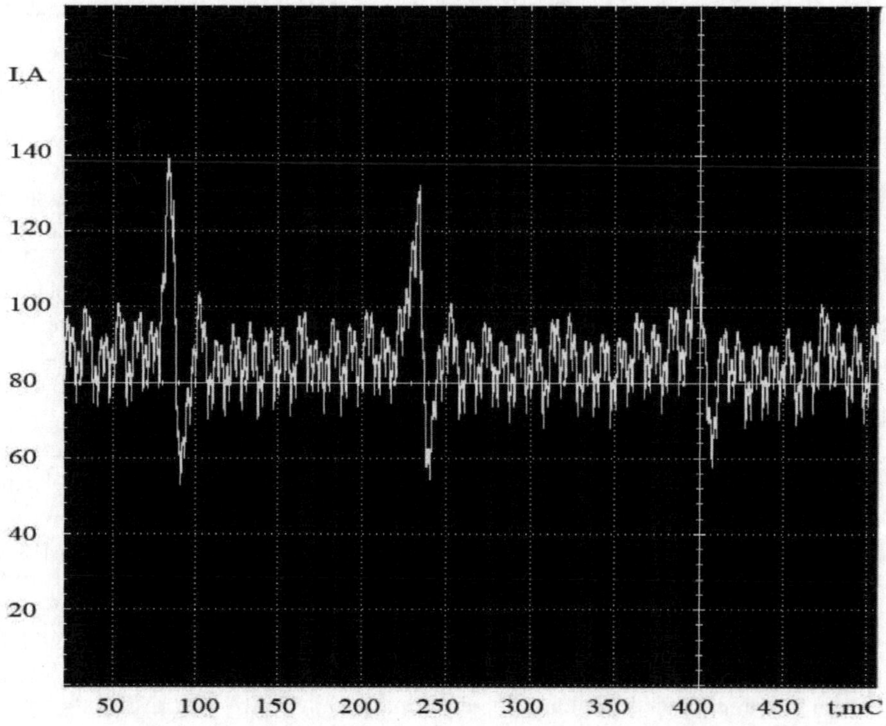

IOP Conf. Series: Materials Science and Engineering **142** (2016) 012007 doi:10.1088/1757-899X/142/1/012007

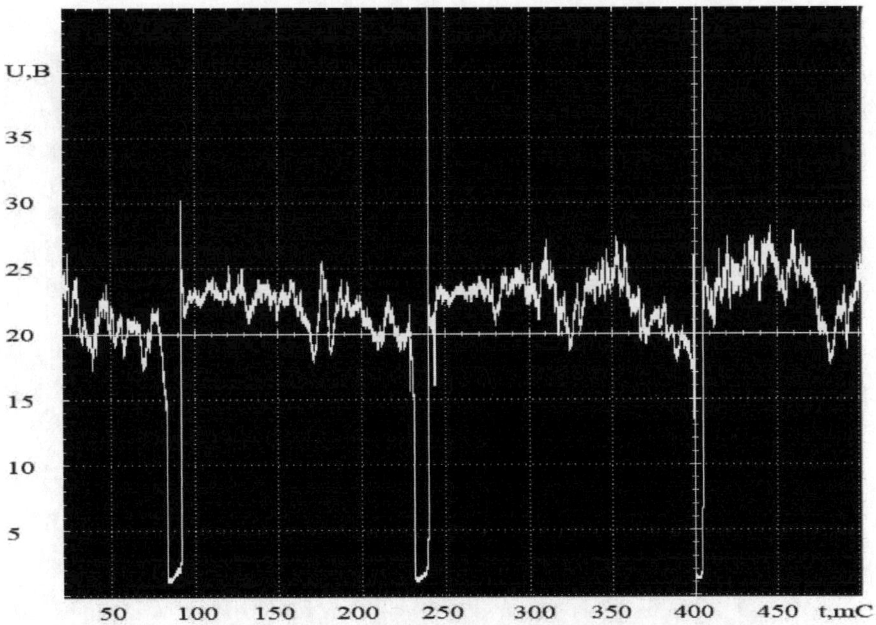

Figure 1 - Waveforms of voltage and current (electrodes LB-52U, 3,2 mm diameter) - diode rectifier

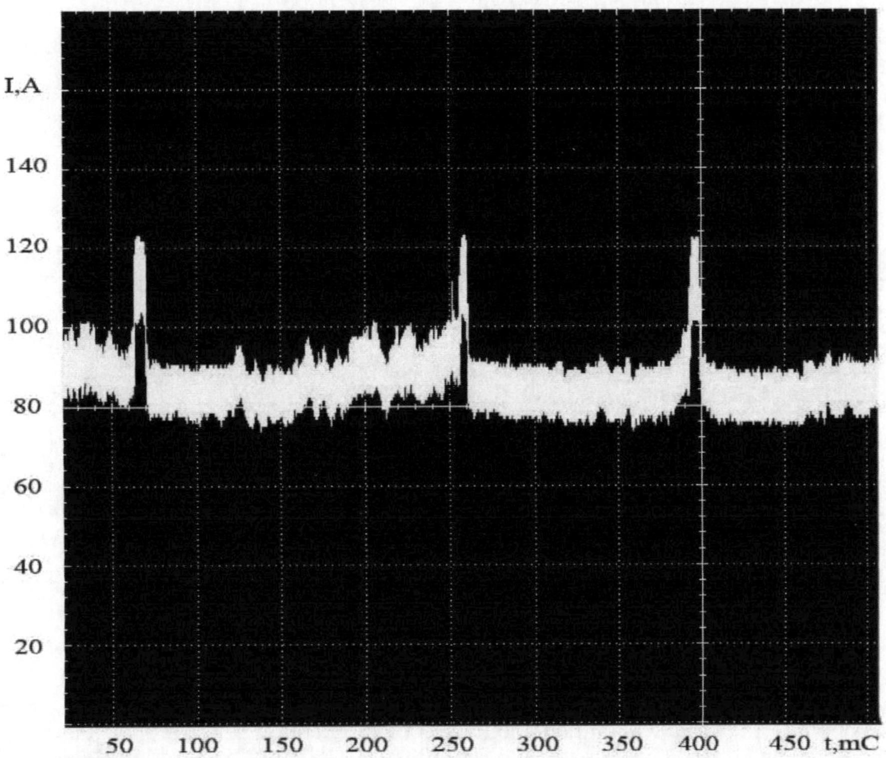

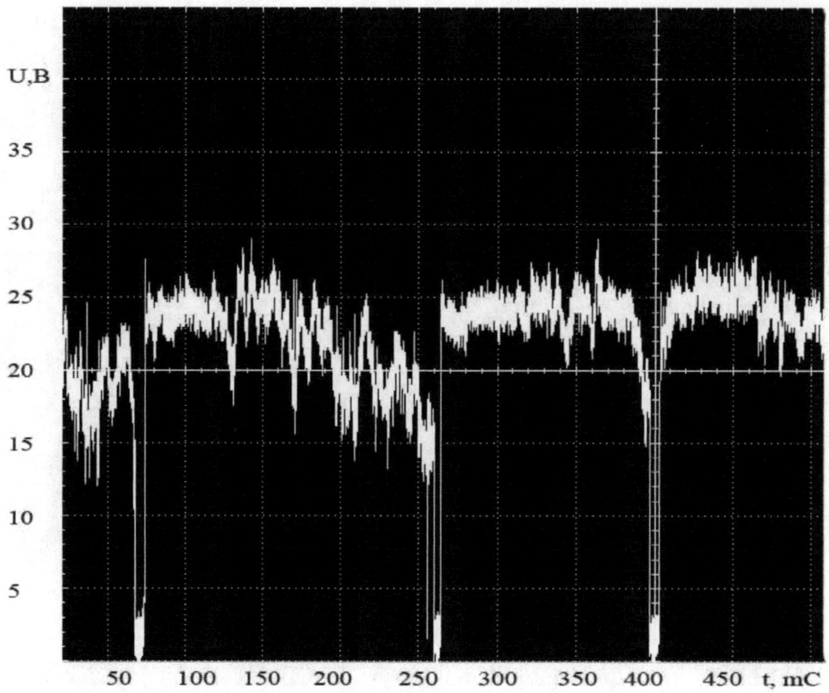

Figure 2 - Waveforms of voltage and current
(electrodes LB-52U, 3,2 mm) - inverter

Table 2 - Chemical composition of weld joints, pipe Ø159 × 6 (09Г2С steel), LB 52U electrodes

Power supply	Element content, mass %							
	C	Si	Mn	S	P	Cr	Ni	Cu
Diode rectifier	0.10± 0.012	0.52± 0.03	1.03± 0.05	0.010± 0.002	0.014± 0.003	0.03± 0.01	0.05± 0.01	0.03± 0.008
Inverter	0.09± 0.005	0.60± 0.03	1.23± 0.05	0.010± 0.002	0.014± 0.003	0.03± 0.01	0.06± 0.01	0.03± 0.008

Table 3 - Average chemical composition of slag

Electrodes	Power supply	CaO, %	SiO_2, %	TiO_2, %	NbO, %	MnO, %	Fe_2O_3, %	Cr_2O_3, %	Al_20_3, %
LB-52U	Diode rectifier	38.66	25.37	9.57	0.10	7.21	13.89	0.17	3.61
	Inverter	36.27	24.187	8.74	0.05	7.48	18.31	0.15	3.66

Analysis of the readings (Tables 2 and 3) shows an increase in mass fraction of the alloying elements: Mn - 0,2%, Si - 0,08% in the weld metal, and a decrease in the proportion of oxides (SiO2, MnO) in the slag phase if using the inverter. This can be explained by the different heat content in molten metal drops while using inverter or diode power supply; that is confirmed by formula 1.

Heterogeneous equilibrium reactions occur between the slag ZnO, SiO_2 and the base metal, which depend on heat content of and temperature of drops (reactions are dissimilar when using different power supplies).

Effect of power supply's energy parameters, on irradiance while welding, was performed with use of radiometer "Кварц-41 – РАТ-2П " (Figure 3.).

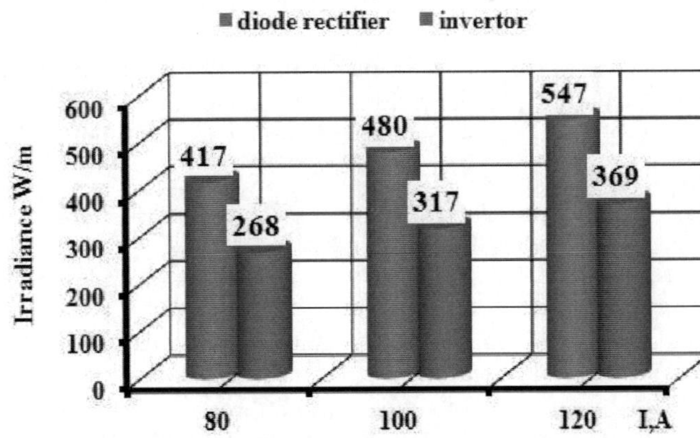

Figure 3 - Dependence of irradiance on welding current (electrodes LB 52U, dia. 3.2 mm)

Test results (Figure 3) show that a lower irradiance is observed when using an inverter power source; therefore, a level of light and heat exposure on workers in the welding area is diminished.

Electrode metal scattering depends on welding process stability. [14] proves a direct relation between scattering and a power of the short-circuit current; that is, if the power supply provides a ratio of $I_{sh.c.}/I_a \rightarrow 1$, it provides less scattering magnitude. Figures 1,2 show this ratio to be 1.38 for the inverter and 1.57 for the diode rectifier . This is also confirmed by earlier studies [7,15-18], resulted in relation between the type of a power supply and the amount of losses of electrode metal scattering. Histogram analysis (Figure 4) shows that Nebula-315 inverter power supply provides a significantly smaller amount of electrode metal scattering (0,3-3,4%), hence the use of coated electrodes is more efficient if compared with a diode rectifier VD-306 (4,4-5,71 %);

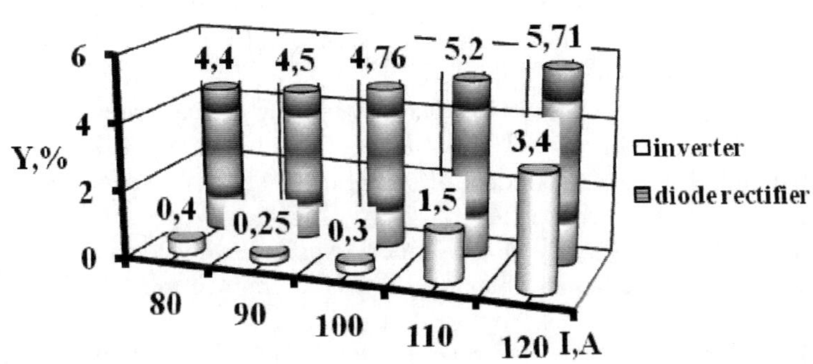

Fig. 4 - Effect of current on electrode metal scattering magnitude while MMA with LB -52U electrodes (base coat), 3.2 mm diameter.

3. Results and conclusion

In order to improve the environmental situation in the areas of welding works, it is recommended to use air trap filters equipped with metal drop pre-purification filters (mechanical dust precipitators - mostly "cyclones") or minimize scattering using inverter power supplies.

Investigations have established that use of inverter rectifiers improves hygienic conditions in welding areas while manual arc welding if compared with a diode rectifier, namely:
1) reducing of quantitative component of welding fumes to 38% and that of manganese to 30%;
2) diminishing of heat emission intensity by 37%.

References

[1] Metal Arc Welding: arc processes in and melting of electrodes / Under. Ed. I.K. Pokhodnya. - Kiev: Naukova Dumka, 1990.-222 p.

[2] I.K. Pokhodnya, Welding fumes - influencing factors, physical properties, methods of analysis / Pokhodnya I.K., Yavdoschin I.R., Gubenya I.P. /Automatic Welding. - № 6. -2011. P.31-34.

[3] Levchenko O.G. Protection against optical radiation while welding / O.G. Levchenko, I.N. Kovtun / Welder in Russia. - № 6. - 2010. - P.44-46.

[4] Lazarenko Y.P. Analysis of arc radiation for arc welding monitoring/ Y.P. Lazorenko, E.V. Shipovalov, V.A. Kolyada /Automatic Welding. № 6. - 2011. - P.39-41.

[5] Ecological value of fumes and scattering metal loss while MIG/MAG welding / welding world 2011. N 16 P. 34-37

[6] Moscovich G.N. Features of inverters performance predictions /G.N. Moscovich / Automatic welding - 2011. - № 9. P.21-28.

[7] Ilyaschenko, D.P. Determination of metal losses through splashing using different power sources/ D.P. Ilyashenko, E.A. Zernin// Welding International. 2011. Vol. 25, No 1, 69–72

[8] Ilyaschenko D.P. Effect of power supply to heat- and mass-transfer while manual arc welding / D.P. Ilyaschenko, D.A. Chinakhov / Welding and diagnostics. 2010. № 6. P. 27-30.

[9] Makarenko V.D. Calculation of kinetic characteristics of electrode drops passing through arc gap while welding with coated electrodes /Makarenko V.D., Shatilo S.P. / Welding production. 1999. №12. P.6 -10.

[10] Fed'ko V.T., Chipalyuk A.S. Electrode metal melting and transfer while arc welding with coated electrodes // Welding production. - 2003. № 2 P.3-11.

[11] Novozhilov N.M. Fundamentals of gas-shielded welding Metallurgy. M - Mechanical engineering. - 1979. - 231 p.

[12] Ilyaschenko D.P., Chinakhov D.A. Investigating Effect of Power Supply on Weld Joints Properties and Health Characteristics of Manual Arc Welding/ Materials of Science Forum, 2011, № 12, pp. 704-705.

[13] Ignatova A.M. Mineral formation in particles of welding fume solid component while high-temperature and short-pyrogenic welding processes // Scientific and Technical Gazette of Volga Region. - 2013 - № 5. - P. 166-173.

[14] Fed'ko V.T. Theory, method and facilities for reducing of scattering and processing time while welding in carbon dioxide. Tomsk: Tomsk State University, 1998. - 432p.

[15] Ilyaschenko D.P., Zernin E.A. Amount of scattering losses while manual arc welding with coated electrodes using different power supplies // New industrial technologies. №4. 2009. - P. 50-52.

[16] Yu D Zemenkov, V V Shalay, M Yu Zemenkova Expert Systems of Multivariable Predictive Control of Oil and Gas Facilities Reliability J Procedia Engineering 113 (2015) 312-315

[17] Mamadaliev R A, Kuskov V N, Popova A A, Valuev D V Alloying elements transition into the weld metal when using an inventor power source J IOP Conf. Series: Materials Science and Engineering 127 (2016).

[18] Valuev D V, Malushin N N, Valueva A V, Dariev R S, Mamadaliev R A Manufacturing component parts of mining equipment with application of hardening technologies J IOP Conf. Series: Materials Science and Engineering 127 (2016).

IOP Conf. Series: Materials Science and Engineering **142** (2016) 012008 doi:10.1088/1757-899X/142/1/012008

Use of Refractory Nanoparticles as a Component of Welding Materials in Welding and Surfacing With Coated Electrodes and Flux Cored Wires

D S Kartsev[a], E A Zernin[b], PhD in Engineering, associate professor

Yurga Institute of Technology, TPU Affiliate
652055, Kemerovo Region, Yurga, 26, Leningradskaya

E-mail: dmitkarcev@mail.ru[a], yuti_sp@bk.ru[b]

Abstract. The authors address to the issue of mainstream directions and application fields of nanostructured coated electrodes and flux cored wires, their distinctive advantages and shortcomings. Some consideration is given to use of refractory nanoparticles and their influence on the structure and properties of metal when welding and surfacing with flux cored wires and coated electrodes. The results of research carried out in this sphere are analyzed.

1. Introduction

The beginning of the 21st century is distinguished by the rapid development of nanotechnologies and nano-materials, which are used in the most important fields of human activity and in all advanced economies of the world. The importance of nanotechnologies and nano-materials is highlighted by such contributions to this field as a great number of scientific papers, whose authors deal with the specifics of nano-materials, the intensity of adopting fundamental and applied engineering products, and the growing amount of investments. These activities will be amongst the basic factors of defense, scientific and economic development of the countries in the next decades [1].

The number of scientific papers focused on nano-materials and nano-technologies is quite impressive, since approximately 15-25.000 articles, whose authors address to this range of problems, have been published over the last few years; it is definitely a very significant factor [1].

Mechanical engineering is one of the main industries applying nano-materials. The technologies of synthesizing wear resistant coatings are adopted in this industry; a lot of attention is paid to nanostructured products and equipment for processing workpieces with nanometer accuracy. What is more important, the quality can be improved via applying nano-additives to the certain technological procedures or appropriate welding regimes [2].

At present the technologies of manual and mechanized arc welding are widely used in construction and mechanical engineering. Manual arc welding is the most accepted welding technique, which is always in demand. This welding process meets principal technological criteria and requirements, for instance: exact high quality weld joints with high parameters of strength and functionality are produced on the base of this process. Flux cored wires are becoming widely applied these days for mechanized welding and surfacing. Use of flux cored wires made it possible to eliminate difficulties of welding and surfacing operations when mounting in open workshops and in field conditions, the processes have become twice to five times more efficient, providing high quality weld joints, as against welding with the solid wire. The number of labor-intensive manual operations to remove metal splatter from structures has been reduced as well [3].

Welding on the base of flux cored wires is currently a promising procedure of mechanized welding capable to improve labor efficiency significantly in comparison with manual arc welding and mechanized welding in shielding gases [2].

It differs from other types of mechanized welding because it combines advantages of both manual arc welding – simplicity, mobility – and those of mechanized welding in shielding gases – high efficiency and quality of produced weld joints [1, 2].

It is interesting to study possibilities of using nano-materials for welding and surfacing with flux cored wires and with coated electrodes simultaneously, since physical characteristics of these processes are similar.

2. Research overview

A flux cored wire is a highly efficient welding material. The advantage of the flux cored wire over the solid wire is its feed rate: the flux cored wire is supplied approximately 20-40% faster as against the solid wire at equal welding currents. Flux cored wires are divided into milled and seamless ones according to their structure [3].

The advantages provided by flux cored wires are highlighted [3]:

• appropriate formation of a weld joint (some grades support backhand welding);

• minimal spattering in proper welding conditions;

• micro-alloying components, ensuring special properties of weld metal, are used in a filler;

• efficiency of welding is at least 1.2 times higher in the flat position and at least 3 times better in vertical and in overhead positions in comparison with the solid wire [4].

In view of weldpool protection from exposure to the atmosphere there are two types of flux cored wires – gas-shielded flux cored wires and self shielded flux cored wires [4].

Flux cored wires are in great demand on the market of welding materials despite their relatively high cost. Since welding and surfacing technologies based on application of flux cored wires are widely accepted in industrial countries, their use in production and building is constantly growing, although short-termed declines are also possible in world economic crises [4].

Special characteristics listed below cause increasing demand for flux cored wires:

1. High efficiency:

— faster speeds of welding and surfacing;

2. Convenience of application:

— broad tolerance limits of welding conditions;

— possibility of fine electrode metal transfer;

3. Low risk of failures when welding:

— slag and shielding gas provide double protection of a weldpool;

— minimal risk of weld porosity, even when site welding;

— low risk of insufficient penetration depth;

4. Reducing total production costs:

— welding and idle times are shortened;

5. Field of application:

— possible to use when underwater welding [5].

The configurations of weld penetration are different when using flux cored and solid wires for consumable electrode welding in shielding gases.

Flux cored wires support more gradient junction of weld beds when multilayer welding or surfacing. Provided that welding technology is used in a proper way one can avoid such defects as overlapping gaps between weld beds, pores or slag impurities [4].

Rutile flux cored wires are used in various branches of industry because of their properties, the main of which is high processability [5]:

• welding can be carried out in difficult conditions and requires no special skills of a welder;

• a high quality weld joint is produced in all spatial positions due to quickly hardening rutile slag;
• high surfacing factor is ensured in all spatial positions, especially in a vertical one when bottom-to-top welding (arc current up to 280 A);
• appropriate micro-alloying is possible, which improves viscosity and mechanical properties of a weld joint provided that concentration of alloying elements is minimal;
• welding in CO_2 with little splashing is possible;
• when welding in a gas mixture (82%Ar+18%CO_2) splashing is almost avoided;
• a root weld can be made on a ceramic substrate;
• automatic orbit welding in all spatial position is possible;
• slag coverage is easy removable [4].

New rutile flux cored wires for welding various inter alia thick-walled constructions are developed due to the high factor of surfacing performed in a vertical position from the bottom to the top [4].

Using a flux cored wire with the charge mixture, containing certain materials, provides modification of a surfaced metal with the components of the charge mixture, so its structure is a fine grained one, and excess phases are distributed uniformly, making the surfaced metal wear resistant. Alloying of the surfaced metal by the components of the flux cored wire results in formation of an oversaturated aluminum solid solution, as the consequence, the properties of the surfaced layer are better than those of the substrate [15].

The effect of refractory materials on mechanical properties of the surfaced metal is an urgent issue requiring deeper investigation in the present day mechanical engineering. It is proposed to add nano-disperse refractory compounds to the filler of a flux cored wire used for surfacing components subject to the abrasive wear, for welding and surfacing thermally stable alloys. These compounds can provide synthesizing the surfaced metals with certain properties. Heat stability tests of the surfaced metal show that nano-disperse refractory compounds used in flux cored wire fillers improve thermal stability of the surfaced metal – the number of cycles "heating - cooling" is increased till a crack network appears and propagates [10, 11]. Abrasive wear resistance tests have revealed that samples produced using the flux cored wire with a definite charge composition are more wear-resistant as against the samples made of the standard flux cored wires [11].

The development of new welding materials with a high-resistant structure, supporting their utilization in aggressive conditions with thermal and force impact, is a burning issue of the present day mechanical engineering. It is a well-known fact that surfaced metals and cast alloys modified by nano-particles of refractory chemical compounds have improved service and process characteristics. Therefore, welding and surfacing technologies based on adding nano-particles of refractory compounds to the weldpool are in a progress [7].

The authors [6] have emphasized that the fine-grain structure of a weld or the surfaced layer is formed due to adding the certain quantity of nano-particles of refractory compounds, so such properties as hardness and wear resistance are improved this way.

Nano-particles can be added in different ways when mechanized or automatic welding or surfacing:
— nano-particles are injected into a shielding gas;
— nano-particles are a component of a flux cored wire;
— a wire can be coated with nano-particles.

As manual arc welding is the most common welding procedure, the study is to be focused on welding and surfacing by coated electrodes containing nano-materials.

The powders with ultradisperse particles can be added to the coated electrodes in the following ways:
- powders are added to the mixed envelope before coating the rod;
- modifying powders are added to the liquid glass [7].

There are certain results available in research into the effect of nano-particles of refractory compounds on the structure and properties of metal.

For instance, in the work by G.N. Sokolov titanium carbonitride powders TiCN with particles 80 to 500 nm are used as nano-components added to fillers of flux cored and compounded wires when surfacing. Microparticles (60 μm) of nickel powder (99.9%) being the component of the wire mixture are used to transport nano-particles into the weldpool. Nano-particles TiCN are added to Ni particles when their co-processing in a planetary-type mill, resulting in obtaining composite nickel granules. The share of nano-particles in composite nickel granules is 30 wt. % for TiCN. These powders are used to produce flux cored and compounded wires with diameter 3 mm and steel 08 KP envelope, which contain 0.1 – 0.6 wt. % refractory nano-particles. A layer of the surfaced metal on the base of iron: heat-resistant nitrogen-containing alloy 15Cr15Ni4AlMn3 is made with these wires when electroslag hard-facing with the fluoride flux ANF-6 and argon-arc surfacing [7].

The study based on the electron microscopy has revealed that the number of spherical intermetallic particles 0.5 to 1.5 μm grows, as well as their distribution is more homogenous in the austenite-martensite structure of the alloy 15Cr15Ni4AlMn3 produced when argon-arc surfacing by the flux cored wire with refractory nano-particles TiCN. Here, the average grain in the metal is 2.5 times smaller as against that of the initial structure, and its strain and cracking resistance gets better when thermal fatigue testing. [7].

The improved mechanical properties of alloys surfaced by refractory nano-particles are possible due to the forming composite structure with spherical intermetallic particles as its basic elements. Electron microscopy analysis of the sections of these intermetallic particles, prepared via ion etching has shown a lot of aggregates of 5 to 50 nm nano-particles. The space between them is 20 to 150 nm. Taking into consideration the dimensions of intermetallic particles we can assume that the number of nano-particles can vary from some hundreds to thousands [7].

The purpose of the work [8] by A.A. Artemyev and G.N. Sokolov is to study the importance of nano-particles TiCN, added to the flux cored wire for formation of the man-made composite structure and properties of the wear resistant layer of the electroslag surfaced metal. The flux cored wire is added to the slag ANF-6 through a hollow electrode. Modifying the surfaced metal requires adding up to 2 wt. % composite powder to the mixture of experimental flux cored wires. The powder contains 70 wt. % nickel micro-powder with grains less 60 μm. The particles of nickel contain 30 wt. % implanted ultradisperse refractory particles TiCN with grains less 80 nm [8].

The metallographic study has identified that structure and phase composition of the added metal based on matrix 20Cr7Mn12Ni2 is different for various concentration of the boride added to the flux cored wire. 28 wt. % concentration of TiB_2 is the cause of a hypoeutectic structure. Boride eutectic based on α-Fe and residual austenite has the microhardness ranging from 12 to 15.5 GPa [8].

Reduced mass share of powder TiB_2 in the wire filler to 18 wt. % is the cause of the eutectic structure with low microhardness (6.7 – 7.5 GPa) and a considerable dispersion. Hardness of the surfaced metal is also decreased and approximates to 45 – 48 HRC [8].

Decreasing the concentration of TiB_2 to 8 wt. % influences on the structure of the surfaced metal, so it consists of alloyed austenite grains with microhardness of 4 – 5 GPa surrounded by separated boride eutectic. Its hardness is low 17 – 20 HRC because of the little eutectic in soft and lamellar austenite matrix [8].

Abrasive wear resistance tests have identified that the 3.5 – fold rise of TiB_2 concentration in the mixture of the flux cored wire results in 2.5 – fold increase of wear resistance of the surfaced metal [8].

Artemyev and Sokolov [8] have revealed dissolution and coagulation of some nano-particles TiCN when electroslag surfacing. Then colonies of coarse irregular particles 1-3 μm are formed. Relying on the atomic-force microscopy the authors have detected that there are impurities 15 to 50 nm in the layer of the surfaced metal. Some particles TiCN fail to get dissolved in the melt and are the centers of crystallization for other titanium compounds. The relative wear resistance is improved more than twice as compared with the reference alloy, its value exceeds fivefold the one of the best industrial wear resistant alloys, surfaced with the standard wire PP Np-170M (15Cr15B3T2) [8].

The authors of the paper [9] A.M. Levchenko, S.G. Parshin and I.S. Antipov faced the problem of synthesizing the flux cored wire for mechanized underwater welding of steels. This wire is manufactured via plastic deformation of a band of steel 08kp according to GOST 3560–73, the mixture is filled, followed by wire removal as soon as the required diameter 1.6 mm is obtained. The mixture consists of ore mineral and chemically pure compounds with a regular granulometric composition, having the coefficient of flux cored wire filling 30 – 35%. Nano-composite material is surfaced electrochemically through colloid sulphate electrolytes, containing nano-disperse particles of halogenide salts and oxides [9].

The authors [9] have pointed at modified structure of the surfaced metal provided that the concentration of TiCN particles in the wire filler exceeds 0.2 wt. % - the grains get smaller 2.0-2.5 times on the average. We can assume that thermodynamically high-stable nano-particles of titanium carbonitride, being slightly dissolved in a metal melt, pass from the mixture of the flux cored wire into the weldpool, and influence the crystallization kinetics of the surfaced metal.

The research into structures has revealed that the increased (more 0.5 wt. %) concentration of the nanostructured powder TiCN in the filler of the flux cored wire is the cause of relatively coarse (up to 2μm) titanium carbonitride impurities, forming in the modified structure of the surfaced metal. These impurities are quite natural and their origination is regarded as melting smaller (less 30 nm) particles of nanostructured powder, their melting temperature can be below 2400 °C. This process occurs at the stage of melting and transfer of electrode metal drops, the temperature of drops is 2500 °C [9].

It is found out in the paper that coarse (more 2 μm) impurities of titanium carbonitride are stress raisers, which deteriorate its fatigue strength considerably. Therefore, the concentration of source nanostructured powder in the mixture of the flux cored wire can not exceed 0.5 w. % [9].

The research into welding and process properties of flux cored wires has revealed that the wire ПП-ПС is distinguished by the uniform melting of envelope and rode, atomized transfer of the electrode metal, it provides welds of a good quality when low, horizontal and vertical underwater welding [9].

S.G. Parshin provides the description of applying nano-disperse particles of activating fluxes and nanostructured electrode materials [12 – 14]. These works are focused on the improvement of welding efficiency, characteristics of electrode metal drop transfer and quality of the weld joints of steel. It is found out that nanostructured fluxes make it possible to control the arc energy properties, improve the transfer of electrode metal drops and quality of the weld joints of steel [12, 13]. The use of the developed nanostructured electrode wires coated with the micro-composite mixture containing halogenide particles in the metallic matrix allows of developing the forced MIG technology – welding of steels, distinguished by the improved efficiency of the process up to до 68.5 %. The weld joints made according to this new technology are characterized by stable formation under deep penetration of the plate rolled metal, even in critical and supercritical conditions at high current densities [14, 25].

The paper [16] by V.B. Litvinenko – Arykov et al is concentrated on the influence of titanium carbonitride nano-powder TiCN, put into the mixture of flux cored wires, on the structure of the surfaced metal and its properties. The composition of the metal is similar to that of high strength austenite-martensite alloys on the base of iron. Flux cored wires with the diameter of 2.6 mm are used for surfacing; the coefficient of filling the mixture is 46 - 47 %. The wires are produced on the base of a well-known technology. The powder of nitrated chromium is added to the filler of the flux cored wires to make a layer of the surfaced metal. For the purpose of the research a composite micro-powder, produced in the Institute of Metallurgy and Materials by A.A. Baykov, is used to provide uniform distribution of TiCN nano-particles in the filler of the flux cored wire. The results have demonstrated that 0.2 – 0.5 (wt. %) of TiCN nanostructured powder added to the mixture of the flux cored wires cause 2.0 – 2.5 fold grinding of grains. Nano-dimensional and micro-disperse TiCN particles in the metal, which is subjected to surfacing support the resistance to plastic deformation at rather high temperatures 750 to 950 ^{0}C.

The works [17] by G. N. Sokolov and I.V. Lysak et al are focused on modifying the structure of the surfaced metal by the nanostructured tungsten carbides. The material to transport the particles of tungsten carbide is nano-disperse powder of nickel (99.9 % Ni) with particles 50 – 80 μm. As the result nickel granules are synthesized, which contain approximately 30 wt. % nano-carbides. The shielding coating of the electrode is made of the mixture containing these granules and potassium-sodium liquid glass. For the experiment they use electrodes OK 43.32 – rutile electrodes for welding low carbon steels (ESAB Group Limited) and UTP 67 S – basic electrodes applied for surfacing the wear resistant alloy (UTP Sweissmaterial GmbH). These electrodes provide the following chemical composition of the surfaced metal (wt. %) C – 0.07; Si – 0.4; Mn – 0.5 and C – 0.5; Cr – 9.0; Si – 3.0; Mn – 0.5, respectively, iron is the base with impurities making the rest. The rods with the diameter 2 mm are coated by the mixture containing nickel carbide granules. The rods are made of wire Sv-06X19H9T. In the process of experiments steel 20 plates are surfaced. First, the structure of the layer surfaced by coated electrodes OK 43.32 is a ferritic pearlite mixture. The addition of nickel and tungsten nano-carbides results in the change of metal, which becomes a modified sub-dispersed solid solution on the base of α-Fe and residual austenite on the grain boundaries. The number of non-metallic impurities, distributed unevenly in the metal and having indefinite boundaries is decreased by 15-20 %. The other nonmetallic impurities are distributed more uniformly and have a globular form. This type of metal structure is to further the improvement of its plasticity and reliability at low temperatures and under cyclic loads. In the process of research into the metal structure, surfaced by electrode UTP 67 S with alloying coating no significant changes of its structure and hardness are revealed. This phenomenon can be referred to the specifics of poorly studied and complex processes of carbon diffusion, occurring in the metal with carbide nano-particles. More efficient influence on the structure of the surfaced metal requires the production of electrodes and wires, containing nano-disperse particles of other refractory materials.

Welding with flux-covered wires and manual arc welding with coated electrodes are somehow similar. Both procedures require the use of powdered components. In the flux-covered wire there are shielding components in the metallic envelope, but on the rod as a coating when manual arc welding. Therefore, it is interesting to study the influence of nano-particles when welding and surfacing both by flux cored wires and coated electrodes.

The paper [18] by S.O. Gordin, A.N. Smirnov and V.L. Knyazkov provides the description of adding 2.0%. titanium carbonitride powder to the coating of electrodes (T-590), which contains ferrochromium, ferroboron, marble, ferrosilicon, fluorspar, ferromanganese, graphite, and potash, to improve the wear resistance of the surfaced coating, operated in conditions of the abrasive wear. When producing the electrodes for the purpose of experiments the nanostructured powder of titanium carbonitride with particles of 60 to 200 nm is added to the coating. The coating is put on the steel Sv-08A rods with the diameter of 4.0 mm. This chemical composition improves the plasticity of the coating mixture. The hardness of the surfaced alloy is up to 66 HRC, with improved wear resistance coefficient and long service life of the remanufactured products. The particles of titanium carbonitride are a refractory compound ($T_{ml} \sim 3100^{o}C$), they improve the toughness of the liquid weld pool metal, increase the rate of metal crystallization and allow of synthesizing a solid solution with martensite-carbide or dendrite structure, reinforced by hard wear resistant phases.

The wear resistance coefficient of the coating made by electrodes containing the nano-powder of titanium carbonitride in the envelope is higher by 57% than that of the coating made by serial electrodes. To identify the thermal stability of the surfaced layer produced by serial and experimental electrodes, hot samples are cooled in water after making a layer. No cracks and delamination are detected when testing the surfaced layers reinforced by carbonitride [18].

The purpose of the work [19, 20] by G.N. Sokolov, A.A. Troshkov et al is to study the influence of nickel micro-granules with nano-disperse tungsten carbide (WC) particles, added to the coatings of welding electrodes, on the structure, impact strength and hardness of the surfaced layer of low carbon metal. The experimental electrodes are manufactured via adding granules of a composite powder to the base coating of industrial electrodes UONI – 13/45 with the diameter of 3 mm. In the process of metallographic study on the surfaced metal and the solidified slag it is revealed that some composite nickel micro-granules get stuck in the viscous slag when passing through the gas-slag environment of the welding reaction zone, but most of them get into the molten metal and alloys it. Tungsten, nickel and carbon are distributed quite uniformly among components of the surfaced metal structure. Their concentration in the local micro-volumes of the alloy depends on the dimensions of the composite nickel granules, quantity and distribution of WC carbide nano-particles added to them. Therefore, addition of composite nickel micro-granules containing nano-disperse tungsten carbide to the base coating of the welding electrodes, is an efficient procedure to modify the low carbon surfaced metal. The concentration of micro-granules Ni + WC (approximately 3 wt. %) in electrode coating supports the improved impact strength of the weld metal as against the samples made by standard electrodes UONI – 13/45. [21, 23, 24]

3. Conclusions

Having analyzed the studies above the results of adding nanostructured powders to the weldpool when arc welding are highlighted:

- the surfaced layer of metal gets modified;
- such characteristics as impact strength and hardness of the weld metal are improved;
- grains in the heat impact area get smaller;
- improvement of arc properties;
- improved formation of the weld;
- increased efficiency of welding.

The results of the literature review demonstrate that nanostructured particles of refractory compounds added to the weldpool when welding and surfacing further the formation of fine grain structure of metal due to crystallization centers developed by refractory materials particles.

Refractory particles added when surfacing have a positive effect on the abrasive wear resistance of the surfaced layer.

Flux cored wires and coated electrodes with nano-particles of refractory compounds are distinguished by satisfactory process characteristics.

One of the prospects of welding and surfacing procedures based on nano-materials is the development of new types of composite wires. Newly developed welding materials are to meet both the process requirements and economic reasonability of use in a particular product. The achievement of the purpose requires solution of the following problems: to design a device and manufacturing procedure of a composite flux cored wire with a complex chemical composition; to determine welding and process characteristics of the developed wires; to reveal the character of influence of new welding materials on the structure and properties of the surfaced metal; to estimate the economic efficiency of adoption new flux cored wires in the industry.

References

[1] S. V. Makarov and S. B. Sapozhkov Use of complex nanopowder (Al_2O_3, Si, Ni, Ti, W) in production of electrodes for manual arc welding // World Applied Sciences Journal 22 (Special Issue on Techniques and Technologies). 2013. P. 87-90.

[2] S. V. Makarov and S. B. Sapozhkov Production of electrodes for manual arc welding using nano-disperse materials // World Applied Sciences Journal. 2014. 29 (6). P. 720-723.

[3] Afanasyev V.K., Smirnov A.N., Gladyshev S.A., Blumenstein V.Yu., Tokarev A.O., Chibryakov M.V., Nikitenko S.M., Kolba A.V., Zyryanov A.P., Martynov M.A., Maslyaev M.V., Sochnev A.V., Tolstogusov V.N. Surfacing alloys: Study guide. Ed. by V.K. Afanasyev. – Kemerovo: “Poligraph”, 2005 – 243 p.

[4] Mazur A.A., Makovetskaya S.V., Pustovoit S.V., Brovchenko N.S. Flux cored wires on world and regional markets of welding materials. Automatic Welding 5 - 6 2015., P. 68 – 74.

[5] Vinogradov V.M., Cherepakhin A.A., Shpunkin N.F. Fundamentals of welding production – M.: “Academia” Publishing, 2007.

[6] Zusin V.Ya. Research into modifying the metal surfaced by flux cored wire with aluminum coating // Vestnik of Azov State Technical University. №2. - 2011, P. 180 – 183.

[7] Sokolov G.N., Zorin I.V., Artemyev A.A., Litvinenko-Arkov V.B., Dubtsov Yu.N., Lysak V.I., Kharlamov V.O., Samokhin A.V., Tsvetkov Yu.V. Specifics of the structure and properties of the surfaced alloys under the action of nano-particles of refractory compounds // Physics and Chemistry of Processing the Materials №2. - 2014, P. 38 – 47.

[8] Artemyev A.A., Sokolov G.N., Lysak V.I. The effect of micro-particles of titanium boride and carbonitride on the structure and properties of the surfaced metal // Metal science and heat treatment of materials. №12. - 2011, P. 32 – 37.
[9] Levchenko A.M., Parshin S.G., Antipov I.S. Welding flux cored wire with nano-composite coating for underwater mechanized welding // Proceedings of the Week of Science at Saint-Petersburg State Polytechnic University December, 2 – 7, 2013., P. 169 – 175.
[10] Ryabtsev I.A., Kondratyev I.A The effect of ultradisperse carbides in flux cored wires on the properties of heat resistant surfaced metal // Automatic Welding. 2009. №6. P. 13-15.
[11] Lukashov A. S., Zernin E. A., Kuznetsov M. A. Application of inorganic nanopowders in welding, surfacing and spraying (review) // Applied Mechanics and Materials. - 2015 - Vol. 770. - P. 299-303. doi: 10.4028/ www.scientific.net/ AMM.770.299
[12] Parshin S.G. Use of ultradisperse particles of activating fluxes to improve the efficiency of MIG/MAG – welding of steels // Welding Production. 2011. №6. P. 16-21.
[13] Parshin S.G. MIG – welding of steel with nanostructured electrode materials // Welding Production. 2011. №10. P. 27-31.
[14] Parshin S.G. Efficiency improvement of WIG – welding of steels and aluminum alloys when adding ultradisperse particles of activating fluxes // Welding Production. 2012. №3. P. 7-11.
[15] [15] Klimpel A., Kik T. Erosion and abrasion wear resistance of GMA wire surfaced nanostructural deposits // Archives of Materials Science and Engineering. 2008. №2. P. 121-124.
[16] Litvinenko – Arkov V. B., Sokolov G. N., Kyazymov F. A. The structure and properties of the heat proof metal surfaced by flux cored wires with nano-particles TiNC // Bulletin of Volgograd State Technical University. 2012. №6. P. 194 – 197.
[17] Sokolov G.N., Lysak I.V., Troshkov A.S., Zorin I.V., Goremykina S.S., Samokhin A.V., Alekseev N.V., Tsvetkov Yu.V. Modifying the surfaced metal structure by nano-disperse tungsten carbides // Physics and chemistry of processing materials. 2009. №6. P. 18 – 25.
[18] Gordin S.O., Smirnov A.N., Knyazkov V.L. The composition of the coating for wear resistant surfacing // Vestnik of Kuzbass State Technical University. №2. 2015. P. 106 – 108.
[19] Sokolov G.N., Troshkov A.S. The effect of nano-disperse carbides WC and nickel on the structure and properties of the surfaced metal // Welding and diagnostics. 2011. №3. P. 36 - 38
[20] The effect of nickel on the structure and mechanical properties of the weld made by electrodes with basic coating / I. K. Pokhodnya, V. D. Makarenko, A. O. Korsun, S. S. Milichenko // Automatic welding. 1986. № 2. P. 1 – 5.
[21] Kuznetsov M.A., Kolmogorov D.E., Zernin E.A., Lukashov A.S. Ways of modifying the surfaced metal by nanostructured powders to further the operating life and reliability of mining equipment components // Mining Informational and Analytical Bulletin. Special issue № 3. Mining engineering. 2012. pp. 306 – 309.
[22] Hunt, W.H. Nanomaterials: Nomenclature, novelty, and necessity / W. H. Hunt // J. of the Minerals, Metals and Materials Society. – 2004. – Vol. 56. – № 10. – P. 13–18.
[23] Nanotechnology. Integrated Processing Systems for Ultra-Precision and Ultra-Fine Products/ Ed. Taniguchi N. – Oxford : Oxford University Press, 1996. – 406 p.
[24] Makienko V.M., Grigorenko V.G., Stroitelev D.V., Romanov I.O. Features of elaboration of welding – deposition flux-cored electrodes powder wires from the mineral raw material of the Far Eastern region / Joint China-Russia symposium on advanced materials processing technology, Harbin, P.R. China. August 21-22, 2006.
[25] Makienko V.M., Stroitelev D. V., Baranov E. M., Romanov U.O. Effect of components of the slag system of flux cored wires on the technological features of the surfacing process and the mechanical properties of the deposited metal / Welding international 2007 Volume 21 Number 3, P. 231-238.

Laser-TIG Welding of Titanium Alloys

G Turichin, I Tsibulsky, V Somonov[1], M Kuznetsov, A Akhmetov

Peter the Great St.Petersburg Polytechnic University
195251, St. Petersburg, Politechnicheskya st, 29, ph. number (812)5529843

[1]e-mail: vlad@ltc.ru

Abstract. The article presents the results of investigation the technological opportunity of laser-TIG welding of titanium alloys. The experimental stand for implementation of process with the capability to feed a filler wire was made. The research of the nature of transfer the filler wire into the welding pool has been demonstrated. The influence of distance between the electrode and the surface of the welded plates on the stability of the arc was shown. The relationship between welding velocity, the position of focal plane of the laser beam and the stability of penetration of plates was determined.

1. Introduction

Titanium and its alloys are indispensable construction materials because they have unique physical properties: high specific strength comparable to steel. However, they have a density in two times less than steels. The titanium alloys have a high melting point and corrosion resistance in aggressive chemical environments, atmosphere and water.

Technology of welding titanium has the distinctive features. The titanium interacts actively with atmospheric gases, such as oxygen, hydrogen, nitrogen during the heating more than 350^0C; because of this, the mechanical properties of joint weld significantly decrease. Titanium alloys are sensitive to thermal cycle due to a heavy increase of grains during heating and cooling in the area of β phase. Therefore, the welding should be done in the atmosphere of inert gas with minimum heat input. For the formation of joint weld from titanium and its alloys in the world practice using various welding methods: TIG welding, friction welding, electron beam welding. However, each of these methods has its own significant imperfection. TIG welding have a low productivity. The friction welding has a narrow field of application, as the dimensions of the welding parts are limited. The electron beam welding has a limitation on the overall dimensions of the weld workpieces and it is a labour intensive process due to there is the necessity of creating a vacuum in the working chamber. The welding of titanium by laser radiation is the most promising. Analysis of the articles of other researchers revealed a wide range of researches in this direction. There are researches with using of laser radiation of solid-state laser in a pulse mode [1], with the aim of reducing residual welding deformations, and in continuous mode for welding thick sheets of titanium (16mm) [2]. In addition, there are researches with using of laser radiation of CO_2 laser. Separate work are devoted to the description of results of the research of laser welding with filler wire [3, 4], other about laser welding of titanium sheets overlap [5]. However, the most advantageous welding process of titanium alloy is laser-arc welding. The main advantages are minimal welding deformation (comparable to the deformations formed at laser welding), high welding speed, high depth of penetration, the capability of welding with gap, and the capability of varying the chemical composition of the weld metal [6-8].

The aim of this work was to evaluate the technological opportunity of laser-TIG welding (LTW) of titanium alloys Ti-1.5Al-1.0Mn and Ti-6Al-4V with a thickness of 5 mm.

Researches in this area are mainly represented with using of electron-beam welding, but also there is the interest to the use of laser welding by a CO_2 laser [9, 10] or solid-state laser [11].

The influence of the parameters of LTW on the formation of the joint weld was investigated during research; also, the analysis of the microstructure and mechanical properties of joint weld was produced.

2. The experimental researches

2.1. Welding equipment

The experimental researches of laser -TIG welding have been carried out with using a laser-arc technological complex with fiber ytterbium laser LS-5 (IPG). It works in continuous mode and has wavelength of radiation of 1070 nm and maximum output power of 5 kW.

The laser beam has been focused with a focal length of 300 mm in a spot with a diameter of 200 microns. The TIG welding source AC/DC TRITON 220 AC/DC Power Sinus (EWM) with the maximum welding current and floating voltage 220A and 45V, respectively, has been used as arc source. The feeder PDGO-601 («ITW») got the filler wire.

2.2. Welding materials

Plates from titanium alloys grade Ti-1.5Al-1.0Mn и Ti-6Al-4V with dimensions 200x400x5 mm were used in the experiments. The welding wire Ti-1.5Al-1.0Mn with a diameter of 1.6 mm has been used as the filler wire.

Chemical composition of titanium alloys are presented in table 1.

Table 1 - Chemical composition of titanium alloys

	Fe	C	Si	Mn	V	N	Ti	Al	Zr	O	H	others
Ti-1.5Al-1.0Mn	< 0.3	<0.1	< 0.15	0.7 - 2		< 0.05	94.138 - 98.3	1 - 2.5	< 0.3	< 0.15	< 0.012	0.3
Ti-6Al-4V	< 0.3	< 0.1	< 0.15		3.5 - 5.3	< 0.05	86.485 - 91.2	5.3 - 6.8	< 0.3	< 0.2	< 0.015	0.3

The argon and helium have been used as shielding gases for prevention of oxidation of the welding pool and metal, heated to a temperature more than 350^0C

2.3. Methods of quality control of the welds, test methods, test equipment

All welds have been subjected to visual inspection for existence of defects. For further study of geometry, internal defects and microstructure of joint weld were made microsections. Researches of the nature of transfer the filler wire into the welding pool and stability of the arc have been carried by high-speed video camera CENTURIO C100, with the video frequency of 3000 frames/sec.

3. Experimental results and discussions

LTW have been carried in a lower position in one pass. The process is performed with maximum output power of laser radiation of 5 kW for getting maximize productivity.

3.1 Melting of a filler wire

During the LTW process filler wire was applied along the seam from the side of the arc at the smallest possible angle relative to the surface plate 10-15^0, figure 1.

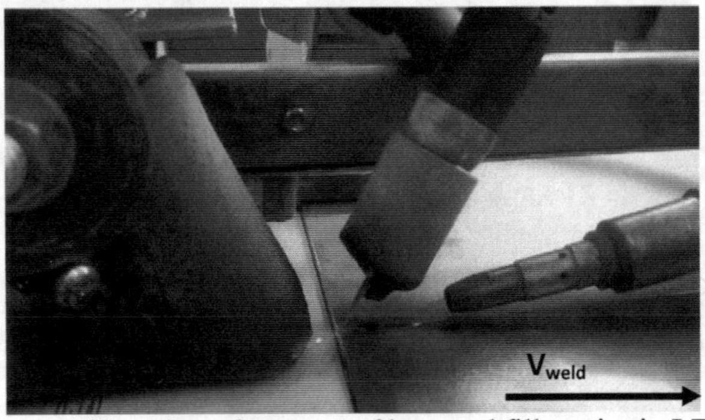

Figure 1 – Scheme of location of sources of heat and filler wire in LTW

For researches of the nature of melting of filler material was used a high-speed video. The frames of high-speed video are shown in figure 2.

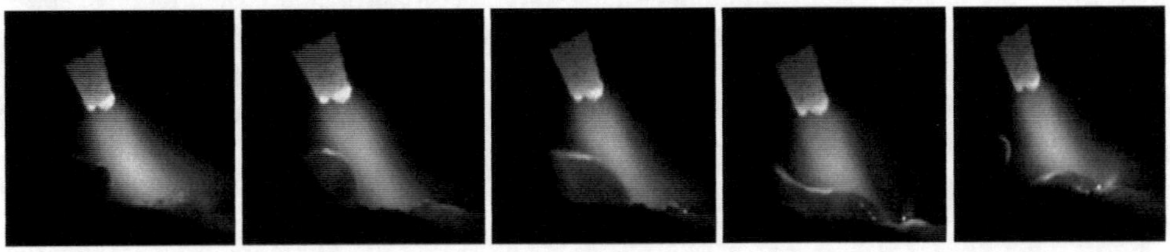

a b c d e

Figure 2 – Transferring of filler material in the welding pool: laser output power (P_L) - 5kW, velocity of LTW (V_{weld}) - 0.9 m/min, current of arc (I) - 140A, arc voltage (U_{arc}) to 18V, filler wire feed speed (V_{wire}) - 1.1 m/min; the distance from the end of the electrode to the surface of the sheet (h) - 3 mm.

As can be seen from figure 2, the process of melting the filler wire can be divided into several stages. The heating of wire in front of by the arc and melting with the formation at the expense of forces of surface tension of a spherical droplet at its end is shown in figure 2a. The growth of a spherical droplet due to the melting of new portions of the filler wire and hold it on the end of the wire due to the forces of surface tension are shown in figure 2b. The amount of molten metal filler wire and the mass of droplet increase and due to the action of gravity, the drop changes its shape (figure 2c). The drop is stretched out (figure 2d) and is separated from the end of filler wire and flowing into the welding pool (figure 2e) when it approaches to the weld pool and increasing the pressure of the arc on the surface.

The cyclical nature of this process provides a uniform transfer of molten filler metal to the welding pool. The melting of the filler wire by arc component of LTW helps to use of costly laser radiation on the penetration of the plates. Arc also provides pre-heating of the plates, increasing their absorption of laser radiation.

3.2. The influence of the distance between the electrode and the surface of the welded plates on the stability of arcing

LTW has been carried with the distance between the electrode and the surface of the plates in the range from 2 to 3 mm. It helps to determine the optimal location of the electrode. The process of LTW

IOP Conf. Series: Materials Science and Engineering **142** (2016) 012009 doi:10.1088/1757-899X/142/1/012009

has been recorded by high-speed video camera. The frames of high-speed video are presented in figure 3.

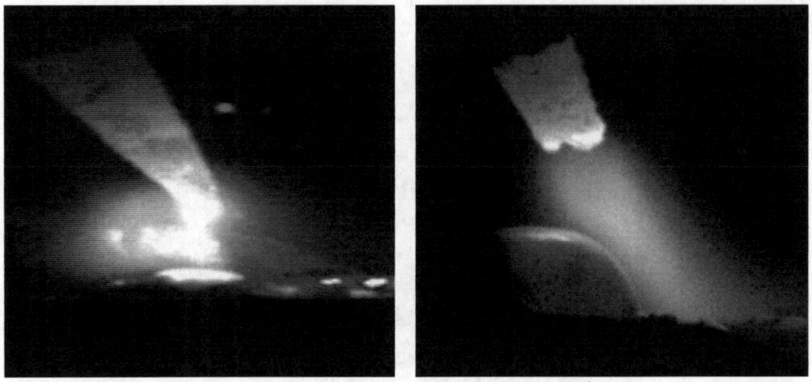

a b

Figure 3 – Influence the distance between the electrode and the surface of the plates h on the stability of arcing: a – h=2 mm, b – h=3 mm, (P_L=5kW, V_{weld}=0.9 m/min, I=140A, U_{arc}=18V, V_{wire}=1.1 m/min)

According to the results of experimental studies of the nature of melting of filler wire (figure 2) the melt droplet during its growth increases in volume with keeping a spherical shape. At some point of time the diameter of the spherical drop is increased to 2 mm, sufficient for the circuits of the arc gap (figure 3a). At the time of short circuit the arc goes out. The shape of drops changes because size of drops, its mass increase, and the pressure of arc on the surface also increase. Thus, the melt droplet at the end of filler wire has not time to reach a diameter of 3 mm and flows into the molten pool via the mechanism described in the previous section (figure 3b), because of this, a short circuit is not formed, and the arc burns steadily.

3.3. The influence of welding speed on the formation of a welded joint

The investigation of the influence of speed of LTW on the formation of a joint weld has been performed in the range of 0.72 to 1.5 m/min. The electrode was positioned at a 30^0 angle from the vertical. The distance between the axis of the laser radiation and the electrode on the surface of the plates was 3 mm. LTW has been carried without filler wire and without the protection of the weld root. The top side of the welding pool and the weld metal were protected by argon with flow rate of 30 l/min. The welding has been carried out at the following mode settings: laser power: 5 kW, embedding of the focal plane of laser beam is 1 mm, the power arc current: 160A, arc voltage: 18V. The appearance of the result of the experiment with stable full-penetration of plates are presented in figure 4.

Front side Backside

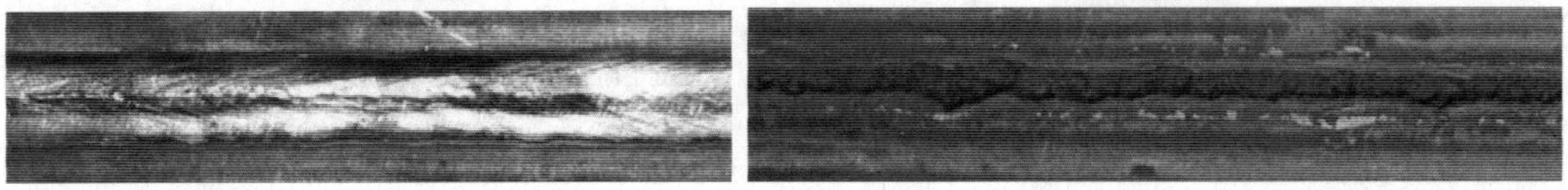

Figure 4 – The appearance of joint weld performed at LTW with welding velocity V_{weld} = 0.72 m/min

The results of welding performed at high speeds and, consequently, low heat input had not stable full penetration. The joint weld formed at a velocity of 0.72 m/min and the maximum welding heat input had the highest stability of full penetration. The results of this study it was decided to perform further experiments of LTW at a welding velocity of 0.72 m/min.

3.4. The influence of deepening the focal plane of laser radiation on the formation of a joint weld

LTW with an embedding of focal plane Δf from 1 to 7 mm relative to the surface of the plates has been performed with the aim of getting a stable full penetration. The welding has been carried out on selected in the previous sub-section velocity with the same values of the other parameters of the mode.

The welding process was carried out without filler wire and without the protection of the weld root. The electrode was positioned at a 30^0 angle from the vertical. The distance between the axis of the laser radiation and the electrode on the surface of the plates was 3 mm. The top side of the welding pool and the weld metal were protected by argon with flow rate of 30 l/min. Appearance of joint weld with stable penetration is presented in figure 5.

Front side Backside

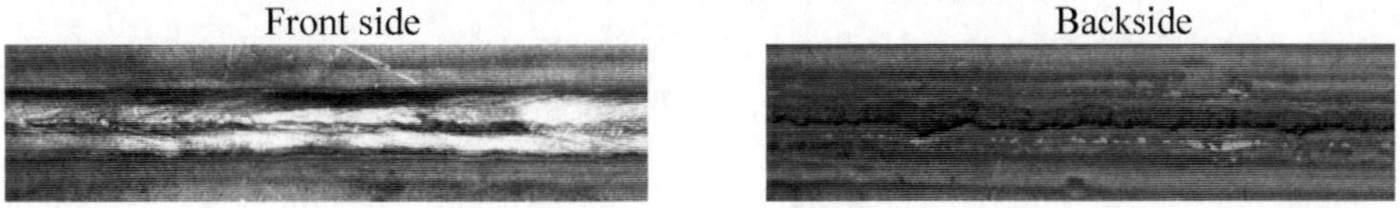

a d

Figure 5 – Appearance of joint weld, Δf =1 mm

The top weld beads had annealing colours (Figure 5a). It shows about to the poor protection during the welding process. Bottom weld beads had a rough surface and a brown color (Figure 5b). It shows about the absence of protection of the weld metal. The reduction of the width of the bottom weld bead and increase the width of the top weld bead at the embedding of the focal plane of the laser beam was found during of visually control of geometry of joint weld.

The stable formation of a bottom weld bead (full penetration) was detected at the deepening of the focal plane relative to the top surface of the plate by 1 mm.

Then LTW with the filler wire was performed. The mode of parameters are presented in table 2. The welding pool and the metal is heated to a temperature above 350^0C were protected by argon. The appearance of welded joints is presented in figure 6.

Table 2 – The mode of parameters of LTW

№	P_L, kW	V_{weld}, m/min	Δf, mm	I, A	U_{arc}, V
8	5	0.72	-1	160	18

Front side Backside

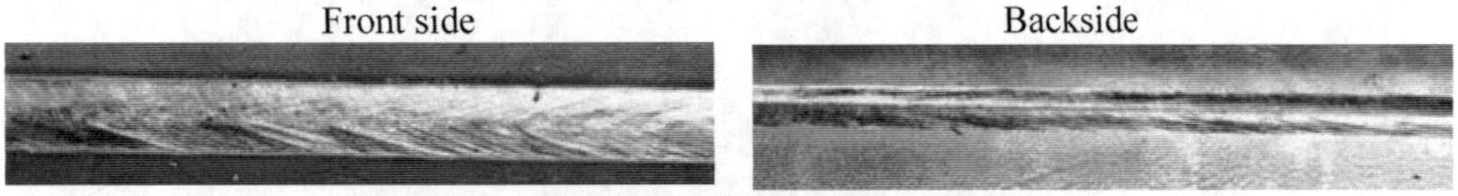

Figure 6 – Appearance of joint weld with filler wire

Any visual defects have been not detected by visual control of joint weld. The joint weld had a stable full penetration. The colour of top weld bead changed from light silver to light straw. The bottom weld bead had a light silver color. It indicates about a uniform quality protection.

3. Metallographic tests

During visual control of the cross section of a joint weld by optical microscope, the increase in the width of weld in the upper and lower parts has been found. The main contribution to end-to-end penetration introduces laser radiation; the energy of the arc is used for melting the filler material and the upper layers of the plate.

3.1. The microstructure of the joint weld

The titanium alloys investigated in this work belong to different groups: Ti-1.5Al-1.0Mn - pseudo-α alloy containing β phase and Ti-6Al-4V – the alloy of the martensitic type, similar to the pseudo α alloy.

The alloys Ti-1.5Al-1.0Mn and Ti-6Al-4V have been welded between each other with application of welding wire with a chemical composition corresponding to Ti-1.5Al-1.0Mn and the resulting structure of the weld metal composed of martensite α' – phase.

Three characteristic areas were found in the HAZ of the joint weld:

the area with big size of grains (overheating), where the temperature varies from the melting temperature to 1300°C;

the area with full recrystallization, where upon cooling, the structures are similar to those that occur at the area with big size of grains;

the area of incomplete recrystallization (figure 7).

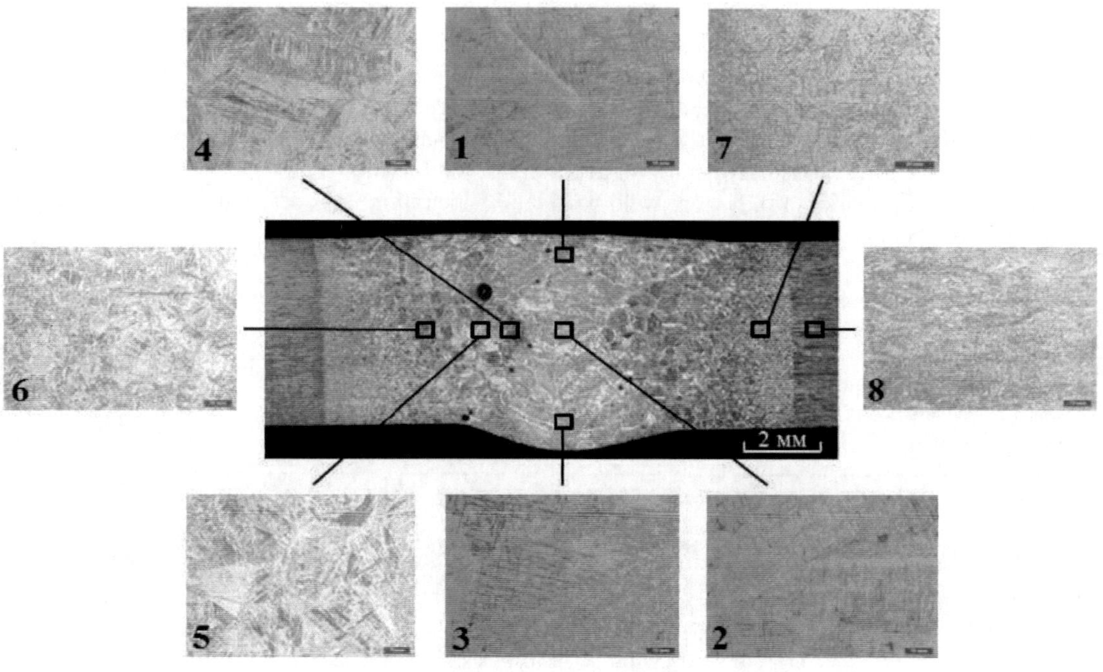

Figure 7 – Microstructure of the joint weld

4. Mechanical tests

4.1. Static tension

From the weld joint were manufactured samples for tests on static stretching according to GOST 1497-84 "Methods of tensile tests". According to test results, the value of temporary resistance ranged from 653 to 661 MPa, and the fracture of specimen occurred in the base metal. This proves that the parameters of the mode of LTW provide a generation a joint weld with the strength simillar to base metal.

IOP Conf. Series: Materials Science and Engineering **142** (2016) 012009 doi:10.1088/1757-899X/142/1/012009

4.2. The microhardness tests

The microhardness tests have been carried out according to the scheme presented in figure 8.

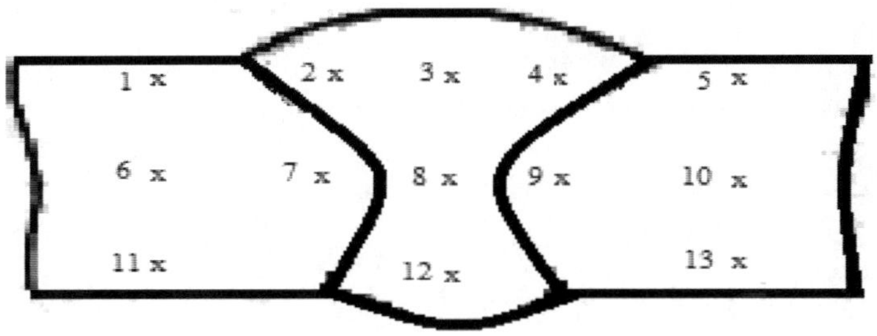

Figure 8 - Scheme of microhardness measurement.

The results of measurement of microhardness are presented in table 3.

Table 3 – The results of measurement of microhardness

№	1	2	3	4	5	6	7	8
HV	283.3	282.9	253.9	272	231.1	272.4	273.8	264.4
	9	10	11	12	13			
	211	213.2	264.6	247.9	214.2			

The weld seam had a hardness middle between Ti-6Al-4V (numbers 1, 6, 11) and Ti-1.5Al-1.0Mn (numbers 5, 10, 13) about 260 HV.

The study of the chemical composition of the weld metal detected that it is formed of filler material Ti-1.5 Al-1.0Mn. The presence of V is not detected in the composition of the weld metal. It indicates about a minimum amount of Ti-6Al-4V in the composition of the weld metal.

Conclusion

During the research, the nature of transport of metal filler wire into the weld pool has been investigated. The melting of it in front of the weld pool by arc with a uniform drop transport mechanism has been detected. The minimal distance between the end electrode and the surface of the welded plates, providing steady-state combustion of arc is 3 mm. The devices providing reliable protection of the molten metal and metal heated over a temperature of 350^{0}C, from the atmosphere have been made. The relationship between welding velocity, the position of focal plane of the laser beam and the stability of penetration of plates was determined.

According to the results of research the technology laser-TIG welding with the filler wire of titanium alloys Ti-1.5Al-1.0 Mn and Ti-6Al-4V with a thickness of 5 mm has been developed. The welded joints with satisfactory mechanical properties were obtained.

References

[1] K. Richter, W. Behr, U. Reisgen, 2007, Low Heat Welding of Titanium Materials with a Pulsed Nd:YAG Laser, Materialwissenschaft und Werkstofftechnik, Volume 38, Issue 1, pp 51–56.

[2] Andre Schneider, Andrey Gumenyuk, Marco Lammers, Andreas Malletschek, Michael Rethmeier, 2014, Laser beam welding of thick titanium sheets in the field of marine technology, Physics Procedia 56 (2014), pp 582 – 590.

[3] Nikolai Kashaev, Volker Ventzke, Manfred Horstmann, Stefan Riekehr, Grigory Yashin, Lennart Stutz, Werner Beck, 2014, Microstructure and Mechanical Properties of Laser Beam Welded Joints between Fine-Grained and Standard Ti-6Al-4V Sheets Subjected to Superplastic Forming, Advanced Engineering Materials, 2015, 17, No. 3, pp 374-382.

[4] Bergmann, J.P., 2005, Mechanical behaviour of overlap joints of titanium, Science and Technology of Welding and Joining, Volume 10, Issue 1, February 2005, pp 50-60.

[5] Liu L.M., Hao X.F., Du X., 2008, Microstructure characteristics and mechanical properties of laser-TIG hybrid welding joint of TA15 titanium alloy, Materials Research Innovations, Volume 12, Issue 3, September 2008, pp 114-118.

[6] Turichin, G. , Velichko, O., Kuznetsov, A., Pevzner, J., Grinin, O., Kuznetsov, M., 2014, Design of mobile hybrid laser-arc welding system on the base of 20 kW fiber laser, Proceedings - 2014 International Conference Laser Optics, LO 2014.

[7] Tsibulskiy, I., Kuznetsov, M., Akhmetov, A. Effect of welding position and gap between samples on hybrid laser-arc welding efficiency, Applied Mechanics and Materials, Volume 682, 2014, pp 35-40.

[8] Wang S.Q., Liu J.H., Chen D.L. 2013, Strain-controlled fatigue properties of dissimilar welded joints between Ti–6Al–4V and Ti17 alloys, Materials and Design, Volume 49, pp 716–727.

[9] Liu P. S., Baeslack III W. A., Hurley J. 1994, Dissimilar Alloy Laser Beam Welding of Titanium: Ti-6AI-4V to Beta-C™, Welding Research Supplement, pp 175–181.

[10] Chiu C.Y., Lu M.Y., Tsay L.W., 2011, Dissimilar laser welding of Ti-6Al-4V to Ti-6Al-6V-2Sn, Advanced Materials Research, Volume 295-297, pp 2353-2357.

[11] Chiu C.Y., Hsieh C.T., Tsay L.W., 2014, Microstructure and notched tensile fracture of Ti–6Al–4V to Ti–4.5Al–3V–2Fe–2Mo dissimilar welds, Materials and Design, Volume 63, pp 14–19.

IOP Conf. Series: Materials Science and Engineering **142** (2016) 012010 doi:10.1088/1757-899X/142/1/012010

A Plasma Reactor for the Synthesis of High-Temperature Materials: Electro Thermal, Processing and Service Life Characteristics

G V Galevskiy[1,a], D. Tech.Sc., Prof., V V Rudneva[1], D. Tech.Sc., Prof., S G Galevskiy[2,b], Cand. Ec. Sc, Assoc. Prof., K I Tomas[3,c], Cand. Tech. Sc, Assoc. Prof., M S Zubkov[3], student of group 10A42

[1]Siberian State Industrial University
654007, Novokuznetsk, ul. Kirova, 42
[2]National Mineral Resources University (University of Mines)
199106, Saint Petersburg, 21- th Line Vasilievsky Island, 2
[3]Yurga Institute of Technology (Affiliate) National Research Tomsk Polytechnic University,
652055, Kemerovo obl., Yurga, ul. Leningradskaya, 26, tel.+7 384 51 777 66

e-mail: [a]kafcmet@sibsiu.ru, [b]nirs@spmi.ru, [c]tki@tpu.ru

Abstract. The three-jet direct-flow plasma reactor with a channel diameter of 0.054 m was studied in terms of service life, thermal, technical, and functional capabilities. It was established that the near-optimal combination of thermal efficiency, required specific enthalpy of the plasma-forming gas and its mass flow rate is achieved at a reactor power of 150 kW. The bulk temperature of plasma flow over the rector of 12 gauges long varies within 5500÷3200 K and the wall temperature within 1900÷850 K, when a cylinder from zirconium dioxide of 0.005 m thick is used to thermally insulate the reactor. The specific electric power reaches a high of 1214 MW/m^3. The rated service life of electrodes is 4700 hours for a copper anode and 111 hours for a tungsten cathode. The projected contamination of carbides and borides with electrode-erosion products doesn't exceed 0.0001% of copper and 0.00002% of tungsten.

This work has been done at Siberian State Industrial University in the framework of government contract No. 11.1531/2014/k with the Ministry of Education and Science RF.

1. Introduction

Processing capabilities currently used for producing and utilizing refractory metals and their chemical compounds, including those in nano- and ultra-dispersed sizes specified, are expanding [1 - 3]. Considerable experience is gained at Siberian State Industrial University relating to research and utiliza-

IOP Conf. Series: Materials Science and Engineering **142** (2016) 012010 doi:10.1088/1757-899X/142/1/012010

tion of three-jet direct-flow plasma reactors with nitrogen-operated arc plasma torches (plasmatrons), and their application in the synthesis of refractory carbides, borides and their compositions, as summarized in the works in reference [4 - 9]. Thus, it is shown that the optimum reactor design should be a structure with evenly spaced plasma torches around the circumference, wherein their plasma jets have an angle of 30-45 degrees to the axis of the reactor; and thermal protection with intensely cooled walls, ensuring a maximum life of the mixing chamber, high and uniform radial temperature distribution and rates with a minimal loss of thermal energy. In fact, the research, by using the sectional calorimetric measurement of energy balances in reactors with an inner diameter of 0.046 m and mixing chambers with different angles of plasma jets, shows [8] that, when an angle of plasma jets inclining to the reactor axis decreases from 90 to 30 degrees, the density of heat flow in the mixing chamber is reduced almost three times that results in its longer service life. However, along with that in both cases, at the initial section of 4-5 gauges long, the heat flow to the walls is characterized by high density that greatly reduces the reactor's capacity in terms of dispersed material heating and evaporating. Heat losses in the reactor can be reduced by lining its channel with a heat insulating material. The applied silica-based skull lining allows the wall temperature to rise, the value of which depends on the power of arc discharge [8]. Thus, with an arc discharge power of 50 kW, the wall temperature at a length of 8 gauges varies from 1600 to 900 K, and in the absence of heat insulation - from 970 to 400 K. Building a "hot" wall helps reduce the heat flow intensity by 15-20% in the average and the temperature factor by 100%, and increase the bulk temperature of the exchange gas by 13% in the mostly heat-stressed zone of the reactor.

Heat transfer in the channel of the plasma reactor has been investigated at values of the Reynolds numbers equal to 700-1500 [4 - 9], taking into account the effect of induced turbulence in the plasma flow, the outflow of heat from plasma to the dispersed raw material and the formation of skull lining or the use of lining on the channel walls in the reactor. For the reactor channel, it is established that the intensity of heat exchange is high within the initial section of up to 4 gauges long, typical for turbulent conditions; criteria-related dependences are obtained for calculating heat-exchange coefficients for the conditions of raw material introducing into the flow.

However, the summarized in [4 - 9] results have been obtained for a laboratory plasma reactor with a capacity of 30-50 kW, information on its thermal, technical and processing characteristics is not complete, the findings do not have process design proposals and recommendations with reference to the scaling up to the industrial level. In this connection, this paper is dedicated to the following issues:

1) Specifying a power of the three-jet reactor to be used at the industrial level, and structural and technical conditions for achieving the target;

2) Determining a bulk temperature of plasma flow;

3) Determining a specific electric capacity of the mixing chamber;

4) Determining a service life of plasma torches;

5) Evaluating the contamination of refractory compounds with erosion products from plasma torch electrodes.

2.Results and discussion

Specifying a power of the three-jet reactor to be used at the industrial level

The following limitations are taken into account when dealing with this issue:

- the specific enthalpy of plasma flow at the reactor inlet should be 7.5-8.5 MJ/kg for efficient dispersed material processing [10];
- the initial speed of plasma flow should not exceed 60-65 m/s [11];
- for plasma generating, the plasma torches should be used with an electric arc that stabilized by gas vortex, since solenoids are structurally and technically difficult to install;
– the plasma torches should operate on a mixture of nitrogen and hydrogen.

The main performance characteristics of plasma torches, potentially suitable for three-jet reactors are shown in Table 1. Comparison of characteristics shows that the plasma torches of EDP-104A and EDP- 114 types are mostly appropriate. A growth in power of these plasma torches is achieved, as a rule, by increasing in the arc current and voltage under conditions of a simultaneous increase in the consumption of plasma-forming gas. It should be noted that there is a clear intention to use technology and ensure the reactor operation at the minimum permissible flow of the plasma-forming gas, usually no more than 50-65% of the maximum possible. This condition combined with higher values of current predetermines a decrease in the thermal efficiency of plasma torches with increasing the power.

Table 1. Main performance characteristics of plasma torches, potentially suitable for three-jet- reactors [11 - 12]

Performance Characteristics	Types of plasma torches		
	EDP-104A	EDP-119	EDP-114
Gas	Nitrogen,	Nitrogen, hydro-	Nitrogen,
	mixed nitrogen and	gen, mixed nitrogen	hydrogen, mixed
	hydrogen.	and hydrogen.	nitrogen and hy-
Gas consumption kg/s		$(6 \div 8)10^{-3}$	drogen
Thermal efficiency	$(1 \div 5)10^{-3}$	$0.5 \div 0.7$	$(6 \div 8)10^{-3}$
Current, A, max	$0.5 \div 0.8$	400	$0.5 \div 0.7$
Arc voltage, V, max	200	500	400
Power, kW, max	250	200	500
Gas heating temperature, K, max	50	5000	200
Nitrogen	5500		5500
Stabilization of electric arc		magnetic	
Tungsten cathode service life, h	Gas vortex, mag-	100	Gas vortex, mag-
Copper anode service life, h	netic	500	netic
Overall length, m	100	0.25	100
Mass, kg	500	4.80	500
	0.21		0.30
	1.45		5.50

Figure 1 gives the values of thermal efficiency for reactors with a power of 30, 50, 80, 120 150 kW and the EDP-104A plasma torches, and reactors with a power of 200, 250 kW and the EDP-114 plasma torches. The values were obtained experimentally by using the calorimetrical measurements of sections and projected by calculating in accordance with recommendations in [9]. As it is seen from the table, the thermal efficiency of the plasma torches is close to 0.50 when a power of 150-250 kW is specified for commercial scale reactors. The specific enthalpy values are achieved with the plasma torches having a power of 80-150 kW, which is required for plasma metallurgy to produce refractory compounds. With this, an increase in power from 80 to 150 kW results in a doubling of mass of the generated gas with reference to the required level of enthalpy.

Thus, the near-optimal combination of thermal efficiency, specific enthalpy of the plasma-forming gas required and its mass flow rate is achieved with the reactor of 150 kW in power. Further increasing the power level is irrelevant, because the designs of plasma torches available for selection do not provide the desired specific enthalpy level of plasma flow.

Determining a bulk temperature of plasma flow

To protect the walls of the reactor channel and reduce the size of a near-wall low temperature zone, the reactor channel was lined using zirconia cylindrical inserts with a wall thickness of 0.005 m and an outside diameter of 0.064 m, reducing its diameter to 0.054 m.

IOP Conf. Series: Materials Science and Engineering **142** (2016) 012010 doi:10.1088/1757-899X/142/1/012010

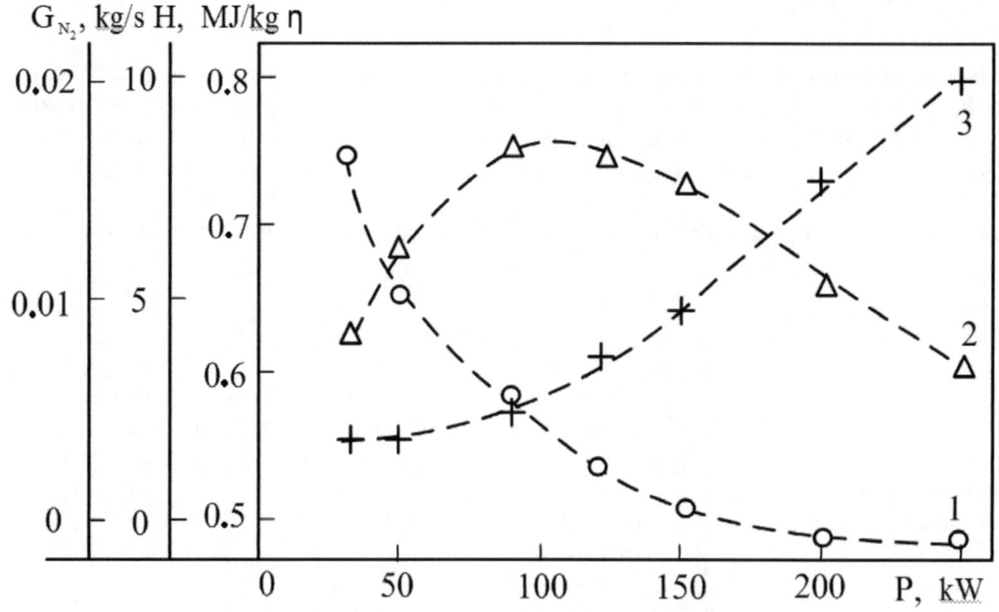

Figure 1 - Dependence of the thermal efficiency (η, 1), the specific enthalpy of plasma flow at the reactor inlet (H, 2), the mass flow of the plasma-forming gas (G_{N_2} , 3) on the reactor power

Selection of zirconia as a heat insulating material is determined by its properties such as refractory quality (a melting point of 2963K), low thermal conductivity (the coefficient of thermal conductivity is 0.42-2.63 Вт/(м·K) at a temperature of 2173 K) and high chemical resistance under the conditions of reactions taking place in the plasma reactor.

We investigated the longitudinal distribution of heat flow, the bulk temperature, and internal surface temperatures of the lined and unlined reactor walls to evaluate experimentally how the thermal protection shield made of zirconium dioxide is effective in protecting the lined and unlined walls of the reactor with a 150 kW power at a mass flow of nitrogen equal to 0.009 kg/s, using the sectional calorimetry. The diameter of the unlined reactor channel was 0.054 m. The heat flow in each section was determined in terms of heat balance by measuring the heating of cooling water. The inaccuracy of values calculated by using calorimetry is ± 10% and caused by errors of flow meters calibration (± 3%), the heat flow determination (± 4%) and calculation (± 3%).

The distribution of bulk temperature in the plasma flow along the reactor length was calculated in terms of a value of bulk enthalpy. In this case, the enthalpy of the gas flow at the section outlet was less than at the inlet by the amount of energy the gas transferred to the walls of this section. The bulk enthalpy was calculated as the arithmetic mean of the input and output values of enthalpy. The internal surface temperatures of lining and the unlined reactor walls were calculated using the experimentally observed values of heat flow.

The main results of research are presented in Figures 2 and 3. It can be seen that the use of lining from zirconia provides the increased flow bulk temperature and wall temperature: along the reactor of 12 gauges, the bulk temperature changes from 5500 K to 3200 K in the case (1) and 2650 K in the case (2), and the wall temperature changes within 1900-850 K and 800-350 K respectively.

Determining a specific electric capacity of the mixing chamber

The specific electric capacity, defined as the ratio of power supplied to the reaction zone to its volume, is an essential characteristic of modern electric and thermal equipment and varies between a value of

0.2 MW/m^3, which is representative for the majority of electric furnaces, and a very high value of 3750 MW/m^3, which is specific for high pressure reaction chambers in the diamond synthesis [13]. The specific electric capacity is defined as

$$W_{sp.} = \frac{P}{V_{r.z.}} = \frac{P_o \cdot \eta}{0.785 \cdot D_r^2 \cdot L_{r.z.}}, \quad (1)$$

Where $W_{sp.}$ is the specific electric capacity, MW/m^3;
P_o is the total power, kW;
D_r is the internal diameter of the reactor channel, m;
$L_{r.z.}$ is the length of the reaction zone, m;
P is the power supplied to the mixing chamber, kW;
$V_{r.z.}$ is the volume of the reaction zone, m^3;
η is the thermal efficiency of the plasma torches.

At the same time, the thermal efficiency of the plasma torches was determined based on the results of calorimetry; the length of the zone, where the dispersed material evaporated, the pyrolysis of hydrocarbons occurred and the reacting mixture of a given composition formed, (the so-called "reaction zone") was assumed to be a 0.5 gauge according to the results of simulating interaction between the plasma and material flows [9]. The values of specific electric capacity are 625, 688, 813 and 1214 MW/m^3 that calculated for the reactor channel with a diameter of 0.054 m, insulated with zirconium dioxide, and plasma torches with a power of 85, 100, 125, 150 kW respectively, which are significantly higher than for traditional electric and thermal equipment.

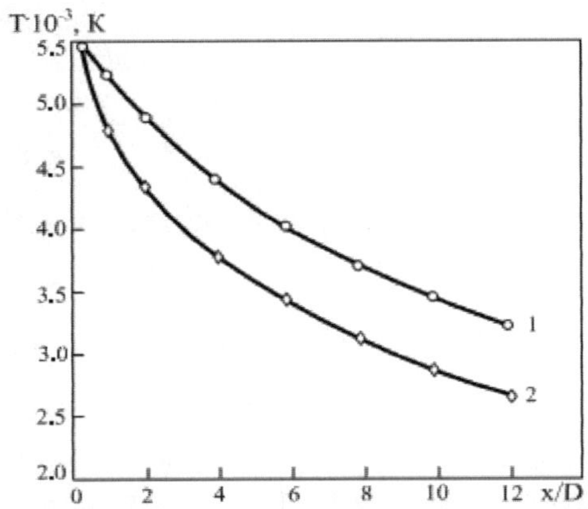

Figure 2 - (T(1) = (5500 ÷ 3200) ± (100 ÷ 60); T(2) = (5500 ÷ 2650) ± (100 ÷ 55)). Longitudinal distribution of bulk temperature for the reactor lined with zirconium dioxide (1) and the unlined reactor (2)

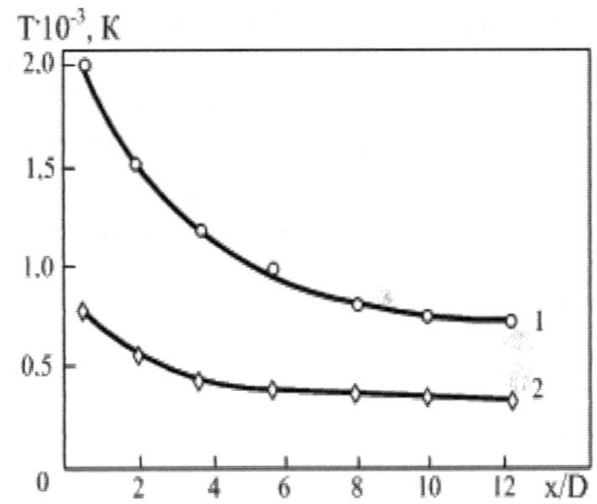

Figure 3 - (T(1) = (1900 ÷ 850) ± (35 ÷ 15); T(2) = (800 ÷ 350) ± (10 ÷ 5)). Longitudinal temperature distribution over the iternal surface of zirconia lining (1), unlined walls of the reactor (2)

IOP Conf. Series: Materials Science and Engineering **142** (2016) 012010 doi:10.1088/1757-899X/142/1/012010

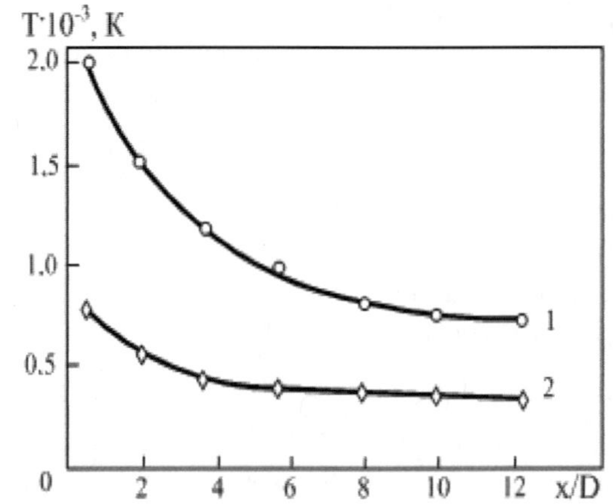

Figure 3 - (T(1) = (1900 ÷ 850) ± (35 ÷ 15); T(2) = (800 ÷ 350) ± (10 ÷ 5)).
Longitudinal temperature distribution over the iternal surface of zirconia lining (1), unlined walls of the reactor (2)

Determining a service life of plasma torches

Service life of a plasma torch defined by erosion of electrode materials is its most important feature. Cathode and anode spots of arcs ignited with cold electrodes are characterized by the extremely high density of heat flow that reaches 104-105 MW/m^3 [12]. Any of known materials is not able to withstand such thermal loads in steady-state conditions. In order to ensure an acceptable level of electrode erosion, near-electrode areas of the arc are moved over the electrode surface by applying aero or electro dynamic forces to those areas. When moving the arc spot, the electrode surfaces is subjected to cyclic thermal shocks, causing the appearance of cracks in the electrode material, resulting in its mechanical failure and reduction in thermal and electrical conductivity. Therefore, the rate of electrode erosion is dependent on physical processes, occurring in the near-electrode areas of the arc discharge, on the electrode surface and inside a crystal lattice of the metal the electrode is made of. It is determined by such transient processes as large- and small-scale arc bridging, the influence of external magnetic field on an arc column, aerodynamics of gas flow in the plasma torch. The cathode erosion also depends on its diameter and design, a composition of plasma gas, amperage, cycles of operation (quantity of plasma torch switching on), quality of thermal contact between a tungsten rod surface and copper cathode holder [12]. Erosion of a copper anode is determined by such factors as conditions of arc cooling, amperage, magnetic strength, and protection of the anode surface with inert or natural gas [12]. An integral characteristic of the electrode erosion processes is specific erosion, measured in kg /C that, of course, does not reveal details of micro processes occurring and their influence, oxide films on the surface of electrodes, thermal stresses in the metal, and features of motion of the arc spots. The specific erosion value varies within $(2 \div 5) \cdot 10^{-12}$ kg/C for the tungsten cathode in nitrogen at the atmospheric pressure and a concentration of oxygen up to 0.5% and a current of 250-300 A. When the concentration of oxygen increases up to 1.5%, the specific erosion value is as high as $(2 \div 4) \cdot 10^{-8}$ kg/C. The specific erosion value of the cylindrical copper anode is also largely determined by oxygen presenting in the working gas. Thus, with a current of 180 A in high purity nitrogen containing about 0.001% oxygen, the specific erosion is $10^{-12} \div 10^{-11}$ kg/C, and it increases by more than an order of magnitude in technical purity nitrogen (0.5% of oxygen).

Service life of a plasma torch is normally assumed as equal to a minor value of service life specified for one of the electrodes. Determining a service life of electrodes experimentally is time-consuming and expensive. Therefore, the service life of the anode and cathode was estimated by calculation. For this purpose, we assumed that a convoluted profile of the worn out anode in place of arc bridging behind the shoulder is similar to a triangle (Figure 4 a), and the diameter of a crater formed on the cathode surface by the arc evaporation process is equal to the diameter of the arc bridging place (Figure 4 b).

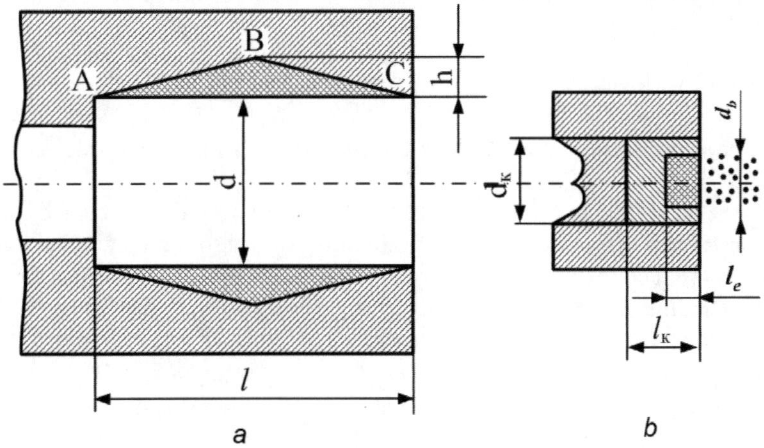

Figure 4 - Geometric models for calculating the service life of anodes (a) and cathodes (b)

The volume of the anode material removed by erosion is defined as follows in m^3

$$V_e = 0,5 \cdot h \cdot \pi \cdot l(d + h) . \tag{2}$$

It is assumed, that $l = 5.0 \cdot 10^{-2}$ m; $h = 0.2 \cdot 10^{-2}$ m; $d = 1 \cdot 10^{-2}$ m, then

$V_e = 0.5 \cdot 0.2 \cdot 10^{-2} \cdot 3.14 \cdot 5 \cdot 10^{-2}(1 \cdot 10^{-2}+0.2 \cdot 10^{-2}) = 1.88 \cdot 10^{-6}$ m^3.

The time of continuous operation for the copper anode is

$$t_a = \frac{\rho_a \cdot V_a}{G_a \cdot I}, \tag{3}$$

Where ρ_a is the density of the anode material, $\rho_a = 8.9 \cdot 10^{-3}$ kg/m^3;

G_a is the specific erosion of the anode, $G_a = 5 \cdot 10^{-12}$ kg/C;

I is the arc amperage, A. $I = 200$ A.

$$t = \frac{8.9 \cdot 10^{3} \cdot 1.88 \cdot 10^{-6}}{5 \cdot 10^{-12} \cdot 200} = 4700 \text{ h}$$

For the cathode, it is assumed that the diameter of the tungsten insert equals $d_к = 0.3 \cdot 10^{-2}$ m, the length of the insert $l_к \le d_к$ is $0.3 \cdot 10^{-2}$ m, the permissible depth of the cathode erosion is $l_e = 0.3\ l_к$. Then $l_e = 0.3 \cdot 0.3 \cdot 10^{-2} = 0.1 \cdot 10^{-2}$ m.

It is assumed that the diameter of the crater, formed on the cathode as a result of arc effect, is equal to the diameter of arc bridging

$$d_b = B \cdot I^{0.5}, \tag{4}$$

Where B is the coefficient; for nitrogen it is equal to $1.6 \cdot 10^{-4}$ [7].

Then $d_b = 2.6 \cdot 10^{-4} \cdot 200^{0,5} = 2.3 \cdot 10^{-3}$ m.

The volume of the material removed by erosion is

$$V_e = \frac{\pi d_b^2 \cdot l_e}{4} = \frac{3.14 \cdot (2.3 \cdot 10^{-3})^2 \cdot 0.1 \cdot 10^{-2}}{4} = 0.42 \cdot 10^{-8} \text{ m}^3 .$$

Taking into account $\rho_к = 19.34 \cdot 10^3$ kg/m^3 for tungsten and assuming $G_к$ $1 \cdot 10^{-12}$ kg/C, we obtain

$$t_k = \frac{19.34 \cdot 10^3 \cdot 0.42 \cdot 10^{-8}}{1 \cdot 10^{-12} \cdot 200} = 4.01 \cdot 10^5 \text{sec.} = 111 \text{ hr.}$$

Consequently, the continuous operation of the plasma torch is determined by the service life of the cathode and exceeds 100 hours that corresponds to the specification. However, it should be noted that the specific erosion values relative to anodes and cathodes are taken for the conditions when a natural gas is added to the plasma gas (technical purity nitrogen) to bind oxygen and protect electrodes, and this contributes to the reasonability to employ this technological method with the purpose of increasing the service life of electrodes.

Evaluating the contamination of refractory compounds with erosion products from torch electrodes

Currently, in the related technical literature there is an opinion widely accepted that the arc plasma torches have their limited use in plasma reactors because of the strong contamination of the produced materials with the products of the electrode-related erosion, i.e. tungsten from cathodes and copper from anodes [12]. In this connection, the study was performed to evaluate possible contamination of refractory compounds with electrode erosion products for the following conditions: the carbide or boride reactor output is 3.0 kg/h, the specific erosion of the copper anode is $5 \cdot 10^{-12}$ kg/C, the specific erosion of the tungsten cathode is $1 \cdot 10^{-12}$ kg/C, the service life is 4700 and 111 hours for the anode and cathode respectively. Then the mass of erosion products formed in an hour makes, for the copper anode:

$$m_{Cu} = \frac{8.9 \cdot 10^3 \cdot 1.88 \cdot 10^{-6}}{4700} = 3.6 \cdot 10^{-6} \text{ kg},$$

For the tungsten cathode:

$$m_W = \frac{19.34 \cdot 10^3 \cdot 0.42 \cdot 10^{-8}}{111} = 7.2 \cdot 10^{-7} \text{ kg},$$

that corresponds to the content of copper $\frac{0.0036}{3000 + 0.0036} \cdot 100$ %=0.0001 %, and tungsten $\frac{0.00072}{3000 + 0.00072} \cdot 100$ %=0.00002 % in the carbide.

According to the obtained values, there is no actual threat of contaminating refractory compounds with tungsten and copper impurities and reducing their useful properties in this regard.

3. Conclusions

The three-jet direct-flow plasma reactor with a channel diameter of 0.054 m was studied in terms of service life, thermal, technical, and functional capabilities. It was established that the near-optimal combination of thermal efficiency, required specific enthalpy of the plasma-forming gas and its mass flow rate is achieved when the reactor has a power of 150 kW. The bulk temperature of plasma flow over the rector of 12 gauges long varies within 5500÷3200 K and the wall temperature within 1900÷850 K, when a cylinder from zirconium dioxide of 0.005 m thick is used to thermally insulate the reactor. The specific electric power reaches a high of 1214 MW/m^3. The rated service life of electrodes is 4700 hours for a copper anode and 111 hours for a tungsten cathode. The projected contamination of carbides and borides with electrode-erosion products doesn't exceed 0.0001% of copper and 0.00002% of tungsten.

Reference

[1] Kuznetsov M.A., Zernin E.A., Danilov V.I., Kolmogorov D.E., Zoubenko L.N. Corrosion and mechanical properties of austenic steel weld joints. // IOP Conference Series: Materials Science and Engineering. – 2015. – №91

[2] Kuznetsov M. A. , Zernin E. A. , Lukashov A. S. , Zhuravkov S. P. Influence of Molybdenum Nanopowder Concentration on Welded Joint Microstructure // Applied Mechanics and Materials. - 2014 - Vol. 682. - p. 438-442.

[3] Zhuravkov S. P. , Pustovalov A. V. , Kuznetsov M. A. , Zhantuarov S. R. Influence of conditions of copper wires electric explosion on dispersity of produced nanopowders // Applied Mechanics and Materials. - 2014 - Vol. 682. - p. 324-326.

[4] Massye P.F. Distribution of a heat transfer and parameters of a laminar layer in an internal subsonic stream of gas at temperatures up to 7700 °C / P.F. Massye, L.K. Beck, E. Roshke // Heat transfer: Works of the American society of mechanical engineers. – 1989. - No. 1. – Pp. 77-85.

[5] Gleiter H. Nanostructured materials: basis concepts and microstructure // Acta Materialia. – 2000. – V. 48. - № 1. – P. 1 – 29.

[6] Kear B.H. Chemical processing and applications for nanostructured materials / B.N. Kear, P.R. Strutt // Nanostruct. Mater. – 1995. – V. 6, № 1 – 4. Pp. 227 – 236.

[7] Galevskii G.V. Nanomaterials and nanotechnologies: assessment, tendencies, and forecasts / G.V. Galevskii, V.V. Rudneva, E.K. Yurkova // Russian journal of non-ferrous metals. – 2007. – Vol.48. – No 2. – P. 157 – 159.

[8] Rudneva V.V. Thermal characteristics of three-jet plasma reactor / V.V. Rudneva, G.V. Galevskii, S.G. Galevskii, E.K. Yurkova // Steel in Translation. – 2007. – Vol. 37. – No. 2. – P. 115 – 118.

[9] Rudneva V.V. Effective processing of disperse raw materials in a plasma reactor / V.V. Rudneva, G.V. Galevskii, S.G. Galevskii, E.K. Yurkova // Steel in Translation. – 2007. – Vol. 37. – No. 5. – P. 425 – 428.

[10] Nozdrin I.V. Chrome carbide – nanotechnology, properties, application: monograph / I.V. Nozdrin, L.S. Shiryaeva. – Saarbruecken (Germany): LAP LAMBERT Academic Publishing, 2014. – 297 p.

[11] Galevsky G.V. Titanium carbide: nanotechnology, properties, application / G.V. Galevsky, V.V. Rudneva, A.K. Garbuzova, D.V. Valuev // JOP Conf Series : Materials Science and Engineering. – 91(2015) 01.2017.

[12] Zhukov M.F. Thermal plasma in technology of new materials: monograph / M.F. Zhukov [etc.]. – Cambridge, 1994. – 580 p.

[13] Yarnell A. The many facets of manmade diamonds. – Chemical and engineering news. – 2004. – V. 82, № 5. – Pp. 26 – 31.

Microconstituents of the Modified Surface Layer of Austenitic Steel With Nanofibres of Aluminium Oxyhydroxide

M A Kuznetsov [1a], E A Zernin [1b], Cand. Sc. in Engineering, V I Danilov [1,3c], Doct. Sc. in Physics and Mathematics, S P Zhuravkov [4d], Cand. Sc. In Chemistry S V Dementyev [1e]

[1]Yurga Institute of Technology (branch) National research Tomsk Polytechnic University, Russia, Kemerovo region, Yurga
[3]Institute of strength physics and materials science SB RAS, Russia, Tomsk
[4]Institute of High Technology Physics of National research Tomsk Polytechnic University, Russia, Tomsk

e-mail: [a]kyznechik_85@mail.ru, [b]yuti_sp@bk.ru, [c]dvi@ispms.tsc.ru, [d]zhursp@yandex.ru, [e]hedg_h@mail.ru

Abstract. In the paper the authors provide the results of experimental study of the effect caused by introduction of nanostructured fibres of aluminium oxyhydroxide into the surface layer of austenitic steel upon its microconstituents. The authors show that, due to introduction of given fibres dendrite size is reduced and equilibrium structure is formed.

1. Introduction

Under various methods of metal surface layer formation the crystalline structure of the given layer influenced by the conditions of liquid-solid transition is one of the factors determining the quality and the properties of this layer [1].

It is known that the process of nucleation can be spontaneous and induced [2]. In the build-up surface layer induced crystallization usually occurs at the fusion boundary where high-melting phases and microconstituents of the base metal may become nucleation centers [1]. In practice high-melting particles are deliberately introduced into the molten metal to increase the amount of induced nucleation centers which results in refinement of grains as the metal solidifies [1, 2].

Currently there are several promising ways used to modify surface layers with powders to obtain brand new properties of the surface: plasma, laser, ultrasonic, treatment with electric arc.

Electric arc was chosen as the means of surface activation as it is widely used in industry and allows introduction of nanodisperse particles after only slight modification of standard equipment.

The electric arc can be applied in three different ways depending on the place and method of application: when remelting the surface, when building-up the surface and for high-temperature treatment of the surface with application of consumable electrode [3, 4, 5].

The aim of the given research work was controlling the process of structure formation of the built-up layer produced from steel (chemical composition of the wire: 0.12% carbon, 18% chromium, 9% nickel; 1-15% titanium).

Relevance of the problem is determined by the fact that the structure directly conditions (specifies) the mechanical and service properties of the surface, i.e. insufficient service properties result in a larger amount of equipment failures and significant time and material losses caused by excessive repairing work [6, 7].

Thus, solution of the problem of properties improvement directly depends upon providing theoretical grounds and experimental verification of homogeneous fine-grained surface structure formation mechanism after introducing nanostructured powder modifiers into the surface layer.

2. Methods of research

To conduct the research built-up modified surface layers were produced. Two types of samples were used: №1 – modified with nanostructured aluminium oxyhydroxide fibers; №2 – without modification. For the purposes of the research the steel (chemical composition of the wire: 0.12% carbon, 18% chromium, 9% nickel; 1-15% titanium) samples were built-up by MIG-welding (welding conditions: current intensity 240-260 A; voltage 28-30 V) in argon atmosphere with electrode wire 12H18N9T 1.2 mm in diameter.

To study the microstructure sections were prepared. To prepare the sections the following methods were applied: mechanical grinding, mechanical polishing with diamond paste ASM 10/7 NVL and chemical etching (hydrochloric acid 75% HCl + concentrated nitric acid 25% HNO_3). The microsections were studied with application of optical metallography method using Neophot-21 microscope and digital camera Genius VileaCam for image recording.

In the given work we applied aluminium oxyhydroxide fibres AlO(OH) (Figure 1) of the following size: diameter of 5 nanometer, length 150 nanometer; specific surface area 150 m^2/g. For AlO(OH) fiber synthesis aluminium conductor electric explosive-produced powders were applied with specific surface area 7.5 m^2/g. Aluminium powders were made from aluminium wire “AM” Ø 0,35 mm. Then nanostructured Al powder was subjected to thermal hydrolysis to produce Al oxyhydroxide nanofibers (AlO(OH)). Aluminium electro-explosive powder was placed into distilled water which had been previously heated up to 60^0C. The aluminium powder reacted with the water under the given temperature for 25-30 minutes. The color of the aqueous suspension changed from dark-grey to white. 10 minutes after the suspension changed its color the container with the reaction mass was taken out of the constant-temperature bath. After cooling to the room temperature suspension was filtered, the deposit was washed with several portions of distilled water (3x300 ml) until its reaction was neutral (pH 5.5-6). Then the powders were dried under 110-115^{0}C. Microphotographs of aluminium oxyhidroxide were presented in Figure 1 [8, 9, 10, 11].

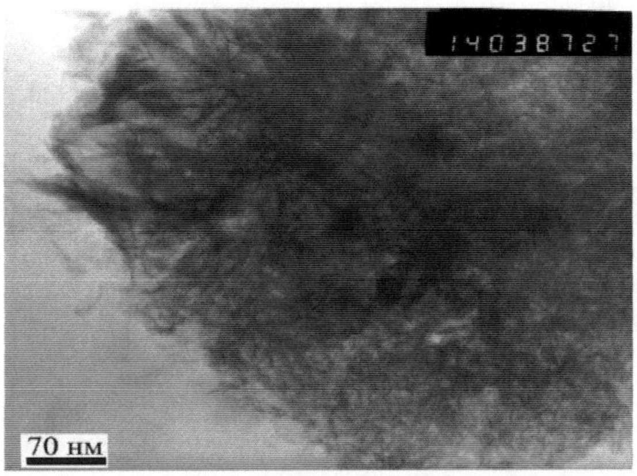

Figure 1 – Transmission electron microscopy: photograph of aluminium oxyhydroxide fiber

Aluminium electric explosive-produced powders on UDP-150 device have spherical form, the size of particles varies within (145-150) nanometer, specific surface area varies within (15-16) m^2/g [8, 10, 12].

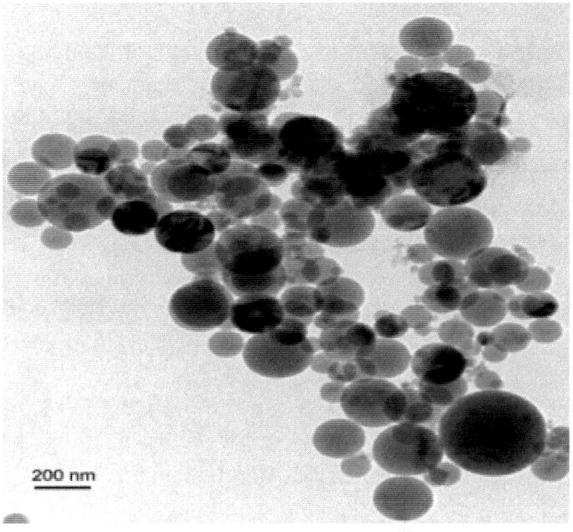

Figure 2 - Transmission electron microscopy of electric explosive-produced powders: aluminium

3. Results of the research and their discussion

Examination of the sections that had not been etched showed that the surface layers built-up with application of two different methods do not have macro- and microdefects.

The samples were studied according to the diagram presented in Figure 3.

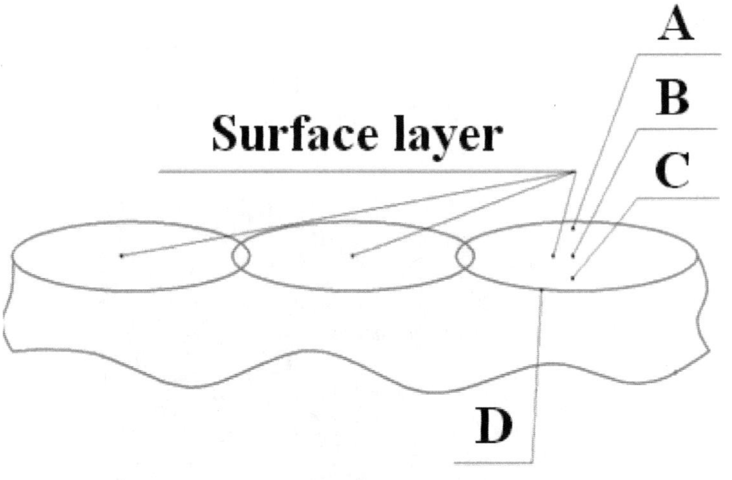

Figure 3 – Examination of the microstructure of the surface layer: A – upper underlayer, B – central underlayer, C – bottom underlayer, D – area of transition from the surface layer to the base one, E – base metal

Point D shows metal structure at the border of fusion and heat-affected zone (HAZ) (Fig. 4). Structure of heat-affected zone does not differ much from that of the base metal. The sample with aluminium oxide has smaller heat-affected zone which corresponds to less heat input and fast cooling.

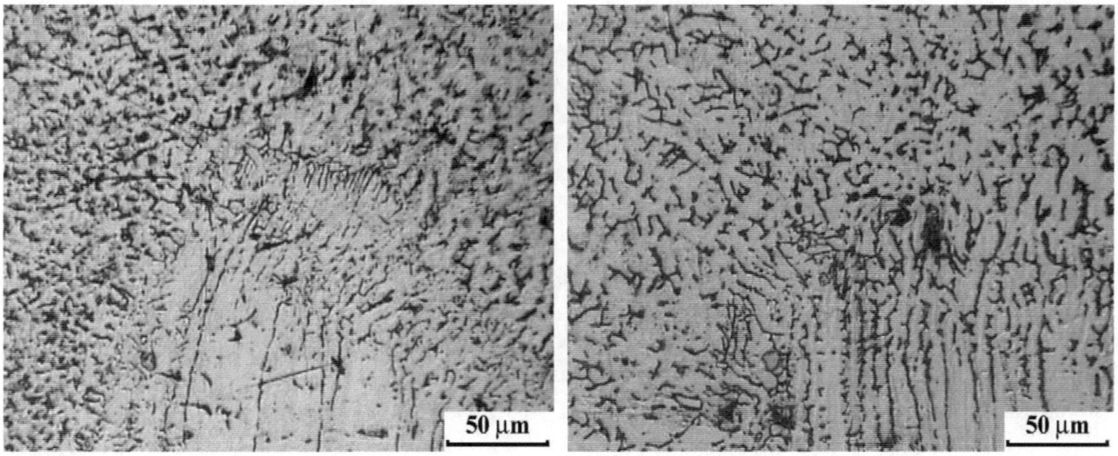

a b

Figure 4 – Structure of the metal at the border between the fusion and heat-affected zone: a – sample №1 modified with nanostructured fibers of AlO(OH); b – sample №2 without the modifier

The grain structure of the heat-affected zone transits to surface layer structure with pronounced dendrites.

The first layer with polyhedral grain structure. In this underlayer polyhedral grains of austenite are observed together with dendrites. This underlayer is not well pronounced in sample №2 produced without the modifier (Fig. 5 b). Its thickness is 20% of the general thickness of surface layer.

The "grain" underlayer is better pronounced in sample № 1 modified with nanostructured fibers of aluminium oxyhydroxide (Fig. 5 a). Here the polyhedral morphology grains can be well seen as they alternate with nonoriented dendrites. The thickness of the considered underlayer is over 30% of the general thickness of the surface layer. The specific feature of the given underlayer is that the grains may contain short and strongly ramified dendrites.

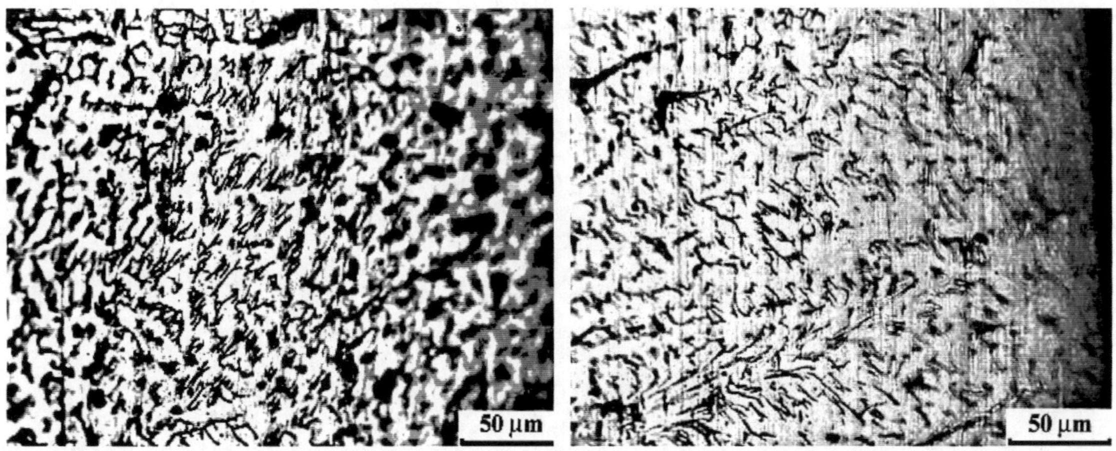

a b

Figure 5 – Microstructure of the underlayer of polyhedral grains: a – sample №1 modified with nanostructured fibers of AlO(OH); b – sample №2 without the modifier

The microstructure of the next underlayer is characterized by relatively short, smaller, strongly ramified nonoriented dendrites. In sample №1 modified with nanostructured fibers of aluminium oxyhydroxide the given underlayer is more pronounced (Fig. 6 a).

The dendrites in the given layer are relatively small as their axes do not coincide to the heat flow direction and they stop growing.

Thickness of the given underlayer in sample № 1 modified with nanostructured fibers of aluminium oxyhydroxide is 34% of general one. The same underlayer in sample № 2 without the modifier is less pronounced. In sample № 2 without the modifier relatively short strongly ramified nonoriented dendrites form a continuous pattern (Fig. 6, b) and in sample № 1 modified with nanostructured fibers of aluminium oxyhydroxide we can observe areas of free surface where, at the same time, grain boundaries can be seen. Thickness of weekly oriented dendrites layer in sample № 2 is 30% of general surface.

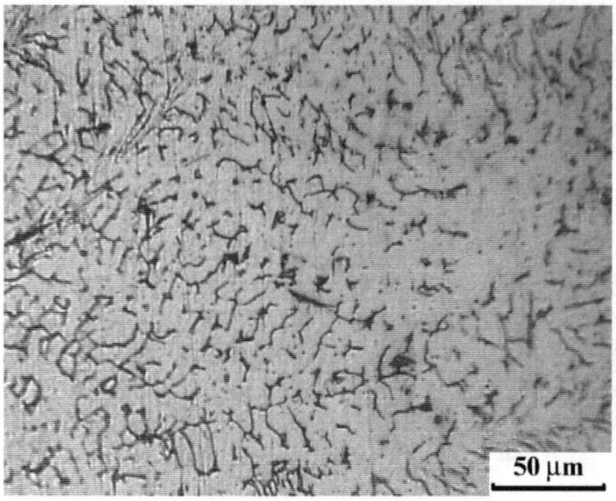

a

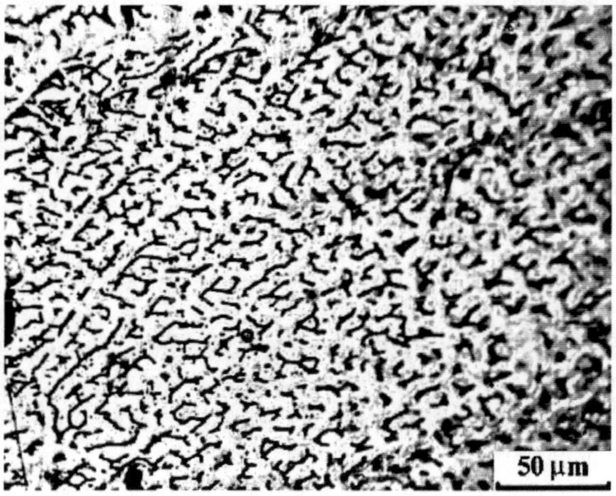

b

Figure 6 – The microstructure of the underlayer of nonoriented dendrites:
a – sample № 1 modified with nanostructured fibers of AlO(OH);
b – sample № 2 without the modifier

The microstructure of the following underlayer is made up of oriented long dendrites (Fig. 7 a, b)

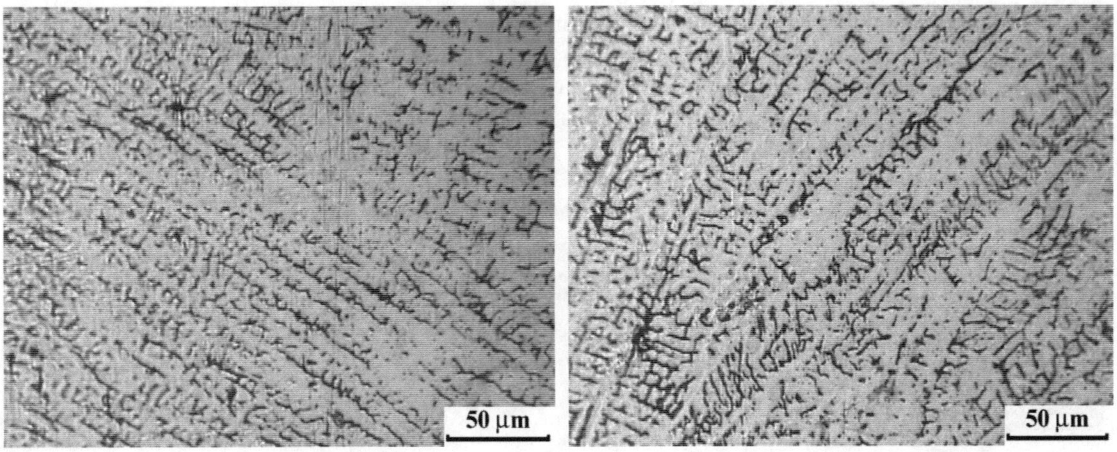

a b

Figure 7 – The microstructure of the underlayer of oriented dendrites: a – sample № 1 modified with nanostructured fibers of ALO(OH); b – sample № 2 without the modifier

Their orientation axis is normal towards the fusion border and is oriented along thermal current direction from the surface layer to the base metal. Strongly ramified thick dendrites are observed in sample № 2 without modifier (Fig. 7 b). Thickness of dendrites is 1.5-2 micrometer and their width (distance between the ends of opposite branches) is 20-25 micrometer. Thickness of dendrites in sample № 1 modified with nanostructured fibers of aluminium oxyhydroxide 0.8-1 micrometer and their width (distance between the ends of opposite branches) is 7-10 micrometer.

Orderly orientation of long dendrite axes is broken right before the fusion border and one more underlayer of weakly oriented dendrites is formed (Fig. 8).

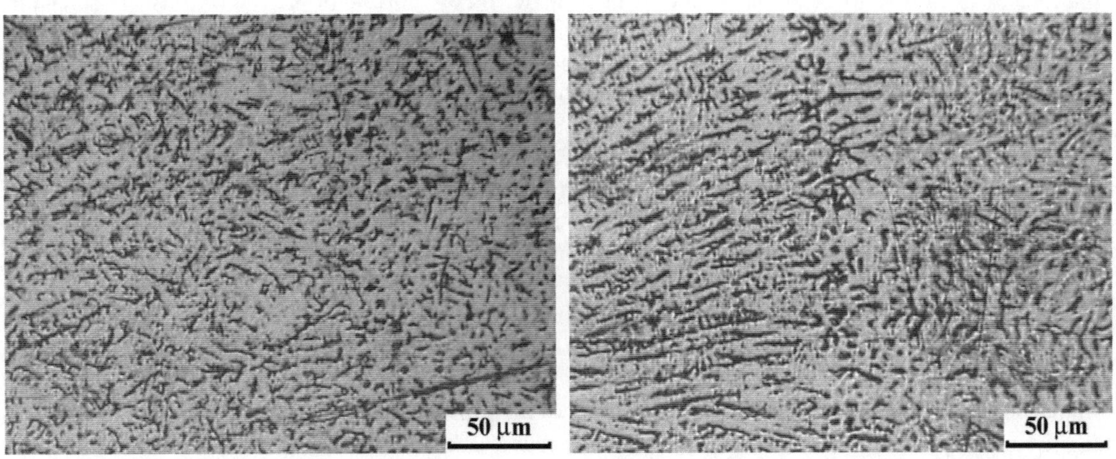

a b

Figure 8 – Microstructure of the underlayer of weakly oriented dendrites:
A – sample № 1 modified with nanostructured fibers of AlO(OH);
B – sample № 2 without the modifier

The given underlayer is weakest pronounced in sample № 2 № without the modifier (Fig. 8 b). Sample № 1 modified with nanostructured fibers of aluminium oxyhydroxide (Fig. 8 a) the border between oriented and nonoriented dendrites is about 25 micrometer thick.

The microstructure of the underlayer can be estimated as intermediate between areas A and B. In sample № 1 modified with nanostructured fibres of aluminium oxyhydroxide a clearly pronounced "grain" layer is observed again (Fig. 9 a).

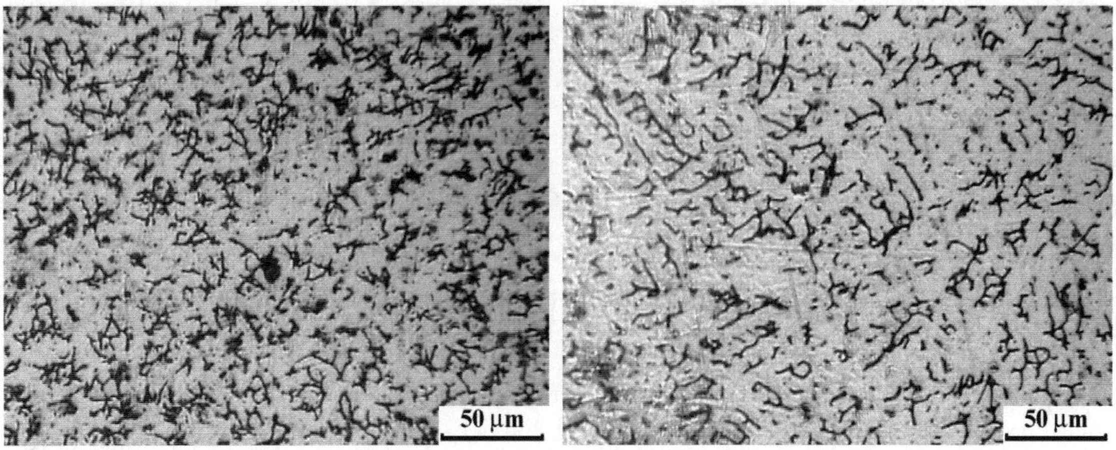

a b

Figure 9 – Microstructure of the underlayer of nonoriented dendrites:
A – sample № 1 modified with nanostructured fibers of AlO(OH);
B – sample № 2 without the modifier

The situation in sample № 2 without the modifier is different (Fig. (b) as the structure of the underlayer is closest to the microstructure in B in terms of morphology. Here we mainly observe chaotically situated nonoriented dendrites with insignificant amount of polyhedral morphology grains. In sample № 1 modified with nanostructured fibres of aluminium oxyhydroxide the underlayer of grain structure was ≈ 4 mm and in sample № 2without the modifier the layer of nonoriented dendrites was ≈ 3 mm.

All studied samples are rather perfect both in terms of their structure and quality of the surface layer. Application of nanostructured fibers of aluminium oxyhydroxide allows regulating the structure of the surface layer. It was established that the surface layer has stratified structure determined by changing the heat removal conditions from the surface inside the molten pool. Heat removal is weak near the free surface that is why crystallizion results in forming polyhedral grain, dendrites do not have time to get formed. Most clearly this process is formed in the sample modified with nanostructured fibers of aluminium oxyhydroxide. In the rest part of surface layer typical dendrite crystallization took place. Approximately half of surface layer volume is taken by oriented dendrites. In the sample modified with nanostructured fibers of aluminium oxyhydroxide it is less that a half and in the sample produced without the modifier it exceeds half of surface layer volume. There is also the underlayer of nonoriented dendrites. This underlayer is especially wide in the sample modified with nanostructured fibers of aluminium oxyhydroxide. According to generally accepted view the more dendrite structure reveals itself, the larger are dendrites the worse are the service characteristics of the given layer [13, 14, 15]. From this point of view the surface layer produced without the modifier is worse than that of the sample modified with nanostructured fibers of aluminium oxyhydroxide. The microstructure of the surface layer modified with nanostructured fibres of aluminium oxyhydroxide is fine-grained and homogenious.

4. Conclusion

1. It has been established that application of nanostructured powders allows controlling the microconstituents of the surface layer.

2. Most clearly the layer of polyhedral grains is observed in the sample modified with nanostructured fibres of AlO(OH). The layer of oriented dendrites in the sample modified with nanostructured fibres of AlO(OH) occupies less than half of the built-up surface layer in terms of thickness. The layer of nonoriented dendrites is the widest in the sample modified with nanostructured fibres of AlO(OH).

3. In terms of dendrite size more equilibrium structure is produced when modifying the surface layer with nanostructured fibres of AlO(OH).

References

[1] Livshits L.S. Metal science for the welders (welding steels). – M.: Mashinostroyeniye, 1979. – 253 p.

[2] Material science / Arzamasov B.N., Mukhin G.G., Makarova V.N., Ryzhov N.M. – M.: MSTU named after N.E. Bauman press, 2004. – 648 p.

[3] Il'yaschenko D.P., Chinakhov D.A., Danilov V.I., Schlyakhova G.V. and Gotovshchik Yu M 2015 IOP Conference Series: Materials Science and Engineering Physical Nature of the Processes in Forming Structures, Phase and Chemical Compositions of Medium-Carbon Steel Welds 91 pp http://iopscience.iop.org/article/10.1088/1757-899X/91/1/012006/pdf.

[4] Il'yaschenko D.P., Chinakhov D.A., Danilov V.I., Schlyakhova G.V. and Gotovshchik Yu M 2015 IOP Conference Series: Materials Science and Engineering Increasing Strength and Operational Reliability of Fixed Joints of Tubes by MMA Welding 91 pp http://iopscience.iop.org/article/10.1088/1757-899X/91/1/012007/pdf.

[5] Pavlov N.V., Kryukov A.V., Zernin E.A. and Gritsenko V.V. 2015 IOP Conference Series: Materials Science and Engineering Systems to Control Molten Metal Transfer in Arc Welding 91 pp http://iopscience.iop.org/article/10.1088/1757-899X/91/1/012021/pdf.

[6] Jian-Yih Wang, Tien-Chan Chang, Li-Zen Chang and Shyong Lee 2006 Materials Transactions Effect of Al and Mn Content on the Mechanical Properties of Various ECAE Processed Mg-Li-Zn Alloys47 4 pp 971-976.

[7] Chen Cuixin, Peng Huifen, Liu Ran, Li Yuanyuan, Zhao Pu. J. 2012 Comput. and Theor. Nanosci Research on inclusions in low alloy steel welds with nano alumina addition 9 pp 1533-1536.

[8] Zhuravkov S., Pustovalov A., Lobanova G., Kvashnina O., Yavorovsky N. Proceedings of 2014 International Conference on Mechanical Engineering, Automation and Control Systems, MEACS 2014. Production of aluminium oxyhydroxides by various methods Tomsk Polytechnic University, 15 December 2014,Tomsk; Russian Federation. (Conference Paper) Article number 6986954.

[9] Zhuravkov S.P., Lobanova G.L., Martemiyanov D.V., Nadeina L.V. 2015 IOP Conference Series: Materials Science and Engineering. Study of sorption properties of aluminium oxides and hydroxides powders obtained by electro-impulse methods 81 pp http://iopscience.iop.org/article/10.1088/1757-899X/81/1/012057/pdf.

[10] Yavorovsky N, Pustovalov A., Lobanova G., Zhuravkov S. Proceedings - 2012 7th International Forum on Strategic Technology. Investigation of the characteristics of aluminum powder obtained in argon with addition of oxygen. (Conference Paper): Article number 6357559. IFOST 2012, Tomsk; Russian Federation.

[11] Zhuravkov S.P., Lobanova G.L., Kornev J.I., Lutoshkina O.S., Khristunova Ye.P. 2016 Key Engineering Materials The study of properties characterizing aluminum hydroxides produced by electric discharge 685 pp 680-684.

[12] Sun J., Simon S.L. 2007 Thermochimica Acta The melting behavior of aluminum nanoparticles 463 pp 32–40.

[13] Surowiec Z., Wiertel M., Gac W., Budzynski M. 2014 Acta physica polonica The influence of reduction process on the iron-molybdenum nanoparticles in modifie MCM-41 silica 125 4 pp 846-849.
[14] Golubtsov V.A., Shub L.G., Ryabchikov I.V., Demin K.Y., Demin Y.S. 2009 Steel in Translation Improving wheel steel by means of barium-based modifiers 39 12 pp 1078-1083.
[15] Golubtsov V.A., Ryabchikov I.V., Yarovoi K.I., Milyuts V.G., Levagin E.Y., Mizin V.G. 2013 Steel in Translation Effectiveness of barium-bearing ferroalloys in steel smelting 43 8 pp 511-514.

Developing and Studying the Methods of Hard-Facing with Heat-Resisting High-Hardness Steels

N N Malushin [1], A P Kovalev [2], D V Valuev[3], E A Shats[4], I F Borovikov[5]

[1,2] 654000 Kemerovo region, Novokuznetsk, Cosmic highway 16, OJSC Evraz ZSMK,
[3,4] 652050, Kemerovo region, Yurga, Leningradskaya str.26 Yurga Technological Institute branch of Tomsk Polytechnic University
[5] The Moscow State Technical University of N.E. Bauman, Moscow, 5, 2nd Baumanskaya

E-mail: valuev@tpu.ru

Abstract: The authors develop the methods of hard-facing of mining-metallurgic equipment parts with heat-resisting high-hardness steels on the base of plasma-jet hard-facing in the shielding-alloying nitrogen atmosphere.

Introduction

Alloying materials developed on the base of chrome-tungsten heat-resisting high-speed steels are widely applied for strengthening mining-metallurgical equipment parts [1, 2]. The most common problem appearing when alloying with chrome-tungsten steels is formation of cold (tempering) cracks in the process of and after alloying. Crack formation is reduced with increase of the temperature of martensite transformation and slowing down the cooling rate. The easiest and the most efficient way to prevent cracking is to regulate the thermal cycle through choosing proper methods and conditions of alloying as well as through application of preheating if necessary. To prevent cracking, reduce internal stresses and obtain enough plastic structure of the hard-faced metal compulsory preheating of the parts up to 350–400°C is applied when they are hard-faced with highly-alloyed electrode wire PP-3H2V8 or solid-drawn EI-701 wire. To eliminate cracking, when hard-facing the high-speed steels the workpieces are to be preheated and additionally heated up to the temperature of 500–700°C [1,2]. The traditional technology of bimetallic products manufacturing (hard-facing with preheating – annealing – hardening – tempering) is unreasonably time and labor consuming. Compulsory application of high-temperature heating, retarded cooling and, as a result, of further hardening of the hard-faced part is a significant disadvantage of the technology. Retarded cooling of the parts which is recommended to prevent cold cracking generates a need for further hardening of the bimetallic item which, by itself, is a difficult task due to the difference between the properties of hard-faced highly alloyed metal and low alloy center. Hardening of some large-sized parts is almost impossible to complete. Current technological processes of tool steels hard-facing do not allow full use of the hard-faced metal properties [1-3]. That is why the problem of developing new methods of chrome-tungsten steels hard-facing is relevant as the new methods allow improving the mentioned above disadvantages of the traditional technology and taking the full advantage of high hardness and wear resistance of chrome tungsten steels as they have been designed by the metallurgists and metal physicists. It is practical to combine the processes of metal hard-facing and hardening as it allows producing a hard-faced layer which state is close to that of a hardened metal. It is also necessary to avoid high-temperature heating or at least reduce its temperature applying new methods of cold cracking elimination.

The work objective is developing and studying the methods of hard-facing with heat-resisting steels high-hardness steels ensuring obtaining hard-faced metal which state is close to the state of the hardened metal and which has no cracks.

Hard-faced metal characterized by high resistance to cold cracking and high mechanical properties can be obtained immediately after the process of hard-facing is finished through regulation of the thermal cycle of hard-facing. The suggested thermal cycle for the multiple-bead deposit of chrome-tungsten hardening steels is shown in Figure 1 [3, 4].

The specific feature of the suggested methods of hard-facing is application of low-temperature preliminary and additional heating. To obtain hard-faced metal with reduced ability for crack formation we have to regulate the level of temporary stresses during the hard-facing process by their partial relaxation due to kinetic plasticity effect which appears at the moment of martensite or bainitic transformation. Relaxation of temporary stresses within the temperature interval of martensite or bainitic transformation prevents cracking during the process of multiple-bead hard-facing.

Cooling of the hard-faced layers and isothermal hardening within the interval of martensite transformation leads to thermal stabilization of austenite. It is worth noting that 70 – 90% of the metal hard-faced in the process of depositing is stabilized austenite. When the amount of austenite increases the stresses reduce, plasticity grows, volumetric changes, deformations and crack sensitivity reduce. The duration of isothermal tempering depends upon the chemical composition of hard-faced metal and can be determined experimentally for each grade of steel or according to the diagram of undercooled austenite transformation.

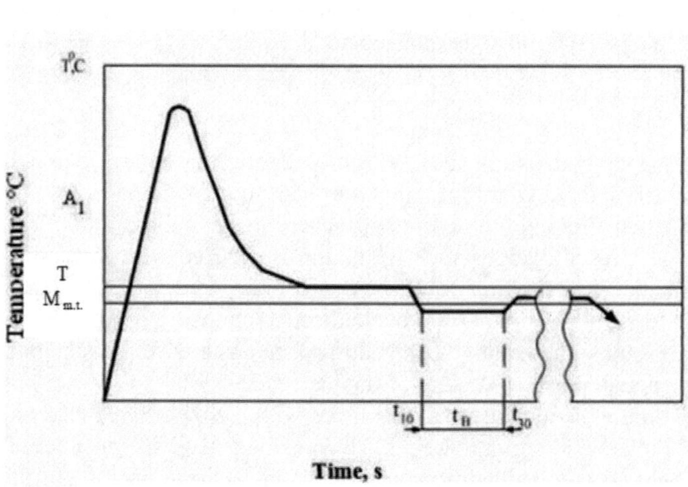

Figure 1. Diagram of the thermal cycle when hard-facing with heat-resisting steels

The level of relaxation of temporary stresses is determined by the time of isothermal hardening t_B under the temperatures below $M_{\text{the beginning of the martensitic transformation}}$, which is found from correlation 1:

$$t_{10} < t_B < t_{30}, \qquad (1)$$

where t_{10}- time required formation of 10% of martensite; t_{30} – time required for formation of 30% of martensite.

If the duration of isothermal hardening is smaller than the duration required to form 10% of martensite no significant relaxation of temporary stresses occurs leading to temporary stress accumulation in the process of multiple-bead hard-facing which, in its turn, may result in crack formation. Besides, relaxation of stresses is important to stop the martensite transformation and to stabilize the austenite. That is why hardening with the time $t_в$ is needed for partial relaxation of temporary stresses in the process of hard-facing and obtaining stabilized austenite. When hardening time exceeds $t_в$ the metal ductility is significantly deteriorated due to formation of a large amount of martensite which results in crack formation. Besides, longer hardening increases the duration of the hard-facing and, thus, reduces the productivity of the process. If the amount of martensite formed in the process of hard-facing exceeds 30% more intense breakdown of martensite occurs. During the process of martensite breakdown carbides are precipitated, that is why high hardness cannot be achieved during further hardening. To obtain homogeneous metal structure of all hard-faced layers cooling and hardening are completed after depositing every single layer.

The special features of the suggested thermal cycle of the hard-facing also include limited time of the hard-faced metal exposure to high temperature heating, high rates of metal cooling in the area of minimal austenite stability, keeping the metal in the austenitic state until the end of the hard-facing process through application of preheating with the temperature $M_{\text{the beginning of the martensitic transformation}}$ +(50-100)°C (Figure 1). To reduce the temporary stresses accumulated in the austenite region we suggest short-term lowering of the preheating temperature by (20-100)°C below $M_{\text{the beginning of the martensitic transformation}}$. Preheating temperature reduction in the process of multiple-bead hard-facing leads to partial relaxation of temporary stresses which happens mainly due to superplasticity effect which happens in the moment of martensite of bainitic transformation. Relaxation of temporary stresses within the temperature range of martensite or bainitic transformation prevents formation of cracks in the process of multiple-bead hard-facing. The level of temporary stresses relaxation is ensured by hardening time $t_в$ under the temperatures below $M_{\text{the beginning of the martensitic transformation}}$

Plasma-jet hard-facing with non-current-carrying filler flux cored wire was chosen as the method of hard-facing. Choosing plasma-jet hard-facing as the basic method of depositing wear-resistant coating to produce new and renew the worn-out parts is explained by a number of advantages laser-jet hard-facing has in comparison to other hard-facing methods such as: high productivity, wide opportunities of alloying the hard-faced metal, opportunity of applying various alloying metals. The wide range of regulation of the heat input into the base and the hard-faced metal allows implementation of the suggested thermal cycle of the hard-facing. The enumerated advantages of plasma-jet hard-facing methods not only significantly increase the technological opportunities of their application but also allow substantial economic effect due to hard-facing of layers with minimal penetration and retaining original physical and mechanic properties of the hard-faced metal. The specific feature of a constricted arc as a heat source is also the opportunity of easy regulation of its thermal and gas-dynamic characteristics within a wide range. Application of a constricted arc with reverse polarity allows avoiding labor consuming operations of preliminary surface preparation complicating the process of hard-facing. The surface of the hard-faced part is cleaned directly in the process of hard-facing due to cathode spattering effect, thus, ensuring the necessary conditions of wetting the surface of the part with the hard-faced metal and producing a defect-free deposit layer. Under plasma-jet hard-facing with reverse polarity we also observe less dilution of the hard-faced metal with the base one [3-7].

The most efficient method for solving a number of process tasks when hard-facing parts of metallurgic equipment like bodies of rotation (rollers, forming rolls, shafts) is application of plasma-jet hard-facing with reverse polarity in shielding-alloying nitrogen atmosphere with non-current-carrying filling flux cored wire. Nitrogen used as a shielding gas as compared to argon allows not only reducing the costs of the hard-facing but also efficient alloying the hard-faced metal with nitrogen from the gas phase immediately in the process of hard-facing, thus, significantly improving its (hard-faced metal) hardness and wear-resistance. Porosity of the hard-faced metal can be prevented by introducing elements with affinity for the nitrogen and fixing it within stable nitrides into the flux cored wire. Introduction of certain amounts of aluminium into the mixture for the flux cored wire has a positive effect on the hardness improvement of the hard-faced metal and porosity prevention. In the process of hard-facing the deposit metal is alloyed by the nitrogen directly from the gas phase which allows additional increasing of the working layer hardness. Aluminium is introduced into the mixture for the flux cored wire to avoid porosity. Aluminium fixes the excessive nitrogen into compounds which cannot be solved in the molten metal and can provide additional strength due to fine disperse nitride particles formation. Additional improvement of the properties of the hard-faced highly alloyed metal as well as ensuring favorable stressed state can be achieved by application of high temperature tempering. Thus, hardness of metal after hard-facing with flux cored wires PP-R18YuN is HRC 52-57 and, in terms of composition its structure is close to that of high-speed steel of R18 type in its tempered state and consists of martensite (about 70%), carbides (up to 20%) and retained austenite (up to 10%). 3-4-time high-temperature tempering under 580°C increases the hardness of the hard-faced metal up to HRC 62-64. Improvement of hardness is explained by transformation of retained austenite into martensite and effect of precipitation hardening. Hardness of the hard-faced metal after tempering for secondary hardness reaches HRC 64-66 [3 -9].

The typical structure of the hard-faced metal after multiple-bead laser-jet hard-facing and high-temperature tempering is shown in Figure 2 (microscope OLYPUS GX – 51, etching solution HNO_3, 3% spirit, magnification ×500).

To complete plasma-jet hard-facing with non-current-carrying flux cored wire we applied a unit assembled from serial equipment.

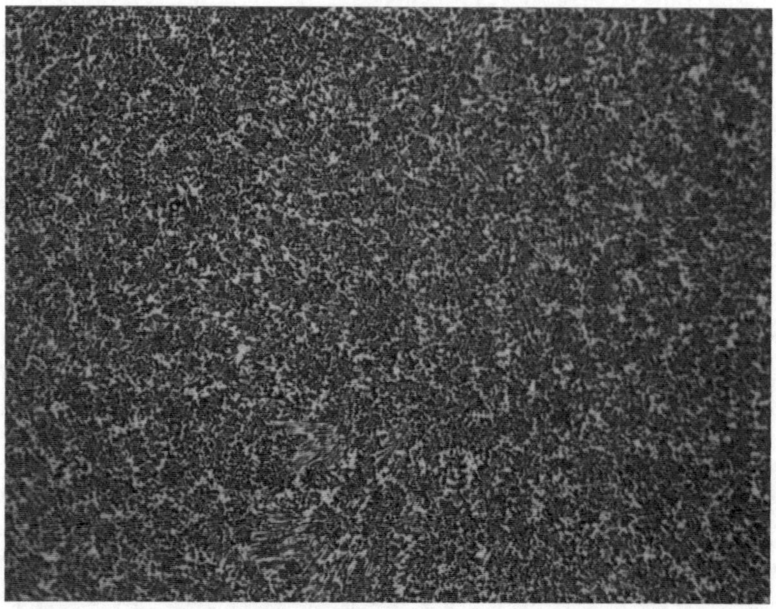

Figure 2. Structure of the hard-faced metal after multiple-bead plasma-jet hard-facing and high-temperature tempering, ×500

The suggested methods are efficient for hard-facing chrome-tungsten steels with low temperatures of phase transformation start ($M_{\text{the beginning of the martensitic transformation}}$ =180° – 380°C). This allows significant reduction of the preheating temperature for R18 steel: from 500–600°C to 230–280°C and even to 80–160°C according to the suggested methods and by itself simplifies the technology of chrome-tungsten steels hard-facing. The methods of hard-facing with heat-resisting steels ensure high resistance of the hard-faced metal to heat-treatment cracking and producing a hardened in the process of hard-facing layer with high hardness and high wear resistance.

Conclusion

1. The traditional technology of chrome-tungsten steels hard-facing with application of high-temperature preheating and retarded cooling is labor consuming, lacks efficiency and does not allow full use of high service properties of the hard-faced metal.

2. The methods of multiple-bead hard-facing were developed basing on following a certain thermal cycle. The specific feature of the suggested methods of hard-facing is application of low-temperature preheating and additional heating ($T_{heating}$ = 230 – 280°C). to produce hard-faced metal with low ability for crack formation we can regulate the level of temporary stresses in the process of hard-facing through their partial relaxation due to the effect of kinetic plasticity in the moment of martensite or bainitic transformation.

3. Reverse polarity plasma-jet facing with non-current carrying flux cored wire in shielding-alloying nitrogen atmosphere was chosen as the process realizing the suggested above method of hard-facing.

References

[1] Frumin I I Automated electric arc welding Metallurgizdat 1961 421
[2] Pimnev A L Zemenkova M Yu Zemenkov Yu D Il'yashchenko D P Mechanical properties of the assembly welded joint of the oil transportation tank after a long-term service J IOP Conference Series: Materials Science and Engineering. 127 (2016)
[3] Malushin N N Valuev D V Ensuring the quality of metallurgic equipment at all stages of their life cycle through plasma-jet hard-facing with heat-resistant steels Tomsk Polytechnic University Press, 2013 358
[4] Fedorov N S Zubkov N S Krepysheva L B Thermal treatment effects upon hardness and phase composition of the hard-faced high-speed steels J Newsletter of Higher Educational institutions. Ferrous metallurgy 8 (1983) 59–63
[5] G V Galevskiy V V Rudneva S G Galevskiy D P Il'yashchenko D S Kartsev Nanosized borides and carbides for electroplating. metal-matrix coatings: specifications, performance evaluation J IOP Conference Series: Materials Science and Engineering. 125 (2016)
[6] Chinakhov D A Chinakhova E D Gotovschik Y M Grichin S V Influence of welding with two-jet gas shielding on the shaping of a welding joint J IOP Conference Series: Materials Science and Engineering. 125 (2016)
[7] Malushin N N Osetkovsky V L Osetkovsky I V Hard-facing of metallurgic equipment parts with heat-resisting with low-temperature preheating J Blank production in engineering 10 (2014) 6 – 10
[8] Gizatulin R A, Senkus V V, Valueva A V, Baldanova A S, Borovikov I F A technology of wastewater sludge treatment J IOP Conf. Series: Materials Science and Engineering 127 (2016)
[9] Galevsky G V Rudneva V V Garbuzova A K Valuev D V Titanium carbide: nanotechnology, properties, application, J. IOP Conference Series: Materials Science and Engineering. 91 (2015)

IOP Conf. Series: Materials Science and Engineering **142** (2016) 012013 doi:10.1088/1757-899X/142/1/012013

Specifics of Pulsed Arc Welding Power Supply Performance Based On A Transistor Switch

N Yu Krampit [1a], PhD in Technical Sciences, T S Kust [1b], PhD in Technical Sciences, M A Krampit [1c]

[1]Yurga Institute of Technology, TPU Affiliate, Russia, Kemerovo Region, Yurga

e-mail: [a]nkrampit@mail.ru, [b] kust@tpu.ru, [c]krampitm@tpu.ru

Abstract. Specifics of designing a pulsed arc welding power supply device are presented in the paper. Electronic components for managing large current was analyzed. Strengths and shortcomings of power supply circuits based on thyristor, bipolar transistor and MOSFET are outlined. As a base unit for pulsed arc welding was chosen MOSFET transistor, which is easy to manage. Measures to protect a transistor are given. As for the transistor control device is a microcontroller Arduino which has a low cost and adequate performance of the work. Bead transfer principle is to change the voltage on the arc in the formation of beads on the wire end. Microcontroller controls transistor when the arc voltage reaches the threshold voltage. Thus there is a separation and transfer of beads without splashing. Control strategies tested on a real device and presented. The error in the operation of the device is less than 25 us, it can be used controlling drop transfer at high frequencies (up to 1300 Hz).

1. Introduction

Switching high currents occurs when designing pulsed arc welding power supply devices, provided that processes are changed rapidly, but weight and size characteristics of these devices are modified minimally. Ascertainment of a rational method for designing pulsed arc welding power supply devices is possible on condition that solutions of problems below are found:

- an appropriate power switch to control large-scale load is selected;
- a control circuit of a power switch is identified;
- a device based on the circuit above is designed and tested.

2. Results and Discussion

Thyristor [1], bipolar transistor [2] and MOSFET [3] are used as power switches in pulsed arc welding power supply devices. Each circuit has its specifics, as well as strengths and shortcomings.

The following drawbacks affecting the power supply device are to be taken into account when designing a thyristor-based power supply device [4]:

- low operating speed because of a long turn-off time;
- restricted controllability: low-power control circuit is used for thyristor activation, whereas power circuit – for its blanking;
- blanking is possible via short voltage impulses, decreasing interference immunity of thyristor-based devices.

A power supply device can be based on a bipolar transistor. There are three main connection circuits of a bipolar transistor available: common base transistors, common emitter transistors and common collector transistors. Each of them has its shortcomings [5].

Defects of the common base connection circuit are as follows:

- low current gain (below 1);
- low input resistance;
- two different voltage sources for power supply.

The common emitter connection circuit has worse temperature and frequency characteristics than the common base connection circuit, so it is its main drawback.

A common collector connection circuit can be considered imperfect because its coefficient of voltage gain is below 1.

The paper presents a voltage feedback algorithm to control drop transfer.

MOSFET is a voltage-controlled transistor in comparison to the bipolar one. It is currently the best possible low-energy circuit in the static quiescent mode [6]. These transistors are of a small size but their capacity amounts to hundreds of Watt. They are also distinguished by high operating speed and good interference immunity.

Pulsed arc welding is a reasonably complex process. Electric current pulse-up is to be at the rigorously defined moment in order to support electrode metal drop transfer to a weldpool. As the result of a couple of milliseconds error the main condition of process control "one impulse – one drop" is broken, as the consequence, electrode metal splashing increases.

As there are a lot of factors affecting the process of welding, it's rather difficult to predict it and set all welding parameters beforehand. Any current network disturbance or slip of a wire supply mechanism can lead to the increase in an inter-pulse period. This problem is rationally solved by means of arc voltage feedback because the arc is shortened; consequently, arc voltage gets diminished when balancing an electrode drop on the wire end (Figure 1). A stable welding process i.e. without electrode metal splashing is possible via regulating the threshold voltage, followed by a pulse pile-up. The control algorithm is presented in Figure 2.

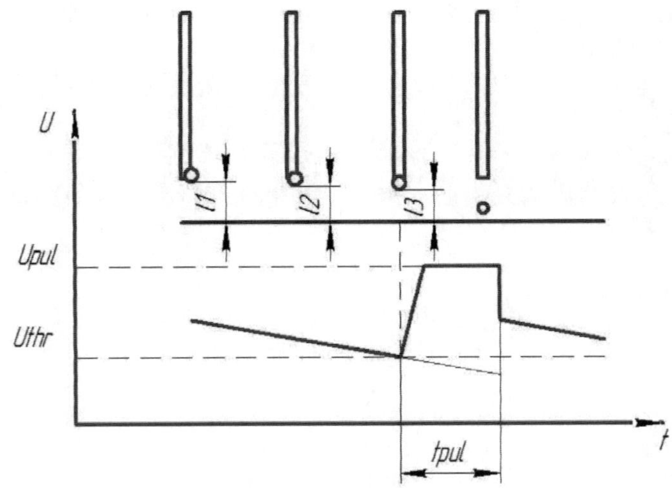

Figure 1 - Control strategy of voltage feedback drop transfer

IOP Conf. Series: Materials Science and Engineering **142** (2016) 012013 doi:10.1088/1757-899X/142/1/012013

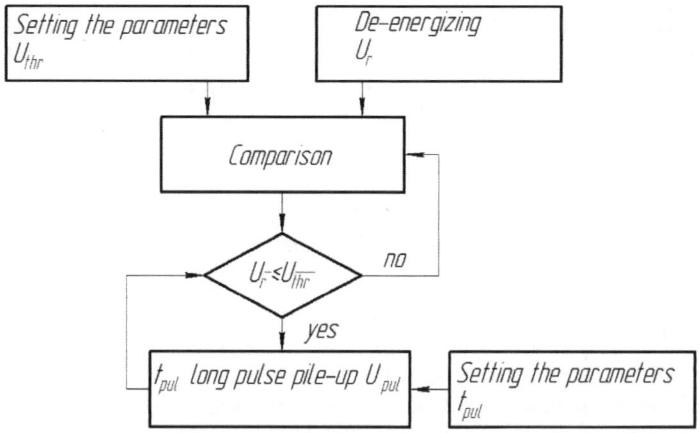

Figure 2 - The algorithm to control voltage feedback drop transfer

Microcontroller Arduino UNO, having 6 analog inputs and 14 digital inputs and outputs, operated at frequency 16 MHz was used to control the transistor. 5 V digital input voltage was sufficient to turn on the MOSFET. They also can be used as pulse-width modulation outputs (PWM), i.e. an irregular shape of the impulse can be set [7, 8].

The algorithm above was tested on pulsed arc power supply device. Oscillograms of the process are shown in Figure 3. They demonstrate that at the frequency less 1300 Hz time inaccuracy is irrelevant and does not exceed 25 us.

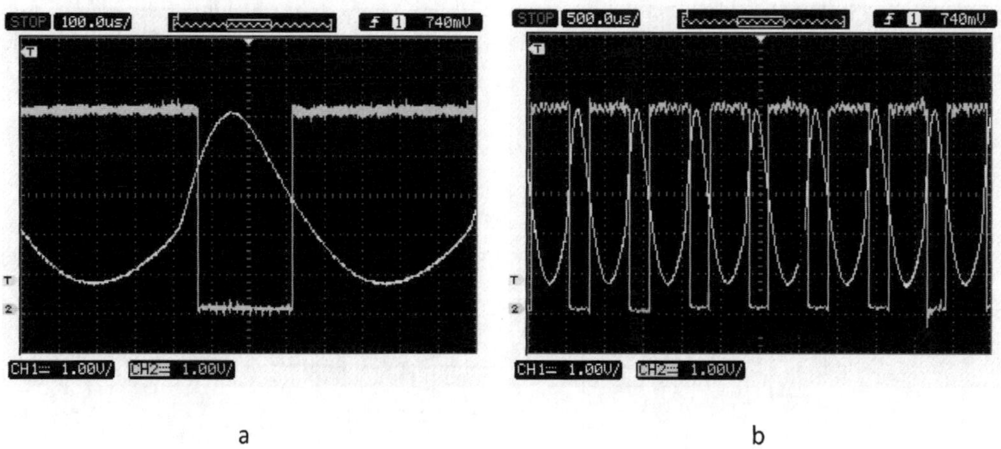

Figure 3 Pulse pile-up at Uthr
f=1300 Hz; Uthr=3 V

The authors [9] suggest a pulsed arc welding mechanism, where a wire section is heated by flowing inter-pulse current. However, overheating is possible. The wire can get crumpled up when entering a terminal clamp owing to limpness caused by heating over the defined temperature. The current flowing through the wire diminishes to the inter-pulse current. These parameters are to be adjusted when welding; and wire overheating is to be avoided.

A thermocouple is to be placed into a terminal clamp in order to de-energize the arc and remove electric current intensity from it. The data are converted in the microcontroller, signaling to the servodrive. The servodrive moves the second terminal clamp, changing the length of a wire section. Its resistance is changed; therefore, inter-pulse current strength and wire heating temperature vary.

One of the most frequent failure reasons of MOSFET-based electronic devices is exceeding the allowed value of drain-to-source voltage. [10]. For instance, voltage spike occurs when switching the inductive load; and its amplitude exceeds the maximal allowed voltage of a MOSFET. As the result, avalanche breakdown of a semi-conductor and transistor damage are possible. A fuse diode VD1 is placed between drain and source in the circuit for MOSFET protection (Fig. 4). Transient MOSFET processes are frequently possible due to the electrostatic discharge. Suppressor VD2 placed between drain and source makes it possible to protect the transistor from input transient processes. A protective diode is mounted with the reverse voltage value, exceeding the MOSFET input value (see Fig. 4).

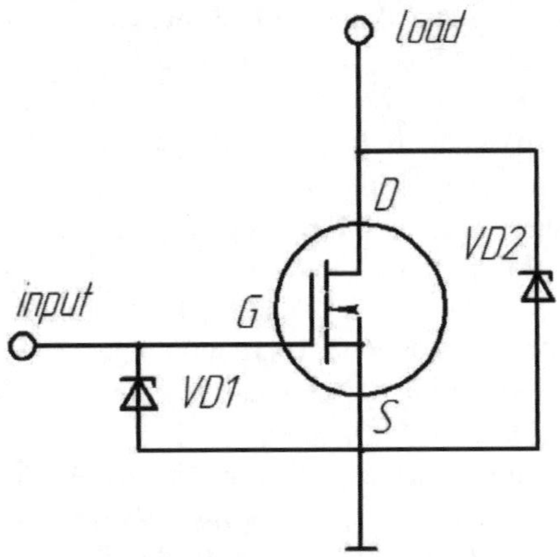

Figure 4 - A diagram of transistor protection

A modulator comprising a microcontroller and a MOSFET was designed and tested as the device base for electrode extension heating by inter-pulse current [11]. Computations of the process and carried out research reveal high energy efficiency. Quality of weld joints at diminished energy input meets the standards (Fig. 5).

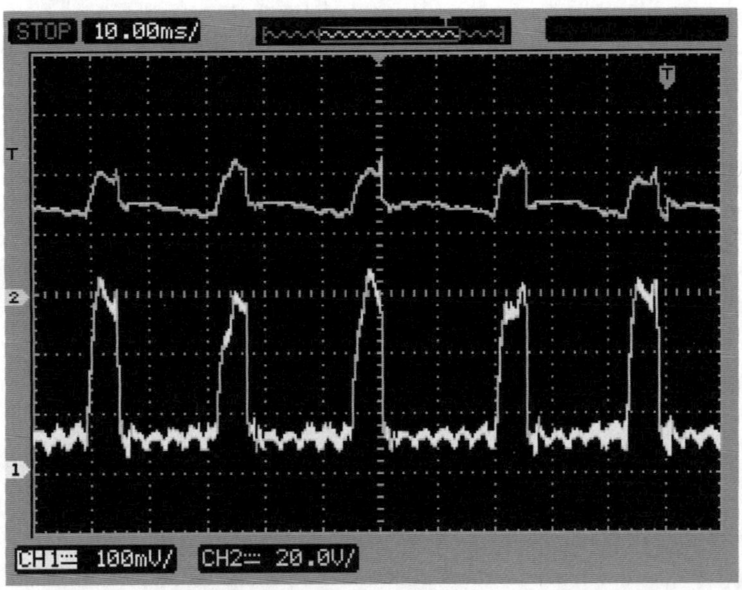

Figure 5 - Electric current and voltage oscillogram of pulsed arc welding based on heating electrode extension by inter-pulse current

3. Conclusions

1. Optimal construction of devices for pulsed arc welding based on MOSFET. Microcontroller with voltage feedback serves control the transistor.
2. A voltage feedback microcontroller facilitates a controlled transfer when pulsed arc welding at frequencies below 1300 Hz.
3. This device allows to create power supplies for pulsed arc welding on the basis of any welding rectifiers at low cost price

Reference

[1] Hao W L, Luo F and Ieee 2009 Development of Constant Voltage Control System for Hot-melt Pipe Welding Iccsse 2009: *Proceedings of 2009 4th International Conference on Computer Science & Education* 491-4

[2] Dudrik J and Trip N D 2010 Soft-Switching PS-PWM DC-DC Converter for Full-Load Range Applications *Ieee Transactions on Industrial Electronics* **57** 2807-14

[3] Zhang Z J, Li H M, Peng Y L and Ieee 2006 Application of MOSFET solid. state high frequency power supply in pipe welding. In: *IEEE Region 10 Conference (TENCON 2006),* (Hong Kong, PEOPLES R CHINA: Ieee) pp 9-11

[4] Shammas N Y A, Eio S and Chamumd D 2009 Electronic Circuits for Switching-Time Reduction of Bipolar Semiconductor Devices *Proceedings of the 13th Wseas International Conference on Recent Advances in Circuits* 28-42

[5] Kuo Y and Lei Y 2004 Fabrication and characterization of hydrogenated amorphous silicon bipolar thin film transistor (B-TFT). In: *Symposium on Amorphous and Nanocrystalline Silicon Science and Technology held at the 2004 MRS Spring Meeting,* (San Francisco, CA: Materials Research Society) pp 709-14

[6] Stepins D 2016 An Improved Control Technique of Switching-Frequency-Modulated Power Factor Correctors for Low THD and High Power Factor *Ieee Transactions on Power Electronics* **31** 5201-14

[7] Panigrahi P K and Sahoo S 2015 Path Planning and Control of Autonomous Robotic Agent Using Mamdani Based Fuzzy Logic Controller and ARDUINO UNO Micro Controller *Proceedings of the 3rd International Conference on Frontiers of Intelligent Computing: Theory and Applications (Ficta) 2014, Vol 1* **327** 175-83

[8] Choi S J and Kang B G 2014 Prototype Design and Implementation of an Automatic Control System Based on a BCI *Wireless Personal Communications* **79** 2551-63

[9] Krampit N.Yu., Krampit M.A. 2015 Pulsed-arc welding with preheating of electrode extension in the break period In: *Welding International.* **29** 210-212

[10] Bodeker C and Kaminski N 2015 Investigation of an overvoltage protection for fast switching silicon carbide transistors *Iet Power Electronics* **8** 2336-42

[11] Knyazkov, A.F., Krampit, N.Yu., Krampit, A.G., Knyazkov, S.A. 2005 Device and method of energy dosing at the pulse-arc CO_2 welding *Avtomatizatsiya i Sovremennye Tekhnologii* **10** 19-21

IOP Conf. Series: Materials Science and Engineering **142** (2016) 012014 doi:10.1088/1757-899X/142/1/012014

Influence of the Introduction of Carbon-Fluorine Additive to the Slag of the Production of Silicomanganese on the Weld Joint Quality[1]

N A Kozyrev[1,a], R E Krukov[1], O E Kozyreva[1], E A Zernin[2,b], D S Kartsev[2]

[1]Federal State Budgetary Educational Institution of Higher Professional Education «Siberian State Industrial University», 654007, Novokuznetsk, 42, Kirov str.,
[2]Yurga Institute of technology, TPU Affiliate, 652055, Yurga, 26 Leningradskaya str.

e-mail: [a]kozyrev_na@mtsp.sibsiu.ru, [b]yuti_sp@bk.ru

Abstract. In the paper a possibility in principle is outlined to use a silicomanganese by-product ladle slag and a gas purification dust of aluminum production for manufacturing welding fluxes. The appropriate component concentration in welding fluxes is determined. The results of metallographic research are provided. The use of carbon-fluorine additive makes it possible to reduce the level of non-metallic impurities in the weld joint.

1. Introduction

Various factors, such as types and methods of arc and weldpool shielding [1, 2]; systems and methods of impulse supply of the welding arc [3, 4]; methods and means of weld metal modification [5, 6]; chemical composition of welding materials and fluxes [7, 8] etc. influence on the quality of weld joints.

At present researchers focus on the issue of developing new fluxes and additives. The use of metallurgical production wastes as components of welding fluxes allows both considerable reducing the cost price of manufactured fluxes and efficient using the production wastes [6-14]. The authors of the paper outline the results of using the non-utilizable silicomanganese by-product ladle slag and gas purification dust of aluminum production as a welding flux, they also consider the effect of these fluxes on the weld joint quality.

2. Results and Discussion

The experiments are carried out using the flux made of the silicomanganese by-product ladle slag smelted in the ore-melting furnaces via the continuous carbothermic process and gas purification dust of aluminum production (Table 1).

The gas purification dust is mixed with the liquid glass and used for manufacturing carbon-fluorine additive FD-UFS on the base of technology covered by a patent of the Russian Federation [15, 16]. The procedure of flux production includes grinding, sieving, sifting the 0.45-2.5 mm faction of silicomanganese by-product slag and mixing with the carbon-fluorine additive taken in various percentage concentration (sample 1 – without the additive FD-UFS, samples 2 to 5 contain 2, 4, 6, 8 %

[1] – The work is performed in Siberian State Industrial University as a project part of the State Order by the Ministry of Science and Education of the Russian Federation № 11.1531.2014/к. The equipment of the Collective Use Center «Materials science» of Siberian State Industrial University and scientific and production center «Welding processes and technologies» is used when testing, examining and measuring

FD-UFS, respectively). The welding modes are tested by the welding tractor ASAW-1250 using welding electrodes Sv-08GA with the diameter of 4 mm. The process of welding 500×75 mm plates with the thickness of 16 mm is butt two-side welding, beveling under the flux layer is not necessary. The samples are welded in the following conditions: Iw=700 A; Ua=30 V; Vw=35 m/h.

The metallographic analysis is performed with the microscope OLYMPUS GX-51 in the bright field and zoom ranging ×100 to ×1000 after etching in 4 % nitric acid solution. The grain size is determined according to GOST 5639-82 when zooming in ×100. The samples are tested to detect non-metallic impurities in terms of GOST 1778-70. The polished surface is examined when zooming in ×100 with the microscope OLYMPUS GX-51.

Table 1 – Chemical composition of the components

Component	MnO	SiO_2	CaO	MgO	Al_2O_3	FeO	Na_2O	K_2O	F	S	P	C
Silicomanganese by-product slag	8.01	46.46	22.85	6.48	9.62	0.38	0.36	0.62	0.76	0.17	0.01	
Gas purifying dust of aluminum production	0.6	2.33	2.1	0.8	43.27	2.1	10.6	0.8	23.6	0.38	0.10	12.5

The welds are shown in Figure 1.

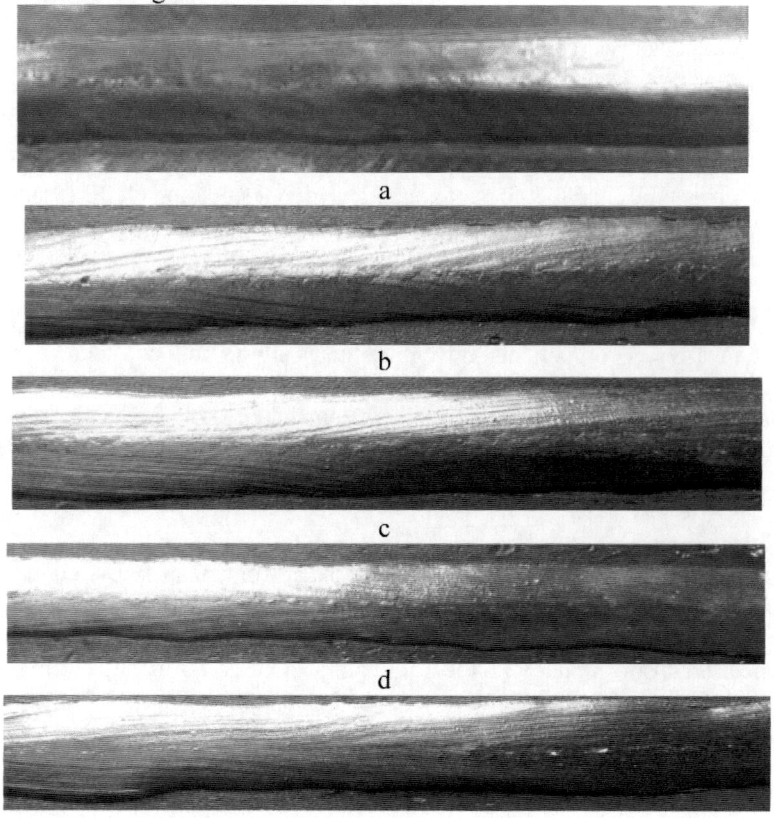

Figure 1 – Weld bead of the samples: a – sample 1, b – sample 2, c – sample 3, d – sample 4, e - sample 5

The welds of the samples under consideration have a Widmanstatten ferritic pearlite structure, which is distinguished by some needle-like ferritic zones, as the experiments have revealed. In some areas ferritic pearlite structure has a lamellar character (Figure 2).

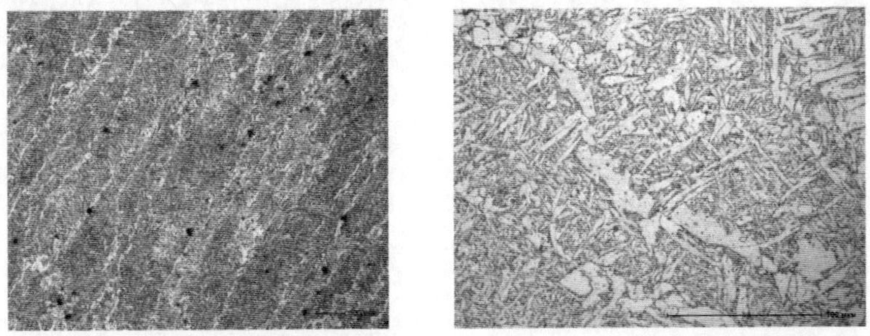

a

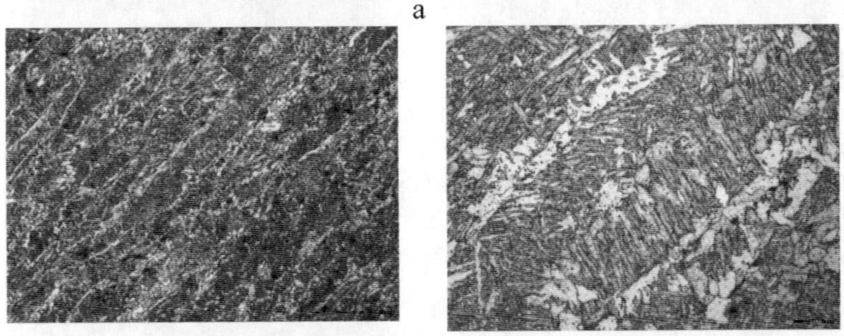

b

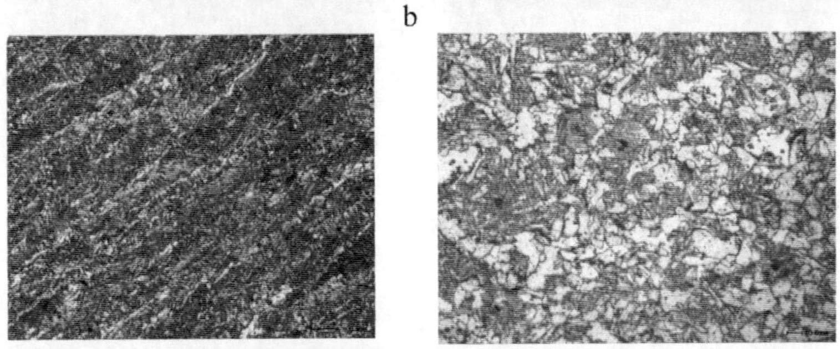

c

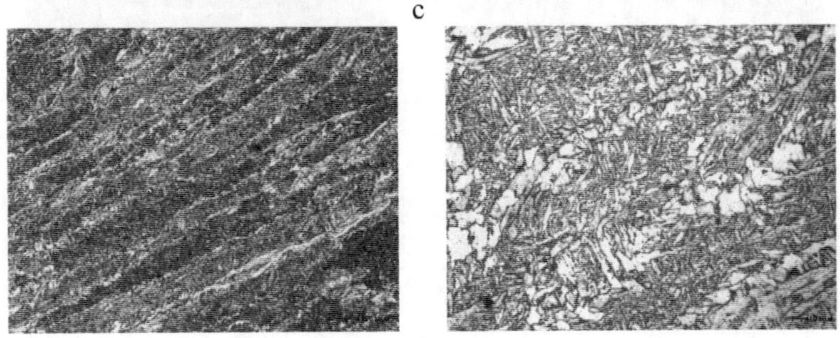

d

e

× 100 × 500

Figure 2 – Weld microstructure of the samples under consideration: a – sample 1, b – sample 2, c – sample 3, d – sample 4, e – sample 5

The grain size in the weld structure of sample 1 meets № 4, 5 according to the graininess scale. The grain size in the weld structure of samples №2 and №3 corresponds to №5, 6 and 7. The grain size of samples №4 and №5 answers № 5, 6, 7. Therefore, it has been revealed that increasing the FD-UFS concentration in the fluxes under consideration furthers the weld grain refinement.

The research into the character of non-metallic impurities has revealed non-deforming, brittle silicates, spot and stitched oxides in the weld area of the samples under consideration (Figure 3).

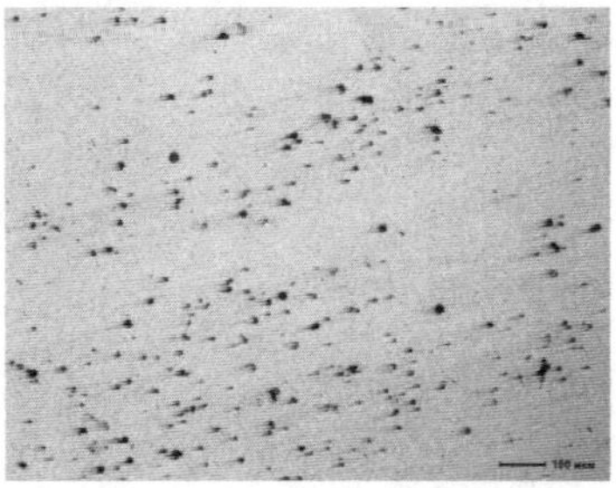

a

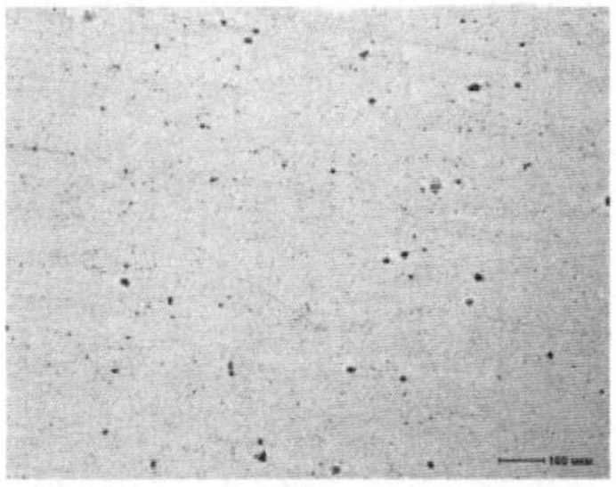

b

c

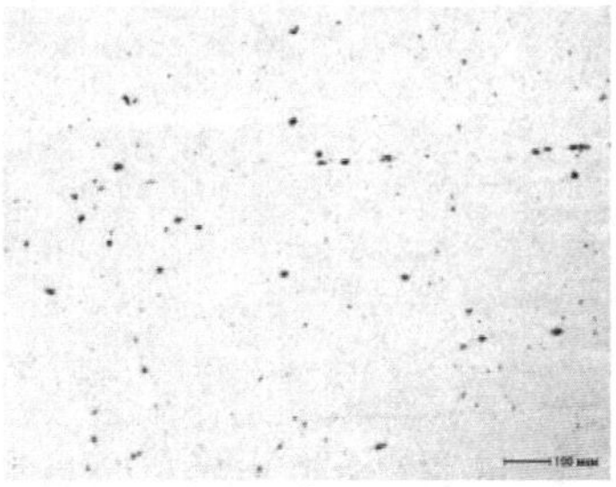

d

e

Figure 3 – Non-metallic impurities in the weld area of the samples under consideration

In the weld area of sample 1 spot oxides of grade 1 (a) are detected, non-deforming silicates, generally, grades 4 (b) and 3 (b), less frequently grade 4 (a), brittle silicates of grade 3 (b) are rare. In the weld area of samples 2 and 3 there are non-deforming silicates of grades 2 (b), 4 (b) and spot oxides of grade 1 (a). In the weld area of sample 4 non-deforming silicates of grades 2 (b) and 1 (a) and spot oxides of grade 1 (a) are detected. In the weld area of sample 5 there are non-deforming silicates of grade 2 (b) and spot oxides of grade 1 (a). Using the carbon-fluorine additive (FD-UFS) in the flux effects positively on the weld, reducing the level of non-metallic impurities, their number and sizes (Figure 3).

The research described above is a foundation of welding fluxes manufacturing based on silicomanganese by-product slag and carbon-fluorine additive FD-UFS [17, 18].

3. Conclusions

1. The conducted experiments make it possible to demonstrate a possibility in principle to use a silicomanganese by-product slag for manufacturing welding fluxes.
2. The use of carbon-fluorine additive FD-UFS together with a silicomanganese by-product slag results in grain refinement.
3. The use of carbon and fluorine containing additive makes it possible to reduce the level of non-metallic impurities in the weld, their sizes and number.

References

[1] Chinakhov, D.A., Vorobyov, A.V., Tomchik, A.A. Simulation of active shielding gas impact on heat distribution in the weld zone // Materials Science Forum. Volume 762, 2013, Pages 717-721.

[2] Chinakhov, D.A. Calculation of gas-dynamic impact of the active shielding gas on the electrode metal drop in gas jet shielded welding // Applied Mechanics and Materials. Volume 379, 2013, Pages 188-194.

[3] Krampit, A.G., Krampit, N.Y A Method for the determination of the geometrical dimensions and area of the welded joint // Welding International. Volume 27, Issue 10, October 2013, Pages 834-836.

[4] Krampit, A.G., Krampit, N.U., Krampit, M.A. Mechanical properties of welded joints in welding with pulsed arcs // Applied Mechanics and Materials. Volume 379, 2013, Pages 195-198.

[5] Kuznetsov, M.A., Zernin, E.A., Danilov, V.I., Karzev, D.S. Application of nanostructured powders to control characteristics of electrode metal transfer and the process of weld structurization // Applied Mechanics and Materials. Volume 379, 2013, Pages 199-203.

[6] Kuznetsov, M.A., Danilov, V.I., Zernin, E.A., Kolmogorov D.E., Zoubenko L.N. Corrosion and Mechanical Properties of Austenic Steel Weld Joints // IOP Conf. Series: Materials Science and Engineering 91 (2015) 012010. doi:10.1088/1757-899X/91/1/012010

[7] Using a new general-purpose ceramic flux SFM-101 in welding of beams/ Yu. S. Volobueva, O. S. Volobueva, A. G. Parkhomenkob, E. I. Dobrozhelac & O. S. Klimenchukd//Welding International.– 2012.- Volume 26. - №8. - p. 649-653

[8] Study of the relationship between the composition of a fused flux and its structure and properties/ Amado Cruz Crespoa, Rafael Quintana Puchola, Lorenzo Perdomo Gonzáleza, Carlos R. Gómez Péreza, Gilma Castellanosa, Eduardo Díaz Cedréa & Tamara Ortíza / Welding International. – 2009.- Volume 23. - №2. - p. 120-131

[9] Special features of agglomerated (ceramic) fluxes in welding/V. V. Golovko & N. N. Potapov//Welding International. – 2011.- Volume 25. - №11. - p. 889-893.

[10] New carbon-fluorine additives for welding fluxes/ N.A. Kozyrev, N.E. Krukov, R. Ev. Krukov, A.V. Roor, L.P. Bashchenko, U.I. Lipatova/ Steel in Translation.-2015. -T. 45. -№ 4. - P. 251-253.

[11] Gasik M.I, Lyakishev N.P. Theory and technology of ferrous alloys electrometallurgy – M.: Intermet engineering, 1999. – 764 p.

[12] Ryss M.A. Ferrous alloys production/ M.A. Ryss. – M.: Metallurgy, 1985. – 344 p.
[13] Potapov N.N. Welding materials for arc welding: reference book in 2 volumes. V.1. Shielding gases and welding fluxes : reference book / N. N. Potapov, B. P. Konishchev, S. A. Kurlanov et al ; edit. by N. N. Potapov. – Moscow : Mechanical engineering, 1989. – 544 p.
[14] Podgaetsky, V. V. Fluxes for automatic and semi-automatic welding / V.V. Podgaetsky, D.M. Rabkin. - Kiev : Publishing AS USSR, 1954. —56 p.
[15] Pat. 2484936 RF, MPK ^{8} B23 K35/362 Ceramic flux-additive/ Kozyrev N.A., Igushev V.F., Krukov R.E., Goldun Z.V.; FSBEI HPO «Siberian State Industrial University.- №2012104939/02(007484), Appl. 13.02.2012.
[16] Pat. 2564801 RF, MPK ^{8} B23 K35/362 Flux-additive / N. A. Kozyrev, R. E. Krukov ; OJSC «Novokuznetsk plant of tank metal constructions by N.E. Krukov». – № 2013144914/02(069340); appl. 07.10.2013 ; pub. 10.10.2015
[17] Decision of patent issue № 2014122996 Flux for welding / Krukov N.E. Krukov E.N., Kozyrev N.A., Krukov R.E., Kozyreva O.E.; OJSC «Novokuznetsk plant of tank metal constructions by N.E. Krukov», Appl. 05.06.14 .
[18] Decision of patent issue № 2014123002 Flux for mechanized welding and surfacing of steel / Krukov N.E. Krukov E.N., Kozyrev N.A., Krukov R.E., Kozyreva O.E.; OJSC «Novokuznetsk plant of tank metal constructions by N.E. Krukov», Appl. 05.06.14.

IOP Conf. Series: Materials Science and Engineering **142** (2016) 012015 doi:10.1088/1757-899X/142/1/012015

Distribution of Liquid Flow Rates in the Process of Bubbling with Gas Through Gas-Permeable Inserts

R A Gizatulin[1], D V Valuev[2], R S Dariev[3], V A Trifonov[4], I F Borovikov[5]

[1,2,3,4] 652050, Kemerovo region, Yurga, Leningradskaya str.26 Yurga Technological Institute branch of Tomsk Polytechnic University
[5] The Moscow State Technical University of N.E. Bauman, Moscow, 5, 2nd Baumanskaya

E-mail: valuev@tpu.ru

Abstract. The authors studied the distribution of the vertical components of the rate in the ascending gas-liquid flow when blowing through the bottom nozzle at two levels under three modes of neutral gas supply. It was estimated that under the intensities of gas (nitrogen) of 2 and 4 L/min·t the type of rates distribution in both cross-sections does not differ from the generally accepted one and practically does not depend upon the intensity of gas supply.

Introduction

The duration of the stirring operation is significantly determined by the time required for bath homogenization in terms of chemical composition and temperature which, in its turn, depends upon the rate gas-ascending flow (current stream). The easiest way to estimate the duration of the blowing through is by simulation. At the same time it is practically impossible to meet all the requirements of exact similarity of the model due to the diversity of the factors effecting the results of the blowing through and the researchers have to give up the idea of absolute similarity of the original to the model. This may be the explanation why the results of the studies we know concerning simulation of the blowing through in the ladle [1-4] do not agree with the results of direct ladle experiments [5, 6].

For this reason we estimated the profile of the rates depending on gas flow and different distances from the nozzle. On the transparent model we studied the distribution of the vertical components of the rate of the liquid in the ascending gas-liquid flow when blowing with bottom nozzle at two levels under three modes of neutral gas supply.

The experiments were carried out in a transparent container with the cross-section 600×600 mm and 1200 mm high. The basic drawing of the unit is shown in Figure 1. A metal rod was plunged into the container filled with water. Two platinum sensors were fixed at one end of the rod at a distance (vertical) of 50 mm from each other to measure the electric resistance of water. Nitrogen was supplied through the nozzle with a hole 1.3 mm in diameter situated in the center of the bottom of the container. Nitrogen consumption was measured by rotameter, its pressure before the nozzle – by spring manometer.

After the container is filled with water up to 1100 mm we opened the gas cylinder valve and settled the fixed gas rate with a rotameter. A certain amount of brine solution was supplied to the gas nozzle with a syringe pump. After bringing in the indicator we turned on the mechanism of aerial film rewind (sensitivity 1300 un.) in the light-beam oscillograph. The rate of the aerial film rewind was 640 mm/sec, accuracy of time marking was 0.005 ± 0.0025 sec.

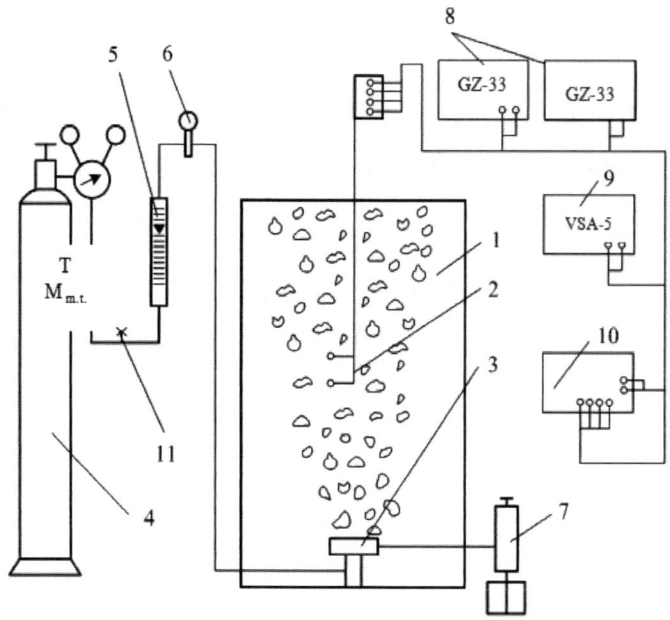

Figure 1. The diagram of the unit to study the liquid flow rates distribution 1 – container with water; 2 – sensors; 3 – nozzle; 4 – gas cylinder; 5 – rotameter; 6 – manometer; 7 – pump; 8 – audio frequency generator; 9 – rectifier; 10 – light-beam oscillograph; 11 – control valve

The indicator does up with the liquid drawn by gas bubbles. When the brine solution reaches the first sensor, water resistance reduces within the gap between the circular and the point electrode and the electric current intensity increases in the sensor circuit. The electric current intensity increases in the second sensor circuit and this is recorded in the aerial film. The change of electric current intensity on the second sensor was controlled by the tester which marked the rush of current in the circuit. After the rush of current the mechanism of aerial film rewind was turned off. After the experiment was over a three-minute pause was made under continuous gas supply until complete homogenizing of the bath. Before the start of the next experiment standard electric current intensity was set for each circuit by changing the output generator voltage. After the adjustment the next experiment was carried out according to the same plan. The time records on the oscillograms allowed determining the interval between the electric current intensity change in the first and in the second sensor circuits. The rate of the liquid flow (ϖ – m/s) under bubblering was found from the formula:

$$\varpi = \frac{l}{\tau}, \tag{1}$$

where l = 50 mm – distance between the sensors;

τ – the time when the salt solution travels from one sensor to the other, sec.

The rate of the flows was measured in the horizontal planes situated at the heights H making 570 and 800 mm from the nozzle exit, along radius R at a pitch of 2 cm starting from the plume centre to its perimeter. The measurements were carried out under the intensities J of gas supply 2, 4 and 8 L/min which corresponded to the intensity of gas supply from 4.3 to 17.2 m^3/h in a 100-tonne ladle. In every coordinate under the given blowdown mode the measurements of the rate were completed (depending on the how the measurements vary) from 6 to 21 times. Altogether about 500 measurements were completed.

The distribution of the vertical component of liquid flow rate distribution in the ascending gas-liquid flow obtained in the course of experiments is shown in Figure 2.

From Figure 2 we can see that under all blowdown modes the rate decreases as we move away from the center irrespectively to the distance from the nozzle. The rates of the flows at 570 mm under the intensities of gas supply of 2 and 4 L/min at corresponding coordinates are practically the same. When gas flow increases up to 8 L/min· flow rate at the given height increases.

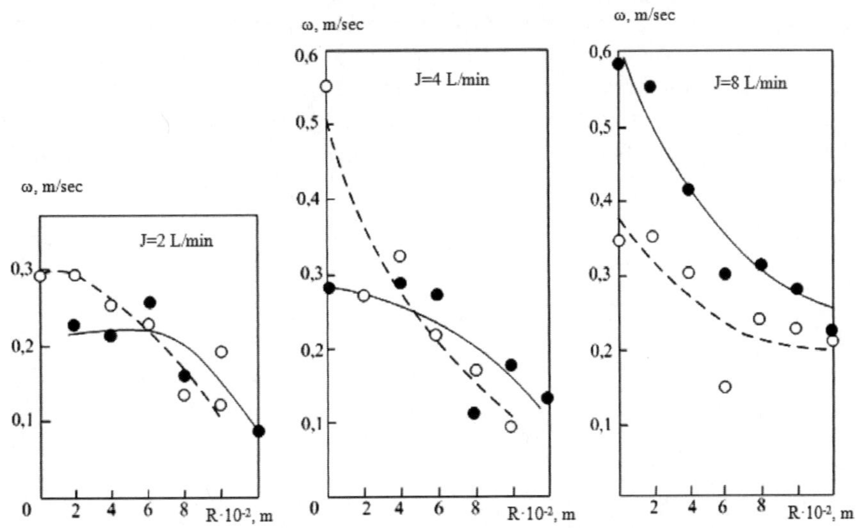

Figure 2. Rate distribution in the ascending gas-liquid flow

At the height H = 800 mm under small gas flows (2 and 4 L/min) the rates, as well as at 570 mm are close to each other in corresponding coordinates. Under maximal in our experiments gas flows (8 L/min) the flow rates at the given height also increase.

The flow rate at the upper level is higher than that at the lower level. This disagrees to the data provided in papers [7, 8]. At the same time under small intensities of gas supply the results of our experiments are similar to the data in [9-10].

The curves in the diagrams are regressions of the following type

$$\varpi = A + BR + CR^2, \qquad (2)$$

where R – distance from the axis of the flow, cm.

The values of the coefficients of the regression and coefficients of the pair correlation are presented in the table.

Table 1 – Values of the coefficients of regression $\varpi = A + BR + CR^2$, and coefficients of correlation (g)

J, l/min	H, mm	A	B	C	g
2	570	0.294	$-0.117\cdot10^{-3}$	$-0.192\cdot10^{-2}$	–0.929
	800	0.187	+0.0225	$-0.254\cdot10^{-2}$	–0.767
4	570	0.504	–0.0687	$+0.230\cdot10^{-2}$	–0.906
	800	0.290	$-0.555\cdot10^{-2}$	$-0.758\cdot10^{-3}$	–0.818
8	570	0.375	–0.0337	$-0.172\cdot10^{-2}$	0.747
	800	0.619	–0.0584	$0.233\cdot10^{-2}$	–0.878

In Figure 3 we provide the results of our calculations of the liquid flow according to the zones of the stream cross-section and the cumulative sum. The liquid flow for every zone was calculated according to the formula

$$V = \omega_i F_i, \quad (3)$$

where $F_i = \pi R^2$ –area of the central zone with the radius of 2 cm; $F = \left(R_i^2 - R_{i-1}^2\right)$ – area of the rest ring zones of the cross-section;

ω_i – flow rate of the liquid at the outer generating line of the ring.

Such method of calculation results in exaggerated amount of the liquid but the type of the dependence does not differ from the dependence obtained if applying the average rate for each zone.

When completing the calculations we did not take into consideration the area of the cross-section taken by the gas phase as the amount of real volume content of the gas (relation of the section part taken by the gas phase to the whole section of the ascending flow $\varphi = F_g/F$ under our intensities of gas supply for both heights is less than 0.01. The exact value is impossible to determine as the boundaries of the flow are fuzzy.

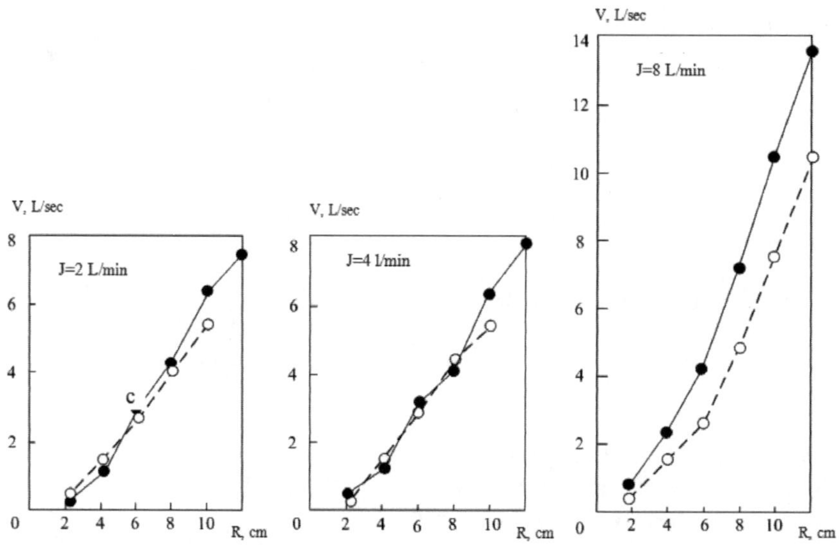

Figure 3. Liquid flow for the zones of the section as a cumulative sum

From Figure 3 we can see that under small gas flows (2 – 4 L/min between 570 and 800 mm) there is no suction of water into the ascending flow. As the zone of bubblering expands new amounts of the liquid are merely involved into the ascending motion. A different situation is observed under J = 8 L/min. Here we can speak about suction of surrounding water into the gas-liquid flow as, starting from R = 4 cm the volume of the liquid passing through every next zone is always larger at 800 mm than at 570 mm.

Conclusion

We applied a transparent model to study the distribution of the vertical components of the rate in the ascending gas-liquid flow in the process of blowing through the bottom nozzle at two levels under three modes of neutral gas supply. It was estimated that, under the intensities of gas (nitrogen) supply of 2 and 4 L/min the type of the rates dependence in both sections does differ from the generally accepted one and does not practically depend on the intensity of gas supply. Under the gas supply of 8 L/min rate distribution differs from the generally accepted distribution both at the height and in the section of the flow.

We applied the cold model to study the distribution of the gas phase in three horizontal sections of the ascending gas-liquid flow blown from the bottom according to the method of remote laser probing. It was established that distribution of the gas flow density in the horizontal plane is described by the exponential dependence; distribution of the gas flow over the main part in the cross-section is similar, i.e. it does not depend upon the intensity of gas supply and the distance between the cross-section and the nozzle. The real volume gas content of the flow is maximal at its axis and decreases exponentially as we move away from it. The expansion angle of the gas flow in general is determined by the density of the gas flow at the nozzle edge; as the flow ascends its value decreases. The results of the study were used for simulating the bubbling stirring of the molten metal.

References

[1] Kadukov V G Kogan A E Fomin N A Stirring intensity effect upon the duration of blowing J Izv. Vuzov 6 (1986) 46 – 57
[2] Chinakhov D A Gas dynamic control of properties of welded joints from high strength alloyed steels J China Welding. Vol 23 3 (2014) 27-31
[3] Saprykina N A Saprykin A A Borovikov I F Sharkeev Y P Influence of layer-by-layer laser sintering conditions on the quality of sintered surface layer of products J Source of the Document IOP Conference Series: Materials Science and Engineering 91 (2015)
[4] Galevskiy G V Rudneva V V Galevskiy S G Il'yashchenko D P Kartsev D S Nanosized borides and carbides for electroplating. metal-matrix coatings: specifications, performance evaluation J IOP Conference Series: Materials Science and Engineering. 125 (2016)
[5] Chinakhov D A Chinakhova E D Gotovschik Y M Grichin S V Influence of welding with two-jet gas shielding on the shaping of a welding joint J IOP Conference Series: Materials Science and Engineering. 125 (2016)
[6] Il'yaschenko D P Chinakhov D A Danilov V I Sadykov I D Pysical nature of the processes in structure forming, phase and chemical composition of pipe permanent joints when MMA welding J IOP Conference Series: Materials Science and Engineering. 125 (2016)
[7] Pimnev A L Zemenkova M Yu Zemenkov Yu D Il'yashchenko D P Mechanical properties of the assembly welded joint of the oil transportation tank after a long-term service J IOP Conference Series: Materials Science and Engineering. 127 (2016)
[8] Saprykin A A Ibragimov E A Babakova E V Engineering and technology based on rapid prototyping J Applied Mechanics and Materials 770 (2015) 622-627
[9] Dmitrienko V I Rozhikhina I D Nokhrina O I Aizatulov R S Platonov M A Reduction of barium and strontium in the ladle treatment of steel J Steel in Translation Vol 42 4 (2012) 305 – 307
[10] Chinakhov D A Increasing reliability of welded joints of hydraulic cylinders in heading machines J Journal of Iron and Steel Research International 17 (2010) 4-7

IOP Conf. Series: Materials Science and Engineering **142** (2016) 012016 doi:10.1088/1757-899X/142/1/012016

Technology Of MIG-MAG Welds Strength Enhancement

S A Solodskiy[1,a] , PhD, Ass. Prof., Yu N Saraev [2,b] Prof., A G Malchik [1,c], Ass. Prof., S E Korotkov [1,d], student

[1] Yurga Institute of Technology, TPU affiliate, Russia, Kemerovo region, Yurga, Leningradskaya str. 26
Institute of Strength Physics and Materials Science of Siberian Branch Russian Academy of Sciences, Tomsk, pr. Academic 2/4

E-mail: serdgio80@inbox.ru [a], litsin@ispms.tsc.ru[b], ale-malchik@yandex.ru [c], korotkovsergej@inbox.ru [d]

Abstract. A new technology of MIG MAG welding control is developed. Authors introduce use of power AC and pulse feed of welding wire in the arc zone, that downsizes the heat affected zone, stabilizes formation of electrode metal droplets, as external magnetic field's effect on the arc is reduced. Principal criteria for electrode metal transfer control, when powered by AC sources, are specified.

Introduction

Conditions of the arc, if power AC supply, differ from conditions of the arc, if DC power supply. The active spot on the cathode changes voltage hundred times per second when a sinusoidal voltage frequency is 50 Hz. At the end of each half-cycle, sinusoidal voltage drops to zero; this diminishes the intensity of arc distance's electrical field. Consequently the gas ionization in the arc column diminishes and near-electrode spatial charge diffuses; that does not comply with the terms of arc stability [1]. However, alternating current arc welding has several advantages, i.e. absence of external magnetic field effect on the arc ensures its spatial stability; cathode's sputtering degrades the oxide layer during half-cycles when a product becomes the cathode. AC arc gives the best temperature fields distribution, which affects quality of microstructure and residual stresses of welded joints. In world practice, AC is used only when non-consumable electrode welding of aluminum metals and alloys, as well as arc welding with special purpose electrodes. The authors present a new technology of mechanized gas metal arc AC welding with a pulse wire feed for welding of steel structures in different spatial positions. This problem was considered by Paton B.E., Lebedev V.A., Mikitin Ya.I.

Implementation

As mentioned above, MIG MAG AC welding has several key advantages if compared with DC welding [2]. However, MIG MAG welding is limited in use because of disadvantages, caused by alternation of current itself. Regular charge recovery, if AC, is a precondition for arc stability while gas shield welding; consequently source of power with high open-circuit voltage is required for agitation of AC arc. Arc current curve becomes distorted; and DC component occurs in the welding circuit.

DC component, in its turn, creates a constant magnetic field the transformer core and the welding choke coil; that diminishes power and stability of the arc. One of the major limitations of AC welding use is arc distance short circuit while of electrode metal transition into the weld pool. Elimination of disadvantages would create an energy-efficient weld technology and increase quality of welding joints due to better thermal cycle and more intense deoxidation and refining of metal. [3]

The proposed welding technology is as follows. Welding is performed in shielding gas, powered by alternating current, with simultaneous use of two pulse sources: mechanical and electrical. Power AC is used synchronized with cycles of pulsed wire feed; procedure is controlled through the feedback paths (Figure 1). Wire melting occurs in alternating (sinusoidal) cycle of T1 current increase, correlated with wire feed cycle pause. Electrode metal transfer occurs when arc distance short circuit, in the cycle of decrease and alter of T2 current polarity due to wire feed pulse. Arc initiation is forced by wire move from the weld in the cycle of arc current increase. To do this, a pulse wire feeder is used (Figure 2) [4].

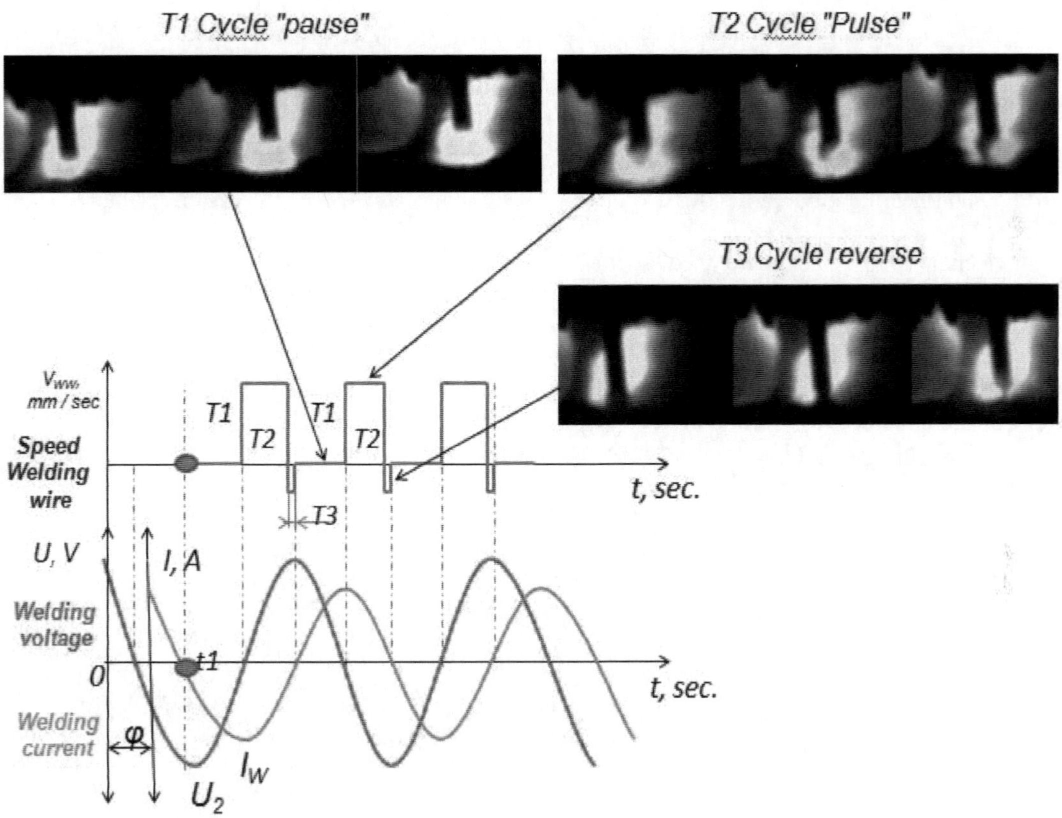

Figure 1 - Sequence diagram of welding process

Technology features. Melting of welding wire and initial formation of electrode metal droplets occur in the moment of no wire feed when arc polarity is reversed. When arc polarity reverses, the anode is hotter than the cathode. Melting rate increases during a pause in welding wire supply; thus a drop of electrode metal is formed. Further, impulse of welding wire feed occurs at time T1, when the arc voltage sinusoid drops to zero, and phase-shifted arc current gradually tends to zero. Accordingly, liquid metal drops transfer through the electrode arc distance occurs with minimum impact of arc forces. Droplet transfer to the weld pool occurs in the moment of short-circuit of arc distance, while the arc current alters from reverse polarity to direct, and at the end of wire feed pulse cycle T2. Thus, arc current has a low value; that diminishes gas dynamic force while droplet detachment, and facilitates smooth transition of electrode metal into the product. Arc initiation is improved and time of liquid wire metal attachment from the weld pool is diminished by controlled wire move upward the product [5].

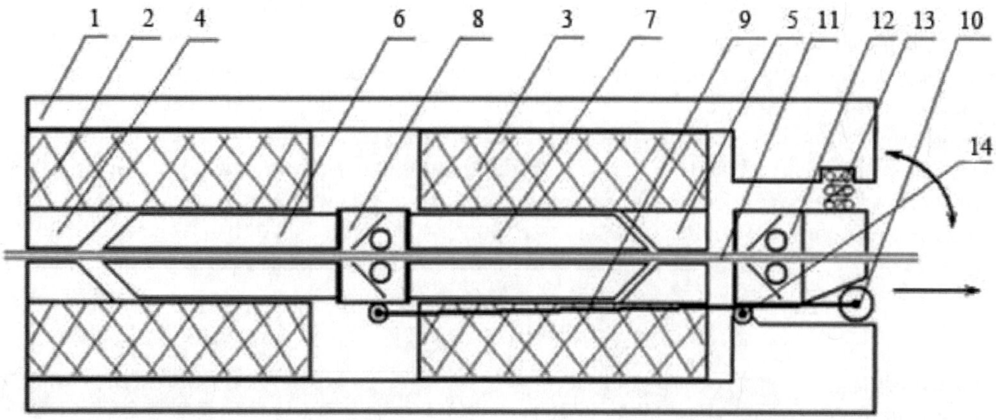

Figure 2 - Design of pulse feeder: 1 - body, 2, 3 - electromagnetic coil; 4, 5 - core; 6 - armature; 7, 8, 12 - ball-bearing guide; 9 - rod; 10 - roller; 11 - welding wire; 13 - compression spring.

Mechanism of pulse wire feed operates as follows. The armature (7) is pulled into of the electromagnetic coil (3) and pulls the armature (6) when electricity is supplied to the coil (3). At the same time the guide (8) is jammed and pulls the wire (11), which is sliding through the guide (12). After switching to the coil (2) reverse action occurs, i.e. the armature (6) is pulled into the coil (2) and pulls the armature (7); the guide (12) is jammed, and the guide (7) is sliding along the wire (11). While return of the armature (6) through the rod (9), roller (10) displacement occurs, which, while moving, ups the front guide edge (13) fixed in the hinge (14). Reverse of the guide (12) drives the wire (11) back from the welding zone, which leads to forced attachment of liquid metal between the wire and the weld pool. The spring (13) holds the guide (12) relative to the roller (10). In this way, alternate switching of coils provides wire movement into the weld zone (11) (shown by arrow), wherein the guide (12), vibrates relative the axis (14) (shown by arrow) and breaks liquid metal contact between the wire and the weld pool while reverse movement of the armature (7).

Besides, to produce high-quality welds, power circuit, controlling feeder's operation, should involve inductive nature of load (electromagnet or motor). Based on the foregoing, mechanism of pulse wire feed together with electric circuit and arc is to be an automated control system with feedback being the key element. Such a control circuit (Fig.5.6) can work with any actuator (electromagnet or motor) of pulse wire feed.

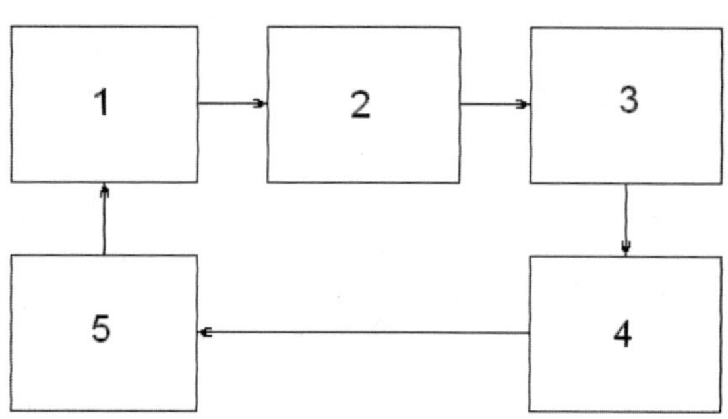

Figure 3 Diagram of pulse feeder control: 1 - correction block; 2 - pulse generator, 3 - power circuit; 4 - pulse feeder; 5 - welding arc;

The control unit monitors variations of arc parameters and corrects welding wire track adequately, at the stage of metal droplet formation and transfer to the weld pool. The unit has voltage or current feedback; the circuit diagram is shown in Figure 4.

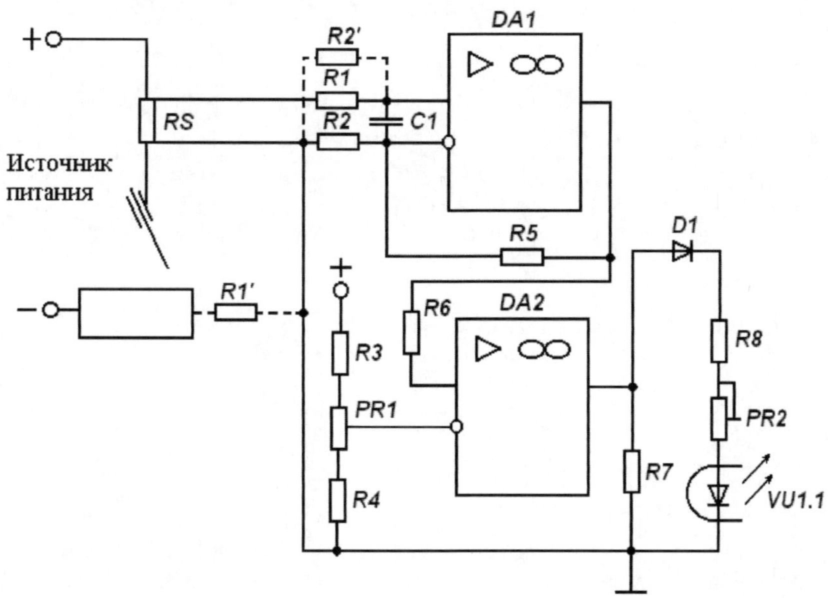

Figure 4 Welding wire pulse feed control unit

The device unit comprises: welding current or arc voltage sensor *(DA1)* and controller for synchronization of welding parameters and control unit input impedance; feedback is a current or voltage (voltage feedback is dotted). The device compares predetermined voltage with a feedback signal (DA2) and, depending on potentiometer's indications (PR1), gives a correction signal to the pulser. To synchronize operation of the correction unit and the pulser, as well as exclude their electrical interconnection, galvanic isolation (transformer or optoisolator) *(VU1)* is to be implemented.

Materials and Methods

Laboratory tests confirmed industrial applicability of the proposed technology. Mechanical properties of welded joints, made of 30ХГСА steel with Св-08Г2С welding wire in carbonic gas environment, were tested. It has been found that if using the proposed technology, impact toughness enhances [6].Results of mechanical tests of welded steel samples (multi-layer weld of articulated pipe joint) [7, 8]

Welding technique	Tensile strength, MPa	KCV Impact toughness, J/cm^2 at 20°C	Weld hardness, HB	Hardness 3TB, HRC (HB)
Conventional DC	635	141	167	22(198)
Recommended technique	634	167	153	23(207)

Metallographic examination of welded joints (Figure 3) showed a decrease in heat-affected zone to 15% due to cyclic rise and decline of arc current with alternation of polarity; if compared with mechanized DC welding in shielded gases. Videography and oscillography showed that electrode metal transfer was more stable, since no axial deviation of the arc; electrode metal scattering diminished to 4%.

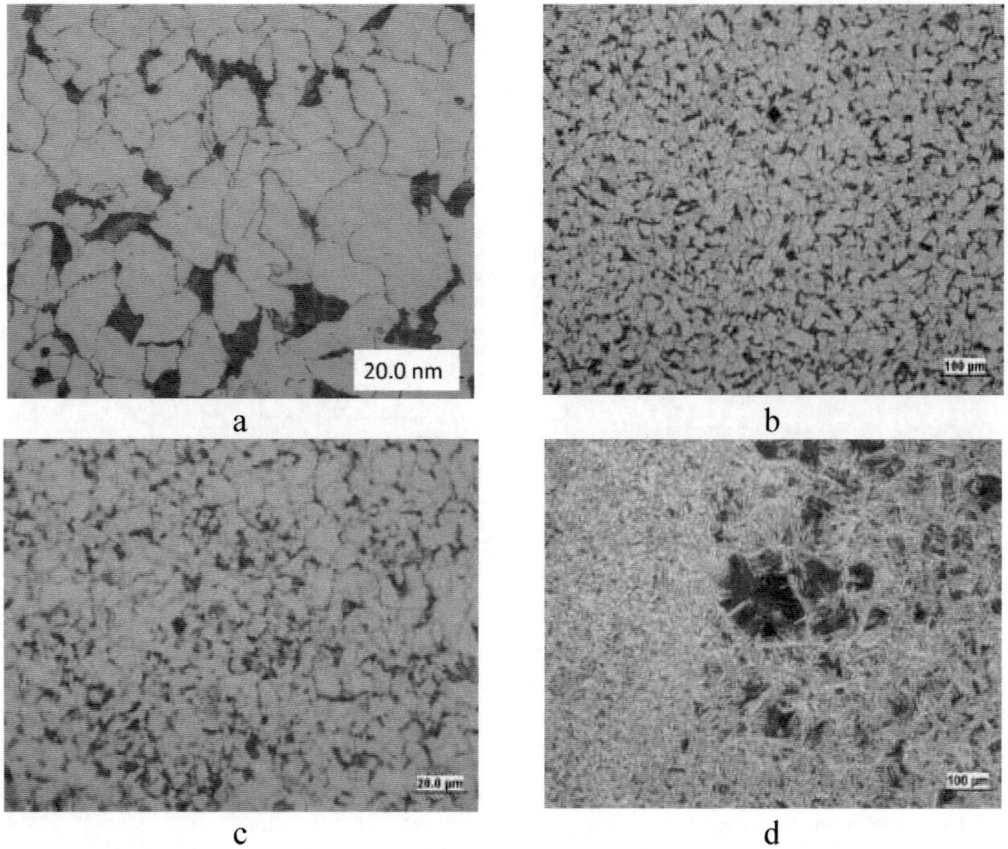

Fig. 3 Weld microstructure: a, b - 09G2S steel microstructure; c - HAZ microstructure (AC arc welding); d - HAZ microstructure (conventional DC welding).

This work was supported by the Russian Science Foundation, the project №16-19-10010.

Conclusions

The technology developed has advantages as follows:

a) Electrode metal transfer is synchronized with polarity alternation of power AC by pulse wire feed, when voltage sine wave decreases and approaches to zero and arc pressure forces are minimal. Stabilization of electrode metal transfer to the weld pool occurs by the aid of automated control system feedback. Transfer is performed on power current of 50 Hz.

b) Volume of the transferred metal droplet is diminished because of lower arc current and no arc pressure during the transfer, thereby reducing both electrode metal heat capacity and alloying elements burnout; impact toughness of weld joints enhances.

c) Transfer rate of the droplet into the weld pool enhances, as short circuit and transfer occur simultaneously in the moment of current zero and polarity alternation from reverse to direct. This minimizes gas-dynamic impact; spattering losses are reduced to 4%

d) Width of HAZ is reduced to 15% as resulted by sinusoidal variations of current intensity and polarity; cyclic electron emission from the cathode provides equilibrium fine-grained microstructure of weld joints.

e) Power consumption is diminished by use of welding AC transformers.

f) HAZ downsize and weld metal dilution occur without use of costly high-tech SMPS.

References:

[1] Saraev Y. Adaptiv pulse-arc welding methods for construction and repair of the main pipelines.

Proceedings of The 2nd South-East European IIW International Congress "Welding – HIGH-TECH Technology in 21st century". Sofia, Bulgaria, October 21st-24th 2010, p. 174 – 177.

[2] Lebedev V.A. Creating mechanized arc-welding equipment with pulsed electrode supply. Russian engineering research. 2009. т. 29. № 2. c. 131-135.

[3] Chinakhov D.A., Agrenich E.P. Computer simulation of thermo-mechanical processes at fusion welding of alloyed steels // Materials Science Forum. – Vols. 575-578 (2008). – Pp. 833-836.

[4] Brunov, O.G., Solodskii, S.A. Physico-mathematical modelling of the transfer of electrode metal droplets into the weld pool. Welding International. 2009. 23 (12), pp. 930-933

[5] Paton B.E, Lebedev V.A, Pichak B.G, S.I Poloskov Evolution pulsing systems electrode wire length welding and surfacing. Welding and diagnostics. Number 3. 2009. p. 46-50.

[6] Brunov, O.G.,Solodskii, S.A.,Zelenkovskii, A.A. Conditions of arc ignition in welding in shielding gases/Welding International 2012 26 (9), pp. 710-712

[7] 7.Chinakhov D. A. Study of thermal cycle and cooling rate of steel 30ХГСА single-pass weld joints // Applied Mechanics and Materials. – Vols. 52-54. – 2011. – p. 442-447. - Mode of access: http://www.scientific.net/AMM.52-54.442

[8] Valuev D. V. , Danilov V. I. Reasons for Negative Formation of Structures in Carbon Steel Processing of Pressure // 7th International Forum on Strategic Technology (IFOST - 2012): Proceedings: in 2 vol., Tomsk, September 18-21, 2012. - Tomsk: TPU Press, 2012 - Vol. 2 - p. 151-154.

The Study of Complex (Ti, Zr, Cs) Nanopowder Influencing the Effective Ionization Potential of Arc Discharge When Mma Welding

S B Sapozhkov [a], E M Burakova

Yurga Institute of Technology TPU Affiliate
Leningradskaya 26, Yurga, Russia Federation, 652055

E-mail: wh13@bk.ru

Abstract. Strength is one of the most important characteristics of a weld joint. Mechanical properties of a weld metal can be improved in a variety of ways. One of the possibilities is to add a nanopowder to the weld metal. Authors of the paper suggest changing the production process of MMA welding electrodes via adding nanopowder Ti, Zr, Cs to electrode components through liquid glass. Theoretical research into the nanopowder influence on the effective ionization potential (Ueff) of welding arc discharge is also necessitated. These measures support arcing stability, improve strength of a weld joint, as the consequence, ensure quality enhancing of a weld joint and the structure on the whole.

Introduction

The 21st century is distinguished by the breakthroughs in development of nano-materials and nanotechnologies. Apparently, technological advance can hardly be ensured via using and combining conventional materials, as well as better understanding their functions is impossible without nano-materials. Sciences and technologies, studying nano-materials, provide unique opportunities for development of future-oriented combinations of materials [1]. Up to date technologies are used in the most important spheres of human activities (industry, defense, radio-electronics, health care system, power engineering, information systems, transport, biotechnology) for manufacturing solar batteries, superconductors, composite and ceramic materials, magnet and colouring pigments, solders, additives to lubricants etc.

At present, technologies enhancing the quality of metalware, resisting to maximal operational loads, are in focus of engineers and researchers. In welding production strength is one of the most important characteristics of a weld joint [2].

The purpose of applying the nanopowders is to achieve a fine-grain structure of a weld metal without any influence on chemical composition of an alloy; these additives cause grain refinement under crystallization, as the result, mechanical properties are improved. It is also perspective to use nano-materials to stabilize arc discharge when fusion welding.

Electric welding arc is a stable and longtime electrical discharge, arising in the atmosphere between hard and liquid electrodes, provided that current density is high, and followed by considerable heat release. Electric discharge in the atmosphere is electric current, flowing through it due to free electrons, and negative and positive ions movable among electrodes under the action of an applied electric field (potential difference between electrodes).

The process involving transformation of neutral atoms and molecules into positive and negative ions is termed ionization. Ionization, arising in a certain volume of the atmosphere, is volume ionization. Volume ionization, resulting from the high-temperature heating of gas, is termed thermal ionization.

At high temperatures energy of the considerable part of neutral gas molecules is sufficient for their splitting into ions as they collide with each other, furthermore, the temperature growth results in the increase of the total number of collisions between gas molecules. At ultra-high temperatures irradiation of gas and incandescence electrons influence on the process of ionization. Ionization is possible at conventional temperatures too, provided that high speeds are imparted by electric field to electrons and ions of the gas. Having a high energy, these particles can break neutral atoms and molecules into ions. Moreover, light or ultraviolet beams, X-rays or emission of radioactive substances can lead to ionization.

Air like all gases has quite a weak electroconductivity in normal conditions due to low concentration of free electrons and ions. That is why, air gap (or other gaseous medium) between electrodes is to be ionized to generate a high electric current, i.e. an electric arc. Ionization can result from applying a rather high voltage to electrodes; electric filed accelerates a few free electrons or ions of the gas; as the result of acquiring a high energy they can break neutral atoms or molecules into ions.

High voltages are not allowed by safety engineering regulations of welding. Therefore, thermal electron and field electron emissions are preferred to carry out research into this process. A lot of free electrons, occurring in metal and having a high kinetic energy, pass to the gaseous medium of the interelectrode space, causing its ionization.

Thermal electron emission involves "vaporization" of free electrons from the surface of metal due to high temperatures. The number of free electrons, acquiring energy sufficient for passing through the energy barrier in the surface layer and leaving the metal is dependent directly on the temperature. The point of field electron (cold) emission is that an outer electric field is developed, which transforms the energy barrier near the surface of metal and supports liberation of electrons with energy sufficient for passing through this barrier.

Gaseous medium ionization is characterized by the degree of ionization, i.e. relation of charged particles in the certain volume to the primary number of particles (before ionization) [3]. Welding is more stable and easier if the ionization potential is low.

Mechanical properties of a weld metal can be improved in a variety of ways. Enhancing strength and operational characteristics of a weld metal is possible via adding a nano-material to it. We consider adding metal nanopowders to liquid glass via ultrasonically induced cavitation by machine UIP 1000hd when manufacturing electrodes for MMA welding. Ultrasonic processor UIP1000hd is widely used in biology, health care, chemistry, and various fields to carry out the following operations: dispersion, homogenization, de-agglomeration, grinding, acoustochemistry, emulsification, bacteria splitting, ultrasonic treatment of liquids in running or periodic mode, constant mixing of various mediums, degassing of liquids, and research into acoustic characteristics of materials [4, 5].

There is not much information on application of nano-materials in welding production available in literature. However, more studies focused on this issue are being carried out these days [2, 6, 7, 8].

Operational characteristics, as well as physical and mechanical properties of metal structures depend both on chemical composition of alloys, they are produced, and on the graininess of their structural elements. The finer the structure is, the better mechanical characteristics of metal structures are. Modification performed at the same time as crystallization is the most typical procedure of grinding structural elements. The efficient method of grinding the structure and enhancing metal characteristics is nanopowder modification of heat-resistant chemical compounds (nitrides, carbides, borides, oxides etc.). However, these powders influence on physical and chemical characteristics of welding alongside with modification.

This paper is aimed at selecting nanostructure elements, which can stabilize the process of welding and regulate crystallization of the weld metal (activator-modifier).

Main part

Various methods of adding nanopowder to the weldpool are available (through welding flux, via spraying in the shielding gas, submerging the electrode into a mixture of nano-materials, welding over the thin layer of nanopowder scattered on metal), but we suppose it is efficient to add a nanopowder to liquid glass when manufacturing electrodes, since losses of a nanopowder are minimal (a nanopowder doesn't get into the weldpool because of splattering, which is not high when MMA welding); this method of adding nano-modifiers is harmless for human health (nanopowder particles can't be breathed into respiratory organs and do not store in lungs); nanopowder is spread smoothly over the weld joint; the process of electrode manufacturing does not get complicated via adding many processing steps.

Substances used for manufacturing electrode coating are very important for welding [9].

Moreover, nanopowders are distinguished by the following characteristics: low liquid metal wettability, it is extraordinary important for performing crystallization centers; high chemical activity; oxidation starts at relatively low temperatures, so nano-dispersed particles can get deactivated in high temperature zone of a weld pool [10].

Nanopowders are powders with specific nano-dimensions. They differ from compact specimens in minute elements of the structure. Nano-materials can be singled out in a certain class of materials due to their dimensions. The diameter of a nano-particle can not exceed 100 nm according to the definition. The diameter of quite a half of available nanopowders is less 30 nm. The diameter of 9% powders in the nano-group is more 100 nm. Nanopowders of the most of producers have particles with diameter 5 to 100 nm. Prices are fixed according to the size of particles, purity and homogeneity of a powder. Activation energy of numerous processes, which nano-particles take part in, is affected by deformed crystal lattice of nano-particles; consequently, their conventional flow and sequence are changed. The system has an important specific feature: potential response speed of objects is increased, i.e. the smaller the object size is, the faster the process is which this object takes part in.

Almost 80% industrially produced nanopowders consist of metal oxides. Nanopowders of fine metals have a considerable and constantly increasing share of the output. Nearly all hard metal elements are produced as nanopowders of fine metals [11].

Modification based on adding refractory particles and aimed at improving mechanical and operational characteristics of the overlay is used more frequently when welding. Grains in the alloy structure can get smaller due to making use of modifiers with 60-100 nm particles; it is possible because the initial radius of crystallization center is minimal and modifier particles distribute evenly in the molten layer, whereas their mass concentration is low.

This paper presents theoretical research into the effect, which complex (Ti, Zr, Cs) nanopowder has on the efficient ionization potential (U_{eff}) of welding arc discharge.

The energy of atom ionization characterizes the particle and it is independent on the method of ionization, whereas ionization potential characterizes the first historical method of ionization.

The energy of atom ionization, measured in eV (electron-volts) is similar numerically to atom ionization potential, measured in V (volts).

The energy of ionization is an important atom characteristic and influences on the nature and bound strength of chemical compounds made by the atom. The energy of atom ionization has a strong effect on deoxidizing properties of the corresponding elementary substance.

Ionization potential is a relation of electron work function of the substance atom to the electron charge:

$$U = \frac{W}{e}, \qquad (1)$$

where U — ionization potential, V; W — electron work function, J; e — electron charge, C.

Complex atoms, consisting of a lot of electrons, have several ionization potentials. The first ionization potential meets electron removal from the shell of the atom, having the weakest bonds with it. The removal of other electrons, which are closer to the atomic nucleus and have stronger bonds, requires more work. Therefore, the second and further ionization potentials, corresponding to removal of the second and further electrons are higher [3].

Table. First ionization potentials (U) of elements

Elements	Cs	K	Na	Ca	Zr	Ti	Mg	C	H	O
U	3.88	4.30	5.11	6.11	6.63	6.8	7.64	11.22	13.53	13.56

It is seen from the Table above that cesium, potassium, sodium, calcium, zirconium, titanium etc. have the lowest ionization potentials. Therefore, these substances are added to electrode coatings or fluxes to support stability of welding arc burning.

We should know how elements affect the weldability of a product to avoid difficulties and eliminate problems when welding.

Titanium is used due to its high corrosion resistance. It increases strength, impact strength of steel, improves its weldability, supports grain refinement in conditions of metal crystallization. It combines with carbon when welding, preventing from synthesis of chromium carbides on grain boundaries and intercrystalline corrosion of chromium-bearing steels of a weld joint.

Zirconium additives improve the ultimate strength, increase compression strength, impact strength and plasticity of steels for general engineering purposes [12]. Weldability of an alloy and weld plasticity get better due to structure refinement [13].

Cesium additive enhances enormously heat resistance of magnesium and aluminum, e.g. 0.3 – 0.4 % cesium additive increases impact strength of magnesium three times and its corrosion resistance is improved considerably. In recent years metal cesium has been widely used in industry as an additive to tungsten when manufacturing electrodes of high-capacity arc-discharge lamps and electrodes for argon, helium and hydrogen welding of aluminum, magnesium, titanium, cerium, stainless steel and other active alloys. This additive makes easier arc initiation and burning at low voltages.

The efficient potential of ionization is potential of a homogeneous gas, producing the same number of charged particles as in the mixture.

As there are several elements in the arc atmosphere formula [6] of U_{eff} is used:

$$U_{eff} = -\frac{T}{5800}\ln\sum_{i-1}^{i=k}\left(C_i^{1/2}\exp\left\{-\frac{5800}{T}U_i\right\}\right) \quad (2)$$

where C_i – gas concentration of gas *i*; U_i– ionization potential of each gas.

Calculating the efficient ionization potential of vapors of ferrum (Fe), calcium (Ca), sodium (Na), potassium (K), magnesium (Mg) with Ti (Figure 1), zirconium Zr (Figure 2), cesium Cs (Figure 3), we have drawn the diagrams.

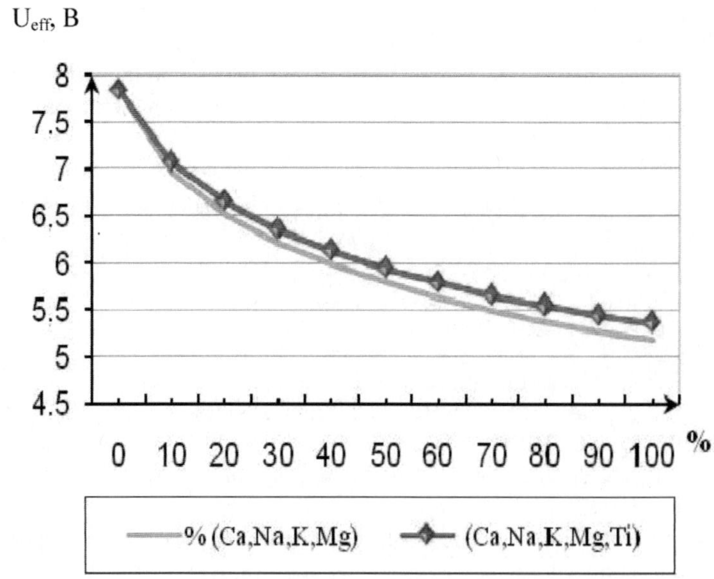

Figure 1 - Change in the efficient potential of ionization for mixed vapors Fe, Ca, Na, K, Mg and Fe, Ca, Na, K, Mg with added Ti.

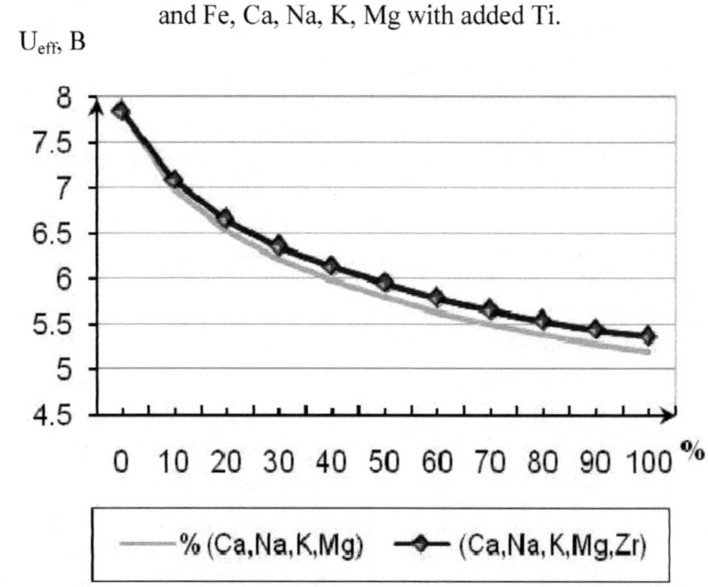

Figure 2 - Change in the efficient potential of ionization for mixed vapors Fe, Ca, Na, K, Mg and Fe, Ca, Na, K, Mg with added Zr.

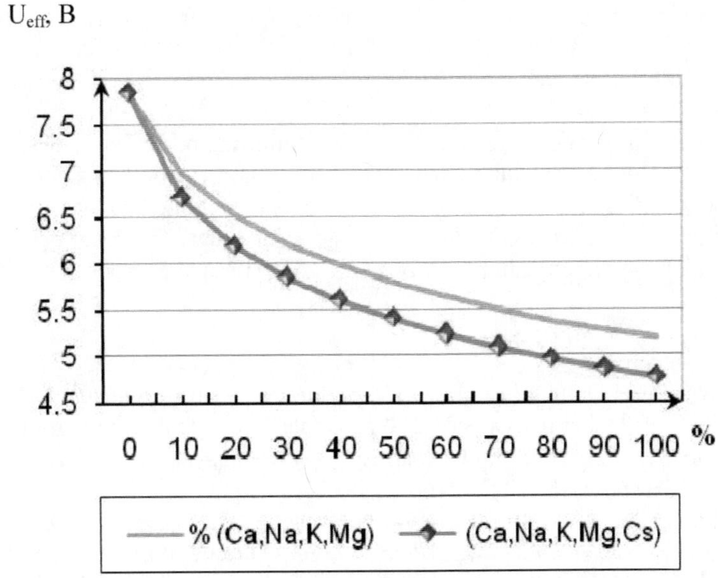

Figure 3 - Change in the efficient potential of ionization for mixed vapors Fe, Ca, Na, K, Mg and Fe, Ca, Na, K, Mg with added Cs.

As one can see from the diagrams, the effective ionization potential is hardly affected via adding Ti, Zr, and it is decreased 6.52 to 6.18 eV provided that mixture containing 20% Cs (mass share) is added.

Having added nanopowder (Ti, Zr, Cs) in equal shares we obtained dependences, depicted in Figure 4.

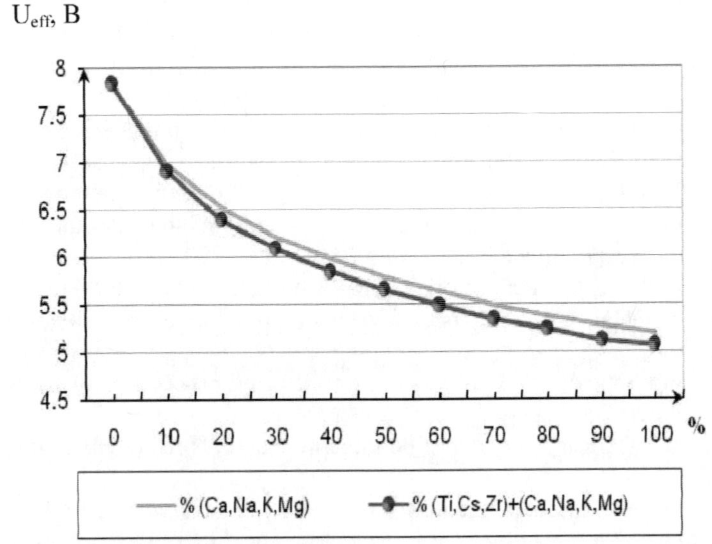

Figure 4 - Change in the efficient potential of ionization for mixed vapors Fe, Ca, Na, K, Mg and Fe, Ca, Na, K, Mg with added Ti, Zr, Cs.

The efficient potential of ionization gets decreased via adding the mixture on the whole. It is necessary to add a little component with a low ionization potential to reduce the ionization potential of the mixture and support burning of the arc. The authors of papers [6, 7] detected a positive effect of a complex nanopowder, containing Ti, Cs, Zr on operational characteristics of the weld joints.

Conclusion

Nowadays nanopowders are used in various fields, e.g. information technologies, chemical industry, medicine and pharmacology, material science, electronics and welding production. However, it is worth emphasizing, the information on using these powders in welding production in not sufficient.

The theoretical study revealed reducing of the general efficient potential of ionization via adding complex nanopowder, containing Ti, Cs, Zr. Therefore, the process of welding gets more stable.

References

[1] Vaia R A and Wagner H D 2004 Frame work for nanocomposites *Materials today* pp 32-37.
[2] Krushenko G G and Reshetnikova S N 2008 Nanotechnologies of strengthening the surface of metal structures *Reshetnikovskie chtenia: proceedings of the XII International Scientific Conference* Krasnoyarsk: SibSAU pp 213–214.
[3] Electric welding arc [Electronic resource] Mode of access: *http://soedenimetall.ru/e-lektricheskaya- svarochnaya-duga/#ixzz3r9JfpMQ3* (date of application: 24.11.15)
[4] Belomestnykh V N and Tesleva E P 2012 Acoustic, elastic, and anharmonic properties of $Sm_{1-x}R_xS$ solid solutions with trivalent impurities (R = Y, La, Tm) *Russian Physics Journal* Vol 55 P 488–494.
[5] Soboleva E G, Igisheva A L and Krit T B 2015 Reduction of longitudinal ultrasonic oscillations in $NaClO_3$ crystals *IOP Conference Series: Materials Science and Engineering* Vol 91 Article number 012032 pp. 1-7.
[6] Makarov S V and Sapozhkov S B 2012 Production of electrodes using complex nanopowders (Zr, Si, Ni, Ti, Cr) *Proceedings of the VIII International scientific and practical conference «Formation of the modern science –* 2012» Prague: Publishing «Education and Science» pp 88–91.
[7] Sapozhkov S B and Makarov S V 2013 Use of complex nanopowder (Al_2O_3, Si, Ni, Ti, W) in production of electrodes for manual arc welding *World Applied Sciences Journal* (Special Issue on techniques and Technologies) V 22 pp 87-90.
[8] Kuznetsov M A and Zernin E A 2012 Nanotechnologies and nanomaterials in welding production (review) *Welding International* Volume 26 Issue 4 p 311-313.
[9] Rodrıguez A G and Perez C R G 2009 Criteria for some methodologies of operative characterization of coated electrodes in welding in short-circuit metallic transference mode *Welding International* Vol 23 No 4 p 282– 290.
[10] Kobernik N V, Mikheev R S, Pankratov A S and Linnik A A 2015 Modern views on modification of welded metal and weld metal by nanoparticles (Review) *Wlding and Diagnostics* pp13-18.
[11] Makarov S V and Sapozhkov S B 2013 Manufacturing procedures of nano-materials *Global scientific potential* pp 67-71.
[12] Fedorchenko I M 1977 Encyclopedia of nonorganic materials *Kiev. Main edition of USE* V 2 pp 734.
[13] Alieva S G, Altman M B and Ambartsumyan S M et al *Industrial alloys of aluminum* (M.: Metallurgy) p 187
[14] Volchenko V P, Yampolskiy V M, Vinokurov V A et al Edited by Frolov V V 1988 *Theory of welding processes*: Study guide on specialty «Equipment and technology of welding production» pp 44-45.

IOP Conf. Series: Materials Science and Engineering **142** (2016) 012019 doi:10.1088/1757-899X/142/1/012019

Three-Stage Character Of Molten Metal Drop And A Hard Substrate Contact Interaction

S B Sapozhkov [a], E M Burakova, E P Tesleva

Yurga Institute of Technology TPU Affiliate
Leningradskaya 26, Yurga, Russian Federation, 652055

E-mail: wh13@bk.ru

Abstract. The paper presents the results of theoretical and experimental studies of the interaction between a molten metal drop and a solid metal substrate and justifies measures to avoid or to weaken this interaction.

Introduction

Welding in carbon dioxide is a key process in all branches of industry both in our country and abroad [1]. A significant drawback of welding in CO2 is high metal splashing, as the result; the surface of the product to be welded is covered with metal spatter. The mechanism of metal spray cohesion with the surface is studied in works [2, 3, 4, 5]. However, further developing and supplementing the theory of spray interaction with the surface support improvement of the shielding means and the quality of products, in particular when manufacturing welded structures of low and medium alloy steels, producing mining equipment, main pipelines etc. Drops of various sizes are splashed when welding in CO2.

Small spatter ~ 0.02 mm leaves the zone of welding with the speed ~ 40 m/s. The angle of its direction and the perpendicular to the surface of the product to be welded does not usually exceed 25°. The bigger spatter (diameter > 1 mm) move slower with bigger opening angle. As the rule, the surface of the products to be welded is covered with big and small spatter, while the nozzle and the neck of the torch with small spatter only.

Main part

The drop contacts with the surface to be welded over three stages: 1) physical contact appears, id. atoms of contacting surfaces are approaching each other, so physical or slight chemical interaction is possible; 2) activation of the contact surfaces (generation of active centers); 3) bulk interaction.

The third stage is distinguished by the growth of interaction both in the plane of the contact attended by generation of strong chemical bonds and around the contact zone. This process occurs in the active centers, which are fields of elastic lattice distortion, or defects of the structure (dislocations, vacancies etc) and is over under coalescence of discrete areas of interaction. The third stage of molten metal drop – product surface interaction is completed in conditions of recrystallization, causing generation of common grains in the contact zone.

The quantity of heat in a drop with the diameter D_d, is equal to q when escaping from the arc space. The drop loses some heat q_{TO}, used for convection and radiation when moving to the surface of the product. At the moment of contact the quantity of heat in the drop is q_k. Temperature distribution ΔT, depending on heat content of the drop is registered in the area where a drop of the diameter D contacts with the surface (the diameter of interaction D_x).

A high speed of the drop (40 m/s and above) is the cause of high pressure in the contact zone. It is known that the drop leaving the arc space is over-heated; its temperature can come up to 2500^0C. High temperature of the contact (T_{cont}) and pressure are important factors of physical and chemical interaction, providing strong drop-surface joining. There is a physical contact between them due to deformation and drop spreading. As the temperature of drops is high and their surface gets deformed, their atoms in the contact zone are excited and tend to form active centers of chemical interaction.

As the result of colliding with the surface, the drop gets deformed, furthering their physical contact. Pressure P is typical for the zone of collision, being component of the total pressure P_t (or dynamic component) and impact pressure P_i, arising due to the effect of a hydraulic shock.

First, liquid volume of the drop is elastically deformed in the zone of drop colliding with the surface. In a time interval $t_y=d/c$ (c – sound speed in the melt, m/s), equal 0.1 microsecond for small and 1.0 microsecond for bigger drops, and sufficient for the shock wave front (moving from the place of its colliding with the product) to reach the free surface of the drop, a thin and flat liquid layer of spreading drop is formed due to its cushioning. Then the drop gets deformed uniformly. The maximal impact pressure is calculated according to the formula below:

$$P_y = \frac{\mu}{2} \cdot \rho_0 V, \quad (1)$$

where μ – stiffness coefficient of the drop, based on the relaxation capacity of the drop liquid, depending on its impact velocity and form; ρ – density of the melt, kg/m^3; V – speed of the drop at the moment of impact, m/s.

The difference of the impact pressure for different diameters of drops is revealed on the base of calculations done for the case above: it is 30 MPa for small drops and 10 MPa for bigger drops, respectively. The total pressure of the drop, calculated by Bernuolli's equation $P_н = \rho \cdot V^2$, is 10 MPa for small drops and 7 MPa for bigger ones.

Provided that the height of the drop decreases evenly from d to h along its axis over the period of deformation, having the speed V of drop movement at the point of impact, the total pressure depends on the time of drop deformation $t_н = \frac{d-h}{V}$, which is 10 microseconds for small and 4 microseconds for bigger drops.

A high impact pressure furthers cleaning of the product surface at the point of impact and causes a physical drop – product contact. The total pressure, acting over the complete period of deformation and hardening of the drop, supports its stronger cohesion with the surface.

The drop deformation is caused by the kinetic energy of movement when hitting against the surface. The crystallization front extends over the surface of the colder product simultaneously with drop spreading. The drop deformation is completed when the crystallization front meets the free surface of the drop. As the process of deformation is over the diameter of drop interaction with the surface is maximal.

The temperature in the contact (temperature of the contact T_{cont}) of the surface (substrate) with a liquid, fast-spreading drop is somewhere between the temperature of the substrate T_s and temperature of the drop T_k ($T_k > T_{cont} > T_s$), being constant for the period t_0, which is equal to the period of drop crystallization.

The drops, welded completely on the surface are hard-to-remove, since their removal requires special tools. The criterion of hard removability is a specific force of shear P_{sh}, which is necessary for removal of such drop.

Components with various degrees of surface roughness are used in manufacturing welded products. Further theoretical and experimental research requires the use of products with following degrees of roughness: as delivered, R_a=1.69 μm; sand jet processed surface, R_a=7.73 μm, and the surface processed with a sanding disk, R_a=16.4 μm.

For the purpose of research a bead joint is made on the samples of steel St3 (GOST 380-85) with various degrees of roughness. The conditions of experiment are as follows: welding in CO_2, a semi-automatic device ПДГ–508 (semi-automatic arc welding in active gas), power source WRV –506, wire Св-08Г2С (GOST 2246, composition: C – 0.08%, Si – 1%, Mn – 2%) with the diameter of 1.6 mm, current 300…320 A, arc voltage 30…32 V.

The most hard-to-remove spatter is detected on the surface processed with a sanding disk, while the surface as delivered is covered with the least spatter. However, the biggest diameter of interaction D_x is identified on the surface as delivered. In the process of carrying out the research into the spatter cohesion with the pre-heated metal surface a dependence of its strength on T_s and T_{cont} (Fig. 1) is revealed, the interaction diameter D_x of a drop (diameter D) with the surface is increased. The worth variant is described by the equation $\frac{D_x}{D} = 1$.

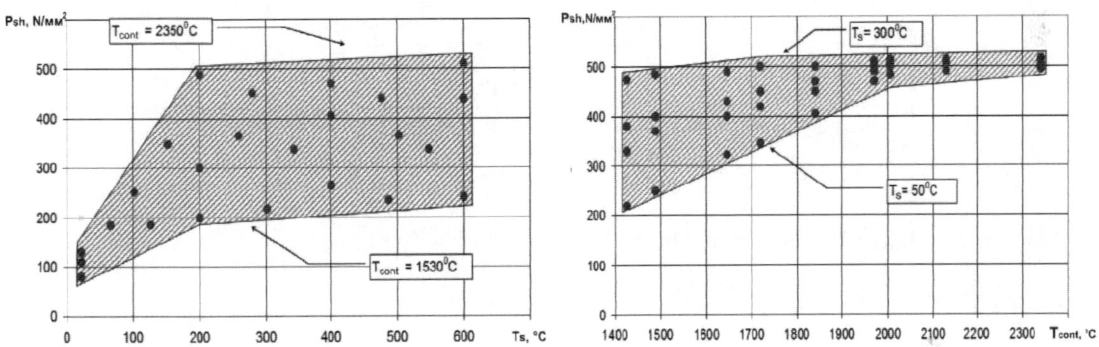

Figure 1- Influence of the substrate temperature T_s and temperature of the contact T_{cont} on the shear pressure P_{sh}

However, theoretical and experimental data on the kinetics of drop spreading over the surface of the product to be welded, obtained in terms of the methods available [1, 6, 7, 8], point at the smallest interaction diameter (D) of molten metal drops with the surface processed by a sanding disk, and at the biggest one with the surface as delivered. The most hard-to-remove drops are detected on the surface processed by a sanding disk; therefore, the cohesion strength of a drop with the surface does not depend on the diameter of their interaction only. This fact can be referred to the processes of contact interaction of molten metal drops with the surface to be welded.

A metallographic analysis of microstructures in the interaction zone is carried out to determine the character of this interaction. The main component of the oxide film on the surface of the drops is FeO. When drops contact with the dross-coated surface of metal, it melts. The oxide film of spatter contacts with the dross of the welded surface in the melt, so cohesion is less strong (P_{av} amounts to 20-50 N/mm^2). The space between the weld and the contact zone influences P_{av} in particular, and since the product is not heated evenly when welding, so one can affirm that cohesion strength of a drop with the surface depends on T_s.

Having studied the cohesion of bigger drops it is revealed their fusion with the surface is possible provided that oxide films are removed or the contact zone is absolutely free of them. The fusion process is regarded to be over when metallic bond is formed totally over the interaction zone, and P_{av}

gets constant ($P_{av} \approx 500$ N/mm^2 – see Fig. 1). However, one should take into consideration cohesion strength depends on the intensity of heat removal, contact and surface temperatures, impact pressure at the moment of contact and oxides on the surface. All these factors further complete or partial fusion of interacting drop and surface. P_{av} varies 60 to 500 N/mm^2. Moreover, rising temperature of the surface is the cause of a longer drop liquid state; then the temperatures get equal due to the intensity of heat removal (Fig. 2).

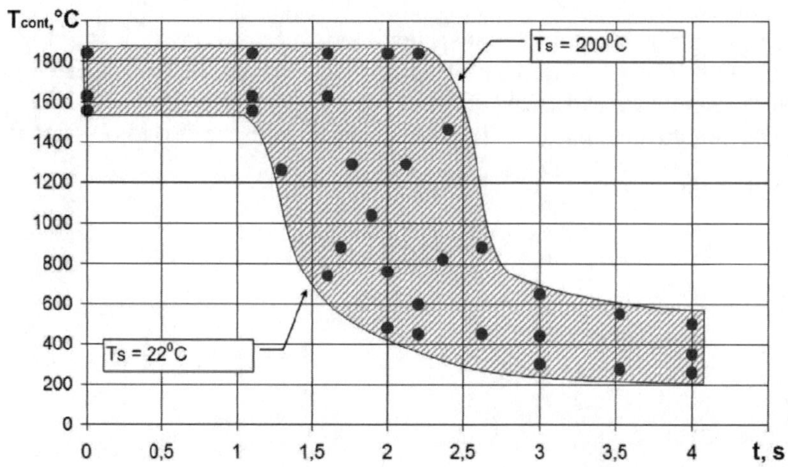

Figure 2 - The contact thermal cycle regarding the substrate temperature T_s

The strength of drop cohesion with the surface depends both on its cleanness and roughness. Higher roughness increases, first, the temperature of contact under the spatter, since the temperature of roughness peaks is high because of insufficient heat removal, second, the total interaction area of drop and surface atoms.

Welded samples made of steel Ст3 with various degrees of roughness are tested to confirm theoretical backgrounds. Macro- and micro-sections are made on the point of spatter and surface contact, optical microscope MIM -7 is used for the purpose of research, providing 100 – 500 x magnification. The microstructures are shown in Figure 3. The results of processing experimental data are given in Table 1.

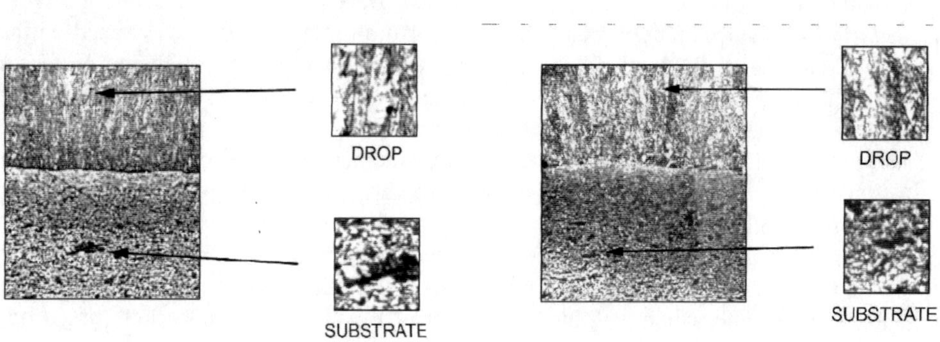

a) b)

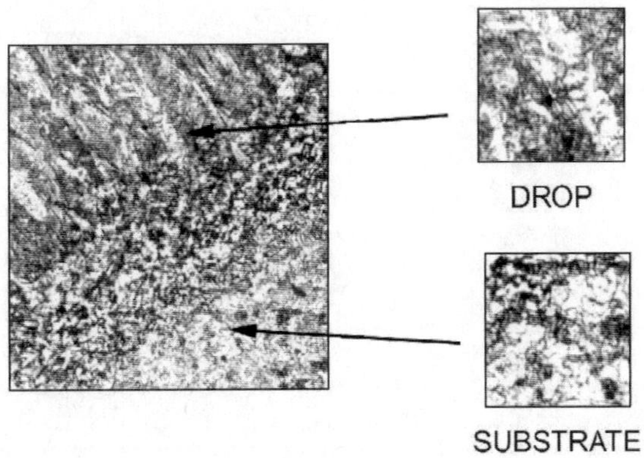

c)

Figure 3 - The microstructure of drop and surface contact zone: a) – the sample as delivered, b) – sample processed in a sand jet chamber, c) – sample processed with a sanding disk (x150, leader – computer magnification x300)

Table 1 - Microstructure analysis of drop and surface contact zone

Figure	Description of the structure
3 a	In the contact zone there is a layer of oxides and non-metallic impurities without obvious layer growth. The nature of interaction is fusion of drop oxides with the dross of the surface (P_{av} – 20-40 N/mm^2).
3 b	In the interaction zone there is a boundary of the contact over the total length. The nature of interaction is partial fusion (P_{av} – 100-150 N/mm^2, $l/D_x \approx 0.2$). On the left there are oxides and non-metallic impurities.
3 c	In the contact zone there are atoms of a drop fused with the substrate (P_{av} - 500 N/mm^2, $l/D_x \approx 1$).
Note:	In all cases under consideration the base metal consists of ferrite and perlite. The drop has a dendrite structure. The growth of dendrites is directed from the contact zone to the free surface of the drop.

Interaction is assessed relying on a non-dimensional quality l/D_x (l – length of fused sections), which influences P_{av} of this drop. Taking into consideration, that P_{av} is measured according to the drop with the area $S_{1/2} = \frac{\pi D_x^2}{8}$, the cohesion strength σ_{cs} is calculated by the formula: $\sigma_{cu} = \frac{8P_{av}}{\pi D_x^2}$. The results are outlined in Figure 4.

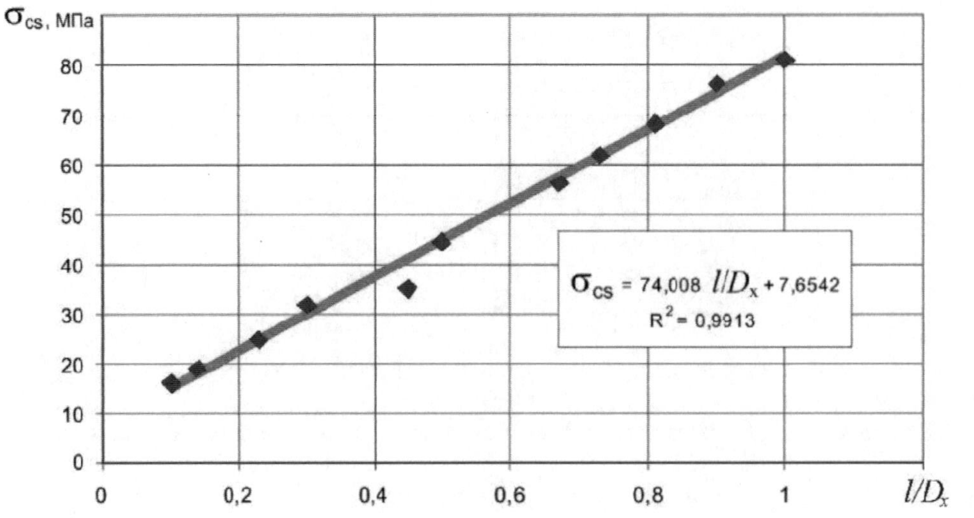

Figure 4 - l/D_x vs. cohesion strength σ_{cs}

Having analyzed the data of experiments it is revealed that the thickness of drop to surface interlayer influences P_{av} of this drop, being a direct proportion, when the drop contacts with the dross-coated surface as delivered.

The data of experiments are computer processed, a trend line is drawn and equation of the dependence of interlayer thickness on P_{av} is determined. It is found out the thicker is the interlayer, the smaller P_{av}. Is P_{av} tends to 0 when thickness of the interlayer consisting mainly of FeO is 12 μm. The conditions reducing or eliminating harmful effect of factors, providing strong cohesion of molten metal drops with the surface, are required for its shielding.

Conclusions

1. Key factors are identified on the base of experiments, which influence the character of spatter (drops) interaction with the surface. The cohesion strength is affected by the intensity of heat removal, temperature of the contact (T_{cont}), temperature of the surface (T_s), force impact of the drop, and oxides on the surface and its roughness.
2. Having analyzed microstructures in the drop – surface contact zone, fusion of the spatter oxide film with the dross on the surface to be welded is revealed, and the least strong cohesion (P_{av} is 20-50 N/mm^2) is formed when drops contact with the metal surface. Fusion of bigger drops with the surface is possible provided that oxide films are removed from the point of contact. The fusion process is regarded to be over when metallic bond is formed totally over the interaction zone, and P_{av} gets constant ($P_{av} \approx 500$ N/mm^2. However, the strength of cohesion is affected by multi-factorial character of contact interaction, resulting in partial or complete drop – surface fusion. P_{av} varies in that range 60 - 500 N/mm^2. Interaction is assessed on the base of a non-dimensional quality l/D_x (l – length of the fused zones). The strict dependence of cohesion strength σ_{cs} on l/D_x is revealed.

References

[1] Fedko V T, Sapozhkov S B, Sokolov P D and Yastrebov A P. 2004 Elements of theory and technology of anti-spatter surface shielding when welding in carbon dioxide Monograph. Tomsk: TSU Publishing 2004 138 p.
[2] Solonenko O P, Kudinov V V, Smirnov A V, Cherepanov A N, Popov V N, Mikhalchenko A A

and Kartaev E V 2005 Micro-Metallurgy of Splats: Theory, Computer Simulation and Experiment JSME International Journal Series B, Vol 48, No 3, 2005

[3] Knotek O and Elsing R. Monte Carlo Simulation of Lamellar Structure of Thermally Sprayed Coatings Surf. Coat. Technol 1987 № 32 P 261 – 271.

[4] Ghafouri-Azar R, Mostaghimi J, Chandra S and Chamchi M A Stochastic Model to Simulation of a Thermal Spray Coating Journal of Thermal Spray Technology 2003 V 12 № 1 P 53-69.

[5] Steinke T, B ker M. Monte Carlo Simulation of Thermal Sprayed Coatings Proc. ITSC'06 Seattle USA May 15 – 18, 2006 P 329 – 334.

[6] Kuznetsov M A and Kolmogorov D E 2015 Corrosion and Mechanical Properties of Austenic Steel Weld Joints IOP Conference Series: Materials Science and Engineering 2015 Vol 91Yurga, Russia May 2015 8 p.

[7] Filonov A V, Kryukov A V and Gusarov D E 2015 Systems to Control Molten Metal Transfer in Arc Welding IOP Conference Series: Materials Science and Engineering Vol 91 Yurga Russia May 2015 6 p.

[8] Chinakhov D A 2015 Dependence of Silicon and Manganese Content in the Weld Metal on the Welding Current and Method of Gas Shielding Applied Mechanics and Materials : Scientific Journal 2015 Vol 756 P 92-96.

IOP Conf. Series: Materials Science and Engineering **142** (2016) 012020 doi:10.1088/1757-899X/142/1/012020

Assignment of Appropriate Conditions for Synthesizing Tungsten Nanopowder by Electric Explosion of Conductors

S P Zhuravkov[1a], A V Pustovalov[1b], M A Kuznetsov[2c], I S Rosliy[1d], E A Zernin[2e]

[1] Institute of High Technology Physics, National Research Tomsk Polytechnic University, 2A, Lenin Ave, Bld. 11, 634028, Tomsk, Russia
[2]Yurga Institute of Technology, TPU Affiliate, Yurga, Kemerovo region, Russia

e-mail: [a]zhursp@yandex.ru, [b]pustovalov@tpu.rum, [c]kyznechik_85@mail.ru, [d]ilya1000@bk.ru, [e]yuti_sp@bk.ru

Abstract. The paper provides the results of experimental research into properties of tungsten nanopowders synthesized by electric explosion of a conductor in argon at various energies, put into the conductor when exploding. The authors have studied how the conditions of synthesizing the tungsten nanopowder influence on the average size of particles.

1. Introduction

Powders of tungsten are used to synthesize high strength, dead-hard heat resistant alloys, high-speed tool steels and welding electrodes. At present special attention is paid to the possibilities of improving the technologies of synthesizing the tungsten powders and their physical and mechanical properties as well. One of the most promising procedures to produce the powders is electric explosion of a conductor (EEC) [1]. This procedure as a method of synthesizing powders is distinguished by simplicity, low prices for the equipment, a relatively high efficiency, and low power inputs. Its principal advantage is a possibility to synthesize powders consisting of nanometer particles with a quite restricted range of their size distribution. These powders have a rather big specific surface area, demonstrating, therefore, their better reactivity. Nowadays tungsten powders produced this way are used for synthesizing tungsten disulphide when manufacturing solid-film lubrications on its base [2], as an active component in ceolite catalysts of non-oxidizing methane conversion into aromatic hydrocarbons [3], and as refractory modifying additives of a weld [4].

Physical properties of metallic tungsten, inter alia its low thermal conductivity, high melting and steaming temperatures complicate the assignment of appropriate conditions for EEC synthesizing the metal tungsten powder.

The purpose of this research is to study the influence of EEC initial conditions on the properties of the synthesized tungsten powder and determine appropriate EEC conditions for manufacturing the best quality powder.

2. Experimental

The research into electric explosion of the tungsten conductor and synthesizing the powders is carried out by the facilities, the general layout and operating principles of which are well-described in papers [5, 6]. Tungsten wire ВА-П-А with the diameter of 0.25 mm is used for EEC. The length of the exploded conductor is varied 40 to 100 mm. The capacitor bank capacity is varied 1.12 to 3.31 mkF, and its charge voltage is changed in the range from 20 to 30 kV. EEC is conducted in argon at the pressure of 2.5 atm.

The energy put into the conductor, energy, emitted at the arc stage of electrical discharge, current density and period of the EEC process are calculated according to the current oscillograms on the base of methodology in [7].

The properties of synthesized powders are studied via scanning electron microscopy (microscope JSM-7500F), X-ray diffraction analysis (difractometer Shimadzu XRD-7000), the specific surface area is measured via thermal nitrogen desorption (analyzer "Sorbtometr-M").

3. Discussion of Results

Typical oscillograms of electric exploding the tungsten conductor are presented in Figure 1. Initial EEC conditions and some properties of the synthesized powders are given in Table 1.

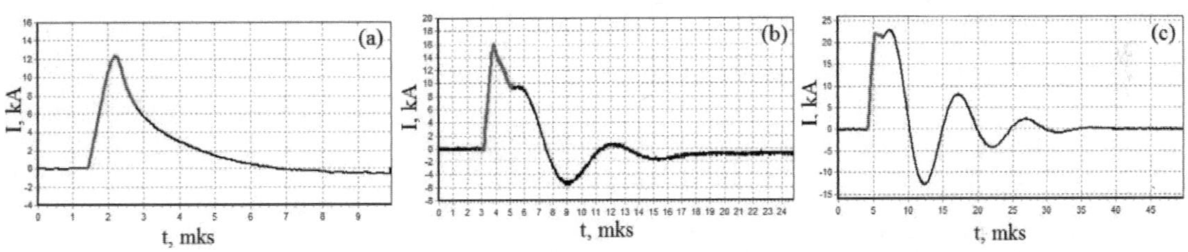

Figure 1 – Current oscillograms of tungsten EEC. a) sample W-1; b) sample W-2; c) sample W-3

In Figure 1 one can see that deceasing length of the exploded conductor for other conditions kept constant results in the increase of the maximal passing current from 12 to 23 kA, but the time of explosion is practically the same about 1.5 ms. It should be noted that the complete tungsten EEC process is at the arc stage of electrical discharge. It is possible due to physical properties of the metal to be exploded: low ionization energy (82 J/mm^3) is the cause of bridging the early conductor electrical explosion by the arc stage of electrical discharge. Moreover, it blocks the further energy input into the conductor, so its higher values fail to be achieved as against EEC of other metals [1].

Having analyzed EEC modes with various initial conditions (Table 1) we revealed that shortening the conductor to be exploded at other conditions of the explosion kept constant causes a slight growth of the specific energy put into the conductor, (W-1 – W-2 or W-4 – W-5), but the energy emitted during the arc stage of electrical discharge increases far quickly. Furthermore, the rise of energy, put into the conductor when exploding, is possible in a certain range only. When exploding the conductor shorter 50-60 mm, decreasing e/e_s and considerable increasing e_{arc}/e_s are registered, being the consequence of the early stage of arc discharge.

Table 1. Conditions of synthesizing tungsten powders and their properties.

Sample	U, kV	C, mkF	L, mkH	l, mm	e/e_s	e_{arc}/e_s	$(e + e_{ars})/e_s$	S, m^2/g	A_S, nm
W-1	20			90	0.4	0.1	0.5	1.1	280
W-2	20			75	0.5	0.2	0.7	1.2	260
W-3	20	1.12	0.81	40	0.5	0.6	1.1	1.4	220
W-4	26			100	0.5	0.2	0.7	1.2	250
W-5	26			80	0.6	0.4	1	1.5	210

W-6	26			50	0.5	1.1	1.6	1.6	190
W-7	30			100	0.5	1.5	2	1.5	210
W-8	30	2.24	0.72	75	0.7	2.1	2.8	1.9	160
W-9	30			60	0.5	2.9	3.4	2	150
W-10	30			40	0.4	4.5	4.9	2.2	140
W-11	30	3.31	0.76	90	0.7	3.8	4.5	2.6	120
W-12	28			50	0.4	5.1	5.5	2.7	110

U – voltage of the capacitor charge; *C* – capacitor capacity; *L* – circuit inductance; *l* – length of the conductor to be exploded; *e* – specific energy put into the conductor in the EEC process; e_{ars} – specific energy emitted at the arc stage of electric discharge; e_S –sublimation energy of tungsten (88.7 J/mm^3); *S* – specific surface area of powder; A_S – average size of particles calculated according to specific surface area on the base of equation – $A_S = 6/(S \cdot \rho)$, where ρ – compact density of tungsten (g/cm^3).

The study on synthesized powders has detected that specific surface area of powders is increased due to the growing energy put into the conductor when exploding and energy emitted at the arc stage of electrical discharge. The relation of A_S to e/e_s and e_{arc}/e_s can be described by the function $A_S = f(e + e_{arc})/e_s$ (Figure 2).

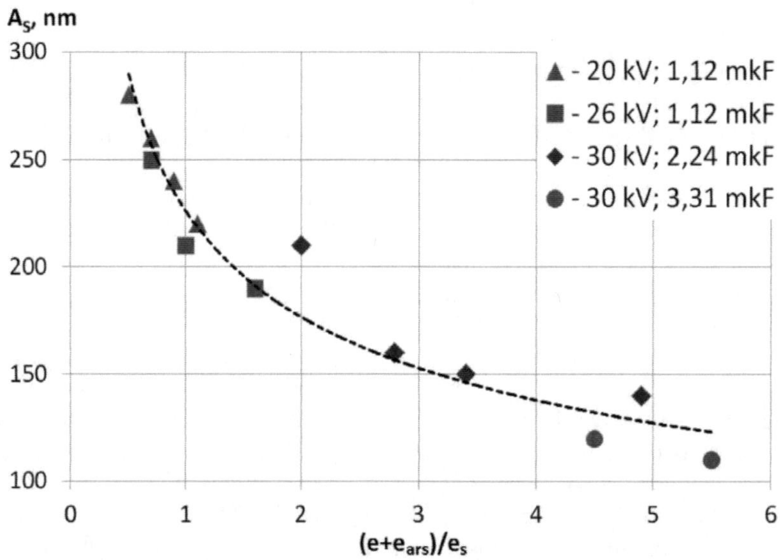

Figure 2 – The average size of particles vs. the sum of energy put into the conductor when exploding and energy, emitted at the arc stage of electrical discharge.

One can see in the plot, given in Figure 2, that the decrease in the average size of particles is typical for the whole interval under consideration. The significant decrease of A_S from 300 to 150 nm is possible in conditions of growing total energy from 0.5 to 3.5 e_S, the increase in energy up to $5.5e_S$ is the cause of the average size loss of particles to 110 nm.

Therefore, the appropriate conditions of EEC synthesizing tungsten powders are modes with the total energy, put into the conductor and energy, emitted at the arc stage of electrical discharge, approximating to $3.5e_S$. The further increase of the energy has only a slight effect on the average size of particles.

The data obtained by the scanning electron microscope demonstrate that the sample synthesized in appropriate EEC conditions (sample W-9) consists of two fractions. The first fraction includes micron particles (Figure 3), the second one – manometer particles (Figure 4). In accordance with the data presented in literature [8] two various fractions of particles are formed due to the specifics of the conductor destructing process under electrical explosion. Micron particles are formed from the bubbling drops when splashing the conductor. The nanometer particles are formed under vapor condensation.

Having processed the obtained data a conclusion is drawn that the size of micron particles ranges from 1 to 3 μm, and maximum of their distribution is 1.5 μm. The size on nanometer particles varies in the range 20 to 200 nm, the maximum is 40-50 nm.

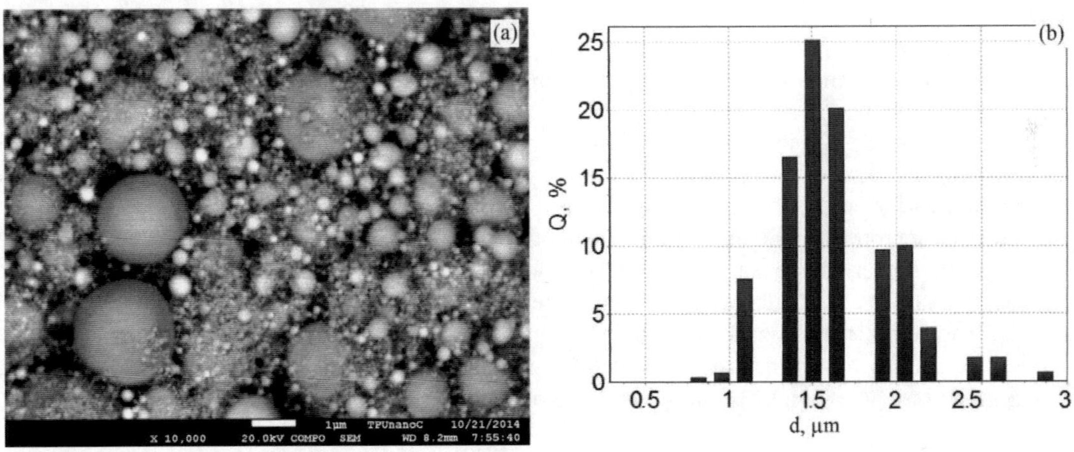

Figure 3 – a) Photo of the particles, sample W-9; b) size distribution of micron particles.

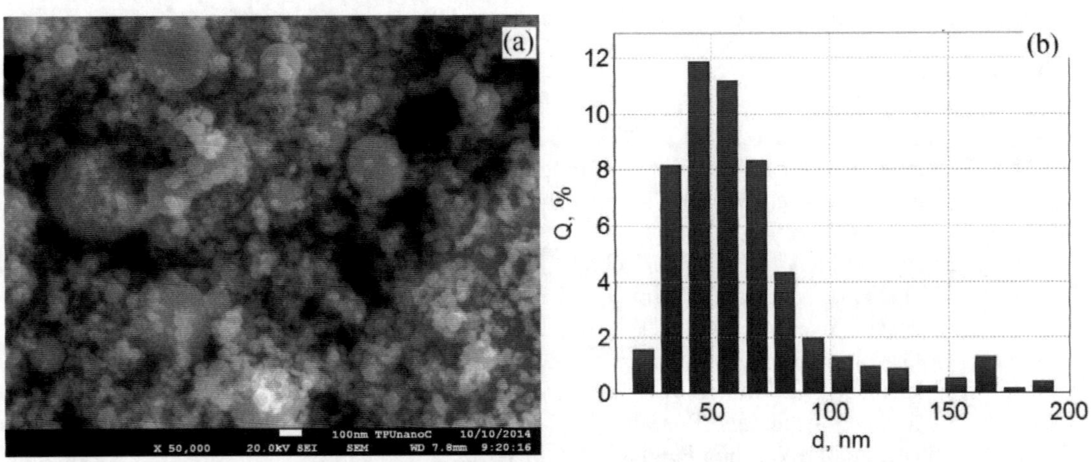

Figure 4 – a) Photo of particles, sample W-9; b) size distribution of nanometer particles.

The powder under consideration consists of two phases αW – 68.8% and βW – 31.2% (Figure 5) according to the data of X-ray diffraction analysis.

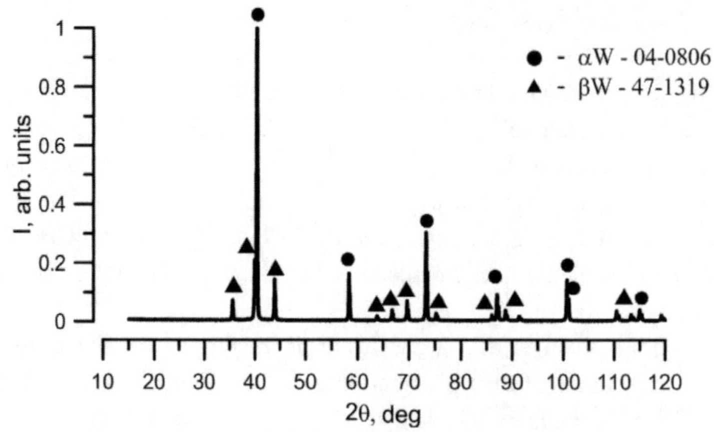

Figure 5 – X-ray diffraction pattern of sample W-9.

4. Conclusions

The research into the electric explosion of tungsten conductors in various initial conditions allows making the following conclusions:

1. The average size of particles calculated on the base of the specific surface area of powder is decreasing because of the growth of energy, put into the conductor when exploding, and energy, emitted at the arc stage of electrical discharge.

2. The appropriate EEC conditions are modes with the total energy, put into the conductor and energy, emitted at the arc stage of electrical discharge, approximating to $3.5e_S$. The further increasing of energy causes a slight decrease of the average size of particles and is not reasonable.

3. The tungsten powder synthesized in appropriate EEC conditions consists of two fractions: a micron one, having the maximum of distribution 1.5 μm and nanometer one with the distribution maximum of 40 - 50 nm. This tungsten powder is a blend of two phases of α- and β- tungsten.

Acknowledgements

The study was performed using the equipment provided by Scientific and Analyzing Centre of TPU, Nano-materials and Nanotechnologies Centre of TPU and Laboratory No. 12 of IHTP TPU.

References:

[1] Yavorovskiy N.A. 1994 Russian Physics Journal Production of ultradispersed powders by the method of electric explosion 4 pp. 114-136.

[2] Vladimir An, Farabi Bozheyev, Franck Richecoeur, Yuri Irtegov 2011 Materials Letters Synthesis and characterization of nanolamellar tungsten and molybdenum disulfides 65 pp. 2381–2383.

[3] Vosmerikov A.V., Echevskii G.V., Korobitsina L.L., Arbuzova N.V., Kodenev E.G., Velichkina L.M. and Zhuravkov C.P. Nonoxdative 2007 Kinetics and Catalysis Methane Conversion into Aromatic Hydrocarbons on Tungsten-Containing Pentasils 48 3 pp. 409-413.

[4] Kuznetsov M.A., Zhuravkov S.P., Zernin E.A., Kolmogorov D.E., Yavorovsky N.A. 2014 Advanced Materials Research Influence of Nanostructured Powder Modifiers on the Structure of a Welding Bead 872 pp. 118-122.

[5] Yavorovsky N, Pustovalov A., Lobanova G., Zhuravkov S. Proceedings 2012 7th International Forum on Strategic Technology. Investigation of the characteristics of aluminum powder obtained in argon with addition of oxygen. (Conference Paper).

[6] Zhuravkov S.P., Pustovalov A.V., Yavorovskiy N.A., Korshunov A.V., Vlasyuk M.N., Nadeina L.V. Key Engineering Materials Property investigation of Pt-Rh alloy nanopowders obtained

IOP Conf. Series: Materials Science and Engineering **142** (2016) 012020 doi:10.1088/1757-899X/142/1/012020

by conductor electric explosion method 685 pp 596-600.
[7] Kvartskhava I.F., Bondarenko V.V., Plyutto A.A. et al. 1956 Journal of experimental and theoretic physics Oscillographic determination of the energy of electric explosion of wires 31 5, pp. 745-751.
[8] Kotov Yu. A. 2009 Nanotechnologies in Russia The Electrical Explosion of Wire: A Method for the Synthesis of Weakly Aggregated Nanopowders 4 7–8 pp. 415–424.

IOP Conf. Series: Materials Science and Engineering **142** (2016) 012021 doi:10.1088/1757-899X/142/1/012021

The Accounting for Materials with Strength Differential Effect in Calculating the Durability in Order to Raise the Exploitation Effectiveness

U E Yakubovskiy, N A Krivchun, V I Kolosov and O L Umanskaya

Tyumen State Oil and Gas Uuniversity 38, volodarskogo St., Tyumen, 625000 Russia.

E-mail: olgaumanskaya @yandex.ru

Abstract. The article is devoted to the mathematical model of bending of machines and constructions. The fact that they are made of multimodulus materials is taken into consideration. The priority is given to the multilayer qualities of such materials. The peculiarity of their deformation under real conditions due to the reciprocal slip of layers is taken into account. The redistribution of forces between the layers is achieved through the variable joint rigidity which is also represented in the given article. The construction of the mathematical model is given in details. The calculation method of bending of composite plates with layers of material with strength differential effect is described. While calculating the stress-strained state, the influence of rigidity of the multilayer bonding was taken into consideration.

1. Introduction

To decrease the material consumption and to preservethe durability and safety at the same time is an important task of raising effectiveness while creating any constructions and mechanisms, which can be solved due to the use of composite materials. Multilayer plates and shells are widely used at the construction sites in the oil and gas industry, in the chemical industry and in the atomic energetics, in creating complex fortified transport mechanisms. The use of such elements is stipulated by their high durability and rigidity while their mass is relatively low, excellent thermal and sound insulation qualities. Their high radiation protection should be mentioned separately.

New and traditional construction materials in use demonstrate different mechanical qualities depending on the main stress signs. In the majority of cases these materials possess different multimodulus qualities, that is they have different elasticity modules when they are stretched and compressed, for example concrete. Its multimodulus degree depends on the size of the specimen, the kind of its filler, humidity and other factors. For example, the multimodulus degree (Young's modulus) of fine-grain concrete on quartz sand and portlandcement equals $E^{+}=0.65\cdot 10^{3}$ MPa, $E^{-}=1.82\cdot 10^{3}$ MPa. The multimodulus quotient (Young's modulus) of heavy-weight concrete (E^{+}/E^{-}) may vary from 1.07 to 1.82. The multimodulus quotient (Young's modulus) of light-weight concrete may be higher, lower or may equal 1.

When constructing block-boxes, multilayer plates, in which the concrete layer is connected with steel sheets with the help of anchors and glue, are used. This bonding has the finite shear rigidity.

IOP Conf. Series: Materials Science and Engineering **142** (2016) 012021 doi:10.1088/1757-899X/142/1/012021

Thus, it is necessary to solve the task of durability of such constructions taking into account the interlayer bonding qualities and multimodulus qualities of the layers.

2. Materials and Method

As in [1] we understand two or more plates when we use the term "a composite plate". Every single plate is an i-layer, the number of which equals n+1.The number of intervals between them (joints) equals n. We will adopt the upper-case numeration. The layers in the composite construction are connected between themselves with elastic-compliance connections, which allow the shear of one layer towards another one. The cross-binding bonds are absolutely rigid. In accordance with it, all the layers have the identical deflection (buckling) W(x,y). The Kirchhoff –Love theory of plates is applied to one separate layer, not for the whole packet of layers. The packet load is directed along the normal and is distributed along the surface in a general-independent way. It is important to take into consideration the rigidity of the interlayer connections and multimodulus qualities of separate layers to solve the task of defining the stress-strained state of such constructions.

The integral characteristics of rigidity for the i-layer of the composite multi-layer plate should be written with the account of the multimodulus qualities of the material. The main force sign is the most important factor in the multimodulus elasticity theory [2]. The points and areas of the body where all main forces have the same signs (that is the body is compressed and stretched along the main forces) will be called the points and areas of the first kind, and we will not concentrate our attention on them in the given article. Those points and areas of the body where one of the main forces has a different sign from the other two forces, will be called the points or areas of the second kind. The physical proportions defining the dependence of the material deformation on the force size, will be written only for the area of the second kind [3].

To represent the integral characteristics of rigidity in Cartesian coordinates, it is necessary to make a transition from the direction of the main platforms to the x-y coordinate system. As being different from other known formulae, we suggest performing this transition through the transformation of the compliance matrix [4].

We will write the compliance matrix in the coordinate system, defining the position of the main platforms for the planar-stressed state as:

$$[a]=\begin{bmatrix} a_{11} & a_{12} & 0 \\ a_{21} & a_{22} & 0 \\ 0 & 0 & a_{66} \end{bmatrix}, \qquad (1)$$

Where the elements a_{ik} are calculated according to the proportions [5].

As being different from other existing models, it is necessary to possess a complete set of the rigidity characteristics in the main coordinate system to make a transition from the coordinate system of the main platforms to the rigidity matrix in Cartesian coordinates. At the same time, when$\sigma_1 > 0$, $\sigma_2 < 0$ at the two orthogonal platforms we have characteristics of the material which correspond to stretching and compressing; that is why it is necessary to differentiate between the shear compliance (shear modulus) $1/G^+$and $1/G^-$. There sultings hear compliance value was calculated by averaging the values which were present at one point on two orthogonal X, Y platforms:

$$a_{66}=\frac{G^{+}+G^{-}}{2G^{+}G^{-}}=\frac{E^{-}(1+\nu^{+})+E^{+}(1+\nu^{-})}{E^{+}E^{-}}, \qquad (2)$$

Using [6], let us make a transition from the compliance matrix (1) in the coordinate system of the main platforms to the matrix in the X-Y coordinate system:

$$A_{11}=a_{11}\cos^4\varphi+a_{22}\sin^4\varphi+(2a_{12}+a_{66})\sin^2\varphi\cdot\cos^2\varphi;$$

$$A_{22} = a_{11}\sin^4\varphi + a_{22}\cos^4\varphi + (2a_{12} + a_{66})\sin^2\varphi\cdot\cos^2\varphi;$$
$$A_{12} = (a_{11} + a_{22} - a_{66})\sin^2\varphi\cdot\cos^2\varphi + a_{12}(\sin^4\varphi + \cos^4\varphi); \quad (3)$$
$$A_{16} = [2(a_{22}\sin^2\varphi - a_{11}\cos^2\varphi) + (2a_{12} + a_{66})(\cos^2\varphi - \sin^2\varphi)]\sin\varphi\cdot\cos\varphi;$$
$$A_{26} = [2(a_{22}\cos^2\varphi - a_{11}\sin^2\varphi) - (2a_{12} + a_{66})(\cos^2\varphi - \sin^2\varphi)]\sin\varphi\cdot\cos\varphi;$$
$$A_{66} = (4a_{11} - 8a_{12} + 4a_{22})\sin^2\varphi\cdot\cos^2\varphi + (\cos^2\varphi - \sin^2\varphi)^2 a_{66}$$

where φ - is an angle, defining the position of the main platforms towards the X, Y coordinates.

Let us write the forces through the deformation components in the following form [7]:

$$\sigma_x = b_{11}\varepsilon_x + b_{12}\varepsilon_y + b_{16}\gamma_{xy}; \quad \sigma_y = b_{21}\varepsilon_x + b_{22}\varepsilon_y + b_{26}\gamma_{xy};$$
$$\tau_{xy} = b_{61}\varepsilon_x + b_{62}\varepsilon_y + b_{66}\gamma_{xy}, \quad (4)$$

where $b_{11} = \dfrac{A_{22}A_{66} - A_{26}^2}{\Omega}$; $b_{22} = \dfrac{A_{11}A_{66} - A_{16}^2}{\Omega}$; $b_{12} = \dfrac{A_{16}A_{26} - A_{12}A_{66}}{\Omega}$ and so on.;

$$\Omega = (A_{11}A_{22} - A_{12}^2)A_{66} + 2A_{12}A_{16}A_{26} - A_{11}A_{26}^2 - A_{22}A_{16}^2.$$

The forces in the i-layer of the plate with the account of its multimodulus qualities are written in the accordance with [8]. Having written the deformations within the Kirchhoff-Love theory of plates, we have:

$$\sigma_x^i = b_{11}^i\left(\varepsilon_x^i - Z^i\frac{\partial^2 w}{\partial x^2}\right) + b_{12}^i\left(\varepsilon_y^i - Z^i\frac{\partial^2 w}{\partial y^2}\right) + b_{16}^i\left(\gamma_{xy}^i - 2Z^i\frac{\partial^2 w}{\partial x\partial y}\right);$$
$$\sigma_y^i = b_{21}^i\left(\varepsilon_x^i - Z^i\frac{\partial^2 w}{\partial x^2}\right) + b_{22}^i\left(\varepsilon_y^i - Z^i\frac{\partial^2 w}{\partial y^2}\right) + b_{26}^i\left(\gamma_{xy}^i - 2Z^i\frac{\partial^2 w}{\partial x\partial y}\right); \quad (5)$$
$$\tau_{xy}^i = b_{61}^i\left(\varepsilon_x^i - Z^i\frac{\partial^2 w}{\partial x^2}\right) + b_{62}^i\left(\varepsilon_y^i - Z^i\frac{\partial^2 w}{\partial y^2}\right) + b_{66}^i\left(\gamma_{xy}^i - 2Z^i\frac{\partial^2 w}{\partial x\partial y}\right).$$

Here b_{mn}^i –is a rigidity characteristic at the point of the i-layer material. $Z^i \in (-h^i/2, +h^i/2)$; ε_x^i, ε_y^i, ε_z^i –are the deformations in the mid-surface plane of the i-layer.

In contradistinction to the existing expressions, summands, which take into consideration the multimodulus qualities of the material, have appeared in (5)-(6).

Integrating in the thickness, let us move to the recording of the linear force $(N_x^i,\ N_y^i,\ S^i)$ and moments $(M_x^i, M_y^i,\ M_{xy}^i)$ in the i-layer from the forces at the arbitrary point integrating in the thickness:

$$N_x^i = \int_{-h/2}^{h/2}\sigma_x^i dz; \quad N_y^i = \int_{-h/2}^{h/2}\sigma_y^i dz; \quad S^i = \int_{-h/2}^{h/2}\tau_{xy}^i dz;$$
$$M_x^i = \int_{-h/2}^{h/2}\sigma_x^i z dz; \quad M_y^i = \int_{-h/2}^{h/2}\sigma_y^i z dz; \quad M_{xy}^i = \int_{-h/2}^{h/2}\tau_{xy}^i z dz, \quad (6)$$

where $-h/2, h/2$ –is the greatest distance from the mid-surface plane to the outermost fibres of the i-layer of the plate.

Plugging the expression for the forces (5) in (6), after integrating we get an expression for the forces and moments in the i-layer:

$$N_x^i = B_{11}^i\varepsilon_x^i + B_{12}^i\varepsilon_y^i - C_{11}^i\frac{\partial^2 w}{\partial x^2} - C_{12}^i\frac{\partial^2 w}{\partial y^2} + B_{16}^i\gamma_{xy}^i - 2C_{16}^i\frac{\partial^2 w}{\partial x\partial y}$$

$$N_y^i = B_{21}^i\varepsilon_x^i + B_{22}^i\varepsilon_y^i - C_{21}^i\frac{\partial^2 w}{\partial x^2} - C_{22}^i\frac{\partial^2 w}{\partial y^2} + B_{26}^i\gamma_{xy}^i - 2C_{26}^i\frac{\partial^2 w}{\partial x\partial y};$$

$$S^i = B_{61}^i\varepsilon_x^i + B_{62}^i\varepsilon_y^i - C_{61}^i\frac{\partial^2 w}{\partial x^2} - C_{62}^i\frac{\partial^2 w}{\partial y^2} + B_{66}^i\gamma_{xy}^i - 2C_{66}^i\frac{\partial^2 w}{\partial x\partial y}; \tag{7}$$

$$M_x^i = -D_{11}^i\frac{\partial^2 w}{\partial x^2} - D_{12}^i\frac{\partial^2 w}{\partial y^2} + C_{11}^i\varepsilon_x^i + C_{12}^i\varepsilon_y^i - 2D_{16}^i\frac{\partial^2 w}{\partial x\partial y} + C_{16}^i\gamma_{xy}^i \quad ;$$

$$M_y^i = -D_{21}^i\frac{\partial^2 w}{\partial x^2} - D_{22}^i\frac{\partial^2 w}{\partial y^2} + C_{21}^i\varepsilon_x^i + C_{22}^i\varepsilon_y^i - 2D_{26}^i\frac{\partial^2 w}{\partial x\partial y} + C_{26}^i\gamma_{xy}^i$$

$$M_{xy}^i = -D_{61}^i\frac{\partial^2 w}{\partial x^2} - D_{62}^i\frac{\partial^2 w}{\partial y^2} + C_{61}^i\varepsilon_x^i + C_{62}^i\varepsilon_y^i - 2D_{66}^i\frac{\partial^2 w}{\partial x\partial y} + C_{66}^i\gamma_{xy}^i$$

Here the integral characteristics of stretching, compression, shear, bendingandtorsionarecalculatedaccordingtothefollowingformulae:

$$(B_{mn}^i)_x = \int_{-\frac{h}{2}}^{\frac{h}{2}}(b_{mn}^i)_x dz \quad (C_{mn}^i)_x = \int_{-\frac{h}{2}}^{\frac{h}{2}}(b_{mn}^i)_x z dz \quad (D_{mn}^i)_x = \int_{-\frac{h}{2}}^{\frac{h}{2}}(b_{mn}^i)_x z^2 dz \quad , \tag{8}$$

where m, n =1,2,6.

On the other hand, the forces existing in the mid-surface plane of the i-layer can be written in accordance with [9]:

$$N_x^i = \frac{\partial^2\varphi^i}{\partial y^2} - \int_0^x \tau_x^i dx + \int_0^x \tau_x^{i-1} dx;$$

$$N_y^i = \frac{\partial^2\varphi^i}{\partial x^2} - \int_0^y \tau_y^i dy + \int_0^y \tau_y^{i-1} dy; \tag{9}$$

$$S^i = -\frac{\partial^2\varphi}{\partial x\partial y}.$$

Let us define $\int_0^x \tau_x^i dx = T_x^i$, $\int_0^y \tau_y^i = T_y^i$, where T_x^i, T_y^i are the shear forces in the interlayer bonding of the i-joint.

Equating the forces N_x^i, N_y^i, S^i in accordance with (7) and (9), let us write the deformation values in the mid-surface plane of the i-layer for the multimodulus materials in the following form:

$$\varepsilon_x^i = B_{11}^{*i}\frac{\partial^2\varphi^i}{\partial y^2} - B_{12}^{*i}\frac{\partial^2\varphi^i}{\partial x^2} - B_{11}^{*i}(T_x^i - T_x^{i-1}) + B_{12}^{*i}(T_y^i - T_y^{i-1}) - B_{16}^{*i}\frac{\partial^2\varphi^i}{\partial x\partial y} - \varepsilon_1^i(w);$$

$$\varepsilon_y^i = B_{22}^{*i}\frac{\partial^2\varphi^i}{\partial x^2} - B_{21}^{*i}\frac{\partial^2\varphi^i}{\partial y^2} - B_{22}^{*i}(T_y^i - T_y^{i-1}) + B_{21}^{*i}(T_x^i - T_x^{i-1}) - B_{26}^{*i}\frac{\partial^2\varphi^i}{\partial x\partial y} - \varepsilon_2^i(w); \tag{10}$$

$$\gamma_{xy}^i = B_{61}^{*i}\frac{\partial^2\varphi^i}{\partial y^2} + B_{62}^{*i}\frac{\partial^2\varphi^i}{\partial x^2} - B_{61}^{*i}(T_x^i - T_x^{i-1}) - B_{62}^{*i}(T_y^i - T_y^{i-1}) - B_{66}^{*i}\frac{\partial^2\varphi^i}{\partial x\partial y} + 2\gamma_{12}^i(w).$$

where $B_{11}^{*i}=\left[B_{22}^{i}B_{66}^{i}-(B_{26}^{i})^{2}\right]/(B_{0}^{i})^{3}$; $B_{12}^{*i}=B_{21}^{*i}=(B_{12}^{i}B_{66}^{i}-B_{16}^{i}B_{62}^{i})/(B_{0}^{i})^{3}$;
$B_{16}^{*i}=B_{61}^{*i}=(B_{12}^{i}B_{26}^{i}-B_{16}^{i}B_{22}^{i})/(B_{0}^{i})^{3}$; $B_{22}^{*i}=\left[B_{11}^{i}B_{66}^{i}-(B_{16}^{i})^{2}\right]/(B_{0}^{i})^{3}$;
$B_{26}^{*i}=B_{62}^{*i}=(B_{16}^{i}B_{21}^{i}-B_{26}^{i}B_{11}^{i})/(B_{0}^{i})^{3}$; $B_{66}^{*i}=\left[B_{11}^{i}B_{22}^{i}-(B_{12}^{i})^{2}\right]/(B_{0}^{i})^{3}$;
$(B_{0}^{i})^{3}=(B_{11}^{i}B_{22}^{i}-(B_{12}^{i})^{2})B_{66}^{i}+2B_{12}^{i}B_{26}^{i}B_{16}^{i}-B_{11}^{i}(B_{26}^{i})^{2}-B_{22}^{i}(B_{16}^{i})^{2}$.

The values $\varepsilon_1^i(w)$, $\varepsilon_2^i(w)$ and $\gamma^i(w)$ are calculated according to the formulae:

$$\varepsilon_1^i(w)=\overline{C}_{11}^{i}\frac{\partial^2 w}{\partial x^2}+\overline{C}_{12}^{i}\frac{\partial^2 w}{\partial y^2}+2\overline{C}_{16}^{i}\frac{\partial^2 w}{\partial x\partial y};$$

$$\varepsilon_2^i(w)=\overline{C}_{21}^{i}\frac{\partial^2 w}{\partial x^2}+\overline{C}_{22}^{i}\frac{\partial^2 w}{\partial y^2}+2\overline{C}_{26}^{i}\frac{\partial^2 w}{\partial x\partial y}; \tag{11}$$

$$2\gamma_{12}^i(w)=\overline{C}_{61}^{i}\frac{\partial^2 w}{\partial x^2}+\overline{C}_{62}^{i}\frac{\partial^2 w}{\partial y^2}+2\overline{C}_{66}^{i}\frac{\partial^2 w}{\partial x\partial y}.$$

We showed in details how to define the $\overline{C}_{mn}^{i}$ values in [1].
Bearing in mind the expressions (7), (10) and (11), let us write the formulae for the moments:

$$M_x^i=-D_{11}^{*i}\frac{\partial^2 w}{\partial x^2}-D_{12}^{*i}\frac{\partial^2 w}{\partial y^2}-2D_{16}^{*i}\frac{\partial^2 w}{\partial x\partial y}+m_1^i(\varphi,T_x,T_y);$$

$$M_y^i=-D_{21}^{*i}\frac{\partial^2 w}{\partial x^2}-D_{22}^{*i}\frac{\partial^2 w}{\partial y^2}-2D_{26}^{*i}\frac{\partial^2 w}{\partial x\partial y}+m_2^i(\varphi,T_x,T_y); \tag{12}$$

$$M_{xy}^i=-D_{61}^{*i}\frac{\partial^2 w}{\partial x^2}-D_{62}^{*i}\frac{\partial^2 w}{\partial y^2}-2D_{66}^{*i}\frac{\partial^2 w}{\partial x\partial y}+m_{12}^i(\varphi,T_x,T_y).$$

Here $D_{11}^{*i}=D_{11}^{i}+C_{11}^{i}\overline{C}_{11}^{i}+C_{12}^{i}\overline{C}_{21}^{i}-C_{16}^{i}\overline{C}_{61}^{i}$; $D_{21}^{*i}=D_{21}^{i}+C_{21}^{i}\overline{C}_{11}^{i}+C_{22}^{i}\overline{C}_{21}^{i}-C_{26}^{i}\overline{C}_{61}^{i}$, and so on.
The values $m_1^i(\varphi,T_x^i,T_y^i)$, $m_2^i(\varphi,T_x^i,T_y^i)$, $m_{12}^i(\varphi,T_x^i,T_y^i)$ are calculated according to the formulae:

$$m_1^i(\varphi,T_x^i,T_y^i)=C_{11}^{*i}\frac{\partial^2\varphi^i}{\partial y^2}+C_{12}^{*i}\frac{\partial^2\varphi^i}{\partial x^2}-C_{11}^{*i}(T_x^i-T_x^{i-1})-C_{12}^{*i}(T_y^i-T_y^{i-1})+C_{16}^{*i}\frac{\partial^2\varphi^i}{\partial x\partial y};$$

$$m_2^i(\varphi,T_x^i,T_y^i)=C_{21}^{*i}\frac{\partial^2\varphi^i}{\partial y^2}+C_{22}^{*i}\frac{\partial^2\varphi^i}{\partial x^2}-C_{22}^{*i}(T_y^i-T_y^{i-1})-C_{21}^{*i}(T_x^i-T_x^{i-1})+C_{26}^{*i}\frac{\partial^2\varphi^i}{\partial x\partial y}; \tag{13}$$

$$m_{12}^i(\varphi,T_x^i,T_y^i)=C_{61}^{*i}\frac{\partial^2\varphi^i}{\partial y^2}+C_{62}^{*i}\frac{\partial^2\varphi^i}{\partial x^2}-C_{61}^{*i}(T_x^i-T_x^{i-1})+C_{62}^{*i}(T_y^i-T_y^{i-1})+C_{66}^{*i}\frac{\partial^2\varphi^i}{\partial x\partial y},$$

where $C_{11}^{*i}=C_{11}^{i}B_{11}^{*i}-C_{12}^{i}B_{21}^{*i}+C_{16}^{i}B_{61}^{*i}$; $C_{12}^{*i}=-C_{11}^{i}B_{12}^{*i}+C_{12}^{i}B_{22}^{*i}+C_{16}^{i}B_{62}^{*i}$;
$C_{16}^{*i}=-C_{11}^{i}B_{16}^{*i}-C_{12}^{i}B_{26}^{*i}-C_{16}^{i}B_{66}^{*i}$; $C_{21}^{*i}=C_{21}^{i}B_{11}^{*i}-C_{22}^{i}B_{21}^{*i}+C_{26}^{i}B_{61}^{*i}$;
$C_{22}^{*i}=-C_{21}^{i}B_{12}^{*i}+C_{22}^{i}B_{22}^{*i}+C_{26}^{i}B_{62}^{*i}$; $C_{26}^{*i}=-C_{21}^{i}B_{16}^{*i}-C_{22}^{i}B_{26}^{*i}-C_{26}^{i}B_{66}^{*i}$;
$C_{61}^{*i}=C_{61}^{i}B_{11}^{*i}-C_{62}^{i}B_{21}^{*i}+C_{66}^{i}B_{61}^{*i}$; $C_{62}^{*i}=-C_{61}^{i}B_{12}^{*i}+C_{62}^{i}B_{22}^{*i}+C_{66}^{i}B_{62}^{*i}$;
$C_{66}^{*i}=-C_{61}^{i}B_{16}^{*i}-C_{62}^{i}B_{26}^{*i}-C_{66}^{i}B_{66}^{*i}$.

Using the equation of equlibrium, following the methodology [9] for the composite plate, the proportions (12) taking into consideration (13), having summed in all the layers, we will get the equation which will be represented in the operator form:

$$L_1(D^*,w)=q-L_2(D^*,w)+\Lambda(\varphi,T_x^i,T_y^i)+L(c,T_x^i,T_y^i); \qquad (14)$$

$$L_1(D^*,w)=D_{11}^*\frac{\partial^4 w}{\partial x^4}+2(D_{16}^*+D_{61}^*)\frac{\partial^4 w}{\partial x^3\partial y}+(D_{12}^*+$$

$$+D_{21}^*+4D_{66}^*)\frac{\partial^4 w}{\partial x^2\partial y^2}+2(D_{26}^*+D_{62}^*)\frac{\partial^4 w}{\partial x\partial y^3}+D_{22}^*\frac{\partial^4 w}{\partial y^4}; \qquad (15)$$

$$L_2(D^*,w)=\frac{\partial^2 D_{11}^*}{\partial x^2}\frac{\partial^2 w}{\partial x^2}+2\frac{\partial D_{11}^*}{\partial x}\frac{\partial^3 w}{\partial x^3}+\frac{\partial^2 D_{12}^*}{\partial x^2}\frac{\partial^2 w}{\partial y^2}+$$

$$+2\frac{\partial D_{12}^*}{\partial x}\frac{\partial^3 w}{\partial x\partial y^2}+2\frac{\partial^2 D_{16}^*}{\partial x^2}\frac{\partial^2 w}{\partial x\partial y}+2\frac{\partial D_{16}^*}{\partial x}\frac{\partial^3 w}{\partial x^2\partial y}+$$

$$+\frac{\partial^2 D_{21}^*}{\partial y^2}\frac{\partial^2 w}{\partial x^2}+2\frac{\partial D_{21}^*}{\partial y}\frac{\partial^3 w}{\partial x^2\partial y}+\frac{\partial^2 D_{22}^*}{\partial y^2}\frac{\partial^2 w}{\partial y^2}+$$

$$+2\frac{\partial D_{22}^*}{\partial y}\frac{\partial^3 w}{\partial y^3}+2(\frac{\partial^2 D_{26}^*}{\partial y^2}\frac{\partial^2 w}{\partial x\partial y}+2\frac{\partial D_{26}^*}{\partial y}\frac{\partial^3 w}{\partial x\partial y^2})+$$

$$+2(\frac{\partial^2 D_{61}^*}{\partial x\partial y}\frac{\partial^2 w}{\partial x^2}+\frac{\partial D_{61}^*}{\partial x}\frac{\partial^3 w}{\partial x^2\partial y}+\frac{\partial D_{61}^*}{\partial y}\frac{\partial^3 w}{\partial x^3})+$$

$$+2(\frac{\partial^2 D_{62}^*}{\partial x\partial y}\frac{\partial^2 w}{\partial y^2}+\frac{\partial D_{62}^*}{\partial x}\frac{\partial^3 w}{\partial y^3}+\frac{\partial D_{62}^*}{\partial y}\frac{\partial^3 w}{\partial x\partial y^2})+$$

$$+4(\frac{\partial^2 D_{66}^*}{\partial x\partial y}\frac{\partial^2 w}{\partial x\partial y}+\frac{\partial D_{66}^*}{\partial x}\frac{\partial^3 w}{\partial x\partial y^2}+\frac{\partial D_{66}^*}{\partial y}\frac{\partial^3 w}{\partial x^2\partial y});$$

$$\Lambda(\varphi,T_x^i,T_y^i)=\sum_{i=1}^{n+1}(\frac{\partial^2 m_1^i}{\partial x^2}+\frac{\partial^2 m_2^i}{\partial y^2}+2\frac{\partial^2 m_{12}^i}{\partial x\partial y});$$

$$L(c,T_x^i,T_y^i)=\sum_{i=1}^{n}(\frac{\partial c^i}{\partial x}\frac{\partial T_x^i}{\partial x}+\frac{\partial c^i}{\partial x}\frac{\partial T_y^i}{\partial y})+\sum_{i=1}^{n}c^i(\frac{\partial^2 T_x^i}{\partial x^2}+\frac{\partial^2 T_y^i}{\partial y^2});$$

$$D_{11}^*=\sum_{i=1}^{n+1}D_{11}^{*i}.$$

The summands $m_1^i(\varphi,T_x^i,T_y^i)$, $m_2^i(\varphi,T_x^i,T_y^i)$, $m_{12}^i(\varphi,T_x^i,T_y^i)$ in the operator form $\Lambda(\varphi,T_x^i,T_y^i)$ are found from (13).

Let us add a continuity deformation equation for the i-layer of the plate to the equation of equilibrium [9].Let us write the deformation continuity equations for the mid-surface plane of each layer using the expressions for the deformations (10) and calculating the following derivatives:

$$L_1^i(B^*,\varphi)+L_2^i(B^*,\varphi)-L_1^i(B^*,T_x^i,T_y^i)-L_2^i(B^*,T_x^i,T_y^i)-\Lambda^i(c,w)=0; \qquad (16)$$

$$L_1^i(B^*,\varphi)=B_{22}^{*i}\frac{\partial^4\varphi^i}{\partial x^4}+2(\frac{1}{2}B_{66}^{*i}-B_{12}^{*i})\frac{\partial^4\varphi^i}{\partial x^2\partial y^2}+B_{11}^{*i}\frac{\partial^4\varphi^i}{\partial y^2}-2B_{26}^{*i}\frac{\partial^4\varphi^i}{\partial x^3\partial y}-2B_{16}^{*i}\frac{\partial^4\varphi^i}{\partial x\partial y^3};$$

$$L_2^i(B^*,\varphi)=2\left[\frac{\partial B_{22}^{*i}}{\partial x}\frac{\partial^3\varphi^i}{\partial x^3}+\frac{\partial}{\partial x}(\frac{1}{2}B_{66}^{*i}-B_{12}^{*i})\frac{\partial^3\varphi^i}{\partial x\partial y^2}\right]+$$

$$+2\left[\frac{\partial B_{11}^{*i}}{\partial y}\frac{\partial^3\varphi}{\partial y^3}+\frac{\partial}{\partial y}(\frac{1}{2}B_{66}^{*i}-B_{12}^{*i})\frac{\partial^3\varphi^i}{\partial x^2\partial y}\right]+(\frac{\partial^2 B_{22}^{*i}}{\partial x^2}-$$

$$-\frac{\partial^2 B_{12}^{*i}}{\partial y^2})\frac{\partial^2\varphi^i}{\partial x^2}+(\frac{\partial^2 B_{11}^{*i}}{\partial y^2}-\frac{\partial^2 B_{21}^{*i}}{\partial x^2})\frac{\partial^2\varphi^i}{\partial y^2}+\frac{\partial^2 B_{66}^{*i}}{\partial x\partial y}\frac{\partial^2\varphi^i}{\partial x\partial y}-$$

$$-\frac{\partial}{\partial y}(2B_{16}^{*i}+B_{61}^{*i})\frac{\partial^3\varphi^i}{\partial x\partial y^2}-\frac{\partial}{\partial x}(2B_{26}^{*i}+B_{62}^{*i})\frac{\partial^3\varphi^i}{\partial x^2\partial y}-$$

$$-(\frac{\partial B_{61}^{*i}}{\partial x}\frac{\partial^3\varphi^i}{\partial y^3}+\frac{\partial B_{62}^{*i}}{\partial y}\frac{\partial^3\varphi^i}{\partial x^3})-(\frac{\partial B_{16}^{*i}}{\partial y^2}+\frac{\partial B_{26}^{*i}}{\partial x^2})\frac{\partial^2\varphi^i}{\partial x\partial y}-$$

$$-(\frac{\partial^2 B_{61}^{*i}}{\partial x\partial y}\frac{\partial^2\varphi^i}{\partial y^2}+\frac{\partial^2 B_{62}^{*i}}{\partial x\partial y}\frac{\partial^2\varphi^i}{\partial x^2});$$

$$L_1^i(B^*,T_x^i,T_y^i)=B_{12}^{*i}\frac{\partial^2(T_y^i-T_y^{i-1})}{\partial y^2}-B_{11}^{*i}\frac{\partial^2(T_x^i-T_x^{i-1})}{\partial y^2}-B_{22}^{*i}\frac{\partial^2(T_y^i-T_y^{i-1})}{\partial x^2}+B_{21}^{*i}\frac{\partial^2(T_x^i-T_x^{i-1})}{\partial x^2}-$$

$$-B_{61}^{*i}\frac{\partial^2(T_x^i-T_x^{i-1})}{\partial x\partial y}-B_{62}^{*i}\frac{\partial^2(T_y^i-T_y^{i-1})}{\partial x\partial y};$$

$$L_2^i(B^{*i},T_x^i,T_y^i)=(T_x^i-T_x^{i-1})(\frac{\partial^2 B_{21}^{*i}}{\partial x^2}-\frac{\partial^2 B_{11}^{*i}}{\partial y^2}-\frac{\partial^2 B_{61}^{*i}}{\partial x\partial y})-(T_y^i-T_y^{i-1})(\frac{\partial^2 B_{12}^{*i}}{\partial x^2}-\frac{\partial^2 B_{22}^{*i}}{\partial y^2}-\frac{\partial^2 B_{62}^{*i}}{\partial x\partial y})-$$

$$-\left[\frac{\partial(T_x^i-T_x^{i-1})}{\partial y}(\frac{\partial B_{11}^{*i}}{\partial y}+\frac{\partial B_{61}^{*i}}{\partial x})-\frac{\partial(T_x^i-T_x^{i-1})}{\partial x}\cdot(\frac{\partial B_{21}^{*i}}{\partial y}+\frac{\partial B_{61}^{*i}}{\partial x})+\frac{\partial(T_y^i-T_y^{i-1})}{\partial y}(\frac{\partial B_{12}^{*i}}{\partial y}+\frac{\partial B_{62}^{*i}}{\partial x})-\right.$$

$$\left.-\frac{\partial(T_y^i-T_y^{i-1})}{\partial x}(\frac{\partial B_{22}^{*i}}{\partial y}+\frac{\partial B_{62}^{*i}}{\partial x})\right]\cdot 2.$$

$$\Lambda^i(c,w)=\frac{\partial^2\varepsilon_1^i}{\partial y^2}+\frac{\partial^2\varepsilon_2^i}{\partial x^2}+2\frac{\partial^2\gamma_{12}^i}{\partial x\partial y}.$$

The values $\varepsilon_1^i(w)$, $\varepsilon_2^i(w)$, $\gamma^i(w)$ are calculated according to (11).

The equations describing the joint layer work adjoining the i-joint loop the system with regard to the sought functions W, φ^i, T_x^i, T_y^i.

Let us single out a separating plane in each joint; longitudinal displacements for the i-layer and (i+1)-layer exist to the both sides of the separating plane. The substractions of the longitudinal displacements Δu^i and Δv^i to the both sides of the separating plane can be written as [9]:

$$\Delta u^i=u^{i+1}-u^i+\frac{\partial w}{\partial x}c^i;\quad \Delta v^i=v^{i+1}-v^i+\frac{\partial w}{\partial x}c^i, \tag{17}$$

where $u^i(x,y)$, $v^i(x,y)$, $w^i(x,y)$ - are longitudinal and transversal displacements of the points of the mid-surface plane of the i-layer; $c^i(x,y)$ - is the distance between the mid-surface planes of the adjacent layers with the account of the variable layer thickness.

The connection between Δu^i, Δv^i and the shear strain in the i-joint will be written as:

$$\tau_x^i=\eta_x^i\Delta u^i;\quad \tau_y^i=\eta_y^i\Delta v^i, \tag{18}$$

where $\eta_x^i(x,y)$, $\eta_y^i(x,y)$ - is the shear stiffness coefficient depending on the shear value of one layer towards another one. The shear strain is recorded through the functions T_x^i и T_y^i accounting for the work of the i-joint:

$$\tau_x^i = \frac{\partial T_x^i}{\partial x}; \qquad \tau_y^i = \frac{\partial T_y^i}{\partial y}. \tag{19}$$

Let us express Δu^i and Δv^i from (19) plugging them in (17).Let us differentiate each equation in x, the second one in y. Let us write the work of the joint equation in the following way:

$$\frac{1}{\eta_x^i}\frac{\partial^2 T_x^i}{\partial x^2} + \frac{\partial}{\partial x}\left(\frac{1}{\eta_x^i}\right)\frac{\partial T_x^i}{\partial x} = \frac{\partial}{\partial x}\left(c^i \frac{\partial w}{\partial x}\right) + \varepsilon_x^{i+1} - \varepsilon_x^i;$$

$$\frac{1}{\eta_y^i}\frac{\partial^2 T_y^i}{\partial y^2} + \frac{\partial}{\partial y}\left(\frac{1}{\eta_y^i}\right)\frac{\partial T_y^i}{\partial y} = \frac{\partial}{\partial y}\left(c^i \frac{\partial w}{\partial y}\right) + \varepsilon_y^{i+1} - \varepsilon_y^i. \tag{20}$$

The obtained system of equations enables to solve thet asks of bending of the multilayer plate whose layers are composed of materials with strength-differential effect considering the influence of the rigidity of the interlayer bonding on the stress-trained state of the construction and this system includes the differential equations of eqilibrium (14), continuity (16) and the work of the joint (20). Ifthenumberoflayersinaplateisn+1, then we have one equation of equilibrium for the whole packet, n+1 compatibility of deformation equations or equations of continuity for each layer separately and 2n equations considering the compatibility of the work of the layers due to the finite rigidity of the bonding connecting them. The total number of equations is 4n. The number of the sought functions is also 4n: the buckling\bending function $W(x,y)$, the same for the whole packet, n+1 functions of forces $\varphi^i(x,y)$ and 2n – shear functions $T_x^i(x,y), T_y^i(x,y)$, and that means that the obtained system is looped.

The mathematical model of bending of composite multilayer plates made of materials with strength-differential effect (SD) includes not only the system of differential equations, but also borderline conditions.

As a result of integrating (14), (18), (20) the sought functions have to satisfy borderline conditions, which correspond to the specific outline fixing. The fact that differential equations are high-order (high-degree) ones allows to take into consideration complex and different kinematical and static conditions of fixing the plate layers.[10]. The borderline conditions for the function which reflects the conditions connected with the outer static indeterminacy, have to be written for every layer, for the function responsible for the inner static indeterminacy of the construction, for every joint. Two groups of borderline conditions were singled out in the calculations: on the outline support (for the whole packet) and on the end faces of the packet (for every i-layer). Based on it, the conditions on the outline are formulated independently from the conditions on the end faces of the layers and joints.

3. Results and Discussion

The authors have worked out the algorithm and methodology of solving tasks devoted to the bending of the composite multilayer plates whose layers are made of multimodulus materials. Hinge support based on the rectangular outline of the packet of the composite construction was considered to be borderline conditions. The end faces are tied with the band, which is absolutely rigid in its plane and absolutely flexible outside the plane, that means we have y=0, b on the boundary; along the X axis the following conditions were fulfilled:

$$W = 0; \quad M_x = 0; \quad T_x^{\,i} = 0; \quad N_x = 0; \quad V^i = 0. \tag{21}$$

The sought functions have been approximated by Fourier two-fold series satisfying the borderline conditions (21).

The transition from the system of the differential equations to the system of algebraic ones has been realized with the help of Galerkin orthogonality. The system of the algebraic equations has been solved with the help of the Gauss method. The integral characteristics of rigidity of the layers and their derivatives have been calculated with the help of the numerical differentiation method. To achieve this aim, we introduced a net with the number of junctions I×J= 21×21on the plate surface. The rigidity values b^i_{mn} have been defined according to formulae (5) at the net corners and on the surface of the composite plate and on five layers in the thickness of the i-layer. Integrating according to (9), the transition to the integral values of rigidity B^i_{mn} C^i_{mn} D^i_{mn} has been realized. The integral values have been acquired according to Simpson's quadrature formula. The derivatives of the integral values of rigidity, stretching, compressing, bending and so on have been recorded in accordance with the finite-difference scheme.

4. Conclusion

The given calculations have shown that the consideration of the finite shear rigidity specifies greatly the stress-strained state of the construction for the three-layer plate, where the middle layer possesses the strength-differential effect. If a multilayer plate is treated as a single-layer plate, then we will get underestimated values of bending and strains. If we consider the case when the layers slide on each other without any impediment, we will get overestimated bending values and, what is more important, overestimated strain values, that means that we unreasonably underestimate the structural capability of the construction. Multimodulus qualities of the material have the greatest influence on the stress-strained state when the rigidity of the interlayer bonding is low. Besides, the change of the stress-strained state with the increase of the coefficient, defining multimodulus qualities, is shown only to some extent.

Reference

[1] Rzhanitsin A 2008 Composite rods and plates (Moscow: Stroiizdat)
[2] Ambartsumyan S 1982 The multimodulus elasticity theory (Moscow: Science)
[3] Ambartsumyan S 1987 The anisotropic plate theory (Moscow: Science)
[4] Wright Howard D and Evans H Ray 1990 A review of composite slab design Resent Res. and Dev. Cold – Form. Steel Des.and Constr.:10th Int. Spes. St. Louis, Mo, Oct. 23 – 24, 1990. – Rolla (Mo)
[5] Krivchun N A and Yakubovskiy U E and Babshanov M T 1999 The bending of the multimodulus composite plate. Oil and Gas. Construction and oil fields arrangemen, 3
[6] Lewinski T 1991 Effective stiffnesses of cilindrical shells of periodic structure Mech. Res.Commun **5**
[7] Young Steven Easterling and W Samuel 1990 Strength of composite slabs Resent. Res. and Dev. Cold – Form., Steel Struct., St. Louis, Mo, Oct. 23 – 24, 1990. – Rolla (Mo)
[8] Reddy J N and Chao W C Nonlinear bending of bimodular material plates Int. J. Solids and Stract 3
[9] Yakubovskiy U E and Kolosov V I 1990 The non-linear bending of the composite plate University press. Construction and architecture **7**
[10] Seide P 1980 An improved approximated theory of the bending of laminated plates. In.: Mech. Todey. Vol. 5.Oxford.

IOP Conf. Series: Materials Science and Engineering **142** (2016) 012022 doi:10.1088/1757-899X/142/1/012022

Innovative Technologies And Modern Facilities In Beekeeping

V A Gaga[1], V N Esaulov[2]

[1]doctor of economic sciences, professor
[2]candidate of economic sciences, associate professor
[1]National Research Tomsk State University
[2]Yurga Technological Institute (branch)
634050, Tomsk, Lenina str, 36, TSU, phone (3822)52-95-20
652055, Kemerovo region, Yurga, Leningradskaya str, 26, phone (38451) 7-77-62

e-mail: gaginov@gmail.com[1], elaresy@mail.ru[2]

Abstract. This article discusses the features of the application of innovative technologies in beekeeping. The authors, based on years of personal experience in beekeeping and learning experience of the best apiaries in Russia and abroad, summarized materials on the topic and offered advanced modern technology in beekeeping to apply, which was successfully tested in other apiaries.

1. Introduction

Russian agro-industrial complex (AIC) includes industries producing agricultural products and involved in its processing and marketing. Agriculture is the foundation of the AIC, because agricultural products are created here. This industry produces 48% of fully ready-to-eat products; there are more than 68% of fixed assets, 60% of the total number of AIC[1]. The main objective of the AIC is full provision of citizens with quality food and consumer goods. People's quality of life is directly dependent on the level of development the AIC and its equipment and on the state of the related sectors of the economy.

Country's transition to a market economy has made changes in the activity of agriculture and it becomes even more important to optimize product prices of APC both intraindustry and in the final product. The competitiveness of the agricultural products is equally important. Besides the price for the average consumer product quality, production environment, compliance with state standards and, of course, taste, appearance, and other are important. Agriculture is particularly dependent on the climate and weather conditions. For Russia's dependence on the climate is particularly great, since more than half of its territory is located in the North zone. The effective area, according to Jean Elisee Reclus, a prominent geographer, is the one that is no higher than 2000 m sea level and where the average annual temperature is below -2^{0}C [1]. In Russia, only 1/3 of the land area is considered to be effective, while it is the coolest in the world.

---

[1] Moshkov Y. Composition and structure of Russian agro-industrial complex. [Electronic resource] URL: http://ru-news.ru/sostav-i-struktura-agropromyshlennogo-kompleksa-rossii/ (date of appeal 02.03.2016)

The third part of all the products that are consumed by man, made with the help of bees, 84% of the plants necessary for human life are pollinated by bees [2].

Beekeeping is a type of agriculture. Currently, it is experiencing serious difficulties associated with a significant reduction in the production of natural bee products [3]. Basically individual beekeepers or small business representatives are engaged in this activity. Therefore, the use of innovative technologies in modern beekeeping is seen as an exception. Part of bee products, without any further processing, is the finished product, for example honey. Other products require processing, such as wax. At the same time large bee farms are in need of the dealer services and beekeeping products from large apiaries (up to 30 beehives) are sold directly by the beekeepers.

Beekeeping in Siberia is particularly difficult: the organization of wintering bees, when the temperature drops to -50^{0}C; cold and rainy spring; short time of honey picking – 2-3 weeks; increasing death of bees in the world; reducing obtain of natural honey; growing adulteration, which is harmful to health; organizational and legal issues unresolved.

The research goal is the studying of modern beekeeping experience in other countries, of innovation in this area of agriculture and the use of the modern technical facilities. On the basis of the received material and of many years of experience of the authors the goal is in developing recommendations for more effective work of beekeepers in Siberia.

To accomplish this goal it is important to solve the following tasks step by step:

1. To study and systematize the experience of modern foreign beekeeping.
2. To establish the reasons adversely affecting the Russian beekeeping.
3. To analyze the possibility of using modern Russian and foreign beekeeping on the territory of Siberia.

Background of this study comes from the fact that the quality of life of citizens determined the development level of the AIC. In our country, research and publications about bees and beekeeping and their invaluable role in human life are very rare. The article analyzes the threatening situation of the death of honey bees around the world, from which not only the level of human life, but also the human life itself on Earth depends directly [4]. The article also studied the growing problem of falsification of honey.

The importance of this paper is in the fact that beekeeping has a strong relationship with other sectors: animal husbandry and plant growing. [5] Essential that the article is considering the dependence of productivity of entomophilous plants from pollination by bees exclusively.

The novelty of the article is that there is considered in the complex not only features and beekeeping problems in Russia and in the whole world, but also offered to the introduction the innovative technologies and modern technical equipment for more efficient apiaries, both in individual farms and industrial scale.

2. Theory

The theoretical basis of this work is based on researches of the famous beekeepers professionals, well-known scientists working in this direction, both Russian and foreign, as well as on the personal many-years experience of beekeepers authors.

For scientific research is applied a universal dialectical materialist method of cognition, which is applicable in agriculture. Beekeeping as a form of agriculture is considered in the development, in cooperation with other branches of the AIC and with related industries. In the studying of this topic the comparative analysis used, comparing the different levels of beekeeping, beekeepers working methods and more, as well as methods of induction and deduction, systematic method.

Humankind started beekeeping more than 15,000 years ago, honey in jars was found in the tomb of Tutankhamen. People were engaged in beekeeping in ancient Greece, in ancient China. Honey picking from the wild bees is considered one of the most ancient forms of human activity, and has survived to this day in some areas. The scientific study of the life of honeybees started from the XVIII century by the first scientists: Swammerdam, Rene Antoine Ferchault de Reaumur.

IOP Conf. Series: Materials Science and Engineering **142** (2016) 012022 doi:10.1088/1757-899X/142/1/012022

The construction of hives has undergone significant changes over the long period of time. In each country, the hives are arranged differently, but basically differ only in size, they all have a frames, vary in size and number too, which are placed in the hive, typically - 12, 16 or 24 frames, can be a different number. Langstroth L L is considered as the founder of the current practice of beekeeping. Francesco De Hruschka invented in 1865 the first centrifugal honey extractor. For protection from bee stings are commonly used light-colored costumes, headpiece with protective netting, gloves and smoker.

In the hive there are 3 types of bees: the queen, which is able to defer for its life, usually up to 3 years, more than 500,000 larvae - the future of bees; worker bees - 30000-50000 in the hive; male bees. Life in the hive is in a certain order: in the spring, thanks to the brood, bee family in the hive increased, usually reaching maximum to honey picking. When swarming the old queen accompanied by young bees leave the hive, the beekeeper must catch the swarm, and then put it into another hive to create a new bee family. In the old hive is still a young barren queen.

The ability to incubate the new queens beekeeper uses to expand the apiary, dividing bee family of one beehive in two. Life of bees exposed too many dangers: fungal diseases, viruses, poisons, pesticides, parasites[1].

Innovate technologies used in modern beekeeping

Innovative technologies increase the efficiency of the apiary. The modern beekeeping is characterized by particularistic specialization. In the absence of queens it is possible to access in special bees nurseries, where large farms are engaged in the withdrawal of queens and for themselves, and for the implementation. The large apiary serving greenhouses, kept some beehives for collection and storage of pollen, which is necessary for feeding the brood, and part of it is forwarded to the working beehives serving greenhouses. Pollen can be purchased in such farms, if necessary.

Currently, there are scientific-production associations and beekeeping complexes. So the good conditions for intensive technologies and the use of modern technical facilities are formed. The large beekeeping farms are equipped with lifting equipment, tractors, multi-frame electric extractors, special centrifuges to produce wax, furnaces for melting wax, and machines for packing honey.

More and more often we can see the multihull hives in modern apiaries; at the same time a significant increasing of the number of frames with honeycomb brings significant honey returns from each hive. It should be noted that such hives are only suitable for strong families, but in Siberia bee family not always has time to prepare to the main honey picking, i.e. grow sufficient amount of brood.

Modern beekeeping should also be maintained by the respective staff. Such personnel are trained at specialized schools, technical schools, universities; working beekeepers are retraining in special courses. Basically training programs are aimed at industrial beekeeping. At present, Russia has more than 1 million apiaries of the amateur beekeepers. Modern amateur beekeeping has half of all bee families in personal property, providing the consumers of these products by honey, wax and pollen. Bees-pollinators are significantly increase productivity of entomophilous plants. The yield of these crops increases and the sum of the implementation of additional products exceeds in 10 times the sum of realization of bee products[2].

In the modern beekeeping is becoming increasingly widespread the use of mobile apiaries. The values of this method are as follows:

1.Delivery of the hives to the place of honey picking increases the harvest of honey; the bees do not spend time and energy on flights from the fields to the hive.

2.Compact placement of hives.

---

[1] Andryushchenko B. Modern beekeeping. [Electronic resource] URL: http://bochkameda.net/o-pchelovodstve/sovremennoe-pchelovodstvo.html (date of appeal 12.03.2016)

[2] Komlatsky V.I. Innovative technology of keeping bee colonies and manufacture of bee products. [Electronic resource] URL:http://www.kaicc.ru/content/maintenance_of_the_family_of_bees (date of appeal 02.03.2016)

3.Easy to inspection frames.

4.A beekeeper can transport trailer with hives, to work with the bees without the smoker and masks.

5.Mobile apiary improves the pollination of crops, increasing their productivity.

This method has many good sides; it is also common in beekeeping of Canada. In Siberia, this method is effective in large apiaries, but if there are several dozen hives in a private farm and the fields are around, the cost for the purchase of equipment and its maintenance, fuel costs and other implementations would exceed the resulting of surplus product – honey.

Most believe that the swarming brings damage to beekeeping, because usually bees lost during it. Swarming also says that the hive is full; it is all right, that part of bees and the old queen fly out of the hive, to create a new bee family. Maurice Maeterlinck describes the preparation of the swarm to fly as the most beautiful time in the hive the day before the abdication, as the old queen voluntarily gives way to young queen [6].

First of all, we must decide what we want from the apiary - honey or apiary expansion. Depending on the decision we proceed to its implementation. We believe that should not fight with a swarm, it should be used for the effectiveness of apiaries, applying modern innovative technologies.

Swarming is one of the most complex and labor-intensive processes in the apiary. The solution to this problem in small apiaries, gardens, dachas takes a lot of time. The literature describes well how to work with the hive after swarming. This work takes 2-3 hours, and if there is a second hull, even more. It should be noted that beekeeper has protective clothing, a mask, and some glasses, the air temperature is + 30^0C, the sun is still up, and work in such conditions is very difficult.

Many years of experience allows us to make some innovation in this important on the apiary process, which requires little or no time. In this case all the problems in the hive the bees decide for themselves, nature regulates everything itself. The main precept of the beekeeper, "as little as possible to disturb the beehive", is running.

At the end of May in Siberia hives with frames are already full, so we need to increase the hive, to give work for bees - to put or little hull, or a second hull. Usually we practice to install a second hull with 7 frames. It should be noted that in Siberia at the end of May and in early June the extension of the hive can not be done, since the cool weather is often at this time of the year in Western Siberia: in the morning it is often no higher than 2^0C and during the day – 6^0-10^0C. To expand the hives it is unfavorable weather, too cold. If you put a little hull or a second hull to the hive, the cool and rainy weather will brood disease, fungal diseases, most often askosferoz. The causative agent is the fungus Ascospharea apis, physical-chemical treatment is not very effective for this kind of infection. The spores of this fungus are very resilient and can remain on the honeycombs, in a beehive, at apiary inventory for up to 4 years, always ready, under favorable conditions, in this case the cold, hit defenseless larvae. We come to the conclusion that it is impossible to expand the hive, it is problematic.

At this time in the hive work is humming, bees are too closely and bees begin to prepare for the swarm. They begin to delay the queen cells, queen seeds eggs inside. After a few days the larvae appears, bees are feeding them by royal jelly during 6 days, then the bees seal the queen cells and after 12 days there is first queen. Due to her whistling the old queen with her retinue are preparing to fly from the hive and in a nice, warm weather the swarm with the old queen comes. Basically swarms output going from 20 June to 10 July, in other areas, these terms can be shifted.

The next day after swarming most beekeepers take apart the hive, carefully inspect each frame, all the queen cells are cut and placed in the box. Number of queen cells on the one frame, on the authors' experience, may be up to 60. At the same time there is always a chance that the beekeeper can not see the queen cells and after 9 days – all over again: the swarm, taking apart, etc. With the second hull on the hive the probability of error increases significantly. If you do not get rid of queen cells, the swarming happens continuously, after 2-3 days, until there are a handful of bees in the hive. This bee family will not bring any honey since 15 August the honey picking in Siberia ends. The death of this family is predetermined: no bees, no honey, the family will die of hunger and cold.

An innovative technology for experimental apiary is as follows. After the release of the first swarms from different hives ("premiers") beekeeper must decide how to use them. Or it, as usual, will be a new family or it is necessary to boost weak family or to connect two "premiers" from different hives together, creating thus a strong bee family. We do not taking apart the hives after the first swarm, we wait for the second swarms ("secondaries"). The connection of the first two swarms from different hives is usually used when the main honey harvest begins in a few days. In Siberia, this time runs usually from 20 July to 5 August, with a shift in one direction or another, depending on the climatic conditions, for 5 days. Further, if after the first swarms hives were not taking apart, as described above, or the queen cell was missed in process, the second swarms are out after 9 days. If the main honey harvest is close or have already started, beekeepers usually use the following known technology, so-called "medovik": the hive from which the "secondary" came transferred to a new location at least 3 meters from the place where stood the old hive. At the old place they put a new hive, which has 6 frames with honeycomb, 1-2 frames with honey and 3-4 frames of brood, cutting the queen cells. Depending on the number of bees they put a second hull with dry frames if it was in the previous, the old hive, from which the "secondary" came; or 2 little hulls. Old beehive, which was moved to a new location, complemented by a pair of honeycombs and dry frames, all queen cells are removed, except the best one.

All flying the bees fly to the old place and evening, after sunset, the beekeeper emptied into new hive a swarm that flew exactly from that place from the old hive. In the old hive in the autumn will be a good new family, and "medovik" will give no less honey than good not swarmed bee family. This is due to the fact that in the "secondary" there is a young barren queen and while she flying around and begin to seed brood, 10-12 days will be passed. Active "medovik" family will be occupied only by harvesting honey these days, as there is nobody to feed at this time.

Creating of "medovik" occurs when the beekeeper missed queen cell, not cut it down and got the "secondary". Described creation of "medovik" involves considerable time and beekeeper forces.

Innovative technology implies do not consciously take apart the hive after the departure of the first swarm and just wait the "secondary". After the "secondary" fly out, we put it in a cool dark place and wait the "secondary" from another hive. Thus swarm can be kept in such conditions up to 4 days. At this time, the hives already have the second hull or little hull. When the "secondary" comes from another hive, put it in a cool place, and after the dusk pour the first "secondary" in the second hive, and the second "secondary" in the first hive. Also we proceed with the third and fourth and subsequent "secondaries" as they departures. This does not apply any of the lattices, and it does not need to seek queens and destroy them in the "secondaries". By morning, the bees will understand which queen is the most powerful: or own, which remained in the hive, or pour into a hive "secondary". At the same time the rest of the queens and the queen cells will be destroyed. The next day at 10 a.m. we can see wonderful flying around both of bee families. Queens in both hives are young, barren purifying flying around and begin to seed, the rest of the bees are active in collecting, receiving, drying of honey and other important matters. It is only necessary to monitor the controlled weighing-machines and to put additional little hulls just in time.

Lost in time is only one night, while the bees do not understand with queens and queen cells and will put things in order in the hive, beekeeper's time spent on these treatments is 10 minutes. Compared with traditional on apiaries methods of working with swarms when during the period of the main honey harvest we lost for two days, and the hive does not get about 10 kg of honey, besides beekeeper spends about 2 hours for work on arrangement of "medovik". If the apiary has 30-40 hives, and the beginning of June the cold, then after June 20 there will be 1-2 swarm each day to the end of July.

With this new technology, some remnants of queen cells we cut off when pumping honeycombs to avoid confusion in the spring with the last year and new queen cell. Only the "secondaries" are important in this technology, at the "premiers" trials, this method has not approached. It is because in the "premier" bees kill the old queen and swarm flies again. In the "secondaries", in the course of our field research, an alien queen wins as a rule. In the study of 28 bee families in 26 cases won alien queen. In the practical application of this technology during 4 seasons, none swarm flew away, bee families were very strong; in Siberian conditions for short honey picking we received at least 70 kilograms of marketable honey from each hive.

Our research results and know-how were discussed at beekeeping department of the Novosibirsk State Agrarian University (NSAU) in May 2015. The scientists of the University are also involved in similar studies, preparing thesis, after defending this innovative technology will be freely available. Expert-scientists (in NSAU) agreed with the results of our research.

Weather anomalies in Siberia are almost every year. Many years were to the point where it was necessary to make a decision: to extend the family with risks of freezing brood and have askosferoz as a consequence. Another way - to wait for the heat and as a result, swarms. Now the climatic conditions do not interfere, apiary is younger, because the "secondaries" always nave young queens.

Research conducted on personal apiary with 30-40 bee families for 4 years. An average per season 70 kg of honey received from each hive. Beekeepers do not spend 2-3 hours to inspect the hive after swarming, do not disturb the bees, and the result is well described in the literature "medovik", but a swarm is from alien family.

By modern technical means used in beekeeping can be attributed frameless honey production. Annually bees are building their new honeycombs without a frame that prevents infectious diseases. In this method beekeeper is loaded minimally. Bees used not a frame, but only the upper bracket, which has a longitudinal groove on the underside, into which wax is poured. These grooves will guide the bees in the construction of new honeycombs. Control Varroa mites is carried out mechanically, for this purpose all the brood honeycombs removes. Mites are no longer there, and bees bring more nectar, as they do not have anyone to feed.

New honeycomb, which are easy to break away from the strip, cut into small pieces, and honey flows almost completely through a large sieve. You can use pressure when the honey picking is big. Quality of the pressed honey is much better; it is more fragrant than that obtained using the honey extractor.

Positive aspects of this method: the bees do not swarm, there are no costs for honeycombs, no time is wasted on the honeycomb stitching to the frame to the honeycomb opening, on the honey pumping by the honey extractor, do not need a place for the storage of the dry frames. [7]

A good and cost-effective method – "Backyard Beekeeping" – is really good in the conditions of the Czech Republic, where the honey picking lasts 3-4 months, and the main harvest last 1-1.5 months. In Siberia this exotic method is not suitable. Siberian bees can not spend time and effort to build their own honeycombs; they need help of the beekeeper. Summer in Siberia is short and climatic conditions not always contribute to the honey picking, and a major harvest lasts only 3 weeks.

Recently appeared in Russian plastic honeycombs and plastic frames belong to the modern technical tools. This equipment is very common among beekeepers in Canada and the United States.

The use of plastic honeycombs has many positive:

1.You do not have to weld honeycombs to the frame; stretch the wire on the frame. Wax is inserted into the frame cuts or fastened with staple gun.

2. Plastic combs during honey pumping withstand the greater load, resulting in the large quantity of pumped out honey.

3. Losses from moths are smaller.

4. Honeycells have more regular shape and do not change it under the weight of honey.

5.These honeycombs are reusable.

6. Negative points include the need for disinfection and high cost[1].

[1] Plastic honeycombs in Russia. [Electronic resource] URL: http://www.pchelsar/com/staty_19/html (date of appeal 16.03.2016)

IOP Conf. Series: Materials Science and Engineering **142** (2016) 012022 doi:10.1088/1757-899X/142/1/012022

Innovative technology for the conservation of populations of honey bees

Along with the introduction of new technologies in the beekeeping industry even more vital topic arises - the preservation of honey bees. Bees are dying all over the world: in the United States since 2007 beekeeping annually loses up to 40% of bee families; in the UK annually lose is up to 10-30 % of the bee families; in Austria - 25%[1]. Process of reduction of bees in Europe has lasted for two decades. In Russia lost 20% of honeybees in general[2]. The causes of the continuous losses bees are not fully explored; and this phenomenon is called the Colony collapse disorder (CCD). It is assumed that the cause is a dangerous disease varroatosis, which is transferred by Varroa mite. There are other opinions that bees are dying from the chemicals that are used intensively in some farms, or form fungal diseases. The migratory beekeeping has many advantages, but it is stressful for bees. Ongoing studies have confirmed that the electromagnetic radiation from antennas and mobile devices disorient bees, and they cannot find their hive. The viability of bees is also affected by a sharp change in temperature. In our opinion, all of these causes exist and all together give a horrendous effect - the mass death of bees. It should be noted that the greatest extinction of bees is noticed in industrialized countries.

The quote belongs to A. Einstein: “Man can live without oxygen for three minutes, without water for three days, and without bees for four years”[3]. Pollination is the main task of the bees as the most powerful pollinators. If the bees disappear, pollinate crops in the world will disappear too, seeds for crops will disappear. Food will not be enough for everyone. Selection moves forward, currently self-pollinated crops are created, which now need pollinators. However, it is known the desired properties are preserved only at created in the first line hybrids.

To preserve the bees we must limit the application of pesticides which kill not only the pests but also bees. Farmers should be able to purchase environmentally-friendly means of protecting plants from pests, to get a natural product without harmful impurities. In America, in the state of Alabama, in 1996 excellent results using organic insecticides were obtained in a pilot plant. [8] Unfortunately, this experience has not received wide acceptance, and as a result we have seen the greatest death of bees in the United States.

The problem of protection of honey bees has grown to global scale. [9] The European Commission has sent 4 million euros to study of the causes of extinction of honeybees[4]. Russia is still in a stable situation, but also there are cases of mass death of bees. Honeybees in our country are protected by the fact that there is much less use of chemicals. It does not come from any particular respect for environmental standards, but because of the abandoned, nowadays uncultivated millions of hectares of arable land, lack of the necessary resources to handle the cultural planting pesticides for farmers.

Modern greenhouses are starting to use domesticated bumblebees. Bumblebees show themselves well as pollinators, but they are not taught to collect honey, after honey picking it has been collected about 50 g of honey[2]. They live in a special house, feed by sugar syrup. Bumblebees work during the day and in the evening returned to the house.

Another substitute of bees can become a man. Due to the application of pesticides more than 20 years ago in the Chinese province of Sichuan bees have disappeared. But the hard-working Chinese people continue to be one of the largest suppliers of pears from the province. They manually pollinate flowers and a result is in the high yields pears [10].

---

[1] Platonova T. The results of the international monitoring of the death of bees in the winter 2014/2015 years. [Electronic resource] URL: http://www.apiworld.ru/1438623225.html (date of appeal 15.03.2016)
[2] If the bees disappear - thoughts on the subject. (date of appeal 17.03.2016): http://federal-pchelovod.ru/o-federacii/novosti/esli-icheznut-pchely/ (date of appeal 16.03.2016)
[3] No bees - no pollination - no food - no man. [Electronic resource] URL: http://opchelkah.ru/category/новости/ (date of appeal 15.03.2016)
[4] 4 million euros will be spend on the search for the causes of the disappearance bees. [Electronic resource] URL: http://www.fumigaciya.ru/news/472 (date of appeal 16.03.2016)

The death of the bees in the winter reaches 30% or more. Proper organization of wintering of bees in the conditions of Western Siberia, allows you to fully preserve the bee colonies to the spring – the experience in the studied apiary for many years gives that result. In the literature on beekeeping they usually write about the preparation of bees for winter the following: after the taking apart of hives and honey pumping you should start to feed the bees by sugar syrup. Such an early feeding of bees is explained by the fact that the old bees have to process the syrup and the young one, who will go in the winter, should not be tired.

In Western Siberia honey picking usually ends up to 5-8 August and many beekeepers start to put hives in the feeder with sugar syrup in 5-10August. The hives at this time still have about 5-6 frames with brood and have nowhere to place the syrup in the required amount for the winter period. An inexperienced beekeeper decides that if the syrup remains, the bees did not eat it, they have rather food. This previously-feeding also stimulates massive theft among bees. To save the hive from invasion of alien bees, beekeeper should close the tap hole on the hive during the day and open at night. If this is not done, in the result of an attack on a hive by alien bees they will take away honey and kill the queen, thus you can lose 3-5 bee families for the winter.

Hence the rule:

Firstly, while preparing for winter it is better to feed bees by sugar syrup because it has 33% of moisture, and honey has15-18%.

Secondly, you should start to feed bees only after the hive will have almost no brood. Normally, this work is carried out from September 5-7, at the same time the number of frames is reduced to 9-10. The syrup is processed by both old and young bees, which will spend the winter. Even if from this work part of the young bees will die, it is still better than losing the whole family, since it did not have enough food. At this time there are no attacks on the hives, because the morning temperature is between -3^{0}C and +2^{0}C.

Thirdly, in October 5-10, when the bees are already formed in compact ball, we remove frames with the syrup and without bees on it in the storage - in the spring of this frames will be useful. When the temperature at noon is between 0^{0}C and + 2^{0}C we treat every street in the hive by “Bipin” only once for saving from the mite. In November 10-15 we put hives in “omshanik” (the special place for hives for wintering) and do not disturb the bees until the end of March. In April 1-10 we put the hives at an apiary.

This method of preparation bees for wintering is used for many years - the safety of bee colonies after wintering is 100%. There was a case of loss of the queens, but in this case we joined two families in one hive.

Innovate technologies for the future of beekeeping

Harvard University conducted experimental research on the creation of RoboBee (robotic bee). This is the smallest UAV, it can fly; can flap the wings very quickly, like bees. The project is scheduled for completion in 10 years[1].

The European Union (EU) has granted 1.4 million Euros to initiators of the project on the development of devices for the remote control of bee’s health; in the media it was called an accelerometer. The use of this device in the future will allow the bee-keeping in the EU and in other countries to move up to higher technology.

The accelerometer will release hundreds of thousands of beekeepers from regular inspections of bee colonies to determine their state of preparation for swarming. Millions of bee colonies will be relieved from stress due to the examinations. Beekeeper will receive information about the state of the hives in the apiary on the Internet and make the necessary actions timely. The use of this instrument will contribute to a deeper study of the biology of bees[1].

While the cause of the mass death of bees is not defined, the scientists in several countries are working together on a project, the implementation of which the cause of death of the bees will be identified. To do this, they create special micro sensors, which will be "put on" by 2.5 million of bees. Thus the scientists will receive information how bees do behave with some changes in a stressful situation. In addition 1,000 of hives all over the Earth will be placed in special Edison Intel boards to get information about the life of the bees inside the hive[2].

Modern beekeeping and growing falsification of honey

Modern beekeeping for the first time faced with a very serious problem - the falsification of honey. Fake honey in different countries reaches 40-80 percent[3].

GOST 25629-2014 Beekeeping. Terms and definitions. The article 153 provides a definition of falsified honey: "Honey, which contains objects that are not characteristic of its natural composition". Article 172 defines the falsification of bee products: "Deliberate action, which resulted in the bee product loses its inherent natural qualities". Firstly, it is difficult to agree with the Article 153 of the State standard, where the product which is not peculiar to the natural composition of honey, forgery, called the honey. Secondly, we do not share the optimism of the author of the article "The scales of honey falsification" A.S. Ponomorev, posted on the website "The World beekeeping" that this standard "will help to clear the market from surrogate"[4], if the definition of adulterated honey itself is not given correctly.

But we support the author of this article in the fact that the blocking the path of honey falsification on the Russian market can be shared with the state organizations with their authority to: "tax authorities, the Federal Antimonopoly Service, the Ministry of Agriculture, the Ministry of Industry and Trade, The Federal Service for Veterinary and Phytosanitary Surveillance, the Federal Service for Supervision of Consumer Rights Protection and Human Welfare".

Adulterated honey is not only knocks down prices, because of what diligent beekeepers are forced to leave the market and keep the bees only for themselves. The main thing is that the consumer pays for the product, which is in 3 times cheaper than natural honey, has incomprehensible content, and harmful to the human health[5]. The greatest number of adulterated honey which is dangerous to health comes from China. Health of the nation depends on the quality of food consumed. The national security depends on food safety.

---

[1] Remote monitoring of bees health in the EU. [Electronic resource] URL: http://federal-pchelovod.ru/o-federacii/novosti/kontrol/ (date of appeal 16.03.2016)

[2] Arslanova A. Microchips will put on the bees. [Electronic resource] URL: http://www.russian7.ru/post/na-pchel-odenut-mikrochipy/full/ (date of appeal 18.03.2016)

[3] Ponomarev A.S. Apimondia intends to fight the falsification of honey. [Electronic resource] URL: http://www.apiworld.ru/1444903289.html (date of appeal 17.03.2016)

[4] Ponomarev A. S. The scales of falsification of honey. [Electronic resource] URL: http://www.apiworld.ru/1435960819.html (date of appeal 17.03.2016)

[5] The round table in the State Duma of the Federal Assembly of the Russian Federation on the issues of beekeeping [Electronic resource] URL: http://federal-pchelovod.ru/o-federacii/novosti/kruglyystol/ (date of appeal 17.03.2016)

Festures of modern beekeeping in foreign countries

Argentina is famous for its honey; its price increases in the world market, the main importers are the USA and Germany. Honey production in recent years has decreased since honey fields plowed under cultivation of corn and soybeans.

In the United States, in good weather in the Midwest, honey bees produce clover honey. This honey is highly prized and expensive.

In Canada the migratory beekeeping predominant, bees actively perform their main task – the pollination of crops, increasing yields in crop production.

In Brazil, honey picking has a very long lasting. In this country, they produce a lot of organic honey, which has a high demand and high prices. Organic beekeeping production depends on environmental in gathering honey place, conditions of bees' detention, storage, transporting of production and of a number of other conditions. Also Argentina, Mexico, Turkey, the EU and Italy produced organic honey. In Russia, there is no regulatory framework on organic honey.

The light amber honey is produced in Vietnam, in Europe it is in demand, increasing of the prices of that honey is projected.

An extra light amber honey is produced in India, the price of honey in this country have grown in recent years[1].

In Serbia, there are excellent spring conditions for the development of bees; honey picking in spring is provided by a variety of flowering gardens. The bees spend the winter outside, as the weather allows, for this period they make 3-10 purifying flying around.

The honey quality in the world market is determined by Codex Alimentarius.

Conclusion

Innovative technologies and modern technical facilities in beekeeping improve the efficiency of beekeeping, on the other hand, if you do not switch to organic products on farms, than or the bees will die, or there will not be migratory beekeeping for pollination of agricultural fields; it is possible to forecast a huge loss in crop production, in the connection with the loss of crop pollinated by bees cultures.

Innovative methods of work described in the paper, facilitate greatly the hard work of the beekeeper especially in small, backyard apiaries. Practical four-year study aimed at improving the work with swarms allowed to develop the know-how, a special technique, which is set out above. Practical recommendations for feeding and wintering bees without loss of bee colonies in Siberia, based on years of experience, are given.

There is also an overview of emerging innovative technologies on beekeeping in the future, as well as the problem caused by the falsification of honey bringing great harm to public health is designated. The problem of legal insecurity of beekeepers and consumers of bee products from counterfeiting is also shown in this article [11].

References

[1] Reclus J E *The Earth and People.* Universal geography. The first book. Volume 1 and 2. SPb. – Ed. Public benefit. 1898 p. 1020
[2] Alison B Why are bees important? *The Guardian*. 17.06.2015
[3] Krivtsov N.I. Beekeeping Russia: state and place in the world *Advances in science and agribusiness technology* 2011 № 09 pp 15-16.
[4] Potts S G and Biesmeijer J C Global pollinator declines: trends, impacts and drivers Trends in *Ecology & Evolution* 2010 Vol 25 issue 6 pp 345-353

[1] Prospects of the global market of honey. [Electronic resource] URL: http://federal-pchelovod.ru/o-federacii/novosti/perspektivy-mirovogo-rynka-meda/ (date of appeal 17.03.2016)

IOP Conf. Series: Materials Science and Engineering **142** (2016) 012022 doi:10.1088/1757-899X/142/1/012022

[5] Ovsyannikov D. A. The u– se of ozonizers in beekeeping during the spring build families *KubGAU Scientific Journal* 2012 № 8 (06) pp 1-13
[6] Maeterlinck M *The Life of th- e Bee*. Dover Publications 2006 176 p.
[7] Vaclav S Without partition walls, without frames, without honey extraction *Beekeeping* 2013 №3 pp 82-83
[8] Briggs T H, Zehnder G W, Witt and James B Organic insecticides not as effective as synthetics, but still useful for vegetable insect control *Highlights of agricultural research* 1996 vol.43 N 3 pp 16-18
[9] Aizen M A, Harder L D The global stock of domesticated honey bees is growing slower agricultural demand for pollination *Current Biology* 2009 Vol 19 № 11pp 915-918
[10] Allison B and McCaallum B *A world without bees*. 2009 Pegasus Publishing 1 edition 336 p.
11. Krivtsov N.I. Status and Development Strategy of Russia beekeeping *Advances in science and agribusiness technology 2008* № 10 pp 27-29

IOP Conf. Series: Materials Science and Engineering **142** (2016) 012023 doi:10.1088/1757-899X/142/1/012023

Enhancing the Lifetime of Welded Joints Using the Protective Coating and Causes the Formation of Defects in Them*

V P Bezborodov[1, 2], Ph. D., associate professor, Yu N Saraev[1], Ph. D., professor

[1]Institute of strength physics and materials science SB RAS, Academichesky Avenue, 2/4, Tomsk, 634055, Russia phone: 8(3822)-492942
[2]National research Tomsk Polytechnic University, 634050, Tomsk, Lenin Avenue, 30

E-mail:litsin@ispms.tsc.ru

Abstract. The paper shows the efficacy of eutectic Nickel coatings for protecting welded joints of steels against corrosion. It increases with increasing content of the eutectic and chemical compounds at the grain boundaries γ-solid solution based on Nickel. Refinement of the structure of the coatings and reduce their heterogeneity allows to increase the protective properties and durability of welded joints against corrosion.
It is established that the probability of formation of cracks in the coatings increases with their thickness, intensity of heating and cooling after reflow, and also with increasing content of the eutectic and chemical compounds at the grain boundaries γ-solid solution based on Nickel. Shown to prevent the formation of cracks in the coatings when they are melt you can use optimal modes of heating and cooling of the composition. Modification of the structure of the coatings by grinding them and reducing heterogeneity allows to improve the resistance to crack formation and protective properties of welded joints against corrosion.

Introduction

The service life of welded joints is determined both by the strength of their material and surface properties [1-4]. Practice technical operation of the equipment shows that the majority of their products working in the conditions of corrosive environments and loads varying in magnitude and sign. Therefore, the reliability and durability of such products is determined by the corrosion resistance of their working surfaces, strength and endurance of welded joints.
The problem of increasing the strength and reliability of welded joints of pipes of low-alloy steels is connected with the need to ensure the complex physico-mechanical properties and equal the strength of the connection zones, the prevention of cold cracking, as well as structures that reduce the resistance of connection delayed fracturing.

The increased service life of products with welded joints by protecting their surfaces against corrosion is an actual scientific-technical task. One of the effective ways to increase the service life of these products is the use of protective coatings. To predict the reliability and durability of welded joints with protective coatings under operating conditions allows the evaluation of their performance in terms of corrosion.

The use of coatings of alloys based on Nickel to protect welded joints against corrosion is limited by their susceptibility to cracking and the formation of pores.

IOP Conf. Series: Materials Science and Engineering **142** (2016) 012023 doi:10.1088/1757-899X/142/1/012023

The purpose of the work. The study of the causes of defects such as cracks and pores, reducing protective properties. Evaluation of possibilities of use of the coatings of eutectic Nickel alloys to protect the surfaces of welded joints against corrosion, increase the service life of products.

Research methods.
Welding of pipes from steels 10Г2С was performed using the power source inverter type MP 2400 company "KEMPPI". For coating used powder materials of Ni-Cr-B-Si-alloys. The substrates were low carbon steel. The coating was applied by welding or argon-nitrogen plasma, subsequent melting. Used various methods of reflow coatings: flame, plasma, laser, electron beam. In research of influence of processes of surfacing and melting coatings on the structure and properties of the compositions used data on surface heating steel in a gas flame, plasma jet, laser and e-beams [5-8]. Certification of the structure of the coatings and the zone of welded joints was carried out on an optical microscope MIM - 9. Conducted qualitative assessment of the impact of melting on the structure of the coatings.

The results of the study
In the production of high activity of technological media exacerbates the total value of wear parts and equipment. The intensity of wear depends on flow rate, chemical conditions, duration of work, presence of abrasive inclusions, and other factors. In terms of the dynamic effects of corrosion and corrosion - abrasive media service life of product is drastically reduced.
A lot of products, working in conditions of corrosive environments, pitting at the same time exposed or threat intercrystalline corrosion and intensive abrasive wear, which is produced in the working surface, accompanied by failure.

In this paper, we propose a combined technology of surface protection of large-size products by combining the operations of welding, electric arc and electron beam welding and melting coatings of Nickel alloys [5-9]. This method allows to solve simultaneously two tasks - to improve the performance of the products and prolong their service life [5, 9]. Such problems can be solved by combination of two operations to replace worn-out fragments of the goods by cutting them and welding of inserts, for example, bypasses the tubes with the operations applied to the weld joint area of protective coatings and their subsequent melting (figures 1, 2). This approach is implemented in the protection of surfaces chemical containers.

Figure 1 - Structure of the transition zone "of the deposited coating PG-12N-02 – a welded joint of steel 10Г2С"

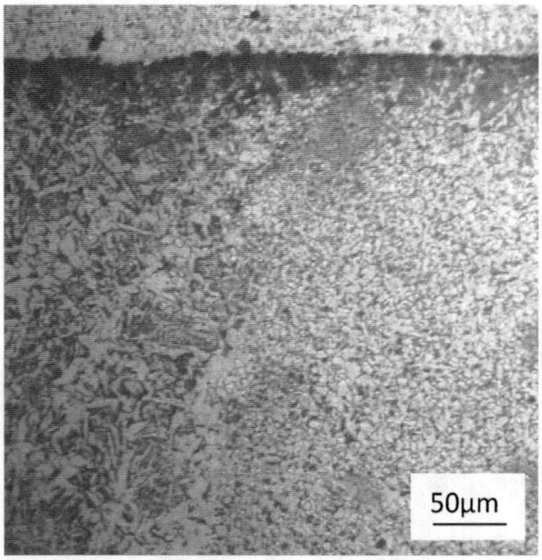

Figure 2 - Structure of the composition "melted coating PG-12N-02 – a welded joint of steel 10Г2С".

In metals, when the erosion occurs through the formation of craters and ductile fracture their edges, the grain size does not play a role if they are larger than the diameter of the crater. However, if these sizes become smaller than the diameter of craters, the erosion resistance increases dramatically. On the basis of qualitative indicator of the corrosion resistance (the change in appearance of the surface) evaluated coatings: "with firmness" - "unstable"; "fit" - "unsuitable" (table).

Table - Chemical resistance of steels and coatings

Of korrozion-tion environment	A temperature of medium, °C	Materials				
		Carbon steel	PG-10N-01	PG-10N-03	PT-19N-01	PS-12H-VK
H_2SO_4	60	D	B	D	B	D
HNO_3	50	D	C	D	C	C
HCl	30-70	D	D	D	D	D
H_2CO_3	120	C	B	B	B	B
NaOH	30-40	B	D	D	B	D
NaCl	30-50	C	B	D	B	B
H_2O	20-140	C	A	B	B	A
Na_2SO_4	0-50	C	B	B	C	B
NH_3	40-120	C	A	D	B	C
CH_2O	30-70	B	B	B	B	B

Note: A, B, C and D groups of resistance of materials against corrosion, correspond to corrosion rates (mm/year): 0,00 - 0,05; 0,05 - 0,1; 0,1 - 0,5; > 0,5

The change in appearance was also assessed by conventional scales scores (according to GOST 9.076-77). At the final stage of the process of corrosion wear of the coating lead to their cracking and subsequent failure.

The performance of the coatings is determined by their structural features. A large heterogeneity of the coatings on the dispersion and distribution of the phases determines the unevenness of processes of wear leads to the formation in the areas of defects of cracks with the subsequent chipping of fragments, which decrease the performance of the stock bearing capacity.

In the formation of coatings with crushed structure and more uniform distribution by volume of carbide and boride phases decreases the size of the collapsing fragments and increase performance properties. The smaller the coating structure, more local nature of the centres of destruction. High resistance of fused coatings is determined by their high strength and structure of supersaturated γ-solid solution Nickel-based distributed fine-grained inclusions of carbide and boride phases, the eutectic of γ-Ni+Ni_3B and alloy γ-Ni+$Cr_{23}C_6$.

In the process of cooling of welded joints with protective coatings [5] at temperatures close to the solidus temperature of the alloy, the coatings are formed by "hot" (crystallization) cracks [6, 9-10]. Their appearance caused by the process of heating and crystallization during cooling of the composition, when heat and are unevenly distributed temperature in the cross section, and, as a result of heating, the coating is compressed, and upon cooling will occur tensile stresses. This occurs under the influence of the processes of lengthening and subsequent shortening of the base metal composition.

When the degree of deformation exceeds the deformation of the composition of the crystallizing ability of the coating, it becomes possible appearance of hot cracks (figure 3).

The probability of their formation increases with increasing content of the eutectic and chemical compounds, crystallizing the grain boundaries of the Nickel γ-solid solution at low temperatures and decreasing the ductility of coatings [10-14]. With the increase of coating thickness, the intensity of their heating and cooling after reflow, the possibility of cracking also increases.

Because the deformation ability of coatings of Nickel alloys is relatively low, occurring during cooling of the compositions after reflow relaxation of residual stresses can lead to the formation of a "cold" brittle cracks.

These cracks occur most often at temperatures around 200 °C due to the occurrence of structural transformations in the coatings and are connected, for example, with the formation in the steel base of martensite as the result of quenching [11, 13, 14].

To prevent the process of formation of cold cracks occurring in the investigated coatings, it is necessary for each of the compositions is to carefully select the modes of heating and cooling when fusing the coatings.

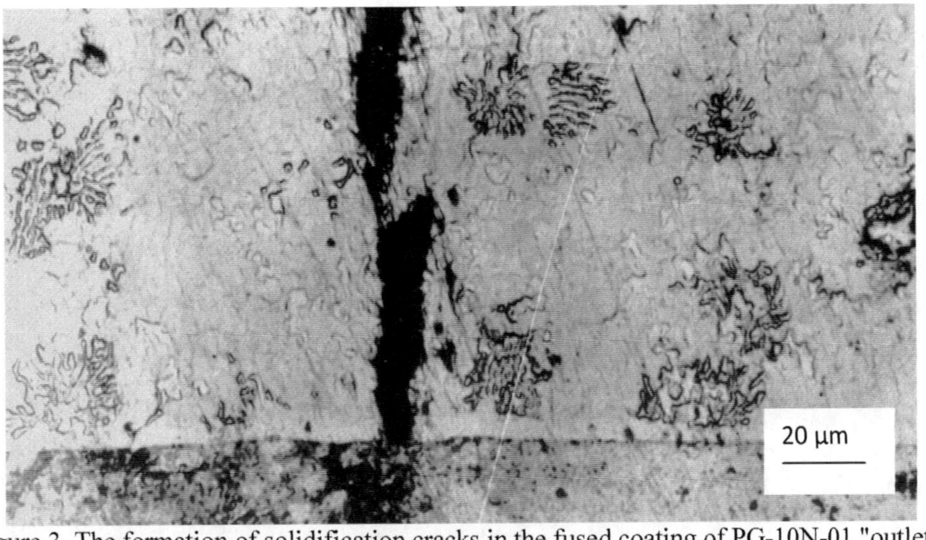

Figure 3. The formation of solidification cracks in the fused coating of PG-10N-01 "outlets" alloy γ-Ni-$Cr_{23}C6$.

Therefore, the use of high-speed and high-energy methods of heating such as laser and electron beams that have many advantages, but producing local intense heating fundamentals, may lead to cracking of the coatings.

This is because the speed of heating and cooling due to uneven local thermal effect and heat sink there are such a high voltage, which lead to the formation of cracks. Thus, in laser and electron beam melting methods because of the very high speed of crystallization of the molten coating often formed "hot" (crystallization) cracks caused by high levels occurring residual stresses and their non-uniform distribution by volume. This is because in such methods, the reflow occurs, in fact, almost complete or, sometimes, the complete dissolution of hardening carbide and boride phases in Nickel γsolid solution. During the subsequent rapid cooling crystallizing coating dissolved in it the elements do not have time to stand and there is a strong local supersaturation plots γ-solid solution. On the borders between these areas and the Nickel matrix, there is tension, leading to the formation of "hot" (crystallization) cracks. The centers of origin can also be and borides and carbides of lamellar eutectic γ-Ni+Ni_3B and alloy γ-Ni +$Cr_{23}C_6$.

The application of volumetric methods of heating (furnaces) and deep warming up the foundations in the flame melting with subsequent relatively slow cooling avoids the formation of cracks in the coatings.

Cause cracking of coatings formed on a steel substrate, the cooling which is sometimes accompanied by martensitic transformation, is that the coating in the cooling period shrinkage, and base metal expansion. To prevent the process of cracking of the coatings is possible by carrying out slow cooling with the furnace, sand, asbestos, etc. or isothermal aging at a temperature of intensive transformation of austenite. Enhancing the resilience of the material of the coatings to the formation of cold cracks contributes to the change of their structure in various ways, accompanied by a decrease in the degree of chemical and structural inhomogeneity of these coatings. In laser and electron beam melting, ultrasonic treatment is a strong refinement of the structure of coatings that reduce the probability of formation of cold cracks. However, at the same time, increase speed of heating and cooling in laser and electron-beam exposure causes the increase in the level of internal stresses, and, as a consequence, the probability of formation of cold cracks in the coatings.

The tendency of the coatings to the formation of hot cracks increases with the growth of the volume content of carbides $Cr_{23}C_6$ and Cr_7C_3, [11, 13, 14].

Stereometric analysis of the coatings performed in the work showed that their porosity, including pore size, number, shape and volume distribution strongly depend on the modes of reflow. Coating, melted in a furnace at temperatures $\leq$ 1050 °C and short times of exposure ($\cong$ 1 min), have a significant porosity (10 - 12 %, vol.). The pores are predominantly spherical in shape, have dimensions of 20 μm and is uniformly distributed over the volume of the coating. At higher temperatures (1050 - 1150 °C) and exposure times $\cong$ 1 - 5 min cake mix low-porous coatings are formed ($\cong$ 4 - 6 %, vol.). Pores (with sizes up to 10 μm) are observed mainly near the surface of the base (figure 4). The exception is if the reflow of the sputtered coatings with friable (loose) structure.

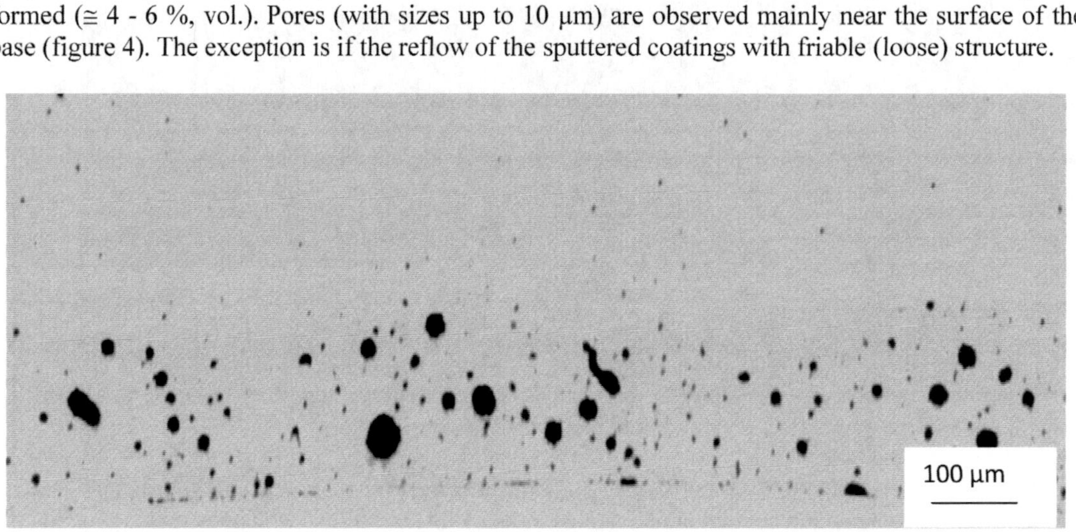

Figure 4. The concentration of pores in the boundary layer of coating at high-speed methods of reflow.

Summary

1. The probability of formation of cracks in the coatings used to protect the welds from corrosion increases with increase in the thickness of these coatings, the intensity of their heating and cooling after reflow, and also with increasing content of the eutectic and chemical compounds, crystallizing the grain boundaries of the Nickel γ-solid solution.

2. To prevent the formation of cracks in the fusing the coatings is possible only with selection of modes of heating and cooling of the composition of each species or isothermal aging at a temperature of intensive transformation of austenite.

3. To increase the resistance of the coatings to cracking, and protection of welded joints against corrosion is possible by modifying the structure of these coatings: grinding and the reduction of heterogeneity.

4. The use of eutectic Ni-Cr-B-Si-alloys can be effective in creating protective coatings for welded joints of steels.

5. The increase in the content of alloying elements in Ni-Cr-B-Si alloys leads to an increase of the volume content of the eutectic of γ-Ni+Ni_3B, alloy γ-Ni+$Cr_{23}C_6$, carbides and boride phases, the increase of strength and corrosion resistance melted and deposited protective coatings and welded joints of steels in General

References

[1] Welded construction. Mechanics of destruction and criteria of serviceability) / V. A. Vinokurov, S. A. Kurkin, G. A. Nikolaev; - ed B. Paton. - M.: Mashinostroenie, 1996, 576 p

[2] Fracture toughness and mechanical properties of structural materials technical systems / V. V. Moskvichev, N. And. Makhutov N., A. P. Chernyaev and others. - Novosibirsk: Nauka, 2002, 334 p.

[3] Lebedev B. F., Pashin G. A., Dudko M. S. the Dependence of impact strength HAZ of welded joints of low alloy steels of the specific energy at welding // Automatic welding. - 1987. - No. 4. - P. 7 - 9.

[4] Saraev Yu. N., Sleptsov O. I., Bezborodov V. P., Nikonova I. V., A. V. Tutev. Analysis of fatigue fracture, the structure and properties of welded joints of pipelines operating in conditions of Siberia and the far North/Installation and special works in construction. - 2005. - No. 8. - P. 18.

[5] Saraev Yu. N., Bezborodov V. P., Selivanov Y. V., Nikonova I. V. The Influence of modes of surfacing coatings on corrosion resistance in acidic environments of welded joints of austenitic steels//metal finishing. - 2007. - № 2. - P. 33 - 36.

[6] Gryazev A. N., Poltev G. L., Kondrat'ev, I. P. Features of the formation of cracks during laser cladding of nickelchromium alloys// Welding production, 1989, №. 9, P. 10 - 12.

[7] Dubasov, A. M., On thermal processes during solidification of the coating particles in plasma spraying//Welding production, 2000, № 2, P. 25 - 28.

[8] Berdnikov A. A., Filippov M. A., Studenok, S. E. Structure of hardened carbon steels after plasma surface heating// materials Science and heat treatment of metals, 1997, № 6, P. 2 - 4.

[9] Larionov V. P., Sleptsov O. E., Saraev Yu. N., Bezborodov V. P. New approaches to the development of modern technologies of welding and coating to ensure the operational reliability of metal structures and products, operating in conditions of Siberia and the far North/Bulletin of the Academy of military Sciences. - 2008. - № 3. - P. 67.

[10] Zbriger A. G. Structure and properties of coatings of tamifluswine alloys//Metallography and heat treatment of metals, 1988, № 4, P. 42 - 44.

[11] Doroshenko L. K., Borisova A. L., Grigorenko G. M. and others. The Processes of melting and crystallization of coatings from tamifluswine Nickel alloys//Automatic welding, 1990, Vol. 10, P. 22 - 23.

[12] Oshchepkov, Y. P., Oshchepkova N. In. Metallographical study tamifluswine carbide powders//Automatic welding, 1976, № 11, P. 32 - 35.

[13] Klinskaya - Rudenskaya N. A., Kopysov V. A. Features of composite coatings based on Ni-Cr-

Si-alloy. The study of the microstructure of coatings//Physics and chemistry of materials processing, 1995, № 1, P. 69 - 81.
[14] Antonova E. A., Andrushchenko N.S., Sinai L. M. Interaction of the coating Ni-Cr-Si-B with steel in the process of fused deposition modeling // Protection of metals. 1971. - T. 7. - Vol. 2. - P. 137 - 142.

Work performed under the project of the Russian science Foundation No. 16-19-10010.

The Influence of Modes of Deposition of Coatings on the Corrosion Resistance of Welded Joints of Steels in Acidic Media*

Yu N Saraev[1], Ph. D., professor, V P Bezborodov[1, 2], Ph. D., associate professor, Y V Selivanov[1], engineer

[1]Institute of strength physics and materials science SB RAS, Academichesky Avenue, 2/4, Tomsk, 634055, Russia phone: 8(3822)-492942.
[2]National research Tomsk Polytechnic University, 634050, Tomsk

E-mail:litsin@ispms.tsc.ru.

Abstract. In this work, effect of welding on corrosion of welded joints of austenitic steel 12KH18N10T. It is shown that the use of pulsed - arc welding steel 12KH18N10T allows you to create a protective coating with dispersed structure with less thermal impact on the zone of the welded joint. Coating is of such structure allows 1.5 to 6 times to reduce the corrosion rate of welded joints of steel 12KH18N10T in active chemical environments. Pulse the process of deposition of coatings on welded joint of steels can be effectively used for the protection against corrosion in the repair of equipment of chemical industry. The results obtained can be recommended for use when welding a protective corrosion - resistant coatings on working surfaces of equipment of chemical productions.

Introduction

Method of welding allows increasing the productivity and quality of products thanks to the strengthening and restoration of worn working surfaces in contact with corrosive environments [1]. However, coatings formed by this method, characterized by the appearance of various defects. The increase in the depth of melting increases the area of sensitization and lack of fusion leads to incomplete fusion of the formed coating from the protected surface [2]. It is necessary to conduct search for new ways of addressing issues of eliminating emerging gaps.

When welding melts the coated electrode and partial melting of the processed surface. On this surface is formed by surfacing a bath, and as you move the heat source relative to the surface is the crystallization of the molten metal in the deposited bead, the geometric size of which is largely determined by the processes occurring in the bath. When welding in pulsed mode chart changes in the welding current (figure 1) differs significantly from the chart using the DC.

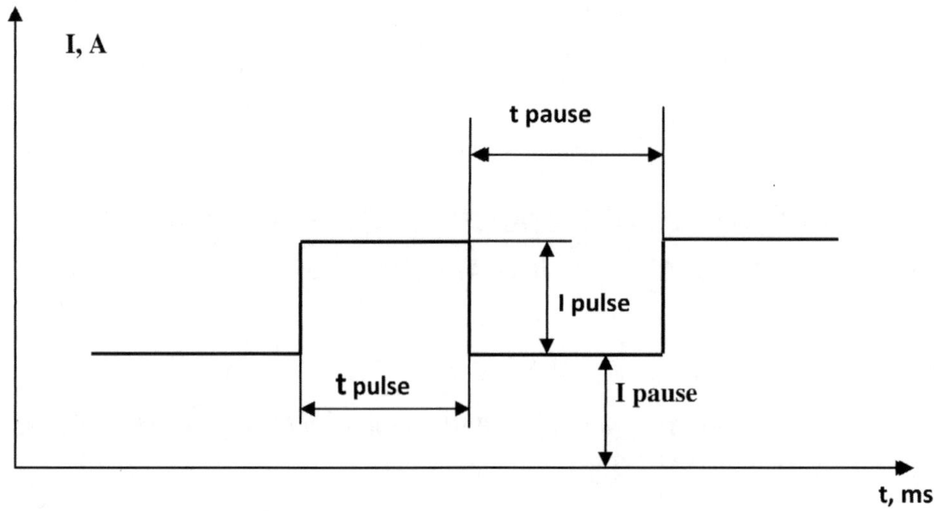

Figure 1. The scheme of change of current surfacing: AI - a working pulse current; IB - base current; t - time of the pulse; tπ - the pause time.

Implementation for surfacing this chart allows the periodization of the physical processes proceeding in the weld pool. As a result of alternating pressure of the arc during formation of the deposited coating metal bath surfacing performs reciprocating motion [3-5]. Such a flow of the process allows surfacing to ensure the circularity of physical processes at the stages of formation of the fused bath and metal crystallization from the melt, which promotes active mixing. The active stirring of weld pool contributes to the alignment of its heat content and establishes required amount of molten metal under the arc to the beginning of the current pulse thereby reducing the depth of melting. Analysis of the factors affecting the formation of deposited coatings shows that they can be divided into two groups. The first group is expressed by the indicator, which in the process of deposition cannot be changed, - the spatial position of the surface to be protected. The second group is associated with parameters of surfacing mode that can be modified during the manufacturing process. In this case, it becomes possible to change the geometrical dimensions of the deposited coating: the height and width of the cushion cover.

New opportunities to control parameters of the second group of factors inherent in the method of overlaying coating by modulated current. In this scenario, you receive and more control over the geometric dimensions of the deposited coating than for welding on direct current with a limited number of adjustable settings.

Additional adjustment parameters - the amplitude of the oscillations of the pulse and pause duration, and repetition rate enables control over the size of surfacing the bath, change the nature of melting and transfer of metal, which contributes to the stability of the formation of the deposited coating. In this case the forces acting at surfacing tub, periodically vary from the maximum (on the interval of the pulse) to a minimum (on the interval of the pause) that provides reciprocating motion of the molten metal.

Thus, movement of the surfacing metal in the bath can effectively control the variation in amperage on the welding interval pulse. With the increase of the current pulse increases the pressure of the arc. As a result, the molten metal is more actively displaced from under the arc in the caudal part of the fused bath. On the interval pause when reducing the arc pressure is proportional to the current pause weld pool tends to return under the arc. It promotes active periodic movement of the metal in the weld pool, the alignment of its heat content and a more uniform distribution of alloying elements throughout the volume of metal deposited. By changing the parameters of the process of surfacing directly in its course, can affect the geometrical dimensions of the formed coating: the surface curvature, the height and width of the cushion.

Corrosion tests of steels used for the manufacture of the equipment working in the conditions of application of acid media, were initiated by the perceived need for replacement of out of order fragments and thereby extend the life of the equipment as a whole. Especially prone to rust and corrosion welded joint of the walls of these units are made of stainless steel. To date the industry uses items of process equipment with welded joints, obtained by manual arc welding by consumable electrodes and mechanized welding. By the aforementioned processes of transfer of deposited metal is performed by a series of short circuits. In this case, the droplet of molten metal from the end face of the electrode and its transfer to the weld pool occurs at high welding current. This leads to the instability processes in the arc gap and the increased splashing of electrode metal, which ultimately worsens operational properties of welded joints. Experience has shown that during long-term operation of the specified equipment operating in an active acidic media, corrosion of metal is about 0.1 mm/year. Corrosion destruction of welded connections reach 4 - 5 mm/year. They are usually expressed in the form of dangerous varieties of intergranular corrosion knife, observed in the HAZ of the base metal and the weld metal. In some cases, corrosion leads to the through destruction of the wall units. Currently rabotosposobnosti the restoration of damaged welded joints of steels at the operating equipment is manual TIG welding with non-consumable electrode. Refurbishment of corroded sections of welded joints is to apply to them a method of overlaying a protective coating. This technological process is accompanied by a large heat input in the weld joint area, reduces the protective effect of the coating deposited coating contributes to the corrosion damage in this zone knife corrosion.

For the purpose of increase of operational properties of welded joints of steels in the work was used the method of pulsed arc welding in the weld joint. It is known that this method of surfacing dramatically reduces the heat input in the weld joint area, affects the structure and properties of welded joints of different steels [6-9]. However, in the literature there are no data on the structure of the deposited metal coating and welded joints of stainless steels and their corrosion properties in nitric acid media.

The aim of this work is to analysed the influence of the used processing methods and pulse coating method for increasing corrosion resistance of welded joints of stainless steels in nitric acid media.

Materials and methods study

Were used a welded joint of plates of rolled austenitic steel 12KH18N10T sizes 300×150×10 mm. Formation of welded joints was carried out using manual arc welding electrodes RLA - 36 with a diameter of 3 mm in three passes welding machine MAGMA 315.

A protective coating was applied on the welded connection by welding - direct current and pulse. Estimation of corrosive wear of the investigated welded joints were conducted according to the accelerated method. Such tests are recommended for quality control of welded joints and evaluation of its tendency to stab corrosion. Test environment served as an aqueous solution of 10 % HNO3 + 3 % HF. The tests were performed at ambient temperature of 95° C. In these conditions, increases the mixing speed of the solution and facilitates the supply of the oxidizing components to the surface of the samples.

The test consisted of 8 cycles of 6 hours each. In real operation conditions of the welded joint corrosion products constantly diverted from the working solution, so in laboratory tests in each cycle was used freshly prepared solution. Corrosion tests of samples of size 80×15×10 mm with welded seams was performed in the equipment operation conditions of partial immersion of samples of each welded joint in fluids at their impact in the area of the waterline. Conducted corrosion tests of samples of welded joints after removal of the deposited coatings by mechanical means. This ensured maximum contact welds with a corrosive medium. Test duration was 30 hours. The corrosion resistance was evaluated by weight change of the samples using an analytical balance type VLR - 200. Registration of corrosive destruction of the surface layers of the samples was conducted on the change in the macrostructures of welds and heat-affected zone (HAZ). The research of structures of welded joints and deposited surfaces was assessed using an optical microscope Axiovert 200 MAT and camera AxioCam HRc Carl Zeiss, were recorded by AxioVision Rel 4.4 software. Structure of welded joints and deposited surfaces was revealed using the reagent: 10 ml of nitric acid, 20 ml of hydrochloric acid, 20 ml of glycerol and 10 ml of 30 % hydrogen peroxide.

Results and discussion

In consideration of the reasons for the resistance of welds to corrosion, it is necessary to highlight the nature of the components of the welded joint as a major factor determining its phase composition and structure. The passivity of chromium steel 12KH18N10T due to the formation on the surface of a thin film of Cr2O3 oxide, soluble in acid and water. As is known, the magnitude of sensitization depends on the amount of heat input during the formation of the welded connection.

Studies of the microstructure of the initial samples showed that the investigated steel has a austenitic structure (figure 2).

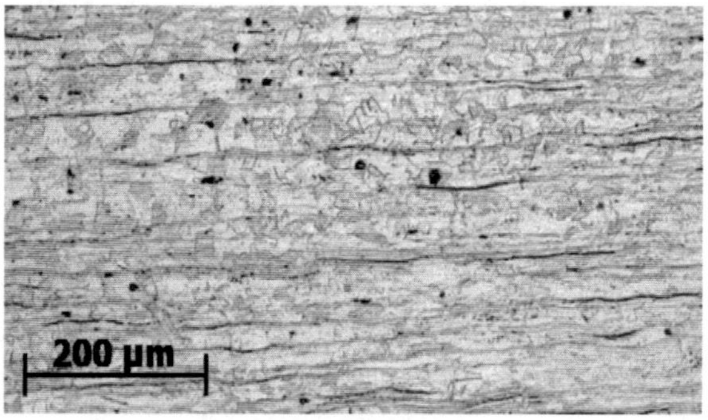

Figure 2. The microstructure of the steel 12KH18N10T.

Ferritic phase is present as separate lines and as a thin layer on the boundaries of polyhedral grains of austenite. The structure of the weld metal is a columnar dendrites (figure 3).

The width and length of the dendrites can reach 20 µm and 500 µm, respectively. The weld metal of samples of samples of welded joints, performed at a constant current, during corrosion tests showed relatively low resistance (figure 4).

Analysis of the dependency of the corrosion rate of the zone of fusion to the time of the tests showed that the corrosion of the samples, deposited at a constant current (figure 6). These samples greatly affected knife corrosion in HAZ of deposited coatings and in the weld metal (figure 7). Corrosion damage occur mainly in the HAZ metal of welded joints. The use of pulsed mode reduces up to 6 times the corrosion rate of connections compared to DC. Analysis of the results of corrosion tests of samples of welded joints after removal of the deposited coatings also showed improved corrosion resistance of the seam area in the coating, made of a pulsed arc method. Most of the corrosion destruction of welded joints of steel are localized in the HAZ (see figure 7).

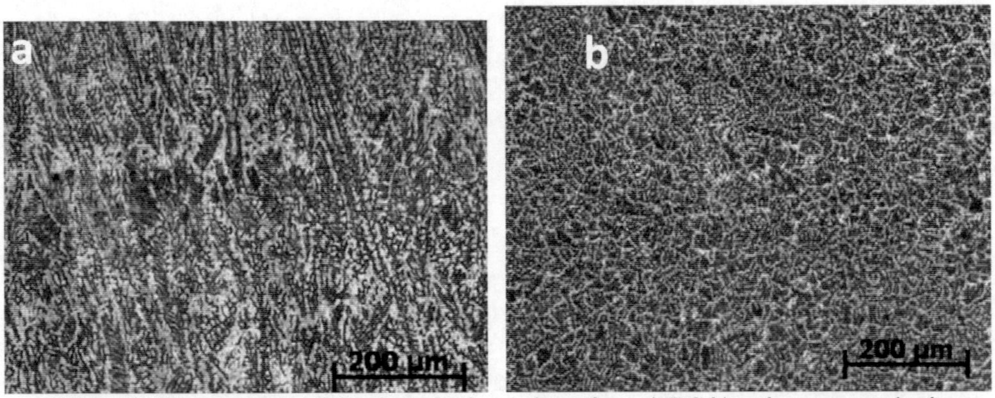

Figure 3. The structure of the deposited metal coating: a) DC b) pulse - arc method.

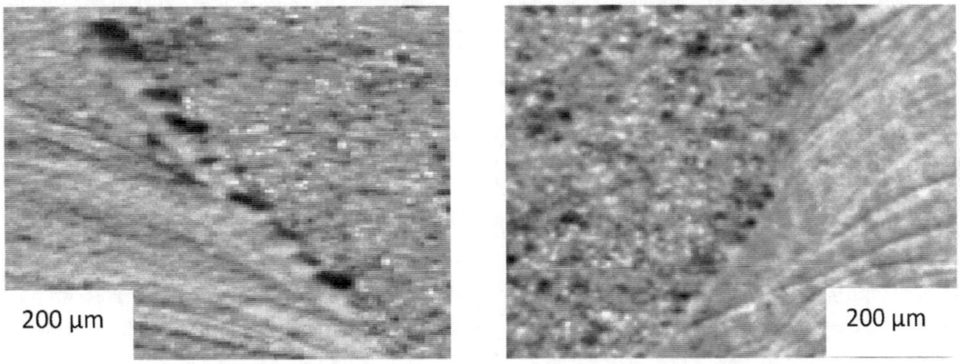

a b

Figure 4. Character of the corrosion destruction of welded joints of steel 12KH18N10T, are obtained: a) DC; b) pulsed arc method.

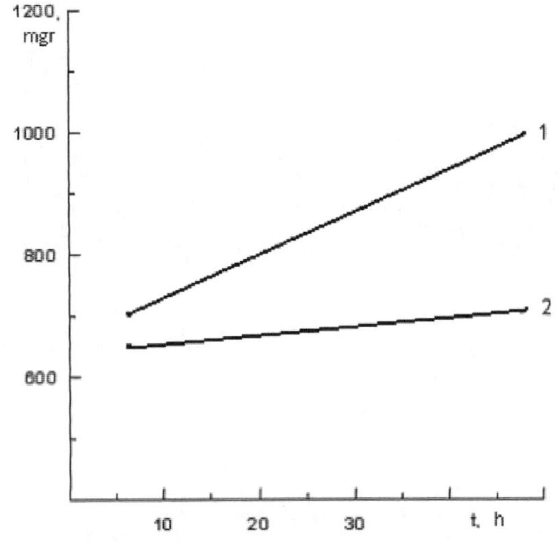

Figure 5. The dependence of the corrosion rate from the time of trial after surfacing coatings: 1 (○) DC, 2 (+) pulsed arc method

The samples welded by the method of adaptive pulse - arc welding of corrosion pattern of destruction changes. Knife-corrosive. Corrosion in the form of pitingo, localiza-bathrooms on the border of the weld.

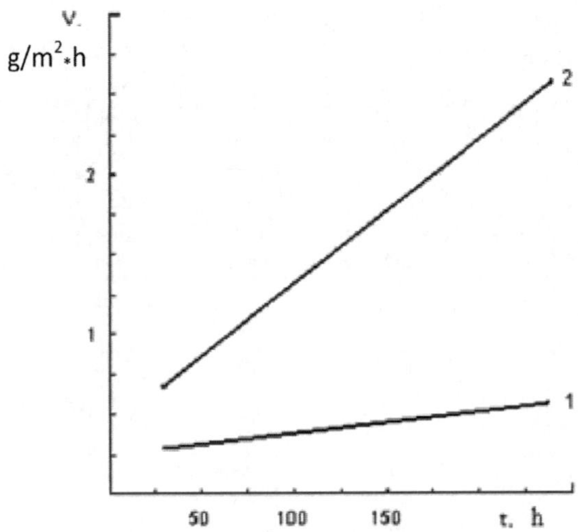

Figure 6. The dependence of corrosion rate of deposited coatings from the time of trial after: 1) direct current 2) pulsed arc method.

The samples welded by the method of adaptive pulse - arc welding of corrosion pattern of destruction changes. Knife-corrosive. Corrosion in the form of pitingo, localiza-bathrooms on the border of the weld.

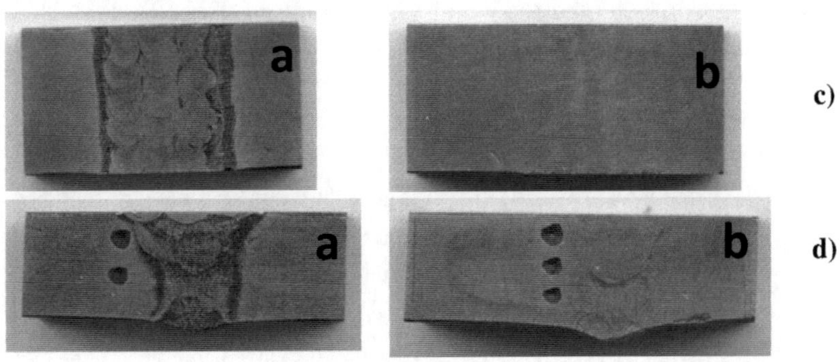

Figure 7. The macrostructures of welds under coatings deposited: a) DC; b) the pulsed - arc method, c) longitudinal section, d) cross-section.

Conclusions

1. Using the technological process of pulse - arc welding steel 12KH18N10T allows for less heat exposure for joint and HAZ of welded joints of steels to form a protective coating with dispersed structure.
2. Surfacing of coatings with dispersed structure obtained by pulsed - arc method, allows in 1.5 - 6 times to reduce the corrosion rate of welded joints of steel 12KH18N10T in active chemical environments.

3. The results of these studies can be recommended for use when welding a protective corrosion - resistant coatings on working surfaces of equipment of chemical productions. Pulsed technological process of deposition of coatings on welded joint of steels can be effectively used for the protection against corrosion in the repair of equipment of chemical industry.

References

[1] Fominykh V. P., Yakovlev A. P., Manual arc welding. Seventh edition revised and enlarged. - M.: Higher school. - 1986. - 366 p.

[2] Ishchenko, Yu. s., Bocharov V. A., Pishchik V. G. Estimation of fusion when welding non-rotating joints//Welding production. - 1981. - No. 3. - P. 10 - 11.

[3] Chernyshov G. G. Influence of force action of the arc on the formation of weld/strength of materials and technologies in welding//Proceedings of the Moscow higher technical school. - 1981. - No. 363. - P. 92 - 101.

[4] Livshits A. S., Greenberg N. And., Kourkoumelis E. Principles of alloying of deposited metal. - M.: Mashinostroenie, 1969. - 188 p.

[5] Hasui A., Moriguchi O. Surfacing and spraying. - M.: Mashinostroenie. - 1985. - 240 p. 6. Kakhovsky N. I., Fartushny, V. G., Yushchenko A. K. Electric Arc welding of steels. - Guide. - Kiev. - Naukova Dumka. - 1974. - 160. p.

[6] Saraev Yu. N., Bezborodov V. P., Selivanov Y. V., Nikonova I. V. The Influence of modes of deposition of coatings on corrosion resistance in acidic environments of welded joints of austenitic steels//metal. - 2007. -№. 2. - P. 33 - 36.

[7] Saraev Yu. N., Bezborodov V. P. The influence of the energy of welding parameters on structure and properties of welded joints of low-alloy steel//Welding production. - 1012. - № 8. - P. 3 - 5.

[8] Robert J K Wood, Tribo-corrosion of coatings: a review / Journal of PHYSICS D: APPLIED P HYSICS, IOP P UBLISHING. 40 (2007) 5502 - 5521, doi:10.1088/0022-3727/40/18/S10.

Work performed under the project of the Russian science Foundation № 16 - 19 - 10010.

IOP Conf. Series: Materials Science and Engineering **142** (2016) 012025 doi:10.1088/1757-899X/142/1/012025

Use of Technogenic Silica Fume and Brown Coal Semi-Coke in the Technology of Silicon Carbide

A E Anikin[a], G V Galevskiy[a], V V Rudneva[a], S G Galevskiy[a], D P Il'yaschenko[b]
[a]Siberian State Industrial University, Novokuznetsk
654007, Novokuznetsk, ul. Kirova, 42
[b]Yurga Institute of Technology (Affiliate) National Research Tomsk Polytechnic University
652055, Kemerovo obl., Yurga, ul. Leningradskaya, 26, tel.+7 384 51 777 66

E-mail: mita8@rambler.ru
Abstract. The paper describes thermodynamic experiments to determine the optimal temperature and time modes for the carbide production process from the briquette charge comprising silica fume and brown coal semi-coke, conditions for chemical enriching of silicon carbide, its phase, chemical and granulometric compositions and particle morphology.

1. Introduction

Production of silicon carbide is one of the most important in the structure of modern metallurgy. About 800 thousand tons are made annually. Silicone carbide is commonly used in steelmaking (about 45% of global demand), abrasive (up 30%) and refractory materials (25%) production. Depending on thermo-mechanical, electro-physical, physicochemical properties, silicon carbide is targeted for the following applications in both traditional and potential fields [1-4]: bonded refractory, structural and functional ceramics, composite materials and coatings, surface and volume modification of materials.

In manufacturing today, silicon carbide is mostly synthesized with using the Acheson method. According to this method, silicon carbide is produced by the carbothermic reduction process in electric furnaces with a capacity of 4.5 - 5.5 MVA at a temperature of 2500 - 2900 K. The silicon carbide material resulting from such process has a crystal structure, where the crystals grow together to form the so-called druses (pieces); after electric furnaces dismantling, the material goes through crushing [5], chemical enriching and size grading. The commercial product of abrasive manufacturers is the carbides in granular forms ranging in size: abrasive grit (2500 - 160 microns), microgrit (160 - 28 microns), micropowders (63 - 10 microns), super thin powder (10 - 1 microns). There are three types of silicon carbide made today: black, green and electro technical.

For quite a long time, the demands of metal making, refractory and ceramic industries for silicon carbide were met by using grinding materials intended for abrasive applications, causing an unnecessary increase in the cost of refractory materials and ceramics; in some cases, the use of silicon carbide was technologically and economically unfeasible in the metallurgical industry. The situation was aggravated by constantly rising energy prices and tightening control to follow ecology regulative requirements of nations. In this connection, in the 80s, the major manufacturers of silicon carbide: Saint - Gobian (France), Exolon - ESK (US - Germany), Carborundum Co. (USA), H. Starck Co. (Germany) and other companies developed and introduced the technology of making the so-called non-ground silicon carbide by synthesis from finely dispersed charges in a furnace usually in the form of micro powders, followed by its chemical enrichment. In the same period, a number of activities were carried out in this field in Russia that demonstrated positive science-based and application-oriented results: a decrease in the temperature of synthesis within 2500 - 2900 by 600 - 1000 ^{o} down to 1900 K, the time required for the synthesis from 26 - 29 hours to 1h, and electricity consumed by a factor of 2 - 4; the content of silicon carbide reached as high as 90.0% by mass in the synthesized products; silicon carbide was made in the form of micro powders of 1 - 5 microns in size; non-commercial and technology-related by-product materials were proved as available for the production of silicon carbide, such as low-quality sand, schungite rock, silica fume, coke fines and dust, and semi-coke; a continuous process for manufacturing silicon carbide was demonstrated as possible, using e.g., tunnel-type, tubular rotary, and shaft-type furnaces. In this regard, it was decided to introduce the developed technologies in a number of enterprises of Minstankoprom. But the socio-economic changes taking place at the turn of the 80 - 90s in the country did not allow these plans to be implemented.

Therefore, providing a scientific background and technological justification for the non-ground silicon carbide production remains an urgent task, which is of great practical importance. In this regard, the aim of this work was to study the processes of carbidizing technogenic silica fume with using semi-coke from brown coal mined at the Berezovsky deposit at the Kansk-Achinsk Coal Basin.

Thermodynamic modeling of the carbide forming processes in Si – O – C and Si – O – C – H systems

Thermodynamic modeling was performed in order to forecast the optimal parameters for making silicon carbide (a ratio of compounds, temperature and pressure in the system), determine equilibrium constants parameters of the reaction (a conversion degree of reactants in carbide, compositions of gaseous and condensed products), and evaluate how independent reactions contributed to the carbide forming processes, with these reactions ensuring the effective processing of raw materials in the course of experiments. The targets of research were the Si - O – C and Si - O - C - H systems, where brown coal semi-coke with a content of volatile products up to 10% was used as a reducing and carbide forming agent.

Values of equilibrium constants were calculated to determine the balanced composition of the Si – O – C and Si – O – C – H systems. The computer simulation software “PLASMA” (Institute of Solid State Chemistry and Mechanochemistry SB RAS), with an in-built database of interaction products for oxide-, boride-, carbide- and nitride-forming systems and designed for modeling high-temperature complex chemical equilibria, was used for calculations. For making calculations, the range of temperature was 800 - 2000 K at a pressure varying within 0.1 to 0.0001 MPa in the systems. The calculations were performed in increments of 100°. The equilibrium constants for reactions to form compounds from the given elements were taken as initial data. In modeling, it was taken into account the possibility of forming gaseous Si, Si^{+}, Si^{-}, Si_2, Si_3, O, O_2, C, C^{+}, C_2, C_3,

C_4, C_5, SiO, SiC, SiC_2, Si_3C, CO, CO_2 in the Si - O - C system and Si, Si^+, Si^-, Si_2, Si_3, O, O_2, C, C^+, C_2, C_3, C_4, C_5, H, H^+, H_2, CH_2, CH_3, CH_4, C_2H, C_3H, C_4H, C_2H_2, C_2H_4, SiO, SiC, SiC_2, Si_3C, CO, CO_2 in the Si - O - C – H, and condensed Si, SiO_2, C, SiC in both systems. The amount of reacting substances and their ratios in the systems under study were determined by using stoichiometry on the basis of the silicon carbide forming reaction and compositions of charged materials. The relevance of this approach to the study of carbide–forming processes was confirmed earlier in works in reference [3-6].

The calculated equilibrium compositions of the Si – O – C and Si – O – C – H systems are shown in Figures 1 and 2 and Tables 1 and 2.

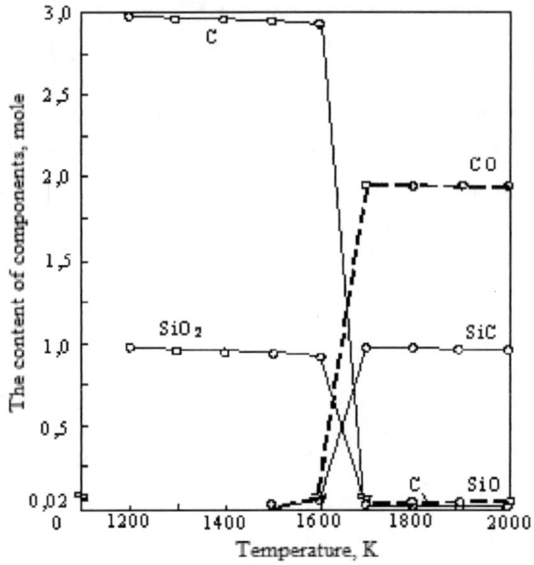

Figure 1 – The temperature dependence of the equilibrium composition of the Si - O - C system at a ratio of Si: O: C = 1: 2: 3 and a pressure P = 0.1 MPa (— condensed and – – – gaseous products)

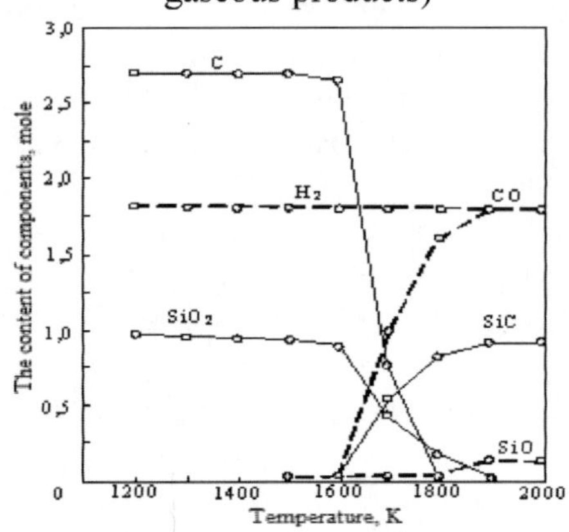

Figure 2 – The temperature dependence of the equilibrium compositions of the Si – O – C – H system at a ratio of Si:O:C:H = 1: 2: 2.7: 3.6 and a pressure P = 0.1 MPa (— condensed and – – – gaseous products)

Table 1 - Equilibrium concentrations (%) of the condensed reaction products in the Si - O - C system for a ratio of Si:O:C = 1:2:3

T, K	P = 0,1 MPa	P = 0,01 MPa	P = 0,0001 MPa

	SiO_2	C	SiC	SiO_2	C	SiC	SiO_2	C	SiC
800	62,5	37,5	0	62,5	37,5	0	62,5	37,5	0
900	62,5	37,5	0	62,5	37,5	0	62,5	37,5	0,02
1000	62,5	37,5	0	62,5	37,5	0,01	62,2	37,3	0,52
1100	62,5	37,5	0	62,4	37,4	0,2	0	0,3	99,71
1200	62,5	37,5	0	61,0	36,7	2,3	0	0,4	99,60
1300	62,5	37,5	0	0	0,2	99,8	0	0,6	99,40
1400	61,8	37,1	1,1	0	0,2	99,8	0	1,0	99,0
1500	61,0	36,8	2,2	0	0,3	99,7	0	1,3	98,7
1600	53,7	32,5	13,8	0	0,4	99,6	0	2,0	98,0
1700	0	0,6	99,4	0	0,6	99,4	0	2,8	97,2
1800	0	0,8	99,2	0	0,9	99,1	0	3,7	96,3

Table 2 - Equilibrium concentrations (%) of the condensed reaction products in the Si - O - C - H system for a ratio of Si:O:C:H = 1:2:2,7:3,6

T, K	P = 0,1 MPa			P = 0,01 MPa			P = 0,0001 MPa		
	SiO_2	C	SiC	SiO_2	C	SiC	SiO_2	C	SiC
900	64,94	35,06	0	64,94	35,06	0	64,94	35,06	0
1000	64,94	35,06	0	64,94	35,06	0	64,94	35,06	0,001
1100	64,94	35,06	0	64,94	35,06	0,002	64,84	35,00	0,16
1200	64,94	35,06	0	64,74	36,04	0,04	62,54	33,86	3,60
1300	64,94	35,06	0	64,54	35,94	0,30	12,47	0	87,53
1400	64,92	35,06	0,03	62,84	34,53	2,63	8,95	0	91,05
1500	63,97	34,51	1,52	51,27	26,61	22,12	0	0	100,00
1600	60,75	32,15	7,10	11,35	0	88,65	0	0	100,00
1700	25,57	8,89	65,54	0	0	100,00	0	0	100,00
1800	9,33	0	90,67	0	0	100,00	0	0	100,00
1900	0	0	100,00	0	0	100,00	0	0	100,00

It can be seen that the process of forming carbides is dominant in both systems

$$SiO_{2(\text{Т})} + 3C_{(\text{Т})} = SiC_{(\text{Т})} + 2CO_{(\text{Г})}. \quad (1)$$

With a stoichiometric ratio of reactants in the charge, the maximum content of silicon carbide in the products is achieved at 1700 K, and at 1900 K when there is a 10% deficiency of carbon. In fact, hydrogen adding to the systems has no influence on the carbide forming process due to the extremely low (less than 0.001 mol) content of hydrocarbons and hydrocarbon radicals within the temperature range of carbide formation. With the Si - O - C system, the equilibrium degree of silicon conversion into carbide does not exceed 0.97% that corresponds to the content of gaseous silicon monoxide in an amount of 0,02mol. As a consequence, it is impossible to obtain homogeneous silicon carbide without free carbon when using the charge (SiO_2 + 3C) of the stoichiometric composition (Table 2). This can be avoided by using a charge with a certain deficiency of carbon (~10%), acting as a reducing agent (see Table 2). In this case, there is a temperature range where both silicon carbide and excessive silica co-exist in the system. And the higher the pressure in the system, the wider this temperature range of both compounds' occurrence. Above these temperatures, silicon carbide is a reducing agent for silica, i.e. the following reaction is possible

$$2SiO_{2(\text{Т})} + SiC_{(\text{Т})} = 2SiO_{(\text{Г})} + CO_{(\text{Г})}. \quad (2)$$

In this reaction, silicon carbide is purified and excessive silica is removed. This reaction has a greater effect when takes place in vacuum.

2. The experimental technique for carbidizing silica fume

The process for carbidizing silica fume was analyzed using the discrete method, including heat treatment of the charge in briquettes at specified temperatures and time, quenching of carbidized products with subsequent chemical analyzes.

The charges were heated in a graphite electric resistance furnace with an internal diameter of 0.075 m (the Tamman type furnace). The temperature inside the furnace was controlled using a tungsten-rhenium thermocouple BP 5/20. Cylindrical briquettes with a mass of about $2{,}5 \cdot 10^{-3}$ kg and an outside diameter of 0.015m were placed in graphite crucibles. Three crucibles were loaded simultaneously inside the working chamber of the heated electric furnace, then heated for 3 - 4 min to a specified temperature, and held therein during a specified period of time within 5 to 90 minutes. After holding at the specified time, the graphite crucibles were removed from the electric furnace and placed in corundum containers, which were tightly plugged, and put for 10 minutes in cold water for rapid cooling of the products formed. This enabled us to prevent oxidation of the products with oxygen and atmospheric moisture. After cooling, the products formed in the reaction were extracted from the crucibles without losses in their amount, then weighed and placed in platinum containers for their subsequent filling with a solution of nitric acid, with a concentration of 7, 5 - 12, 5M, mixed with hydrofluoric acid, with a concentration of 2,0 - 3M, taken at a ratio of 1: (3 ÷ 5), and boiled during a period of 2.5 - 3.0 hours. Treating silicon carbide with a mixture of nitric and hydrofluoric acids ensures the removal of impurities of aluminum, magnesium, calcium, iron, manganese, nickel, chromium, copper, sodium, potassium, silicon and their oxides, and is recommended for the silicon carbide powder of a grain size from 5 to 200 microns [5] . In the presence of nitric acid, metallic impurities and their oxides form soluble nitrate compounds, while using hydrofluoric acid only may cause the formation of insoluble fluorides. After cooling, the insoluble residue in the form of silicon carbide admixed with free carbon was collected with using a paper filter, then the collected on the filter residue was washed with distilled water and acetone, dried to obtain a constant mass, and weighed. Following that, according to the recommendations in [1-4], the insoluble residue was heat treated at a temperature of 1073 - 1123 K for 1.5 - 2.0 hours to remove free carbon by oxidation. The content of free carbon and silicon carbide in the products of carbidization was calculated in the context of their loss in mass. The percentage yield of silicon carbide was calculated as the ratio of actual yield, being the average amount of silicon carbide produced with the three crucibles, to theoretically possible yield.

3. Experimental carbidization of silica fume - brown coal semi-coke, measurement and discussion of results

The following materials were used to carry out the laboratory works: (i) silica fume, collected as a by-product in the production of silicon (MK-Kr) and high-silica ferrosilicon (MK-FS), with a content of silicon dioxide equal to 93.92 and 93.00 % by mass respectively and a specific surface of 25000 m^2/kg [6-10], and (ii) brown coal semi-coke (BPK), produced at the Berezovsky deposit at the Kansk-Achinsk Coal Basin, containing, %: carbon - 81.9, volatile matter - 9.5, ash - 8.6, a specific surface of 264000 m^2/kg [1-5]. In all cases, the charges had a stoichiometric composition to obtain carbides.

In terms of time and temperature conditions for the carbidization process, the single-component briquette charge comprising silica fume MK-Kr, MK-FS-BPK were investigated at temperatures of 1883, 1923, 1973 K and during a period of time of 5, 10, 15, 20, 25, 30 min. The results of experiments are shown in Figures 3 and 4.

In both cases, the process of forming carbides begins from the first minute, and at the beginning of the fifth minute the yield of carbide produced reaches 5.23 - 7.21% at a temperature of 1873 K, 17.06 - 22.30 % at a temperature of 1923 K, and 20.20 - 23, 51 % at a temperature of 1973 K.

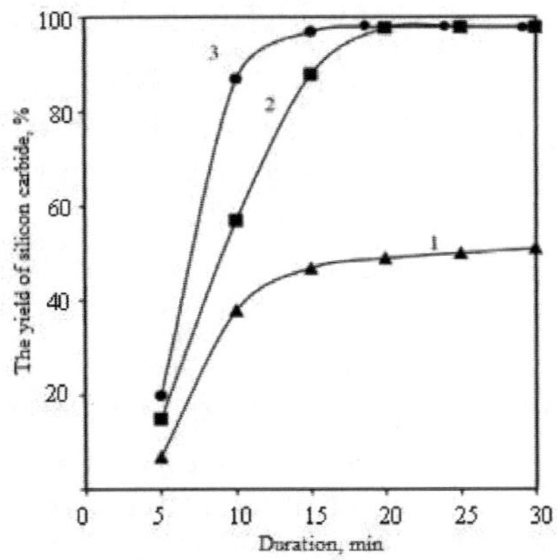

Figure 3 – Relationship between the percentage yield of silicon carbide, temperature and duration of heat treatment for the MK-FS-BPK charging mixture
(1 – 1873 K (Δ) = (7,21 ÷ 53,13) ± (0,20 ÷ 1,13);
2 – 1923 K (□) = (17,06 ÷ 96,82) ± (0,55 ÷ 1,35);
3 – 1973 K (○) = (20,20 ÷ 97,01) ± (0,61 ÷ 1,45))

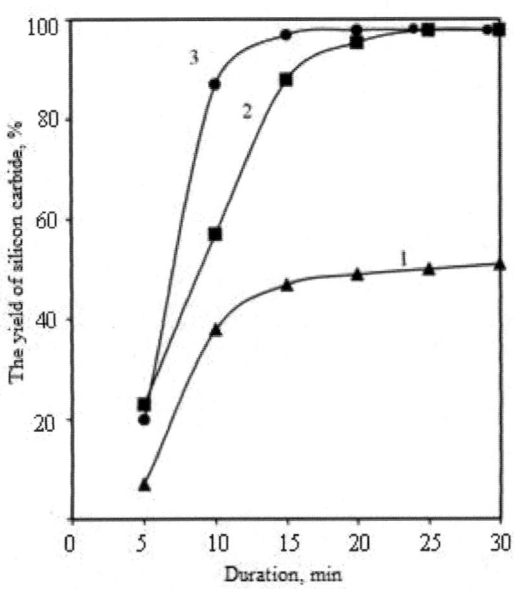

Figure 4 - Relationship between the percentage yield of silicon carbide, temperature and duration of heat treatment for the MK-Kr-BPK charging mixture
(1 – 1873 K : (Δ) = (5,23 ÷ 46,26) ± (0,19 ÷ 1,11);

$$2 - 1923\ \text{K} : (\square) = (22{,}30 \div 96{,}78) \pm (0{,}59 \div 1{,}33);$$
$$3 - 1973\ \text{K} : (\circ) = (23{,}51 \div 97{,}62) \pm (0{,}67 \div 1{,}38))$$

The carbidization process ends at a temperature of 1973 at the beginning of the fifteenth minute and 1923 K at the twentieth minute. The mass percentage of carbide is 92.50 for the MK-FS charge and 94.90 for the MK-Kr charge in the products after heat treatment. These time and temperature conditions for forming carbide are confirmed by the study of phase compositions of products after heat treatment. Thus, it was established that, in both cases, the cubic crystal structure (β-SiC) of silicon carbide prevailed by the start of the tenth minute, and partially modified to form the hexagonal structure (α-SiC_{II}), when heated to a temperature of 1923 to1973 K and held for 50 - 90 minutes. Alpha ferrite accompanied the process of forming carbide in silica fume of the MC-FS-BPK mixture. There was a glassy phase present in all test samples, which were formed, presumably, by silicates of calcium, magnesium and iron.

As a result of chemical processing, the SiC content exceeded 90% in carbide, i.e. the rate specified for abrasive micro powders of a grain size of 1 - 2 microns. The rate of enrichment was sufficiently high, standing for 87 - 95% in the presence of impurities: oxides and iron. Silicon carbide was characterized by a high content of silicon dioxide that amounted to more than 7%, and this allowed us to consider it as a promising material for the production of silica-bonded silicon carbide refractories, normally containing 84 – 89 % of SiC and 6 – 12% of SiO2.

Basic characteristics with their ranges and optimal conditions for obtaining and processing silicon carbide are given in Table 3 for a series of five parallel experiments. It can be seen that the examined technology options enable us to produce silicon carbide with reproducible phase, chemical and granulometric composition, and they can be regarded as a technological basis for developing an industrial production of non-ground silicon carbide from briquette finely dispersed mixtures of silica fume – brown coal semi-coke.

Table 3 – Optimal conditions for producing and enriching silicon carbide and its properties

Conditions for producing and enriching silicon carbide and its main properties	SiC (MK–FS+BPK)	SiC (MK–Kr + BPK)
Charge mixture, %:		
- silica fume (MK)	58,80	58,04
- brown coal semi-coke (BPK)	42,20	41,96
Specific surface of mixed components, m^2/kg		
- silica fume	20000 - 22000	
- brown coal semi-coke	255000 - 265000	
Temperature of carbidization, K	1923 - 1973	
Duration of carbidization, min	25 - 20	
Chemical enrichment of products		
- Concentration of hydrochloric acid, %	35	
- Solid-to-Liquid Ratio	1:2	
- Temperature, K	383	
- Duration, h	3	
Phase composition	β - SiC	
Chemical composition, %:		
SiC	90,42 - 90,86	90,94 - 91,18

Al_2O_3	0,10 -0,25	0,10 -0,15
CaO	0,32 - 0,53	0,37 - 0,49
MgO	0,12 - 0,18	0,03 - 0,10
Fe	0,08 - 0,12	0,05 - 0,11
SiO_2	7,32 - 7,94	6,97 - 7,13
free C	0,41 - 0,54	0,63 - 0,81
free Si	0,13 - 0,17	0,23 - 0,27
Yield of silicon carbide, %	96,84 - 97,12	97,31 - 97,63
Fineness and morphology of silicon carbide particles:		
- Specific surface, m^2/kg	8000 - 9000	
- Range in size, micron	0,2 - 1,0	
- Shape	Irregular	

4. Conclusions

Experimental works were performed to study and analyze the thermodynamic processes of deoxidizing industrial silica fume with brown coal semi-coke to produce silicon carbide.

It is established that the process of forming carbide is dominant in both Si - O - C and Si - O - C - H systems chosen for the thermodynamic analysis. With a stoichiometric ratio of reactants in the charges, the maximum content of silicon carbide in the products after the reduction reaction is achieved at 1700 K, and when there is a 10% deficiency of carbon - at 1900 K. The second charge composition is more technically advantageous due to obtaining carbide without impurities.

The single-component briquette charges of silica fume (MK-FS and MK-Cr) and brown coal semi-coke were processed at temperatures of 1883, 1923, and 1973 K with a range of time from 5 to 30 min to produce carbide. The following time-temperature conditions were determined as the optimal: a temperature of 1923 - 1973 K during a period of 20 - 15 minutes.

The feasibility, optimal conditions and parameters for the chemical enrichment of the formed products were defined as follows: when hydrochloric acid treating with a concentration of minimum 35% at a temperature of 353 K and a S:L ratio = 1: 2, and duration of 3 hours, the carbide content in the products is higher than 90%, impurities of iron and oxides are removed by 87 - 95%.

The particle size distribution, the morphology of particles, the phase and chemical properties were studied relating to both processing techniques employed to produce silicon carbide. It is established that the cubic crystal structure (β-SiC) of silicon carbide is formed in the process, which is partially modified to form the hexagonal structure (α-SiC_{II}) when heated to a temperature ranging from 1923 to 1973 K and maintained for a period of time between 50 and 90 minutes. The content of carbide after enrichment is 90, 42 - 90, 86 % when using the MK-FS silica fume and 90, 94 – 91,18 % when using the MK-Kr silica fume. In both cases, the silicon carbide produced is in the form of micro powders with a specific surface of 8000 – 9000 m^2/kg, and the particles have an irregular shape and a size of 0.2 - 1.0 microns.

This work has been done at Siberian State Industrial University in the framework of government contract No. 11.1531/2014/K with the Ministry of Education and Science RF.

References

[1] Rudneva, Viktoriya Vladimirovna/ Sovershenstvovanie plazmometallurgicheskoj tehnologii proizvodstva nanoporoshkov karbida kremnija/ dissertacija ... doktora tehnicheskih nauk [Improvement of plasma metallurgy for the production of chromium carbide nanopowders. Doctoral Thesis in Tech, Sc.]: 05.16.06 Novokuznetsk 2009 (in Russian)
[2] Yakushevich N.F. Vzaimodeistvie ugleroda s oksidami kal'tsiya, kremniya, alyuminiya [Interaction of carbon with oxides of calcium, silicon, aluminum]. Novokuznetsk: izd. SibGIU, 1999. 250 p. (in Russian)
[3] Zel'berg B.I., Chernykh A.E., Elkin K.S. Shikhta dlya elektrotermicheskogo proizvodstva kremniya [Charge for electroheat silicon production]. Chelyabinsk: Metall, 1994. 320 p. (in Russian)
[4] Garshin A.P., Shumyacher V.M., Pushkarev O.I. Abrazivy i materialy konstruktsionnogo naznacheniya na osnove karbida kremniya [Abrasives and materials of constructional materials on the basis of silicon carbide]. Volgograd: izd. VolGASU, 2008. 189 p. (in Russian)
[5] Shiryaeva, Lyudmila Sergeevna. Razrabotka nauchnyh i tehnologicheskih osnov plazmometallurgicheskogo proizvodstva karbida hroma. Dissertacija ... kandidata tehnicheskih nauk [Development of scientific and technological foundation for plasma metallurgy for the chromium carbide production. Candidate's Thesis in Tech, Sc.]: 05.16.02 Novokuznetsk 2013 (in Russian)
[6]. Nozdrin I.V., Galevskiy G.V., Shiryaeva L.S., Terent'eva M.A. Sintez i jevoljucija dispersnosti boridov i karbidov vanadija i hroma v uslovijah plazmennogo potoka [Synthesis and evolution of dispersion in vanadium and chromium borides and carbides in conditions of plasma flow]// Journal Izvestiya. Ferrous Metallurgy. 2011. No.10. pp. 12 – 17 (in Russian)
[7] Kryukov R.E., Kozyrev N.A., Galevsky G.V., Bendre Y.V., Goryushkin V. F., Valuev D.V. Some aspects of oxidation-reduction reactions under carbon-bearing flux welding, J. IOP Conference Series: Materials Science and Engineering. 91 (2015)
[8] Valuev D V, Malushin N N, Valueva A V, Dariev R S, Mamadaliev R A Manufacturing component parts of mining equipment with application of hardening technologies J IOP Conf. Series: Materials Science and Engineering 127 (2016).
[9] Valuev D V, Malushin N N, Valueva A V, Dariev R S, Mamadaliev R A Traditional technology of chromium-tungsten steels facing, its disadvantages and suggestions for their eliminations J IOP Conf. Series: Materials Science and Engineering 127 (2016).

IOP Conf. Series: Materials Science and Engineering **142** (2016) 012026 doi:10.1088/1757-899X/142/1/012026

Manganese Content Control in Weld Metal During MAG Welding

D A Chinakhov[1], E D Chinakhova[2], A S Sapozhkov [1]

[1]Yurga Institute of Technology of National Research Tomsk Polytechnic University, 652055, Leningradskaya st., 26, Yurga, Kemerovskaya region, Russia
[2]Siberian State Medical University, 634055, Moskovsky trakt 2, Tomsk, Russia

e-mail: chinakhov@tpu.ru

Abstract. The influence of the welding current and method of gas shielding in MAG welding on the content of manganese is considered in the paper. Results of study of the welded specimens of steels 45 when applying welding wire of different formulas and different types of gas shielding (traditional shielding and double-jet shielding) are given. It is found that in MAG welding the value of the welding current and the speed of the gas flow from the welding nozzle have a considerable impact on the chemical composition of the weld metal. The consumable electrode welding under double-jet gas shielding provides the directed gas-dynamics in the welding area and enables controlling the electrode metal transfer and the chemical composition of a weld.

Introduction

At present welding is widely applied to obtain permanent joints from a wide range of metallic and non-metallic materials and structural composites in the Earth's atmosphere, in the World Ocean and in the Space. In spite of constantly increasing application of light alloys, polymer materials and composites in constructions and articles, steel is still used as the main engineering material. The global market of welding equipment and services is expanding proportionally to the growth of world steel consumption.

Welding processes develop according to complex physical and chemical laws under high temperature. The complex of various factors and phenomena influences the quality of joint welds. This influence is particularly significant in welding hardenable items. New welding technologies are developed and the early ones are steadily improved. Welding techniques, methods, weld properties and configuration control are also being improved [1–7].

The main role in the formation of welded joints belongs to the processes occurring in a drop of molten metal and the electrode metal in the weld pool [1, 2, 3, 13]. The speed and completeness of the physical and chemical processes in the electrode metal droplet formation determine the quality of the weld and its chemical composition [12, 13]. Welding current, arc voltage, chemical composition of the electrode wire, size and droplet growth time, the frequency of droplet transfer from the electrode to the weld pool, the composition of the shielding gas and its rate of flow from the torch [12, 13,] influence the processes taking place in a drop of molten electrode metal significantly.

IOP Conf. Series: Materials Science and Engineering **142** (2016) 012026 doi:10.1088/1757-899X/142/1/012026

Methodology

This work is aimed at the study of the influence of active shielding gas (CO_2) and welding current consumption on Mn content in weld metal in consumable electrode welding.

When welding steels with consumable electrode in shielding gas, the composition of the weld metal depends on the share of the electrode and the base metal, the intensity of the reactions in the droplet and the weld pool, the loss of elements of the electrode and the base metal [1, 2, 7], as well as the flow rate and the composition of the shielding gas [1, 2, 6, 7].

We conducted a comparative calculation of the speed of the shielding gas flow from the nozzle of the traditional single-jet nozzle (cone type) and from the designed double-jet shielding nozzle (Table 1). The shielding gas is CO_2, gas density b is 1.97 kg/m^3, gas flow rate Q is 20 l/min.

The results of the calculations show that in double-jet shielding the speed of the gas flow from the nozzle is 2.1 times higher than in one-jet shielding, the gas flow rate being the same [16, 17]. To check the adequacy of the calculated data we conducted an experiment. Under laboratory conditions, using a hot-wire anemometer Dwyer Series 471, we measured the speed of shielding gas flow at the nozzle exit at different flow rates (Table. 1).

Table 1.

The speed of flow of gas through a nozzle and the power of the shielding gas jet

Method	Sectional area S, mm^2	Gas rated speed, м/c	Experimentally measured speed of gas flow, м/c
Single jet	173	1.93	1.83
Double jet	82	4.05	3.95

Comparing the calculated and experimental data showed that the relative error does not exceed 10%, and developed a model for calculating the shielding gas flow through nozzle provides a satisfactory accuracy [16, 17].

Results and Discussion

To assess the impact of the shielding gas flow through nozzle speed on the transfer of electrode metal droplets to the weld pool we made a high speed recording of waveforms of current and voltage (Fig. 1) with a digital oscilloscope Agilent Technologies DSO1012A.

We made a weld bed with welding wire Sv-08G2S diameter of 1.2 mm in CO_2 shielding gas The weld bed was made using a single-jet (cone type nozzle) and a double-jet gas shieldong. Welding conditions: I = 200 A, arc voltage U = 26 V, welding wire extension L = 12 mm, the shield gas flow Q = 20 liters/min. The experiments were performed under the same conditions and the same modes. The power supply sources were VS-300B, welding machine was VD - 1500.

Analysis of the waveforms has shown, that with increasing the speed of the shielding gas though nozzle (double-jet), gas-dynamic effect on the droplet of the electrode metal is also increased, the frequency of droplet transfer from the electrode to the weld pool (an average of 1.6 times as compared with the single jet gas shield), and the intensity of the metallurgical processes on the surface of the drop increases too, the droplet size decreases as the droplet transfers to the weld pool (Fig. 1).

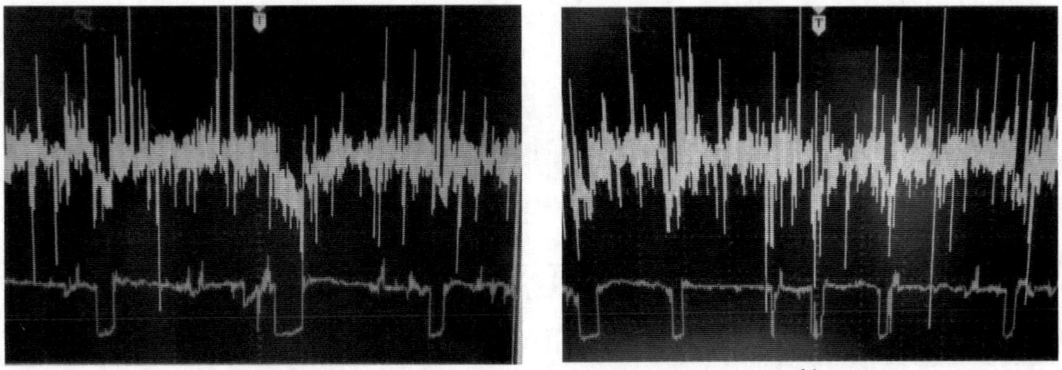

a) b)

Fig. 1. Waveforms of current and voltage: a) Traditional Single-jet gas shielding; b) double-jet gas shielding

Mn content in weld metal decreases as shielding gas CO_2 flow rate rises in open-arc welding the diameter of wire is 1.2 mm [6, 7]. The experiments carried out on multi-layer 45 steel welded samples with welding wires of various composition (Sv-08G2S, Sv-08GSMT) and various methods of gas shielding (the traditional one and double-jet shielding) demonstrated the dependence of Mn level in weld metal on welding current (Tables 2–4). The obtained results demonstrate the influence of welding current and gas outflow rate on the chemical composition of weld metals and, consequently, their performance characteristics.

From 250A (U = 27V, l = 12 mm), when the arc becomes submerged, Mn in weld metal is getting practically stabilized, moreover, its level is much higher than that in weld metal when in open-arc welding (140–220 A).

Table 2

Chemical composition of the base and materials

Material	Weight fraction of elements, [%]				
	C	Mn	Si	Cr	Ni
According to the steel grade guide [9]					
Steel 45	0.42–0.5	0.5–0.8	0.17–0.37	≤0.25	≤0.3
Sv-08G2S	0.05–0.11	1.8–2.1	0.7–0.95	≤0.2	≤0.25
Sv-08GSMT	0.08–0.12	1.0–1.3	0.4–0.7	≤0.3	≤0.3
Averaged results of experiments					
Steel 45	0.44	0.67	0.31	0.19	0.11
Sv-08G2S	0.07	1.80	0.78	0.06	0.07
Sv-08GSMT	0.12	1.12	0.54	0.15	0.14

In traditional open-arc gas shielded welding with Sv–08G2S wire in the current range 140–190 A 30–35 % manganese is burnt out (0.58–0.6 % the weight fraction), and in welding at 280 A about 20% manganese (0.32–0.35 % the weight fraction) is burnt out when consumption of the shielding gas (Tabl. 3, fig. 1, a) held equal. The less manganese burnout is possible as welding wire feed rate is increased, transfer of a drop into the weld pool is getting faster, and the time of drop overheated condition is reduced. In the same conditions in double-jet gas shielding welding the burnout of manganese increases and amounts to about 36–41 % (0.7–0.74 % the weight fraction) and 24 % manganese (0.4–0.44 % the weight fraction), consequently. This fact demonstrates the growing gas-dynamic impact on drop of molten metal and more intensive metallurgical processes.

Table 3

Averaged research results of chemical composition of weld metal in 45 steel samples, welded with Sv-08G2S wire in CO_2 ($U = 27V$, $l = 12$ mm, $Q = 25$ l/min, $V_w = 20$ cm/min)

Welding current, [A]	Weight fraction of elements, [%]				
	C	Mn	Si	Cr	Ni
	Traditional single-jet gas shielding				
140	0.120	1.205	0.430	0.080	0.100
190	0.140	1.214	0.450	0.090	0.100
230	0.152	1.264	0.461	0.089	0.098
280	0.150	1.461	0.556	0.096	0.104
	Double-jet gas shielding				
140	0.145	1.066	0.342	0.082	0.092
190	0.133	1.104	0.379	0.098	0.103
230	0.135	1.243	0.444	0.080	0.097
280	0.165	1.363	0.534	0.108	0.092

The situation is similar in welding with Sv–08GSMT wire (Tabl. 4, Fig. 1, b). In traditional open-arc gas shielded welding in the current range 140–190 A 30–40 % manganese is burnt out (0.33–0.44 % the weight fraction), and in welding at 280 A about 20% manganese (0.21–0.23 % the weight fraction) is burnt out when the consumption of the shielding gas held equal. In double-jet gas shielding welding the burnout of manganese is 39–42 % (0.45–0.47 % the weight fraction) and 29 % (0.32–0.34 % the weight fraction), respectively.

It's determined that the percentage of manganese burnout while transfer from a wire into a weld stays nearly similar and amounts to 35% at the current up to 200 A, and about 20% at the current up to 280 A, but the weight fraction of the burnt out manganese changes depending on the initial level in a welding wire. The manganese level in weld metal differs from that of the initial level in a welding wire and amounts to about 5% in the traditional single-jet and double-jet gas shielding.

Table 4

Averaged research results of the chemical composition of weld metal of 45 steel samples, welded with Sv-08GSMT wire in CO_2 ($U = 27V$, $l = 12$ mm, $Q = 25$ l/min, $V = 20$ cm/min)

Welding current, [A]	Weight fraction of the elements, [%]						
	C	Mn	Si	Cr	Ni	Mo	Ti
	Traditional single-jet gas shielding						
140	0.184	0.682	0.221	0.120	0.122	0.16	0.012
190	0.124	0.796	0.268	0.130	0.143	0.17	0.017
230	0.136	0.864	0.328	0.125	0.126	0.17	0.019
280	0.144	0.901	0.341	0.167	0.150	0.17	0.019
	Double-jet gas shielding						
140	0.206	0.652	0.212	0.135	0.131	0.16	0.015
190	0.161	0.669	0.229	0.126	0.127	0.15	0.01
230	0.151	0.739	0.235	0.148	0.137	0.15	0.012
280	0.169	0.792	0.280	0.147	0.134	0.15	0.018

In open-arc welding the magnitude of the welding current and active gas outflow rate from the welding nozzle have a considerably influence on the chemical composition of weld metal. In double-jet gas shielding with increase in the impact of active shielding gas jet on a drop of electrode molten metal the intensity of metallurgical processes on the drop surface rises [10].

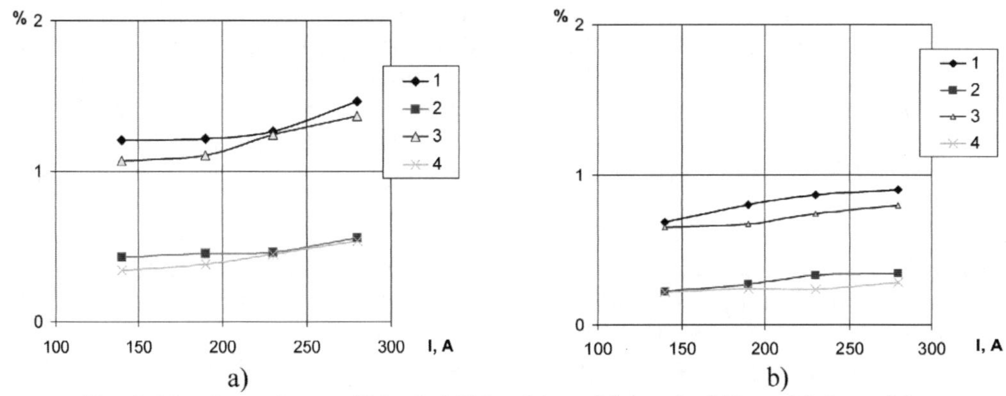

Fig. 2. The dependence of Mn and Si level in weld metal of 45 steel joint welds: a) welding with Sv–08G2S wire; b) welding with Sv–08GSMT wire. Traditional single-jet gas shielding: line 1 – Mn, line 2 – Si; Double-jet gas shielding: line 3 – Mn, line 4 – Si

Many scientists in their works [1–7, 11] note the more advantageous conditions of interaction between metal and gas and slays at the stage of a drop as against the weld pool. Work [2] states that the specific surface area of molten metal drops is 5–22 times (in accordance to the drop size) bigger than the specific surface of a weld pool and their specific oxidation speed is about 39 times higher. A conclusion can be made, that the chemical composition of weld metal determines basically the chemical composition of electrode molten metal drops.

The analysis of the results of research [1, 2, 6–8, 10] demonstrated, that the chemical composition of weld metal, especially Mn and Si levels depends on a lot of factors: the temperature of a drop (welding current strength, voltage, gas shielding method), the time of interaction between a drop and shielding medium (welding current strength, consumption and gas outflow rate – they increase and the time reduces), the drop size, welding wire chemical composition, base metal chemical composition. With increase in the welding current Mn level grows in weld metal, and with increase in the voltage its level falls. The decrease in Mn and Si level in weld metal occurs as the consumption of the shielding gas (CO_2) increases in open-arc welding.

Among all alloying elements that are parts of welding wires manganese has the lowest boiling temperature and the heat of vaporization (Tabl. 5). This fact causes its intense vaporization and oxidation on the surface of molten metal drop, which temperature is about 3000 K [2]. Increasing electrode metal droplets heating temperature rises manganese evaporation from its surface and the manganese content the in the weld metal reduces.

Table 5

Thermal-dynamical properties of elementary substances which are parts of the chemical composition of a welding wire (in initial conditions) [12]

Item	Substance							
	Mn	Si	Cr	Ni	Fe	Ti	Mo	C Graphite
Vaporization temperature, [kJ/mole]	221	383	342	378.6	340	422.6	590	–
Boiling temperature, [K]	2235	2623	2945	3005	3134	3560	4885	5100
Melting temperature, [K]	1517	1688	2130	1726	1812	1933	2890	3820
Density, [g/cm^3]	7.21	2.33	7.19	8.9	7.87	4.54	10.22	2.25

On the other hand, when the current increases in mechanized welding with consumable electrode, the manganese content in the weld metal is increased. Stable welding process with increasing welding current provides for increased wire feed rate, i.e. it speeds up the transfer of droplets into the weld pool and reduces the time of its stay in the superheated state.

Double-jet shielding causes the increase in the force of active shielding gas impact on the filler metal drop and surface of the weld pool. When varying the gas-dynamic impact (consumption and rate of gas flow), transfer of filler metal drops, stability of the welding process, as well as chemical composition of weld metal, thermal and other processes in consumable electrode welding can be controlled, and required properties of joint weld can be formed [13-15].

Conclusion

It was found that in MAG welding in CO_2 the welding current magnitude and the gas flow rate though nozzle (gas consumption, method of gas shielding, electrode extension) are of significant importance for the chemical composition of weld metal. The consumable electrode welding under double-jet gas shielding provides the directed gas-dynamics in the welding area and enables controlling the electrode metal transfer and the chemical composition of a weld, stabilizing the welding process.

Work performed under the project of the Russian science Foundation No. 16-19-10010.

References

[1] Beresovsky B.M. Mathematical models of arc welding: in 7 v. V. 4. Fundamentals of thermal processes in welded workpieces. Chelyabinsk: YuUrGU Press. 2006. 547 p.
[2] Novozhilov M.N. Fundamentals of metallurgy of arc welding in gases, Moscow, Machinebuilding. 1979.
[3] Welding and welded materials: In 3 v. V. 1. Weldability of materials: reference book, edited by E.L. Makarov, Moscow, Metallurgy, 1991.
[4] Chinakhov D.A. Gas Dynamic Control of Properties of Welded Joints from High Strength Alloyed Steels. China Welding. Vol. 23 No.3 (2014) pp. 27-31.
[5] Brunov O.G., Solodskii S.A, Zelenkovskii A.A. Conditions of arc ignition in welding in shielding gases. Welding International. 2012. V. 26. 9. P. 710-712
[6] Chinakhov D.A. Calculation of Gas-dynamic Impact of the Active Shielding Gas on the Electrode Metal Drop in Gas Jet Shielded Welding, Applied Mechanics and Materials. Vol. 379 (2013), pp. 188–194.
[7] Chinakhov D.A., Zuev A.V., Filimonenko A.G. Gas-dynamic Impact of a Shielding Gas Jet on the Drop Transfer When Welding with a Consumable Electrode. Advanced Materials Research Vol. 1040 (2014) pp. 850-853.
[8] Grebenchuk V.G., Karasev M.V., Rabotinsky D.N., Karaseva S.M., Rozert R. The influence of mechanized welding with poVVER BRIDGE 60M metal cored wire on molten metal properties when bridge conduits welding, Welding and diagnosis, 1 (2009) 19–24.
[9] Steels and alloys guide, edited by V.G. Sorokin, Moscow, Mashinebuilding, 1989.
[10] Chinakhov D.A., Vorobyov A.V., Tomchik A.A. Simulation of active shielding gas impact on heat distribution in the weld zone, Materials Science Forum, Vol. 762 (2013) 717–721.
[11] Lenivkin V.A., Dyurgerov N.G., Sagirov H.N. Technological properties of a welding arc in shielding gases, Moscow, Machinebuilding, 1989.
[12] Bigeev A.M. Steel metallurgy. Study guide, 2nd edition, Moscow, Metallurgy, 1988.
[13] Krampit, A.G., Krampit, N.U., Krampit, M.A. Mechanical properties of welded joints in welding with pulsed arcs, Applied Mechanics and Materials, Vol. 379 (2013) 195–198.
[14] Novikov O.M., Rad'ko E.P., Ivanov E.N., Ivanov N.S. Development of a new technology of shielding gases arc welding on the basis of gas flows pulsations and ionization potentials. Welder-professional. 2006. 6. P. 10–13, 16.

[15] Chinakhov D.A., Chinakhova E.D., Gotovschik Y.M., Grichin S.V. Influence of welding with two-jet gas shielding on the shaping of a welding joint. IOP Conf. Series: Materials Science and Engineering 125 (2016) 012013. doi:10.1088/1757-899X/125/1/012013.

Modelling the Spectrum of Potency a Stationary Random Process in the Form of Spline First Order at the Random Number of Data in Instant of Time

I Ustinova[1], A Kozlova[2]

[1] Associate Professor, *National Research Tomsk Polytechnic University*, Lenin Avenue 30, Tomsk 634050, Russia
[2]Student, *National Research Tomsk Polytechnic University*, Lenin Avenue 30, Tomsk 634050, Russia

E-mail: igu@sibmail.com, Chibi.moon@mail.ru

Abstract. The spectrum of potency as same the function of correlation is one of the most important characteristic of the second process random order. The spectrum of potency allows to judge about, that structure of process gives opportunity to take estimation of spectrum composition of useful signals and hindrances, allows to produce synthesis (reconstruction) the signals and to build filters and to obtain the estimates filtration. The purpose of this presenting work is modeling of process random stationary potency spectrum by random dates number in measurement moments. Using by probability theory methods and mathematical statistics was derived unbiased estimator of the potency spectrum in the form of spline first order, and were researched statistic characteristics estimation.

1. Introduction

One of the modern methods digital processing of signals is digital the spectrum analysis. It is based on modeling and studying the spectrum potency of signals, which is accepted as a some random process [1]. Density function the spectrum of potency defines the distribution of the dispersion a random process the frequency [2]. There are three basic ways of obtaining function the density of the spectrum:

1. Determination density of the spectrum $S(\omega)$ a given correlation function $R[\tau]$. Computation $S(\omega)=\dfrac{1}{2\pi}\int_0^\infty R[\tau]\cos(\omega\tau)d\tau$ difficulty, because this integral not always computed in elementary function, and the correlation function known, as rule, on the discrete set of the variable τ [3–7].
2. Estimation the density of the spectrum with the help of procedure using fast Fourier transforms. Such approach for at the spectrum analysis virtual and, as rule, to provide deriving acceptable results. However, this approach has a meaningful constraint, to wit, limiting the frequency resolution, i.e. ability to distinguish the frequency lines two or more signals [4].

3. To classic methods the spectrum of potency modeling applies periodogram methods, which in the discrete Fourier transformation conform to direct for succession, acquired realization quantization of a random process. However, as shows in [8], periodogram is not consistent estimation the spectrum of potency.
4. In the works [9-11] set, that there is a relation the spectrum of potency with a fractal characteristics a random process. Aforementioned a relation expressed as following: $S(\omega) = \frac{c}{\omega^{5-2D}}$, where c – the scaling constant, dependent on the amplitude of the signal, D – the fractal dimension, moreover this intercommunication carry estimating disposition, because she is asymptotically solution of integral equation. [12-15].
5. If about the investigated process notorious some plurality of the information, allowing pick acceptable approximation of the process, in this occurrence can get more precise the spectrum estimation, previously defined axes, chosen the model by result dimension [5].

The aim of this work is modeling the spectrum of potency a random fixed process in the form of spline first order at random number data at the time of determined as follows [16-18]. The function $S(t)$, is defined and continuous on the interval $[a, b]$, is denominated a polynomial spline of order m with clusters $x_j \in \{a = x_0 < x_1 < x_2 < \ldots < x_n = b\}$,if on each of the interval $[x_{j-1}, x_j)$, $j = \overline{1,n}$ $S(t)$ there is an algebraic polynomial of power m.

2. Statement of problem

Let the values of the process $y(t)$ are measured on the interval at time $[0;T]$. The moments of measure t_i, $i = \overline{1,N}$ are knows precisely. We will assume, that $y_i = y(t_i) + \xi_i$, $\xi(t)$ – the random function, the presence of which is caused by measurement errors, external hindrance etc. Supposed, that hindrance measurement $\xi_i = \xi(t_i)$, i=1, 2, … – independent, identically distributed a random expressions with mathematical waiting $M[\xi_i] = 0$ and dispersion $D[\xi_i] = \sigma^2$. We knows the sequence of values $y_1, y_2, \ldots, y_N$, where $y_i = \frac{y_{i1} + y_{i2} + \ldots + y_{in_i}}{n_i}$. Here n_i – independent a random expression, distributed by the Poisson law with argument λ and $M[y(t_i)] = 0$, $M[y(t_j)y(t_i)] = R[t_j - t_i]$. By results of observations required to construct estimation $\hat{S}(\omega)$ the spectrum of potency $S(\omega)$ in the form of spline first order. A similar problem is considered in the work [19], however, it used a different approach to obtaining a spline, to wit, approach in which first searched estimation the correlation function, then, is estimation the spectrum of potency, moreover coefficients of the spline are estimated everything at once.

3. Solution of the problem

We divide the whole frequency axis ω on the interval $[0\,;\Omega]$, $[\Omega\,;2\Omega]$, $[2\Omega\,;3\Omega]$, …. We will consider the statistics,

$$Q = \frac{1}{\lambda^2 \pi T} \sum_{i,j;i \neq j} n_i n_j y_i y_j \varphi(t_j - t_i), \tag{1}$$

where $\varphi(\tau)$ for the function evenly condition:

$$\varphi(\tau) = \int_{-\infty}^{+\infty} \Phi(\omega) \cos \omega\tau d\omega, \tag{2}$$

which in the integral $\int_{-\infty}^{+\infty} \Phi(\omega)\cos\omega\tau d\omega$ converges. By averaging (1) by the expression n_i, we obtain the expression $\overline{Q} = \frac{1}{\pi T}\sum_{i,j;i\neq j} y_i y_j \varphi(t_j - t_i)$.

We seeing, that $M[\overline{Q}] = \frac{1}{\pi T}\sum_{i,j;i\neq j} R[t_j - t_i]\varphi[t_j - t_i]$. May assume, that the following holds approximation: $M[\overline{Q}] \approx \frac{1}{\pi T}\int_0^T\int_0^T R[u-v]\varphi(u-v)dudv$. After the change of variable, we will get $\int_0^T\int_0^T R[u-v]\varphi(u-v)dudv = \begin{vmatrix} u-v=z \\ du=dz \end{vmatrix} = \int_0^T dv \int_{-v}^{T-v} R(z)\varphi(z)dz$. We interchange the order of integration, by using the range of integration ,on the figure1.

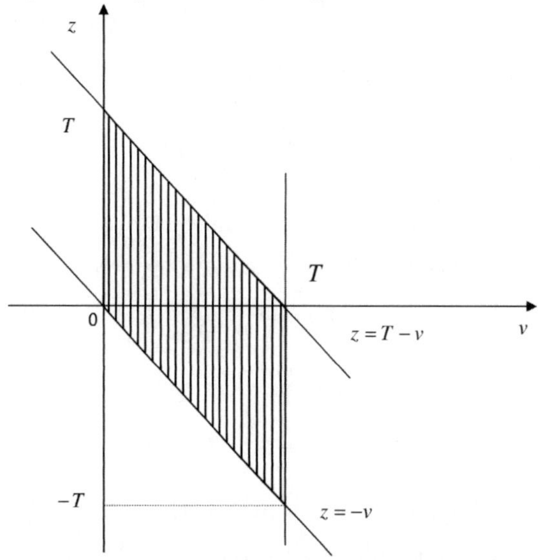

Figure 1. The range of integration.

This will allow from the original double integral go to the single integral

$$\int_{-T}^{0} R[z]\varphi(z)dz\int_{-z}^{T} dv + \int_{0}^{T} R[z]\varphi(z)dz\int_{0}^{T-z} dv = \int_{-T}^{0}(T+z)R[z]\varphi(z)dz + \int_{0}^{T}(T-z)R[z]\varphi(z)dz =$$

$= T\int_{-T}^{T}\left(1-\frac{|z|}{T}\right)R[z]\varphi(z)dz$. In the asymptotic case, when $T\to\infty$, the expression $\frac{|z|}{T}\to 0$ and, accordingly, $\int_{-T}^{T}\left(1-\frac{|z|}{T}\right)R[z]\varphi(z)dz \approx \int_{-\infty}^{\infty} R[z]\varphi(z)dz$. Thus

$M[\overline{Q}] \approx \frac{T}{\pi T}\int_{-\infty}^{+\infty} R[z]\varphi(z)dz = \frac{1}{\pi}\int_{-\infty}^{+\infty} R[z]\varphi(z)dz$. By using the expression (2), we will get

$M[\overline{Q}]\approx\frac{1}{\pi}\int_{-\infty}^{+\infty}R[\tau]\int_{-\infty}^{+\infty}\Phi(\omega)\cos\omega\tau d\omega d\tau=\frac{2}{\pi}\int_{-\infty}^{+\infty}\Phi(\omega)d\omega\int_{0}^{+\infty}R[\tau]\cos\omega\tau d\tau$. By using knows the ratio of Wiener – Kninchin [1], we will get

$$M[\overline{Q}]\approx\int_{-\infty}^{+\infty}\Phi(\omega)S(\omega)d\omega. \quad (3)$$

For construct estimation $\hat{S}(\omega)$ the spectrum of potency $S(\omega)$ in the form of spline first order, we will assume, that on the interval $[(k-1)\Omega\,;k\Omega]$ the function $S(\omega)$ presented in the form $S(\omega)=S_{k-1}\frac{k\Omega-\omega}{\Omega}+S_k\frac{\omega-(k-1)\Omega}{\Omega}$. Consider the integral $I=\int_{(k-1)\Omega}^{k\Omega}S(\omega)\left(A\frac{k\Omega-\omega}{\Omega}+B\frac{\omega-(k-1)\Omega}{\Omega}\right)d\omega$, computation of this integral leads to the expression $\frac{\Omega}{6}\{S_{k-1}(2A+B)+S_k(A+2B)\}$. Noticing that when $A=2$ и $B=-1$ $I=\frac{\Omega}{2}S_{k-1}$. Then

$$\Phi(\omega)=\begin{cases}\frac{2}{\Omega}\left(2\frac{k\Omega-\omega}{\Omega}-\frac{\omega-(k-1)\Omega}{\Omega}\right), & \omega\in[(k-1)\Omega;k\Omega],\\ 0, & \omega\notin[(k-1)\Omega;k\Omega]\end{cases}$$

By using the expression (2), we find $\varphi(\tau)=\int_{(k-1)\Omega}^{k\Omega}\frac{2}{\Omega}\left(3k-1-3\frac{\omega}{\Omega}\right)\cos\omega\tau d\omega=$

$$=\begin{cases}\frac{2}{\Omega}\left\{\frac{3}{\Omega\tau^2}(\cos(k-1)\Omega\tau-\cos k\Omega\tau)-\frac{\sin k\Omega\tau}{\tau}-\frac{2\sin(k-1)\Omega\tau}{\tau}\right\}, & \omega\in[(k-1)\Omega,k\Omega],\\ 0, & \omega\notin[(k-1)\Omega,k\Omega]\end{cases}$$

We assume that $k=1$ $\varphi(\tau)=\varphi_0(\tau)$, at $k=2$ $\varphi(\tau)=\varphi_1(\tau),\ldots$. Substituting into expression (1) found of values $\varphi(\tau)$, we find the consistently $S_0, S_1,\ldots$, which we join by straight line segments, that gives on estimate $S(\omega)$ in the form of spline first order. From constructing estimates of the coefficients spline it follows, that the resulting estimates of knots of the spline are unbiased.

4. The estimate of the dispersion obtained estimates the spectrum of potency

Now we find an asymptotic estimate of the dispersion $D[Q]$ the statistic Q at $T\to\infty$. For this consider $Q^2=\frac{1}{\lambda^4\pi^2T^2}\sum_{j=1}^{N}\sum_{i=1}^{N}\sum_{l=1}^{N}\sum_{k=1}^{N}n_jn_in_ln_ky_jy_iy_ly_k\varphi(t_j-t_i)\varphi(t_l-t_k)$. By averaging of this expression by values n_i, we will get $\overline{Q}^2=\frac{1}{\pi^2T^2}\sum_{j=1}^{N}\sum_{i=1}^{N}\sum_{l=1}^{N}\sum_{k=1}^{N}y_jy_iy_ly_k\varphi(t_j-t_i)\varphi(t_l-t_k)$. By averaging by realization of the process $y(t)$ at fixed moments of measurement, we will get

$$M[y_jy_iy_ly_k]=M[y(t_j)y(t_i)y(t_l)y(t_k)]=R[t_j-t_i]R[t_l-t_k]+R[t_j-t_l]R[t_i-t_k]+$$
$$+R[t_j-t_k]R[t_i-t_l]$$

When

$$M\left[\overline{Q}^2\right]=\frac{1}{\pi^2T^2}\sum_{j=1}^{N}\sum_{i=1}^{N}\sum_{l=1}^{N}\sum_{k=1}^{N}\varphi(t_j-t_i)\varphi(t_l-t_k)\left(R\left[t_j-t_i\right]R\left[t_l-t_k\right]+R\left[t_j-t_l\right]R\left[t_i-t_k\right]+\right.$$
$$\left.+R\left[t_j-t_k\right]R\left[t_i-t_l\right]\right) \quad (4)$$

Next to consider by averaging over the moments of measurement each term contains a factor of the form $R\left[t_j-t_i\right]R\left[t_l-t_k\right]$ expressions (4). Here possible the following variants:

1. All index j,i,l,k various. Making the change t_j on u, t_i on u', t_l on v, t_k on v' we get $\frac{1}{\pi^2T^2}\int_0^T\int_0^T\int_0^T\int_0^T R\left[u-u'\right]R\left[v-v'\right]\varphi\left(u-u'\right)\varphi\left(v-v'\right)dudvdu'dv'$, the computation which gives as $T\to\infty$ the square of the expectation the statistic Q, to wit, $\frac{1}{\pi^2}\left(\int_{-\infty}^{+\infty}R[z]\varphi(z)dz\right)^2$, which is subtracted when finding the dispersion.

2. Two indexes are equal, to wit, $i=k$, i.e. $t_i = t_k$, when considering summand of the expression(4) will have the form $\frac{1}{\pi^2T^2}\int_0^T\int_0^T\int_0^T R\left[u-u'\right]R\left[v-u'\right]\varphi\left(u-u'\right)\varphi\left(v-v'\right)dudvdu'$, which after the change of variables and changes the order of integration may be reduced to the form $\frac{2}{\pi^2T^2}\left\{T\int_0^T R[z]\varphi(z)dz\int_{z-T}^{T}R[t]\varphi(t)dt+\int_0^T R[z]\varphi(z)dz\int_{z-T}^{T-z}tR[t]\varphi(t)dt\right\}$, which shows, that attached to $T\to\infty$ the main contribution to the dispersion will add summand $\frac{1}{\pi^2T}\left(\int_{-\infty}^{+\infty}R[z]\varphi(z)dz\right)^2$. In case equality of indices: j and k, i and l, j and l.

3. Two pairs of indexes are equal, to wit, $j=l$, $i=k$, when considering summand of the expression (4) will have the form $\frac{1}{\pi^2T^2}\int_0^T\int_0^T R\left[u-u'\right]R\left[u-u'\right]\varphi\left(u-u'\right)\varphi\left(u-u'\right)dudu'$. Making the substitution and by changing the order of integration this the integral reduced to the form $\frac{1}{\pi^2T^2}T\int_{-T}^{T}\left(1-\frac{|z|}{T}\right)R^2[z]\varphi^2(z)dz$, in which $T\to\infty$ seeks to $\frac{1}{\pi^2T}\int_{-\infty}^{+\infty}R^2[z]\varphi^2(z)dz$.

Having considered also the other two expressions (4), after the conducted research it can be argued, that the dispersion statistic attracted to Q at $T\to\infty$ decreases as $\frac{1}{T}$.

5. The simulation modeling obtained estimated

All calculation and construction of the graph performed in the program Mathcad 2000. Now we present some results the simulation modeling obtained estimated with a shot discussion of the results. Below presented the result of modeling realization a stationary random process, which described by difference equation the first order

$$x(n)=0.8x(n-1)+\nu(n), \quad (6)$$

where $v(j)=\frac{1}{j}\sum_{i=1}^{j}\xi_i$, the values j - a random values, distribution by the Poisson law with arguments $\lambda=3$, ξ_i - normal distribution a random values with $M[\xi_i]=0$ and $D[\xi_i]=\sigma^2$.

As a first element generates realization $x(0)$ was taken the value 1.5. The modeling result of the realization a stationary random process, present on the figure 2.

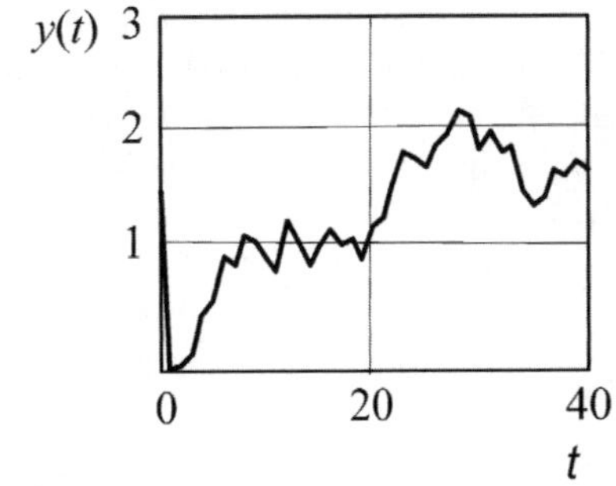

Figure 2. A model realization a stationary random process.

On the following pictures constructed estimation the spectrum of potency, the correlogram method and in the form of spline first order.

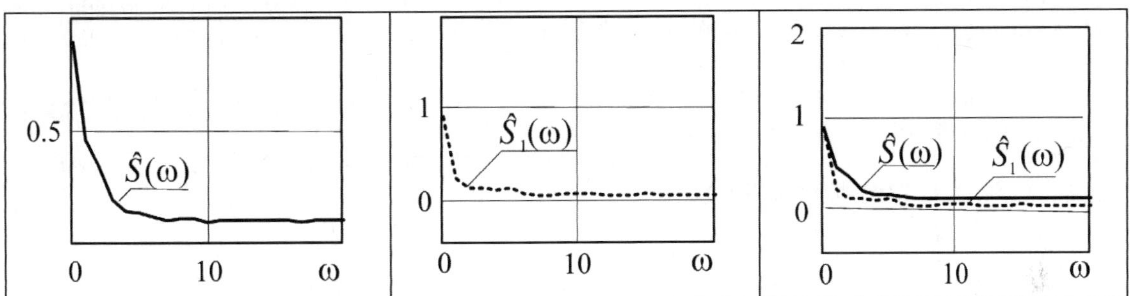

Figure 3. Presents estimated: $\hat{S}(\omega)$ the spectrum of potency $S(\omega)$, the correlogram method, $\hat{S}(\omega)$– in the form of spline first order for realization a random process figure 2

To evaluate the significance obtained estimates the spectrum of potency. For quality studies obtained a model, we use the Fisher – Snedecor. For this we will find the observed value of the criterion

$$F=\frac{\dfrac{\sum_{i=1}^{n}\left(\hat{S}_{1i}-\bar{S}\right)^2}{m-1}}{\dfrac{\sum_{i=1}^{n}\left(S_i-\hat{S}_{1i}\right)^2}{n-m}}$$

, here m –the numbers of estimated parameters, n – sample size, $\bar{S}$ – sample mean, $S_i, i=\overline{1,n}$ – values the spectrum for the model realization on a random process (6),which has the form: $S_i=\frac{0.09}{0.41-0.4\cos i}$ and $\hat{S}_{i1}$, $i=\overline{1,n}$ – evaluated values the spectrum of potency in the form of spline. The observed value of the criterion was equal to 23. 684, that to surpassed by far table value $F_{\alpha,m-1,n-m}$= 3.6 the criterion Fisher-Snedecor's at the significance level $\alpha=0.05$ and the number of degrees of freedom $k_1 = m - 1$ and $k_2 = n - m$. As $F > F_{\alpha,k_1,k_2}$, then obtained evaluation the spectrum of potency in the form of spline first order is important. The index of correlation

$$R_{\hat{S}\omega} = \sqrt{1 - \frac{\sum_{i=1}^{n}\left(S_i - \hat{S}_{1i}\right)^2}{\sum_{i=1}^{n}\left(S_i - \bar{S}\right)^2}}$$

=0.997 shows, that the estimates obtained the spectrum in the form of spline first order closely related with analytical form the spectrum a random process (6).

Thus, the spline evaluation right shows a form the spectrum of potency.

6. Conclusion

Was performed the calculation of the coefficient of spline first order in the annex to evaluation the spectrum of potency a stationary random process, when at each instant of time is a random number of measurement.

It is shows, that the obtained estimates are unbiased. It is established, that the dispersion of the estimates asymptotically behaves as $1/T$, where T – time of observation.

References

[1] Marple S Lawrence, Jr 1990 *Digital spectral analysis with applications* (Moscow: Mir) p 584
[2] Ventzhel E S, Ovcharov L A 2000 Theory of stochastic processes and its engineering applications (Moscow: Vysshaya shkola) p 383
[3] Romanenko A F, Sergeev G A 1968 *Issues applied analysis of random processes* (Moscow: Sovetskoe radio) p 256
[4] Krivosheev V I 2006 Modern methods of digital signal processing (digital spectral analysis) (Nizhniy Novgorod: NNGU) p 117
[5] Sergienko A B 2003 *Digital signal processing* (St. Peterburg: Piter) p 608
[6] Bendat Julius S, Piersol A 1989 *Applied analysis of random data* (Moscow: Mir) p 540
[7] Bendat Julius S 1965 Principles and applications of random noise theory (Moscow: Nauka) p 463
[8] Rabiner L, Gould B 1978 Theory and application of digital signal processing (Moscow: Mir) p 848
[9] Person B N J, Albohr O 2005 On the nature of surface roughness with application to contact mechanics, sealing, rubber friction and adhesion *J. of Physics: Condensed Matter* **17** 1 – 62
[10] Majumdar A, Bhushan B 1999 Characterization and modeling of surface roughness and contact mechanics. Handbook of Micro/Nanotribology. (CRC Press LLC) p 861
[11] Fedotov A 2010 The power spectrum as a characteristic of the surface roughness *J. Photonics* **6** 18–21
[12] Peitgen H O, Saupe D 1988 *The Sciens of Fractal Images* (New York: Springer-Verlag) p 312
[13] Flandrin P 1989 On the Spectrum of Fractional Brownian Motions *IEEE Transactions on Information Theory* **35** 197 – 199
[14] Mandelbrot B B, van Ness J W 1968 Fractional Brownian motions, fractional noises and applications *SIAM Review* **10** 422 – 437
[15] Berry M V 1979 Diffractals *J. Phys. A: Math. Gen.* **12** 781 – 797
[16] Bakhvalov N S 1975 *Numerical methods* (Moscow: Nauka) p 632
[17] Korneychuk N P 1984 *Splines in approximation theory* (Moscow: Nauka) p 352
[18] Shevelitskiy I V 2003 Interpolation splines in problems of digital signal processing *J. Exponenta Pro. Matematika v prilozheniyakh* **4** 42–53
[19] Terpugov A F, Konstantinova I G 2000 Estimation of the power spectrum of a stationary random process splines of the first order with a random number of measurements *J. Vestnik of Tomsk state University* **269** 85 – 89

Method of Electrolyte-Plasma Surface Hardening of 65G and 20GL Low-Alloy Steels Samples

Bauyrzhan Rakhadilov[1,a], **Laila Zhurerova**[1,b], **Alexander Pavlov**[2,c]

[1]D. Serikbaev East Kazakhstan state technical university, Serikbaev str. 19, 070010, Ust Kamenogorsk city, Kazakhstan
[2]S. Amanzholov East Kazakhstan state university, 30 Gvardeiskoi divisii str. 34, 070002, Ust Kamenogorsk city, Kazakhstan

E-mail: [a]bor1988@mail.ru, [b]leila_uka@mail.ru, [c]Alexandr_Pavlov_1988@mail.ru

Abstract. This work is devoted to formation of modified surface layers in 65G and 20GL steels which using for the manufacture of railway transport parts, as well as the study of influence of the parametersof electrolyte-plasma surface hardening methodon the changes in structural-phase states, improving of wear-resistance.
The process of electrolyte-plasma surface hardening of 65G and 20GL steels samples conducted in the electrolyte from water solution of 20% sodium carbonate, in the mode ~850°C – 2 seconds, ~1200°C - 3 seconds. It is established that in the initial state 20GL steel has ferrite-pearlite structure, and the 60G steel consists of pearlite and cement structure. After application of electrolyte-plasma surface hardening is observed the formation of carbides particles and martensite phase components in the structure of 20GL and 60G steels. It is determined that after electrolyte-plasma surface hardening with heating time - 2 seconds, the abrasive wear-resistance of 65G and 20GL steels increased to 1.3 times and 1.2 times, respectively, and the microhardness is increased to 1.6 times and 1.3 times, respectively.

Introduction.

In connection with the increasing requirements for durability and reliability of railway transport parts such as spring beams, carts, pivots and coupler of a freight car, and become a significant priority problem of surface hardening and improving the operating properties of these parts. For the manufacture of railway transportparts which is operating in conditions of dry friction, high contact and impact loads leading to wear, low-alloy steels are often using. Analysis of studies showed that after prolonged use of these parts, the wear of their working surfaces is accompanied by a reduction in operating properties, which in particular causes deterioration in the quality of manufactured parts or is connected with their frequent replacement [1, 2].

The traditional methods of hardening of railway transport parts areused for a long time and largely exhausted its possibilities. However, the improvement of operational performance of the working surface in conditions of increased wear, evaluates mainly to the structural phase changes, as well as improving mechanical and tribological properties of surface layer.

Recently in the industry resource-saving technologies of surface hardening in the low-temperature plasma is one of the most promisingmethods.This method give the ability to change the structure and properties of working surfaces and, as a consequence, to improve their operating properties.

Therefore, the aim of this work is to provide a method of electrolyte-plasma surface hardening (EPSH) for railway transportparts, which allows to obtaining a modified hardened surface layer with high wear-resistance.

Results and Discussion

The 65G and 20GL steelssampleswere used as a research material.The samples subjected to standard thermal process under the following modes for steels: 65G (quenching 830°C, oil, the release at 470 °C, air) and 20GL (quenching 880 - 900°C, the release at 600 - 650°C)and electrolyte-plasma quenching (~850°C - 2s, ~1200°C-3 s). The chemical composition of 65G and 20GL steels presented in table 1.

Table 1 – Chemical composition of 65G and 20GL steels

Steel grade	C	Mn	Si	Cr	Ni	Cu	S	P
65G	-	0.9 0-1.20	0.17- 0.37	~ 0.25	0.25	~ 0.20	~ 0.035	~ 0.035
20GL	0.15 - 0.25	1.2 - 1.6	0.2 - 0.4	-	-	-	~ 0.04	~ 0.04

Structural studies of 65G and 20GL steels samples conducted at the D. Serikbayev East Kazakhstan State Technical university, "Nanotechnology and new materials" Research Institute, by X-Ray diffraction method by XPertPRO X-Ray diffractometer in monochromatization CrKα radiation (λ=2.2897 Å), elemental surface analysis after electrolyte-plasma modification at the FEI XL 30 FEG scanning electron microscope, metallographic analysis on the NEOPHOT 21 and AXIOPHOT-2 optical microscopes, mechanical testing of microhardness was performed at the PMT-3M according to State standard 9450-76 and measurement of wear-resistance when rubbing against loosely fixed abrasive particles according to State standard 23.208-79. 65G and 20GL steels samples after mechanical grinding and polishing using diamond pastes, subjected to selective chemical etching to reveal the microstructure of steel surfaces [3].

Electrolyte-plasma surface hardening (EPSH) of the steels samples were carried out at the installation that are designed and manufactured at the "Nanotechnology and new materials" Research Institute [4]. The installation consists of a power source, electrolyte-plasma processing chamberand the personal computer [4, 5]. EPSH process of 65G and 20GL steels samples conducted in the electrolyte from water solution of 20% sodium carbonate in the following modes: supply voltage between anode and sample during heating to the quenching temperature – 320 V, the heating time - 2 and 3 seconds, and the samples were heated up to a temperature of ~850, ~1200°C, respectively.

Metallographic analysis showed that the 20GL steelsurface in the initial state has a ferritic-perlitic structure, 60G steel consists of pearlite and cementite (Fig.1-a, d). After EPSH for 2 swe observing the formation of the martensitic phase constituting in the structure of 65G and 20GL steels (Fig.1-b, e), with increasing heating time up to 3 s there is a coarsening of martensite grains.

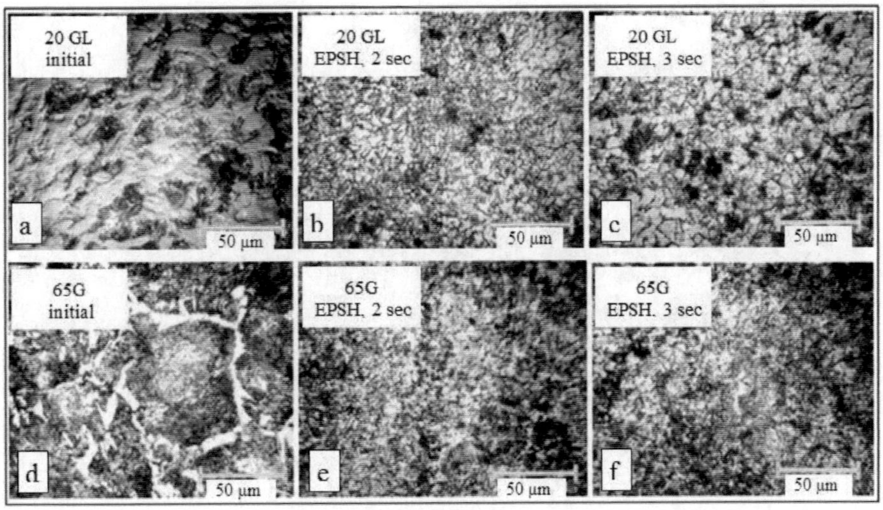

Figure 1 - Microstructure of 65G and 20GL steelsbefore and after EPSH
a, b, c - 20GL steel; d, e, f - 65G steel

Figure 2 shows images of 20GL and 65G steels microstructure in thecross section after electrolyte-plasma surface hardening with heating duration - 2 s. From figure 2 it is seen that the cross section structure of 20GL and 65G steels after electrolyte-plasma surface hardening is divided into 3 zones: on the surface there is a 1-zoneteletravail hardened layer with martensitic structure; 2-zone - layer thermal influence; 3-zone- matrix. The thickness of the modified layer of 65G and 20GL steelsafter electrolyte-plasma surface hardening is ~500-550 µm.

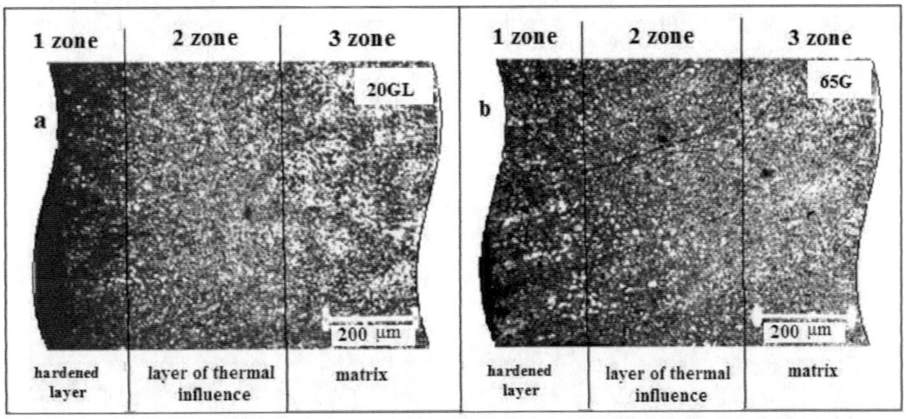

Figure 2 - Microstructure of the cross section of 65G and 20GL steels after EPSH

In order to study in detail changes in the morphology and structure of 65G and 20GL steels samples, electron-microscopic analysis of the surfacewas conducted. Figure 3 shows SEM-images of 20GL (Fig.3, d-f) and 65G (Fig.3, a-e) steel surfaces before and after EPSH. The structure of 65G and 20GL steels in the initial state consists of ferrite and lamellar pearlite, and cementite. After EPSH the formation of martensite at the boundaries is observed, which is introducing as a small particles. It is assumed that the detected fine particles are the carbides of alloying elements. The wear-resistance of 20GL steel is improved, possibly due to the formation of these small particles. Since, it is known [6] that the dispersed separation prevents the volume of grains relatively mild matrix from abrasion.

IOP Conf. Series: Materials Science and Engineering **142** (2016) 012028 doi:10.1088/1757-899X/142/1/012028

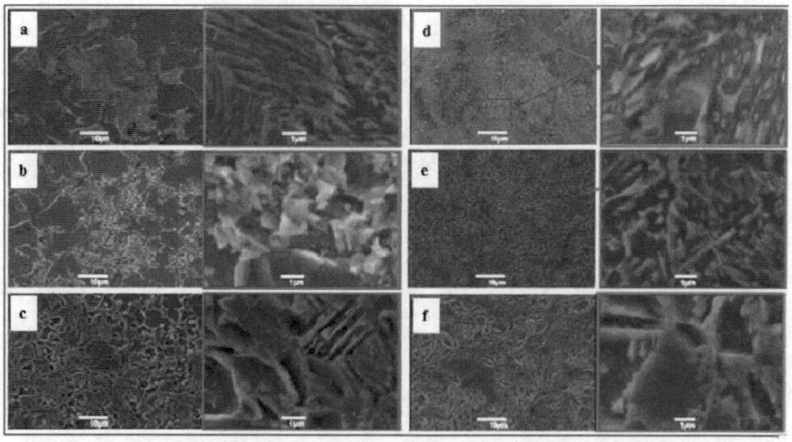

Figure 3 – SEM images of the surface of 65G and 20GL steels
a, d - before processing; b, e – after EPSH with heating time 2 sec;
c, f – after EPSH with heating time 3 sec

For the purpose of identification of change of element structure of a surface, the power dispersive analysis of a surface of samples of steel 65G and 20GL before and after EPSH has been carried out. In figures 4-5 results of the power dispersive analysis are shown. Analyses have shown that after EPSH with heating during 2 sec. on a surface of steel 20GL oxides are observed. Nevertheless, after EPSH considerable changes of element structure of a surface of 65G and 20GL steels isn't observed.

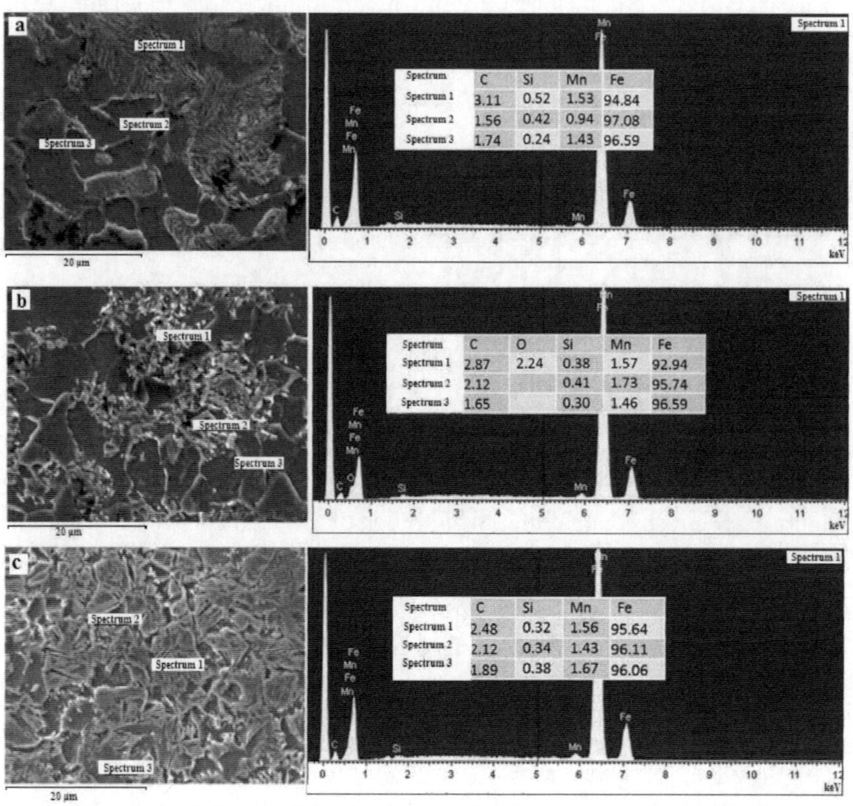

Figure 4 – Results of the SEM analysis of a surface of steel 20GL
a, b – after EPSH heating duration of 2 sec.; c– after EPSH heating duration of 3 sec

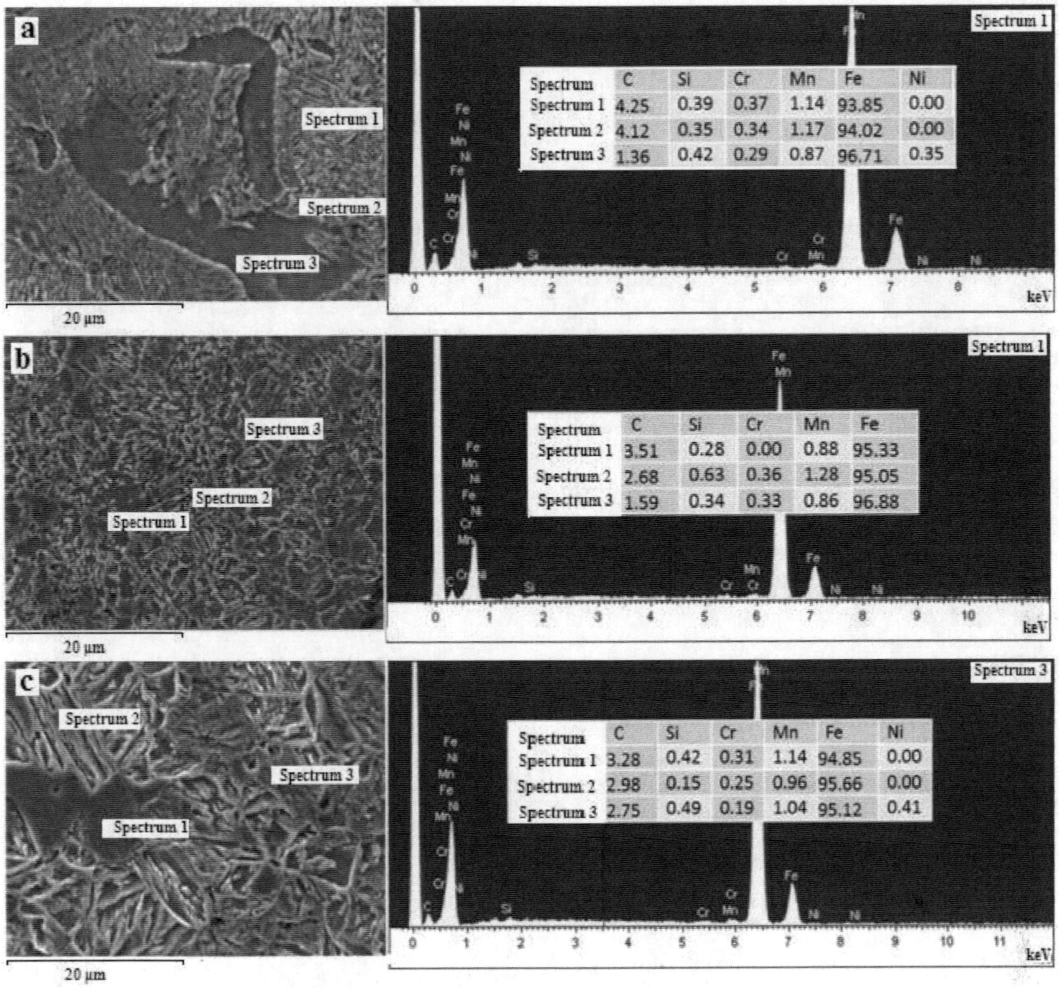

Figure 5 - Results of the SEM analysis of a surface of 65G steel
a, b – after EPSH heating duration of 2 sec.; c – after EPSH heating duration of 3 sec

Figure 6 shows the wear rate (mm^3/Nm) of the samples of 65G and 20GL steels before and after EPSH. Tests were conducted by the "ball-disk" scheme. It is seen that all the processed samples show a significant decrease in wear-rate in comparison to the original material. The wear-rate of the 65G and 20GL steels and after EPSH during 2 s is reduced to 30%, indicating a significant increase of steels wear-resistance.

Abrasive wear-resistance was evaluated by comparing the mass loss of 65G and 20GL steels samples before and after EPSH. The relative wear-resistance of the steel samples was determined by the formulas according to State standard [7]. As we can see from figure 6, mass loss of samples steels 65G and 20GL and after hardening EPSH less than the samplesin the initial state, indicating increased resistance to abrasive wear of 65G and 20GL steels after surface hardening. After EPSH the abrasive wear-resistance of 65G and 20GL steels increased to 1.3 times and 1.2 times, respectively.

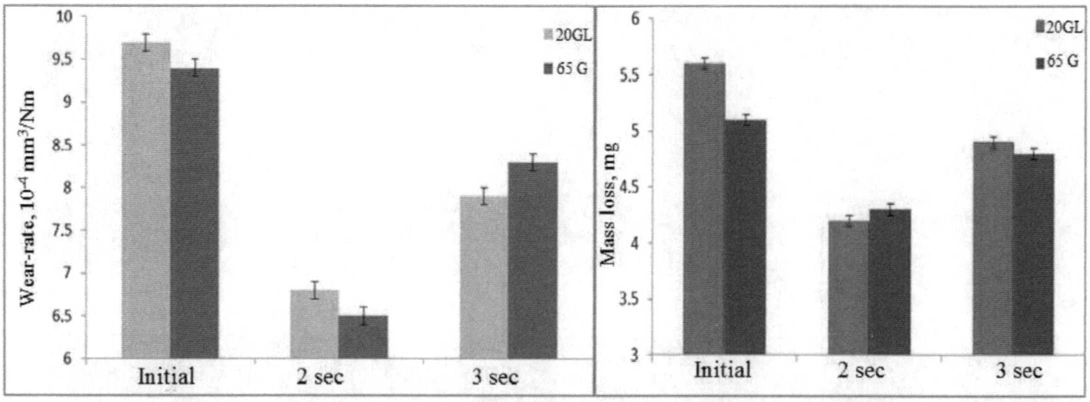

Figure 6 – Evaluation of wear-resistance of 65G and 20GL steels samples before and after EPSH

Figure 7 shows the dependence of the microhardness of 65G and 20GL steels from duration of electrolyte plasma processing. The microhardness of 65G and 20GL steels in the initial state is 2430 and 1690 MPa, respectively. It is established that microhardness of 20GL steel after EPSH (heating time of 2 s) was increased to 1.6 times, and the hardness of 65G steelafter EPSH (heating time of 2 s) was increased to 1.3 times, comparing to the initial state.

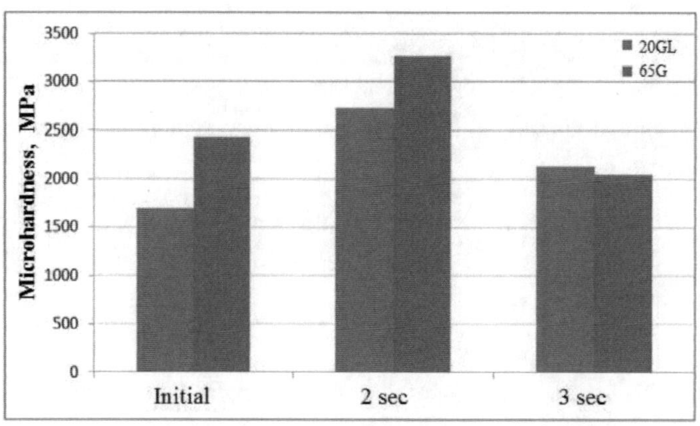

Figure 7 - Microhardness of 65G and 20GL steels

Conclusion

Thus, based on the analysis of the obtained experimental results of investigation of modified surface layers of 65G and 20GL steels after EPSH, we can draw the following conclusions:

– the EPSH methodis developed and optimal modes of surface hardening of low-alloy steels 65G and 20GL in plasma electrolyteis determined, which allows to obtain a modified surface layer with ~500-550 μm thickness and with improved performance characteristics;

– it is determined that the abrasive wear-resistance of 65G and 20GL steels increased to 1.3 times and 1.2 times, respectively;

– it is established that the microhardness of 65G and 20GL steels after EPSH (heating time - 2 s) was increased to 1.6 times and 1.3 times, respectively, depending to the initial state.

– it is revealed that after electrolyte-plasma surface hardening morphological structure of 20GL and 60G steels consists of martensite grains, with the small carbide particlesin the boundaries;

– it is discovered that the structure of the cross section of 65G and 20GL steelsafter EPSH has a zonal characteristic: on the surface there is a 1-zone teletravail hardened layer with martensitic structure; 2-zone – layer thermal influence; 3-zone - matrix.

Thus, studies have shown perceptivity and expediency of application of the developed method for improvement of operational properties of railway transportdetails, operating in conditions of friction and wear.

This work was financially supported by Committee of Science of MES RK by the “Development of resource-saving technologies of surface hardening of steel parts for railway transport” project under the contract № 63 of 12 February 2015.

References

[1] S. A. Turtaev, Regularities of structure formation in plasma surface processing of metal products and the development on this basis of resource-saving industrial technologies / Doctor of Sciences / National research technological University "MISIS". Moscow, 2013, p. 42.

[2] V. S. Nagovitsin, M. Y. Veshkurtsev, V. G. Shakhov, A. N. Golovash, Technical diagnostics of rolling stock // Herald of transport, 2002, № 5, pp. 19-22.

[3] L.V. Baranova, E. L. Demina, Metallographic etching of metals and alloys. Guide. – M.: Metallurgy, 1986, p. 256.

[4] M. K. Skakov, B. K. Rakhadilov, D. B. Zarva, V. A. Gulkin, Installation of electrolyte-plasma treatment. Kazakhstan Patent № 29978, 2015.

[5] B. K. Rakhadilov, M. N. Bekkaliev, A. J. Minasov, Z. H. Aringazina, M. K. Rakhadilov, S. R. Kozhanova, Experimental studies of the energy characteristics of electrolyte-plasma discharge // Bulletin EKSTU, 2015, № 1, pp. 62-68.

[6] Y. I. Ustinovshikov, The Allocation of the second phase in solid solutions. –M.: Nauka, 1988, p. 172.

[7] State standard 23.20879 Maintenance of wear-resistance of products. Method of testing materials for wear-resistance in friction against loosely fixed abrasive particles.

IOP Conf. Series: Materials Science and Engineering **142** (2016) 012029 doi:10.1088/1757-899X/142/1/012029

Elasticity of the $Sm_{1-x}Y_xS$ alloy Based on Ultrasonic Measurements

EG Soboleva[1,a], AL Igisheva[1,a] and KW Wojciechowski[2]

[1] Yurga Institute of Technology, Tomsk Polytechnic University, Leningradskaya st., 26, 652055, Yurga, Russia
[2]Institute of Molecular Physics, Polish Academy of Sciences, ul. Mariana Smoluchowskiego 17, 60-179, Poznan, Poland

[a]email: sobolevaeno@mail.ru

Abstract. The elastic moduli, sound velocities, Grüneisen parameter, Poisson's ratios and brittleness-plasticity criterion ratios are studied for the $Sm_{1-x}Y_xS$ alloys. Their dependence on the concentration of alloy components including a valence transition from semiconductors into the metal phase is presented. Auxeticity (negative Poisson's ratio) is found for some concentrations.

1. Introduction

In recent years there has been increasing interest observed in studies of materials with negative Poisson's ratio ($\sigma< 0$) which are called "auxetics" [1]. Scientists have studied mechanisms and have established criteria of negative Poisson's ratios appearance in isotropic and anisotropic solids [1,2]. The extreme values $\sigma = 0.5$ and $\sigma = -1$ for ideal elastic continuous media were analyzed from the point of view of physical acoustics as its methods were widely used in experimental studies of elastic properties of various materials [3]. Previously we completed a brief analysis of sound velocities relations in the context of limit values of Poisson's ratios for 94 elastically isotropic elements and compounds including four auxetics [4]. In a number of cases there was a disagreement in the values of Poisson's ratios for one and the same substance when calculating with various relations of sound velocities although the initial formulas of elasticity theory for σ through elastic moduli were equivalent. We used the data for elastic parameters of polycrystals obtained from rigidity constants in accordance with constants of monocrystals in Voigt-Royce-Hill approximation [5].

In this work we study the interrelation of sound velocities and Poisson's ratios relations by the example of only one system $Sm_{1-x}Y_xS$ but in more detail than in [4]. The choice of this object is determined by the fact that constant c_{12} according to experimental results [6] takes positive or negative values depending on yttrium concentration in samarium sulfide. The given fact allows one for more unambiguous treatment of the limits of sound velocities relations when the hybrid system transits into the auxetic state. Besides, $\sigma < 0$ in $Sm_{1-x}Y_xS$ is registered near valence transition $Sm^{2+} \rightarrow Sm^{3+}$ [7] and study of materials with the intermediate valence in relation to their simultaneous demonstration of anomalous properties is of special interest.

2. Calculations

The known formulas for the Poisson's ratios of isotropic bodies [3, 8] could be applied to approximate the formulas for σ expressed through the sound velocities:

$$\sigma_x = \frac{x^2 - 2}{2x^2 - 2}, \quad x = \frac{\upsilon_L}{\upsilon_t} \tag{1}$$

$$\sigma_y = 0.5y^2 - 1, \quad y = \frac{\upsilon_\ell}{\upsilon_t} \tag{2}$$

$$\sigma_{xy} = 0.5 - \frac{y^2}{6x^2 - 8} \tag{3}$$

$$\sigma_{z_{1,2}} = \frac{-\left(z^2 - 1\right) \pm \sqrt{\left(z^2 - 1\right)^2 + 8z^2\left(z^2 - 1\right)}}{4z^2}, \quad z = \frac{\upsilon_L}{\upsilon_\ell}, \tag{4}$$

where υ_t – velocity of propagation for transverse elastic waves, υ_ℓ – velocity of longitudinal waves propagation in the plug, υ_L – velocity of propagation longitudinal elastic waves.

However, no experimental values are known for elastic moduli of $Sm_{1-x}Y_xS$ polycrystals and sound velocities in them. Therefore in the present work we provide calculations of bulk and shear moduli based on rigidity constants c_{11}, c_{12} and c_{44} of this system cubic monocrystals measured by the pulse ultrasonic method [6]. When establishing the given moduli we applied the approximations of Voigt (B_V, G_V), Royce (B_R, G_R), Voigt-Royce-Hill (B_{VRH}, G_{VRH}) [5], G. Peresada (G_{Per}) [9], K.S. Alexandrov (G_{Al}) [10]. Poisson's ratios along the specific crystallographic directions, $\sigma_{<hk\ell>}$ were calculated according to the formulas presented in the references [2, 6].

Velocities of purely longitudinal and traverse waves propagation for isotropic elastic bodies and in three specific directions of cubic monocrystals of hybrid system $Sm_{1-x}Y_xS$ were found through the known relations of elasticity theory and physical acoustics [5, 8, 11, 12].

Grüneisen parameter γ, a nonharmonic measure of interatomic oscillations and non-linearity of interatomic interrelation forces, was calculated according to the previously established formula [12]:

$$\gamma = \frac{3}{2}\left(\frac{3\upsilon_L^2 - 4\upsilon_t^2}{\upsilon_L^2 + 2\upsilon_t^2}\right). \tag{5}$$

3. Results and their discussion

Changes of velocities of elastic waves propagation in crystallographic directions <100>, <110> and <111> of cubic monocrystals $Sm_{1-x}Y_xS$ depending on their composition are presented in Figure 1 (a, b, c). The obtained data provide us with a number of basic conclusions: functions υ_L (x), υ_t (x) are linear in both phases of the studied system; dependence υ_ℓ (x) is linear only in semiconductor phase $Sm_{1-x}Y_xS$; with the growth of yttrium concentration at the initial stage ($0 < x \le x_c$, where x_c – critical concentration) velocities of traverse waves propagation decrease and those of longitudinal waves increase; acoustic anisotropy is qualitatively equal for longitudinal waves ($\upsilon_{L\langle 100\rangle} > \upsilon_{L\langle 110\rangle} > \upsilon_{L\langle 111\rangle}$, $\upsilon_{\ell\langle 100\rangle} > \upsilon_{\ell\langle 110\rangle} > \upsilon_{\ell\langle 111\rangle}$) and is different for traverse waves ($\upsilon_{t_2\langle 110\rangle} > \upsilon_{t\langle 111\rangle} > \upsilon_{t\langle 100\rangle} = \upsilon_{t_1\langle 110\rangle}$); at the isostructural transition all sound velocities change abruptly (in a discontinuous way) but differently in terms of quality and quantity: velocities of longitudinal waves decrease, velocities of traverse waves increase. The data on the mentioned fact are given in Table 1 (here "+"corresponds to the discountinuous growth of sound velocity at the alloy transition from the semiconductor phase into the metallic one and "–"corresponds to the discountinuous decrease).

IOP Conf. Series: Materials Science and Engineering **142** (2016) 012029 doi:10.1088/1757-899X/142/1/012029

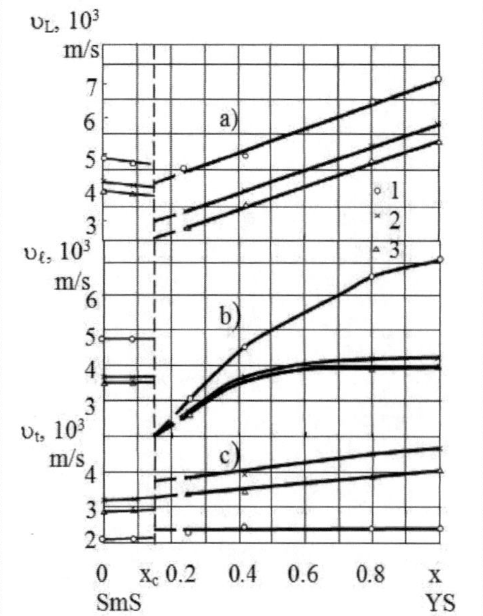

Figure 1. Concentration dependence of propagation velocities for longitudinal (υ_L, υ_ℓ) and traverse (υ_t) elastic waves in specific directions of $Sm_{1-x}Y_xS$ monocrystals.

1 – <100>, 2 – <110>, 3 – <111>, x_c – critical concentration, $\upsilon_{t_1\langle 110\rangle} = \upsilon_{t\langle 100\rangle}$.

Table 1. Abrupt changes of sound velocities and their relations at valence transition in the $Sm_{0.85}Y_{0.15}S$ (%) alloy

Direction in the monocrystal	$\Delta\upsilon_L/\upsilon_L$	$\Delta\upsilon_\ell/\upsilon_\ell$	$\Delta\upsilon_t/\upsilon_t$	$\frac{\Delta(\upsilon_L/\upsilon_t)}{(\upsilon_L/\upsilon_t)}$	$\frac{\Delta(\upsilon_\ell/\upsilon_t)}{(\upsilon_\ell/\upsilon_t)}$	$\frac{\Delta(\upsilon_L/\upsilon_\ell)}{(\upsilon_L/\upsilon_\ell)}$
polycrystal	-22.5	-45.2	+6.7	-29.0	-50.0	+26.5
<100>	-12.8	-57.4	+8.3	-4.6	-53.6	+42.9
<110>	-25.0	-45.2	+11.6**	-13.8*	-45.5*	+23.9
				-35.4**	-31.8**	
<111>	-32.0	-42.9	+9.8	-40.0	-47.4	+14.3

Note: * $\upsilon_t = \upsilon_{t_1}$, ** $\upsilon_t = \upsilon_{t_2}$, signs "+","–" see explanations in the text

Relations of sound velocities υ_L/υ_t, υ_ℓ/υ_t and υ_L/υ_ℓ for specific directions in the monocrystal and for the polycrystal depending upon the composition of the mixed system are presented in Figure 2 (a, b, c). Concentration dependences of these parameters are different from functions υ (x) in Figure 1 and can be reduced to the following: in the semiconductor phase of the $Sm_{1-x}Y_xS$ alloy (x < 0,15) with the growth of yttrium concentration all relations of sound velocities decrease linearly; in the metallic phase these relations are non-linear with the change of x, and functions υ_L/υ_ℓ (x) have minimums near x ≈ 0.4 for three directions in the monocrystal and for the $Sm_{1-x}Y_xS$ polycrystal; values of sound velocities relations in the polycrystal take the intermediate position between the values of corresponding relations in the monocrystal; during the transition from the semiconductor into metallic phase ($x = x_c$) υ_L/υ_t and υ_ℓ/υ_t decrease abruptly and υ_L/υ_ℓ increases. The values of changes of these relations are shown in Table 1 and approximately amount to 0.5% - 50%; the important condition for the acoustic problem of auxetics: relations $\left(\upsilon_L/\upsilon_{t_2}\right)_{\langle 110\rangle}, \left(\upsilon_L/\upsilon_t\right)_{\langle 111\rangle}$ in the interval of concentration values $0.15 < x < 0.40$ become less than one; $\left(\upsilon_\ell/\upsilon_{t_2}\right)_{\langle 110\rangle} < 1$ in the whole metallic phase of the alloy and in the pure yttrium sulfide (YS); the curve $\left(\upsilon_\ell/\upsilon_t\right)_{\langle 111\rangle}(x)$ almost repeats the curve $\left(\upsilon_\ell/\upsilon_{t_2}\right)_{\langle 110\rangle}(x)$ for $x > x_c$ and only slightly exceeds the one in the interval $0.42 < x < 0.80$; near the valence transition from the metallic phase and other sound velocities relations (υ_ℓ/υ_t) decrease to values lower than one.

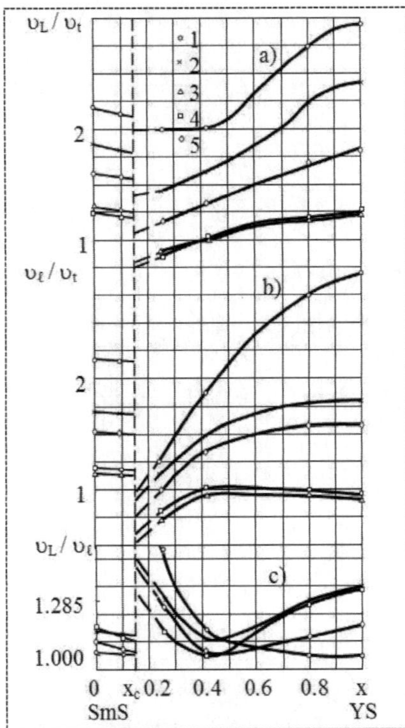

Figure 2. Relations of propagation velocities of longitudinal and traverse elastic waves depending upon the $Sm_{1-x}Y_xS$ alloy composition. 1, 2, 3, 4 – monocrystal, 5 – polycrystal. Directions in the monocrystal: 1 - <100>, 2 – <110.001>, 3 – <110.1$\bar{1}$0>, 4 – <111>

In Table 2 we present anisotropic Young's moduli, shear moduli, Poisson's ratios of $Sm_{1-x}Y_xS$ alloy monocrystals and also jumps of the mentioned elastic parameters at the valence transition from the semiconductor into the metallic phase ($x = x_c$). Anisotropy of modules $E_{\langle 100\rangle} > E_{\langle 110\rangle} > E_{\langle 111\rangle}$, $G_{\langle 100\rangle} < G_{\langle 110\rangle} < G_{\langle 111\rangle}$ is retained for all compositions of the studied alloy monocrystals including the pure primary components. The established inequalities between the elastic moduli are characteristic of cubic ion monocrystals with lattices of the NaCl type. In a number of cases anisotropy of elastic properties is characterized by relation of Young's moduli $E_{\langle 100\rangle}/E_{\langle 110\rangle}$ for two crystallographic directions. If we analyze results of Table 2 from this point of view the relation of the given moduli will make: SmS ($E_{<100>}/E_{<110>} = 1.62$), $Sm_{0.91}Y_{0.09}$ ($E_{<100>}/E_{<110>} = 1.62$), $Sm_{0.75}Y_{0.25}$ ($E_{<100>}/E_{<110>} = 1.29$), $Sm_{0.58}Y_{0.42}S$ ($E_{<100>}/E_{<110>} = 1.54$), $Sm_{0.80}Y_{0.20}S$ ($E_{<100>}/E_{<110>} = 2.36$), YS ($E_{<100>}/ E_{<110>} = 2.63$). This way, we can state that compositions of $Sm_{0.75}Y_{0.25}S$ and $Sm_{0.58}Y_{0.42}S$ alloys with negative Poisson's ratios have decreased anisotropy of elastic properties.

IOP Conf. Series: Materials Science and Engineering **142** (2016) 012029 doi:10.1088/1757-899X/142/1/012029

Anisotropic Poisson's ratios of $Sm_{1-x}Y_xS$ monocrystals (Table 2) fulfil inequality characteristic for cubic monocrystals (except for $Sm_{0.75}Y_{0.25}S$ and $Sm_{0.58}Y_{0.42}S$): $\sigma_{\langle 110.001\rangle} < \sigma_{\langle 100\rangle} < \sigma_{\langle 111\rangle} < \sigma_{\langle 110.1\bar{1}0\rangle}$. Alloy compositions with yttrium concentration of $0.15 \le x \le 0.29$ have negative Poisson's ratios in all three specific crystallographic directions. When extrapolating the results for the metallic phase, minimum values of Poisson's ratios are observed at critical yttrium concentration ($x = x_c$) and Poisson's ratio equals $\sigma_{\langle 100\rangle}(x_c) = -0.98$. for direction <100>.

Table 2. Anisotropic Young's, shear (GPa) moduli, Poisson's ratios and their relative jumps (%) at the valence transition in the monocrystal of the $Sm_{1-x}Y_xS$ alloy

	$E_{<100>}$	$E_{<110>}$	$E_{<111>}$	$G_{<100>}$	$G_{<110>}$	$G_{<111>}$	$\sigma_{<100>}$	$\sigma_{<110.001>}$	$\sigma_{\langle 110.1\bar{1}0\rangle}$	$\sigma_{<111>}$
YS	240.62	91.54	75.87	27.70	44.32	55.41	0.086	0,033	0.652	0.457
SmS	124.92	77.22	68.50	26.90	38.61	41.69	0.086	0,053	0.435	0.273
$Sm_{0.91}Y_{0.09}S$	126.61	78.57	69.74	28.00	38.38	43.80	0.038	0,024	0.403	0.245
$Sm_{0.75}Y_{0.25}S$	58.55	45.28	42.10	32.00	47.07	55.84	-0.671	-0,519	-0.293	-0.342
$Sm_{0.58}Y_{0.42}S$	123.23	80.15	71.79	34.70	49.76	58.18	-0.299	-0,195	0.154	0.034
$Sm_{0.20}Y_{0.80}S$	218.15	92.48	77.59	29.00	45.40	55.95	0.044	0,019	0.595	0.338
$Sm_{0.85}Y_{0.15}S$		ΔE/E			ΔG/G				Δσ/σ	
	-82.4	-78.1	-83.3	-7.0	+14.5	+16.7	-19800	-3750	-247.5	-385.7

Table 3 presents density (ρ), elastic (B, G, E, σ) properties, acoustic (υ_L, υ_ℓ, υ_t, $\bar{\upsilon}_{sq}$, $\bar{\upsilon}$) properties, plasticity-fragility criterion (B/G) and Grüneisen parameter (γ) of $Sm_{1-x}Y_xS$ alloy polycrystals. Here we provide values and signs of relative jumps of all Table 3 parameters at the critical concentration of alloy components ($Sm_{0.85}Y_{0.15}S$). All alloy compositions are characterized by greater resistance to unilateral pressure deformation in comparison to uniform compression (E > B), and sound velocities make "regular" string $\upsilon_L > \upsilon_\ell > \upsilon_t$ and $\bar{\upsilon}_{sq} > \bar{\upsilon}$. SmS is more fragile than YS but at the same time it can be easily turned into an "absolutely fragile" alloy (B/G → 0) with a comparatively small addition of the second component (x = 0.15). This condition is likely to prevent us from studying the mechanical properties of the $Sm_{0.85}Y_{0.15}S$ alloy close to the critical point. Grüneisen parameter depending upon the alloy composition demonstrates behavior similar to function B/G (x) and at critical concentration equals γ (0.15) ≈ 0. If we take into consideration the fact that at the electronic phase transition Poisson's ratio is close to the limit value σ = -0.82, we observe a demonstration of three important factors: one of the lowest known Poisson's ratio values, record of absolute fragility and the almost complete harmonization of the interatomic oscillations. The given combination requires further understanding. Abrupt changes of the elastic and acoustic parameters of the $Sm_{0.85}Y_{0.15}S$ alloy polycrystal presented in Table 3 are quite large and can be related to similar changes for the monocrystal of the same alloy composition.

Values of sound velocities and Poisson's ratios relations obtained by formulas (1) – (4) for elastically isotropic polycrystals $Sm_{1-x}Y_xS$ are given in Table 4. From this table we can clearly see that for positive σ_x, σ_y or σ_{xy} a positive value of the σ_{z_1} root is obtained and, vice versa, when σ_x, σ_y and σ_{xy} are negative a negative value of σ_{z_2} is obtained.

Table 3. Density, elastic moduli and their relation, Poisson's ratio, sound velocities, Grüneisen parameter of the $Sm_{1-x}Y_xS$ polycrystal and changes of the given characteristics at the critical concentration of components (%)

	ρ, 10^3 kg·m^{-3}	B (GPa)	G (GPa)	E (GPa)	σ	υ_L (m·s^{-1})	υ_ℓ (m·s^{-1})	υ_t (m·s^{-1})	$\bar{\upsilon}_{sq}$ (m·s^{-1})	$\bar{\upsilon}$ (m·s^{-1})	B/G	γ
YS	4.830	96.95	48.88	125.54	0.284	5794	5098	3181	4235	3546	1.983	1.679
SmS	5.690	50.33	36.50	88.18	0.208	4171	3937	2533	3174	2800	1.379	1.316
$Sm_{0.91}Y_{0.09}S$	5.780	45.67	38.28	89.76	0.172	4090	3941	2573	3161	2832	1.193	1.186
$Sm_{0.75}Y_{0.25}S$	6.090	8.33	48.05	49.32	-0.487	3448	2846	2809	3037	2969	0.173	0.222
$Sm_{0.58}Y_{0.42}S$	5.820	25.68	50.36	91.36	-0.093	3994	3962	2942	3330	3169	0.510	0.597
$Sm_{0.20}Y_{0.80}S$	5.183	79.67	48.88	121.74	0.245	5286	4846	3071	3950	3407	1.630	1.477
$Sm_{0.85}Y_{0.15}S$	+8.2	-100.0	+13.8	-92.2	-647.6	-21.9	-43.8	+6.7	-7.3	+2.6	-100.0	-78.2

Table 4. Relations of sound velocities and calculated relations of Poisson's ratios with their application by formulas (1) – (4) for the polycrystals of the $Sm_{1-x}Y_xS$ system

	x*	y	z	σ_x	σ_y	σ_{xy}	σ_{z_1}	σ_{z_2}
YS	1.8214	1.6026	1.1365	0.284	0.284	0.284	0.284	-0.397
SmS	1.6467	1.5543	1.0594	0.208	0.208	0.208	0.208	-0.262
$Sm_{0.91}Y_{0.09}S$	1.5896	1.5317	1.0378	0.173	0.173	0.173	0.172	-0.208
$Sm_{0.75}Y_{0.25}S$	1.2275	1.0132	1.2115	-0.487	-0.487	-0.487	0.327	-0.487
$Sm_{0.58}Y_{0.42}S$	1.3576	1.3467	1.0081	-0.093	-0.093	-0.093	0.085	-0.093
$Sm_{0.20}Y_{0.80}S$	1.7213	1.5780	1.0908	0.245	0.245	0.245	0.245	-0.325

Note: * sign $x = \upsilon_L/\upsilon_t$

4. Conclusions

1. Anisotropy of sound velocities has been determined for pure components and various compositions of alloy $Sm_{1-x}Y_xS$ monocrystals: $\upsilon_{L\langle 100\rangle} > \upsilon_{L\langle 110\rangle} > \upsilon_{L\langle 111\rangle}$, $\upsilon_{\ell\langle 100\rangle} > \upsilon_{\ell\langle 110\rangle} > \upsilon_{\ell\langle 111\rangle}$, $\upsilon_{t\langle 100\rangle} < \upsilon_{t\langle 111\rangle} < \upsilon_{t\langle 110\rangle}$. Anisotropy of longitudinal wave propagation velocities is reflected in the anisotropy of Young's modulus and that of traverse waves – in the shear modulus anisotropy.
2. The concentration dependences of sound velocities, their ratios, elastic moduli and Poisson's ratios for mono- and polycrystals of the $Sm_{1-x}Y_xS$ alloy were studied. All these characteristics change their values abruptly (in a discontinuous way) at the critical concentration of yttrium sulfide in samarium sulfide $x_c = 0.15$ (transition from the semiconductor phase of the alloy into the metallic one). Significant changes are demonstrated by Poisson's ratios (for example, $\Delta\sigma/\sigma_{<100>} \approx 2\cdot 10^4$ %).
3. It has been confirmed that the $Sm_{1-x}Y_xS$ alloys with concentration of components $0.15 < x < 0.75$ are partial auxetics ($\sigma_{<100>}$ of the monocrystal < 0), and in the narrower interval of concentrations ($0.15 < x < 0.50$) are (complete) auxetics (σ of the polycrystal < 0).

References

[1] Alderson KL, Alderson A, Grima JN, Wojciechowski KW 2014 Phys. Status Solidi (b) 250 263.
[2] Goldstein RV, Gorodtsov VA, Lisovenko DS 2013 Phys. Status Solidi (b) 250 2038.
[3] Lipsett AW, Beltzer AI 1988 J. Acoust. Soc. Am. 84 2179.
[4] Belomestnykh VN, Soboleva EG 2012 7th International Forum on Strategic Technology (IFOST - 2012) 1 499.
[5] Tesleva E, Belkova T 2014 Applied Mechanics and Materials 682 519.
[6] Hailing Tu, Saunders GA and other 1984 J. Phys. C: Solid State Phys. 17 4559.
[7] Belomestnykh VN, Tesleva EP 2012 7th International Forum on Strategic Technology (IFOST - 2012) 2 381.
[8] Frantsevich IN, Voronov FF and Bakuta SA 1982 Elastic constants and elastic moduli of metals and non-metals Kiev 286.
[9] Peresada GL 1971 Phys. Status Solidi A4 K 23.
[10] Alexandrov KS 1967 Reports of Academy of Sciences USSR 176 295.
[11] Soboleva EG, Igisheva AL and Krit TB 2015 IOP Conf. Series: Materials Science and Engineering 91 1-7
[12] Belomestnykh VN Tesleva EP and Soboleva EG 2008 Technical Physics Letters 34 867.

Methodology of strength calculation under alternating stresses using the diagram of limiting amplitudes

V V Konovodov[1], A V Valentov[2], I S Kukhar [1], O Yu Retyunskiy[2], A S Baraksanov[2]

[1] Novosibirsk State Agrarian University
147 Nikitin St, 630039 Novosibirsk, Russia; phone: (383) 267-35-07
[2] Yurga Institute of Technology, affiliate of Tomsk Polytechnic University
26 Leningradskaya St, 652050 Yurga, Russia; phone: (38451)-6-05-37

E-mail: valentov@mail.ru

Abstract. The work proposes the algorithm to calculate strength under alternating stresses using the developed methodology of building the diagram of limiting stresses. The overall safety factor is defined by the suggested formula.
Strength calculations of components working under alternating stresses in the great majority of cases are conducted as the checking ones.
It is primarily explained by the fact that the overall fatigue strength reduction factor ($K_{\sigma g}$ or $K_{\tau g}$) can only be chosen approximately during the component design as the engineer at this stage of work has just the approximate idea on the component size and shape.

1. Object of research

Let us consider defining safety factor under uniaxial stress state and under pure shear.

Pure shear occurs in the points of bar of circular cross section working in torsion. In most cases safety factor is defined by the suggestion that working stress cycle occurring in the component under study while its operation is similar to the limit cycle. It means that skewness coefficients R and parameters ρ of working and limit cycles are the same [1].

Safety factor can be most simply defined in case of symmetric cycle of stress variation, as fatigue endurance limits of material by these cycles are usually known. In cases with lack of experimental data on fatigue endurance limits of materials one can apply empirical relations between fatigue endurance limits by symmetric cycles and ultimate strength limit (σ_β) under tension.

Safety factor is a ratio of fatigue endurance limit defined for a component to the reference value of the maximum stress [1,2], occurring in the component critical point. Reference value is the value of stress defined by the basic formulas of materials strength, i.e. not considering the factors influencing the value of fatigue endurance limit (stress concentration, etc.).

Methods:

Thus, to define safety factor under symmetric cycles we obtain the following dependencies:
under bending:

IOP Conf. Series: Materials Science and Engineering **142** (2016) 012030 doi:10.1088/1757-899X/142/1/012030

$$n = \frac{\sigma_{-1g}}{\sigma_{\max}} = \frac{\sigma_{-1}}{K\sigma_g \cdot \sigma_{\max}}; \quad (1)$$

under tension - compression:

$$n = \frac{\sigma_{-1pg}}{\sigma_{\max}} = \frac{\sigma_{-1p}}{K\sigma_g \cdot \sigma_{\max}}; \quad (2)$$

under torsion:

$$n = \frac{\tau_{-1g}}{\tau_{\max}} = \frac{\tau_{-1}}{K\tau_g \cdot \tau_{\max}} \quad (3)$$

There are several ways to build schematized diagram of limiting amplitudes [4,11]. However we need additional experimental data for that purpose. The only way where one can deal without the mentioned data is given in Fig. 1.

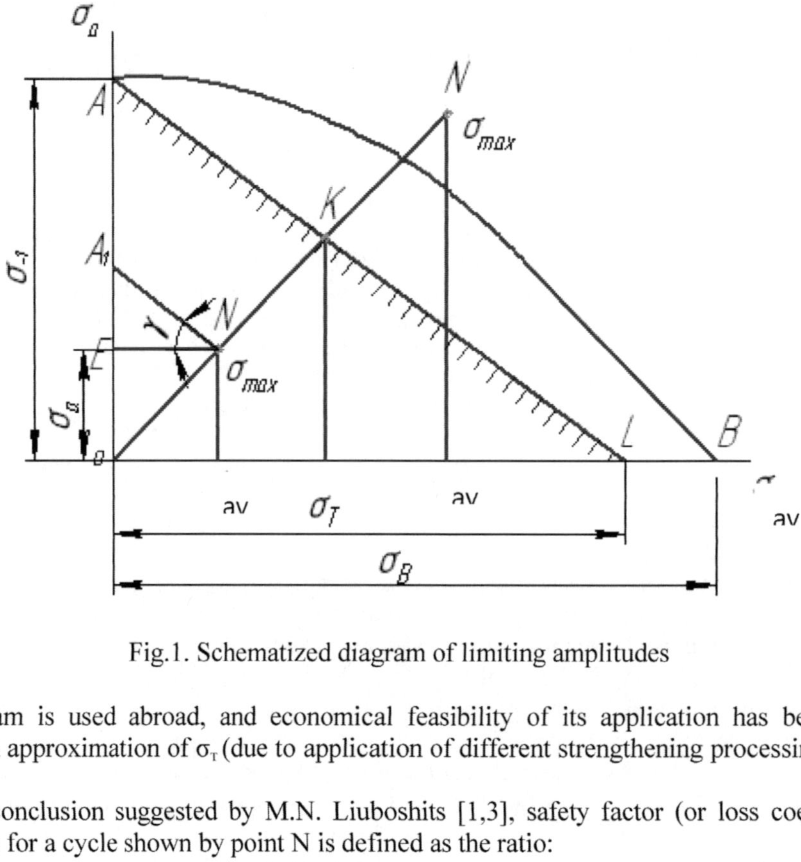

Fig.1. Schematized diagram of limiting amplitudes

Such diagram is used abroad, and economical feasibility of its application has been recently increasing with approximation of σ_T (due to application of different strengthening processing methods) to σ_B.

Using the conclusion suggested by M.N. Liuboshits [1,3], safety factor (or loss coefficient) by fatigue fracture for a cycle shown by point N is defined as the ratio:

$$n = \frac{OK}{ON} = \frac{OA}{OA_1} \quad (4)$$

Following from Fig.1:

$$OA_1 = OE + EA_1 = \sigma_a + \sigma_{av} \cdot tg\gamma = \sigma_a + \psi_\sigma \cdot \sigma_{av},$$

where $tg\gamma = \psi_\sigma$.

Then we put the obtained values of OA and OA_1 into the ratio (4):

$$n = \frac{\sigma_{-1}}{\sigma_a + \psi_\sigma \cdot \sigma_{av}} \quad (5)$$

The similar case is with variables of shear stresses:

$$n = \frac{\tau_{-1}}{\tau_a + \psi_\sigma \cdot \tau_{av}} \quad (6)$$

The values ψ_σ and ψ_τ depend on the accepted calculation of schematized diagram of limiting amplitudes and on the component material.

To define safety factor for a definite component we need to take into account the influence of fatigue strength reduction factor $K\sigma_g$ ($K\tau_g$).

According to scientists [2,3,5,8] stress concentration, size effect and surface condition influence only the values of limiting amplitudes and show almost no influence on the values of limiting mean stresses. Thus, as accepted in practice of calculations fatigue strength reduction factor is referred only to amplitude cycle stress. Then the final formulas to define the safety factor by fatigue fracture can be as follows:

under bending:

$$n = \frac{\sigma_{-1}}{K\sigma_g \cdot \sigma_a + \psi_\sigma \cdot \sigma_{av}} \quad (7)$$

under torsion:

$$n = \frac{\tau_{-1}}{K\tau_g \cdot \tau_a + \psi_\tau \cdot \tau_{av}}. \quad (8)$$

Under tension-compression we should use the formula (7), but instead of σ_{-1} one should put fatigue endurance limit σ_{-1p} under symmetric cycle tension-compression.

Formulas (7) and (8) are true in all ways of schematization of diagram of limiting amplitudes, only the values of coefficients ψ_σ (ψ_τ) are changed.

Formula (7) is obtained for cycles with positive mean stresses ($\sigma_{cp} \geq 0$): for cycles with negative (compressive) mean stresses ($\sigma_{cp} < 0$) it should be assumed that $\psi_\sigma = 0$, i.e. proceed on the assumption that in compression zone the line of limiting stresses is parallel to X-axis.

Along with safety factor by fatigue fracture one should define safety factor by fluctuating stresses.

Under bending (or under tension-compression):

$$n_{\sigma_T} \frac{\sigma_T}{\sigma_{max}} = \frac{\sigma_T}{\sigma_a + \sigma_{av}} \quad (9)$$

under torsion:

$$n_{\tau_T} = \frac{\tau_T}{\tau_{max}} = \frac{\tau_T}{\tau_a + \tau_{av}} \quad (10)$$

The lowest of the safety factors defined by the formulas (7), (8) or (9), (10) should be accepted as the reference one.

In some simple cases it is possible to calculate fatigue resistance by admissible stress $[\sigma_R]$, corresponding to the specified characteristic of a cycle (ρ_σ or ρ_τ) [6,7]. Let us derive a formula for the admissible normal stress by the cycle with the characteristic ρ_σ. Suggesting that in formula (7) n=[n], $\sigma_a = [\sigma_a]$; $\sigma_{av} = [\sigma_{av}]$, we obtain:

$$[n] = \frac{\sigma_{-1}}{K\sigma_g[\sigma_a] + \psi_\sigma[\sigma_{cp}]} = \frac{\sigma_{-1}}{[\sigma_a](K\sigma_g + \psi_\sigma \frac{[\sigma_{av}]}{[\sigma_a]})} = \frac{\sigma_{-1}}{[\sigma_{av}](K\sigma_g \frac{[\sigma_a]}{[\sigma_{av}]} + \psi_\sigma)}$$

Thus we get:

$$[\sigma_a] = \frac{\sigma_{-1}\rho_\sigma}{[n](K\sigma_g \cdot \rho_\sigma + \psi_\sigma)},$$

$$[\sigma_{cp}] = \frac{\sigma_{-1}}{[n](K\sigma_g \cdot \rho_\sigma + \psi_\sigma)},$$

but

$[\sigma_R] = [\sigma_a] + [\sigma_{av}]$,

and, therefore,

$$[\sigma_R] = \frac{\sigma_{-1}(\rho_\sigma + 1)}{[n](K\sigma_g \cdot \rho_\sigma + \psi_\sigma)} \tag{11}$$

The same way for calculation of admissible shear stress:

$$\tau_R = \frac{\tau_{-1}(\rho_\tau + 1)}{[n](K\tau_g \cdot \rho_\tau + \psi_\tau)} \tag{12}$$

We shall record without justification a dependence to define safety factor while bar operation under joint action of bending with torsion or torsion with tension (compression) or bending with torsion and tension (compression) [9,10,11,12], i.e. for those cases when plane stress state occurs in the component critical point. The overall safety factor in the mentioned cases is defined from the equation:

$$\frac{1}{n^2} = \frac{1}{n^2_\sigma} + \frac{1}{n_\tau^2} \tag{13}$$

Where n – the overall safety factor, n_σ – safety factor by normal stresses; n_τ – safety factor by shear stresses.

The same way we define the overall safety factor by yield stress, one just needs to substitute n_σ and n_τ with $n_{\sigma T}$ and $n_{\tau T}$ accordingly.

Formula (13) is applicable in case where normal and shear stresses in the verified point of a component are changed simultaneously, i.e. they reach their maximum and minimum values at the same time.

Results

Using system approach within the conducted studies enabled to formulate the fundamental principles of how to provide reliability of components and increase their fatigue strength.

Strength and fatigue properties of material show to some extent the fatigue properties of the ready components however they cannot serve the reliable basis to define fatigue limit of a definite kind of product operating under some dynamic conditions.

Thus to define safety factor one needs to have experimental data on components fatigue strength, including the ones by symmetric cycle, as the last parameter is included in the majority of dependencies connecting the limiting stresses under different coefficients of asymmetry.

In order to obtain fatigue characteristics by asymmetric cycles of loading, fatigue testings are conducted under different asymmetries of a cycle. Diagram of limiting amplitudes is built on the basis of testing results. The diagram characterizes the dependency between maximum and minimum limiting stresses $\tau_{maa,}\tau_{\min}$ (along the Y-axis) and the mean cycle stress (X-axis).

The limits of component fatigue strength by different coefficients of asymmetry r can be defined from diagram of limiting amplitudes, as well as fatigue safety factor, as the ratio of ultimately admissible stress τ_r, when the spring bears the basic number of loading cycles Nb to the maximum stress, acting in the spring under dynamic loading [8,9]:

$$n = \frac{\tau_r}{\tau_{\max}} \tag{14}$$

Conclusion

All the known ways of calculating fatigue strength of components come to defining fatigue safety factor. In this case the value of safety factor when calculating its fatigue strength depends on the accuracy of defining forces and stresses, the level of technology to recover components and other factors that can find application in calculation. Safety factors can vary from 1.3 to 3 due to the mentioned reasons.

Special research devoted to studies of strength in conditions of cyclic loading, and practice of machinery operation proves that the state of the surface layer of components defines their durability to a large extent as due to environmental influence and imperfect processing the surface has damages and micro roughnesses playing the role of stress concentrate. The largest influence of the surface layer condition on the fatigue strength is explained by the fact that the surface layers undergo the largest stresses while working under torsion and bending.

Thus understanding the existing knowledge of the influence of design and engineering and dynamic factors on the fatigue strength of machinery components one can conclude that applying processing procedures increasing the material quality and the state of surface layer can be an effective way to increase their fatigue resistance and service life. In order to predict the service life of the recovered machinery components it is necessary to develop methodology to evaluate their operational life depending on the recovery technology considering the fatigue characteristics of metal in the definite conditions of operation and define the influence of dynamic factors on the fatigue strength of a product.

References

[1] Darkov A.V. Strength of Materials /A.V. Darkov, G.S. Spiro. - M .: Higher School, 1975. - 654 p.

[2] Lasukov, A.A., Mokhovikov, A.A. Influence of tool modified layer on stress - Strain state of cutting wedge, Proceedings - 2012 7th International Forum on Strategic Technology, IFOST 2012

[3] Retyunsky O. Y. , Kurilin A. A. Stipulation of conditions of CMF with required allocation of tool materials // Applied Mechanics and Materials. - 2014 - Vol. 682. - p. 474-479.

[4] Putilova U.S., Nekrasov Yu.I., Lasukov A.A. Loading of Manufacturing System Elements at Unsteady Cutting and the Models of their Arrangement Deviations, Applied Mechanics and Materials, Vol. 682 (2014), pp. 192-196

[5] Valentov A.V., Konovodov V.V., Agafonov E.V. Prediction of residual and operational stress during cutting tools soldering for machining of weld surfaces / Bulletin of the Novosibirsk State Agrarian University. 2013. № 1 (26). Pp 107-110.

[6] Retyunsky O.Y. Method of making performs for composite reversible cutting plates // Applied Mechanics and Materials. - 2013 - Vol. 379. - p. 24-27.

[7] Retyunsky O.Yu., Valentov A.V., Solomatin A.A. Statistical analysis of results of experimental studies on turning of surfaces of mining machines parts, recovered by surfacing / Mining informational and analytical bulletin (scientific and technical journal). 2012. № S3. Pp 250-254.

[8] Lasukov V. V. , Rozhkova S. V. , Abdrashitova M. O. , Ilkin E. E. , Novoselov V. V. Mathematical Model of the Information Factor of the Evolution of the Lemaitre–Friedmann Primordial Atoms in Superspace-Time // Russian Physics Journal. - 2016 - Vol. 58 - №. 9. - p. 1265-1270.

[9] Valentov A.V., Konovodov V.V., Agafonov E.V. Prediction of residual and operational stress during cutting tools soldering for machining of weld surfaces / Bulletin of the Novosibirsk State Agrarian University. 2013. № 1 (26). Pp 107-110.

[10] Gubaydulina R. K. , Petrushin S. I. , Galeeva A. A. Selecting an Economical Variant of the Manufacturing Method of Engineering Product Fabrication under Current Conditions // Applied Mechanics and Materials. - 2014 - Vol. 682. - p. 613-616
[11] Petrushin S. I. , Gubaydulina R. K. , Grubiy S. V. , Likholat A. V. On the Problem of Wear Resistant Coatings Separation From Tools and Machine Elements // IOP Conference Series: Materials Science and Engineering. - 2015 - Vol. 91, Article number 012048. - p. 1-7
[12] Petrushin S. I. , Gubaydulina R. K. , Grubiy S. V. Optimization of Products Life Cycle // Applied Mechanics and Materials. - 2015 - Vol. 770. - p. 662-669

IOP Conf. Series: Materials Science and Engineering **142** (2016) 012031 doi:10.1088/1757-899X/142/1/012031

On Accuracy Order of Fourier Coefficients Computation for Periodic Signal Processing Models

I V Korytov[1] and S E Golosov[2]

[1] Associate Professor, Department of High Mathematics, Tomsk Polytechnic University, 634050, Russia, Tomsk, Lenin Avenue, 30
[2] Student, Institute of Power Engineering, Tomsk Polytechnic University, 634050, Russia, Tomsk, Lenin Avenue, 30

E-mail: korytov@tpu.ru

Abstract. The article is devoted to construction piecewise constant functions for modelling periodic signal. The aim of the paper is to suggest a way to avoid discontinuity at points where waveform values are obtained. One solution is to introduce shifted step function whose middle points within its partial intervals coincide with points of observation. This means that large oscillations of Fourier partial sums move to new jump discontinuities where waveform values are not obtained. Furthermore, any step function chosen to model periodic continuous waveform determines a way to calculate Fourier coefficients. In this case, the technique is certainly a weighted rectangular quadrature rule. Here, the weight is either unit or trigonometric. Another effect of the solution consists in following. The shifted function leads to application midpoint quadrature rules for computing Fourier coefficients. As a result the formula for zero coefficient transforms into trapezoid rule. In the same time, the formulas for other coefficients remain of rectangular type.

1. Introduction

Fourier series is applied to expand function that describes a waveform of periodic signal. As soon as the function is represented with its values obtained at discrete points, we have to substitute it with another function. The process of discretization of continuous wave is known as a time sampling. The new function is usually a step one. Its values in each sampling interval are equal to the values at the left limit of the interval. Trigonometric polynomial as a partial sum of series is a tool for modelling initial waveform [1], [2], [3]. One problem concerning truncated Fourier series consists in increasing amplitude of oscillations at jump discontinuities. The problem is known as the Gibbs phenomenon. There are various points of view to discuss the issue and resolve it [4], [5], [6], [7]. In the paper we consider continuity and discontinuity in points where data are obtained. To avoid the maximum of amplitude at the points of observation we suggest to shift the step function. As a result, the function becomes continuous at these points. In any case, when step function is chosen for sampling, an algorithm of calculation of Fourier coefficients appears as a certain quadrature rule. Suggested shift of step function leads to distinctions between used quadrature rules. As previous works of one of the authors are devoted to numerical integration and its error estimation [8], [9], we emphasize computational aspect of the issue. According to stated, in the article, we consider quadrature rules of order 0 and 1.

This paper is organized as follows.

Firstly, we set initial conditions for the two models. This way, we construct two piecewise constant functions matching the two models. Also we consider formulas for computation Fourier coefficients concerning the two cases.

Secondly, we consider weighted quadrature rules of low algebraic order. At first, we give the rules, that are accurate on all constants, then we perform rules, that are accurate on all linear functions. At the same time, we consider elementary and composite weighted quadrature rules.

Lastly, we show that each formula of Fourier coefficient matches certain quadrature rule. In addition, we discuss the order of approximation.

2. Two models for waveform

In this section, we introduce two step functions. One of them contains points of observed values as jump discontinuities. Another contains points of observed values as points of its continuity.

To begin with, we set the function of signal waveform. Suppose $f^*(t)$ is an unknown function which describes a periodic signal waveform, and $T = 2l$ is its period. Denote by $[a,b]$ the fundamental period of $f^*(t)$, where $b = a + 2l$.

Further, we describe general features of approximating function as a step function. Suppose $f(t)$ is a step function which approximates $f^*(t)$. Let t be an argument that presents a time variable ($t > 0$), and let t_k be the points where observed values of $f^*(t)$ are obtained. Denote by N the number of sampling intervals, therefore the number of points of observation is $N + 1$. By y_k denote the observed values: $y_k = f^*(t_k)$, $k = 0,\dots,N$. Let t_k be equidistant: $t_{k+1} - t_k = h$, $k = 0,\dots,N-1$, so $t_k = a + kh$, $k = 0,\dots,N$. In particular, $t_0 = a$, $t_N = b$.

We give more details of setting step functions in subsections 2.1 and 2.2 in order to specific cases.

2.1. Top-left corner model

In this subsection, we introduce model with piecewise constant function, where the points of observed values are jump discontinuities.

Suppose $f(t) = y_k$, $t \in [t_k, t_{k+1}] = [a + kh, a + (k+1)h]$, $k = 0,\dots,N-1$. Then Fourier coefficients for the step function are

$$a_0 = \frac{1}{l}\int_a^{a+2l} f(t)\,\mathrm{d}t = \frac{1}{l}\sum_{k=0}^{N-1}\int_{a+kh}^{a+(k+1)h} y_k\,\mathrm{d}t, \tag{2.1}$$

$$\begin{Bmatrix} a_n \\ b_n \end{Bmatrix} = \frac{1}{l}\int_a^{a+2l} f(t)\begin{Bmatrix} \cos\frac{\pi n t}{l} \\ \sin\frac{\pi n t}{l} \end{Bmatrix}\mathrm{d}t = \frac{1}{l}\sum_{k=0}^{N-1}\int_{a+kh}^{a+(k+1)h} y_k \begin{Bmatrix} \cos\frac{\pi n t}{l} \\ \sin\frac{\pi n t}{l} \end{Bmatrix}\mathrm{d}t. \tag{2.2}$$

Integrating right members of equation (2.1) and equation (2.2) by t, we obtain

$$a_0 = \frac{h}{l}\sum_{k=0}^{N-1} y_k, \tag{2.3}$$

$$\begin{Bmatrix} a_n \\ b_n \end{Bmatrix} = \frac{1}{\pi n}\sum_{k=0}^{N-1} y_k \begin{Bmatrix} \sin\frac{\pi n(a+(k+1)h)}{l} - \sin\frac{\pi n(a+kh)}{l} \\ \cos\frac{\pi n(a+kh)}{l} - \cos\frac{\pi n(a+(k+1)h)}{l} \end{Bmatrix}. \tag{2.4}$$

In this case, sampling interval coincides with interval of integration. The step function is discontinuous at points t_k and continuous at other points of the interval.

2.2. Midpoint model

In this subsection, we consider a model with a piecewise constant function, where the points of observed values are points of continuity. We shift the function introduced in the subsection 2.1 on half-step leftwards. This yields that length of the very left and the very right intervals decreases on the half. Thus the points of observed values turn out to be the middle points of interval of continuity.

Suppose

$$f(t)=\begin{cases} y_0, & t\in\left[t_0,t_0+\frac{h}{2}\right], \\ y_k, & t\in\left[t_{k-1}+\frac{h}{2},t_k+\frac{h}{2}\right], \quad k=1,\dots,N-1, \\ y_N, & t\in\left[t_{N-1}+\frac{h}{2},t_N\right] \end{cases}$$

According to definition, Fourier coefficients for this function are

$$a_0=\frac{1}{l}\left(\int_a^{a+\frac{h}{2}} y_0\,\mathrm{d}t+\sum_{k=1}^{N-1}\int_{a+\frac{h}{2}+(k-1)h}^{a+\frac{h}{2}+kh} y_k\,\mathrm{d}t+\int_{a+\frac{h}{2}+(N-1)h}^{b} y_N\,\mathrm{d}t\right), \tag{2.5}$$

$$\begin{Bmatrix} a_n \\ b_n \end{Bmatrix}=\frac{1}{l}\left(\int_a^{a+\frac{h}{2}} y_0\begin{Bmatrix}\cos\frac{\pi nt}{l} \\ \sin\frac{\pi nt}{l}\end{Bmatrix}\mathrm{d}t+\sum_{k=1}^{N-1}\int_{a+\frac{h}{2}+(k-1)h}^{a+\frac{h}{2}+kh} y_k\begin{Bmatrix}\cos\frac{\pi nt}{l} \\ \sin\frac{\pi nt}{l}\end{Bmatrix}\mathrm{d}t+\int_{a+\frac{h}{2}+(N-1)h}^{b} y_N\begin{Bmatrix}\cos\frac{\pi nt}{l} \\ \sin\frac{\pi nt}{l}\end{Bmatrix}\mathrm{d}t\right). \tag{2.6}$$

Then, integrating right members of equation (2.5) and equation (2.6) by t, we have

$$a_0=\frac{h}{l}\left(\frac{y_0}{2}+\sum_{k=1}^{N-1}y_k+\frac{y_N}{2}\right), \tag{2.7}$$

$$\begin{Bmatrix} a_n \\ b_n \end{Bmatrix}=\frac{1}{\pi n}\left(y_0\begin{Bmatrix}\sin\frac{\pi n\left(a+\frac{h}{2}\right)}{l}-\sin\frac{\pi na}{l} \\ \cos\frac{\pi na}{l}-\cos\frac{\pi n\left(a+\frac{h}{2}\right)}{l}\end{Bmatrix}+\sum_{k=1}^{N-1}y_k\begin{Bmatrix}\sin\frac{\pi n\left(a+\frac{h}{2}+kh\right)}{l}-\sin\frac{\pi n\left(a+\frac{h}{2}+(k-1)h\right)}{l} \\ \cos\frac{\pi n\left(a+\frac{h}{2}+(k-1)h\right)}{l}-\cos\frac{\pi n\left(a+\frac{h}{2}+kh\right)}{l}\end{Bmatrix}\right.$$
$$\left.+\,y_N\begin{Bmatrix}\sin\frac{\pi nb}{l}-\sin\frac{\pi n\left(a+\frac{h}{2}+(N-1)h\right)}{l} \\ \cos\frac{\pi n\left(a+\frac{h}{2}+(N-1)h\right)}{l}-\cos\frac{\pi nb}{l}\end{Bmatrix}\right). \tag{2.8}$$

In this case the interval of integration is shifted left towards the sampling interval. The step function is continuous at points t_k.

3. Quadrature rules of low order

In this section, we consider quadrature rules based on interpolating polynomials of degree 0 and 1. We assume that an integrand is given by tabulating its arguments and values. In this case tabulated arguments match the nodes of interpolation.

Moreover, we consider so called weighted quadrature rules which contain given tabulated function as a factor within integrand. The integrand as a product consists of the given function and some another function called a weight one. In general, the weight function is arbitrary, known, and defined in interval of integration. In certain case, the weight function is given by specific formula.

Each subsection in the section we begin with general form of weighted quadrature rule. Then we give examples with unit and trigonometric weight. This concerns either constant or linear functions.

3.1. Elementary weighted quadrature rules

To begin with, we define the general weighted quadrature rule. As soon as quadrature sum is a linear combination of integrand values, its coefficients are called weights. We should tell the difference between the 'weight function' and the 'weights'. The weight function is factor within integrand, and weights are numerical coefficients in quadrature sum. Denote by M a number of the very right node such that $M+1$ is a number of nodes, by x_k nodes, by A_k weights, and by $p(x)$ weight function of quadrature rule. Let $f(x)$ be a tabulated function, and $[a,b]$ interval of integration. Then a weighted quadrature formula is defined as following

$$\int_a^b p(x)f(x)\,\mathrm{d}x \approx \sum_{k=0}^{M} A_k f(x_k).$$

Let m be a degree of interpolating polynomial. As known the degree depends on the number of nodes of interpolation as $m = M$. Therefore, quadrature rule accurate on interpolating Lagrange polynomials $L_m(x)$ of degree m is

$$\int_a^b p(x)L_m(x)\,\mathrm{d}x = \sum_{k=0}^{m} A_k L_m(x_k).$$

In following subsubsections we give the quadrature rules with $m = 0.1$.

3.1.1. Accuracy on constant functions. Now suppose that the quadrature rule is accurate on all constant functions $f(x) = L_0(x) = \alpha$, where α is arbitrary real number. So, elementary on $[a,b]$ weighted quadrature rule accurate on polynomials of degree 0 is

$$\int_a^b p(x)\alpha\,\mathrm{d}x = A_0\alpha \;\Rightarrow\; \int_a^b p(x)f(x)\,\mathrm{d}x \approx A_0 f(x_0),\quad A_0 = \int_a^b p(x)\,\mathrm{d}x. \tag{3.1}$$

There are three cases for x_0. The following methods are called as below depending on location of the node within interval of integration: if $x_0 = a$ the rule is called top-left corner method; if $x_0 = \frac{a+b}{2}$ the rule is called midpoint method; if $x_0 = b$ the rule is called top-right corner method. The cases are important for our purpose.

3.1.2. Accuracy on linear functions. Now we turn to the quadrature rule accurate on all linear functions $f(x) = L_1(x) = \beta x + \gamma$, where β and γ are arbitrary real numbers. Then

$$\int_a^b p(x)(\beta x + \gamma)\,\mathrm{d}x = \sum_{k=0}^{1} A_k(\beta x_k + \gamma).$$

Nodes of the formula are set: $x_0 = a$, $x_1 = b$. As the rule is accurate on polynomials of degree 1 and less the following system should be solved to find weights A_k

$$\begin{cases} \int_a^b p(x)\alpha\,\mathrm{d}x = A_0\alpha + A_1\alpha \\ \int_a^b p(x)(\beta x + \gamma)\,\mathrm{d}x = A_0(\beta x_0 + \gamma) + A_1(\beta x_1 + \gamma) \end{cases}$$

Thus, elementary on $[a,b]$ weighted quadrature rule accurate on polynomials of degree 1 with weights A_0 and A_1 is

$$\int_a^b p(x)f(x)\mathrm{d}t \approx A_0 f(x_0) + A_1 f(x_1),\quad A_0 = \frac{B-aA}{b-a},\quad A_1 = \frac{Ab-B}{b-a}, \tag{3.2}$$

$$A = \int_a^b p(x)\,\mathrm{d}x,\quad B = \int_a^b xp(x)\,\mathrm{d}x.$$

To simplify expressions for A_0 and A_1 we assume that $P(x)$ is antiderivative for the weight function $p(x)$. Then integrating A and B we have

$$A_0 = P(b) - \tfrac{1}{b-a}\int_a^b P(x)\,\mathrm{d}x,\quad A_1 = \tfrac{1}{b-a}\int_a^b P(x)\,\mathrm{d}x - P(a). \tag{3.3}$$

This is the quadrature rule with the weight in general form. If the weight is unit, the rule becomes classical trapezoid rule.

3.2. Composite weighted quadrature rules

The interval of integration is sectioned on N equal subintervals. An elementary quadrature formula is used on each subinterval. The rule as a result of summing the parts is known as composite quadrature rule. We begin with introducing general forms of lower order rules.

Suppose $h = \frac{b-a}{N}$. Then, applying the rule represented in equation (3.1) to each subinterval, and summing integrals over all subintervals, taking in account we have

$$\int_a^b p(x)f(x)\,\mathrm{d}x = \sum_{k=0}^{N-1}\int_{a+kh}^{a+(k+1)h} p(x)f(x)\,\mathrm{d}x \approx \sum_{k=0}^{N-1} A_{0k}\begin{Bmatrix} f(a+kh) \\ f\left(a+kh+\frac{h}{2}\right) \\ f(a+(k+1)h) \end{Bmatrix}. \tag{3.4}$$

Weights in equation (3.4) are $A_{0k} = P(a+(k+1)h) - P(a+kh)$, $k = 0,\ldots,N-1$. Double index in A_{0k} means that weight A_0 from equation (3.1) is applied to subinterval whose left limit is numbered by k. Thus, equation (3.4) represents quadrature formula accurate on constant functions.

Furthermore, applying equation (3.2) with specifics (3.3) to subinterval and summing, we have

$$\int_a^b p(x)f(x)\,\mathrm{d}x \approx \sum_{k=0}^{N-1}\left(A_{0k}f(a+kh) + A_{1k}f(a+(k+1)h)\right) = B_0 f(a) + \sum_{k=1}^{N-1} B_k f(a+kh) + B_N f(b). \tag{3.5}$$

Weights in equation (3.5) are

$$B_0 = P(a+h) - \tfrac{1}{h}\int_a^{a+h} P(x)\,\mathrm{d}x,\quad B_N = \tfrac{1}{h}\int_{b-h}^{b} P(x)\,\mathrm{d}x - P(b-h),$$

$$B_k = P(a+(k+1)h) - \tfrac{1}{h}\int_{a+kh}^{a+(k+1)h} P(x)\,\mathrm{d}x + \tfrac{1}{h}\int_{a+(k-1)h}^{a+kh} P(x)\,\mathrm{d}x - P(a+(k-1)h). \tag{3.6}$$

Expression for weights B_k in equation (3.6) is true for all $k = 1,\ldots,N-1$. Thus, equation (3.6) represents quadrature formula accurate on linear functions.

3.2.1. Unit weight. Now we turn to specific weight functions. We begin with unit weight function. Let $p(x) = 1$, then $P(x) = x$, and $\int_a^b P(x)\mathrm{d}x = \frac{b^2-a^2}{2}$. The expressions are called unit weight parameters. Substituting weights in equation (3.4) by the parameters, we have classical rectangular rule with $A_{0k} = h$, $k = 0,\ldots,N-1$

$$\int_a^b f(x)\,\mathrm{d}x \approx h\sum_{k=0}^{N-1}\begin{Bmatrix} f(a+kh) \\ f\left(a+kh+\frac{h}{2}\right) \\ f(a+(k+1)h) \end{Bmatrix}. \tag{3.7}$$

In other words, this is a composite unit weighted quadrature formula exact on constant functions. Further, substituting weights in equations (3.5) and (3.6) by unit weight parameters we have classical trapezoid rule with $B_0 = B_N = \frac{h}{2}$, and $B_k = h$, $k = 0,\dots,N-1$

$$\int_a^b f(x)\,\mathrm{d}x \approx h\left[\frac{f(a)}{2} + \sum_{k=1}^{N-1} f(a+kh) + \frac{f(b)}{2}\right] \tag{3.8}$$

Thus we have two composite quadrature rules with unite weight.

3.2.2. Cosine weigh. We continue with trigonometric weight function. Now we turn to cosine weight. Let $p(x) = \cos \alpha x$, then $P(x) = \frac{1}{\alpha}\sin \alpha x$, and $\int_a^b P(x)\,\mathrm{d}x = \frac{1}{\alpha^2}(\cos \alpha a - \cos \alpha b)$. Here α is an arbitrary real number. Substituting weights in equation (3.4) by this, we have

$$\int_a^b \cos \alpha x f(x)\,\mathrm{d}x \approx \sum_{k=0}^{N-1} A_{0k} \begin{Bmatrix} f(a+kh) \\ f\left(a+kh+\frac{h}{2}\right) \\ f(a+(k+1)h) \end{Bmatrix}. \tag{3.10}$$

Weights in equation (3.10) are $A_{0k} = \frac{1}{\alpha}[\sin \alpha(a+(k+1)h) - \sin \alpha(a+kh)]$, $k = 0,\dots,N-1$. Further, substituting weights in equations (3.5) and (3.6) by cosine weight parameters, we have

$$\int_a^b \cos \alpha x f(x)\,\mathrm{d}x \approx B_0 f(a) + \sum_{k=1}^{N-1} B_k f(a+kh) + B_N f(b). \tag{3.11}$$

Weights in equation (3.11) are

$$B_0 = \frac{1}{\alpha}\sin \alpha(a+h) + \frac{1}{h\alpha^2}(\cos \alpha(a+h) - \cos \alpha a),$$

$$B_N = -\frac{1}{h\alpha^2}(\cos \alpha b - \cos \alpha(b-h)) - \frac{1}{\alpha}\sin \alpha(b-h),$$

$$B_k = \frac{1}{\alpha}\sin \alpha(a+(k+1)h) + \frac{1}{h\alpha^2}(\cos \alpha(a+(k+1)h) - 2\cos \alpha(a+kh) + \cos \alpha(a+(k-1)h))$$
$$-\frac{1}{\alpha}\sin \alpha(a+(k-1)h). \tag{3.12}$$

Expression for weights B_k in equation (3.12) is true for all $k = 1,\dots,N-1$. Thus we have two composite quadrature rules with cosine weight.

3.2.3. Sine weight. Finally we turn to sine weight. Suppose $p(x) = \sin \alpha x$, then $P(x) = -\frac{1}{\alpha}\cos \alpha x$, and $\int_a^b P(x)\,\mathrm{d}x = \frac{1}{\alpha^2}(\sin \alpha a - \sin \alpha b)$. Here α is an arbitrary real number. Substituting weights in equation (3.4) by this, we have

$$\int_a^b \sin \alpha x f(x)\,\mathrm{d}x \approx \sum_{k=0}^{N-1} A_{0k} \begin{Bmatrix} f(a+kh) \\ f\left(a+kh+\frac{h}{2}\right) \\ f(a+(k+1)h) \end{Bmatrix}. \tag{3.13}$$

Weights in equation (3.13) are $A_{0k} = \frac{1}{\alpha}[\cos \alpha(a+kh) - \cos \alpha(a+(k+1)h)]$, $k = 0,\dots,N-1$. Further, substituting weights in equations (3.5) and (3.6) by sine weight parameters, we have

$$\int_a^b \sin \alpha x f(x)\,\mathrm{d}x \approx B_0 f(a) + \sum_{k=1}^{N-1} B_k f(a+kh) + B_N f(b). \tag{3.14}$$

Weights in equation (3.14) are

$$B_0=\tfrac{1}{h\alpha^2}\left(\sin\alpha(a+h)-\sin\alpha a\right)-\tfrac{1}{\alpha}\cos\alpha(a+h),\quad B_N=\tfrac{1}{\alpha}\cos\alpha(b-h)-\tfrac{1}{h\alpha^2}\left(\sin\alpha b-\sin\alpha(b-h)\right),$$

$$B_k=\tfrac{1}{\alpha}\cos\alpha(a+(k-1)h)+\tfrac{1}{h\alpha^2}\left(\sin\alpha(a+(k+1)h)-2\sin\alpha(a+kh)+\sin\alpha(a+(k-1)h)\right)$$
$$-\tfrac{1}{\alpha}\cos\alpha(a+(k+1)h). \qquad (3.15)$$

Expression for weights B_k in equation (3.15) is true for all $k=1,\dots,N-1$.Thus we have two composite quadrature rules with sine weight.

4. Fourier coefficients

In this section, we compare formulas for Fourier coefficients of piecewise constant function and quadrature rules of lower orders.

4.1. Case of discontinuity

Let us remember that $y_k=f^*(t_k)$, $k=0,\dots,N$. Comparing (2.3) and (3.7), we have

$$a_0=\frac{1}{l}h\sum_{k=0}^{N-1}f^*(a+kh)\approx\frac{1}{l}\int_a^{a+2l}f^*(t)\,\mathrm{d}t, \qquad (4.1)$$

Here $\alpha=\frac{\pi n}{l}$. Thus, computing formula for zero Fourier coefficient matches composite unit-weighted quadrature rectangular rule for tabulated function $f^*(t)$ in $[a,a+2l]$. Comparing (2.4) and (3.10) (3.13), we have

$$\begin{Bmatrix}a_n\\b_n\end{Bmatrix}=\frac{1}{l}\frac{l}{\pi n}\sum_{k=0}^{N-1}f^*(a+kh)\begin{Bmatrix}\sin\frac{\pi n(a+(k+1)h)}{l}-\sin\frac{\pi n(a+kh)}{l}\\ \cos\frac{\pi n(a+kh)}{l}-\cos\frac{\pi n(a+(k+1)h)}{l}\end{Bmatrix}\approx\frac{1}{l}\int_a^{a+2l}\begin{Bmatrix}\cos\frac{\pi nt}{l}\\ \sin\frac{\pi nt}{l}\end{Bmatrix}f^*(t)\,\mathrm{d}t. \qquad (4.2)$$

Thus, computing formulas for nth Fourier coefficients match composite cosine- and sine-weighted quadrature rules accurate on constants for tabulated function $f^*(t)$ in $[a,a+2l]$.

4.1.1. Case of continuity. Comparing (2.7) and (3.8), we have

$$a_0=\frac{1}{l}h\left[\frac{f^*(a)}{2}+\sum_{k=1}^{N-1}f^*(a+kh)+\frac{f^*(b)}{2}\right]\approx\frac{1}{l}\int_a^{a+2l}f^*(t)\,\mathrm{d}t. \qquad (4.3)$$

Thus, computing formula for zero Fourier coefficient matches composite unit-weighted quadrature trapezoid rule for tabulated function $f^*(t)$ in $[a,a+2l]$. In other words, in comparison with the top-left corner model the midpoint model provides computation with quadrature rule of higher order.

Looking at (2.8) and (3.8), we see that

$$\begin{Bmatrix}a_n\\b_n\end{Bmatrix}=\frac{1}{l}\frac{l}{\pi n}\left(f^*(a)\begin{Bmatrix}\sin\frac{\pi n(a+\frac{h}{2})}{l}-\sin\frac{\pi na}{l}\\ \cos\frac{\pi na}{l}-\cos\frac{\pi n(a+\frac{h}{2})}{l}\end{Bmatrix}+f^*(b)\begin{Bmatrix}\sin\frac{\pi nb}{l}-\sin\frac{\pi n(a+\frac{h}{2}+(N-1)h)}{l}\\ \cos\frac{\pi n(a+\frac{h}{2}+(N-1)h)}{l}-\cos\frac{\pi nb}{l}\end{Bmatrix}\right.$$

$$\left.+\sum_{k=1}^{N-1}f^*(a+kh)\begin{Bmatrix}\sin\frac{\pi n(a+\frac{h}{2}+kh)}{l}-\sin\frac{\pi n(a+\frac{h}{2}+(k-1)h)}{l}\\ \cos\frac{\pi n(a+\frac{h}{2}+(k-1)h)}{l}-\cos\frac{\pi n(a+\frac{h}{2}+kh)}{l}\end{Bmatrix}\right). \qquad (4.4)$$

Comparing, we see that, although expressions have similar structure, weight coefficients do not provide higher order of quadrature rule. So, we have to write

$$\begin{Bmatrix} a_n \\ b_n \end{Bmatrix} = \frac{1}{l}\left[B_0 f(a) + \sum_{k=1}^{N-1} B_k f(a+kh) + B_N f(b)\right]$$

$$\approx \frac{1}{l}\left[\int_a^{a+\frac{h}{2}} \begin{Bmatrix} \cos\frac{\pi n t}{l} \\ \sin\frac{\pi n t}{l} \end{Bmatrix} f^*(t)\,\mathrm{d}t + \int_{a+\frac{h}{2}}^{b-\frac{h}{2}} \begin{Bmatrix} \cos\frac{\pi n t}{l} \\ \sin\frac{\pi n t}{l} \end{Bmatrix} f^*(t)\,\mathrm{d}t + \int_{b-\frac{h}{2}}^{b} \begin{Bmatrix} \cos\frac{\pi n t}{l} \\ \sin\frac{\pi n t}{l} \end{Bmatrix} f^*(t)\,\mathrm{d}t\right]. \quad (4.5)$$

Expression in square brackets before approximate equality sign means following. The first part is an elementary top-left corner quadrature rectangular term, the middle part is a composite midpoint quadrature rectangular sum, and the last part is an elementary top-right corner term. Expression in square brackets after approximate equality sign matches trigonometric-weighted definite integrals of tabulated waveform function. We see that order of quadrature rule remains the same in this case.

5. Conclusion

Taking the set of tabulated function values as an initial data for calculating Fourier coefficients, we have a task to compute approximately some definite integrals. The way to complete the task is use of appropriate quadrature rules. If the modelling function is a piecewise constant, the rules appear to be composite of rectangular or trapezoid weighted type.

Acknowledgments

This work is funded within the framework of realization of Strategic Programme on National Research Tomsk Polytechnic University Competitiveness Enhancement in the Group of Top Level World Research and Academic Institutions.

References

[1] Rairan D and Garzon M H 2015 Period estimation for adaptive reconstruction of (quasi-) periodic signals *Int. J. Adapt. Control Signal Process* **29(1)** pp 64–80

[2] Ohlsson H Eldar Y C Yang A Y and Sastry S S 2014 Compressive shift retrieval *IEEE Trans. Signal Process.* **62(16)** pp 4105–4113

[3] Schimmel J 2012 Audible aliasing distortion in digital audio synthesis *Radioengineering* **21(1)** pp 56–62

[4] Adcock B and Hansen A C 2012 Stable reconstructions in Hilbert spaces and the resolution of the Gibbs phenomenon *Appl. Comput. Harmonic Anal.* **32(3)** pp 357–388

[5] Rim K S and Yun B I 2013 Gibbs phenomenon removal by adding Heaviside functions *Adv Comput. Math.* **38(4)** pp 683–699

[6] Zhu H, Ding M and Gao D 2014 Resolution of the Gibbs phenomenon for fractional fourier series *IEICE Trans. Fund. Electron. Commun. Comput. Sci.* **E97-A(2)** pp 572–586

[7] Berriochoa E, Cachafeiro A and Díaz J 2015 Gibbs phenomenon in the Hermite interpolation on the circle *Appl. Math. Comput.* **253** pp 274–286

[8] Korytov I V 2013 Function representing error functional of a cubature formula in Sobolev space *Bulletin of the Tomsk Polytechnic University. Mathematics and Physics. Mechanics* **323 (2)** pp 21–25

[9] Korytov I V 2011 The extreme function of a linear functional in the weighted Sobolev space *Vestnik Tomskogo Gosudarstvennogo Universiteta. Matematika i Mekhanika* **2** pp 5–15

IOP Conf. Series: Materials Science and Engineering **142** (2016) 012032 doi:10.1088/1757-899X/142/1/012032

Compulsory Checking of Nuclear Power Engineering Materials by Direct and Eddy Current

V V Larionov, A M Lider, D A Sednev, Shupeng Xu

Tomsk Polytechnic University, 30 Lenin Avenue, Tomsk, Russian Federation

E-mail: lvv@tpu.ru

Abstract. The testing technology of copper parts designed for dry storage of spent nuclear fuel with application of direct and eddy current has been developed. Measurements results of flaw quantity caused hydrogenation and oxidation processes are presented. Evolution of copper M 001 flaw structure during hydrogenation from gaseous medium is analyzed. It has been demonstrated that the dependence of copper ρ electrical resistance on number of flaws in its structure has dome shaped character and changes with eddy current frequency change. Number of flaws formed by hydrogen depends on direction (100) or (200) of the crystal structure of copper lattice.

1. Introduction

Copper containers are widely used for dry storage of spent nuclear fuel [1]. Hydrogen plays a significant role during their production and operation as it dissolves in melting during welding metal. Hydrogen ingresses into metal from air which contains water vapour, from moisture covering electrodes, from oxides located on the surface of metal product and electrodes. Under high temperature moisture fumes and dissociates with heat absorption. Hydrogen is contained in electrode coatings and in metal. As a result of components production, gasing under operating conditions, especially during welding, when hydrogen inevitably precipitates, «hydrogen disease» occurs under high temperature in copper. For express diagnostics of welded joints in particular, various methods of nondestructive testing and their combinations are used [2].

During operation by hydrogenation complex structural changes are frequently observed in the surface layer and generally in metal. Saturated layers, characterized by the presence of hydrogen implementation phases and changes of the parameters of the crystal lattice at different depths of the metal, are formed. As a result, there are structural defects. Despite the diffusion effects, which are characteristic of titanium, hydride formation, reducing the mobility of hydrogen, there is hydrogen distribution in the depth of the titanium sheet. The analysis of copper is of great importance. In particular, in the manufacture of electro-vacuum devices, OFC with oxygen of not more than 0.0005% is used. When oxygen in copper is excessive the "hydrogen disease", caused by the presence of hydrogen in copper, occurs. The quality of copper is determined by measuring the number of bends of copper strips, pre -annealed in a hydrogen atmosphere. This process is not technological. Thus, the analysis proves the relevance of finding alternative non-destructive methods for determining the concentration of hydrogen in metals. Eddy current transducers are used for a variety of measurements of properties and for detail and material control. However, in this case there is a following problem. If the controlled samples are sufficiently thin, the signal measured with an eddy-current transducer is dependent not only on the material properties of the sample, but also on its thickness. Due to the fact that the conductivity is a function of hydrogen concentration in metals, the eddy-current method can be used to measure the hydrogen content in metals. However, in all cases, the topical problem is the accuracy and sensitivity of the measurements.

In the present paper the evolution of copper M 001 flaw structure during oxygen and hydrogen saturation is analyzed. Principal physics for application of current methods for these purposes is change in ρ copper electrical resistance dependant on the presence of gases and flaws in copper. Defects of crystal lattice influence specific resistance of metals. Change in electrical resistance in different frequencies of eddy currents leads to deformation of obtained dependence.

The goal of the work is to develop the technology for diagnostics of copper containers with spent nuclear fuel using the method of current analysis.

2. Materials and Methods. Experimental Procedure

Copper M 001 having the following composition [%wt.]: 0.01 Fe; 0.1 Si; 0.07 C; 0.024 O; 0.04 N was used for studies. Hydrogenation was performed according to Sievert's method [3] under the temperature up to 1000 ^{0}C and oxidation equal to 600 ^{0}C. Measurements of hydrogen concentration were performed using hydrogen analyser RHEN602 by LECO. Density of dislocation in titanium was determined by broadening of X-ray lines using diffractometer Shimadzu XRD-6000 in radiation Cu-Kα. Analysis of phase composition was performed using PCPDFWIN and PDF-4+ data base and software for full profile analysis POWDER CELL 2.5. Hydrogen depth distribution was performed by means of magnetic spectral analyser (3MA, Germany). Resistance measurement of titanium samples in direct current was performed by 4 probe method [4] using hard- and software by «KEIHLEY INSTRUMENTS».

3. Results

Hydrogen interacts with the present structural defects as well as it induces new defects formation and dislocations appearance [5, 6]. This leads to changes in electrical resistance. Change in specific resistance of dislocations, related to the unit of their density N_d in the unit of volume in dependence to hydrogen concentration in metal was estimated using the formula [7]:

$$D = \frac{\rho_d}{N_d} = \frac{\hbar k_F \Omega_a Q}{n_s e^2}, \quad (1)$$

where k_F – volume of wave vector on Fermi level, Ω_a – atomic volume, Q – transport cross-section of electrons scattering, n_s — number of current carriers per atom, e – volume of electron charge, n_s – number of charge carriers.

Dislocations density N_d was determined using the formula [8] based on X-ray diffraction analysis data

$$N_d = \pi\beta^2 ctg^2\Theta / 16b^2, \quad (2)$$

where β – broadening of X-ray lines conditioned by lattice microdeformation, θ – the angle which corresponds to the maximum of X-ray line, b – Burgers vector. For each weight concentration of hydrogen in copper, specific resistance of samples was measured using probe method [4].

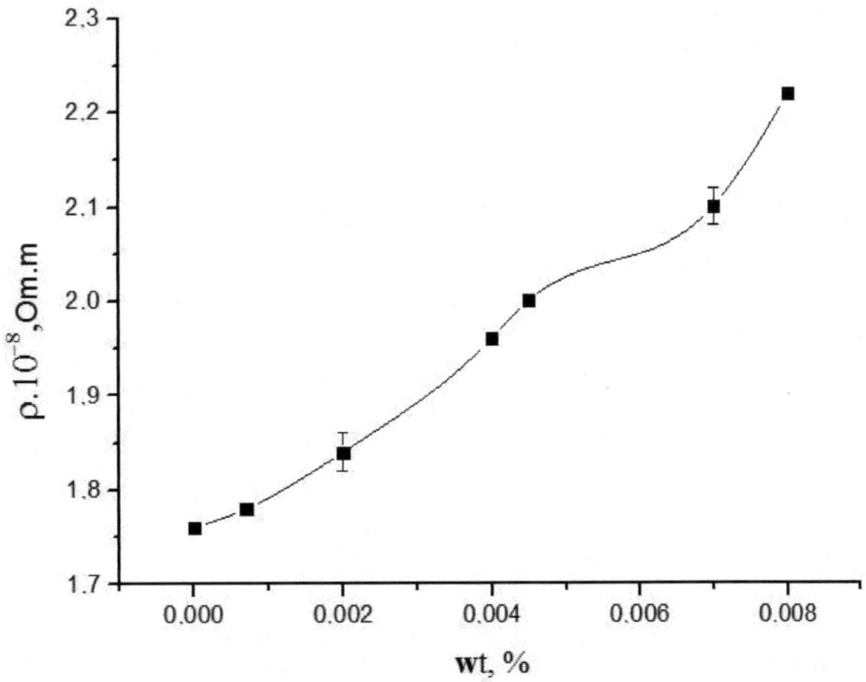

Figure 1. Dependence of Specific Electrical Resistance on Hydrogen Concentration in Copper

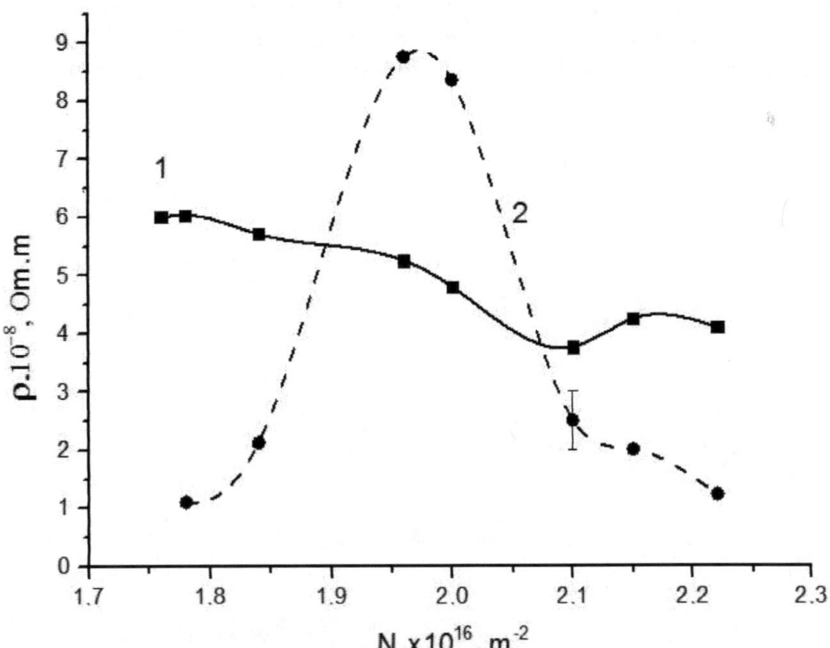

Figure 2. Dependence of Specific Electrical Resistance on Defects Concentration in Hydrogenated Copper (1- direction (100), 2 – (200))

IOP Conf. Series: Materials Science and Engineering **142** (2016) 012032 doi:10.1088/1757-899X/142/1/012032

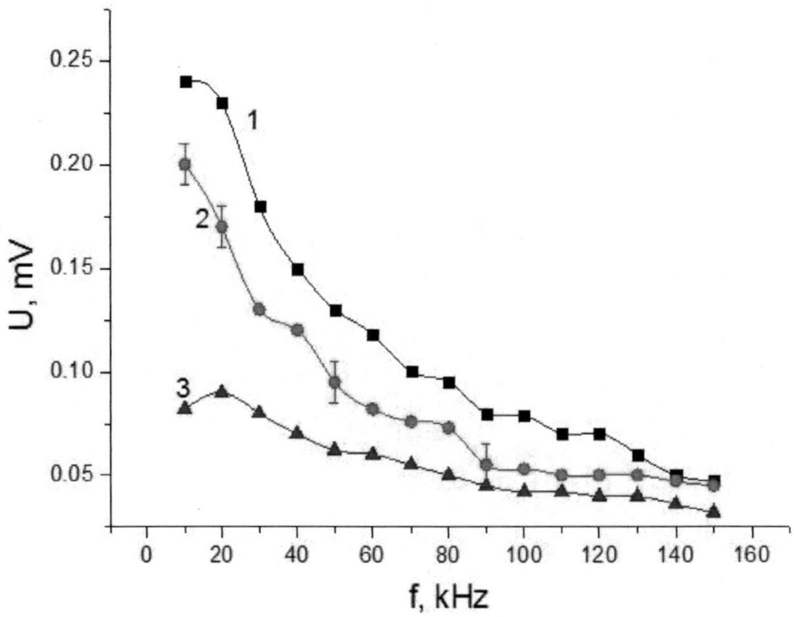

Figure 3. Dependence of Magnetic Spectrometer Indications on Eddy Current Frequency, 1 – Hydrogenated and then Oxygenated Copper, 2 – Sample Hydrogenated up to Oxidation, 3 – Initial Sample

The change of parameter P can be caused by different area of a sample covered by eddy current at different depth of its penetration. The hydrogenation can change both electric conductivity and the area of a layer located at depth δ. Since the experimental separate determination of the area value is impossible, the relative change of parameter P during hydrogenation allows evaluation of metal structure change extent, which enables the analysis of hydrogen content at different depth of a material. This is demonstrated by a plot in fig 4. The change of parameter P at frequencies varying from 50 to 100 kHz testifies not the change of thickness, but a prevailing change of metal structure caused by the hydrogenation.

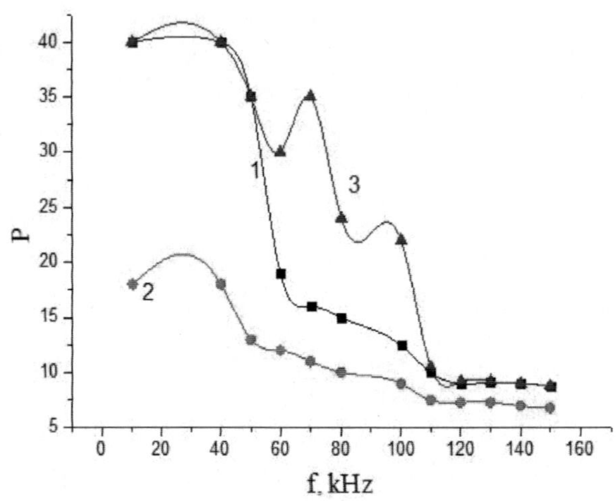

Figure 4. Dependence of parameter P on frequency of 3MA magnetic spectrometer eddy current (1 – Δd/d=0.05/1.7); 2 – Δd/d=0.05/1.75; 3 is hydrogenated metal)

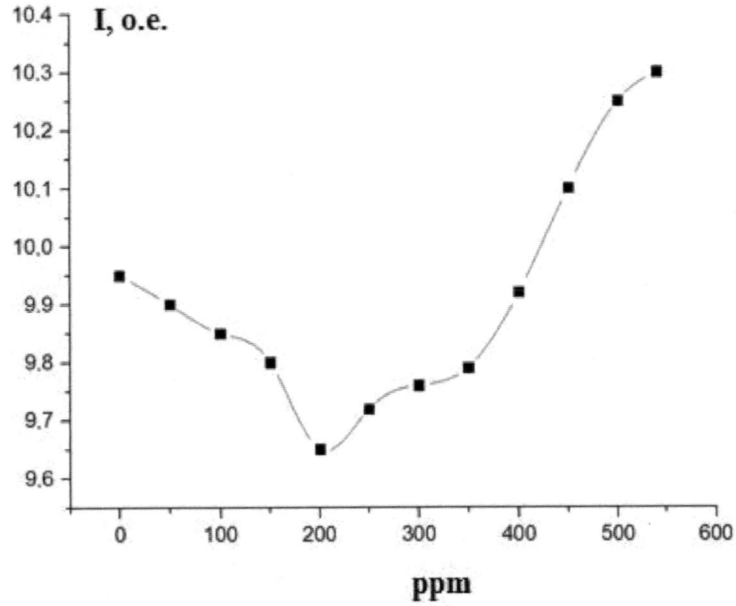

Figure 5. Dependence of active component of eddy current (frequency of 200 kHz) on hydrogen content in VT1-0 titanium

The dependence of eddy current transducer signal on hydrogen concentration in VT1-0 titanium for different frequencies is depicted in fig. 5. The dependence can be specifically used as calibration curves for the analysis of hydrogen content in every particular case.

Thus, the study results allow creation of nomogram atlas for determination of the condition of exposed products [9]. Alternatively, application of nomograms provides an intelligent control in the conditions of personnel limited access to continuously exposed objects.

4. Discussion

Figures 1-3 represent the results of experimental studies of copper samples. In all cases the electrical resistivity and half-width of X-ray spectrum were measured. Within the studied hydrogen concentrations in copper the virtually linear dependence of electrical resistance increase on hydrogen concentration in copper is observed (Fig.1). Virtually decreasing dependence of electrical resistivity on number of defects calculated using the formula (1), is characteristic for the direction (100) of crystal lattice of copper (Fig.2). Equally the dependence of specific resistance of hydrogenated copper samples for direction (200) of crystal lattice possesses dome-shaped character. The difference between the curves 1 and 2 in Fig. 2 on directions of crystal lattice is probably related to change in the number of hydrogen atoms introduced to copper lattice and changes of dislocations character. Dependence of the parameter $D = \rho d/Nd$ (formula 1) on hydrogen concentration in copper in the first case carries decreasing character, in the second case this value virtually does not change. This indicates the decrease of total transport cross-section of electrons scattering on defects caused by hydrogen (formula 2).

Fig. 3 shows the results of copper samples measurements performed by eddy currents. The difference between the curves is conditioned by the following defects. Under regulated by standards the oxygen concentration in copper is up to 0.065 % on the mass, some of oxygen within the limits of solubility is located in copper crystal lattice. The other part in the form of oxide particles is stochastically located in copper matrix. Voids occurrence is possible in the proximity of copper oxides particles in oxygen-containing copper samples [10], due to hydrogen and oxygen interaction.

Under low frequencies the maximum difference for initial hydrogenated and oxidized samples is observed. In this case the penetration depth of eddy current in sample is maximum (Fig. 3). The difference is decreasing with increase of eddy current frequency. This is conditioned by the fact that electrical conductivity on the copper samples surface is for the main part performed by impurities rather than hydrogen and oxygen. It should be underlined that the behavior of the curves within the limits of ± 0.02 mV depends on the degree of initial samples preparation for studies.

5. Conclusion

The potential of using eddy current method for testing of nuclear power engineering materials is mainly provided by maintenance personnel safety. In research, certain interest is represented by the comparison of electrical resistance measurements performed under direct and eddy current along with X-ray diffraction analysis in order to identify the conditions of current propagation on defects of hydrogenated metals. Imperfection of hydrogenous copper structure is evident with the presence of electrical resistance dependences on directions 100 and 200 of copper crystal lattice and is identified by measurements of eddy current of different frequencies.

Acknowledgements

This study was partly financed by the Government of RF (Nauka grant no. № 1524, theme 0.1325. 2014 and Federal Target Programme «Research and Development in Priority Fields of Russia`s Technological Complex Development for 2014 - 2020». Project unique identifier RFMEFI57514X0048.

References

[1] Larsen E, Watkins A., McJunkin T., Pace D., Bitsoi R. 5th International Conference on NDE in Relation to Structural Integrity for Nuclear & Pressurized Components, San Diego, May 2006.
[2] Larionov V V, Shupeng Xu, Kun Shi, Krening M X 2015 Adv. Mater. Res. **1084** 21
[3] Kudiiarov V N, Lider A M, S.Y. Harchenko S Y 2014 Adv. Mater. Res. **880** 68
[4] Kobeleva S P 2007 Industrial laboratory **73** 60
[5] Laptev R S, Bordulev Y S, Kudiiarov V N, Lider A M, G.V. Garanin G V 2014 Adv. Mater. Res. **880** 134
[6] Gabis I E, Voyt A P, Evard EA, Zaika Yu V, Chernov I A, Yartis V A 2005 J. Alloys Comp. **404-406** 312
[7] Karolik 2011 A S Mater. Sci. (Materialovedenie in Russian) **4** 15
[8] Malychin D G, Korneeva V V 2010 Bulletin of the Kharkov University. Series Physical **887** 115
[9] Larionov VV, Lider A M, Garanin G V 2015 Adv. Mater. Res. **1085** 335
[10] [Loginov Yu N, Demakov S L, Illarionov A G, Ivanova M A, Karabanalov M S 2013 Physical Mesomechanics **16** 99

IOP Conf. Series: Materials Science and Engineering **142** (2016) 012033 doi:10.1088/1757-899X/142/1/012033

Mathematical Modeling of Non-Stationary Hydraulic Process Occurring in the Gas Centrifuge Cascade During the Separation of Multicomponent Isotope Mixtures

A A Orlov[1], A A Ushakov[2] and V P Sovach[2]

[1] National Research Tomsk Polytechnic University, 634050, Lenin av, 30, Tomsk, Russia
[2] SC "PA Electrochemical Plant", 663690, Pervaya promyshlennaya ul, 1, Zelenogorsk, Russia

E-mail: ushakovaa2015@sibmail.com

Abstract. This article presents results of development of the mathematical model of nonstationary separation processes occurring in gas centrifuge cascades for separation of multicomponent isotope mixtures. This model was used for the calculation parameters of gas centrifuge cascade for separation of germanium isotopes. Comparison of obtained values with results of other authors revealed that developed mathematical model is adequate to describe nonstationary separation processes in gas centrifuge cascades for separation of multicomponent isotope mixtures.

1. Introduction

Gas centrifuge (GC) method of isotope separation is industrial method of uranium enrichment. High efficiency and long-term experience of gas centrifuge operation enabled to extend the application domain of this separation method and devise the industrial plant for separation stable and radioactive isotopes of various chemical elements.

Process gases are used for isotope separation by GC method. The molecules of this gases consists of the atoms of one or several chemical elements (for example, Kr, Xe, SiF_4, WF_6, $Fe(CO)_5$, $Cd(CH_3)_2$). Therefore, process gas is called a multicomponent isotope mixture (MCIM).

Technological processes of MCIM separation in GC cascade goes continuously in the stationary mode. In the operation of cascade there are nonstationary hydraulic processes, characterized by time-variable values of MCIM pressure and flows in cascade stages. For example a filling of GC cascade by process gas is nonstationary hydraulic process.

During nonstationary processes it is necessary to ensure safety of the equipment and to minimize losses of cascade productivity. The GC cascades for MCIM separation have insignificant gas content, and, as a consequence, low inertia. It leads to increased influence of nonstationary processes on the cascade efficiency.

In this regard, actual task is full-scale research of nonstationary process.

This problem had been discussed in the papers [1-7], authors of which had developed of dynamic stability principles, analytical and numerical solution methods of nonstationary hydraulic equations and mathematical models of nonstationary processes in free cascades. These processes was researched in the cascades for uranium isotopes separation. The presented models have different object details.

In paper [2] there was the description of the nonstationarity's influence during multicomponent isotope mixture separation processes. The papers about developed mathematical models of nonstationary hydraulic processes in these cascades don't exist.

We have developed of mathematical model of nonstationary hydraulic processes in gas centrifuge cascade for separation of multicomponent isotope mixtures and realized it as program software.

2. The description of mathematical model of nonstationary processes

2.1. Basic parameters of stage and cascade

MCIM separation takes place into a cascade (figure 1). A cascade consists of S separation stages numbered by i index ($i=\overline{1,S}$) and connected by counter-current type [1]. The cascade has three flows: incoming feed flow F, output light fraction flow P and heavy fraction flow W.

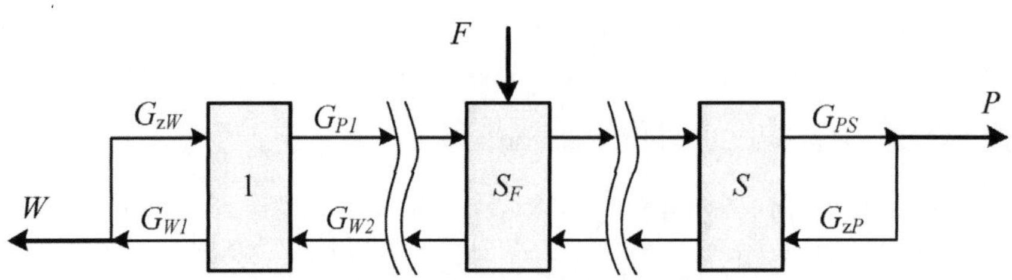

Figure 1. Separation cascade.

Separation stage consists of n parallel-connected GC. The stage has three flows: incoming feed flow G_F, output light fraction flow G_P and heavy fraction flow G_W (figure 2).

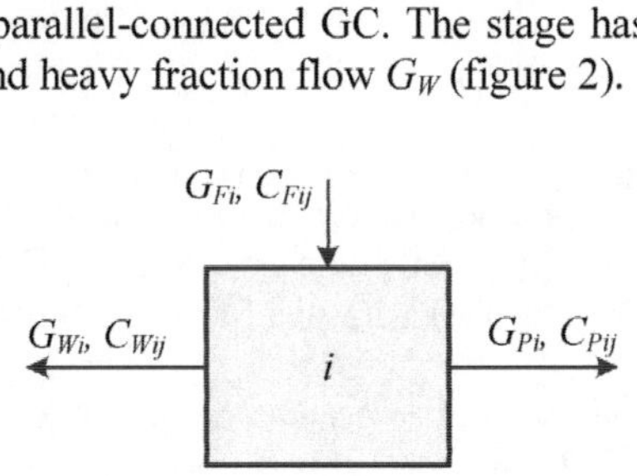

Figure 2. Separation stage.

During stationary hydraulic process the flows and concentrations of components in stage are interrelated by material balance equations and balance equations of each component (in the absence of material losses):

$$G_{Fi} = G_{Pi} + G_{Wi}, \tag{1}$$

$$\theta_i = \frac{G_{Pi}}{G_{Fi}}, \tag{2}$$

where θ_i is cut of the i-th stage.

The flows of cascade and concentrations of the components are interrelated by balance equations:

$$F = P + W. \tag{3}$$

1.2. Capacities of model

A method of calculation non-stationary separation processes is based on the following points:

1. Each stage is composed of finite number of capacities. The value of process gas into cascade (the holdup) is distributed in these capacities.
2. The condition of process gas into capacity is characterized by pressure value.
3. Basic equation of mathematical model is balance equation of process gas and some components in each capacity.

Each capacity has the following properties: the volume V; the pressure of process gas P, concentration of the j-th component C_j, physical properties of process gas (molar mass μ, dynamic viscosity η, the temperature T); the quantity of process gas into capacity (holdup) H.

In the regards, the holdup of cascade is distributed in $4S$ capacities: feed manifold, GC, heavy fraction manifold and light fraction manifold (figure 3).

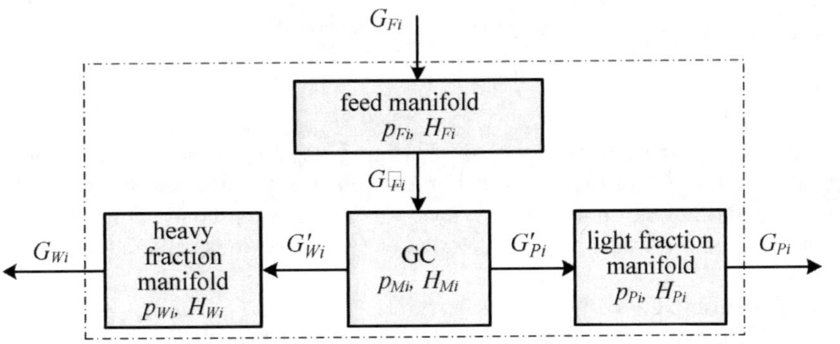

Figure 3. Capacities of i-th stage.

The incoming flow G_{Fi} enters to feed manifold. The flow G'_{Fi} comes in GC and have been separated to flows G'_{Pi} and G'_{Wi} which left to light fraction manifold and heavy fraction manifold. The flows G_{Pi} and G_{Wi} left out from the stage.

The value of concentration C_{Fij} is given by:

$$C_{Fij} = \frac{1}{G_{Fi}} \cdot \begin{cases} \delta_i F C_{Fj} + G_{zW} C_{Wij} + G_{Wi+1} C_{Wi+1j}, & i = 1 \\ \delta_i F C_{Fj} + G_{Pi-1} C_{Pi-1j} + G_{Wi+1} C_{Wi+1j}, & i = \overline{2; S-1} \\ \delta_i F C_{Fj} + G_{Pi-1} C_{Pi-1j} + G_{zP} C_{Pij}, & i = S \end{cases}, \tag{4}$$

$$\delta_i = \begin{cases} 1, & i = S_F \\ 0, & i \neq S_F \end{cases}, \tag{5}$$

Such hydraulic parameters of a stage as values of its pressure and flows are interconnected by following equations:

$$G'_{Fi} = f\left(P_{Fi}\right) \quad (6)$$

$$G'_{Pi} = f\left(P_{Pi}\right) \quad (7)$$

$$G_{Wi} = f\left(P_{Wi}\right) \quad (8)$$

The frequency rotation of the GC rotors and the temperature of equipment operation mode are assumed constant.

1.3. Basic equations describing the nonstationary processes

Basic equations of nonstationary hydraulic processes are balance equations of process gas in all capacities:

$$\begin{cases} \dfrac{dH_{Mi}}{dt} = G'_{Fi} - G'_{Pi} - G'_{Wi} \\ \dfrac{dH_{Fi}}{dt} = G_{Fi} - G'_{Fi} \\ \dfrac{dH_{Wi}}{dt} = \begin{cases} G'_{Wi} - W - G_{zW}, & i = 1 \\ G'_{Wi} - G_{Wi}, & i = \overline{2;S} \end{cases} \\ \dfrac{dH_{Pi}}{dt} = \begin{cases} G'_{Pi} - G_{Pi}, & i = \overline{1;S-1} \\ G'_{Pi} - P - G_{zP}, & i = S \end{cases} \end{cases} \quad (9)$$

The task of calculation the parameters of nonstationary separation processes is reduced to solution differential equation system (9) of first order with initial conditions (values of pressure in the initial time). The values of flows and holdups are arbitrary functions of the time; the system (9) is nonlinear. These circumstances make impossible the analytical solution of the task set in general case.

2. The solution algorithm of nonstationary separation equations

We consider the capacity with the holdup H. Total flow G_{in} feds to the capacity, the flow G_{out} left out from the capacity.

Differential equation from the system (9) describing non-stationary hydraulic processes in this capacity is written in the general form as:

$$\frac{dH}{dt} = G_{in} - G_{out}. \quad (10)$$

The value of capacity holdup has been calculated by following equations:

$$H = a \cdot P, \quad (11)$$

$$a = \frac{\mu V}{RT}, \quad (12)$$

where a is specific capacity holdup, kg·Pa^{-1}; P is pressure of process gas into the capacity, Pa; R=8,314 J·mole^{-1}·K^{-1} is universal gas constant.

Differentiating the equation (12) we have had:

$$\frac{dH}{dt} = a\frac{dP}{dt}. \quad (13)$$

The changes of capacity holdup are directly proportional to changes of pressure. It depends on values of flows Gin and Gout which are proportional to the pressure P:

$$a\frac{\mathrm{d}P}{\mathrm{d}t}=G_{\mathrm{in}}-G_{\mathrm{out}}. \qquad (14)$$

Replacing derivatives to difference equations by backward Euler method [8] and obtaining the equations (10), (11), (14) on the k-th time layer (at the point in time $t^{(k)}$), we have received:

$$a\frac{P^{(k)}-P^{(k-1)}}{\Delta t}=G_{\mathrm{in}}\left(P^{(k)}\right)-G_{\mathrm{out}}\left(P^{(k-1)}\right), \qquad (15)$$

$$\Delta t=t^{(k)}-t^{(k-1)}, \qquad (16)$$

where Δt is time step; k is the index of time layer.

Knowing pressure on previous (k-1)-th time layer (at the point in time $t^{(k-1)}$), we have solved derived nonlinear algebraic equation and have found values of pressure in the point of time $t^{(k)}$. The calculation have been realized as iterative process: on each iteration cycle calculation of pressure and flows have been done from first stage to S-th stage. The initial approximation on k-th time layer is the values of pressure and flows on previous (k-1)-th time layer.

Thus the calculation of nonstationary hydraulic processes reduces to replacement differential equations of first order by difference equations, solution of derived nonlinear algebraic equations and iterative evaluation of pressures in each capacity and flows on every time layers satisfying balance equation of process gas into cascade:

$$\left|\frac{\left(F^{(k)}-P^{(k)}-W^{(k)}\right)\cdot\Delta t}{H'^{(k)}-H'^{(k-1)}}-1\right|\le\sigma, \qquad (17)$$

$$H'^{(k)}=\sum_{i=1}^{S}\left[H_{Mi}^{(k)}+H_{Fi}^{(k)}+H_{Wi}^{(k)}+H_{Pi}^{(k)}\right], \qquad (18)$$

where H' is the cascade holdup, kg; σ – the condition of iteration process end.

We have realized the algorithm described herein as software product, developed on Delphi programming language using Embarcadero Delphi XE2.

The developed mathematical model has been tested for modeling of nonstationary hydraulic processes in cascades for separation of Si, Xe, Ni and W isotopes. The deflections between estimated and real values are 7.5 %. As the result we have ascertained that developed model is universal for the calculation of nonstationary hydraulic processes in gas centrifuge cascade for separation of multicomponent isotope mixtures using different process gases.

3. Conclusion

1. The result of research is creation of mathematical model of nonstationary hydraulic processes in gas centrifuge cascade for separation of multicomponent isotope mixtures, representing the differential equation system of first order with initial conditions.
2. The solution algorithm of equation system described nonstationary separation processes is developed and realized as software product.
3. We have made the calculation parameters of nonstationary hydraulic processes in GC cascade for separation of Si, Xe, Ni and W isotopes.
4. We have ascertained that developed model is suitable for the calculation of nonstationary hydraulic processes in gas centrifuge cascade for separation of multicomponent isotope mixtures using different process gases.

IOP Conf. Series: Materials Science and Engineering **142** (2016) 012033 doi:10.1088/1757-899X/142/1/012033

References

[1] Cohen K 1951 *The Theory of Isotope Separation as Applied to the Large-Scale Production of U-235* (New York: McGraw Hill)

[2] Sulaberidze G A *et al* 2011 *Theory of cascades for separation of binary and multicomponent isotope mixtures* (Moscow: MEPhI) p. 368

[3] Vlasov V A, Butov V G, Goldobin D N, Orlov A A, Timchenko S N Modeling of nonstationary processes in gas centrifuge cascades 2007 *Tomsk Polytechnic University news* **310** 103

[4] Vlasov V A, Orlov A A, Butov V G, Timchenko S N Modeling of nonstationary hydraulic processes in gas centrifuge cascades for uranium isotopes separation 2009 *Tomsk Polytechnic University news* **315** 94

[5] Orlov A A, Timchenko S N, Sidorenko V S Mathematical Model of Non-Stationary Hydraulic Process Occurring in Gas Centrifuges for Uranium Enrichment 2015 *Advanced Materials Research* **1084** 673

[6] Filimonov S V, Skorynin G M, Goldobin D N Modeling of nonstationary processes in automatic control system of industrial gas centrifuge cascade for uranium enrichment 2004 *Universities news. Physics* **47** 173

[7] Filimonov S V, Skorynin G M, Orlov A A, Goldobin D N Modeling of nonstationary hydraulic processes in industrial gas centrifuge cascade for uranium enrichment 2006 *Tomsk Polytechnic University news* **309** 85

[8] Pirumov U G 2003 *Numerical methods* (Moscow: Drofa)

Thermodeformational Behavior of Cubic Crystals of Sodium Cyanide

[1]E P Tesleva, [1]L B Gil, [1]A V Solovyan

[1]Yurga institute of technology of National Research Tomsk Polytechnic University
Yurga, Kemerovskaya obl., Russia

E-mail: tesleva@mail.ru

Abstract. In the paper the authors provide results of the study of anisotropic and isotropic acoustic (elastic waves propagation velocities), dimensional (elastic moduli) and nondimensional (Poisson's ratios) elastic characteristics of mono- and polycrystals of sodium cyanide within the temperature range 283.7÷473 K.

1. Introduction

For most solids it is considered normal to decrease under temperature growth according to the linear law and small dependence (or no dependence) under low temperatures of elastic waves, elastic constants and elastic moduli propagation rates [1-4]. Any deviation from the given functional changes is to be treated as anomalous. In this context the results of experimental study of temperature dependences of stiffness constants $c_{ij}(T)$ of alkali metals cyanides monocrystals completed by Haussühl and others [5-9] afford ground for such conclusion. The characteristics of thermoelastic behavior of high-temperature disordered cubic phases of cyanides MeCN (Me = Na, K, Rb, Cs) are as follows:

a) Velocities of elastic waves propagation increase with temperature growth within significant temperature ranges.

b) One of two shear constants (in NaCN stiffness constant c_{44}) tends to zero when temperature approaches critical value of T_c (T_c – temperature of structural phase transition NaCN I → NaCN II, T_c = 283.7 K [7], 288.1 K [10]) according to dependence $T_{44} = d(\log c_{44})/dT = 1/(T \log T/T_0)$, where T – temperature in K, and T_0 – specific constant (for NaCN T_0 = 255.4 K [7]).

c) Volume compressibility and thermal expansion under temperature change demonstrate dependences typical for stable solids.

The theory of emolliating of stiffness constants and phonons in ion molecular crystals on the example of alkali cyanides is considered in [11]. The authors suggest correlations describing renormalization of phononic frequencies due to interaction between translational and rotational degrees of freedom. According to the correlations under some temperatures stiffness constants c_{11}, c_{44} may go to zero. In alkali cyanides, as temperature grows, constant c_{44} is the first to go to zero, in superoxides – c_{11}. This difference is determined by the correlation between the short-range repulsive forces and the interaction force of quadrupole moment of molecule with electric field gradient fluctuating due to phonons. Dependence of c_{11}, c_{44} on temperature is also influenced by anharmonicity and direct interaction of molecules. Due to $c_{44} = 0$ for NaCN the calculated critical temperature of T_0. The difference is associated with application of mean field approximation when calculating.

Information on elastic moduli (Young modulus E, shear modulus G) for three specific crystallographic directions <100>, <110>, <111> of the cubic phase ($T > T_c$) of NaCN monocrystal, values of these moduli of sodium cyanide polycrystals as well as anisotropic ($\sigma_{<hkl>}$) and isotropic (σ) Poisson's ratios of the given compound from the group of cyanic crystals are not found in reference literature. In the given paper the authors determine and discuss the mentioned above elastic characteristics of cubic NaCN and some properties of cyanide associated with its elasticity, particularly anharmonic.

2. Theoretical basis

The lattice of isomorphic sodium, potassium, rubidium cyanides under room temperature is of KCN-type, a distorted type of rock salt lattice (NaCl, type B_1, spatial group Fm-$3m$) which formally is impossible as the crystal lattice of the given substances is made up of three different atoms (Fig. 1) [12]. But as temperature decreases the structures of their lattices become less symmetrical (Table 1) and then positions of carbon and nitrogen atoms can be distinguished: the distance between them is 0.105 nanometer. In the cubic phases of alkali metals cyanides ion CN^- behaves as if its symmetry was spherical. This is only statistically average due to rotation of completely-bound ("dumbbell") C and N atoms around the common centre of gravity and, thus, imitate spherical symmetry.

Figure 1. Elementary cell of the cubic phases of KCN-type crystals: large spheres – CN^-, small – Na^+; K^+, Rb^+ [12]

The mechanism which is alternative to the rotational mechanism of disorder in the lattices of cyanides is orientation of CN^- ions in a limited number of positions [13]. Excess entropy $\Delta S \approx Rln2$ under III → II transition in salts NaCN and KCN quite evidently show that in orthorhombic forms II of these crystals CN^- ions are randomly distributed between two orientations. From excess entropy under NaCN II → NaCN I transition follows that it results in significant further disorder orientation in high-temperature cubic phase: $\Delta S \approx Rln4$ and corresponds to the transition to eight orientations of cyanide ions. Analysis of the results of heat capacity, nuclear magnetic resonance of inelastic spectrum in NaCN I showed that in the cubic phase barriers of CN^- ions reorientation are relatively small (8 12 kJ/mole) [13]. A rather common characteristic of disordered cubic forms of alkali metals cyanides is that motion of ions of one type notably affects orientational motion of ions of another type (cyanide ions).

IOP Conf. Series: Materials Science and Engineering **142** (2016) 012034 doi:10.1088/1757-899X/142/1/012034

Table 1. Thermodynamic and structural data on cyanides of alkali metals [13]

Cyanides	III ↔ II T_c (K), $\Delta S_c/R$	II ↔ I T_c (K), $\Delta S_c/R$
NaCN	171.9; 0.61 (λ) monoclinic	288.1; 1.41 (λ) orthorombic NaCl
KCN	 monoclinic	168.3; 1.01 (λ) orthorombic NaCl
RbCN		110.3; 0.695 (F) orthorhombic NaCl
CsCN		193.1; 1.31 rhombohedrical CsCl

Note: T_c – temperature of transition, ΔS_c – entropy change, R – universal gas constant, (λ) – smooth transition, (F) – abrupt transition.

Results and discussion

The given work is based on experimental data on stiffness constants c_{ij} of monocrystal NaCN I in the interval 283.7 – 473 K obtained with ultrasonic resonance at the frequency of 15 mHz with limiting error for c_{11}, c_{44} and c_{12} accordingly of 0.4%, 0,8% and 1.0% [7]. Experiment conducted at higher frequencies (order of units – dozens of gigahertz, method of Brillouin scattering) brings to values of c_{11}, c_{12} which are lower by 6-7% and c_{44} which is by 14% lower for NaCN under 20°C [14]. This fact for c_{44} was checked and discussed in [15]: there is no c_{44} dispersion and loss of stability of NaCN I lattice under $T \rightarrow T_c$ continues due to mode frequency of transversal acoustic phonons vanishing. Elasticity of low-temperature forms of pure sodium cyanide was not studied.

Table 2. Anisotropic sound velocities and Gruneisen parameters of NaCN monocrystal

T, K	Crystallographic direction										
	[100]			[110]					[111]		
	υ_L	υ_t	γ	υ_L	υ_{t1}	υ_{t2}	γ_1	γ_2	υ_L	υ_t	γ
1	2	3	4	5	6	7	8	9	10	11	12
283.7	3967	385	4.360	3731	385	1816	4.340	2.089	3396	1500	2.396
293	3981	454	4.310	3556	454	1846	4.263	1.873	3403	1530	2.340
313	4004	557	4.220	3569	557	1898	4.151	1.790	3412	1583	2.243
333	4020	645	4.133	3582	645	1936	4.043	1.733	3423	1624	2.171
353	4031	720	4.050	3592	720	1967	3.942	1.688	3433	1659	2.112
373	4040	793	3.964	3602	793	1994	3.838	1.650	3443	1691	2.060
393	4046	860	3.879	3612	860	2016	3.737	1.621	3455	1719	2.016
413	4049	920	3.797	3620	920	2033	3.641	1.599	3466	1743	1.981
433	4049	973	3.723	3626	973	2047	3.556	1.580	3474	1763	1.949
453	4045	1025	3.647	3630	1025	2059	3.469	1.563	3480	1782	1.920
473	4039	1073	3.572	3632	1073	2067	3.385	1.552	3486	1798	1.896

All calculated ratios used in the given paper are provided in [16, 17]. Results of current research are provided in Tables 2-4 and in Figure 2. Table 2 provides data on anisotropic rates of propagation of longitudinal (υ_L), transverse (υ_t) elastic waves, Gruneisen parameters γ in directions <100>, <110>, <111> of NaCN monocrystal in the interval from 283.7 K to 473 K. Except longitudinal sound velocity along face diagonal <110> all other propagation velocities increase non-linearly with temperature growth. Maximal temperature dependence is demonstrated by transverse wave velocity along cube edge (<100>, increase by 1.79). Anisotropic Gruneisen parameters along three specific directions of NaCN monocrystal significantly decrease with temperature growth, especially $\gamma_{2<110>}$ (by 26%), which is not typical for ion crystals, for example, for NaCl. At the same time, for all studied temperature range Gruneisen parameters maintain the inequation: $\gamma_{<110>} > \gamma_{1<110>} > \gamma_{<111>} > \gamma_{2<110>}$.

Table 3. Elastic anisotropy factor *A*, Couchy relation Δ, bulk moduli *B*, Young *E* and shear *G* of NaCN monocrystal

T, K	*A*	Δ	*B*	$E_{<100>}$	$E_{<110>}$	$E_{<111>}$	$G_{<100>}$	$G_{<110>}$	$G_{<111>}$
				hPa					
283.7	0.045	61.51	18.2	14.45	0.93	0.71	0.24	0.46	0.66
293	0.061	43.76	18.1	14.86	1.28	0.98	0.33	0.62	0.88
313	0.086	28.40	17.9	15.56	1.90	1.47	0.50	0.91	1.27
333	0.111	20.83	17.7	16.05	2.51	1.96	0.66	1.19	1.62
353	0.134	16.43	17.6	16.45	3.08	2.42	0.82	1.45	1.94
373	0.158	13.32	17.4	16.78	3.67	2.91	0.99	1.71	2.26
393	0.182	11.16	17.2	17.03	4.25	3.40	1.16	1.96	2.55
413	0.205	9.59	17.0	17.22	4.81	3.87	1.32	2.20	2.82
433	0.226	8.46	16.8	17.35	5.30	4.30	1.48	2.41	3.05
453	0.248	7.51	16.6	17.44	5.79	4.74	1.63	2.61	3.27
473	0.269	6.75	16.4	17.47	6.26	5.15	1.78	2.80	3.47

Table 4. Density ρ, thermal-expansion coefficient α, sound velocities, elastic moduli, thermodynamic γ_D and acoustic γ_a Gruneisen parameters of NaCN polycrystal

T	ρ, 10^3	α, 10^{-6}	υ_L	υ_t	*E*	*G*	*B/G*	γ_a	γ_D
K	kg/m^3	K^{-1}	m/s		hPa				
283.7	1.602	58	3527	908	3.87	1.32	13.75	3.62	1.10
293	1.599	58	3538	953	4.25	1.45	12.44	3.55	1.31
313	1.593	59	3555	1029	4.91	1.69	10.60	3.42	1.50
333	1.587	59	3572	1094	5.50	1.90	9.33	3.32	1.49
353	1.582	60	3586	1151	6.05	2.10	8.37	3.22	1.51
373	1.576	60	3600	1207	6.60	2.30	7.56	3.12	-
393	1.570	60	3614	1258	7.12	2.49	6.92	3.04	-
413	1.564	61	3627	1305	7.59	2.66	6.39	2.96	-
433	1.558	62	3635	1345	8.01	2.82	5.97	2.89	-
453	1.552	63	3642	1384	8.41	2.97	5.60	2.82	-
473	1.546	64	3647	1419	8.78	3.11	5.27	2.76	-

Elastic anisotropy factor and Cauchy relations (Table 3) are indicative of NaCN crystal being strongly anisotropic where the forces of interaction between the ions are also acentral. For example, for NaCl under normal temperature $A = 0.72$ and $\Delta = 1.02$ (are close to one) [15]. Anisotropic Young $E_{<hkl>}$ and shear $G_{<hkl>}$ moduli as opposed to bulk modulus increase with temperature growth and the following inequations are maintained: $E_{<100>} > E_{<110>} > E_{<111>}$, $G_{<100>} < G_{<110>} < G_{<111>}$.

Elastic and some other characteristics of NaCN polycrystal are presented in Table 4 (note: for cubic lattices B of mono- and polycrystals coincide). Here we would like to pay attention to two moments.

1. Relation of B/G moduli in NaCN is large especially near T_c. Such values of B/G are characteristic of superplastic materials. To compare let us provide typical values of B/G for malleable gold (6.03), fragile diamond (0.83) and NaCl which is on the border of fragility-plasticity (1.73).
2. Bad agreement between acoustic and thermodynamic Gruneisen parameters. However heat capacity of NaCN was measured only in low-temperature region (5-345 K [10, 18]) and γ_a and γ_D are compared for small overlapping temperature range bordering with T_c which value differs due to two methods of recording (see Introduction).

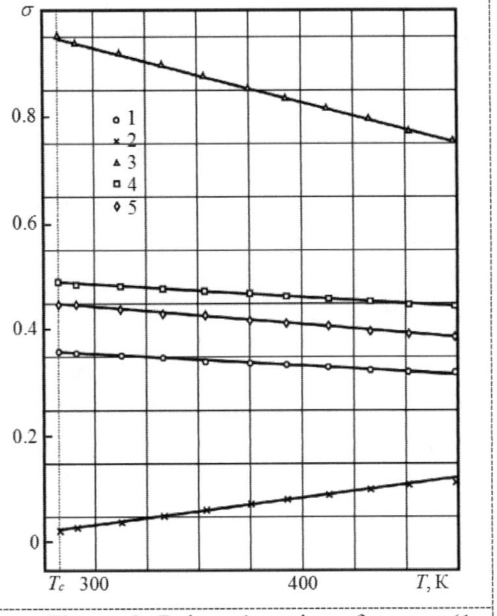

Figure 2. Poisson's ratios of mono- (1, 2, 3, 4) and polycrystals (5) of NaCN dependence on temperature 1 $-\sigma_{<100,001>}$, 2 $-\sigma_{<110,001>}$, 3 $-\sigma_{<110,1\bar{1}0>}$, 4 $-\sigma_{<111,111>}$

Dependence of Poisson's ratios of sodium cyanide mono- and polycrystals on temperature is presented in Fig. 3. Longitudinal-transverse deformation of NaCN along direction <110> is diametrally opposite: value of $\sigma_{<110,001>}$ is minimal and its function to temperature is positively sloped, $\sigma_{\langle 110,1\bar{1}0\rangle}$ is maximal, near T_c its value tends to one and $d\sigma_{\langle 110,1\bar{1}0\rangle}/dT < 0$. In point of T_c transition into partially ordered orthorhombic phase NaCN II isotropic Poisson's ratio approaches upper limit 0.5 (σ = 0.454) and under further temperature decrease a break is obvious on dependence $\sigma(T)$. All Poisson's ratios of the cubic phase of NaCN I crystal are positive within the studied temperature range and are expected to stay positive until melting (T_{melting} = 835 K). Only Poisson's ratio $\sigma_{<110,001>}$ can reach zero in phase NaCN II near the temperature of 250 K (extrapolation from T_c). This value of temperature well coincides with experimentally found parameter T_0 = 255.4 K applied for describing temperature dependence of shear stiffness $c_{44}(T) = 2.144(T - T_0)/T_0$ hPa [7]. If $\sigma_{<110,001>}$ of NaCN II takes negative values under $T < T_c$ and, thus, rhombic sodium cyanide turns into non-axial auxetic still remains open to question.

Conclusion

On the base of data on stiffness constants the authors studied temperature dependences of anisotropic and isotropic acoustic and elastic characteristics of sodium cyanide mono- and polycrystals within the range 283.7÷473 K. It has been established that as temperature grows we observe anomalous increase of sound velocities (except longitudinal velocity along <110>). Uncharacteristic temperature dependence is observed for anisotropic Gruneisen parameters – along three specific directions – with temperature growth they significantly decrease. NaCN crystal is a strongly anisotropic object where forces of interaction between ions are significantly non-central. Relation of B/G moduli in NaNC indicates superplasticity of the crystal near T_c. Temperature dependences of anisotropic moduli (Young, shear, bulk) and Poisson's ratios are linear.

References

[1] Sorokin, B.P., Glushkov, D.A., Aleksandrov K.S. 1999 *Physics of the Solid State* **41** 208
[2] Belomestnykh V N, Tesleva E P, Soboleva E G 2008 *Technical Physics Letters* **34** 867
[3] Belomestnykh V N, Soboleva E G 2014 *Applied Mechanics and Materials* **682** 170
[4] Belomestnykh V N, Soboleva EG 2012 *7th International Forum on Strategic Technology (IFOST - 2012)* **1** 499
[5] Haussühl S. 1973 *Solid State Communic.* **13** 147
[6] Kasser W., Buchenau U., Haussühl S. 1976 *Solid State Communic.* **18** 287
[7] Haussühl S., Eckstein J., Recker K., Wallrafen F. 1977 *Acta Cryst.* 33 847
[8] Haussühl S. 1979 *Solid State Communic.* **2** 181.
[9] Loidl A., Haussühl S., Kjems J.K. 1983 *Z. Phys.* **50** 187
[10] Matsuo T., Suga H., Seki S. 1968 *Bull. Chem. Soc. Japan.* **41** 583
[11] Sahu D., Mahanti S.D. 1982 *Phys. Rev. B: Condens. Matter.* **26** 2981.
[12] Stokes H. T., Decker D. L., Nelson H. M., Jorgensen J. D. 1993 *Phys. Rev.* **47** 11082
[13] Parsonage N.G. Staveley L.A.K. Disorder in Crystals, Clarendon Press, Oxford, 1978.
[14] Satija S.K., Wang C.H. 1977 *J. Chem. Phys.* **66** 2221
[15] Rehwald W. 1982 *Phys. Letters.* **87** 245
[16] Belomestnykh V.N., Tesleva E.P. 2012 *Russian Physics Journal.* **55** 488
[17] Tesleva E. P., Belkova T. A. 2014 *Applied Mechanics and Materials* **682** 519
[18] Messer Ch.E., Ziegler W.T. 1941 J. *Amer. Chemic. Society.* **63** 2703

Effect of Intensive Plastic Deformation on Microstructure and Mechanical Properties of Aluminum Alloys

Bauyrzhan Rakhadilov[1,a], Gulzhaz Uazyrkhanova[1,b], Alexandr Myakinin[1,c], Zhuldyz Uazyrkhanova[1,d]

[1]D.Serikbayev East Kazakhstan State Technical University, Serikbayev Str., 19, 070010, Ust-Kamenogorsk, Kazakhstan

E-MAIL: [a]Bor1988@mail.ru; [b]GUazyrhanova@mail.ru; [c]miakininsania@mail.ru; [d]zhuldyzuaz@mail.ru

Abstract-In work it was studied the influence of intensive plastic deformation on structure and mechanical properties of aluminum alloys. Intensive plastic deformation was carried out by using equal-channel angular extrusion. It is shown that the most efficient angle of intersection of the channels is the angle of Φ=120°, which ensures defect-free parts at the highest possible level of accumulated strain (e=8). It is established that the intensive milling grain structures in aluminum alloys AMG6 and AMC occurs at ECAE-12 passes, while the intersection angle of the channels of 120°. After ECAE-12 in aluminum alloys the grain refinement reaches to the size of ~1.0-1.5 μm. It is determined that as a result of equal channel angular pressing, the microhardness of alloy AMG6 increases almost 4 times in comparison with the initial state, the microhardness of alloy AMC increases by almost 4.5 times in comparison with the initial state. It is shown that ECAE-12 mass loss is reduced to 5.4 and 5.6 mg, which shows an increase in wear-resistance of aluminum alloys AMG6 and AMC 13-14 %.

Introduction

One of the main tasks of modern materials science is to obtain high-strength materials with increased strength and technological characteristics. The solution of this problem is an effective way to give metals and alloys with submicrocrystalline (QMS) and nanocrystalline (NC) structure. [1-5]. The most promising method of obtaining the QMS and NDT of materials is intensive plastic deformation (IPD) [1-3,6]. Intensive plastic deformation – is a complex physico-chemical process, in which, while changing the shape and structure of the initial state, it changes its physico-chemical and mechanical properties. With various types and modes of plastic deformation in crystalline materials with different type of crystal lattice is observed the basic phenomenon of fragmentation, i.e. the deformation refinement of structure of materials [3,5,7]. Obtained by using the method of IPD ultrafine-grained materials possess high strength and significant impact resistance compared to traditional compositions of materials, and these materials are currently the subject of extensive research throughout the world. The most widely used method of IPD, allowing providing high intensity and a more homogeneous strain state is equal channel angular pressing (ECAE). Method ECAE has several advantages over other methods of IPD that are associated with the implementation of the process of plastic deformation of the simple shear scheme. ECAE allows with minimal energy consumption to accumulate a large and uniform strain over the loop, without changing the cross section of samples [9,10].

In connection with the foregoing, the aim of this work was to study influence of ECAE regimes on the microstructure and mechanical properties of aluminum alloys AMG6 and AMC.

Material and Experimental Procedure

As research material was selected as aluminum alloys AMG6 and AMC. These alloys are widely used in modern aviation, aircraft, shipbuilding, railway transport, road transport, construction, petroleum and chemical industries. Chemical composition of aluminum alloys: AMC – 96.35-99% Al; 1-1.5% Mn; 0.6% Si; 0.7% Fe; 0.05-0,2%Cu; 0.1% Zn, AMG6-91.1-93% Al; 0.5-0.8% Mn; 0.4% Si; 0.4% Fe; 0.1%Cu; 0.2% Zn; 5.8-6.8% Mg.

Experimental studies of the structure and mechanical tests were carried out in laboratories of D. Serikbayev East Kazakhstan State Technical University, "Nanotechnologies and new materials" Scientific Research Institute, "IRGETAS" Regional laboratories of engineering profile by optical microscopy methods on the microscope ALTAMI-МетlM, X-Ray analysis, X-Ray diffractometer X'pert PRO and microhardness testing on a PMT-3M and abrasive wear. Shooting diffraction patterns was performed using CuKα - radiation (λ=2.2897 A^o) with a voltage of 35 kV. Measurement of the samples microhardness was carried out according to Vickers method with a load on the indenter P=1N and holding time 10 sec. Tests of durability were carried out on experimental setup for testing the abrasion by friction are not rigidly attached to the abrasive particles according to the "rotating roller – flat surface" scheme in accordance with the State Standard 23.208-79 that corresponds to the American standard ASTM S 6568.

In the present work to produce ultrafine-grained structure in aluminum alloys was developed and manufactured universal equipment for equal channel angular pressing (ECAE) (figure 1). For aluminum alloys typical are not high temperature processing. They can be processed at room temperature, in which they have sufficient deformability. Accordingly, for pressing in such circumstances, there is reliable operation of equipment and the use of special techniques to ensure manufacturability of the process. A distinctive feature of this tooling is that it allows ECAE at different angles of intersection of channels. The snap-in has two stamps with angles of channels intersection - 90 and 120^0. Through the use of stamps, crossing the different angles you can vary the intensity of deformation effects in the loop. Given that during multi-cycle processing, the resource of plasticity of the sample is reduced, you may need to change the level of a single strain per cycle of treatment. The principle of operation of the snap-in is as follows. The sample is fed to the vertical channel of the molding and pressed into the punch in a horizontal channel. By subsequent pressing the second sample, or simulator (harvested from specially selected material) of the first sample is removed (ejected) from the horizontal channel. Processing cycles are repeated by the planned number of process cycles and route selected of pressing.

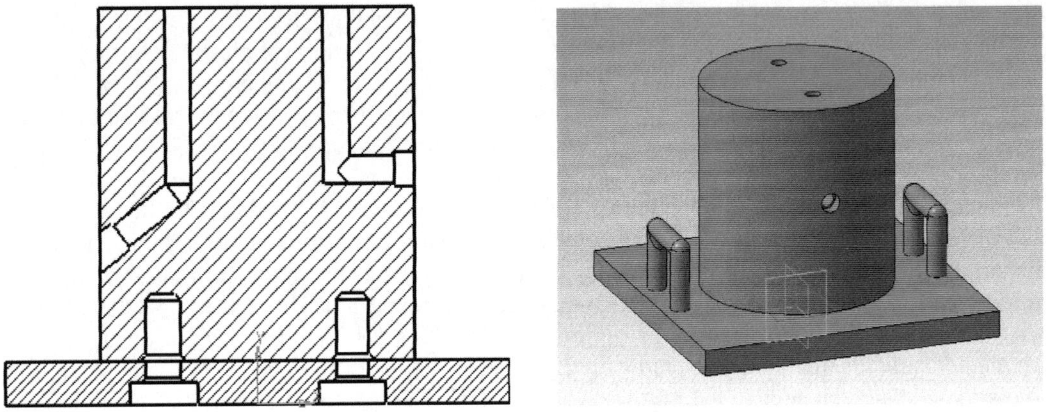

Fig. 1. The equipment for ECAE

The study used an initial billet of aluminum alloy AMC and AMG6 with diameter -12 mm and length -50 mm. The diameters of the extrusion channels were as follows: the input (vertical) of 13 mm, output 12 mm. The angle of intersection of channels pressing were respectively 90 and 120°. Used routes pressing: B (90° rotation) and C (rotated 180°). ECAE was performed using tooling that was installed on a hydraulic press brand KNWP 30M. The pressed samples were produced with pressure up to 30 kg/sm^2, at the pressing time up to 40 sec. at each pass.

Results and Discussion

One of the objectives of this work is the determination of the ECAE, which provides a solid defect-free bulk billets with fine-grained structure and study the influence of process parameters on the forming structure. An important parameter of ECAE is the angle of intersection of channels - Φ. It defines the degree of deformation for one cycle of treatment (e) and accumulated strain (e_N) for several cycles. Study of the effect of intersection angles and channels on deformability is presented in table 1.

Table 1 - Influence of the intersection angles of channels in ECAE of AMG6 alloy at room temperature in the deformed state during multicycle processing

The intersection of angles channels, Φ°	Amount of deformation per one cycle, e	The maximum number of cycles to failure sample (N)	The maximum accumulated deformation (e_N)
90	1.15	4	4.6
120	0.67	12	8

Also studies have been conducted on the effect of the angle of intersection of channels in the deformation of AMG6 alloy. As a result of researches it is revealed that when the angle of intersection of the channels equal to 90 (e=1.15) billet destroyed already after the fourth cycle of compression. Optimal results were obtained by ECAE with an angle of intersection of the channels 120°, in which the number of processing cycles has reached 12, with e=8. In such regimes have the opportunity to obtain defect-free billet with a more uniform structure.

Research and comparative analysis showed that the most efficient angle of intersection of the channels is the angle of Φ=120°, which ensures defect-free parts at the highest possible level of accumulated strain (e=8).

The results of optical microscopy showed that aluminum alloys AMG6 and AMC in the initial state is mainly characterized by equiaxed shape of the grains (figure 2).

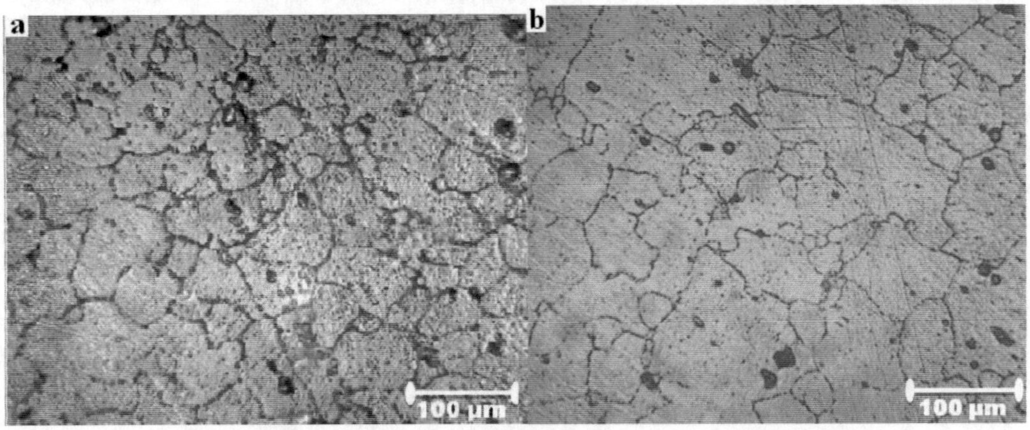

Fig. 2. Microstructure of aluminum alloys in the initial state:
(a) AMG6; (b) AMC.

Determined that ECAE leads to a significant refinement of the aluminum alloys structure, depending on the cycles number. In figures 3 and 4 show microstructure alloys AMG6 and AMC after ECAE with different modes. From the presented results it is possible to highlight data corresponding to two passes ECAE. After ECAE-8 (120) and ECAE-12 (120) in aluminum alloys is the grain size of the starting material to the size of ~ 1.0-1.5 μm. These results show that the average grain size for the alloy AMC decreased by almost 13 times compared to the initial state, and for AMG6 alloy decreased by almost 17 times.

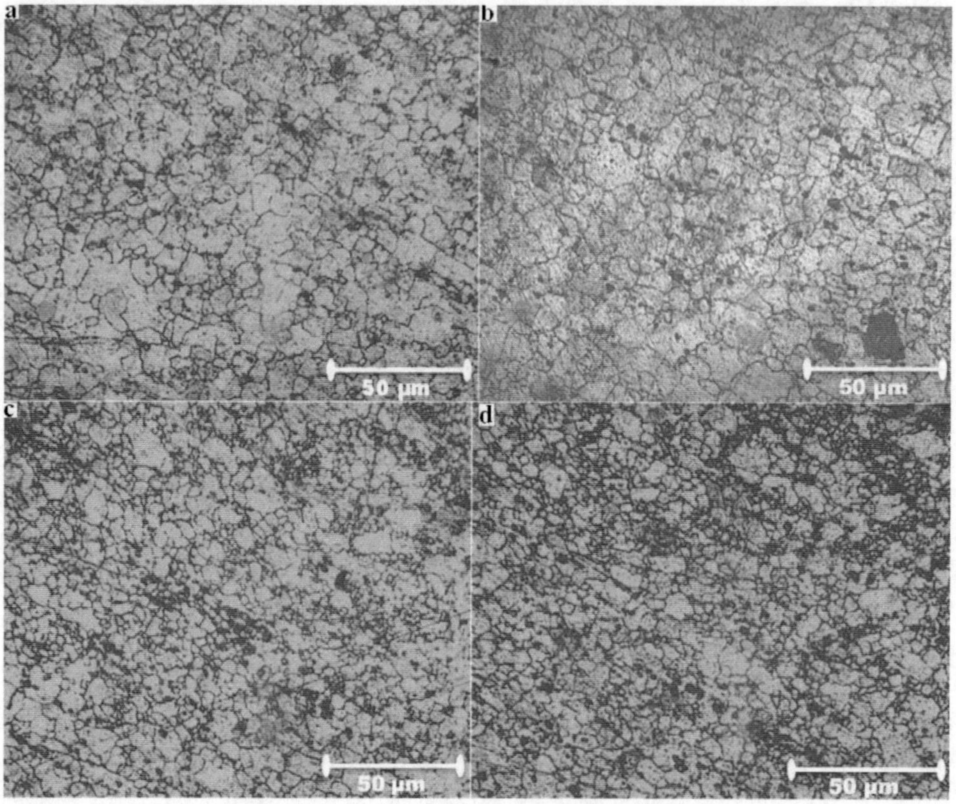

Fig. 3. The microstructure of the alloy specimens after ECAE AMC-4 (90) (a), ECAE-4 (120) (b), ECAE-8 (120) (c) and ECAE-12 (120) (d)

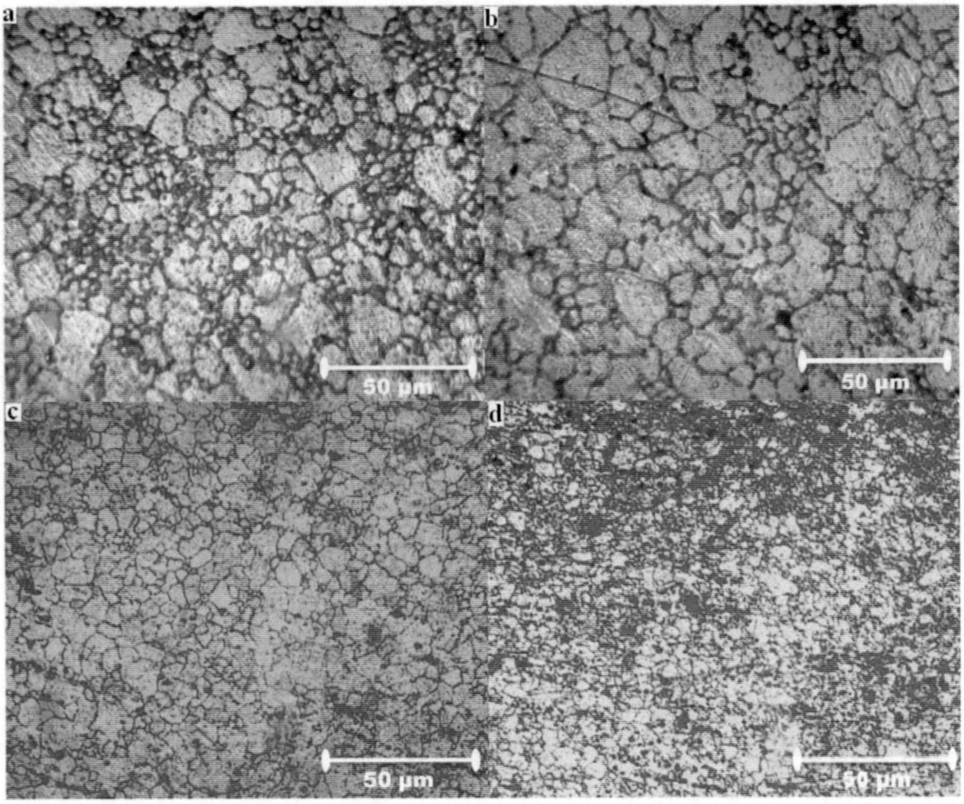

Fig. 4. The microstructure of alloy specimens after ECAE AMG6-4 (90) (a), ECAE-4 (120) (b), ECAE-8 (120) (c) and ECAE-12 (120) (d)

As shown by the results obtained, refinement of the structure at ECAE is accompanied with the increase of the microhardness of aluminum alloys AMG6 and AMC. Figure 5 shows histograms of the microhardness according to Vickers AMG6 and AMC aluminum alloys. After ECAE, an increase of the microhardness depending on the number of passes. While samples processed by ECAE when the angle of intersection of channels 900 microhardness higher than in samples processed by ECAE when the angle of intersection of channels 1200. However, ECAE with an angle of intersection of the 900 channels were not given the opportunity get blanks with defect-free structure with aisles greater than 4. But during ECAE with angles intersection of channels 1200 have the opportunity to obtain the billet with homogeneous defect-free structure when the number of passes to 12. Determined that as a result of ECAE (12 passes), the microhardness of AMG6 alloy increases to 1640 MPa, which is almost 4 times higher than in the initial state (Fig.5). As a result, the ECAE (12 passes), the microhardness of AMC alloy increases to 1820 MPa, which is almost 4.5 times higher than in the initial state.

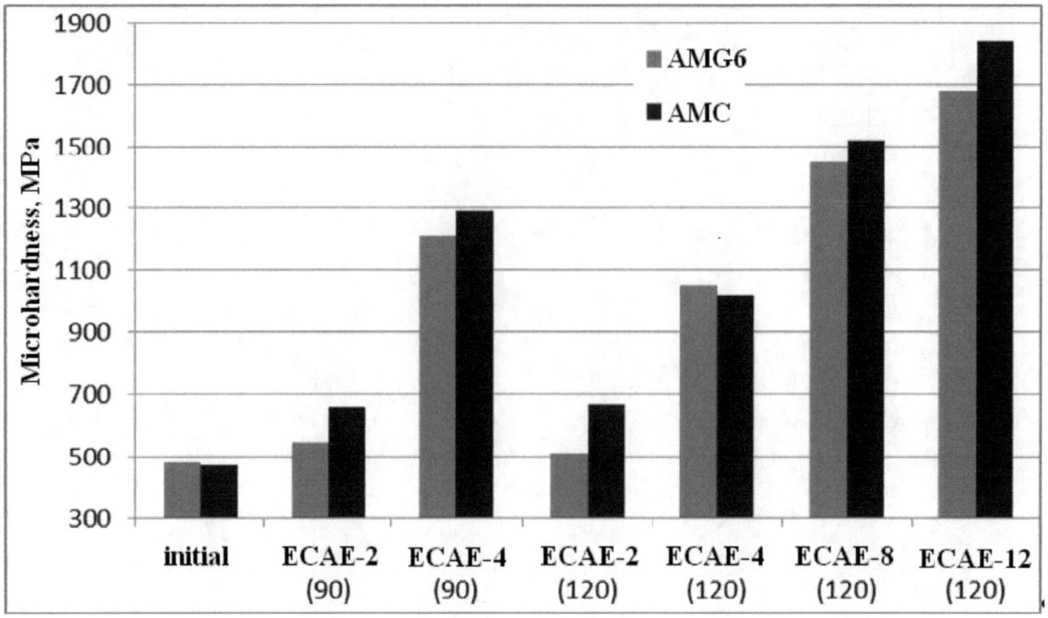

Fig. 5. The microhardness of the AMG6 and AMC alloys before and after ECAE

Figure 6 shows the values of the mass loss of AMG6 and AMC alloys samples before and after ECAE under different conditions. It is seen that the change in mass loss after ECAE are not significant at passages 2 and 4. Significant change is only observed after ECAE, 8 and 12 passes. After ECAE-12 mass loss is reduced to 5.4-5.6 ppm, which shows an increase in wear resistance of aluminum alloys AMG6 and AMC 13-14 %.

Properties of materials are determined by the pattern formed in the processing structure. We carried out a systematic study on the change of structure and mechanical properties (microhardness and wear resistance) of aluminum alloys, depending on the number of passes and the angle between channels pressing ECAE. It is shown that by varying the number of passes to obtain different grain sizes and different values of the mechanical characteristics.

Figure 7 shows a graph of the microhardness of the average grain size of AMG6 and AMC aluminum alloys. From the picture you can see the inverse dependence of the microhardness from the size of the grains. I.e. with the decrease of grains, the microhardness of the alloys increases. It is associated with a particular state of a crystal lattice in the grains and grain boundaries structure.

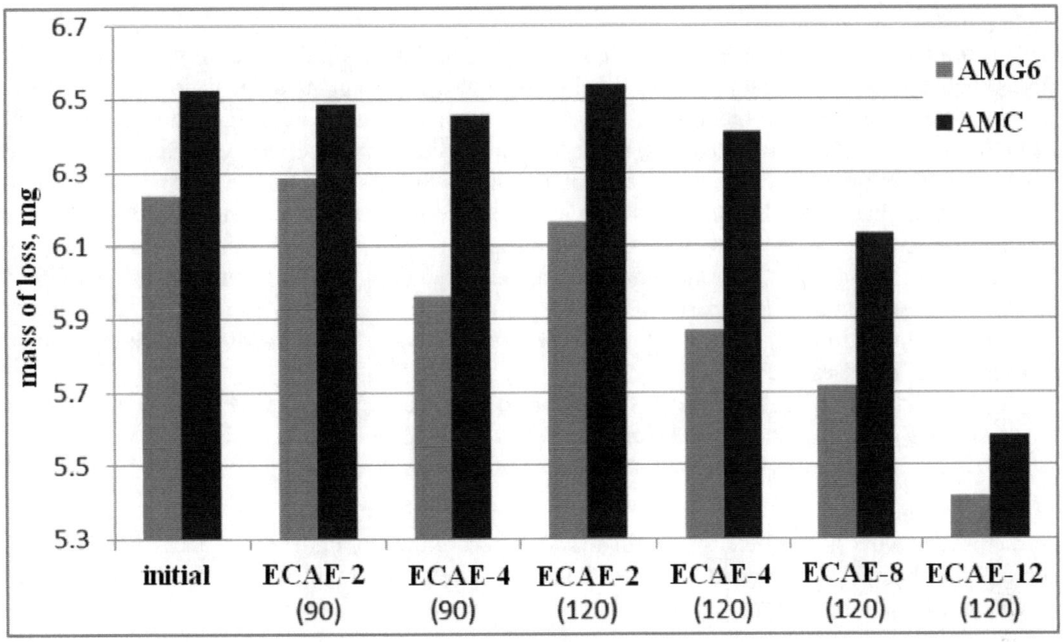

Fig. 6.The mass loss of AMG6 and AMC alloys samples before and after ECAE under different conditions.

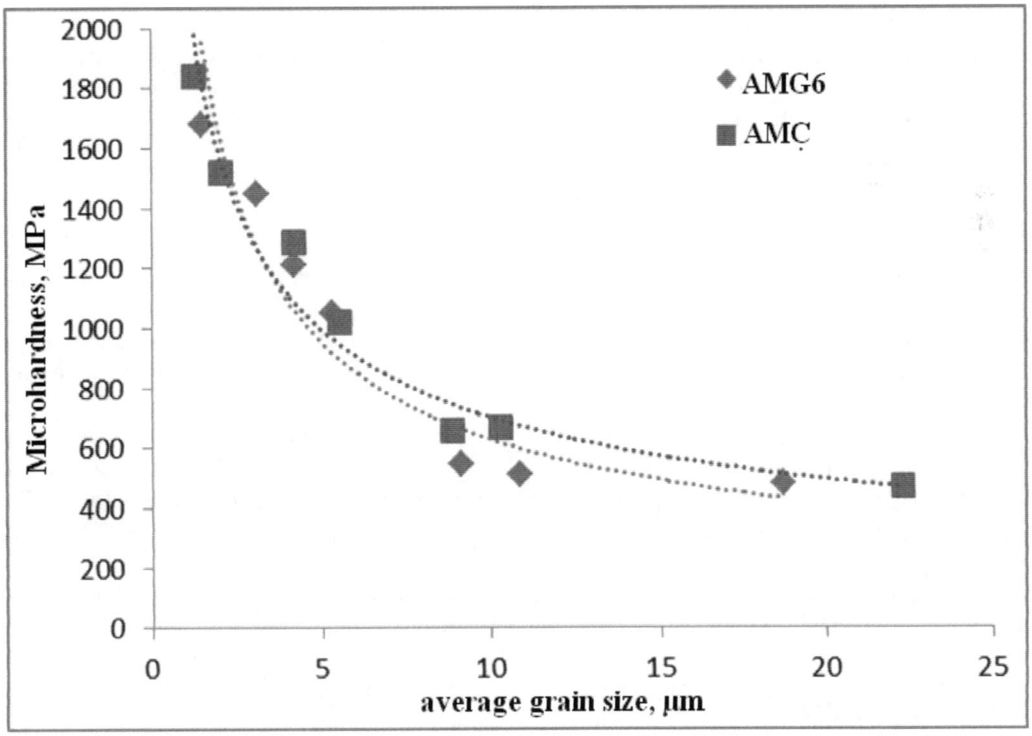

Fig. 7. A graph of the microhardness from the average grain size of AMG6 and AMC aluminum alloys

Conclusions

Thus, on the basis of the received research results about intensive plastic deformation influence on microstructure and mechanical properties of aluminum alloys we can make the following conclusions:

- it is established that the most intensive grinding grain structures in AMG6 and AMC aluminum alloys occurs at ECAE with 12passes, while the intersection angle of the channels of 120^0. After ECAE-12 in aluminum alloys the grain refinement reaches to the size ~ 1.0-1.5 μm;
- it is shown that during ECAE-12 with an angle of intersection of the channels 120° at e=8 allows to obtain a defect-free billet with a more uniform structure;
- determined that result in equal-channel angular pressing, the microhardness of AMG6 alloy increases almost 4 times in comparison with the initial state. Microhardness of AMC alloy increases by almost 4.5 times in comparison with the initial state. It is shown that with decrease in grain size - microhardness increases.
- it is shown that ECAE-12 mass loss is reduced to 5.4 and 5.6 mg, which shows an increase in wear resistance of AMG6 and AMC aluminum alloys to 13-14 %.

Acknowledgments

The present work is executed by the financial support of the Committee of Science MES RK on the theme "The obtaining submicrocrystalline aluminum alloys with high physico-mechanical properties by intensive plastic deformation" under contract № 327 by 13 may 2016.

References

[1] R. Z. Valiev, I. V. Alexandrov, Bulk nanostructured metallic materials: preparation, structure and properties. Moscow: IKC "Akademkniga".- 2007. -398 p.

[2] N.A.Tereshchenko, I.L.Yakovleva, T.A.Zubkova, M.V. Chukin, and N.V. Koptseva, Structure Levels of Pearlite Deformation in Carbon Steel of Eutectoid Composition // The Physics of Metals and Metallography.-2013. -Vol.114, №5. -P.430-439.

[3] M. Calcagnotto, Y. Adachi, D. Ponge, D. Raabe, Deformation and fracture mechanisms in fine- and ultrafine-grained ferrite/martensite dual-phase steels and the effect of aging // Acta Materialia. -2011. -№59. -P.658-670.

[4] R. Song, D. Ponge, D. Raabe, J. Speer, D. Matlock, Overview of processing, microstructure and mechanical properties of ultrafine grained bcc steels //Materials science and Engineering A. -2006. -№441. -P.1-17.

[5] L. Storojeva, D. Ponge, D. Kaspar, D. Raabe, Development of microstructure and texture of medium carbon steel during heavy warm deformation // Acta Materialia. -2004. -№52. -P.2209-2220.

[6] M. Skakov, G. Uazyrkhanova, N. Popova, M. Scheffler: Influence of deformation on the phase structure of a 30CrMnSi steel. // Materials testing. – 2013. -Vol.55. -№1. -P.51-54

[7] D. Raabe, R. Kumar, Tensile deformation characteristics of bulk ultrafine-grained austenitic stainless steel produced by thermal cycling // Scripta Materialia. -2012. -№66. -P.634-637.

[8] R.Z. Valiev, G.L. Terence, Principles of equal-channel angular pressing as a processing tool for grain refinement // Progress in Materials Science -2006, Vol. 51, Issue 7, P.881–981

[9] Segal V. M. Development of material processing intensive shear deformation / V. M. Segal // Metals. – 2004. – No. 1. – P. 5-14.

[10] V. I. Elagin, Structural nanostructured alloys based on aluminum // Technology of light alloys - 2008. - No. 2. - P. 6-20.

The Simulation of the Trend of the Time Series in the form of the Spline of Third-Order With a Random Number of Data at the Moments of Measurement

I Ustinova, E Pakchomova

Tomsk Polytechnic University, Tomsk, Russia

E-mail: igu@tpu.ru, peg@tpu.ru

Abstract. The possibility of simulation of trend of a time series as a spline of third-order with a random number of data at the moments of measurement is discussed. Estimations of coefficients of the spline are obtained in the explicit form. The statistical characteristics of the received estimations are studied in details.

1. Introduction

One of the crucial problems of the analysis of the functioning of complicated technical systems is the simulation of the trend of telemetric parameters. Telemetric parameters are characterized the state of the system in some discrete moments of time [1]. The choice of a mathematical model and the definition of a scheme of an observation of the process are the main aspect of solve of this task. Mathematical model should describe the trend (a systematic component) of the observed values of a random process adequately. Here are the general ways of the simulation of the trend of the time series:

1. In the form of the polynomial p-degree of time t. Taking p large enough, one can get a fairly accurate representation of the trend of the time series. The value of the parameter p is selected separately in each case.
2. The curves which reflect "the net growth" can be described by exponential functions.
3. The curves of the population growth can be described by the logistic-curve in the form $f(t)=k/(1+e^{-\lambda t})$ [2].
4. As a linear combination of known functions.
5. In the form of splines.

In the classical theory of the time series the measurements are made thru equal time intervals and only one measurement is made at each moment time [2, 3]. However, there is another scheme of measurements in which at each time is produced a random number of measurements [4, 5]. Note that the simulation of the trend in a polynomial form whose order is higher than four is impractical due to the big error in the estimation of coefficients of the polynomial. On the other hand if the number of observations is large, the polynomial of the low-order can describe the true trend unsatisfactorily. The solution is in the estimation in the form of the spline of the trend of the time series. Generally, the spline is determined as follows [6-8]. The function $S(t)$ defined and continuous on the interval $[a,b]$

called a polynomial spline of the order m with nodes $x_j \in \{a = x_0 < x_1 < x_2 < \ldots < x_n = b\}$ if at each of the intervals $[x_{j-1}, x_j)$, $j = \overline{1,n}$, $S(t)$ is an algebraic polynomial of degree m.

2. The statement of the problem

Let there is the time series $y(t) = f(t) + \xi(t)$ representing the sum of some determined function $f(t)$ which is the trend of the process $y(t)$ and $\xi(t)$ is the stochastic function which emergence is defined by errors of measurements, hindrances etc. We will believe that hindrances of measurements $\xi_i = \xi(t_i)$, i=1, 2,..., are the independent, equally distributed random variables with the mathematical waiting $M[\xi_i] = 0$ and dispersion $D[\xi_i] = \sigma^2$. We know the sequence of values $y_1, y_2, ..., y_N$, where $y_i = \dfrac{y_{i1} + y_{i2} + ... + y_{in_i}}{n_i}$. Here n_i are independent a random expression, distributed by the Poisson law with an argument λ. We assume that the time points t_i are known to us exactly. Concerning the trend it is supposed that it represents the spline of the third order. In this case, the entire period of the observation $[0; T]$ is divided into parts of equal length: $[0; T_0]$, $[T_0; 2T_0]$, ..., $[(n-1)T_0; nT_0]$. On each such interval, the trend is estimated in the form of the polynomial of the third degree. On the borders of the segments, these polynomials are sewn together so as to obtain a continuous curve. Such the piecewise polynomial curve is called by the spline [5-9].

3. The solution of the problem

Let on an interval number k the trend can be presented in the form of the polynomial of the third order $f_k(t) = a_k + b_k(t/T_0) + c_k(t/T_0)^2 + d_k(t/T_0)^3$. We will believe that on each time interval a counting of time is conducted from the beginning of this interval therefore for a sewing together of polynomials on the ends of intervals the condition $f_{k-1}(T_0) = f_k(0)$ should be satisfied, that is $a_k = a_{k-1} + b_{k-1} + c_{k-1} + d_{k-1}$. By results of measurements of values $y_i^{(k)}$ on the interval number k it is necessary to construct estimates of coefficients a_k, b_k, c_k and d_k, moreover, there is a condition:

$$\hat{a}_k = \hat{a}_{k-1} + \hat{b}_{k-1} + \hat{c}_{k-1} + \hat{d}_{k-1} . \tag{1}$$

Estimates of missed coefficients for the interval number k we find by the method of least squares, based on the conditions:

$$Q_k = \sum_{i=1}^{N} n_i \left[y_i^{(k)} - \left(\hat{a}_k + b_k (t_i/T_0) + c_k (t_i/T_0)^2 + d_k (t_i/T_0)^3 \right) \right]^2 \Rightarrow \min_{b_k, c_k, d_k} , \tag{2}$$

N is the number of measurements at the interval of number k. Differentiating the equation (2) on variables b_k, c_k, d_k, we will receive system of the equations

$$\begin{cases} (b_k/T_0)\sum_{i=1}^{N} n_i t_i^2 + (c_k/T_0^2)\sum_{i=1}^{N} n_i t_i^3 + (d_k/T_0^3)\sum_{i=1}^{N} n_i t_i^4 = \sum_{i=1}^{N} n_i \left(y_i^{(k)} - \hat{a}_k\right) t_i, \\ (b_k/T_0^2)\sum_{i=1}^{N} n_i t_i^3 + (c_k/T_0^3)\sum_{i=1}^{N} n_i t_i^4 + (d_k/T_0^4)\sum_{i=1}^{N} n_i t_i^5 = \frac{1}{T_0}\sum_{i=1}^{N} n_i \left(y_i^{(k)} - \hat{a}_k\right) t_i^2, \\ (b_k/T_0^3)\sum_{i=1}^{N} n_i t_i^4 + (c_k/T_0^4)\sum_{i=1}^{N} n_i t_i^5 + (d_k/T_0^5)\sum_{i=1}^{N} n_i t_i^6 = \frac{1}{T_0^2}\sum_{i=1}^{N} n_i \left(y_i^{(k)} - \hat{a}_k\right) t_i^3. \end{cases}$$

from which we find the explicit form of the estimates of the missing parameters:

$$\hat{b}_k = \frac{30T_0}{\lambda N(N^2-1)(N^2-4)(N+3)(N^2+N+3)} \cdot \left\{ \frac{5(4N^6+12N^5+4N^4-12N^3-11N^2-3N+2)}{2N+1} \cdot S_1 - \right.$$
$$-10N(3N^3+6N^2-N-4)\cdot S_2 + \frac{7(6N^4+12N^3+3N^2-3N+2)}{2N+1}\cdot S_3 -$$
$$-\left[\frac{5(4N^6+12N^5+4N^4-12N^3-11N^2-3N+2)}{2N+1} \cdot \sum_{i=1}^{N} n_i t_i - 10N(3N^3+6N^2-N-4)\cdot \sum_{i=1}^{N} n_i t_i^2 + \right.$$
$$\left. \left. + \frac{7(6N^4+12N^3+3N^2-3N+2)}{2N+1} \cdot \sum_{i=1}^{N} n_i t_i^3 \right] \hat{a}_k \right\},$$
$$\hat{c}_k = \frac{60T_0^2}{\lambda N(N^2-1)(N^2-4)(N+3)(N^2+N+3)} \cdot \left\{ -5N(N+1)(3N^2+3N-4)\cdot S_1 + 6(8N^3+12N^2-2N-3)\cdot S_2 \right.$$
$$-35N(N+1)\cdot S_3 - \left[-5N(N+1)(3N^2+3N-4)\cdot \sum_{i=1}^{N} n_i t_i + 6(8N^3+12N^2-2N-3)\cdot \sum_{i=1}^{N} n_i t_i^2 - 35N(N+1)\cdot \right.$$
$$\left. \left. \cdot \sum_{i=1}^{N} n_i t_i^3 \right] \hat{a}_k \right\}, \qquad (3)$$
$$\hat{d}_k = \frac{210T_0^3}{\lambda N(N^2-1)(N^2-4)(N+3)(N^2+N+3)} \cdot \left\{ \frac{6N^4+12N^3+3N^2-3N+2}{2N+1} \cdot S_1 - 10N(N+1)\cdot S_2 + \right.$$
$$+ \frac{5(3N^2+3N+2)}{2N+1} \cdot S_3 - \left[\frac{6N^4+12N^3+3N^2-3N+2}{2N+1} \cdot \sum_{i=1}^{N} n_i t_i - 10N(N+1)\cdot \sum_{i=1}^{N} n_i t_i^2 + \right.$$
$$\left. \left. + \frac{5(3N^2+3N+2)}{2N+1} \cdot \sum_{i=1}^{N} n_i t_i^3 \right] \hat{a}_k \right\},$$

where $S_1 = \sum_{i=1}^{N} n_i y_i^{(k)} t_i$, $S_2 = \sum_{i=1}^{N} n_i y_i^{(k)} t_i^2$, $S_3 = \sum_{i=1}^{N} n_i y_i^{(k)} t_i^3$. Replacing in (3) evaluation a_k by the expression (1), we obtain the system for determining the coefficients of the spline on the interval number k. That is $\hat{a}_k = \hat{a}_{k-1} + \hat{b}_{k-1} + \hat{c}_{k-1} + \hat{d}_{k-1}$,

$$\hat{b}_k = \frac{30T_0}{\lambda N(N^2-1)(N^2-4)(N+3)(N^2+N+3)} \cdot \left\{ \frac{5(4N^6+12N^5+4N^4-12N^3-11N^2-3N+2)}{2N+1} \cdot S_1 - \right.$$
$$-10N(3N^3+6N^2-N-4)\cdot S_2 + \frac{7(6N^4+12N^3+3N^2-3N+2)}{2N+1}\cdot S_3 -$$
$$-\left[\frac{5(4N^6+12N^5+4N^4-12N^3-11N^2-3N+2)}{2N+1} \cdot \sum_{i=1}^{N} n_i t_i - 10N(3N^3+6N^2-N-4)\cdot \sum_{i=1}^{N} n_i t_i^2 + \right.$$
$$\left. \left. + \frac{7(6N^4+12N^3+3N^2-3N+2)}{2N+1} \cdot \sum_{i=1}^{N} n_i t_i^3 \right] \left(\hat{a}_{k-1} + \hat{b}_{k-1} + \hat{c}_{k-1} + \hat{d}_{k-1}\right) \right\}, \qquad (4)$$

$$\hat{c}_k = \frac{60T_0^2}{\lambda N\left(N^2-1\right)\left(N^2-4\right)\left(N+3\right)\left(N^2+N+3\right)} \cdot \Big\{-5N\left(N+1\right)\left(3N^2+3N-4\right)\cdot S_1 + 6\left(8N^3+12N^2-2N-3\right)\cdot S_2$$
$$-35N\left(N+1\right)\cdot S_3 - \Big[-5N\left(N+1\right)\left(3N^2+3N-4\right)\cdot\sum_{i=1}^{N} n_i t_i + 6\left(8N^3+12N^2-2N-3\right)\cdot\sum_{i=1}^{N} n_i t_i^2 - 35N\left(N+1\right)\cdot$$
$$\cdot\sum_{i=1}^{N} n_i t_i^3\Big]\left(\hat{a}_{k-1}+\hat{b}_{k-1}+\hat{c}_{k-1}+\hat{d}_{k-1}\right)\Big\},$$

$$\hat{\mathrm{d}}_k = \frac{210T_0^3}{\lambda N\left(N^2-1\right)\left(N^2-4\right)\left(N+3\right)\left(N^2+N+3\right)} \cdot \Big\{\frac{6N^4+12N^3+3N^2-3N+2}{2N+1}\cdot S_1 - 10N\left(N+1\right)\cdot S_2 +$$
$$+\frac{5\left(3N^2+3N+2\right)}{2N+1}\cdot S_3 - \Big[\frac{6N^4+12N^3+3N^2-3N+2}{2N+1}\cdot\sum_{i=1}^{N} n_i t_i - 10N\left(N+1\right)\cdot\sum_{i=1}^{N} n_i t_i^2 +$$
$$+\frac{5\left(3N^2+3N+2\right)}{2N+1}\cdot\sum_{i=1}^{N} n_i t_i^3\Big]\left(\hat{a}_{k-1}+\hat{b}_{k-1}+\hat{c}_{k-1}+\hat{d}_{k-1}\right)\Big\}.$$

$$\hat{\mathrm{d}}_k = \frac{210T_0^3}{\lambda N\left(N^2-1\right)\left(N^2-4\right)\left(N+3\right)\left(N^2+N+3\right)} \cdot \Big\{\frac{6N^4+12N^3+3N^2-3N+2}{2N+1}\cdot S_1 - 10N\left(N+1\right)\cdot S_2 +$$
$$+\frac{5\left(3N^2+3N+2\right)}{2N+1}\cdot S_3 - \Big[\frac{6N^4+12N^3+3N^2-3N+2}{2N+1}\cdot\sum_{i=1}^{N} n_i t_i - 10N\left(N+1\right)\cdot\sum_{i=1}^{N} n_i t_i^2 +$$
$$+\frac{5\left(3N^2+3N+2\right)}{2N+1}\cdot\sum_{i=1}^{N} n_i t_i^3\Big]\left(\hat{a}_{k-1}+\hat{b}_{k-1}+\hat{c}_{k-1}+\hat{d}_{k-1}\right)\Big\}.$$

$$\Big\{\frac{6N^4+12N^3+3N^2-3N+2}{2N+1}\cdot S_1 - 10N\left(N+1\right)\cdot S_2 +$$
$$+\frac{5\left(3N^2+3N+2\right)}{2N+1}\cdot S_3 - \Big[\frac{6N^4+12N^3+3N^2-3N+2}{2N+1}\cdot\sum_{i=1}^{N} n_i t_i - 10N\left(N+1\right)\cdot\sum_{i=1}^{N} n_i t_i^2 +$$
$$+\frac{5\left(3N^2+3N+2\right)}{2N+1}\cdot\sum_{i=1}^{N} n_i t_i^3\Big]\left(\hat{a}_{k-1}+\hat{b}_{k-1}+\hat{c}_{k-1}+\hat{d}_{k-1}\right)\Big\}.$$

Now we investigate statistical characteristics of the received estimates, namely, we will prove their asymptotic unbiasedness and sustainability. Because $E[y_i^{(k)}] = a_k + b_k(t/T_0) + c_k(t/T_0)^2 + d_k(t/T_0)^3$ ($E[x]$ is the mathematical expectation of a random variable x)

$$E[S_1|n_i] = \frac{\lambda N(N+1)}{2}\left\{a_k + \frac{2N+1}{3T_0}b_k + \frac{N(N+1)}{2T_0^2}c_k + \frac{(2N+1)(3N^2+3N-1)}{15T_0^3}d_k\right\}$$

$$E[S_2|n_i] = \frac{\lambda N(N+1)}{2}\Big\{\frac{2N+1}{3}a_k + \frac{N(N+1)}{2T_0}b_k + \frac{(2N+1)(3N^2+3N-1)}{15T_0^2}c_k +$$
$$+\frac{N(N+1)(2N^2+2N-1)}{6T_0^3}d_k\Big\},$$

$$E[S_3|n_i] = \frac{\lambda N(N+1)}{2}\left\{\frac{N(N+1)}{2}a_k + \frac{(2N+1)(3N^2+3N-1)}{15T_0}b_k + \frac{N(N+1)(2N^2+2N-1)}{6T_0^2}c_k + \right.$$
$$\left. + \frac{(2N+1)(3N^4+6N^3-3N+1)}{21T_0^3}d_k\right\},$$

$$b_k = \frac{15T_0}{(N-1)(N^2-4)(N+3)(N^2+N+3)}\left\{\frac{10(4N^6+12N^5+4N^4-12N^3-11N^2-3N+2)}{\lambda N(N+1)(2N+1)}M[S_1|n_i] + \right.$$
$$+ \frac{20(3N^2+3N-4)}{\lambda}M[S_2|n_i] + \frac{14(6N^4+12N^3+3N^2-3N+2)}{\lambda}M[S_3|n_i] -$$
$$\left. - \frac{486N^6+1458N^5+895N^4-640N^3-691N^2-128N+60}{6(2N+1)}a_k\right\},$$

$$c_k = \frac{15T_0^2}{(N-1)(N^2-4)(N+3)(N^2+N+3)}\left\{\frac{20(3N^2+3N-4)}{\lambda}M[S_1|n_i] + \frac{24(4N^2-1)(2N+3)}{\lambda N(N+1)}M[S_2|n_i] - \right.$$
$$\left. - \frac{140}{\lambda}M[S_3|n_i] - \left(59N^4+118N^3-13N^2-72N-12\right)a_k\right\},$$

$$d_k = \frac{15T_0^3}{(N-1)(N^2-4)(N+3)(N^2+N+3)}\left\{\frac{14(6N^4+12N^3+3N^2-3N+2)}{\lambda N(2N^2+3N+1)}M[S_1|n_i] - \right.$$
$$\left. - \frac{140}{\lambda}M[S_2|n_i] + \frac{70(3N^2+3N+2)}{\lambda N(N+1)(2N+1)}M[S_3|n_i] - \frac{7(N-1)(N^2-4)(N+3)}{6(2N+1)}a_k\right\}.$$

We will designate: $\Delta a_k = \hat{a}_k - a_k$, $\Delta b_k = \hat{b}_k - b_k$, $\Delta c_k = \hat{c}_k - c_k$, $\Delta d_k = \hat{d}_k - d_k$ By averaging over the random variables n_i we get the following system:

$$\Delta\bar{a}_k = \Delta\bar{a}_{k-1} + \Delta\bar{b}_{k-1} + \Delta\bar{c}_{k-1} + \Delta\bar{d}_{k-1},$$

$$\Delta\bar{b}_k = \frac{-15T_0}{(N-1)(N^2-4)(N+3)(N^2+N+3)}\left\{\frac{5(4N^6+12N^5+4N^4-12N^3-11N^2-3N+2)}{(2N+1)} - \right.$$
$$\left. - \frac{10N(3N^3+6N^2-N-4)}{3}(2N+1) + \frac{7(6N^4+12N^3+3N^2-3N+2)N(N+1)}{2N+1}\right\}\cdot$$
$$\cdot\left(\Delta\bar{a}_{k-1} + \Delta\bar{b}_{k-1} + \Delta\bar{c}_{k-1} + \Delta\bar{d}_{k-1}\right),$$

$$\Delta\bar{c}_k = \frac{-30T_0^2}{(N-1)(N^2-4)(N+3)(N^2+N+3)}\left\{-5N(N+1)(3N^2+3N-4) + 2(8N^3+12N^2-2N-3)\cdot\right.$$
$$\left.\cdot(2N+1) - \frac{35N^2(N+1)^2}{2}\right\}\cdot\left(\Delta\bar{a}_{k-1} + \Delta\bar{b}_{k-1} + \Delta\bar{c}_{k-1} + \Delta\bar{d}_{k-1}\right),$$

$$\Delta\bar{d}_k = \frac{-105T_0^3}{(N-1)(N^2-4)(N+3)(N^2+N+3)}\left\{\frac{6N^4+12N^3+3N^2-3N+2}{2N+1} - \right.$$
$$\left. - \frac{10N(N+1)(2N+1)}{3} + \frac{5(3N^2+3N+2)N(N+1)}{2(2N+1)}\right\}\cdot\left(\Delta\bar{a}_{k-1} + \Delta\bar{b}_{k-1} + \Delta\bar{c}_{k-1} + \Delta\bar{d}_{k-1}\right).$$

For asymptotically unbiased of the estimates it is necessary that this system was stable. This is possible if the roots of its characteristic equation (5) will be on the module less units.

$$\begin{vmatrix} 1-\lambda & 1 & 1 & 1 \\ a & a-\lambda & a & a \\ b & b & b-\lambda & b \\ c & c & c & c-\lambda \end{vmatrix} = 0 \qquad (5)$$

In this equation the following designations are entered:

$$a = \frac{-15T_0}{(N-1)(N^2-4)(N+3)(N^2+N+3)} \left\{ \frac{5(4N^6+12N^5+4N^4-12N^3-11N^2-3N+2)}{(2N+1)} - \right.$$
$$\left. -\frac{10N(3N^3+6N^2-N-4)}{3}(2N+1) + \frac{7(6N^4+12N^3+3N^2-3N+2)N(N+1)}{2(2N+1)} \right\},$$

$$b = \frac{-30T_0^2}{(N-1)(N^2-4)(N+3)(N^2+N+3)} \left\{ -5N(N+1)(3N^2+3N-4) + 2(8N^3+12N^2-2N-3)\cdot \right.$$
$$\left. \cdot(2N+1) - \frac{35N^2(N+1)^2}{2} \right\},$$

$$c = \frac{-105T_0^3}{(N-1)(N^2-4)(N+3)(N^2+N+3)} \left\{ \frac{6N^4+12N^3+3N^2-3N+2}{2N+1} - \right.$$
$$\left. -\frac{10N(N+1)(2N+1)}{3} + \frac{5(3N^2+3N+2)N(N+1)}{2(2N+1)} \right\}.$$

Three roots $\lambda_1, \lambda_2, \lambda_3$ of the equation (5) are equal to zero and $\lambda_4 = 1+a+b+c$, that is $\lambda_4 = 1 - \frac{5T_0}{2(N^2+N+3)} \left\{ \frac{7T_0^2}{2N+1} - 6T_0 + \frac{6N^2+6N+5}{2N+1} \right\}$. Its value is determined by the expression N/T_0, in which N is the number of points in the interval $[(k-1)T_0; kT_0]$. For recursive estimation of the coefficients of the spline of the third order requires at least three points on each partial interval. By solving the inequality $|\lambda_4| < 1$, we see that it is true for any $N/T_0 > 0$ (figure 1). Thus, the recursive estimates (4) of parameters of the spline are sustainable, and therefore asymptotically unbiased.

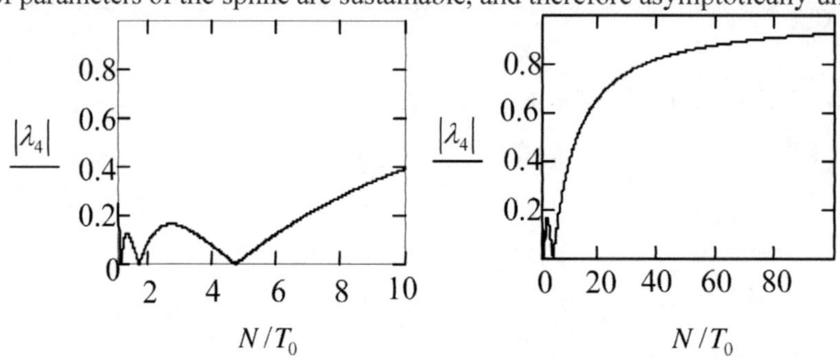

Figure 1. The dependence of the fourth root of the characteristic equation from N/T_0.

4. Conclusion

As a result of the conducted research are obtained by the recurrent evaluation of the coefficients of the spline of the third order in the modelling of the trend of the time series with random number of data at the moments of measurement. The stability of the estimates of the parameters of the spline and their asymptotic unbiased are installed.

References

[1] Trivozhenko B E 1989 *Identification of Trends of Time Series and Flows of Events* (Tomsk: Tomsk state University).
[2] Kendall M G, Stuart A 1966 *The advanced theory of statistics. Design and analysis and time-series* (London: Published by Griffin & Company).
[3] Anderson T W 1971 *Statistical analysis of time series* (New York: Wiley and Sons).
[4] Idrisov F F, Konstantinova I G 1999 Identification of trends of time series for a random number of measurements *Russian Physics Journal* **42** 379 384.
[5] Ustinova I G , Pakhomova E G 2015 Spline estimate of the time series trend for a random number of data at measurement instants *Vestn. Tomsk. Gos. Univ. Mat. Mekh.***1(33)** 20 36.
[6] Bakhvalov N S 1975 *Numerical methods* (Moscow: Nauka).
[7] Korneychuk N P 1984 *Splines in approximation theory* (Moscow: Nauka).
[8] Shevelitskiy I V 2003 Interpolation splines in problems of digital signal processing *J. Exponenta Pro. Matematika v prilozheniyakh* **4** 42 53.
[9] March L C, Cormier D R 2002 *Spline Regression Models* (London: SAGE PUBLICATION Ltd).

IOP Conf. Series: Materials Science and Engineering **142** (2016) 012037 doi:10.1088/1757-899X/142/1/012037

Simulation of Tasks Distribution in Horizontally Scalable Management System

D Kustov[1], A Sherstneva[2], I Botygin[3]

[1] E-learning Institute, National Research Tomsk Polytechnic University, Lenin Avenue, 30, Tomsk, 634050, Russia
[2]Department of Higher Mathematics, Institute of Physics and Technology, National Research Tomsk Polytechnic University, Lenin Avenue, 30, Tomsk, 634050, Russia
[3]Department of Computer-Aided Designed System, Institute of Cybernetics, National Research Tomsk Polytechnic University, Lenin Avenue, 30, Tomsk, 634050, Russia

Abstract. This paper presents an imitational model of the task distribution system for the components of territorially-distributed automated management system with a dynamically changing topology. Each resource of the distributed automated management system is represented with an agent, which allows to set behavior of every resource in the best possible way and ensure their interaction. The agent work load imitation was done via service query imitation formed in a system dynamics style using a stream diagram. The query generation took place in the abstract-represented center – afterwards, they were sent to the drive to be distributed to management system resources according to a ranking table.

1. Introduction

Currently, there have been significant changes in automated systems infrastructure for various complicated organization-technical complexes. Constant hardware modernization, production number increase and cost reduction are not the only reasons behind it. The difficulty of data processing itself has increased (especially in scientific research applications), while applied problems are solved in a distributed manner.

The amount of stored and processed data is yet another factor that actively influences the research and development of the new information-calculation structures. The modern management systems are often using a territorially distributed structure, formed of tens of thousands computers under the concepts of meta-computing and grid-technologies. Sketchily, such structure includes computing nodes functioning as calculators and data storages, communication network, middleware management software multi-layered architecture, various targeted domain-functional modules and systems. While I do not claim it as an in-depth description, the main features of such infrastructures are their large scale defined by the territorial distribution and decentralized management due to non-deterministic process management type.

Even those features listed allow to conclude that it's difficult, if not impossible, to use distributed automated management systems for strict formal mathematical methods.

Such circumstance largely defines the special role of the imitational modeling during the design of distributed automated management systems (DAMS) [1]. Only an imitational model allows to adequately represent DAMS primary components, their interaction and the dynamically changing system condition.

Therefore, it is possible to replay all the DAMS functioning process over time, accounting for its designed logic structure, various management factors, random events and limitations. Let's point out the prospect of using imitational models for integration with decision support systems. In this case, it is possible to search for optimal variants of distributed automated management system at every step of development.

Various concepts, technologies and implementation methods are used to solve technological and methodological issues of DAMS development, for planning and implementing the rational use of its resources [2]. One of the approaches that provides the necessary DAMS functioning efficiency, is, definitely, load balancing of its software-hardware components. The solution for this problem is targeted towards providing the necessary technological interaction between integral components in DAMS functioning process. The load balancing (equalization) ensures optimized calculations for distributed architecture of informational and computational resources of complex systems, and increases their reliability as well.

2. Simulation model of load balancing

The current work researches the load balancing algorithm [3] developed by authors. The testing was done in a model experiment imitating the distributed automated management system functioning with a more than 500 interacting components. Regarding the model itself, DAMS was represented as a generalized mass-service system with a multiple channels, service errors, non-homogenous service query streams, with a multiphase service query processing, unlimited queues, open-type system with limited reliability. All the DAMS hardware-technical and software-informational infrastructure components are acting as the generic model objects.

The imitational model development uses universal programming languages (C, C++, C#, Java, Fortran etc.), specialized languages (GPSS/H, GPSS World, SIMSCRIPT, SIMPLEX3,GASP, SIMULA, SLAM etc.) [4-15], specialized modeling environments (MicroSaint, iGrafx Process, SIMUL8, VisSim, Extend, Arena, AnyLogic, Enterprise Dynamics etc.) [16-22]. Using the universal programming languages most certainly allows to receive the models most adequately resembling the real objects. Specialized modeling environments, however, allow to significantly reduce the model development time and easily modify them if necessary.

As the modeling environment, the authors used the domestic object-oriented imitational modeling environment AnyLogic [23, 24]. The combine approach, including agent modeling and system dynamics, was used as a modeling method. The system dynamics was used for distributed system components abstraction, and agent model was employed to adequately represent the functional software-hardware components. In other words, for the studied problem, every distributed automated management system component was represented as an agent, which is the best way to set a behavior for each resource, to provide a proper way for their interaction.

The resource (agent) load imitation was modeled via service query generation formed with the system dynamics style using a stream diagram to create a determined and alternative operations passing algorithm. The query generation is done via abstract-view center; then, queries are delivered to the drive (distributor) that distributes queries to DAMS resources. The query distribution is executed according to the following principle:

Step 1. If any service queries remain in the drive, the execution priority is given to the resources that aren't processing any queries at the moment, i.e. are currently free.

Step 2. If there are no free resources at the moment, the busyness coefficient of every node is calculated $L = P*(Q+1)$, where P is resource performance, Q the number of queries in queue for this resource. The query is sent to the resource with the lowest busyness coefficient. Specifically, it is sent to queue. Afterwards, the busyness coefficient for this resource is recalculated accordingly.

Step 3. If the query has finished being processed, it is deleted.

For this model to resemble a real system more closely, the following randomly generated events have been included:

- The possibility of DAMS resource refusing to accept query;

- The possibility of resource failure at any moment of time;
- The possibility of resending processed request to another DAMS resource for its further processing.

The AnyLogic Statechart class object were chosen as the primary blocks of the imitational model. Those objects can determine a state diagram that describes system functioning as a set of states with different readiness coefficients and state transitions. Additionally, every transition has a percentage rate assigned statically or dynamically calculated during modeling. Every resource used one of the following states:

- Resource initial state – «FreeState» (DAMS resource is ready to receive service queries);
- «BusyState» – DAMS resource is busy processing the incoming queries;
- «BreakState» – DAMS resource is not available (break/failure).

«FreeState» is pretty simple and always relates to an initial state for any distributed automated management system resource. In such condition DAMS resource is always ready for incoming queries.

After receiving a request, resource enters the «busyState» and starts processing. The processing duration can be calculated as a query processing time multiplied by resource components performance. By default, the query processing time is equal to a single unit of modeling time. DAMS resource performance is determined as generalized characteristics performance for every component that belongs to this resource. Resource performance is correlated by its load and can be defined by dimensionless coefficient between 0 and 1. The processing duration may vary from 1 up to 10 modeling time units. After query processing is finished, there is a chance to redirect query to another random system resource (the probability is 0.05 by default). Otherwise, the query is deemed as processed. The processed query counter increases by 1, resource enters «freeState».

The «breakState» is necessary for modeling breaks/failures of distributed automated management system. The chance to enter break/failure state is defaulted to 0.1 and calculated for every unit of modeling time. Entering this state will return the query in process to the query pool. The duration of break/failure state varies between 1 and 100 modeling time units. After break/failure state expires, the resource enters the initial «freeState».

The following parameters were chosen as DAMC imitational model input data:

- The total number of DAMC resources (agents).
- Service query generation intensity.
- The time required to process a single service query.
- The resource break/failure rate, defined by Bernoulli distribution.
- The resource break/failure duration, determined as a random value within minimum and maximum boundaries.
- The chance to resend a service query from one resource to another.

All of the default data above can be set for imitational modeling with model interface. The work end condition was time limit. The experiment used time limit of 200 modeling time units, which is sufficient to receive a complete work load data for each resource and system as a whole.

There was a series of experiments with different input data values. The varying parameters were: query generation intensity and system resource number. The query stream intensity varied from 1 to 150 per modeling time unit. The number of DAMC resources was determined by the following array a = {5, 10, 20, 30, 50, 75, 100, 125, 150, 200, 250, 300, 350, 400, 500}.

As a result of DAMC imitational model work, the following numerical criteria were chosen for estimation:

- The total number of queries processed by DAMC.
- The total number of DAMC breaks/failures.
- The total DAMC load coefficient.
- The performance of every DAMC resource.
- The number of processed queries for each DAMC resource.

IOP Conf. Series: Materials Science and Engineering **142** (2016) 012037 doi:10.1088/1757-899X/142/1/012037

- The number of failures for each DAMC resource.
- The load coefficient for each DAMC resource per unit of modeling time.
- The number of queries in queue for every DAMC resource per unit of modeling time.

All the received data was recorded in a structured manner in an external file. The imitational experiments have shown that the researched load balancing algorithm allows to determine the optimal amount of queries per unit of time to equally load all DAMC resources for any component infrastructure (up to 500 agents), component load or query routing. Moreover, after the received data analysis, the DAMC functioning quality can be estimated.

It is based upon such general information-communicational environment parameters as flow coefficient, average query processing time, average query processing delay, failure rate etc. The recommendations for increasing DAMC resilience have been created, which allows to reduce time and computation costs. Finally, the criteria in use allowed to get a complete statistic on distributed automated management system model functioning.

References

[1] Devyatkov V.V. Metodologiya i tehnologiya imitatsionnyh issledovaniy slozhnyh sistem: sovremennoe sostoyanie i perspektivy razvitiya. - M.: INFRA-M, 2013. - 448 p.
[2] Sherstnev V. S. , Sherstneva A. I. , Botygin I. A. , Kustov D. A. Distributed information system for processing and storage of meteorological data // Key Engineering Materials . - 2016 - Vol. 685. - p. 867-871.
[3] Botygin I.A., Tartakovsky V.A., Sherstneva A.I. Algoritm balansirovki nagruzki v raspredelennyh sistemah upravleniya // Informatizatciya i sviaz. – 2015. – № 1. – P. 28-31.
[4] Jefferson D.R. Virtual Time / Jefferson D.R. // Assoc. Comput. Mach. Trans. Programming Languages and Systems, 1985. – № 7. P. 404-425.
[5] Computer Simulation // Minuteman Software, available at http://www.minutemansoftware.com, 2015.
[6] GPSS World // Elina-Computer, available at http://elina-computer.ru/static/gpss-world.html, 2015.
[7] GPSS/H Software Products // Wolverine Software, available at http://www.wolverinesoftware.com/GPSSHProducts.htm, 2015.
[8] SIMSCRIPT III // CACI Advanced Simulation Lab, available at http://www.simscript.com/products/products.html, 2015.
[9] Jalote P. Fault Tolerance in Distributed Systems // P. Jalote. USA: Prentice Hall, New Jersey, 1994. – 448 p.
[10] What is Simplex3? // SOI-Software, available at http://www.simplex3.net/Body/Introduction/English/indexAbstract.html, 2015.
[11] Philip J. Kiviat, A. Colker A General Activity Simulation Program. – RAND Corporation, Document Number: P-2864, 1964. – 9 p.
[12] Sklenar J. Introduction to OOP in SIMULA, available at http://staff.um.edu.mt/jskl1/talk.html, 2015.
[13] Birtwistle, G.M., O.-J. Dahl, B. Myhrhaug and K. Nygaard: SIMULA Begin, AUERBACH Publishers Inc, 1973.
[14] Pooley, R.J.: An Introduction to Programming in SIMULA, Oxford, Blackwell Scientific Publications, 1987.
[15] Kirkerud, B.: Object-Oriented Programming with SIMULA, Addison-Wesley, 1989.
[16] About Micro Saint Sharp // Alion Science and Techology, available at http://www.microsaintsharp.com, 2015.
[17] iGrafx Process 2015 // iGrafx, LLC, available at http://www.igrafx.com/products/process-modeling-analysis/process, 2015.
[18] SIMUL8 // SIMUL8 Corporation, available at http://www.simul8.com, 2015.
[19] VisSim. A graphical language for simulation and model-based embedded development // Visual

Solutions, available at http://www.vissim.com, 2015.
[20] Simulation with ExtendSim // Imagine That Inc., available at http://www.extendsim.com/prods_overview.html, 2015.
[21] Arena // Arena Simulation Software, available at https://www.arenasimulation.com, 2015.
[22] Enterprise Dynamics // INCONTROL Simulation Solutions, available at http://www.incontrolsim.com/en/enterprise-dynamics/enterprise-dynamics.html, 2015.
[23] AnyLogic // AnyLogic, available at http://www.anylogic.ru/overview, 2015.
[24] Kelton W., Law A. Simulation Modeling and Analysis. McGraw Hill Higher Education, 3rd edition, 2000 – 784 p.

IOP Conf. Series: Materials Science and Engineering **142** (2016) 012038 doi:10.1088/1757-899X/142/1/012038

Enhancement Effects of Transition and Vavilov-Cherenkov Radiation Mechanisms Under Grazing Interaction of Fast Electrons With a Thick Substrate Applied by Thin Layer

E Irribarra[1], A S Kubankin[2,3,4], A Sotnikov[2], R M Nazhmudinov[2,4], T Fam[2], A P Starovoytov[2]

[1] Escuela Politécnica Nacional, Quito, Ecuador
[2] Laboratory of Radiation Physics, Belgorod State University, Belgorod, Russia
[3] National Research Tomsk Polytechnic University, Tomsk, Russia
[4] Lebedev Physical Institute, Moscow, Russia

E-mail: kubankin@bsu.edu.ru

Abstract. The paper presents the results of a theoretical study and a mathematical model of radiation processes occurred during the grazing interaction of fast electrons with semi-infinite targets applied on a thin amorphous layer. The developed model considers Vavilov-Cherenkov and transition radiation mechanisms and predicts the possibility to enhance the angular radiation density under grazing incidence of fast electrons on the layer. The characteristics of possible extreme vacuum ultraviolet and soft X-ray sources are estimated.

1. Introduction

An important spectral range of electromagnetic radiation are the extreme vacuum ultraviolet (EUV) and soft X-ray regions which are used in various fields of science and technology [1]. Nowadays many EUV sources have been developed, the most powerful of which are synchrotrons and free electron lasers. The main drawback of these sources is the high cost of their production and operation which limits their availability. A promising direction for the development of EUV sources lies in the use of radiation mechanisms occurred during the interaction of fast charged particles with matter [2–4].

The coherent mechanisms of X-ray generation by relativistic electrons in periodic structures like a multilayer mirror [5,6] or a crystal [7-9] can be used to obtain high radiation fluxes for different applications. But the existence of a strong radiation absorption in the target limits the use of the mentioned mechanisms to develop EUV and soft X-rays sources. It is important to mention also that the coherent mechanism yield is substantially decreased when the energy of the emitting particle decreases [10,11].

The photoabsorption problem can be solved when radiation is generated during contactless interaction of the charged particle with the structured target, for example the Smith-Purcell radiation. In addition, multiple passes of the emitting particles through targets mounted inside cyclical accelerators can be used to decrease the influence of photoabsorption on the radiation yield in EUV and soft X-rays regions [12-14]. Among the drawbacks of this mechanism one can underline the necessity of a small value of the impact parameter of the interaction between the incident particle and the target structured surface and a small size of the transverse beam to obtain a high efficiency of the source. In this context, transition radiation (TR) and Vavilov-Cherenkov (VC) mechanisms are convenient and can be used to develop EUV and soft X-ray sources. An important feature of TR and VC is the significant yield in the soft spectral range for relativistic electrons. The possibility to use TR and VC as a radiation source was shown in [15,16]. Nevertheless, the yield of TR decreases substantially when the energy of the emitting particle decreases and the VC mechanism is limited by photoabsorption. The possibilities to enhance the yield of these mechanisms are relevant task. An approach to solve this task was proposed in [17,18], where the grazing interaction of fast electrons with a flat semi-infinite amorphous target is considered. The main result of these works is the predicted possibility to increase the angular radiation density by more than then times in comparison to the usual case.

This paper presents the calculation results of TR and VC characteristics obtained during the interaction between fast electrons and a target which consists of a thin amorphous layer applied on the surface of a semi-infinite amorphous substrate (the thickness of the substrate layer is much more than the absorption length of the considered photons). Such structure of the target can be easily produced to develop EUV and soft X-ray radiators. The high quality of the surface is an indispensable condition for the realization of the considered effects. The relativistic system of units used for the calculations presented below is $\hbar = c = 1$.

2. Theory

Let us consider the radiation produced by a relativistic electron moving with a constant velocity $\vec{V}$ in a target consisting of a thin amorphous layer with thikness L applied on a substrate with thickness much more than the photoabsorption length of the considered radiation. The semi-infinite model for the substrate can be used as it is presented in Fig.1 where $\vec{n}$ is propagation direction of a radiated photon, $\chi_1(\omega)$ and $\chi_2(\omega)$ are the dielectric susceptibility of the layer and substrate respectively (ω is the energy of the radiated photon). The presented geometry of the radiation process represents real targets, which can be produced with different methods of film coating.

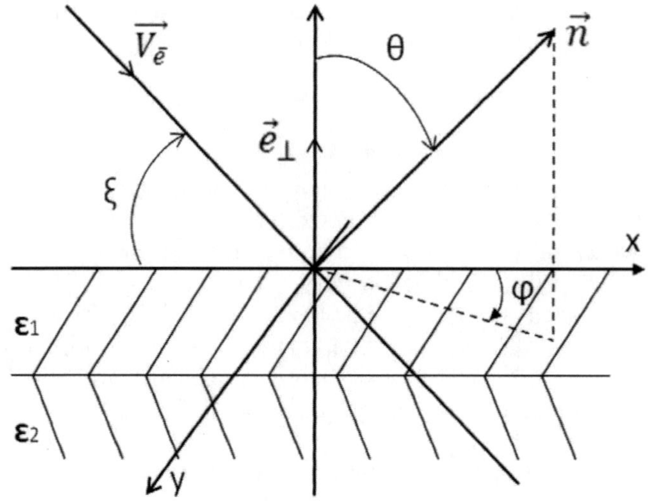

Figure 1. The geometry of radiation process

The radiation characteristics can be calculated from Maxwell equations and ordinary boundary conditions for the electromagnetic field. Using the Fourier-transform of electric and magnetic fields one can obtain the following equations for the Fourier-components of the electric field in vacuum – I, the layer – II and the substrate – III

$$\vec{E}^{I}_{\omega\vec{k}} = \frac{i\omega e}{2\pi^2}\frac{\delta\left(\omega-\vec{k}\vec{V}\right)}{k^2-\omega^2}\left(\vec{V}-\vec{k}\frac{\vec{k}\vec{V}}{\omega^2}\right)+\vec{a}_{k_\parallel}\delta\left(k_\perp-\sqrt{\omega^2-k_\parallel^2}\right), \tag{1}$$

$$\vec{E}^{II}_{\omega\vec{k}} = \frac{i\omega e}{2\pi^2}\frac{\delta\left(\omega-\vec{k}\vec{V}\right)}{k^2-\omega^2\varepsilon_1}\left(\vec{V}-\vec{k}\frac{\vec{k}\vec{V}}{\omega^2\varepsilon_1}\right)+\vec{b}_{k_\parallel}\delta\left(k_\perp-\sqrt{\omega^2\varepsilon_1-k_\parallel^2}\right)+\vec{c}_{k_\parallel}\delta\left(k_\perp+\sqrt{\omega^2\varepsilon_1-k_\parallel^2}\right), \tag{2}$$

$$\vec{E}^{III}_{\omega\vec{k}} = \frac{i\omega e}{2\pi^2}\frac{\delta\left(\omega-\vec{k}\vec{V}\right)}{k^2-\omega^2\varepsilon_2}\left(\vec{V}-\vec{k}\frac{\vec{k}\vec{V}}{\omega^2\varepsilon_2}\right)+\vec{d}_{k_\parallel}\delta\left(k_\perp+\sqrt{\omega^2\varepsilon_2-k_\parallel^2}\right). \tag{3}$$

The vectors $\vec{a}_{k_\parallel}$, $\vec{b}_{k_\parallel}$, $\vec{c}_{k_\parallel}$ and $\vec{d}_{k_\parallel}$ determine the radiation in the layer and inside the substrate. Calculating the vector $\vec{a}_{k_\parallel}$ is of great interest because it represents the radiation from the target in vacuum. It is convenient to introduce the basis of orthogonal unit vectors

$$\vec{e}_\perp, \quad \vec{e}_\parallel = \frac{\vec{k}_\parallel}{k_\parallel}, \quad \vec{e}\,' = \frac{[\vec{k}_\parallel, \vec{e}_\perp]}{k_\parallel}. \tag{4}$$

The unit vectors $\vec{e}_\parallel$ and $\vec{e}\,'$ lie in the surface of the target. The unknown coefficients for the electric fields in (2) can be determined using ordinary boundary conditions. The coefficient $\vec{a}_{k_\parallel}$ determines the radiation field in vacuum and it can be represented in the following form $\vec{a}_{k_\parallel} = \left(\vec{e}_\perp - \vec{e}_\parallel k_\perp/k_\parallel\right)a_\perp + \vec{e}\,'a'$.

Calculating the Fourier-integral in wave-zone by the stationary phase method, one can obtain the following expression for the emission amplitude

$$\vec{E}^{Rad}_\omega = \int d^3k\,\vec{a}_{\vec{k}_\parallel} e^{i\vec{k}\vec{r}} \rightarrow -2\pi i\omega n_\perp \vec{a}_{\vec{k}_\parallel=\omega\vec{n}_\parallel}\frac{e^{i\omega r}}{r} = \vec{A}_\omega \frac{e^{i\omega r}}{r},$$

$$\omega\frac{dN}{d\omega\, d\Omega} = \left|\vec{A}\right|^2 = \left|\vec{A}_\perp\right|^2 + \left|\vec{A}'\right|^2 = 4\pi^2\omega^2 n_\perp^2\left(\left|\frac{a_\perp}{n_\parallel}\right|^2 + \left|a'\right|^2\right). \tag{5}$$

The resultant expression for the calculation of the spectral-angular distribution of radiation in considered conditions can be represented in the next form

$$\left|\vec{A}_\perp\right|^2 = \frac{e^2}{\pi^2}\frac{n_\perp^2}{V_\perp^2 n_\parallel^2}\frac{1}{\left|\left(\varepsilon_1\gamma+\varepsilon_2\beta\right)\left(\beta+\varepsilon_1 n_\perp\right)e^{-i\omega L(\beta-p)}+\left(\varepsilon_1\gamma-\varepsilon_2\beta\right)\left(\beta-\varepsilon_1 n_\perp\right)e^{i\omega L(\beta+p)}\right|^2}\times$$

$$\left|\frac{2\beta\left(\chi_1-\chi_2\right)}{p-\gamma}\frac{n_\parallel^2\left(\varepsilon_1 V_\perp - p\right)-\gamma\left(\varepsilon_1\vec{n}_\parallel\vec{V}_\parallel - n_\parallel^2\right)}{p^2-\beta^2}+\right.$$

$$+\frac{\chi_1}{p^2-n_\perp^2}\left(\frac{\varepsilon_1\gamma+\varepsilon_2\beta}{p-\beta}\left(\beta\left(\vec{n}_\parallel\vec{V}_\parallel-n_\parallel^2\right)+n_\parallel^2\left(p-V_\perp\right)\right)e^{-i\omega L(\beta-p)}+\right.$$

$$\left.\left.+\frac{\varepsilon_1\gamma-\varepsilon_2\beta}{p+\beta}\left(\beta\left(\vec{n}_\parallel\vec{V}_\parallel-n_\parallel^2\right)-n_\parallel^2\left(p-V_\perp\right)\right)e^{i\omega L(\beta+p)}\right)\right|^2, \tag{6a}$$

$$|A'|^2 = \frac{e^2}{\pi^2}\frac{V_{\parallel}^2 n_{\perp}^2 n_y^2}{V_{\perp}^2 n_{\parallel}^2}\frac{1}{\left|(\gamma+\beta)(\beta+n_{\perp})e^{-i\omega L(\beta-p)}+(\gamma-\beta)(\beta-n_{\perp})e^{i\omega L(\beta+p)}\right|^2}\left|2\beta\frac{\chi_1-\chi_2}{p-\gamma}\frac{1}{p^2-\beta^2}-\right.$$

$$\left.-\frac{\chi_1}{p^2-n_{\perp}^2}\left(\frac{\gamma+\beta}{p-\beta}e^{-i\omega L(\beta-p)}-\frac{\gamma-\beta}{p+\beta}e^{i\omega L(\beta+p)}\right)\right|^2, \quad (6b)$$

where $p=\left(1-\vec{n}_{\parallel}\vec{V}_{\parallel}\right)/V_{\perp}$, $\beta=\sqrt{n_{\perp}^2+\chi'_1+i\chi''_1}$, $\gamma=\sqrt{n_{\perp}^2+\chi'_2+i\chi''_2}$, $\chi''>0$.

The components of vectors $\vec{n}$ and $\vec{V}$ can be presented in a convenient form for the calculation of spectral-angular distribution of emitted photons

$$V_{\perp}=V\sin(\xi),\ V_{\parallel}=V_x=V\cos(\xi),$$
$$n_{\perp}=\cos(\Theta),\ n_x=\sin(\Theta)\cos(\varphi);\ n_y=\sin(\Theta)\sin(\varphi).$$

The presented expressions (6a) and (6b) can be used for the calculation of radiation spectral-angular characteristics for both cases when emitting electron flight in the target ($V_{\perp}<0$) and flight out from one ($V_{\perp}>0$).

3. Calculations

An important case for calculations is the radiation produced in conditions of grazing interaction of fast electrons with the target surface because of the prediction of radiation effects enhancement for TR [17] and VC [18]. These effects were predicted for the radiation generated during the flight out of the radiating electron from the semi-infinite amorphous target. The semi-infinite target is not convenient because of strong multiple scattering of the electron inside the target under grazing incidence conditions. The expressions (6) allow calculating the possibility of the enhancement realization as in [17,18] when the electrons cross the layer moving from vacuum. In accordance with [17] the angular density of TR can increase in ten times. This effect is explained by the combined action of two mechanisms of the formation of maxima in the angular distribution of TR. The first one corresponds to the case of total external reflection, when the maximum of the angular distribution is observed near the angle of total external reflection. The second mechanism is occurs due to the reconstruction of the Coulomb field of the charged particle when the velocity increases. The angular density can increase substantially when both mechanisms take place. The results of TR angular distribution calculations are presented in Fig.2 for the Si substrate and C layer. The real and imaginary parts of the dielectric susceptibilities for the layer and the substrate were calculated according to [19]. One can see the growing of TR angular density at the angle of incidence $\xi=\pi/8$ (curve 3). The curve 3 illustrates the manifestation of TR enhancement as described in [17].

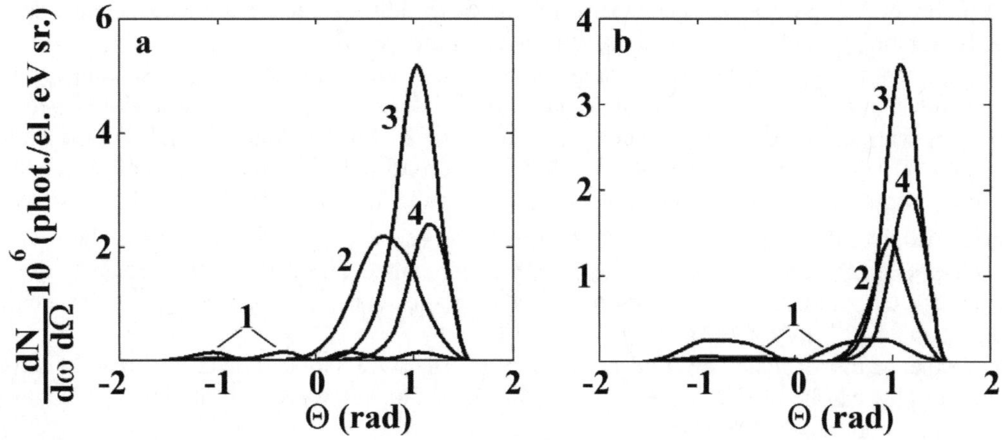

Figure 2. The effect of TR angular density enhancement for 50 eV photons under grazing interaction of 300 keV electrons with the target consisting of C layer 300 nm thickness on Si substrate. Figures "a" and "b" are the radiation under incidence of electrons from layer and vacuum sides respectively. Curves: 1 - $\xi = \pi / 2$; 2 - $\xi = \pi / 4$; 3 - $\xi = \pi / 8$; 4 - $\xi = \pi / 18$

A second case of interest for calculations is the VC radiation produced under conditions when the emitting electrons interact with the target provided that the incidence angle is less than the VC cone angle. The radiation angular density can grow under that condition in ten times [18]. The theoretical results of VC angular distribution are presented in Fig.3. One can see the angular density growing and the cone structure transformation. The oscillations manifested near the peaks are explained by the interference effects occurred the photons reflection at the layer surfaces.

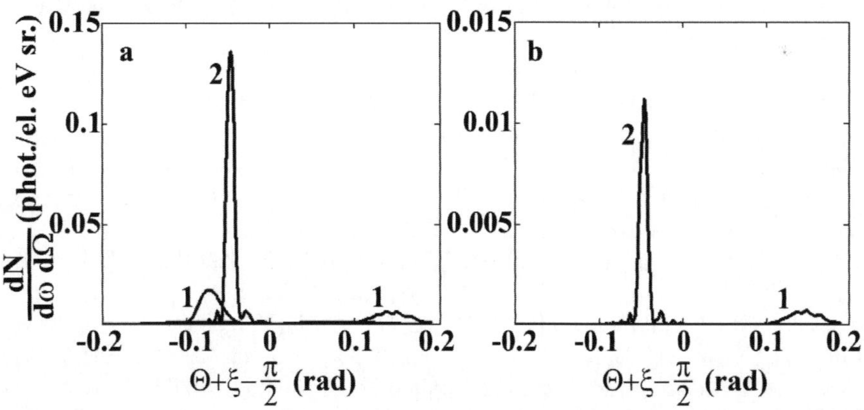

Figure 3. The effect of VC angular density enhancement for 284.1 eV photons under grazing interaction of 10 MeV electrons with the target consisting of C layer 300 nm thickness on Si substrate. Figures "a" and "b" are the radiation under incidence of electrons from layer and vacuum sides respectively. Curves: 1 - $\xi = \pi / 15$; 2 $- \xi = \pi / 60$

4. Conclusions

The model of TR and VC radiation is developed for the grazing incidence geometry of the interaction between fast electrons and a target consisting of thin amorphous layer applied on a thick amorphous substrate. The performed calculations predict the possibility to enhance the angular density of TR and

VC radiation under the grazing incidence of electrons on the layer from the vacuum side. The mechanism of radiation enhancement is the same as described in [18].

It is important to highlight the simple target construction and the radiation process geometry which facilitates the application of the mentioned effect. The EUV and soft X-ray radiation can be generated by electrons with an energy of the order of 100 keV and 10 MeV for the TR and VC mechanisms respectively. The TR mechanism allows to produce wideband radiation in a wide spectral range, VC mechanism allows to produce a quasimonohromatic radiation in contrast with TR. The typical radiation yield for TR generated under grazing interaction of electrons with the target is about of 10^{-6} photons per (electron eV sr.) and for the VC is about of 0.1 photons per (electron eV sr.). The difference for the realization of the presented characteristics consists in the energy of the incidence electrons – about 10 MeV for VC and 0.1 MeV for TR.

The results of the work can be used to develop laboratory radiation sources. A compact setup as described in [20] can be used to develop the VC source. The TR source can be built using electron guns with an energy order of 100 keV.

Acknowledgments

The work was supported by a program of the ministry of education and science of The Russian Federation for higher education establishments, project №14.578.21.0063 (RFMEFI57814X0063).

References

[1] Attwood D 1999 Soft X_Rays and Extreme Ultraviolet Radiation: Principles and Applications
[2] Ter-Mikaelian M 1972 High Energy Electromagnetic Processes in Condenced Media, Wiley, New York
[3] Rullhusen P, Artru X and Dhez P 1999 Novel Radiation Sources Using Relativistic Electrons, Word Scientific, Singapore
[4] Potylitsyn A P, Ryazanov M I, Strikhanov M N and Tishchenko A A 2011 Diffraction Radiation from Relativistic Particles, Springer-Verlag Berlin Heidelberg
[5] Nasonov N, Kaplin V, Uglov S, Piestrup M and Gary C 2003 Phys. Rev. E **68** 03654
[6] Kaplin V V, Uglov S R, Sohoreva V V, Bulaev O F, Voronin A A, Piestrup M, Gary C and Fuller M 2009 Nucl. Instrum. Methods Phys. Res. B **267** 777
[7] Adishchev Y N, Verzilov V A, Potylitsyn A P, Uglov S R and Vorobyev S A 1989 Nucl. Instrum. Methods Phys. Res. B **44** 130
[8] Scandale W, Arduini G, Assmann R, Gogolev A S et al. 2011 Physics Letters B **701** 180
[9] Takahashi Y, Hayakawa Y, Kuwada T, Tanaka T, Sakae T, Nakao K, Nogami K, Imagaki M, Hayakawa K and Sato I 2012 X-Ray Spectrometry **41** 210
[10] Nasonov N N, Kubankin A S, Zhukova P N, Goldstein M, Williams D L, Piestrup M A and Park H 2007 Nucl. Instr. and Meth. in Phys. Res. B **254** 259
[11] Baryshevsky V G, Batrakov K G, Feranchuk I D et. al. 2007 Phys. Lett. A **363** 448
[12] Kaplin V V, Uglov S R, Bulaev O F, Goncharov V J, Voronin A A, Piestrup M A, Gary C K, Nasonov N N and Fuller M K 2002 Applied Physics Letters **80** 3427
[13] Gary C K, Kaplin V V, Kubankin A S, Nasonov N N, Piestrup M A, and Uglov S R 2005 Nucl. Instrum. Methods Phys. Res. B **227** 216
[14] Kaplin V V, Uglov S R, Bulaev O F, Goncharov V J, Piestrup M A and Gary C K 2001 Nucl. Instrum. Methods Phys. Res. B **173** 3
[15] Piestrup M A, Kephart J O, Park H, Klein R K and Pantell R H 1985 Phys. Rev. A **32** 917
[16] Knulst W, van der Wiel M J, Luiten O J and Verhoeven J 2003 Appl. Phys. Lett. **83** 4050
[17] Kubankin A S 2008 Tech. Phys. Lett. **34** 927
[18] Gary C, Kaplin V, Kubankin A, Nasonov N, Piestrup M and Uglov S 2005 Nucl. Instr. and Meth. in Phys. Res. B **227** 95
[19] Henke B, Gullikson E and Davis J 1993 At. Data Nucl. Data Tables **54** 181
[20] Uglov S R, Zabaev V N, Kaplin V, Kuznetsov S 2012 Journal of Physics: Conference Series **357** 012012

IOP Conf. Series: Materials Science and Engineering **142** (2016) 012039 doi:10.1088/1757-899X/142/1/012039

Simulation of the Sample Alignment Process for the White Beam Tomography

A H Ozdiev[1], Dr. S N Liventsov[2]

[1]Department of General Physics, Institute of Physics and Technology, Tomsk Polytechnic University, Russia, Tomsk, 30 Lenin Avenue, 634050
[2]Department of Electronics and Automation of Nuclear Plants, Institute of Physics and Technology, Tomsk Polytechnic University, Russia, Tomsk, 30 Lenin Avenue, 634050

E-mail: ozdiev@tpu.ru

Abstract. Nowadays growth of synthetic crystals is a very prospective direction of research activity. However, it is well known that in synthesized crystals there are still many defects and the main idea is to synthesize crystal of ideal form and structure. But in purpose to remove defects first you have to characterize them. This paper describes the method for the characterization of such defects and mainly concentrates on the problem of sample alignment procedure, offering the way to avoid misalignments.

1. Method description

The idea is to use combination of topography and tomography (topo-tomography) with application of the white beam.

1.1. Topography and tomography

In order to characterize defects X-ray topography can be used. Topography is a 2D imaging of some area of analyzed object in purpose to find the defects. It can be implemented with using of monochromatic or white beam. Tomography is 3D imaging of scanned object with possibility to analyze it slice by slice. Combining those two methods allows not only find the defects, but presents them in 3D-view. In case of dislocations [1] described method helps to know how they are located inside crystal. This method already was implemented with monochromatic beam [2], but in current work the white beam have been used.

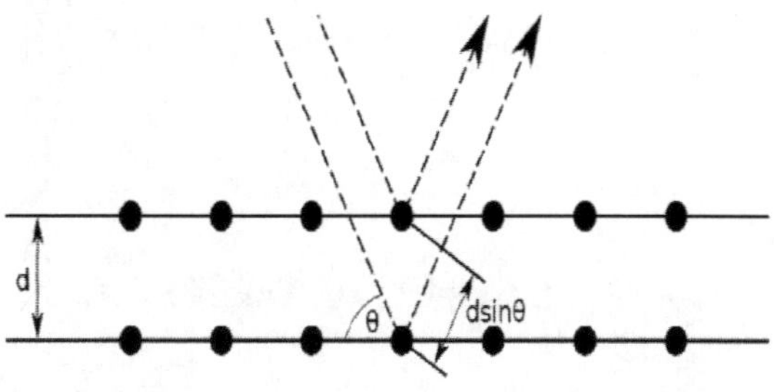

Figure 1. Schematic of Bragg's condition

X-ray topography is a method of obtaining images of defects in crystals by X-ray diffraction. This is the non-destructive method for characterizing and monitoring of the real structure of the material. Due to its high sensitivity to defects in the crystal lattice, this method makes it possible to analyze sample of quite big thickness and area. The sensitivity of X-ray topography to the presence of defects in the crystal lattice connected with the local deformation of the lattice in the location of the defect, for example, dislocations. The reflection of the beam from the deformed region of the crystal is different from the reflection of the beam from perfect part of crystal. This difference allows finding the defects.

To get the diffraction, the Bragg condition (Figure 1) has to be fulfilled. Bragg diffraction occurs when electromagnetic radiation or subatomic particle waves with wavelength comparable to atomic spacing's are incident upon a crystalline sample, are scattered in a specular fashion by the atoms in the system, and undergo constructive interference in accordance to Bragg's law [3]. For a crystalline solid, the waves are scattered from lattice planes separated by the interplanar distance d.

$$2d \times sin\theta = n\lambda \tag{1}$$

where n is an integer determined by the order given, and λ is the wavelength. A diffraction pattern is obtained by measuring the intensity of scattered waves as a function of scattering angle. Very strong intensities known as Bragg peaks are obtained in the diffraction pattern when scattered waves satisfy the Bragg condition.

Tomography (Figure 2) is a generating of three-dimensional image of the inside of an object from large series of two-dimensional X-ray images or projections taken around a single axis of rotation. In the end from tomographic reconstructions obtains 3D volume of object, which is possible to analyze slice by slice [4].

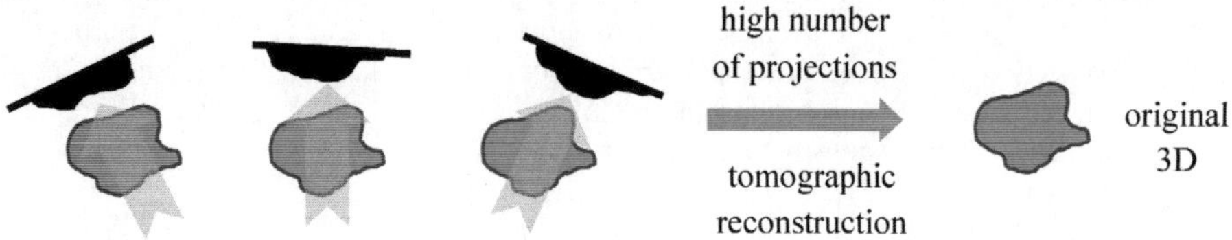

Figure 2. Basic principle of tomography

1.2. Principle of the method

In contrast to the monochromatic case, in which accurate sample adjustment is often necessary in order to reach diffraction conditions, the Bragg equation is always and automatically fulfilled in case of a white X-ray beam: whatever the angle at which the beam hits a specific lattice plane, there is always one wavelength in the incident

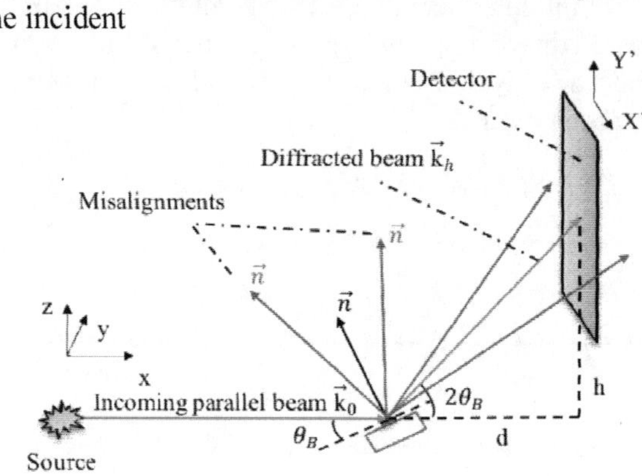

Figure 3. Misalignments

spectrum for which the Bragg angle is fulfilled just at this precise angle (on condition that the spectrum is wide enough).

So, Bragg's law works in each rotation angle, it means that in each rotation angle possible to get 2D-projection. But the problem is to adjust the orientation of the sample without any misalignments (Figure 3). Because if the sample is not aligned, the projections move during rotation, which does not allow 3D reconstruction. Misalignment in our case means that the diffraction normal to lattice plane of sample is not parallel to rotation axis. This is the main problem – problem of neutralizing of misalignments [5].

To solve the problem was used the experimental setup. This setup allows adjusting position and orientation of the sample in the space. Setup consist of several main and important parts namely source of beam, tilts, which change the orientation of the sample and the detector, which allows to find diffracted beam. Directions of position changing of sample for tilts 1 and 2 are shown on Figure 4. Directions of position changing of sample for tilts 3 and 4 are parallel and perpendicular, accordingly.

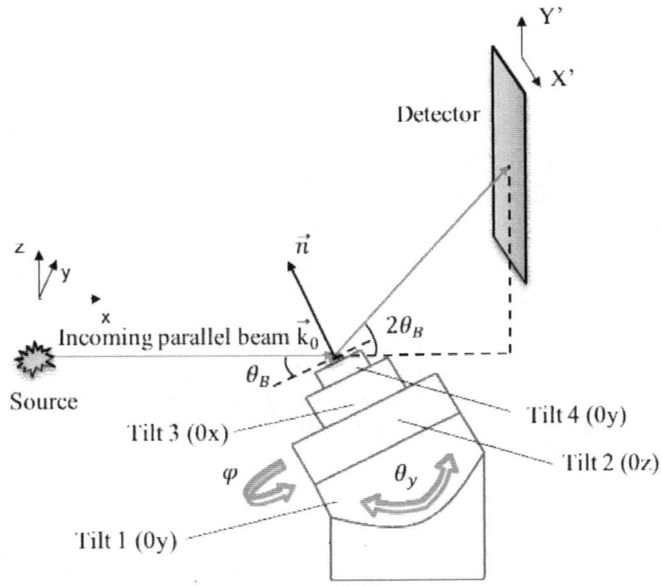

Figure 4. Misalignments

2. Simulation

To simulate the position of diffracted beam on the detector the coordinates of hit has to be calculated. Necessary calculations were made in Wolfram Mathematic 8.0.

2.1. Mathematical model

For calculation were used following formulas and mathematical expressions:

Tilts: rotation around 0y, angle θ_y; rotation around 0z, angle φ; rotation around 0y, angle β for $\theta_y = 0$, $\varphi = 0$; rotation around 0x, angle α for $\theta_y = 0$, $\varphi = 0$. $n' = (R'_y \cdot R'_x) \times n_0$, where $R'_y(\beta)$ – matrix of rotation around 0y, tilt 3 and $R'_x(\alpha)$ – matrix of rotation around 0x, tilt 4. Initial normal:

$$n_0 = \begin{pmatrix} 0 \\ 0 \\ 1 \end{pmatrix} \quad (2)$$

IOP Conf. Series: Materials Science and Engineering **142** (2016) 012039 doi:10.1088/1757-899X/142/1/012039

Misalignments includes in the system by random initialization of α and β. Such as: α = (a1-a2), where a1 is random part and a2 is adjustable part. β = (b1-b2), b1 –random, b2 – adjustable. The normal for the whole system:

$$n = (R_y(\theta_y).\, R_z(\varphi)).n' \tag{3}$$

where n' – normal for tilts 1 and 2; R_y – matrix of rotation around 0y, tilt 1; R_z– matrix of rotation around 0z, tilt 2. Diffracted beam:

$$k_n = k_0 - 2n(k_0.\, n) \tag{4}$$

Then from X, Y and Z components of diffracted beam S-parameter and hit coordinates can be calculated:

$$S = \frac{d}{k_{nx}} \tag{5}$$

d – distance to the detector; $X = S \times k_{ny}$ - X coordinate of hit; $Y = S \times k_{nz}$ - Y coordinate of hit.

On the figure 5 presented a user interface, which is simulating the experimental adjusting procedure. The green point is the diffracted beam, the white field is the detector, the scale bars are in cm. By changing of α and β angles user can change incline orientations of the sample. By changing of angle φ changes the rotation angle. Changing of θ_y angle allows changing Bragg angle. Also it possible to change position of detector by adjusting of X' and Y' coordinates in detector CS. Corresponding script is presented in figure 6.

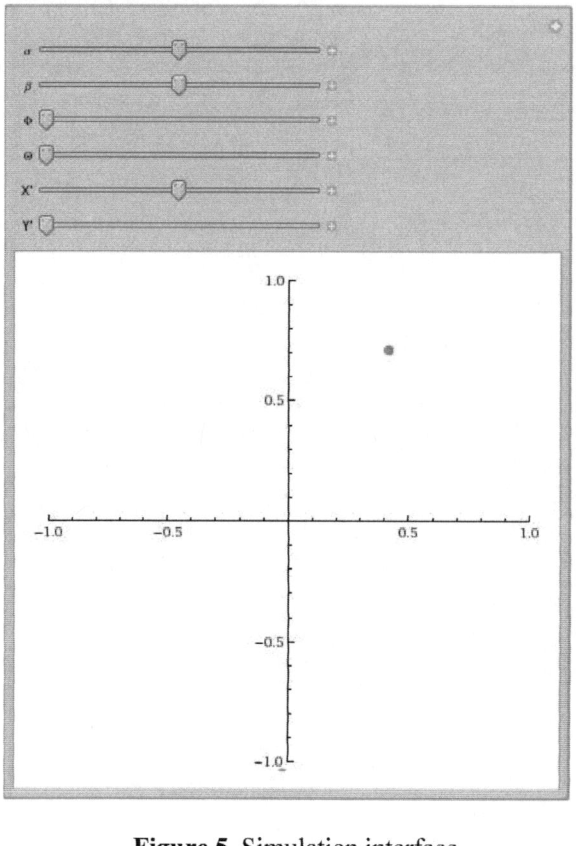

Figure 5. Simulation interface

```
α=alpha; (*Tilt angle*)
β=beta; (*Tilt angle*)
φ=phi; (*Rotation angle*)
Subscript[θ, y]=theta; (*Tilt angle*)
d=10; (*Distance to the detector*)
Subscript[k, 0]={1,0,0}; (*Initial beam*)
Print[Style["Source beam:",Blue, 14]]
Print["Subscript[k, 0]=",MatrixForm[Subscript[k, 0]]]
Ry1={{1,0,0},{0,Cos[α],-Sin[α]},{0,Sin[α],Cos[α]}}; (*Matrix of rotation aroung Ox for tilt #4*)
Rx1={{Cos[β],0,Sin[β]},{0,1,0},{-Sin[β],0,Cos[β]}}; (*Matrix of rotation around Oy for tilt #3*)
Print[Style["Matrix of rotation around 0x for tilt 4:",Blue, 14]]
Print["Subscript[R^', x]=",MatrixForm[Ry1]]
Print[Style["Matrix of rotation around 0y for tilt 3:",Blue, 14]]
Print["Subscript[R^', y]=",MatrixForm[Rx1]]
n2={0,0,1};
n=(Ry1.Rx1).n2; (*Normal n^'*)
Print[Style["Normal for system of tilts 3 and 4:",Blue, 14]]
Print["Normal n=",MatrixForm[n]]
Rz={{Cos[φ],-Sin[φ],0},{Sin[φ],Cos[φ],0},{0,0,1}}; (*Matrix of rotation around Oz for tilt #2*)
Print[Style["Matrix of rotation around 0z for tilt 2:",Blue, 14]]
Print["Subscript[R, z]=",MatrixForm[Rz]]
Ry={{Cos[Subscript[θ, y]],0,-Sin[Subscript[θ, y]]},{0,1,0},{Sin[Subscript[θ, y]],0,Cos[Subscript[θ,
y]]}}; (*Matrix of rotation around Oy for tilt #1*)
Print[Style["Matrix of rotation around 0y for tilt 1:",Blue, 14]]
Print["Subscript[R, y]=",MatrixForm[Ry]]
Subscript[n, L]=(Ry.Rz).n; (*Normal n*)
Print[Style["Normal for whole system:",Blue, 14]]
Print["Subscript[n, L]=",MatrixForm[Subscript[n, L]]]
Subscript[k, n]=Subscript[k, 0]-2 Subscript[n, L] (Subscript[k, 0].Subscript[n, L]); (*Diffracted
beam*)
Print[Style["Diffracted beam:",Blue, 14]]
Print["Subscript[k, n]=",MatrixForm[Subscript[k, n]]]
Print[Style["Components of diffracted beam:",Blue, 14]]
Print["X-component Subscript[k^x, n]=",Subscript[k^x, n]=Subscript[k, n][[1]]]; (*X component of
diffracted beam*)
Print["Y-component Subscript[k^y, n]",Subscript[k^y, n]=Subscript[k, n][[2]]]; (*Y component of
diffracted beam*)
Print["Z-component Subscript[k^z, n]=",Subscript[k^z, n]=Subscript[k, n][[3]]]; (*Z component of
diffracted beam*)
Print["Parameter s=",s=d/Subscript[k^x, n]]; (*Parameter S*)(*Coordinates of hit*)
Print[Style["Hit coordinates:",Blue, 14]]
Subscript[d, x]=s Subscript[k^y, n]; (*X coordinate*)
Subscript[d, y]=s Subscript[k^z, n]; (*Y coordinate*)
Print["X=",FullSimplify[Subscript[d, x]]];
Print["Y=",FullSimplify[Subscript[d, y]]];
(*Random missalignment angles*)
a12=RandomReal[{-10,10}];
a1=a12*Pi/180;
b12=RandomReal[{-10,10}];
b1=b12*Pi/180;
(*ρ=φ, a2=α, b2=β, σ=θ*)
```

```
(*X coordinate function*)
X[ρ_,a2_,b2_,σ_]:=-((20 (-Cos[b1-b2] Cos[ρ] Sin[a1-a2]+Sin[b1-b2] Sin[ρ]) (Cos[ρ] Cos[σ] Sin[b1-b2]+Cos[b1-b2] (Cos[σ] Sin[a1-a2] Sin[ρ]-Cos[a1-a2] Sin[σ])))/(1-2 (Cos[ρ] Cos[σ] Sin[b1-b2]+Cos[b1-b2] (Cos[σ] Sin[a1-a2] Sin[ρ]-Cos[a1-a2] Sin[σ]))^2));
(*Y coordinate function*)
Y[ρ_,a2_,b2_,σ_]:=-((20 (Cos[a1-a2] Cos[b1-b2] Cos[σ]+(Cos[ρ] Sin[b1-b2]+Cos[b1-b2] Sin[a1-a2] Sin[ρ]) Sin[σ]) (Cos[ρ] Cos[σ] Sin[b1-b2]+Cos[b1-b2] (Cos[σ] Sin[a1-a2] Sin[ρ]-Cos[a1-a2] Sin[σ])))/(1-2 (Cos[ρ] Cos[σ] Sin[b1-b2]+Cos[b1-b2] (Cos[σ] Sin[a1-a2] Sin[ρ]-Cos[a1-a2] Sin[σ]))^2));
Manipulate[Graphics[{PointSize->Large,Green,Point[{X[ρ*Pi/180,a2*Pi/180,b2*Pi/180,σ*Pi/180]-x,Y[ρ*Pi/180,a2*Pi/180,b2*Pi/180,σ*Pi/180]-y(*-d*Tan[(Pi/90)σ]*)}]},Axes->True,AxesOrigin->{0,0}, PlotRange-> {{-1,1},{-1,1}}],{{a2,0,"α"},-90,90,.1},{{b2,0,"β"},-90,90,.1},{{ρ,0,"Φ"},0,360,90},{{σ,0,"Θ"},0,20,1},{{x,0,"X'"},-5,5,.1},{{y,0,"Y'"},0,10,.1}]
```

Figure 6. The script of the simulation of the alignment procedure

3. Conclusion

Since the method is a combination of topography and tomography, the alignment procedure aims to fulfil two conditions. First, Bragg's law has to work for each tomographic rotation angle of the mounted sample with a fixed Bragg angle (Figure 7). This is complicated by the fact that by using white beam Bragg's law is always fulfilled. The second conditions requirement of tomography, because to reconstruct the 3D volume of the sample, the image on the detector has to stay in the same position for all view angles of the sample.

Using slits the beam size was reduced so that diffracted beam can be displayed on the detector as a point. Misalignments can be neutralized by changing of incline orientation of the sample. It is possible to check by changing of rotation angle. If there are no misalignments changing of rotation angle will not cause changes in the hit coordinates on the detector. It was reached by turning always +/- 180° and trying to make the diffracted beam (point) stay at the same position on the detector, then the same operation was repeated for the 90° rotated case. And then back and iteratively the adjustment is improved. Developed algorithm was tested at the ANKA synchrotron facility [6].

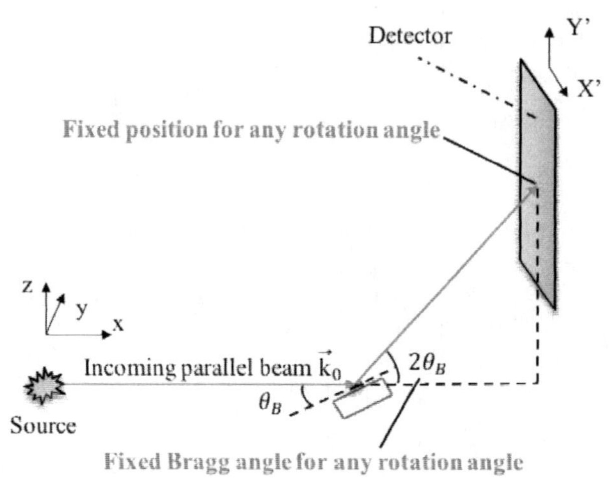

Figure 7. Alignment procedure

References

[1] James Shackelford (2009). Introduction to Materials Science for Engineers (7th ed.)
[2] Wolfgang L. (2010) *Characterization of polycrystalline materials by X-ray diffraction tomography* (New York: Springer US)
[3] Bragg W. L., (1914) «The Diffraction of Short Electromagnetic Waves by a Crystal», Proceedings of the Cambridge Philosophical Society, 17, 43
[4] A. C. Kak and Malcolm Slaney, Principles of Computerized Tomographic Imaging
[5] Helfen L (2011) *Synchrotron and neutron laminography for three-dimensional imaging of devices and flat material specimens* (Berlin: Carl Hanser Verlag GmbH & Co. KG.)
[6] 2013 *ANKA instrumentation book* (Karlsruhe: ANKA Synchrotron Radiation Facility, Karlsruhe Institute of Technology, a member of the Helmholtz Association. IPS, Institute for Photon Science and Synchrotron Radiation)

Experimental Simulation of the Radionuclide Behaviour in the Process of Creating Additional Safety Barriers in Solid Radioactive Waste Repositories Containing Irradiated Graphite

A O Pavliuk[1], S G Kotlyarevskiy[1], E V Bespala[2], E V Zakarova[3], N I Rodygina[3], V M Ermolaev[3], I M Proshin[3], A Volkova[3]

[1]JSC « Pilot and Demonstration Center for Uranium-Graphite Nuclear Reactor Decommissioning», Seversk, Tomsk region, Russia
[2]Tomsk polytechnic university, Tomsk, Russia
[3]Frumkin Institute of Physical Chemistry and Electrochemistry RAS, Moscow, Russia

E-mail: bespala_evgeny@mail.ru

Abstract. Results of the experimental modeling of radionuclide behavior when creating additional safety barriers in solid radioactive waste repositories are presented. The experiments were run on the repository mockup containing solid radioactive waste fragments including irradiated graphite. The repository mockup layout is given; the processes with radionuclides that occur during the barrier creation with a clayey solution and during the following barrier operation are investigated. The results obtained confirm high anti-migration and anti-filtration properties of clay used for the barrier creation even under the long-term excessive water saturation of rocks confining the repository.

1. Introduction

The majority of the at-reactor repositories of the shutdown Industrial Uranium-Graphite Reactors (IUGR) were in operation for 35÷40 years. They have played their positive part in providing the radiation safety of the personnel, population, and the environment in accordance with the regulatory requirements in effect in 50s-70s of the previous century. There are several dozen at-reactor repositories at the IUGR sites of the Russian Federation where irradiated graphite-containing solid radioactive waste (SRW) of the reactor origin is currently located.

To date the nuclear legacy at-reactor repositories do not entirely pass strict modern requirements imposed on the storage of radioactive materials. Those requirements could be only met if additional engineered safety barriers are created in the existing at-reactor repositories so that in combination with the repository confining rocks (natural barrier) they make a multi-barrier protective system. The developed system of barriers should provide the SRW disposal conditions for the radionuclide migration to be possible merely by means of the diffusion processes. The radionuclide diffusion coefficients in the barrier material should be as low as to ensure the level of activity emerging out of the boundaries of the repository to be within the intervention level, i.e. the regulated parameters defining the boundary between the radioactive and non-radioactive medium.

At the JCS Pilot and Demonstration Center for Uranium-Graphite Nuclear Reactor Decommissioning (JSC «DC») an option of creating the additional protective barriers by means of filling the free repository space by a clayey solution is being considered as a promising solution for the problem [1,2]. The barrier is installed by pumping the clayey solution through the perforated pipes. The pipe location grid size is determined so as to distribute the pumped clayey mass through the entire repository volume. After the clay is dried, the formed clay monolith contains the embedded solid radioactive waste. The implementation of the technique in question is the most relevant for the irradiated graphite-containing repositories as to date the ultimate scenario of handling this waste is not defined yet [3]. The approach suggests a delayed solution of the problem for a period of time long enough to make a decision on whether the SRW is to be retrieved or not.

To prove the efficiency of the technique proposed for the creation of engineered clay barriers, a repository mockup has been fabricated at the JSC «DC». It was used to conduct a series of runs aimed at the simulation of the radionuclide behavior when creating additional protective barriers in the existing repositories.

2. The procedure of experimental runs

The repository mockup was fabricated in the shape of a combined metal vessel. The schematic layout of the mockup cross-section is shown in figure 1; the locations of SRW fragments, engineered clay and natural rock barriers are indicated on the layout. The rock for the mockup was drawn nearby the existing at-reactor SRW repository.

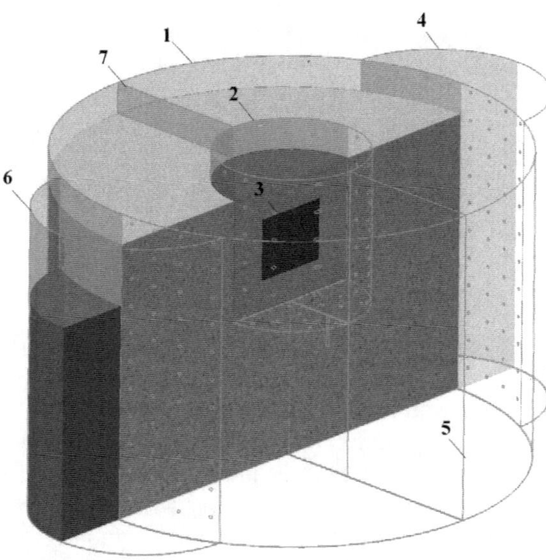

Figure 1. The mockup layout: 1 – the external vessel loaded with the rock; 2 – the internal vessel with the holed wall, initially loaded by the clayey solution; 3 – SRW fragments in a basket; 4 – the right tank; 5 – the partition wall; 6 – the left tank with water; 7 – the partition wall.

The mockup consists of the two cylinder-shaped nested steel vessels. The external vessel of 1.5 diameter and 0.5 m height is divided with a solid partition into two parts loaded with the layer-by-layer compacted rock. Walls of the internal vessel are holed such as their permeability is significantly higher than that of the concrete walls of the actual repository. SRW fragments of steel, alloys, and irradiated graphite are positioned into this vessel in the metal basket. Taking into account the gamma-activity of the samples and the mass of the SRW fragments the gross gamma-activity of the waste amounted ~6.2×10^{8} Bq and was defined primarily by ^{137}Cs and ^{60}Co radionuclides. The activity of the alpha-emitting trans-uranium (TRU) elements did not exceed 1.0×10^{5} Bq.

The gross activity of radionuclides contained in SRW loaded into the internal mockup vessel are given in table 1.

Table 1. Gross activity of SRW positioned in the internal mockup vessel, Bq.

^{137}Cs	^{90}Sr	^{60}Co	^{134}Cs	^{152}Eu	^{154}Eu	^{155}Eu	TRU
5.9×10^8	3.0×10^8	1.6×10^7	2.9×10^6	2.5×10^4	1.0×10^6	2.3×10^5	$<1.0\times10^5$

The internal vessel and the basket with SRW fragments were filled with clay of the Zavarzinskoe Deposit containing 10% of sand. Clays of the Zavarzinskoe Deposit (located in the Tomsk Region) belong to the illite-kaolinite type characterized as follows: the content of particles sized below 0.01 mm varies from 55 thru 65%; the components in the descending order are illite, kaolinite, and montmorillonite.

To get the maximum filling of free space between SRW fragments we used the clayey solution containing sandy-clayey mix and water at the ratio of 1/1. Once in a while water was pumped out of the internal vessel. Redistribution of the residual moisture between the clay and the rock resulted in the clay drying and eventually in the formation of the clayey monolith with embedded SRW fragments. During the run the mockup was covered by the polyethylene film to prevent the moisture variations in the surface layer.

The tanks located on each side of the mockup were separated from the external vessel by the holed partition. After the clayey monolith was formed, the left tank was filled by the underground water taken from the well located nearby the at-reactor repository. The water level in the tank was higher than the bottom of the basket with SRW fragments positioned into the internal vessel and was maintained constant during the whole run. Thus, a directional water flow was simulated.

The right tank was used for monitoring. In case there was no filtration of water though the clay monolith during the run, the right tank should have remained dry. Conditions in the left tank were simulating the situation when the rocks in the bottom part of the repository were below the ground water level. On the contrary, the conditions in the right tank were simulating the situation when the repository bottom was located higher than the ground water level.

The radionuclide behavior studies were performed by sampling the clay and the rock periodically over a particular time interval. The measurement of γ-emitting ^{60}Co, Eu isotopes, ^{137}Cs, and ^{241}Am radionuclides was performed using the gamma-spectrometer equipped with GMX25P4 (Ortec) ultra-pure germanium detector. The content of β -emitting ^{90}Sr in the samples was measured using the Tri-Carb 3180 TR/SL (Perkin Elmer) liquid scintillation spectrometer. The specific activity of ^{239}Pu was analyzed using the extraction-chromatographic technique followed by the electrolytic depositing of plutonium and measuring samples on the Alpha Aria (Ortec) low-background alpha-spectrometer [4].

The speciation of radionuclides adsorbed by clays was determined by the selective desorption technique. The species are subdivided into groups based upon the distinctions in the solubility thereof. The selective desorption techniques are widely used in the radionuclide behavior studies in the natural and natural-technogenic systems, i.e. the protective barriers [5, 6]. Using the results of the selective desorption (see table 2) one can evaluate the content of radionuclides existing as mobile species (Stages I and II), potentially mobile species (Stage III), and the species fixed on rocks (Stages IV and V).

Table 2. The successive desorption procedure used to determine the sorption mechanism (S/L=1/10).

Stage	Composition of the desorbing solution	Desorbed species
I	Distilled water, 20°C	Water soluble
II	NH_4Ac, 1 mole/L, pH 4,8 (HAc), 20°C	Exchangeable
III	1 mole/L HCl, 20°C	Surface complexation
IV	6 mole/L HCl, 20°C	Soluble in acids

Stage	Composition of the desorbing solution	Desorbed species
V	6 mole/L HNO_3, 85°C	Soluble in acids

3. The experimental modeling of the radionuclide behaviour - results

In 8 months of filling the internal and external vessels the rock surrounding the internal vessel was sampled as follows: IB-1 ÷ IB-4 and IC-1 ÷ IC-5 samples were taken in the vicinity of the central SRW containing vessel and IIC-1 ÷ IIC-5 samples were taken at 100 mm distance from it. The sampling was performed using the piston sampler. The samples were taken along the flow, IB-1 ÷ IB-4 samples before the internal SRW containing vessel and IC-1 ÷ IC-5 samples - after it.

The sampling layout is shown in figure 2. The number assignment in the sample labels corresponds to the downward direction from top to bottom. The radionuclide content in the samples is given in tables 3-5.

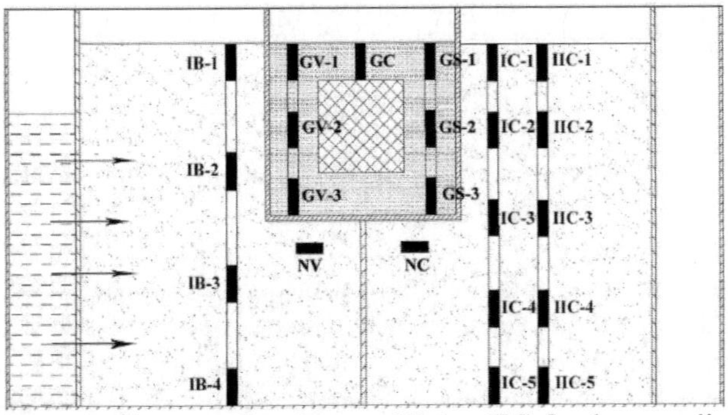

Figure 2. The sampling layout: IB-1 ÷ IIC-5 - the sampling points; → -the direction of the water flow through the holed wall of the left tank.

Table 3. The radionuclide content in the rock samples taken in the vicinity of the central SRW containing vessel and at 100 mm distance from it, Bq/g.

S. No.	Sample	^{137}Cs	^{90}Sr	TRU
1.	IB-1	0,6	<0,02	<0,02
2.	IB-2	0,6	<0,02	0,02
3.	IB-3	0,2	<0,02	<0,02
4.	IB-4	<0,02	<0,02	<0,02
5.	IC-1	1,5	<0,02	<0,02
6.	IC-2	0,4	<0,02	<0,02
7.	IC-3	0,4	<0,02	0,03
8.	IC-4	<0,02	<0,02	<0,02
9.	IC-5	<0,02	<0,02	<0,02
10.	IIC-1	<0,02	<0,02	<0,02
11.	IIC-2	<0,02	<0,02	<0,02

There were no Eu isotopes, ^{60}Co, and ^{90}Sr radionuclides detected in the samples. From the data obtained it follows that a negligible quantity of ^{137}Cs and TRU radionuclides has been detected in the immediate vicinity of the SRW containing vessel. There is a possibility the sample was contaminated in the process of sampling. At the 100 mm distance from the wall of the vessel the concentration of all radionuclides was below the detection limit.

If in the samples taken from the rock in the external vessel there was practically no radionuclides

IOP Conf. Series: Materials Science and Engineering **142** (2016) 012040 doi:10.1088/1757-899X/142/1/012040

detected, in the internal vessel a transition thereof into the clay monolith was observed after the contact of clay with the SRW fragments (see table 4). The process of the radionuclide desorption from the SRW and sorption thereof on clay occurred with the participation of water remained after the installation of the clay barrier. The radionuclide distribution along the height of the internal vessel is irregular; probably, it is attributed to the different level of the contamination of various SRW fragments located in the basket. As compared to other radionuclides, the content of ^{137}Cs in clay is considerably higher.

Table 4. The radionuclide content in clay samples taken from the internal vessel simulating the SRW repository and in the rock under the vessel, Bq/g.

S.No.	Sample	^{137}Cs	^{60}Co	^{90}Sr	TRU
1.	GS-1	$5.7 \cdot 10^2$	0.8	30	4.5
2.	GS-2	$7.8 \cdot 10^2$	0.3	19	2.0
3.	GS-3	$4.7 \cdot 10^2$	0.2	32	1.6
4.	NV	0.9	<0.01	0.5	8.3
5.	NC	<0.01	<0.01	<0.02	<0.02

In the rock under the internal vessel (NV sample) the radionuclide activity level is lower, TRU being the only exception.

In 10 months one more sampling has been conducted; hence the total duration of the run was 18 months. During this time the rock in the left part of the vessel was saturated with water so that water appeared on the rock surface; however there was no penetration of water into the right part of the vessel. In the rock samples taken in the vicinity of the internal vessel the presence of radionuclides was not detected, the only exception being IV-1 sample with 1.3 Bq/g of ^{137}Cs.

The difference in the content of ^{137}Cs in the samples taken at various times is associated with the fact that at the initial stage the distribution of the excessive water arisen from the barrier installation process due to the contact with SRW caused the non-uniform contamination of the rock. Besides, the samples in the course of time were not taken from the same point rather from the various mockup areas. During the second sampling the set clay monolith barrier prevented the filtration of the directed water flow through the contaminated zone.

The character of the radionuclide distribution in clay was unchanged since the first sampling, the only difference being the higher contamination level (see tables 4 and 5).

Table 5. The radionuclide content in the clay in the internal vessel simulating the SRW repository, Bq/g.

S.No.	Sample	^{137}Cs	^{60}Co	^{90}Sr	TRU
1.	GV-1	$7.5 \cdot 10^3$	1.8	40	0.6
2.	GV-2	$6.3 \cdot 10^3$	1.0	20	0.5
3.	GV-3	$0.2 \cdot 10^3$	1.3	1.0	1.15
4.	GC	$52.6 \cdot 10^3$	3.2	$0.7 \cdot 10^3$	2.4
5.	GS-1	$0.7 \cdot 10^3$	1.1	16	5.7
6.	GS-2	$1.4 \cdot 10^3$	1.0	10	1.4
7.	GS-3	$0.6 \cdot 10^3$	0.05	16	1.5

The highest content of ^{137}Cs and ^{90}Sr was found in GC sample of clay taken from the top part of the internal vessel. Probably it is associated with the SRW fragments positioned in the top part of the basket and possessing the high level of the surface contamination by ^{137}Cs and ^{90}Sr.

The third sampling was conducted in 10 months, the total duration of the run by then was 28 months. During all that time the soil was watered, but there was no water in the right part of the vessel.

The samples were taken along the flow in the vicinity of the central SRW containing vessel and at 100 mm distance from it. In the rock samples before the internal vessel the content of ^{137}Cs increased as follows: IB-1 – 1.5; IB-3 – 0.5; IB-4 – 0.5 Bq/g. In the samples taken along the flow after the internal vessel the content of ^{137}Cs did not exceed 0.4-0.6 Bq/g. Other fission products were not detected also. In two samples, namely, IC-2 and IC-3 the content of TRU was 0.05 and 0.07 Bq/g. In the samples taken at the 100 mm distance from the internal vessel the content of ^{137}Cs was either below its detection limit or slightly higher. Along the way there was no marked variation in the activity level of clay. As before, the maximum contamination of samples was with ^{137}Cs.

When evaluating the protective properties of the barrier material, the important factors are, along with its sorption properties, the radionuclide speciation on the barrier material and the trend of the redistribution thereof in time taking into account the barrier has to keep its protective properties for a long period of time. The results of the selective radionuclide desorption from the clay samples taken during the run are given in figure 3. The results were obtained in the study of the GS-1 samples taken in the vicinity of the internal vessel along the flow.

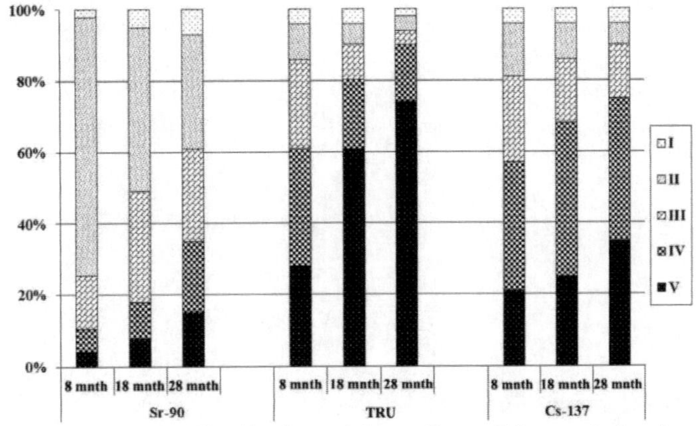

Figure 3. The distribution of the radionuclide speciation in clay (I – water soluble; II – exchangeable; III – mobile; IV and V – strongly fixed) depending on the time of the radionuclide interaction with the barrier material.

As it follows from the results of radionuclide speciation studies, in 8 months from the beginning of the run merely 10% of ^{90}Sr existed in the strongly fixed species (Stages IV and V). The corresponding values for ^{137}Cs and TRU are 60%. As the interaction time increases, the radionuclide speciation is redistributed towards the growth of the strongly fixed species content. In 28 months of interaction of clay with radionuclides the content of the strongly fixed species for ^{90}Sr amounted 35%, and for ^{137}Cs and TRU – 75% and 90%, respectively.

4. Conclusion

1. In order to evaluate the efficiency of the protective barrier installation in the existing SRW repositories the radionuclide behavior has been investigated in the combined barrier made of the illite-kaolinite clay and the repository confining rocks. The experiments were run for 28 months using the repository mockup containing solid radioactive waste fragments including irradiated graphite.

2. In the internal vessel of the mockup simulating the SRW repository the contamination of clay has occurred on the barrier installation stage, when the clayey solution used for the installation desorbed the radionuclides from the surface of SRW fragments. The following formation of the clay monolith effectively prevented the further distribution of ^{137}Cs, gamma-emitters, ^{90}Sr, and alpha-emitting TRU radionuclides through the rock. For the most readily desorbed ^{137}Cs radionuclide the difference in its concentration between the clay and the rock in the vicinity of the repository simulating vessel reached four orders of magnitude.

3. Thus, during more than two years of observation no significant distribution of radionuclides in the rock has occurred, even though under the maintained conditions of the constant wetting. It proves the high anti-migration and anti-filtration properties of clay used for the barrier material.

4. References

[1] Izmestiev A, Pavliuk A, Kotlyarevskiy S 2015 *Advanced Materials Research* **1084** pp 613–619
[2] Yushitsin K V, Kotlyarevsky S G, Izmestiev 2012 *Nuclear Technology and Environmental Safety* **2** pp 100–105
[3] Tsyganov A A, Komarov E A, Kotlyarevsky S G, Pavlyuk A O, Shamanin I V, Nesterov V N 2007 *Izvestiya of Tomsk Polytechnic University*. **310/2** 94–98
[4] Phillips G W, Marlow K W 1976 *Ibid* **94/2** p 349
[5] Tesser A, Campbell P G C, Bisson M 1979 *Anal.Chem.* **51** 844–851
[6] Pavlotskaya F I, Goryachenkova T I, Kazinskaya I E 2003 *Radiokhimiya* **45/5** 471–478

IOP Conf. Series: Materials Science and Engineering **142** (2016) 012041 doi:10.1088/1757-899X/142/1/012041

Data and Knowledge Base on the Basis of the Expanded Matrix Model of Their Representation for the Intelligent System of Road-Climatic Zoning of Territories

A Yankovskaya[1, 2, 3, 4, 5, a], **D Cherepanov**[1, b], **O Selivanikova**[3, c]

[1]Tomsk State University of Architecture and Building, Russia, 634003, Tomsk, 2 Solyanaya Sq. Tel. (3822) -65-07-82
[2]Tomsk State University, Russia, 634050, Tomsk, 36 Lenin Ave.
[3]Tomsk Polytechnic University, Russia, 634050, Tomsk, 2 Lenin Ave.,
[4]Tomsk State University of Control Systems and Radioelectronics, Russia, 634050, Tomsk, 40 Lenin Ave.
[5]Siberian State Medical University, Russia, 634050, Tomsk, 2 Moscovskiy Trakt

E-mail: [a]ayyankov@gmail.com, [b] d_n_ch@mail.ru, [c] selov@tpu.ru

Abstract. An extended matrix model of data and knowledge representation on the investigated area, as well as a matrix model of data representation on the territory under investigation, are proposed for the intelligent system of road-climatic zoning of territories (RCZT) – the main information technology of RCZT. A part of the West Siberian region has been selected as the investigated territory. The extended matrix model of knowledge representation is filled out by knowledge engineers with participation of highly qualified experts in the field of RCZT. The matrix model of data representation on the territory under investigation is filled out by persons concerned in RCZT of the motor-roads management system.

1. Introduction

The existing motor-roads network in the Russian Federation currently does not allow to fully meet the needs of the economy and the competitiveness of the international transportation of goods by motor-road throughout the territory of Russia. The importance of motor-road transportation cannot be overestimated in preservation the territorial integrity of the Russian Federation, its geopolitical influence, and its competitiveness. [1] In this connection, the development of modern information technologies in the motor-road management system assumes a special prominence.

Road-climatic zoning of territories (RCZT) includes the allocation on the territory of the Russian Federation of road-climatic zones characterized by one or another combination of heat and moisture, which stipulates, within the limits of a single zone, the formation of interrelated types of soil and vegetation [2 – 7]. However, other factors that significantly affect the condition of motor-roads can vary significantly within each zone [3 – 5]. In this regard, approaches to RCZT have to be refined.

Consideration of specificities related to natural-and-climatic conditions of West Siberian and other regions of Russia, in addition to the existing design standards in the motor-road industry [2], allows to divide road-climatic zones into homogenous territories [3 – 5] characterized by typical and peculiar geocomplexes (climate, geology, topography, and other conditions). Within each homogeneous territory the homotypic road structures, in the first instance especially roadbed and pavement design, are characterized by approximately the same strength and stability. Such division into homogeneous territories should prevent significant financial and economic losses throughout the life cycle of motor-roads, associated with their maintenance to the desired safety standards. In addition, it is necessary to develop a modern methodology of RCZT for clarification of the boundaries of homogeneous territories that had been allocated on the basis of previously used approaches. Thus, the development of new information technologies in the field of RCZT is beyond question.

The proposed in the publication [8] intelligent information technology of road-climatic zoning of territories (IIT RCZT) will allow to attribute the area where a motor-road is planned to be built to a known type of the homogeneous territory. For this purpose, it is necessary to use matrix way of data and knowledge representation in the intelligent system (IS) of RCZT, which serves as the basis of IIT RCZT consisting of an expanded matrix model of data and knowledge representation on the investigated area of the West Siberian region and a matrix model of data representation on the region under investigation, as described is in the publication [8].

Below is given the matrix way of data and knowledge representation used for RCZT; structuring of data and knowledge; characteristic and classification features and their values; formation of the database and knowledgebase in IS RCZT.

2. Matrix way of data and knowledge representation

In the publication [8] it is proposed to use an extended matrix model of data and knowledge representation for RCZT. Integer values of characteristic features, including group features (features each of which is divided into another features that take integer values corresponding to certain intervals) as well as compulsory features are elements of the expanded description matrix ($\mathbf{Q}^e$)..

Integer matrix of descriptions ($\mathbf{Q}^e$) defines a description of objects in the space of characteristic features $z_1, z_2, \ldots, z_m$. The column of the matrix $\mathbf{Q}^e$ corresponds to each characteristic feature. The row of the matrix $\mathbf{Q}^e$ corresponds to each reference point at which the values of characteristic features are defined. Thus, the value of an integer characteristic feature (including the compulsory feature corresponding to one of the reference points) is the element of the matrix $\mathbf{Q}^e$ [8]. It shall be noted that compulsory features are not involved in the definition of regularities, but serve only to map the position of zones, subzones, and road districts.

Moreover, the description matrix $\mathbf{Q}^e$ is complemented by rows that represent the knowledge of highly qualified experts. It shall be noted that if the value of the characteristic feature in the additional row is not known or not indicated by highly qualified experts, this fact is designated by a dash ("–") in the respective element of the matrix $\mathbf{Q}^e$.

$$
\begin{array}{c}
\\ \mathbf{Q}^e =
\end{array}
\begin{array}{c}
\begin{array}{ccccccc} z_1 & z_2 & z_3 & z_4 & z_5 & z_6 & z_7 \end{array} \\
\begin{bmatrix}
3 & 4 & 8 & 9 & 5 & 1 & 3 \\
2 & 5 & 7 & 9 & 4 & 1 & 5 \\
3 & 5 & 7 & 8 & 3 & 0 & 8 \\
4 & 2 & 6 & 8 & 4 & 1 & 5 \\
3 & 3 & 8 & 8 & 4 & 1 & 5 \\
2 & 3 & 8 & 8 & 4 & 1 & 3 \\
1 & 2 & 8 & 8 & 3 & 2 & 3 \\
1 & - & 7 & 8 & 4 & 0 & 6 \\
2 & 6 & 7 & 9 & 5 & 1 & 3 \\
1 & 4 & 8 & 8 & 3 & 0 & 4
\end{bmatrix}
\end{array}
\begin{array}{c} \\ 1 \\ 2 \\ 3 \\ 4 \\ 5 \\ 6 \\ 7 \\ 8 \\ 9 \\ 10 \end{array}
\quad
\mathbf{R}^e =
\begin{array}{c}
\begin{array}{ccc} k_1 & k_2 & k_2 \end{array} \\
\begin{bmatrix}
1 & 1 & 1 \\
1 & 1 & 2 \\
1 & 2 & 1 \\
1 & 2 & 1 \\
1 & 1 & 1 \\
2 & 1 & 1 \\
1 & 2 & 2 \\
2 & 2 & 1 \\
2 & 1 & 2 \\
2 & 2 & 2
\end{bmatrix}
\end{array}
$$

Fig. 1 Matrices $\mathbf{Q}^e$ and $\mathbf{R}^e$

Integer matrix of distinctions ($\mathbf{R}^e$) defines the partition of objects into equivalence classes for each classification mechanism. Integer values of classification features are elements of the distinctions matrix $\mathbf{R}^e$. We shall restrict ourselves to the distinctions matrix of the diagnostic type where each subsequent column parts the previous one into equivalence classes. In accordance with the practice of RCZT [3 – 5] we shall use three classification features of the diagnostic type, which correspond to: 1st – zones, 2nd – subzones, 3d – road districts.

Some additional rows of the distinctions matrix, filled out by highly qualified experts, may differ from the previously filled on the basis of experimental data, because in the learning sample some combinations of values of classification features can be absent.

Information on the objects under investigation is represented only by the matrix $\mathbf{Q'}^e$, similar to the matrix $\mathbf{Q}^e$, and determined the description of the territory under investigation subjected to RCZT.

An example of a description and distinctions extended matrices for ten reference points is given on fig. 1. Rows formed on the basis of the knowledge of highly-qualified experts are absent in matrices $\mathbf{Q}^e$, $\mathbf{R}^e$.

3. Structuring data and knowledge

Formation of the data and knowledge base is carried out on the basis of structurization data and knowledge given in the article [8]. To create IIT RCZT it is advisable to use the following three types of characteristic features proposed in [8]: characteristic features (CF), group CF, and compulsory CF.

We shall list 6 CF: type of vegetation; terrain relief; calculated soil moisture (CSM); content of sand particles; evaporation from the surface of the roadbed in summer months; type (type and subtype) of soil.

The following are 7 group CF: Selyaninov hydrothermal coefficient (SHC); plasticity index (PI); size of soil particles (SSP); average annual air temperature (AAT); snow cover depth (SCD); soil freezing depth (SFD); relative frost heave (RFH).

The following four features are used as compulsory features that do not participate in the identification of particularities but serve only to map the position of road-climatic zones, subzones, and road areas: name of the zone; name of the subzone; name of the road district; name of the reference point.

In addition to characteristic features there are three classification features: zone; subzone; road district.

4. Characteristic and classification features and their values

The knowledge base is formed on the basis of description and distinctions matrices [9]. The values of elements of the description matrix are determined based on the information database. The values of characteristic features or intervals or their values are encoded by integers. Intervals of partitioning of feature values are determined on the base of expert knowledge. The result is a set of integer values of characteristic features.

The following are integer values encoding characteristic features, associated to columns of the description matrix:

z1 – vegetation type: 1 – tundra vegetation; 2 – forest-tundra vegetation; 3 – forest vegetation (northern taiga, with propagation of permafrost soils); 4 – forest vegetation (middle taiga); 5 – forest vegetation (southern taiga); 6 – forest-steppe vegetation; 7 – steppe vegetation; 8 – desert and desert-steppe vegetation;

z2 – terrain relief: 1 – flat with a relative elevation of the relief (RER) up to 25 m; 2 – hilly with RER from 25 m up to 200 m; 3 – mountainous (low mountains area) with RER from 200 m up to 500 m, and with a prevailing slope gradient (PSG) from 5° up to 10°; 4 – mountainous (mid-mountain terrain) with RER from 500 m up to 1000 m, PSG from 10° up to 25°, and elevation above sea level of about 1000-2000 m; 5 – mountainous (highland area) with RER from 1000 m, PSG more than 25 °C, and elevation above sea level more than 2000 meters;

z3 – calculated soil moisture (CSM) – the ratio of the natural soil moisture (obtained in the calculated – spring period of the year) to the moisture at the liquid limit: 1 – low soil moisture with CSM up to 0.4; 2 – normal soil moisture with CSM from 0.4 up to 0.6; 3 – increased soil moisture with CSM from 0.6 up to 0.8; 4 – waterlogged soil with CSM from 0.8 up to 1;

z4 – content of sandy particles (expressed as percentage): 1 – sandy loam [50, 100); 2 – silty loam [0; 50); 3 – light loam, sandy [40; 100); 4 – light loam, silty [0; 40); 5 – heavy loam, sandy [40; 100); 6 – heavy loam, silty [0; 40); 7 – light clay, sandy [40; 100); 8 – light clay, silty [0; 40); 9 – heavy clay;

z5 – evaporation from the surface of the roadbed during summer months: 1 – extremely low, from 100 mm up to 150 mm (arctic deserts); 2 – very low evaporation, from 150 mm up to 200 mm (Siberian tundra provinces); 3 – low evaporation, from 200 mm up to 400 mm; 4 – average evaporation, from 400 mm up to 600 mm (taiga, Central and Central Black Earth regions of Russia, Krasnodar region); 5 – increased evaporation, from 600 mm up to 700 mm (mixed forests); 6 – high evaporation, from 700 mm up to 800 mm; 7 – very high evaporation, from 800 mm up to 900 mm (steppes); 8 – extremely high evaporation, from 900 mm up to 1000 mm (semi-deserts and deserts);

z6 – kind (type and subtype) of soil: 1 – light loam, large; 2 – light loam; 3 – silty loam; 4 – silty loam, heavy; 5 – light loam; 6 – light loam, silty; 7 – heavy loam; 8 – heavy loam, silty; 9 – sandy clay; 10 – silty clay; 11 – fat clay.

The following are integers encoding group characteristic features, associated to columns of the description matrix:

z7 – Selyaninov hydrothermal coefficient (SHC): 1 – excess moisture of soils with SHC from 1.4 up to 5; 2 – significant soil moisture in respective years with SHC from 1 up to 1.4; 3 – lack of soil moisture with SHC from 0.5 up to 1; 4 – arid areas with SHC up to 0.5;

z8 – plasticity number (PN, expressed as percentage): 1 – incoherent soils (sand, etc.) with PN from 0 up to 1; 2 – loam with PN from 1 up to 7; 3 – light loam with PN from 7 up to 12; 4 – heavy loam with PN from 12 up to 17; 5 – light clay with PN from 17 up to 27; 6 – heavy clay with PN from 27;

z9 – soil particle size (SPS, in mm): 1 – clay fraction with SPS up to 0.005; 2 – silty fraction with SPS from 0.005 up to 0.05; 3 – sandy fraction with SPS from 0.05 up to 0.5; 4 – large-sized fraction (coarse sand, gravel, boulders) with SPS from 0.5 up to 250;

z10 – average annual air temperature (AAT, in Celsius): 1 – extremely low temperature with AAT from -15.5 up to -10; 2 – very low temperature with AAT from -10 up to -6; 3 – low temperature with AAT from -6 up to -2; 4 – average temperature with AAT from -2 up to 2; 5 – high temperature with AAT from 2 up to 6; 6 – very high temperature with AAT from 6 up to 10; 7 – extremely high temperature with AAT from 10 up to 14.2;

z11 – snow cover depth (SCD, in cm): 1 – snowless areas with SCD up to 30; 2 – little-snow areas with SCD from 30 up to 50; 3 – moderate-snow areas with SCD from 50 up to 70; 4 – snow areas with SCD from 70 up to 100; 5 – extremely snowy regions with SCD from 100 up to 290;

z12 – soil freezing depth (SFD, in cm): 1 – shallow depth with SFD from 50 up to 180; 2 –average depth with SFD from 180 up to 220; 3 – great depth with SFD from 220 up to 260; 4 – very great depth with SFD from 260 up to 300; 5 – extremely great depth with SFD from 300 up to 600;

z13 – relative frost heave (RFH, %): 1 – non-heave soils with RFH less than 1; 2 – weakly-heave soils with RFH from 1 up to 4; 3 – heaving soils with RFH from 4 up to 7; 4 – strongly heaving soils with RFH from 7 up to 10; 5 – excessively heaving soils with RFH more than 10;

The following are integers encoding classification features, associated to columns of the distinctions matrix:

k1 – zone (1 – zone I, 2 – zone II, 3 – zone III, 4 – zone IV, 5 – zone V);

k2 – subzone (1 – subzone F, 2 – subzone H, 3 – subzone M);

k3 – road district (1 – district 1, 2 – district 2, 3 – district 3, 4 – district 4, 5 – district 5).

Frameworks of the article do not allow to present a representative fragment of description and distinctions matrices ($\mathbf{Q}^e$ and $\mathbf{R}^e$).

5. Data and knowledge base

The basis for the construction of the data and knowledge base for IIT RCZT is the above-described expanded matrix description of data and knowledge, as well as the description matrix of investigated territories. The data and knowledge base for IIT RCZT, which contains all characteristic features and their values, both in regards to the investigated territories and territories under investigation, as well as classification features of the investigated territories, is formed taking into account the peculiarities of natural-and-climatic conditions and the refined regularities of the influence of a geographic complex on the flow of the water-and-thermal regime of road structures.

To create IIT RCZT, the intelligent instrumental tool IMSLOG (previously created in the laboratory of intelligent systems of TSUAB [10]) is being further advanced. IMSLOG is designed to reveal various types of regularities, including fault-tolerant diagnostic tests; decision-making and justification with the use of cognitive tools.

The development of IIT IMSLOG is carried taking into account the specificity of the road-climatic zoning of territories.

In difference of the previously used concept "knowledge base - scheme of investigation – knowledge", the extended version of IIS IMSLOG uses the concept "project – scheme of investigation – data and knowledge base – knowledge". The modified subsystem of investigation templates is an interactive scheme of the investigation that includes: control elements (buttons, drop-down lists, text fields, etc.); display elements of graphic and text information (schedules, information windows, dynamic animation, cognitive graphics tools, etc..); report templates; summary, representing tables displaying the current status of calculations according to a presigned scheme of investigation; subsystem of security objects; file storage subsystem. Elements of the investigation template subsystem play a module of plug-ins in the previous version of IIT IMSLOG.

Methods of linear, nonlinear, and object-oriented programming, as well as technologies OLEDB and COM, have been used in the development of software modules. Software modules of the investigation scheme elements are stored in the library of algorithms, and are made in the form of plug-in dynamic modules. Each implements a given set of algorithmic and interface methods or methods of operation with data and knowledge. In difference from the unnamed input and output parameters of template elements used in previous versions, in the advanced version IIS IMSLOG each element of the scheme has a set of named parameters used to exchange data between modules. Software modules have been developed in C ++ in the environment Embarcadero RAD Studio XE.

6. Conclusion

It is proposed to construct a database and knowledgebase for the intelligent information technology of road-climatic zoning of territories on the basis of the proposed extended matrix description of data and knowledge, as well as on the basis of the description matrix of territories under investigation.

The structurization of data and knowledge on the road-climatic zoning of territories, necessary for the development of IIT RCZT, based on test methods of pattern recognition and cognitive graphics tools is given in the paper. The structurization was fulfilled with the use of reference data as well as data collected in the course of long-term field and laboratory studies in the territory of Western Siberia [3 – 5].

For the first time, intervals of values of characteristic features have been separated and their integer values used in the description matrices have been given.

Formation of the data and knowledge base on the base of the new concept "project – investigation scheme – data and knowledge base – knowledge" proposed in the developed IIT IMSLOG that will be used to construct of the intelligent system RCZT, which is the base of creating IIT DKRT, has been given.

The developed IIT IMSLOG can be used in the construction of intelligent systems to solve a wide range of diagnostic, prognostic, and organizational-administrative tasks from various problem and interdisciplinary areas. Reasonability of its use for the diagnostics of the stability of structural materials, including complex composite materials, to various types of deforming effects is beyond question. For these purposes, first of all, it is necessary to carry out structurization of data and knowledge in regards to deforming effects according to the matrix model of data and knowledge representation, used in intelligent systems based on test methods of pattern recognition, support of the diagnostic decision-making, and tools of cognitive graphics for justification of decision making results [9]. Further, it is necessary to create a data and knowledge base, as well as to construct an application intelligent system on deforming effects on the basis of IIT IMSLOG.

7. Acknowledgment

We are grateful to proff. V.N. Efimenko and associate proff. S.N. Efimenko for providing data on road-climatic zoning of territories.

The work is supported by Russian Foundation for Basic Research, projects №14-07-00673-a, № 16-07-00859 (partly)

References

[1] The national program of modernization and development of automobile roads of the Russian Federation until 2025. Project from March 25, 2004 / Ministry of Transport and Communications of the Russian Federation. – Moscow, 2004.

[2] 34.13330.2012. Set of rules. Automobile roads. The updated edition of SNIP 02.05.02 – 85 *. Automobile roads. Gosstroy of the USSR. – Moscow: CISD of Gosstroy, 2004 (approved by the Order of the Ministry of Regional Development of Russia, dated 30.06.2012, No. 266).

[3] Efimenko V. N., Efimenko S. V., Sukhorukov A. V. Technology for Assignment of Predicted Behavior of Subgrade Soils in the Design of Road Pavements. Key Engineering Materials. Vol. 683 (2016). pp. 250-255.

[4] Efimenko S., Efimenko V., Sukhorukov A. Peculiarities of strength and deformability properties of clay soils in districts of Western Siberia. ADVANCED MATERIALS IN TECHNOLOGY AND CONSTRUCTION (AMTC-2015): Proceedings of the II All-Russian Scientific Conference of Young Scientists «Advanced Materials in Technology and Construction». Vol. 1698 (2016). 070020.

[5] Efimenko V. N., Efimenko S. V., Sukhorukov A. V. Accounting for natural-climatic conditions in the design of roads in western Siberia // Sciences in Cold and Arid Regions. Vol. 7. Issue 4. August. 2015. pp. 307-315.

[6] Russam K., Coleman J.D., The Effect of Climatic Factors on Subgrade Moisture Conditions // Geotechnique, XI, 1961, № 1, P. 22–28.

[7] Zapata C.E., Houston W.N., Calibration and validation of the enhanced integrated climatic model for pavement design. – Washington, D.C.: Transportation Research Board, 2008. – 62 p.

[8] Yankovskaya A., Efimenko S., Cherepanov D. Structurization of data and knowledge for the information technology of road-climatic zoning // Applied Mechanics and Materials. Vol. 682. 2014. – P. 561-568.

[9] Yankovskaya A.E. Logic tests and tools of cognitive graphics. Publishing house: LAP LAMBERT Academic Publishing, 2011. – 92 p.Yankovskaya A.E., Gedike A.I.,

[10] Ametov R.V., Bleikher A.M. IMSLOG-2002 Software Tool for Supporting Information Technologies of Test Pattern Recognition // Pattern Recognition and Image Analysis. Vol. 13. 2003, № 4. – P. 650-657.

Structural Modification of La_2O_3-TiO_2-B Mixture After Mechanical Activation

O Yu Dolmatov, V V Zakusilov, M S Kuznetsov, A O Semenov

National Research Tomsk Polytechnic University, Tomsk, Russia

E-mail: kms@tpu.ru

Abstract. Accelerators require development of new and more powerful installations, generating the streams of the charged particles, affecting the development of machine building industry, energy power industry and other fields of science and technology. Great attention is paid not only to the modernization of the installations, in whole, but also to their individual elements. Cathode is the most important element of the installations, generating the stream of charged particles.

Using lanthanum hexaboride as a cathode material allows getting maximum efficiency of cathode assembly due to the thermodynamic and emission properties of lanthanum hexaboride. This paper studies the properties changes of lanthanum oxide mixture, titanium and boron oxides under the influence of mechanical activation to obtain the final product, based on lanthanum hexaboride and titanium diboride, corresponding to the high performance properties of the emitter. The study resulted in determination of time and frequency of mechanical activation to obtain the particles with the smallest average size of 6.3 microns for the mixture of lanthanum oxide, titanium and boron oxides powders.

1. Introduction

Nowadays, the world faces with an acute problem of creating electron flows generating devices used in charged-particle accelerators, electron tubes, microscopes and other devices with operating principle based on the physical phenomenon – the electron emission.

Operating consistency of the emission cathode depends on the material from which it is made. The use of rare earth metals hexaborides as cathode materials ensures the greatest efficiency of the cathode assembly. Among the compounds of the borides with the rare earth metals lanthanum hexaboride stands out, because of the possibility to obtain a number of the advantages over traditional materials used in cathode technology: minimal work function of the electrons (~ 2.7 eV); high melting point (2800 K); ability to retain the emission properties at low vacuum and under intense ion bombardment; high density of obtained emission current ($\approx 2 \cdot 10^3$ A/cm^2, at a temperature of 2000 K); no poisoning in the air; high performance [1, 2].

Powder metallurgy methods, which are currently used to obtain the lanthanum hexaboride compound, have a number of drawbacks: poisoning of the final product as a result of the synthesis process; technologically sophisticated equipment; large energy losses due to prolonged annealing and synthesis of the products. One of the latest technologies for the production of lanthanum hexaboride is a self-propagating high-temperature synthesis (SHS), which allows obtaining materials with desired properties. SHS method is based on local initiation of the reaction, resulting in moving of the chemical transformations along the mixture in the combustion wave mode (self-propagation) to form solid products.

SHS is possible to control during three stages of implementation: preparation of the batch components, conducting self-propagating synthesis and cooling of the final products. Preparation stage is considered to be more sophisticated, since changing the parameters of the batch, it is possible to control the final properties of synthesized materials, for example, the use of mechanical activation of the initial components of the mixture [3]. The solid body adopts the energy, acquired during the activation, in the form of point and line defects, and the accumulated excess energy is released during the synthesis process [4, 5].

2. Experiment

During the research the experiment was carried out to study the influence of mechanical activation on the morphological properties of the mixture of chemically pure (99.9 wt. %) powders of lanthanum oxide, titanium oxide and boron mixed at the stoichiometric coefficients for the following reaction:

$$La_2O_3 + 3TiO_2 + 24B \rightarrow 2LaB_6 + 3TiB_2 + 3B_2O_3 \quad (1)$$

Titanium dioxide was added to obtain titanium diboride, which provides structural properties improvement and reduces the work function of the final product.

Starting reagents were thoroughly mixed in KB-15S pin type cubic mixer using ERWEKA AR 403S universal drive for 30 minutes. The powder mixture was pre-dried in a muffle furnace at 100 °C for 5 hours at atmospheric pressure. Mechanical activation was carried out in a planetary ball mill of AGO-2C type; grinding was carried out using metal balls with a diameter of 4 mm; the ratio of the balls mass to the mass of the processed mixture was 10/1.

In the first part of the experiment, the dependence of the average particle size of the mixture from the rotational rate of activating mill was studied. Rotational rate varied in the range of 10 to 40 Hz at a pitch of 10 Hz at a constant rotation time of the activating mill of 10 minutes.

3. Results and discussion

The pictures of the powders before mechanical activation, made by a scanning electron microscope Philips SEM 515, clearly show large particles (Figure 1, a). After mechanical activation (Figure 1, b) the size of the particles decreases significantly as a result of the shock-abrasive interaction of the grinding balls with the particles in a planetary mill.

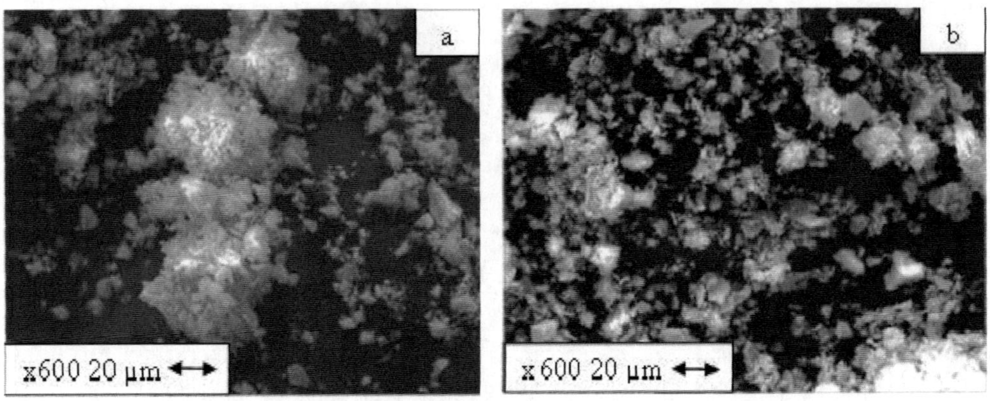

Figure 1. Pictures of the powders mixtures ($La_2O_3 + TiO_2 + B$):
a – before mechanical activation; b – after mechanical activation

To measure the surface area of the mixture components, the initial samples were BET-analyzed using surface area analyzer META SORBI-M. The results are shown in Table 1.

IOP Conf. Series: Materials Science and Engineering **142** (2016) 012042 doi:10.1088/1757-899X/142/1/012042

Table 1. Measurement of the surface area of the mixture components ($La_2O_3 + TiO_2 + B$) (activation time is 10 minutes).

Mill rotation rate (Hz)	Centripetal acceleration, made by the balls (m/s^2)	Measured Value $A_{sp.}$ (m^2/g)
Without mechanical activation		1.19±0.02
10	27.03	2.00±0.01
20	54.05	3.16±0.04
30	81.08	5.14±0.05
40	108.11	5.03±0.05

It may be noted that the largest surface area for the mixture ($La_2O_3 + TiO_2 + B$) is observed at a frequency of 30 Hz. This is the threshold value and increase of the rotational rate leads to decrease of the specific surface area.

In order to determine the particle size distribution, the components of the initial batch were subjected to the grain-size analysis using a laser particle size analyzer SALD-7101.

Figure 2 shows the particle size distribution before and after mechanical activation at a frequency of 10 Hz.

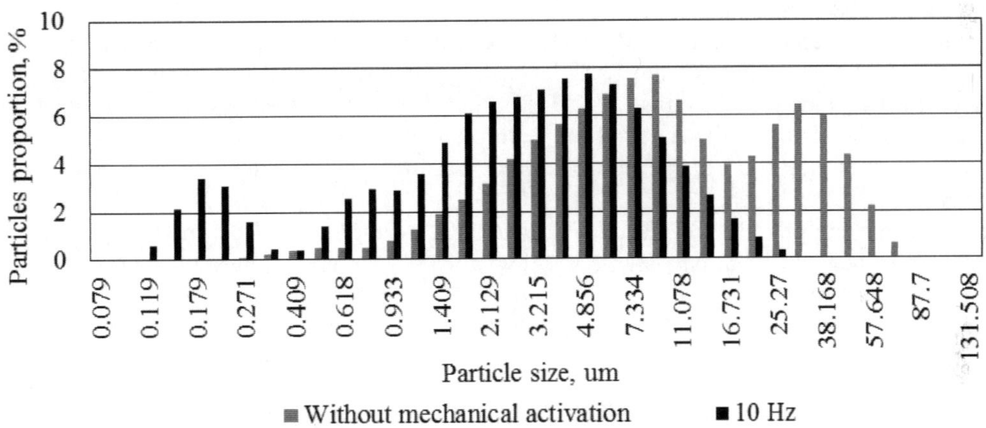

Figure 2. Grain-size composition of the batch before and after mechanical activation with a frequency of 10 Hz

Comparison of the size distributions shows that the particle size of the mixture after mechanical activation significantly reduces due to the distribution displacement to smaller sizes area.

At increasing the activation frequency up to 20 Hz and 30, the distribution displacement to smaller sizes area continues. However, at increasing the frequency up to 40 Hz, the proportion of the particles with a large size significantly increases. Figure 3 shows the particle size distribution after mechanical activation with frequency of 30 and 40 Hz.

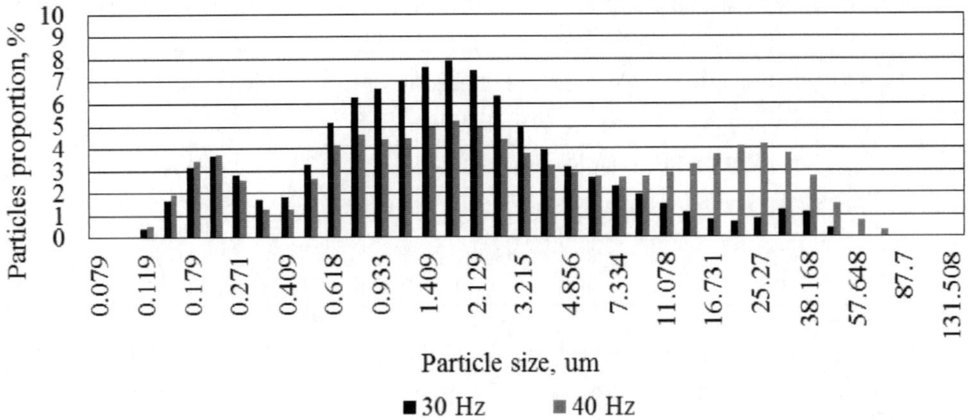

Figure 3. Grain-size composition of the batch after mechanical activation with frequency of 30 and 40 Hz.

Comparing the distribution of particle size after mechanical activation with frequency of 30 and 40 Hz, it can be seen that with an increase of the activation frequency, the maximum particle size increases, the proportion of the particles with a large size also significantly increases. Such negative effect of mechanical activation at increasing the rotational rate is due to the fact that the particles under the action of the shock-abrasive forces reach their minimum size, begin to "glued" to one another under the action of mechanical forces, arising in mechanical activation mill, and form agglomerates.

For complete analysis it is necessary to compare the average size of the mixture particles before and after mechanical activation. The average particles size is a universal characteristic of all mechanical activation modes of the batch components, so, the obtained results could be used in all installations, where mechanical activation is able to be implemented. Calculation of average particle size of the mixture is made as using grain-size analysis. The obtained values are presented in Table 2.

Table 2. Measurement of average particle size

Mill rotation rate (Hz)	Centripetal acceleration, made by the balls (m/s^2)	Average particle size of the mixture, um
Without mechanical activation		15.05
10	27.03	9.51
20	54.05	7.31
30	81.08	3.67
40	108.11	8.49

The data presented in Table 2 and in Figures 2 and 3 show the reduction in particle size, indicating an increase of the surface contact area. According to [6, 7], the higher the surface contact area, the lower the temperature gradient, which high values negatively influence on obtaining the final product. However, comparing the particles size, obtained at rotation rate of 30 Hz and the rotation rate of 40 Hz (see Figure 3), it can be seen that the maximum particle size increases, the proportion of the particles with a large size significantly increases.

In the second part of the experiment the effect of activation time on average particle size in the mixture was studied. The activation time was varied in the range from 5 to 30 minutes at a pitch of

5 minutes and at a frequency of 30 Hz, provided the smallest size and the greatest average specific surface area of the particles.

As a result of the surface area analysis, the following experimental data were obtained and presented in Table 3. The highest specific surface area of the particles is observed for 25 minutes of activation time.

Table 3. Measurement of the surface area of the mixture components ($La_2O_3 + TiO_2 + B$).

Activation time, minutes	Measured value A_{sp}, m^2/g
5	13.650±0.053
10	9.938±0.052
15	9.942±0.079
20	9.4765±0.057
25	36.840±0.313
30	4.063±0.036

Conducting grain-size analysis of the batch composition shows that at short-time activation (5 minutes) and at rotation frequency of 30 Hz of activation mill, there is a significant decrease of the average particle size of the mixture in comparison with the initial value of 15.05 um (Figure 4).

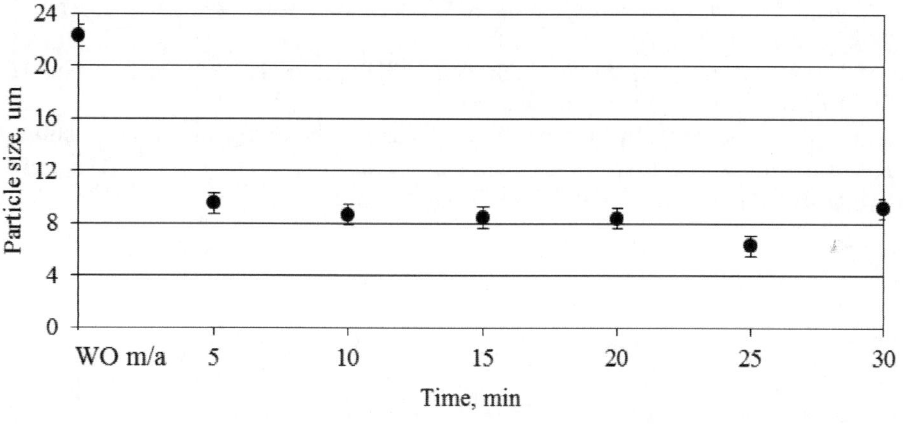

Figure 4. Dynamics of the changes of average mixture particle size ($La_2O_3 + TiO_2 + B$) depending on the activation time in a planetary ball mill.

Also, it should be noticed that the average particle size at mechanical activation ranging from 5 to 20 minutes slightly decreases, but at 25-minute activation a strong decrease of the average size is observed and, then, the average size increases in the case of 30-minute activation. Activation time of 25 minutes is a threshold, at increasing time there is an agglomeration of the particles, because of riveting one particle on the other to form agglomerates.

4. Conclusion

The study revealed that because of conducting reagents mechanical activation, there is a decrease of average particle size, leading to an increase of the contact area of the particles. In turn, this increases the reaction rate and improves the uniformity of the combustion wave flow, providing obtaining homogeneous and high-quality samples.

For the system ($La_2O_3 + TiO_2 + B$) the dependence of the average particle size on the activation frequency and time in a planetary mill is experimentally proved; the values of the rotation frequency and time of the activating mill are determined, at which particles with the smallest dimension of 6.3 mm – 30 Hz and 25 minutes, respectively, are formed. Further uncontrolled increase of the activation frequency or time could lead to the "saturation".

Thus, starting with certain values, mechanical activation will have a negative impact on obtaining the desired product due to reducing the particles contact area, which leads to a decrease of the reaction energy yield and the nonuniformity of SH-synthesis flow. Therefore, for each reagent structure the values of frequency and time of mechanical activation must be individualized, because they have a strong influence on obtaining of the final product.

References

[1] Kresanov V.S. et all. Highly efficient electron emitter based on lanthanum hexaboride: M., "Energoatomizdat", 1987. – 152 p.

[2] Wang L. Electronic structures and properties of lanthanide hexaboride nanowires / L. Wang, G. Luo, D. Valencia etc. // Journal of Applied Physics, — 2013. № 14. – P. 76-81.

[3] Merzhanov A., Mukasyan A. Tverdoplamennoe gorenie. - M.: Torus Press, 2007. - 336 p.

[4] Fundamental'nye osnovy mekhanichekoi aktivatsii, mekhanosinteza i mekhanolhimiicheskikh tekhnologii (Basic Principles of Mechanical Activation, Mechanosynthesis, and Mechanochemical Processing), Avvakumov, E.G., Ed., Novosibirsk: Izd. Sib. Otd. RAS, 2009.

[5] Rogachev, A.S., Shkodich, N.F., Vadchenko, S.G., Baras, F., Chassagnon, R., Sachkova, N.V., and Boyarchenko, O.D., Reactivity of mechanically activated powder blends: Role of micro and nano structures, Int. J. Self-Propag. High-Temp. Synth., 2013, vol. 22, no. 4, pp. 210–216.

[6] Bernard, F. and Gaffet, E., Mechanical alloying in SHS research, Int. J. Self-Propag. High-Temp. Synth., 2001, vol. 10, no. 2, pp. 109–132.

[7] Kochetov N. A. Mechanoactivation of Ni-Al blends in a ball mill: Influence of ball size / N. A. Kochetov, I. D. Kovalev // International Journal of Self-Propagating High-Temperature Synthesis. — 2014. № 3. — P. 171-173.

IOP Conf. Series: Materials Science and Engineering **142** (2016) 012043 doi:10.1088/1757-899X/142/1/012043

A Computer Model of the Evaporator for the Development of an Automatic Control System

K A Kozin, E V Efremov, O P Kabrysheva and M I Grachev

Institute of Physics and Technology, Tomsk Polytechnic University, 30 Lenin Avenue, Tomsk, 634050, Russian Federation

E-mail: efremov@tpu.ru

Abstract. For the implementation of a closed nuclear fuel cycle it is necessary to carry out a series of experimental studies to justify the choice of technology. In addition, the operation of the radiochemical plant is impossible without high-quality automatic control systems. In the technologies of spent nuclear fuel reprocessing, the method of continuous evaporation is often used for a solution conditioning. Therefore, the effective continuous technological process will depend on the operation of the evaporation equipment. Its essential difference from similar devices is a small size. In this paper the method of mathematic simulation is applied for the investigation of one-effect evaporator with an external heating chamber. Detailed modelling is quite difficult because the phase equilibrium dynamics of the evaporation process is not described. Moreover, there is a relationship with the other process units. The results proved that the study subject is a MIMO plant, nonlinear over separate control channels and not self-balancing. Adequacy was tested using the experimental data obtained at the laboratory evaporation unit.

1. Introduction

Implementation of a closed nuclear fuel cycle requires substantiation of spent nuclear fuel (SNF) technology through studies at the experimental and pilot equipment. At the same time creation of an effective radiochemical production with regard to its high radioactive, nuclear and environmental hazards is impossible without the use of automatic control systems (ACS) [1].

For conditioning solutions in spent fuel reprocessing technologies the method of evaporation, which is conditioning of liquid waste and concentration of the desired products, is often applied. Therefore, the effective continuous operation of the technological process will depend on the operation of the evaporation equipment. Its essential difference from similar devices is a small size due to a relatively small-scale production for safety assurance [1].

Such devices are MIMO plants, nonlinear over separate control channels and they are not self-balancing. Therefore, to study various ACS using a one-effect evaporator with an external heating chamber for evaporation of uranyl nitrates solution (see figure 1), it is necessary to create a mathematical model describing the dynamic relationships between key process variables. At the same time, according to a general engineering rule, for the synthesis of ACS 10% accuracy is quite acceptable, due to the approximate formulas for analytical calculations of the controller parameters [2].

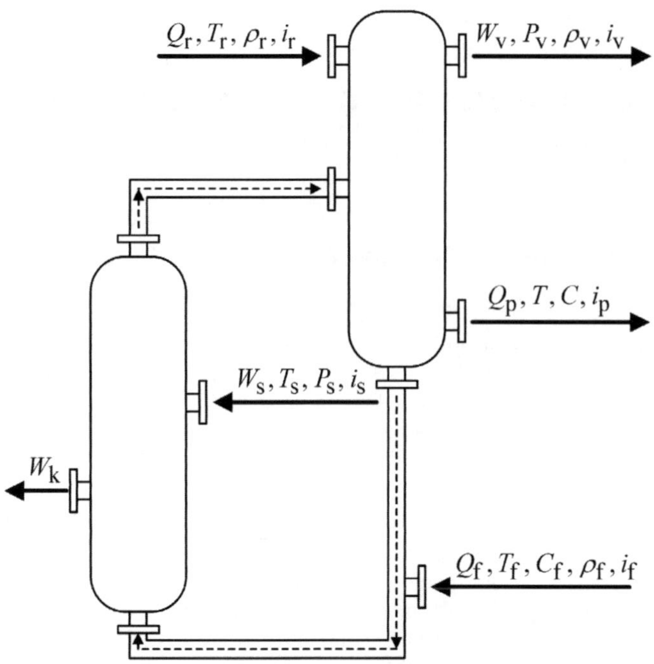

Q_r, Q_p, Q_f – the volume flow of reflux, the evaporated uranium re-extract and feed solution of uranium re-extract, respectively, m^3/h;
T_r, T, T_f, T_s – the temperature of reflux, solution in the unit, feed solution of uranium re-extract and heating steam respectively, °C;
$\rho, \rho_v, \rho_r, \rho_f$ – the density of the solution in the unit, secondary steam, reflux, feed solution of uranium re-extract respectively, kg/m^3;
W_k, W_s, W_v – the mass flow of the condensate, heating and secondary steam respectively, kg/h;
P_s, P_v – the pressure of the heating and secondary steam respectively, kPa;
$i, i_s, i_f, i_r, i_v, i_p$ – the enthalpy of the solution in the unit, the heating steam, the feed solution, reflux, secondary steam and the evaporated uranium re-extract respectively, J/kg;
C, C_f – the concentration of the evaporated uranium re-extract and the feed solution of uranium re-extract respectively, kg/m^3

Figure 1. Evaporator scheme

2. Mathematical model

A mathematical model of uranium re-extract evaporation should represent the following:

- time history of the solution level in the unit according to the flow of feed solution of uranium re-extract, evaporated uranium re-extract, reflux and heating steam flow;
- time history of concentration / density of evaporated solution, depending on the flow rate and the concentration / density of the feed solution and the reflux, flow of the evaporated uranium re-extract and heating steam;
- time history of the solution temperature in the unit depending on the temperature of feed solution, reflux, heating steam and its pressure.

In addition, the model should take into account mutual influence of the controlled parameters, mentioned above.

As a rule, in analytical models of chemical units, mathematical formulation comes to generation of equations of mass and heat balance of the system, based on the fundamental laws of nature: the laws of conservation of mass and energy [3–6].

Mathematical formulation had to be simplified due to the lack of data on the thermodynamic properties of the evaporated product, physical and thermodynamic parameters of the heating and juice steam, as well as the necessary data for the description of the heat lost to the environment, etc.

As a result, time history of the material, element and heat flows in the unit is described by the following system of equations (1), where A – cross sectional area of the evaporator, m^2; c_f – specific thermal capacity of the feed solution, kJ/(°C·kg); ρ_{w} – water density, kg/m^3 and the mass flow of secondary steam W_{v} is calculated by the following algebraic expression (2).

$$\begin{cases} \dfrac{dh}{dt} = \dfrac{1}{A}\left(Q_{\text{f}} + Q_{\text{r}} - Q_{\text{p}} - \dfrac{W_{\text{v}}}{\rho_{\text{w}}}\right) \\ \dfrac{d\rho}{dt} = \dfrac{1}{Ah}\left(W_{\text{v}}\left(\dfrac{\rho}{\rho_{\text{w}}} - 1\right) - Q_{\text{f}}\rho_{\text{f}}\left(\dfrac{\rho}{\rho_{\text{f}}} - 1\right)\right) \\ \dfrac{dT}{dt} = \dfrac{\left[W_{\text{s}}(i_{\text{s}} - i_{\text{c}}) + Q_{\text{f}}\rho_{\text{f}}(i_{\text{f}} - i) + Q_{\text{r}}\rho_{\text{r}}(i_{\text{r}} - i) + W_{\text{v}}(i_{\text{v}} - i) - L(T - T_{\text{ar}})\right]}{\rho_{\text{f}}c_{\text{f}}Ah} \end{cases}, \quad (1)$$

$$W_{\text{v}} = \frac{Q_{\text{f}}\rho_{\text{f}}c_{\text{f}}T_{\text{f}} - Q_{\text{p}}\rho cT + Q_{\text{r}}\rho_{\text{r}}i_{\text{r}} + W_{\text{s}}i_{\text{s}}}{i_{\text{v}}}. \quad (2)$$

Enthalpy of heating steam, secondary steam, and dew is determined using the expression obtained by fitting the tabular data:

$$\begin{aligned} & i_{\text{s,v}} = 2.5 \cdot 10^6 + 1813 \cdot T_{\text{sat}} + 0.417 \cdot T_{\text{sat}}^2 - 0.11 \cdot T_{\text{sat}}^3 + 2090 \cdot \left(T_{\text{s,v}} - T_{\text{sat}}\right) \\ & T_{\text{sat}} = \frac{2147}{\left(10.76 - \lg(P_{\text{s}})\right)} - 273.2 \\ & i_{\text{c}} = \left(-0.0051 \cdot T_{\text{s}}^2 - 1.5595 \cdot T_{\text{s}} + 2467.1\right) \cdot 10^3 \end{aligned} \quad (3)$$

The developed approach and mathematical description of the evaporator as a controlled object was implemented as a computer model in the MATLAB / Simulink package. For testing qualitative adequacy of the model, research results of the evaporators presented in [7] were used. One of the study subjects there was a two-stage vacuum evaporator "Edinstvo" [8].

3. Simulation results

Figure 2 shows the transition processes of the solution level in the second device of "Edinstvo" evaporator and in the developed model of uranium re-extract evaporator. The flow of the feed solution $\Delta S_{\text{f}} = 1700$ kg/h to the unit was changed up to 62%.

As it is seen from the graphs, the behavior of the level in both cases is identical, and the channel of the controlled object can be regarded as an astatic link.

Figure 3 shows the transient responses of the concentration in the first device of "Edinstvo" evaporator and the appropriate transitional process in the developed computer model of the evaporator. The fluid flow from the unit $\Delta S_{\text{p}} = -250$ kg/h was changed up to 30%.

The above graphs show that transient responses of concentration obtained at the evaporator and the simulation results are qualitatively equal. In addition, this channel of the controlled object can be regarded as a first order aperiodic link.

Figure 4 shows the transient responses of the temperature of the solution in the first device of "Edinstvo" evaporator and the appropriate transitional process in the developed computer model of the evaporator. The heating steam flow from the unit $\Delta S_{\text{p}} = 480$ kg/h was changed up to 17%.

The above graphs show that the transient responses of temperature, obtained at the evaporation unit and the simulation results are qualitatively equal.

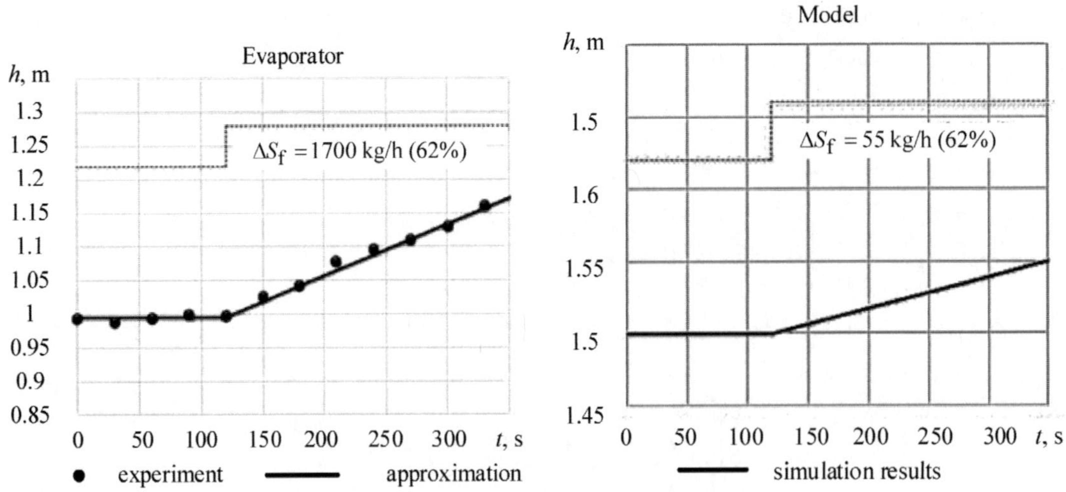

Figure 2. Transition process in the level

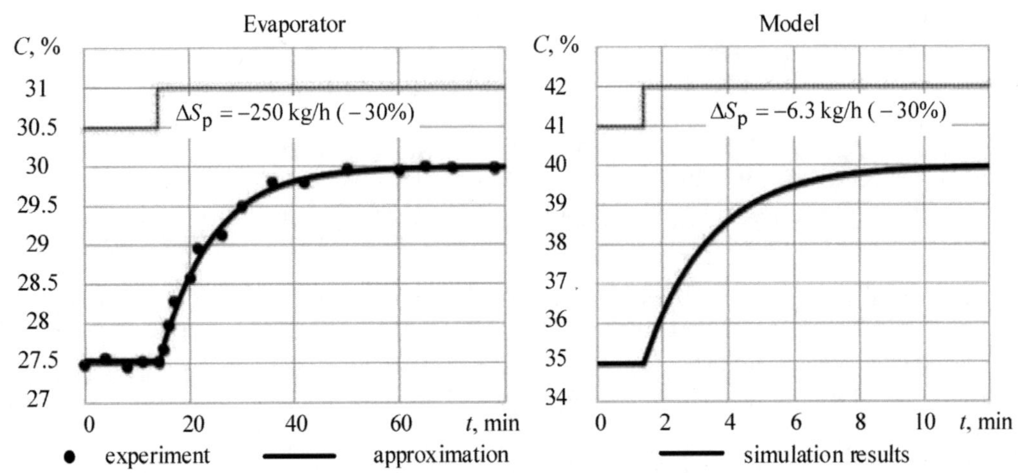

Figure 3. Transition process in concentration

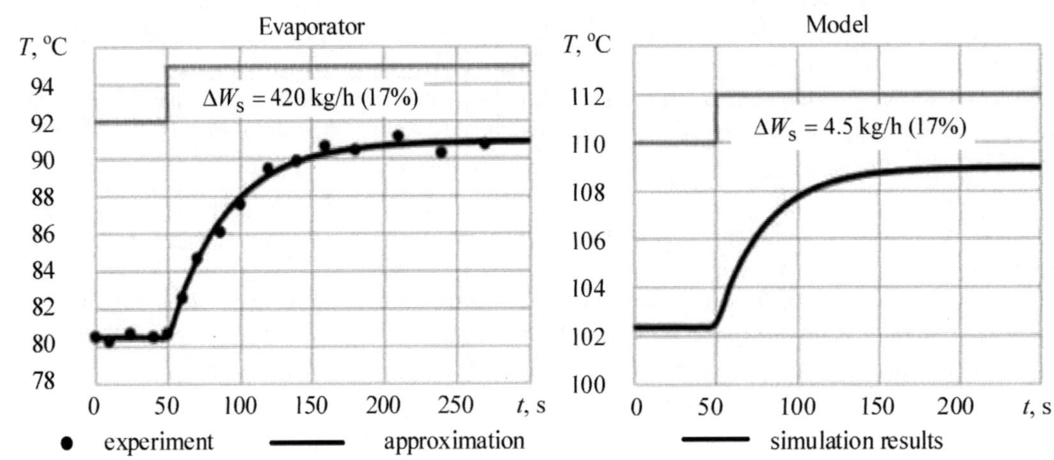

Figure 4. Transition process in temperature

Furthermore, similar to the transient responses of concentration, it can be seen that the object of the present channel can be regarded as a first order aperiodic link. Comparison of simulation results with the data [9] shows their qualitative agreement.

The analysis of the given above transient responses shows that the time constant of the evaporators for concentration / density is substantially higher than time constant for the level and temperature. This fact should be considered when developing ACS of an evaporation unit.

Qualitative adequacy of the developed model was tested using the experimental data obtained at a laboratory evaporator at V.G. Khlopin Radium Institute, St. Petersburg.

This unit is equipped with an automated control system, based on two control algorithms.

The first algorithm is designed for evaporation mode with a required density. The level of the solution in the still is maintained constant by controlling the flow of feed solution. In the transient mode (to achieve the desired density) the solution in the still was not drained. In the continuous mode, the solution was drained in proportion to the average flow of the feed solution.

The second algorithm provides required density of the solution at the output of the evaporation unit in a continuous mode. The desired density of the solution in the still was maintained by the flow of feed solution, while the level was maintained by draining the solution in the still.

The following experiment was carried out at the laboratory evaporator. In the starting mode, after initial filling with the feed solution and beginning of the evaporation process control was maintained by the first algorithm in the transient mode. After reaching a required value of density $\rho = 1400\,\text{kg/m}^3$, that is, tapering off to a steady-state regime, control was maintained by the second algorithm. If the measured density of the solution in the still was greater than ρ, the maximum flow rate of feed solution was set. If it was less than that, the minimum flow rate was set. That is, the control was maintained according to the relay rules.

After tapering off to a steady-state regime, transition process was recorded while set point of density was changed from $\rho = 1400$ kg/m^3 to $\rho = 1500$ kg/m^3.

A similar experiment was carried out on the developed computer model of the evaporation unit. The experimental data and simulation results are shown in figure 5.

Another experiment was carried out at the laboratory evaporator. After reaching a required density value of $\rho = 1315\,\text{kg/m}^3$, it was decreased to $\rho = 1285\,\text{kg/m}^3$. The results of the experiment are presented in figures 6b–6d.

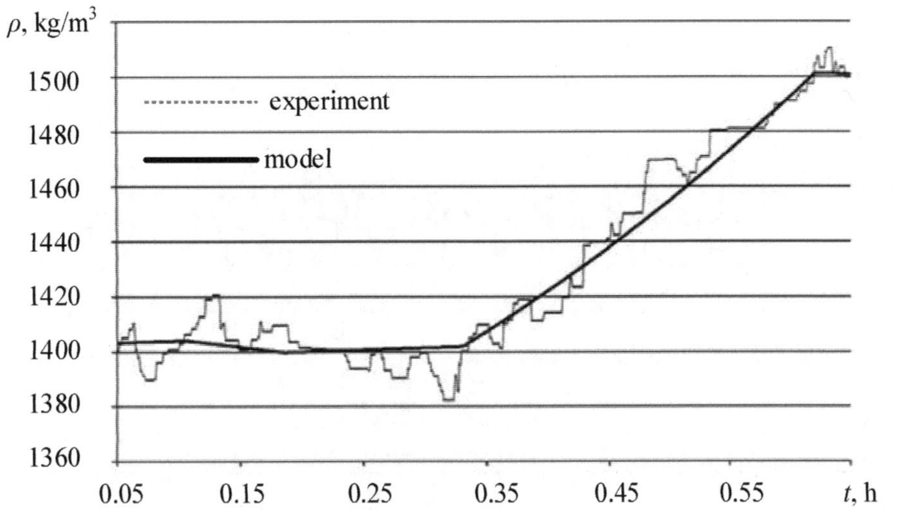

Figure 5. Transition process in density

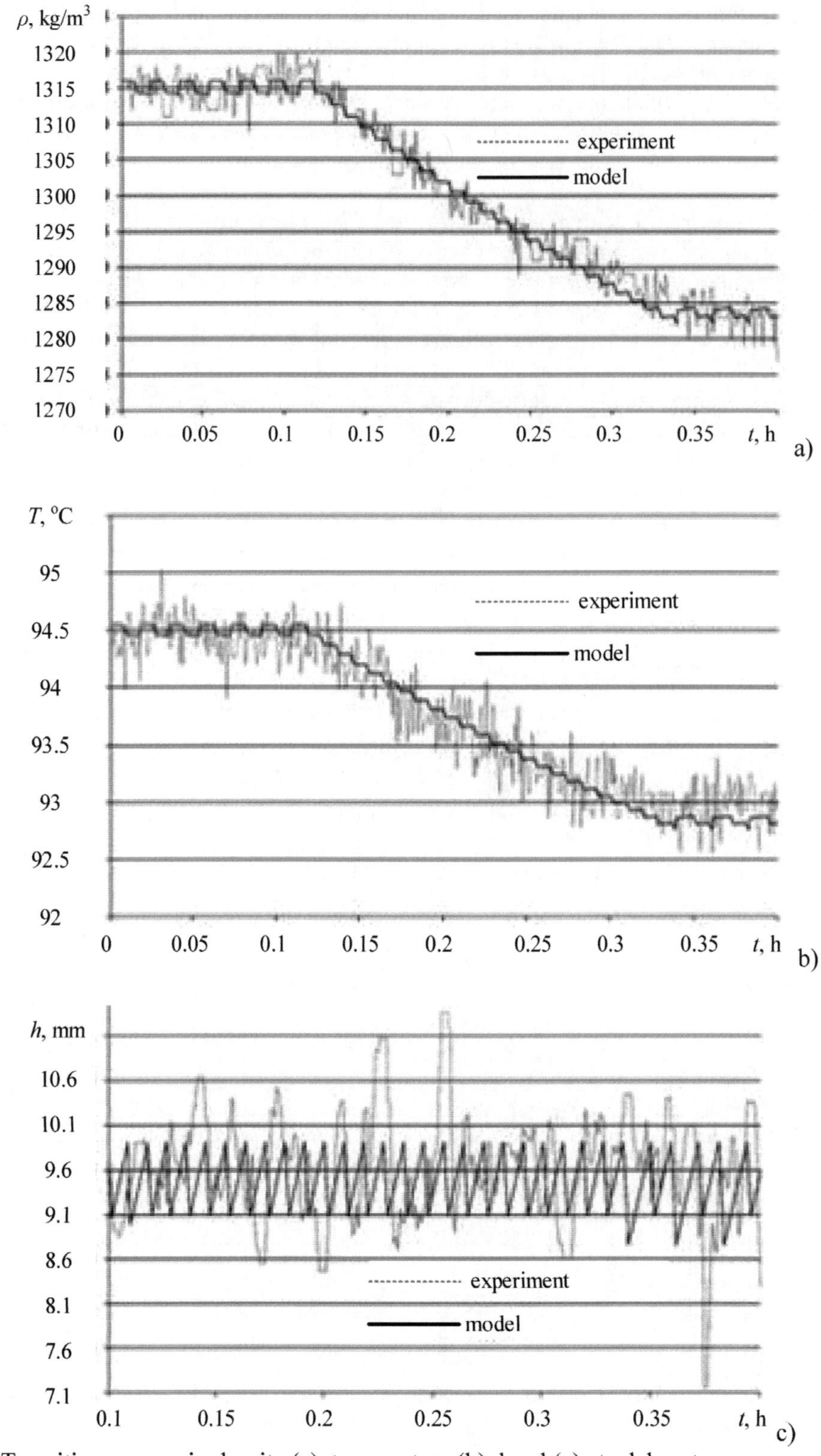

Figure 6. Transition process in density (a), temperature (b), level (c) at a laboratory evaporator and in the computer model

4. Conclusion

The relative standard errors of simulation in the level, density and temperature of the solution were less than 9%, 5% and 7%, respectively, which satisfy the required specification.

Thus, in the course of this work the qualitative and quantitative adequacy of the developed computer model was proved, which allows its application for the development of ACS an evaporation unit.

References

[1] Sheviakov I Y, Kozin K A, Goryunov A G, Hoffman F E, Zilberman B J and Ryabkov D V 2014 *Proc. Int. Conf. "Physical and technical problems of nuclear science, energy and industry"* (Tomsk: Tomsk Polytechnic University Press) p 52.
[2] Sovetov B J and Yakovlev S A 1998 *Modelling of systems: Textbook for universities* (Moscow: Visshaja shk.)
[3] Gafarov V V and Glebov M B 1991 *Mathematical modeling of the basic processes of chemical industry* (Moscow: Visshaja shk.)
[4] Lutsenko V A and Finyakin L N 1984 *Mathematical modeling of chemical-technological processes on analog computers* (Moscow: Chemistry)
[5] Lutsenko V A and Finyakin L N 1979 *Analog computers in chemistry of chemical technology* (Moscow: Chemistry)
[6] Samarskii A A and Mikhailov A P 2001 *Mathematical modeling: the Ideas. Methods.Examples* (Moscow: FIZMATLIT)
[7] Taubman E I 1970 *Calculation and simulation of evaporators* (Moscow: Chemistry)
[8] Egorova O V, Liventsova N V, Efremov E V and Grachev M I 2014 *Procedia Chemistry* **11** 139–46
[9] Kiew M. Kam, Prabirkumar Saha, Moses O. Tadel and Rangaiah G P 2002 Models of an Industrial Evaporator System for Education and Research in Process Control *Developments in Chemical Engineering and Mineral Processing* **10** 105–27.

IOP Conf. Series: Materials Science and Engineering **142** (2016) 012044 doi:10.1088/1757-899X/142/1/012044

Modelling of free positron states in TiH_x

O.N. Imas[1], I.Yu. Karataeva[2], K. B. Fedorov[1]

[1]Department of Higher Mathematics, Tomsk Polytechnic University, 30 Lenin Avenue, Tomsk, 634050, Russia
[2]Department of Theoretical Physics, Tomsk State University, 36 Lenin Avenue, Tomsk,634050, Russia

E-mail: onm@tpu.ru, karin@phys.tsu.ru

Abstract. Electron energy structure, positron spectrum and positron characteristics of α-Ti and α-$TiH_{0.125}$ were calculated. Self-consistent calculations of the band structure were performed by the linear muffin-tin orbital method in the atomic sphere approximation. Modelling has been made on low content of hydrogen into α-Ti with expanded close-packed hexagonal cell inclusive 8 titanium atoms. Variation of sphere radiuses permitted to consider anisotropy and spherical symmetry of potential. Positron potential and positron wave function were calculated on a base of self-consistent electron density. Then positron probability of existence into TiH_x lattice and lifetime were founded. Theoretical calculation indicated a satisfactory agreement of positron characteristics absolute values with the experimental data is achieved but the tendency of values with hydrogen defects increasing is not. The reason of divergence is discussed. On the basis of experimental data and theoretical calculations it was shown that different hydrogen atom states demonstrate the different influence in the lifetime spectra.

1. Introduction

Metals are the most important structural material. In many branch there is actual thread of hydrogen corrosion of metals. Well operation of oil and gas, operation of equipment in chemical and nuclear industry are the most risk because hydrogen and hydrogen containing medium makes up a large share of the working environment. Modification of the physical and mechanical properties of metals and alloy materials by hydrogen is significant problem. The used materials must combine the resistance to high stresses with acceptable high temperature deformation. However hydrogen influence on it strength characteristics depends from material analysis. At the same time visible surface change could be unobserved. But sample even having super difficult configuration requires precise diagnostic which would not destroy construction of sample.

Nondestructive and contactless methods, possibility to measure parameters of samples under different temperatures coupled with fast response to local change of electron density defines positron diagnostics as one of the unique method among traditional analysis of structure of matter. For metals and allows positron annihilation technique reveals the electron momentum distribution and Fermi level energy that largely determines their mechanical, electrical and magnetic properties [1]. Moreover positron spectroscopy is sensitive to define structure, cause and concentration of spots and extended defects, to investigate the disrupt blanket on metals, alloys, semiconductors and ionic crystals [2]. In that case positron annihilation method is useful as for nature and concentration of impurities analysis as for study of electron structure modification because of any factors. Now it is the defect spectroscopy tool. In [3, 4] the annihilation of positrons has been studied by *angular* distribution on annihilation photons method (ADAP) and lifetime has been measured in metal hydrides. In [5] hydrogen influence on the dynamic of defect formation (crater and crack) in α-titanium has been investigated by sight with scanning microscopy and with average positron lifetime measurement.

Ab initio electron-positron structure and positron characteristics were calculated mainly for metals and alloys which has simple crystal structure. More compound one require improved calculation methods of electron density which is the basis for construction of the positron spectrum. Even less there are first principle calculation of hydrogen-containing compounds. In [6] electron and positron characteristics of tungsten changing with impact of hydrogen and helium were calculated by augmented plane wave method. Quasi-free positron lifetime *ab initio* for titanium hydride (TiH_2) was obtained by [7]. In this work we study influence of the atomic hydrogen on electron structure of titanium and its positron characteristics. Absorbing hydrogen at the standard pressure α-Ti crystal lattice extends, ratio c/a of the hexagonal closely packed (HCP) lattice parameter decreases. Dilution of hydrogen in metal shows uneven distribution from surface to volume. That explains differing degree of destruction of α-Ti on a surface and in a volume. Thus, after 360 minutes electrolytic hydrogenation of titanium samples "the traces of destruction" in the form of increased dislocation concentration were emerged [5]. However the whole crystal fracture within atomic hydrogen $11 \cdot 10^{-5}$aт% H was not observed.

In this paper we calculate electron structure, positron spectrum and positron characteristics of α-Ti and α-$TiH_{0.125}$ and compare them with experimental one. Comparing calculated electron structure results with different models crystal potential of pure titanium and titanium with a hydrogen impurity we try to make more exact the positron annihilation characteristics. An attempt to interpret the thermalized positron behavior in crystal by *ab initio* method was made.

2. Method of calculation

Band structure of metal hydrides was calculated by the self-consistent linear muffin-tin orbital method in the atomic sphere approximation (LMTO-ASA) adjusted for overlap. Ceperley and Alder approximation was used for the exchange-correlation potential [8]. The self-consistency procedure was applied at 90 k-pontes in the irreducible part of the Brillouin zone for HCP lattice. The self-consistency was regarded to have been achieved by the variation energy eigenvalues from iteration to iteration did not exceed 3 mRy and the presser which was calculated at each iteration by Pettifor formula [9] was less than 1 kbar. Wigner–Seitz (WZ) sphere radii were identical and equal to average WZ-radius. The projection of the expanded HCP cell on the XOY plate is in figure 1.

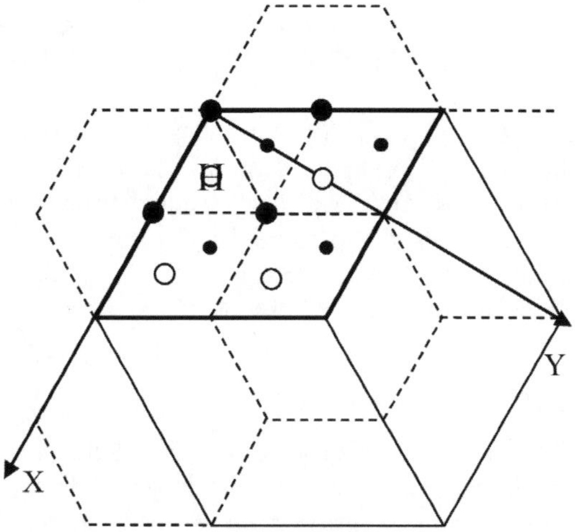

Figure 1. Project of expanded HCP cell to XOY.
● – Ti in z=0 plane,
• – Ti in z=c/2 plane,
○ – E in z= c/4 and z= 3c/4 planes,
H – H-defect in oct-pore (1/4;√3/12;c/4).

IOP Conf. Series: Materials Science and Engineering **142** (2016) 012044 doi:10.1088/1757-899X/142/1/012044

Crystal lattice was simulated by repeating hexagonal extended cells which contain 8 titanium atoms (lattice parameter of titanium: a=0.2952 x 2 nm, c=0.4684 nm [10]). Hydrogen atoms placed in octal pore with coordinate (1/4;√3/12;c/4). The others 7 octal pores were filled up empty spheres having zero electron density in order to consider the crystal potential anisotropy. The atoms positions of the expanded cell are in table 1.

Table 1. Coordinates of atoms in expanded cell per unit lattice parameter *a*.

atom	X	Y	Z
Ti	0.000000	0.000000	0.000000
	0.500000	0.000000	0.000000
	-0.25000	0.233020	0.000000
	0.250000	0.433020	0. 000000
	0.000000	0.288680	0.396828
	-0.25000	0.721680	0.396828
	0.250000	0.721680	0.396828
	0.500000	0.288680	0.396828
H	0.250000	0.144340	0.198414
Empty sphere (E)	0.750000	0.144340	0.198414
	0.000000	0.289000	0.198414
	0.500000	0.289000	0.198414
	0.250000	0.144340	0.595242
	0.750000	0.144340	0.595242
	0.000000	0.289000	0.595242
	0.250000	0.289000	0.595242

Used method of electron structure calculation demands to input addition empty spheres because LMTO assumes spherical symmetry of potential within the atomic spheres and zero potential out of spheres. That admission is not true for the most of crystals especially for dielectrics and semi-conductors. However, it could be corrected by putting the additional atomic spheres in void of cell which possess high symmetry enough. Thus we can solve some problems: to decrease the overlap, to increase spherical symmetry of potential, to decrease the volume between the spheres. The calculation of the hydrogenizes titanium in HCP lattice was carried out with two variants of atomic spheres radii which were equal to each other's (R_{Ti}=R_H=R_E=2.4233 a.u.) and increased titanium radius (R_{Ti}=2.8000 a.u., R_H=R_E=1.8670 a.u.).

The positron states problem have been solved on basis of two-component density functional theory [11, 12]. Positron potential and positron wave function were calculated on a base of self-consistent electron density. It was considered the approach of low positron density, positron-electron correlation potential was depending on electron density only and was not perturbed by positron. Using this potential single-particle Schrödinger equation was solved by LMTO-ASA method and positron wave function was found. It was used calculating probability of positron distribution to atomic spheres, annihilation rate and lifetime.

3. Discussion of results

Electron band structure and lower positron band of $TiH_{0.125}$ with equaled atomic sphere radii are in figure 2. Filled levels of the valence band is here only. The presence of hydrogen defect leads to formation of the separated lower band. The Fermy energy is upper and the occupied part of the valence band is increased. There is the band gap. Band structure of α-Ti and $TiH_{0.125}$ is similar in every

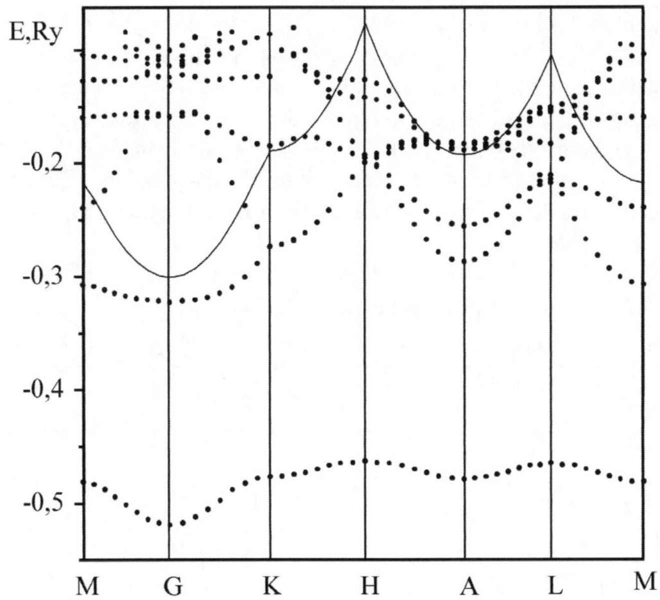

Figure 2. Electron structure (dashed lines) and the lowest energy positron band (solid line) of $TiH_{0.125}$.

respect of α-Zr and $ZrH_{1.5}$ [13]. But numerical and scale of characteristics differs.

The positron band is similar to the lowest valence electron band that agrees well with theoretical representation about quasi-free positron state in solid. Positron energies at G point are -0.200, -0.301 and -0.211 Ry for Ti, $TiH_{0.125}$ (equal atom spheres) and $TiH_{0.125}$(different atom spheres), respectively. Electron band structure and positron band peculiarity is kept when the ratio of the atomic spheres is changed. It should be noted that increasing of the atomic sphere titanium radius and decreasing of the hydrogen and empty spheres radii insignificantly reduces the atom spheres overlap. But the values of characteristics (bottom and width of the valence band, energy values of the symmetrical points) shifted by an amount greater than the accuracy of the calculation. Comparative energy data are in the table 2.

Table 2. Electron band structure values of pure and hydrogenize α-Ti. Fermy energy E_F (Ry), the occupied part of the conduction band E_F-Γ_1 (Ry), terms Γ_1 and Γ_4 (Ry).

α-TiH_x	α-TiH R_{Ti} =3.0532 a.u.	α-$TiH_{0.125}$	
		R_{Ti}=R_H=R_E=2.4233 a.u.	R_{Ti}=2.8000 a.u.; R_H= R_E=1.8670 a.u.
E_F (Ry)	-0.0526	0.0680	0.0046
E_F-Γ_1 (Ry)	0.4744	0.5978	0.6393
Γ_1 (Ry)	-0.5270	-0.5198	-0.6347
Γ_4 (Ry)	-0.2680	-0.3006	-0.2112

The calculated positron characteristics and experimental the first component of positron lifetime is in table 3. The calculated value of thermalized positron lifetime agrees with the experimental one. However, the calculated trend of lifetime decrease with defects increase in α-titanium disagrees with the experiment [14]. In this connection the band structure and positron lifetime was calculated with corrected potential. Titanium radius of the atomic sphere was enlarged and hydrogen radii of the atomic spheres were diminished. The radii change significantly affected on the charge redistribution in atoms. The charge increased in titanium spheres and decreased in defect sphere and void (table 3). As a result probability of positron distribution in titanium sphere and annihilation increased from there. It was caused lessening of positron lifetime in the hydrogenated titanium. The other reason of disagreement of the experimental trend with theory would be swelling of titanium sample at its hydrogenation which was obtained by experiment. It could not be considered by theory.

Table 3. Positron lifetime τ(ps), probability of positron distribution W (%) to atomic spheres (S), electron charge Q (el.) in atomic spheres and probability of electron-positron annihilation ω (%).

	α-Ti	α-$TiH_{0.125}$			α-$TiH_{0.125}$			SIC-$TiH_{1.0}$		
τ(ps) [14]	150	154 ($TiH_{0.01}$)								
τ(ps)	158.1	154.5			149.8			163.5		
atom spheres radii (a.u.)	3.0532	$R_{Ti}=R_H=R_E=2.4233$			$R_{Ti}=2.800$ $R_H= R_E=1.867$			$R_{Ti}=R_H=R_E=2.0656$		
sphere	Ti	Ti	H	E	Ti	H	E	Ti	H	E
W (%)	100	32.25	4.90	62.85	59.87	1.93	38.20	10.05	23.72	66.23
ω(%)	100	42.19	5.35	52.46	64.79	2.24	32.97	16.68	27.97	55.35
Q (el/at)	22	20.47	2.48	1.48	21.28	1.35	0.75	19.68	1.64	0.84

In order to take account of the volume cell increasing by hydrogenation electron structure and positron lifetime were calculated for $TiH_{1.0}$ in cubic phase and lattice parameter a=0.4397 nm. Elementary cubic cell with 16 atoms per cell was designed like HCP and discussed in [7]. The calculation showed that positron redistributed into hydrogen trap a lot, although charge of hydrogen sphere changes lightly.

Note that experimental positron lifetimes of the first and the second components showed the slight increase of the first component τ_1 together with decrease of the corresponding intensity I_1. It could be explained that density of free electrons is not changed a lot by hydrogenation. For the hydrogen charged samples the second component decreases together with increase of the corresponding intensity compared with the original titanium samples: τ_2=369 ps, 362 ps, 321 ps, 331 ps for 0 min, 60 min, 120 min, 360 min hydrogen charged titanium, respectively [14]. It could be explained that positron annihilation increases from region of the vacancy-type defects which are trapping hydrogen atoms. In that case electron density of vacancy defect increases. It have to lead to the lifetime decreasing of positron trapped by the defects.

4. Conclusion

Thus, in the present work the detail investigations of electron band structure of α-Ti and hydrogen charged titanium α-TiH_x to x=1 were performed. The positron states and annihilation characteristics were calculated in frames of LMTO-ASA method, which allows us to analyze the contributions of the different electron states in electron-positron annihilation process in titanium samples as well. There is the satisfactory agreement with experimentally measured the delocalized positron lifetime, but not the tendency. It can be connected with either experiment or calculation method precision. To analysis the localized positron it is necessary to use the method which allows to model the soft core.

IOP Conf. Series: Materials Science and Engineering **142** (2016) 012044 doi:10.1088/1757-899X/142/1/012044

References

[1] Belyaev V, Grafutin V, Phuntikov Yu, Khmelevskii N, Shtockii Yu, Htut M, Aung Ya 2008 *Voprosy Atomnoi Nauki i Tekhniki. Seriya: Fizika Yadernykh Reaktorov.* **1** 35.

[2] Grafutin V and Prokop'ev E 2002 *Phys.Usp.* **45** 59.

[3] Savin V, Andrievskii R, Gorbachev V, Tsyganov A 1972 *Physics of the Solid State* **14**, 1120

[4] Tsyganov A, Andrievskii R, Gorbachev V, Savin V 1974 *Inorganic Materials* **10**, 6, 1030

[5] Chernov I, Lider A, Cherdancev Yu, Garanin G, Nikitenkov N, Krening M, Surkov A 2000 *Physocal Mesomechanics* **3** 6 97.

[6] Troev_T, Popov E, Nankov N and Yoshiie T 2010 *Journal of Physics: Conference Series* **207** 1.

[7] Kul'kova S, Muryzhnikova O, Beketov K 1996 *International Journal of Hudrogen Energy* **21** 11/12, 1041.

[8] Ceperley D, Alder B 1980 *Phys. Rev. Lett.* **45**, 7, 566.

[9] Pettifor D 1976 *Commun. Phys.* **1**, 5, 151.

[10] Papaconstantopoulos D. 1986 *Handbook of the band structure of elemental solids*. Plenum Press, New York. 329.

[11] Boev O, Puska M, Nieminen R 1987 *Phys. Rev.* **B36**, 7786.

[12] Puska M, Jepsen O, Gunnarsson O, Nieminen R 1986 *Phys. Rev.* **B34**, 2695

[13] Imas O 2005 *Proc. 9th Russian-Korean Int. Symp. on Science and Technology, KORUS-2005.* (*Novosibirsk*), p 156

[14] Arefev K, Boev O, Imas O, Lider A, Surkov A, Chernov I 2003 *Physics of the Solid State* **45** 3.

IOP Conf. Series: Materials Science and Engineering **142** (2016) 012045 doi:10.1088/1757-899X/142/1/012045

Efficiency Assessment of Using Flammable Compounds from Water Treatment and Methanol Production Waste for Plasma Synthesis of Iron-Containing Pigments

Anastasia P Shekhovtsova, Alexander G Karengin

National Research Tomsk Polytechnic University, Lenina Avenue, 30, Tomsk, 634050, Russia

E-mail: karengin@tpu.ru

Abstract. This article describes the possibility of applying the low-temperature plasma for obtaining iron-containing pigments from water purification and flammable methanol production waste. In this paper were calculated combustion parameters of water-saltorganic compositions (WSOC) with different consists. Authors determined the modes of energy-efficient processing of the previously mentioned waste in an air plasma. Having considered the obtained results there were carried out experiments with flammable dispersed water-saltorganic compositions on laboratory plasma stand. All the experimental results are confirmed by calculations.

1. Introduction

Water pollution which caused by the industrial development is a problem of modern society. Conventional water treatment technologies are using chemicals, which makes the process expensive [1,2]. Nonthermal plasma found applicability in various field such as sterilization [3], cancer cell control [4], surface treatment [5], air treatment [6] and biomedical applications [7] due to high performance and low cost [8-10]. When considering the implementation of the technology, a small power supply will be required [2,11].

Scientific and technical progress is intended to improve the quality of human life and to protect people from the many natural and anthropogenic factors, but an inevitable consequence of this progress is rapidly increasing negative impact on the environment. Industrial waste is one of those factors. On the other hand, polluting wastes may be used in industry and usage of them is economically profitable [12].

Tomsk underground water intake (city of Tomsk, Russia) annually produces after water treatment about 3000 tons of water purification waste (WPW) in form of precipitates having following consist [13,14]: Fe (30.3 %), Mn (4.5 %), Si (4 %), Ca (1 %), Mg (0.26 %), Al (0.1 %), Cu (0.05 %) and H_2O (all the rest, ≈ 60 %).

A similar problem is at the wastewater treatment plants of many other localities in different countries all over the world where population uses underground water with high iron and manganese content. It should be noted, that at some regions (for example in China) underground water is the only source of unleavened water and that is why its purification is the special task [15].

Nowadays, actual problem is energy-efficient and environmentally safe obtaining of iron-containing pigments from WPW and wide application of them in the metallurgy (iron-containing raw), in the manufacture of paving slabs, curbs, decorative tiles, glass composite materials, heat-resistant and protective coatings, etc. [16].

However, conventional thermal methods of obtaining pigments from precipitations require significant energy and labor costs.

A significant reduction of energy consumption for the process can be achieved using plasma recycling. In this case precipitations undergo recycling in form of inflammable water-saltorganic compositions (WSOC) with optimal consist having adiabatic combustion temperature about 1200 °C [17].

Thus it is of interest to apply direct plasma obtaining of pigments from WPW for different purposes.

2. Calculation of combustion parameters of WSOC consist of WPW

Tomsk Company «Sibmethachim» occupies a leading position in Russia in the methanol production. It produces about 800 000 tons annually. At the same time it produces about 3 500 tons of inflammable methanol production waste (MPW) consisting of: CH_3OH (50–70 %);
C_2H_5OH (2.6–5.3 %); C_3H_7OH and other higher alcohols (2.5–4.7 %) and H_2O (all the rest percentage) [18].

In present those waste is burned and emitted into the atmosphere. Such MPW utilization technology is environmentally unsafe and leads to a significant loss of thermal energy.

That is why it is of interest to jointly recycle WPW and inflammable component (in this case MPW) in air plasma in forms of WOC. It should be noted that MPW provides more efficient recycling of these wastes.

Liquid combustible compositions are compositions with a net calorific value of more than 8.4 MJ/kg [19]. Net calorific value of inflammable composition should calculate as net calorific value of liquid waste Q_l:

$$Q_l = \frac{(100 - W - A)\cdot Q_d}{100} - \frac{2.5W}{100}, [MJ/kg], \tag{1}$$

where Q_d is net calorific value of dry waste [MJ/kg]; W and A are respectively content of water and solids [%]; 2.5 is value of latent heat of vaporization of water at 0 °C [MJ/kg].

That is why it is of interest to jointly recycle WPW and inflammable component (in this case MPW) in air plasma in forms of WSOC. It should be noted that MPW provides more efficient recycling of these wastes.

Recommendation to consider liquid compositions with $Q_l \geq 8.4$ MJ/kg as inflammable is overstated for many industrial compositions containing combustible components with low values of Q_d.

In contrast to the work [19], we proposed a more objective measure of flammability of liquid compositions in the form of an adiabatic combustion temperature Tad:

$$T_{ad} = \frac{Q_l + C_{was}t_{was} + \alpha \vartheta^0_{ox} C_{ox} t_{ox}}{\upsilon C}, [\circ C], \tag{2}$$

where C_{was} is average mass heat capacity of liquid waste [kJ/kg]; t_{was} is temperature of liquid waste [°C]; α is oxidant flow coefficient; ϑ^0_{ox} is theoretical flow of oxidant (air) [m^3/m^3]; C_{ox} is average heat capacity of an oxidizer [kJ/(m^3·°C)]; t_{ox} is temperature of an oxidizer [°C]; υ are combustion gaseous products of inflammable components of liquid waste [m^3/kg]; C is average volumetric heat capacity of combustion products of liquid waste [kJ/(m^3·°C)].

Figure 1 shows influence of content of WPW and MPW on the WSOC adiabatic combustion temperature. WSOC have different consists.

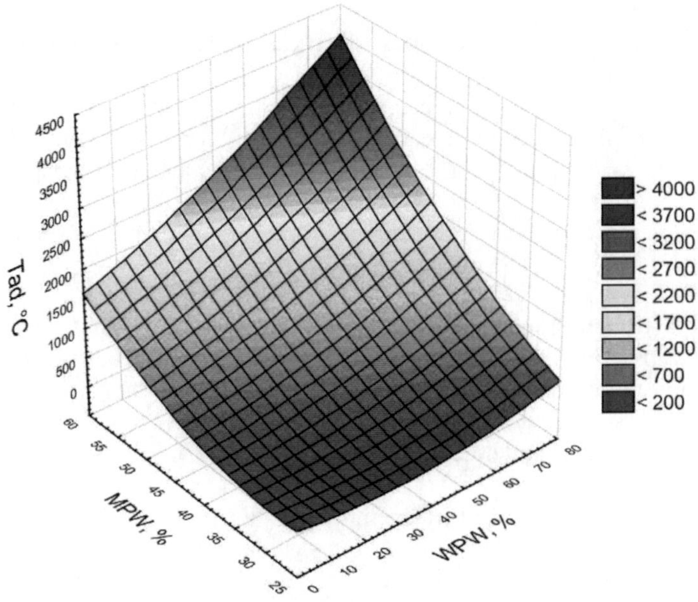

Figure 1. Influence of WPW and MPW content on the WSOC adiabatic combustion temperature

Considering obtained results for practical application may be recommended optimal consist of inflammable WSOC1 (with maximum content of waste): 50% WPW : 50 % MPW.

3. Calculation of equilibrium consists of products formed during WSOC recycling

To determine optimal modes of the plasma recycling there were calculated equilibrium consists of gaseous and condensed products of joint WPW and MPW recycling for recommended (optimal) WSOC 1. Licensed PC program TERRA was used to perform the calculations with following parameters: atmospheric pressure (0.1 MPa), temperature range (300–4000 K), mass fraction range of air heat transfer agent (0.1–0.95).

Figure 2 shows characteristic equilibrium consists of gaseous (a) and condensed (b) products of optimal WSOC1 recycling in plasma. Air mass fraction is 79 %.

IOP Conf. Series: Materials Science and Engineering **142** (2016) 012045 doi:10.1088/1757-899X/142/1/012045

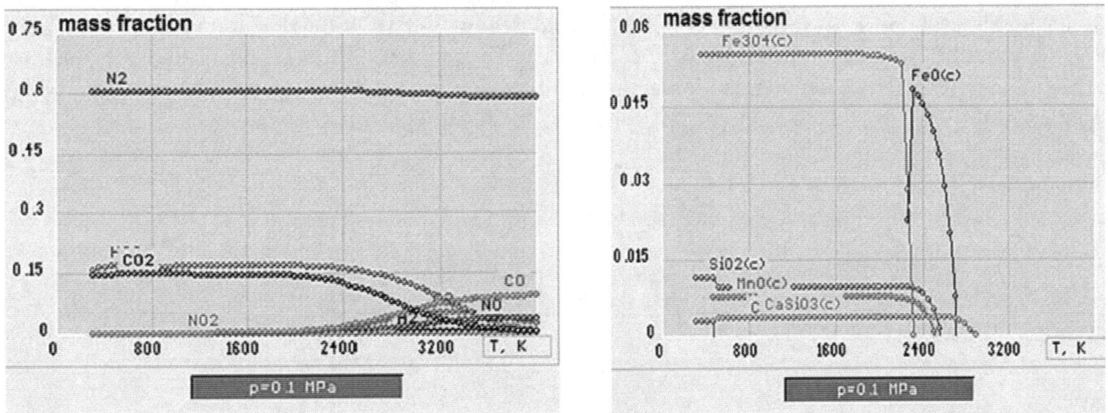

a) b)

Figure 2. Equilibrium consists of gaseous (a) and condensed (b) products of optimal WSOC1 recycling in plasma (79 % air : 21 % WSOC1)

The analysis of equilibrium consists (figure 2) shows main gaseous products of WSOC1 recycling at temperature lower 1500 K and 79 % of air mass fraction: N_2, H_2O and CO_2. Condensed products include simple and complex metal oxides, comprising magnetic iron oxide Fe_3O_4(c). This fact indicates a possibility to apply magnetic sedimentation. Absence of soot C(c) and insignificant amount of gaseous CO, NO and NO_2 indicates that the plasma recycling of optimal WSOC1 occurs optimally.

Air mass fraction increasing up to 80 % (figure 3) does not influence on consist of gaseous products, but changes consist of condensed ones. The last include simple and complex metal oxides, comprising nonmagnetic iron oxide Fe_2O_3(c).

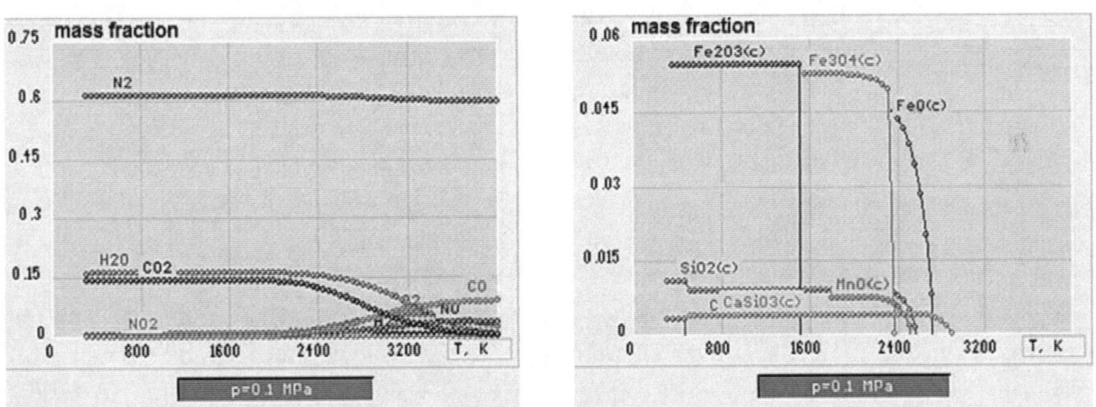

a) b)

Figure 3. Equilibrium consists of gaseous (a) and condensed (b) products of optimal WSOC recycling in plasma (80 % air : 20 % WSOC1)

Considering obtained results for practical application may be recommended optimal modes of process:

- WSOC consist:50 % WPW : 50 % MPW;
- air mass fraction: 79 % air, 21 %WSOC1;
- temperature range: (1500 ± 100) K.

4. Experimental confirmation of recommended modes for WPW and MPW recycling in plasma

Plasma stand «Plasma unit on the basis of high-frequency generator VChG 8-60/13-01» was used to experimentally verify recommended process conditions of joint plasma recycling wastes in high-frequency air plasma torch discharge. Figure 4 shows the scheme of this plasma unit.

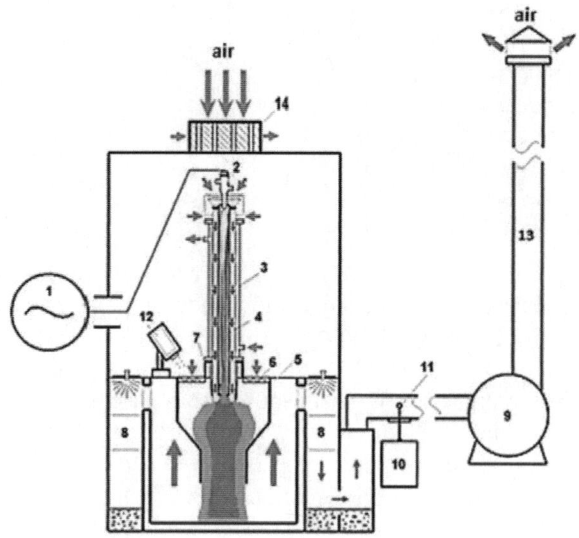

Figure 4. Scheme of laboratory plasma unit: high-frequency generator (1); copper electrode (2); high-frequency torch plasmatron (3); high-frequency torch discharge (4); reactor (5); reactor impeller (6); dispersant (7); unit for water scrubbing of exhaust gases (8); exhaust fan (9); gas analyzer (10); sampling device (11); pyrometer (12); gas duct (13); heater (14)

Plasma unit includes high-frequency generator VChG 8-60/13-01 (oscillatory power up to 60 kW, the operating frequency of 13.56 MHz). High-frequency energy from generator via coaxial output 6 is supplied to the water-cooled copper electrode 2 of high-frequency torch plasmatron 3. Plasmatron generates air plasma jets with temperature to 4000 K. Exhaust fan 9 provides a constant air flow through the plasmatron 3 and reactor 5.

Input unit of plasma gas (air) is provided with an impeller (input area S_{imp} = 25 cm^2). It is placed at the entrance of the plasmatron and is designed for input and swirling the air heat transfer agent into the discharge chamber of the plasmatron. Discharge chamber is made of quartz tube with a diameter of 50 mm and a length of 860 mm.

Reactor impeller 6 is designed for swirling the air heat transfer agent entering the reactor and allows adjusting the input area of the impeller within 0–1650 cm^2.

Gas analyzer 10 and sampling device 11 are made for determining the consist of the exhaust gases after unit for water scrubbing 8 and for measurements with Pitot tube of air heat transfer velocity through plasmatron and the reactor.

Pyrometer 12 is designed for non-contact temperature measurement (through absorption of produced carbon dioxide) of the plasma combustion of inflammable dispersed WOC into reactor.

Experiments on plasma recycling of WPW and MPW were carried out to confirm the calculated results. Wastes recycled in form of model inflammable WOC consisted of WPW and ethanol (instead of poisonous methanol). Ethanol and methanol have close combustion parameters (Table 1).

Table 1. Combustion parameters of MPW components

Name of fuel component	Net calorific value, MJ/kg	Flash-point, °C	Ignition point, °C	Self-ignition point, °C
methanol	19.95	6	13	440
ethanol	26.80	13	18	400

Analysis of combustion parameters of MPW components shows that it need an air jet temperature $T_{jet} \geq 400$ °C for reliable WOC ignition in the reactor.

Figure 5 shows the influence of the anode current of the generator and water content on the operating temperature of the process plasma recycling model dispersed compositions «WPW-ethanol» (flow ≈ 1000 liters per hour).

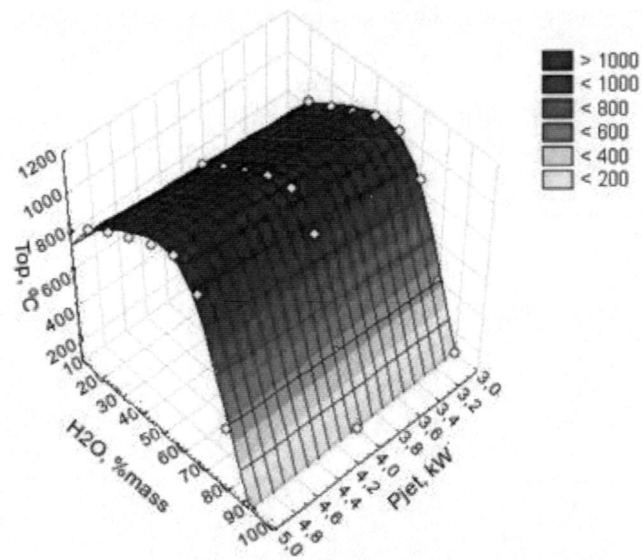

Figure 5. Influence of the anode current of the generator and water content on the operating temperature of the process plasma recycling model compositions «WPW-ethanol»

Analysis of the resulting graphical dependence suggests that the operating temperature in the reactor reaches a maximum temperature T≈1100 ° C at a water content in the composition is ≈40%. The results are consistent with the calculated values. Reject water content in a smaller or larger side reduces the operating temperature of the combustion model of the composition in the reactor.

5. Conclusion

WPW and MPW plasma recycling in form of optimal WOC can significantly reduce the specific energy consumption for the process as well as to apply magnetic separation for efficient extraction of iron-containing pigments being plasma recycling products. All the obtained calculated and experimental results may be used for creating of industrial equipment for plasma obtaining of iron-containing pigments from WPW and MPW.

References

[1] Zheng X, Blais J, Mercier G, Bergeron M, Drogui P 2007 *Chemosphere* **68** 1143–52
[2] Shimizu K, Masamura N, Blajan M 2013*J. Phys.: Conf. Ser.* **441** 012005
[3] Tanino M, Wang X, Tahashima K, Katsura S, Mizuno A 2005 *IEEE Trans. on IAS* **2** 784–8
[4] Iseki S, Nakamura K, Hayashi M, Tanaka H, Kondo H, Kajiyama H, Kano H, Kikkawa F, Hori M 2012 *Appl. Phys. Lett.* **100** 113702
[5] Ceyter N D, Morent R, Leys C, Gengembre L, Payen E 2007 *Surf. Coat. Tech.* **201** 7066–75
[6] Shimizu K, Blajan M, Kuwabara T, 2011 *IEEE Trans. on Ind. Appl.* **47** 2351–58
[7] Kong M G, Kroesen G, Morfill G, Nosenko T, Shimizu T, Dijik J V, Zimmermann J L 2009 *New J. Phys.* **11** 115012
[8] Handa T, Minamitani Y 2009 *IEEE Trans. on Plasma Sci.* **37** 179–83
[9] Shih K Y, Locke B R 2011 *IEEE Trans. on Plasma Sci.* **39** 883–92
[10] Zhao J L, Zhou, Su J, Guo H C, Wang X S, Gong W M 2007 *AIChE J.* **53** 3204–09
[11] Kuwahata H, Kimura K, Ohyama R I 2010 *J. Surf. Sci. Nanotech.* **8** 381–83
[12] Bobovich B B 1999 *Recycling of industrial waste* (Moscow: SP Intermet Engineering)
[13] Usova N T, Kutugin V A, Lotov V A, Lukashevich O D 2011*Bull. Tom. Pol. Univ.* **319,3** 36–39
[14] Lisetskiy V N, Bryukhantsev V N, Andreychenko A A 2003 *Capture and recycle of sediment of water treatment at city of Tomsk* (Tomsk: NTL Publ.)
[15] Zhen H L 2011 *Appl. Mech. Mater.* **94–96** 706–9
[16] Yakovlev S V 1999 *Treatment and disposal of precipitation of industrial wastewater* (Moscow: Khimiya)
[17] Karengin A G, Tundeshev N V, Novoselov I Yu, Oreshkin E A 2015 *Adv. Mater. Res.* **1084** 327–32
[18] Karengin A A, Karengin A G, Vlasov V A 2015 *Rus. Phys. J.* **58,2.2** 195–200
[19] Bernadiner M N, Shurygin A P *Flame processing and disposal of industrial waste* (Moscow: Chemistry)

IOP Conf. Series: Materials Science and Engineering **142** (2016) 012046 doi:10.1088/1757-899X/142/1/012046

Control of the Physical and Technical Properties of Water in Technological Processes

V D Klopotov[1], N P Gorlenko[2], Yu S Sarkisov[2], A K Kulchenko[3], A A Klopotov[2,3]

[1]National Scientific Research Tomsk Polytechnic University
[2]Tomsk State University of Architectureand Building
[3]Surgut State University
[4]National Scientific Research Tomsk State University

E-mail: vdklopotov@mail.ru

Abstract. The physical and technical properties of water activated by the electrochemical treatment in a two-chamber electrolizer are investigated. The regularities of changes inthe values of acidity, redox potential, ionic composition, concentration of oxygen, structural organization of catholyte and anolyte are revealed. The possibility of controlling the properties of the liquid for more efficient extraction of polymetallic minerals by flotation is described.

1. Introduction

Currently, the most common method for controlling the properties of water is its treatment with external physical impacts or introduction of chemical additives of both organic and inorganic nature. In this case, in water and in aqueous solution there are changes that occur in the structural organization of the entire system, caused by generation of hydrate formations of dissolved ions and gas hydrates, which leads to the disruption of the interaction among the individual structural components of water. In particular, the quantitative ratio of associated and individual dipoles of water molecules changes, as well as its physical and technical properties. The abovementioned processes have a significant impact on the course of technological processes where water is used as a solvent or a dispersed phase. Therefore, for optimum results, in particular, in the flotation technology, it is necessary not only to control its quality but to purposefully regulate its properties as well.

During the electrochemical treatment of industrial water in a diaphragm electrolyzer the following takes place: change in the ionic composition, pH, Eh of water, saturation with finely dispersed electrolytic gases (hydrogen, oxygen), and its division into two products: alkaline water – catholyte, and acidic water – anolyte. Moreover, the final ionic composition depends on the initial composition of water and the quantity of the electricity used for 1 m^3 of solution. The electrochemical treatment of water allows to change the pH by (4 – 5) units, Eh by (300 – 500) mV to the region of negative or positive values, as well as to reduce the content of hardness salts by 50 – 85%. These processes have a significant effect on the flotation of refractory ores. For example, in all cases, the reduction in the redox potential (RP) to the negative direction by more than 200 mV results in an increased recovery of some (zinc) and decreased recovery of other (copper, iron, tin) valuable components [1-4].

The aim of the work is to study the changes in physical and technical properties of water and aqueous solutions during its treatment with the use of the electrochemical method, and to assess the impact of the activated liquid on the effectiveness of polymetallic ores flotation.

2. Materials and methods

The electrochemical treatment of water was carried out using the device shown in Figure 1.

Plates made of stainless steel X18H9T were used as electrodes. The diaphragm was made of a filter cloth. The electrode spacing was 100 mm. The power was supplied by the rectifier VSA-5. The surface density of the current was varied in the range of (16-24) A/m^2, the volume current density – (0.1 – 0.5) A/l, and the power consumption – 6 kWh/m^3.

The control of the acidity (pH) and the redox potential (Eh) was carried out using anionometer I-120. The quantity of the concentration of the dissolved oxygen was measured using an electrochemical analyzer.

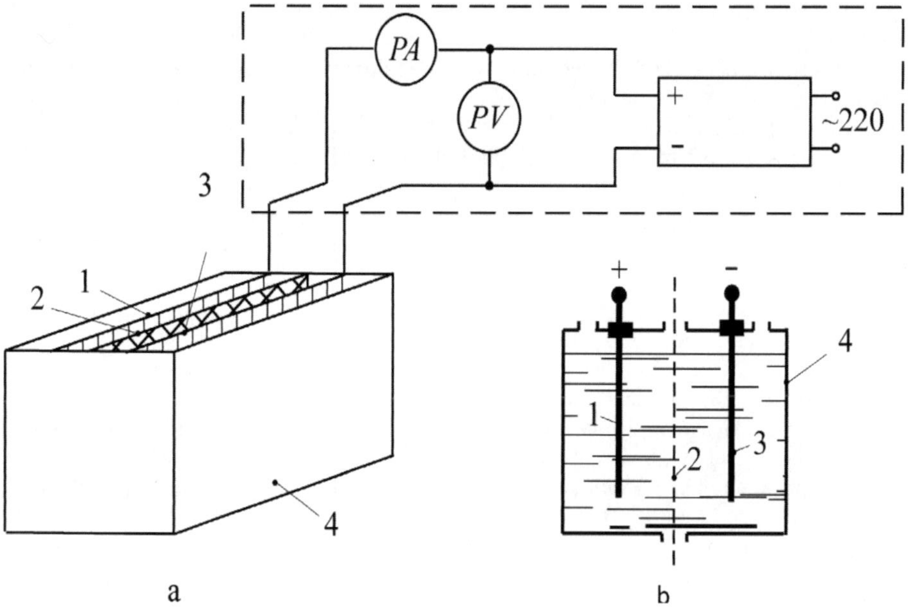

Figure 1. Principal (a) and schematic (b) images of the electrolizer: 1 –anode; 2 – diaphragm; 3 – cathode; 4 – case

The electrical properties of water were controlled using the dielectrometry method [5]. The measuring cell comprises a cylindrical glass vessel with a diameter of 8 cm, in which the tested liquidin the amount of 500 ml and the two plates (electrode) of the capacitor made of a nonmagnetic material (area 64 cm^2 each) located on opposite sides of the vessel are placed (Figure 2).

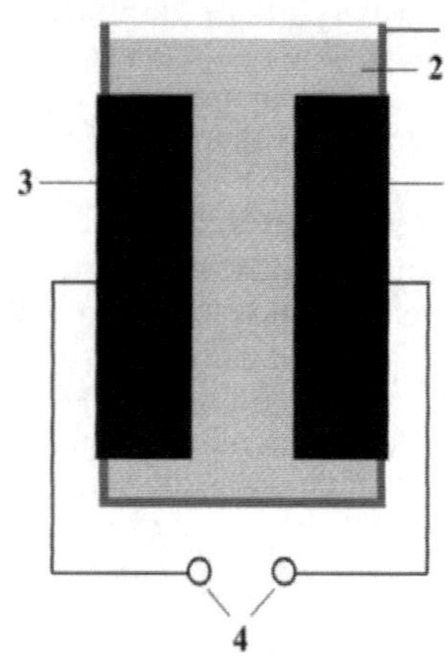

Figure 2. The arrangement of the capacitor plates in the measurement cell: 1 – liquid container, 2 – test liquid, 3 – capacitor plates, 4 – terminals for connecting the signal from the sine wave generator.

Results and discussion

An installation for measuring liquid parameters allows to register the change in the electric capacity of liquids (based on the change in the value of the reactive current and upon deducting the capacity of the measuring cell without the liquid from the total capacity of the capacitor). The voltage to the measuring cell is applied from a sine wave generator FSG-2007, the cell signal is amplified by an instrumentation amplifier based on the circuit INA 217 and is measured on an oscilloscope PDS5022S. The value of the inductance is measured using a device LCR-9063.

The paper (Figure 3-4, Table 1) shows the results of the research on changes in the physical and technical properties of water passed through a DC electrode pair.

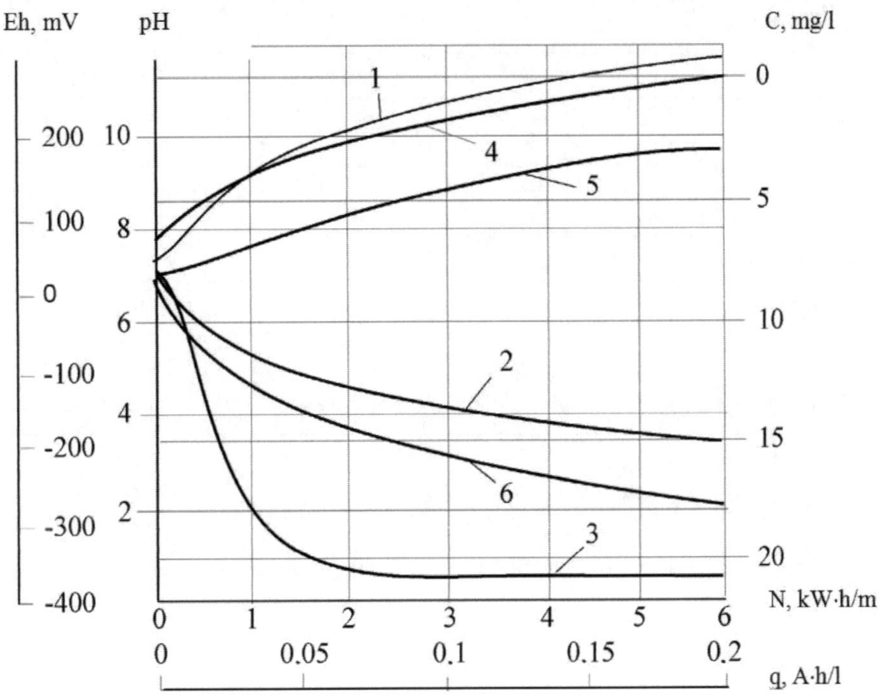

Figure 3.Changes in pH (1, 2), Eh (3, 4), and concentration of the dissolved oxygen (C) (5, 6) depending on the cost of electricity (N) and the quantity of electricity (q) in catholyte (1, 2, 3) and anolyte (2, 4, 6).

As can be seen from figure 3, pH and Eh of process water change significantly even when the electricity consumption reaches0.5 kWh/m^3 and the amount of electricity reaches 0.017 A·h/l. In the future, changes in pH and Eh are exponential. The catholyte is characterized by the values pH=11.4; Eh=460 mV (Figure 5, curves 1 and 3); Anolyte– pH=3.45; Eh=260 mV (curves 2, 4) at the electricity cost 6 kWh/m^3 and the amount of electricity 0.2 A·h/l.

The ionic composition of process water significantly changes in both the anolyte and the catholyte. (Table 1). This is especially important for hardness salts, as its reduction leads to an intensification of flotation processes [6].

The total hardness of water during the electrochemical treatment decreases with an increase in its acidity and alkalinity. In the anolyte, at pH 3,45the total hardness reduces by 4.6 times, and in the catholyte at pH 11.38 by 2.3 times. The experimentally obtained optimum hardness of water is in the range of 1-2 mg-eqv/l. At this value, the extraction of fine particles of minerals with a grain size of (10÷5) micron increases due to the strengthening of the mineralization of air bubbles. With an increase in the water hardness above 3 mg-eqv/l the extraction of slurry particles significantly reduces (Figure 4). It has been revealed that anions of hardness salts hardly affect the flotation.

The indicated values of the content of calcium and magnesium salts are optimal, because further decrease in water hardness requires additional power consumption. The decrease in the total hardness can be attributed to the transition of calcium and magnesium ions from the anolyte to the cathode space and to the formation of low-solubility precipitations in the catholyte.

It is known that the change in the structure of water solutions may also influence the course of various processes [6]. Figure 5 shows that after carrying out the electrolysis of distilled and technical water there is a significant change in its electrical capacitance, measured at frequencies of (1-300) kHz, which indirectly indicates a change in the structural organization of the solution, in particular, in the field of ion hydration, associative formations of water dipoles via hydrogen bonding and other.

Table 1. The ionic composition and the physical-chemical parameters of process water and products of its electrochemical treatment

Indicators		Process water	Anolyte	Catholyte
pH		7.23	3.45	11.38
Eh, mv		+95	260	-450
Total hardness, mg-eqv/l		6.45	1,4	2.75
Solids, mg/l		597	487	608
Cations mg/l	Ca^{2+}	68.0	10.0	4.0
	Mg^{2+}	24.0	17.1	54.2
	NH_4^+	1.8	2.3	1.9
	Fe_{tot}	0.0625	1.05	0.0
Anions mg/l	Cl^-	39.2	52,27	24.5
	CO_3^{2-}	18.0	0.0	168.2
	HCO_3^-	426.0	0.0	180.0
	NO_3^-	0.0	0.31	0.0
	NO_2^-	1.29	1.08	0.47
	S^{2-}	9.14	7.85	5.77
	SO_4^{2-}	58.65	79.85	38.29
	$S_2O_3^{2-}$	7.26	7.26	7.26
	SO_3^{2-}	7.21	3.2	2.4
Oxygen content, mg/l		6.8	22.2	3.1

Thus, the electrochemical treatment of water leads to substantial changes in pH, redox, ionic composition, concentration of dissolved oxygen in water, structural organization of the solution, which can be controlled by specifying the quantity of electricity and electricity costs. When implementing technological processes that are sensitive to these parameters it is possible to achieve their intensification. In particular, flotation of fine-grained mineral ores relates to such technologies [7].

The paper includes an assessment of the effectiveness of the flotation of polymetallic ores with the use of electrochemically activated process water. In particular, it has been revealed that the catholyte significantly activates the flotation of tin: thus, the recovery of this metal into a concentrate of the primary cleaning of the tin-lead flotation increases by more than 40 times (from 1.34% in the basic mode up to 64.45%). In the anolyte, the enrichment effect is significantly lower and it depends on the concentration of the dissolved oxygen.

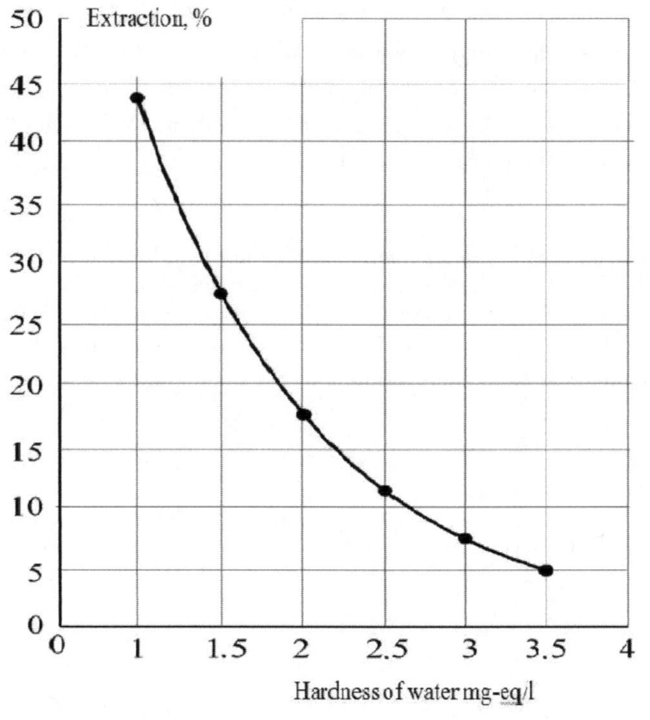

Figure 4. The effect of water hardness on the extraction of dispersed particles (10-15 microns) of minerals.

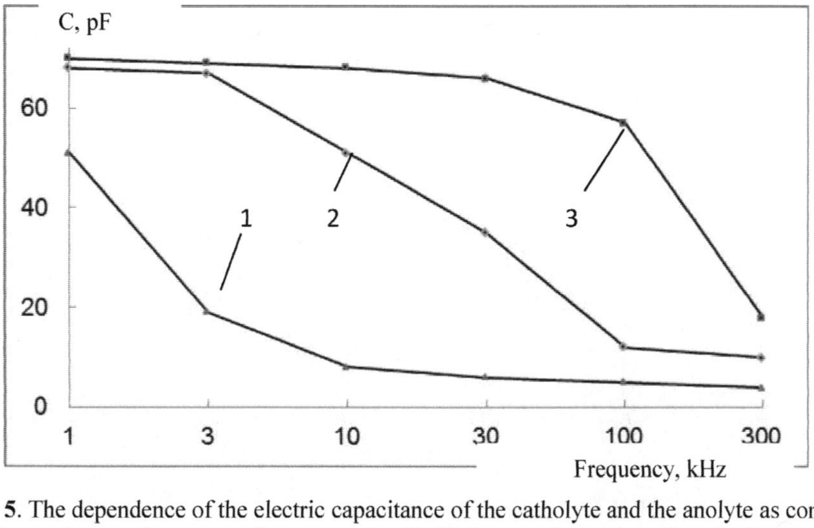

Figure. 5. The dependence of the electric capacitance of the catholyte and the anolyte as compared to distilled water on the reactive current frequency: 1 – distilled water, 2 –anolyte; 3 –catholyte.

Conclusion

1. It has been shown that during the electrochemical activation of liquids in a two-chamber cell the physical and technical properties of water significantly change. These parameters can be adjusted by setting the amount of electricity and the cost of electricity.

2. It has been established that the use of the catholyte in the flotation process of refractory finely-disseminated oxidized tin-lead ores leads to a significant increase in the extraction degree of tin mineral particles. The effect of the process intensification of fine particles extraction from aqueous media is caused by formation of electrolysis gases, reduction in the water hardness indicator, and formation of a new structural organization of the solution.

References

[1] Glembotskiy V A, Klassen V I. 1981 *Flotation methods of enrichment* (Moscow, Nedra).
[2] Chanturia V A, Zelentsov V I. 1984 *Electronic Materials Processing*. – N**2** 80
[3] Chanturia V A, Lunin V D. 1979 *News of Universities. Non-ferrous metallurgy*. N**3** 38
[4] Chanturia V A, Lunin V D. 1983 *Electrochemical methods for intensification of the flotation process* (Moscow, Nauka)
[5] Laptev B I, Sidorenko G N, Gorlenko N P, Sarkisov Yu S 2014 *Water and ecology. Problems and Solutions* **76** 21
[6] Klassen V I, Shcherbakov S V. 1965 *The structure of water and the possibility of changing it with the use of enrichment processes* (Moscow, Nauka)
[7] Chanturia V A, Filippov V V. 1979 The electrochemical method of water treatment during the flotation of polymetallic ores of the complex material composition *The 13th International Congress on Mineral Processing* vol 2 (Warsaw) pp. 5-29.

IOP Conf. Series: Materials Science and Engineering **142** (2016) 012047 doi:10.1088/1757-899X/142/1/012047

Passive Identification is Non Stationary Objects With Closed Loop Control

Valeriy F Dyadik[1], Igor S Nadezhdin[1], Alexey G Goryunov[1], Flavio Manenti[2]

[1] Tomsk Polytechnic University, Institute of Physics and Technology, Department of Electronics and Automation of Nuclear Plants, Russia, Tomsk, Lenina St 30, 634050
[2] Politecnico di Milano, Dipartimento di Chimica, Materiali e Ingegneria Chimica «Giulio Natta» Italy, Milano, Piazza Leonardo da Vinci 32, 20133

E-mail: kun9@list.ru

Abstract. Typically chemical processes have significant nonlinear dynamics, but despite this, industry is conventionally still using PID-based regulatory control systems. Moreover, process units are interconnected, in terms of inlet and outlet material/energy flows, to other neighbouring units, thus their dynamic behaviour is strongly influenced by these connections and, as a consequence, conventional control systems performance often proves to be poor. However, there a hybrid fuzzy PID control logic, whose tuning parameters are provided in real time. The fuzzy controller tuning is made on the basis of Mamdani controller, also exploiting the results coming from an identification procedure that is carried on when an unmeasured step disturbance of any shape affects the process behaviour. This paper presents procedure for identifying technological object control in a closed loop, i. e. that operates the automated control system. The variation in the controlled variable, caused by the change of the non-measurable disturbance, is considered the initial signal for the identification procedure. The parameters of the control object are found by optimization method Levenberg-Marquardt.

1. Introduction

Nowadays, in chemical and nuclear industries (i.e. integrated separations, extractions crystallization processes to purify U and Pu from other fusion side-components) any processing step has a high level of automation. On the one hand, all processes are high responsibility technology (HRT), i.e. high performance technology with respect of safety level. On the other hand, they are also complex dynamic systems. Experience shows that the usually the conventional PID controller is not effective for complex dynamic systems [1]. A number of researchers have conducted studies to develop a hybrid fuzzy PID control logic, whose tuning parameters are provided in real time. The fuzzy controller tuning is made on the basis of Mamdani controller, also exploiting the results coming from an identification procedure. The problem of identifying has attracted and attracts a lot of research interests and many methods are available for this problem in literature [2].

A priori information about the dynamics of the controlled plant is required for the synthesis of PID-based fuzzy logic controller. Hammerstein and Wiener models may be used to describe complex dynamics real-life processes [3, 4]. Hammerstein and Wiener models are methodologies constituted by the combination of a static nonlinearity (N) and a linear system (L), respectively in the N-L and L-N form. The problem of identifying N and L from input-output data has attracted and attracts a lot of research interests and many methods are available for this problem in literature [5]. The nonlinear dynamic system can be approximated by a linear dynamic system near the operating point, which is sufficient for PID tuning. It is not a simple task to define the parameters of the linear dynamic model approximation in the closed-loop system. In [6, 7] active methods of identification are proposed; here sine waves in input are used to excite the Wiener continuous-time system and frequency methods are used to determine the unknowns. Unknown additive disturbances create problems for closed-loop identification. Good results can be obtained by using MATLAB system identification toolbox for the identification of the parameters of the process with the use of ARX, ARMAX, BJ state space, polynomial models and others [8].

2. Material and methods for passive identification

The proposed method employs algorithms for the plant identification coupled with fuzzy systems such as Mamdani controllers [9]. The layout of a generic automatic control systems plant is presented in Fig. 1 while a scheme of an adaptive fuzzy controller is shown in Fig. 2.

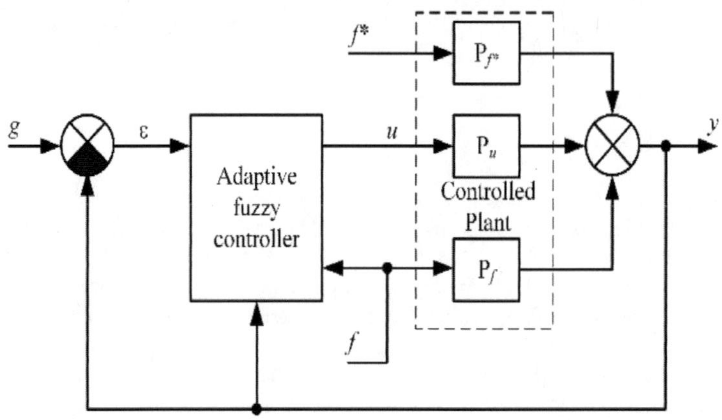

Figure 1. Fuzzy adaptive control system. g : reference signal; f^* : non-measurable disturbance; f: measurable disturbance; P_u : plant control channel; P_f : plant disturbance channel; P_{f*} : plant non-measurable disturbance channel; y: controlled variable; ε : control error is defined as $\varepsilon = g - y$.

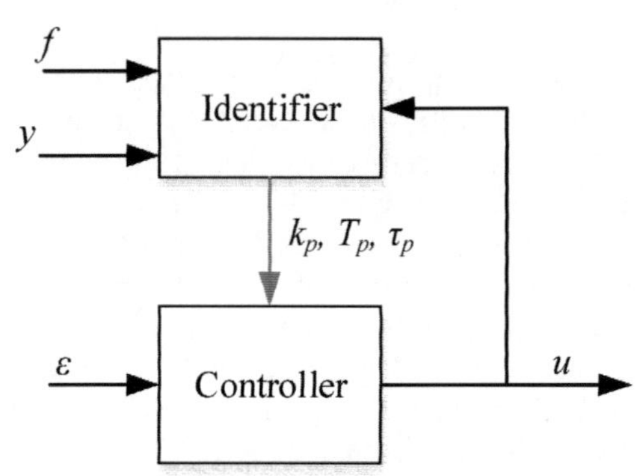

Figure 2. Adaptive fuzzy controller for automatic control systems.

The operating principle of identifier (Fig. 2), is shown in Figure 3. In this paper use the step disturbance variables, but is not known value and duration of these disturbances.

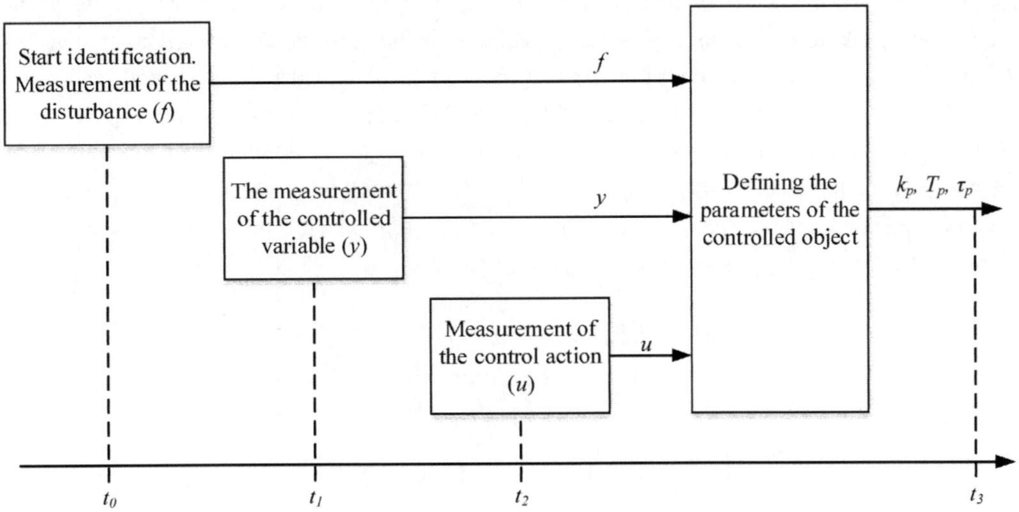

Figure 3. The principle of operation identifier.

The identification is performed in the closed-loop system in those operating conditions where the edge of the transient is reached (see Fig. 4).

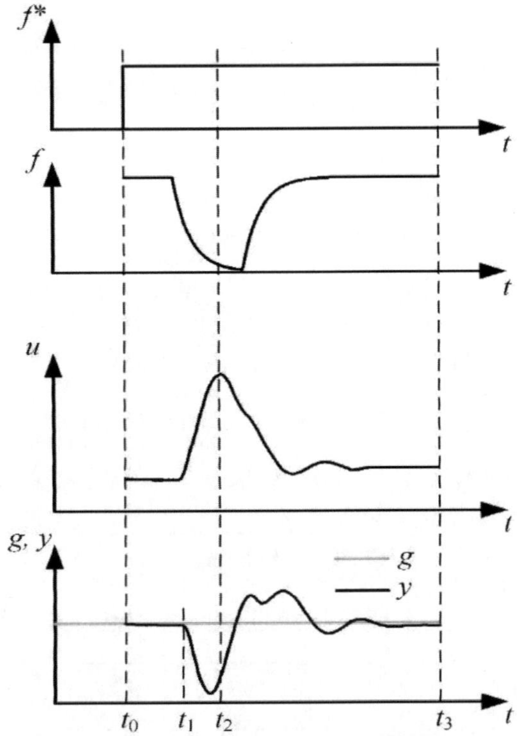

Figure 4. Transient of the automatic control systems for the identification of plant parameters in a closed loop.

The variation in the controlled variable, caused by the change of the non-measurable disturbance f^*, is considered the initial signal for the identification procedure. The time instant t_0 where the non-measurable step disturbance f^* undergoes a step change is unknown. The time instant t_1 is defined by the deviation threshold of y from g by $\Delta y > y_g$, where y_g is the required control accuracy, and t_2 is the time instant of the y variable sign change. The parameters of the plant are defined by Levenberg-Marquardt optimization method. In this case, the measured disturbance f, the control action u and the controlled variable y (see Fig. 1 – 4) are supplied to the identifier input in the time interval whose lower and upper bounds are, respectively, t_2 and t_3. The parameters of the plant (k_p, T_p, τ_p) and those related to the disturbance (k_p^f, T_p^f, τ_p^f) are optimized, in accordance with the reported statements, by the iterative prediction-error estimation method (pem) by Matlab:

$$\begin{cases} y_j^m = y_j^u\left(u_j, k_p, T_p, \tau_p\right) + y_j^f\left(f_j, k_p^{\,f}, T_p^{\,f}, \tau_p^{\,f}\right) \\ S = \sqrt{\sum\limits_{j=1}^{n}\left(y_j^m - y_j\right)^2 \Big/ (n-1)} \\ S \to 0 \end{cases} \tag{1}$$

where y_j^u, y_j^f are a digital representation of the plant model with lagging for channels u and f, S is the optimized functional, n is the dimension of access. The dynamics of the plant is described as a linear system with the transfer function $W_p^u(s)$ that represents channel u and the transfer function $W_p^f(s)$ that stand for channel f:

$$W_p^{\,u}(s) = \frac{k_p}{T_p \cdot s + 1} \cdot e^{-\tau_p s}, \; W_p^{\,f}(s) = \frac{k_p^f}{T_p^f \cdot s + 1} \cdot e^{-\tau_p^f s} \tag{2}$$

3. Results of the passive identification with closed loop control

The proposed the procedure for identifying controlled object was used in the identifier, which is used in the development of automated control system with fuzzy logic. The identifier and the automatic control system was are implemented in the package MATLAB/Simulink.

Realized identifier makes determination of the parameters of the transfer functions of the control object on control and disturbance (2). Table 1 shows the results of identification of parameters of control object after making the system disturbances. Into the system was introduced a step disturbance with unknown characteristics.

Table 1. The results of the identification.

Object parameters	The parameter values of the transfer function of the object		The results of the identification	
	Control	The perturbation	Control	The perturbation
k_p	4.3	4	4.7	4.4
T_{p1}	90	90	107.8	107.1
T_{p2}	25	-	-	-
T_{p3}	2.5	-	-	-
τ_p	12	30	62.6	26

The controlled object is represented transfer function the third order. As a result, the identification of the control object was described by the transfer function of the first order with delay. The discrepancy of transient processes after identification was 0.3%, the transient processes are presented in Fig. 5.

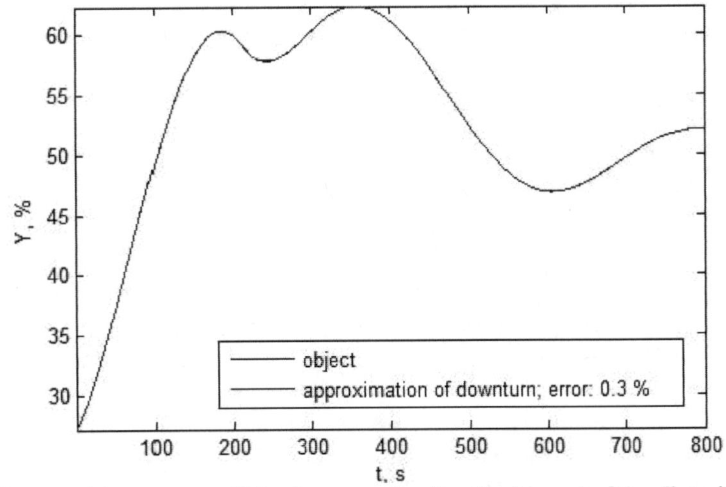

Figure 5. The transition process ("the downturn", after the impact of the disturbance) for the identification parameters of the controlled object

It was also investigated the work identifier at entering the system step disturbances with noise. As the noise interference used additive white Gaussian noise. The results of identification are shown in Table 2.

Table 2. The results of the identification.

Object parameters	The parameter values of the transfer function of the object		The results of the identification	
	Control	The perturbation	Control	The perturbation
k_p	4.3	4	10	8.6
T_{p1}	90	90	239.3	241.7
T_{p2}	25	-	-	-
T_{p3}	2.5	-	-	-
τ_p	12	30	41.8	28.3

Fig. 6 shows the transition process that indicate that an error in the identification of 1.2 %.

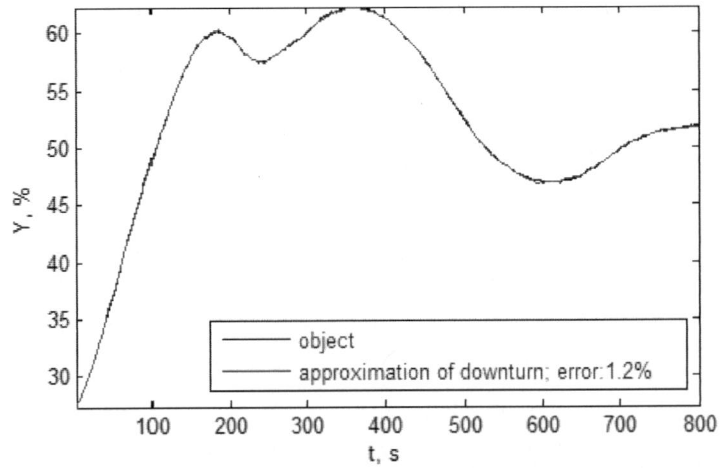

Figure 6. The transition process ("the downturn", after the impact of the disturbance with noise) for the identification parameters of the controlled object

The use of white noise explained by its wide distribution in nature and technology. In practice, most of the signals contain the noise. The use for research step disturbance with noise, will allow more realistically simulate the identification procedure. As seen from the transient processes, the proposed identifier performs its functions, even at disturbance with noise.

4. Conclusions

As a result of the work was implemented identifier in structure of system control with fuzzy logic. The work was performed in the package MATLAB/Simulink. Identification of the parameters of the control object is carried out in a closed loop, with step disturbances of unknown magnitude and duration. Error occurred while identify the parameters of the object was 0.3 %. In studying of the proposed procedures the identification, on the disturbance was superimposed noise. As the interference was used additive white Gaussian noise. Error occurred while identification of parameters of the object at step disturbance with white noise was 1.2 %. The results demonstrate the possibility of determining the parameters of the object in a closed loop, using the proposed identification procedure.

5. Acknowledgements

This work was funded as part of the Federal government-sponsored program «Science» by Tomsk Polytechnic University.

References

[1] Gao Z., Kong D., Gao C., Chen M. 2013. Modeling and control of complex dynamic systems. Journal of Applied Mathematics. Article number 151372. DOI: 10.1155/2012/869792
[2] Ikhouane F., Giri F. 2014. A unified approach for the parametric identification of SISO/MIMO Wiener and Hammerstein systems. Journal of the Franklin Institute. 351(3) 1717–27.
[3] Biagiola S.I., Figueroa J.L. 2009. Wiener and Hammerstein uncertain models identification. Mathematics and Computers in Simulation. 79(11) 3296–313.
[4] Dempsey E.J., Westwick D.T. 2004. Identification of Hammerstein models with cubic spline nonlinearities. IEEE Transactions on Biomedical Engineering. 51(2) 237–45.
[5] Zhou L., Li X., Pan F. 2013. Gradient based iterative parameter identification for Wiener nonlinear systems. Applied Mathematical Modelling. 37(16-17) 8203–09.
[6] Giri F., Rochdi Y., Chaoui F.Z. 2009. An analytic geometry approach to wiener system frequency identification. IEEE Transactions on Automatic Control. 54(4) 683–96.
[7] Giri F., Rochdi Y., Radouane A., Brouri A., Chaoui F.Z. 2013. Frequency identification of

nonparametric Wiener systems containing backlash nonlinearities. Automatica. 49(1) 124–37.

[8] Rabbani M.J., Hussain K., Khan A.U.R., Ali A. 2013. Model identification and validation for a heating system using Matlab system identification toolbox. In: Proceedings of the 1st International Conference on Sensing for Industry, Control, Communications, and Security Technologies, ICSICCST 2013, Karachi, Pakistan: June 2013. IOP Conference Series: Materials Science and Engineering. 51(1) DOI: 10.1088/1757-899X/51/1/012022.

[9] Xu X.Q. 2014. The application of MATLAB for fuzzy control system simulation. Applied Mechanics and Materials. 494-495 1306–9.

IOP Conf. Series: Materials Science and Engineering **142** (2016) 012048 doi:10.1088/1757-899X/142/1/012048

Fuzzy Adaptive Control System of a Non-Stationary Plant

Igor S Nadezhdin[1], Alexey G Goryunov[1], Flavio Manenti[2]

[1] Tomsk Polytechnic University, Institute of Physics and Technology, Department of Electronics and Automation of Nuclear Plants Russia, Tomsk, Lenina St 30, 634050
[2] Politecnico di Milano, Dipartimento di Chimica, Materiali e Ingegneria Chimica «Giulio Natta» Italy, Milano, Piazza Leonardo da Vinci 32, 20133

E-mail: kun9@list.ru

Abstract. This paper proposes a hybrid fuzzy PID control logic, whose tuning parameters are provided in real time. The fuzzy controller tuning is made on the basis of Mamdani controller. In addition, this paper compares a fuzzy logic based PID with PID regulators whose tuning is performed by standard and well-known methods. In some cases the proposed tuning methodology ensures a control performance that is comparable to that guaranteed by simpler and more common tuning methods. However, in case of dynamic changes in the parameters of the controlled system, conventionally tuned PID controllers do not show to be robust enough, thus suggesting that fuzzy logic based PIDs are definitively more reliable and effective.

1. Introduction

Nowadays the conventional proportional-integral-derivative (PID) controllers are the most widely used for process control in most of the industrial plants. The success of PID control logic can be attributed to the achievement of simple structures of automatic control systems (ACS) and its effectiveness for linear systems [1, 2]. There is a wide variety of PID controllers tuning rules: the Ziegler-Nichols rule [3], the magnitude optimum method [4], the direct synthesis methods [5], the Internal Model Control methods [6], the minimum error integral criteria [7], the iterative feedback tuning method [8], the virtual reference feedback tuning method [9], the approximate M-constrained integral gain optimization method [10], AMIGO method [11] and others. The required quality of a PID control system can be achieved by means of a variety of tuning rules once a linear model of the controlled system and a criteria for the assessment of the control performance are chosen.

Usually the conventional PID controller is not effective for complex dynamic systems. The complex dynamic systems are those systems with non-linear static characteristics, i.e. those systems that are described by differential equations with time-varying parameters. This feature essentially complicates the design and analysis of PID-based control systems and decreases their control performance.

A number of researchers have conducted studies to combine a conventional PID controller with a fuzzy logic controller (FLC) in order to achieve a better control quality in ACS rather than the one guaranteed by conventional PID controllers. The idea of using fuzzy sets is successfully applied, for the first time, in the control of a dynamic plant developed by Mamdani and Assilian. Currently, there are different types of FLC, but a PID-based FLC is the most common and practical for applications to

IOP Conf. Series: Materials Science and Engineering **142** (2016) 012048 doi:10.1088/1757-899X/142/1/012048

ACS [12]. Such FLC is equivalent to a conventional PID controller for the input-output structure [13]. PID-based FLC may be constructed by sequentially incorporating FLC and PID controllers or paralleling PID and FLC (PID with an adapter based on FLC). Moreover, the use of FLC logic makes it easy to add nonlinearities and additional input signals to the control law, that, in turn, allows to apply PID-based FLC to complex dynamic systems.

The purpose of the research is to develop a method of synthesis for low-level ACS (relative to HRT), which will provide the required control performance also in the presence of a significant change in the process parameters and several step disturbances with unknown amplitudes and durations. A Low-level ACS must fulfill the following limitations: control in the tight real-time mode should be performed with hot standby of the controllers; applied controllers have limited computation abilities which do not allow an extension of the mathematical support functions; for the purpose of control, conventional PID controllers should be employed.

2. Material and methods for the fuzzy adaptive control of a generic plant

The proposed method employs algorisms for the plant identification coupled with fuzzy systems such as Mamdani controllers [14]. The layout of a generic ACS plant is presented in Fig. 1 while a scheme of an adaptive fuzzy controller is shown in Fig. 2.

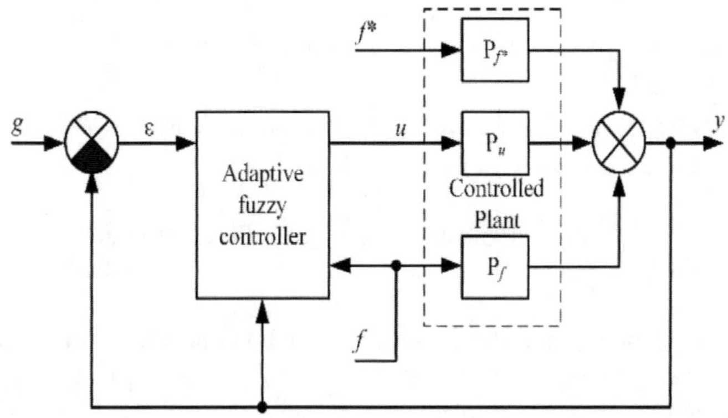

Figure 1. Fuzzy adaptive control system. g : reference signal; f^* : non-measurable disturbance; f : measurable disturbance; P_u : plant control channel; P_f : plant disturbance channel; P_{f^*} : plant non-measurable disturbance channel; y: controlled variable; ε : control error is defined as $\varepsilon = g - y$.

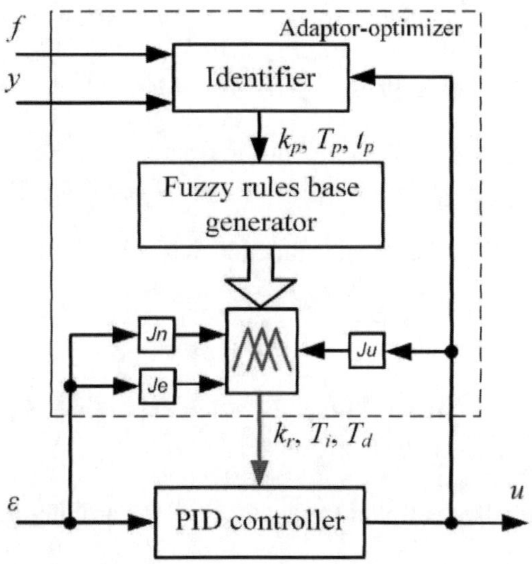

Figure 2. Adaptive fuzzy controller for an ACS.

The optimization problem consists of maximizing or minimizing a functional which plays the key role from the viewpoint of the design of adaptive and optimal control systems. It is addressed here in the following form:

$$\min\left(Je_k + Ju_k + Jn_k\right) \quad (1)$$

where

$$Je_k = \sqrt{\frac{\sum_{j=k}^{k+he}\left(\varepsilon_j\right)^2}{he-1}} \quad (2)$$

$$Ju_k = \sqrt{\frac{\sum_{j=k}^{k+hu}\left(u_j - u_k\right)^2}{hu-1}} \quad (3)$$

Jn_k – the number of control error oscillations in the interval *he*, (2)

where $k = 1,2,\ldots\infty$, ε_j – the control error, u_j – the manipulated variable, *he* – the control error interval, *hu* – the control interval, *j* – the index of time sampling.

The adaptor-optimizer of the suggested ACS (see Fig.2) includes the following blocks: an identifier, a fuzzy rules base generator, a Mamdani fuzzy output controller and *Jn*, *Je* and *Ju* terms calculation engines.

The procedure for calculating the parameters of the PID controller exploits the Mamdani controller with those fuzzy rules previously obtained by minimizing the functional (Eq. 1). Membership functions are generated based on two groups of parameters of the PID, calculated by two different methods.

The first group of PID controller parameters (k_r^{OM}, T_i^{OM}, T_d^{OM}) is calculated by magnitude optimum method. The second group of PID controller parameters is calculated by Ziegler-Nichols. The third group of PID controller parameters is calculated by AMIGO method. In addition, the PID-regulator parameters were determined by means of the method developed in Mikhalevich et al. [15].

The next group of PID controller parameters is calculated by means of the amplitude-phase frequency characteristic (APFC) of the open-loop system $W_{open}(j\omega)$. The PID controller parameters (krmax, Timax, Tdmax), corresponding to the closest to the imaginary axis left eigenvalue of the matrix of the open-loop system Wopen(jω), which is in the range of $-1\cdot(1\div0{,}8)$, are determined.

As a result, by applying two groups of PID controller parameters, the base of the fuzzy rules of Mamdani controller is generated. This basis set of rules and the algorithm required for their evaluation is generated with Matlab.

The PID controller is described by the transfer function Wr(s) reported in expression (4):

$$W_r(s) = k_r + \frac{k_r}{T_i \cdot s} + k_r \cdot T_d \cdot s \quad (4)$$

where k_r – transfer coefficient, T_i – integral time, T_d – derivative time. k_r, T_i and T_d parameters are calculated by means of the Mamdani controller (Fig. 2).

The digital realization of the PID controller (4) is described via a finite-difference form:

$$\begin{cases} u_j = k_r \cdot \varepsilon_j + C_j + \dfrac{k_r \cdot T_d}{\Delta t} \cdot \left(\varepsilon_j - \varepsilon_{j-1}\right) \\ C_j = C_{j-1} + \dfrac{k_r \cdot \Delta t}{T_i} \cdot \left(\varepsilon_j + \varepsilon_{j-1}\right) \end{cases} \quad (5)$$

where Δt – sampling time.

An adaptive fuzzy controller (see Fig. 2) can be implemented into the industrial software of the process controller (PLC). Adaptor-optimizer can be implemented as a DLL user-library for SCADA systems. It does not require a tight real-time mode and operates asynchronously with a PID controller.

3. Results of the closed-loop transients coming from the application of the ACS

Fig.3 shows the APFC of the plant and the open-loop system. The PID controller is tuned by the magnitude optimum method («Magnitude optimum» line) and the frequency method («Maximum» line). The PID controller is tuned by means of the adaptor-optimizer in the case in which the graph of the APFC of the open-loop system is between «Maximum» and «Magnitude optimum» curve (Fig. 3).

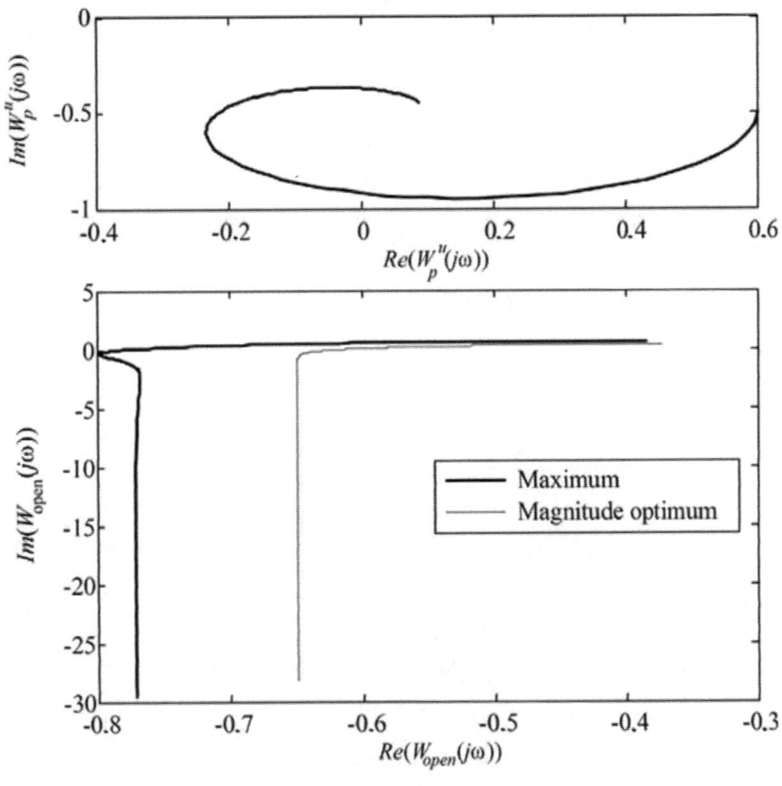

Figure 3. APFC of the plant and the open-loop system

Fig. 4 shows the transient for a 15% step disturbance. The settling time in the suggested ACS (named LF PID in the text) is 10% smaller than in the ACS based on a PID controller tuned by the magnitude optimum method (MO). Moreover, the LF PID method is also compared, in terms of performance, with PID controllers configured using AMIGO and Ziegler-Nichols methods along with the new method proposed in Mikhalevich et al. As Fig. 11 demonstrates, the worst performance is

IOP Conf. Series: Materials Science and Engineering **142** (2016) 012048 doi:10.1088/1757-899X/142/1/012048

observed in the process controlled with the PID tuned with AMIGO method while the best control quality is obtained when the process is controlled with the regulator set up with the help of fuzzy logic in real time. This visual conclusion is confirmed by the calculated integral quality indicators presented in Table 1.

A dynamic coefficient of control *Rd* is used for the evaluation of the decrease in dynamic error:

$$Rd = \frac{\Delta Y_c^{\max}}{\Delta Y_o^{\infty}} \qquad (6)$$

where $Y_c^{\max}$ is the maximum amplitude of one controlled variable in the closed-loop system and Y_o^{∞} is the amplitude of the same controlled variable in the open-loop system.

Fig. 5 shows the transient of the ACS for a 30% step disturbance *f* and a 30% parameters change in W_p^u. The new plant parameters have been calculated at t = 0.48h (see Fig. 5) and LF PID has provided a decrease in *Rd* by 10% by this time in comparison with MO PID. The online adaptation of the PID controller tuning parameters in both the ACSs has made it possible to decrease the settling time by 2, and *Rd* by 3 times in comparison with the non-adaptive system (see Fig. 6). The controlled systems equipped with PID controllers tuned by different methods are analyzed, in terms of dynamic response, under the effect of the same aforementioned perturbation and a change of 30% in the parameters of the process. In this case, the best dynamic response was shown by the system with the regulator tuned with Ziegler-Nichols method, the worst result was shown by the system with the regulator tuned with AMIGO method. This conclusion is also confirmed by the calculated integral quality indicators presented in Table 1.

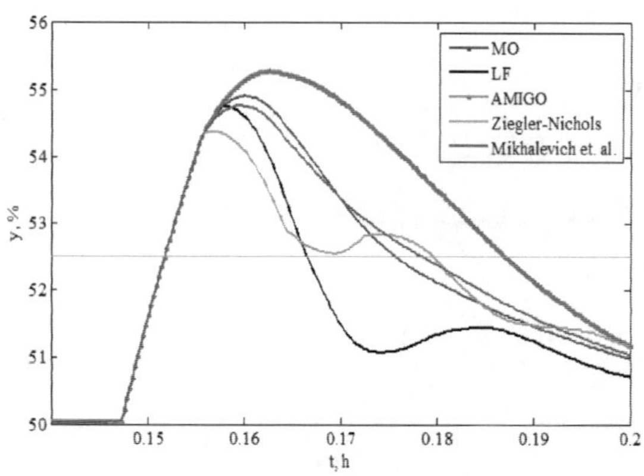

Figure 4. The transient of the ACS for a 15% disturbance.

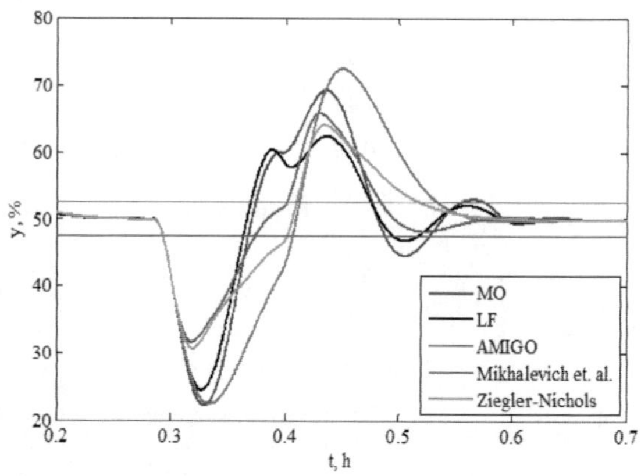

Figure 5. The transient of the ACS for a 30% step disturbance *f* and a 30% plant parameters change.

Table 1. Integral quality indicators

PID controller tuning method	15% disturbance		30% step disturbance on *f* and 30% plant parameters change	
	IAE	ISE	IAE	ISE
LF	0.13	0.31	2.30	32.00
MO	0.16	0.47	2.91	46.81
AMIGO	0.20	0.72	3.69	68.91
Z-N	0.15	0.37	2.20	25.32
Mikhalevich et. al.	0.17	0.47	2.59	34.23

A study on the controlled system stability was conducted too. The process, controlled with a PID regulator tuned by means of either Ziegler-Nichols method or Mikhalevich et al. strategy or by means of fuzzy logic in real time, is selected for the stability test. The achieved transients are presented in Fig. 7. The system, whose controller is tuned using fuzzy logic in real time, proves to possess the best stability. The calculated integral quality indicators assess the previous statement (Table 2).

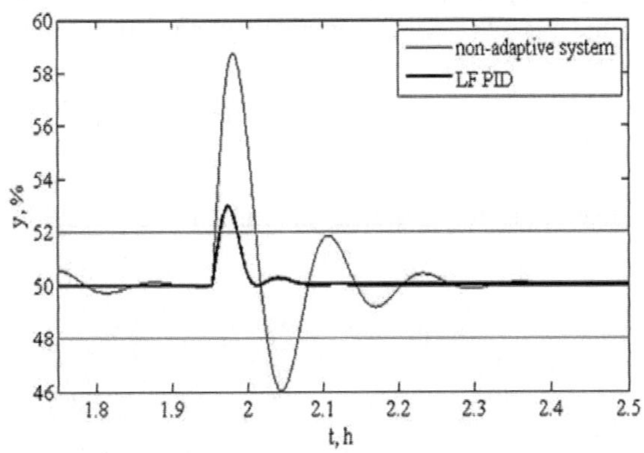

Figure 6. Transient of the ACS for a 15% disturbance and a 30% plant parameters change.

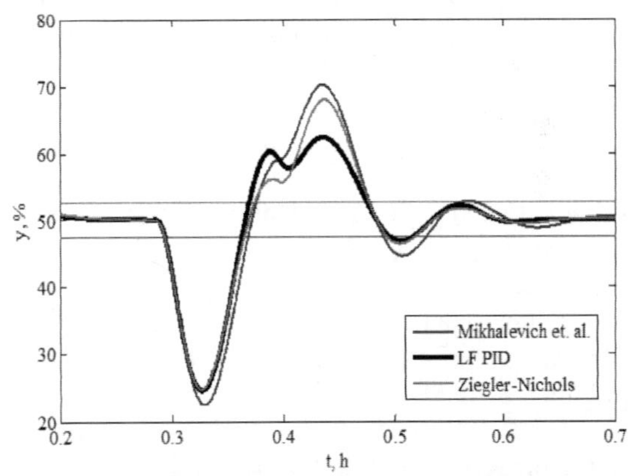

Figure 7. Checking the stability of the control system.

Table 2. Integral quality indicators

PID controller tuning method	30% step disturbance on *f* and 30% plant parameters change	
	IAE	ISE
LF	2.30	32.00
Z-N	2.45	36.19
Mikhalevich et. al.	2.99	47.94

In the near future the research on the suggested method is planned to be carried out in presence of disturbances in more complicated plant models, for example, in models of chemical processes [16, 17], water-treatment networks [18], power control channels of nuclear reactors and others. In addition, a comparison with the model predictive control (MPC) [19] will be investigated.

4. Conclusions

The suggested method of automatic control system synthesis, which is based on the application of a fuzzy adaptor-optimizer, allows developing automatic control systems for generic plants, which provide low sensitivity to the process parameters instability. A continuous process the adaptation of the controller parameters, in cases of significant instability, make the settling time and the dynamic coefficient of control 2-3 times smaller in comparison with a non-adaptive system. The application of the suggested adaptor-optimizer makes it possible to increase the control quality, when a significant error in the plant parameters identification might occur, by 10-15 % in comparison with the adaptive control system based on a PID controller where the controller parameters are calculated by magnitude optimum method, Ziegler-Nichols method, AMIGO method and others. In some rare cases, also the standard tuning methods show good results. However, fuzzy-logic based real-time tuning policies are suitable to be applied in any circumstance, thus being also more flexible and general than conventional offline tuning strategies.

5. Acknowledgements

This work was funded as part of the Federal government-sponsored program «Science» by Tomsk Polytechnic University.

References

[1] Åström K., Hägglund T. 2001. The future of PID control. Control Engineering Practice. 9(11) 1163–75.
[2] El-Bardini M., El-Nagar A.M. 2014. Interval type-2 fuzzy PID controller for uncertain nonlinear inverted pendulum system. ISA Transactions. 53(3) 732–43.
[3] Skogestad S. 2004. Simple analytic rules for model reduction and PID controller tuning. Modeling, Identification and Control. 25(2) 85–120.
[4] Umland J.W., Safiuddin M. 1990. Magnitude and symmetric optimum criterion for the design of linear control systems: what is it and how does it compare with the others. IEEE Transactions on Industry Applications. 26(3) 489–97.
[5] Panda R.C. 2008. Synthesis of PID tuning rule using the desired closed loop response. Industrial and Engineering Chemistry Research. 47(22) 8684–92.
[6] Shamsuzzoha M., Lee M. 2007. IMC-PID controller design for improved disturbance rejection of

time-delayed processes. Industrial and Engineering Chemistry Research. 46(7) 2077–91.

[7] Huang H.P., Jeng J.C. 2003. Identification for monitoring and autotuning of PID controllers. Journal of Chemical Engineering of Japan. 36(3) 284–96.

[8] Hjalmarsson H., Gevers M., Gunnarsson S., Lequin O. 1998. Iterative feedback tuning: theory and applications. IEEE Control Systems Magazine. 18(4) 26–41.

[9] Kansha Y., Hashimoto Y., Chiu M.S. 2008. New results on VRFT design of PID controller. Chemical Engineering Research and Design. 86(8) 925–31.

[10] Åström K., Hägglund T. 2004. Revisiting the Ziegler-Nichols step response method for PID control. Journal of Process Control. 14(6) 635–50.

[11] Panagopoulos H., Åström K.J., Hägglund T. 2002. Design of PID controllers based on constrained optimization. IEE Procedings – Control, Theory and Applications. 149(1) 32-40.

[12] Chen K.Y., Tung P.C., Tsai M.T., Fan Y.H. 2009. A self-tuning fuzzy PID-type controller design for unbalance compensation in an active magnetic bearing. Expert Systems with Applications. 36(4) 8560–70.

[13] Saad N., Arrofiq M.A. 2012. A PLC-based modified-fuzzy controller for PWM-driven induction motor drive with constant V/Hz ratio control. Robotics and Computer-Integrated Manufacturing. 28(2) 95–112.

[14] Xu X.Q. 2014. The application of MATLAB for fuzzy control system simulation. Applied Mechanics and Materials. 494-495 1306–9.

[15] Mikhalevich S.S., Baydali S.A., Manenti F. 2015. Development of a tunable method for PID controllers to achieve the desired phase margin. Journal of Process Control. 25 28–34.

[16] Goryunov A.G., Liventsov S.N., Rogoznyi D.G., Chursin Yu A. 2011. A dynamic model of a multicomponent no equilibrium extraction process in a pulsating column. Radiochemistry. 53(3) 278–83.

[17] Goryunov A.G., Mikhaylov V.S. 2012. The automatic control system of a multi-component nonequilibrium extraction process in the pulse column. Journal of Process Control. 22(6) 1034–43.

[18] Nadezhdin I.S., Goryunov A.G., Manenti F., Ochoa Bique A.O. 2016. Mathematical Modeling of EDM Method of Water Purification. Lecture Notes in Engineering and Computer Science: Proceedings of The International MultiConference of Engineers and Computer Scientists 2016. 254-258.

[19] Manenti F. 2011. Considerations on nonlinear model predictive control techniques. Computers and Chemical Engineering. 35(11) 2491–509.

IOP Conf. Series: Materials Science and Engineering **142** (2016) 012049 doi:10.1088/1757-899X/142/1/012049

Simulation of Uranyl Nitrate Crystallization Process in Linear Crystallizer Using Simsar Software

A Ochoa Bique[1], A Gozhimov[1], Yu Chursin[1] and O Schmidt[2]

[1]Tomsk Polytechnic University, 30 Lenina Street, Tomsk 634050, Russia
[2] A. A. Bochvar All-Russian Scientific Research Institute for Inorganic Materials, 5a Rogovaya Street, Moscow 123098, Russia

E-mail: anthonob@tpu.ru

Abstract. The paper deals with simulation of linear crystallizer work process for the research of technic operating modes and searching the most effective for material's nano-purity achievement. The model is realized by using SimSar software. Importance of device's geometry and process variables are marked. The model was included in the complex's composition of closed nuclear fuel cycle.

1. Introduction

Development of modern technology for nuclear waste handling is focused on getting of material in the pure state (nano-purity). In chemical industry, extraction [1] and crystallization [2] processes are used for this thing. The crystallization process is a complex process involving solid phase in the crystal form. Phase transformations and heat and mass transfer processes forming the conditions at the interface play a key-role in the formation of the material properties. To properly grow uranyl nitrate hexahydrate (UNH) crystals and operate uranium crystallization processes, it is important to carry out comprehensive studies of the phenomena involved in. For these reasons, crystallization of uranyl nitrate was studied in the papers of scientists from Russia and Japan [3-7]. A prominent role here is given to the theoretical studies using different mathematical model. The relevance of the task is related to the modeling of the crystallizer work process and detection by the method of the virtual experiment of unsteady regimes of its behavior. The virtual experiment implementation is due to the complexity of setting real experiments and a lack of knowledge about the process. At the present time a stand for refining of extraction and crystallization treatment process of spent nuclear fuel (SNF) is developing in Russia in the framework of "Proryv" project stream. In needs a computation-intensive for the model developing of all stand's parts and concurrent operation. It is expected of chemical process modeling and hardware modeling at the same time. The scope of the paper is the modeling of UNH crystals growth in linear crystallizer. SimSar software developing for modeling of closed nuclear fuel cycle (CNFC) production is used for end of purpose.

2. Mathematical Model of UNH Crystals Growth in Linear Crystallizer

A liquid nitric-acid uranyl nitrate solution (liquid solution) is ideally considered as a three-component liquid:

$$C_{UN}^{m} + C_{HNO_3}^{m} + C_{H_2O}^{m} = \sum_{i=1}^{NC} C_i^{m} = 1 \tag{1}$$

where C_i^m is a mass content of the i-th component in the liquid solution. Mass content values of uranyl nitrate (UN) and nitric acid (HNO_3) are evaluated by initial conditions of uranium (U) and HNO_3 concentrations.

The dependence between C_{UN}^m and $C_{HNO_3}^m$ can be described [7]:

$$C_{UN}^{m} = A - \psi \cdot C_{HNO_3}^{m},\ A = 1/(1+\alpha) = 0.785 \tag{2}$$

where the parameter α is the ratio of six molar mass of water to the molar mass of UN (it needs six molecules of water per the one molecule of UN to get the one crystal of uranyl nitrate hexahydrate). $\psi = (A - C_{UN}^{m,int}) \cdot C_{HNO_3}^{m,int\ -1}$ - the slope of the line, determined by the composition of the initial melt.

The growing rate of the crystal radius R is stated as follows:

$$\frac{dR}{dt} = k(C_{UN}^{m} - C_{UN}^{m,sat}) \cdot \frac{L}{U_s} - R \tag{3}$$

where k - the growth rate of the crystal phase inclusions (in the general case k is the function of temperature and liquid phase composition); U_s - a speed of solid phase movement; $C_{UN}^{m,sat} = C_{UN}^{m,sat}(T,\psi)$ – mass fraction of UN saturation in liquid solution (the function of temperature and composition).

Liquid and solid phases in the crystallizer active volume move with different velocities owing to their mass densities difference. The velocity of solid phase movement $U_s(t)$ can be written by the following expression considering Stokes law:

$$U_s(t) = U_{liq} + \frac{2}{9}\frac{g}{\upsilon}R^2\left[\frac{\rho_s}{\rho_{liq}} - 1\right] \tag{4}$$

where U_{liq} - liquid velocity; g – gravity; υ - liquid phase kinematic viscosity (function of the temperature and composition); ρ_s and ρ_{liq} – mass densities of solid and liquid phases, respectively.

The density of the liquid phase [5]: $\rho_{liq} = [1.023 + 2.936\cdot10^{-2}\cdot C_{HNO3} + 1.313\cdot10^{-3}\cdot C_U - (4.681\cdot10^{-4} + 3.475\cdot10^{-5}\cdot C_{HNO3})\cdot T]\cdot 10^3$.

The conservation principle applied for UN and HNO_3 leads to the Eq. (5) and Eq. (6).

$$\frac{dC_{UN}^{m}}{dt} = \frac{Q_0 \cdot C_{UN}^{m,int} - U_{liq}\cdot S\cdot(1 - 4/3\cdot\pi R^3(t)n)\cdot C_{UN}^{m} - U_s\cdot S\cdot 4/3\cdot\pi R^3(t)n\cdot A}{U_{liq}\cdot S\cdot(1 - 4/3\cdot\pi R^3(t)n)} \tag{5}$$

$$\frac{dC_{HNO3}^{m}}{dt} = \frac{Q_0 \cdot C_{HNO3}^{m,int} - U_{liq}\cdot S\cdot(1 - 4/3\cdot\pi R^3(t)n)\cdot C_{HNO3}^{m}}{U_{liq}\cdot S\cdot(1 - 4/3\cdot\pi R^3(t)n)} \tag{6}$$

where Q_0 is an initial volumetric flow rate; S – sectional area; n - amount of crystallization centers.

The energy balance for the active volume of the crystallizer is stated as follows and provide the temperature variation due to enthalpic bulk contributions, internal heat exchange and external (cooling jacket) heat exchange contributions:

$$\frac{dT}{dt} = T_{in} - T - \frac{\pi\cdot D\cdot\chi\cdot(T - T_{cj})}{\rho_{liq}\cdot C_{liq}\cdot Q_0\cdot\left(1 - 4/3\cdot\pi R^3(t)n\right) + 4/3\cdot\pi R^3(t)n\cdot C_s\cdot\rho_s\cdot S\cdot U_s} \tag{7}$$

where D – diameter of the cross section of the crystallizer active volume; C_s and C_{liq}– specific heat of solid and liquid phases; χ – heat conductivity of the media (solid + liquid phases) in the considered crystallizer working volume section; T_{cj}- temperature of the cooling jacket.

A resulting model of crystallization process was adapted for modeling of a concrete device:

- achieved differential equations is represented in Cauchy's form;
- device's geometrics are taken into account;
- a device's work process is implemented for all source and target faucet conditions;

- a heat exchange model is implemented.

3. Virtual Experiments

A resulting model is implemented in SimSar software. Figure 1 and 2 show a model general form. Rate is 1 l/h, temperature of cooling jacket in the 1 and 2 zones is 2 and 1.5°C properly.

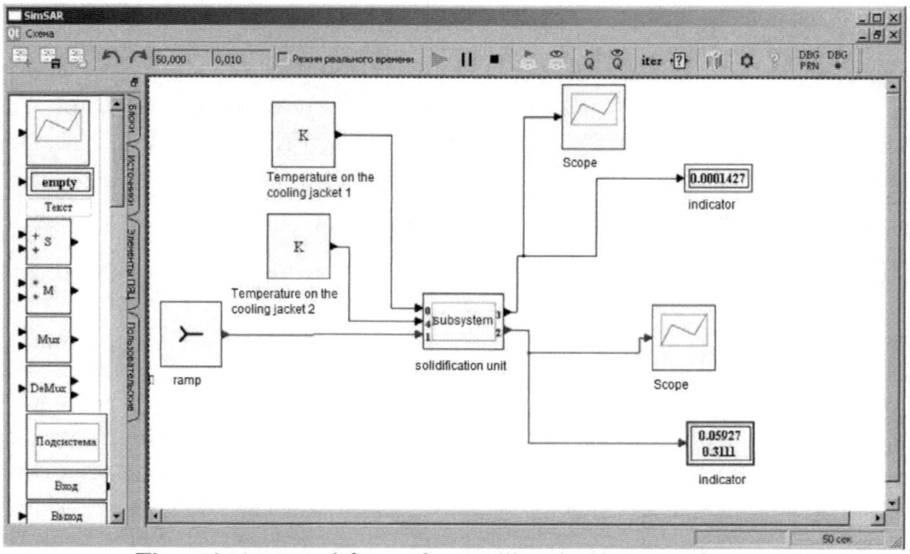

Figure1. A general form of crystallizer in SimSar software.

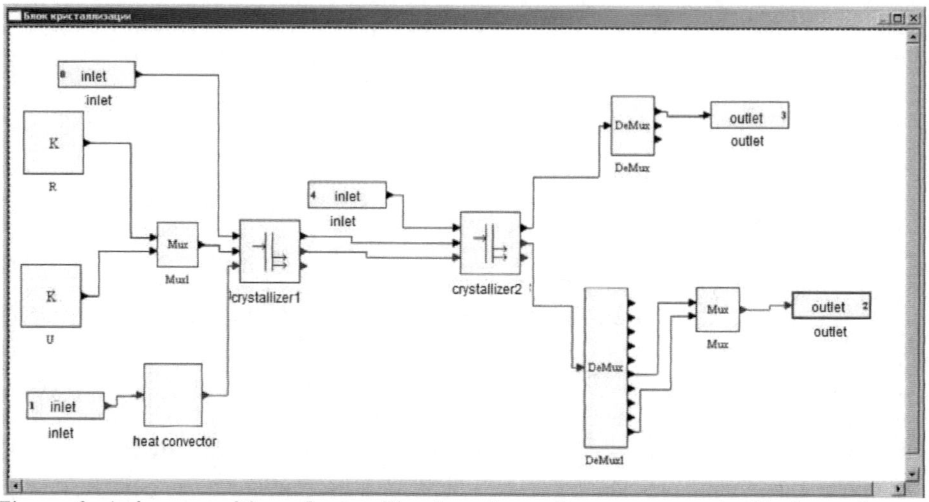

Figure 2. A decomposition of crystallizer's active volume on zones in SimSar software.

Results of simulation are showed in Figures 3 and 4. Figure 3 presents the trends of UN and HNO_3 mass fractions in the liquid phase % depending on time. A step-function response is not monotonous due to dependence between UN and HNO_3 concentrations. Figure 4 presents trend of UNH crystal growth.

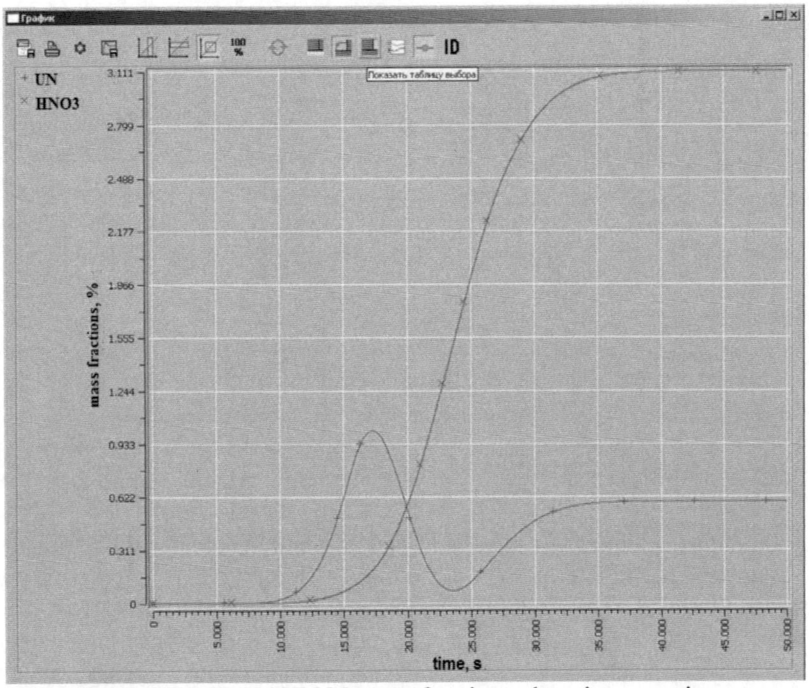

Figure 3. UN and HNO3 mass fractions changing over time.

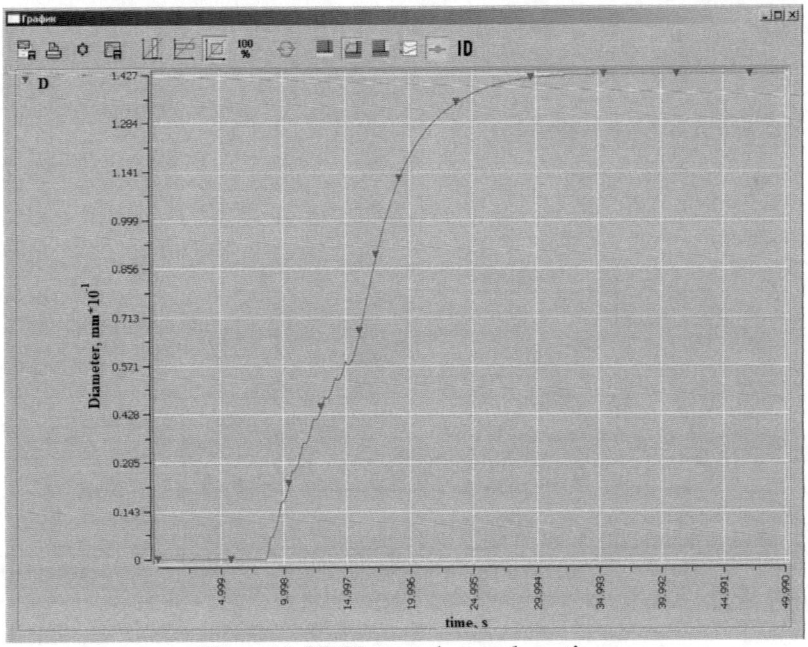

Figure 4. UNH crystal growth on time.

Model of linear crystallizer was implemented in Matlab Simulink for validity check of SimSar. (Figure 5)

IOP Conf. Series: Materials Science and Engineering **142** (2016) 012049 doi:10.1088/1757-899X/142/1/012049

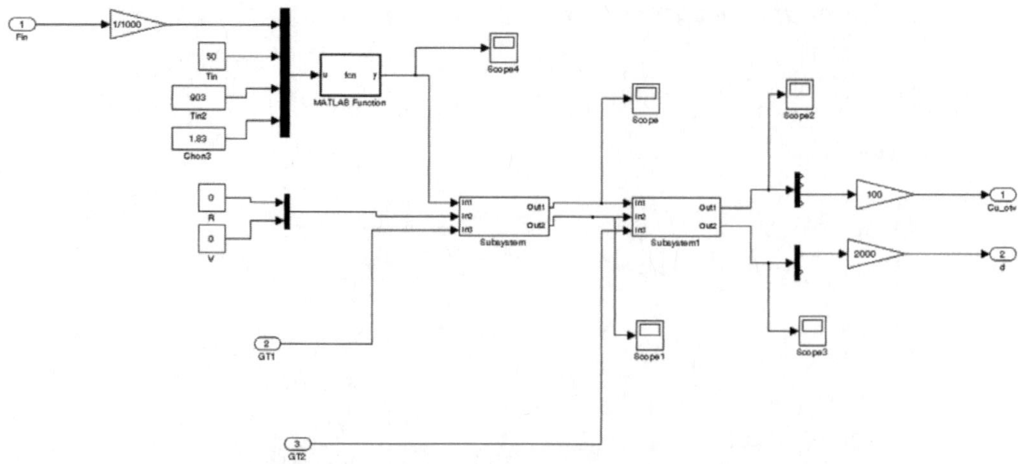

Figure 5. A model's implementation in Matlab Simulink.

Results of simulation in SimSar software and Matlab Simulink are equal (Figure 6). There are used the same ode1 equation solver and the one of simulation step. SimSar's rate of computing 26.2 times faster than Matlab (Table 1). The model of linear crystallizer in SimSar is developed on C++, but in Matlab Simulink it is developed by s-function. The model is failed verification due to deficiency of the real device, but results of simulation are equal with results getting in A. A. Bochvar All-Russian Scientific Research Institute for Inorganic Materials.

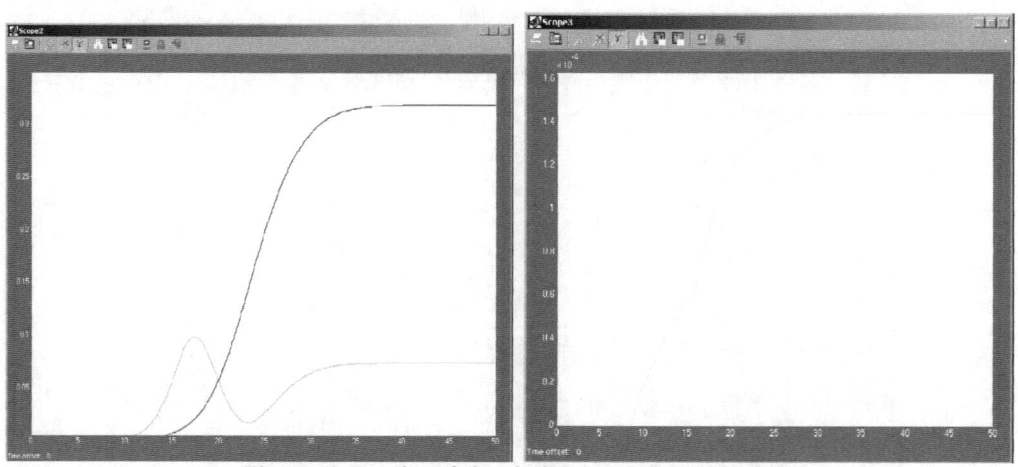

Figure 6. Results of simulation in Matlab Simulink.

Table 1. Inlet and outlet simulation characteristics in different software

	SimSar	Matlab Simulink
Simulation time (s)	50	50
Equation solver	ode1	ode1
UN mass fraction is coming out in outlet (%)	0.006	0.006
HNO_3 mass fraction is coming out in outlet (масс. %)	0.031	0.031
Diameter of UNH crystals (mm)	0.14	0.14
Rate of computing (s)	0.678	17.76

4. Conclusion

The model analyzing different temperature regimes of crystallization process allows choosing the most effective operating modes of linear crystallizer for material's nano-purity achievement. Results of simulation are equal with results getting in A. A. Bochvar All-Russian Scientific Research Institute for Inorganic Materials. Simulation of crystallizer work process in SimSar software faster than Matlab Simulink, thus this fact allows to integrate the model of linear crystallizer in calculation complex of CNFC production. Further, it is planned to test model against real data.

5. References

[1] Goryunov A G, Liventsov S N, Rogoznyi D G and Chursin Yu A 2011 A dynamic model of a multicomponent nonequilibrium extraction process in a pulsating column *Radiochemistry* **53(3)** 278-283

[2] Goryunov A G, Dyadik V F *et al.* 2011 *Math modeling of water-extraction nuclear fuel reprocessing processes* (Tomsk: TPU) p 237

[3] Yano K 2009 Research and development of crystal purification for product of uranium crystallization process *Proc. Int. Conf. GLOBAL 2009* pp 9093

[4] Shibata A, Kaji N, Nakahara M, Yano K, Tayama T, Nakamura K, Washiya T, Myochin M, Chikazawa T and Kikuchi T 2009 Current status on research and development of uranium crystallization system in advanced aqueous reprocessing of FaCT project *Proc. Int. Conf. GLOBAL 2009* pp 9154.

[5] Chikazawa T, Kikuchi T, Shibata A, Koyama T and Homma S 2008 Batch crystallization of uranyl nitrate *J. of Nucl. Sci. Technol.* **45(6)** pp 582-587.

[6] Homma S, Ishii J, Kikuchi T, Chikazawa T, Shibata A, Koyama T, Koga J and Matsumoto S2008 Flowsheet study of U-Pu co-crystallization reprocessing system *J. Nucl. Sci. Technol.* **45(6)** pp 510-517.

[7] Veslov S, Volk V, Kasheev V, Podimova T and Posenitsky E 2015 Mathematic simulation of crystallization refining process of spent nuclear fuel reprocessing desired products in linear crystallizer *Advanced Materials Research* **1084** pp 666-672.

IOP Conf. Series: Materials Science and Engineering **142** (2016) 012050 doi:10.1088/1757-899X/142/1/012050

Inaccuracy Determination in Mathematical Model of Labsocs Efficiency Calibration Program

M Kuznetsov, T Nikishkin, S Chursin

National Research Tomsk Polytechnic University, Tomsk, Russia

E-mail: chursinss@tpu.ru

Abstract. The study of radioactive materials quantitative inaccuracy determination caused by semiconductor detector aging is presented in the article. The study was conducted using a p-type coaxial GC 1518 detector made of a high-purity germanium produced by Canberra Company and LabSOCS mathematical efficiency calibration program. It was discovered that during 8 years of operation the efficiency of the detector had decreased due to increase of the dead layer of the germanium crystal. Increasing the thickness of the dead layer leads to 2 effects, which influence on the efficiency decrease: the shielding effect and the effect of reducing the active volume of the germanium crystal. It is found that the shielding effect contributes at energies below 88 keV. At energies above 88 keV the inaccuracy is connected with the decrease of the germanium crystal active volume, caused by lithium thermal diffusion.

1. Introduction

Gamma spectrometry is one of the most common methods of non-destructive analysis of nuclear materials and radioactive substances in the field of nuclear non-proliferation, as well as in the field of accounting and control of nuclear materials. Germanium detectors came into widespread acceptance in the gamma spectrometry because of their high resolution and efficiency. Eventually, the outer layer of the germanium crystal undergoes some changes because of the influence of various external factors.

These changes may be mechanical, chemical and radioactive. Mechanical changes mainly refer to micro-scratches and abrasions that occur during the operation process. Chemical changes are usually meant the detector outer surface oxidation, which leads to increase of the germanium crystal dead layer.

Also, there is an accumulation of irreversible radiation damage under the influence of ionizing radiation. Various forms of ionizing radiation interaction with the detector material are the reason of the defects in its crystal lattice structure, and, in most cases, these injuries are increasing proportionally to the radiation dose. The larger the detector, the earlier the degradation of its energy resolution begins when it is exposed by ionizing radiation. As a result of these changes, the detector changes its characteristics. In particular, its own detection efficiency changes, as well as the effectiveness of gamma-ray absorption by the detector. [1,5,6]

Efficiency calibration and measurement technique significantly affect the accuracy of measurements. There are 2 methods of efficiency calibration:

- Mathematical – using mathematical models and calculations.
- Experimental – using special calibration sources.

2. Mathematical method of calibration

Mathematical methods make unnecessary the calibration sources. However, these methods require certain mathematical models to be created. There are many methods that offer the simulation algorithms for the profile function of the detection system energy and spatial response. Simulating such functions is a challenge, so the mathematical packages are developed to make the user free from self-calculation.

In most cases, the manufacturer creates its own mathematical model of the profile function of the detection system energy and spatial response and the user, with help of the software package, creates a geometric model of the experiment. Then, the software package calculates the detection efficiency of detectors radiation using a mathematical model developed by the manufacturer and the geometric one created by the user.

3. Experiment

The aim of the work is to determine the inaccuracy of mathematical model of LabSOCS (Laboratory Sourceless Object Calibration Software) calibration program, developed in June, 17, 2007 for a coaxial high-purity GC1518 germanium detector produced by Canberra for 2015.

All the measurements and analysis presented in this paper were performed using the LabSOCS program of mathematical efficiency calibration in Genie-2000 software environment. The use of the program provides the most accurate method to conduct qualitative and quantitative gamma-analysis.

The software package is based on a mathematical model of the detector, which describes the profile function of the detector system spatial and energy response to ionizing radiation. Such model is created when the detector is in a particular state, so when we change the detector efficiency it is necessary to correct a mathematical model, because the profile function of the detector system spatial and energy response to ionizing radiation varies over time due to the detector parameters changes.

GC1518 detector is made of high purity germanium. While the detector is produced, the cylindrical sample external surface is sputtered with lithium that diffuses into the sample. The properties of lithium embedded in germanium allow creating sufficiently large areas (with the thickness of more than 1 cm) of almost full compensation and, consequently, the areas with the conductivity close to its own. It is connected both with extremely high mobility of lithium ions into tetravalent crystals and a low energy of its ionization [2].

When the detector is regularly kept at a room temperature, deep lithium thermal diffusion into the germanium crystal actively occurs. As a result, the detector dead layer has increased during 8 years and the active volume of the germanium crystal has decreased. The experimental part presents 2 measurements series of the gamma radiation standard samples in the source-detector geometry using high-purity germanium coaxial detector. 3D source-detector geometric model was created (Figure 1) for each series of the experiments in LabSOCS software package.

Figure 1. 3D geometric model of HpGe detector with the source

In the first series of measurements, standard gamma-radiation samples were in geometry of point source at a distance of 0 mm from the center of the detection unit housing normally. In the second series of measurements, the same standard gamma radiation samples were at a distance of 100 mm from the center of the detection unit housing normally. In particular, the following point sources of gamma radiation were measured: Mn-54, Co-57, Co-60, Zn-65, Cd-109, Sn-113 and Cs-137, Am-241. The measuring time for each source was 300 seconds. This spectrum range time is sufficient, because during this time the number of the counts in the peak is more than 10 thousand. Further, the measured samples spectra were processed using existing mathematical model of the detector, developed in June, 17, 2007, in the energy range from 50 keV to 2000 keV in LabSOCS software package. Obtained activity of measured sources was recalculated at the date of certification to compare it with certified values.

Table 1. Experimental results

Source	E (w), keV (%)	Certified activity at the date of certification, kBq	Measured activity at the date of certification (distance 0 mm), kBq	Inaccuracy (distance 0 mm), %	Measured activity at the date of certification (distance 100 mm), kBq	Inaccuracy (distance 100 mm), %
^{241}Am	59.0 (36.3)	104.0	89.7	13.7	74.1	28.7
^{109}Cd	88.0 (3.7)	118.0	114.9	2.6	98.7	16.4
^{57}Co	122.0 (85.5) 136.0 (10.6)	131.0	124.7	4.8	117.2	10.5
^{113}Sn	391.0 (64.9)	155.0	141.8	8.5	143.9	7.2
^{137}Cs	661.6 (85.1)	101.9	95.1	6.7	91.7	10.0
^{54}Mn	834.0 (99.9)	134.0	125.9	6.1	119.5	10.8
^{65}Zn	1115.0 (50.7)	135.0	116.9	13.4	111.9	17.1
^{60}Co	1173.2 (99.8) 1332.5 (99.9)	90.9	68.2	24.9	81.7	10.1

The table shows that obtained active source values are different from their true certified values with an inaccuracy of more than 5%, indicating that mathematical model is not in accordance with actual profile function of the detector system spatial and energy response to ionizing radiation.

Am-241 high measurement inaccuracy at source-detector distance of 100 mm could be explained by the shielding effect arising from the detector dead layer increase. The shielding effect is usually observed at low energies. Therefore, for gamma rays energies of above 88 keV, this effect makes a small contribution, as it is shown in the Table.

Co-60 high measurement accuracy at the source-detector distance is due to the cascade summation effect. To avoid it, it is necessary to carry out measurements at the distance of, at least, 10 cm from the source.

The Table shows that at energies above 88 keV the measurement inaccuracy is no more than 15%. This inaccuracy is connected with the decrease of the germanium crystal sensitive volume, caused by lithium thermal diffusion.

Thus, the detector has changed its properties since a mathematical model of the detection system was developed. As a result, the effectiveness of gamma-ray detection by the detector has changed.

4. The shift of the calibration curve on semiconductor detector efficiency

Figure 2 and Figure 3 show the efficiency curve of the detector detection in the point geometry for 2007 and 2015 at the source-detector distance of 0 mm and 100 mm normally, respectively.

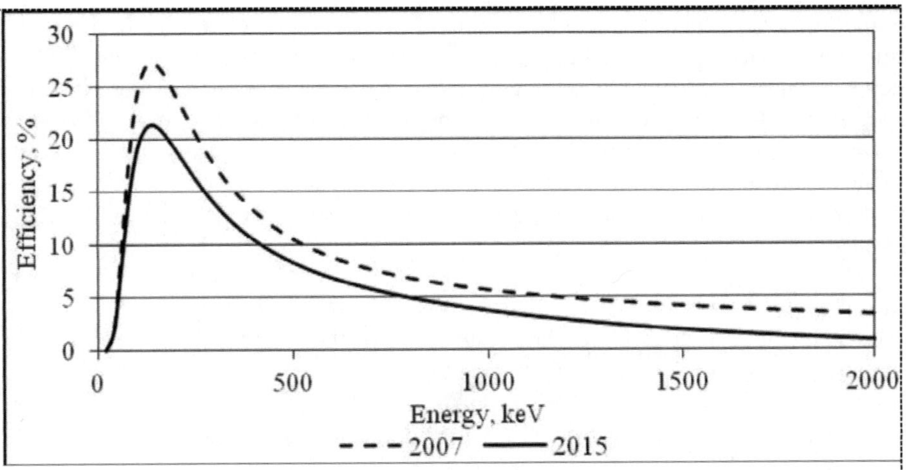

Figure 2. Detector efficiency at the source-detector distance of 0 mm

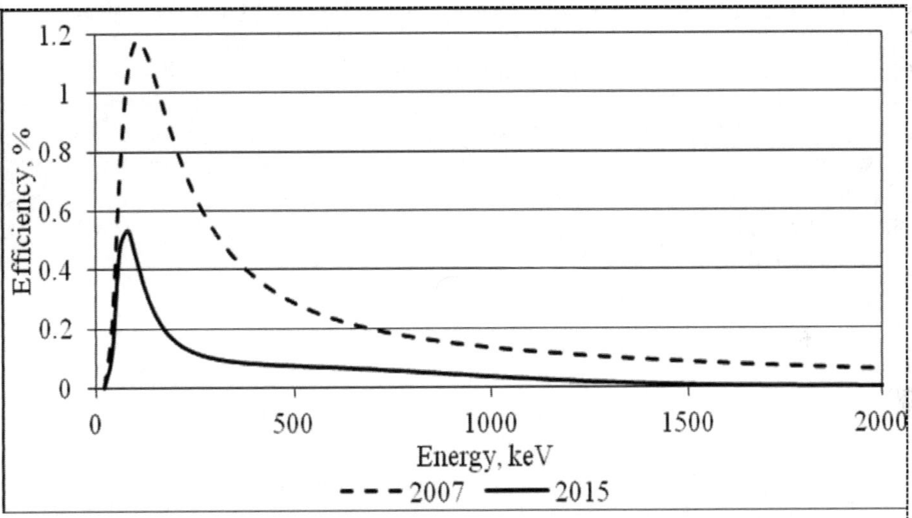

Figure 3. The detector's efficiency at the source-detector distance of 100 mm

The figures show that the efficiency of the detector has changed. There is a general decrease of the absolute detection efficiency of the detector in the whole energy range, which is connected with reducing the sensitive area of the detector. In addition, there is a greater reduction in the detection efficiency of the detector in the low energy region, where the shielding effect occurs due to the increase of germanium crystal dead layer.

Thus, the necessity to create an actual mathematical model of the detector that meets the real function of the detector system spatial and energy response is obvious. Also, it should be noted that it is necessary to increase samples measuring time, since the absolute detection efficiency has decreased, which will affect the rapidity of analysis.

5. Conclusion

As a result of the study, the following conclusions are made. The effectiveness of gamma-ray detection by the detector has changed during 8 years. There is a significant reduction in the detection efficiency of the detector at absolute value over the entire range of energies, but the decrease is stronger in low-energy region than in the high-energy one. This is due to the dead layer thickness increase, and, consequently, to the decrease in the sensitive volume of the detector [3].

Measurements inaccuracy of more than 5% occurs in almost the entire area. With such a high inaccuracy it is not recommended to conduct quantitative analysis of nuclear materials and radioactive substances, especially in the field of accounting and control of nuclear materials, as in this field the mass of nuclear materials must be determined up to the gram. [4] Consequently, for further detector use to obtain the most accurate measurements, it is necessary to develop its actual mathematical model corresponding to the actual profile function of the detector system spatial and energy response to ionizing radiation.

References

[1] Boyko V I, Silaev M E 2011 Methods and instruments for measuring the of nuclear and other radioactive materials *ISTC* 356

[2] Moskvin A V, Semenova V V and Teplykh V F 2005 The new chemist's and technologist's handbook. Radioactive substances. Harmful substances. Hygienic standards St. Petersburg *Professional* 1141

[3] Ngo Quang Huy 2010 The influence of dead layer thickness increase on efficiency decrease for a coaxial HPGe p-type detector *Nuclear Instruments and Methods in Physics Research Section A: Accelerators, Spectrometers, Detectors and Associated Equipment* Volume 621 Issues 1–3 pp 390–394

[4] Federal rules and regulations in the field of nuclear energy use NP-030-05 «Basic rules of accounting and control of nuclear materials»

[5] Tatarnikov D. A., Godovykh A. V. Radiation Detection System. Advanced Materials Research : Scientific Journal. – 2014. – Vol. 1040 : High Technology: Research and Applications 2014 (HTRA 2014). – p. 980-984

[6] Reilly D., Ensslin N., Smith H., Kreiner S. Passive Nondestructive Assay of Nuclear Materials. NUREQ/CR-550, LA-UR-90-732

IOP Conf. Series: Materials Science and Engineering **142** (2016) 012051 doi:10.1088/1757-899X/142/1/012051

The Influence of Meteorological Parameters and Other Factors on Soil Radon Dynamics

V S Yakovleva[1], P M Nagorsky[2], A G Kondratyeva[1], N V Mishina[1]

[1]National Research Tomsk Polytechnic University, 30 Lenin Ave., Tomsk, 634050, Russia
[2]Institute of Monitoring of Climatic and Ecological Systems (IMCES) Siberian Branch of the Russian Academy of Sciences, 10/3 Academichesky avenue, Tomsk, 634055, Russia

E-mail: vsyakovleva@tpu.ru

Abstract. The paper presents the results of the research in the degree of the effect of space weather meteorological parameters and factors on the dynamics of soil radon levels and α- and β-radiation flux densities in a seismically passive region. The cross-correlation analysis showed a significant correlation of β-radiation flux density with temperature in summer, and no correlation in winter. A significant relation between α- and β-radiation flux densities and pressure within the intra-annual range was not observed. The investigation of the high-intensity precipitation effect on radon volumetric activity and α- and β-radiation flux densities showed their abnormal increase. The dependence of the anomaly duration on the depth was revealed. The abnormal jumps in α- and β-radiation flux densities data series occur in the snow-melting periods as well. Low-intensity precipitations significantly violate the standard "diurnal variations" of α- and β-radiation soil fluxes and radon volumetric activity. Fourier analysis showed the diurnal (24 hours) and semidiurnal (12 hours) harmonics for the observed radiation values at a depth of 0.5 m. The obtained results can be used for interpretation of the data on the soil radon monitoring in order to predict earthquakes, etc.

1. Introduction

Zones of strains and compressions occurring in the lithosphere cause changes in the intensity of the gas flow (including radon) from the earth interior into the atmosphere [1–10]. Thus, variations of soil radon are an indicator of changes in stresses in the lithosphere and they are used as a precursor of earthquakes. The interpretation of the monitoring data on radon in seismic areas is a complex problem since the variations related to the change in stresses in the lithosphere are affected by "background" variations due to meteorological or other factors. These "background" variations are interferences and the spectrum and the factors affecting them are far from being fully studied. These background variations should be explored in seismically passive areas [11–12]. Therefore, the main research aim was to study the degree of the effect of meteorological parameters and space weather factors on the dynamics of soil radon in a seismically passive region.

2. Materials and methods

The investigation of the influence of meteorological parameters and other factors on soil radon dynamics was started from the end of 2008 at Tomsk Observatory of Radioactivity and Ionizing Radiation [13, 14]. Volumetric activity (VA) of soil radon was measured with help of highly sensitive intellectual scintillation α- and β-radiation detectors BDPA-01 and BDPB-01 (ATOMTEX, Republic of Belarus), respectively, with high time resolution of data series (10 min). Ionizing radiation (IR) detectors were mounted into boreholes of 0.5 and 1 m depths. The distance between boreholes was about 1.5 m. Radon isotopes and its decay products (DPs) radiometer EQF 3200 (SARAD, Germany) was used for calibration to convert the measured α- and β-radiation flux densities (FD) into soil radon VA [15].

In order to determine the degree of external factors influence the monitoring of meteorological (air temperature, pressure, air relative humidity, wind speed and direction, surface temperature and soil temperatures), actinometrical and atmospheric-electrical values was performed via automated information measuring system of IMCES. The precipitation data were taken from http://rp5.ru/8218/ru.

3. Results and analysis

The analysis of seasonal variations of α- and β-radiation FDs showed that, on the whole, seasonal changes by both types of IR measured at 2 depths are very close. However, variations of β-radiation FD in soil have more dispersion compared to the α-radiation FD. The dispersion of variations at a depth of 0.5 m for both types of radiation is greater than that at a depth of 1 m.

The following seasonal pattern has been identified. After melting of the snow cover all the IR detectors recorded a significant increase in α- and β-radiation FDs: at a depth of 1 m, it was ~2-fold, and at a depth of 0.5 m, it was 2.5–3-fold. This is due to the filling of the top layer pores with the soil water, which reduces the exhalation of soil radon into the atmosphere and causes it to accumulate in the soil.

It is well known that pressure and temperature have a significant impact on the variation of soil radon. A decrease in temperature is accompanied by gas compression, whereas, an increase is followed by its extension, therefore, when the temperature of the soil decreases, new portions of the atmospheric air enter the soil, and quite the opposite process occurs when heated. These processes have distinct diurnal variations.

To study the dependence of the α- and β-radiation FDs on meteorological parameters a cross-correlation analysis was performed using MATLAB. The example of cross-correlation between β-radiation FDs at depths of 1 and 0.5 m and the temperature in winter and summer of 2013 is shown in Fig. 1 a–d. It shows that in summer, there is a correlation of the β-radiation FD with temperature, and in winter, no correlation is observed. An interesting pattern can be observed when considering different depths in summer: the correlation with temperature at a depth of half a meter is significantly greater than that at a depth of 1 meter. Moreover, at a depth of 0.5 m, the correlation is positive with an average coefficient of 0.6, and at a depth of 1 meter, the correlation is negative with the same coefficient. The obtained result is important both to interpret the data on soil radon monitoring and to improve the models of radon transport.

The example of cross-correlation between α-radiation FDs at the depths of 1 m and 0.5 m and the temperature in winter and summer is shown in Fig. 2 a-d. The relation of α-radiation FDs and the temperature of the atmosphere can be observed only at the depth of half a meter during the summer months. The average value of correlation coefficient is about 0.6. The relation between alpha-field at a depth of 1 m and the temperature was not significant. In the winter months, the effect of the atmosphere temperature on the dynamics of the soil alpha and beta fields was not observed.

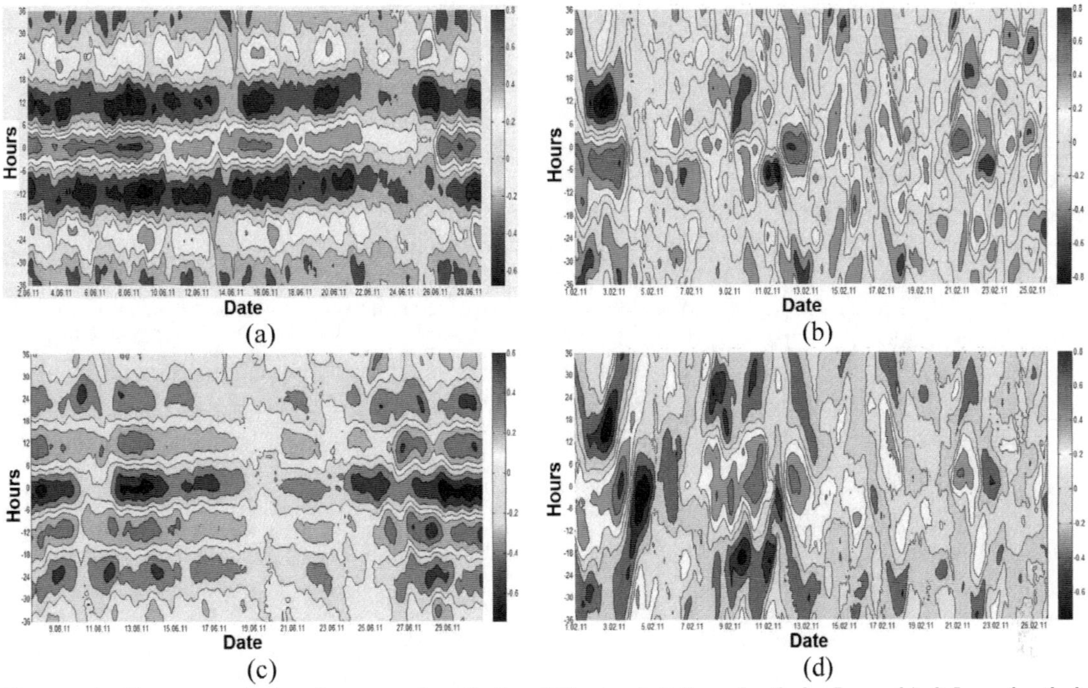

Figure. 1. Cross-correlation between β-radiation FD at: a) 0.5 m depth in June; b) 0.5 m depth in February; c) 1 m depth in June; d) 1 m depth in February and air temperature

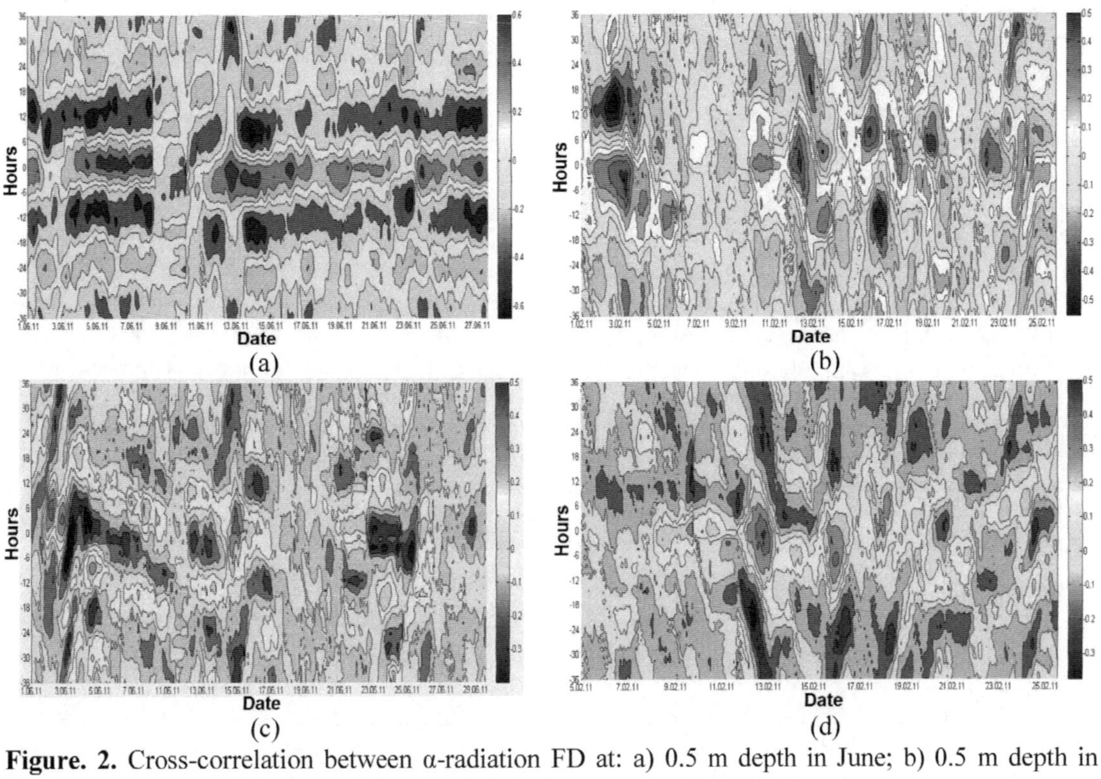

Figure. 2. Cross-correlation between α-radiation FD at: a) 0.5 m depth in June; b) 0.5 m depth in February; c) 1 m depth in June; d) 1 m depth in February and air temperature

IOP Conf. Series: Materials Science and Engineering **142** (2016) 012051 doi:10.1088/1757-899X/142/1/012051

The cross-correlation analysis also did not confirm a significant relation between α- and β-radiation FDs and pressure within the intra-annual range.

Consider the influence of precipitations on the variations of soil radon VA and α-, β-radiation FDs. Fig. 3 presents variations of radon VA at a depth of 1 m measured by the radiometer EQF 3200 (Rn 1 m), the series of β-radiation FD at 1 m depth and α-radiation FDs at 0.5 m and 1 m depths, recovered through multiplication by calibration coefficients. All radiation values showed abnormal abrupt increase for the precipitation intensity of 47 mm (13 of July, 2011). The abnormal increase in the radon VA was ~4.5-fold, α- and β-radiation FDs at the depth of 1 m increased by a factor of ~2.5 and ~1.6, respectively; α-radiation FD at the depth of 0.5 m went up by a factor of ~2.8. It can also be seen that the anomaly duration depends on the depth. The flux of α-radiation at a depth of 0.5 m recovered to its original value after 2 days, and the flux of α-radiation, as well as the flux of β-radiation and radon VA, at a depth of 1 m recovered only after 5 days. It is reasonable to assume that the anomaly can be even greater, and its duration can be shorter at the depths up to 50 cm. These abnormal jumps could be misinterpreted as the earthquake precursors in the seismic zone.

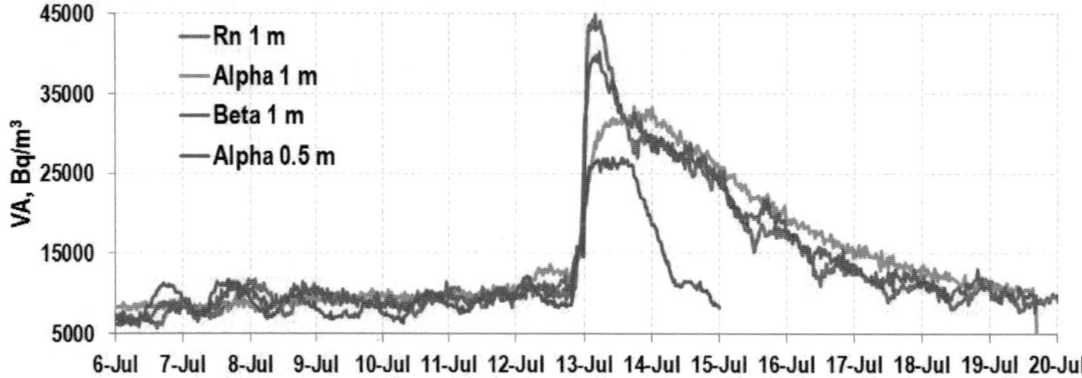

Figure. 3. The influence of high-intensity precipitations on dynamics of soil radon volumetric activity, α- and β-radiation FDs

It was found that low-intensity precipitations do not cause the abnormal jumps but significantly violate the standard "diurnal variations" of α- и β-radiation soil fluxes at a depth of 0.5 m, leading to its insignificant increase.

Consider the results of Fourier analysis for the time series of α- and β-radiation FDs at different depths. The examples are presented in Fig. 4. In the series of α- and β-radiation FDs at half a meter, the diurnal harmonic (24 hours) associated with solar tides can be clearly observed. The peak of the solar tide is caused by non-uniform heating of the atmosphere and the ground surface by solar radiation. The semidiurnal harmonic (12 hours) for β-radiation FD, which is also caused by the influence of the solar radiation reaching the earth's surface, can be clearly observed. It is difficult to distinctly identify the periodicals corresponding to the solar and lunar tides in the series recorded at a depth of 1 m. This fact confirms the minor effect of the tides on the radon dynamics at a depth of 1 m.

4. Conclusion

The analysis of the study results on the influence of meteorological parameters and space weather factors on the soil radon dynamics in a seismically passive region allows making the following conclusions:

- It was revealed a significant correlation of the β-radiation flux density with temperature in summer, but in winter, no correlation is observed. The cross-correlation analysis also did not confirm a significant relation between α- and β-radiation FDs and pressure within the intra-annual range.

- When processing the results of monitoring in the earthquake zones, it is essential to consider that high-intensity precipitations of more than 20 mm may cause abnormal jump of the soil α- and β-background and radon concentration at a depth of up to 1 m. The shape and duration of the anomalies are similar to the geochemical precursors of earthquakes.
- The low-intensity precipitations significantly violate the standard "diurnal variations" of the soil α- and β-radiation flux densities at a depth of 0.5 m, and lead to radiation increase.
- The abnormal jumps caused by high soil moisture can also occur in the snow-melting periods.
- Fourier analysis revealed the diurnal (24 hours) and semidiurnal (12 hours) harmonics in the time series of α- and β-radiation flux densities at a depth of 0.5 m.

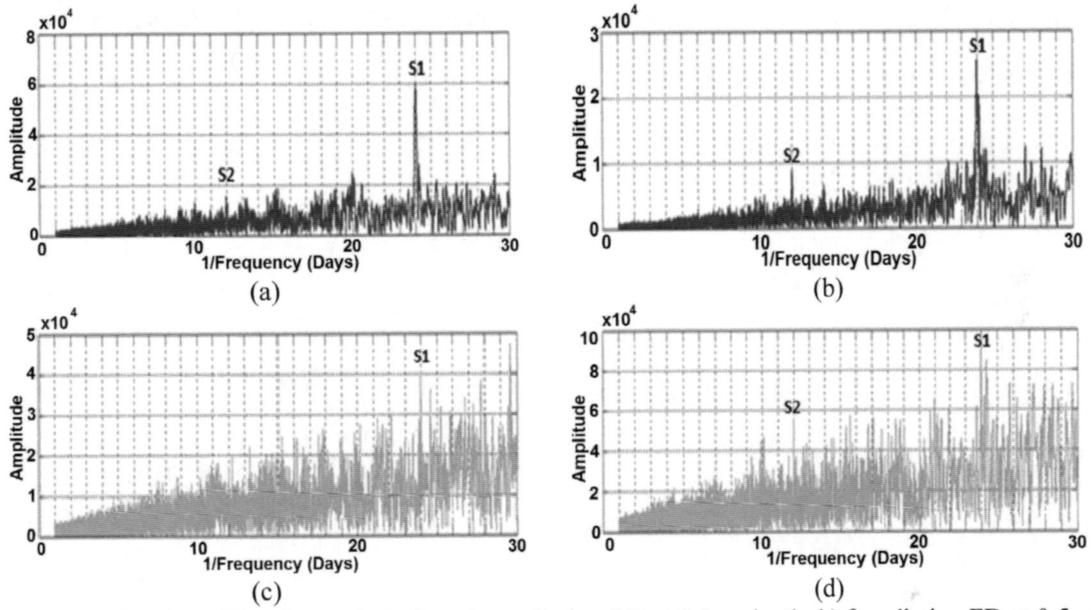

Figure. 4. Results of Fourier analysis for: a) α-radiation FD at 0.5 m depth; b) β-radiation FD at 0.5 m depth; c) α-radiation FD at 1 m depth; d) β-radiation FD at 1 m depth

The obtained results could be useful for interpretation of soil radon monitoring data with an aim of prediction of the stress-deformed state of the Earth's crust.

Acknowledgments
This work is funded by Russian Federal Target Program № 14.575.21.0105 (RFMEFI57514X0105) of Ministry of Education and Science of Russian Federation.

References
[1] Durrani S A and Ilić R 1997 *Radon measurements by etched track detectors: applications in radiation protection, earth sciences and the environment* (Singapore: World Scientific) p 387
[2] Firstov P P, Yakovleva V S, Shirokov V A, Rulenko O P, Filippov Yu A and Malysheva O P 2007 *Annals of Geophysics* **50** 547–556
[3] Virk H S, Sharma A K, Walia V 1997 *Current Science* **72** 656–663
[4] Segovia N, Mena M, Pena P, Tamez E, Seidel J L, Monnin M and Valdes C 1999 *Radiat. Meas* **31** 307–312
[5] Zmazek B, Vaupotic J, Zivcic M, Premru U, Kobal I 2000 *Fizika B (Zagreb)* **9** 111–118
[6] Monnin M M 2001 Radon over volcanic and seismic areas *Radionuclides and heavy metals in Environment* ed M V Frontasyeva et al. (Kluwer Academic Publishers) pp 319–330

[7] Outkin V I, Yurkov A K and Krivasheev S V 2001 Dynamics of radon concentration in soil for predicting earthquakes *Third Eurosymp. on Protection against Radon: Proc.* 137–141
[8] Steinitz G, Begin Z B and Gazit-Yaari N A 2003 *Geology* **31** 505–508
[9] Yakovleva V S and Karataev V D 2007 *Journal of Volcanology and Seismology* **1** 67–70
[10] Isaev A V, Ilguzin D A and Yakovleva V S 2005 Peculiarities of radon transfer in zone of dynamic influence of breaking *11th International Scientific and Practical Conference of Students, Postgraduates and Young Scientists "Modem Techniques and Technologies": Proceedings* 172–174
[11] Yakovleva V S and Karataev V D 2004 *Radiatsionnaya Biologiya. Radioekologiya* **44** 216–220
[12] Yakovleva V S 2005 *Annals of Geophysics* **48** 195–198
[13] Yakovleva V S, Karataev V D, Vukolov A V, Ippolitov I I, Kabanov M V, Nagorsky P M, Smirnov S V, Firstov P P and Parovik R I 2009 *ANRI* **4** 55–60
[14] Yakovleva V S and Nagorsky P M 2015 *Bulletin of KRAUNTS Physical and mathematical sciences* **1** 65–71
[15] Yakovleva V S and Nagorsky P M 2015 *Bulletin of KRAUNTS Physical and mathematical sciences* **1** 54–64.

IOP Conf. Series: Materials Science and Engineering **142** (2016) 012052 doi:10.1088/1757-899X/142/1/012052

Segmentation Effect on Inhomogeneity of [110]-Single Crystal Deformation

D V Lychagin [1,2,3], E A Alfyorova [1], V P Nesterenko [1]

[1]National Research Tomsk Polytechnic University, Lenin Avenue, 30, Tomsk, 634050, Russia
[2]National Research Tomsk State University, Lenin Avenue, 36, Tomsk, 634050, Russia
[3]Tomsk State University of Architecture and Building, Solyanaya Square, 2, Tomsk, 634003, Russia

E-mail: dvl-tomsk@mail.ru , katerina525@mail.ru

Abstract: This work presents a detailed analysis of segmentation process in FCC single crystals with compression axis [110] and side faces $(\bar{1}10)$ and (001) considering effect of octahedral shear crystal-geometry and basic stress concentrators. Sequence of meso-band systems formation on side faces is determined. Macro-segmentation patterns are specified, that are common to the FCC single crystals under investigation. It is proved that rectangular shape of highly compressed crystals, elongated in direction of operating planes, is conditioned by orientation symmetry of compression axis, single crystal side faces and shears directions, which are characteristic for the given orientation. The specified patterns are characteristic only for the samples with initial height-to-width ratio equal to 2. When varying sample height relative to the initial one, segmentation patterns will also vary due to crystal geometry variations.

1. Introduction

Deformation topography of FCC metals and alloys varies in relation to stacking-fault energy, crystallographic orientation, deformation degree and conditions [1-5]. Deformation topography differs in different areas of crystal faces and within polycrystalline grains. Crystallite partition into areas with different orientations, number and density of sliding traces occur during plastic deformation. [6]. In each separate area deformation is realized by sliding systems, which differ from those in adjacent areas of the crystal. This phenomenon is known as segmentation. The areas can be regarded as deformation domains. They have both characteristic sliding systems and specific shear procedures. In the latter case, shear domains are studied, where sliding is realized with formation of macro- and meso-clots, macro- and meso-bands of deformation. Clots' distinguishing feature is realization of shears along parallel sliding planes in crystal interior layers. This tendency is enhanced with diminishing of stacking fault energy. Bands deformation is characterized by noticeable distortion of shear traces within them; so deformation mechanism requires separate examination. [6-8].

Magid K.R. et al. [9] has been investigated heterogeneity of copper single crystal deformation. The findings demonstrate shear macro-segmentation in a sample. End and central areas of the crystal have distinctions; however, [9] does not examine this phenomenon in detail.

In polycrystals of FCC metals a single grain is divided into domains with different lattice orientations [10]. Uneven grain deformation causes sub-structural features, which occur on the crystal surface as various deformation topographies. Authors [11] suggests a crystal plasticity model, which considers the single crystal as a stack of domains. Individual domains are deformed homogeneously and correlate velocity and tracing continuity with adjacent domains. All domains co-accommodate behavior of the imposed deformation. The model predicts crystal lattice reorientation, distribution of sliding traces, localization of deformation. Effect of crystal-geometric orientation on segmentation character and type of deformation topography structural elements FCC single crystal are examined. It is established that segmentation nature relates to crystallographic orientation of both compression axis and side faces. The principal is orientation of octahedral shear planes referred to end apexes and edges, which are areas of higher strains [7, 12].

Thus, segmentation process is an integral part of crystal plastic deformation and has its own characteristics and laws, which are related to internal physical-chemical and crystallographic material parameters and to exposure conditions. Segmentation results in formation of certain topography structural elements on the surface, causes distribution of local deformations and facilitates deformation uniformity. Transfer of Al single crystal segmentation behavior to nickel crystals of the same orientation, as well as examination of their evolution during plastic deformation, are of scientific interest.

This work submits an analysis of segmentation of Al and Ni FCC single crystal with compression axis orientation [110] and lateral faces $(\bar{1}10)$ and (001); sequences of deformation domains based on crystallography and geometry of octahedral shear as well as a role of basic stress concentrators are examined.

2. Materials and methods of research

Object of the study is FCC single crystals with compression axis orientation [110] and lateral faces $(\bar{1}10)$ and (001). Slipping is performed along crystal systems <110> {111}. Two octahedral planes, each with two shear directions, are equally loaded for single crystals with compression axis [110]. Schmid factor for uniaxial stress state scheme is equal to 0.41. If sample's height to width ratio is 2, it is possible to distinguish a volume, wherein octahedral shear planes have free exposure to free side faces (Fig. 1). When such crystallographic orientation, shear along octahedral planes is initiated by linear stress concentrator. In the case under examination, there are only two active edges, which are in the plane (001); arrangement of shear planes, relative to base stress concentrators, facilitates each of eight apexes to activate at least one shear system. This crystal-geometric configuration affects formation of topography and segmentation order. Deformation meso-bands being formed on all side faces is a characteristic element of deformation topography of single crystals with compression axis [110]. This paper presents the experimental results obtained from single crystals compressive deformation tests. The deformation was performed with Instron ElectroPuls E10000 testing machine, at rates $1.4{\cdot}10^{-3}$ c^{-1}, at room temperature. To reduce friction force graphite grease was applied. Deformation topography was examined under optical microscope Leica DM 2500P.

3. Experimental results and discussion

Figure 1a shows crystallographic images of crystals under examination. Obviously, on parallel edges $(\bar{1}10)$ two inclined shear traces are to be developed, which will be parallel to exposures of planes (111) and $(11\bar{1})$. The shear along these planes, on the faces (001), is to develop a single system of horizontal traces parallel to sample ends. Figure 1b shows 3-D orientation of one octahedral plane family in the single crystal with side faces $(\bar{1}10)$ and (001). In the central part of the single crystal there is a volume, wherein octahedral shear planes have exposures at all free side faces.

Both octahedral shear plane families share the single crystal into seven volumes: A, A_1, D, D_1, E, E_1, K (Fig. 1b). Area K corresponds to the area wherein both octahedral shear plane families have exposures to all free side faces. Areas D, D_1 and E, E_1 correspond to the areas wherein only one octahedral shear plane family has exposure to the face. Uniaxial compression scheme is realized in these areas. The planes, crossing hindered deformation areas with non-uniform multidirectional compression, have exposure to areas A and A_1. Therefore, these areas have predicted lower density of shear traces. The face (001) is divided into E, K and E_1, wherein horizontal shear traces are to be formed (Fig. 1c). In E and E_1 they will be predominantly formed according to one principle, and in K – to two.

The octahedral shear planes geometry is such that one of the planes of each family goes through end edge, being an area of maximum shear stresses [7, 12]. In this case, the shear plane is effectively activated by the linear stress concentrator [7, 12]. Different arrangement of shear domains (marked areas), relatively to testing machine plungers, causes equidistant shear in both directions along the close-packed direction in the central area. In other areas shears will be conducted with testing machine plungers. Now consider macro-segmentation of shear deformation in FCC single crystals with compression axis orientation [110], and faces $(\bar{1}10)$ and (001).

IOP Conf. Series: Materials Science and Engineering **142** (2016) 012052 doi:10.1088/1757-899X/142/1/012052

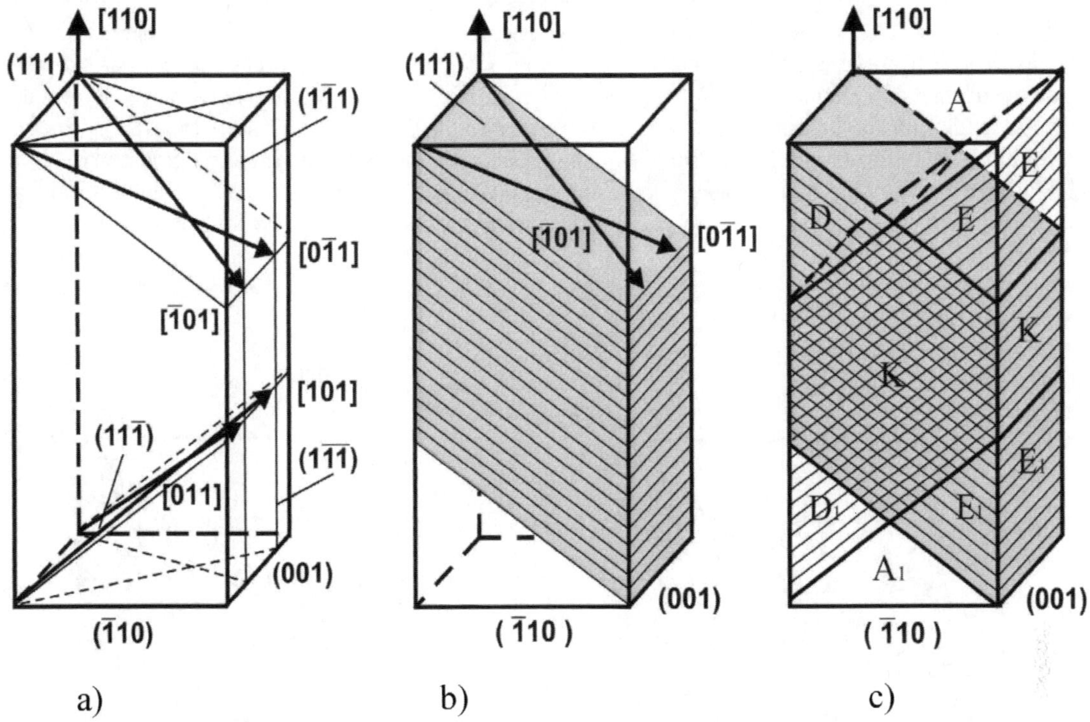

Figure 1 Crystal-geometry of single crystal with compression axis orientation [110] and side faces ($\bar{1}$10) and (001): plane orientation scheme {111} (a); scheme of one plane family orientation, with exposure to all free edges (b); for two families of equally loaded planes (c) [7]

The octahedral shear planes geometry is such that one of the planes of each family goes through end edge, being an area of maximum shear stresses. In this case, the shear plane is effectively activated by the linear stress concentrator [12]. Different arrangement of shear domains (marked areas), relatively to testing machine plungers, causes equidistant shear in both directions along the close-packed direction in the central area. In other areas shears will be conducted with testing machine plungers. Now consider macro-segmentation of shear deformation in FCC single crystals with compression axis orientation [110], and faces ($\bar{1}$10) and (001).

Analysis of aluminum and nickel deformation topography shows similarity of shear geometry in the samples being examined. Trace analysis of deformation topography on the side faces proves that most of the traces result from shear planes (111) and (11$\bar{1}$). On faces ($\bar{1}$10) two intersecting trace systems are formed, on faces (001) - one horizontal trace system. On faces (001) vertical shear traces are observed in local volumes near the ends. Distribution of shear traces is non-regular both in any system and on the faces. Fig. 2 shows a scheme of shear traces distribution in aluminum single crystal. Nickel single crystals deformation topography has minor differences. Thus, crystallographic factor, which specifies regular geometry of shear planes in the crystal volume, markedly contributes to the segmentation pattern. Fig. 2a is a scheme of primary macro-segmentation of shear deformation. Domains A and A_1 (Fig. 1c), with the lowest density of active shear planes, are formed in the end areas. There are small areas wherein a macro-level share is not observed. Obviously, in these domains the smallest magnitude of local shear deformations is to be predicted. As noted above, deformation in these areas is hindered because of uniform compression. Trace density is significantly higher near the lower operating plunger if compared with the upper fixed one. Domains D and D_1, E and E_1 are the areas wherein asymmetrical shear is realized along equally loaded planes. In any domain - D, D_1, E and E_1 - the shear along one plane, either (111) or (11$\bar{1}$) , prevails. In the central domain K symmetrical shear occurs along both planes. It is proved by comparable densities of shear traces along planes (111) and $(11\bar{1})$ on the faces $(\bar{1}10)$; aggregate pattern of shear traces in this area is a grid with rhomb or parallelogram elements. The face (001) is filled with horizontal shear tracks. Morphology of experimentally identified primary domains correlates with that, theoretically predicted by crystallographic analysis. With enhancing of deformation magnitude, additional shear domains occur in aluminum and nickel single crystals (Fig. 2c). From the very beginning of deformation, higher trace density is observed at the boundaries of domains A and A1, at sample apexes. They are marked as separate domains *B* and B_1, *C* and C_1,(Fig. 2a). Such high density of shear traces in these domains is caused by stress concentrations in sample apexes and edges, near testing machine plungers.

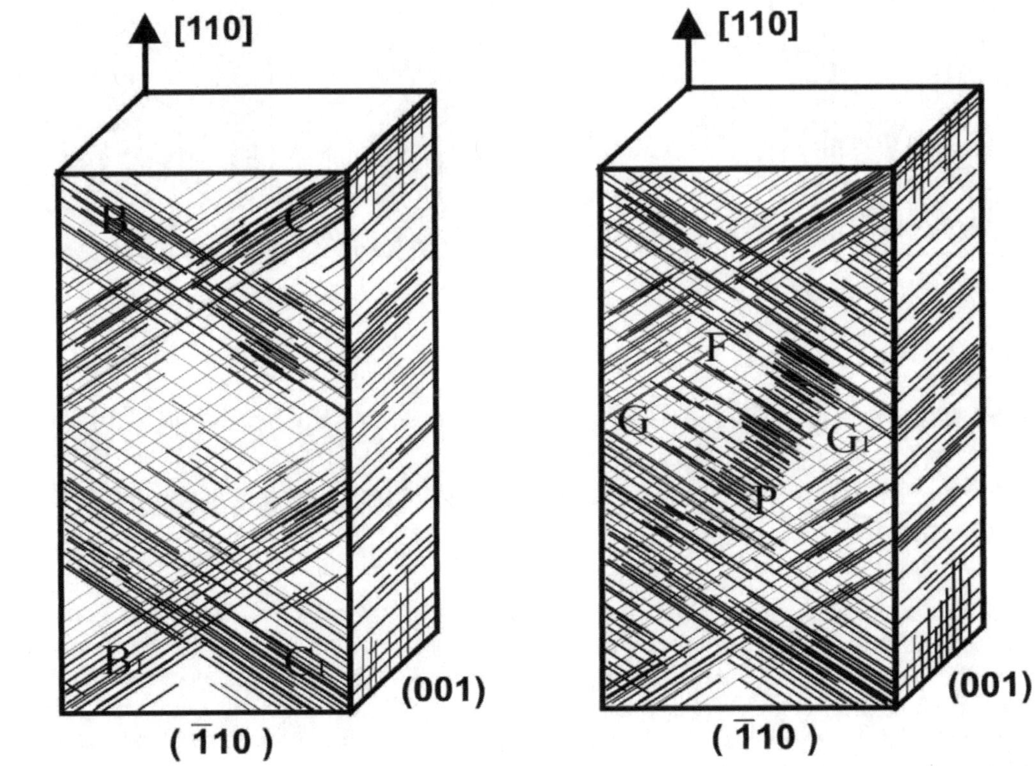

a) b)

Figure 2 Shear pattern on faces $(\bar{1}10)$ and (001) aluminum single crystal with compression axis orientation [110] after deformation to $\varepsilon = 0.03$ (a) and $\varepsilon = 0.06$ (b) [7]

In the course of plastic deformation, primary domains shear structure evolves and secondary macro-segmentation of deformation develops. Deformation band systems are formed on the face. Deformation band systems are clearly visible in the single crystals of nickel at $\varepsilon = 0.05$. In the central domain K four characteristic areas are observed: domains *G*, G_1, *F* and meso-band system *P* (fig. 2*b*). The first three are shear deformation domains, wherein face shear $(\bar{1}10)$ forms a grid with 0.5 ... 50 mc elements. They are areas of the primary domain K, wherein meso-bands are not formed. The meso-band *P* is oriented to any face diagonal. Examination of shear pattern on parallel faces $(\bar{1}10)$ proves that the diagonal band is a volumetric defect. The band is oriented at an angle of approximately 30^0 to the compression axis within a single crystal volume. Diagonal band boundaries are indistinct; however, if to draw a line through the middle of the band on the face $(\bar{1}10)$, this line will be curved relative to the direction $[\bar{1}10]$. Another diagonal band, ambidextrous, is less pronounced, i.e. shear symmetry is broken while diagonal bands formation (Fig. 3a).

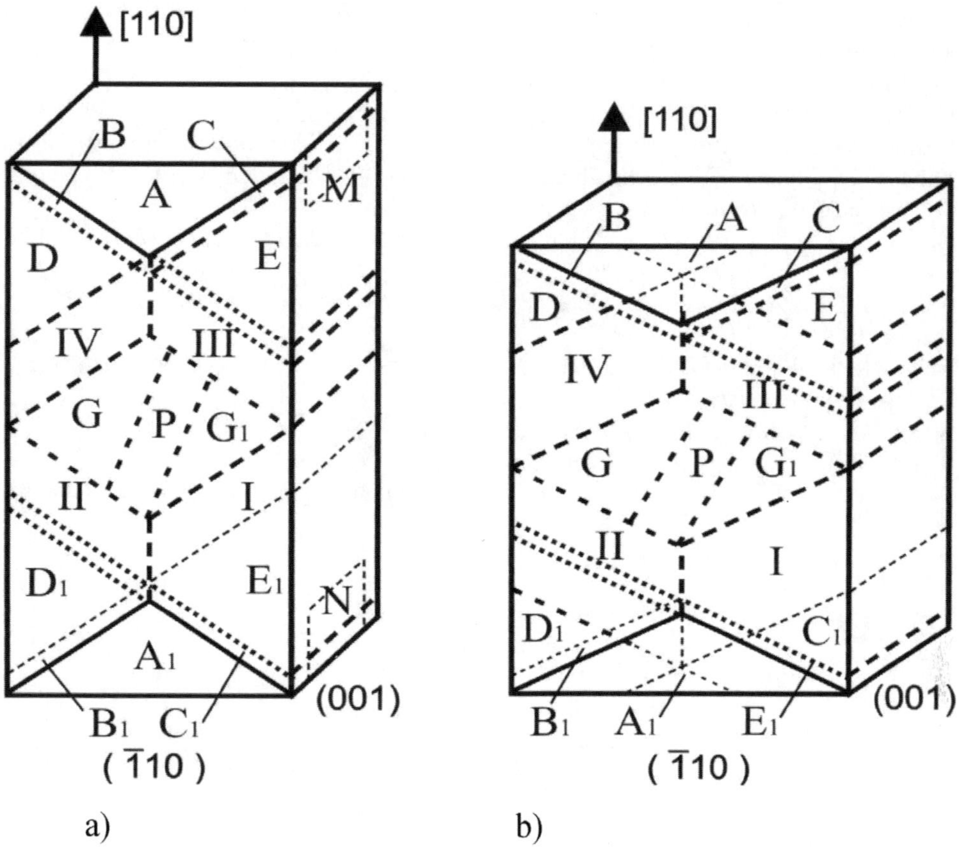

Figure 3 Scheme of secondary fragmentation in [110]-single crystals: deformation $\varepsilon \approx 0.05$ (a); $\varepsilon \approx 0.16$ (b) [7]

Meso-band system developments on side faces $(\bar{1}10)$ occur in a certain order. They fill area K and adjacent areas *D*, D_1, *E*, and E_1 in a specific pattern (Fig. 3) in the following sequence: I→II→III→IV. Meso-band systems initiate at the operating plunger (lower end of the sample), and then - at the fixed plunger. These are principal deformation domains; some four - eight domains (two - four at each end on the face (001)), are observed near the plungers of the testing machine. Such domains are formed by shear traces from unloaded octahedron according to uniaxial compression schemes only. These domains are expected when scheme of complex strain state, caused by end friction, is realized. This is verified by joint analysis of traces on all single crystal faces, including the end face. Studies confirm that this phenomenon is a general one. Such aggregate of deformation domains is characteristic only for samples with initial height-to-width ratio equal to 2. When varying sample height, segmentation pattern will also vary.

4. Conclusions

Thus, macro-segmentation regularities are analogous in the investigated FCC single crystals. Orientations of stress axis, single crystal side faces and shear directions are symmetrical, causing initial symmetry of shear spatial organization in single crystals with compression axis [110] and side faces $(\bar{1}10)$ and (001). With deformation enhancing, shear symmetry is broken to some extent, but prevailing effect of two equally stressed planes continues until large-scale deformation. This results in rectangular shape of highly compressed crystals, elongated in direction of operating planes.

Acknowledgements
The reported study was funded by RFBR, according to the research project No. 16-32-60007 mol_a_dk.

IOP Conf. Series: Materials Science and Engineering **142** (2016) 012052 doi:10.1088/1757-899X/142/1/012052

References

[1]. Kuhlmann-Wilsdorf D and Wilsdorf H 1953 *Acta Metallurgica* **1 4**.
[2]. Robert J. Asaro 1983 *Advances in Applied Mechanics* **23** 1.
[3]. Hughes D A, Hansen N 1991 *Mater, Sci. Technol.* **7** 544.
[4]. Gurald Franz, Farid Abed-Meraim, Marcel Berveiller 2013 *International Journal of Plasticity* **48** 1.
[5]. Zaitsev K, Klimenov V, Loshchilova M, Polovnikov C 2015 IOP Conference Series: Materials Science and Engineering **91 1** Article number 012056.
[6]. Bay B, Hansen N, Hughes D A, Kuhlmann-Wilsdorf D 1992 *Acta Metall. Mater.* **40 2** 205.
[7]. Lychagin D V 2006 *Phys. Mesomech.* **9** 3-4 95.
[8]. Lychagin D V, Tarasov S Yu, Chumaevskii A V, Alfyorova E A 2015 *International Journal of Plasticity* **69** 36.
[9]. Magid K R, Florando J N, Lassila D H, LeBlanc M M, Tamura N and Morris J W 2009 *Philosophical Magazine* **89 1** 77.
[10]. Kumar M A, Mahesh S 2013 *International Journal of Plasticity* **44** 95.
[11]. Kumar M A., Mahesh S 2012 *International Journal of Plasticity* **36** 15.
[12]. Alferova E, Lychagin D, Chernyakov A 2014 *Applied Mechanics and Materials* **682** 485.

IOP Conf. Series: Materials Science and Engineering **142** (2016) 012053 doi:10.1088/1757-899X/142/1/012053

Development of Misorientation in FCC Single Crystals Under Compression at Different Scales

D V Lychagin [1,2,3], **E A Alfyorova** [1], **D V Soprunov** [1]

[1]Tomsk Polytechnic University
634050, Russia, Tomsk, pr. Lenina, 30
[2]Tomsk State University
634050, Russia, Tomsk, pr. Lenina, 36
[3]Tomsk State University of Architecture and Building,
634003, Russia, Tomsk, Salt sq., 2

e-mail: dvl-tomsk@mail.ru , katerina525@mail.ru

Abstract. This article presents analysis of FCC single crystals local areas reorientation under compression. The reorientation process was examined at different scales, from sample size scale to that of dislocation substructure. It has been found that disorientation in meso and macro levels is determined by accumulation of misorientation at the level of dislocation subsystem. Our research allows quantifying of accumulated misorientation magnitude. The results of this study illustrate interrelation of rotational and translational deformation modes both at the same scale level and various scale levels.

1. Introduction

During plastic deformation, there are two constantly interacting processes that specify translational and rotational components of plastic deformation. At early stages of experimental investigation of plastic deformation phenomenon, only one process was observed and described because of imperfections of research methods and facilities [e.g. 1,2].

X-ray analysis made it possible to monitor crystals' reorientation, and development of optical and electron microscopy - to observe glide dislocations. It has been established that both processes are closely related, but the reorientation is the most characteristic for large-scale plastic deformations and higher dislocation densities [3-5]. While deformation starting, changes in a defect-free crystal are performed because of dislocation glide. However, in some cases, the reorientation process can be initiated at an early stage of deformation. This phenomenon is known as twinning mechanism, which occurs if a certain crystallographic orientation, low stacking fault energy and temperature are involved.

All facts aforesaid refer to the entire range of crystalline materials, which are investigated not only by physics and chemists, but also geologists and other scientists and engineers. This approach is applicable to the study of deformation processes in single crystals with FCC lattice. Despite its apparent simplicity (no grain boundaries) single crystals are sufficiently complex objects because of anisotropy properties and aspects of symmetry. Therefore, the plastic deformation is traditionally considered at different scale levels; that enables to explore certain aspects of the phenomenon.

Investigations of plastic deformation in single crystals is important for basic science and for practical application, if extrapolating the data to a single grain of a polycrystalline aggregate or a small set of single crystals. It determines the relevance of this research. In this way, the purpose of the work is examination of misorientation development at various scale levels in relation to shearing strain processes. This phenomenon is examined in FCC samples of pure metals (aluminum, nickel, copper).

2. Materials and method

FCC single crystals (copper, aluminum, nickel) were examined. Stacking fault energy (SFE) is large enough in aluminum and nickel, so, in these materials, cross dislocation glides are rather frequent. SFE of copper is lower; however, it also has a cellular dislocation substructure within the second stage of strain hardening curves. Deformation of these materials is performed at room temperature along octahedral gliding planes. Compressive strain was performed by Instron ElectroPuls E10000 testing machine at a rate $1.4 \cdot 10^{-3}$ sec^{-1} at room temperature. Graphite lubricant was used to reduce friction force. Strain relief pictures were examined under Leica DM 2500P optical microscope and Tescan Vega II LMU raster electron microscope. To determine the misorientation of local areas electron backscatter diffraction (EBSD) attachment to a microscope Tescan Vega II LMU was used.

3. Experimental results and discussion

Let's begin examination of misorientation in single crystals with laws' descriptions starting with a macroscopic level. Crystallographic orientation changes are closely related to samples' shape changes while plastic deformation. The latter depends on the strain pattern and on work-piece shape to which deformation is applied. Laboratory experiments enabled us to determine a load pattern (compressive strain) and a sample form (tetragonal prism). On the one hand, factor of shape's effect on compressive strength was identified and eliminated (sample is stable with the ratio of height to width equal to two). On the other hand, deformation with friction is complicated by difference in stress condition patterns in the end faces (uneven hydrostatic compression) and the central part of the sample (uniaxial compression). Let's examine the results of experiments that were performed on samples - tetragonal prisms, subjected to compression with mechanical friction. Gliding crystallography was examined as if developing according to the systems {111}<110>. Single crystals were oriented relative to compression axis along corner lines of a standard stereographic triangle with a set of various side faces.

Previously, the authors performed systematization of strain relief structural elements, depending on crystallographic orientation of compression axes and side edges. Proportions of strain relief structural elements were determined; that clarified how zones of these elements, at macro- and meso-level, were involved in plastic deformation of investigated single crystals [6]. Magnitude of plastic deformation heterogeneity of [111] nickel single crystals was determined experimentally for examined crystallographic orientations of various side faces at various scales, considering strain relief formation and compressive stress distribution [7]. It was found that localization zones do not occur within deformation domains. Development of inner structures occurs in such a way to reduce the strain heterogeneity and approximate an average deformation value to that of local deformation. Interface areas of neighboring deformation domains and end-face deformation areas are deformation localizations areas. The domain interface area has different glide systems; in end-face areas deformation is increased because of material transfer while mechanical friction. The role of stress distribution nature in plastic deformation inhomogeneity is shown in [6,7].

We consider single crystals with a ratio of height to width equal to two. Samples with such an aspect ratio show the greatest stability under compression in the case of isotropic materials. In the case of single crystals shear anisotropy has its own individual impact. As part of the problem statement, let's examine how compression axes' and lateral faces' crystallographic orientation affect sample stability while being compressed. Various crystallographic orientations of compression axis along the long sample axis determine various resistances to deformation, depending on orientation of close-packed planes and shear directions in the volume [6]. It's been found that the greatest stability of the

IOP Conf. Series: Materials Science and Engineering **142** (2016) 012053 doi:10.1088/1757-899X/142/1/012053

sample under compression is achieved when the sample symmetry axis, relative to the compression axis, coincides with gliding systems arrangement symmetry, relative to this axis. This feature depends on the crystallographic orientation of the side faces. Besides orientation of lateral sides affects development of single crystal volume deformation and causes its discontinuity.

The most detailed investigations of deformation heterogeneity were carried out on nickel [7].

It's been found that at the macro level in single crystals with compression axis $[\bar{1}11]$ deformation occurs more inhomogeneous for samples - tetragonal prisms. In this case, deformation axis orientation significantly reduces possibility of symmetric shear relative to side faces and macro-stripes forming large deformation domains. Intense folding deformation occurs in neighboring domains.

Development of compressing deformation contributes to changing in single crystals' $[\bar{1}11]$ orientation; that is expressed at the macro level in sample shape curving and crystallographic reorientation of its parts.

Studies have shown that the sample body is divided several reoriented fragments [8]. A lattice orientation is less changed in the central fragment. In adjacent fragments reorientation is more pronounced. Crystal lattice orientation $[\bar{1}10]$ is heavily biased. The lattice rotation occurs about the axis [110]. In end-face fragments changes of crystal lattice orientation are not observed.

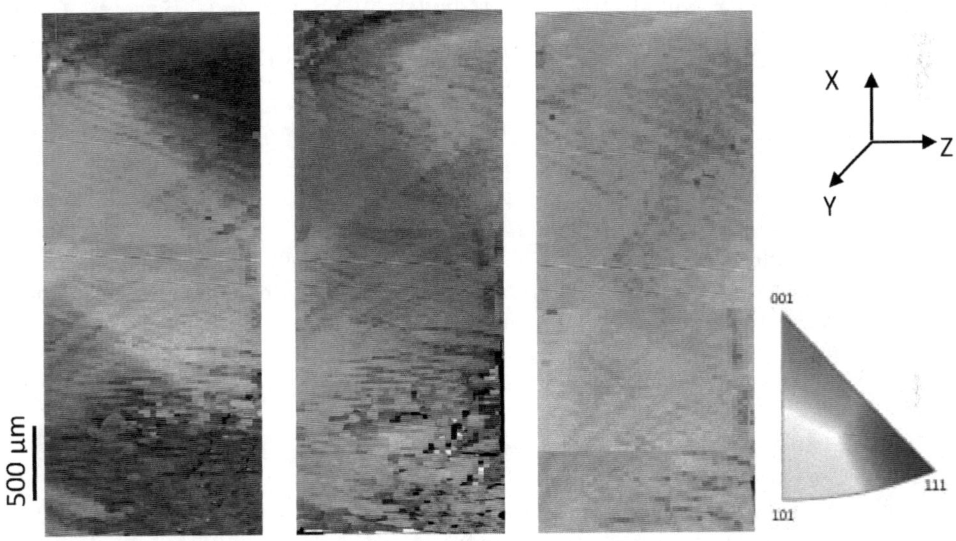

a) b) c) d)

Figure 1. Formation of reorientation areas relative to axes X, Y, Z (a-c), standard stereographic triangle (d), nickel e = 22%

With further increase of strain (22%), reorientation areas' macro-fragmentation is not changed (Fig. 1). Fragments' reorientation degree increases uniformly. Crystal lattice orientation remains practically unchanged in end-face fragments. Central fragment's lattice reorients in the direction [101] relative to the deformation axis X. It's been noted that the largest reorientation areas are observed relative to the compression axis X, and almost are not observed relative to the axis Z. Such nature of reorientation in different areas of the single crystal is conditioned by several reasons. Original orientation in end-faces is permanent, if the scheme is implemented non-uniform compression due to friction, which hinders the development of deformation processes in these areas. Fragments, located on borders of regions with different stress state patterns, are reoriented the most intensively with less deformation. This is caused by force-moment effect on a selected body, as values of stress tensor components vary in neighboring areas of this body. Reorientation is lagged in the central fragment. Reorientation of areas within single crystals of nickel, copper and aluminum occurs likewise.

Trusov's investigations [9-11] show that non-uniform stress field activates various sets of glide systems. In a local area, only one possible glide system operates in a certain time. Meanwhile, according to Taylor criterion, crystal body deformation can occur only under effect of five independent glide systems. In practice, usually, effects of more than three or four glide systems in local material volumes are not observed. In general, this is insufficient for any deformation. Therefore, this disadvantage should be compensated by reorientation of crystal areas. Based on this approach, the authors of [12] proposed a crystalline solid flow model, where disadvantage of active glide systems is complemented by rotary modes. We observe the same phenomenon in the above experiment on copper and nickel.

Compressive strain is not effective while deforming by six still loaded with shear systems in FCC single crystals with orientation of compression axis $[\bar{1}11]$ [7]. In such crystals deformation has a number of distinctive features. First, deformation occurs by shear with formation of macroscopic deformation bands systems. Second, the deformation bands systems are localized in certain areas of the crystal, which do not provide its complete deformation. Third, number and location of still loaded with shear systems do not provide orientation stability while deformation. Because of this, the single crystal is divided for disoriented fragments.

Primarily, a single crystals central area reorientation occurs. Reorientation in this body starts a mechanism of rotary deformation and also involves additional shear systems, being not active before. This is an illustrative interrelationship of translational and rotational modes of deformation on the same scale level.

Experiments prove that deformation domains should be sorted as translational and rotational ones [13]. Boundaries between of translational and rotational domains differ from grain boundaries of a polycrystalline sample. The difference is in the length and width of the boundaries, as well as in continuous transfer from one strain domain to another. In [14] authors identify these boundaries as a separate element of structural deformation with their geometric and structural parameters. Depending on domain orientation relative to applied stress, displacement deformation vectors have multidirectional or reverse directions in adjacent domains. Prevailing shearing orientation in the single crystal is always directed towards free side faces.

Areas of different deformation domains are clearly observed in the picture of displacement vectors' fields [14]. Comparison of deformation domain areas and spatial distribution of strain tensor components (shear and rotary) indicates an increase of all tensor components at the boundaries of deformation domains. This proves that increase of shear strain provides increase of rotational component. Thus, resulting deformation is higher at domains' boundary, if compared with that within a separate domain.

Data obtained with EBSD-analysis indicate accumulation of misorientation in domain boundaries. Our studies have found that misorientation increases when approaching the deformation domain boundary and significantly increases in the boundary.

Accumulation of misorientation in deformation domain depends on deformation mechanism at a meso-level. Previous studies [6] show that, depending upon single crystals' orientation, domain deformation occurs as a shearing along parallel shear planes with formation of slip bands in the form of meso- or macroscopic deformation bands. If moderate deformation degree, macroscopic deformation bands are shearing and do not lead to reorientation of mesoareas within the domain. With increase of strain degree, accumulation of dislocations' like-sign over-density occurs in local areas of the crystal, which contributes to development of misorientation. In addition, in the surface area, misorientation effect is manifested as deformation folds. Fig. 2 a-b represents the area of folds formation. EBSD-analysis is carried out in cross-section. It allows you to explore the disorientation with distance from the surface.

In this case we observe crystal local areas' reorientation in places of folds' forming. Geometric image of boundaries (Fig. 2a) illustrates distribution and magnitude of disorientation boundaries in the areas occupied by various types of structural relief elements. Color corresponds to value of misorientation angles (Fig. 2b). Here it is necessary to make a certain topical excursus. The boundaries are drawn with the program, which implies a certain disorientation interpolation on a segment, perpendicular to the boundary line. Thus, the boundaries, interpreted with the program, are diffused in a certain domain, determined by the area of interpolation. We can see that, for a given degree of deformation, disorientation angles do not exceed 5°...10°. There is disorientation magnitude to 5 ° in reoriented stripes; and larger angles along their boundaries (Fig. 2d). Consequently, folding contributes to formation of new boundaries within the single crystal.

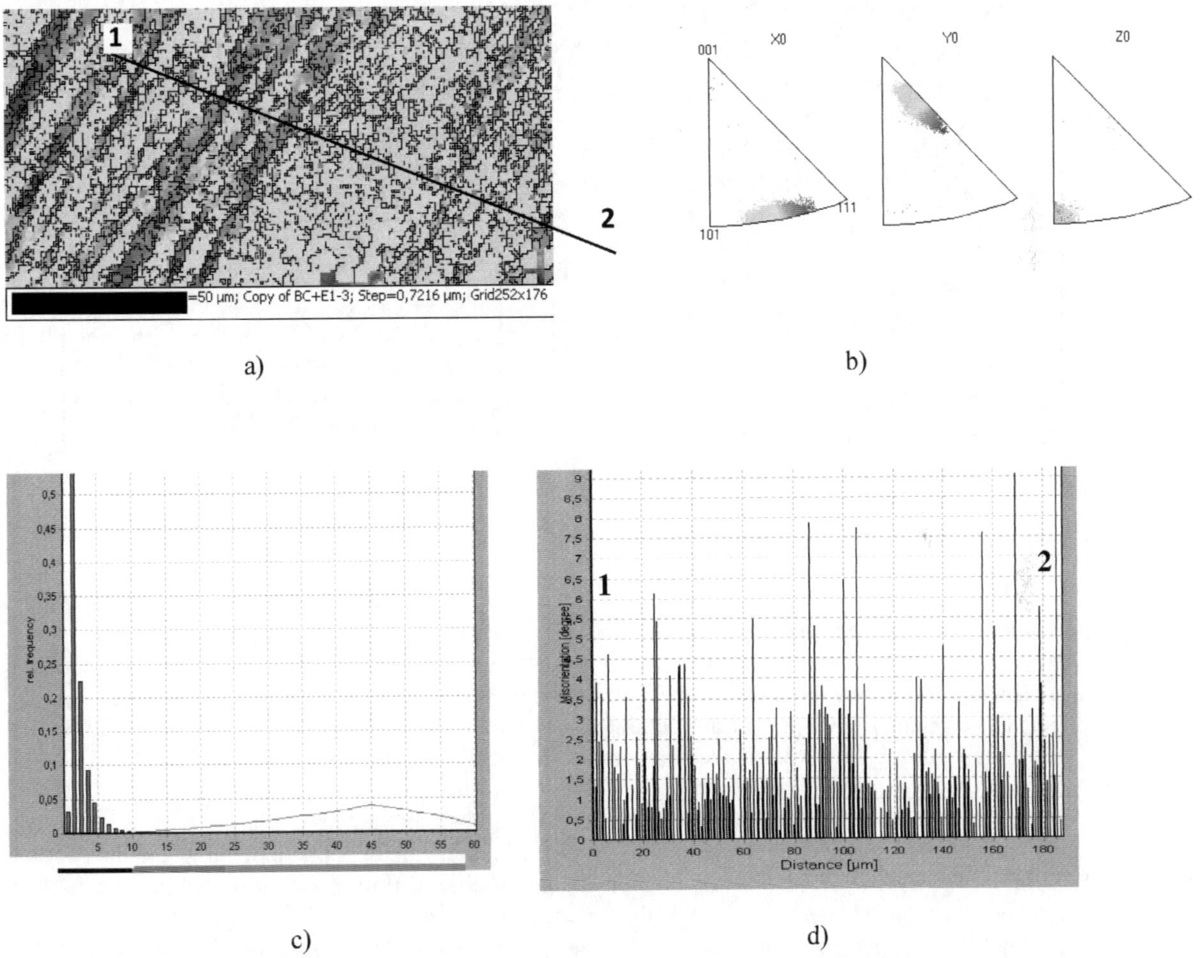

Figure 2. Geometric image of boundaries (a), standard stereographic triangle (b), frequency of angles misorientation distribution (c), transverse angles misorientation distribution 1-2 (d), Nickel, e = 16%

With use of EBSD-analysis observed disorientations were compared to disorientations of dislocation subsystem's level. EBSD record was conducted with a certain minimum grade value of 1 pixel = 5.5 microns. So, if disorientation boundary, after processing, goes along the boundary between adjacent pixels, then disorientation is at a distance of 5.5 microns. Nickel dislocation structure has the form of dislocation cells (Fig. 3a) in the initial degree of deformation. While increase of strain, dislocation over-density is accumulated in the boundaries of dislocation cells as is evidenced by change in contrast between adjacent cells or groups of cells (Fig. 3). Contrast changes usually become apparent when disorientations' magnitudes are 0.5°. This can be verified either with electron microscope analysis or direct measuring with the goniometer, locating the tilt angle at a certain orientation of the tilt axis in the foil plane. Size of the dislocation cell is 0.5 microns at the deformation degree being examined [15]. That is, 11 dislocation cells are placed in one pixel, which give the accumulated disorientation 5.5° observed in the experiment. Thus, EBSD-defined misorientation magnitude agrees with that of disorientation, accumulated during deformation in cellular dislocation substructure.

a)

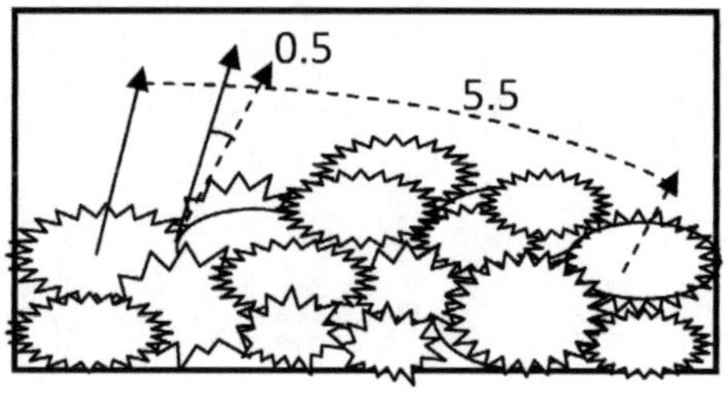

b)

Figure 3. Cellular dislocation substructure (a) and its misorientation accumulation diagram (b)

Over-density accumulation varies in cell walls of finite thickness. Depending on like-sign dislocations' distribution at the boundary, it has various magnitudes of disorientation while transitioning across the boundary; and transforms sequentially into blocked, fragmented and then sub-grain structure. Volumes, bounded with disorientations, constitute structure elements which can be subjected to reorientation under certain conditions. In fact, it occurs, in particular, under conditions of material super-plastic flow, when the grain boundary gliding or microporosity formation facilitates rotation of grains in the boundary.

At meso-level, individual rotational defects of very significant magnitudes can be observed, mainly those of disclination type. There are many studies that interpret phenomena in the dislocation structure as manifestations of disclination formations. By now, findings of investigations, obtained while experimenting with metallic materials, are incomplete. It is our belief that a disclination loop, experimentally obtained (by authors), is rather valuable; the loop was formed in iron-nickel ordered alloy with compression axis orientation [001] [16]. Magnitude of azimuthal disorientation, determined by splitting of reflexes on an electron diffraction pattern, is 10 °. In this case, disclination loop formation facilitated "knife-edge" shear boundary formation in the rest of the crystal volume, which resulted in a shift of dislocation cell wall by 0.2 ... 0.4 micron. Earlier authors classified this phenomenon as a mechanism for destruction of stable cellular structure of the ordered alloy. The experimental results also illustrate interrelation of translational and rotational deformation modes at the level of mesodefects of the dislocation-disclination subsystem. Deformation mechanism is a parallel action of two "knife-like" edges, which share cell boundaries in opposite directions. After the strong shear ceases, within the crystal a large-scale disclination loop is formed; which has a large misorientation angle between loop interior part and surrounding matrix.

4. Conclusions

Thus, analysis of scientific papers and authors' own results show interrelation of disorientation processes at different scale levels. Accumulation of misorientation at dislocation subsystem's level (lower scale level) leads to accumulation of misorientation at meso- and macro-levels. Our research allows quantifying of accumulated misorientation magnitude and monitoring of crystallographic orientation changes in deformation elements at all scales. Furthermore, the results of this study illustrate interrelation of rotational and translational deformation modes both at the same scale level and various scale levels.

Acknowledgements

The reported study was funded by RFBR, according to the research project No. 16-32-60007 mol_a_dk
The investigations were carried out using facilities of the centre for collective use "Analytical Centre of Geochemistry of Natural Systems" at Tomsk State University, Russia.

References

[1] Honeycombe R W K 1984 *The Plastic Deformation of Metals*. Edward Arnold & American Society of Metals, London.
[2] Higashida K, Takimura J, Narita N 1986 *Mater. Sci. Eng.* **81** 239.
[3] Ke-Shen Cheonga, Esteban P Busso 2006 *Journal of the Mechanics and Physics of Solids* **54** 671
[4] Koneva N A, Zhukovskiy S P et al. 1985 *Physics of Metals and Metallography* **60** **1** 157.
[5] Hatherly M 1982 *Strength Met. and Alloys. In: Proc. 6th Int. Conf., Melbourne*, Oxford, **2** 1181.
[6] Lychagin D V 2006 *Phys. Mesomech.* **9** 3-4 95.
[7] Lychagin D V, Alfyorova E A, Starenchenko V A 2011 *Phys. Mesomech*. 14 (1-2), 66.
[8] Lychagin D V, Tarasov S Yu, Chumaevskii A V, Alfyorova E A 2015 *International Journal of Plasticity* **69** 36.
[9] Trusov P V, Volegov P S, Shveykin A I 2013 *Computational Materials Science* **79** 429.
[10] Trusov P V, Shveykin A I, Nechaeva E S, Volegov P S 2012 *Physical Mesomechanics* **15** **3-4** 155.
[11] Trusov P V, Shveykin A I 2013 *Physical Mesomechanics* **16** **1** 23.
[12] Arul Kumar M, Mahesh S 2012 *International Journal of Plasticity* **36** 15.
[13] Teplyakova L A, Kozlov E V 2006 *Physical Mesomechanics* **9** **1-2** 53.
[14] Kibitkin V V, Solodushkin A I et al. 2011 *Optoelectronics, Instrumentation and Data Processing* **47** 4 83
[15] Starenchenko V A, Shaekhov R V et al. 1999 *Russ. Phys. J.* **42** **7** 653.
[16] Teplyakova L A, Koneva N A et al. 1988 *Russ. Phys. J.* **31** **2** 18.

Analysis of Structure Destroyed Metal after Diffusion Heat Treatment

A M Apasov[1], E V Kozlov[2], S N Fedoseev[1]

[1]Yurga Institute of Technology, National Research
Tomsk Polytechnic University Affiliate,
26 Leningradskaya str., Yurga, Kemerovo region, RUS
[2]Tomsk State University of Architecture and Building,
2 Solyanaya Square, Tomsk, Tomsk region, RUS

E-mail: mchmyti@rambler.ru, sfedoseev@tpu.ru

Abstract. It was accomplished research of the structure steel which carbonitriding and subsequent heat treatment was exposed for its cause's destruction to discover. For measure quality field of metal were used methods optical, appearing electronic microscopy and X-ray diffraction. Therefore one of the principal problems were research phase composition, grain and dislocation structure of a metal the gear teeth. Mechanism of rising hear cracks in the gear teeth on different stages her making and their trajectories of evolution were determined.

Introduction

Uninterrupted operation of the conveyor in the development of powerful of underground layers black coal is associated with a high degree of reliability guarantee [1] drive of reducer, providing continuous feed conveyor belt and prevents blockages. This in turn can only be achieved by using the manufacturing process quality toothed gears working in shear, shear and buckling.
In connection with this problem development defect-free manufacturing technology toothed gear is very important. One method of increasing wears resistance and surface strength of a thermo chemical treatment, particularly nitro carburizing.

The study of the structure of nitro cemented of metal layer of the tooth gear is dedicated to the work [2], in which the established that with increasing distance from the surface nitro carburized layer is changed that is graded structure formed in the material. Work [3] was the logical continuation of the cycle of research, the results of which were published in the article [2]. At the same time the main objective is to continue the study of the gradient structure-phase states that are formed in the constructional alloyed steel 20H2N4A martensite class during the high temperature carbonitridation and subsequent heat treatment methods of X-ray diffraction and optical microscopy. At the same time been installed depending a structurally-phase states steel 20H2N4A on the distance from the surface carbonitridation. It is shown that low nitro carburizing and temper have a significant impact on the parameters of the gradient structurally-phase states.

In the article [4] are shown data that allow establishing the mechanism of nucleation of microcracks and to determine the trajectory of their development. Nevertheless, until now fabrication stages remain unknown toothed gear, in which cracks arise and further features of their formation. The solution of this important problem and is dedicated to this research.

Methodology Research

It is known that one of the basic metal imperfections subjected to carburizing and, in particular, nitriding are internal residual stresses. Thus in the surface layer compressive stresses are created in the mid – stretching. They can reach significant values and, especially, in a complex and variable configuration of parts, assemblies and products. Arising large internal stresses are sources of nucleation, formation and development of micro cracks, which then (under certain conditions) are transformed into a trunk to form crack-free surfaces, eventually leading to destruction of the product as a whole.

Carbonitriding work piece surface (gear teeth) of steel 20H2N4A was carried out at a temperature of 920 °C. Thus the thickness layer was nitro cemented 1500 microns. Thermal treatment of the samples after carbonitridation consisted of high tempering at 620 °C (1 hour), 820 °C by hardened in oil and tempering at a low 180 °C for 1 hr (Table. 1).

Table 1. Procedure of sampling for investigation

Number of the sample	Sample status	The distance from the surface of tested samples
Original	Condition on delivery	The Centre sample
Sample 8	Cutting the gear teeth	1) Surface; 2) 1.6 mm from the surface of the tooth
Sample 12	Nitro cementation 920 °C	1) Surface; 2) 2.07 mm from the surface of the tooth
Sample 13	Tempering at 620 °C, 1 hour	1) Surface; 2) 2.17 mm from the surface of the tooth
Sample 14	Hardening from 820 °C the oil	1) Surface; 2) 2.0 mm from the surface of the tooth
Sample 15	Tempering at 180 °C, 1 hour	1) Surface; 2) 2.1 mm from the surface of the tooth

For carrying out optical microscopy on the electric spark machine tooth portions were cut from the gear intact. They are represented in Figure 1, and thin sections are made in parallel (section 1, Figure 1a) and perpendicular (section 2, Figure 1b) nitro carburized plane relative to surface of the tooth. Grinding conducted to remove traces of spark erosion. To create a mirrored surface sample is subjected to electrolytic polishing. Wherein:

1. For detection of microcracks applied electrolytic etching in a supersaturated solution of chromic anhydride in H_3PO_4.
2. The grain structure is determined by electrolytic etching in a 10% HCl solution.
3. The structure set Carbide resulting samples electrolytic etching in 50% solution of hydrochloric acid.

Subsequently, the prepared samples were investigated by metallographic microscope MIM-10 at magnifications of 50, 100, and 380 times. The increase to 1500-fold was necessary for the manufacture of microstructures pictures.

Based on optical microscopy and solutions structure nitro cemented zones were identified layer, the shape and dimensions of the real and hereditary grain size and location coarse carbonitrides, the presence of microcracks and their trajectory.

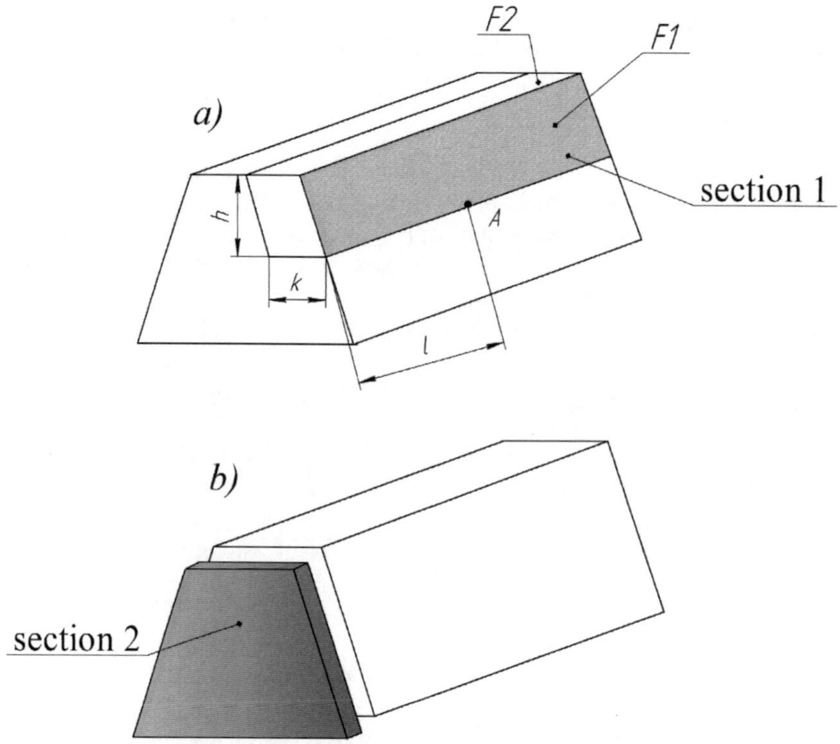

Figure 1. Scheme of cutting samples from the gear tooth for metallography, electron microscopy and X-ray analysis (section 1) and intaglio (section 2) in the electric spark machine.
A – area of research material (by electron microscopy and X-ray analysis) arranged on different surfaces of the tooth: h = 4,5 mm; l = 15 mm; k = 0 (tooth surface – F1) and 2 mm (F2)

X-ray diffraction (XRD) sample made by X-ray diffractometer DRON-1.5 automatic scanning X-ray beam angles in the range from 19 to 150 degrees in the filtered Fe-Kα-radiation. Recording was carried out in automatic mode on a chart tape rate 1/2 degree per minute. The X-ray indicates the reflexes to determine the main phase components. Using the main diffraction maxima of α-Fe and γ-Fe phase using known techniques allowed the calculation of lattice parameters (extrapolation cos2α function), internal microdistortions and microtension type II, as well as hold on to texture analysis (reference data taken from [5, 6]).

From the analysis of the broadening of diffraction maxima of were obtained information about microdistortions and sizes of coherent scattering of the crystal lattice of the main phase units. The broadening of diffraction maxima of due to three factors: the finite width of the spectral line (Fe-Kα), geometric shooting conditions and the state of the test sample [7]. The contribution to the broadening of reflex due to the imperfection of the crystal, called the true physical broadening of the X-ray line β (2θ). It is obtained by comparing the width of the reflex of the crystal with the width of the reflex from the reference defect-free sample. As a reference was used single crystal of chemically pure of the annealed iron.

Performed by X-ray phase analysis was qualitative and quantitative. Filing [5, 6] were used in conducting the qualitative phase analysis. As indicate the x-ray reflexes to determine the main phase components. Quantitative phase analysis was characterized on the basis of the integrated intensity of the diffraction lines of values. Having defined the texture relative intensity of the reflection (110) α-Fe to the intensities of the reflexes (200), (211) and (220) of the same phase, followed by comparison with reference intensity ratio of α-Fe. The intensity of the reflection (110) α-Fe is taken as 100%.

Transmission electron microscopy (TEM) was performed on thin foils. The thin metal plates were cut from the different layers of the tooth (Figure 1a) parallel to the surface of the nitro cemented layer.
Cut off to the electric spark machine metal plate thickness of 0.25 mm ~ previously been chemically cleaned by cutting traces in H_2O_2 solution with the addition of a few drops of HF. Then realized their thinning electrolytic ally in a supersaturated solution of chromic anhydride in N_3PO_4 to a thickness of ~ 0.17 m and viewing electronic microscope type EM-125 and EM-125K at an accelerating voltage of 125 kV. In this operation an increase in the electron microscope column ranged from 8.000 to 25.000 times. A final increase was achieved by using photographic printing.
To identify the phases present in the metal diffraction analysis was applied using the darkfield technique. The phase analysis carried out in accordance with the equilibrium diagrams of Fe-C-N, Cr-C-N and Ni-C-N and other iron alloys ternary diagram [8, 9]. When the diffraction pattern analysis verified the possible existence in the nitro carburized layer of the order of 30-35 table phase.
Images of thin metal structure used for:

- Classification of its morphological features;
- Determining the size, volume fraction and places of localization of secondary phases and discharge;
- The definition of scalar ρ and $\rho_\pm$ excess dislocation density;
- Establishing the value of the amplitude of the curvature-torsion æ and torque stress τ.

It has been viewed by a significant area of the foil and captured 30 to 40 micrographs and micro diffraction paintings to them for each test sample of a particular layer. The scalar dislocation density ρ measured by the cross-section [10]. Excessive density of dislocations in the local areas of the sample is determined by the curvature-torsion of the crystal lattice [11-13]:

$$\rho_\pm = \frac{1}{b} = \frac{d\varphi}{d\ell}, \qquad (1)$$

where – the Burgers vector, dφ/dℓ – the gradient of disorientation, i.e. orientation change with the movement of a particular plane on the model, φ – angle, ℓ – distance. Direct measurements found that the width of the Bragg reflection for the elements of the first period is about 10 [14].
Therefore, by measuring the width of extinction contour curvature-torsion Δℓ, you can define dφ/dℓ from the condition:

$$\frac{d\varphi}{d\ell} = \frac{\Delta\varphi}{\Delta\ell} = \frac{1^0}{\ell}. \qquad (2)$$

Results and Discussion

Final thermal processing gear (hardening and subsequent low tempering) crushed ferrite grains in the study began. The size and orientation of the grains are extremely important for the problem of formation of microcracks, as well as for stress control material strength.
Shredding grain size increases the metal fracture [15], but it does not solve the whole problem.
The study 20H2N4A steel structural in all states except the original contains cracks. Their number varies according to the heat treatment and the distance from the surface of the gear tooth. These cracks are divided into three types:

1. the cleavage fracture;
2. quasi-cleaved fracture,
3. crack ductile fracture.

It was found that the most common fractures quasi-cleaved and ductile fracture.
Structure of cracks in the test steel 20H2N4A [4] is fundamentally different from the classical structure of the metal circuit with crack [4], which generates ahead of itself a small plastic zone with a

certain dislocation density. This area is surrounded by the elastic stress unreformed metal. Away from stress cracks in the metal are absent.

According to scheme [4], all the metal previously is deformed and has high dislocation density. At the mouth of the crack forming a special zone with a high level of elastic stress fields and increased the scalar dislocation density. In this zone, the elastic deformation much greater than metal plastic. In the rest of the metal, both the deformation contribution commensurate. In particular, the study steel 20H2N4A crack length ranges from 1.0 to 4.0 microns, and the length of the zone ahead of it – 0.5–7.0 mm. The density of dislocations in elastic stress zone may exceed the average density of dislocations in 1.5–2 times.

It is established that after the first nucleation microcracks occurring when cutting the teeth, their parameters increase after each subsequent heat treatment. It can be noted that the metal embrittlement, begun after cutting of gear teeth on the gear cutting, developed after each subsequent heat treatment.

Measurement of internal stress fields was carried out by X-ray analysis and transmission electron microscopy. It has been shown that the resulting gradual voltages of heat treatment processes increase, without exceeding 1000 MPa. This value critical and indicates the start of the fracture process because tensile strength of the test steel is approximately 1000 MPa. Since there is heterogeneity of internal stress fields throughout the volume of metal gears, it is the peak (maximum) voltages will serve as real sources of nucleation of microcracks.

In the zone located ahead of the front propagating the crack, the maximum stress level equal to 6500 MPa, reached after the nitrocarburizing process. Later in subsequent thermal treatments, this value decreases, remaining nevertheless significant ~ 4000 MPa.

This implies that only a high-strength metal in the way of propagating cracks will prevent the gear from damage.

Figure 2 shows more detailed information about the structure of the highly loaded areas [4] in the crack head. This figure shows the values of the amplitudes of the crystal lattice curvature-torsion χ in two cases: 1 – directly at the crack tip; 2 – highly stressed at the boundary of zones 1 and zone 2 [4]. In zone1 curvature-torsion elastic component of the crystal lattice (χ resilient) is always greater than the plastic components (χ ductile) that is provided by dislocations. In the zone 2, both components are comparable. The internal stress field in any of the areas is proportional to the total value of x, which is also given in this figure. The amount of stress in zone 1 for most types of heat treatment, either increasing or decreasing slightly. The maximum stress field are either after nitrocarburizing (sample 12), or after high-temperature tempering (sample 13), or after hardening (sample 14). Their numerical values are sufficient for the emergence of new cracks, as well as for the further development of already formed. The degree of hardening zones 1 and 2 is that the plastic of high stress relaxation difficult. Therefore, the main mechanism of stress relaxation data are the nucleation the processes of formation and development of cracks.

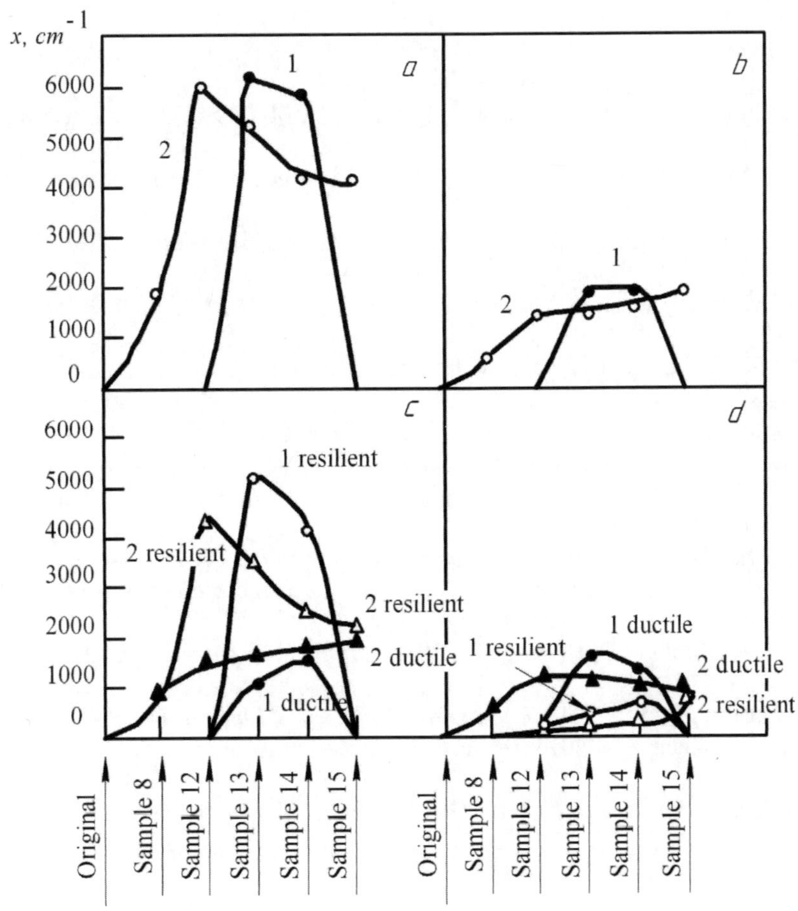

Figure 2. The change of curvature of the crystal lattice amplitude torsion χ, measured by electron microscopy, depending on the heat treatment on the tooth surface of gears (1) and at a distance of ~ 2 mm from the surface of the tooth (2):
a, b - the crack in the head; b, d - in elastically deformed zone ahead of the crack;
a, b - the average value of the curvature-torsion χ amplitude;
c, d - resilient (1 resilient, 2 resilient) and ductile (1 ductile, 2 ductile) components of the curvature-torsion amplitude

Conclusion

As a result of studying the structure of the tooth nitro cemented layer of metal gears:

1. The quantitative characteristics of internal stress fields and microcracks.
2. The structure of the area ahead of the front of developing cracks.
3. It is shown that microcracks are generated already with cutting teeth and reach the most active state after nitrocarburizing, quenching and tempering high.
4. It was found that in all cases on the metal surface relaxation processes developed stronger. Therefore, activity of cracks inside the tooth above the metal, and the length of microcracks and their "sharpness" increases as the distance from nitrocarburizing layer into the metal.
5. It was determined that in all investigated conditions the metal has a significant amount of microcracks and the amplitude of such internal stress fields, which predicts a high probability of failure in the operation of the gears in the drive gear of the coal conveyor.

References

[1] Apasov A M Galevsky G V 2008 Methods of research, testing, analysis and control in metallurgy and materialovedinii Tomsk 488

[2] Kozlov E V, Popov N A, Malinovskaya V A, Apasov A M 2005 *J. Izvestia Ferrous Metallurgy* **8** 26-29.

[3] Kozlov E V, Malinovskaya V A, Popova N A, Horseshoe V P Apasov A M 2006 *J. Izvestia Ferrous Metallurgy* **2** 33-36.

[4] Abbasov A M, Apasov A A, Kozlov E V 2013 *J. Bulletin of the Tomsk Polytechnic universiteta* **2** 25-32/

[5] Hu W, Yunxia W, Chenshuo M, Zhaolin Z 2014 *J. Applied Mechanics and Materials* **456** 406–410

[6] Babcsan N, Leitlmeier D, Degischer H.P, Banhart J 2004 *J. Advanced Engineering Materials* **6** 421–428

[7] Mao D, Edwards J R, Harvey A 2006 *J. Chemical Engineering and Science* **61** 1836–1845

[8] Banhart J, Bellmann D, Clemens H 2001 *J. Acta Materialia* **49** 3409–3420.

[9] Rodzevich A P, Kuzmina L V, Gazenaur E G, Krasheninin V I 2014 *J. AIP Conference Proceedings* **1623** 519-522

[10] Yang C C, Nakae H 2003 *J. Materials Processing Technology* **141** 202–206.

[11] Ibragimov E A, Saprikin A A, Babakova E V 2014 *J. Advanced Materials Research* **1040** 764–767

[12] Nokhrina O I, Rozhihina I D, Hodosov I E 2015 *J. IOP Conference Series: Materials Science and Engineering* **91** 12-17

[13] Ferro P, Lazzarin P, Berto F 2012 *J. Materials Science and Engineering* **554** 122–128

[14] Fedoseev S N, Mukhtar Z M 2015 *J. IOP Conference Series: Materials Science and Engineering* **91** 1–7

[15] Nakae H, Fukami M, Kitazawa T, Zou Y 2011 *J. China Foundry* **8** 96–100

IOP Conf. Series: Materials Science and Engineering **142** (2016) 012055 doi:10.1088/1757-899X/142/1/012055

Calculation of the Intensity of Adhesive-Fatigue Wear of Cutting Tools

V Bibik[1], N Ivushkina[2], D Arhipova[3]

[1]Assistant Professor, Yurga Institute of Technology of Tomsk Polytechnic University 652055, Russia, Kemerovskaya oblast, Yurga, Leningradskaya str., 26.
[2]Assistant, Yurga Institute of Technology of Tomsk Polytechnic University 652055, Russia, Kemerovskaya oblast, Yurga, Leningradskaya str., 26.
[3]Student of Yurga Institute of Technology of Tomsk Polytechnic University 652055, Russia, Kemerovskaya oblast, Yurga, Leningradskaya str., 26.

E-mail: bibik@tpu.ru

Abstract. On the base of kinetic equation of strength the authors suggest the method carbide tools wear intensity calculation allowing comparing wear resistance basing on tool material thermal diffusivity coefficient. The authors obtain equations of tool wear resistance dependences upon its thermal diffusivity which show close correlation between these parameters. The results of wear intensity calculations correspond well to the experimental results under the cutting rates corresponding to the region of adhesive-fatigue wear. The inserts with low thermal diffusivity coefficient are characterized by lower wear rate at the inial and normal stages of wear.

1. Introduction

Modern unmanned NC machine tools are very sensitive to the homogeneity of the cutting properties of inserts. Modern engineering production also makes wide use of split-design tools, such as face milling cutters which apply several inserts at one time when cutting. If these inserts have highly disperse wear resistance properties, then, some of them may fail prematurely which can lead either to defects or to equipment failure.

At the same time, due to their composition and technology of fabrication, inserts are characterized by great dispersion of wear resistance. Wear resistance for the inserts of the same brand but of different batches may differ by dozens of times, within one batch – by several times, for the various points of the insert – by 1.5 – 3 times [1].

2. Results and Discussion

Analysis of carbide cutting tools operation under factory conditions showed that in most cases the tools are operated at cutting rates corresponding to the temperatures of up to 900°C in the cutting area, i.e. under cutting rates that are smaller than the rate corresponding to the break-over point of the curve showing dependence of wear upon the temperature. In that region of the curve wear resistance of the tool increases with temperature growth. This temperature region corresponds to adhesive wear. Adhesive wear occurs when particles of tool material are broken away due to combined forces of adhesion between the tool material and the workpiece. According to many experimental studies [2],

IOP Conf. Series: Materials Science and Engineering **142** (2016) 012055 doi:10.1088/1757-899X/142/1/012055

the contact surfaces of tools are always covered with adhered particles of worked material of varying thickness. It means that destruction of adhesive bonding far more often affects the area of the worked material than that of tool material. That is why the areas of actual contact of tool material appear to be loaded with cyclic forces and their destruction occurs due to fatigue. Thus, it is important to estimate the wear rate of carbide tools under temperatures below 900°C with consideration to the fatigue nature of adhesive wear.

According to many research papers the process of fatigue damage accumulation and development is associated with micro- and macroflow of material which is determined by the processes of nucleation, motion, generation and disappearing of linear defects – dislocations [3].

Starting to talk about hard alloy destruction we should note the peculiarities of hard alloy structure. It contains grains of tungsten carbide cemented together with cobalt bound and according to many researchers its strength is due to its carbide "skeleton". When the hard alloy is destroyed the fracture may develop along the following directions: across tungsten carbide grains, along the boundary between carbide grains, along WC-Co phase boundary and, finally, across the cobalt phase. The major amount of destruction affects the boundary between the cementing phase and the carbide grains, i.e. the fracture goes round carbide grains and the view of the breaking shows well-defined contours of tungsten carbide grains.

Assuming that carbide grain is exposed to constant shear load and accepting the simplification that carbide grain is spherical, let us use the kinetic equation of strength [4] to calculate the time until cobalt bound destruction and carbide grain detachment, replacing normal stress σ with tangential stress $\tau*$.

$$\tau = \tau_0 \exp\frac{U_0 - \gamma\tau_*}{kT}, \qquad (1)$$

where T – test temperature; k – Boltzmann constant; U_0 – energy of interatomic bond; τ_0 – period of thermal oscillation of atoms; γ - activation volume.

In paper [4] it is established that energy of interatomic bond

$$U_0 = \varepsilon_* \frac{C_a}{\alpha}, \qquad (2)$$

where ε_* - ultimate strain of the interatomic bond; C_a- atomic heat; α - linear expansion coefficient.

Activation volume [4], i/e/ the volume where the initial crack is formed can be determined as

$$\gamma = \frac{C_a}{\alpha E}\chi, \qquad (3)$$

where E – modulus of elasticity; χ - overload factor.

Overload factor [4]

$$\chi = \Lambda/a_0, \qquad (4)$$

where Λ - free path length of phonons; a_0 – atomic size.

Free path length of phonons can be found from Debuy equation

$$\lambda = \frac{1}{3}cS\Lambda\rho, \qquad (5)$$

where λ - thermal conduction; c– specific thermal capacity.

Consequently

$$\Lambda = \frac{3\lambda}{cS\rho}. \qquad (6)$$

Thermal conduction is expressed through thermal diffusivity (a)

$$\lambda = c\rho a. \qquad (7)$$

After combining (7) and (6) we obtain

$$\Lambda = 3a/S. \qquad (8)$$

Applying (8) to (4) and then, applying the obtained expression to (3) we obtain

$$\gamma = \frac{3C_a a}{\alpha E S a_0}, \tag{9}$$

where a – thermal diffusivity

Thus, durability of carbide grain

$$\tau = \tau_0 \exp \frac{\frac{C_a}{\alpha}\left(\varepsilon_* - \frac{3a\tau_*}{ESa_0}\right)}{kT}. \tag{10}$$

Under the real conditions the carbide grain is affected by cyclic saw-like loading

$$\tau_* = 0.5\tau_{\max} - \frac{\tau_{\max}}{\pi}\left(\sin\omega t + \frac{1}{2}\sin 2\omega t + \frac{1}{3}\sin\omega t + _{...}\right), \tag{11}$$

where ω=2π/p – cyclic frequency; p - period.

Considering every contact point being exposed to 10^3-fold shear stress impact [2] and knowing the cutting rate we can determine the period of loading.

According to the theory of damage summation and according to Bailey criterion [5] during the time period Δt durability reduce $\Delta t/\tau_{i,}$ occurs if the destruction rate during this period was $1/\tau_{i,}$. The cobalt bound will be completely destructed under [5]

$$\sum\left(\frac{\Delta t}{\tau_i}\right) = 1. \tag{12}$$

Or in the integrated form

$$\int_0^\tau \frac{dt}{\tau(t)} = 1 \tag{13}$$

Replacing in formula (13) $\tau(t)$ with the durability formula (10) we obtain

$$\int_0^{t_p} \frac{dt}{\tau_0 \exp\frac{U_0 - \gamma(a)\tau_*(t)}{kT}} = 1. \tag{14}$$

Applying the form of loading (11) into formula (14), knowing the experimental value of thermal diffusivity and solving the equation by iteration with reference to t_p, we can find the durability of carbide grain. Taking the simplification that carbide grain is spherical we find the volume of carbide grain (V), wear intensity

$$J = \frac{V}{t_p}. \tag{15}$$

Set by the wear criterion the worn part of the tool is found. For the front surface [6]

$$V_w = \frac{2}{3} a b h_0, \tag{16}$$

where a – width of the crater; b- length of the crater; h_0 – depth of the crater.

For the back surface [6]

$$V_w = \frac{tg\alpha h_K^2 b}{2(1 - tg\gamma tg\alpha)}, \tag{17}$$

where a – width of the chip; α- back clearance angle; γ- face angle; $h_з$ – wear of the back surface. And so, we can find the tool durability

$$T = \frac{V_w}{J}. \tag{18}$$

In equation (14) activation volume γ is the function of thermal diffusivity, so to compare the intensity of wear of a batch of inserts we need to measure the thermal diffusivity of every insert and calculate the approximate intensity of wear.

To measure the thermal diffusivity the method of "flash" [7] was chosen. This method allows finding the thermal diffusivity within several seconds. The test unit included the following equipment: helium-neon laser LG-78, pulse neodymium-glass laser GOS-301, a unit for noncontact temperature measuring TAU-4 (Tomsk Polytechnic University), pulse generator G5-56, oscillograph S8-14 [8].

Two techniques were applied for the thermal diffusivity measuring. The first one was to expose the face of a flat specimen which initial temperature was maintained at the same level to uniformly distributed pulse heating. At the back of the specimen chronological thermogram was registered (the two-sided technique). The second technique was to heat the specimen face with the chronological diagram of this face being registered (the one-sided technique).

The specimen was heated by the neodymium laser impulse (wave-length 1.06 micrometer, pulse duration 4 ms, energy 1.3÷14 Joule, the diameter of the heated area on the specimen surface – 4÷9 mm). The temperature measurements of the heated areas were carried out with TAU-4 (Tomsk Polytechnic University) and registered by an oscillograph with a storage tube. The pulse generator was used to delay the oscillograph for the time of the neodymium lasering, (for the one-sided technique). The visible helium-neon laser is necessary for optical channel mastering. To separate the thermal radiation of the specimen (wave length 2-8 micrometer) from the reflected laser impulse (wave-length 1.06 micrometer), one-sided thermal diffusivity measurement technique being used, a germanium filter placed at TAU-4 input port was applied

For the two-sided technique, we employed the Parker formula Eq. 19, [7] to evaluate thermal diffusivity.

$$a = 0.139L^2/\tau_{0.5} \tag{19}$$

where $\tau_{0.5}$ - half of the time needed to achieve the maximum temperature on the specimen, L – specimen thickness.

For the one-sided technique, thermal diffusivity was calculated from a theoretical equation describing the cooling stage of the specimen in the cylindrical coordinates Eq. 20 [9]

$$T(x,r) = \frac{W}{c\rho L}\left[1 + 2\sum_{n=1}^{\infty}\cos n\pi x \cdot \exp(-n^2\pi^2 F_0)\right] \times \left[1 + 2\sum_{m=1}^{\infty}\frac{J_1(Z_m R_H/R) J_0(Z_m r/R)}{(Z_m R_H/R) J_0^2(Z_m)}\exp\left(-Z_m^2\frac{L^2}{R^2}F_0\right)\right], \tag{20}$$

where J_0, J_1, - Bessel first-class zero and first order function; Z_m - positive roots of a characteristic equation: $J_1(Z)=0$; W - absorbed energy; c - heat capacity; ρ- density; $F_0=a\,\tau/L^2$ - Fourier criterion (no dimensional time); a - thermal diffusivity; τ - time; R - the specimen radius; R_H - - the heated area radius.

To try the suggested method of adhesive-fatigue wear calculation we calculated the intensity under structural steel 1040 machining by a cutter produced from hard alloy MR7 (9% Co, 6% Tic, 85% WC) for one of the inserts.

The volume of the worn part of the tool was calculated according to formula (16). The calculated wear intensity was 0.12 mm^3/min. Intensity of wear obtained in the course of wear testing was 0.01 mm^3/min.

To analyze the thermal diffusivity influence upon the tool wear resistance the following types of reversible cemented carbide cutting plates were chosen: MP7, P35 (9% Co, 91% WC, three-layer coating TiC-Ti(CN)-TiN), produced in Sweden.

All the plates were thoroughly examined for fractures, crumbling and other defects with an optical microscope. Resistance of MP7 grade plates was tested when steel-turning (1040 steel) under the following cutting conditions V=200 m/min, S=0.18 mm/rev t=1 mm. The resistance criterion – wear crater width 0.8 mm. P35 grade plate resistance was taken from [10].

The plate thermal diffusivity was estimated according to the method described above.

In the conclusion the correlation between the cemented carbide cutting inserts wear resistance and their thermal diffusivity coefficients was estimated. The corresponding correlation diagrams and graphical correlations are represented in fig. 1-3. The regression equations and correlation coefficients are shown in table 1.

Table 1. Correlation coefficients

Tool material	Technique	Regression equation	Correlation coefficients
MP7	One-sided	T=9.38 - 0.23a	-0.63
MP7	Two-sided	T=11.65 - 0.12a	-0.62
P35	One-sided	T=51.18 - 2.92a	-0.66

Diagram 4 shows MP7 grade plates wear curves having the smallest, the medium and the largest thermal diffusivity coefficient values. It is evident that the plates with the lower thermal diffusivity coefficient are characterized by lower wear factor at the initial and normal wear stages.

3. Conclusion

Under the cutting rates corresponding to adhesive-fatigue wear, wear of cutting tool occurs due to accumulation of defects in the crystal lattice in the cobalt bound. When the density of dislocations in the cobalt bound achieves its critical values, crack initiation and its further growth occur resulting in detachment of carbide phase grain. On the base of kinetic equation of strength the authors suggest a method for calculating carbide tool wear intensity. The given method allows completing comparative analysis of wear resistance on the base of tool material thermal diffusivity coefficient. The intensity of tool wear is inversely related to thermal diffusivity coefficient.

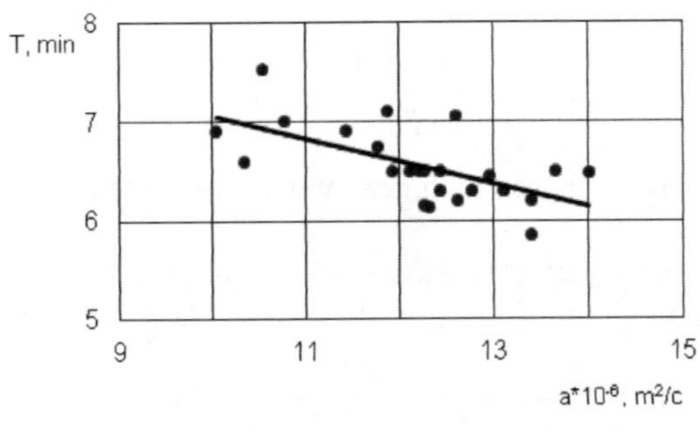

Figure 1. Correlation diagram and T (a) function for MP7 (two-sided technique)

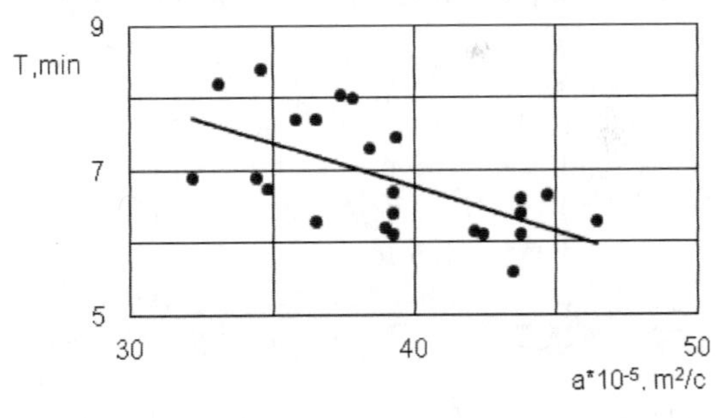

Figure 2. Correlation diagram and T (a) function for MP7 (one-sided technique)

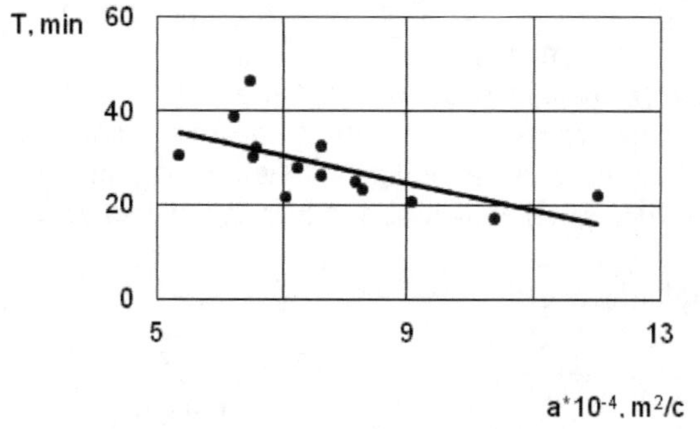

Figure 3. Correlation diagram and T (a) function for P35 (one-sided technique)

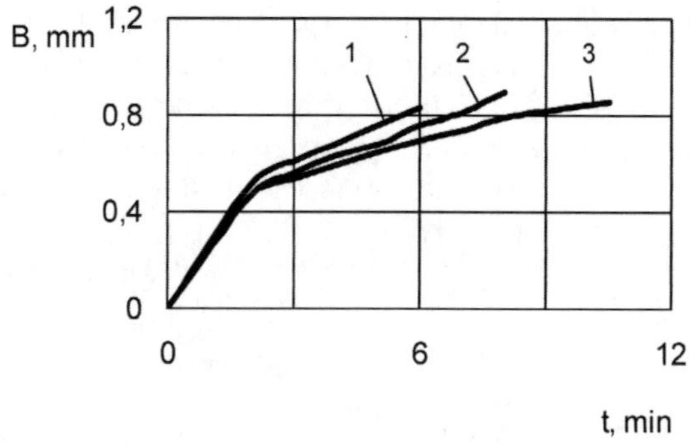

Figure 4. Plate wear curves for different thermal diffusivity values for MP7 (1 – 43,5·10-5 m^2/c; 2 – 39,2·10-5 m^2/c ; 3 – 34,6·10-5 m^2/c)

References

[1] Etin A.O.Wilson A.L. Reliability of the cutting tools with modern cutting material inserts. // Machine tools and the cutting tool. - 1983, No 7. pp. 21-23 Development of the science dealing with metal-cutting./ Under the editorship of Bobrov V.F. and others - M.: Mashinostroyeniye, 1967. - 416 p.

[2] Ivanova V.S. Fatigue degradation of metals. – M.: Publishing house for ferrous and non-ferrous metallurgy, 1963. – 272 p.

[3] Zhurkov S.N. Dilaton mechanism of solid bodies strength // Physics of strength and plasticity. – L.: Nauka, 1986. – P. 5-11.

[4] Regel V.R., Slutsker A.I., Tomashevsky E.E. The kinetic nature of solids. – M.: Nauka, 1974. – 560 p.

[5] Petrushin S.I. Study of the cutting properties of inserted cutters when machining gray cast iron. Ph.D. thesis in Engineering Science: 05.03.01. – Pres. on 18.02.80 : Appr. 25.02.80. – M.: 1979. – 149 p.

[6] Loladze T.N. Wear of cutting tools. – M.: Publishing house for technical and scientific literature. 1958. – 356 p.

[7] Parker W.J., Jenkins R.J., Buttler C.P., Abbot G.L. Flash Method of Determining Thermal Diffusivity, Heat Capacity and Thermal Conductivity. - J. Appl. Phus., vol.32, No.9, Sept. 1961. - pp. 1679-1684.

[8] Bibik V., Petrova E. Methods of Forecasting Wear Resistance of Cutting Tools Based on the Properties of Their Materials. Applied Mechanics and Materials Vol. 682 (2014) pp 491-494. Trans Tech Publication, Switzerland.

[9] Nagorski Z. Podstawy teoretyczne impulsowych metod badania wspotczynnika wyrownywania temperaturi materialow modeli elementow konstrukcynych // Archiwum termodynaviki. - 1984. - Vol.5. - № 3/4.

[10] Vavilov V.P., Pushnykh V.A., Shipulin A.V. Forecasting of tools wear resistance using their thermal diffusivity. // Problems of ensuring high quality of engineering products. / Proeedings of international scientific and technical conference. – Krasnoyarsk, 1994. – P. 223-240.

IOP Conf. Series: Materials Science and Engineering **142** (2016) 012056 doi:10.1088/1757-899X/142/1/012056

Controlling Explosive Sensitivity of Energy-Related Materials by Means of Production and Processing in Electromagnetic Fields

A P Rodzevich[1,a], E G Gazenaur[2,b], L V Kuzmina[2,c], V I Krasheninin[2,c] and P N Sokolov[1,a]

[1]Department of Ferrous Metals, Yurga Institute of Technology of Tomsk Polytechnic University, Russia
[2]Department of Chemistry of Solids, Kemerovo State University, Russia

e-mail: [a]nimez@tpu.ru, [b]gazenaur@kemsu.ru, [c]specproc@kemsu.ru

Abstract. The present work is one of the world first attempts to develop effective methods for controlling explosive sensitivity of energy-related materials with the help of weak electric (up to 1 mV/cm) and magnetic (0.001 T) fields. The resulting experimental data can be used for purposeful alternation of explosive materials reactivity, which is of great practical importance. The proposed technology of producing and processing materials in a weak electric field allows forecasting long-term stability of these materials under various energy impacts.

Introduction

Modern constructional materials are often subjected to extreme impacts (separate or combined action of electro-magnetic fields, laser radiation, light, temperature) which cause various physical and chemical transformations in the materials and making their application areas limited. From the range of impacts, great interest is attracted by electro-magnetic fields of uncontrolled origin (different manifestations of static electricity, radiation from powerful electric and radio devices, other man-made radiation sources). The impact of such radiations on explosive materials can initiate quickly developing processes in them and cause fires and explosions. Due to that, it is important to develop effective methods of controlling explosive sensibility and stability of energy-related materials. Among other areas such control is greatly demanded by coal industry.

This research is one of the world first attempts to develop effective methods of controlling explosive sensitivity of initiating explosives (on the example of silver azide whiskers) using the technology of production and processing of energy-related materials in superweak electric (up to 1 mV / cm) and magnetic (up to 0.001 T) fields. The resulting experimental data can be used for purposeful alternation of explosive reactivity which is of great practical importance.

Silver azide whiskers were chosen as a test object for this research. They represent a traditional model object used to create a theory of fast reactions in solids. Under the influence of external factors of different nature they are capable to undergo irreversible transformations generating inert end products (molecular nitrogen and metal), which can be easily analyzed by conventional physical and chemical methods. Physicochemical properties of silver azide have been studied in detail, the band structure and the parameters of the crystal lattice have been determined [1-4]. There are two practically significant applications of silver azide whiskers: they serve as explosives with high initiating ability [1, 5], and as sensitive sensors measuring very low electric and magnetic fields.

Technology of inorganics crystallization in weak electric fields suggested in [6-8] allowed obtaining samples with improved performance (monodispersity, the minimum number of defects, increased shelf life, the size controlled by varying the electric field in crystallization, etc.). Samples of silver azide obtained at automated installation for synthesis and crystallization described in [7] contain the minimum number of defects. The content of impurities of positive metal ions Zn^{2+}, Cu^{2+}, Fe^{3+}, Al^{3+}, Pb^{2+}, assessed by atomic emission and X-ray fluorescence analysis methods, makes up $2 \cdot 10^{-5}$ mole percent in average; dislocation density determined by etch pits [9] is $2 \cdot 10^{2}$ cm^{-2} in average.

The results of our studies [10] showed that the process of solid-phase decomposition of silver azide can be effectively controlled by a non-contact electric field. Thus, this research focuses on application of weak electromagnetic fields simulating real conditions of storage and transportation of explosives and on development of technology to control sensitivity of energy-related materials.

Experimental technique

Crystallization in a weak non-contact electric field was carried out by the method described in [6] at the apparatus shown in Fig. 1.

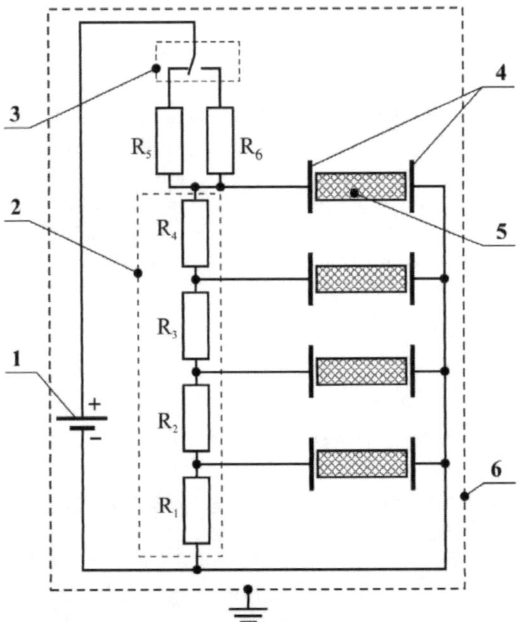

Fig. 1. Installation for crystallization in the electric field: 1. power supply; 2. switch of the voltage adjustment range; 3. block of the voltage divider; 4. electrodes; 5. crystallizer; 6. grounded shield housing.

The crystallizer is a cylindrical weighing bottle partially filled with a solution of the corresponding salt. To increase the reliability of the experimental results n crystallizers were connected to various steps of a voltage divider (Scheme $n = 4$). An element of 1.5V is used as the power supply. It provides a weak constant non-contact electric field without additional hardware and eliminates interferences. As seen in Figure 1, the proposed apparatus allows varying the voltage of the electric field in the experimental cell in the range from 10^{0} to 10^{-6} (from 10^{0} V to 10^{-3} V when switched on via resistance R5=1.1 MMOhm, and from 10^{-3} V to 10^{-6} V when switched on via resistance R6 = 1.1 GOhm). The voltage divider unit includes a serially connected resistances R1 = 2.2 kOhm, R2 = 20kOhm, R3 =

200kOhm, R4 = 2mOhm with terminals at each stage. Crystallization in the electric field was carried out until the solvent completely evaporated (evaporation rate was 0.407 g / 24h at the temperature of 23 ± 2 °C).

Figure 2 shows the technology of producing samples in a weak magnetic field. The weighing bottle with a solution of silver azide was placed in a uniform and non-uniform magnetic field. The gradient of the magnetic field was created by permanent magnets. Thus assembled structures were mounted in a grounded shield housing. Magnetic field distribution was measured by milliteslametre in points at the distance of 0.5 cm (instrument error at the measurement area of 200 mT was about ± 2%).

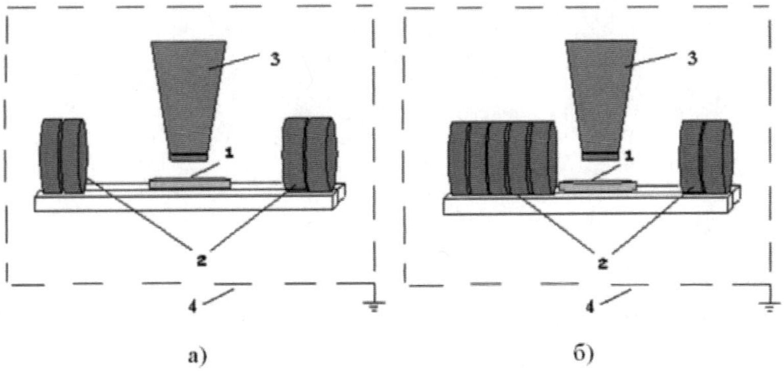

Fig. 2. Installation diagram for crystallization in a uniform (a) and non-uniform (b) magnetic fields 1 - crystallizer; 2 - permanent magnets 3 - microscope eyepiece; 4 - grounded shield housing.

The crystallization process was observed through the microscope, "Biolam" (ocular zoom × 120), photo and video shooting were performed.

Control of the defect structure of the materials was carried out by X-ray fluorescence (research was conducted at the installation JEOL JSM-6390 LA, combined PMA-SEM) and atomic emission (using atomic emission spectrometer with inductively coupled plasma iCAP 6500) analysis methods.

Since in silver azide whiskers, reaction regions (RR)coincide with dislocations emergences, the study of dislocation structure of crystals was carried out by etching holes method [8]. Contrasting etch pits were obtained by etching the crystals in 10 % aqueous solution of sodium thiosulfate. Individual edge dislocations runs, local and overall dislocation density of the samples were measured using an optical microscope.

The test samples were stuck at both ends onto a mica substrate pre-defatted with ethanol with the help of BF-6 glue which has high electrical resistance.

Explosive sensitivity of the samples was determined as the explosion delay time at which the explosion of the sample can be registered judging by a flash or audible signal with 50% confidence. To study the decomposition patterns we used a scheme of direct experiment: exposure - response. The explosive decomposition of silver azide whiskers was initiated in two ways:

1. by an impulse of solid-state YAG: Nd^{3+}laser LDPL10M 10 ns in duration at a wavelength of 1064 nm (impulse energy being up to 1.1J) and by an impulse of ytterbium quasi-continuous laser from 200 ms to 20 ms fiber in duration at a wavelength of 1070 nm (pulse energy being up to 20J) [4,10]. Initiating exposure on the surface of the sample was determined using a pyroelectric head PE50BF-DIF-C (OphirPhotonics) and controlled by a signal of calibrated photodiode. A luminescence signal was recorded using a photomultiplier Hamamatsu H11526-20-NN. It provided a time reference of the signal moment caused by the sample explosion with the initiating laser impulse. It also allowed assessing the luminescence signal increase rate and recording the time when the luminescence caused by the explosion products arose. Only the central part of the laser beam was used to initiate the explosion ensuring a sufficiently homogeneous distribution of the excitation intensity on the sample surface. The spread of initiating impulse energies did not exceed 3%. Approximation of the

experimental data was performed based on the dependence of explosion probability on initiation exposure within micro model of chain reaction of energy-related materials explosive decomposition proposed in [11].

2. by constant contact electric field. The electric field 300 kV / m in strength was created using the DC power supply (Fig. 3). The test samples were stuck at both ends onto a mica substrate pre-defatted with ethanol with the help of BF-6 glue which has high electrical resistance. Gallium was used for electrical contacts (inter-electrode distance being 1 mm). Under these conditions monopolar injection of holes which promote a chemical reaction in reaction sites takes place [9].

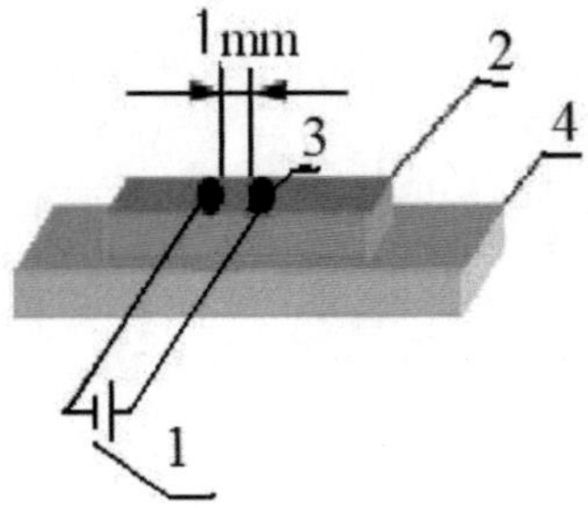

Fig. 3. Experimental setup for the investigation of explosive sensitivity in electric-field decomposition: 1 - power supply; 2 - crystal; 3 - gallium contacts; 4 - mica substrate.

Explosive sensitivity was defined as an explosion delay time. Measuring the explosion delay time has a number of features due to a probabilistic nature of the explosive decomposition especially at low strength areas. The measured value of the explosion delay time contains a probabilistic component, and experiments have shown that it depends on a sample biography. Explosion delay time at contact electric field intensity of 300 kV / m is adopted as a reference value (this field intensity corresponds to the explosion delay time of 360 ± 20 on average with the probability of 80-95%).

Application of the technology of energy-related materials treatment in noncontact electric field for controlling the reactivity of the materials on the example of silver azide whisker is analyzed in paper [9]. Figure 4 shows an experimental installation for a research into explosive sensitivity of silver azide samples in the non-contact electric field of different spatial configurations in a wide range of electric field strengths from 10^{-3} V/m to 10^{4}V /m.

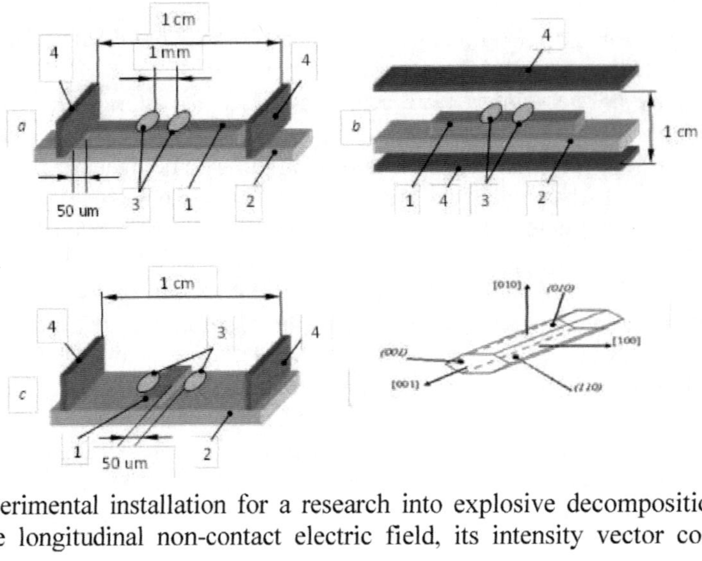

Fig. 4. An experimental installation for a research into explosive decomposition of silver azide samples: a) in the longitudinal non-contact electric field, its intensity vector coinciding with the

crystallographic direction [001]; b) and c) in cross the non-contact electric field, its intensity vector coincides with the crystallographic directions [010] and [100]; 1 - ATM crystal, 2 - mica substrate, 3 - gallium contacts, 4 - electrodes.

Results and Discussion

Dependence of explosion probability after laser initiation with the above characteristics on the initiation exposure showed that silver azide whiskers grown in the electric field have higher initiation threshold (approximately 50% higher than control samples grown without the field) (Fig. 5). Furthermore, the exposure threshold value increases with increasing intensity of non-contact electric and magnetic fields during crystallization.

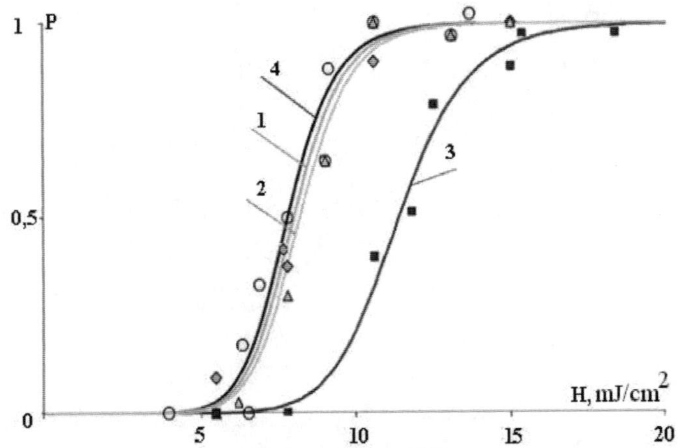

Fig. 5. Dependence of explosion probability (P) of initiation exposure (H) of silver azide samples grown at different electric field intensities during crystallization: curve 1 - 10^{-4} V / m; curve 2 - 10^{-3} V / m; curve 3 - 10^{-1} V / m; Curve 4 - without the field.

To analyze the explosive luminescence of silver azide whiskers we determined the time passed from exposure to the laser impulse to luminescence peak caused by the explosive decomposition for each batch of samples. Within each batch an average time and standard deviation (S) were calculated. Mathematical results of oscillogram processing are summarized in Table 1.

Table 1.
The standard deviation of the time to reach the maximum luminescence of silver azide samples

Electric field intensity during crystallization of (E, V / m)	0	10^{-3}	10^{-2}	10^{-1}	10^{0}
Standard deviation (S)	17.2	3.4	1	8.8	0.57

The standard deviation of the time to reach maximum luminescence with crystals grown in the electric field is less than that with crystals grown by the conventional method (0.57 ÷ 3 depending on the electric field intensity for the samples grown by crystallization versus 17.2 for the samples grown by the conventional method).

When initiating the explosive decomposition by a contact electric field it is experimentally proved that explosive sensitivity of samples grown in the electric field decreases.

Explosion delay time is 6000s in average for the samples obtained in the electric field compared with 360s for the samples grown by conventional methods.

Figure 6 shows experimental results of processing energy-related materials by non-contact electric field of different configurations on the example of silver azide whiskers in order to control explosive materials sensitivity. Electric field intensity varies from 10^{-3}V/m to 10^{4}V/m. To make the figure distinct decimal logarithm of field intensity was plotted along the horizontal axis. The straight line in Figure 6 represents explosive sensitivity of the silver azide samples in the absence of non-contact electric field.

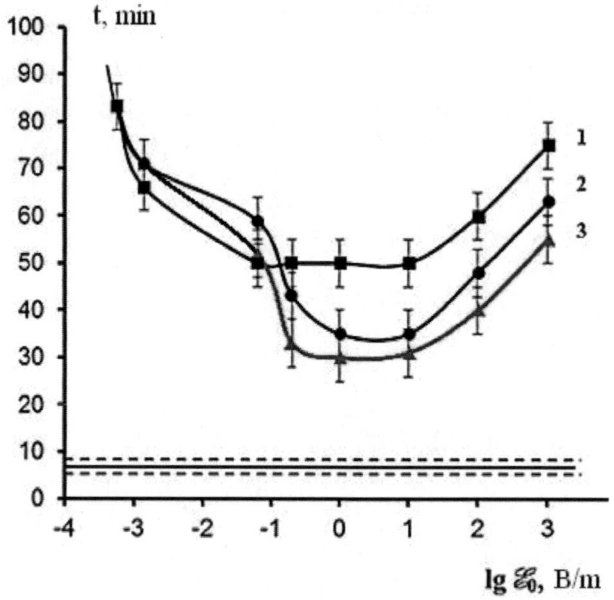

Fig. 6. Dependence of explosive sensitivity of silver azide on the intensity of non-contact controlling: 1, 3 - crross electric field (intensity vector coinciding with the crystallographic directions [010] and [100], respectively); 2 - longitudinal electric field (intensity vector coinciding with the crystallographic direction [001].

When non-contact electric field intensity goes down sensitive of explosive samples decreases sharply. Thus, when the non-contact electric field intensity is less than 10^{-2} V/m explosion, the explosion delay time is more than $6{\cdot}10^{3}$. Note that regardless of the mutual configuration of the electric fields, curves almost coincide passing through the extremum especially at the sites with field intensity less than 10^{-3} V/m. The arrow in Figure 6 shows that the time before explosion is more than 4hours (after experiments stopped).

Conclusions

Thus, with the help of the applied technology of obtaining and processing energy-related materials in a weak electromagnetic field with the specified intensity range we can control their explosive sensitivity changing defective and dispersed structure of these materials. It makes possible to use the experimental results for a targeted control of hazardous materials stability. It should be noted that the time before samples explosion is directly dependent on the intensity of the non-contact constant electric and magnetic fields in the crystallization and on the amount of RR in the sample.

This work was supported by a grant of RFBR №16-03-00313 A "Explosive sensitivity of energy-related materials in external fields."

IOP Conf. Series: Materials Science and Engineering **142** (2016) 012056 doi:10.1088/1757-899X/142/1/012056

References

[1] Walker, B.F. Energetic Materials. Physics and chemistry of inorganic azides. / Ed. by H. D. Fair, B.F. Walker. - New York: Plenum Press, 1977. - Vol. 1. - 503 p.

[2] Saprykin A.A. Influence of mechanical activation of copper powder on physicomechanical changes in selective laser sintering products. / A.A. Saprykin, E.A. Ibragimov, E.V. Babakova, V.I. Yakovlev. // AIP Conf. Proc. 1683: 020199, 2015.

[3] Gordienco, A. B. Electronic Structure of Metal Azides / A. B. Gordienco, Yu. N. Zhuravlev, A. S. Poplavnoy // Phys. stat. sol. (b), 1996. - V. 197. - № 2. - P. 707-719.

[4] Gora, T. Elektronic structure of the azide ion and metal azid / T. Gora, D. S. Downs, P. J. Kemmey, et al. // Energetic materials. - New York: Plenum Press, 1997. - V. 1. - P. 193-250.

[5] Aduev, B.P. Explosive decomposition of heavy-metal azides. / B.P. Aduev, É.D. Aluker, G.M. Belokurov, Yu.A. Zakharov, et al. // Journal of Experimental and Theoretical Physics. - 1999. - T. 89. - № 5. - C. 906-915.

[6] Rodzevich, A.P. On a mechanizm of an electric field influence on physicotecnical specifications of materials / A.P. Rodzevich, E.G. Gazenaur, V.I. Krasheninin // Applied Mechanics and Materials. - 2013. - Vol. 379. - p. 154-160.

[7] Rodzevich, A.P. The Technology of Production and Treatment of Materials in the Electric Field / A.P. Rodzevich, E.G. Gazenaur, G.M. Belokurov. // Applied Mechanics and Materials. - 2014. - Vol. 682. - pp. 206-209.

[8] Rodzevich, A.P. Features of Physical and Chemical Properties of Inorganic Salts, Grown in the Electric Field. / A.P. Rodzevich, E.G. Gazenaur, A.S. Walnukova, L.V. Kuzmina. // The 7th International Forum on Strategic Technology IFOST2012, 2012. pp. 296-299.

[9] Sangwal K. Etching of crystals: Theory, experiment, and application. In S. Amelinckx and J. Nihoul, Eds., Defects in Solids. North-Holland: Amsterdam, 1987. 497 p.

[10] Zakharov, V. Yu. The control of solid phase decomposition of silver azide by noncontact electric field / V. Yu. Zakharov, V. I. Krasheninin, L. V. Kuz'mina, et al. // Solid State Ionics. - 1997. - V. 101-103. - P. 161-164.

[11] Kriger, V.G. The effect of crystal size on initiation of decomposition of heavy metal azides by pulse radiation. / V.G. Kriger, A.V. Kalenskii. // Russian Journal of Physical Chemistry. - 1996. - T. 15. - № 3. - C. 351-358.

[12] Aluker, E.D. Early Stages of Explosive Decompostion of Energetic Materials, in Focus on Combustion Research, ed. by Sung Z. / E.D. Aluker, B.P. Aduev, Y.A. Zakharov, A.Y. Mitrofanov, A.G. Krechetov // Focus on Combustion Research. / Aluker E.D -New York: Novapublishers, 2006. P. 55.

IOP Conf. Series: Materials Science and Engineering **142** (2016) 012057 doi:10.1088/1757-899X/142/1/012057

Research of Influence Modification of Natural Concentrate on Quality Metal

S N Fedoseev, R A Gizatulin, E A Korotkova

Yurga Institute of Technology, National Research
Tomsk Polytechnic University Affiliate
Kemerovo region, Yurga, Leningradskaya str. 26, RUS

E-mail: sfedoseev@tpu.ru

Abstract. Questions of increase of mechanical, technological and service properties of metal at minimum cost to produce it are relevant for the metallurgical enterprises.
Modification of complex steel alloys containing reactive elements is one of the effective ways to improve the quality of steel. At the same time the direct costs for the use of modifiers are 0.2–0.3%, which little effect on the cost of production.
The paper presents the results of the application of natural concentrates as a modifier steel. The effects on the metal quality changes due to the impact of the modification concentrates demonstrate the effectiveness of their application. As a result of modification decreased the content of nonmetallic inclusions and grain size. Reduction of impurity modified metal of was the cause more high plastic properties, especially, impact strength at ordinary and low temperatures of tests. Based on the experimental data evaluated hardening mechanisms that lead to a significant improvement of physic-mechanical properties of the metal workpiece after administration modifier.

Introduction

Metal processing outside the furnaces aimed at reducing the concentration therein of harmful impurities, contamination by nonmetallic inclusions, improved mechanical properties and specific, level of which is revealed during the operation of fabricated metal products (heat resistance, fatigue strength, cold resistance, corrosion resistance, etc.). At the same time when using a secondary treatment not always successfully resolved issues related to the quality of the metal: the elimination of cracks in the bars and theaters, reducing chemical heterogeneity of rolled metal and its products and in the manufacture of steel pipes – ensuring high corrosion resistance and others.

Less contamination of the modified metal is a cause more high plastic properties, especially, impact strength at low temperatures and routine testing.

Significant role may be played modifiers reduce or eliminate the negative impact of located in a metal gas – nitrogen and hydrogen. In the case of high concentrations affects the casting sieve porosity, sinks, cracks and even flakes. To eliminate the defects caused by the nitrogen can be useful modifiers, which include those elements have strong nitride – titanium or zirconium. To neutralize the harmful effects associated with hydrogen, recommended rare earth-containing modifiers that promote the binding of hydrogen in lasting hydrides, thus preventing or slowing its isolation in the solid metal in the form of flakes. Due to this safety can be increased level of hydrogen concentrations, reduced intensity and duration of degassing a process the liquid steel, as well as reduced heat treatment castings anti flakes up to complete elimination of the latter.

Results are determined by modifying primarily contents of active elements (Ca, Ti, and rare earth al.) both in the liquid metal (in the ladle, ingot mold, the mold) and in the finished steel. The magnitude of these concentrations depend on many factors – the steel grade, the extent of it deoxidation, place and method of modifier additives and the presence and composition of slag in the ladle, casting, duration, etc.

Results and Discussion

Modification of steel natural concentrates or minerals in recent years attracted considerable interest in the metallurgists as a source of modifying of steel. Recently there has been a tendency to replace ordinary carbon steels to high-resource-efficient micro-alloyed steel in the world steel production. Which are used to dope the rare and expensive alloy elements such as vanadium, titanium, niobium, rare earth metals, barium, strontium, and others. But significant costs on the use of clean alloy elements or their alloys restrain the development directions of micro-alloying and modifying steels and alloys. Current solutions this problem is the use of natural mineral concentrates as the replacement of expensive pure elements as modifiers for micro-alloying steel. Such mineral concentrates are an excellent substitute for expensive alloying elements with microalloying steel and more available for use.

One such example of the use such concentrates is the use of ilmenite, zircon sands deposit Tugansk in the modification and microalloying steel and cast iron, the chemical composition shown in Table 1.

Table 1. The chemical composition of the concentrates

Components		Ilmenite	Zircon
TiO_2	no less	60.0	4.0
ZrO_2	no less	–	60.0
VO_2	no less	3.0	–
Al_2O_3	no less	5.0	1.0
Fe_2O_3	no less	–	1.0
SiO_2	no less	4.0	–
Humidity, %	no more	0.5	0.5

As a result of a number of studies on the influence of the given deposit concentrates on the changes in the structure and properties of the steel were obtained results presented in the works [1, 2]. As a result of the experimental heats was identified structural changes, in particular the number of non-metallic inclusions was reduced.

The modifying mixture loaded on the bottom of a pouring ladle 3 per kg inoculant per ton of steel. After holding of melt in furnace to made steel casting ladle modifier. After three minutes of exposure in a ladle with a modifier steel began to pour.

Research quality and properties of all metal blanks performed on the developed and concerted the enterprises-manufacturers programs and schemes in the expanded volume of receiving and acceptance and periodic tests. All testing and research were carried out by the standardized methods, which ensures the objectivity and comparability of the results, including the ability to compare the results of earlier conducted research and testing, including testing of serial production.

The test results in the volume of blanks acceptance tests after heat treatment for various models are shown in Table 2.

Table 2. The results of blanks test (heat treatment mode number 1)

№№ of templates	σ_p, MPa	$\sigma_{0.2}$, MPa	$\sigma_в$, MPa	δ, %	ψ, %	$KCU^{+20°C}$ MJ/m^2	KCU $_{-50°C}$, MJ/m^2	KCT $_{-50\,°C}$, MJ/m^2	K_{1C} $_{-50\,°C}$, H/mm$^{3/2}$
1	1187	1295	1413	11.5	35.0	0.29	0.20	0.12	2815
2	1187	1324	1413	9.4	24.0	0.29	0.12	0.11	2384
3	1197	1295	1393	9.4	30.0	0.33	0.20	0.11	2433
4	1207	1334	1432	10.0	32.0	0.27	0.12	0.11	2570
5	1197	1324	1442	10.5	30.0	0.29	0.22	0.10	2286

From the above data it is evident that All Templates in all tested final heat treatment conditions meet the requirements of state standard V5192-78 laid down for the category strength of O-120, in terms of mechanical properties and crack resistance. To achieve the sustained and higher rates of mechanical properties, especially on indicators of contraction ratio and toughness at temperatures below freezing, were carried out more research and testing of quenching and tempered regimes.

Results of chemical composition of the metal blanks control are shown in Table 3. The results of monitoring the content of nonferrous metal impurities are shown in Table 4.

Table 3. The results of monitoring of the chemical composition of the metal blanks

№№ of templates	Content of elements,%									
	C	Mn	Si	P	S	Cr	Ni	Cu	Mo	V
1	0.42	0.26	0.25	0.012	0.003	1.04	3.42	0.22	0.57	0.16
2	0.40	0.31	0.28	0.006	0.009	1.10	3.38	0.17	0.57	0.16
3	0.39	0.30	0.32	0.009	0.007	1.05	3.39	0.21	0.58	0.15
4	0.40	0.30	0.26	0.005	0.004	1.09	3.40	0.24	0.59	0.13

Table 4. The results of monitoring the content of non-ferrous metal impurities in the material blanks

№№ of templates	Content of elements,%							
	Sn	As	Pb	Sb	Zr	Bi	Al	Ti
1	0.009	0.012	< 0.001	0.003	0.022	0.002	< 0.005	0.020
2	0.006	0.012	< 0.001	0.002	0.025	0.001	0.010	0.022

As seen from the data on the chemical composition of the metal blanks fully complies the requirements of state standard V5192-78.

Control of the microstructure and fracture of metal, and the tensile test and impact toughness at + 20 °C - 50 °C were conducted directly at the manufacturing plants; test and crack resistance impact strength of on the with a crack samples at - 50 °C and a metal durability study conducted on samples prepared the enterprises-manufacturers blanks.

Control was carried out on metal macrostructure macro section prepared in accordance with state standard 10243-75. Etching performed in 50% hydrochloric acid solution at a temperature of 60 °C for 10–15 minutes. Fractures were made of the same samples, which held control of the macrostructure of metal. For this purpose, the samples were applied to the cuts macro sections the opposite side, and then performed their destruction on the copra or press. Sample photo metal of the microstructure is shown in Figure 1.

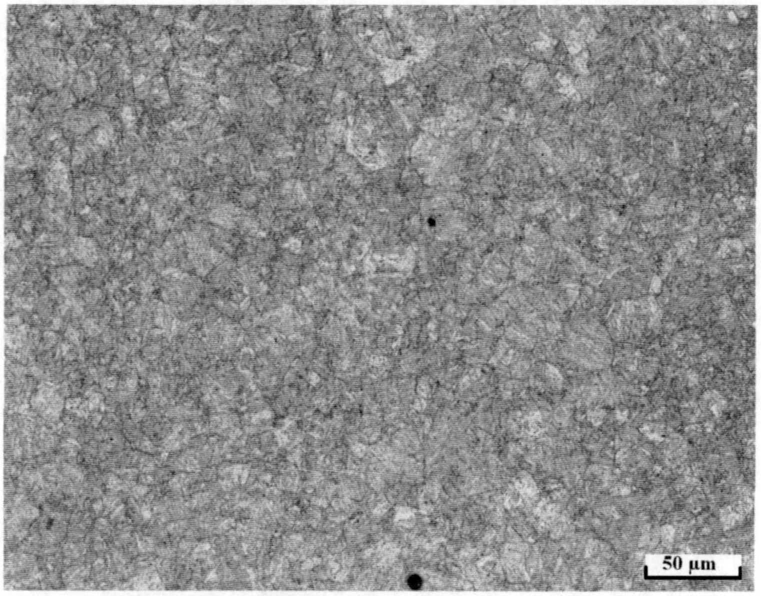

Figure 1. Photography metal of the microstructure

Structure of metal is sorbitol. From the presented data show that the grain size metal parts billet meets the requirements of the Terms of Reference. Assorted metal over the cross section is not blank.

Control content gas (oxygen and nitrogen) in a metal was carried out using an automated analyzer Eltra ON900. To improve the accuracy of determining the gas content of the gas analysis procedure is conducted repeatedly for each sample and the results were subjected to statistical analysis. The results of monitoring the content gas (oxygen and nitrogen) in a metal blanks shown in Table 5.

Table 5. The results of monitoring the gas content in a metal blank

№№ of templates	Oxygen		Nitrogen	
	Concentration	Average deviation	Concentration	Average deviation
1	0.0058	0.0017	0.0062	0.0004
2	0.0088	0.0017	0.0069	0.0004
3	0.0068	0.0011	0.0068	0.0003

From the presented data show that the average content of gases in the metal blanks is quite high, but it fits into the limits set by the requirements of the Terms of Reference (no more than 0.008% of each).

Microsections for the research of non-metallic inclusions were made of halves of the samples after the tensile tests, the impact strength and crack resistance. Analysis of contamination of have become non-metallic inclusions was carried out on non-etched thin sections by ASTM E1245-03 standard and in accordance with state standard 1778-70 using appropriate Thixomet.Pro image analyzer modules.

Micro x-ray spectroscopic studies of the chemical composition of non-metallic inclusions were performed on a scanning electron microscope ZEISS SUPRA 55 VP. The microscope is equipped with a computer-controlled scanning electron beam and digital recording signals and images, X-ray microanalyzer INCA WAVE and INCA X-MAX. Type of inclusions typical encountered in the metal blanks is shown in Figure 2.

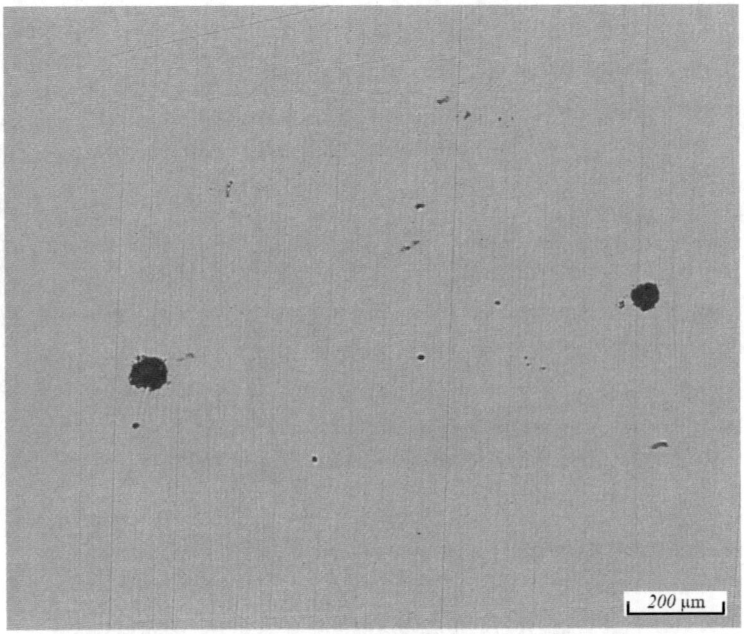

Figure 2. Type of typical inclusions occurring in the blanks

As a result of research, it was found that the reason for the decline of properties blanks is large concentrations of nonmetallic inclusions onto the surface of the fracture. On average, the content of non-metallic inclusions in the metal blanks ranged up to 2.5 points, but some areas were (metal volumes), which was an increase in the content of some types of non-metallic inclusions to 4 points. The main types of inclusions non-deformable, ductile and fragile silicates.

Furnace processing with modifying powder of concentrates showed its process ability and allowed to get steel correspond the set requirements.

Experiments are carried out on improving industrial smelting of technology by using steel as refining and modifying elements titanium, vanadium, identified the following impact modifier, on the characteristics steel components. Experiments are carried out on improving industrial smelting of technology by using steel as refining and modifying elements titanium, vanadium, and et al. identified the following impact modifier, on the characteristics steel components.

Titanium improves the purity of steel by nonmetallic inclusions on oxides silicate fluxing effect renders it makes them more fusible that promotes coagulation of inclusions and the removal of the slag. The total pollution index of steel with a titanium additive to 0.11 % is in the range of 0.016–0.020. A further increase in the titanium content leads to a sharp drop in toughness and wear resistance and an increase in total pollution index to 0.029.

With increasing vanadium content in steel toughness is reduced, and the relative wear increases, this is associated with the processes of carbide-, nitride- and carbonitride formation, which contribute to improving the initial hardness of steel. However, in this case there is a risk of inclusions location carbides, nitrides and carbonitrides of at austenite grain boundaries, which can lead to embrittlement of the metal and reducing the toughness. Study of the microstructure of impact samples at the site of fracture showed that it is a finely dispersed austenite with vanadium carbides, located within and along the grain boundaries.

Zirconium, having a high affinity for oxygen, actively deoxidizes steel and its deoxidizing ability in conditions steelmaking processes above deoxidizing ability aluminum, whereby it prevents the formation of harmful inclusions in the steel aluminous. The formed zirconium de oxidation products easily deformed and have a thermal linear expansion coefficient similar to those of the steel, and thus during heating and cooling of the metal they do not create a tension therein, unlike aluminous inclusions that cause micro cracks. Zirconium also reduces the coefficient of oxygen activity of the melt, thereby increasing the degree of assimilation of titanium them.

Conclusion

On the basis conducted researches implemented steel smelting technology with the use of modifier deoxidizing complex mineral concentrates. In the process enhancing the properties of the castings obtained by grain refinement due to the formation in the melt and in the solid steel additional nucleation (vanadium and titanium nitrides), crystallite growth restriction (zirconium), of modifying the non-metallic inclusions (titanium, zirconium). Due to such possible to effectively manage the processes of both primary and secondary crystallization cast steel technology. In such a way one can effectively influence the processes of absorption by removing of harmful impurities and gases, changes in the character of crystallization and redeployment crystallizing phases and their quantities. Achieved by reducing the size of dendritic crystallites and austenite grains when introduced into the steel elements of the data is accompanied by an increase in the grain boundary surface, decreasing the specific border concentration of impurities.

References

[1] Fedoseev S N, Lychagin D V, Sharafutdinova A S 2014 *J. Advanced Materials Research* **1040** 236–240
[2] Fedoseev S N, Mukhtar Z M 2015 *J. IOP Conference Series: Materials Science and Engineering* **91** 1–7
[3] Lu K, Lu J. 1999 *J Mater Sci Technol* **15** 193
[4] Anand Kumar S, Ganesh Sundara Raman S, Sankara Narayanan T.S.N, Gnanamoorthy R 2012 *J. Surf. Coat. Technol* **206** 442
[5] Hu W, Yunxia W, Chenshuo M, Zhaolin Z 2014 *J. Applied Mechanics and Materials* **456** 406–410
[6] Babcsan N, Leitlmeier D, Degischer H.P, Banhart J 2004 *J. Advanced Engineering Materials* **6** 421–428
[7] Mao D, Edwards J R, Harvey A 2006 *J. Chemical Engineering and Science* **61** 1836–1845
[8] Banhart J, Bellmann D, Clemens H 2001 *J. Acta Materialia* **49** 3409–3420.
[9] Rodzevich A P, Kuzmina L V, Gazenaur E G, Krasheninin V I 2014 *J. AIP Conference Proceedings* **1623** 519-522
[10] Yang C C, Nakae H 2003 *J. Materials Processing Technology* **141** 202–206.
[11] Ibragimov E A, Saprikin A A, Babakova E V 2014 *J. Advanced Materials Research* **1040** 764–767
[12] Nokhrina O I, Rozhihina I D, Hodosov I E 2015 *J. IOP Conference Series: Materials Science and Engineering* **91** 12-17
[13] Ferro P, Lazzarin P, Berto F 2012 *J. Materials Science and Engineering* **554** 122–128
[14] Li K, Xie M Z, Liu H 2009 *J. Material Science and Technology* **25** 777–783
[15] Nakae H, Fukami M, Kitazawa T, Zou Y 2011 *J. China Foundry* **8** 96–100

IOP Conf. Series: Materials Science and Engineering **142** (2016) 012058 doi:10.1088/1757-899X/142/1/012058

Use Zircon-Ilmenite Concentrate in Steelmaking

S N Fedoseev, T N Volkova

Yurga Institute of Technology, National Research
Tomsk Polytechnic University Affiliate
Kemerovo region, Yurga, Leningradskaya str. 26, RUS

E-mail: sfedoseev@tpu.ru

Abstract. Market requirements cause a constant search for new materials and technologies, for their immediate use in increasing requirements for material and energy efficiency, as well as to the quality of steel. In practice, steel production in the tended recently of more stringent requirements for the chemical composition of the steel and its contamination by nonmetallic inclusions, gas and non-ferrous metals. The main ways of increasing of strength and performance characteristics fabricated metal products related to the profound and effective influence on the crystallizing metal structure by furnace processing of the melt with refining and modifying additives. It can be argued that the furnace processing of steel and iron chemically active metals (alkali-earth metals, rare-earth metals, and others.) is an integral part of modern production of high quality products and competitive technologies.
Important condition for development of methods secondary metallurgy of steel is the use of relatively inexpensive materials in a variety of complex alloys and blends, allowing targeted control of physical and chemical state of the molten metal and, therefore, receive steel with improved performance. In this connection the development of modifying natural materials metallurgy technologies presented complex ores containing titanium and zirconium, is a very urgent task.

Introduction

Titanium and zirconium are strategic types of minerals and the development of their raw material base for Russia a pressing need. Russia in demand for titanium and zirconium raw are met by its own production has not more than 2–3%. Placers ore concentrates are the cheapest source of raw materials in the world are the main industrial source of titanium minerals and zircon. By the number of prospected and estimated of stocks of titanium and zirconium resources, Russia is one of the leading places in the world. However, all the known placers of Russian comparison with similar foreign fields characterized by a difficult geological and hydro geological condition for development several worse technological properties of ore sands (smaller size ore minerals, more clay content) and, as a consequence low profitability of their mastering. Therefore, when huge resources discovered placers in Russia, this resource base is not used, and titanium minerals and zircon are imported. Improving technology and increasing use of integrated ores – one of the main ways to increase investment appeal of Russian placer fields.

The major an industrial type of zirconium and titanium field are complex zircon-ilmenite placers of where is concentrated 92% and 52% of world reserves, respectively, for each component. Development this type fields plays a significant role in the economy of the various states, and

IOP Conf. Series: Materials Science and Engineering **142** (2016) 012058 doi:10.1088/1757-899X/142/1/012058

especially Australia, South Africa, India, Brazil. Currently buried coastal-marine placers of zircon and ilmenite are revealed in various regions of Russia. A detailed are prospected three of placer fields, but only one of them, Tugansk (Tomsk Region), prepared for the use, the other two – Central (Tambov region) and Tulun (Irkutsk region) – are currently in need of reassessment. Previously are prospected or evaluated at the exploratory stage 7 complex zircon-ilmenite placer deposits: Georgievskoe, Tarskoe, Ordynskoe, Nikolaevskoe (Western Siberia), Lukoyanovskaya (Nizhny Novgorod region), Kirsanovsky (Tambov region), Bashpagirskoe (Stavropol Territory). At involving these facilities into operation they must be considered not only as a source of ilmenite and zircon concentrates, as well as complex fields of quartz, kaolin, titanium, zirconium, and rare earth elements.

These fields are characterized the fact that in addition to the basic element ores contain, as a rule, a large list of other metals, in the amount commensurate with the principal. In these fields of rare earth metals in the ore are elements: beryllium, gadolinium, gallium, hafnium, germanium, ytterbium, yttrium, lanthanum, lithium, niobium, rubidium, scandium, thallium, tantalum, thorium, uranium, cesium, cerium, and zirconium. All this type of ore have a large set of items, including, manganese, zinc, tin, molybdenum, lead and others. The above data give grounds be concluded that the mineral resources in Western Siberia can serve as a serious basis for the development of the metallurgical industry, the production of ferro-alloys and master alloys, stainless steel, cast iron and non-ferrous metals.

In industrial ores contains 0.5–35% TiO_2, in the magmatic of disseminated ore fields are usually 7–10% TiO_2. Placers are often characterized by low titanium content. However, relatively simple to obtain titanium concentrates from alluvial deposits make them cost effective exploitation. The mined material processed at the processing factory, where receive separate concentrates: ilmenite, rutile, zircon and others. The majority of the obtained of titanium concentrates contain an entire group of impurity elements (Sc, V, Ta, Nb, TR, Ga, and others.), representing the industrial value. Especially valuable among them is costly scandium, which is permanently contains in ilmenite (up 0.02%) and rutile (0.01%).

In Tomsk region by the Company JSC "Tugansk Mining Plant" Ilmenite "" Alexandrovsky of the South area Tugansk is mined placer fields of titanium dioxide of ore sands. Where hereinafter enrich it to yield ilmenite and zircon concentrates, chemical compositions shown in Table 1 and 2.

Table 1. The chemical composition of the concentrates

Product name	Contents of components in the product, %					
	TiO_2	ZrO_2	Fe_2O_3	SiO_2	Al_2O_3	o.i.
Ilmenite	60,0	–	–	4,0	5,0	15,0
Zircon	4,0	60,0	1,0	–	1,0	18,0

Table 2. The distribution of the rare earth elements in the concentrate initial

Sample	Contents of components in the product, grams per ton								
	Th	La	Hf	Sc	Ce	Yb	Tb	Ta	Co
№1	446	1378	2790	122	2715	191	14,6	63,9	54,7
№2	813	1335	2095	88,4	2054	142	30,5	40,6	25,6
№3	649	1454	2348	105	2514	176	28,3	55,7	49,1

As the prevalence in the crust, titanium is ranked seventh among other elements. Titanium in the crust significantly larger than Mg, Zr, Cr, V and other elements. In the form of alloys, mainly iron, titanium is used for alloying, degassing and deoxidation of steel and alloys. In smelting of corrosion-resistant and heat-resistant steel, titanium is used as a stabilizer, linking and prevents the formation of carbides of chromium carbides. Titan intercrystallite reduces the corrosion and improves weld ability of corrosion resistant chromium-nickel steels. Steels processed containing titanium or some of it, have improved mechanical properties.

Zirconium - the only rare metal, world consumption of which hundreds of thousands of tons. More than 85% of the produced zircon feedstock used in the form of a mineral zircon containing 65–66% ZrO_2. Zircon concentrate (98–99% zircon) is widely used in the production of building and sanitary ceramics, refractory, and foundry services. About 10% of zircon is subjected to recycling for zirconium and its various compounds, 5% is metal and alloys. Zirconium dioxide is a valuable alloying additive in ferrous and non-ferrous metallurgy. Zirconium alloys are manufactured medical equipment and implants and thread for neurosurgery.

In industry, initial raw material for the production of zirconium is zirconium concentrates zirconium mass content of not less than 60–65% obtained an enrichment of zirconium ore. The basic methods for producing metallic zirconium concentrate – chloride, fluoride and alkaline processes.

Zirconium is an active a deoxidizer and modifier steel excelling at certain concentrations of deoxidizing aluminum capacity. Zirconium binds nitrogen in durable joints are (ZrN) and sulfur (ZrS), neutralizing their harmful impact on the quality of steel. In combination with other elements Zr increases the viscosity, strength and wears resistance of steel.

Zirconium, having a high affinity for oxygen, actively deoxidizes steel and its deoxidizing ability in conditions steelmaking processes above deoxidizing ability aluminum, whereby it prevents the formation of harmful inclusions in the steel aluminous. The formed zirconium deoxidation products easily deformed and have a thermal linear expansion coefficient similar to those of the steel, and thus during heating and cooling of the metal they do not create a tension therein, unlike aluminous inclusions that cause micro cracks. Zirconium also reduces the coefficient of oxygen activity of the melt, thereby increasing the degree of assimilation of titanium them.

Titanium and zirconium are tied into stable compounds, nitrogen and sulfur, neutralizing their harmful effect on the steel. The emerging in the steel refractory carbides, nitrides and sulfides of titanium and zirconium are used in crystallization additional center nucleation, resulting in a dense granular metal structure. When heating the treated steel with titanium and zirconium data inclusions slow grain growth, thereby preventing the occurrence of a defect such as "superheat", enabling to intensify the processes of forging, stamping, heat treatment, carburizing the metal due to heating to higher temperatures.

Lower limit of titanium content in the alloy caused by the fact that when the content of it less than 30% its expense increases and therefore silicon content in the steel. The increase in the titanium content in the alloy is more than 50% is impractical because it would result to an increase in value due to deterioration of technical and economic parameters of the alloy production.

When the content of zirconium in the alloy is less than 1% of its effects on the treated metal is weakly expressed. Zr content is more than 25% is inexpedient due to increased melting temperature of the alloy and, as a consequence, decreases in the degree of its assimilation melt.

Silicium forms stable compounds with titanium and zirconium, thus reducing the carbon content in the alloy. If silicon content in the alloy is less than 15% increases the solubility of carbon in it. Increasing the silicon content of more than 30% reduction is inexpedient consequence the content of total titanium and zirconiums in the alloy is and increase of its consumption when processing of steel.

Formulation of the problem

One of the urgent problems of metallurgical production is to improve quality of smelted metal at relatively high cost. The need to increase the yield of metal from the smelting and, most importantly, to improve the mechanical properties of the lower cost. For example, an increase in product life of responsible assignment, in most cases, connected with decision of the metal the uniformity increasing problems in both structure and composition, and to reduce pollution steel harmful impurities. Complex alloying is not always provides the desired properties, however, researchers are trying to improve the quality of existing steel brands exposure to crystallizing or recrystallizing the metal during heat treatment and deformation. In all cases the quality and the deformed thermally treated steel depends on the structure and properties of the metal.

One of the most effective methods to improving the quality of steel is to use different methods of modification. Under the modifying understand such effects, in which considerably are modified the structure and properties at practically unchanged number of major components. Modifying of is carried by small additives soluble and insoluble impurities, vacuuming and refining, exposure to ultrasonic and low-frequency vibrations, external fields (electric, magnetic), regulation of the heat sink rate and others.

Most cost effective way to improve the quality of the steel melt modification should be considered as soluble and insoluble additives because it does not require any additional expensive equipment. Introduced into the melt small additions of soluble and insoluble impurities affect crystallization parameters dislocation structure, degassing, nonmetallic inclusions and formation of secondary phases, liquation, change of shrinkage, peel speed deformation and ingot solidification, as well as recrystallization and grain growth. Recently, been great improvements in the use of modifiers to improve the quality of steels and alloys.

Are currently a large number of modifiers mechanisms of action of hypotheses. However, most of them are not sufficiently backed up by experimental data, and those that are based on a comprehensive and in-depth research, apply only to the pure metals and binary and ternary alloys. In relation to the multi-component alloy, smelted in a production environment, this hypothesis is not entirely are valid, because it does not take into account a number of side factors, associated with the steelmaking process, but have a significant impact on the final properties of metals

To address question of the advisability modification the steel or other modifier or a mixture of the components necessary to know some of the characteristics of the melt and especially the work of formation of nuclei of gas bubbles and crystallization centers. The work formation of crystal nuclei determines the propensity to steel super cooled, which is determined by the composition and degree of purity crystallizing steel. However, the choice of fuses and modifiers associated with their melting kinetics, solubility and activity in the investigated steel.

Modifying of steel of natural concentrates or in minerals have recently has attracted considerable interest in the metallurgical steel as a source of modification. Recently, in a world practice of steel production trend replacement of ordinary carbon steels to economical high strength micro alloyed steel. For alloying of which use rare and expensive alloying elements such as vanadium, titanium, niobium, rare earth metals, barium, strontium, and others. However significant costs of the use of clean alloy elements or their alloys restrain the development directions of micro-alloying and modification of steels and alloys. Current solutions this problem is the use of natural mineral concentrates as the replacement of expensive pure elements as modifiers for micro-alloying steel. Such mineral concentrates are an excellent substitute for expensive alloying elements with micro alloying steel and more available for use.

Research on the influence different types of modifiers with varying degrees of perfection of structure and composition are of considerable interest when discussing modifying the mechanism activated or become isomorphic insoluble impurities. Of even greater interest is the work on the influence particle size and distribution of the insoluble impurities on the mechanical properties of the composite alloys, but such data are still very scarce.

Results and Discussion

Placers are often characterized by low contents of the basic element (Ti, Zr, etc.). However, the relatively simple production of concentrates from placers makes them cost-effective exploitation. The mined material is processed at the processing factory, where receive separate concentrates: ilmenite, rutile, and zircon. Which later goes to the consumer to the corresponding production.

As a result of a series of studies on the influence of Tugansk field concentrates on the changes in the structure and properties of the steel been received results showing the characteristics of the data changes.

Steel produced by modifying existing technology using ultra disperse powders of zirconium oxide, titanium, niobium, hafnium, vanadium, aluminum, obtained from natural field Tugansk concentrate. Modifying mixture loaded onto the bottom of a pouring ladle counting 3 kg modifier per ton of steel. After holding of melting in the furnace make steel casting ladle with a modifier. After three minutes of exposure in a ladle with a modifier steel began bottling.

Influence of modifier on the quality and structural characteristics of the samples evaluated for austenite grain size, quantity, shape and distribution of nonmetallic inclusions in the grain boundaries and inside the grains, the presence of carbides.

Microsections for the study of nonmetallic inclusions were made of halves of the samples after the tensile tests, impact strength and crack resistance.

Before the metallographic analysis samples were several stages preparation of micro sections that is selected in accordance with characteristics of the sample material. Preparation of micro sections steps include:

1. Tenderloin samples cutting machine BUEHLER Delta Abrasimet Cutter using abrasive wheel HH type of metal materials and steel hardness 50–60 HRC at a constant water cooling;
2. Hot Mounting Presses in the phenol resin in BUEHLER SimpliMet 1000 automatic press;
3. The grinding and polishing machine to BUEHLER Phoenix 4000 using the following consumables: grinding of papers Buehler CarbiMet 180 grit, Buehler CarbiMet 240 grit, Buehler CarbiMet 600 grit; polishing with diamond paste tissue Metadi particle size 6 microns and 3 microns and suspension MasterPrep 0.05 microns.

Analysis of the microstructure of the metal was carried out using motorized optical microscope Zeiss Axiovert 200 MAT and Nikon Epiphot TME, which are an integral part of the complex analytical Thixomet.Pro.

Analysis of contamination of steel non-metallic inclusions carried out on thin sections not etched by ASTM E1245-03 standard and in accordance with National State Standard 1778-70 using appropriate Thixomet.Pro image analyzer modules.

In assessing the non-metallic inclusions in accordance with National State Standard 1778-70 used automated method, in which separation of oxides on the "non-deformable silicates" after recognition by the color inclusions made, "point oxides", "oxide stringers", "silicate fragile" and "plastic silicates." Assigning points conducted by the measured volume fraction for each type of inclusions on the basis of the known calibration curve "point-volume fraction."

The results of monitoring the content of nonmetallic inclusions in the metal of the samples are presented in Table 3.

Table 3. Results of control the content of nonmetallic inclusions in sample

№ samples	The types of non-metallic inclusions							
	OS	OP	SC	SD	CH	C	HC	HT
1	–	0.0	2.0	2.0	0.5	2.0	–	1.0
2	0.5	0.0	0.0	0.0	4.0	0.5	–	1.0
3	–	0.0	3.0	0.5	0.0	3.0	0.0	1.0
4	–	0.0	1.0	2.5	3.0	2.5	–	1.0
5	–	0.0	1.5	0.0	3.0	1.5	–	2.5

As can be seen from the data, average content nonmetallic inclusions in the samples ranged up to 3 points. The main types of inclusions – non-deformable and brittle silicates and nitrides stitch.

Micro x-ray spectroscopic studies of the chemical composition of nonmetallic inclusions were performed on a scanning electron microscopic ZEISS SUPRA 55 VP.

Type inclusions specific encountered in the samples is shown in Figure 1.

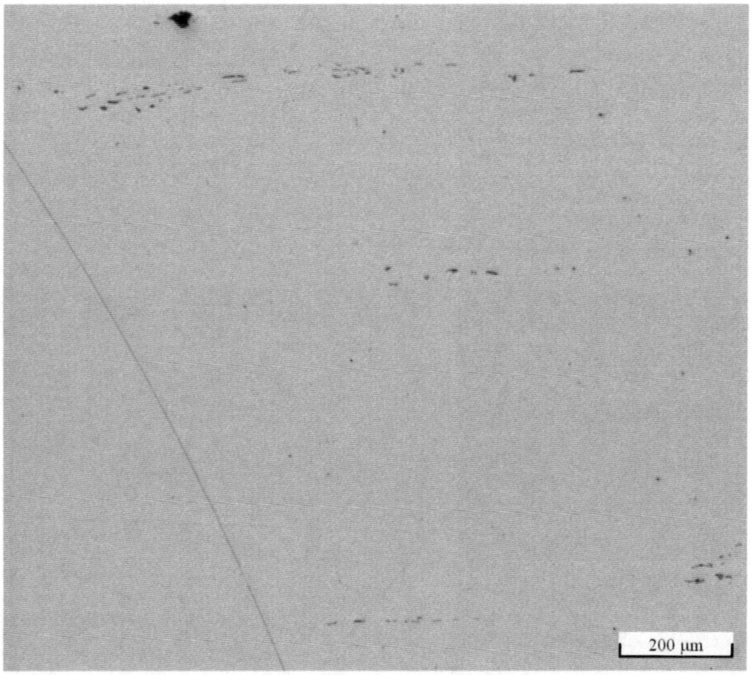

Figure 1. Type specific nonmetallic inclusions encountered in the sample

Detailed study of nonmetallic inclusions in the fracture detected samples for mechanical tests were conducted on selected samples in which unsatisfactory results in terms of metal plastic properties were obtained. As a result of research, it was found that the reason for the decline of properties blanks is large concentrations of nonmetallic inclusions onto the surface of the fracture. Type sample fracture is shown in Figure 2.

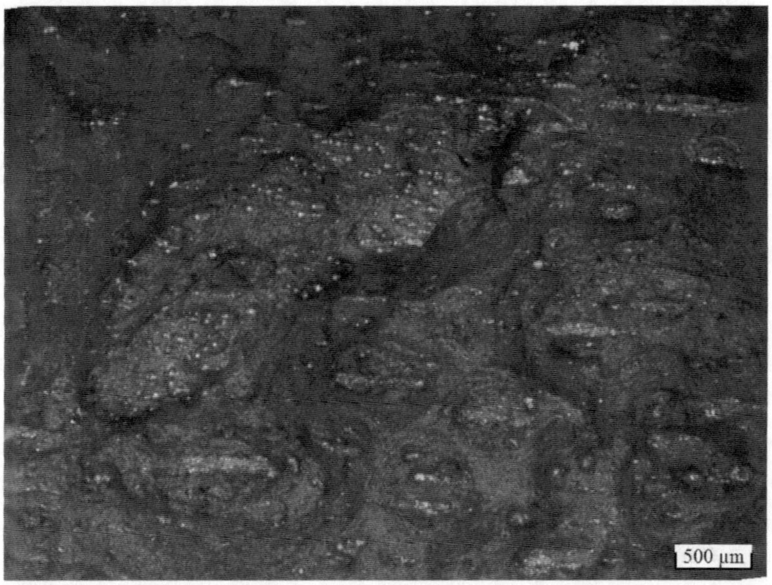

Figure 2. Type of fracture of the sample with lower plastic properties

At of metallographic research on microsections inclusion of both types were in the form of lines. To determine the composition of inclusions was their microprobe analysis was conducted, which resulted in the maps of the distribution of chemical elements contained in the inclusions.

The data obtained show that in the absence of titanium a metal are formed in the inclusions containing large amounts of calcium aluminates, and calcium and manganese sulphides. A characteristic feature of such inclusions is that they are in the form of clusters and layers on the primary grain boundary and lead to a decrease of material properties, primarily plastic. The excess of titanium contrary, leads to the formation of a metal nitride and carbonitride inclusions, which are located in the metal in the form of clusters and chains extended along the direction of deformation.

The characteristic form of the sample metal microstructure is shown in Figure 3.

Figure 3. The characteristic form of the metal microstructure

As can be seen from the data presented, the metal microstructure is austenite. The size of the metal grains of both pieces is within the limits set by the requirements of the Terms of Reference.

According to the study it was found that in the macrostructure of metal and fracture free from defects: friability, bubbles, cracks, foreign objects, captivity, shells, flakes, breaks.

The study revealed microstructure of the steel sample the influence of modifier on the microstructure of the samples compared to melting, obtained by the existing technology, namely: In particular, the steel melt modification significantly improved the uniformity of steel structures, which in its turn contributes to its density and it leads to improved isotropy, i.e. enhances of uniformity of internal stress distribution under load, reduce cracking, chipping and pore formation.

The modified samples no carbides, blowholes, a decrease in the number and size of nonmetallic inclusions on the boundaries and within the grains, as well as visually observed a decrease in grain size.

Conclusion

In the process of modifying steel zircon-ilmenite concentrate was obtained by a slight increase in the mechanical characteristics. The modification leads to reduction degree of contamination of the metal non-metallic inclusions, the fact of the presence of titanium and zirconium contributed to this.

Thus, the use of mineral concentrates perhaps steel a modifier and partial or complete replacement of expensive alloying elements. However, there is more to understand many factors influencing the properties and structure of the steel obtained in the modification.

Works on studying the influence of different types of modifiers with varying degrees of perfection of structure and composition are of considerable interest in the discussion of modifying the mechanism by activated or steel isomorphic insoluble impurities. Also greater interest working on the effect of particle size and distribution the insoluble impurities on the mechanical properties of the composite alloys, but such data is still very scarce.

References

[1] Fedoseev S N, Lychagin D V, Sharafutdinova A S 2014 *J. Advanced Materials Research* **1040** 236–240
[2] Fedoseev S N, Mukhtar Z M 2015 *J. IOP Conference Series: Materials Science and Engineering* **91** 1–7
[3] Lu K, Lu J. 1999 *J Mater Sci Technol* **15** 193
[4] Anand Kumar S, Ganesh Sundara Raman S, Sankara Narayanan T.S.N, Gnanamoorthy R 2012 *J. Surf. Coat. Technol* **206** 442
[5] Hu W, Yunxia W, Chenshuo M, Zhaolin Z 2014 *J. Applied Mechanics and Materials* **456** 406–410
[6] Babcsan N, Leitlmeier D, Degischer H.P, Banhart J 2004 *J. Advanced Engineering Materials* **6** 421–428
[7] Ibragimov E A, Saprikin A A, Babakova E V 2014 *J. Advanced Materials Research* **1040** 764–767
[8] Banhart J, Bellmann D, Clemens H 2001 *J. Acta Materialia* **49** 3409–3420.
[9] Rodzevich A P, Kuzmina L V, Gazenaur E G, Krasheninin V I 2014 *J. AIP Conference Proceedings* **1623** 519-522
[10] Yang C C, Nakae H 2003 *J. Materials Processing Technology* **141** 202–206.
[11] Mao D, Edwards J R, Harvey A 2006 *J. Chemical Engineering and Science* **61** 1836–1845
[12] Nokhrina O I, Rozhihina I D, Hodosov I E 2015 *J. IOP Conference Series: Materials Science and Engineering* **91** 12-17
[13] Li K, Xie M Z, Liu H 2009 *J. Material Science and Technology* **25** 777–783
[14] Ferro P, Lazzarin P, Berto F 2012 *J. Materials Science and Engineering* **554** 122–128
[15] Nakae H, Fukami M, Kitazawa T, Zou Y 2011 *J. China Foundry* **8** 96–100

IOP Conf. Series: Materials Science and Engineering **142** (2016) 012059 doi:10.1088/1757-899X/142/1/012059

Feeding Devices Design for Selective Laser Melting Formation of Heterogeneous Powder Structures

O I Grinin, E A Valdaytseva, I T Lasota[1], Ya B Pevzner, V V Somonov

Peter the Great St.Petersburg Polytechnic University
195251, St. Petersburg, Politechnicheskya st, 29, ph. number (+7812)2979800

[1]E-mail: rgltc@mail.ru; i.lasota@ltc.ru

Abstract. The article presents the principles of selective laser melting technology for manufacturing of polymetallic products. The results of theoretical investigations of heat and mass transfer at the border of materials are shown. Types of feeding devices design have been demonstrated.

1. Introduction

The Additive Manufacturing (AM) industry has undergone tremendous growth in recent years. Although it seems to be a relatively new technology, it has been present over the last twenty years, but it is in the late decade when a big development has taken place [1]. One of the type of additive manufacturing is selective laser melting technology (SLM) [2].

The major AM trends are improvement of quality and shape complexity of manufactured products. SLM technologies allows to produce products of the functionally graded materials (FGM) [3,4], characterized by a continuous change from one layer to another following characteristics: chemical composition, morphology and crystalline structure.

Further SLM-technology development is impossible without the development and manufacturing of specialized equipment for its implementation [5,6]. SLM machines for the manufacture of products from FGM differs from usual SLM machines by feeding devices. These feeding devices provide in layer the formation of heterogeneous powder structure of two different metals/alloys and each of the metal/alloy has its own predetermined plane coordinates.

Heterogeneous powder structures forming methods and equipment development provides an opportunity to combine the materials during the laser sintering process and polymetalic products manufacturing [7,8].

There are some practical applications of development, such as:

- Gradient structures manufacturing;
- Improving products strength characteristics and decreasing weight and size;
- Improving products heat resistance, abrasion resistance, corrosion resistance.

2. Heat and mass transfer at the border of materials during laser synthesis

According to theoretical investigations of heat and mass transfer and blending of materials at the border during laser synthesis the size of transition zone was determined. The size of transition zone in a vertical direction is about tenths of millimeter, and around half millimeter in a horizontal direction. Therefore resolution of powder feeders should be within 1 millimeter range.

The thickness of the powder layer has a size in a few powder fractions [9]. It is necessary to provide melting of material on the whole layer thickness and solid adhesion between the layers for the production of a homogeneous product. In this case, heat problem can be described as a two-dimensional problem. It means that the temperature field is determined similarly to the temperature field created by a linear heat source in the plate. In the case of using different powders (heterogeneous materials) the situation can also be modeled as a result of laser irradiation on the plate composed of two materials (see figure 1).

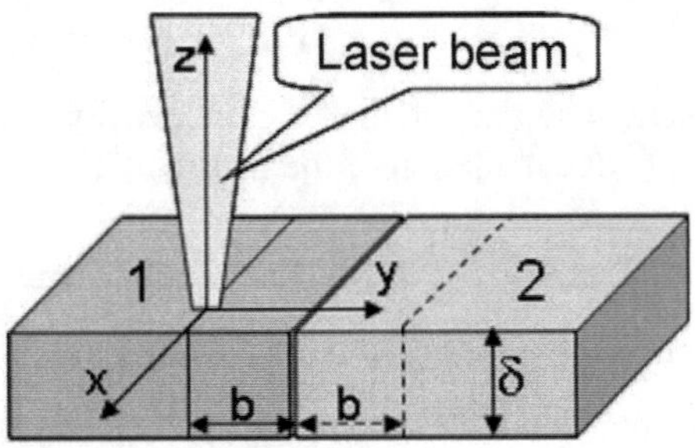

Figure 1. Scheme of the heat source movement

For the calculation of the heat transfer between layers it was accepted that contact area between the powder is negligibly small and change of powder fraction during melting was disregarded. At the border of the two materials continuity conditions of temperature field and of continuous heat flow were applied:

$$T_1\big|_{y=b} = T_2\big|_{y=b} \text{ and } \lambda_1 \frac{\partial T_1}{\partial y}\bigg|_{y=b} = \lambda_2 \frac{\partial T_2}{\partial y}\bigg|_{y=b} \quad (1)$$

Where T1, λ1 and T2, λ2 are temperature field and coefficients of heat conductivity of the 1st and 2nd material.

The fundamental solutions of the heat equation were used for the solving heat problem. Thermal field for the 1st material was determined as superposition of thermal fields from the real heat source and the reflected source (1). Thermal field for the 2nd material was determined as field of source located at a distance "c" from the border.

It is impossible to obtain an exact solution for any value of "x", so area was limited from x = 0 to x = l, where l — value equal to the length of the molten bath, this will allow to minimize an error in the most important area of calculation. Considering conditions above an approximate determination of the required distance has been obtained as a result of serial changes:

$$c \approx 2l\frac{\chi_1 - \chi_2}{\chi_1\chi_2} + b\frac{\chi_2}{\chi_1} \quad (2)$$

Where χ1, χ2 are thermal coefficients of the two materials.

In contrast to the 1st material, solution of the problem of the second material thermal field determination is an approximation suitable for use at the large distances from the heat source. It is necessary to have an exact solution of the heat problem in the second material for determination the temperature when the heat source moves near the border of materials. The linear theory of heat conduction has been used to obtain the solution as before. Thermal field can be described by the equation with boundary conditions:

$$V_0 \frac{\partial T}{\partial x} = \chi \Delta T \qquad T\big|_{y=b} = T(x) = \frac{2\lambda_1}{\lambda_1 + \lambda_2} T_1^{(0)}(x,b,z),\ \mathrm{T}\big|_{y\to\infty} \to 0,\ \mathrm{T}\big|_{x\to\pm\infty} \to 0 \tag{3}$$

T1(0) – thermal field from the real heat source in the first material.

The temperature distribution in the second material was calculated using the following expression. The heat source moves in the first material parallel to the border of materials. The source speed is constant.

$$T = \frac{V\lambda_1 c_2 \rho_2}{\pi\lambda_2(\lambda_1 + \lambda_2)} \int_{-\infty}^{\infty} T_1^{(0)}(x',0,z) e^{\frac{V(x-x')}{2\chi}} \frac{y}{\sqrt{y^2 + (x-x')^2}} K_1\left(\frac{V}{4\chi}\sqrt{y^2 + (x-x')^2}\right) dx' \tag{4}$$

Thermal properties in a molten zone and the melt surface tension were taken in view of percents of the materials in the melt. Example of calculation of the thermal field is shown in figure 2.

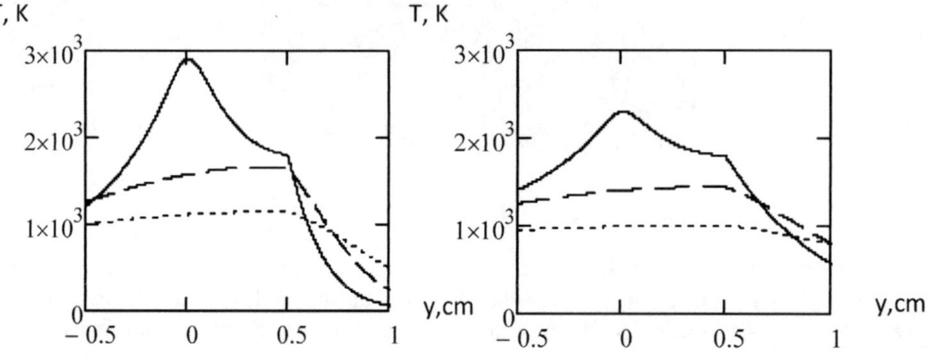

Figure 2. The thermal field in the contact zone (y=0.5 cm – the border of the two materials) The solid line corresponds coordinate "x" of the heat source, dash line – offset back at 1 cm, dotted line – at 3 cm.

3. Feeding methods in laser sintering

Heterogeneous powder structure formation process of two different metals or alloys can be realized in two basic ways: using the multi-nozzle dispensing system and using a controlled feed from the feeding bunkers.

The order of formation is shown in figures 3 and 4.

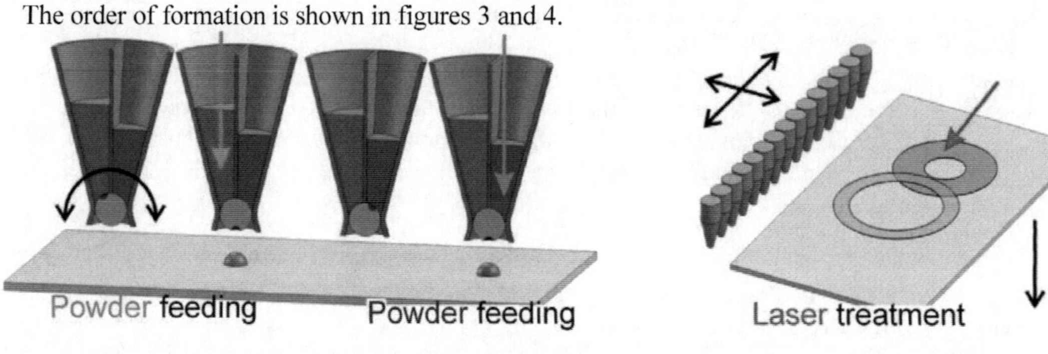

Figure 3. Heterogeneous powder structure formation order using the multi-nozzle dispensing system

Layers formation occurs by moving the multi-nozzle system over the building platform and feed of required powder at each point, then laser treatment starts.

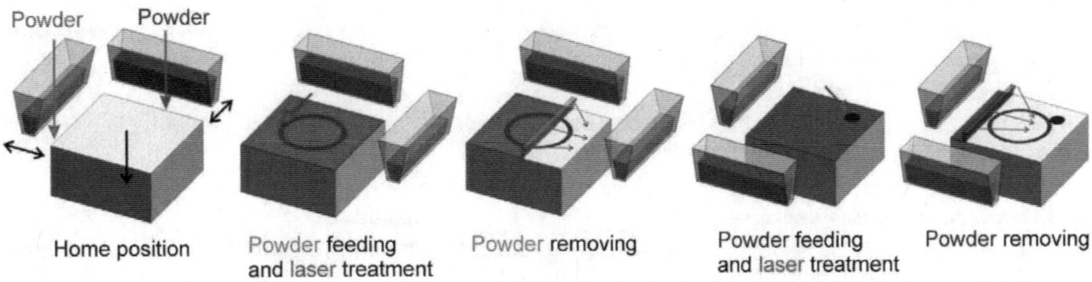

Figure 4. Heterogeneous powder structure formation order using a controlled feed from a feeding bunkers

Process of layer formation with using bunker system divides into two steps:

- deposition of the first powder , laser treatment and removing the untreated powder
- deposition of the second powder , laser treatment and removing the untreated powder

Feeder may include powder collector for the untreated powder removing from the building surface area.

The powder layer removing can be realized by using the «vacuum cleaner» principle. Powder removes through the moving slot nozzles, which are moving at low height (~1mm) above the building plane. But by using such nozzles it is hard to provide stable powder collecting characteristic. This makes necessary to collect the powder with a large margin over the layer thickness, order to avoid the cavities and overhangs formation, which can enter into the new powder layer from another material. This significantly increases an amount of the extracted and fed powder amount.

Alternatively, it is proposed to use a moving rotor as a powder collector working tool. Rotor tube is made from the porous material such as sintered metal powder (see figure 5). There is vacuum in the rotor inner. When the rotor is rolling along the powder plane, there will be the gasostatic force attracting the powder to the collector outside surface. The value of this force and thickness of collected powder layer are proportional (in the first approximation) to the vacuum inside of the rotor (gas pressure falling between the inner volume and the outside rotor surface).

Figure 5. Powder collector rotor (tube from a porous material)

The rotor can roll across the whole powder surface.

The pressure required to hold the powder on the rotor surface, in the first approximation, does not depend on the powder diameter, and is calculated by the formula:

$$\Delta P = 4/3 * \rho * g * \Delta R \quad (5)$$

At $\Delta R >> R$, where:

ΔP – the pressure difference between the processing area and the rotor inner;

ρ – powder density;
g – the free-fall acceleration;
ΔR – the rotor wall thickness;
R – the rotor tube radius.

It should be mentioned that the powder feeding bunker device provides productivity on a few orders higher [10].

4. Structure of feeding devises

The model of the feeding devise was designed for the comparing experiments of two methods described above. The model allows to realize both methods.

The main parts of powder feeder are:

- The moving system
- The feeder

General view of the moving system is shown in Figure 6.

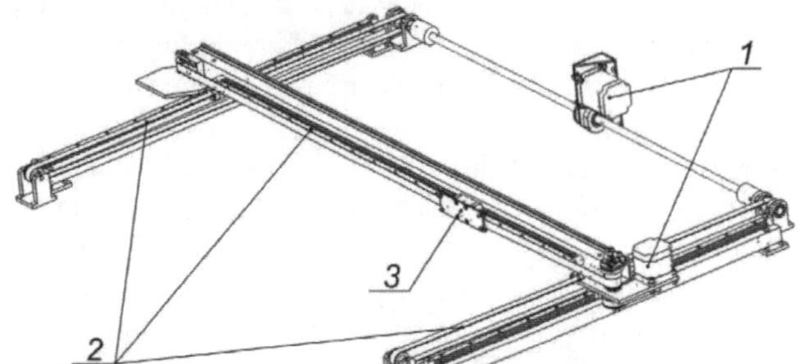

Figure 6. General view of the moving system (1 – drives; 2 – rails; 3 – carriage for fixation of the feeder)

Moving system is designed as a two-axis portal with linear guides, stepper motors and toothed belts as the motion transmission elements.

The multi-nozzle dispensing feeder and the bunker feeding device are shown in figure 7 and 8.

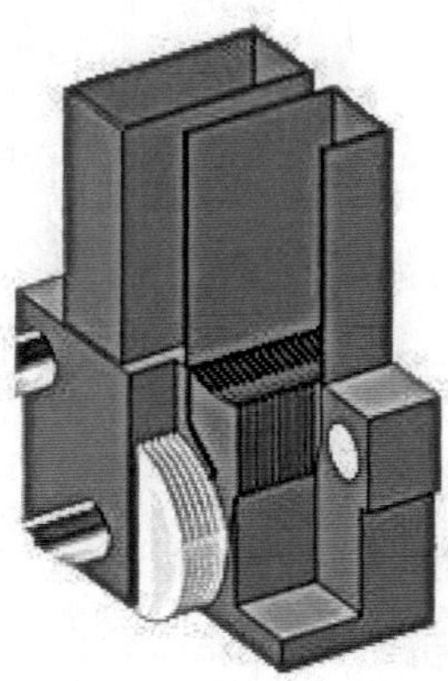

Figure 7. The multi-nozzle dispensing feeder with dispensing roller consisted of a set of thin discs (about 1mm), each of which is equipped with individually controlled drive.

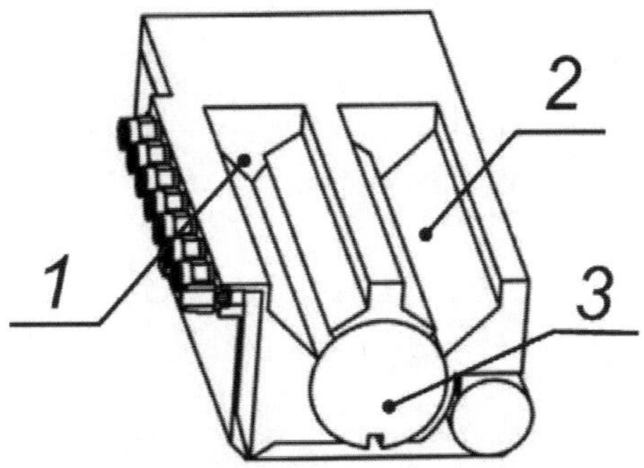

Figure 8. The bunker powder feeding device with the dispensing roller (1-first powder, 2- second powder, 3- dispensing roller)

5. Conclusion
According to theoretical investigations of heat and mass transfer and blending of materials at the border during laser synthesis technical requirements for powder feeding devices for SLM machines were formulated. Models of powder feeding devices were designed for experimental research of heterogeneous powder structures formation processes.

References
[1] Ian Gibson, David W. Rosen, Brent Stucker. Additive Manufacturing Technologies: Rapid Prototyping to Direct Digital Manufacturing. Springer 2010;

[2] Additive manufacturing: opportunities and constraints. A summary of a roundtable forum held on 23 May 2013 hosted by the Royal Academy of Engineering, Royal Academy of Engineering, November 2013, 21p;

[3] Mohammad Vaezi, Srisit Chianrabutra, Brian Mellor, Shoufeng Yang. Multiple Material Additive Manufacturing – Part 1: A Review. 71p;

[4] Yadroitsev I,Bertrand P. Selective laser melting in micro manufacturing. Annals of DAAAM for 2010 & Proceedings of the 21st International DAAAM Symposium, Volume 21, No. 1, 2010, pp

[5] David Espalin, Jorge Ramirez, Francisco Medina, Ryan Wicker. Multi-Material, Multi-Technology FDM System. 2012. SFF Symposium. Austin, Texas, USA, pp828-835;

[6] E.L. Doubrovski, A. Cencen, S. M. van Bennekom, J.C.Verlinden, J.M.P. Geraedts. Challenges in Design & Digital Fabrication of Embedable Electronics and Freeform Surface Interfaces, 9p;

[7] Wei Gao, Yunbo Zhang, Devarajan Ramanujan, Karthik Ramani, Yong Chen, Christopher B. Williams, Charlie C.L. Wang, Yung C. Shin, Song Zhang, Pablo D. Zavattieri. The status, challenges, and future of additive manufacturing in Engineering. Computer-Aided Design, Volume 69, December 2015, pp 65-89;

[8] Konrad Bartkowiak, Sven Ullrich, Thomas Frick, Michael Schmidt. New Developments of Laser Processing Aluminium Alloys via Additive Manufacturing Technique. Physics Procedia 12 (2011) pp 393–401;

[9] I. Gibson I D., W. Rosen I B., Stucker. Additive Manufacturing Technologies. Rapid Prototyping to Direct Digital Manufacturing, 2010, p116;

[10] M. Ott, M. F. Zaeh. Multi-Material Processing in Additive Manufacturing SFF Symposium 21 Proceedings International SFF Symposium, Austin, Texas, USA (2010), pp195-203

IOP Conf. Series: Materials Science and Engineering **142** (2016) 012060 doi:10.1088/1757-899X/142/1/012060

Analysis of the Lifecycle of Mechanical Engineering Products

R H Gubaydulina[1], S V Gruby[2,2], G D Davlatov[1,3]

[1,3]Yurga Institute of Technology of National Research Tomsk Polytechnic University Affiliate, Tomsk, Russia
[2]Moscow State Technical University by N.E. Bauman, Moscow, Russia

E-mail: [1,3] victory_28@mail.ru. [2] gru@bmstu.ru.

Abstract. Principal phases of the lifecycle of mechanical engineering products are analyzed in the paper. The authors have developed methods and procedures to improve designing, manufacturing, operating and recycling of the machine. It has been revealed that economic lifecycle of the product is a base for appropriate organization of mechanical engineering production. This lifecycle is calculated as a minimal sum total of consumer and producer costs. The machine construction and its manufacturing technology are interrelated through a maximal possible company profit. The products are to be recycled by their producer. Recycling should be considered as a feedback phase, necessary to make the whole lifecycle of the product a constantly functioning self-organizing system. The principles, outlined in this paper can be used as fundamentals to develop an automated PLM-system.

1. Introduction

“Product lifecycle” is a complex index of product lifetime that has been introduced recently and is thought of as its lifetime on the whole, including engineering design, manufacture, service and disposal of manufactured products. It is considered that a global automated information system entirely covering this cycle and abbreviated to PLM (Products Lifecycle Management) is to be developed urgently [1,2]. Prior to the development of PLM-system it is necessary to carry out a comprehensive analysis of the entire process of mechanical engineering product life cycle control [3].

2. Results and Discussion

In terms of PLM concept a product appears primarily in the head of the developer as a general idea of the structure. Then this guiding idea is implemented in the set of needed engineering design documents. While engineering design planning the first problem of optimal design is resolved, as well as its criteria and aims of optimization are set on the ground of requirements for both manufacture and service of the product. [3-6]

The next phase of product development involves a great number of human and material recourses, as the result the phase of machine manufacture is implemented as a certain serial production. Hence, the second problem of optimization arises, which generally focuses on overall reducing of production costs and on timely transition to manufacturing a completely new product.

After the product is sold at the market the consumer starts its service. This phase requires for consideration of the third optimization problem concerning the consumer’s interests. Among them are minimization of operational costs and determination of an optimal term of maintenance, followed by product replacement by a new one and its disposal [7].

The disposal of the machine is a final phase of product life cycle. An open cycle is usually under consideration so it is assumed that a product subsequently getting through the phases of marketing investigations, engineering design, manufacture, maintenance and disposal completes its lifetime, therefore, meets some private or public requirements. At the same time we know that such open function chains viewed from the standpoint of the theory of automated regulation and control are not efficient. The system is to be closed to improve its controllability and stability through introduction of a feedback. It is suggested to use the phase of disposal as such a feedback [8,9].

Fig. 1 shows a variant of closed PLM-system functioning for purposes of mechanical engineering enterprise. The entire control is concentrated in the block of marketing. Alongside with investigation and analysis of the market it has other tasks. The main focus of the marketing phase is on development of feasibility analysis and preliminary specifications (TDA) for product design. At the same time its functions are to be as follows: determination of economically substantiated lifetime of the machine T_E, obsolescence period of this machine structure T_M, optimal program of product manufacture $N_{\max}$, estimated product price K_P, limits of current expenses and capital investments in manufacture, optimal cycle time of product manufacture, estimated profit of the enterprise and a number of other macroeconomic indexes. T_E, determined at this phase, is to be a guideline for optimal machine engineering design according to the criterion of reliability margin and a reference for the user in the technical certificate of the product maintenance. The price of the product and its changes caused by the current situation based on the principles of fair price for both producer and consumer is given on the market and constantly monitored via a feedback. The block of marketing is also responsible for strategic planning of manufacture phase determining the time of transition to manufacture of a new product structure through the obsolescence period. This block obtains timely information from the structure dealing with disposal to estimate current expenses on product maintenance, this information is used for T_E and K_P assessment [5,10].

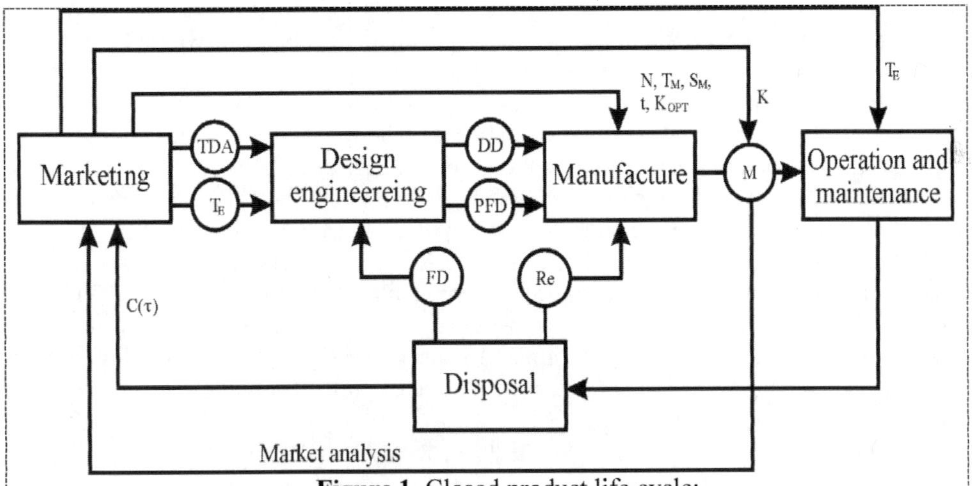

Figure 1. Closed product life cycle:
TDA – preliminary specifications; DD – design documents;PFD – process control documents; M – market; FD – fault detection; Re – renovation.

The product is utilized in order to reduce the burden on the environment and to optimize the life cycle of the product, moreover, to close the cycle according to a positive feedback. Here two problems of this phase are to be distinguished, fault detection of the depreciated machine and renovation of its parts [3,7].

Fault detection of the depreciated machine is aimed mainly at obtaining of source information on performance deterioration of the product as the whole and its components, aggregates, units, and couplings followed by statistic processing the results of such monitoring. As these data significantly interest designer and manufacturer of the product a special service responsible for put-back and disposal of over-aging machine is to be established at mechanical engineering enterprises.

Faults are to be controlled according to specially developed methods involving a wide range of up to date flaw detection facilities. This control is to be attended by sorting machine components aimed at their further renovation and re-use in new products or at recycling to source material [5,6].

For efficient implementation of the phase under consideration the structure of any machine is to be improved to provide a complete diagnosis of its current and final condition. First of all, it is to be equipped with microchips providing information on the intensity of machine maintenance, concerning both the period of service and modes of its operation. It is the data that make it possible to precise the regularities of physical depreciation of the product, therefore, determine the economically estimated lifetime, and a planned price of the product.

A number of over-aging assemblies and components of the machine might be of quite high quality after fault detection is completed. Among them are, as a rule, base and frame components of the product designed with a significant safety factor and service life because their faults often cause untimely failure of the whole machine. Generally speaking, the more components like those mentioned above are in the structure of the machine the less optimal this structure is. At the same time these components of the machine can be reconstructed and applied either for assembling similar machines unless they are still produced or as repair parts in technical support service. A great number of repair mechanical engineering technologies has been recently developed, which enable complete restoring of the product lifetime [10].

Three mentioned above issues of product lifetime optimization are interconnected, moreover, inequivalent. This is because of the so-called consumer dictate typical for the most market relations involving the manufacturer of the product and the consumer of it. The consumer determines the reasonability of purchasing any product following his needs and financial resources. On the other hand, producer dictate is also possible, especially in monopolistic economy, where the consumer has no choice when purchasing a product. As the biggest economies of the world have antimonopoly laws the latter is rather rare and not typical for well-developed market relations. Therefore, the third problem of optimization (optimization of service phase) is the main one and determines development of optimal industrial economy; the first two problems are inferior ones [10].

When servicing the product the consumer gets a positive effect from its use, on one hand, on the other hand, has input and current operational costs. Input costs of the consumer include value of the product and costs arising from assembling and start-up of the product. They are of the same importance as capital investments at the phase of manufacture. Current costs depend on the purpose and structure of the product and can embrace costs on electricity, fuel, technical service, repair components and so on. Their purpose is similar to the cost price in manufacturing process. The following formula is recommended to apply for calculation of product servicing [5,10]:

$$C = C_K \cdot \tau + \frac{K_P}{\tau + 1}, \qquad (1)$$

where C_K is a coefficient with the dimension (standard unit of value)/(standard unit of time)^2; τ - current term of product maintenance, standard unit of time; K_P - product price, standard unit of value.

Figure2 provides the calculation of the dependence (1) for various values K_P, $C_K = 200$ is constant.

IOP Conf. Series: Materials Science and Engineering **142** (2016) 012060 doi:10.1088/1757-899X/142/1/012060

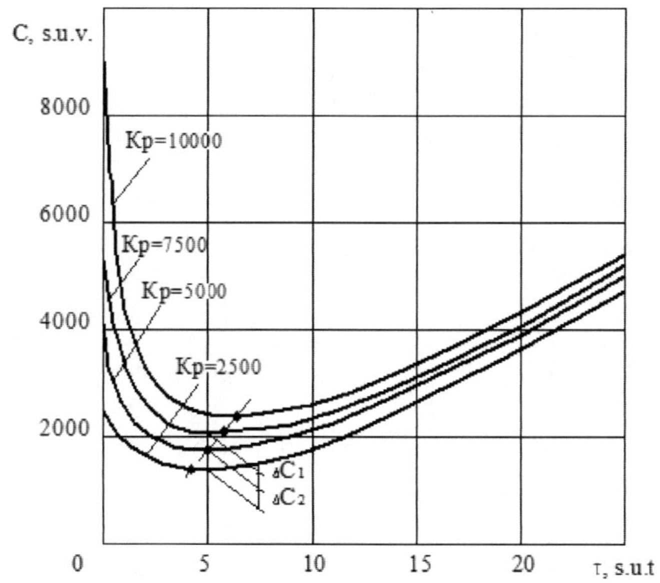

Figure 2. The dependences of current reduced expenditure on maintenance on various prices of its selling.

Figure demonstrates that all curves have a global minimum of specific reduced expenditure of the consumer for a certain period of product servicing. Let us denote the duration of the lifetime as T_E and call it an "economically substantiated lifetime", which is further thought of as a period of product maintenance and after its completion the specific reduced expenditure of the consumer are the minimum ones. Further maintenance of the product is unprofitable; the product is to be replaced by a more up to date one.

The value of an "economically substantiated lifetime" can be calculated from the equation (1), by taking a derivative with respect to the time τ and equating it to zero:

$$T_E = \sqrt{K_P / C_K}\,, \qquad (2)$$

In terms of (2) the value of "economically substantiated lifetime" doesn't depend on the absolute values of input costs of product K_P purchasing and on the level of specific current costs of its maintenance C_K, but on their correlation. The higher the correlation of price to the current costs is the higher the optimal term of its maintenance is to be and vice versa. It is drawn attention to the fact that reducing price of the product causes reduction of its "economically substantiated lifetime" (see the dashed line in Fig.1), that is, a cheap product is to have shorter optimal lifetime. However, the fact of a rigid functional relation between the price of the product and its "economically substantiated lifetime" is a more important conclusion drawn from mentioned above dependences[11,12].

It makes itself evident in the following way: for known values C_K and T_E the price of the product K_P can't be determined at random, but it must have a quite determined optimal value (fair price). We analyze this fact in more details on the examples.

Let us take as a basis the curve corresponding with the input expenditure on purchasing a product K_P=5000 standard units of value. The coordinates of optimal point A for this correlation are as follows (see Fig.2): T_E=5 standard units of time; $C_{\min}$=1833 standard units of value. We consider two variants of rough setting the price of the product, agreeing with K_P=7500 standard units of value and K_P=2500 standard units of value. In the first variant we have a cooperative financial advantage of product manufacturer and seller at the rate ΔC_1= 417 standard unit of value (see Fig.1), and in the second one – the advantage of the consumer at the rate ΔC_2=416 standard units of value. In other words, the first variant agrees with the dictate of the manufacturer, and the second one – with the dictate of the consumer of the product. To meet the requirements of both sides when purchasing and selling the price of the product is to agree with its optimal value, calculated according to the equation below:

$$K_P = C_K \cdot T_E^2. \quad (3)$$

Therefore, only planned and fact specific current expenditure on product maintenance matching exactly cause a balance of economic interests of both manufacturer and consumer and provide an optimal result for them [10].

On the basis of carried out analysis the following principle of any mechanical engineering product maintenance can be stated: to minimize the cooperative costs of consumer and manufacturer a product with certain structure and of certain quality is to be utilized within its economically substantiated lifetime, the latter depends on the values of input expenditure on purchasing and current costs on service of this product.

On the other hand, if optimal lifetime of the product is set and specific maintenance costs are determined, to minimize the overall expenditure of the manufacturer and consumer price of the product is to be rather certain, its excess or reducing cause the losses of either consumer or manufacturer of the product. This price is optimal (fair) both for the phase of manufacturing and maintenance of the product and must be a guideline for the manufacturer and a seller. That is why, in contrast to the notorious market uncertainty, determining the price of the product on the basis of competition, pricing in terms of economically substantiated lifetime adds a planned nature to the activity of a mechanical engineering company both when starting manufacture of a new product and in resolving other industrial problems [11,12].

For the foregoing reasons the phase of engineering design of a mechanical engineering product is to focus primarily on providing its economically substantiated lifetime and needed current expenditure on maintenance. Here, the methods of machine design are to foresee obtaining equal operating life for all its components. A structure is thought of as an optimal one which is designed on the basis of uniform strength, wear and other particular objective functions providing a uniform distribution of service properties (uniform inflexibility, resistance to corrosion etc.). When developing an engineering design of the machine it is necessary to search for the most optimal form of the components, units and the machine as the whole, as well as for optimal distribution of physical and mechanical properties of the construction material. A perfect product can be designed by simultaneous optimizing form and material only [3]. These principles of design enable forced restriction of product lifetime by the value of economically substantiated lifetime. Then, a consumer will face a problem of purchasing a new, more up to date machine, and a manufacturer will be able to plan volumes and development outlooks of the manufacture through constant relations with the users.

The problem of an optimal design project of the product viewed generally is to have the following solution sequence [5].

1. Calculation of an optimal operational life of the component on the basis of economically substantiated lifetime of the machine.
2. In terms of the uniform strength criterion calculation of the range of optimal forms of the product from certain material, subjected to either concentrated or distributed force and thermal loads.
3. According to the principle of uniform strength determination of the optimal structure of material for a particular form of the product.
4. Synthesis of the range of optimal solutions for form and variants of projects with optimal distribution of physical and mechanical properties of product material according to uniform strength. This phase is to be performed by the method of successive approximations, that is, after the first solution of full-strength form is made, a complying with it distribution of elasticity modulus of the component material is studied, afterwards, an optimal profile in terms of the changed VAT is

developed by a numerical method, and so forth. It is calculated till the form of the product matches the structure of its material with a set degree of approximation.

5. Calculation of optimal projects range of rubbing surfaces form of the product, manufactured from a certain material in accordance with the uniform wear.

6. Determination of tribological properties optimal distribution on product rubbing surfaces and development of projects of selective structured wear resistant coating.

7. Synthesis of optimal projects in terms of form and intensity of rubbing surfaces wearing providing a mode of uniform wear.

8. Optimization and synthesis of product projects with uniformly distributed properties in terms of other criteria of optimization (uniform inflexibility of the structure, equal corrosion resistance etc.).

9. Synthesis of the range of full-strength and uniformly wearing forms of the product with optimized structure of its material.

The final phase of optimal design consists in selecting from all technically and technologically possible variants the most economical one through comparing the term of its service life with the economic stability, determined in section 2. This methodology results in obtaining an optimal project of a perfect product with needed strength, wear resistance and efficiency for given conditions [11-15].

For stable functioning of the main production corresponding facilities of auxiliary production are necessary, the key purpose of the latter includes design and manufacture technological facilities required for new machine production. This industry has a single-piece and small-scale manufacture, so the principles of flexible technology and equipment with numerical control are required. This tool-making and experimental production is to be the third basis of the contemporary mechanical engineering enterprise and guarantee it a high efficiency.

Summary

- An economically substantiated lifetime of the product conforming to the minimal overall expenditure of the manufacturer and consumer of the machine is the basis of optimal organization of mechanical engineering production.
- The structure of the machine and its manufacturing technology are economically interrelated through the maximum profit of the enterprise.
- Disposal of the products is to be implemented by the enterprise-manufacturer and play a part of a feedback link to make the complete lifetime of the product a stable functioning self-organizing system.
- Principles stated in this paper can be used as the basis of automated PLM-system development.

References

[1] Zheng Qing-chun, Hu Ya-hui, Lv Hui-juan Research on management for manufacturing information and integrated technology oriented to PDM/PLM // Zuhe jichuang yu zidonghua jiagong jishu = Modular Machine Tool and Automatic Manufacturing Technique.– 2008, Vol.7.– C. 84 – 86, 89.

[2] Trifonov D. Implementation of PLM is, first and foremost, restoring order at the facility. // CAD/CAM/CAE information-analytical PLM – log. -2008, Vol.7.– pp. 19–21.

[3] Petrushin S.I.. The choice of optimal technology of manufacturing of engineering products. Tomsk: TPU Publishing, 2013.- 182p.

[4] Swift K.G., Bukker J.D. Selecting a process. From development to production.M.: publishing house "Tekhnologii", 2006.

[5] Petrushin S.I., Gubaydulina R.H. Organization of mechanical engineering products lifetime. Tomsk: TPU Publishing, 2012.- 200p.

[6] Blashchuk, M. Y., Kazantsev, A. A., Chernukhin, R. V. Capacity Calculation of Hydraulic Motors in Geokhod Systems for Justification of Energy-Power Block Parameters. // *Applied*

*Mechanics and Materials.*Vol. 682.- 2014.- pp. 418-425.

[7] Petrushin S.I., Gubaydulina R.H. Disposal as the closing stage of the product life cycle engineering // Vestnik machinostroyeniya. 2012. Vol.9. pp. 82 - 85.

[8] Petrushin S.I., Gubaydulina R.H. New principles of mechanical engineering organization//The7[th]international Forum on Strategic Technology IFOST 2012 September 17 – 21, 2012. Tomsk polytechnic University. VOLUME II pp.129 – 133.

[9] Petrushin S.I., Gubaydulina R.H. Optimization of the maintenance phase of mechanical engineering products. Vestnik machinostroyeniya. 2010. Vol.7. pp. 68 - 72.

[10] Gubaidulina R.H., Petrushin S.I. Economically sound operation of engineering products. Technical officer, M.: "Economica i financy" Theoretical and research and practical periodical Vol.3. (2010) pp.75 – 78.

[11] Petrushin S.I., Gubaydulina R.H. Optimization of the transition to new mechanical engineering products manufacture. // Vestnik machinostroyeniya. (2011), Vol 12. pp. 80 - 82.

[12] Konovodov V. V., Valentov A. V., Kukhar I. S. Analysis of the influence of warming on the quality of soldered instruments // IOP Conference Series: Materials Science and Engineering. - 2015 - Vol. 91, Article number 012053. -pp. 1-5.

[13] Saprykina N A, Saprykin A A, Borovikov I F, Sharkeev Y P, Influence of layer-by-layer laser sintering conditions on the qu ality of sintered surface layer of products, //IOP Conf. Series: Materials Science and Engineering.- 2015.- Vol.91. Article number 012031.- pp.1-6.

[14] Saprykina N. A., Saprykin A. A. Improvement of surface layer formation technology for articles produced by layer-by-layer laser sintering // Applied Mechanics and Materials. Vol. 379. – 2013. pp. 56-59.

[15] Petrushin S. I., Gubaydulina R. K., Grubiy S. V., Likholat A. V. On the Problem of Wear Resistant Coatings Separation From Tools and Machine Elements // IOP Conference Series: Materials Science and Engineering. - 2015 - Vol. 91, Article number 012048. - p. 1-7.

IOP Conf. Series: Materials Science and Engineering **142** (2016) 012061 doi:10.1088/1757-899X/142/1/012061

Modeling the Temperature Fields of Copper Powder Melting in the Process of Selective Laser Melting

A A Saprykin[1], E A Ibragimov[1], E V Babakova[1]

[1]Yurga Institute of Technology, TPU Affiliate

E-mail: Ibragimov@tpu.ru

Abstract: Various process variables influence on the quality of the end product when SLM (Selective Laser Melting) synthesizing items of powder materials. The authors of the paper suggest using the model of distributing the temperature fields when forming single tracks and layers of copper powder PMS-1. Relying on the results of modeling it is proposed to reduce melting of powder particles out of the scanning area.

1. Introduction

A lot of process variables are taken into account when SLM synthesizing products. More than a hundred of variables are outlined in the literature, influencing somehow the SLM process [12], however, the researchers point at some key variables, inter alia: scanning speed; laser emission power; diameter of the laser; thickness of the powder layer; gaseous medium; the strategy of scanning. For instance, the authors [2–4, 8] have considered in details the relevance of speed of scanning, laser emission power, and powder thickness for synthesizing products of copper powder PMS-1. In papers [5–7] the impact of a gaseous medium on forming the single layers of copper powder is studied, as well as the effect of laser beam diameter on the thickness of the layer to be sintered is examined. The papers [9, 10] are focused on the importance of laser beam scanning strategy for development of thermal stresses and strains of the sintered samples.

The most appropriate process conditions for sintering products of copper powder PMS-1 are determined on the base of research results outlined in paper [2]. Cubic samples with the sides 10×10×10 мм (Figure 1) are produced in these conditions. The mechanical compressive strength of the samples exceeds 105 MPa, and the porosity approximates to 15%. Further investigations require the production of samples in the form of a 6×6×30 mm bar in process conditions № 1 and № 2 (Figure 1).

All the samples produced in process conditions № 1 appear not to have any significant defects and meet the pre-set configuration (Figure 1, a). No visible defects are detected on the cubic sample produced in process conditions № 2, however, some defects occur on the lateral faces when synthesizing a bar (Figure 1, b). The defects are melted particles of powders, located on the lateral faces out of the scanning area. The defects originate mainly during the second phase of sintering the layer.

A 3D thermal model concentrated on sintering a single track and a single layer according to a pre-set strategy is developed to detect the causes of defects. The model problems are solved by a mathematical package COMSOL Multiphysics, which is an efficient interactive environment for FEM simulating and calculating a great number of scientific and technical issues based on differential equations in partial derivatives (PDE). A lot of works are available, the authors of which deal with simulating the interaction between the laser beam and substances by means of this mathematical package [11-16].

Process conditions № 1: W=30 W; V=3000 mm/min; S=0.1 mm; t=300 °C
a)

Process conditions № 2: W=30 W; V=2000 mm/min; S=0.1 mm; t=300 °C
b)

Figure 1. SLM produced samples.

2. Methods of research

In the SLM process the powder is distributed, as a rule, over the whole operating area of the machine, and the section of sintering is far smaller (Figure 2)

IOP Conf. Series: Materials Science and Engineering **142** (2016) 012061 doi:10.1088/1757-899X/142/1/012061

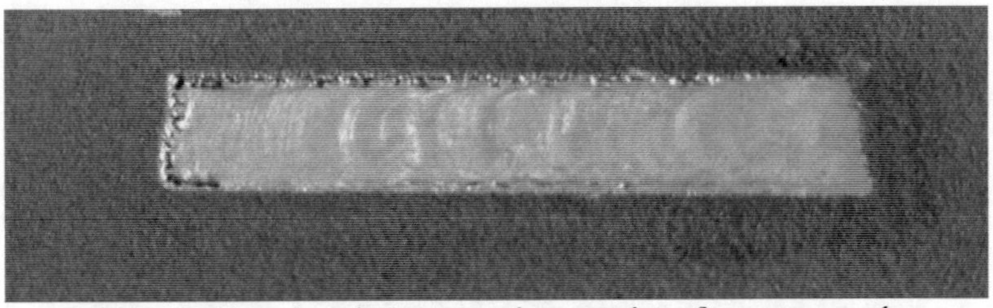

Figure 2. Sintering the rectangular samples of copper powder.

Therefore, for the purpose of modeling the geometry of powder in a freely poured condition can be limited by some millimeters relatively the section to be sintered.

In Figure 3 there is a model structure of sintering a single track on the powder layer. And in Figure 4 there is a model presented to sinter a single layer of powder poured on the previously sintered layers according to the pre-set strategy.

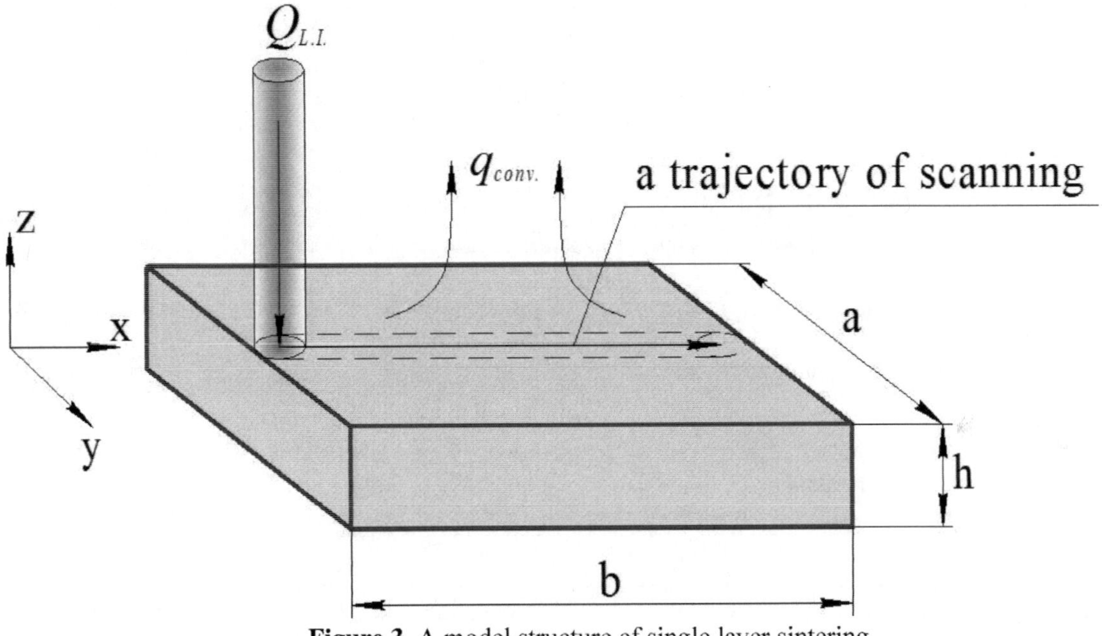

Figure 3. A model structure of single layer sintering.

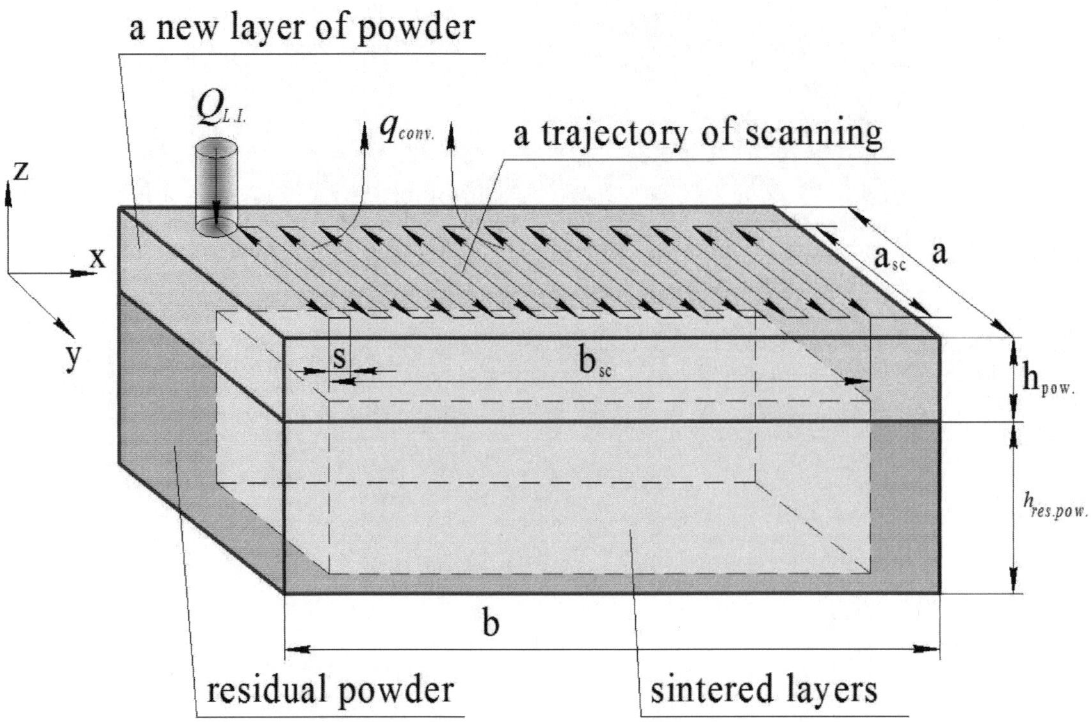

Figure 4. A model structure of sintering a single layer.

In the proposed model heat is distributed in the sample due to thermal conductivity only. A mathematical model describing the heat transfer by means of thermal conductivity is expressed by the thermal conductivity equation given below:

$$\rho C\frac{\partial T}{\partial t} - \nabla\cdot(k\nabla T) = Q, \quad (1)$$

where T – temperature; ρ – density; C – thermal capacity; k – thermal conductivity factor; Q – originating or absorbed heat.

Heat-transfer properties of powdered materials (ρ, C, k) differ considerably from those of solid (monolithic) materials, they are identified experimentally and given in Table 1.

Laser impact is determined as a volumetric source of heat, the intensity of which depends on laser impact at various depths of the powder layer. The equation to calculate the laser impact heat is as follows:

$$Q(x,y,z) = Q_0(1-R_c)\cdot\frac{A_c}{\pi\sigma_x\sigma_y}e^{-\left[\frac{(x-x_0)^2}{2\sigma_x^2}+\frac{(y-y_0)^2}{2\sigma_y^2}\right]}\cdot e^{-A_c z}, \quad (2)$$

where Q_0 – laser emission power; R_C – reflection coefficient; A_C – absorption coefficient; $e^{-\left[\frac{(x-x_0)^2}{2\sigma_x^2}+\frac{(y-y_0)^2}{2\sigma_y^2}\right]}$ – 2D Gauss distribution of emission power over the sample surface in the plane x, y; $e^{-A_c z}$ – exponential decay of power over the layer depth of a sample (Bouguer law).

The following assumptions are to be taken into consideration when implementing the model:

- reflection and absorption coefficients are constant;
- thermal effects of phase transformations are not taken into account;
- the surface of powder layer, along which the laser beam is moved, is parallel to the plane x-y of the system of coordinates;
- the upper plane of the powder layer is smoothed out according to z=0, consequently, the effect of power absorption can be expressed as follows: *exp(-Ac·abs(z))*;
- the center of laser beam can be displaced via changing the variables x_0 and y_0;

- the diameter of laser beam is calculated on the base of standard deviation parameters σ_x and σ_y.

Free convection with the upper surface of powder layer is taken into account as the boundary conditions. In Table 1 there are designations and values of variables provided, which we rely on when solving the problem.

Table 1. Designations and values of variables

A variable	Model for 1 track	Model for 1 layer
The laser spot radius relatively the coordinate axes σ_x/σ_y, mm	0.1/0.1	0.1/0.1
Laser power Q_0, W	15	30
Power distribution in the laser spot	Gauss distribution	Gauss distribution
Speed of scanning v, mm/s	50	33.3
Scanning pitch s, mm	-	0.1
Temperature of powder pre-heating, °C	25	400
Ambient temperature, °C	25	25
Dimensions of a new powder layer a/b/$h_{pow.}$, mm	2.5/10/1	12/36/0.1
Dimensions of residual powder a/b/$h_{res.pow.}$, mm	-	12/36/0.4
Scanning area dimensions $a_{sc.}/b_{sc.}$, mm	10	6/30
Dimensions of sintered layers $a_{sc.}/b_{sc.}/h_{res.pow.}$	-	6/30/0.4
Thermal conductivity coefficient of powder, W/(m·°C)	0.36	0.36
Thermal capacity of the powder, J/(m^3·°C)	6.1	6.1

Heat-transfer properties of copper powder differ significantly from those of solid copper. The following properties of copper powder, necessary for modeling are identified experimentally: bulk density $\rho_{pow.}$=1.86 g/cm^3, thermal conductivity coefficient $k_{pow.}$= 3.6×10^{-3} W/(m·°C) and thermal capacity $C_{pow.}$=6.1 J/(m^3·°C).

The results of modeling single tracks are given in Figure 5.

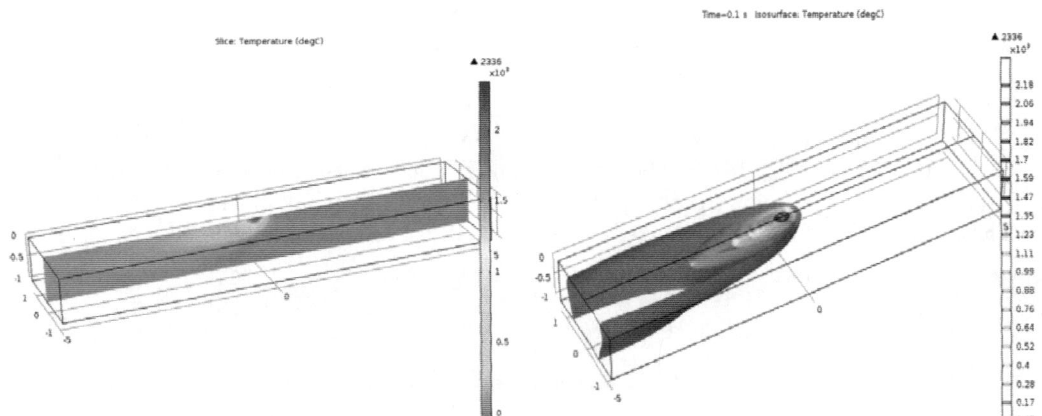

Figure 5. Temperature distribution when sintering 1 track.

The results of modeling a single track demonstrate that the depth of the zone with temperature above that of metal melting approximates to 0.2 mm. Therefore, the thickness of the sintered track is about 0.2 mm. These results meet the data of experiments outlined in [1] for powder PMS-1.

IOP Conf. Series: Materials Science and Engineering **142** (2016) 012061 doi:10.1088/1757-899X/142/1/012061

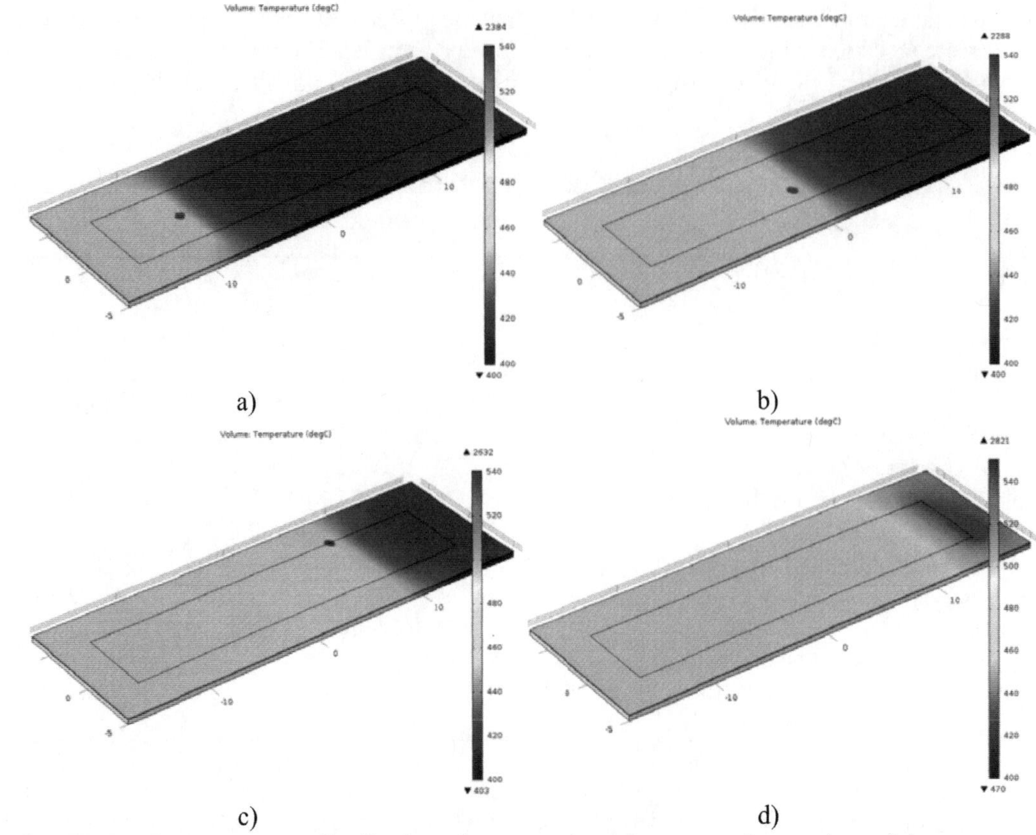

Figure 6. Temperature distribution when scanning a layer at various points of time: 10 s (a); 25 s (b); 40 s (c); 54 s (d).

When modeling the process of a single layer sintering, it is of interest how the temperature is distributed through the sample at various points of time. Since the temperature in the laser spot is much higher than that of a sample, temperature scaling is carried out for appropriate representation of temperature fields through the sample.

The results of modeling a single layer are given in Figure 6.

In Figure 8 one can see the plots of changing powder temperature 0.3 mm from the scanning area. The area under consideration is shown in Figure 7. Each plot demonstrates the distribution of temperature along the area to be studied with a 0.5 mm step at the points of time when a laser beam approaches the end line of the scanning area.

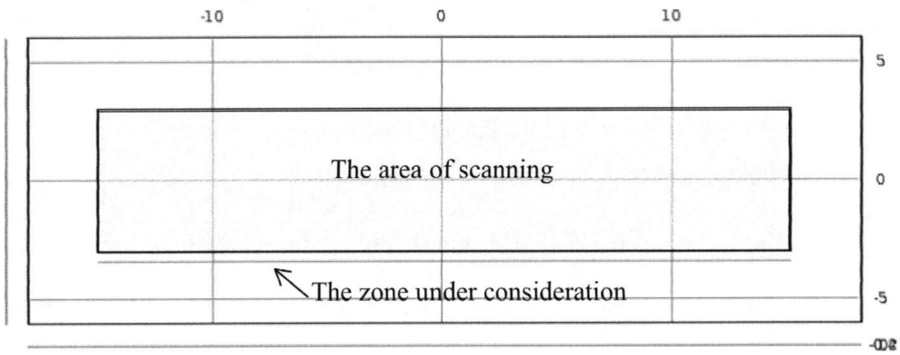

Figure 7. The zone to be examined along the area of scanning.

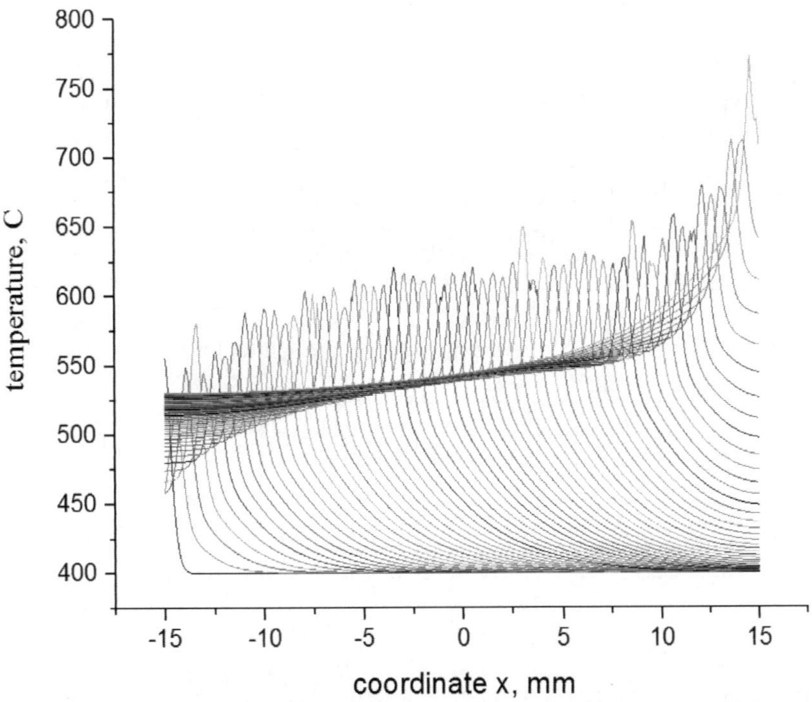

Figure 8. Powder temperature variation along the side line of the sintering area.

3. Results and discussion

Analyzing the data presented in Figure 6 and 8 one see that powder located on each side of a laser scanning area is heated when sintering a layer. The total time of scanning the surface is 54 seconds in the model. In the first 20 seconds the not-sintered powder is heated a little (Figure 6, a). In the period 20 to 40 seconds powder gets more heated, but then its temperature does not change much (Figure 6, b, c). During the end period 40 to 54 seconds the temperature of powder on the sides and in front of the scanning area goes up very fast (Figure 6, d).

During the second and third periods of heating the powder on the scanning area sides conditions are provided for self-melting of powder particles out of the scanning area under a slight energy impact (Figure 1, b). Non-uniform density distribution in the powder layer is also very important for self-melting of powder particles. This is possible because of different sizes and dendritic shape of powder particles when making a new layer of powder. These are the causes of non-uniform distribution of powder particles over the surface.

Decreasing the effect of powder heating on the sides is possible via changing the strategy of scanning (Figure 8).

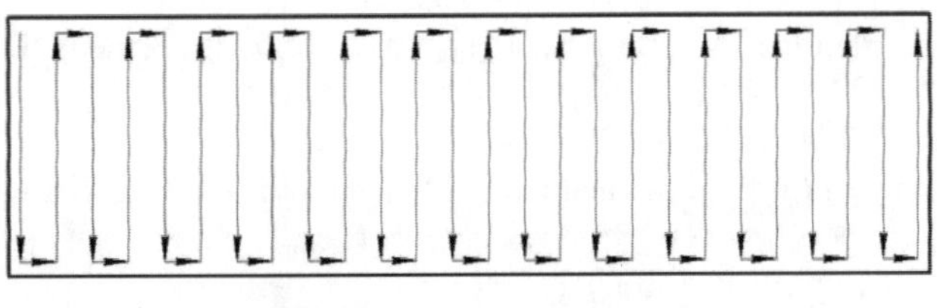

a)

The sequence of sintering the sections

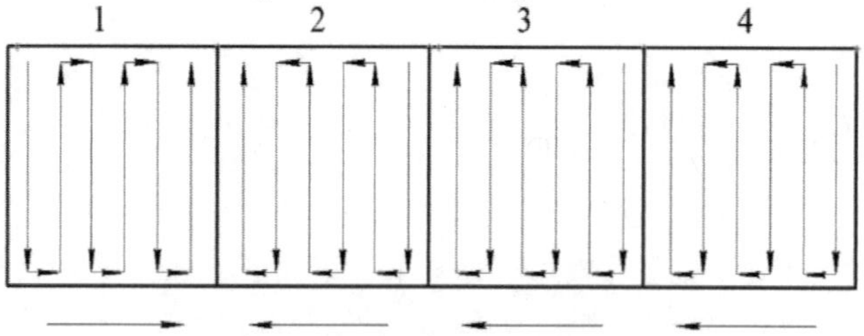

Direction of the scanning front

b)

The sequence of sintering the sections

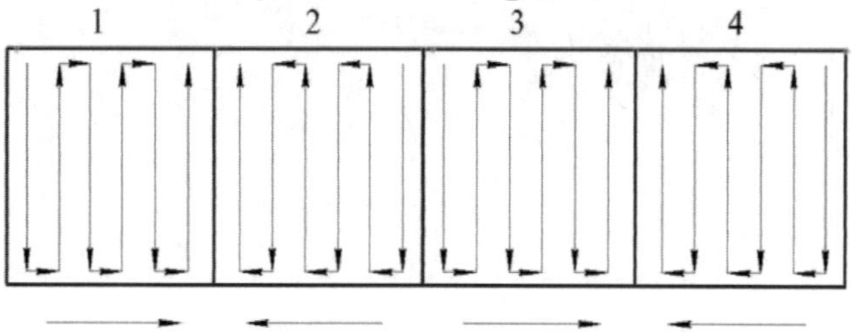

Direction of the scanning front

c)

Figure 9. Strategies of scanning: a – complete line-oriented zigzag; b – "close to each other"; c – "away from each other".

It is suggested to divide the area of scanning into sections and scan them "close to each other" (Figure 9, b) or "away from each other" (Figure 9, c). In these conditions the powder on the scanning area sides can get cooled down due to convection and thermal conductivity.

4. Conclusion

The mathematical model describing the distribution of heat fields proposed in the paper makes it possible to determine the sections on the surface providing the most appropriate conditions for uncontrolled self-melting of powder particles. Alongside with key process variables of SLM synthesis one should consider the configuration of the layer to be scanned. It is proposed in this study to divide extended spaces of the surface into the sub-areas and scan them so that the free powder out of the scanning area get cooled down. This technological decision makes it possible to reduce the number of defect products.

References

[1] Saprykin, A.A. Saprykin, N. A 2012. Engineering Support for Improving Quality of Layer-by-layer Laser Sintering *J. IFOST2012* **129–132**

[2] Mani M., Lane B., Donmez A., Feng S., Fesperman R. Measurement Science Needs for Real-time Control of Additive Manufacturing Powder Bed Fusion Processes: Web: http://dx.doi.org/10.6028/NIST.IR.8036.

[3] Saprykin A. A., Ibragimov E. A., Babakova E. V., Yakovlev V. I. 2015 Influence of mechanical activation of copper powder on physicomechanical changes in selective laser sintering products *J. AIP Conf. Proc.* **1683**, 020199

[4] Saprykin A A, Gradoboyev A V, Yakovlev V I, Ibragimov E A, Babakova E V 2015 Comparative analysis of processes to activate copper powder ПМС-1 for SLS synthesizing products *J. Metal processing* **82–88** 3 (68)

[5] Ibragimov E A, Saprykin A A, Babakova E V 2014 Influence of Laser Beam Machining Strategy at SLS Synthesis *J. Advanced Materials Research* **764–767** 1040

[6] Saprykina N A, Saprykin A A, Borovikov I F, Sharkeev Y P, Influence of layer-by-layer laser sintering conditions on the quality of sintered surface layer of products 2015 *J. IOP Conf. Series: Materials Science and Engineering* **91** 012031.

[7] Saprykina N A, Saprykin A A, Matrunchik M S, Formation of Surface Layer of Cobalt Chrome Molybdenum Powder Products with Differentiation of Laser Sintering Modes 2014 *J. Applied Mechanics and Materials* **294-298** 682

[8] Galevsky G.V., Rudneva V.V., Garbuzova A.K., Valuev D.V., Titanium carbide: nanotechnology, properties, application 2015 *J. IOP Conference Series: Materials Science and Engineering.* **1-7** 91

[9] Saprykin A. A., Babakova E. V., Ibragimov E. A. Fiber-Reinforced Composites in Rapid Prototyping Technologies 2015 *J. Applied Mechanics and Materials* **185-188** 770

[10] Li C., Fu C.H., Guo Y.B., Fang F.Z. Fast Prediction and Validation of Part Distortion in Selective Laser Melting 2015 *J. Procedia Manufacturing.* Volume **1–11** 30.

[11] Thijs L Kempen K, Kruth J-P ,Van Humbeeck J Fine- structured aluminium products with controllable texture by Selective Laser Melting of pre-alloyed AlSi10Mg powder 2013 *J. Acta Materialia* **1809-1819** 61

[12] Tomashchuk I., Jouvard J.-M., Sallamand P., Numerical modeling of copper-steel laser joining 2007 Excerpt from the Proceedings of the COMSOL Users Conference 2007 Grenoble [electronic resource] access mode: www.researchgate.net/publication/266443725_Numerical_modeling_of_copper-steel_laser_joining.

[13] Biancol N., Manca O., Nardini S. Tamburrino S. Transient Heat Conduction in Semi-Infinite Solids Irradiated by aMoving Heat Source 2007 Excerpt from the Proceedings of the COMSOL Users Conference Grenoble [electronic resource] access mode: www.researchgate.net/ publication/234006400_Transient_Heat_Conduction_in_Solids_Irradiated_by_a_Moving_Heat_Source

[14] Karbasi H. COMSOL Assisted Simulation of Laser Engraving / Excerpt from the Proceedings of the COMSOL Conference 2010 Boston [electronic resource] access mode:http://www.comsol.com/ paper/comsol-assisted-simulation-of-laser-engraving-7574

[15] V. Bruyere, C. Touvrey, P.Namy, Comparison between Phase Field and ALE Methods to model the Keyhole Digging during Spot Laser Welding / Excerpt from the Proceedings of the COMSOL Conference 2013Rotterdam [electronic resource] access mode:www.comsol.com/paper/download/181647/bruyere_paper.pdf

[16] M. Darif, N. Semmar, Numerical Simulation of Si Nanosecond Laser Annealing by COMSOL Multiphysics / Excerpt from the Proceedings of the COMSOL Conference 2008 Hannover [electronic resource] access mode: http://www.comsol.com/paper/numerical-simulation-of-si-nanosecond- laser-annealing-by-comsol-multiphysics-5586

[17] Pryor, Roger W. Multiphysics modeling using COMSOL : a first principles approach / Roger W. Pryor. 2010.

[18] Kumar S, Kruth J-P 2009 Composites by Rapid Prototyping Technology *J. Materials & Design* **1–23.**

[19] Murali K, Chatterjee A N, Saha P, Palai R, Kumar S, Roy S K, Mishra P K, Choudhury A R 2003 Direct selective laser sintering of iron-graphite powder mixture *J. Journal of Materials Processing Technology* **179–185** 136

IOP Conf. Series: Materials Science and Engineering **142** (2016) 012062 doi:10.1088/1757-899X/142/1/012062

Laser Treatment on the Coating Surface Having Been Performed by Means of Plasma Surfacing With Powder Made of M2 Steel*

A A Khaidarova [1], S A Silantiev [2]

[1,2]Tomsk Polytechnic University, Lenina Str., 30 Bld., Tomsk, 634050

E-mail: [1]haydarova@tpu.ru; [2]silantev-sergei@mail.ru

Abstract. In this study researches were carried out about the impact of pulsed laser irradiation on the surface of M2 steel which had been surfaced beforehand on steel 20 by poweder-pasma surfacing technique. The surface treatment was performed by the single point laser action and by successive imposition of overlapped impulses. During the surfacing the average irradiation power was varied from 15.8 to 21.0 W. The impulse duration in all points was constant and it was 7 msec. In this study it was researched the influence of laser-beam power change on the depth of penetration, changes in the structure and microhardness of treated areas.

Introduction

There are different ways how to solve some issues connected with working surfaces of machinery parts and machines with high mechanical and tribotechnical characteristics, the following are: the coating application with overlaying or sputtering, the thermo treatment, the surface treatment with highly concentrated sources of energy such as: an electronic beam, concentrated arc plasma spray or laser. Special preferences are given to the laser as the means that allows you to change the structure of the metal surface layer at high speed and this means does not require special protection from the ambient air unless the treated material requires it.

With the help of laser treatment researchers are able to solve such tasks as surface texturing and modification [1-3], the increase of wearing capacity, the hardness, the strength and the fatigue resistance of surface parts [1-14]. It is all possible due to the high thermal gradient which arises in the area of laser irradiation. It is possible to achieve austenitic-martensitic transformations in the smallish volume of material due to the rapid heating and cooling processes during the laser treatment thus changing and strengthening separately taken areas [7.8].

The choice of laser irradiation parameters especially influences the structural and characteristic change of the material to be treated. In works [4,9] it is shown that the essential factors affecting the depth of the laser penetration are the following: the wave length of an emitter, the average power, the energy and the radiation energy density and the laser speed under the continuous-wave irradiation. In recent years, the processes of structural changes of steel under the impact of short-term radiation treatment with laser pulses are being actively studied [1-4,7,9]. During the pulse treatment it is necessary to take into account such parameters as: the peak power density, duration and pulse repetition rate. In the work [12] it is shown that the laser treatment in pulsed mode allows you to get greater action depth due to the higher peak power density than the treatment in the continuous-wave mode. Besides the use of pulsed laser treatment allows you to control the surface characteristics with its minimal geometry distortions.

The most efficiency of laser treatment is achieved when it is applied to the steels, which are able to harden themselves and undergo austenitic-martensitic transformations [8]. In this regard, fast-cutting steels as M2, R18, R12 are of interest. In works [13,14] it was shown that coatings on the basis of such kind of steels which were made with electron-beam and plasma surfacing techniques are stable in the conditions of abrasive wear. It might be supposed that the additional treatment of such kind of coatings with laser pulses will considerably enlarge the field of their application by means of the surface texturing, which has proved itself in works [15,16].

Technical publications do not show enough information about the influence of the laser beam power on the structure and hardness formation of tool steels after laser treatment. That is why it is necessary to carry out a research on the influence of pulsed laser irradiation parameters on the penetration depth, structure and characteristics of fast-cutting steels like M2.

Materials and research methods

A pre-polished surface coating made of M2 steel with roughness of Ra 0.08 was treated with laser irradiation. The coating was treated beforehand with plasma powder surfacing technique at current strength of 200 A and surfacing speed of 0.17 cm/s.

The laser welding machine HTF-50T with aluminum-yttrium solid-state laser (YAG: Nd3 +) was used as the radiation source.

For the needs of the comparative analysis four single treated areas have been done and also it has been done one linear area which represents a number of impulse treatment points applied against each other with partial overlapping. Modes of surface treatment were remarkable for power pulses of rectangular shape (Table 1). The laser irradiation treatment was performed in the open air without protection for the surface coating against oxidation.

Table 1. Mode parameters for a laser irradiation action on the structure and characteristics of surfaced layer made of M2 steel.

Mode №	Peak power (qp), KW	Average power (qA), W	Pulse energy, J	Pulse duration, ms	Pulse repetition rate, Hz
1	1.50	15.8	10.50	7	Single pulse
2	2.00	21.0	14.00	7	Single pulse
3	2.42	25.4	16.94	7	Single pulse
4	4.05	42.5	28.35	7	Single pulse
5	2.00	21.0	14.00	7	1.5

The studies about treated metal were carried out in the surface layer and in the cross-section along the central axis of the track and points. The study of macro and microstructure was carried out by means of optical and electron metallography. Optical metallography was conducted with the help of a microscope OLYMPUS GX51 and an application module SIAMS 700 in the bright field mode and in the differential interference contrast. Electron microscopy was done with a scanning electron microscope Hitachi S-3400N, equipped with an attachment for elemental analysis EDS, WDS.

The structure identification of the treated metal was accomplished by dipping the pre-polished surface into reagents with the following compositions: HNO3(ml):C2H5OH(ml)=4:100; HCl(ml):CuCl2·H2O(g):H2O(ml):C2H5OH(ml)=20:1:15:65. Metal holding time in the solutions ranged from 20 to 60 seconds.

Microhardness was measured with HVS-1000 device with 100 microns depth in each step at 5H load.

Results and discussions

Figure. 1 represents the microstructure of the single points surface, the microstructure was figured out with optical metallography only after the surface polishing and etching. All points have three distinctive areas: reflow zone (RZ) - the central part of the spot; hardened zone (HZ) - unevenly etched circle which is remarkable for the structural heterogeneity; heat-affected zone (HAZ) - bright circle which is ringed with the dark, well-etched circle.

Thermocapillary and capillary phenomenon that occur in melting points as well as the martensitic transformations both result in the surface deformation due to the surface tension and due to the cracks formation in the center of the spot. At the same time, the greater the coefficient of penetration shape is (k=h/d), which increases along with power growth of laser irradiation, the larger deformations and cracks become (Figure. 1,d).

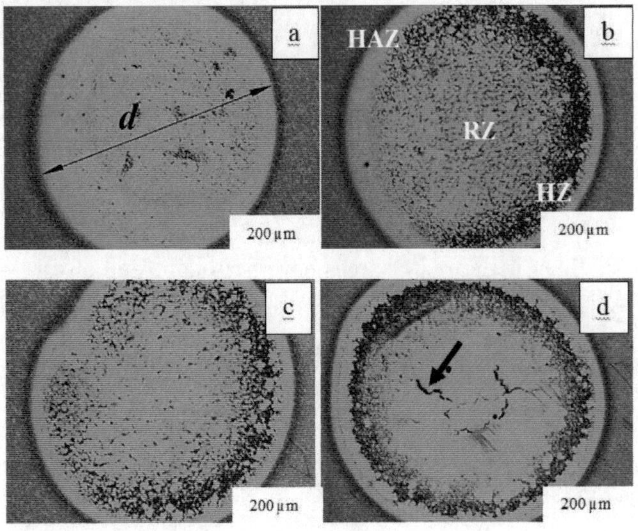

Figure 1. The microstructure of the point surface which was achieved with average laser irradiation power: a – 15,8 W, b – 21.0 W, c – 25,4 W, d – 42,5 W

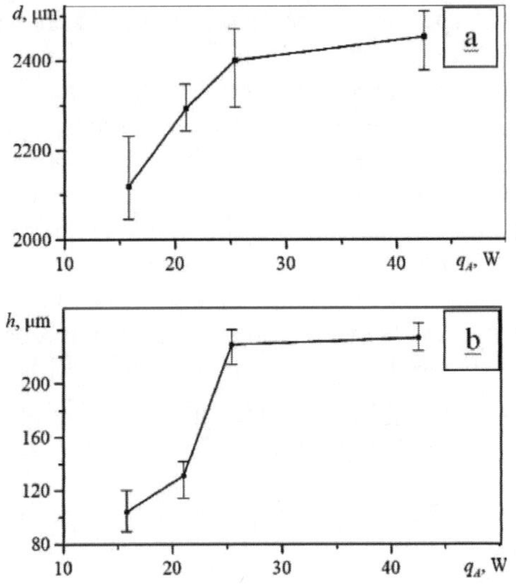

Figure 2. The graph of the spot irradiation diameter (a) and the penetration depth (b) by laser power growth

The total length of the laser heating zones was accepted as the total depth of laser action zone, while structural and phase changes occurred in laser heating zone with respect to the original steel structure. The penetration depth increases along with the growth of power or pulse energy (Figure. 2b), however, these changes are of irregular character. The heating depth is 100 ... 130 µm when an average power laser pulse is 15.8 ... 21.0 W. The further power increase up to 25.4 W leads to the step like increase of the penetration depth up to 230 µm with a simultaneous increase of the molten metal layer on the walls of a steam-gas channel, as a result it occurs the slowdown of penetration even when the power is increased up to 42.5 W. The Form of laser action points in the cross section is close to hemispherical.

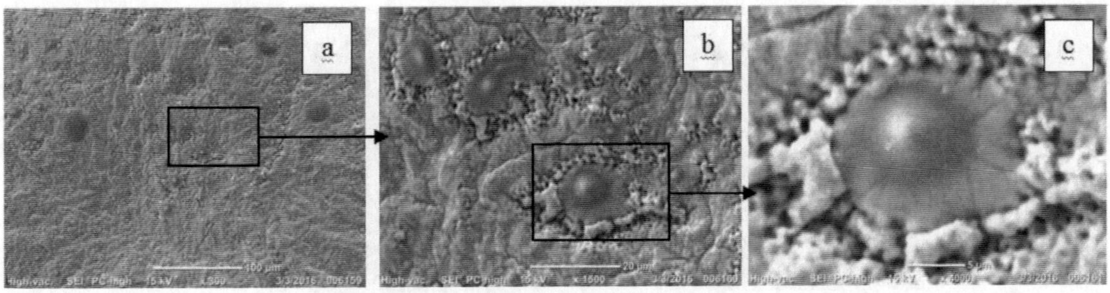

Figure 3. Microstructures of reflow zone.

The protection absence of the treated melt surface during the laser air reflow results in the gases absorption from the ambient under the influence of the convection flows in a thin surface layer. This facilitates the metal oxidation on the whole surface of the points, and this is proved by a large amount of oxygen discovered with microroentgen spectral analysis (Figure. 4).

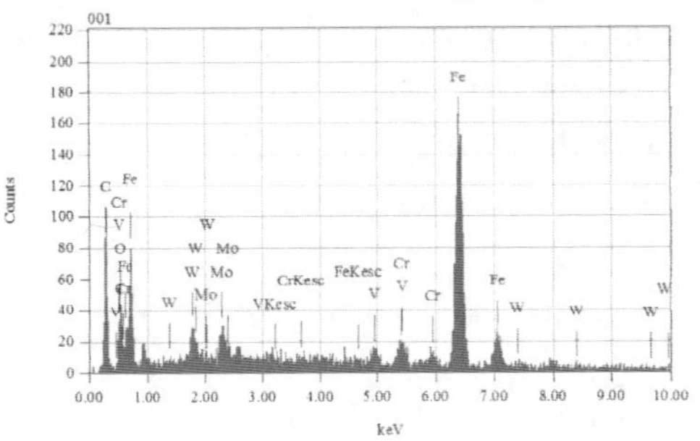

Figure 4. Microroentgen spectral analysis carried out on the reflow zone.

Element	Wt%	At%
C	04.18	15.16
O	07.69	20.93
Mo	03.32	01.51
V	01.95	01.66
Cr	03.69	03.09
Fe	71.76	55.87
W	07.5	01.78

The hardened zone offers the austenitic-martensitic matrix with a grid of eutectic carbides which are situated along the grains boundaries (Figure. 5a). Unlike the lamellar structure of eutectic carbides, which are in the initial state in the steel M2, carbides in the hardened zone are offered as dispersed formations which are situated closer to the reflow zone, and the rods which are situated closer to the heat-affected zone. At the same time the amplification of eutectic phases branching occurs close to the HAZ. It is observed a large number of dispersed precipitates with size of ~0.5 μm inside grains. Resting on the results of X-ray structural analysis carried out in works [17,18], it can be assumed that the precipitates present carbides which compositions include W, Mo, Cr, V and Fe.

Heat-affected zone is well etched in the reagent HNO3(ml):C2H5OH(ml)=4:100 (Figure. 5,b.). The photomicrograph clearly shows coarse-acicular martensitic structure with a grid of eutectic carbides at the grain boarders. Fine-dispersed precipitates are not observed in the heat affected zone while they are typical for the hardened zone. The HAZ width does not depend on the laser beam power and it can range from 80 to 110 μm.

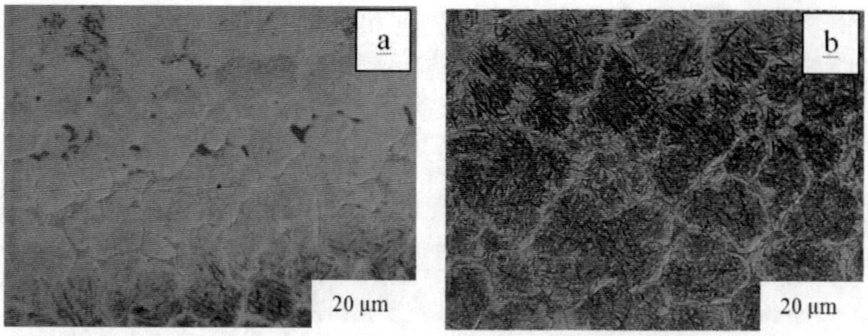

Figure 5. Microstructure of a hardened zone (a) and heat-affected zones (b).

The metal structural changes after laser reflowing allows for a change in hardness in some areas of laser action. Distribution of microhardness across the reflow points was measured after the surface polishing when roughness reached Ra 0.08 μm.

Heat-affected zone represents the most weakened area of the laser action points. The level of hardness in the HAZ decreases by 25 ... 30% in comparison with the hardness of the initial material which has not been treated (Figure. 6). This is due to the tempering of the martensitic boundary which is close the hardened zone, this process happens in a very short period of time in the case of the metastable structure heating [4].

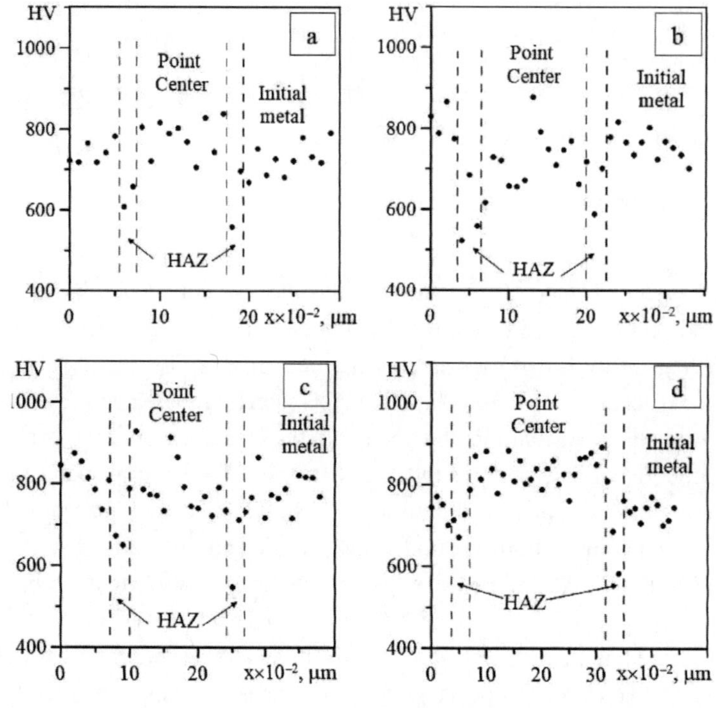

Figure 6. The length wise distribution of microhardness over laser action points.

It is observed an uneven distribution of microhardness due to the formation of structural heterogeneity in the central part of points and in the hardened zone, this is particularly evident during the laser beam treatment with the power from 15.8 to 25.4 W (Figure. 6, a-c). The average hardness of these zones is comparable with the hardness of M2 steel which is in its condition after surfacing (~790... 800HV).

IOP Conf. Series: Materials Science and Engineering **142** (2016) 012062 doi:10.1088/1757-899X/142/1/012062

When the laser power is increased up to 42.5 W, so the average level of the treated metal hardness is increased by 10 ... 15% mainly due to the widening of the hardened zone which contains a large amount of eutectic and secondary carbides (Figure. 6,d). Alignment around a heating spot is facilitated by the increase of the laser action power, along with the increase of the average level of the treated metal hardness.

While the surface is being treated by overlapped laser points, it is necessary to consider the presence of narrow weakened areas under pulsed laser action on M2 steel.

When the surface of M2 steel is treated by successive imposition of intersecting laser pulses, so it happens a uniform penetration of the initial metal to the depth of ~130 ... 140 μm, what corresponds to the penetration depth for a single pulse with power of 21.0 W (Figure. 7,a).

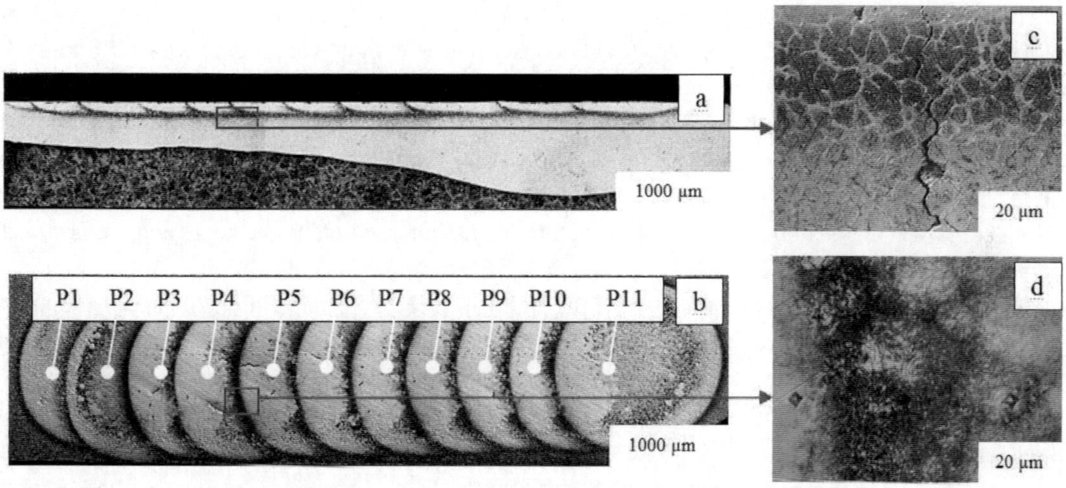

Figure 6. The microstructure of the cross section (a) and the path surface (b) done by a series of laser irradiation pulses and the HAZ microstructure in cross section under the point P.5 (c).

It is observed the cracks formation on the surface of central points (P.3 P.6 ... in Figure. 6,b), and also below them in the HAZ (Figure. 6c).

It is formed a narrow heat-affected zone in areas where points are overlapped (Figure. 6. a,b). This area having been formed on the pre-reflowed metal area has a fine-acicular martensite structure (Figure. 6d). The hardness of this area is 20 ... 30% lower than the hardness of the reflow zone and the hardened zone (Figure. 7). In general, the average level of metal hardness that has been treated with a series of pulses is ~800HV and it is at the level of coating hardness that has been made of M2 steel immediately after surfacing.

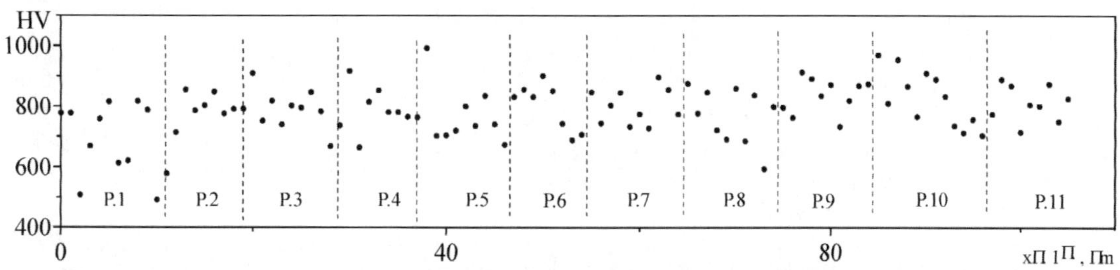

Figure 7. The microhardness distribution in the surface layer of the laser path action which is directed from the first point to the last one.

Thus, thermal cycling which occurs when M2 steel is consecutively treated by pulse laser irradiation does not allow considerably change the hardness of the treated layer in comparison with the hardness of M2 steel after surfacing.

Conclusion

The increase of the average laser pulse power from 15.8 W up to 25.4 W results in the twofold depth increase of laser action on the M2 steel coating. Further increase in the average power up to 42.5 W facilitates the growth of the molten metal layer on the walls of the steam-gas channel, which leads to the decrease of penetration depth.

The three zones formation under laser treatment facilitates the irregular distribution of microhardness on the surface of the treated material. It does not occur the considerable strengthening of the surface when M2 steel is treated by single laser pulses with power ranging from 15.8 to 25.4 W. The average hardness of the treated area with laser irradiation is within the limits of the M2 steel hardness after plasma surfacing. The hardness increase of the treated area is facilitated by laser irradiation power increase up to 42.5 W and this is 10 ... 15% higher in comparison with the hardness of M2 steel after surfacing. Mode with an average power of 42.5 W and pulse energy of 28.35 J is considered to be the best in respect to the structure formation, the hardness change and the laser action depth.

The treatment of the steel surface with the help of successive imposition of laser irradiation in overlapped manner facilitates the initiation and development of cracks in the areas under the laser action, at the same time the structures that are being formed do not allow to increase the hardness of the M2 steel with the help of periodical weakening of the treated material in the HAZ while the points are overlapped.

Work performed under the project of the Russian science Foundation № 16 - 19 - 10010.

References

[1] D.H. Kam, S.Bhattacharya, J.Mazumder 2012, *J.Micromech.Microeng.* **22** 105019–105025.

[2] S. Hammouti, A. Pascale-Hamri, N. Faure, B. Beaugiraud, M. Guibert, C. Mauclair, S. Benayoun, S. Valette 2015. *Applied Surface Science*. 357 1541–1551.

[3] Bizi-Bandoki P., Benayoun S., Valette S., Beaugiraud B., Audouard E. 2011, *Applied Surface Science*. **257** 5213–5218.

[4] Akhtar S.S., Yilbas B.S. 2014, *Comprehensive Materials Processing.* **9** 25-46.

[5] V. Oliveira, N.I.Polushkin, O.Conde, R.Vilar 2012, *Opt. Laser Technol.* **44** 2072–2075.

[6] M. Tang, V.Shim, Z.Y.Pan, Y.S.Choo, M.H.Hong. 2011, *Laser Micro/Nanoeng.* **6** 6–9.

[7] Abdellah L., Bruno C., Frederic J., Henry A. 2011, *Optics & Laser Technology*. **43** 1257-1263.

[8] Ho Jun Shin, Young Tae Yoo. Microstructural and hardness investigation of hot-work tool steels by laser surface treatment. Journal of materials processing technology 201(2008) 342–347.

[9] Zhihui Zhang, Luquan Ren, Ti Zhou, Zhiwu Han, Hong Zhou, Li Chen, Yu Zhao 2010, *Journal of Bionic Engineering,* **7** 67–76.

[10] Linjian Chaia, Baofeng Chena, Shuyan Wang, Ning Guo, Can Huang, Zhimin Zhou, Weijiu Huang 2016, *Applied Surface Science*. **364** 61–68.

[11] Abdellah Laazizi, Bruno Courant, Fre´de´ ric Jacquemin, Henri Andrzejewski 2011. *Optics & Laser Technology,* **43** 1257–1263

[12] Eurico Assuncao, Stewart Williams 2013, *Optics and Lasers in Engineering*. **51** 674–680.

[13] S.F. Gnyusov, A.A. Ignatov, V.G. Durakov, S.Yu. Tarasov 2012, *Applied Surface Science*. **263** 215–222.

[14] S.F. Gnyusov, S.Yu. Tarasov 2013, *Surface & Coatings Technology*. **232** 775-783
[15] A. Dunn, K.L. Wlodarczyk, J.V. Carstensen, E.B. Hansen, J. Gabzdyl, P.M. Harrison, J.D. Shephard, D.P. Hand 2015, *Applied Surface Science*, **357** 2313–2319
[16] S.Razi, Kh.Madanipour, M. Mollabashi 2016, *Optics&LaserTechnology*, **80** 237–246.
[17] Zhou Bin, Shen Yu, Chen Lun , Cui Zhen-shan 2011, *Journal of iron and steel research, International*, **18** 41-48.
[18] Zhou Xuefeng, Yin Xiaoyan, Jiang Jianqing, Zhu Wangl 2012, *Journal of rare earths*, **30** no. 10 1075.

IOP Conf. Series: Materials Science and Engineering **142** (2016) 012063 doi:10.1088/1757-899X/142/1/012063

Development of a Metal Cutting Tool Fase in Order to Create the Conditions of Ringed Chips Wrapping

M Korchuganova [1], A Syrbakov [1], T Chernysheva [1], G Ivanov [2], M Korchuganov[1]

[1] Yurga Technological Institute of Technology, TPU affiliate
Kemerovo Region, Yurga, 26 Leningradskaya RUS
[2] The Moscow State Technical University by N.E.Bauman
Moscow, 5, 2and Baumanskaya, RUS

E-mail: kma77@list.ru , kormar@tpu.ru

Abstract. When processing ductile metals with high cutting speed, there is a need to take additional measures for a comfortable and safe formation and removal of chips. In the conditions of large-scale manufacture, it is recommended to produce flow chips in the form of short fragments, while in the conditions of small-lot and single-piece manufacture, it is reasonable to wrap the chips spirally with a rather small turn radius. Such way of chips formation reduces the time of its removal from the working area as well as facilitates its transportation and processing. In order to solve the problem of chip wrapping and breakage, almost all modern manufacturers of tools with replaceable many-sided plates (RMSP) followed the way of complication of tool faces and determination of the areas of effective chip breaking. On the one hand, the suggested solution turns out to be effective; however, as showed the analysis of recommended cutting modes for complex forms of RMSP made by leading manufacturers, they all correspond to the definite cross section of the cut-layer $S/t=0.1$.

1. Introduction.

In the process of metal working, predominantly flow chips are developed which are represented by fragments of different length and form. It was noticed, that during the efficient life, even in constant duty of one and the same tool, the chips of various forms are created which, in its turn, negatively affects the stability of the process. Here are the most important reasons for this [1,5,6,7,8]:

1) The variability of the forces that affect the chips. In the process of formation, the mass of the chips is constantly growing, which provokes the shift of its gravity center. The chips-forming element, which constitutes an obstacle in the motion path of the chips, causes an additional resistance force. These forces create a flexure moment and define the instability of stress distribution at the root of the chips. In the process of cutting, the root of the chips is situated in the zone of plastic flow, that is why even a small change of stress distribution makes the form of the chips change. The form is changing in order to decrease flexure moment and resistance force, i.e. the chips is able to adapt to external forces and obstacles in its way.

2) Tool geometry modification. It is practically impossible to sustain a constant geometry of the cutting edge in the process of cutting because of wear and spoiling, which also leads to the form change of chips.

3) The inhomogeneity of the processed material by its chemical composition and structure, which causes the change of the shearing angle and chips form. Mostly, the cut-layer is hardened to some extent by the previous processing which affects the form of the created chips as well.

The flow chips of the general type, spirally wrapping, are usually hardly breakable, because the chip formation zone constitutes a "plastic hinge" which easily turns under the influence of deviations from chip breakers, thus bypassing these obstacles [1]. That is why the special concern is made here about extreme cases of chips wrapping: the so called flat wrapping when the chip slides on the face, and the ringed wrapping in the sectional plane of chip flow [5]. The second variant is more spread in practice and easier applied in the chip breaking.

Considering the process of restricted cutting with a cutter with the grade angle of the main cutting edge $\lambda = 0$, regardless of its other geometrical parameters, cutting modes, and the processed material, during the process of formation of flow chip the preconditions of its wrapping are created, simultaneously relative to two mutually transverse axles. It is suggested that one of these axles is orthogonal to the diagonal line of the cross-section of cut-layer, while the other axle is parallel to the reference plane and to the diagonal line of the cross-section of the cut-layer of metal. The basis parameter which determines the wrapping radius of the chip in the plane of the frontal cutter's edge is the direction and the speed differential of the widest distant points of the chip at the beginning of its existence as an independent solid body. The variability of the deformation grade of the metal chip by its width, which determines the speed differential of single elements at the beginning of its formation, in its turn, depends on an average ration of chip deformation and the ration between the width and the depth of the cut-layer. The more the average deformation ratio of the chip is as well as the less the ratio between the width and the depth of the cut-layer is, the more intensive is chip wrapping in the front face. In the case of predominant chip wrapping in the front face, the chips flows in the form of helical spiral.

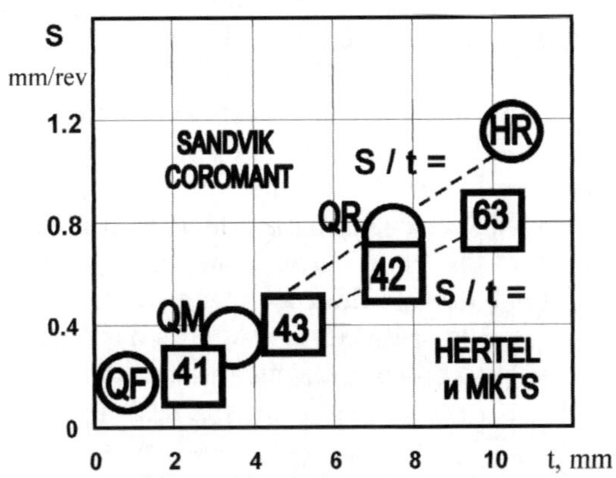

Figure 1. The arrangement of the areas of reliable chip breaking with cutting modes, geometry and the face form of RMSP.

Final appearance of the chip is formed in several steps: 1) its formation; 2) movement on the tool face within the contact length following a certain trajectory; 3) wrapping in the vertical and (or) in the horizontal plane; 4) natural or artificial chip breaking. Most geometrical parameters of the cutting unit and cutting process as a whole have a complex effect on the form and the dimensions of created chips [6]. For instance, the flow direction of the chips on the face depends on the primary φ and the secondary φ_1 angles on the plan [6], on the inclination angle of the main cutting edge λ [5], the ratio between the supply and the cutting depth ***S/t*** (or between the depth and the width of the cut-layer ***a/b***) [1] as well as on the form of the cutting unit on the plan. In case the tool face is plain, the chip wrapping in the sectional plane of its flow is primarily affected by the ratio ***S/t***, while in case it is determined by chip-forming elements [2], the geometry of the plate in the chips flow direction is most important. The form of the chips and its flow direction are correlated [3, 5], however the connection between them is not uniquely defined and depends on the cutting conditions. There is no correlation between the curve radius and the number of chips turns, because their values constitute the results of different chips formation processes: the chips radius characterizes the process of chips wrapping, and the number of turns characterizes the process of chip breaking [1].

Chip breakage is a complex and poorly studied process, because the breaking ability of the chips is determined by a large number of factors connected to physical characteristics of chips material which significantly differ from the characteristics of the source material [4,6]. In general, the chips is broken due to two reasons: in the process of unbending when meeting an obstacle, as well as under the influence of gravity force. While in the first case, in order to break the helical-curve chip into fragments, you need to adjust its hardness by the depth or width of the cut-layer, the second case is hard to describe.

An effective chip breaking, in single cases, can be achieved by means of the choice of appropriate processing modes: delivery S, depth t, and cutting speed V. Many prominent companies manufacturing metal-cutting tools have in their catalogues plates with a plane face and a relatively wide recommended chip breaking zone which matches analogous zones for RMSP with chip-forming elements [2,3].

Within the last years, the stock-list of RMSP for the tools with mechanical mounting has significantly broadened by means of variation of their face form. In the catalogues [2,3], each RMSP form is connected with a certain chip breaking range of cross-sections of the cut-layer. We have analyzed cutting modes which provide with a proper chip breaking without any special plate geometry made by Sandvik Coromant and Hertel [2]. It turned out (Figure 2) that almost all of them lie within the ratio range $S/t = 0.08 - 0.1$, regardless of the form and the dimensions of chip-forming elements. Apparently, it is connected with the fact, that the cross-section of the chips in this area becomes harder, and the chips is more inclined to breaking. By the ratio of $S/t<0.08$, the cross-section of the chips becomes laminar, and the spirally wrapping chip breaks affected by the gravity force or an external force.

In Figure 1, there is a comparison of the areas of chip breaking using turning tools with mechanical RMSP mounting which are sharpened with different face geometries. The line $S/t = 0.1$ divides the area of the cross-sections of the cut-layer into the zones of stable and unstable chip breaking. On the top of it, the centers of the stable breaking zones are marked [1,7] as well as, according to the work [1], the connecting lines of these centers for a plain face, a variable bevel, and a radial impression. According to Figure 1.2, these centers and lines are situated in the zone of stable breaking.

The existence of multiple geometry forms of RMSP faces shows that there are no universal forms with chip breaking flutes which are able to process various materials in a wide range of advances. Usually, for CNC metal-turning lathes, a set of plates is used designed for semi-rough and final turning.

The analysis of the literary sources gives evidence that the existing data are insufficient for the determination of the form and the dimensions of chips using geometrical parameters of cutting units. In case of a complex face topography of RMSP, its influence on the form and the dimensions of created chips either is absent or is given in the catalogues of foreign companies in the form of empirically calculated zones of stable chip breaking.

2. Results and Discussion.

2.1. The conditions of ringed chips wrapping in the sectional plane of its flow.

Having conducted the analysis of literary sources, of publications in Russian and foreign technical periodicals as well as of the reference data concerning the questions of flow chip wrapping and breaking, we were able to make the following conclusions:

1) The simplest and the most approachable method of chip wrapping and breaking is to change the cutting mode; however, it guarantees a proper chip form only within the range restricted by the ratio $S/t > 0.1$.

2) A chip breaker, a ledge, or a radial groove on the face somehow broaden the range of sustainable chip breaking; moreover, the areas of application of chip-forming grooves and ledges on the face are considerably similar to the areas of stable chip breaking using cutting modes.

3) When performing a restricted cutting, the scheme of wrapping and breaking in the sectional plane of chip flow is the most widespread. Here, a problem arises of the determination of the conditions which help achieve this wrapping using cutting modes as well as the geometry of tool faces. This problem has not yet been solved in a general way.

4) The recommendations concerning the choice of tool faces are as a rule restricted by small changing ranges of cutting modes.

Therefore, a model was suggested of a ringed flow chip wrapping in the sectional plane of its flow.

From the point of view of stable cutting processes, processing safety as well as transport and reprocessing conditions, flat-spiraled chips is the most convenient as it is characterized by the largest value of packed density. This chip form is created by means of wrapping in the sectional plane of its flow.

In figure 2, the cutting unit of a tool is depicted with the root of the chip flowing in the direction y'. Here, ***OAB*** is a conditional shear surface [1], ***O'A'B'*** is the cross-section of the chips, the form of which can be deduced by means of a projection of the shear surface on the normal plane to the axis X'. On the power contact spot ***OO'B'B***, normal and tangential loads are distributed according to a certain law between the chip and the tool face. As a first approximation, we can suppose that the stresses on the conditional shear surface and tangential loads are evenly distributed. Then, the points of application c_s and c_f of the cumulative forces R_S and F_f will match the gravity centers of curvilinear figureures ***O'A'B*** and ***OO'B'B***. If the chips does not meet any obstacles in its way, then, having set the chip flow speed on the tool face as constant, we deduce that $R_S = F_f$, i.e. this is the case when the forces that affect the chips from the side of conditional shear surface as well as from the tool face are balanced.

In a general case, the points of application c_s and c_f do not match which causes the flexure moment that conditions the so called "natural" chips wrapping of ***a*** general kind. Hereby, arm l_v (Figure 2) defines the chip bend in the vertical plane towards the tool face. The arm of force l_h conditions the chips wrapping in the face plane. It is possible, that the correlation between arm l_h and arm l_v defines the correlation between the radius and the pitch distance of the helical spiral of the chips. It follows thence that ringed chips wrapping corresponds with the condition $l_h = 0$, when chips forming and friction forces lie in the same vertical plane. This condition is provided when the horizontal coordinate of the gravity center of chip power contact area matches the tool face [7,8].

In order to follow the condition of ringed chips wrapping in the sectional flow plane, it is necessary to know the direction of its flow and the dimensions of the contact mark. In a general case, the cutting edge can be described as a sequence of points (x_i, y_i, z_i), while its previous position is described as $(x_i\text{-}S, y_i, z_i)$, where S is delivery, mm/rev. Let us consider the cross section of the cut-layer which is created by two consequent positions of the cutting blade (figure 4). In case of a curvilinear blade of no particular form, it is always possible to define the coordinates of nodal points of the cross-section of the cut layer for certain values of cutting depth t and delivery S: *B, C* and *O*. Let us assume that angle β_I is constant in the chips flow direction for any point of the working section and let us apply the scheme with a single conditional surface shear (CSS) which, in figure 2, is presented by the projection ***BAO*** on the general plane. The value of area of the projections of the conditional shear surface on the coordinate planes F_x and F_y, the ratio of which defines the initial chips flow angle, is determined by the sum of elementary areas which are restricted with the tool face from below as well as with the conditional shear surface from above:

IOP Conf. Series: Materials Science and Engineering **142** (2016) 012063 doi:10.1088/1757-899X/142/1/012063

$$F_x = \sum (x_i - x_{i+1}) \cdot z_{ci} \,;\; F_y = \sum (y_i - y_{i+1}) \cdot z_i \,, \tag{1}$$

where z_{ci} is chips height, which, taking into account all the consumptions, equals to

$$z_{ci} = a_{ci} / tg\beta_1 \tag{2}$$

where a_{ci} - is the depth of the cut-layer in the flow direction of the chips.

Further chips flow trajectory on the tool face is defined by the tool face form, thus the equation of the chip flow angle shall be as follows:

$$\eta_d = arctg\left(\frac{\sum (x_i - x_{i+1}) \cdot a_{ci}}{\sum (y_i - y_{i+1}) \cdot a_{ci}} \right) \pm arctg\left(\frac{z_B - z_O}{y_B - y_O} \right). \tag{3}$$

where η_d is the dynamic flow angle [1].

According to (3), it follows that angle η_d depends only on cutting depth, delivery, which determine the nodal points of the cross-section of the cut-layer, and on the cutting edge geometry on the plan.

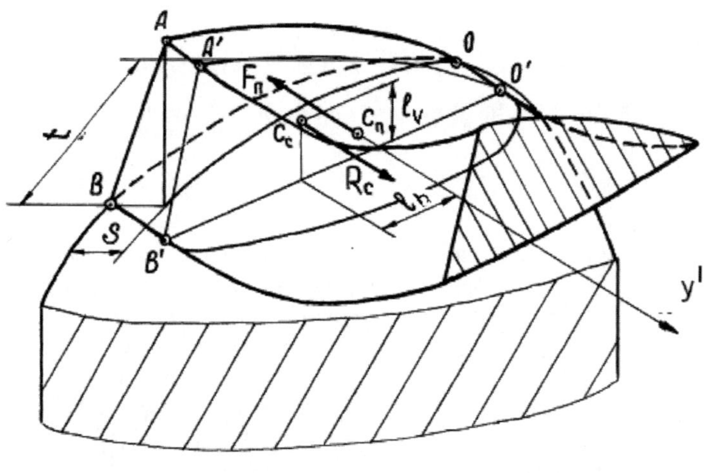

Figure 2. Diagram Curling continuous chip [1].

In the equation (3), the value of the angle of exit is evident in the left part and subtended in the right part, this is why its determination demands the application of the method of successive approximation. There was a computer program developed for the identification of η_d.

This algorithm functions as follows. For the purpose of plotting a conditional shear plane projection in the direction of the initial chips flow, let us define the depth of the cut-layer using the normal line to it a_{ni}. Then, the chips height on the ith section equals to

$$h_i = a_i / tg\beta_1, \tag{4}$$

while the coordinates of the gravity center of shear surface are:

$$x_c = \frac{\sum x_{ci} \cdot \Delta S_i}{\sum \Delta S_i}; \qquad z_c = \frac{\sum z_{ci} \cdot \Delta S_i}{\sum \Delta S_i}, \tag{5}$$

where x_{ci}, z_{ci} - are the coordinates of gravity centers of surface elements which constitute the projections of the conditional shear surface;

ΔS_i - is the area of the surface element of the chips cross-section.

The contact mark of the chips with the tool face is built in the flow direction consequently for each depth of the cut-layer a_i. For this purpose, according to the equation, the plastic length of the chips contact with the tool face is defined and protracted in the flow direction which is determined by the equation (3) so that, in result, there is a contour of the contact mark ***ABOC*** (Figure 2).

If the coordinates of gravity centers of the chips cross-section x_{mi} and contact mark x_{ci} are situated in the same straight line in the flow direction with angle η_s, then the formed chips shall wrap in the vertical plane. If the position of gravity centers does not meet this condition, it is necessary to change the cutting conditions. According to (5), for a positional change of gravity centers by the constant cross-section of the cut-layer, it is necessary to change the coordinates z_O и z_B.

In the conditions of mismatching of gravity centers, the cutting modes can be changed in case of the calculation of recommendations on ringed wrapping, as well as geometrical parameters of the cutting edge can be changed if a new tool is projected.

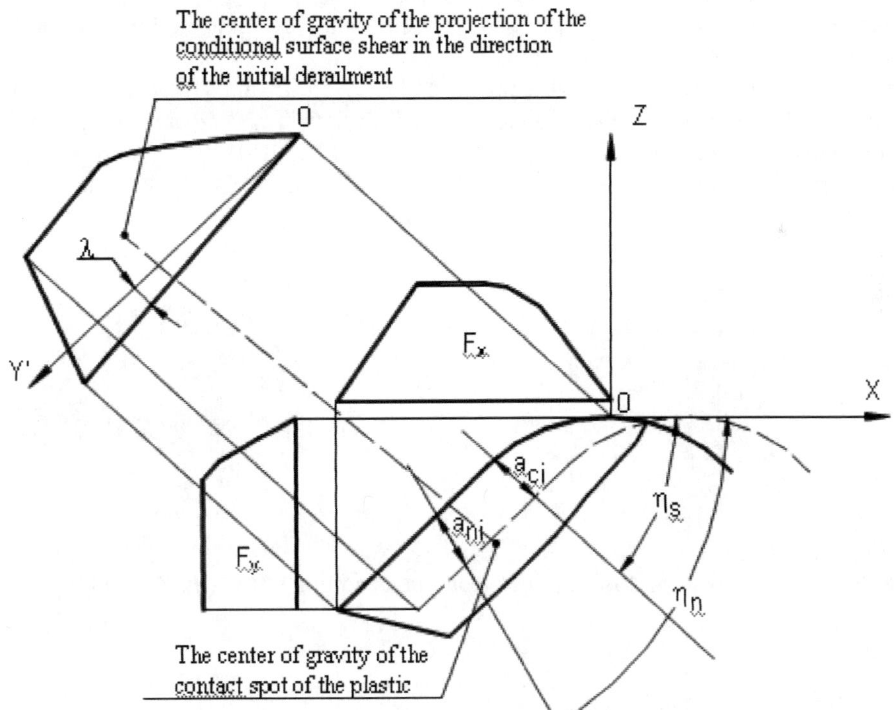

Figure 3. Scheme to the determination of centers of gravity transverse chip, the spots of contact force and angle go chips.

Thus, the developed program, to the specified accuracy, helps calculate the flow angle to the diagonal η_n, the flow angle η_s as a ratio of the area of the projections F_x/F_y as well as the coordinates of gravity centers of the chips height and the plastic contact mark. Moreover, the program is able to define the absolute difference between the coordinates which can be used for the assessment of the form of the created chips. The program is equipped with a function which makes it able to determine the coordinates of any points of the cutting edge profile, the conditional shear surface, and the lengths of the plastic contact mark, as well as to save these values as a text file. The program graphically depicts calculated values and has a zooming function of selected areas.

As follows from the analysis of the algorithm, the front angle, the shear angle, and the angle of slide lines approach on the tool face do not affect the coordinates of gravity centers of the chips cross-section and contact mark; thus, they do not affect the conditions of chips wrapping in the vertical plane as well.

3. Conclusion

For the conditions of restricted cutting, there was the condition of ringed wrapping created in the sectional flow plane. Accordingly, in case of matching forces that affect the chips from the tool face and from the conditional shear surface, a scheme can be realized which allows to create flat-spirally formed chips. The most important parameter defining chips form is the primary angle of its flow on the tool face. For the purpose of determination of the conditions of ringed chips wrapping, there was an algorithm created which helped to conclude that the front angle, the shear angle, and the angle of slide line approach on the tool face do not affect the condition of matching forces which influence the chips from the side of the tool face and the conditional shear surface. It is only the delivery, the cutting depth, and geometrical parameters of the cutting edge that affect the flow angle. This algorithm is helpful when elaborating the recommendations on chips wrapping under changing cutting conditions for a certain geometry of the cutting edge.

The deduced calculated dependences can be used for the designing of chip-forming elements for various materials and processing conditions.

Theoretical determination of effective profile and location ID on the front surface of the instrument should be based on the task of Curling chips, which in turn is inextricably linked with the accepted model of the chip formation. Figure 4 shows the form of the RMSP, the shape of the front surface calculated from the conditions of the circular Curling of the chips. Made plates showed high efficiency and increased bulk density of the chips depending on the cutting conditions 2 to 5 times.

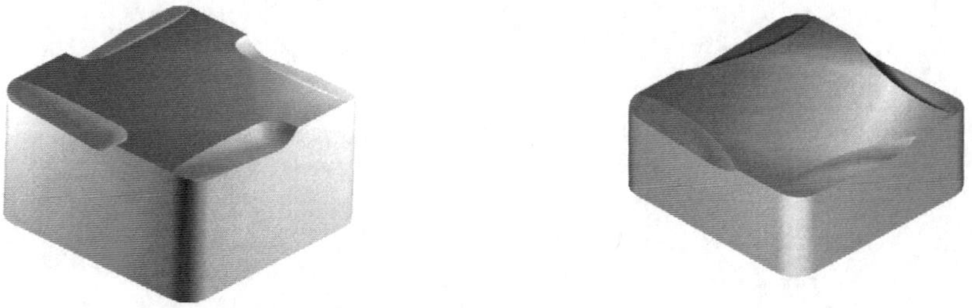

Figure 4. New forms RMSP designed from the condition of the ring Curling.

Reference

[1] Korchuganova M A Design indexable inserts. Design of cutting tools. - Saarbrucken: LAP LAMBERT Academic Publishing GmbH & Co. KG, 173 (2012).

[2] Coromant Turning Tools 2012: Catalogue.- Sandviken: AB Sandvik Coromant, 130 (2012).

[3] Innovations Master Catalog Tooling Sistems 2013. – Kennametall Inc., Latrobe. 1576 (2013)

[4] Nekrasov R Yu, Starikov A I, Lasukov A A Entering the operative correction machining processes CNC. Published under licence by IOP Publishing Ltd • IOP Conference Series: Materials Science and Engineering, Volume 91, conference 1

[5] Nakayama K., Ogawa M. Basic rules on the form of chip in metal cutting // CIRP Ann., 1978. –

27, № 1. – С. 17 – 21.

[6] Proskokov A.V. "Theory of Constrained Cutting: Model of Chip Formation with a Developed Plastic-Deformation Zone", Materials Science Forum, Vol. 762, pp. 782-789, Jul. 2013 http://www.scientific.net/MSF.762.782

[7] Proskokov A. V. , Petrushin S. I. Process Chip Formation with a Single Conditional Shear Surface // Applied Mechanics and Materials. - 2014 - Vol. 682. - p. 304-312 http://www.scientific.net/AMM.682.304

[8] Proskokov A. V. , Filippov A. V. , Gorbatenko V. V. Experimental Study of Plastic Deformation Intensity in Cutting Zone // Applied Mechanics and Materials. - 2015 - Vol. 770. - p. 200-204 http://www.researchgate.net/publication/275547244_Experimental_Study_of_Plastic_Deformation_Intensity_in_Cutting_Zone

IOP Conf. Series: Materials Science and Engineering **142** (2016) 012064 doi:10.1088/1757-899X/142/1/012064

Computer Simulation of Replaceable Many Sider Plates (RMSP) with Enhanced Chip-Breaking Characteristics

M Korchuganova [1], A Syrbakov [1], T Chernysheva [1], G Ivanov [2], E Gnedasch [1]

[1] Yurga Technological Institute of Technology, TPU affiliate
Kemerovo Region, Yurga, 26 Leningradskaya RUS
[2] The Moscow State Technical University by N.E.Bauman
Moscow, 5, 2and Baumanskaya, RUS

E-mail: kma77@list.ru , kormar@tpu.ru

Abstract. Out of all common chip curling methods, a special tool face form has become the most widespread which is developed either by means of grinding or by means of profile pressing in the production process of RMSP. Currently, over 15 large tool manufacturers produce tools using instrument materials of over 500 brands. To this, we must add a large variety of tool face geometries, which purpose includes the control over form and dimensions of the chip. Taking into account all the many processed materials, specific tasks of the process planner, requirements to the quality of manufactured products, all this makes the choice of a proper tool which can perform the processing in the most effective way significantly harder. Over recent years, the nomenclature of RMSP for lathe tools with mechanical mounting has been considerably broadened by means of diversification of their faces

1. Introduction.

The analysis of literature data [1,3] showed that the most efficient chip in the automated manufacture is of the kind "flat spiral" or "cylindrical spiral", and for this, it is necessary that it wraps in the sectional flow plane. Previously [2,3,5], the conditions of circular chip wrapping have been formulated, and a calculation method of dynamic canting angle of the cutting edge has been suggested for the achievement of those conditions. It was established that, for a certain cutting edge form and one combination of depth and feed rate, there is only one angle λ_d which creates the condition of circular wrapping. However, when cutting with varying duty, which is mostly spread in the manufactures, there is a need of such face form of RMSP that can provide with the conditions of circular wrapping for all possible cross-sections of the cut layer within a specified range. It can be achieved by means of changing the canting angle of the cutting edge.

We suggest to split the development process of new RMSP constructions with chip formation elements into three stages. In the first stage, on the basis of theoretical provisions or empirical dependences, we determine the initial data for RMSP designing; in the second stage, we design chip formation elements on working areas of the RMSP face by means of geometric computer simulation; in the third stage, we elaborate the final appearance of the construction of special and specialized RMSP taking into account their spatial orientation.

Methods of simulation of replaceable many-sided plates

The initial data for the choice or constructing of cutting tools include the information about the material and the form of a blank part and a component part, on the basis of which they choose the geometry of tool cutting edge as well as cutting modes. Following parameters must be set:

φ, φ_1 – the main and the auxiliary angle on the plan;
λ – canting angle of the cutting edge;
γ – rake;
r – tip radius, mm;
i – number of profile points of the cutting edge (maximum 255);
l – length limit of the main cutting edge, l $\approx$ 1,2 t;
S – feed rate, mm/rev;
t – cutting depth, mm;
ω_f – angle of emergency of flow lines onto the face;
β_1 – canting angle of the conventional shear plane.

Main angle φ and auxiliary angle φ_1 in the plan depend on the form of work surface; rake γ is designated according to the mechanical characteristics of work material as well as to the processing type [1]; in the process of finishing, tip radius of the cutting edge r depends on the required surface undulation of the component part according to the drawing; cutting depth and feed rate are determined on the basis of allowance, mechanical characteristics of the material, as well as the features of work surface.

General methods of simulation of replaceable many-sided plates were implemented in the computer program, the calculation results of which are the flow angle for determined cutting conditions, the coordinates of gravity centers of chip cross-section and contact mark as well as files including the coordinates of points which describe the profile of cutting edge, chip cross-section, and chip contact mark with the flat face. Additionally, the program visualizes the section of the cut layer, chip cross-section in the direction of its flow, and contact mark.

The condition of circular wrapping can be used in order to solve two tasks: first, to pick cutting modes (*S, t*) for a certain cutting edge geometry (φ, $\varphi 1$, λ, r); second, to pick a λ_d for a certain section form of the cut layer (φ, φ_1, r S, t). In the first case, the geometric parameters of a tool are determined in accordance with cutting modes; in the second case, they receive the necessary data for face form computer simulation.

Using the computer program, they calculate the dependences of the flow angle and the dynamic canting angle of the cutting edge on cutting modes with varying geometric parameters of the plate. In order to observe the condition of circular wrapping, they change angle λ_d up to the moment of gravity centers equality of chip cross-section and contact mark with the face within the accuracy of 0.001 mm.

As long as the dependences of λ_d on the feed rate and the cutting depth are directly-proportional [1,5], we can design the form of chip formation area of the RMSP face as a ruled surface.

In the cutting process, for varying parameters of section of the cut layer, the developed program helps project RMSP analogous to the case of varying cutting depth, varying feed rate as well as their constant ratio *S/t*.

2. Results and Discussion.

2.1. Designing of a chip formation geometry on the RMSP face with varying cutting depth.

When processing with CNC metal-turning lathes, they mostly perform cutting with varying depth and constant feed rate. This case is present when turning blank parts with varying allowance.

The designing of the RMSP face for such processing type is characterized by the fact that all variants of sections of the cut layer have one mutual point C (figure 1) on the main cutting blade the position of which is determined by the feed rate. The starting phase of projecting a chip formation area of the face is the determination of coordinates of nodal points on the main cutting edge which depend on the change interval of cutting depth. In fig. 1, index 1 corresponds with tmin, n with tmax, and i with the current value of cutting depth t. The origin of coordinates is put onto the corner of the cutting blade. From this point, the value of the half feed rate is plotted on axis X to the right, and the coordinate of the nodal point C on axis Y is found. In order to determine the coordinates (x, y) of nodal points of the section on the main cutting blade, they transfer the distances in the direction of axis Y which equal to the cutting depths within the projecting range of the plate. Thus, points B_i are found.

In order to accomplish this, we draw connecting lines through the nodal points and the origin of coordinates, register the mutual nodal point determined by feed rate, lay off angle λ_d (fig. 1), and, on the crossing with connecting lines, we find Z-coordinates of the nodal points for the maximum cutting depth. Then, we determine Z-coordinates of the nodal points for the minimum cutting depth. Through a point on the auxiliary cutting edge corresponding with the half feed rate as well as through the nodal points of the minimal and the maximal cutting depth, we can draw a plane. When this plane crosses the surface that limits the plate in the top area, a transitional curved surface is formed.

For the purpose of following the condition of circular wrapping, it is necessary that angle λ_d is constant along the whole chip-tool contact area. Preliminarily, a full contact mark is built on the tool face. For each section variant of the cut layer, through a reference point on the tool face corresponding with the condition of matching gravity center of the contact mark and cross-section of the chip, we lay off distances in the flow direction from the cutting edge which equal a full contact length.

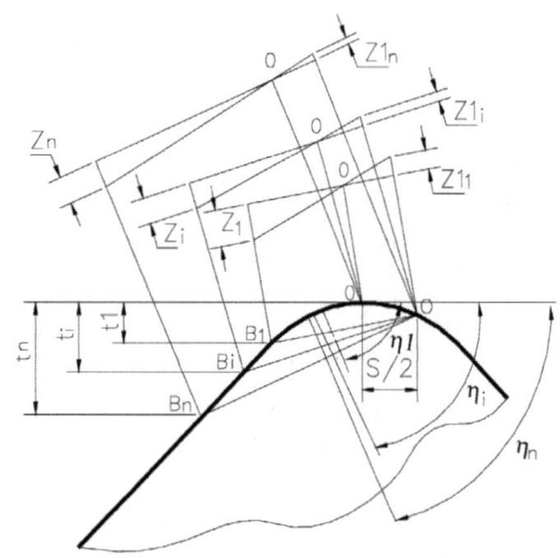

Figure 1. Scheme to the determination of coordinates of chip formation surface for the condition of constant feed rate

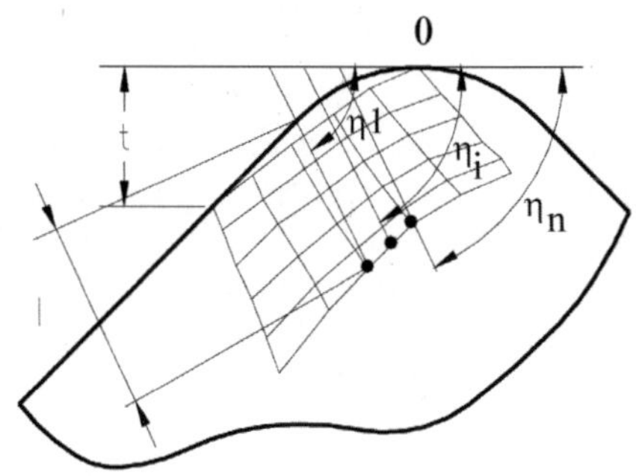

Figure 2. Simulation of the contact mark with varying cutting depth

Connecting the ends of this distances, we create the curve of contact end of the chip with the tool face (fig. 2). A line drawn through the nodal points of this mark must be bent to the basic plane with the same angle λ_d as the cutting edge.

The built area of the tool face will be the main chip formation element; and it is characterized by the fact that reference contact lengths shall be equal — it is only the points on the main cutting blade from which they are laid off that shall be different.

In fig. 3 and 4, there are sketches of a square and a triangular RMSP represented, as well as solid models and project diagrams of chip breaking with varying cutting depth. These plates designated for external linear turning have been formed into 20G cutters with RMSP of hard alloy T15K6 with feed rate S = 0.2 mm/rev and the cutting depth variation range t = 0.5–2 mm. The square plate is designed for the mounting into the cutter body with the entering angle $\varphi = 45^o$, the triangular — $\varphi = 90^o$.

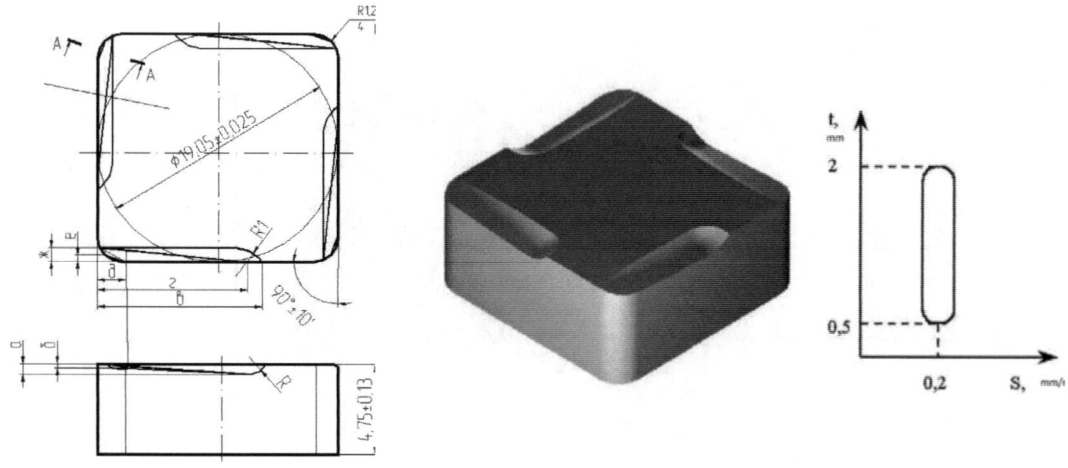

a b c

Fig. 3: Example of a square RMSP construction which follows the condition of circular chip wrapping with varying cutting depth:
a) sketch drawing; b) solid model; c) designed diagram of chip breaking

2.2. Designing of a chip formation geometry on the RMSP face with varying feed rate

A possible case of processing with CNC metal-turning lathes is cutting with varying feed rate and a constant cutting depth. It is especially characteristic for the lathes with adaptive control systems of processing accuracy when the feed rate is considered as a control parameter.

Here, all section variants of the cut layer have one mutual point B on the main cutting blade (fig. 5) the position of which is indicated by constant cutting depth. The form of contact mark (fig. 6) is conditioned by the fact that when changing the feed rate, there is a proportional change of the full contact length.

In fig. 7, there is an example of a triangular RMSP for the processing with t = 1 mm and S = 0.05–0.5 mm/rev (steel 20G — T15K6).

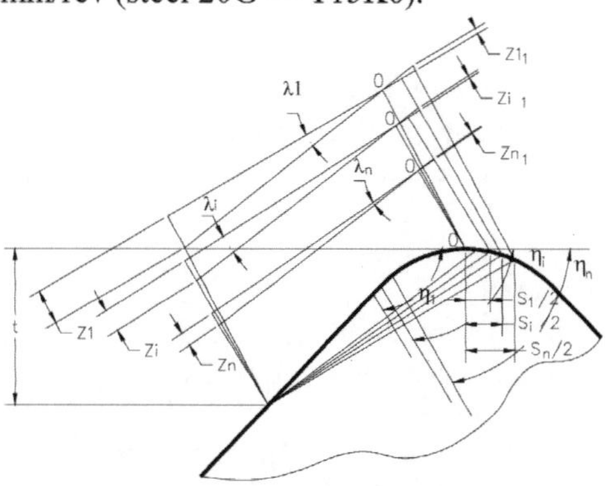

Fig. 5: Scheme to the determination of coordinates of chip formation surface for the condition of constant cutting depth

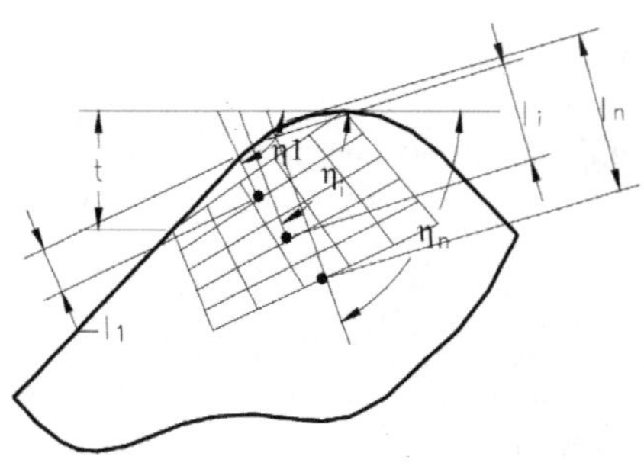

Fig. 6: Simulation of the contact mark with varying feed rate

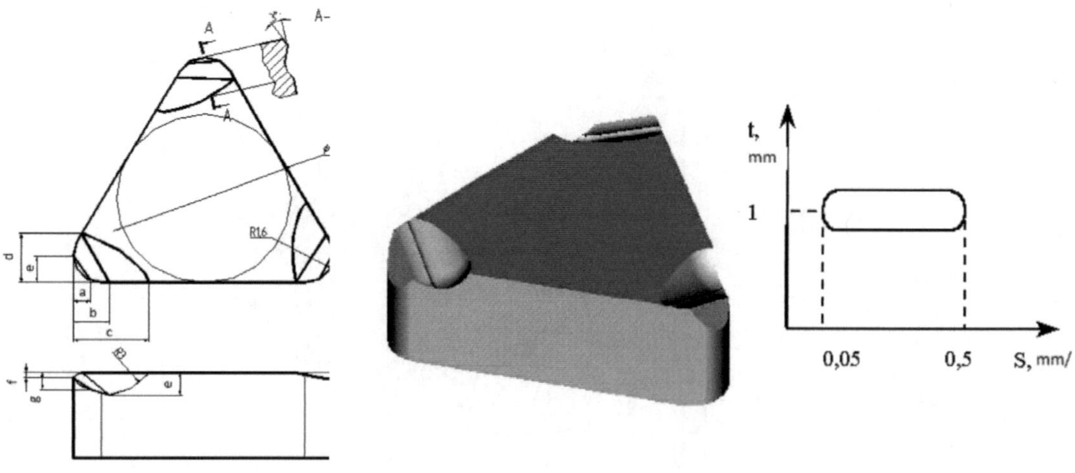

a b c

Fig. 7: Example of a triangular RMSP construction which follows the condition of circular chip wrapping with varying feed rate: a) sketch drawing; b) solid model; c) designed diagram of chip breaking

2.3. Designing of a chip formation geometry on the RMSP face with constant ratio of feed rate and cutting depth

The case of simultaneous change of cutting depth and feed rate is approximate with the variant when value S/t is constant. Besides, it was established [1,2,3] that the centers of chip breaking diagrams of most RMSP for lathe turning are positioned within the range S/t = 0.08...0.1 which corresponds with the area of accurate breaking.

A plate is designed analogous to the above described variants. The scheme of finding the coordinates of chip formation surface is presented in fig. 8. As seen from it, in the process of a proportional change of cutting depth and feed rate, there are no mutual points characterizing the section of the cut layer.

In fig. 9, there are RMSP models represented with improved chip formation features for the following conditions: work material is Steel 20G; RMSP material is T15K6; t = 0.5–2 mm; S = 0.05–0.2 mm/rev.

In order to create a spontaneous variation range of the section of the cut layer, the simulation of the chip formation area of the tool face must be conducted by means of overlapping of a variety of specific solutions. This results in a not ruled form of this surface and demands the improvement of software.

2.4. Designing of additional constructional parts on the RMSP face

The methods of computer simulation of a chip-formation area on the RMSP face elaborated in the previous chapters help to exactly describe the form of this area based on the initial data. At the same time, the RMSP construction is determined by other constructional parts: plate form, thickness, availability or lack of holes, accuracy, and other constructive parameters [2,3]. At this, they consider the stability of the plate [6,7], its wear-resistance as well as the methods of its fixation and orientation in the tool body [5].

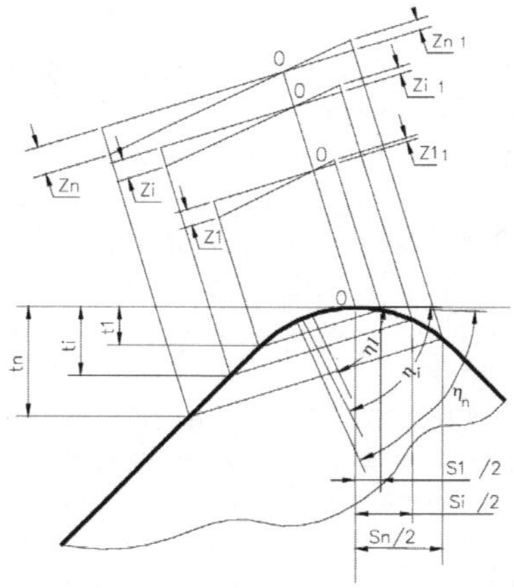

a

Fig. 8: Scheme to the determination of coordinates of chip formation surface for the condition of constant ratio of feed rate and cutting depth

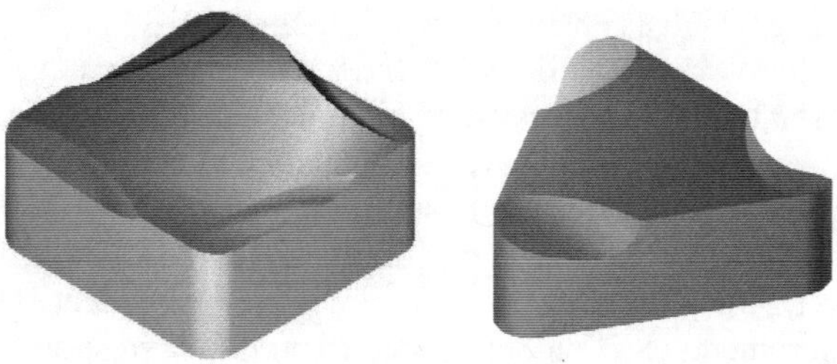

b c

Fig. 9: Examples of solid RMSP models following the condition of circular wrapping when the ratio of cutting depth and feed rate is S/t = 0.1: a) square $\varphi = 45^o$; b) triangular $\varphi = 90^o$

For the ratio S/t > 0.1, the received face form is enough for its secure breaking. At the same time, the observance of the condition of circular wrapping of smooth chip does not guarantee its breaking. In this case, it is necessary to create chip lead grooves and ledges on the face which work according to the principle of forced chip wrapping with its subsequent stop at the cutting surface. We can suppose that it is most rational to put the starting point of chip wrapping radius into the point of contact end of the chip with the face [4,8].

Based on the analysis of RMSP manufactured by Sandvik Coromant, Hertel and Moscow combine of hard alloys (MCHA), we have unveiled that an average distance from the main cutting edge to the step-type chip breaker depends on the feed rate and the cutting depth as follows:

$$l_{av}=4.2\,S^{0.1}$$

$$l_{av}=1.2\,t^{0.5}$$

In general, the recommended designing order of a new RMSP form with improved chip-breaking characteristics consists in the following:

1. Creation of a solid plate model in the rectangular coordinates system established taking into account the orientation angles of the plate in the body (AutoCAD usage is possible).
2. Determination of conditions of circular chip wrapping.
3. Simulation of a chip formation area on the RMSP face.
4. Designing of additional constructional parts on the face.
5. Combination of a solid model of the oriented plate and the modeled face object in order to create a complex topography on the plate face.
6. Positioning of the plate into the rectangular coordinates system for the purpose of realization of its final construction.

3. Conclusion

The suggested method of computer simulation of RMSP with improved chip breaking characteristics helps to design a face area providing the circular wrapping of flow chips.

The face area of a metal-cutting tool indicated to the circular chip wrapping is a complex ruled surface which can be replaced by a plane.

There were RMSP constructions created for a varying cutting depth, varying feed rate, and a constant ratio of feed rate and cutting depth. The elaboration of a RMSP for a general case of changing the section of the cut layer demands a switch from AutoCAD-based software to the software of the type Mechanical Desk Top or ProEngineering.

Apart from the considered approach to RMSP designing according to the chip breaking condition, a full RMSP construction must simultaneously meet the conditions of stability, firmness, and other requirements. It is recommended to put the corresponding constructional parts beside the limits of the full contact length with the face.

Created solid RMSP models can be realized by means of profile molding with subsequent baking, or by means of electrochemical and electrophysical processing methods. At the same time, after a small construction change, they can be used for the finalization of RMSP with flat faces by means of grinding of chip lead grooves.

Reference

[1] Korchuganova M A Design indexable inserts. Design of cutting tools. - Saarbrucken: LAP LAMBERT Academic Publishing GmbH & Co. KG, 173 (2012).

[2] Coromant Turning Tools 2012: Catalogue.- Sandviken: AB Sandvik Coromant, 130 (2012).

[3] Innovations Master Catalog Tooling Sistems 2013. – Kennametall Inc., Latrobe. 1576 (2013)

[4] Nekrasov R Yu, Starikov A I, Lasukov A A Entering the operative correction machining processes CNC. Published under licence by IOP Publishing Ltd • IOP Conference Series: Materials Science and Engineering, Volume 91, conference 1

[5] Nakayama K., Ogawa M. Basic rules on the form of chip in metal cutting // CIRP Ann., 1978. – 27, № 1. – C. 17 – 21.

[6] Proskokov A.V. "Theory of Constrained Cutting: Model of Chip Formation with a Developed Plastic-Deformation Zone", Materials Science Forum, Vol. 762, pp. 782-789, Jul. 2013 http://www.scientific.net/MSF.762.782

[7] Proskokov A. V. , Petrushin S. I. Process Chip Formation with a Single Conditional Shear Surface // Applied Mechanics and Materials. - 2014 - Vol. 682. - p. 304-312 http://www.scientific.net/AMM.682.304

[8] Proskokov A. V. , Filippov A. V. , Gorbatenko V. V. Experimental Study of Plastic Deformation Intensity in Cutting Zone // Applied Mechanics and Materials. - 2015 - Vol. 770. - p. 200-204 http://www.researchgate.net/publication/275547244_Experimental_Study_of_Plastic_Deformation_Intensity_in_Cutting_Zone

IOP Conf. Series: Materials Science and Engineering **142** (2016) 012065 doi:10.1088/1757-899X/142/1/012065

Controlling the type and the form of chip when machining steel

S.V. Gruby[1], A.A. Lasukov[2,1], R.Yu. Nekrasov[3] E.V. Politsinsky[2] and D.A. Arkhipova[2]
[1]Moskow State Technical University by N.E. Bauman, ul. Baumanskaya 2-ya, 5/1, Moscow, 105005
[2]Yurga Institute of Technology, TPU affiliate 26, Leningradskaya street, Yurga, 652050, Russia, tel.+ 7 38451 77761
[3]Tyumen State Oil and Gas University 38, Volodarskogo street, Tyumen, 625000, tel.+ 7 3452 4124 65

[1]E-mail: lasukow@rambler.ru,

Abstract. The type of the chip produced in the process of machining influences many factors of production process. Controlling the type of chip when cutting metals is important for producing swarf chips and for easing its utilization as well as for protecting the machined surface, cutting tool and the worker. In the given work we provide the experimental data on machining structural steel with implanted tool. The authors show that it is possible to control the chip formation process to produce the required type of chip by selecting the material for machining the tool surface.

1.Introduction

The type of the chip influences many factors of production process. Development of high-speed machine tools, machines with increased capacity and strength, introduction of new tool materials lead to significant increase of machining efficiency and its productivity mainly due to increase of cutting speed. From the growth of the cutting speed the need arises for providing safe and rational chip removal. Controlling the type of chip when metal cutting is important for producing swarf chips and for easing its utilization as well as for protection of machined surface, cutting tool, worker, for reducing the energy input when machining materials [1]. With the development of automated machining operations the necessity of chip formation control increased significantly [2, 3]. The ability for controlling the chip formation process is also important in the process of machining the reconditioned machine parts [4]. The practice of structural materials machining has developed one of the most wide-spread methods – cutting with the tool which has a flute on the front surface or steps in the path of chip escape. They ensure satisfactory chip removal. The removed chip in this case is, as a rule, continuous. Machining of materials with the property of discontinuous chip formation is not a factor preventing from cutting and working at high speed. At the same production of special elements for the tool arises certain difficulties and leads to the tool cost increase.

Producing discontinuous chip when machining eases its removal from the cutting area which is especially important when operating automated machines. Zelinsky A.N. [5] notes that cutting with discontinuous chip formation is related to increasing tool wear resistance.

Study of chip formation is the most efficient and inexpensive method to understand the characteristics of materials machining.

According to research works by Kufarev G.L. [6] to ensure chip control the following conditions should be observed: producing chip which is tough enough; curling of the chip due to unhomogeneity of deformation in various layers of metal; directing the flow of chip towards natural obstacle which curves the chip in the opposite direction to curling. Discontinuous chips are usually tough, with varied strength and have weak longitudinal bonds in the areas of deformation. curling of these chips is achieved due to the difference of speeds of separate layers of metal in the plastic domain of chip formation area. The chip flow direction to curve it as needed can be changed by changing (shortening) the radius of curvature of the chip.

Obtaining of necessary form and type of chip strongly depends upon the contact relations at the front surface of the tool. It is obvious that those phenomena are associated with adhesive behavior of machined and tool materials.

The increased adhesive capability of the contacting materials leads to the growth of friction at the front surface. And this results, as has been shown above, in changing of the force of interaction between the materials and formation of continuous chips. Behavior of metals under friction depends upon their mutual solid-phase solubility and ability of forming chemical compounds with each other. Thus, the main factor determining the friction behavior at the front surface is the ability of metals to form solid solutions and intermetallic compounds. That means that the chemical properties of machined materials which determine their mutual solid-phase solubility have the determining influence upon the friction behavior.

If we know under which conditions the continuous chip changes into discontinuous one we can control the chip type without adding special elements to the instrument, thus, making it cheaper [6]. We can control the chip type by changing the properties of the front surface of the tool to change the surface conditions. One of the ways of solving the given problem is ion implantation of the tool with ions of various materials. This, in its turn, leads to significant increase of tool wear resistance when machining titanium, heat-resistant and corrosion-resistant alloys usually associated with severe wear of the tool.

Changing the properties of the front surface of the tool [8, 9] we can control the process of chip formation through transformation of its basic characteristics: contact length, friction ratio, cutting forces, shearing angle, etc. Ion implantation of the cutting tool changes the adhesive contact between the chip and the tool. This leads to changes in the character of chip formation increasing chip fragility in most cases which eases its breaking and removing from the cutting area. Other ways of high-speed steel tool impacting are shown in [10], treatment of inserts with pulsed electron beam is considered in [16].

Choice of material for implantation should be made according to the condition of maximum reduction of probability of seizure between the machined material and the tool.

According to the said above we can make the conclusion that it is possible to control the process of chip formation achieving its required form and type, through changing the characteristics of the front surface. Considering that ion implantation influences the tool efficiency together with changing the chip formation character it is worth applying tools with modified surface for machining.

2. Methods of experimental study

In the given work we provide experimental data on machining structure 30HGSA steel with ion implanted tool. Inserts having quadrihedral shape from carbide alloy T5K10 were used as tool material. Implantation was completed on ION-700 with the following materials: Al, BN, TiB_2, ZrGf and Zr. Besides we applied inserts with TiN coating which were additionally implanted with various materials.

The workpiece was machined at the CNC screw-cutting lathe 16K20F3C32 under the constant supply S=0.21mm/rev and cutting depth t=2mm. The cutting speed changed within rather wide range V=50…200 m/min. Changing of the cutting force and chip formation characteristics were registered at the same time.

3. Results and Discussion

As the experimental data on structure steel machining show the chip formation is rather sensitive to the changes of tool surface properties.

Machining with unimplanted tool within the whole range of cutting conditions leads to formation of continuous chip which enmeshes significantly. Such chip is inconvenient in terms of safe work and, being accumulated in the working area, it disturbs normal work of equipment. Besides this type of chip flows down the machined surface deteriorating its quality [11]. Winding of the chip around the tool may lead to its loss of function. Holding-up to remove this chip leads to equipment downtime and, thus, to productivity reduction. Machining with this kind of tool is associated with high cutting force values which tells upon wear resistance of the tool leading to its deterioration.

In Tables 1 and 2 we present the changes of the chip and its edge according to the change of cutting speed and type of the modified surface.

Changing of chip appearance Table 1

Implanted material	Speed of cutting, m/min.				
	50	80	126	159	202
T5K10 (basic)	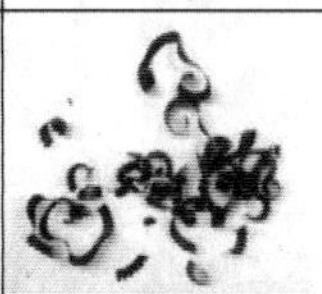				
Al				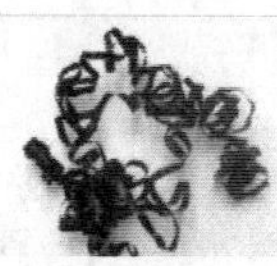	
TiB_2					
Zr			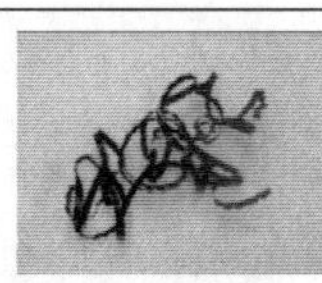		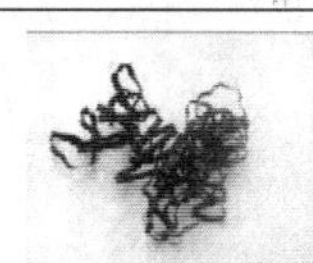
Al with coating					
TiB_2 with coating			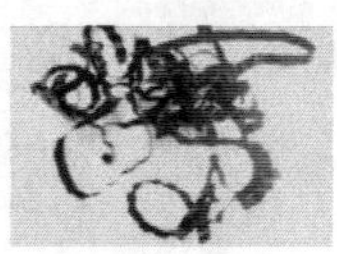		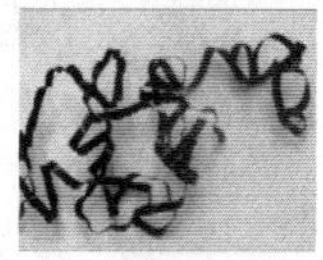
ZrGf with coating		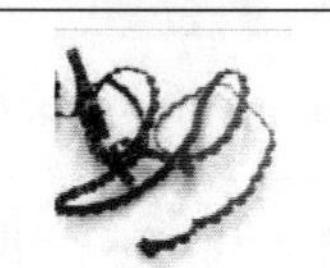	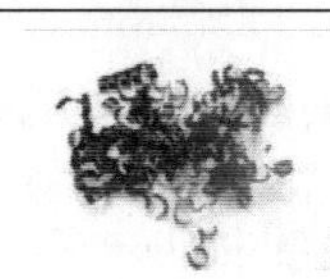		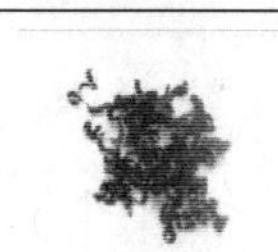

Analysis of the cutting process characteristics showed the changes of the cutting forces, chip shrinkage and shearing angle when machining under the described conditions. It is interesting to note that implantation of the instrument by any of materials applied in our experiments leads to reduction of the cutting force, chip shrinkage and increase of the shearing angle. All that results in more favorable conditions for machining steel as well as in reduction of power consumption when cutting [12].

Chip edge

Table 2

Implanted material	Cutting speed, m/min.				
	50	80	126	159	202
T5K10 (basic)	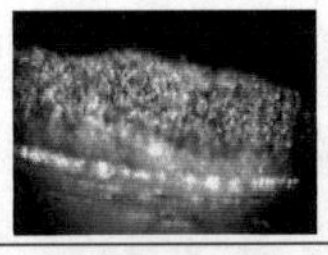	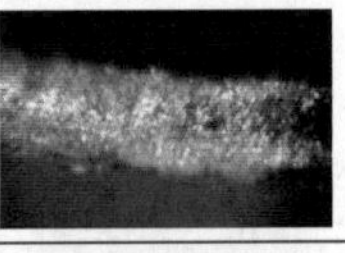	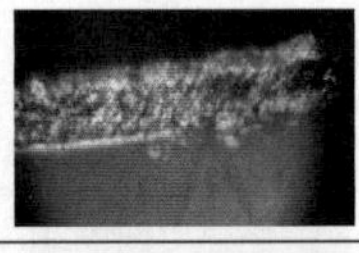	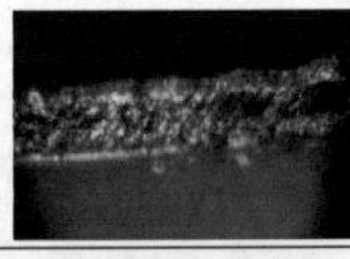	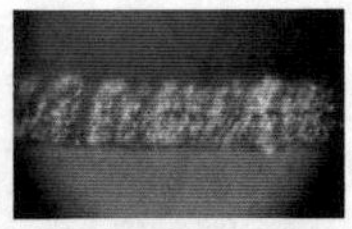
Al		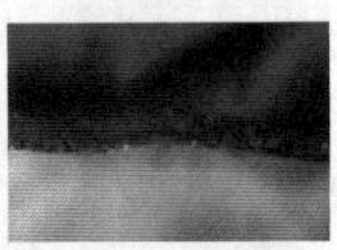	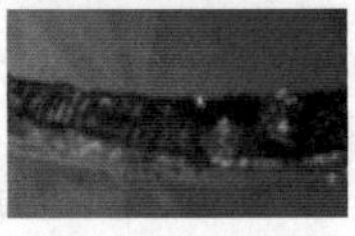	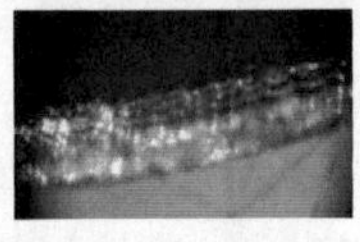	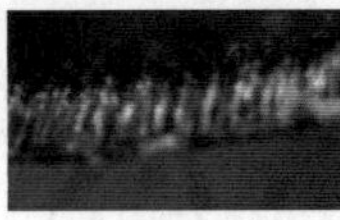
TiB_2				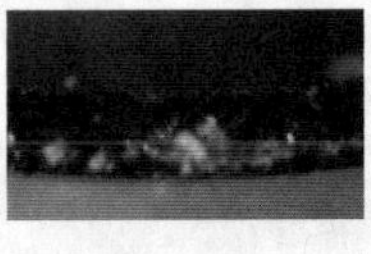	
Zr	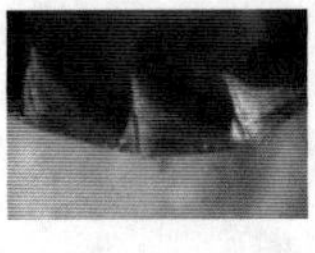		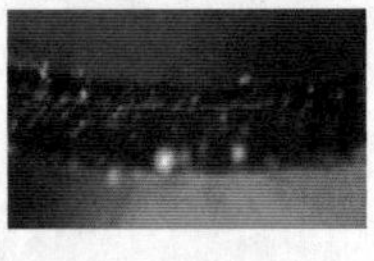	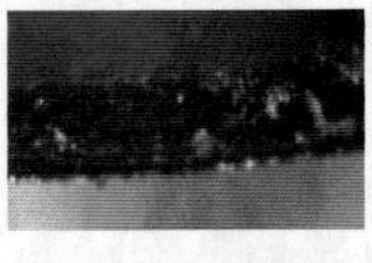	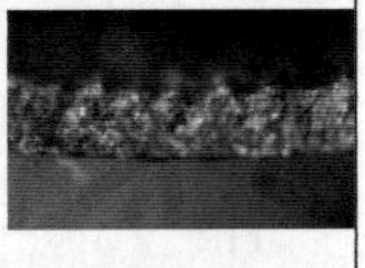
Al with coating	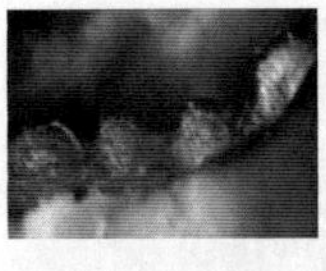	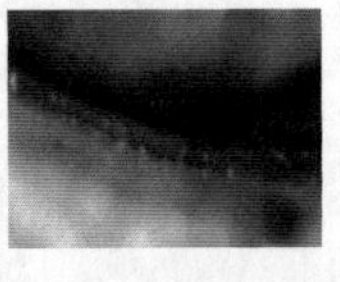	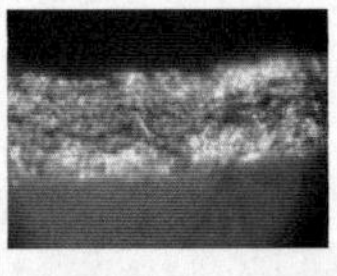	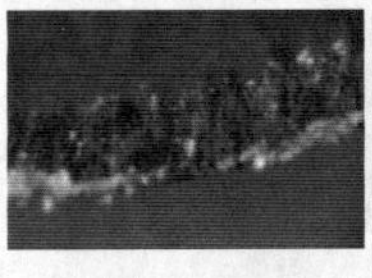	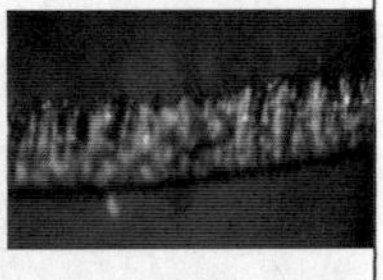
TiB_2 with coating	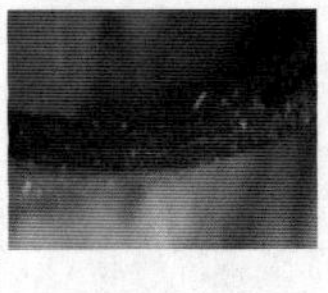	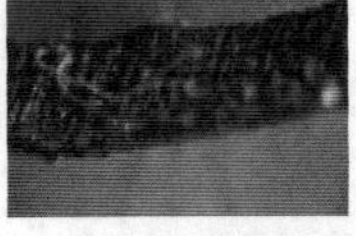		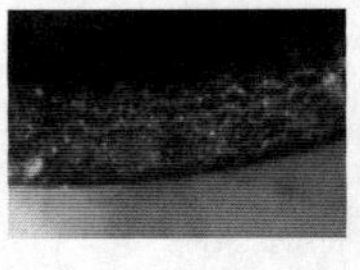	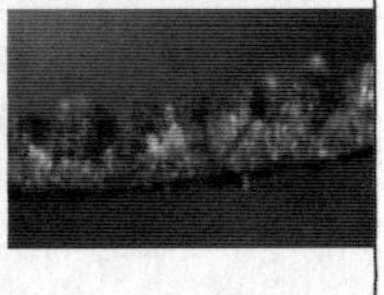
ZrGf with coating	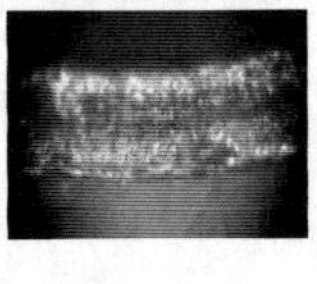	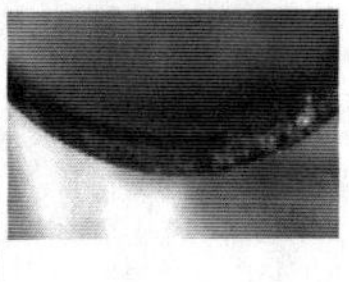		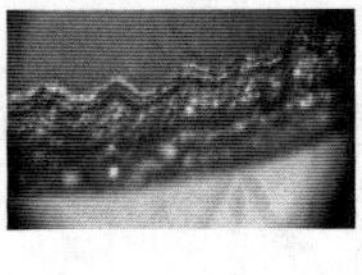	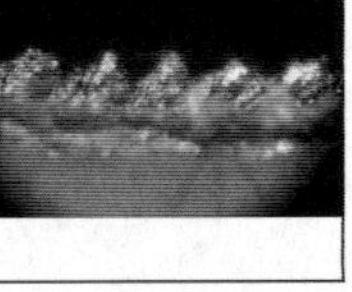

It should be noted that tool coated with TiN and additionally implanted with TiB_2 does not practically impact the process of chip formation when machining 30HGSA steel. As one can see from the presented photographs the form of the chip changes very little, the edge of the chip does not change very much either. At the same time the main component of the cutting force intensively decreases within the studied range of cutting speeds. The shearing angle is large but stays practically unchanged within the studied range of the cutting speeds during the process of machining. Chip shrinkage is low and with the growth of cutting speed changes insignificantly.

When implanting the tool with other materials the chip from being obviously discontinuous under low cutting speeds changes to continuous as the cutting speed grows. The appearance of the chip shows that with speed growth the discontinuous "bit-like" chip turns into tapes and in some cases forms coils which are hard to remove from the cutting zone. This type of chip is difficult to work with under speeds applied in production.

The properties of the tool surface impact chip formation in a different way even if the tool is implanted with similar material. From Table 1 we can see that machining with the tool implanted with aluminum (Al) the chip turns into tapes with the growth of cutting speed. The chip produced un the process of machining with previously applied coating (titanium nitride – TiN) and implanted ions (Al) was continuous under low speeds (the ripples on the edge were not seen) and remains the same, at the first glance, under high-speed cutting. But even a small effort makes it crush into pieces. Thus, the properties of coating together with those of implanted material reduce the forces at the front surface, namely the force of friction, and it leads to formation of such chip.

Machining of steel with TiN-coated inserts implanted with ZrHf ions (20% hafnium) is of practical interest. When machining with tool of this kind we observe a reversed pattern of chip formation. As speed grows the chip becomes discontinuous. As it is shown in Table 1 chip produced under the cutting speeds close to those aaplied at factories (speed of approximately 200 m/min) crushes into separate elements. Such chip is convenient for removing, transportation and recycling. Analysis of the cutting force changes showed that the main component of the cutting force becomes significantly smaller (by up to 300 – 350 N) in comparison to that when cutting with unimplanted tool. The values of the cutting force also appeared to be lower in comparison to those when cutting with the tool implanted with other materials. Chip shrinkage under cutting with such tool is low and does not change significantly with the change of cutting speed. It indicates that the deformation degree decreases when machining with such tool resulting in reduction of cutter load and improving tool efficiency.

These phenomena can be explained by the change of adhesive properties on the front surface of the cutter. When we work with the tool implanted by different materials the force of adhesion of chip and tool surfaces is different. Similar phenomena are observed when coating machine parts [13]. Adhesion causes stagnation: at some part of the contact area adhesion between different metallic surfaces (chip-cutter) becomes larger than slip resistance of the contact layer of the chip, the thin layer slows down and further sliding occurs inside the chip. The given layer, evidently, prevents the chip from finally breaking into pieces under high-speed cutting. Intensity of adhesion is mainly determined by the ability of the contacting metals for mutual solution [14].

The characteristics of such interaction of the tool implanted with ZrHf ions can be supposedly explained by little interaction of these elements with iron. The adhesion at the front surface reduces, so does the cutting force, the chip becomes discontinuous. This points to the fact that the degree of deformation varies depending upon the properties of tool surface.

In work [15] the authors analyze the resultant of the cutting force and the radial stresses in the tool. The value of the resulting force in all cases of working with implanted tool is smaller than that when working with usual tool. The value of the resulting force is the smallest when working with TiN+ZrHf coated tool. The authors also establish reduction of stresses when working with implanted tool. According to the authors the stresses in the tool are reduced due to reduction of the resultant of the cutting force and reduction of the angle between the resultant of the forces and the bisector of the angle of wedge ψ.

When working with TiB_2 implanted tool the given angle [15] increases within the whole range of cutting conditions under study. it is associated with low values of forces in the horizontal plane. When working with TiN+ZrHf coated tool the given reduces and has rather low values. The resultant approaches the bisector of the cutting edge angle. The photographs of the chips show that under tool implantation with TiB_2 the chips become matted as speed of cutting grows. If the tool is treated with

TiN+ZrHf the chip turns into its discontinuous form as the speed grows and becomes convenient for removing from the cutting zone.

4. Conclusion

The research shows that ion implantation of the cutting tool evidently leads to change of the tool resistance and of machined surface quality. Thus, we can say that it is possible to control the process of chip formation, improve tool efficiency, reduce energy input, ensure safe chip removal from the cutting zone through changing the properties of tool surface with ion implantation.

Analysis of experimental data allows recommending application of T5K19 carbide alloy inserts with TiN coating implanted with ZrHf ions (20% hafnium) for producing discontinuous chip in the process of 30HGSA steel machining at automated lines and CNC tools without implementing additional elements for chip crushing.

References

[1] Aco Anti, Petar B Petrovi, Milan Zeljkovi, Borut Kosec and Janko Hodoli 2012 The influence of tool wear on the chip-forming mechanism and tool vibrations *Materials and technology* vol. 46 Issue 3 p.p.279–285.

[2] Zhang X D, Lee L C and KHW Seah 1995 Knowledge Base for Chip Management System, Journal of *Materials Processing Technology* vol. 48 pp. 215-221.

[3] Nekrasov R Yu, Starikov A I, Lasukov A A 2015 Entering the operative correction machining processes CNC *IOP Conference Series: Materials Science and Engineering* vol. 91 (1) 012042/

[4] Valentov A V, Konovodov V V and Agafonova E V 2013 Forecasting residual and operating stress in soldering cutting tools with tungsten-free hard alloy inserts *Applied Mechanics and Materials* vol. 379 pp. 28-31.

[5] Zelinsky A N 1974 Improving the productivity of machining through controlling the processes of the cut layer material deformation: author's *abstract ... Candidate of Technical Sciences* Tula.

[6] Kufarev G L, Okenov K B and Govorukhin V D 1970 Chip formation and the quality of machined surface under restricted cutting. Frunze: Mektep 169 p.

[7] Lasukov A A 2015 Selection of machining conditions in terms of the temperature dependence of chip formation *Russian Engineering Research* vol.35 Issue 9 pp. 679-681.

[8] Vasin S A, Vereshchaka A S and Kushner V S 2001 Metal cutting. Thermomechanical approach to the system of interrelations when cutting: Textbook for technical higher educational institutions. M.: MSTU named after N.E. Bauman publishers 448 p. ill.

[9] Petrushin S. I., Gubaydulina R. K., Grubiy S. V., Likholat A. V. 2015 On the Problem of Wear Resistant Coatings Separation From Tools and Machine Elements *IOP Conference Series: Materials Science and Engineering* vol. 91, Article number 012048 pp. 1-7

[10] Petrova V A, Bakanov A A and Walter A V 2014 Crack Resistance of Weld Seals, Hardening Methods and Composite Tools Quality *Applied Mechanics and Materials* vol. 682 pp. 431–437.

[11] Guosheng Sua, Zhanqiang Liuc, Liang Lie and Bing Wangc 2015 Influences of chip serration on micro-topography of machined surface in high-speed cutting *International Journal of Machine Tools and Manufacture* vol. 89 pp. 202-207.

[12] Astakhov V P, Shvets S V and Osman M O M 1997 Chip structure classification based on mechanics of its formation *Journal of Materials Processing Technology* vol. 71 pp. 247 257.

[13] Klimenov V A, Kovalevskaya Z G, Ul'yanitskii V Y, Zaitsev K V and Borozna V Y 2010 Effect of ultrasound treatment of the substrate on the formation of the coating in detonation spraying *Source of the Document Welding International* vol. 24 Issue 10 pp. 803-807.

[14] Brukhov V V Improving the tool wear resistance with ion implantation method 2003 Tomsk: STL publishers 120 p.

[15] Lasukov A A and Mokhovikov A A 2012 Influence of modified layer of tool on stress - Strain state of cutting wedge *Proceedings - 20127th International Forum on Strategic Technology, IFOST 2012* 6357720.

[16] Ovcharenko V E 2012 Evolution of the structure of plasma metal-ceramic coating under pulsed electron-beam treatment *Inorganic Materials: Applied Research* vol.3 pp. 210-215.

IOP Conf. Series: Materials Science and Engineering **142** (2016) 012066 doi:10.1088/1757-899X/142/1/012066

Synthis and Phisical And Chemical; Properties of $SiO_2 - B_2O_3$ and $SiO_2 - P_2O_5$ Thin Film Systems and Powders

A G Mal'chik[1,a*], S V Litovkin[1,b], V I Seregin[2,c], P V Rodionov[1,d], S O Kryuchkova[1,e]

[1]652055, Yurga, ul. Leningradskaya 26, Yurga Institute of Technology (Affiliate) National Research Tomsk Polytechnic University, Russia
[2]105005, ul. Baumanskaya 2-ya, 5/1, Moscow, Bauman Moscow State Technical University, Russia

E-mail: [a]ale-malchik@yandex.ru, [b]protoniy@yandex.ru,[c]vyacheslavseregin@bk.ru, [d]rodik-1972@yandex.ru, [e]Lizakup@bk.ru,

Abstract. The $SiO_2 - B_2O_3$ and $SiO_2 - P_2O_5$ films were synthesized by using film forming solutions having a P_2O_5 content of up to 30% and B_2O_3 up to 40%. Properties of the film-forming solutions and binary oxides were examined. The physical and chemical processes occurring in the solution during the heat treatment of films were examined. The conditions for producing films of different thicknesses were determined. The kinetic parameters were calculated.

1. Introduction

Currently, a lot of studies are devoted to inorganic silicate materials with a focus on their applications and functional uses. Available results relating to the synthesis and study of physical and chemical properties of materials allow us to design and develop thin-film based materials and devices for use in various industries, including mechanical engineering. A substance of thin films in the solid phase is a special state, and its properties are different from those of the same substance in the bulky state. The difference lies in the fact that such systems are characterized by a range of particle dimensions that dominates most of the physical and chemical properties [1,2]. The properties of thin-film materials sharply contrast with the properties of materials in the bulky state due to their specific features of macrostructure, which is very dependent on processes occurring during the formation of film systems. For various applications in modern machinery and tool manufacturing industries, products surfaces made of glass, ceramics, porcelain or other materials are often required to be hydrophobic and have wetting resistance to acidic or saline solutions and organic liquids. For this purpose it is often sufficient to apply a thin film of a specific chemical nature. Such films are designed for applications where it is important to maintain transparency, expand functionality of devices under particular sever weather conditions, and extend their service life. A variety of organic-silicone compositions are the most frequently used as hydrophobic coatings [3,4].

The aim of this work is to obtain thin films of $SiO_2 - B_2O_3$ and $SiO_2 - P_2O_5$ systems and study physical and chemical processes behind the synthesis as well as their properties. In the course of this work, the properties of film-forming solutions based on tetraethoxysilane, ethyl alcohol, orthophosphoric and boric acids were investigated; ranges of stable film forming properties were studied; The physical and chemical processes of forming oxides during the heat treatment were examined; the optimal conditions for the formation of $SiO_2 - B_2O_3$ and $SiO_2 - P_2O_5$ films were determined; the physical and chemical properties of the obtained films were studied.

2. Experimental part

To obtain thin films of a specified composition, we used film forming solutions (FFSs), prepared on the basis of 96% ethanol, tetraethoxysilane, boric and orthophosphoric acids. The films were prepared on silicon and glass substrates. Two methods were used for obtaining: the processes of centrifugation at 3000 - 5000 rev/min and stretching at 1 - 5 mm/s. Samples were thermostatically controlled at a temperature of 333K for 30 minutes. The final formation of the film was made at temperatures of 473-873K for 1 hour. A glass viscometer (d = 0,99 nm) was used to measure viscosity of solutions. The dried FFSs were analyzed by a NETSCHSTA 499C (in air; standard - calcined $\alpha - Al_2O_3$) in a temperature range of 25 to 1000 °C. The refractive index and thickness of films were measured at 5 points of each sample using laser ellipsometers LEF-3M and "SE400advanced". The morphology of the film surface was examined by a scanning electron microscope (SEM) HITACHI-3000.

3. Results and Discussion

The process of forming double oxides is complex and its processing history is influenced by conditions of preparing FFSs. For this purpose, during the first phase we studied the influence of adding boric and orthophosphoric acids on properties of FFSs. Since the main processes of forming FFSs are hydrolysis and polycondensation, a criterion for film-forming ability is viscosity of solutions. A detailed study of the system shows that the viscosity of the solution with orthophosphoric acid added is higher at the initial moment of time. However, the time to solution aging remains the same as the time to aging solutions without adding H_3PO_4, i.e. for two days. Subsequently, the viscosity in the orthophosphoric acid – added system changes more significantly than in the plain system, and this points to the fact that H_3PO_4 accelerates hydrolysis and polycondensation, reducing a solution's expiration time.

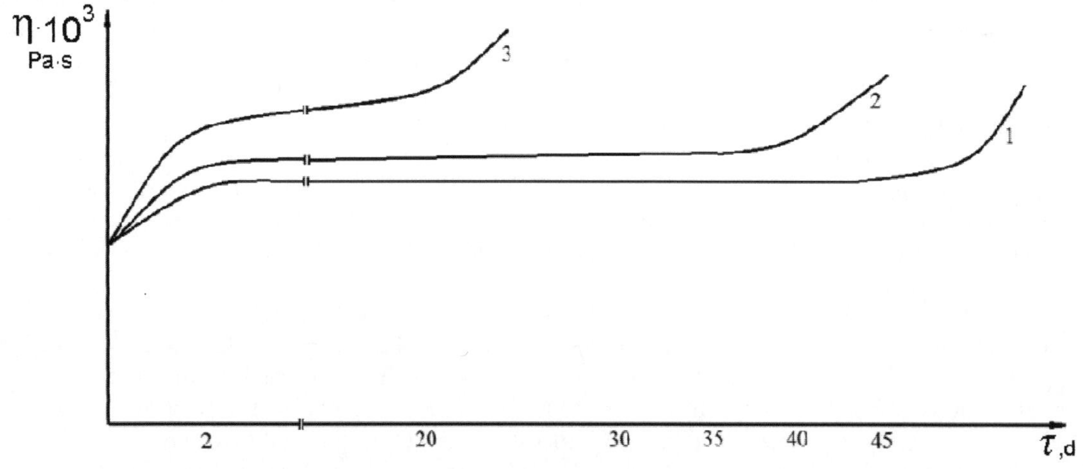

Figure 1. Time – dependent changes in viscosity of the FFSs with orthophosphoric acid added: 1 P_2O_5 – 10% by wt.; 2 P_2O_5 – 20% by wt.; 3 P_2O_5– 30% by wt.

FFS without adding H_3PO_4 becomes a gel after 45 days after its preparation; FFS with adding orthophosphoric deteriorates after 35 days, moreover, the higher concentration of H_3PO_4, the more the solution is subject to the aging process. The dependency of FFS viscosity on time is shown in Figure 1 [5, 6].

The film-forming properties of FFS with adding boric acid were measured by a viscometric method. The addition of boric acid has a less effect on the plain system than orthophosphoric acid, as H_3BO_3 (pK_I=9,2) is much weaker than H_3PO_4 (pK_I=2,1). The time to solution aging remains the same, i.e. two days. FFSs with boric acid added deteriorates in 43-44 days, and in this case, a concentration of boric acid does not cause a substantial change in viscosity, consequently, the time to expiration of the solution (Figure 2).

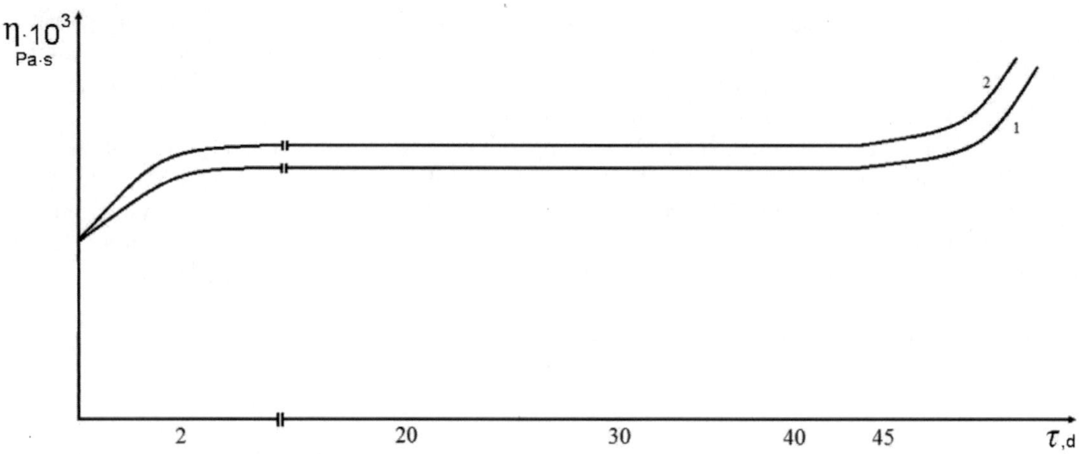

Figure 2. Time-dependent changes in viscosity of the FFSs with boric acid added: 1 B_2O_3- 10% by wt.; 2 B_2O_3- 30% by wt.

As a result, we established limits for a range of stabilized film-forming properties with regard to FFSs with H_3BO_3 and H_3PO_4 additions. The maximum content of B_2O_3 in the film reaches 40 % by weight. A further increase in boric acid concentration is limited by poor solubility in ethanol. The maximum content of P_2O_5 in the film is 90 % by weight. However, FFSs are not self-sustaining; the time to their aging varies 5 to 10 days at high concentrations of orthophosphoric acid in FFSs. The hydrolysis process rapidly flows at high concentrations of H_3PO_4, resulting in the formation of glass-like polymer that is not suitable for producing films. Accordingly, FFSs with a content of P_2O_5 up to 30% in the film is classified as most technically possible.

The process of forming $SiO_2 – P_2O_5$ oxides is complicated and undertakes several phases. When comparing thermal micrographs of powders in the SiO_2 and $SiO_2 – P_2O_5$ systems, it can be noted that the addition of phosphoric acid to FFS reduces a decomposition temperature to 873K in comparison with the plain system – 973K. The formation process proceeds in three steps, as evidenced by the presence of three peaks in the DTA - curve (Figure 3). The first phase takes place at a temperature of up 433K and corresponds to the process of polycondensation. In this case, the polycondensation process is faster than in the plain SiO_2 system, this is indicative regarding a smaller value of activation energy (Table 1). The second phase starts at 473K and ends at 523K. This results in the decomposition process of polyorganosiloxanes, with the simultaneous processes of phosphoric acid decomposition and P_2O_5 oxide formation. Energetically, these processes are constrained, and the energy of activation is the highest at this phase (Table 1). During the third phase, an intermediate is formed according to equation 1 with subsequent break-down to $SiO_2 – P_2O_5$.

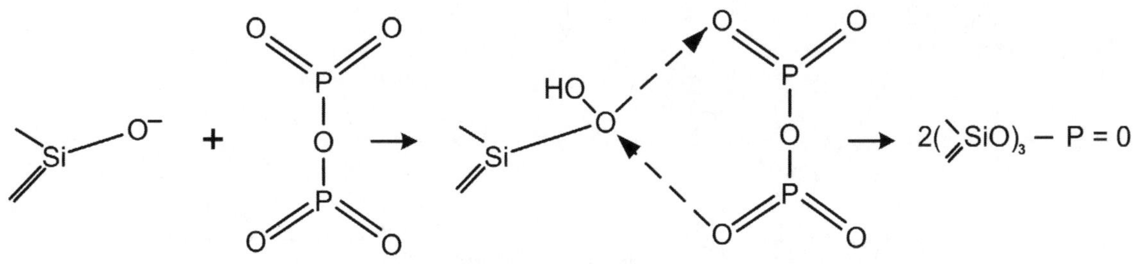

Table 1 Energy of Activation for the processes of obtaining powders

System	Phases	T, K	E_a, kJ/mol
SiO_2	1	333 – 433	48,9
	2	433 – 973	70,4
$SiO_2 - P_2O_5$	1	333 – 433	42,6
	2	433 – 523	81,9
	3	523 – 873	32,0
$SiO_2 - B_2O_3$	1	333 – 433	54,2
	2	433 – 823	37,9

From scientific literature it is known that the energy required for breaking bonds P - O and Si - O have very high values of 597 kJ/mol and 408 kJ/mol respectively [7,8,9]. But the energy of activation in the third phase is not high, and it can be assumed that this occurs due to the formation of an intermediate, reducing the temperature of decomposition by 100K. We can also assume that the formation of $SiO_2 - P_2O_5$ oxides in the thin film state will proceed in the same way, in three stages. Therefore, a three step-wise process is required to ensure the formation of uniform structured films: first, drying at 333K for 15-20 minutes, then annealing at 523K for 30 min and final annealing at 873 K.

If we compare the thermal micrographs of SiO_2, $SiO_2 - P_2O_5$ and $SiO_2 - B_2O_3$ powders [10], it can be observed that adding boric acid to FFSs reduces a temperature of decomposition to 823K, which is 50 K lower than the temperature of decomposition when orthophosphoric acid is added. The process of forming oxides in the $SiO_2 - B_2O_3$ system develops in three phases as well, there are only two peaks observed in the DTA-curve (Figure 4). At a temperature of 433K two processes take place simultaneously: the decomposition of boric acid and polycondensation, both having the highest value of activation energy (Table 1). During the next phase, an intermediate is formed according to equation 2, which decomposes to $SiO_2 - B_2O_3$.

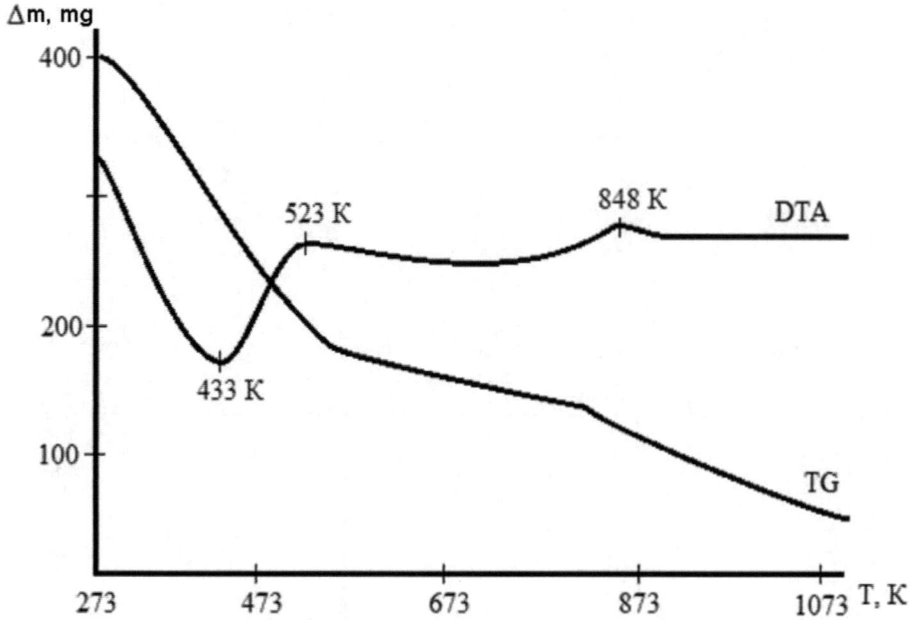

Figure 3. Thermogram of the $SiO_2 - P_2O_5$ formation by decomposing FFS based on tetraethoxysilane and orthophosphoric acids

The formed intermediate leads to a reduction in the energy of activation, as compared with the plain system, although it is higher than the energy of activation in the orthophosphoric acid added system. This apparently relates to the fact that the energy of B – O bond breaking is as high as 807 kJ /mol, so the process of breaking the intermediate takes place with difficulty and followed, accordingly, by an increase in the energy of activation. We can assume that the formation of $SiO_2 - B_2O_3$ oxides in the thin film state will proceed similarly in three stages. Therefore, a step-wise process is required to ensure the formation of uniform structured films: first, drying at 333K for 15-20 minutes, then annealing at 433K for 30 min and final annealing at 823K.

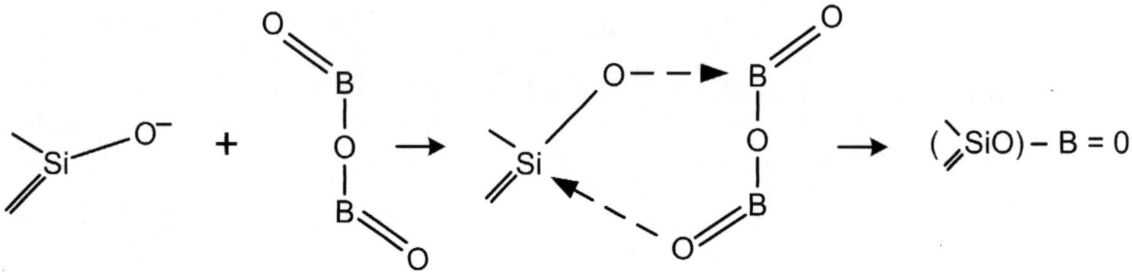

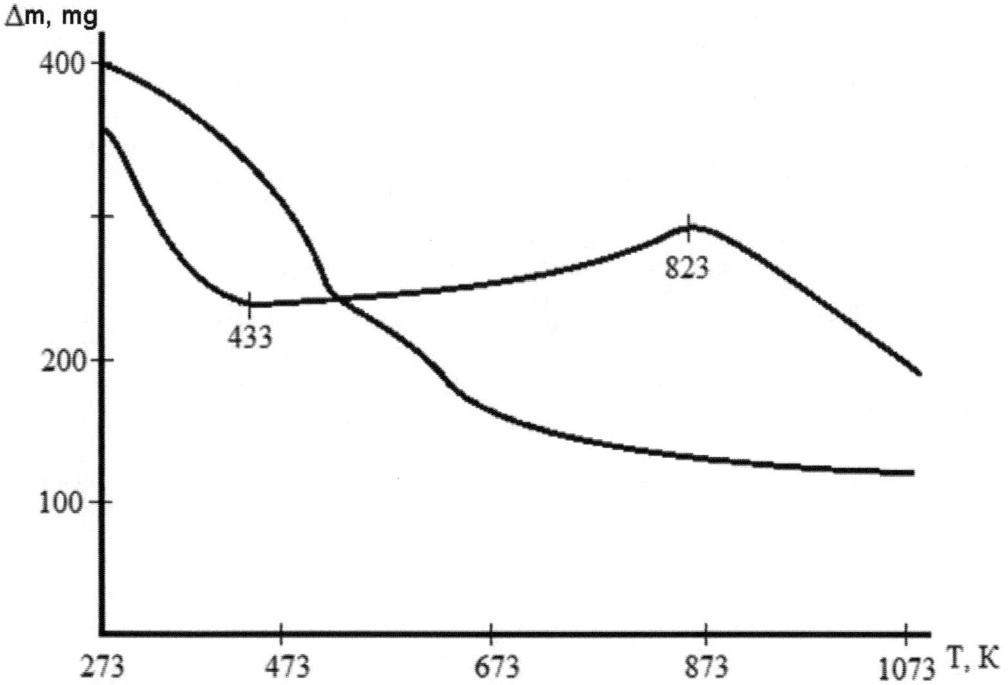

Figure 4. Thermogram of the $SiO_2 - B_2O_3$ formation by decomposing FFS based on tetraethoxysilane and boric acid

The study of physical and chemical properties of thin films allows us to make conclusions on possible uses of films with a specified composition. The $SiO_2 - P_2O_5$ films received from FFSs are nonporous, homogeneous and have a good adhesion to the substrate (F = 0,993 kg/mm^2) (Figure 5). The values of refractive index are shown in Table 2 depending on the time to FFS aging, the rate of stretching and centrifugation.

Table 2. Dependencies of refractive indices and $SiO_2 - P_2O_5$ film thicknesses on the film forming processes

Property	Time to solution aging, day			A rate of stretching, mm/s			Centrifugation, rev/min		
	2	10	20	1	3	5	3000	4000	5000
n	1,378	1,476	1,536	1,463	1,474	1,431	1,479	1,473	1,415
d, nm	55,28	107,36	274,14	224,63	286,4	198,92	167,77	156,56	196,60

Figure 5 A micrograph of the $SiO_2 - P_2O_5$ film

Figure 6 A Micrograph of the $SiO_2 - B_2O_3$ film

It should be noted that the film thickness and refractive index grow with increasing the time to solution aging. Apparently, the hydrolysis and polycondensation processes are faster when H_3PO_4 is added, causing the viscosity of the solution to increase that results in the increased thickness and refractive index.

The obtained $SiO_2 - B_2O_3$ films are nonporous and have a good adhesion (F = 0,981 kg / mm^2) (Figure 6). The optical properties of the $SiO_2 - B_2O_3$ films are presented in Table 3.

Table 3. Dependencies of refractive indices and $SiO_2 - B_2O_3$ film thicknesses upon the film forming processes

Property	Time to solution aging, day			A rate of stretching, mm/s			Centrifugation, rev/min		
	2	10	20	1	3	5	3000	4000	5000
n	1,431	1,464	1,473	1,415	1,438	1,436	1,467	1,434	1,456
d, nm	198,92	224,63	286,4	167,77	178,32	193,24	198,87	177,83	181,77

In contrast to the $SiO_2 - P_2O_5$ films, the refractive index for the $SiO_2 - B_2O_3$ films does not increase noticeably with the time to FFS aging. This is explained by the fact that the viscosity of FFS with boric acid added does not change significantly.

4. Conclusions

Based on the results of research, the following conclusions can be made: using the viscometry methods, the influence of boric and orthophosphoric acids additives on the FFS properties were analyzed; the ranges of the solutions' stabilized film-forming properties were established. Application of H_3BO_3 is limited to 40% by wt. because of poor solubility of boric acid in ethanol, and H_3PO_4 is limited due to the fact that an increase in adding orthophosphoric acid reduces the expiry time of the solution up to 10 days. Therefore, the technologically usable solutions are those with acid additives up to 30% by wt. The mechanism for the formation of complex $SiO_2 - B_2O_3$ and $SiO_2 - P_2O_5$ oxides was studied; the optimal conditions for the formation of films are as follows: for $SiO_2 - P_2O_5$: 15-20 min at 333K, 30 min at 523K and 30 min at 873 K, and for $SiO_2 - B_2O_3$: 15-20 min at 333K, 30 min at 433K and 20-25 min at 823K.

The structure and physicochemical properties of the films were studied. The refractive index for the $SiO_2 - P_2O_5$ films varies from 1.38 to 1.53; the film thickness depends on the time to solution aging. The refractive index for the $SiO_2 - B_2O_3$ films varies from 1.41 to 1.47; the film thickness is practically independent of the aging time. The films with the specified compositions can be recommended as protective coatings in microelectronic and optic application.

References

[1] Borilo L.P. Tonkoplenochnye neorganicheskie nanosistemy [Thin-film inorganic nanosystem]/ Ed. Prof., D. in Eng.Sc. V.V. Kozik. – Tomsk: Publ. TGU, 2003. – 134 p.

[2] Petrovskiy G.T., Shashkin V.S., Yakhkind A.K. Osnovnye napravlenija zol'-gel' sinteza stekloobraznyh materialov dlja optiki iz kolloidnyh form kremnezema [Main processes of the sol-gel synthesis for glass-like materials from silica colloid forms // Fizika i himija stekla [Physics and chemistry of glass]– 1997. – Vol. 23. – No. 1. – pp. 43–53.

[3] Synthesis and properties of films in the SiO_2-Bi_2O_3 system Borilo L.P., Kuznetsova S.A., Kozik V.V., Zabolotskaya A.V., Mal'chik A.G. Russian Journal of Inorganic Chemistry. 2014. T. 59. No.10. pp. 1065-1068.

[4] Shabanova N.A., Popov V.V., Sarkisov P.D. Himija i tehnologija nanodispersnyh oksidov [Chemistry and technology of nano dispersed oxides]. – Moscow: ICC «Akademija», 2006. – 309 p.

[5] Kozik V.V., Borilo L.P., Brichkova V.Yu. Sintez i izuchenie svojstv tonkoplenochnoj i dispersnoj sistemy SiO_2-P_2O_5 [Synthesis and properties of the thin dispersed films in the SiO_2-P_2O_5 system]. – Tomsk: Publ. TPU - 2004. Vol. 307. No. 6. pp. 92-96.

[6] Borilo L.P., Petrovskaya T.S., Lyutova E.S., Spivakova L.N.Sintez i fiziko-himicheskie svojstva tonkoplenochnyh i dispersnyh funkcional'nyh silikofosfatnyh materialov [Synthesis and physicochemical properties of the thin dispersed films from silicophosphate materials]. - Tomsk: Publ. TPU. 2011. Vol. 319. No. 3. pp. 43-47.

[7] Kharitonov N.P., Ostrovskiy V.V. Termicheskaja i termookislitel'naja destrukcija poliorganosiloksanov [Thermal and thermo oxidation destruction of polyorganosilica]. – Leningrad: Nauka [Science]. - 1982. – 207 p.

[8] Gurvich L.V., Karachevtsev G.V., Kondrat'ev V.N. Jenergija razryva himicheskih svjazej. Potencialy ionizacii i srodstvo k jelektronu [Energy of bond breaking. Ionizing potential and electron accepting ability]. – Moscow: Nauka [Science]. - 1974. – 351 p.

[9] Fialko M.B. Neizotermicheskaja kinetika v termicheskom analize [Non-isothermal kinetics in thermal analysis] - Tomsk: Publ. TPU, 1981. – 110 p.

[10] Physicochemical investigation of formation of SiO_2 powders and films from film-forming solutions. Mal'chik A.G., Borilo L.P., Kozik V.V.Russian Journal of Applied Chemistry. 1996. T. 69. No. 2. pp. 198-201.

IOP Conf. Series: Materials Science and Engineering **142** (2016) 012067 doi:10.1088/1757-899X/142/1/012067

Direct Alloying of Steel with Nickel Concentrate

O I Nokhrina[1], I D Rozhikhina[2], I E Proshunin[2]

[1]Yurga Institute of Technology, National Research
Tomsk Polytechnic University Affiliate,
26 Leningradskaya str., Yurga, Kemerovo region, RUS
[2]Siberian State Industrial University,
42 Kirova str., Novokuznetsk, Kemerovo region, RUS

E-mail: kafamsf@sibsiu.ru

Abstract. A technology of alloying steel with nickel reduced from nickel concentrate is analysed and developed. Limits of reduction concentration areas are defined. An optimal composition of nickel concentrate pellets and a method of feeding them into the furnace are deduced from experiments. It is proved that when pellets made of nickel concentrate and coke are added into the charge during steel smelting by the technology of alloyed scrap remelting, nickel recovery achieves 92–95%. The technology was tested by smelting DSP-40 steel.

Introduction

Nickel is one of the most important and widely used alloying elements. Alloyed steels and complex iron-nickel alloys production takes from 60 to 65% of the nickel produced per year. The rest 35–40% are used in non-ferrous metallurgy and chemical industry, first of all for making electroplates [1–5].
Steel is usually alloyed with nickel in electric arc furnaces by introducing nickel or ferronickel into the furnace charge. There are no any showings of using nickel oxides or nickel concentrates in domestic metallurgical industry. However, a technology of direct alloying of steel with manganese, chromium and vanadium is successfully used. Ad-vantages of this technology consist in throughout recovery of an alloying element which means eliminating ferroalloy smelting stage and thus leads to cost saving [6, 7]. Nickel production is a resource- and energy-consuming process, which is why application of nickel concentrate for direct alloying of steel can be advantageous. Nickel concentrate consisting of 42–45 % Ni, 2–3 % Mn, 1–1.5 % Fe, 0.3–0.5 % can be produced by complex beneficiation of ferromanganese nodules and polymetallic manganic ferrous materials [5, 8 – 10].
The aim of this work is a scientific justification, research and development of a technology for producing nickel alloy steel using nickel concentrate.

Methods and Materials

When melting steel in arc furnaces, carbon and silicon dissolved in liquid steel can serve as reducing agents at different melting stages.
To determine the conditions under which nickel is reduced from nickel-containing oxide materials we used thermo-dynamic modeling methods based on calculation of equilibrium states in model thermodynamic systems. Calculation of thermodynamic equilibrium reveals a possibility of obtaining

different substances which are important for estimation of the limit state and determines acceptable values. To carry out the thermodynamic modeling we used “Terra” software complex.
Nickel reduction was studied in a three-component system Ni–O–C represented by a set of substances *l*NiO – *n*C, where *l* – a number of NiO moles, *n* – a number of carbon moles. Values of *l* and *n* were set as a reference composition of the system.

The research procedure involved the following stages: calculating possible compositions and determining thermodynamic conditions required for nickel reduction; determining limits of reduction concentration areas; finding input flow parameters which provide the best composition of the system under equilibrium conditions.

Possible compositions which could result from nickel reduction in thermodynamic system consisting of Ni–O–C and Ni–O–C–Fe were determined by varying the number of carbon moles in the system. It helped us to find the limits of reduction concentration areas. The reference composition of the model system Ni–O–C was the following: 1mole of NiO and *n* moles of carbon. The calculations were made in the temperature interval from 573 to 1873 K which corresponds to melting temperature.

Kinetic laboratory tests of nickel reduction from oxide at the temperature range 973–1473 K were carried out by the method of continuous weighing by a common technique [7].

For treating steel in a ladle, briquettes were composed. They consisted of the nickel concentrate obtained by beneficiation of polymetallic manganiferous materials (% mass. Ni – 45.0, Mn – 2.3, Fe – 1.4, Co – 0.5, Cu – 0.1, P – < 0.015, SiO_2 – other impurities – 2.82) and coke produced by “EVRAZ-ZSMK” Ltd with chemical composition shown in Table 1.

Table 1. Chemical composition of coke by “EVRAZ-ZSMK” Ltd

	A^d	V^{daf}	W^p	Ash composition, % mass									
				SiO_2	Al_2O_3	MnO_2	MgO	CaO	Fe_2O_3	P_2O_5	K_2O	Na_2O	CO_2
Coke	13.6	2	2	51.1	23.3	0.16	1.58	1.2	17.46	0.5	1.2	0.2	74.4

Proportion of the nickel monoxide and the coke was calculated by the reaction:

$$NiO + C = Ni + CO.$$

The coke and the nickel concentrate were mixed thoroughly. Liquid glass having a density of 1.4 g/cm^3 and amounting 5 % of the mass was used as a bonding agent. Briquettes were pressed by a hydraulic press having a tonnage of 10 tons. The briquettes produced were of cylinder shape, 40 cm high and 25 mm in diameter. The briquettes were place in an alundum pot which in its turn was installed into a graphite pot and suspended on a special device in a preheated resistance furnace.

Isometric soaking was carried out at the temperatures of 1073, 1173, 1273 and 1473 K. The thermocouple BP 5.20 was used for measuring temperatures. Electronic weighing scales recorded changes in the sample weight every 60 seconds during reducing nickel from the oxide. The briquettes were soaked in the furnace until the loss of sample weight stopped. The resulting products were subjected to X-ray phase analysis.

A degree of nickel reduction at different temperatures after 15, 20 and 25 minutes of soaking was calculated by an equation:

$$CB(\%) = \frac{G \cdot \left[\frac{M_1}{M_1 + M_2}\right]}{M_3} \cdot 100 = \frac{\frac{G \cdot 16}{28}}{M_3} \cdot 100 = \frac{G \cdot 0.57}{M_3} \cdot 100,$$

Where G – loss of sample weight;
$M_1 = 16$ and $M_2 = 12$ – atomic weights of oxygen and carbon;

M_3 – weight of oxygen in the original sample

Treatment of steel in the ladle was studied under laboratory and workshop conditions. In the laboratory, steel was melted in a laboratory arc furnace, 10 tons capacity. A method of loading of nickel oxide into the arc furnace was experimentally determined and tested. Pallets 20–30 mm in diameter were made from the nickel concentrate, less than 0.5 mm fraction, obtained by beneficiation of polymetallic manganiferous ores and small coke.

The pallets were loaded into the furnace in two ways:

I – into the charge;

II – during the recovery period onto the heel of metal before building up slag.

The quantity of pallets was calculated to produce steel containing 1 % of nickel.

Experimental smeltings were carried out by conventional two-slag technology: the furnace was charged with a fusion mixture consisting of scrap metal, pellets, coke, and lime carbonate required to generate slag. When nickel concentrate pellets were used, a percentage of coke in the fusion mixture was increased providing for the amount of carbon necessary to reduce nickel from oxide state and for carbon loss. Pellets were placed closer to the slopes. The scrap metal used for smelting had the following composition: %: C – 0.275, Si – 0.267, Mn – 0.423, Cr – 0.175, Ni – 0.1, S – 0.027, P – 0.028, Fe – the rest.

The composition of the fusion mixture is shown in Table 2.

After the furnace was charged, the electrodes were dipped and the electric arc was lightened. The smelting time was 20 min.

By the end of the smelting period, slag was sintered in the furnace and some slag samples were taken for chemical analysis. Iron ore was loaded in batches to initiate a decarburation reaction. After five minutes of soaking, the slag was flushed out; metal and slag samples were collected for chemical analysis.

After removing oxidizing slag preliminary deoxidization with ferrosilicon was carried out in order to obtain a required percentage of silicon in the finished steel. After deoxidizers were added, lime carbonate and quartzite amounting 3 % of the metal weight were loaded into the furnace as slag-forming materials. After melting slag-forming materials and getting liquid slag the later was deoxidized by coke powder and ground ferrosilicon. Then slag and metal samples were collected for a chemical analysis.

The metal was tapped into the ladle. The ingot thus obtained was cut height wise into three equal parts. The resulting metal was analysed.

Table 2. Fusion materials for experimental smelting

Materials	Smeltings					
	1	2	3	4	5	6
Weight of briquettes	0.757	0.756	0.758	0.677	0.678	0.676
Composition of briquettes, % weight:						
Nickel concentrate	85	85	85	95	95	95
Coke	10	10	10	–	–	–
Bondings	5	5	5	5	5	5
Scrap metal weight, kg	9	9	9	9	9	9

For industrial tests of direct alloying of steel with nickel, steel was smelted in arc furnace DSP-40. Pellets for direct alloying were made of nickel concentrate and coke in stoichiometric proportion. The metal for test smelting had been certified according to the procedure adapted at "Stal NK" Ltd.

Results and Discussion

Thermodynamic calculations at different temperatures for Ni–O–*n*C system showed that the greatest reduction of nickel at the temperatures of 1073 K was obtained with 0.5 mole of carbon per 1 mole of NiO, and at the temperatures of 1573 and 1873 K the same result was achieved with 0.2 mole of carbon per 1mole of NiO. Complete reduction of nickel from its oxide in Ni–O–C–Fe system at the temperature of 1873 K takes 0.5 mole of carbon. If the amount of carbon in the system is increased over these values, iron begins to reduce from the oxide alongside with nickel (Figure 1).

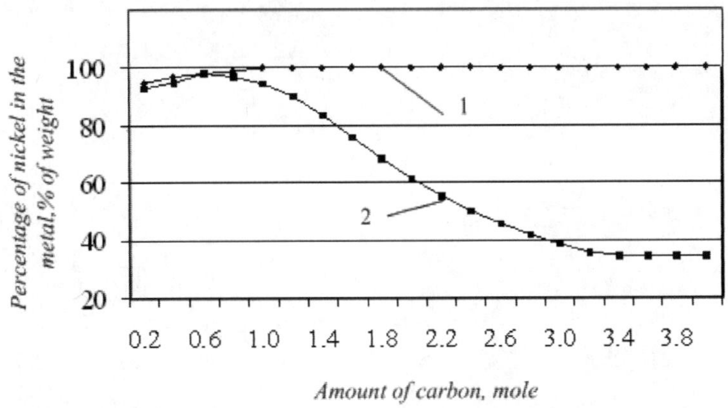

Figure 1. Dependence of nickel percentage in the metal on the amount of carbon in Ni–O–*n*C system at the temperature of 1073 K (1) and in Ni–O–*n*C–Fe system at the temperature of 1873 K (2)

Thermodynamic simulation showed that in Ni–O–*n*C system in Ni–O–*n*C system nickel can be cpmpletely reduced from the oxide at the temperature of 1073 K and in Ni–O–*n*C–Fe system it happens at the temperature of 1873 K.

A kinetic research and an X-ray phase analysis (Table 2) proved that at the temperatures of 1173 – 1473 K the complete reduction of nickel from its oxide takes 20–30 min, while at 1073 K the reduction process takes a longer time (70 min).

Samples which were soaked at the temperatures of 1173 – 1473 K are different in structure and color, and incorporate sintered bits with characteristic metallic luster. Dependences of nickel reduction rate on soaking time under such conditions demonstrate a more explicit increase which proves that the process becomes more active at a higher temperature (Figure 3). The X-ray phase analysis also proves this conclusion. The amount of the nickel oxide and the reducing agent decreases and the amount of nickel considerably increases.

Table 2. X-ray phase analysis

Soaking temperature, K	Phase composition	
	A lot	Some
1073	NiO, C	Ni*
1173	Ni	NiO, C
1273	Ni	NiO*, C*
1473	Ni	NiO*, C*

* – Found in a small quantity

Figure 2 shows samples after twenty minutes of soaking in the furnace at different temperatures.

a – 1073 K; b –1173 K; c – 1273 K; d – 1473 K

Figure 2. Samples after 20 min soaking at different temperatures

According to the X-ray phase analysis at the temperature of 1473 K the percentage of nickel in the sample is greater and the amount of nickel oxide and carbon is lower than in the samples soaked at other temperatures.

Figure 4 shows the values of nickel reduction from its oxide. As we can see nickel reduction values greatly depend on the temperature of soaking nickel in the furnace.

The tests proved that the reduction of nickel from its oxide by solid carbon in the resistance furnace takes place at as low as 1073 K, but the process is slow. At 1173 K and above the reduction of nickel from its oxide is more intensive and takes less time.

Thus, when smelting steel in the electric arc furnace at the temperature range of 1173–1473 K, complete reduction of nickel from its oxide can take 20–30 min.

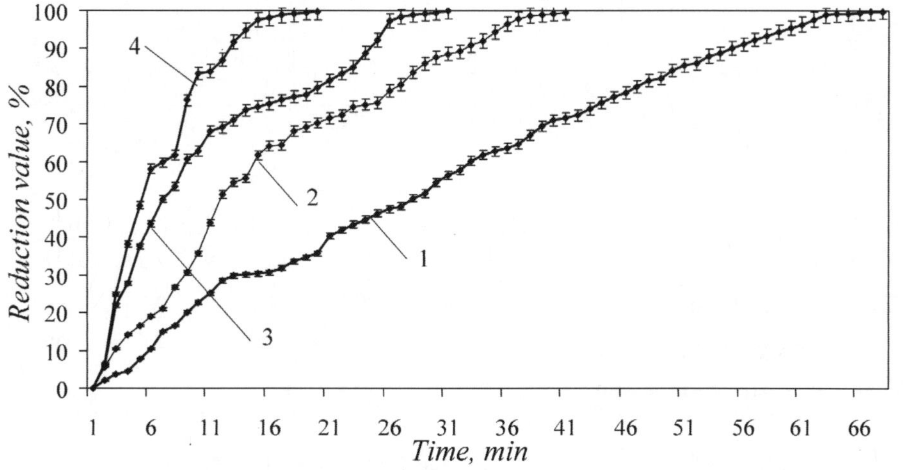

Figure 3. Dependence of nickel reduction values on isometric soaking time at the temperatures of 1073 K (1), 1173 K (2), 1273 K (3) and 1473 K (4)

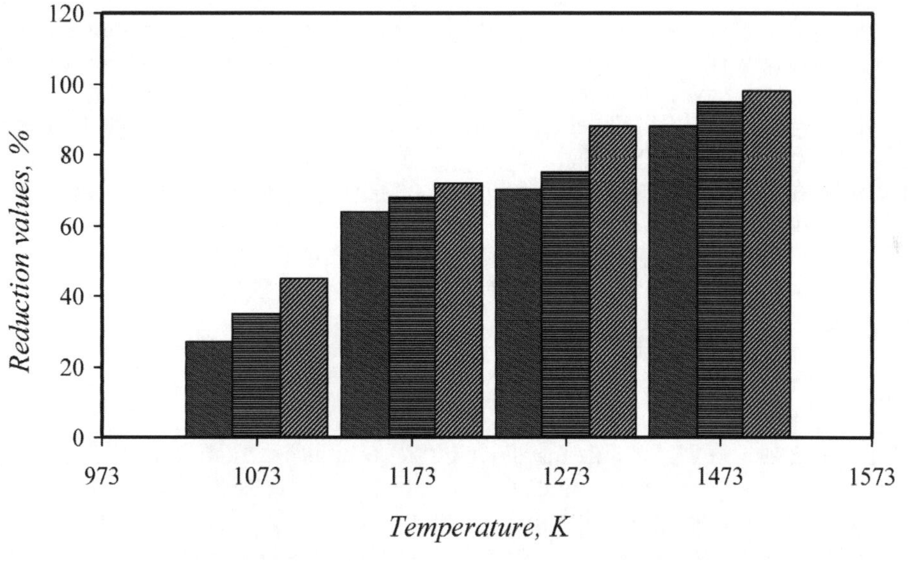

Figure 4. Dependence of nickel reduction values on temperature (soaking time – 15 min, – 20 min, – 25 min)

Laboratory tests resulted in finding the most effective method of loading the nickel oxide into the electric arc furnace. For smeltings 1 – 3, pellets made of nickel concentrate were used, for smeltings 4 – 6, pellets of nickel concentrate mixed with coke was applied. It was experimentally found out that the reduction of nickel is completed during smelting. Nickel output with pellets made of nickel concentrate and coke (smeltings 4, 5, 6) is as high as 97–98 %, while when pure nickel pellets are used (smeltings 1, 2, 3) it makes up 93–95 % (Table 3).

Table 3. Resulting data of alloying steel with nickel

Parameter	Values for smeltings					
	1	2	3	4	5	6
Ni percentage, %:						
– in scrap metal	0.1	0.1	0.1	0.1	0.1	0.1
– at the end of a smelting	2.85	2.79	2.83	2.91	2.94	2.93
– in the finished metal	2.85	2.79	2.83	2.91	2.94	2.93
– in the slag at the end of a smelting	0.144	0.207	0.165	0.084	0.057	0.063
Ingot weight, kg	9.96	9.97	9.95	9.98	9.97	9.95
Nickel reduction, % of weight	95.2	93.1	94.5	97.2	98.1	97.9

Nickel reduction produced by pellets with coke addition is higher and exceeds the values obtained with pellets of pure nickel concentrate by 3 – 5%.

When nickel is added at the beginning of the reduction process its reduction decreases. This is evidently caused by its partial evaporation in the arc zone as boiling temperature of nickel is relatively low.

Based on the experimental results obtained from thermodynamic stimulation of nickel reduction in the electric arc furnace we have developed a technology of direct alloying of chrome-nickel stainless steel 08(12)X18H10T. Smelting was performed by means of alloyed scrap remelting with complete oxidation. The furnace was charged with pellets made of nickel concentrate and coke. This technology has been tested in a steel smelting workshop of "Stal NK" Ltd. Industrial application of the developed technology of alloying steel with nickel using nickel concentrate showed that nickel reduction in the workshop conditions can reach 97 – 98 %.

The metal obtained in test smelting by the technology of direct alloying has been certified according to the procedure adopted at "Stal NK" Ltd, and it fully meets the requirements.

It is estimated that the use of nickel concentrate produced by hydrometallurgical beneficiation of polymetallic manganese ores and ferro-manganese nodules instead of pure nickel will reduce the cost of an alloying agent.

Conclusion

The theoretical and experimental research proved that nickel concentrate produced by beneficiation of polymetallic manganese ores is suitable for alloying steel with nickel in electric arc furnaces. The tests showed that direct alloying of steel is more effective when pellets consisting of nickel concentrate and small coke are fed into the charge, the optimal ratio of nickel oxide and reducing agent being 7:1. The resulting nickel reduction values obtained during the steel smelting is as high as 92–95 %.

References

[1] Duan Yuping, Zhang Yahong, Wang Tongmin, Gu Shuchao, Li xin, Lv Xingjun 2014 *J. Materials Science and Engineering* **185** 86-93.

[2] Santiago Sosa Haudet, Martín A. Rodriguez, Ricardo M. Carranza 2015 *J. Procedia Materials Science* **8** 21-28.

[3] Nokhrina O I, Rozhikhina I D, Kichigina O Y, Goryushkina Y V, Rozevich A P 2012 *J. Steel in Translation* **42** 802-805.

[4] Yan-li An, Hua-yun Du, Ying-hui Wei, Ning Wang, Li-feng Hou, Wan-ming Lin 2013 *J. Materials & Design* **46** 627-633.

[5] Anand Kumar S, Ganesh Sundara Raman S, Sankara Narayanan T S N, Gnanamoorthy R 2012 *J. Surf. Coat. Technol* **206** 442.
[6] Hu W, Yunxia W, Chenshuo M, Zhaolin Z 2014 *J. Applied Mechanics and Materials* **456** 406–410.
[7] Babcsan N, Leitlmeier D, Degischer H.P, Banhart J 2004 *J. Advanced Engineering Materials* **6** 421–428.
[8] Fedoseev S N, Mukhtar Z M 2015 *J. IOP Conference Series: Materials Science and Engineering* **91** 1–7
[9] Mao D, Edwards J R, Harvey A 2006 *J. Chemical Engineering and Science* **61** 1836–1845.
[10] Nokhrina O I, Rozhikhina I D, Proshunin I E, Khodosov I E 2015 *J. Steel in Translation.* **45** 295–300.
[11] Rodzevich A P, Kuzmina L V, Gazenaur E G, Krasheninin V I 2014 *J. AIP Conference Proceedings* **1623** 519-522.
[12] Yang C C, Nakae H 2003 *J. Materials Processing Technology* **141** 202–206.
[13] Ibragimov E A, Saprikin A A, Babakova E V 2014 *J. Advanced Materials Research* **1040** 764–767
[14] Nokhrina O I, Rozhihina I D, Hodosov I E 2015 *J. IOP Conference Series: Materials Science and Engineering* **91** 12-17.
[15] Ferro P, Lazzarin P, Berto F 2012 *J. Materials Science and Engineering* **554** 122–128.
[16] Fedoseev S N, Lychagin D V, Sharafutdinova A S 2014 *J. Advanced Materials Research.* **1040** 236-240.
[17] Zhang W, Singh P, Muir D 2002 *J. Hydrometallurgy* **63** 127 – 135.

IOP Conf. Series: Materials Science and Engineering **142** (2016) 012068 doi:10.1088/1757-899X/142/1/012068

Manufacturing and Application of Metalized Ore-Coal Pellets in Synthetic Pig Iron Smelting

O I Nokhrina[1], I D Rozhikhina[2], I E Khodosov[2]

[1]Yurga Institute of Technology, National Research
Tomsk Polytechnic University Affiliate,
26 Leningradskaya str., Yurga, Kemerovo region, RUS
[2]Siberian State Industrial University,
42 Kirova str., Novokuznetsk, Kemerovo region, RUS

E-mail: kafamsf@sibsiu.ru

Abstract. The article presents research data on manufacturing and application of metalized ore-coal pellets in synthetic pig iron smelting. A technology of pellets metallization by means of solid-phase reduction of iron from oxides using hematite-magnetite iron ore and low-caking coal as raw materials is described. Industrial testing of replacing 10, 15, and 20% of waste metal by the metalized ore-coal pellets in the coreless induction furnace IST-1 is described. Optimal temperature and time conditions of feeding the metalized pellets into the furnace in smelting pig iron of SCh-40-60 grade are determined.

Introduction

Iron and steel companies suffer from a scarcity of burden materials and thus they have to find new sources of raw materials. Production and application of metallized iron products is a promising solution of this problem [1]. In Russia as well as all over the world direct-reduced iron (DRI) or hot briquetted iron (HBI) produced in shaft furnaces in Midrex® and HYL that uses converted natural gas as a reducing agent [2]. The use of natural gas provides obtaining materials with high percentage of metallic iron (more than 85%) and low percentage of gangue material. But in this process, gas consumption is as high as 300 – 400 m^3 per a ton of the metalized material. It hampers adoption of this technology in the regions lacking natural gas [3]. At the same time alternative reducing gases such as hydrogen, coke gas, gases released by coal gasification as well as solid carbon reducing agents are coming into use. Metalized materials reduced by solid agents are produced in revolving furnaces in SL/RN process and in rotary-hearth furnaces in Fastmelt® and ITmk3® processes [4].

Coal and its derivative products are the most common solid reducing agents. There are a number of researches focusing on solid-phase reduction of iron from oxides using coal but a commonly accepted scientifically based technology has not been developed. Experts disagree on their characterization of solid-phase reduction of iron from iron ores oxides using coal because of different physical and chemical properties of coal types and iron-bearing materials, complexity of the system, and simultaneous reduction processes involving solid carbon and gaseous substances. It should be taken into consideration that reactions of direct and indirect (gaseous) reduction of iron from iron ore oxides take place simultaneously.

Based on the research data having been obtained recently we determined conditions for producing metalized materials using coal of different grades as a reducing agent. Our experiments consisted in thermodynamic simulation of solid-phase reduction of iron from iron ore oxides, high-temperature experiments, research into properties of raw and finish materials, as well as mathematical analysis of the obtained data.

This work resulted in a developed technique of producing metalized coal-ore pellets (Figure 1).

In Russia and abroad the metalized raw materials are used for smelting free of non-ferrous impurities steels in arc furnaces [6]. The metalized material can also be used for smelting low-phosphorus and low-sulfur impurity-free n on-ferrous metals of synthetic pig iron. Such pig iron is smelted in induction furnaces in on-site foundries at machine-building plants. Cast iron, recycled scrap, and waste metal.

Remelting metalized materials in induction furnaces has a number of evident advantages. In induction remelting the metal in the furnace are constantly mixed in the melting bath. Induction smelting conditions provide degassing of the liquid melt and allow obtaining metal with low nitrogen content because the induction furnace has no high temperature areas unlike which happen under the arcs of the electric arc furnace.

When using metalized materials in burden charge it is necessary to mind their differences from conventional materials. The following characteristics of the metalized materials make them different from waste metal and cast iron: lumps size, apparent density, specific surface, oxide percentage. Table 1 shows characteristics of the metalized pellets, ore, briquettes, and waste metal. It also should be accounted that the metalized materials obtained with coal as a reducing agent contain some amount of solid carbon.

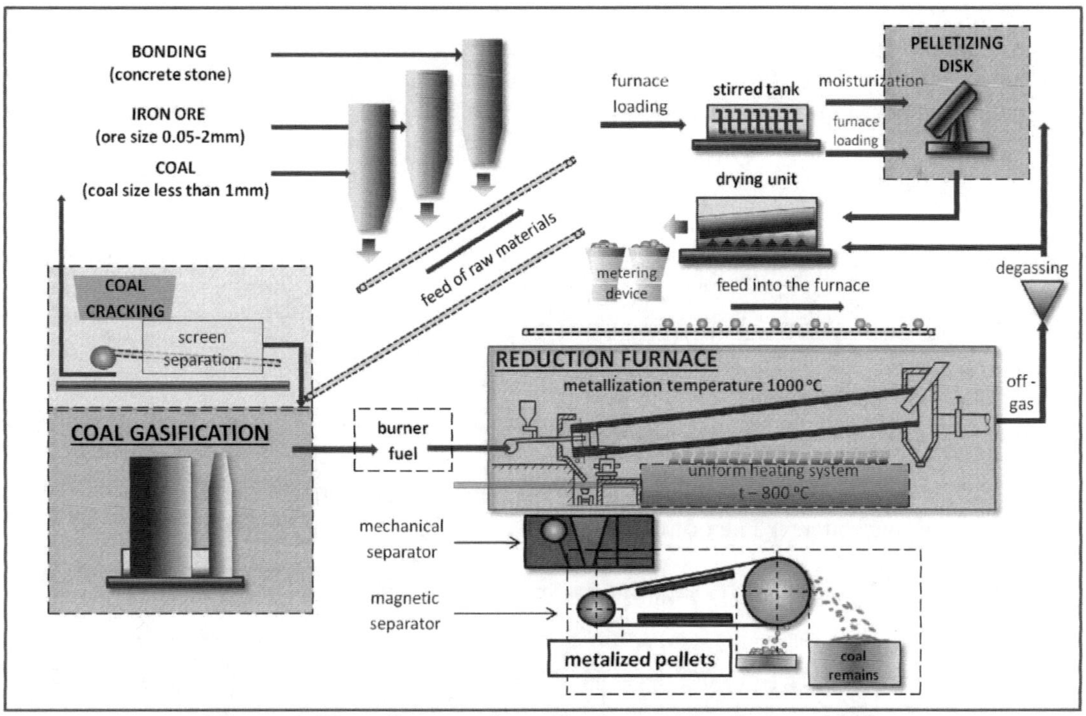

Figure 1. Flow diagram of metalized ore-coal pellets production

IOP Conf. Series: Materials Science and Engineering **142** (2016) 012068 doi:10.1088/1757-899X/142/1/012068

Table 1. Physical properties of metalized materials and waste metal

Material	Properties			
	Lump size, mm	Bulk nass, t/m^2	Apparent density, g/sm^3	Total porosity, %
Metalized pellets	3–20	1.6–2.1	2.7–3.3	45–70
Metalized ore	3–20	1.8–2.1	3.5–3.9	45–70
Briquettes:				
– cold briquetting	Up to 70	2.8	4.0	25–30
– hot briquetting		3.2	5.5	15–20
Waste metal	6–2500	0.5–2.9	–	–

Summarizing available data of using metalized materials in induction furnaces we concluded that the metalized materials applied in induction smelting must have the greatest possible percentage of metal and the lowest percentage of gangue materials. To provide conditions for recovery of light metalized furnace-charge by the molten metal it is important to prevent slag cooling. It is also necessary to remove slag from the metal surface. Working with acid slag is possible when acid bottom and lining is used and percentage of phosphorus in the metalized material is low.

Conditions of induction smelting are good for the use of metalized materials containing solid carbon which in its turn can take part in postreduction of iron oxides and act as a carburizing agent. Metalized materials produced using coal as a reducing agent can be efficiently applied as furnace-charge in melting synthetic pig iron.

Thus, development of a technique for using metalized ore-coal pellets in induction smelting is an important task for researchers.

Methods and Materials

To produce a pilot lot of metalized pellets we used iron-ore concentrate (by Abagursk ore-processing plant, city of Novokuznetsk), hematite-magnetite ore (by Tashtagol ore mine), and low-caking coal of CC grade (by Bachatsk coal mine). Table 2 shows properties of the materials in use.

Table 2. Properties of the materials in use

Material	Chemical composition, %							Moisture content
	$Fe_{общ}$	SiO_2	Al_2O_3	P_2O_5	S	MgO	CaO	W^r, %
Iron ore	52.3	14.24	2.51	0.072	0.42	0.76	0.94	5.78
Iron ore concentrate	61.2	6.75	2.2	0.02	0.41	2.31	1.96	8.27
Coal ash	2.6	46.6	16.3	0.37	0.02	1.58	2.5	-
Coal properties	C^{fix}–70.2 %; V^{daf}–20.6 %; A^d–6.41 %;							6.7

Table 3 shows the composition of raw ore-coal mixture. Produce 100 kg of metalized pellets took 84 kg of iron-ore concentrate, 28 kg of iron ore, 27.5 kg of CC grade coal, 7 kg of concrete stone.

Table 3. Composition of metalized ore-coal pellets

Material	Chemical composition, %							Moisture content,%	C^{fix}, %
	$Fe_{общ}$	SiO_2	Al_2O_3	P_2O_5	S	MgO	CaO		
Ore-coal pellets	42.36	9.53	2.65	0.03	0.30	1.48	1.35	~10	16.4

The metalizing process was carried out in the batch furnace. Raw pellets together with small-sized coal (0.5 mm) were placed into the furnace. After that the furnace was locked and hermetically sealed with asbestos cord. The small-sized coal was used to maintain the reduction conditions in the furnace.

The furnace was operating as the following: during the first 20 min the temperature did not exceed 200 °C to remove moisture from the materials and prevent pellets cracking; then the temperature was raised up to 1000 °C and isometric soaking was held for 80 min. The resulting pellets were separated from the remains of coal. Table 4 shows average data of chemical analysis of the pellets.

Table 4. Chemical composition of metalized ore-coal pellets

Metalized ore-coal pellets	Chemical composition, %				
	Fe_{met}	$Fe_{general}$	SiO_2	CaO	C
	80.20	82.90	11.25	0.30	1.9

Experiment of the technique using metalized ore-coal pellets in the furnace-charge were carried out in coreless induction furnace IST-1. This furnace is applied in a foundry shops of a mechanical plant to produce synthetic pig iron of SCh 40-60 and SCh 44-64 grades.

For experimental smeltings, 10, 15 and 20% of waste metal in the furnace charge were sequentially replaced by metalized ore-coal pellets. The metalized pellets were fed when the melting pot was 60% filled with melted metal which provided the most use of electro-magnetic mixing process in the metal bath. The speed of pellets feed made up 10 kg/min. To each batch of the pellets (10 kg) 0.05 kg of calcium fluoride was added. The pellets having been visually absorbed and the metal surface having been covered with melted slag, the furnace was switched off (Figure 2). After that the metal was tapped into the pouring ladle, metal surface was cleaned from the slag, and the metal from the ladle was poured back into the furnace.

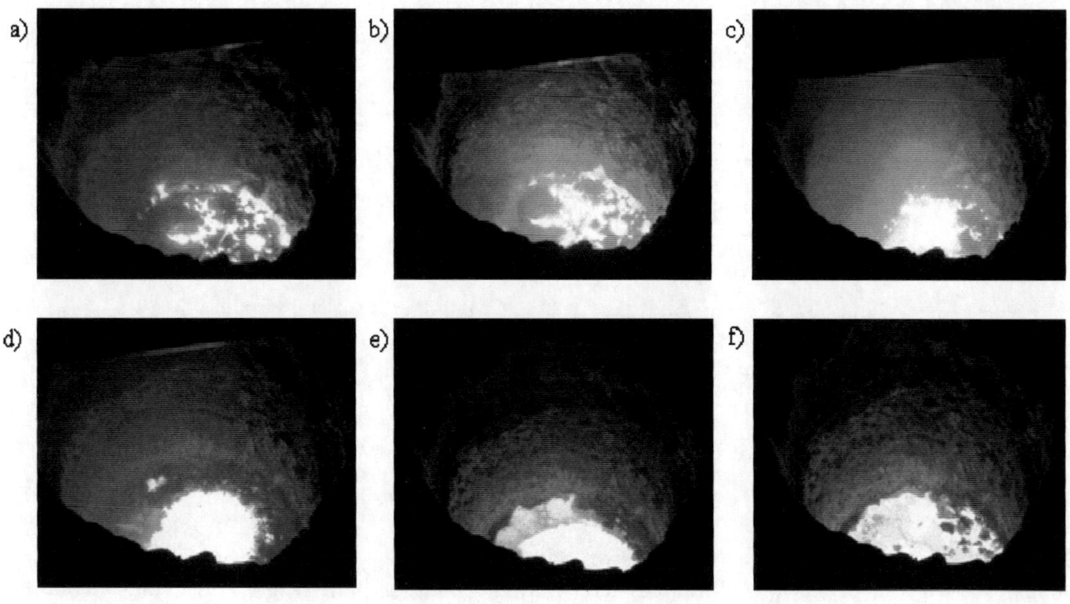

Figure 2. Recovery of 10 kg of metalizes ore-coal pellets during: a – 2 min; b – 2.5 min; c – 3 min; d – 3.5min, e – 4min; f – 5min after feeding into the furnace

Results and Discussion

Characteristics of conventional smelting using metalized ore-coal pellets in the furnace charge are shown in table 5. The smelting mode is shown in Figure 3. Table 6 shows analyses data of metal and slag samples which were collected in experimental smelting.

Table 5. Experimental smelting characteristics

	Conventional smelting	Smelting with metalized pellets		
		10 %	15 %	20 %
Furnace-charge composition, kg/t:				
cast iron	495	495	495	495
metalized pellets	-	115	185	245
recycled scrap	100	100	100	100
waste metal	300	200	150	100
Added materials, kg/t:				
ferrosilicon	4	4.2	4.4	4.5
carbon	5	2.5	2	1.5
Slag quality, kg/t	13	34	72	110
Amount of released gases, kg/t	3	17	21	25
Energy consumption, mJ/t	2124	2358	3437	4100

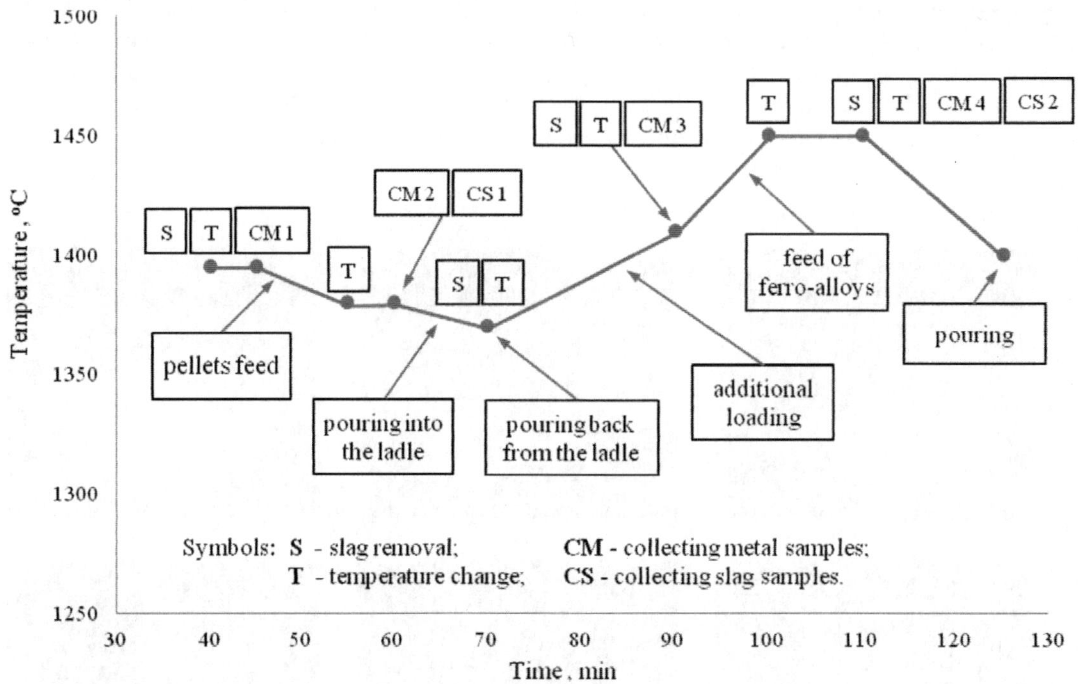

Figure 3. Pig iron smelting using metalized pellets

The experiments show that when batches of pellets are fed into the melting pot filled with melted metal to 60%, the pellets are entrained deep into the melting bath (Figure 3). This phenomenon happens due to the electro-magnetic mixing. Each 100kg of pellets generate 15-20kg of slag. Besides, silicon loss increases since silicon together with carbon evidently takes part in post-reduction of iron oxides found in metalized pellets.

Processes having place during pellets feed can be described as the following (Figure 4). When the pellets are entrained deep into the melting pot and get into the melted metal, a few processes happen simultaneously: post-reduction of FeO accompanied by release of gaseous CO; solving metal phase in the melted metal, carbon recovery and slag formation.

IOP Conf. Series: Materials Science and Engineering **142** (2016) 012068 doi:10.1088/1757-899X/142/1/012068

Table 6. Chemical analysis of samples obtained in experimental smelting using 10% of metalized pellets in furnace-charge

Samples*	Weight content, %							
	C	Si	Mn	Cr	Ni	Cu	S	P
Sample 1	3.04	0,91	0.48	0.05	0.04	0.06	0.076	0.10
Sample 2	3.04	0,87	0.49	0.05	0.02	0.05	0.065	0.09
Sample 3	3.08	0,82	0.47	0.04	0.03	0.05	0.068	0.10
Sample 4	2.96	3,03	0.89	0.05	0.03	0.05	0.060	0.10
Slag samples*		FeO	SiO_2	Al_2O_3	CaO	MgO	S	P
Slag sample 1		18.0	58.42	10.13	4.51	-	0.75	0.029
Slag sample 2		3.5	52.79	14.42	6.33	9.39	0.63	0.019
* - conditions of collecting samples are shown in Figure 2.								

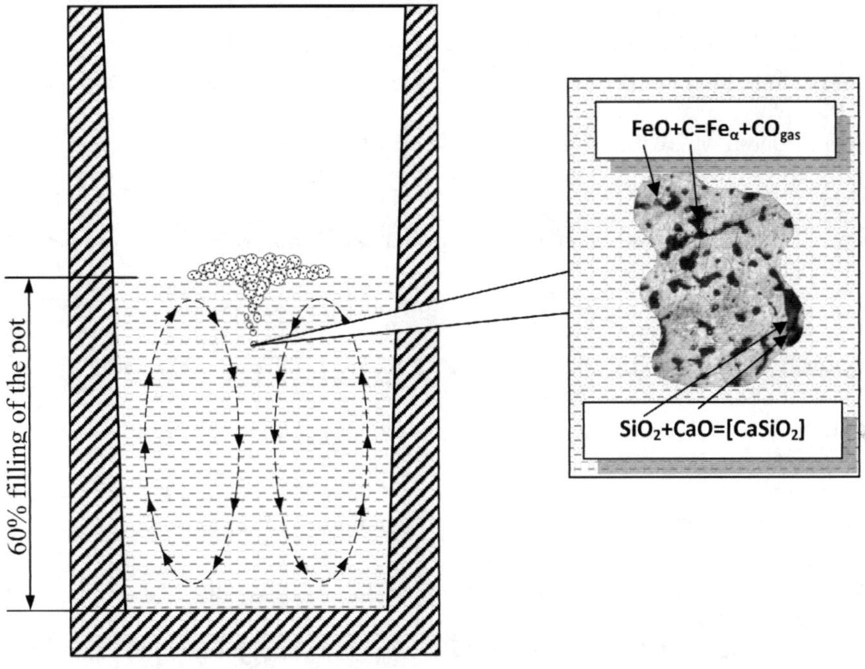

Figure 4. Entraining of pellets deep into the melting pot

The experimental smelting using metalized ore-coal pellets show that the percentage of sulfur in the pig iron does not increase when the quantity of the pellets changes. To minimize a negative effect of slag on the furnace lining it is recommended to implement an intermediate operation: tapping of the slag from the pouring ladle and then loading the metal back into the melting pot.

Conclusion

The resulting data conform to findings of other researches and let us conclude that the metalized ore-coal pellets produced by the developed technology can be applied as furnace-charge in induction smelting of synthetic pig iron. Products cast of the obtained pig iron meet all quality requirements and do not have more defects than cast products made of pig iron smelted with conventional furnace-charge.

Utilization of metalized ore-coal pellets in the furnace-charge provides greater choice of raw materials and technologies for smelting synthetic pig iron in induction furnaces. A continuous process of manufacturing the metalized pellets and feeding them into the furnace at the temperature of 600–800 °C is worth attention. This technique reduces heat consumption and leads to a considerable decrease in capital and operating costs. A preliminary economic estimation of application of metalized materials produced according to the technology developed by the authors showed that if waste metal in the furnace charge is replaced by 15–20% of the metalized ore-coal pellets, the cost of charging materials is reduced by 7%.

References

[1] Duan Yuping, Zhang Yahong, Wang Tongmin, Gu Shuchao, Li xin, Lv Xingjun 2014 *J. Materials Science and Engineering* **185** 86-93.
[2] Santiago Sosa Haudet, Martín A. Rodriguez, Ricardo M. Carranza 2015 *J. Procedia Materials Science* **8** 21-28.
[3] Nokhrina O I, Rozhikhina I D, Kichigina O Y, Goryushkina Y V, Rozevich A P 2012 *J. Steel in Translation* **42** 802-805.
[4] Yingyi Zhang 2013 *J. Advanced Materials Research* **746** 533-53
[5] Fedoseev S N, Mukhtar Z M 2015 *J. IOP Conference Series: Materials Science and Engineering* **91** 1–7
[6] Sharifi E, Barati M 2010 *J. The Minerals* **41**(5) 1018-1024
[7] Yan-li An, Hua-yun Du, Ying-hui Wei, Ning Wang, Li-feng Hou, Wan-ming Lin 2013 *J. Materials & Design* **46** 627-633.
[8] Anand Kumar S, Ganesh Sundara Raman S, Sankara Narayanan T S N, Gnanamoorthy R 2012 *J. Surf. Coat. Technol* **206** 442.
[9] Hu W, Yunxia W, Chenshuo M, Zhaolin Z 2014 *J. Applied Mechanics and Materials* **456** 406–410.
[10] Babcsan N, Leitlmeier D, Degischer H.P, Banhart J 2004 *J. Advanced Engineering Materials* **6** 421–428.
[11] Gizatulin R A, Fedoseev S N, Dariev R S 2016 *J. IOP Conf. Series: Materials Science and Engineering* **125** 012001
[12] Mao D, Edwards J R, Harvey A 2006 *J. Chemical Engineering and Science* **61** 1836–1845.
[13] Nokhrina O I, Rozhikhina I D, Proshunin I E, Khodosov I E 2015 *J. Steel in Translation.* **45** 295–300.
[14] Rodzevich A P, Kuzmina L V, Gazenaur E G, Krasheninin V I 2014 *J. AIP Conference Proceedings* **1623** 519-522.
[15] Yang C C, Nakae H 2003 *J. Materials Processing Technology* **141** 202–206.
[16] Ibragimov E A, Saprikin A A, Babakova E V 2014 *J. Advanced Materials Research* **1040** 764–767
[17] Nokhrina O I, Rozhihina I D, Hodosov I E 2015 *J. IOP Conference Series: Materials Science and Engineering* **91** 12-17.
[18] Ferro P, Lazzarin P, Berto F 2012 *J. Materials Science and Engineering* **554** 122–128.
[19] Fedoseev S N, Lychagin D V, Sharafutdinova A S 2014 *J. Advanced Materials Research.* **1040** 236-240.
[20] Zhang W, Singh P, Muir D 2002 *J. Hydrometallurgy* **63** 127 – 135.

Specific Nature of Application of Various Carbon Components for Mechanochemical Synthesis of Titanium Carbide

V P Reva, V U Yagofarov, D A Gulevskii, A E Filatenkov and Yu N Mansurov

Far Eastern Federal University, Vladivostok, Russia

e-mail: festurvp@mail.ru, vyagofarov@gmail.com, psihomail@mail.ru, 3po91@mail.ru, yubarsmans@gmail.com

Abstract. It has been established that structure of carbon modifications, obtained via pyrolysis of vegetable feed, and ash content of natural graphite are crucial factors during mechanochemical synthesis of titanium carbide under vibratory conditions. The possibility of synthesis of titanium carbide with minimal sulfur content has been shown.

1. Introduction

During last years the scope of applications of titanium carbide in production of composite systems of various purposes both in powder technologies and conventional metallurgy has been increased [1]. One of the prospective methods of titanium carbide synthesis is high temperature mechanochemical [2, 3] synthesis [4 – 11], which allows obtaining products with low content of harmful impurities [6, 8].

Chemical composition of synthesized titanium carbide depends mainly on the purity of applied carbon modifications [7, 10– 12]. Searching of effective substituents of traditionally used ash with higher sulfur content is an important issue in mechanochemistry of carbides. Previously the prospects of using carbon modifications from vegetable feed and natural graphite in mechanochemical synthesis of titanium carbide with low sulfur content were shown. Titanium carbide was synthesized in vario-planetary mill Pulverisette-4 (Fritsch, Germany) [6 – 8]. In present study titanium carbide synthesis was carried out in energy-intensive vibratory mill that allowed revealing new aspects of mechanochemical technology.

This work aimed on investigation of crucial factors of mechanochemical synthesis of titanium carbide conducted under vibratory treatment using carbon modifications obtained from pyrolyzed vegetable feed and also from ash and natural graphite.

2. Experimental

As initial materials various types of renewable vegetable feed were used: moss *Sphagnum fuscum* and *Sphagnum magellanicum* and also stems of gumbo (*Abelmoschus esculentus*) of "Green velvet" sort and sprouts of corn of "Catherine SV" sort. Vegetable feed was preliminary sieved in order to remove foreign inclusions, washed with distilled water, then dried at 100-110 °C and dispersed in disintegrator

DESI-11 (Estonia) down to 100-150 μm. Carbon modifications, which were involved in the synthesis of titanium carbide, with amorphous and crystalline-amorphous structure were fabricated via pyrolytic technology at following temperatures: 950, 11150, 1300 and 1500 °C. Specific surface area (single-point BET) of carbon modifications synthesized from vegetable feed, ranged within 100-220 m^2/g. Ash of PM-15 brand and natural graphite with various ash content were also used as carbon precursors.

Titanium was used as powder of PTES-2 brand with particle size 150-200 μm and purity 99.8%. The stoichiometry $TiC_{0,8}$ was used to calculate amounts of precursors for mechanochemical synthesis of titanium carbide. Activation of precursors and mechanochemical synthesis of titanium carbide were carried out in sealed container (mechanoreactor) of energy intensive vibratory mill, working at 750 min^{-1} frequency of container oscillations and 90 mm amplitude. As grinding bodies the balls (14 mm diameter) of steel of SCh15 brand were used. Grinding intensity (relation between mass of precursors and mass of grinding balls) was 1:20.

Synthesis completion was identified by abrupt temperature change in mechanoreactor measured on its outer wall by IR laser pyrometer C-20.1. After temperature jump (figure1), vibratory treatment was stopped.

Influence of nature of carbon reagents on synthesis conditions was assessed by the time delay of mechanochemical synthesis – time from the beginning of mechanoactivation of the precursors to the moment of temperature jump.

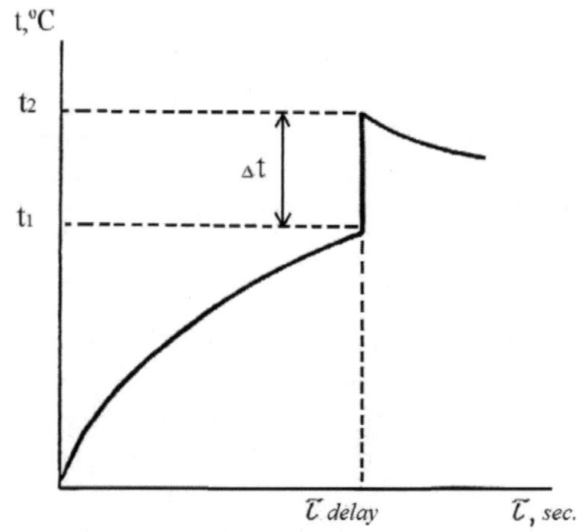

Figure 1. Thermogramm of mechanochemical synthesis of titanium carbide.

Phase composition of carbon modifications and synthesized carbides was determined by XRD on D8 ADVANCE (Germany) device using Cu-Kα source.

Particle size distribution was obtained on laser analyzer "Analyzette 22" (Fritsch, Germany).

Morphology of carbon precursors was studied on scanning electron microscope of high resolution Hitachi S5500 (Japan). Investigation of specific surface area was carried out on sorption analyzer Sorptometer-M, CJSC "KATAKON", (Russia, Novosibirsk), value of specific surface area was determined by thermodesorption of nitrogen.

Sulfur and carbon content in carbon modifications and in carbide powders was measured using sulfur and carbon analyzer CS 600 of "LECO" (USA).

3. Results and Discussion

During the first step, carbon modifications with amorphous and amorphous-crystalline structure were obtained from renewable vegetable feed (agricultural wastes) by pyrolysis at various temperatures: 950, 1150, 1300 and 1500 °C. Carbon precursor formed during after pyrolysis at 950 °C has amorphous structure and doesn't identified by XRD. As can be seen, morphology of amorphous carbon in all cases is predetermined by original structure of certain type of vegetable feed (figure 2).

At 1150-1500 °C pyrolysis leads to amorphous-crystalline structure of the carbon.

Carbon modifications obtained after pyrolysis of vegetable feed, were used as carbon precursors for mechanochemical synthesis of titanium carbide. Ash of PM-15 brand was used to conduct comparative analysis as recommended in [4].

Mechanochemical synthesis involved abrupt temperature change. Temperature of reactor walls, measured by laser IR pyrometer C-20.1 immediately after mechanochemical synthesis of titanium carbide, was 80-130 °C.

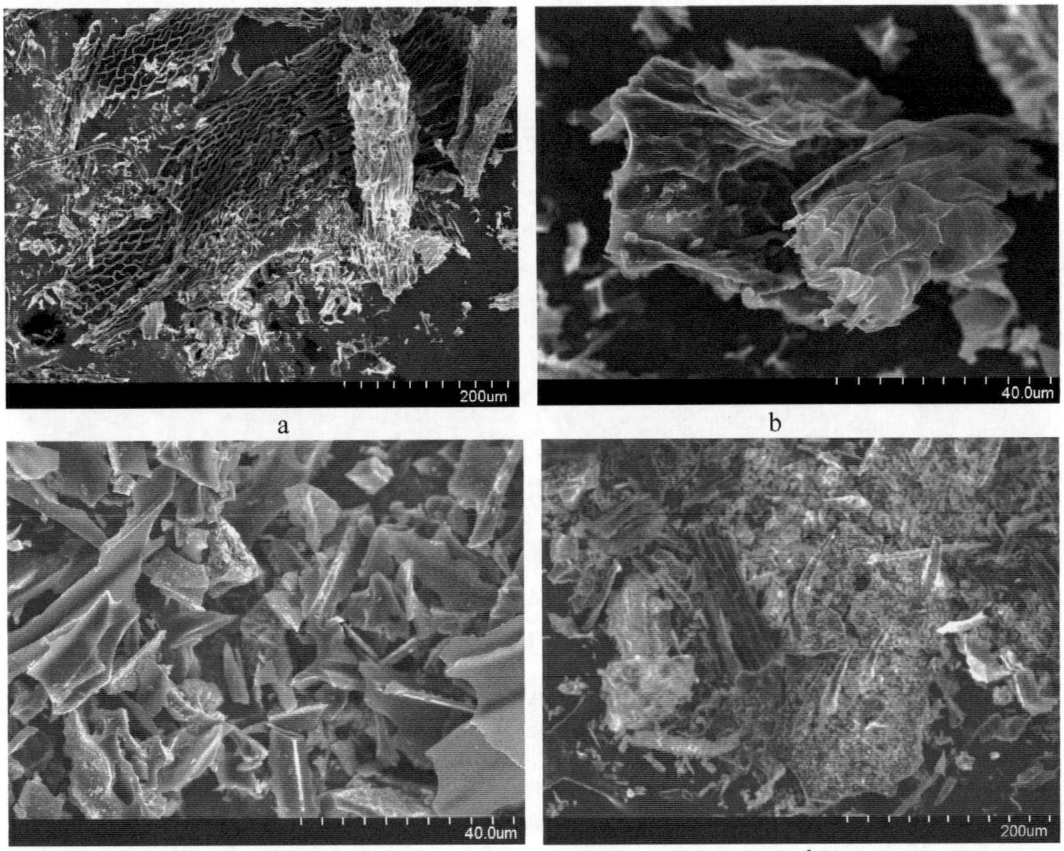

a b
c d

Figure 2. SEM image of amorphous carbon after pyrolysis at 950 °C.
a - *Sphagnum fuscum*, b - *Sphagnum magellanicum*, c - gumbo, d - corn of "Catherine SV" sort.

It was established that structure of carbon modifications, obtained under different conditions of pyrolysis of vegetable feed, influence on time delay of mechanochemical synthesis of titanium carbide (figure 3).

Thus, the higher temperature at which vegetable feed is annealed the faster mechanochemical synthesis is. However, one should note that the higher temperature of pyrolysis, the lower temperature jump $_{\Delta}t$ is ($_{\Delta}t = t_2 - t_1$ – the difference between the temperature of the reactor walls after and before temperature jump).

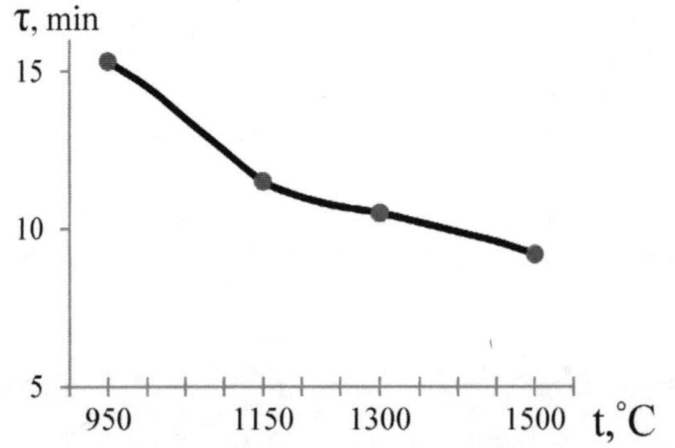

Figure 3. Influence of pyrolysis temperature of sphagnum moss on time delay of mechanochemical synthesis of titanium carbide.

So, if pyrolysis temperature was 950 °C the temperature jump was 40-50 °C, in the case of 1500 °C pyrolysis temperature the jump value was just 7-10 °C (figure 4).

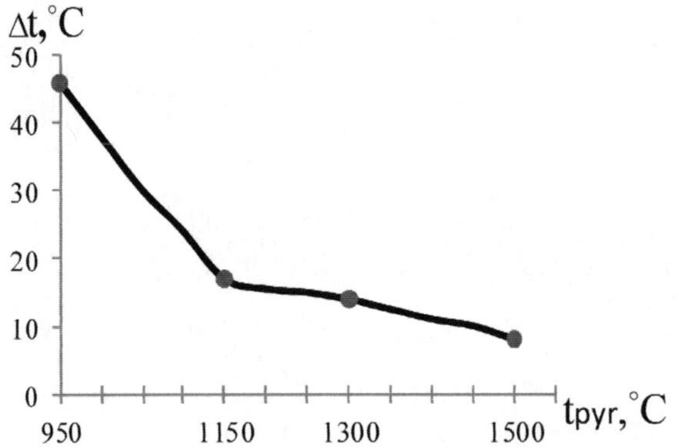

Figure 4. Influence of sphagnum moss pyrolysis temperature on temperature jump during mechanochemical synthesis of titanium carbide.

During visual assessment of the products, obtained using carbon components and under temperatures from 1150 °C and higher, the presence of agglomerates (figure 5a) 3-8 mm in diameter has been revealed.

If carbon material that was obtained by pyrolysis at 950 °C is used, then titanium carbide is formed in the bulk of reacting mixture without formation of agglomerates (figure 5b).

a b

Figure 5. Products of mechanochemical synthesis of titanium carbide. Pyrolysis temperature of sphagnum moss: a – 1150 °C, b – 950 °C.

IOP Conf. Series: Materials Science and Engineering **142** (2016) 012069 doi:10.1088/1757-899X/142/1/012069

According to XRD analysis, mechanochemical synthesis with carbon components obtained after 950 °C pyrolysis results in only titanium carbide. If pyrolysis is increased to 1150 °C, the powder product contains titanium, and agglomerates are a cake of titanium and titanium carbide (figure 6).

Thus, to conduct mechanochemical synthesis of titanium carbide under vibratory conditions one must not use carbon components that have amorphous-crystalline and crystalline structure. Usage of carbon precursor with amorphous structure is the most promising.

Modifications of amorphous carbon from various vegetable feed obtained by pyrolysis at 950 °C were used to conduct mechanochemical synthesis of titanium carbide. The influence of the type of vegetable feed on the time delay of synthesis of titanium carbide is presented (figure 7).

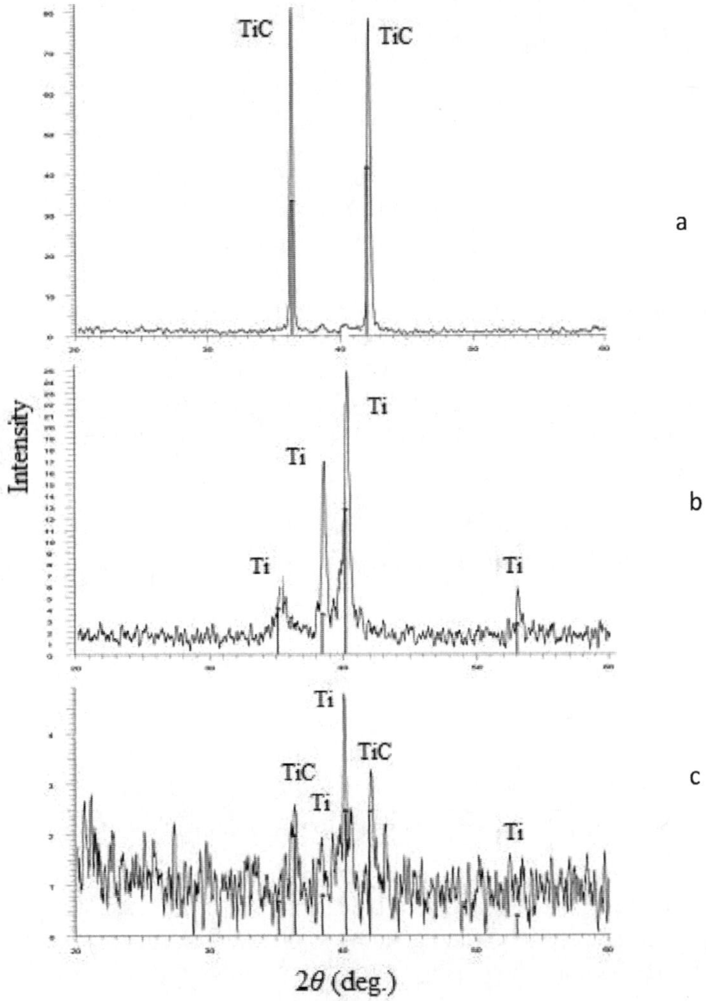

Figure 6. XRD spectra of the products of mechanochemical synthesis of titanium carbide. Pyrolysis temperature: a – 950 °C; b, c – 1150 °C (b – powder, c - agglomerate).

As one can see, the type of vegetable feed influence significantly on the time delay of mechanochemical synthesis if titanium carbide.

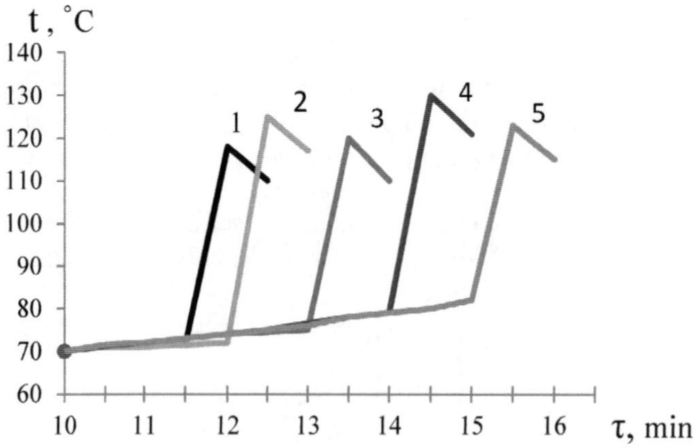

Figure 7. Time delay of mechanochemical synthesis of titanium carbide depending on the type of vegetable feed: 1 – ash PM-15; 2 – magellanicum; 3 – corn of "Catherine SV" sort; 4 – gumbo; 5 – brown sphagnum.

Laser analysis showed that maximal dispersity corresponds to the powder of titanium carbide synthesized using carbon obtained by annealing sphagnum moss, while particle size of titanium carbide is 1-25 μm and 70 % of the particles have size lower than 10 μm.

In order to reveal the influence of the type of vegetable feed, using to obtain carbon precursor, on the chemical composition of synthesized product, the investigations of sulfur and free carbon content in titanium carbide were carried out (table).

Table. Sulfur and free carbon content in titanium carbide

Titanium carbide, synthesized using carbon modification	Chemical compositions, %	
	$C_{free.}$	S
from moss magellanicum	0.29	0.012
from sphagnum moss	0.39	0.014
from corn	0.49	0.085
from gumbo	0.59	0.095
Ash PM-15	0.50	0.19

Temperature of pyrolysis of vegetable feed - 950 °C

Sulfur content in titanium carbide synthesized using carbon modifications obtained by pyrolysis of vegetable feed is lower than that in titanium carbide synthesized using ash PM-15, from 2 to 17 times. The best result is observed for carbon modifications formed from magellanicum moss. Free carbon content in titanium carbide synthesized using investigated carbon modifications is also lower than that using ash PM-15.

The possibility of conducting mechanochemical synthesis of titanium carbide under vibratory conditions using natural graphite as carbon component.

Graphite with ash content from 1 to 5.2% was used. Graphite and ash PM-15 were used to make carbon composites with graphite content varied from 10 to 90 wt. %. Graphite and ash PM-15 were preliminary dried at 150 °C for 3 hours after what the carbon composites with carious graphite content were fabricated.

The influence of ash content of graphite and its concentration in carbon composite on time delay of mechanochemical synthesis of titanium carbide is presented (figure 8).

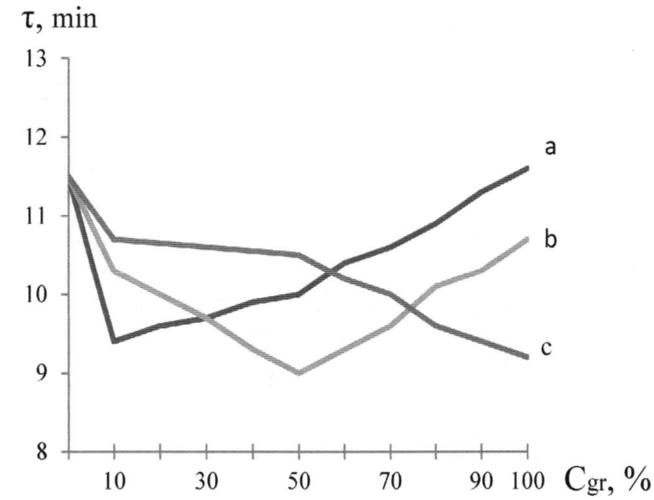

Figure 8. Influence of ash content of graphite and its content in the composite on time delay of synthesis. Ash content of graphite: a – 5.2%; b – 4.8%; c – 1.0%.

Chemical composition of titanium carbide obtained by mechanochemical synthesis using as carbon precursor the composite “ash + graphite” showed, that sulfur content in all of the synthesized products from 2 to 3 times lower than in titanium carbide obtained using ash.

Thus, application of graphite with low ash content as carbon component for conducting mechanochemical synthesis of titanium carbide allows decreasing applied energy while increasing quality of obtained product.

4. Conclusions

The structure of carbon components is the crucial factor in implementation of mechanochemical synthesis of titanium carbide under vibratory treatment conditions. Titanium carbide synthesized using carbon modifications obtained from vegetable feed possess chemical composition with low sulfur and free carbon content that is preferably for implementation in production of dispersion strengthened steel, fireproof materials and also in powder metallurgy. The crucial factor in conducting mechanochemical synthesis of titanium carbide in the presence of graphite is ash content of graphite.

References

[1] Garbuzova A K Rudneva V V Galevsky G V Shiryaeva L S 2014 Analysis of current condition of fabrication and application of titanium carbide *J Reporter of Siberian State Industrial University* **1** p 58 – 66

[2] Takacs L 2014 What Is Unique about Mechanochemical Reactions? *Acta Physica Polonica A* **126** pp 1040– 43

[3] Kajdas C 2015 Mechanical Activation of Chemical Process *Materials Sciences and Applications* **6** pp 60–67

[4] Popovich A A 1993 Microstructure and Properties of Titanium Carbide and Titanium Carbide-Based Alloys Obtained by Tribochemical Synthesis *Journal of Alloys and Compounds* **196** pp 97–104

[5] Popovich A A , Reva V P, Vasilenko V N, Popovich T A and Belous O A 1993 Mechanochemical Method for Obtaining Powders of Refractory Compounds *Poroshkovaya Metallurgiya* Issue 2 pp 37–43

[6] Reva V P, Onishchenko D V, Chakov V V and Voronov B A 2012 Promising Carbon Modifications for the Mechanochemical Synthesis of Titanium Carbide *Doklady Chemistry* **445** Part 2 pp 152–154
[7] Onishchenko D V and Reva V P 2012 Creation of useful carbon materials from plants *Coke and Chemistry* **55** Issue 5 pp 192–196
[8] Reva V P, Onishchenko D V and Kuryavyi V G 2013 Mechanochemical Synthesis Criteria for Titanium and Tungsten Carbides with Participation of Different Carbon Components *Metallurgist* **56** Issue 11 pp 912–918
[9] Dorofeev G A, Ladjanov V I, Lubnin A N, Gilmutdinov F Z , Kuzminykh E V and Ivanov S M 2011 Initial stage of mechanochemical synthesis in the Ti-C exothermic system *Bulletin of the Russian Academy of Sciences: Physics* **75** Issue 11 pp 1427-34
[10] Rahaei M B, Yazdani rad R, Kazemzadeh A, Ebadzadeh T. 2012 Mechanochemical synthesis of nano TiC powder by mechanical milling of titanium and graphite powders. *Powder Technology* **217** pp 369–76
[11] Li B, Cui L, Zheng Y and Xu C 2007 Synthesis of TiC powder by mechanical alloying of titanium and asphalt *Chinese Journal of Chemical Engineering* **15** Issue 1 pp 138–140
[12] Mullins M E and Riley E 1989 The effect of carbon morphology on the combustion synthesis of titanium carbide *Journal of Materials Research* **4** Issue 2 pp 408–411

IOP Conf. Series: Materials Science and Engineering **142** (2016) 012070 doi:10.1088/1757-899X/142/1/012070

The effect of cutting conditions on power inputs when machining

S I Petrushin[1], S V Gruby[2,2], Sh C Nosirsoda[1,3]

[1,3]Yurga Institute of Technology of National Research Tomsk Polytechnic University Affiliate, Tomsk, Russia
[2]Moscow State Technical University by N.E. Bauman, Moscow, Russia

E-mail: [1]psi@tpu.ru., [2] gru@bmstu.ru., [3] Shoni_1997@mail.ru

Abstract. Any technological process involving modification of material properties or product form necessitates consumption of a certain power amount. When developing new technologies one should take into account the benefits of their implementation vs. arising power inputs. It is revealed that procedures of edge cutting machining are the most energy-efficient amongst the present day forming procedures such as physical and technical methods including electrochemical, electroerosion, ultrasound, and laser processing, rapid prototyping technologies etc, such as physical and technical methods including electrochemical, electroerosion, ultrasound, and laser processing, rapid prototyping technologies etc. An expanded formula for calculation of power inputs is deduced, which takes into consideration the mode of cutting together with the tip radius, the form of the replaceable multifaceted insert and its wear. Having taken as an example cutting of graphite iron by the assembled cutting tools with replaceable multifaceted inserts the authors point at better power efficiency of high feeding cutting in comparison with high-speed cutting.

1. Introduction

A lot of attention is paid to the problems of energy saving and efficient use of energy resources all over the world [1,2]. Economical power consumption in the production process is one of key factors of intensifying and improving the production efficiency. It is particularly important for power-consuming industries of economy [3-5]. Let us address to this issue regarding mechanical engineering. Any technological process involving modification of material properties or product form necessitates consumption of a certain power amount. Procedures of edge cutting machining are the most energy-efficient amongst the available forming procedures [6-8]. For instance, energy consumed for removal of a volume when turning is taken as a unity, then this value needed for polishing increases hundredfold, and for electrochemical and electro-physical processing even thousandfold and more [9]. Therefore, when developing up-to-date technologies one should take into account the benefits of their implementation vs. arising power inputs. On the other hand, power costs in traditional technologies depend on the conditions of machining, first of all, on the cutting mode. Let us consider this problem regarding turning ferrous metals by assembly cutting tools with replaced multifaceted inserts fastened mechanically [9-11].

2. Results and Discussion

Power inputs H of machining are thought as electric energy consumed for removal of one kilogram of chips [9]:

$$H = \frac{N \cdot \tau}{W \cdot \lambda} \text{ kW-hour/kg}, \quad (1)$$

where N – power, supplied for cutting, kW;

τ – time of cutting, hour;

W – volume of removed chips over the time of cutting, m^3;

λ – density of the material to be processed, kg/m^3.

Values N and W are in line with the formulae given in [9]

$$N = \frac{P_Z \cdot V}{60 \cdot 1020}; \quad (2)$$

$$W = 60 \cdot 10^{-6} \cdot V \cdot t \cdot S \cdot \tau, \quad (3)$$

where P_Z – tangential component of the cutting force, N;

V – velocity of cutting, m/min;

t – depth of cutting, mm;

S – length-wise supply, mm/turn.

Using expressions (2) and (3) in formula (1) and giving all values in the same numbers of dimension, we obtain:

$$H = \frac{P_Z}{3{,}67 \cdot t \cdot S \cdot \lambda}. \quad (4)$$

We use the data given in [11] for further analysis, considered for outer length-wise turning of graphite iron GI 25 (λ=7.15 kg/m^3) by cutting tools with three-faceted inserts of hard alloy VK 6.

It can be seen in formula (4) the component of the cutting force P_Z is required to calculate the power inputs.

The experiment aimed at detecting the influence of cutting modes on the force of cutting are conducted by the cutting tool VAZ with a three-facetted regular insert, its outside angle is 01331-160308 K6 GOST 19045-80; and geometrical parameters are as follows: $\alpha = 6^0$, $\gamma = 5^0$, $\varphi = 90^0$, $\varphi_1 = 30^0$, $\lambda = 0^0$, r=0.8 mm. Reference parameters of cutting modes: $V = 1{,}0 м/с$, t=2mm, S=0.57 mm/turn. No cutting fluid is required. Each point of experiments is passed at least five times. The speed of cutting is varied 0.33 to 1.67 m/s (Figure 1).

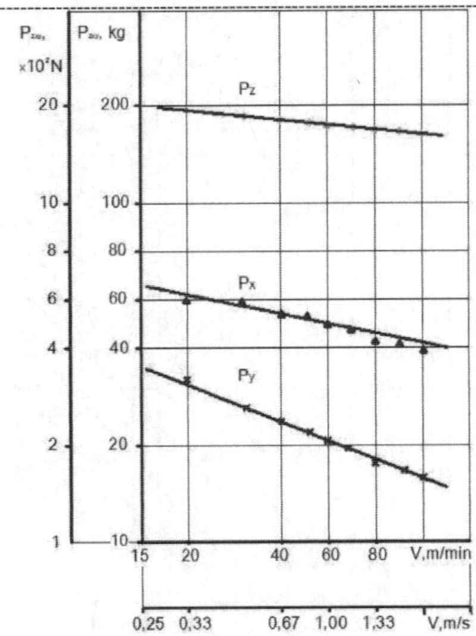

Figure 1. The effect of cutting speed on the cutting force components: GI25-VK6; cutting tool – (VAZ), a rectangular insert with the outside angle, $\varphi = 90^0$; t=2 mm; S=0.57 mm/turn.

The experiments have revealed that increasing speed of cutting results in monotonous decreasing of all its components, while the component P_Z is the slowest one, and P_y diminishes most intensively. Corresponding approximations are as follows:

$$P_Z = 235.6\ V^{-0,07}; \tag{5}$$

$$P_y = 122.4\,V^{-0,44}; \tag{6}$$

$$P_x = 138.0\ V^{-0,27}. \tag{7}$$

The results of experiments focused on identifying the influence of depth of cutting and supply on the force of cutting are presented in Figure 2 a, – for the cutting force component P_Z, in Figure 2 b – for the cutting force component P_y and in Figure 2 c – for the component P_x.

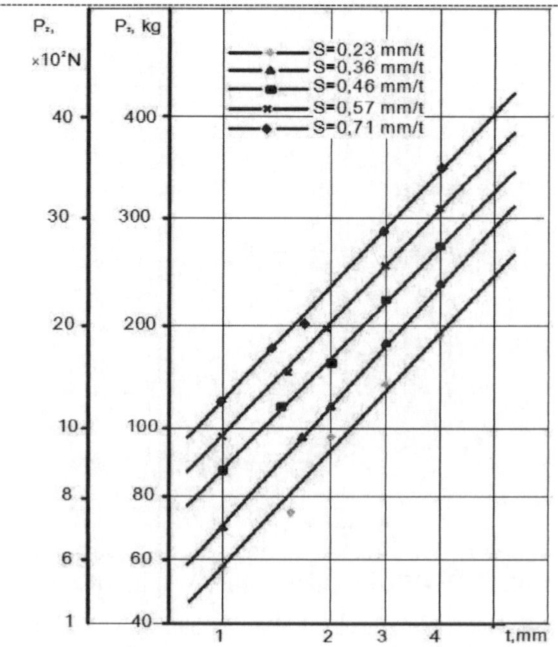

Figure 2a. The effect of the depth of cutting and supply on the component P_Z, GI25-VK6, cutting tool – VAZ, three-faceted insert with the outer angle, V =1.0 m/s

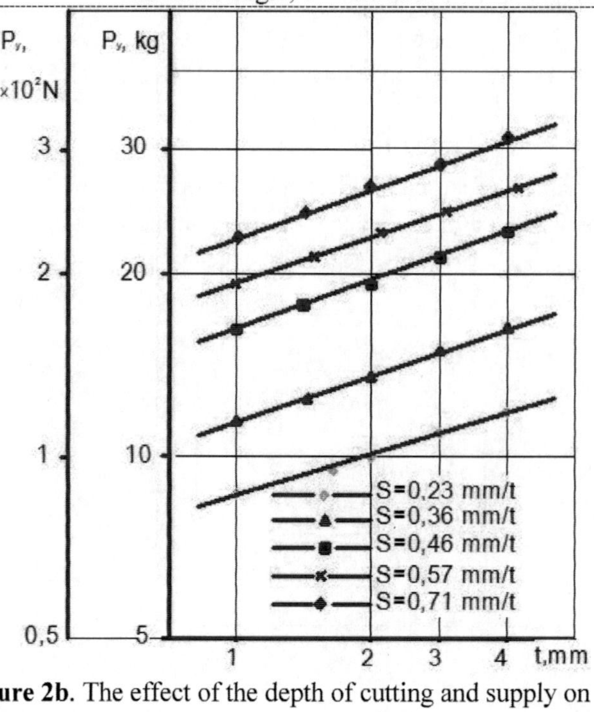

Figure 2b. The effect of the depth of cutting and supply on the component P_y, GI25-VK6, cutting tool – VAZ, three-faceted insert with the outer angle, V =1.0 m/s

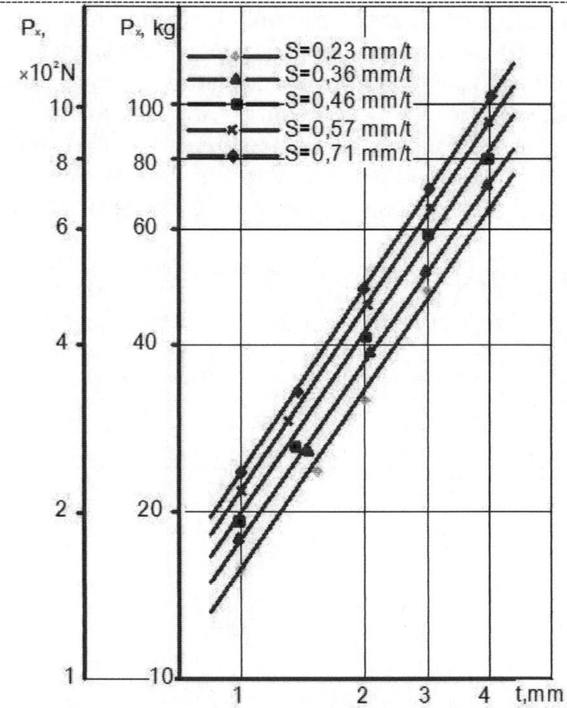

Figure 2c. The effect of the cutting depth and supply on the components P_x, GI25-VK6, cutting tool – VAZ, three-faceted insert with an outer angle, V =1.0 m/s

One can see in the Figures above that supply and depth of cutting are independent on each other and the expression below can be accepted as the mathematical model:

$$P_z = C_{Pz} \cdot t^{X_{Pz}} \cdot S^{Y_{Pz}} ; \tag{8}$$

$$P_y = C_{Py} \cdot t^{X_{Py}} \cdot S^{Y_{Py}} ; \tag{9}$$

$$P_x = C_{P_x} \cdot t^{X_{P_x}} \cdot S^{Y_{P_x}} . \tag{10}$$

Results of experiments processed according to the "classical" method are relevant for deducing the formulae of the depth of cutting and supply:

$$P_z = 145,7 \cdot t^{0,85} \cdot S^{0,68} ; \tag{11}$$

$$P_y = 28,9 \cdot t^{0,29} \cdot S^{0,78} ; \tag{12}$$

$$P_x = 28,9 \cdot t^{1,13} \cdot S^{0,36} . \tag{13}$$

One should note the speed of cutting is not taken into account in the formula (4), however, its effect on the power inputs is incorporated in the component P_Z. Substituting equation (5) in it and numerical values of the constants, we obtain:

$$H = 0,787 \cdot V^{-0,07} \text{, kW-hour/kg.} \tag{14}$$

The dependence of power inputs on the depth of cutting and supply can be obtained the same way by substitution (8) in (4):

$$H = \frac{0,397}{t^{0,15} S^{0,32}} \text{, kW-hour/kg.} \tag{15}$$

The power inputs are decreased when intensifying the cutting mode according to the analysis of (14) and (15). The most efficient method of their reduction is increasing the supply, whereas increasing the depth of cutting is the least efficient method. The growth of cutting speed can save the power somehow, but it affects negatively the lifetime of the tool. So, the cutting speed influences on the power inputs to the extent 0.07, and on the wear resistance – to the extent of 4.2 [11].

And expanded formula given below is deduced to calculate the power inputs, taking into account both the mode of cutting and the tip radius r, the form of the replaceable multifaceted insert and its wear:

$$H = 0{,}0735 \cdot V^{-0{,}07} \cdot t^{-0{,}15} \cdot S^{-0{,}32} \cdot r^{0{,}01} \cdot K_{\phi p_z} \cdot K_{hp_z} . \tag{16}$$

Correcting coefficients of formula (16) are given in Tables 1 and 2. Formula (16) is true in the following ranges of augment variation: V =0.33 – 1.67 m/s; t=1 – 4 mm; S =0.23 – 0.71 mm/turn; r =0.8 – 2.6 mm.

Table 1. Correcting coefficient for the value of wear

Wear h_3, mm	K_{hpz}
0.0	1,00
0.5	1.01
0.8	1.02
1.0	1.04
1.2	1.11
1.5	1.27

Table 2. Correcting coefficient for the insert form

Insert form	K_{fP2}
Regular trihedral with the outer angle	1,00
Regular trihedral insert	1.12
Irregular trihedral with the hole and clearance grooves	1.10
Rectangular with the whole and clearance grooves	1.14
Five-sided with the hole	1.12
Hexahedral with the hole and clearance grooves	1.05
Rhombic with the hole and clearance grooves	1.11
Parallelogram with clearance grooves, left.	0.96

Summary

- Procedures of edge cutting machining part blanks are the most energy-efficient ones as against such up-to-date technologies as physical and technical methods including electrochemical, electroerosion, ultrasound, and laser processing, rapid prototyping technologies etc. [12-14]. Therefore, it is to be taken into account when adopting up-to-date technologies.
- An expanded formula for calculation of power inputs is deduced, which takes into consideration the mode of cutting together with the tip radius, the form of the replaceable multifaceted

insert. These formulae can be used for developing standards of power consumption, which is relevant for production activity of the enterprise.

- Having taken as an example cutting of graphite iron by the assembled cutting tools with replaceable multifaceted inserts the authors point at better power efficiency of high feeding cutting vs. high-speed cutting. Therefore, production profitability can be increased, as well as the competitive ability of the manufactured product can be improved.

References

[1] Dulzon A.A., Ushakov V.Ya., Tchubik P.S. Resource-efficiency – fundamental of the stable civilization development // Proceedings of Tomsk Polytechnic University. – 2012.–V.320. – №6. – pp. 39–46.

[2] Boyett J.H., Boyett J.T. Management-Guide: Die Top-Ideen der Management- Gurus.-Muenchen: Econ, 1999.–399p.

[3] Klimova G.N. Energy saving at industrial enterprises. – Tomsk: TTPU Publishing, 2008. – 181 p.

[4] Ushakov V.Ya. Energy saving and improvement of energy efficiency: social and economic, managerial and legal aspects. Tomsk: TPU Publishing, 2011. – 280 p.

[5] Gubaydulina R. K., Petrushin S. I., Galeeva A. A. Selecting an Economical Variant of the Manufacturing Method of Engineering Product Fabrication under Current Conditions // Applied Mechanics and Materials. - Vol. 682. –. 2014– pp. 613-616.

[6] Petrushin S.I., Proskokov A.V. Theory of Constrained Cutting: Chip Formation with a Developed Plastic Deformation Zone //Russian Engineering Research, 2010. -т. 30 -№ 1 - pp. 45–50

[7] Petrushin S. I. Differential equation for tool wear // Russian Engineering Research. - 2015 - Vol. 34 - №. 12. -pp. 756-762.

[8] Konovodov V. V., Valentov A. V., Kukhar I. S. Analysis of the influence of warming on the quality of soldered instruments // IOP Conference Series: Materials Science and Engineering. - 2015 - Vol. 91, Article number 012053. -pp. 1-5.

[9] Granovsky G. I., Granovsky V. G. Cutting of metals. – M.: Higher school, 1985. – 304 p.

[10] Dulzon A.A. Motivation and energy efficiency // EKOTEK. – 2009. – №4 (33). – pp. 40–41.

[11] Petrushin S. I., Gruby S. V. Processing cast irons and steels by assembly cutting tools with replaceable multifaceted inserts. – Tomsk: TPU Publishing, 2000. – 156 p.

[12] Saprykina N A, Saprykin A A, Borovikov I F, Sharkeev Y P, Influence of layer-by-layer laser sintering conditions on the qu ality of sintered surface layer of products, //IOP Conf. Series: Materials Science and Engineering.- 2015.- Vol.91. Article number 012031. - pp.1-6.

[13] Blashchuk, M. Y., Kazantsev, A. A., Chernukhin, R. V. Capacity Calculation of Hydraulic Motors in Geokhod Systems for Justification of Energy-Power Block Parameters. // *Applied Mechanics and Materials.*Vol. 682. - 2014.- pp. 418-425.

[14] Saprykina N. A., Saprykin A. A. Improvement of surface layer formation technology for articles produced by layer-by-layer laser sintering // Applied Mechanics and Materials. vol. 379. – 2013. pp. 56-59.

Selective Laser Melting of Metal Powder Of Steel 316l

V G Smelov [1,2,3], A V Sotov [1,2], A V Agapovichev[1] and T M Tomilina [2]

[1]Samara State Aerospace University (National Research University), Moskovskoe sh. 34, Samara, 443086 Russia
[2]Institute of Mechanical Engineering named after A.A. Blagonravov of Russian Academy of Sciences, Russia

[3]e-mail: sotovanton@yandex.ru

Abstract. In this article the results of experimental study of the structure and mechanical properties of materials obtained by selective laser melting (SLM), metal powder steel 316L was carried out. Before the process of cultivation of samples as the input control, the morphology of the surface of the powder particles was studied and particle size analysis was carried out. Also, 3D X-ray quality control of the grown samples was carried out in order to detect hidden defects, their qualitative and quantitative assessment. To determine the strength characteristics of the samples synthesized by the SLM method, static tensile tests were conducted. To determine the stress X-ray diffraction analysis was carried out in the material samples.

1. Introduction

Today 3D-printing is a fast developing technology that produces three-dimensional objects directly from digital models through additive process, typically by precipitation and "cured in place" successive layers of polymers, ceramics or metals. Unlike traditional production processes associated with subtraction, for example, a cutting process or shaping techniques, such as stamping, bending and molding, AT connects the material layers to create a final product. Originally, this technology was conceived as a method for producing prototypes, but currently additive manufacturing has improved to the extent that they are increasingly used to obtain the finished product. Therefore there are high demands on the quality of the synthesized material to details grown.

Selective laser melting technology (SLM) is a fast growing technology that allows to produce metal parts in a short time. Interest in the technology is growing from year to year. At the moment, there are several well known works devoted to the study of physical and mechanical properties and structure of the samples obtained by the SLM. For example, in works [1-3] studies on the impact of the mode selective laser melting refractory metal powder on the structure of the material are presented.

The essence of SLM technology is the layer-by-layer manufacturing of a detail in which the heat source is a laser. Schematic diagram of the process is shown in figure 1. Construction of part 1 occurs on the high-platform 2. The platform is a plate approximately 20 mm thick attached by screws to the piston 3. The piston is responsible for moving building platform by an amount equal to the thickness of the layer construction. Accuracy of the piston affects the physical and geometric characteristics of the articles produced and typically is ± 0,005 mm [4]. The metal powder enters the zone via constructing recoater (dispenser) 4. From the main hopper 5, which is a building material, the powder is dosed via a screw recoater. Recoater then, moving horizontally, delivers a new batch of powder to the growing area and levels it with silicone blade 6. Process openings 7 serve to collect metal powder which does not participate in the fusion process, the powder is stored in bottles with the possibility of further sieving. After laying a new layer laser processing is implemented where the laser through a system of mirrors 8 9 scans the powder layer, fusing it with the previous layer. Build process itself takes place in a sealed chamber filled with an inert gas (argon, nitrogen). The process is cyclical and repeats until the product is completely finished.

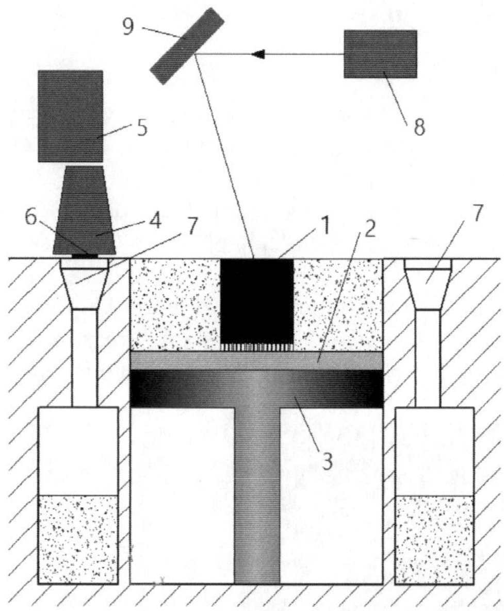

1 - manufactured item; 2 - construction platform; 3 - piston; 4 - recoater (dispenser); 5 - the main hopper; 6 - silicone knife; 7 - technological holes; 8 - laser; 9 - a system of mirrors

Figure 1. Diagram of SLM process.

The aim of this study was to conduct a comprehensive study of the structure of the sample material obtained by selective laser melting (SLM) of powder metal steel 316L. The results of the study of the surface of the powder particle morphology are presented in the article, as well as 3D X-ray quality control of the samples grown, conducted static tests of cylindrical samples in tension.

2. The methodology of the study

The objects of research are samples grown from the metal powder 316L of stainless steel on the 3D machine SLM 280HL. The manufacturer of the device is a German company SLM Solutions. Infrared fiber laser with a wavelength of 1075 nm and a maximum power of 400 watts is used to alloy powder. The laser is operated in a continuous mode. This machine has an open system that allows the use of powders for the manufacture of parts of one's own production. Construction of the detail is carried on a metal platform, the maximum dimensions are up to 280x280 mm. To prevent oxidation and combustion of the powder particles during the process of forming the monolayers working chamber filled with inert gas, wherein the oxygen content does not exceed 0.2%.

Before the process of growing samples on 3D machine as input control, the morphology of the surface of the powder particles was studied, particle size analysis was carried out. Metal powder steel 316L, a Russian analogue is 03H16N15M3 stainless steel was used as a building material . The study of the powder was carried out using a metallurgical microscopes METAM LE-31.

3D X-ray quality control of the samples grown was performed to detect hidden defects, using microfocus X-ray inspection system with CT function on the basis of X-ray machine Nikon Metrology

XT H 225. X-ray system XT H 225 is an X-ray camera, suitable for working with objects in real-time time, as well as for computed tomography objects. In this system, it is possible to create a difference in the X-ray gun potential of up to 225 kW X-ray power to 225 watts. In conjunction with a flat panel high-definition receiver,the system provides high-contrast images.

To determine the strength properties of the samples obtained by selective laser scanning state standard GOST 1497-84 "Tensile Test Methods" was used [5]. This standard establishes methods for static testing of products stretching. For tensile tests were selected samples of type VII number 3. The diameter of the cylindrical part of the working sample to test was 8 mm, the initial sample gauge length 40 mm, working length 52 mm.

3. Results and discussion

In selective laser melting technology strict requirementsare imposed for metal powders on the quality of the surface of the particles, and particle size distribution. Figure 2 shows the results of the study of the surface morphology of 316L powder. It can be seen that the particle surfaces have sufficiently correct spherical powder particles in a small amount of visible growths. The formation of these nodules is associated with powder preparation technology, in this case, the metal powder was obtained by gas atomization of the melt [6-8]. Another important criterion for assessing the validity of the powder for SLM is the size of grains of powder. Figure 2b shows a histogram of particle size distribution of the powder particles of steel 316L. It is seen from the analysis that the average particle diameter is about 15 ... 60 mm, which satisfies the requirements for SLM technology powders.

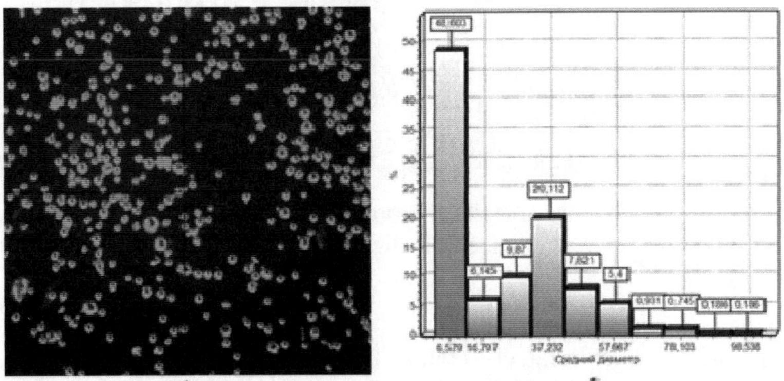

а) б)

a) The surface morphology of the powder particles; b) The size distribution of the powder

Figure 2. Study of metal powder steel 316L.

To study the structure and mechanical features of SLM parts, cylindrical samples have been designed. Then, in a special software product that came with the 3D machine model samples were positioned on the construction platform of 100x100 mm. The smaller size of the platform is justified by decrease of powder flow supplied to the laser beam scanning area. To prevent shear parts, its buckling during printing, as well as for ease of separation of the final product were raised platform construction parameters and configuration support material in the form of thin-wall block designs.

Samples grown on 3D car were sent to the X-ray control of the structure of the resulting material. Scanning of the sample was subjected to a median of about 17 mm in size. Presence of internal pore size of 38 ... 41 mm was found in the manufactured details. Figure 3 shows the results of computed tomography, arrows mark pores formed in the cross section of the samples.

Figure 3. 3D X-ray inspection of the structure of grown material samples.

To determine the strength properties of the components synthesized by the SLM, cylindrical samples according to GOST 1497-84 were designed and manufactured on SLM 280HL. The surface roughness of the grown parts is quite high, requiring the introduction of additional machining on a lathe [9-12]. Diagram of the process tests the force-elongation is shown in figure 4.

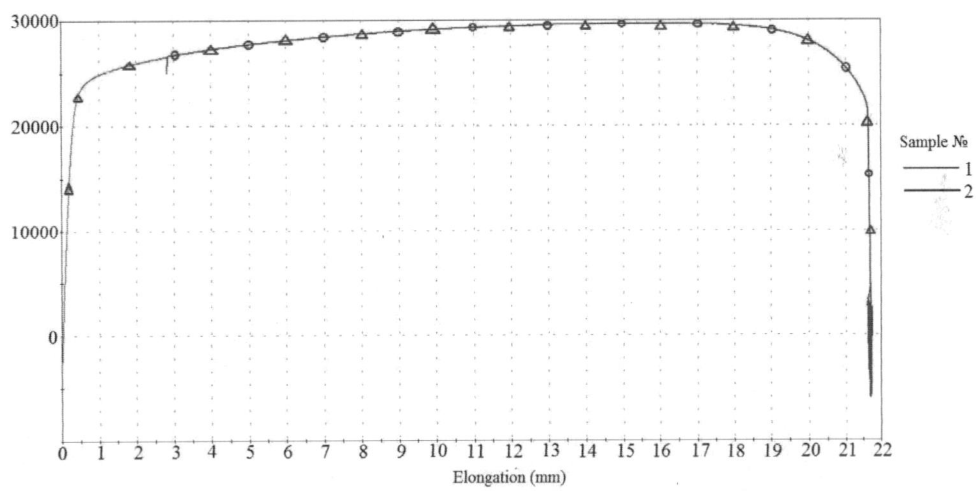

Figure 4. Diagram of the process of tests the force-elongation.

Results of mechanical testing of tensile specimens grown in 316L steel metal powder and comparison with the properties of the samples obtained by casting technology are presented in Table 1.

Table 1. The values of the mechanical parameters of the molded and grown material.

Technology	Strength limit, $\sigma_в$, MPa	Yield strength $\sigma_т$, MPa	Relative elongation, ε, %
SLM	579	455	50
Casting [13]	485	170	40

It is seen that the ultimate strength of SLM samples is 15% above with respect to the tensile strength of cast samples. This result can be explained by the presence of of the borders like those of "micro-welded seams" (MWS-borders). Crystallographically and chemically these boundaries are quite different from the "normal" grain boundaries. MWS-boundary has significant difference in the chemical composition of the base material and has a significant effect on the kinetics of deformation processes. MWS-borders create effective stoppers for the movement of lattice dislocations and an additional contribution to the strengthening of the material [2].

Significantly different is yield strength of the samples from the samples obtained by casting (about 63%). This significant increase in yield strength indicates that plastic deformation will develop in SLM samples at high strain. Ceteris paribus grown items will withstand heavier loads, in contrast to parts obtained by casting.

To determine the internal stress in the material of the grown samples, X-ray diffraction analysis was carried out. The XRD patterns are shown in figure 5, the values are summarized in table 2.

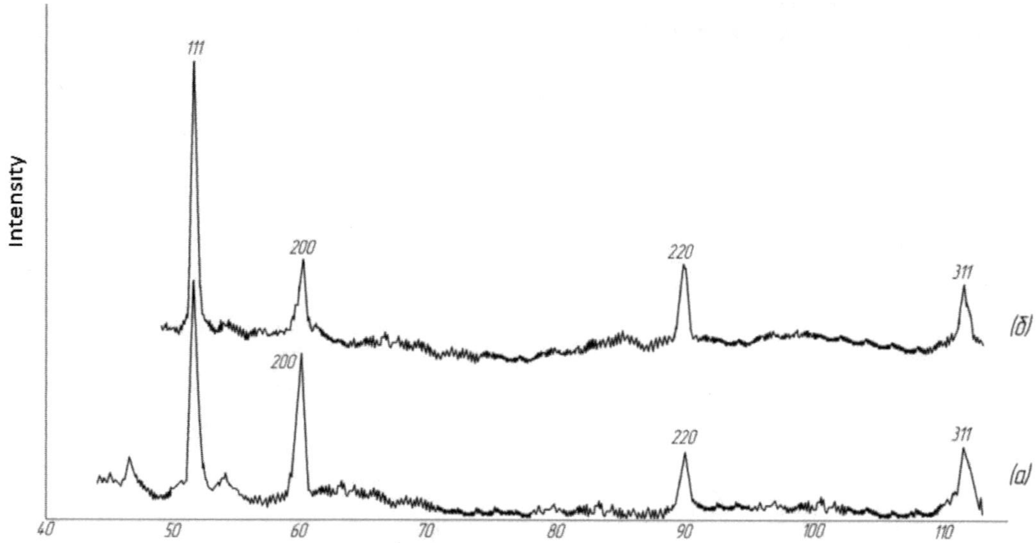

a) in the initial state after fusion; b) after annealing

Figure 5. Steel 316L Radiographs obtained by SLM.

Table 2. Breakdown of diffraction patterns.

HKL		B, mm	B_0, mm	E, kg/mm^2	R, mm	2θ°	θ°	tgθ	σ_{II}, MPa
Deformed state	111	3,0	4,0	21000	180	51,6	25,8	0,483	-603
	200	4,5	6,0			60,3	30,15	0,583	-750
	220	5,0	6,0			89,9	44,45	0,981	-386
	311	7,0	9,5			111,6	55,8	1,471	-436
Initial state	111	4,9	4	21000	180	51,6	25,8	0,483	+543
	200	6	6,6			60,2	30,1	0,581	+301
	220	7	6,5			90	45	1,000	+437
	311	11	9			111,6	55,8	1,471	+396

Table 2 shows that the displacement of the lines has not been identified, i.e. macrostresses are absent. Microstresses of type II are present and have a negative value in comparison with the initial state of the material. Presence of type II microstresses can point to inhomogeneous heating or cooling of the powder during sintering. Tensile and compressive stresses, appearing at the same time lead to plastic shear, which result in residual microtension.

4. Conclusion

1. As an input control, morphology of the surface of the metal powder particles of steel 316L was studied and the particle size distribution analysis implemented. Results showed that the particles have a spherical shape, a small amount of powder particles has visible growths. Histogram of the particle size distribution of the powder is also created and the average size was 37 microns that corresponds to standard.

2. The structure of the samples obtained by selective laser melting metal powder steel 316L was investigated. 3D X-ray quality control of manufactured samples showed that the material has internal pores of 38 ... 41 mm.

3. Mechanical properties of synthesized samples have been determined, where the tensile strength of 579 MPa yield strength of 455 MPa, elongation of 50%. Comparison with a molded sample showed that the tensile strength and yield strength SLM samples are 15% and 63% more respectively.

4. To improve the properties of material of details manufactured by SLM, further studies on the influence of heat treatment on the quality of the material structure and mechanical properties are required.

Acknowledgements
This work was supported by the Russian Science Foundation (project no. 15_19_00284) in the experimental part concerning the investigation of mechanical parameters of the 316L stainless steel manufactured by selective laser melting.
Samples for testing were manufactured on the equipment of CAM technology common use center (RFMEFI59314X0003). This work was supported by the Ministry of education and science of the Russian Federation in the framework of the implementation of the Program «Research and development on priority directions of scientific-technological complex of Russia for 2014-2020».

References

[1] Sufiiarov V Sh, Popovich A A, Borisov E V and Polozov I A 2015 Selective laser melting of heat-resistant nickel alloy *Tsvetnye Metally*. Vol. 2015, Issue 1, pp 79-84
[2] Grjaznov M U, Shoshin S V and Chuvildeev V N 2012 Effect of mesostructural hardening of steel 316L during stratified laser alloying *Bulletin of the Nizhny Novgorod University named after N. Lobachevsky*. №5 (1) pp. 45-50
[3] Grjaznov M U, Shoshin S V and Chuvildeev V N 2014 Physical and mechanical properties and structure of the alloy Inconel 718, obtained by layering laser fusing technology *Bulletin of the Nizhny Novgorod University named after N. Lobachevsky*. №4 (1) pp 99-104
[4] Nazarov A P 2013 Features of machine design for selective laser sintering *Vestnik MSTU "STANKIN"* № 1 pp 76-79
[5] GOST 1497-84 1986 Test methods for stretching *Moscow: Publishing House of Standards*. 36 p
[6] Zlenko M A, Popovich A A and Mutylin I N 2013 Additive technologies in mechanical engineering *SPb .: Publishing house of the Polytechnic University*. 222 p
[7] Riemera A, Leudersa S, Thönea M, Richarda H A, Tröstera T and Niendorfd T 2014 On the fatigue crack growth behavior in 316L stainless steel manufactured by selective laser melting *Engineering Fracture Mechanics*. Vol. 120 April pp 15–25
[8] Smelov V G, Sotov A V and Agapovichev A V 2016 Recovery Technology Features of Aerospace Parts by Layering Synthesis *Key Engineering Materials*. Vol. 684, pp. 316-322

[9] Fangxia X, Xinbo H, Shunli C and Xuanhui Q 2013 Structural and mechanical characteristics of porous 316L stainless steel fabricated by indirect selective laser sintering *Journal of Materials Processing Technology*. Vol. 213 Issue 6 June pp 838–843

[10] Vdovin R A and Smelov V G 2014 Elaboration of a casting defects prediction technique via use of computer-aided design systems *International Journal of Engineering and Technology (IJET)*. Vol. 6 Issue 5 Oct-Nov pp 2269-75

[11] Smelov V G, Sotov AV and Murzin S P 2016 Particularly selective sintering of metal powders by pulsed laser radiation *Key Engineering Materials*. Vol. 685 pp. 403-407

[12] Olakanmia E O, Cochranea R F and Dalgarnoc K W 2015 A review on selective laser sintering/melting (SLS/SLM) of aluminum alloy powders: Processing, microstructure, and properties *Progress in Materials Science*. Vol. 74 pp 401–477

[13] http://goodner.ru/services/info/marks/316l-ti.

Influence of Material Models Used in Finite Element Modeling on Cutting Forces in Machining

Vusal Jivishov, Elchin Rzayev

Azerbaijan Technical University.
H. Javid pr. 25, Az 1073, Baku, Azerbaijan

E-mail: elchin_rz@mail.ru

Abstract. Finite element modeling of machining is significantly influenced by various modeling input parameters such as boundary conditions, mesh size and distribution, as well as properties of workpiece and tool materials. The flow stress model of the workpiece material is the most critical input parameter. However, it is very difficult to obtain experimental values under the same conditions as in machining operations.. This paper analyses the influence of different material models for two steels (AISI 1045 and hardened AISI 52100) in finite element modelling of cutting forces. In this study, the machining process is scaled by a constant ratio of the variable depth of cut h and cutting edge radius rß. The simulation results are compared with experimental measurements. This comparison illustrates some of the capabilities and limitations of FEM modelling.

Introduction

With the rapid increase of computing performance in the last two decades, finite element modeling of the machining process became possible and can be conducted with minimal costs. Finite element methods are appropriate tools for predicting variables such as stresses, strains, temperatures and cutting forces during the machining process. However, these predicted variables are significantly influenced by the simulation input parameters and boundary conditions. The most significant input parameter for the simulation are the material flow stress data, also known as material model. Several researchers studied the discrepancies of different material models for the same material and found out that it has a great influence on the predicted variables in machining [1, 2, 3].

This paper investigates the influence of material model on the modeling of cutting forces. The investigations taking place by a constant ratio of cutting edge radius r_β to depth of cut h. The experiments are carried out using two different steels, AISI 1045 (Ck45) and hardened AISI 52100 (100Cr6), as workpiece material. Uncoated carbide cutting tools with edge radii of 0.01 mm and 0.5 mm are used to cut the AISI 1045 workpiece and CBN cutting tools with edge radii of 0.008°mm and 0.12 mm are used to cut the hardened AISI 52100. The predictions are compared with cutting forces obtained from orthogonal cutting experiments conducted with the same materials and cutting conditions. The edge radii of all of the cutting tools that are used in the experiments are measured preliminary.

DEFORM-2D®, a commercial finite element code that has been validated for cutting simulations [15], is used for the simulations. In this study orthogonal cutting experiments are represented with plane strain model since no side flow was observed. To achieve simulation models close to the reality, large and small workpieces were used. Furthermore small sized mesh elements are used in the simulations especially near the tools' edges (less than 0.001 mm for smaller depth of cut h). This cutting model is used for the simulations using the different material models of each steel for the respective cutting conditions.

Material Models for AISI 1045

Four material models of AISI 1045 are compared in this study, Johnson–Cook [4], Koppka [5], El-Magd [6] and Oxley [7]. The mentioned material models are determined using compression tests, except for the Koppka model. Koppka's model is based on flow stress data obtained from machining tests, and thereupon the Johnson-Cook's equation was modified. All material models are compared using their behavior at averaged values of strain, strain rate and temperature. The cutting forces and chip morphology are obtained from cutting simulation and experiment.

Johnson-Cook's model:

$$\sigma = \left[A + B(\varepsilon_p)^n\right]\left[1 + C\ln\left(\frac{\dot{\varepsilon}_p}{\dot{\varepsilon}_0}\right)\right]\left(1 - \theta^m\right) \quad [1]$$

El-Magd's model:

$$\sigma(\varepsilon, \dot{\varepsilon}, T) = \left(K(B + \varepsilon)^n + \eta\dot{\varepsilon}\right)\exp\left[-\beta_1 \frac{T - T_0}{T_m}\right] \quad [2]$$

Oxley and co-workers used the velocity modified temperature concept to describe material properties as a function of strain rate and temperature.

$$\sigma = \sigma_1 \varepsilon^n \text{ where } n = f(T_{mod}) \quad [3]$$

Thereby T_{mod} is calculated as follows

$$T_{mod} = T\left[1 - \nu \log\left(\frac{\dot{\varepsilon}}{\dot{\varepsilon}_0}\right)\right] \text{ where } \nu = 0.09 \text{ and } \dot{\varepsilon}_0 = 1 \quad [4]$$

increases as the temperature increases and decreases as the strain rate increases. T and $\dot{\varepsilon}$ describe the testing temperature and strain rate, and v and $\dot{\varepsilon}_0$ are material constants for a given material and a range of testing conditions

Koppka's model:

$$\sigma = \left[B\varepsilon^n\right]\left[1 + C\ln\left(\frac{\dot{\varepsilon}}{1000}\right)\right]\left[T^* + ae^{-0.00005(T-700)^2}\right] \quad [5]$$

where

$$T^* = \left[\frac{T_{melt} - T}{T_{melt} - T_{room}}\right]$$

Tab. 1 shows the different constants for the material model equations. The Oxley model does not have any constants.

Tab. 1: Constants for Material Models (AISI 1045).

Johnson-Cook	A = 316 MPa; B = 1067 MPa; n = 0.107; C = 0.0277; m = 0,7
El-Magd	K = 1341 MPa; B = 0; n = 0.17; η = 0.02 MPa·s; β_1 = 1,5; T_m = 1808 K
Koppka	B = 996.1 MPa; n = 0.168; C = 0.097; a = 0.275; T_{melt} = 1480 °C

For the analyses, the flow stress curves of the studied material models for AISI 1045 are compared at the same conditions of strain, strain rate, and temperature. The used constants are defined in tab 1. Fig. 1 illustrates a comparison of true stress-temperature curves at constant values of strain (1 mm/mm) and strain rate (3000 1/sec) that are estimated from simulation. The figure shows that all curves obtained from the different models except Koppka's are located within a range and slope up to a temperature of 400 °C. At higher temperatures, Oxley's and Koppka's models have large shape deviations compared to the other models. Fig. 2 illustrates true stress versus true strain at a temperature of 250 °C and at a strain rate of 3000 1/sec. The figure shows that all material models differ at a temperature of 250 °C and a strain rate of 3000 1/sec. Fig. 3 shows true stress versus strain rate at a temperature of 20 °C and at a strain of 1. It is shown clearly that the material behavior based on the El-Magd model is different from the other models at very high strain rates (> 104 /sec).

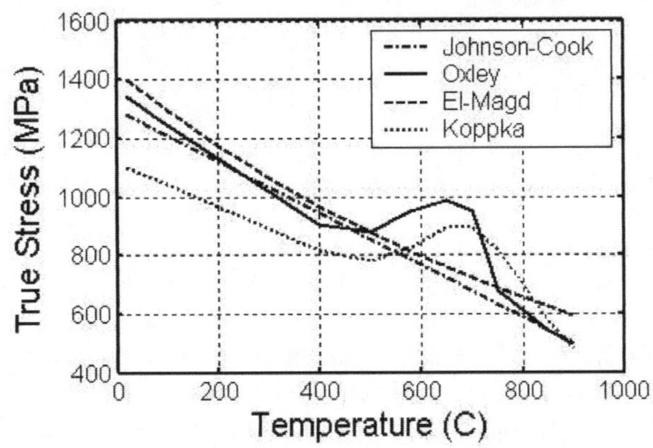

Fig. 1: Stress versus Temperature at True Strain=1 and Strain Rate=3000/sec (AISI 1045)

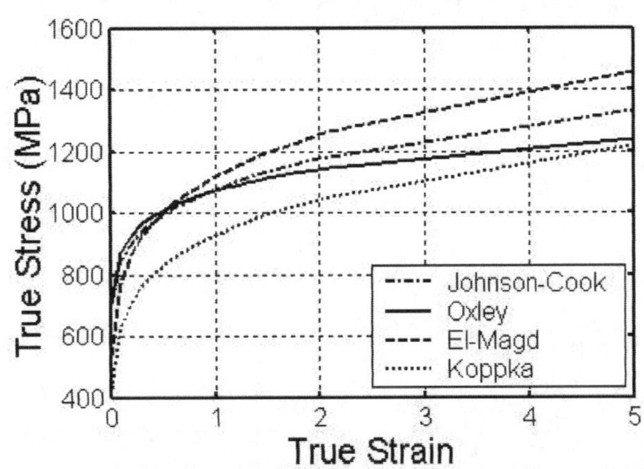

Fig. 2: Stress versus Strain at Temperature=250 °C and Strain Rate=3000/sec (AISI 1045)

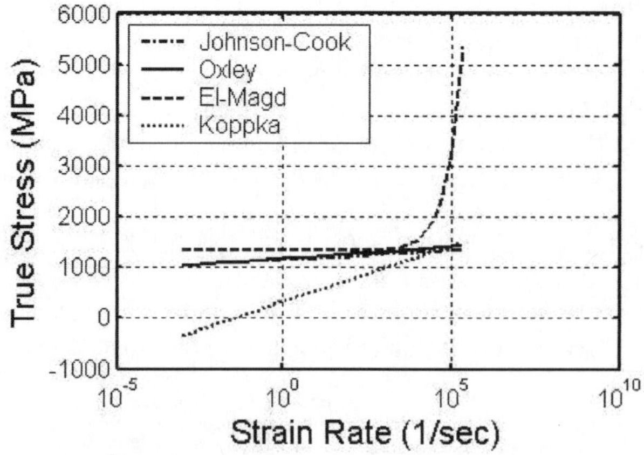

Fig. 3: Stress versus Strain Rate at Strain=1 and Temperature=20° C (AISI 1045)

Although all the models are for the same material there are significant variations between each model. This difference can influence the output variables from machining simulation.

Material Models for AISI 52100

There are several material models for AISI 52100 available in the literature [8-13]. Four different material models of hardened AISI 52100 are considered in this study. The Poulachon model [8], the Poulachon-IEP model with the same equation as the one used by Poulachon but with experimental constants obtained by IEP (Institut of experimenthal Physics, Universty of Magdeburg) [9], the Huang model [11], and Huang's model coupled with Umbrello's model [12]. The equations used in each model are listed below with constants illustrated in tab. 2.

Poulachon's model and Poulachon-IEP's model:

$$\sigma = (A + B\varepsilon^{-n})(1 - CT) \qquad [6]$$

Huang's model:

$$\sigma = (A + B\varepsilon^{n}) \cdot (1 + C \cdot \ln \dot{\varepsilon}) \left| 1 - \left(\frac{T - 25}{1487 - 25} \right)^{m} \right| \qquad [7]$$

Umbrello's model:

$$\sigma = B(T)(C\varepsilon^{n} + F + G\varepsilon)[1 + (\ln(\dot{\varepsilon})^{m} - A] \qquad [8]$$

where

$$F(HRC) = 27.4HRC - 1700.2 \qquad [9]$$

$$G(HRC) = 4.48HRC - 279.9 \qquad [10]$$

Huang's and Umbrello's models are coupled so that the material behaviour at the primary shear zone is represented by Huang's model and the deformation on the remaining workpiece body is represented by Umbrello's model. The flow stress coupled model proved to give better predictions for cutting forces and chip formation compared to orthogonal cutting experiments [14].

Tab. 2 shows the constants of the various material models of AISI 52100.

Poulachon	A = 11.032 MPa; B = 4783 MPa; n = 0.0946; C = 0,00129
Poulachon-IEP	A = 2000 MPa; B = 1600 MPa; n = 0.2; C = 0.0173
Huang	A = 774.78 MPa; B = 134.46 MPa; n = 0.3710; C = 0.0173; m = 3.171
Umbrello	A = 0.0567; n = 0.083; C = 1092 MPa; m = 0.1259

A similar approach to the one used for comparing the various AISI 1045 material models is used for comparing hardened AISI 52100 material models. Fig. 4 shows true stress versus temperature curves at a strain of 1 and at a strain rate of 3000 1/sec.

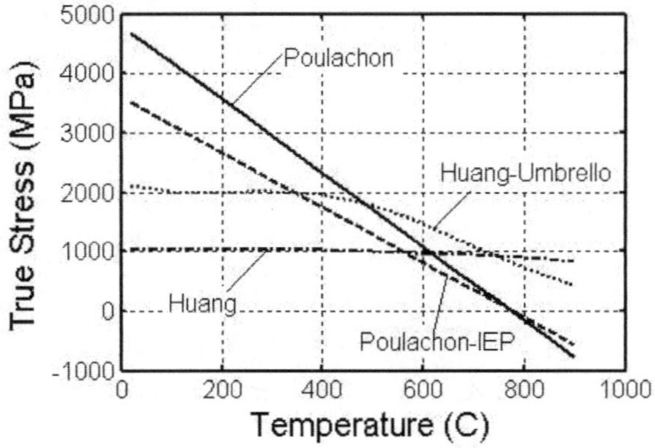

Fig. 4: Stress versus Temperature at True Strain=1 and Strain Rate=3000/sec (AISI 52100)

The curves generated from the presented material models are different. At room temperature, Poulachon's model has the highest value of true stress and Huang's model has the lowest true stress value. In addition, Huang's model is less sensitive to the increase of temperature. Fig. 5 shows true stress versus true strain curves at a temperature of 250 °C and a strain rate of 3000 1/sec. Poulachon's model gives the highest level of true stress and Huang's model gives the lowest, corresponding to what is observed on fig. 4. Finally fig. 6 illustrates a comparison of true stress versus strain rate curves at fixed values of strain and temperature. The figure shows that the Poulachon and Poulachon-IEP models are not dependent on strain rate, whereas the flow stress value based on Huang's and Huang-Umbrello's models increases with the increase of strain rate.

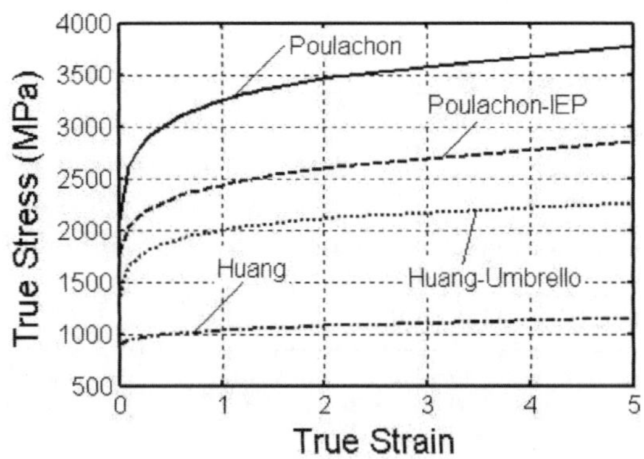

Fig. 5: Stress versus Strain at Temperature=250 °C and Strain Rate=3000/sec (AISI 52100)

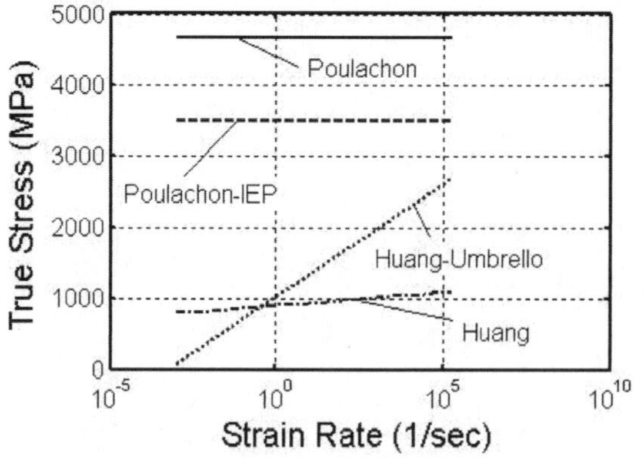

Fig. 6: Stress versus Strain Rate at Strain=1 and Temperature=20° C (AISI 52100)

Results

The cutting conditions used in simulation are identical to those of the experimental investigations. The results show that the influence of material models on cutting forces is significant. In fig. 7 a comparison of the cutting forces obtained from cutting AISI 1045 by a constant ratio of r_β / h = 1 and at a depth of cut h = 0.01 mm is shown. The cutting forces predicted using Oxley's material model are closer to the measured forces then the forces predicted using Koppka's model, the highest error occurs when using El-Magd's model.

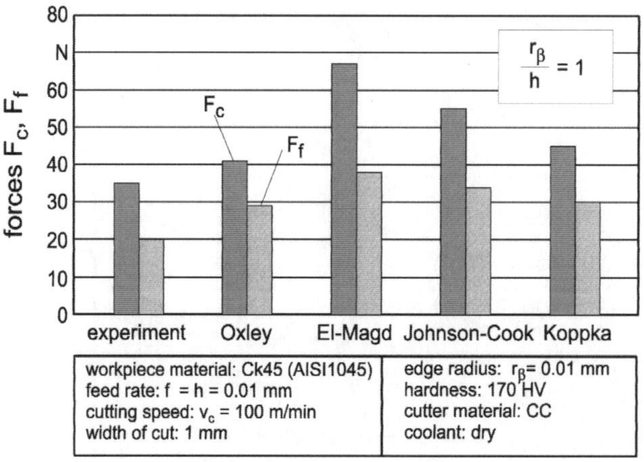

Fig. 7: Forces of AISI 1045 by smaller depth of cut h

Fig. 8 illustrates the cutting forces that are predicted by a constant ratio of r_β / h = 1 and at a depth of cut h = 0.5 mm. In the figure, Koppka's model is adequate to the experimental force of cutting but gives smaller value for the feed force. Therefore Oxley's model is expected to give better predictions of the cutting forces. However, the cutting forces predicted using El-Magd's model are very low compared to the measured forces. In fact, the forces predicted using El-Magd's model are larger than the measured forces by a smaller depth of cut and smaller than the measured forces by a larger depth of cut.

The analysis of the forces that are generated from the various material models of AISI 1045 shows that Oxley's model gives the best predictions when cutting by a constant ratio of r_β / h = 1 and different depth of cut h.

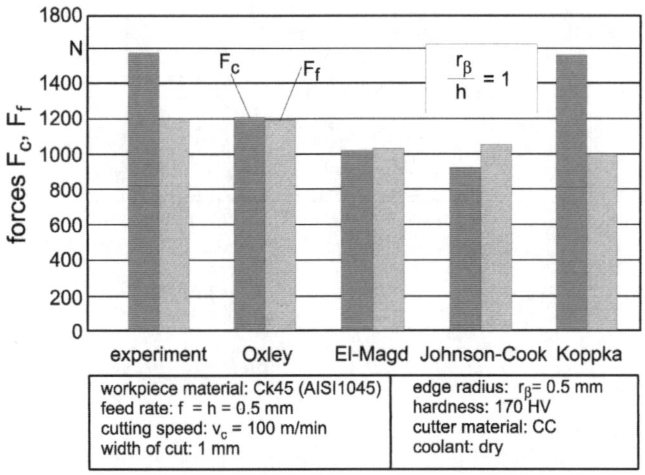

Fig. 8: Forces of AISI 1045 by larger depth of cut h

Due to the difference in hardness and material microstructure, the forces generated from cutting hardened AISI 52100 are larger than the forces generated from cutting AISI 1045. Moreover, the tendency to generate segmented chips is higher when cutting hardened AISI 52100.

Fig. 9 presents the forces obtained from the cutting of hardened AISI 52100 by a constant ratio of r_β / h = 1 and at a depth of cut h = 0.008 mm. Poulachon's and Poulachon-IEP's model lead to a good prediction of forces. Huang's model gives the lowest forces. The forces generated from the Huang-Umbrello coupled model are higher than the forces generated when Huang's model is used alone, but the forces are still very low compared to the measured forces. When comparing the forces by a constant ratio of r_β / h = 1 and at a depth of cut h = 0.12 mm, Poulachon's model also gives the better forces, as in fig. 10. In this case, the difference between Poulachon's model and Poulachon-IEP model in predicting the forces is larger than by the ratio of r_β / h = 1 and at a depth of cut h = 0.008 mm. The forces predicted using Huang's model by larger depth of cut h = 0.12 mm are as low as those by smaller depth of cut h = 0.008 mm.

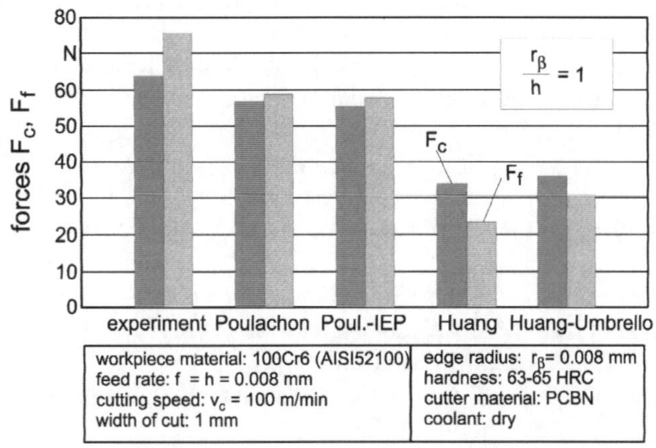

Fig. 9: Forces of AISI 52100 by smaller depth of cut h

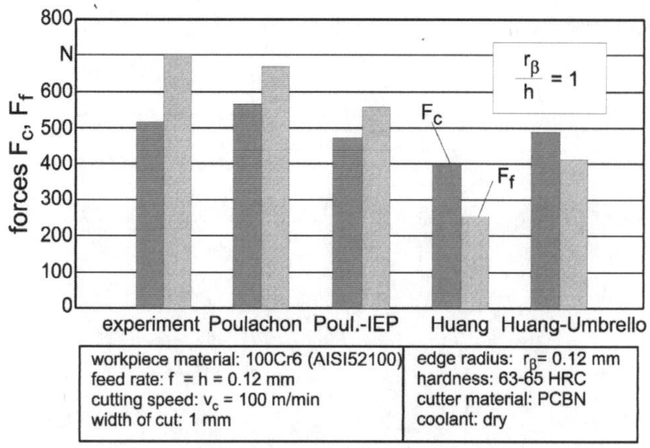

Fig. 10: Forces of AISI 52100 by larger depth of cut h

Summary and Conclusion

The results of this study show that the influence of material model on cutting forces is significant. The forces are different when different material models are used. For AISI 1045, Oxley's flow stress model leads to better results in predicting cutting forces than other models considered in this study. For AISI 52100, the Poulachon's material model shows a better prediction of cutting forces. The forces predicted using Huang-Umbrello's coupled model are much lower than the forces predicted using Poulachon's model. The results give a good understanding of how material models for the same steel are different. This leads to the conclusion that the material model may need to be validated for the specific application before the simulation takes place. Also, this observation gives more emphasis to the need for a material testing method that is capable of generating more robust material models that cover all of the material behaviour during machining.

References

[1] *Meyer, L. W.; Halle, T.; Herzig, N.:* Determination of special material behaviour and devolopment of constitutive equations for numerical simulations of high speed cutting process. Proceedings of the 8th CIRP International Workshop on Modeling of Machining, Chemnitz, Germany (2005), pp. 131-138.

[2] *Shi, J.; Liu, R.:* The Influence of Material Models on Finite Element Simulation of Machining. Journal of Manufacturing Science and Engineering, vol. 126 (2004), pp. 849-857.

[3] Adibi-Sedeh, A. H.; Vaziri, M.; Pednekar, V.; Madhavan, V.; Ivester, R.: Investigation of the Effect of Using Different Material Models on Finite Element Simulations of Machining. Proceedings of the 8th CIRP International Workshop on Modeling of Machining, Chemnitz, Germany (2005), pp. 215-224.

[4] *Johnson, R.; Cook, H.:* A Constitutive Model and Data for Metals Subjected to Large Strains, High Strain Rate, and Temperatures. International Symposium on Ballistics, The Hauge, The Netherlands (1983), pp 1-7.

[5] *Koppka, F.; Sahlan, H.; Sartkulvanich, P.; Altan, T.*: Experimental determination of flow stress data for FEM simulation of machining operations. *ERC Report,* ERC/NSM-01-R-63, 2001.

[6] *El-Magd, E; Treppmann, C.:* Mechanical Behavior of AA7075, Ck45N and TiAl6V4 at High Strain Rates. Materialsweek, Munich (2000), pp 1-8.

[7] *Oxley, P. L. B.:* The mechanics of machining: an analytical approach to assessing machinability. E. Horwood, Chichester, England, 1989.

[8] *Poulachon, G.; Moisan, A.; Jawahir, I.S.:* On modeling the influence of thermo-mechanical behavior in chip formation during hard turning of 100Cr6 bearing steel. *Annals of the CIRP*

(2001), Vol. 50, 1, pp.31-36.

[9] *Clos, R.:* Preliminary report of Project "Groesseneinfluesse beim Hartzerspanen", April, 2004

[10] *Ramesh, A.:* Prediction of process-induced microstructural changes and residual stresses in orthogonal hard machining. Ph. D. Thesis, George W. Woodruff School of Mechanical Engineering, Georgia Institute of Technology, Atlanta, Georgia, 2002.

[11] *Huang, Y.:* Predictive modeling of tool wear rate with application to CBN hard turning. Ph.D. Dissertation, The Ohio State University, Columbus, Ohio, 2002.

[12] *Umbrello, D.; Hua, J.; Shivpuri, R.:* Hardness-based flow stress and fracture models for numerical simulation of hard machining AISI 52100 bearing steel. Journal of Materials Science and Engineering (2004), Vol. A 374, pp. 90-100.

[13] *Caccialupi, A.:* Systems development for high temperature, high strain rate material testing of hard steels for plasticity behavior modeling. Master Thesis, George W. Woodruff School of Mechanical Engineering, Georgia Institute of Technology, Atlanta, Georgia, 2003.

[14] *D'Anna, L.; Al-Zkeri, I; Altan, T.:* Influence of CBN cutting tool design upon tool life and machined surface integrity, experimental and numerical investigation – verification of input data for FE modeling of hard turning. ERC/NSM Report No. HPM/ERC/NSM-05-R-38, The Ohio State University, 2005.

[15] *Yen, Y.C.; Jain, A.; Chigurupati, P.; Wu, W.T.; Altan, T.:* Computer Simulation of Orthogonal Cutting Using a Tool with Multiple Coatings. Journal of Machining Science and Technology (2004), Vol. 8, pp. 305-326.

IOP Conf. Series: Materials Science and Engineering **142** (2016) 012073 doi:10.1088/1757-899X/142/1/012073

Paths of Improving the Technological Process of Manufacture of GTE Turbine Blades

R A Vdovin[1,2], V G Smelov[1], M A Bolotov[1] and N D Pronichev[1]

[1]Samara State Aerospace University (National Research University), Moskovskoe sh. 34, Samara, 443086 Russia

[2]e-mail: vdovin.ssau@gmail.com

Abstract. The article provides an analysis of the problems at manufacture of blades of the turbine of gas-turbine engines and power stations is provided in article, and also paths of perfecting of technological process of manufacture of blades are offered. The analysis of the main systems of basing of blades in the course of machining and the control methods of the processed blades existing at the enterprises with the indication of merits and demerits is carried out. In work criteria in the form of the mathematical models of a spatial distribution of an allowance considering the uniform distribution of an allowance on a feather profile are developed. The considered methods allow to reduce percent of release of marriage and to reduce labor input when polishing path part of a feather of blades of the turbine.

1. Introduction

The modern gas-turbine engines (GTE) and gas-turbine installations (GTI) incorporate several thousands of details and assembly units. However the most loaded, responsible and technologically composite details of any GTE and GTI are blades which define reliability, a resource, cost and labor input of manufacture of the engine.

In the considered work as an object of research turbine GTE rotor blades (figure 1) as the high-loaded figurine details received by a molding method with a directional crystallization from express materials were. Therefore, preparations of blades – non allowances with a high operational properties [1].

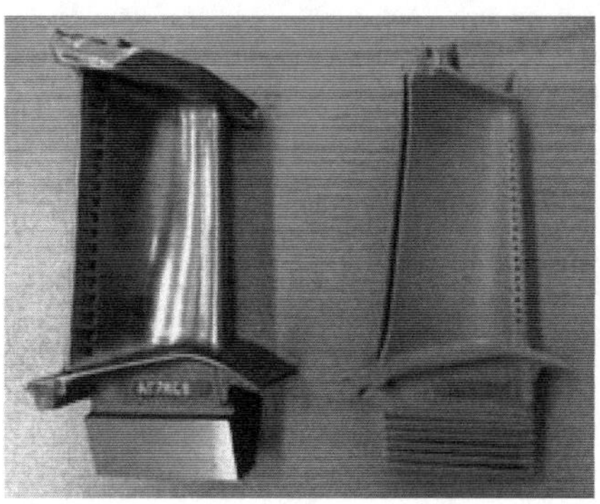

Figure 1. Turbine GTE rotor blades.

The development stage and realization of technological process of manufacture of blades of the turbine is characterized by a number of the research and production problems significantly influencing labor input, efficiency and accessible quality of turbine blades. It is caused by the following features of their manufacture [2, 3]:

1. Blades of GTE are the most responsible details in the engine to a great extent defining reliability and non-failure operation of work of GTE in general. Therefore, the manufacturing techniques and monitoring of blades have to provide stability of quality of their production and exclude a possibility of installation in the engine of blades with deviations by the geometrical sizes, quality of a surface, with metallurgical and other defects.
2. Complexity of geometric forms and requirements of a high precision of manufacture of blades, and also, the increased accuracy of orientation of a feather of a blade in space by relation to the blade lock. Accuracy of manufacture of a feather is in limits 0.05 … 0.15 mm. The latch part by means of which blades fasten to disks is produced with an accuracy of 0.01-0.02 mm.
3. Mass character of manufacture of blades. Even when releasing test pieces of engines manufacture of blades has serial character.
4. Use of expensive and scarce materials for manufacture of blades. In this regard the technological process of manufacture of blades has to guarantee minimum percent of marriage.
5. The poor workability of the materials applied to manufacture of blades. Blades of the turbine produce from the alloys on a nickel basis having rather larger hardness at high viscosity.

The combination of the specified factors also defined specificity of production of blades [4, 5, 6].

2. Description of the models

To judge uniform of distribution of an allowance, it is necessary to have criterion of this uniform of distribution, in other words – criterion of an optimality [7].

Let's provide the formal description of a problem of a spatial distribution of an allowance.

We will explain substance of the developed criterion of an optimality by means of an example of distribution of an allowance at preparation of a turbine blade in space (figure 2).

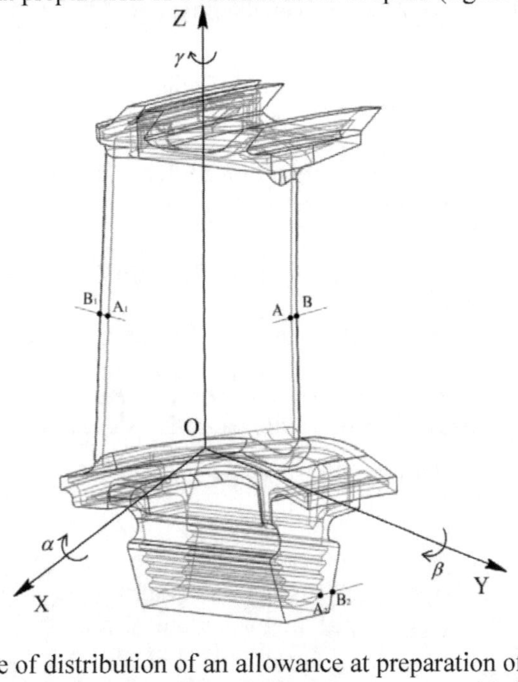

Figure 2. Picture of distribution of an allowance at preparation of a turbine blade.
———— contour of preparation of a blade, ———— blade detail contour.

The analysis of figure 2 allows to draw a conclusion that the contour of preparation of a blade has 6 degree of freedoms: movements on axes X, Y, Z and turn (rotations) on corners α, β, γ rather corresponding axes. The task consists in that by movement on these degree of freedoms of a contour of preparation to reach the most favorable relative positioning of a contour of preparation and a contour of a detail of a blade, i.e. in mathematical statement it is necessary to achieve keeping of identity: AB=A1B1=A2B2.

Three types of criteria of an optimality (K) were developed for the solution of an objective in the considered statement.

Serves as the first criterion of an optimality (K1) equation (1) maximal of the distances taken on a normal to a shovel detail contour. Let's designate this distance through δi. Measurement to be taken in n blade detail contour points. Then, the optimality criterion has an appearance:

$$K_1 = \max_{(i)} \delta_i = \max_{(i)} \delta_i (X,Y,Z,\alpha,\beta,\gamma); \quad (1)$$

$$(i=1, \ldots, n).$$

The problem of optimum installation of a contour of preparation of a blade concerning a contour of a detail of a blade will consist in such choice of values of the X, Y, Z variables, α, β, γ at which the size K1 minimum equation (2) is provided:

$$\min_{(X,Y,Z,\alpha,\beta,\gamma)} K_1 = \min_{(X,Y,Z,\alpha,\beta,\gamma)} \left\{ \max_{(i)} \delta_i (X,Y,Z,\alpha,\beta,\gamma) \right\}; \quad (2)$$

$$(i=1, \ldots, n).$$

As the second criterion of an optimality (K2) equation (3) it can be accepted minimum of deviations δi:

$$K_2 = \min_{(i)} \delta_i = \min_{(i)} \delta_i (X,Y,Z,\alpha,\beta,\gamma); \quad (3)$$

$$(i=1, \ldots, n).$$

The problem of management consists in that change of the X, Y, Z variables, α, β, γ to provide the maximal value of criterion K2 equation (4):

$$\max_{(X,Y,Z,\alpha,\beta,\gamma)} K_2 = \max_{(X,Y,Z,\alpha,\beta,\gamma)} \left\{ \min_{(i)} \delta_i (X,Y,Z,\alpha,\beta,\gamma) \right\}; \quad (4)$$

$$(i=1, \ldots, n).$$

From the practical point of view, the criterion of K2 often is the most successful as at installation of preparation of a blade by this criterion preparation moves away from dangerous provisions in which the allowance can be the negative.

As the third criterion of an optimality (K3) equation (5) the difference between maximal and minimum of deviations can serve:

$$K_3 = K_1 - K_2 = \max_{(i)} \delta_i - \min_{(j)} \delta_j = \max_{(i)} \delta_i (X,Y,Z,\alpha,\beta,\gamma) - \min_{(j)} \delta_j (X,Y,Z,\alpha,\beta,\gamma); \quad (5)$$

$$(i=1, \ldots, n)\ (j=1, \ldots, n).$$

The problem of optimum distribution of an allowance is reduced to minimization of this criterion:

$$\min_{(X,Y,Z,\alpha,\beta,\gamma)} K_3 = \min_{(X,Y,Z,\alpha,\beta,\gamma)} \left\{ \max_{(i)} \delta_i (X,Y,Z,\alpha,\beta,\gamma) - \min_{(j)} \delta_j (X,Y,Z,\alpha,\beta,\gamma) \right\}; \quad (6)$$

$$(i=1, \ldots, n)\ (j=1, \ldots, n).$$

Shortcomings of criterion of this kind is need of formation of K3 in the form of a difference of two rather larger sizes. This difference at an approximation to a minimum becomes small and commensurable with the hindrances complicating a possibility of use of criterion of K3.

3. Results and discussion

The manufacturing techniques of blades are improved, mainly, in the direction of increase of mechanization and automation now. The exception of manual skills (a subjective factor) allows not only to reduce labor input, but also to increase quality of manufacture of blades [8, 9].

Increase of effectiveness of technological process of manufacture of blades of the turbine GTE, first of all, has to solve a problem of increase of economic indexes of process, namely: increases in the material capacity factor (MCF); decrease in labor input of an manufactures; reductions of a production cycle of manufacture of details and decrease of costs of technological preparation of production [10].

Let's carry out a qualitative analysis of the technological process of manufacture of the rod (cooled) turbine GTE blade existing at the enterprise.

Today in mechanical engineering it is accepted to make turbine blades of the preparations received by investment casting with an accuracy, whenever possible, the blade (so-called non allowance molding) excluding the subsequent machining of a profile of a feather. According to the carried-out statistical analysis by production of shovels of the turbine of the 1st step of GTE to 30% it is rejected in connection with existence of pouring defects (blisters, blockages, multidirectional crystalline structure); to 10% makes marriage on discrepancy of geometrical parameters of a blade of design documentation and non-optimum distribution of an allowance in path part of a blade. In spite of the fact that blades of gas-turbine engines make under constant technological conditions, using the same expendables, the geometry of cast preparations of blades of the turbine is not stable. It is bound to individual conditions, a subjective factor and the modes of melting filling of each mold piece [11, 12].

Available geometrical accuracy of blades of the turbine is defined by the terminating finishing operations of polishing realizing a method of trial passes and measurements. In these operations an iterative successive approximation to the given design documentation of geometry of a blade by means of measurement and partial completion is supposed. On a set of serial productions, for measurement of a profile of a feather of a blade the specialized adaptation realizing a sample method of measurement is used (figure 3).

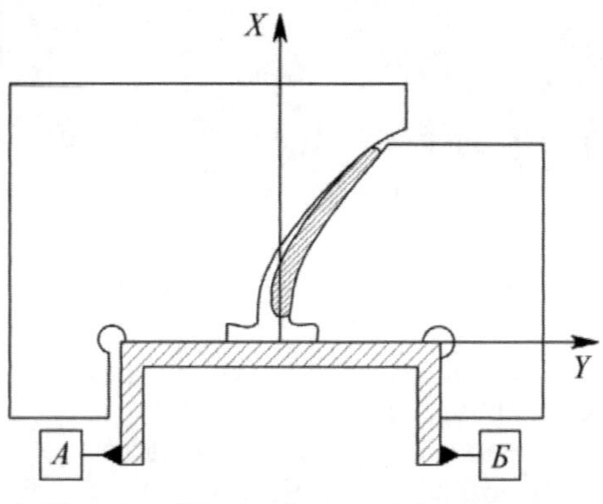

Figure 3. The simplified scheme of the sample appliance.

The part of the blades arriving after molding are often rejected because of shortage of an allowance in certain areas of section of a profile of a feather (have, a so-called, negative allowance). Therefore preparation of a blade is rejected. However, similar preparations can be allowed for further machining as deviations which have blades allow to enter a theoretical profile of a feather by its best placement in limits of tolerance set by design documentation.

Let's carry out the analysis of the control methods existing in machine-building enterprises and checks of a profile of a feather of blades of GTE:

– a sample method (measurement is carried out between a surface of a back and a trough of a feather an express template in this section of a feather of a blade);

– the method based on use of the device of the optical-mechanical monitoring of blades (DOMMB);

– the method based on use of contact measuring TESA, LK cars which compare an actual surface of preparation of a detail to its mathematical model in particular points, the given programmatically;

– the method based on use of noncontact laser "OPTEL" instruments.

From the given control methods of blades of GTE the first two are most widespread in serial production today, and the DOMMB appliance is used for monitoring of a feather of compressor blades, and adaptation of sample type is intended for monitoring of blades of the turbine [13].

The third and fourth control methods of a profile of blades as the most modern and perspective instrument of monitoring only begin to introduce recently on the enterprises of machine-building branch. Their rather sluggish introduction and use is bound, first of all, to the high cost of installations and need of a highly skilled engineering upkeep as installations of a similar class demand express conditions at their placement on temperature, environmental purity, vibration.

Sample method – the most widespread way of monitoring and an assessment of an optimality of distribution of an allowance of a feather of a blade of GTE. It consists in use of express templates, i.e. for each blade there is a particular set of reference templates (figure 4): *(a)* – the 1st step of the turbine GTE; *(b)*– the 4th step of the turbine GTE. . The blade on the technological bases located on half-internal and latch part is installed in the device and check a feather profile by serial combination of templates with surfaces of a back and a trough. In places of a deviation of a profile of a feather from a template carry out mechanical operational development by glazing by the abrasive tool.

(a) *(b)*

Figure 4. Sample equipments for monitoring of preparations of turbine blades.

Now at the enterprise the sample equipments represented in figure 4 for monitoring of geometry of a profile of turbine blades of GTE are used [14].

Essential factor, at a similar method of measurement of a profile of a feather, limiting body height of efficiency and putting into dependence accessible quality on a subjective factor is lack of automatic keeping of the sizes.

It is possible to increase effectiveness of technological process of manufacture of blades of the turbine GTE the next ways: having increased quality of foundry processes (due to use of the reduction mix and a ligature, introduction of systems of computer model operation of foundry processes, etc.); application of original design decisions – use of the gravitational lock devices capable to reduce marriage on mold pieces of blades of the turbine; introduction in technological process of manufacture of blades of the additive technologies and methods of fast prototyping; modernization of operations of monitoring of a profile of a feather of blades. One of essential reserves is increase of level of operations of monitoring of path part of a blade which determine the reached accuracy, considering the method of trial passes and measurements realized in technology. This way is less expensive.

In the course of mechanical operational development of a profile of a feather of a turbine blade one of the most important questions is the choice of optimum system of basing for the uniform distribution of an allowance on a feather. The problem consists in the irregular space shape of a feather, lack of developed surfaces for precise basing when processing and monitoring, a small rigidity of thin-walled blades.

Because sample devices do not consider an optimum spatial distribution of an allowance on a feather of preparation of a blade, often there are situations bound to non-uniformity of distribution of an allowance which special case is existence of the negative allowance as it was said above that, in turn, leads to marriage.

Detailed studying of algorithm of monitoring of blades of the turbine on the sample device allowed to reveal a number of shortcomings of this method:

- ✓ at a similar control method it is very difficult to reveal the real size of shift of a contour of a back and trough in a Y direction – y as the result of an assessment of size of shift is influenced at the same time also by the shift of contours of a back and trough x dimension – x, an error of a corner of a curling of a feather and an error of a form of the checked contour;
- ✓ the small efficiency and poor objectivity caused by lack of a spatial distribution of an allowance on a blade feather and also
- ✓ influence on result of measurement of conditions of installation of measuring templates (a distortion of templates, size of effort of a clip of a template to a blade feather, etc.);
- ✓ existence of a subjective factor bound to a big share of application of manual skills on basing of a blade and the subsequent operations of grinding and glazing of a feather;
- ✓ need of application of unwieldy sample and measuring economy.

Thus, at a stage of monitoring of the received preparations of blades of the turbine there is a problem, the bound to basing of preparations and the subsequent uniform distribution of an allowance on a feather. In this regard, as it was told above, about 10% of preparations of blades are rejected because of existence of the negative allowance. Having resolved an issue of the uniform distribution of an allowance in preparations of blades by technological methods it is possible to increase percent of release of suitable production.

The schemes of basing of blades of GTE applied today can be divided into the following groups on a way of installation of preparation:

1) basing of preparation of a blade immediately on the machine;
2) basing of preparation of a blade, using the equipment readjusted on dimensions;
3) basing of preparation of a blade, using the unified equipment;
4) basing of preparation of the blade which is filled in a briquette.

Basing of preparation of a blade immediately on the machine allows processing only by methods, not the bound to the considerable efforts. Basing of blades with use of equipment or filling in a briquette allow to conduct processing by milling.

According to [15] at the engine-building enterprises apply the following schemes of basing of blades of GTE (figure 5).

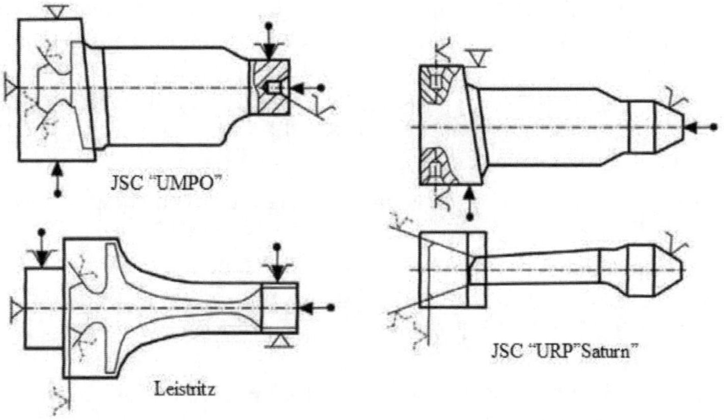

Figure 5. The existing systems of basing of blades in the course of manufacture.
———— technological bases, – – – – – – design bases.

IOP Conf. Series: Materials Science and Engineering **142** (2016) 012073 doi:10.1088/1757-899X/142/1/012073

It is possible to solve a problem of the uniform distribution of an allowance in preparations of blades of the turbine GTE by design methods, in a path of perfecting of a design of the sample device, having added several degree of freedoms (in the form of the correcting screws) by means of which, preparation of a blade falling apart on three axes and gets an opportunity to rotate concerning each of axes.

4. Conclusion

Thus, authors of work made the following recommendations allowing to reduce labor input of operations of monitoring of a profile of a feather of turbine blades and to increase effectiveness of technological process of manufacture of blades:

1. On the basis of production statistics it is necessary to define the required quantity of degree of freedoms in the sample device for an optimum spatial distribution of an allowance in preparation of a blade.
2. To define less expensive design ways of realization of a problem of a spatial distribution of an allowance in industrial equipment.
3. Practical realization and the experimental approbation of the offered recommendations under production conditions the enterprises.

These studies were conducted on the equipment of CAM technology common use center (RFMEFI59314X0003).This work was supported by the Ministry of education and science of the Russian Federation in the framework of the implementation of the Program "Research and development on priority directions of scientific-technological complex of Russia for 2014-2020".

This work was financially supported by the Ministry of Education and Science of the Russian Federation within the project № 14.578.21.0131 (RFMEFI57815X0131) of the Federal Target Program "Research and development on priority directions of scientific-technological complex of Russia for 2014-2020 years".

References

[1] Vdovin R A and Smelov V G 2014 Elaboration of a casting defects prediction technique via use of computer-aided design systems *International Journal of Engineering and Technology (IJET)*. 6 Issue 5 pp 2269-75

[2] Pechenin V A, Bolotov M A, Ruzanov N V and Yanyukina M V 2015 Optimization of measurements of the geometry of parts with complex surfaces *Measurement Techniques*. 58(3) A007 pp 261-268

[3] Vdovin R A, Smelov V G and Golanov S P 2015 Analysis of results of computer simulation of the casting process *Applied Mechanics and Materials (AMM)*. 770 pp 223-228

[4] Khaimovich A I and Balaykin A V 2014 Analysis of titanium alloys plastic properties under severe deformation conditions in machining *Journal of Engineering and Applied Sciences (ARPN)*. 9(10) pp 1828-33

[5] Agapovichev A V, Balaykin A V and Smelov V G 2015 Production technology of the internal combustion engine crankcase using additive technologies *Modern Applied Science*. 9 Iss. 4 pp 335-343

[6] Vdovin R A 2013 Multiproduct manufacture process improvement *Izvestiya of Samara Scientific Center of Russian Academy of Sciences*. 15, № 6(3) pp 612-619

[7] Bolotov M A, Kapenkina E A, Rusanov N V and Pechenin V A 2014 Choice of production measuring instruments based on techno-economic analysis, taking into account the type I error and type II error *Journal of engineering and applied sciences (ARPN)*. 9(10) pp 1834-41

[8] Vdovin R A 2012 3D virtual simulation and casting process optimization for "Swirler of II duct" part of GTE with use of computer technologies *Vestnik of Samara State Aerospace University named after Korolev S.P. (of National Research University)*. 3(3) pp 115-120

[9] Khaimovich A I, Balaykin A V and Kondratiev A I 2014 Methodology of rheological material properties phenoinenological modeling at high speed cutting by reverse analysis *Research Journal of Applied Sciences*. 9(11) pp 753-760

[10] Agapovichev A V, Balaykin A V, Smelov V G and Agapovichev A V 2015 Application of Additive Technologies in the Production of Aircraft Engine Parts *Modern Applied Science*. 9 (4) pp 151-159

[11] Belousov A I and Badamshin I Kh 2015 Conceptual Approach to design of materials for GTE turbine blades *Izvestiya VUZ. Aviatsionnaya Tekhnika*. 58(3) pp 73-78

[12] Kostyshev V A and Khaimovich A I 2015 Analytical modeling of the microstructure evolution of titanium alloys during high speed stamping of the blades of gas turbine engines *Izvestiya VUZ. Tsvetnaya Metallurgiya*. 56(2) pp 49-56

[13] Barvinok V A, Smelov V G, Sotov A V and Kosirev S A 2015 Restoration of the face of a GTE blade tip by pulse laser cladding *Journal of Machinery Manufacture and Reliability*. 44, No. 7 pp 658–662

[14] Bolotov M A, Pechenin V A and Ruzanov N V 2014 Modeling of coordinate measuring geometrical parameters form and location complexity profile of compressor blade GTE *Research Journal of Applied Sciences*. 9(12) pp 1143-48

[15] Zaripova L Z 2009 Controlling the size and geometric orientation of cast turbine blades *Vserossiyskaya nauchno-tekhnicheskaya konferentsiya "Problemy i perspektivy razvitiya aviatsii, nazemnogo transporta i energetiki "ANTE-2009"* [Proc. of the 5th All-Russian Scientific Conference "Problems and prospects of the development of aviation, land transport and power engineering "Ante-2009"] pp 372-376

IOP Conf. Series: Materials Science and Engineering **142** (2016) 012074 doi:10.1088/1757-899X/142/1/012074

Computer Simulation Methods for Crushing Process in an Jaw Crusher

Ilia Il'ich Beloglazov[1], Dmitrii Andreevich Ikonnikov[2]

Department of automation of technological processes and production
National Mineral Resources University (University of Mines)
199106, St. Petersburg, Vasilevsky Island, 21 line,2 Russia

E-mail: [1]filosofem@mail.ru, [2]ArtSciencesD@gmail.com

Abstract. One of the trends at modern mining enterprises is the application of combined systems for extraction and transportation of the rock mass. Given technology involves the use the conveyor lines as a continuous link of combined technology. The application of a conveyor transport provides significant reduction of costs for energy resources, increase in labor productivity and process automation. However, the use of a conveyor transport provides for certain requirements for the quality of transported material. The maximum size of the rock mass pieces is one of the basic parameters for it. The crushing plants applies as a coarse crushing followed by crushing the material to the maximum size of piece which possible to use for conveyor transport. It is often represented by jaw crushers. Modelling of crushing process in jaw crushers allows to maximally optimize workflow and increase efficiency of the equipment at the further transportation and processing of rocks.
We studied the interaction between walls of the jaw crusher and bulk material by using discrete element method (DEM) in this paper. The article examines the process of modeling by stages. It includes design of the crusher construction in solid and surface modeling system. Modelling of the crushing process based on the experimental data received via the crushing unit BOYD. The process of destruction and particle size distribution in the study was done. Analysis of research results shows a comparability of actual experiment and modeling process.

1. Introduction.

Preparation of a rock mass for enrichment includes quite energy-intensive process of crushing. Initially, the rock mass crushing made by drilling and blasting operations. The next stage to reduce the medium size of rock mass pieces used crushing plant for coarse splitting. The next stage to reduce the medium size of rock mass pieces is the use of crushing plant for coarse splitting. It is mostly provided by jaw crushers. Among the various crushing equipment jaw crushers are widely used. The advantages of jaw crushers are simplicity of the design, reliability, small overall dimensions and weight, as well as easy maintenance and repairing. Disadvantages of jaw crushers is a periodical operation mode and availability of large swinging masses.

We studied the interaction between walls of the jaw crusher and bulk material by using discrete element method (DEM) in this paper. Operations via ball mills, autogenous and semiautogenous grinding mill rather widely describes in the scientific literature [1-4]. Whereas modeling of jaw crushers is possible to find only a small amount of articles actually [5,6,7,8]. Modeling of the grinding operation in a jaw crusher is an exciting and actual problem.

The jaw crushers have proliferated among the various crushing equipment. It is used for coarse crushing.

Modeling of crushing process allows to select the most effective mode of equipment operation. Also it allows to study process of breaking visually and "from the inside". DEM is one of the new instruments for research of the destruction process for various materials [8,9,10]. In recent years this approach quickly became a powerful and effective instrument of modeling. The fundamental assumption of this method is that matter is composed of separate discrete particles. This method is based on representation of material as the aggregate of interacting particles (usually it is spherical particles). It is moving according to Newton's classical laws with taking into account the external force fields (there is gravity, friction, momentum and elastic interaction). DEM allows to get comprehensive data of an arbitrary particle at moment of interest: location, operating force, speed, acceleration, direction of movement and energy of the particle.

DEM calculation cycle represents as a step by step calculation per unit time, over which the law of motion is applied to each particle and each contact surface. The boundary conditions are updated in the modeling process. The interaction between the elements formed and stopped automatically. DEM calculation cycle in the general form is shown in Figure 1. During one modeling step calculation is performed with average range from 10^{-8} to 10^{-6} seconds.

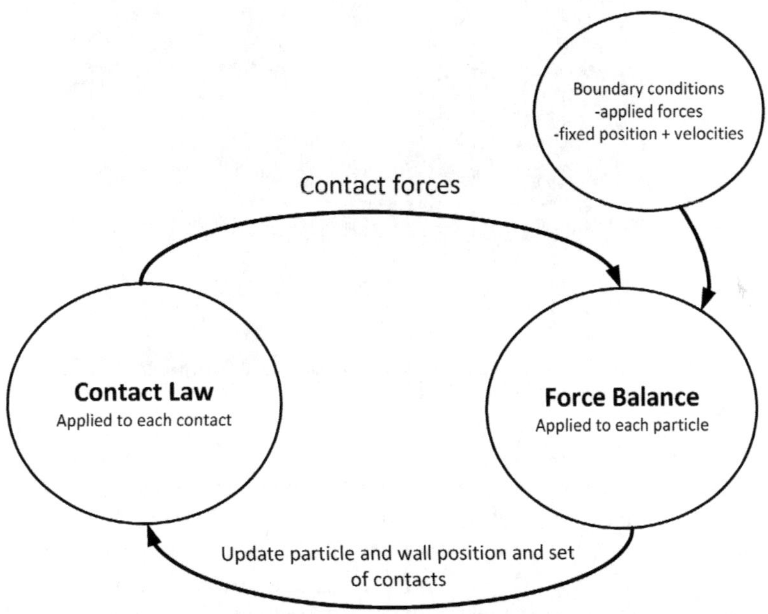

Figure 1. Calculation cycle

Using the DEM requires rather simple, but very high demands on computer performance. Moreover, it is need to use modern software systems for maintenance of system flexibility with the ability to operative change parameters of the model, for parametrical optimization, to adding new properties into the model, to testing number of structures in order to select the most effective solutions.

The Rocky software package [12] has been selected for modeling of technological processes with bulk materials and applicable equipment as such an instrument among variety of different software packages and available integrable platforms [10,11] in existence today. Modeling has been done by using a workstation in a specialized laboratory of the Automation of technological processes and production facilities department of the National Mineral Resources University (University of Mines).

We have made review of the literature and analyzed data of physical experiments in production conditions using crushing plants at the same time.

2. Modeling

Modelling of crushing process is restricted in a jaw crusher work area. It is the area where the grinding of the material and movement of adjacent geometric surfaces (cheek crusher) does. Modelling of the crusher design produced in real scale by used solid and surface modeling system of the SolidEdge [13]. Design details which complicate the process of modeling or not involved in it (for example: the shaft, the frame, connecting rod, etc.) not included in the three-dimensional model. Using the Solid Edge as a three-dimensional modeling environment enables not only to make the construction of the model, but it also allows you to quickly and efficiently make any modification to the design of the model. Subsequently, the created model can easily be imported in the ROCKY (Fig. 2).

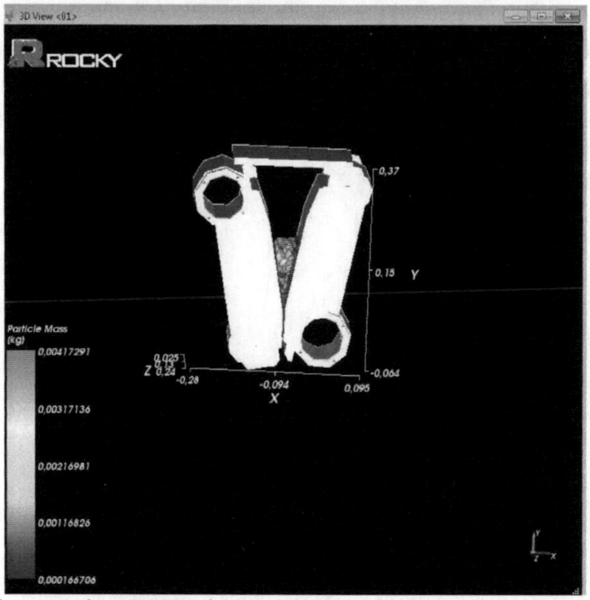

Figure 2. he jaw crusher BOYD in the Rocky, which contains processed material

3. Modelling of material for research

The structural concrete as a material was selected for fragmentation. This material is widely available and its properties are well understood. Rocky Software features allow you to set many parameters generated particles (Table 1). Unlike many other tools for modeling using the MDE, which use either spherical particles or other pre-form [14, 15] in Rocky has an option to select an arbitrary particle geometry. This allows the maximum to take into account features of the material used and the structure at failure.

Table 1 - Parameters of particles for research

Parameter	Value
Rotation of particles	Yes
Rolling resistance	0
Vertical aspect ratio	1,50
Horizontal aspect ratio	0,75
Number of corners	25
Density	1600 kg /m^3
Loading Stiffness	$1*10^8$ N/m^2
The bulk density	not taken into account

The complex polygon shape has been selected as the particles (Fig. 3)

Figure 3. The shape of the processed material particles.

For each particle in contact with other particles and / or geometric surfaces calculated total energy density of this contact. If this energy is greater than the minimum energy of the particle destruction (E_{min}), the E_{min} defined as

$$E_{min} = E_{min.r}(L_r/L) \tag{1}$$

where L - particle size,

$E_{min.r}$ the minimal specific energy for the average particle size for the selected material (this material constant)

L_r - average particle size, minus the minimum energy added to the total energy of the previous contacts E_sum:

$$E_{cum} = E_{cum} + E - E_{min} \tag{2}$$

Particle fracture probability P is calculated by the formula [16]:

$$P = 1 - \exp\left(-S \cdot (L/L_r) \cdot E_{cum}\right. \tag{3}$$

where S - the parameter of the particle breakage strength.

If the particle is destroyed, then fragments are generated by an algorithm Voronoi fracture [17] in accordance with the distribution indicated by the built-in functions T_{10}:

$$T_{10} = M \cdot (1 - \exp\left(-S \cdot (L/L_r) \cdot E_{cum}\right) \tag{4}$$

where M is a constant for the selected material. The size distribution of fragments of the wreckage determined from the value of T_{10}. The distribution of particle size is calculated based on the model Gaudin-Schumann [18], where the mass fraction of particles retained in each calculation interval calculated by an equation with two given constants:

$$W_p = 100 * \left(\frac{x}{k}\right)^m \tag{5}$$

Where Wp - mass yield,%

x - particle size,

k - particle size constant

m - distribution constant

Table 2 – The modeling parameters

Parameter	Value	
Minimum Size	0,005	m
Reference size	0,007	m
Minimum specific energy	0	J/kg
Select function coefficient	1	J/kg
Maximum T10 value	10,00	

Note that in the above Tables 1 and 2 list only a few parameters of the simulation. Functionality of the program covers more of the properties of elements and process parameters.

Furthermore, it is necessary to consider the number of particles participating in the simulation. When splitting the number of particles increases several times compared with the initial state, while significantly increasing the calculation time model. For example, for numerical processing only 1000 particles require about 20 hours of computation on a workstation with 9 cores.

4. Modeling the dynamics of movement of the cheeks

In crushers with a complex movement the movable jaw fixed to the eccentric portion of the main shaft, further acting as a connecting rod. At the bottom of the connecting rod is attached to the spacer plate that can oscillate around a fixed point. With this consolidation, by the rotation of the main shaft, each point of the movable jaw will describe the trajectory of a closed curve in the form of elongated ellipses. At this point the trajectory will have lower longitudinally extending closed curves, and when approaching the top of the trajectory of the movable cheek increasingly will approach the form of a circle. When approaching a stationary cheek occur simultaneously lower the movable jaw down. material other than grinding exposed on the impacts from the stekni cheeks, so that there is a material and pushing it easy abrasion. In other words there is involuntary discharge of the crushed material.

Therefore, with the same overall dimensions of the crusher with a complex motion more productive. It is this feature of crushers with a complex movement of interest for modeling and optimization. The software package allows for complex motion, adjusting the parameters such as frequency and amplitude of the motion.

In addition to movement crusher jaws considered the physical characteristics of the surface and, if necessary, it is possible to control the deterioration of "jaw armor", thus predicting the area exposed to the greatest abrasion

5. Calibration and experiments

For the simulation of the crushing process, first of all, you need to calibrate the static parameters. Estimate the actual state of the particles as a result of crushing, particle size distribution and the number of degrees of freedom.

We create the data for the calibration setting parameters of the material, making a comparison between experiment and simulation. Thus, one should always pay attention to portability of simulation results for process parameters such as particle size distribution and specific energy consumption. The next step is to change the model parameters as long as they will not reflect reality. Maximum use of the actual specific characteristics and process parameters are always gives a good foundation for the rapid and successful simulation. The sequence of the calibration procedure is described in detail in [19, 20].

Table 3 – Parameters of the experiment and the model

Experiment	**Model**
Aggregate	Around 1200 of the particles
ρ=1600 kg/m^3	ρ=1600 kg/m^3
d=0,1…16 mm	d=5…16 mm
E=30…40 kN/mm^2	Stiffness aggregate k=$6 \cdot 10^6$ N/m Stiffness cement k=$5 \cdot 10^6$ N/m
Compression strength $\sigma_в > 35$ N/m	Normal strength $\sigma > 1{,}5 \cdot 10^7$N/m^2
Coefficient of friction concrete-steel $\mu \approx 0.3$	$\mu = 0.3$

Experimental data [13, 20] was used in the subsequent model calibration and verification of simulation results. A large number of experiments on the jaw crusher were carried out to investigate conditions for concrete recycling. The obtained experimental data could also be used for calibrating the DEM models and validating the results. Following process parameters were varied: the feeding rate of the concrete pieces between 4.5 and 13.3 t/h which corresponds to the volume fraction of solids in the process area of 2.6 to 12.4%, the smallest gap width between 20 and 35 mm and the peripheral rotor speed between 25 and 38 m/s. Test samples of the material were only a concrete pieces shaped complex polygons.

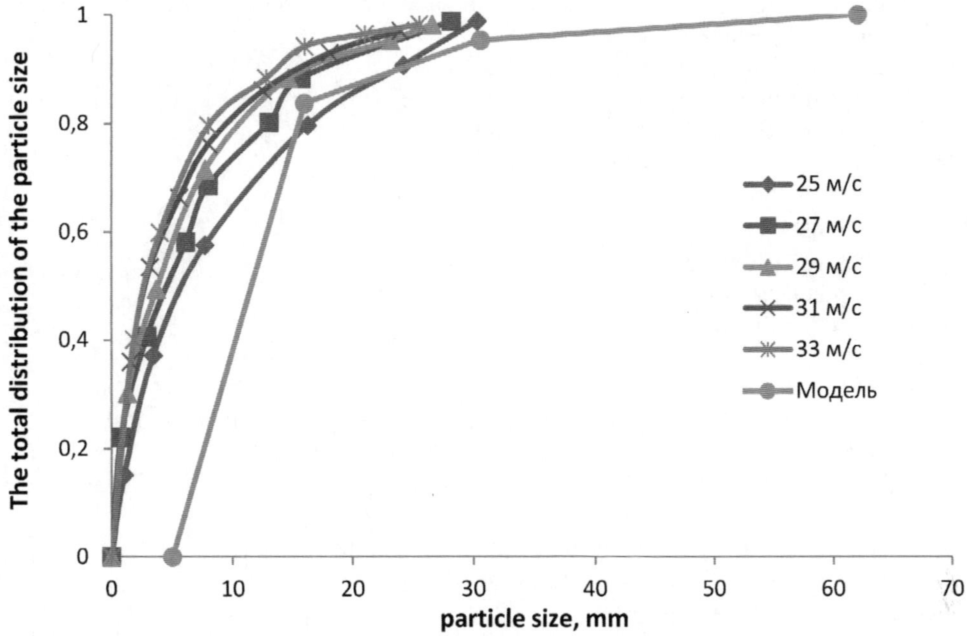

Figure 4. Particle size distribution after the experiment and modeling at various rotor speeds

Figure 4 shows a comparison of experimental data and simulation results. The graph shows the distribution of particle size obtained milled products. Experimental distribution curves shown in the figure, are correlated in order. The higher rotor speed, the smaller the resulting product respectively. simulation curve lies below the rest of the experimental data, since the minimum size of the simulated particle was 5 mm. Selecting this particle size was primarily limited time settlement models.

6. Conclusions and prospects

Simulation of crushing using jaw crusher is only one of many examples of the application of the DEM. The features of this simulation can be the basis of material intended for a more detailed study of the process in view of a more detailed experimental data.

It is noteworthy that the amount of time for calculations increases sharply with the number of particles and largely determines the accuracy of the results. Therefore, new methods must meet the following conditions:

1. Interaction of specialized software allows you to quickly change the design elements and optimize the operation of the unit. The simulation results can be instantly imported for further careful evaluation.

2. In the simulation process, it is impossible to take into account a disproportionately large number of particles generated as a result of crushing, resulting in a significant increase in computation time. Smaller particles in particular occur in vast numbers and make the calculation last much longer. Hence, their interaction with other particles and walls can hardly be simulated. A pressure term, obtained from the Maxwell velocity distribution, replaces their interaction forces so that they do not need to be simulated. It can be predicted that the computing time will be saved remarkably.:

List of accepted designations.

DEM – discrete element method

L – particle size,

$E_{min.r}$ – the minimum specific energy of the average particle size for the selected material (material constant)

IOP Conf. Series: Materials Science and Engineering **142** (2016) 012074 doi:10.1088/1757-899X/142/1/012074

L_r – average particle size, minus the minimum energy added to the total energy of the previous contacts E_{cum}
S – breaking strength of particles.
E_{min} – the minimum energy for particle destruction
P – probability of particle destruction
M – constant for the selected material.
T_{10} – the function for calculating of particle destruction
W_p – mass yield, %
x – particle size,
k – constant of particle size
m – distribution constant

References

[1] Morton, D., Dunstull, S.: Using the Web to increase the availability of DEM-based mill modelling. Minerals Engineering. issue 17 (2004) pp. 1199-1207
[2] Morrison, R. D., Cleary, P. W.: Using DEM to model ore breakage within a pilot scale SAG mill. Minerals Engineering. issue 17 (2004) pp.1117-1124
[3] 14. Mishra, B. K., Murty, C.: On the determination of contact parameters for realistic DEM simulations of ball mills. Pow. Technology 115 (2001) 290-297
[4] Djodjevic, N., Shi, F. N., Morrison, R. D.: Applying discrete element method modelling to vertical and horizontal shaft impact crushers. Minerals Engineering 16 2003. pp. 983-991
[5] Tomas, J., Schreier, M., Gröger, T., Ehlers S.: Impact crushing of concrete for liberation and recycling. Powder Technology. 105 (1999) pp. 39-51
[6] A. Refahi Comparison between bond crushing energy and fracture energy of rocks in a jaw crusher using numerical simulation / J. Aghazadeh Mohandesi, B. Rezai The Journal of The Southern African Institute of Mining and Metallurgy. Vol. 109 (2009) pp. 709-717
[7] Legendre D. Assessing the energy efficiency of a jaw crusher / Zevenhoven R/ Energy Vol. 74 (2014) pp. 119–130
[8] Paul W. Cleary Simulation of particle flows and breakage in crushers using DEM: Part 1 – Compression crushers, / Matthew D. Sinnott Minerals Engineering Vol 74. (2015) pp. 178–197
[9] Arsentev V.A., Blekhman I.I., Blekhman L.I., Weisberg L.A. The methods of particle dynamics and discrete element as a tool for investigatingand process optimization for the processing of natural and man-made materials / Obogashhenie rud. issue 1 (2010) pp. 30–35.
[10] Feoktistov A.Ju, Blehman L.I, Vasil'kov V.B., Ivanov K.S. The application of discrete element method for modeling of processes in mining and metallurgical industry / Zapiski Gornogo instituta, volume 192. №1. 2011
[11] Cundall, P. A., Potynondy, O., Konietzky, H.: Manual of the DEM program system Particle Flow Code in 2 and 3 Dimensions (PFC) Vers. 3.1, Itasca Consulting Group Inc., Minneapolis, Minnesota USA 2005
[12] http://www.rocky-dem.ru/rocky/review/
[13] Anatolij M., Belokopytov C., Lepestov A. The construction of special equipment in the Solid Edge software. CADmaster №5 (78) 2014 p.35-39
[14] Schubert, W.; Jeschke, H.: DEM-Simulation of the Breakage Process in an Impact Crusher. Wissensportal baumaschine.de 4 2005
[15] Khanal, M., Schubert, W., Tomas, J.: Ball Impact and Crack Propagation – Simulations of Particle Compound Material, Granular Matter, 5 (4) 2004 p. 177-184
[16] J. Bruchmuller, B.G.M. van Wachem, S. Gu, K.H. Luo / Modelling discrete, incremental, repetitive and/or simultaneous particle breakage', 7th International Conference on Multiphase Flow ICMF 2010, Tampa, FL, May30-June 4, 2010.
[17] F. Preparata, M. Shejmos. Computational Geometry: Introduction. - M .: Mir, 1989. P. 29515

[18] Stamboltzis G.A. Calculation of Gates–Gaudin–Schuhmann and Rosin Rammler parameters from the size analysis of the coarse part of the distribution. Min Metall Ann 1989;72–73 p. 29–38.
[19] Krivcov A.M. Deformation and fracture of solids with microstructure. - M .: Fizmatlit 2007, 304.
[20] Schubert, W., Khanal, M., Tomas, J.: Impact Crushing of Particle-Particle Compounds – Experiment and Simulation, International Journal of Mineral Processing, 75 (1-2) 2005. p. 41-52

IOP Conf. Series: Materials Science and Engineering **142** (2016) 012075 doi:10.1088/1757-899X/142/1/012075

Physical Modeling of Contact Processes on the Cutting Tools Surfaces of STM When Turning

V A Belozerov, M H Uteshev

Tyumen State Oil and Gas University, 38, Volodarskogo street, Tyumen, 625000, Russia

E-mail: kafedra_tm@tsogu.ru

Abstract. This article describes how to create an optimization model of the process of fine turning of superalloys and steel tools from STM on CNC machines, flexible manufacturing units (GPM), machining centers. Creation of the optimization model allows you to link (unite) contact processes simultaneously on the front and back surfaces of the tool from STM to manage contact processes and the dynamic strength of the cutting tool at the top of the STM. Established optimization model of management of the dynamic strength of the incisors of the STM in the process of fine turning is based on a previously developed thermomechanical (physical, heat) model, which allows the system thermomechanical approach to choosing brands STM (domestic and foreign) for cutting tools from STM designed for fine turning of heat resistant alloys and steels.

1. Introduction

By results of research of process of a struzhkoobrazovaniye when cutting tempered staly by the tool from an elbor – are made P following conclusions: 1) turning by cutters from a composite 01 tempered staly is more expedient, than crude owing to more favorable conditions for deformation of a struzhkoobrazovaniye – when turning soft (not tempered) staly typical drain shaving is formed, when turning tempered staly processes of formation of elements of shift occur more intensively sustavchaty shaving with undivided elements iobrazutsya; microhardness of shaving tempered by staly is much lower than the initial size (material of shaving razuprochnyatsya), and at crude on the contrary; 2) intensity of deformation processes in a zone of contact of shaving with a forward surface of a cutter when turning tempered by staly is much lower, than when turning crude; 3) increase in speed and depth of cutting reduces deformation size; 4) wear of cutters sharply intensifies deformation processes of shift and friction.

From the analysis of literary data and operating experience of cutters the following scopes of composites when turning are recommended tempered staly: 1) composites 01 and 02 – thin and fair turning without blow of details from tempered with the staly hardness of HRC 55-70 with giving to 0.15 mm / about and cutting depths to 1-1.5mm; 2) a composite 05 and his modifications – fair and semi-fair turning without blow of details from HRC 40-60 tempered staly average hardness with giving to 0.3 mm / about and depths of cutting of 0.05-2.0 mm; 3) composites 09 and 10, 10D – thin, fair and semi-fair turning with blow and without blow of details from tempered by staly hardness not above HRC 60 with giving to 0.2 mm / about and cutting depths 0.05-1.5mm.

It is specified in work that the available information on application of cutters from a composite when processing of the hardly processed alloys is insignificant, it is often contradictory and can't be a basis for what - or recommendations.

There are four main groups of the materials which are effectively processed by composites: the bleached cast iron or the white HRC 50-65 cast iron alloyed by nickel, or chrome; some strengthening HRC 38 alloys, some brands of gray HB 200-220 cast iron. When turning without cooling of nickel alloys with plates from a composite the following modes of cutting are recommended: speed of cutting is 80-120 m/min, giving of 0.2-0.3 mm / about, depth of cutting is 1 mm.

Efficiency of processing of metals cutting substantially depends on a right choice of geometrical parameters of the cutting tool. Geometrical parameters have to correspond completely to specifications of the carried-out operation, provide an opportunity to appoint the most economy mode of cutting and consequently, to allow to conduct processing with high efficiency at the minimum expenses of means.

On the basis of the pilot studies conducted in Vniialmaz and in other organizations the geometrical parameters of the lezviyny tool equipped with CTM, providing the high durability and the maximum wear resistance of the tool are determined.

Scopes of composites and ceramics are substantially blocked. Cutters with plates or inserts from a composite 01 possess ability to samozatachivatsya at wear, that is to keep the radius of a rounding off of the cutting edge almost invariable in limits ρ=20-50 micron during the entire period of firmness. It promotes reduction of a roughness of the processed surface and, as a result, increases firmness of cutters by technological criterion at fair processing. Cutters from ceramics don't possess such ability – the edge at them is rounded much more intensively and more. Therefore the composite 01 has advantage in comparison with ceramics when turning of the modes of cutting characteristic of these materials tempered staly high hardness in all range. The ceramics is more preferable when turning preparations from not tempered constructional staly at cutting speeds from above 250m/min composites of all brands are almost disabled.

The Lezviyny tool from STM (composites) at introduction in many cases replaces: 1. The cutters equipped with firm alloys, the cutting ceramics or diamonds on operations of fair and thin turning and boring. 2. Abrasive, elborovy and diamond wheels on grinding operations. The feasibility study on efficiency of the tool from STM in the first and second cases is provided in works.

In the first case the lezviyny tool from STM can fit into the existing technological processes as they are applied on the same equipment, as the replaced tool. However, ensuring higher speeds of cutting and receiving more quality processing leads in some cases to change and simplification of technological processing: operations of operational development, grinding in or a shabreniye are excluded, the necessary number the honingovalnykh of operations, that is in these cases is reduced, the existing technological processing of details undergoes some changes.

In the second case, when replacing abrasive tools and operations of grinding of heat resisting alloys and staly, technological processing of details undergoes considerable changes, the lezviyny tool from STM allows to concentrate operations, for example, on one machine with CNC. One of the basic principles of technology of mechanical engineering – concentration of operations – is carried out, as a rule, due to improvement of the machine and its automation. In this case the high cutting properties of the tool are the cornerstone of this principle. Experience of application the lezviynykh of the STM tools shows that concentration of operations is carried out due to performance of draft and fair operations on this equipment, one tool and simultaneous processing by one tool of details of various hardness. In this case technological processing should be made according to the scheme: exact preparation - heat treatment - full processing by the lezviyny tool from STM.

Cutters from STM are the tool for finishing processing therefore they should be used on the metal-cutting machines of the increased, high and especially high precision possessing necessary vibrostability and rigidity. The most effective is use of the tool from STM the machines having a high top limit of frequency of rotation of a spindle (3000 rpm and more) and a low limit of longitudinal giving (0,005-0,01) $\times 10^{-3}$ m / about.

The main area of effective use of the cutting tool equipped with STM, the automated production on the basis of machines with CNC, multi-purpose machines, automatic transfer lines, special high-speed machines, the flexible production modules (FPM) and the flexible production systems (FPS). Efficiency of such systems is provided with their ability to be reconstructed quickly on release of a new product with the high efficiency reached by round-the-clock work of the equipment in an automatic cycle. Under such circumstances works sharply increase requirements to reliable operation of machines, all industrial equipment and is concrete to reliability and wear resistance of the cutting tool.

When processing on machines with program control and GPM of preparations from tempered staly on many finishing operations by the abrasive tool it is almost difficult to provide the required rates of increase of productivity of processes owing to low dimensional firmness of circles, and also practical impossibility of processing (by the adjusted size) and with a deep water of cutting. Besides, when processing defects as charging of the processed surfaces of preparations and education in their blanket of cracks, prizhog and other defects arise the abrasive tool. Machining of such preparations by the lezviyny tool is more productive and practically excludes emergence of the specified defects. However, application of cutters from firm alloys or quick cutting steels is limited to his low dimensional firmness which is sharply decreasing with increase of hardness tempered by staly. In this regard the most perspective tool materials are STM on the basis of KNB.

It should be noted that reliability of the tool equipped with STM increases at increase in speeds of cutting in strictly recommended limits when processing each concrete material. It is essentially important to consider it when using such tool on the automated equipment.

High microhardness and heat resistance of modifications of nitride of pine forest in combination with fine-grained structure allow to receive optimum prerequisites for processing of the preparations produced from heat resisting and stainless steels and alloys. In comparison, for example, with turning by tools from firm alloys or quick cutting steels, processing by the tools equipped with STM is characterized by smaller forces that positively affects reduction of size of wear and fluctuation of the tool, elastic deformations in technological system. Also the deviation of a form and height of roughnesses on the processed surfaces decreases. These features are very favorable when processing of the basic, sealing, interfaced, step and shaped surfaces, boring of exact openings.

We have developed a thermo-mechanical model of a thin turning heat-resistant alloys and steels STM cutters on CNC machines, flexible manufacturing units (GPM) and machining centers is based on experimental studies. The basis of this model, the system proposed by thermomechanical approach to the selection of domestic and foreign brands to private label cutters for fine turning of heat resistant super alloys and steels [1].

This thermo-mechanical model allows to develop and implement in the course of processing optimization (physical, thermal) process model of fine turning of superalloys, and has become a tool of the STM. Design optimization model (physical optimization) is performed using graphical techniques - charting of the family (nomograph) to determine the contact characteristics and parameters of the cutting process the cutting conditions (t, s, v_{opt}) having a constant value for optimal values [2].

2. Thermomechanical model

Before considering the issues related to the creation and implementation of the optimization model of the process of fine turning of superalloys tools STM need to dwell on the problems of modeling and creating physical models of processes.

The term "modeling" is defined as a substitution of one object (original) to another (the model). Next, you need to study the properties of the model. The substitution is made in order to simplify, reduce the cost, accelerate the study of the properties of the original.

Initially, all models can be divided into two groups: physical (real, real); mathematical (abstract, conceivable).

Thus, process modeling involves making assumptions varying degrees of importance. This must be met a number of requirements to the models: the adequacy of sufficient accuracy, expediency, efficiency.

The need arises in the modeling techniques in different situations for the pilot study, a comprehensive analysis and forecasting of processes of passage. An important area of the simulation is to take management and design solutions in the study process.

Originally all models can be divided into two groups: physical (material, real); mathematical (abstract, imaginable).

Thus, modeling of processes assumes acceptances of assumptions of this or that degree of importance. At the same time a number of requirements to models has to be met: adequacy, sufficient accuracy, expediency, profitability.

Need for methods of modeling arises in various situations for a pilot study, the comprehensive analysis and forecasting of passing of processes. The important sphere of modeling is adoption of administrative and design decisions at research of processes [3,4,5].

Thermomechanical model is the basis for the creation of an optimization model of the process of fine turning of superalloys and steel cutters of the STM, with low tensile strength. Implementation of thermomechanical model of the process of fine turning of superalloys and steel tools from the STM solves the thermal problems with differing from those given in the literature boundary conditions, approximating the calculated according to the actual contact processes fine turning of superalloys and steel cutters of the STM.

We set task is to carry out sufficiently substantiated systematic approach to the selection of private label brands (domestic and foreign) for fine turning heat-resistant alloys on CNC machines, machining centers and flexible performance model, as well as on the universal equipment. Thermomechanical proposed approach to selecting brand STM, consisting in simultaneous account while choosing the thermophysical characteristics STM thermal conductivity λ_{toll} (λ_t) physical and mechanical characteristics of the STM (instrumental characteristics of the material strength) - the tensile strength $\sigma_{stretching}$ (σ_{st}). These boundary conditions when selecting brands STM differ from those given in the literature.

Thermomechanical approach to pilot studies of contact processes at a thin current of heat resisting alloys cutters from STM allows to draw a conclusion that the most effective from the point of view of dynamic durability from three considered STM (composites) - an elbor - P, a composite 05IT and a geksanita - R-is geksanit - P as due to redistribution of average contact tension on forward and back cutter surfaces from a geksanit - P tension of qN, qF on a forward surface increase, and on a back surface tension of qN_1 decreases by 1,1 – 1,5 times and tension of qF_1 a cutter from an elbor - P in the considered range of the modes of cutting decreases by 1,2 - 1,64 times in comparison according to tension of qN_1 , qF_1 . It is connected with influence of size of coefficient of heat conductivity λ_t a geksanita - P on the fact that at a geksanit - P qN, qF increase, and qN_1 , qF_1 decrease in comparison with the corresponding contact tension at an elbor - P at thin turning of heat resisting alloys.

We have developed the principle of choosing the effective use of cutting tools made of composites and STM, manufactured abroad, in the processing of a variety of processed materials. This principle is the realization of thermo-mechanical approach to the choice of domestic brands STM (composites) and foreign STM CBN intended for fine turning of heat resistant steels and alloys tverdostyuHRC35-40, hardened alloy steel HRC35-40 hardness (medium hardness). The principle of choice of brands STM shows that the most effective composites, serially produced for the fine turning of heat resistant steels and alloys are domestic STM: composite 10, 10D composite, composite 05IT, composite 09 (PTNB-IR-1) and foreign STM CBN: vyurtsin (Japan), sumiboronBN200 (Japan), the ICE-50 (England). As the name STM CBN vyurtsin (Japan) is a method for preparation, composition and structure of the crystal lattice of the domestic analog STM CBN - geksanit - P as well, these are STM wurtzite-like hexagonal (wurtzite) modifications of boron nitride. Also in accordance with the above, vyurtsin and geksanit-P have almost the same size thermal characteristics- thermal conductivity (geksanit- P – λ_t=20$\frac{Wt}{м*К}$; vyurtsin – λ_t=21$\frac{Wt}{м*К}$).

Our experiments show that hardened to a high hardness steel to effectively handle the entire range of composites. But superalloys most effectively processed composites 05IT, 10, 10D. This is due to the fact that they have low thermal conductivity in size, which leads to a decrease in the contact stress on the rear surfaces of the incisors and dynamic strength and the durability of the cutters will be higher than that of the composites of cutters 01, 02, 03, kiborita-1, amborita. Therefore, the composites 01, 02, 03, Kiborita-1, Amboro ineffective at HRSA. We have found that, ceteris paribus, along with the characteristics of strength, a determining influence on the resistance and dynamic strength of cutters is the coefficient of thermal conductivity of the selected brand STM. Krome addition, the need to use cutters of a composite 05IT, the composite 09, 10 of the composite, the composite · 10D, vyurtsina , sumiboronaBN200 (Japan), the ICE-50 (England) in fine turning heat-resistant alloys and steels tverdostyu HRC 35-40 (medium hardness) is caused by the necessity of rational use of tools STM CBN, cutters of which are expensive.

It was determined that the thermomechanical approach to choosing brands STM is connected not only with the peculiarities of physical and mechanical and thermal interaction of the processed materials and tools in the process of cutting, but is based on the principles of economic efficiency and expediency of the incisors of the STM.

Thus, the selection of stamps on the basis of STM thermomechanical systemic approach to the study of phenomena in contact relationship during the cutting process, that is, physical interaction analysis instrument and processed materials in fine turning of superalloys incisors and STM relies on the interaction of the physical laws.

System thermomechanical approach when choosing a brand STM for fine turning of heat resistant super alloys and steels provides a selection of brands CTM not only for the existing domestic and foreign instrumentation STM for cutting tools, but also for new tools STM (domestic and foreign), when known values of thermophysical characteristics (coefficient thermal conductivity λ_t and the physical and mechanical characteristics of the STM tool (tensile strength σ_{st}). This is one of the main advantages of the process models created thermomechanical fine turning of superalloys tools from STM.

Identify physical pattern of systemic thermomechanical approach when choosing brands STM for processing high-temperature alloys and steels (based on established thermo-mechanical model of the process of fine turning of superalloys cutters from STM) applies, in our opinion, all the combinations of processed steel and alloys with different types of tool materials treatment process. For all treated steels and alloys in the choice of the brand of instrumental material groups (hard alloys Group VC, hard alloys TC group; hardmetal; cermet; mineral ceramics and other tool materials) necessary to implement the system thermomechanical approach, choosing the tool material grade for the respective brand of processed material (steel and alloy), taking into account the lower value of thermophysical characteristics (thermal conductivity λ_t) and the physical and mechanical characteristics of the tool material (tensile strength σ_{st}). This systematic approach thermomechanical brand when selecting a tool material must provide redistribution of contact stresses and temperatures at the front and rear surfaces of the cutting tool in the direction of reducing the contact stress and temperature on the back surface, and due to this rise period dynamic strength and durability of the cutting tool.

Before creating an optimization model of the process of fine turning of superalloys tools from STM we identified prerequisites of physical contact to optimize performance at the same time on the front and back surfaces of the tool from the STM.

Conducted research and comprehensive analysis of the results leads to the conclusion that the process of interaction between the treated and instrumental materials when cutting the front and rear surface of the instrument, followed by a chip, it is a single pin deformation process, so the chip velocity factor the K_L, defined on the front surface of the tool, affects not only the characteristics of the contact acting on the front surface of the tool [6], but also on the characteristics of the contact acting on the rear surface - forces N_1, F_1, friction coefficient μ_1, contact stresses σN_1, τF_1.

The analysis shows that the most pronounced decrease with the smooth growth factor chip shortening K_L ot 1.14 to 1.48 in the process of fine turning HN62MVKYU alloy - VD, HRC38 "sharp" cutter of geksanit - P contact characteristics on the rear surface of the instrument - the normal pressure force N_1 i average normal contact qN_1 voltage. As is known, the cutting part of the tool is delicate, since the tensile strength limits constitute STM σ_{st} = 240-490 MPa, so one of the main problems of these machining tool materials is simultaneous with the regular mechanical abrasive wear on the rear surface of the destruction of the instrument in the form of crumbling and spalling (sometimes chips) on the back side, which is actually the uncontrolled probabilistic nature [7,8,9]. This proves that you must strive to ensure the constancy of the coefficients of friction μ_1 on the rear surfaces of the incisors of the STM in the process of fine turning of heat resistant super alloys and steels.

3. Results and Discussion. Creation optimizing (physical, thermal) models

The prerequisites for the creation of the optimization model and the physical optimization of the contact characteristics (under development) are the following: 1) Our study forces N_1, F_1, the average contact stress qN_1, qF_1, coefficients of friction $\mu 1$ on the rear surfaces of the incisors of the STM from the changes the velocity factor chips K_L pri thin turning heat-resistant alloys. 2) Research Poletika M.F., which indicates that most clearly affects the chip velocity factor K_L on hand at the back edge of N_1, F_1 and friction coefficient μ_1 at h_{flang} (h_f) = 0.03-0.04mm ("sharp" cutting edge), that is, immediately after the running-in period the tool [10]. Experiments have shown that with increasing power K_L ,N_1, F_1 uvelichivayutsya. 3) We have used the method of physical optimization Makarov A.D. and the cutting speed υ_h^{opt} wholesale, meet the minimum relative wear hopz incisors of the STM. 4) Nekrasov Y.I. came to the conclusion that in the processing of hard materials in a range of cutting conditions "for the build-up" combination of cutting conditions in which the ratio shortening chips $K_L$opt changes ensure consistent power and thermal load of the cutting edges and the regime maintain a constant safety factor of the strength of the cutting tool (n_{opt} = const) [11,12]. It has been proven that physical optimality criteria on the front surface of the tool in the contact zone are both constant optimum ratio shortening chips $K_L$opt wholesale, cutting temperature θopt and the mean normal contact stresses qN opt wholesale. Constant optimal ratio shortening chips K_L opt wholesale determine the optimum cutting speed υ_{KL}^{opt} wholesale. In this case the optimum cutting speed wholesale υ_{KL}^{opt} is equal to the cutting speed corresponding to the relative tool wear on a back surface (on Makarov A.D.) - υ_h^{opt} wholesale, the wholesale est $\upsilon_{KL}^{opt} = \upsilon_h^{opt}$ (as provided by the constancy K_L = const , θopt = consti, qNopt = const).

At a position where $K_L$opt = const, θopt = const, q Nopt = const, there is physical optimization of the process of cutting the front surface of the cutting tool for a given combination "processed material - Tool material", that is, the provision is actually physical law in the interaction of the specific tool and processed materials. At the same time at a constant depth of cut for different values of the respective innings obtain optimum cutting speed $\upsilon_{KL}^{opt} = \upsilon_h^{opt}$ wholesale.

Constant optimum ratio shortening chips $K_L$opt wholesale and constant optimal temperature cutting θopt are the criteria of the physical interaction of the processed materials and instrumental, that is determined by the voltage-thermal state of the cutting part of the tool at the wholesale vershiny.$K_L$opt is the physical criterion of deformation processes in the cutting zone with chip and θopt - criterion teplofizicheskogo interactions processed and instrumental materials.

Our studies have shown that at constant optimal values of the velocity factor K_L^{opt} chips ~ wholesale, temperature cutting θopt , average normal contact stresses qNopt wholesale on the front surface of the tool of the STM with different combinations of the cutting parameters υ_h^{opt} (υ_h^{opt} - cutting speed corresponding to a relative minimum wear of the cutter hopz STM flank as a result of the physical process optimization method Makarov A.D.) relationship of forces on the rear poverhnosti N_1 / F_1 constant, ie constant ratio $F_1 / N_1 = qF_1 / qN_1 = \mu_1$. In addition, as mentioned above, the cutting speed $\upsilon_{KL}^{opt} = \upsilon_h^{opt}$wholesale.

As a result, detection of the background of the optimization model of the process of fine turning of superalloys tools from STM identifies areas (zones) of physical optimization contact characteristics (and contact processes) simultaneously on the front and rear surfaces of the incisors of the elbor - P composite 05IT, geksanit - P at Alloy turning HN62MVKYU-VD, HRC38. Provides constant optimal values K_L opt wholesale, θopt, qNopt wholesale on the front surfaces of the incisors of the STM and the permanent optimal coefficients of friction μ_1 on the rear surfaces of the incisors of the STM. It is also provided constant optimum cutting speed $\upsilon_{KL}^{opt} = \upsilon_h^{opt}$ for cutting tools from STM (at constant cutting depth t and feed S).

Thanks to the analysis of prerequisites of physical optimization of process the optimizing model of management in the course of thin turning of heat resisting alloys by contact characteristics at the same time on forward and back surfaces of the tool from STM has been created.

Physical optimization of contact characteristics of process of cutting extend not only to a lobby, but also to a back surface of the tool from STM. It occurs because the criterion of a physical optimality of interaction of the processed and tool materials - constant optimum coefficient of acceleration of K_L^{opt}shaving - operates not only contact processes on a forward surface of the tool (together with θ_{opt}° and q_N^{opt}) but also on a back surface. K_L^{opt}= const operates the size of constant optimum coefficient of friction on a back surface μ_1^{opt}=const on which size sizes of friction force F_1, forces of normal pressure of N_1 and average contact tension on a back surface of qF_1, qN_1 depend, that is also management of constancy of ratios $\mu_1^{opt} = f(K_L^{opt}) = \frac{F_{1opt}}{N_{1opt}} = \frac{qF_{1opt}}{qN_{1opt}} = const.$

Thus, the optimal ratio of permanent shortening chips $K_L$opt (chip shrinkage) - This is a combined (complex) characteristic of the interaction of the processed materials and instrumental in a physical process optimization, ie optimization criterion of physical contact processes simultaneously on the front and back surfaces of the cutting tool.

Developed a flowchart of an optimization model control during cutting contact characteristics simultaneously on the front and back surfaces of the tool in a STM physical optimization. Optimization model of contact management processes developed during the design phase operation of the process and is already being implemented at the time of surface treatment detali. Optimizatsionnaya model is presented as a family of graphs related to each other and forming a nomogram control the loading process at the same time on the front and back surfaces of the tool from the STM and for this nomogram determined optimal constant contact characteristics on the front and back surfaces of the STM tool and cutting conditions are optimized parameters for fine turning of the tool superalloy.

In accordance with the classification of physical models of the processes we have developed a nomogram optimization model can be attributed to the circuit (graphic) form a physical model as well as a representation of the model is produced in the graphical language (as a family of plots - nomogram).

4. Conclusion

Contact characteristics of thin turning of heat resisting alloys on a nickel basis by cutters from STM are experimentally determined and on the basis of it dependences of average contact tension on forward and back surfaces of cutters from STM from cutting mode parameters are established. At an intesifikation of these parameters of the modes of cutting in processing on machines about a CNC, flexible production modules and the processing centers in the conditions of real machine-building production, and also in vitro at application of the flexible production systems (FPS).

It is proved and experimentally confirmed that the most effective at thin turning of heat resisting alloys from the point of view of durability from three considered cutters from STM – an elbor - P, a composite 05IT and a geksanita - P – are cutters from geksanita-R as due to redistribution of average normal and tangent contact tension on forward and back surfaces of a cutter from geksanita-R average normal and tangent contact tension on a forward surface increases, and on a back surface average normal contact tension decreases by 1.1-1.5 times and average tangent contact tension decreases by 1.2-1.64 times in comparison respectively with the average normal and tangent contact tension of a cutter from elbora-R in the considered range of the modes of cutting.

The developed optimization model of fine turning of superalloys tools of the STM in a physical optimization has allowed to establish the relationship of contact performance and contact processes simultaneously on the front and back surfaces of the tool from the STM in order to stabilize the contact characteristics with the rear surface of the cutting part of the STM, as STM tools are fragile, that is, have a low value for the tensile strength σ_{st}.

The advantage of the implementation of process control optimization model cutting part loading of STM simultaneously on the front and rear surfaces based on the identified natural physical laws governing the interaction of the specific material being processed and the tool during the cutting process is that in this. There is no need to create complex devices and systems for the diagnosis and control of the parameters of the cutting process (contact characteristics and modes of cutting parameters). The process of contacting the treated material and in cutting tool takes the form of self-contact characteristics simultaneously on the front and back surfaces of the STM tool, based on the established physical contact superalloy patterns of interaction of the tool and in the STM processing.

In optimizing model of management of loading process at the same time on forward and back surfaces of the cutting parts of tools from STM at thin turning of heat resisting alloys and staly, the centers and the flexible production modules (FPM) tempered alloyed staly on machines with CNC, processing in the conditions of economic optimization of process of cutting at a design stage of operation of technological process and during processing of a surface of a detail criteria are involved: economic – the greatest economic efficiency of processing of a surface of a detail (profit) – $Э_{max}$, rub / with; criterion of reliability of the cutting tool – constant optimum probability of no-failure operation - P_3^{opt}, %; criterion of dynamic durability of the tool – constant optimum coefficient of margin of safety of the cutting part of the tool at top -n_{opt}; criterion of physical interaction of the processed and tool materials in the course of cutting (the generalized characteristic of the intense deformed state in a cutting zone) at the same time on forward and back surfaces – constant optimum coefficient of shortening of shaving -K_L^{opt}; criterion of physical interaction of the processed and tool materials in the course of wear and destruction of the cutting part of the tool on a back surface – constant optimum coefficient of friction on a back surface of the cutting part of the tool - μ_1^{opt}; criterion of quality of processing of a surface of a detail of R_a^{opt} (R_z^{opt}) – the constant optimum height of microroughnesses of the processed detail surface – external technological criterion of process of cutting (criterion of quality of processing of a surface) which unlike shrinkage of shaving and temperature of cutting is easier for controlling in the course of cutting by methods of active control when processing on the machine with CNC.

Physical and economic optimization of parameters of the modes of cutting at thin turning of heat resisting alloys with the hardness of HRC by 37 - 39 cutters from STM on machines with CNC, the processing centers and GPM is performed, are established economically reasonable, considering strength opportunities and necessary probability of no-failure operation of tools from STM in the conditions of the automated production.

References

[1] R.Yu. Nekrasov, A.I. Starikov, M.A. Korchuganov. Performance assessment of carbide tooling under termal and loading conditionds. Applied Mechanics and Materials Vol. 682 (2014) pp 414 – 417.

[2] Lasukov A.A. Selection of machining conditions in terms of the temperature dependence of chip formation. Russian Engineering Research, 2015. Vol. 35, Issue 9, September 2015, pp. 679–681.

[3] R.Yu. Nekrasov, A.I. Starikov, I.V. Soloviev. Simulation of technological systems for diagnosis and management machining with CNC. Applied Mechanics and Materials, 2015. Vol.770. pp 617-621.

[4] O.S. Lomova. Mathematical modeling of structural changes in the surfaces of workpieces with thermal perturbations in the process of grinding // Omsk Scientific Bulletin , №2-120, Omsk : Izd.vo VPO OmSTU , 2013. - pp. 95-98.

[5] V.G. Shalamov . Mathematical modeling in metal cutting : a tutorial. - Ufa : Izd YuzhUrGU, 2007 .
[6] M. B. Gordon, Methodology and some results of a study of patterns of distribution of the friction forces and tangential stresses on the contact length of the chip from the front surface of the cutter, Issues in the application of cutting fluids in machining metals, Ivanovo. 1965, pp. 41 - 44.
[7] V. F. Bobrov, Determination of stresses in the cutting part of the cutting tools, High-performance cutting in mechanical engineering, Moscow, 1966, pp. 228-233.
[8] G. S. Andreev, The stress state of the cutting part of the cutter with periodic cutting, Durability of the cutting tool, Moscow, 1969, pp. 28-37.
[9] R.Yu. Nekrasov, A.I. Starikov, A.A. Lasukov. Entering the operative correction machining processes CNC. IOP Conf. Series: Materials Science and Engineering 91 (2015) 012042 doi:10.1088/1757-899X/91/1/012042.
[10] M. F. Poletika, M. H. Uteshev, Study of the processes of cutting the polarization-optical method, Bulletin of the Tomsk Polytechnic Institute, 1964, pp. 114-115.
[11] R.Yu. Nekrasov, U.S. Putilova, A.I. Starikov, D.A. Kharitonov. Diagnostics of cut-layer deformation and rational tool loading in numerically controlled lathes. Russian engineering research, 2014. Vol. 34, No.12, pp. 826–828, Allerton Press, inc. 2014.
[12] R.Yu. Nekrasov, U.S. Putilova, I.V. Soloviev. Laser interferometry method of stress determination in loaded cutting tool parts. Applied Mechanics and Materials Vol. 698 (2015) pp. 537–541.

IOP Conf. Series: Materials Science and Engineering **142** (2016) 012076 doi:10.1088/1757-899X/142/1/012076

Comparative Structural Strength Research of Hardened Carbon Steel and Hot-Rolled Alloy Steel

A V Bogomolov, A N Zhakupov, A T Kanayev, I A Sikach, K K Tugumov

S. Toraighyrov Pavlodar State University, Kazakhstan
140008, Pavlodar, Lomov Street, 64, tel. +7(7182)687572

e-mail bogomolov71@mail.ru

Abstract. Experiments on quantitative evaluation of fatigue strength showed that St5ps and St5sp carbon steels with A400 strength class can be fully applied for erection of constructions and buildings having cyclical loads during operation. Study of corrosion resistance of hardened carbon steel in comparison with hot-rolled alloy steel consists in difference in structures and hence, difference in intensity of electric and chemical processes featuring presence of steel in concrete. Structure of St5sp steel with A400 strength class in surface area has significantly less corrosion rate than ferritic-perlitic structure of 35GS steel with A400 strength class.

1. Introduction

Technology of combined stress and thermal hardening as any other meaningful idea has its perspectives offshoots.Studying microstructure and properties of hardened reinforced and angle section we paid our attention to the fact that sometimes when hardening based on interrupted tempering with subsequent self-tempering, strength mechanical properties of ordinary low-carbon steels are better than strength characteristics of alloy steels in hot-rolled state. For example, thermal hardening of bar section made of low-carbon steel allows to obtain rolled section which conforms to the level of mechanical characteristics of 09Mn2, 09Mn2Si alloy steels in hot-rolled state. Mechanical properties of low-alloy steels of this group after thermal hardening conform to strength class which meets properties of expensive nickel-containing 10CrSiNiCu types and 14Mn2NV steels with carbonitride hardening.

Besides, thermal hardening sharply increases cold resistance and impact strength level at temperature – 60°C exceeds 0.3 MJ/ m^2 [1,2].

As methodological basis of replacement of hot-rolled low-alloy steels by hardened low-carbon steels and expensive alloy steels by thermally hardened low-alloy steels an equal strength principle is accepted combined with analysis of stressed state and operation conditions of reinforced profiles. Study of temperature-time parameters of deformation processing of thermally hardened profiles made of 35GS low-alloy steels shows that there is possibility of its replacement by St3 and St5 low-carbon steels by thermal hardening. Study of chemical composition of 35GS, St5ps and St5sp steels demonstrates that manganese content in 35GS low-alloy steels is higher by 0.25-0.35%, and silicone - by 0.50% than in St5ps and St5sp steels with close content of carbon. Such increase in manganese and silicone content leads to higher resistance of undercooled austenite within temperature range from 400 to 600 ° C and as a consequence to increase of perlite component in the structure and hardening of ferrite by means of dissolving of alloy components in it. It results in improvement of strength properties in these steels by 100 MPa in average with some reduction of its plasticity and weldability.

Experiments show that such strength improvement necessary for transformation of reinforced steel made of 35GS low-alloy steel into reinforced steel made of St5 low-carbon steel can be obtained by hardening thermal treatment without increase of manganese and silicone content [2].

Rapid development of industrial and civil construction in our country, especially in northern regions with severe climatic conditions as well as some cases of breakage of reinforced concrete constructions due to breakage of reinforced bars require to carry out study of structural strength of reinforced bars and work out additional requirements for properties.

2. Material and methods of researches

St5ps, St5sp, 35GS reinforced steels with profile diameter of 14 mm of Karaganda metallurgical complex (ArcelorMittal Temirtau) served as a material for the research. 14 mm diameter rebar was chosen based on the following factors: profile diameter of 14 is one of the popular minimum profiles which are the basis for manufacture of standard impact test sample of maximum section. Moreover, profile diameter of 14 is a middle profile out of rebar grades with diameter of 10-18 mm to be hardened up to A400 strength class. St5ps and St5sp carbon steel grades were hardened up to A-400 strength class using combined stress-thermal treatment in rolling process. Reinforced bars of hot-rolled low-alloy steels were used for comparative research. Mechanical properties ($\sigma_в$. $\sigma_т$ and others) were determined according to standard method. Chemical composition and mechanical properties of steels used for structural strength studies are showed in the tables 1 and 2 below.

Table 1. Chemical composition of studied 35GS, St5ps and St5sp steel grades,%.

Steel grade	State	C	Mn	Si	Cr	S	P
35GS	Hot-rolled	0,31	1,12	0,86	0,19	0,034	0,023
St5sp	Thermal hardening	0,34	0,70	0,21	-	0,023	0,012
St5ps	Thermal hardening	0,35	0,77	0,15	-	0,036	0,014

Table 2. Mechanical properties of studied hot-rolled and thermally hardened steels.

Steel grade	State	$\sigma_в$,, MPa	σ_T MPa	δ,%	Strength class
35GS	Hot-rolled	685	446	23,80	A400
St5sp	Thermal hardening	765	610	24,0	A400
St5ps	Thermal hardening	724	545	21,2	A400

It should be noted that evaluation of mechanical properties of metal materials shall be based on the following evaluation criteria [3].

1) Criteria determined with no reference to structural features and product service conditions.

2) Evaluation criteria of material structural strength. They feature material operation capability in operation conditions. These criteria are in full correlation with product service properties.

Strength and plastic characteristics of test samples with static loads were determined in the following range of temperatures: (+ 20°C) to (-70 °C). As stress risers contribute to brittle failure, especially at low temperatures, sensitivity of reinforced steels to presence of damages was studied. Besides, superficial damages which create stress concentrations were imitated with the help of sharp notches made on samples at the depth of about 0.05 % of profile nominal diameter with bending radius of r=0.2 mm. Given that longitudinal rib is the most protruded element of periodic profile it can be considered as the element exposed to different mechanical damages to the maximum extent. That's why 35GS reinforced steels with A400 strength class were evaluated to sensitivity and longitudinal rib of St5ps and St5sp steels with A400 strength class were evaluated to surface damages with static tension in temperature range of (+ 20° C) (- 60 °C). Notch on one of core longitudinal rib with 0.6 mm depth was made by hardness indentation with 0.15 mm spherical radius of the point.

Cold resistance of reinforced steels was studied based on test results of dynamic bending of prismatic samples with mild (r = 1 mm) notch. Tendency of steels to brittle failure was evaluated based on critical brittle point T_{crit}. determined by planimetering of viscous component in sample fracture.

Periodic profile of bar reinforcement providing sufficient cohesion with concrete creates local stress concentration in reinforcement bar and as a result it leads to reduction of bar fatigue strength. There is a significant impact of fillet radius of transverse projection with bar body on endurance limit of reinforced bars. Evaluation of endurance limit of steels studied was analyzed based on stress diagrams during cyclic tension regime.

Corrosion resistance of reinforced bars was determined not only in concrete and solution $Ca(OH)_2$ imitating liquid phase of cement brick but in environments which allow to toughen test conditions and evaluate impact of aggressive industrial environments with reference to behavior of reinforced bars (sulphates and chlorides). Impact of sulphates on corrosion of reinforced bars was determined using gravimetric method with 72-hours etching of cylindrical samples in 10% solution of sulphuric acid. Impact of chlorides on reinforced bars was determined using electrochemical characteristics of reinforced bars pressed into concrete by periodic moistening with 10 % solution of sulphuric acid. Electrochemical behavior of steel in concrete was studied using potentiometer by taking anodic and cathodic polarization curves and by registration of stationary potentials with time change.

3. Research results

Study of St5ps and St5sp carbon steel samples with A400 strength class and 35GS low-alloy steels with A400 strength class for static tension within the temperature range of (+20^{0}C) - (- 60^{0}C) showed that the values of temporary resistance and yield point rose by 50-75 MPa when temperature fell from - 25 to_- 60.

Besides, yield point for 35GS steel with A400 strength class at - 60 °C reaches 500 MPa and 620-650 MPa for St5ps and St5sp steels with A400 strength class. Temporary resistance reaches values of 750 and 790-830 MPa accordingly. The results obtained show that during tensile test of samples at low temperatures properties of St5ps, St5sp and 35GS steels with A400 strength class change in fact equally. In addition, mechanical properties required by the standards are fully provided. Therefore, it is hardly to specify advantages of any of the above-mentioned steels.

Study of sensitivity of reinforced steels to damages creating stress concentration shows that damage of periodic profile has no impact on mechanical properties both at room and low temperatures while presence of stress riser on core body, especially on its longitudinal rib, causes considerable change of properties, especially at low temperatures. These tests demonstrated that St5ps and St5sp reinforced steels of A400 strength class are low-sensitive to notches on longitudinal rib and relative elongation reduces 1.7-2.0 times for 35GS hot-rolled steels and 1.3-1.5 times for thermally hardened steels.

IOP Conf. Series: Materials Science and Engineering **142** (2016) 012076 doi:10.1088/1757-899X/142/1/012076

It is known that metals and alloys with lattice structure depending on frost fracture threshold may display brittle breakage or viscous breakage. Frost fracture threshold is not a material constant component as it is greatly depends on its structure, stress concentrations, material size etc. Knowing frost fracture threshold and working temperature of material in operation it is possible to evaluate its temperature viscosity limit, i. e. temperature range between frost fracture threshold and working temperature. The more temperature viscosity limit, the less possibility of brittle breakage. As brittle and viscous nature of breakage during impact bending of steel can be identified based on fracture type, frost fracture threshold was determined according to fiber percentage (B %) of opaque, fibrous component in fracture. Number of fibers (B) in fracture was determined as ratio of area of fibrous (viscous) fracture to primary estimated sample section. According to curve with coordinates fiber percentage- test temperature, frost fracture threshold was determined, namely, temperature value at which fracture has 50 % of fiber which corresponds to KCU/2 impact strength.

Experimental data studied signify that application of St5ps and St5sp steels with A400 strength class instead of 35GS steels with A400 strength class for erection of reinforced concrete constructions operating in low temperature conditions of northern Kazakhstan will improve their sustainability.

Study of endurance of reinforced bars made of 35GS steels rolled on right and left mill streams through passes with different output demonstrates that wear of passes and increase of fillet radius of protrusions with core body virtually has no impact on strength characteristics and plasticity during static tension but leads to reduction of endurance limit by 8-10 %.

It should be noted that removal of stress risers (mechanically turned samples) allowed to avoid their impact and endurance limits obtained on mechanically turned samples of reinforced bars with different profile geometry have equal value.

It is established that fatigue strength characteristics of St5ps and St5sp thermally hardened carbon steel are equal to the same characteristics of 35GS A400 low-alloy steel. Based on the said experimental studies it can concluded that St5ps and St5sp carbon steels with A400 strength class can be fully applied for erection of constructions and buildings having cyclical loads during operation.

35GS reinforced steel with A400 strength class is generally used as weld reinforcement for reinforced concrete constructions and has mechanical properties exceeding pre-planned estimated characteristics. However, there were cases when these constructions broke due to corrosion of reinforced steels although strength characteristics met GOST 5781 requirements. Therefore, there is a need to evaluate corrosion resistance of reinforced steels.

Reinforced steels can be regarded as resistant to stress corrosion fracture if during stresses of $0.9\sigma_T$ time to fracture exceeds 100 h. As tests based on these method showed, time to breakage of a sample made of St5 steel with A400 strength class loaded with 390 MPa amounted to 120 hours, i.e. reinforced steels with A400 strength class regarded as corrosion-resistant.

However, such evaluation of tendency to stress corrosion fracture cannot give full picture of steel corrosion resistance, especially in case of exposition to aggressive environments. It is known from practical experience that minor defects on metal surface can start damage of reinforced steels and spot corrosion reduces strength characteristics by 20-30%. Therefore, it is necessary to begin more detail study both of stress corrosion fracture and general corrosion of reinforced steels, work of galvanic couples appearing because of selective affection of metal surface by aggressive environments due to appearance of contacts between reinforced steels and aggressive environment during denuding of reinforced steels.

It is known that concrete is protective medium for reinforced steels and its corrosion starts after destruction of concrete or existence cracks in it. Protective properties depend both on chemical nature of concrete and its permeability for aggressive environments. As it was said, steel corrosion rate should be defined not only in solution $Ca(OH)_2$ imitating liquid phase of cement brick but also in environments which allow to toughen test conditions and evaluate impact of aggressive industrial environments on reinforced steel behavior, in particular, sulphates and chlorides.

Sulphates present in most natural waters as well as in industrial waste waters. Sulphate solutions cause corrosion of concrete and reinforced concrete constructions which are in direct contact with them. Chlorides present in concrete components and often are used as additives to speed up curing of concrete mix. Table 3 shows comparative data on corrosion rate of reinforced bars with diameter 14 mm with A400 strength class in different environments.

Table 3. Corrosion resistance of reinforced steel with A400 strength class.

Environment for corrosion test	Steel grade	Strength class	Corrosion rate, mm/year	Resistance grade according to the scale	Resistance category
H_2SO_4	35GS	A400	0,37	2	Resistant
	St.5sp	A400	0,30	2	Resistant
NaCl	35GS	A400	1,28	3	Low-resistant
	St.5sp	A400	0,77	2	Resistant

As it follows from the above-mentioned data, in studied environment (H_2SO_4, NaCl) index of corrosion resistance of St5sp thermally hardened steel is better than that of 35GS hot-rolled steel. It can be seen that corrosion rate in 3% solution of sodium chloride, chosen as the environment for corrosion tests, in St5sp thermally hardened carbon steel is by 60% lower than that in 35GS silicone-manganese steel.

4. Results review

Combined deformation and thermal treatment of reinforced profiles made of St5sp and St5ps carbon steels provides improvement of strength properties up to level of GOST 5781 requirements for reinforced profiles made of alloy steels with A400 strength class. It follows herefrom that instead of reinforced profiles made of alloy steels with A400 strength class it is more effective and economic to produce equally strong reinforced profiles made of ordinary St5sp and St5ps carbon steels with A400 strength class.

However, presence of equal mechanical properties ($\sigma_в$, $\sigma_т$, $\delta 5$) is not a required and sufficient basis for use of hardened carbon steels in reinforced concrete constructions instead of 35GS, 09Mn2Si alloy steels and others as strength, plastic and viscous characteristics defined by GOST 5781 neither reflect operation conditions and destruction nature during operation nor stressed state of reinforced bars of periodic profile. Therefore, based on mechanical properties defined by GOST 5781, it is actually impossible to establish which of these steels are better in real operational conditions. Hence, there are actuality and academic and research importance of study of structural strength of these steels.

Strength study at low temperatures demonstrate that during tensile tests of samples at low temperatures mechanical properties of hardened carbon and alloy steels with A400 strength class change in fact equally. In addition, values of mechanical properties required by the standards are fully provided. Therefore, it is hardly to specify advantages of any of the above-mentioned steels even at low temperatures [4].

Evaluation of sensitivity of reinforced steel with A400 strength class to surface damages shows that presence of notch does not result in considerable change of strength property values of hardened and hot-rolled steels and change nature with temperature fall from —20°C to -70°C. However, as the experiments demonstrated, during testing of samples with notches, 14 out of 15 samples made of 35GS steel with A400 strength class were broken whereas number of broken samples made of St5sp steels with A400 strength class amounted to only 7 and only 8 for St5ps steels with A400 strength class, i. e. if during testing of samples with notches virtually all samples made of hot-rolled steels were broken then during testing of samples made of thermally hardened steels only half of them were broken.

Presence of notches on samples leads to significant reduction of relative elongation of 35GS hot-rolled steels: from 24-25% to 12-15%. Relative elongation of samples made of hardened steels broken on base metal is equal to elongation of samples without notches. In cases when samples were broken because of notches there was reduction of relative elongation of St5sp steels from 20% to 15%, and to 12% for St5ps steels. According to plastic characteristics, hot-rolled steels in comparison with thermally hardened steels seem to be more sensitive to notch. Samples made of thermally hardened steels given the presence of notches broke with plastic deformation. Test data showed that St5sp and St5ps steels with A400 strength class are low-sensitive to notches on longitudinal rib. Moreover, notches on longitudinal rib cause significant reduction of relative elongation of 35GS hot-rolled steels and less significant one for steels with A400strength class.

During evaluation of structural strength of reinforced bars in reinforced concrete constructions intended for operation at low temperatures the data of dynamic impact bending tests of bar samples were used. However, during evaluation of behavior of reinforced bars at low temperatures, impact tests shall be accepted as quite conditional due to significant difference between stressed state in impact sample and in real reinforced bar.

As working reinforced steel made of reinforced concrete constructions is mainly undergone to tensile and compressing stresses, crack resistance evaluation method shall be accepted as main criterion during evaluation of reinforced steels [5].

Widespread application of reinforced steels in reinforced concrete constructions which undergone repeated loads also generates a need for study of reinforced steels for endurance. Steel sensitivity to stress concentration was studied using effective concentration coefficient which is a ratio of endurance limit of smooth sample to stress limit. While evaluating sensitivity of reinforced steels to stress concentrations it should be concluded that 35GS, St5sp and St5ps steels have similar stress concentration coefficient equal to 1.77; 1.76 and 1.74 accordingly. This is due to presence of close structures in mechanically turned samples made of 35GS hot-rolled steels and St5sp and St5ps hardened steels.

Microfractographic study of fracture surfaces showed that fatigue damages of 35GS low-alloy steels St5sp and St5ps carbon steels have equal nature. The results obtained confirm available data about the fact that the important factors featuring endurance of reinforced bars are geometric parameters of periodic profile of reinforced steel [6,7].

Corrosion is an electrochemical process because anodic (oxidizing) and cathodic (reducing) reactions are referred to processes with electrons transmission [8].

Table 4. Data on galvanic corrosion microcouples.

Galvanic corrosion microcouples	Anode (oxidant)	Cathode (reducer)
	Area with big energy	Area with less energy
Perlite	Ferrite	Cementite
Grain boundaries	Grain boundaries	Grain body
Fine and coarse grains	Fine grains	Coarse grains
Structural imperfections	Defects (dislocations, vacancy defects etc.)	Perfect structure
Deformed areas	Deformed and phase hardening areas	Annealed metal or alloy
Stress areas	Loaded areas	Non-loaded areas

Imperfect structures serve as anodes because adjacent areas have excess deformation energy and process of oxidation of atoms of these adjacent areas to ions requires less additional energy. In case of heterogeneous structures phase compositions as a rule differ greatly. For example, if ferrite composition (carbon content 0.006 %) and cement (6.67 % C) in perlite is not equal, then they have different electrode potentials and form microgalvanic couple.

Improvement of corrosion resistance of thermally hardened steel in comparison with 35GS hot-rolled low-alloy steel consists in difference of structures and hence, in difference of intensity of electric and chemical processes featuring presence of steel in concrete. As it is known, heterogeneous structures are less resistant to corrosion attack than homogenous structures as solid solutions. In heterogeneous structures this is due to the fact that matrix works as an active anode and inclusions or particles of second phase work as a cathode. Besides, cathode activity considerably depends on shape and dispersiveness of impurities. For example, laminated structures have more high resistance to corrosion attack in comparison with grain structures. Troostite structures have minimum corrosion resistance [9,10]. Apparently, it is due to significant dispersiveness of carbide constituent in troostite structures which increases number of anode-cathode microcouples. Increase of tempering temperature leading to coagulation of cementite (carbide constituent) as well as to removal of internal strains originated during quenching reduces tendency to corrode.

Structure of St5sp thermally hardened steel of A400 strength class in surface area (secondary sorbite) has far less corrosion rate than ferritic-perlitic structure of 35GS steel with A400 strength class. Besides, it should be noted that ferritic-perlitic structure of 35GS steel is more electrochemically heterogeneous and rate of anode and cathode processes which determine corrosion resistance is considerably higher than in more equilibrium structure of St5sp steel with A400 strength class having different structure at the surface area and in the core center. Thus, while studying corrosion resistance of hardened steels it should be taken into consideration the difference in microstructure based on core diameter. If there are signs of crack origin on the surface then its growth and distribution depends on corrosion resistance of layers following surface hardened ring. This poorly studied question requires further study.

Increased resistance observed to corrosion of St5sp and St5ps steels in comparison with 35GS steels shows that obtainment of structural states close to balanced ones due to thermal hardening contributes to increase of corrosion résistance.

5. Summary

1.Results obtained demonstrate that while tensile testing of samples at low temperatures, properties of hardened carbon steel and hot-rolled low-alloy steel with A400 strength class change in fact equally. In addition, mechanical properties required by the standards are fully provided. It is established that in fact damage of transversal rib of periodic profile has no impact on mechanical properties both at room and low temperatures while presence of stress riser on core body, especially on its longitudinal rib, causes considerable change of properties, especially at low temperatures.

2. Given the more high values for impact bending, thermal hardening of carbon steels provides shifting of cold-shortness curve towards low temperatures as compared to 35GS hot-rolled steel. Analysis of critical temperatures of cold resistance T_{crit} and notch sensitivity demonstrates that St5sp reinforced steel with A400 strength class has cold-shortness threshold by 60^{0}C less than that of 35GS reinforced steel of the same strength class. Application of St5sp and St5ps steel instead of 35GS steel for manufacture of reinforced concrete constructions operating in low temperature conditions of northern Kazakhstan will improve their sustainability.

3. It is established that fatigue strength values of St5sp and St5ps thermally hardened carbon steel are equal to the same characteristics of low-alloy steel of the same steel strength class. St5sp and St5ps carbon steels with A400 strength class can be fully applied for erection of constructions and buildings having cyclical loads during operation.

4. Corrosion resistance characteristics of thermally hardened steels have become better than that of 35GS hot-rolled steels. Comparative analysis demonstrates that St5sp and St5ps steels have positive potential for reinforced steel, less area of affected surface, low porosity of general corrosion and what is more important – considerably less size of corrosive holes which can initiate crack origin.

References

[1] Lenard, J.G., 2002. Metal Forming Science and Practice. Elsevier Science, pp: 378.

[2] Guo, Z.X., 2005. The Deformation and Processing of Structural Materials. Cambridge: Woodhead Publishing Limited; Boca Raton: CRC Press, pp: 331.

[3] Weng, Y., H. Dong and Y. Gan, 2011. Advanced Steels: The Recent Scenario in Steel Science and Technology Metallurgical Industry Press, Beijing and Springer-Verlag GmbH Berlin, pp: 511

[4] Doege, E., and B. Bernd-Arno, 2007. Umformtechnik. Grundlagen, Technologien, Maschinen. Springer Berlin Heidelberg New York, pp: 913

[5] Lin, J., D. Balint and M. Pietrzyk, 2012. Microstructure evolution in metal forming processes. Cambridge: Woodhead Publishing, pp: 402.

[6] Brandon, D. and Kaplan, W.D, 2008.Microstructural Characterization of Materials. John Wiley & Sons Ltd., Chichester, England. pp: 536.

[7] Saravanan, R. and M.P. Rani, 2012. Metal and Alloy Bonding: An Experimental Analysis. Charge Density in Metals and Alloys. London: Springer, pp:151.

[8] Flewitt, P.E.J. and R.K.Wild, 2003. Physical Methods for Materials Characterisation, Taylor & Francis, pp: 602 .

[9] Kanayev A.T. and A.V. Bogomolov, 2015. Assessment of efficiency of mechanisms of hardening ferritic-pearlitic steels. Cambridge Journal of Education and Science, Cambridge University Press,, Vol.2(14) : 493-499.

[10] Kanaev, A. T., E. N. Reshotkina and A.V. Bogomolov, 2010. Defects and Thermal Hardening of Reinforcement Rolled from Continuous Cast Billet. Steel in Translation, Vol. 40(8): 586–589.

Stability and Process of Destruction of Compressed Plate of Layered Composite Materials With Defects

L A Bokhoeva[1a], V E Rogov[2], A S Chermoshentseva[3], D V Lobanov[4]

[1]East Siberia State University of technology and management, Ulan-Ude, Kluchevskaya str., 40v, 670013, Russia
[2]Baikal Institute of Nature Management Siberian Branch of the Russian Academy of Sciences, Ulan-Ude, Sakhyanovoy str., 6, 670047, Russia
[3]Bauman Moscow State Technical University, Moscow, Baumanskaya 2-ya str., 5/1, 105005, Russia
[4]Bratsk State University, 40, Makarenko St., Irkutsk region, Bratsk, 665709, Russia

e-mail: [a]bohoeva@yandex.ru

Abstract. Interlayer defects in composite materials are a pressing problem, which affecting their performance characteristics. In this research, we considered the problem of the stability and of the fracture process of the compressed thin plate made of laminated composite materials with the interlayer defects. In this research we had got a critical equation for a plate with interlayer defect. The experiment showed the effect and the quantity of nano-dispersed additives on the mechanical properties of composite materials with interlayer defects.

Introduction

Application of composite materials in aircraft and aerospace industries has led to the need to take into account new factors, which affecting their performance characteristics, such as non-linear nature of the deformation; the risk of fracture along the interface between the layers - defects type of delamination [1-10]. In this research, we considered the problem of the stability and of the fracture process of the compressed thin plate made of laminated composite materials with the interlayer defects. Experimentally, we found the effect of nano-dispersed additives on the mechanical properties of composite materials with interlayer defects.

In this research we consider the element construction in the form of a homogeneous plate. This plate has the thickness H, one the width and the length L. The plate has only one thin of delamination, which has the thickness h (h<<H), the length l. The axial force P acts on the plate. The ends of the plate are rigidly fixed (Figure 1).

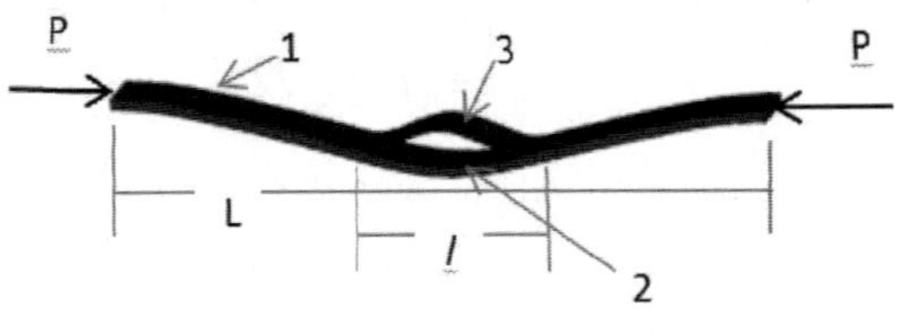

Figure 1. The plate with only one thin of delamination

Different parts of the plate (defect - 3, the lower part of the defect - 2, base plate - 1) are described by differential equations:

$$D^i_{j,k}\frac{d^4 y_i}{dx_i^4}+P_i\frac{d^2 y_i}{dx_i^2}=0 \quad (1)$$

where i = 1,2,3; $P_i, D^i_{j,s}$ - the load and the stiffness at bend i-th part. The stiffness at bend of laminated composite with layers deferent orientation depends from the orientation angles of the fibers in the layers, from the relative content of different orientations of the layers, from the arrangement of layers by thickness of the composite. The calculation of stiffness for each part can be approximately represented in the form:

$$D^i_{j,s}=\sum_{m=1}^{n} C^m_{j,s}\delta_m\left[\frac{\delta_m^2}{12}+\left(\chi_0-\chi_m+\frac{\delta_m}{2}\right)^2\right],$$

where $C^m_{j,s}$ - the characteristics stiffness of m-th layer, dependent on the elastic modulus, shear modulus, Poisson's ratio and the orientation angle of the unidirectional fiber layer $E_1, E_2, G_{1,2}, \mu_{12}, \mu_{21}, \varphi$; χ_0 - the distance from the edge of the area to the neutral line; χ_m - the distance from the edge of the area to the $(m+1)$ layer; n - the quantity of layers of i-th part; δ_m - the thickness m-th layer i-th part; j,s=1,2,6.

The expression is the solution of equation (1) in the first part:

$$y_1=\frac{\theta}{k_1\sin[k_1 b]}\left(1-\cos[k_1(L/2-x)]\right),$$

the second part $0\le x\le l/2$

$$y_2=\frac{\theta}{k_1\sin(k_1\cdot b)}\left\{\frac{k_1\sin(k_1\cdot b)}{k_2\sin(k_2\cdot l/2)}[\cos(k_2\cdot x)-\cos(k_2\cdot l/2)]+1-\cos(k_1\cdot b)\right\},$$

the third part $0\le x\le l/2$

$$y_3=\frac{\theta}{k_1\sin(k_1\cdot b)}\left\{\frac{k_1\sin(k_1\cdot b)}{k_3\sin(k_3\cdot l/2)}[\cos(k_3\cdot x)-\cos(k_3\cdot l/2)]+1-\cos(k_1\cdot b)\right\},$$

where θ - the angle of rotation at the start of the defect; $b=L/2-l/2$; $k_i^2=\frac{P_i}{D^i_{j,s}}$.

Of the three conditions of compatibility we had obtained the equation for determining the critical load:

$$\kappa_{\kappa p}ctg(k_{kp}\cdot b)+(1-\bar{h})^2 k_{\kappa p}ctg\left(\frac{k_{\kappa p}\cdot l/2}{1-\bar{h}}\right)+\bar{h}^2 k_{\kappa p}ctg\left(\frac{k_{\kappa p}\cdot l/2}{\bar{h}}\right)+\frac{3\cdot\bar{h}\cdot(1-\bar{h})}{l/2}=0,$$

where $\bar{h}=h/H$; $k_{kp}^2=\frac{P_{1kp}}{D^1_{j,s}}$.

For research the behavior of postcritical plates with interlayer defect is necessary to solve a system of nonlinear equations:

the equation equilibrium of forces

$$\varepsilon_1=\bar{h}\cdot\varepsilon_3+\left(1-\bar{h}\right)\cdot\varepsilon_2;$$

the equation equilibrium of moments

$$D^1_{j,s}\theta\cdot k_1\cdot ctg(k_1\cdot b)+D^2_{j,s}\theta\cdot k_1\cdot ctg(k_1\frac{l}{2})+D^3_{j,s}\theta\cdot k_3\cdot ctg(k_3\frac{l}{2})=G_{x,y}h(H-h)(\varepsilon_2-\varepsilon_3);$$

the equation compatibility of the change the length of the upper and lower parts of the defective part

$$\varepsilon_3 l-\frac{l\cdot\theta^2[\cos(k_3\cdot l/2)\sin(k_3\cdot l/2)-k_3\cdot l/2]}{4\cdot k_3\cdot l/2\cdot\sin^2(k_3\cdot l/2)}=\varepsilon_2 l-\frac{l\cdot\theta^2[\cos(k_2\cdot l/2)\sin(k_2\cdot l/2)-k_2\cdot l/2]}{4\cdot k_2\cdot l/2\cdot\sin^2(k_2\cdot l/2)}+H\cdot\theta$$

the equation compatibility of the change of the length of the entire plate

$$\varepsilon_0 L=2\varepsilon_1 b+\frac{\theta}{4\cdot k_1}\left[-2ctg(k_1\cdot b/2)+\frac{1}{\sin(k_1\cdot b/2)\cos(k_1\cdot b/2)}+\frac{k_1\cdot b/2}{\sin^2(k_1\cdot b/2)\cos^2(k_1\cdot b/2)}\right]+$$
$$+\varepsilon_2 l-\frac{l\theta^2[\cos(k_2\cdot l/2)\sin(k_2\cdot l/2)-k_2\cdot l/2]}{4\cdot k_2\cdot l/2\cdot\sin^2(k_2\cdot l/2)}+h\theta$$
;

where ε_i - the relative deformation i-th part; $G_{x,y}$ - the shear modulus of the composite. The system of equations holds for $P > P_{kp}$.

To assess the potential growth the defect of delamination, we used Griffith energy criterion. Griffith energy criterion can evaluate the intensity of the release of the elastic energy at the crack tip by the value of axial forces and bending moments, which act in various cross-section. Various cross-section adjacent to the top of the delamination. Postcritical deformation was affecting on the characteristics of the energy prior to the growth of defect. We obtained an analytical expression for the energy release rate.

A number of theories describe a change of characteristics of polymers as a result of introduction of a disperse phase in them. One way of strengthening the interlayer composite materials with defects is the introduction in a matrix the fine particles of submicron size, in the process of forming [11-12]. In such the dispersion-strengthened materials the strength of the particles themselves is not used. These particles serve as a barrier to crack propagation in the matrix. Thus, these particles strengthen the initial phase.

The experiment

For experimental research the control samples (CS) were fabricated without nano-additives, but with the included defect between the layers (10 samples). This paper presents the study of composite materials (CM). Their mechanical properties change depending on the degree of their hydrophobic powder (Tarkosil T-20) filling. As the result we have received the CM test samples with the underlying interlayer defects with the addition of binder nanosized particles (silicone dioxide) with different volumetric concentrations from 0.1% to 0.5%. For research, 30 samples were made with the different mass concentration: C = 0.1%; C = 0.3%; C = 0.5%. For a comparative analysis the control samples (CS) were fabricated without nano-additives, but with the included defect between the layers (10 samples). As nano-modifying additives PCM we chose the nanosized silicone dioxide powder Tarkosil - 20 (T-20). Teflon film was embedded in the laminate. It is simulate the presence of a delamination. Teflon film thickness is 10-100 microns. The data for the samples is given in Table 1.

Table 1. The data for the samples – plate with a through defect

#	Name of nano-particles	Mass concentration of nano-particles, %	L(mm) length of sample	H(mm) thickness of sample	l(mm) length of defect	h(mm) thickness of defect	Number of layers sample	Number of layers in the defective portion	The angle of placing filaments
1	T-20	0.1	230	2	60	0.4	10	2	$(45,-45)_5$
2	T-20	0.3	230	2	60	0.4	10	2	$(45,-45)_5$
3	T-20	0.5	230	2	60	0.4	10	2	$(45,-45)_5$
4	CS	-	230	2	60	0.4	10	2	$(45,-45)_5$

The samples were fabricated of twenty-ply pre-preg (fiberglass or carbon fiber). The industrial grade glass is T-25 (VM) specification 6-11-380-76. The unidirectional material has the following characteristics: fiberglass - $E_1 = 5.4 \cdot 10^4$ MPa, $E_2 = 1.2 \cdot 10^4$ MPa, $G_{12} = 0.5 \cdot 10^4$ MPa, $\mu = 0.28$. In order to determine the tensile strength of the samples we used an electromechanical testing machine Instron 5982 (Figure 2). The loading process is performed according to the automatic locking force applied to the sample P on the axial displacement. Continuous compression of the sample was continued until complete delamination layer from the rest of the defective part or the complete destruction of the sample.

Figure 2. The stand for testing

Some experimental results are shown in Table 2: the critical load of the defective area (P_1), the critical load of the defective plate (P_2) and the load during the fracture of the plate (P_3). We received a graphical the load dependence on the sample displacement (Figure 3).

Table 2. The results of experimental research

Type of sample	P_1 (H)	P_2 (H)	P_3 (H)
CS	486.7	458.3	140
CM with the addition T-20 in mass % 0.3	601.7	535.25	268
CM with the addition T-20 in mass % 0.5	576.5	569.25	283.3
CM with the addition T-20 in mass % 0.1	486.9	391	240

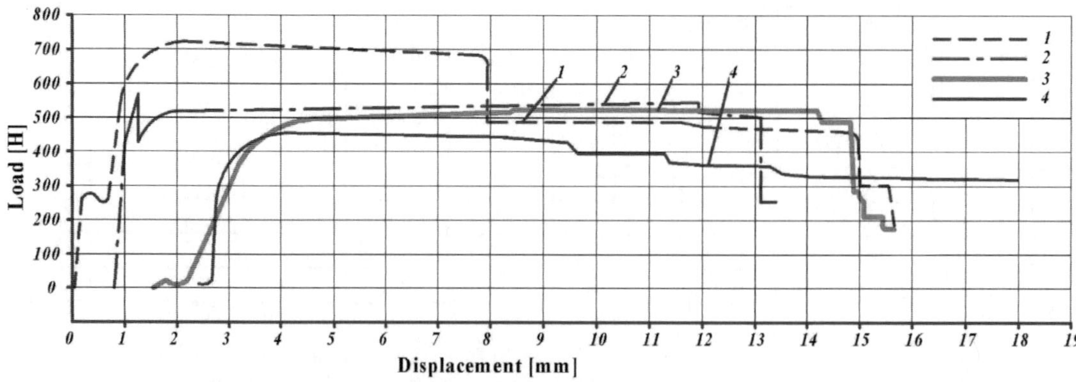

Figure 3. The load versus displacement of composite materials. 1. the composite material with the addition of Tarkosil - 20 in mass % 0.3, 2. the composite material with the addition of Tarkosil - 20 in mass % 0.5, 3. the composite material with the addition of Tarkosil - 20 in mass % 0.1, 4. the control samples

Figure 3 shows: 1. the composite material with the addition of Tarkosil - 20 in mass % 0.3, 2. the composite material with the addition of Tarkosil - 20 in mass % 0.5, 3. the composite material with the addition of Tarkosil - 20 in mass % 0.1, 4. the control samples.

Conclusion

1. In this research we had got a critical equation for a plate with interlayer defect.

2. The experiment showed the effect and the quantity of nano-dispersed additives on the mechanical properties of composite materials with interlayer defects. The analysis of the data showed that the introduction of powder Tarkosil in CM with defects leads to the change in the strength of the mixture made on their basis. The best result of P3 as compared to the control samples was detected in the samples with nanomaterials with T-20 at the concentration of 0.3%.

This research will be used to meet the challenges of strengthening the interlayer defects in laminated composites, the appearance of which is the result of imperfections in manufacturing technology and the effects of operational loads, due to the addition of the nanoparticles in the binder.

Reference

[1] Alfutov N A Essentials of calculation on the stability of elastic systems *Moscow: Engineering* (1991) p 311
[2] Alfutov N A, Zinoviev P A and Popov B G Calculation of multilayer plates and shells made of composite materials *Moscow: Engineering* (1984) p 263
[3] Bolotin V V Delaminations in composite structures its origin, buckling, growth and stability *J. Composites. Part B.* 3 (1996) pp 129-145
[4] Bolotin V V, Zebelyan Z H and Kurzin L A Stability of compressed elements with defects type of delaminations *J. Problems of strength* 7 (1980) pp 3-8
[5] Bolotin V V Defects type of delaminations in constructions of composite materials *J. Mechanics of Composite Materials* 2 (1984) pp 239-256
[6] Bolotin V V and Novikov Yu N The mechanics of multilayered structures *Moscow: Engineering* (1980) p 375
[7] Yin W L, Sallam S N and Simitsees C J Ultimate axial load capacity of a delamindtes bam-plate *J. AIAA* vol 24 1 (1986) pp 123-128
[8] Burt C W and Martindale D L Exact simplified method for research of behavior of thin plates at large deflections *J. Aerospace Engineering* 11 (1988) pp 131-138

[9] Bokhoeva L A, Rogov V E and Chermoshentseva A S Stability of circular defects such as delaminations in elements of construction at condition of shear *J. Modern technologies. System analysis. Modeling* 4 (2014) pp 19-22
[10] Bokhoeva L A, Rogov V E and Chermoshentseva A S Determination of critical loads using the energy criterion of stability for the local circular defect *J. Systems. Methods. Technologies* 4 (2014) pp 32-37
[11] Brusentseva T, Zobov K, Filipov A, Bazarova D, Lkhasaranov S, Chermoshentseva A and Syzrantsev V Introduction of nanopowders and mechanical properties of the materials on epoxy resins *J. Nanoindustry* 3 (2013) pp 24-29
[12] Chermoshentseva A S, Pokrovskiy A M and Bokhoeva L A The behavior of delaminations in composite materials – experimental results *IOP Conference Series: Materials Science and Engineering* vol 116 (2016) p. 012005

Research of Residual Stress After Deformation and Annealing of a Titanium Alloy Ti-5Al-5Mo-5V-1Cr-1Fe Billets

A V Balaykin[1], E A Nosova[1], N V Galkina[1,2]

[1]Samara State Aerospace University, 34, Moskovskoye shosse, Samara, Russia, 443086

[2]E-mail: nata12_92@mail.ru

Abstract. Research of opening angle changes in bent billets of Ti-5Al-5Mo-5V-1Cr-1Fe alloy after bending and annealing at temperature 800°C for 1, 2, 3, 4, 5 hours has been madeto improve the accuracy of the titanium partsmanufacture. The change of surface residual micro- and macrostrains via X-ray analysis was found depending on the duration of annealingduration. It was found that the value of the opening angle between the legs bent billets is defined by the influence of the level of residual microstress. To maintain the geometry of the bent billets it is recommended to anneal them at a temperature of 800°C for 2 hours.

1. Introduction

Residual stress appears in the formed products during manufacturing operations and material processing, such as molding, welding, forging, rolling, stamping, machining and a heat treatment. The occurrence of residual stress in most cases is caused by the following factors: the heterogeneity of a plastic deformation in cross sections of details, uneven temperature distribution in terms of products, irregularity of processes of phase, the structural and physic-chemical transformations in the product material [1].

Typically, residual stresses play a negative role. It is found that the residual stresses decrease the strength of products under varying cycle loads, affect friction wear, stimulates corrosion processes [2 - 4]. There are enough examples of destruction caused by significant technological residual stresses [5 - 9].

At present, there are many methods for measurement of residual stress based on partial damage of the measured product (drilling holes and measuring surface deformation with using special sensors). The most accurate and reliable method for determining residual stresses on the cylindrical surface has been proposed by Sachs. It is based on the product bore in layers of desired value and subsequent measurement of the axial and radial deformation. Further, according to the formulas of theElasticity Theory, a calculation of the principal stresses in the cylinder is implemented. However, this method is very complex, time-consuming and cannot be implemented for thin workpieces and products used in the construction of aircraft engines.

X-ray methods [10] for the study of residual stresses on the surface of parts of any shape that can be used to analyze the structure of thin-walled objects. They allow studying macro-, micro stresses on a 0.5 mm depth. Therefore the X-ray method is available for a thin body part.

In the manufacture of parts and components made of titanium alloys all kinds of machined castings and forgings are used: turning, milling, drilling, grinding, and polishing. From the one hand, these techniques provide a high accuracy of products, thereby reducing the vibration and noise in aircraft engine parts [11-12]. On the other hand, when the decompensation occurs during machining and relaxation of internal and surface residual stresses appears, leading to a distortion of workpieces configuration.

An important feature of the machining of titanium alloys is a need to ensure the resource, especially fatigue, features that are to a great extent caused by the quality of the surface layer formed after cutting treatment.

In mechanical treatment of cutting, when the tool removes a part of the workpiece surface (processing allowance), the material undergoes a significant structure changes and redistribution of internal stresses. The cutting tool is also the source of the stress, because tool heats and deforms a material in a treatment zone. This residual macrostress may have a significant impact on the quality of the finished part, causing warping of the workpiece or cracks on the materialsurface. The ability of residual stresses is more dangerous in the reason stress relaxation occur over time - when processing is already finished, the mechanism's detail suddenly begins to wear out quickly due to the exceeding of size over limits.

Alloy Ti-5Al-5Mo-5V-1Cr-1Feis intended for highly loaded structures and components, continuously operating at temperatures of 350 - 400 ° C. Force parts, parts of management system, fasteners of force bolts are produced of it. The process of manufacturing of parts made of Ti-5Al-5Mo-5V-1Cr-1Fe includes stockpiling manufacturing (casting, forging), annealing, machining [13-14].

The aim of this research was to investigate the change in configuration and level of surface macro- and microstress in flat workpieces of titanium alloy Ti-5Al-5Mo-5V-1Cr-1Fe after bending and heating.

2. Materials and Methods

Samples of size 2x10x30 mm were cut from a commerce rod (AMTU 451-67) by spark erosion, which ensures the structure and properties savingas in the original piece. The chemical composition of the alloy in accordance with Russian Standard OST 90013-81 is presented in Table 1.

Table 1. Chemical composition of Ti-5Al-5Mo-5V-1Cr-1Fe, mass percentage.

Fe	C	Si	Cr	Mo	V	N	Ti	Al	Zr	O	H
0.5 – 1.5	Up to 0.1	Up to 0.15	0.5 - 2	4 – 5.5	4 – 5.5	Up to 0.05	78.485 – 86.6	4.4 – 5.9	Up to 0.3	Up to 0.2	Up to 0.015

After cutting samples were subjected to bending deformation in a tool die with an apex angle of 90 deg. and fillet radius of 10 mm (die scheme is shown in Figure 1). After bending, the samples were annealed in laboratory furnace at 800°C for 1, 2, 3, 4 and 5 hours.

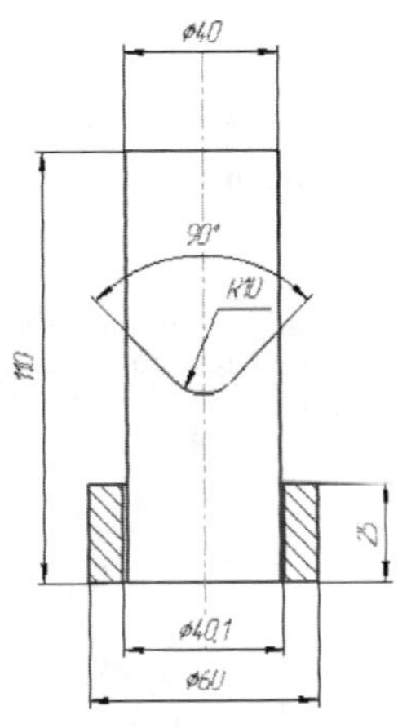

Figure 1. Scheme of the instrumental die.

Change of the geometry of the samples was determined by the change in the angle between the workpiece shelves after annealing. The measurements were made using the software product KOMPAS-3D. Disclosure angles of the scanned samples were measured before and after annealing. The measuring principle is shown in Figure 2.

IOP Conf. Series: Materials Science and Engineering **142** (2016) 012078 doi:10.1088/1757-899X/142/1/012078

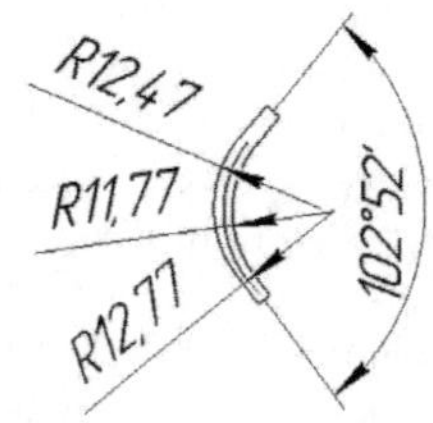

Figure 2. Scheme of measurement of the angle-disclosure.

Residual macro- and microstresses were studied using X-ray diffraction analysis for the convex and concave sides of the samples using focusing by Bragg - Brentano. According to [15], the residual stress was considered for the plane (211), a middle plane since secondary interference lines have the minimum error.

3. Results

According to the calculation results, we noted residual macro and micro stresses curves depending on annealing time of bent samples (Figure 3 and 4).

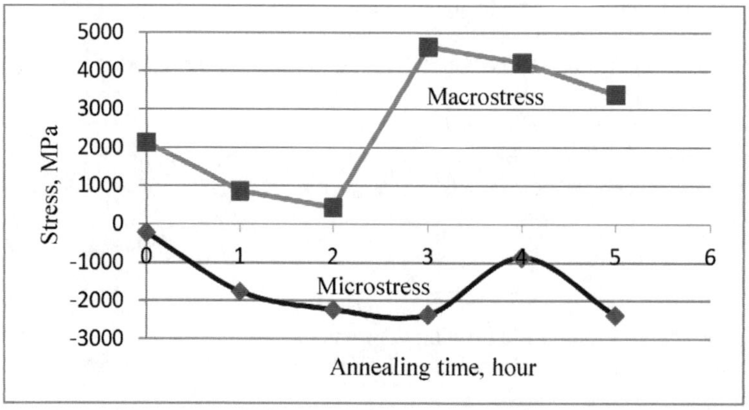

Figure 3. Dependence of residual macro- and micro stresses of annealing time in the outside (stretching zone).

Figure 3 analysis shows that annealing duration less two hours reduces macrostresses nearly 5 times. Longer duration causes a sharp increase in macrostresses almost 10 times. Then with increase of annealing time up to five hours macrostresses were reduced by 20%. It should be noted that annealing does not change macrostresses kind - they remain positive (or stretching). Microstresses have negative values, indicating their compressive origin. They decrease in modulus with increasing of annealing time up to three hours at about 11 times, then decrease in absolute value to 64%. Further, with increase of annealing time microstresses also increase.

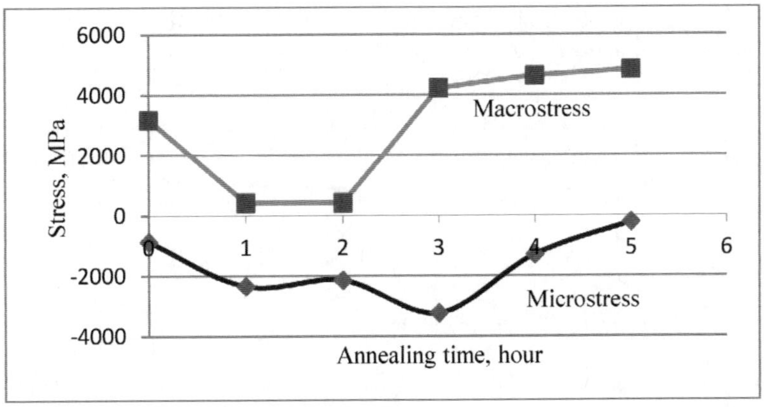

Figure 4. Dependence of macro- and micro residual stress of annealing time in the inner side (compression zone).

Figure 4 shows that annealing up to two hours reduces the macrostresses approximately 7.5 times compared to deformed state. Longer durationcauses a sharp increase in the macrostresses almost 11 times. Micro stress decrease in modulus with increasing annealing time up to three hours to approximately 73%, then reduced modulo 15 times.

Comparison of Figures 3 and 4 shows that the macro- and microstress are higher in the compression zone than in the stretch zone. Macrostresses reducing at the initial stages of annealing at period up to 2 hours can be explained by a decrease in defects in the structure due to polygonise processes, emerging high angle boundaries and division of separate grains into subgrains. Microstresses that are balanced within a few grains, increase due to the emergence of the dislocation walls. Start of a recrystallization process is accompanied by the appearing and growth of new grains with a minimum distortion, which is accompanied by a decrease in microstrain. Perhaps, in this case, the driving force of grain boundaries is the residual microtension. Increase of macro stresses during annealing for more than three hours can be explained by the appearance of fine-grained structure as a result of the primary recrystallization, the length of grain boundaries increases, which causes an increase in residual macrostresses.

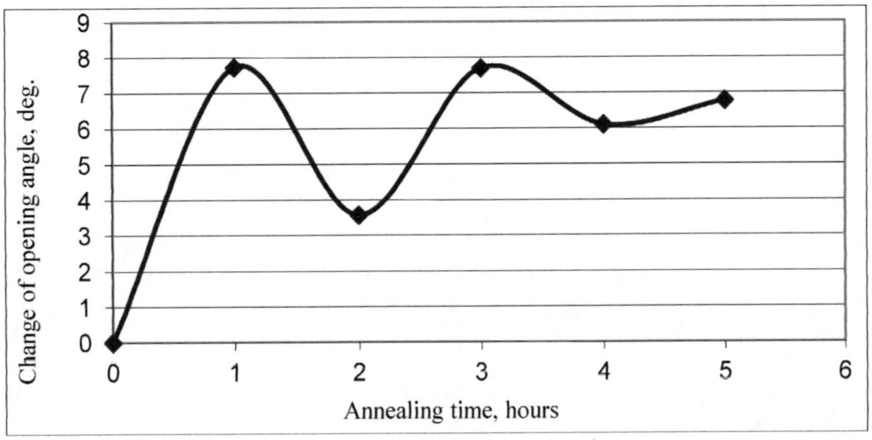

Figure 5. The relative change in the angle of opening between the shelves of bent billets.

Figure 5 shows that the angle after annealing increases. Thus, increasing the annealing time leads to the opening of angle between the bent workpiecesshelves. But this change is nomonotonic, as in the case for a change in the residual macro- and microstresses. It is possible that the change of geometry samples affects the driving force depending on the level of macro- and microstresses.

To determine the influence of the residual macro and micro stresses on the change of the opening angle between the legs of bent billets diagrams were constructed. Since stress is symmetrical in compressive and strain zones relatively neutral layer, angle change curves were constructed according the middle arithmetic valuesof residual stress for strain and compressive zones and shown in Figure 6 and 7.

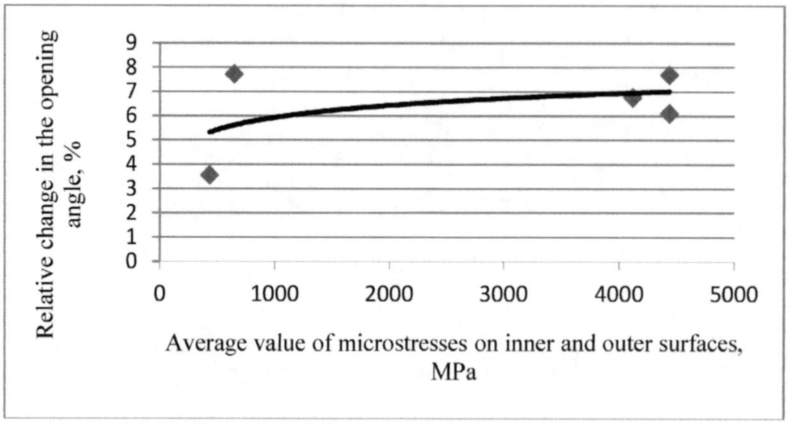

Figure 6. Dependence of the relative change in the angle of opening of the average value of macrostress.

Figure 6 shows that the increase in the average value of macro stress from 500 MPa to 4500 MPa increases the opening angle more than 2 times. But a clear correlation between the change of the macrostresses and the angle of opening billets between shelves is not observed.

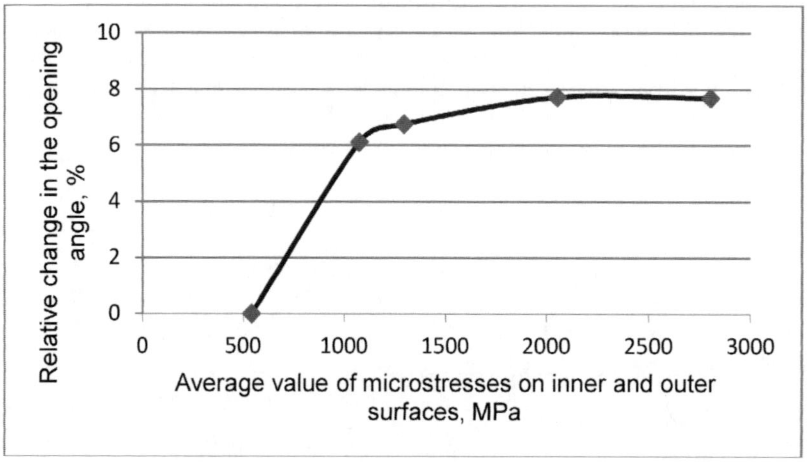

Figure 7. Dependence of the relative change in the angle of opening of the average value of microstress.

Figure 7 shows that an increase in the average value of microstresses from 500 to 3000 MPa results in an increase of more than 8 times in the relative change of angle. At the same time there is a steady monotonic dependence of the relative change of the opening angle between the legs of the micro stress values.

4. Conclusions

1. It was established that during bending samples of alloy Ti-5Al-5Mo-5V-1Cr-1Fe macrostresses are formed in the surface layers, which have a positive sign or stretching character, and microtension have a negative contractive character. Changes in the macro and micro stress on the outer and inner sides of the bent workpieces have similar behavior patterns, but their level of compression zone exceeds similar values measured in the stretch zone.

2. An increase in geometry change is observedwith the increase of macro- and microstress. The driving force of recrystallization processes in the bent billets after heating at 800°C is the residual micro stress.

3. For minimal warpage, shape preserving of cold-deformed bent workpieces during subsequent heating as a result of thermal or mechanical treatment they must be subjected to annealing at a temperature of 800°C for 2 hours. Thus minimum values of macro stresses and opening angle between the legs of bent billets are achieved.

References

[1] KielS. 1992 Experimental determination of residual stresses with the ring-core method and an online measuring system. Exp Tech; 16(5): 17–24.

[2] Zhandano I.M., Gonchar A.K. 1978 Determining the residual welding stresses at depth in metals. Automatic Welding; 31: 22–24.

[3] Totten G. 2005 Handbook on residual stress. SEM, Bethel, ISBN: 978-0871707291; 1: 417.

[4] Makhmutov N.A. 1973 The resistance of structural elements to brittle fracture.M.: Nauka, - 200c.

[5] Newman R.C., Whitney W.R. 2001 Award Lecture: “Understanding the Corrosion of Stainless Steel”, Corrosion, Vol.57, No. 12, pp 1030-104, NACEInternational.

[6] Steklov O.I. 1990 Resistant of materials and constructions to corrosion under stress. -M .: Engineering, 1990. 384 p.21.

[7] Ereklintsev V.V. Pashkov Y.I, Stepanenko A.I. 1990 Analysis of the damage of gas pipes and ways to improve the quality of the pipes. M .: Metallurgy, S. 35-42.

[8] Rosenstein I.M. 1995 Accidents and reliability of steel tanks. M .: Nedra, 254s.

[9] Brennan F.P., Ngiam S.S. and LeeC.W. 2008 An Experimental and Analytical study of Fatigue Crack Shape Control by Cold Working, Engineering Fracture Mechanics, Volume 75, Issues 3-4, Pages 355-363

[10] Gorelik S.S., Skakov Y.A., Rastorguev L.N. 2002 X-ray and electron-optical analysis. Moscow.

[11] Makaryants G.M., Prokofiev A.B.,Shakhmatov E.V.2015 Vibroacoustics Analysis of Punching Machine Hydraulic Piping // 2nd International Conference on Dynamics and Vibroacoustics of Machines, DVM 2014; P. 17-26

[12] Murzin S.P., Shakhmatov E.V. 2015 A Study of Vibration Characteristics and Determination of the Conditions of Nanopores Formation in Metallic Materials during Laser Action // 2nd International Conference on Dynamics and Vibroacoustics of Machines, DVM 2014; P. 266-271

[13] Smelov V.G., Sotov A.V., Agapovichev A.V. 2016 «Recovery technology features of aerospace parts by layering synthesis», Key Engineering Materials Vol. 684 pp 316-322

[14] Vdovin R.A. and Smelov V.G. 2014 Elaboration of a casting defects prediction technique via use of computer-aided design systems. International Journal of Engineering and Technology (IJET), V. 6 Issue 5, pp 2269-75.

[15] Zhuravel L.V. 2012 Radiography metals: laboratory practical / L.V. Zhuravlev, SA Nefedov. - Samara: Publishing House "Samara University", 76 p.

Formation Wear Resistant Coatings on Martensite Steel Hardox 450 by Welding Methods

S V Konovalov[1*], V E Kormyshev[1], S A Nevskii[1], Yu F Ivanov[2,3] and V E Gromov[1]

[1]Siberian State Industrial University, Novokuznetsk, 42 Kirov Street, 654007, Russia
[2]Institute of High Current Electronics Siberian Branch of Russian Academy of Science, Tomsk, 2/3 Akademicheskii av., 634034, Russia
[3]National Research Tomsk Polytechnic University, Tomsk, 30 Lenina av., 634050, Russia

[*]E-mail: konovserg@gmail.com

Abstract. By methods of modern physical materials science the investigations analysis of phase composition, defect substructure, mechanical and tribological properties of Cr-Nb-C-V containing coatings formed in surfacing on martensitic wear resistant steel Hardox 450 were carried out. It was shown that surfacing resulted in the formation of high strength surface layer 6 mm in thinness. This layer had wear resistance 138 times greater than that of the base and friction coefficient 2.5 times less. Having analyzed the deflected mode of the deposited material in conditions of friction, a conclusion is drawn that plastic shear deformation is localized in the surface layer due to the high friction factor, as the result, scratches are formed. The maximum of tangential stress is deflected deep into the material provided that friction factor is low. On the basis of the investigations by methods of X-ray structural analysis and transmission diffraction electron microscopy it was shown that increase strength and tribological properties of surfacing metal were caused by its phase composition and state of defect substructure, namely, availability of interstitial phases (more than 36%) and martensitic type of α-phase structure.

1. Introduction

At present surface and subsurface layers of materials are to meet a lot of requirements because of growing production of high-duty products [1]. These products are used in conditions of enormous loading the surface layers, where most defects are detected. Such processes as wear, corrosion, crack initiation and propagation are the most intense in surface layers; that is the major cause of product performance degradation. Therefore, a surface layer should possess certain characteristics (high hardness and wear resistance, compressive residual stresses etc.). As the consequence, its modification via developing a gradient structure is required [2]. There are some procedures available to synthesize a gradient structure, e.g. electro explosive alloying and electron beam treatment [3], plasma processing [4], various types of weld deposition [5 – 8], laser processing [9]. Weld deposition of wear and corrosion resistant materials is easy-to-implement [7]. This procedure makes it possible to process significant areas of material; that is particularly important for large-dimensional products.

Flux cored wires are applied frequently alongside with electrodes for producing wear resistant deposited layers [10]. However, their wide application is complicated because of insufficient information on the structure, phase composition and hardening mechanisms of the deposited metal in diverse operational conditions. Let us note that the structure and phase composition of material deposited on steel Hardox 400 by electric-arc method have already been studied in our papers [11, 12]. We have found out that secondary phases lead to strengthening, but no detailed consideration of its mechanisms has been provided yet.

Therefore, this paper is focused on the study of synthesizing nanostructural wear resistant coatings and identifying mechanisms of their strengthening.

2. Materials and Methods

Steel Hardox 450 (0.19-0.26 %C, 0.70% Si, 1.60% Mn, 0.25% Cr, 0.25% Ni, 0.25% Mo, 0.004% B, 0.025% P, 0.010% S (Wt. %, ferrum is Bal.))) is used as the base material, weld deposition (1.4 %C, 1.0 %V, 7.0 % Cr, 8.0 % Nb, 1.2 % W(Wt. %, ferrum is Bal.)) is carried out in the shielding gas atmosphere, containing Ar – 98 %, CO_2 – 2 %; welding current is 250–300 A, and welding arc voltage is 30–35 V [11].

X-ray structural analysis and transmission diffraction electron microscopy are used when studying phase composition and defect substructure of the deposited metal. X-ray structural analysis is carried out by difractometers DRON-7 and Shimadzu XRD 6000 (Bragg-Brentano geometry, cobalt and copper k_α-radiation). To determine phase composition, primary and secondary stresses we rely on methods, outlined in [13]. Microscope EM-125 is used for implementation of transmission electron microscopy analysis. The microstructure is analyzed according to the procedure in [14]. The foils are produced via electrolytic thinning of samples, cut out of the deposited metal, of the deposited metal – steel contact area, and of the steel spaced 15 mm from the area of contact. Mechanical properties of the deposited metal and steel are characterized regarding to microhardness [15] (HVS-1000A). Tribological characteristics of the deposited metal are identified via measuring wear resistance and friction factor (tribotester Tribotechnic). Wear is tested in the following conditions: a ball 3 mm in diameter, made of hard Ti (VK8)-based alloy is used as a counterbody. The counterbody is moved along the circle 4 mm in diameter on the surface of the sample at a linear speed of 2 cm/s and basic load 5 N. The total number of counterbody revolutions is 5000. Wear factor V is calculated as the ratio of the cut out material volume [16] to the sample load and friction length (cross-sectional area of wear groove, as well), and friction factor μ.

3. Results and discussion

Figure 1 shows the results of microhardness measurement. Analyzing the data, presented in this figure, it should be noted that a high-strength 6 mm thick surface layer is synthesized under weld deposition; and microhardness of this layer varies from 9.5 to 11.5 GPa. Microhardness of the material diminishes fast in a direction away from the deposited metal surface, reaching a level of 6.5 GPa. Therefore, the deposited layer is twice as hard as the base metal, provided that thickness of the deposited layer is ≈6 mm. Wear factor of the deposited layer is $V = 95.1 \cdot 10^{-6}$ mm^3/N·m, and that of steel is $V = 0.69 \cdot 10^{-6}$ mm^3/N·m as tribological testing the base and deposited metal demonstrates. Friction factors are 0.259 and 0.104 for the deposited metal and base, respectively.

In view of up-to-date physical mesomechanics [17] one should look for an explanation of improving strength properties of the deposited metal on various scale levels of the structure. On a macroscopic level it can be explained due to the changing deflected mode. The approximation below is used to calculate it: normal $\sigma_{yy}(x, y)$ and tangential stresses $\tau_{xy}(x, y)$ are applied to section $-l \le x \le l$, $y = 0$ of an elastic semi-plane $-\infty \le x \le \infty$, $y \le 0$; here $\sigma_{yy}(x,0) = l^2 - x^2$, $\tau_{xy}(x,0) = \mu \cdot \sigma_{yy}(x,0)$, where μ – friction factor. Kolosov and Muskheshvili method is taken into account to solve the problem [18]. Initiation of plastic flow in conditions of a sliding contact is

IOP Conf. Series: Materials Science and Engineering **142** (2016) 012079 doi:10.1088/1757-899X/142/1/012079

detected (provided that yield criterion Tresca [19] is used) relying on the maximal value of the principal tangential stress τ_1. Isolines of stress τ_1 are shown in Figure 2 for $l = 3$, $\mu = 0.3$.

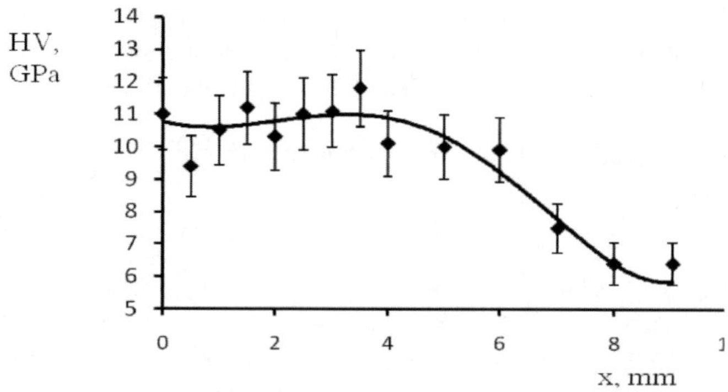

Figure 1. Microhardness profile of the system «deposited metal / steel»

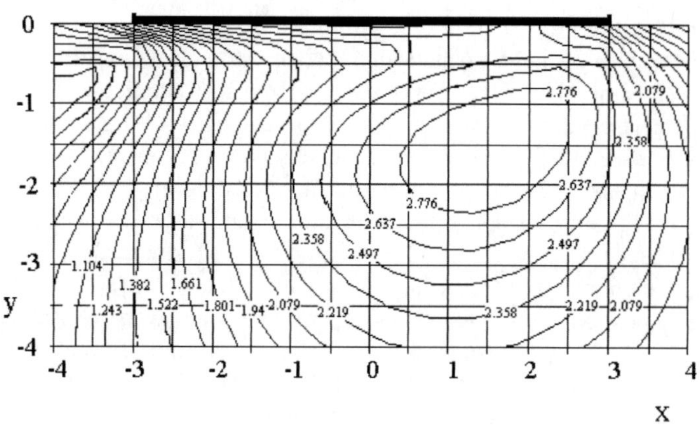

Figure 2. Isolines of maximal tangential stress for l = 3, μ =0.3, y – depth of sample, x – width of sample

Let us consider the relation of tangential stresses on the surface to maximal tangential stresses in the subsurface layer in dependence of friction factor. The maximal tangential stress in the plane of deformation equals to:

$$\tau_1 = \frac{1}{2}\left(\left(\sigma_{xx} - \sigma_{yy}\right)^2 + 4\tau_{xy}^2\right)^{1/2} \tag{1}$$

Numerical computations based on formula (1) are provided in [20], their results are presented in Figure 3. Here, [τ] – stress of shear initiation, τ_1 – maximal tangential stresses, τ(x,0) – stresses on the surface. As one can see in the Figure, there are no shears anywhere at $\mu < \mu_1$, $\tau_1 < [\tau]$, $\tau(x,0) < [\tau]$, as well as no plastic deformation, leading to material wear, occurs. At $\mu_1 < \mu < \mu_2$, $\tau_1 > [\tau]$, $\tau(x,0) < [\tau]$ tangential stresses exceed the ultimate shear stress in the subsurface layer. At $\mu_2 < \mu < \mu_3$ stresses on the surface and in the layer exceed the ultimate shear stress, that is why, shears are possible both on the surface and in the subsurface layer. At $\mu > \mu_3$ isolation of areas is not registered, and stresses diminish gradually deep into the material. This fact implies that shears occur mainly on the surface; that is illustrated by scratches on the surface.

Their presence can be explained the following way. As soon as plastic shear arises in the subsurface layer, normal and tangential stresses increase on the surface, so scratches occur. The area of maximal tangential stresses is displaced close to the surface, as the result of augmenting friction factor (Figure 4).

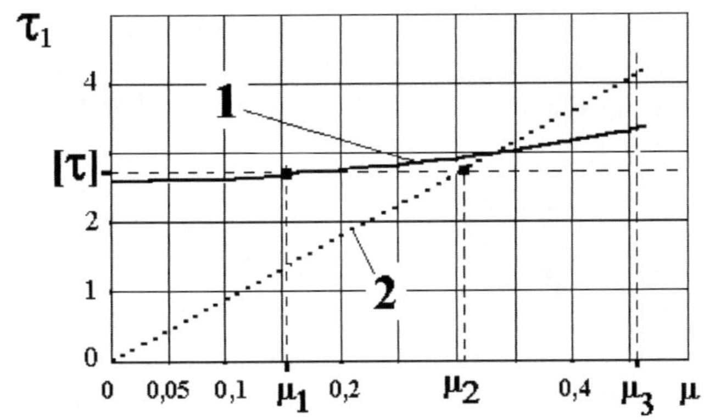

Figure 3. Isolines of maximal tangential stress for $l = 3$, $\mu = 0.3$

The growth of the resistance to shear is caused by changing structure and phase states and defect substructure of the material (meso- and micro-scale levels). The results of X-ray phase analysis are given in Table 1. The data in Table 1 show that a deposited metal is a multiphase material. Basic phases (α-phase is not taken into account), relating to interstitial phases, are ferrum oxide, ferrum and chromium carbides. Analyzing the size of coherent scattering areas we can see that dimensions of the phases are in a nanoscale range. Taking into consideration the calculated primary and secondary stresses, a conclusion can be drawn, that ferrum oxide, having a significant strengthening effect together with the matrix, is the cause of the maximal stresses.

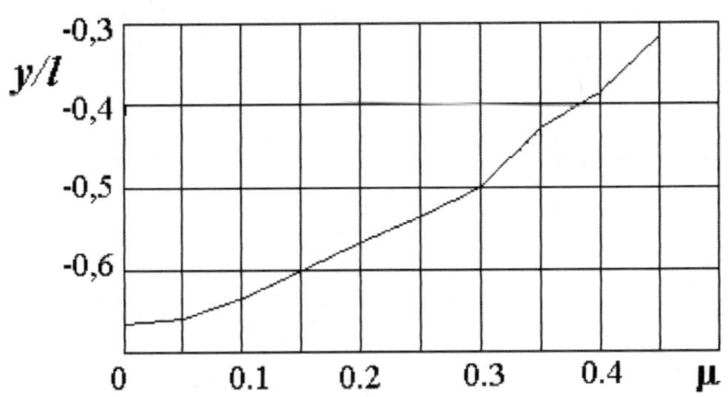

Figure 4. The relation of y depth to the length l of section with shear in dependence on friction factor μ

The study on the defect substructure of steel Hardox 450 has revealed that it is a polycrystalline aggregate based on α-phase. In bulk α-phase grains a lamellar structure dominates, being apparently the result of martensite $\gamma \rightarrow \alpha$ transformation (Figure 5 a). A sub-grain structure is registered infrequently (Figure 5 b).

Secondary phase particles are detected in bulk lamellae (Figure 5 b), on boundaries of lamellae and sub-grains (Figure 5 a). We have determined that these particles are ferrum carbide Fe_3C.

The study outcomes of the defect substructure of the deposited material are in line with the data of X-ray phase analysis. The principal phase is α-phase (Figure 6 a). It has a martensite structure. Indexing micro electron diffraction patterns of the martensite structure reveals retained austenite (Figure 6 a (insert) arrows indicate reflexes of retained austenite), which is located on the boundaries of martensite crystals in form of long inter-layers. Figure 6 b and Figure 6 c show images of oxide and carbide phases. Ferrum oxide Fe_3O_4 and vanadium carbide V_8C_7 are detected. It is quite distinct that particles of special carbides are round, dimensions of the particles vary from one to tens nm. Carbide phase particles are located on the boundaries and in bulk martensite crystals.

The structure of the deposited - base metal contact area is like the base steel structure in a lot of parameters. The martensite structure is identified. In martensite crystals there is a dislocation substructure in form of chaotically distributed dislocations and dislocation balls.

Table 1. The relation of maximal tangential shear stress τ_1 to friction factor. $[\tau]=2.7$; $\mu_1=0.155$; $\mu_2=0.31$; $\mu_3=0.47$

Phase	Phase characteristics				
	ΔV, mass. %	D (CSR), nm	$\Delta d/d$, 10^{-3}	$\sigma_1+\sigma_2$, MPa	$<\sigma>$, MPa
α-Fe	63.2	24.53	2.03	1413	396
Fe_3O_4	28.8	9.85	7.254	6380	1670
Fe_3C	6.0	14.04	3.395	-	-
CrC	2.0	10.55	2.914	5828	1107

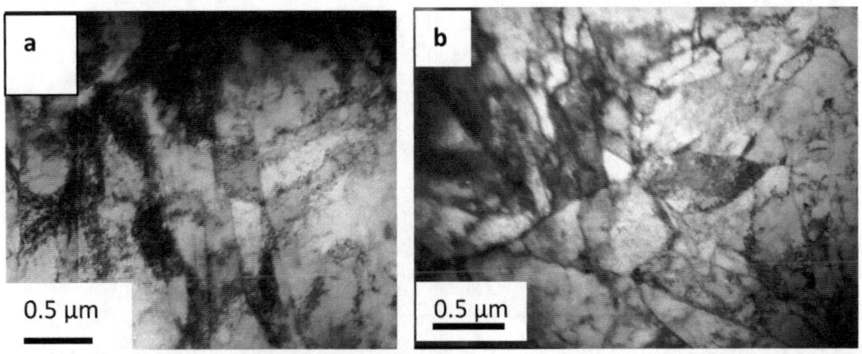

Figure 5. Electron microscope image of steel Hardox 450 structure. Lamellar structure (a), sub-grain structure (b)

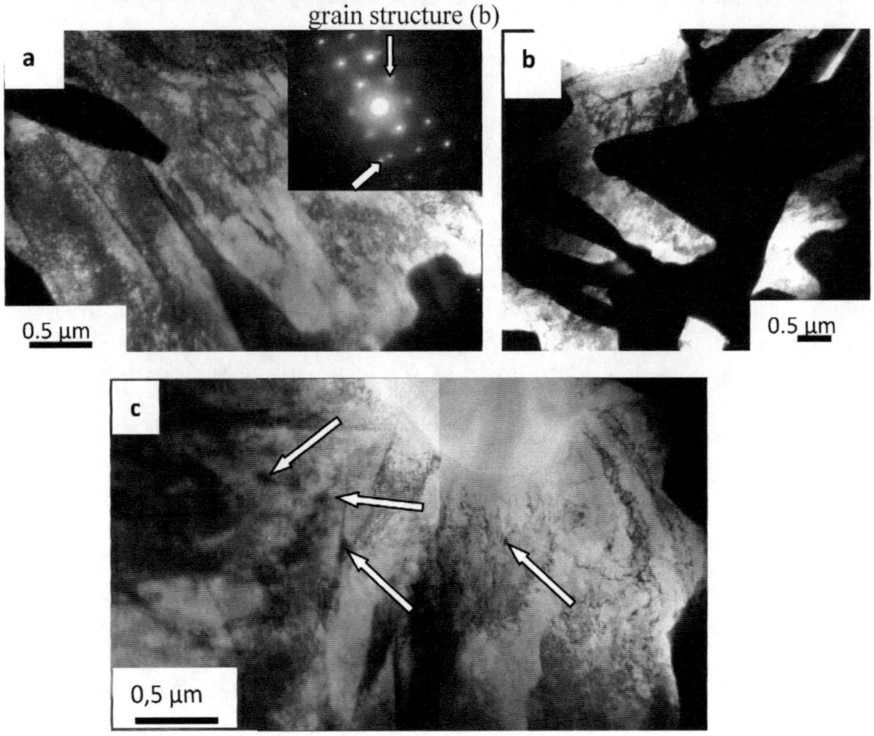

Figure 6. Images of the deposited material α-phase (a), ferrum oxide (b), and vanadium carbide (c) structures made by electronic microscopy

IOP Conf. Series: Materials Science and Engineering **142** (2016) 012079 doi:10.1088/1757-899X/142/1/012079

4. Conclusion

The structure and properties of material deposited on steel Hardox 450 are studied by methods applied in up-to-date material science on various (macro, micro, and nano) scale levels of the structure. It has been ascertained that microhardness and wear resistance of the deposited layer are twice and 140 folds as high as these characteristics of steel Hardox 450, respectively. Friction factor of the deposited material is four times lower than that of the base. The deflected mode of the material, calculated in conditions of friction, shows that plastic shear deformations are localized in its surface layer. This fact is emphasized by changing surface relief (occurrence of scratches). It has been revealed that mechanical properties of the deposited layer are improved due to formation of multiphase submicro and nanostructures; their strengthening is conditional on the martensite structure of α-matrix and high (more 36 %) inclusion volume fraction of oxide and carbide phases.

Acknowledgements

The research is carried out under financial support of Russian Science Foundation (project 15-19-00065).

References

[1] D'Ans Pierre, Degrez Marc 2015 *Surf. and Coat. Tech.* **276**.

[2] Yang X, Ma X, Moering J, Zhou H, Wang W, Gong Y, Tao J, Zhu Yu, Zhu X. 2015 *Materials Science and Engineering: A.* **645** 280.

[3] Sosnin K V , Ivanov Yu F, Gromov V E , Budovskikh E A, D.A. Romanov 2014 *Journal of Surface Investigation. X-ray, Synchrotron and Neutron Techniques.* **8** 1286.

[4] Béjar M A, Henríquez R 2009 *Materials & Design* **30** 1726.

[5] Gnyusov S F, Ignatov A A, Durakov V G, Tarasov S Yu 2012 *Applied Surface Science* **263** 215.

[6] Rao L, Wang Sh, Zhao J, Geng M, Ding G 2014 *Journal of Iron and Steel Research, International* **21** 869.

[7] Słoma J, Szczygieł I, Sachajdak A 2011 *Archives of Civil and Mechanical Engineering* **11** 437.

[8] Chen X, Fang Y, Li P, Yu Z, Wu X, Li D 2015 *Materials & Design* **65** 1214.

[9] Yilbas B S, Shuja S Z 2013 *Laser Surface Processing and Model Studies*, Berlin: Springer.

[10] Senthilkumara B, Kannanb T. 2015 *Measurement* **62** 127.

[11] Kapralov E V, Budovskikh E A, Gromov V E, Ivanov Yu F 2015 *Russian Physics Journal* **58** 471.

[12] Kapralov E V, Raykov S V, Budovskikh E A, Gromov V E, Vashchuk E S and Ivanov Yu.F. 2014 *Bulletin of the Russian Academy of Sciences: Physics* **78** 1015.

[13] Zevin L S, Kimmel G 1995 *Quantitative X-Ray Diffractometry*, Berlin: Springer.

[14] Fultz B, Howe J M 2013 *Transmission Electron Microscopy and Diffractometry of Materials*, Berlin: Springer.

[15] Pelleg J 2013 *Mechanical Properties of Materials*, Berlin: Springer.

[16] Randall N 2002 *CSM Instruments Application Bulletin* **18** 3.

[17] Panin V E, Egorushkin V E, Panin A V 2010 *Physical Mesomechanics* **13** 215.

[18] Muskhelishvili N I 1966 *Nekotorye osnovnye zadachi matematicheskoi teorii uprugosti (Some Basic Problems of Mathematical Elasticity Theory)* Moscow: Nauka.

[19] He Q -C, Vallée C, Lerintiu C 2005 *ZAMP* **56** 357.

[20] Molotkov S G 2006 *Russian Physics Journal* **49** 86.

Fractography of Fatigue Fracture Surface in Silumin Subjected to Electron-Beam Processing

S V Konovalov[1], K V Aksenova[1], V E Gromov[1], Yu F Ivanov[2,3] and O A Semina[1]

[1]Siberian State Industrial University, Novokuznetsk, Russia
[2] Institute of High Current Electronics Siberian Branch of Russian Academy of Science, Tomsk, Russia
[3]National Research Tomsk Polytechnic University, Tomsk, Russia

E-mail: konovserg@gmail.com

Abstract. The surface modification of the eutectic silumin with high-intensity pulsed electron beam has been carried out. Multi-cycle fatigue tests were performed and irradiation mode made possible the increase in the silumin fatigue life more than 3.5 times was determined. Studies of the structure of the surface irradiation and surface fatigue fracture of silumin in the initial (unirradiated) state and after modification with intense pulsed electron beam were carried out by methods of scanning electron microscopy. It has been shown, that in mode of partial melting of the irradiation surface the modification process of silicon plates is accompanied by the formation of numerous large micropores along the boundary plate/matrix and microcracks located in the silicon plates. A multi-modal structure (grain size within 30-50 μm with silicon particles up to 10 μm located on the boundaries) is formed in stable melting mode, as well as subgrain structure in the form of crystallization cells from 100 to 250 μm in size). Formation of a multi-modal, multi-phase, submicro- and nanosize structure assisting to a significant increase in the critical length of the crack, the safety coefficient and decrease in step of cracks for loading cycle was the main cause for the increase in silumin fatigue life.

1. Introduction

One of the most widespread causes of equipment, mechanisms, machines and structures failure is fatigue fracture of parts. In this connection, prevention of fatigue fracture of critical parts and, consequently, increase in their service life is an urgent problem [1]. Process of fracture begins, as a rule, with appearance of microscopic cracks which when developed result in complete fracture of article [2]. Fatigue cracks initiate in near-surface layer in zone of action of stress concentrators that may be inclusions of the second phase, micropores, scratches and dents. Therefore, in the process of fatigue fracture the state of material surface layer and its processing with hardening capable of significant increase in fatigue life of material [2] play the essential part.

Effective method of surface hardening of metals and alloys and, as a consequence, increase in fatigue service life is material's surface processing with high-intensity electron beam of submillisecond duration. This method enables change in structure of tens micrometers thick surface layer by transforming it into multi-modal structure-phase state with practically unchanged structure-phase state of the main volume of the alloy [3, 4].

The use of electron beams as a means of metals and alloys modification results in significant change in structure-phase state of surface layers and, as a consequence, increase in corrosion resistance, wear resistance and microhardness, unattainable in conventional methods of surface processing [5-10].

The purpose of the research is to analyze the modification features of silumin structure with high-intensity pulsed electron beam, detection of mechanisms responsible for fracture of silumin subjected to multi-cycle fatigue tests.

2. Material and methods of research

Test material was Al-Si alloy (silumin) (Al – 89.77, Si – 9.88, Fe – 0.35, wt. %). Fatigue test of 8×14×145 mm samples were done on the special setup according to cyclic asymmetrical cantilever bending scheme [1]. Simulation of crack was performed with half circle notch of 10 mm radius. At the test temperature 300 K frequency of samples bending cyclic amounted to 15 Hz under 10 MPa load.

Irradiation of samples surface for fatigue tests was performed on the setup "SOLO" [2] with the following parameters: electron energy U = 18 KeV; pulse repetition frequency f = 0.3 Hz; pulse duration of electron beam τ = 150 μs; energy density of electron beam E_S = 15 and 20 J/cm^2; number of pulse of action N_1 = 3 and N_2 = 5. The face surface of samples was irradiated i.e. the surface being located above the crack simulating notch. Not less than five samples were tested by each mode of irradiation. Studies of irradiation and fracture surface were performed with methods of scanning electron (setup SEM-515 Philips) microscopy.

3. Results and their discussions

Fatigue life tests of silumin irradiated samples revealed a very wide spectrum of results whose values depend significantly both on material structure and electron-beam irradiation mode and they differ by more than order of magnitude (from 132000 to 517000 cycles up to fracture) [11, 12]. Maximum fatigue life increase (~by a factor of 3.5) provided the irradiation mode with high-intensity pulsed electron beam with parameters 20 J/cm^2; 150 μs; 5 pulses.

The distinctive feature of test silumin in the initial state is presence of a large number of silicon inclusions of platelet shape predominantly. Plates locate chaotically or decorate the grain boundaries of the alloy. Plate sizes of silumin under study vary in the range from units to tens of micrometers. Irradiation of silumin surface with high-intensity pulsed electron beam depending on energy density of electron beam is accompanied by either melting of sample surface (15 J/cm^2; 150 μs; 3 pulses – the first mode) or melting of material surface layer of thickness from units to tens of micrometers (20 J/cm^2; 150 μs; 5 pulses – the second mode).

Electron-beam processing (EBP) according to the mode No.1 initiating the starting stage of silicon inclusions melting results in the running of two interconnected processes: globularization of silicon inclusions (Fig. 1, a) and brittle destruction of silicon plates (Fig. 1, b). Plate destruction process is accompanied by the formation of numerous micropores located along the interface plate/matrix and microcracks located in silicon plates (Fig. 1, b). In the micrograph of fatigue destruction surface it is seen that fatigue crack forms on the sample surface (Fig. 1, c). The large silicon plates located on the surface (Fig. 1, a) and in near-surface sample layer (Fig. 1, f) being critical stress concentrators are responsible for fatigue crack formation. As a result the fatigue tests with 180000 cycles resulted in plate destruction (Fig. 1, d) and formation of extended microcracks (Fig. 1, e). Thus, formation of micropores and microcracks in surface layer weakens the material in electron-beam processing in the mode No. 1. It is the determining factor facilitating only insignificant increase in fatigue life of material.

Electron-beam processing according to the mode No.2 initiating high-speed melting and subsequent high-speed crystallization results in significant structure refinement of silumin sample surface (Fig. 2). According to morphological attributes the surface layer structure differs from structure of initial sample and the sample irradiated by the mode No. 1. A homogeneous structure of granular (cellular) type forms on the surface of irradiation (Fig. 2, a, b). Regions with sizes from 30 to

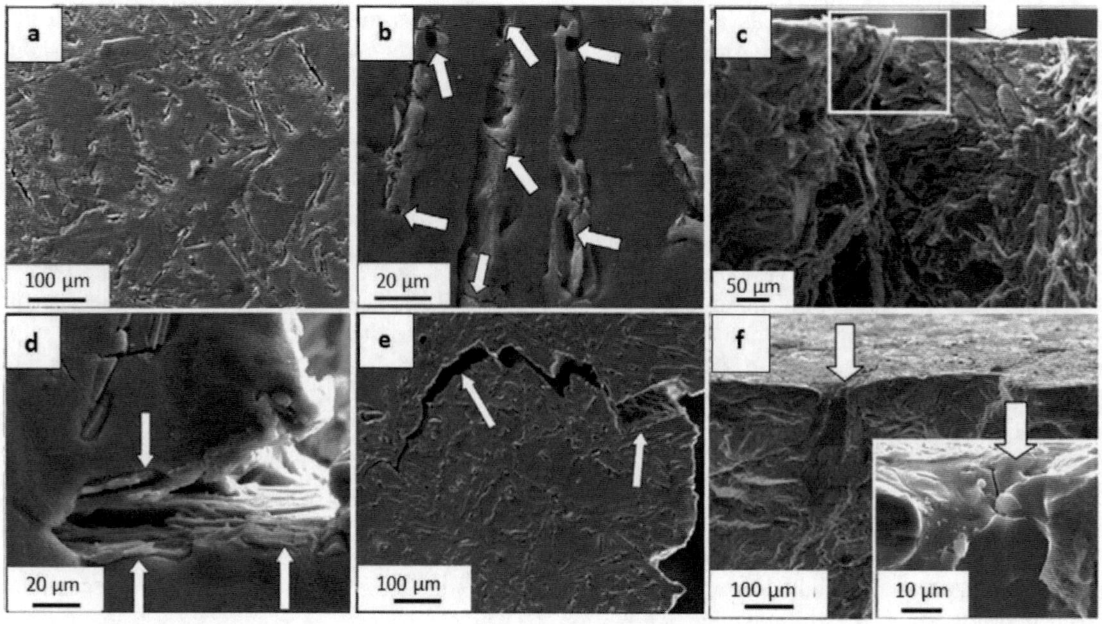

Fig. 1. Surface structure of silumin sample processed with electron beam according to the mode No. 1 and fractured after 180000 cycles. a, b – state prior to fatigue tests; c-f – post-fatigue tests states. Arrows designate: in (b) – micropores and microcracks in silicon plates; in (c) – surface of irradiation (area of fatigue crack formation); in (d) – silicon plates fractured in the process of fatigue tests; in (e, f) – microcracks having been formed in fatigue tests.

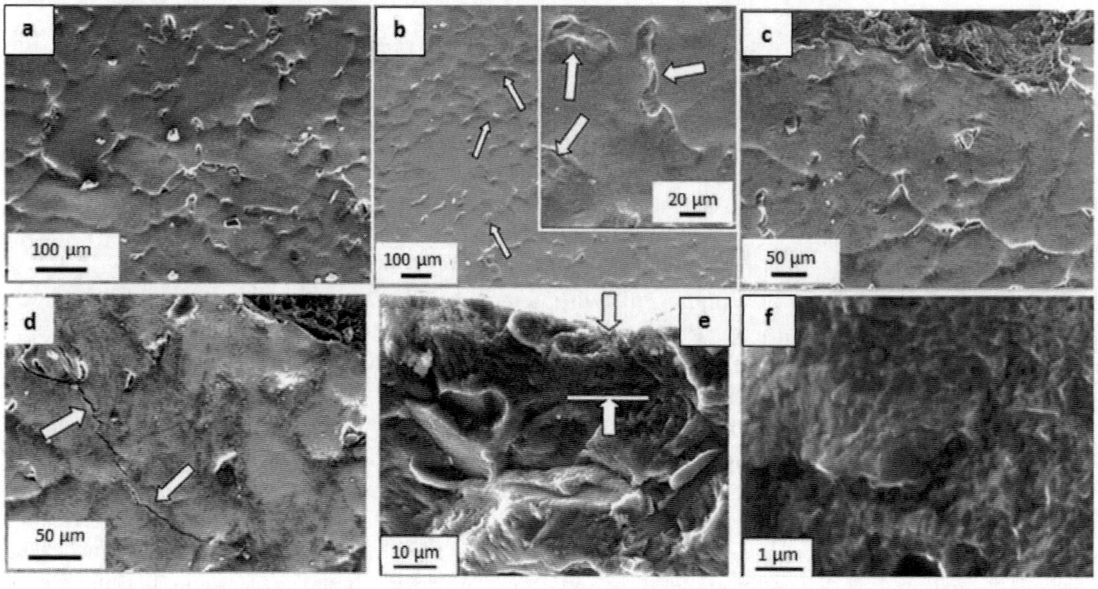

Fig. 2. Surface structure of silumin sample processed with electron beam according to the mode No. 2 and fractured after 517000 cycles. a, b – state prior to fatigue tests; c-f – post-fatigue tests states. Arrows designate: in (b) – silicon particles; in (d) – microcrack having been formed in the process of fatigue tests; in (e) – silumin layer thickness melted with electron beam.

50 μm are separated by silicon interlayers whose transverse sizes do not exceed 10 μm (Fig. 2, b). Thickness of melted layer varies within 20 μm (Fig. 2, e). It is important that stress concentrators capable of being sample fracture sources are not observable on the edges of fracture (Fig. 2, c). The cracks parallel to fracture surface locate at some distance from it (Fig. 2, d). It is evident that the concentrator having been the cause of sample fracture located under the surface, most probably, at the interface of liquid and solid phases.

High-speed crystallization results in two-phase subgrain structure (silicon and aluminium-based solid solution), whose cell sizes vary from 100 to 250 nm (Fig. 2, f). It is the formation of similar submicro- and nanosize multiphase structure in electron-beam processing according to the mode No. 2 that is the determining cause facilitating a multiple increase in its fatigue life.

The results of fractographic investigations allowed the analysis of fatigue fracture mechanisms of silumin samples subjected to electron-beam processing according to different modes. Polycrystalline character of structure (granular structure of aluminium-based solid solution and a large number of comparatively big silicon plates) results in multiple branching of material fracture front (Fig. 3, a-c). A large number of microscopically visible fracture traces located in parallel is formed which is especially characteristic for samples with maximum number of cycles up to fracture (Fig. 3, c). It is determined that in silumin under study a mixed mechanism of fatigue fracture is realized. Microrelief of fatigue fracture is presented by pits of ductile fracture and quasi-cleaved facets. Pits are dominant element of fracture surface structure and they form as a result of cutting of micropores through which aluminium grain fracture occurred (Fig. 3, d). Silicon plates fail according to cleavage mechanism (Fig. 3, e).

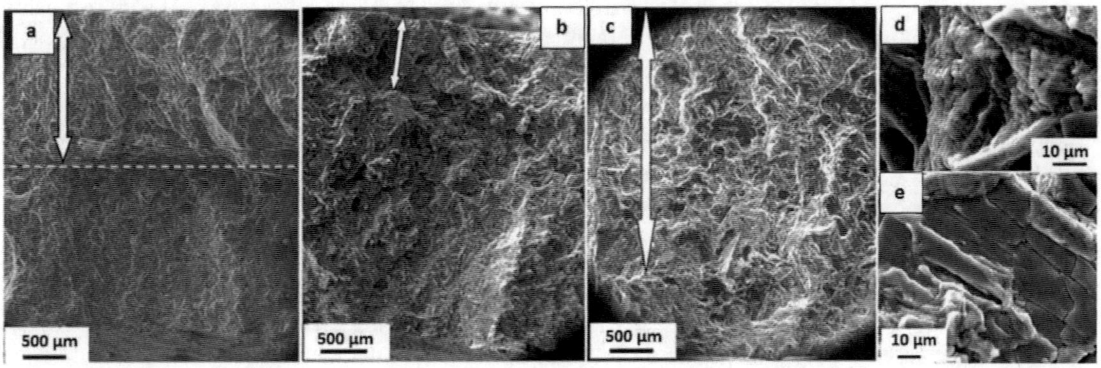

Fig. 3. Fractography of fatigue fraction surface of silumin samples. A, в – initial state, 130000 cycles; b – electron-beam processing according to the mode No.1, 180000 cycles; c – electron-beam processing according to the mode No.2, 517000 cycles. Arrows designate the fatigue crack growth zone.

On the surface of fatigue fracture of silumin samples three distinctive zones are revealed – zone of crack fatigue growth, zone of finish breaking and zone of accelerated crack growth separating them [13]. Deformation processes proceeding in fatigue tests of the material develop, in full measure, in zone of crack fatigue growth and essentially to a lesser degree, in zone of finish breaking. The width of zone of crack fatigue growth in silumin samples under study is correlatively connected with number of cycles up to fracture, i.e. it depends on material irradiation mode with electron beam. Width of zone of crack fatigue growth is taken equal to critical length of crack [1]. When analyzing Table 1 it is seen that electron-beam processing according to the mode No. 2 allows the increase in critical crack length by a factor of 2 thereby increasing the service life of material working capacity.

From the magnitude of ratio of fatigue growth zone area to area of zone occupied by finish breaking it can be judged on the value of safety coefficient of the given material: the less this ratio, the lower the safety coefficient at the same load intensity of fatigue tests [14]. Electron-beam processing according to the mode No. 2 increases safety coefficient (by a factor of 1.6) of material operation (Table 1).

Fatigue striations being a trace of crack displacing for one step in every loading cycle belong to the important indication of fatigue destruction zone of the material [1, 13, 14]. The distance between striations is determined by the capacity of the material to withstand the propagation of fatigue crack: the less distance between striations, the larger the resistance of the material to crack propagation. It is seen from Table 1 that crack step in one cycle of fatigue loading in silumin samples processed according to the mode No. 2 is less by a factor of ~2.7, consequently, the given samples have a markedly higher resistance to fatigue crack propagation.

Table 1. Surface parameters of silumin fatigue fracture.

Mode of EBP	Width of crack fatigue zone, mm	Safety coefficient	Crack step, μm
Without EBP, 130000 cycles	1.80	0.53	0.75
15 J/cm^2; 150 μs; 3 pulses; 180000 cycles	0.75	0.24	0.95
20 J/cm^2; 150 μs; 5 pulses; 517000 cycles	3.50	0.86	0.28

4. Conclusion

It is shown that different states of structure form in silumin surface layer depending on the parameters of electron-beam processing. In electron-beam processing according to the mode No. 1 initiating the starting stage of silicon inclusions melting the formation of highly defective surface layer containing micropores and microcracks weakening the material are observed. In electron-beam processing according to the mode No. 2 initiating a high-speed melting and subsequent high-speed crystallization the structure of cellular type with silicon in the form of extended nanosize interlayers or globular shape inclusions is formed in up to 20 μm thick near-surface layer.

Analysis of fatigue fracture mechanisms of silumin samples subjected to electron-beam processing according to different modes was carried put. It was detected that large silicon plates of micron and submicron sizes being undissolved in electron-beam processing were the sources of fatigue microcracks.

Surface irradiation of silumin samples with electron beam according to the optimal mode No. 2 allows (1) the increase in critical crack length by a factor of 2 thereby increasing the service life of material working capacity; (2) the increase in safety coefficient of material operation by a factor of 1.6; (3) the decrease in crack step in one cycle of fatigue loading by a factor of 2.7 which is indicative of a higher resistance to fatigue crack propagation.

It was determined that the main causes of increase in fatigue life of silumin irradiated with high-intensity pulsed electron beam were the formation of multiphase submicro- and nanosize structure in the surface layer and refining of large silicon plates.

This work has been supported by a grant of the President of the Russian Federation for state support of young Russian scientists – Doctors of Sciences (project MD-2920.2015.8) and state task No 3.1496.2014/K.

References

[1] Kotsan'da S 1990 *Fatigue cracking of metals* (Moscow: Metallurgiya).
[2] Gromov V E, Ivanov Yu F, Vorobiev S V and Konovalov S V 2015 *Fatigue of steels modified by high intensity electron beams* (Cambridge: Cambridge International Science Publishing Ltd).
[3] Engelko V, Yatsenko B, Mueller G and Bluhm H 2001 *Vacuum* **62/2 (3)** 211.
[4] Proskurovsky D I, Rotshtein V P, Ozur G E, Ivanov Yu F and Markov A B 2000 *Surf. and Coat. Technol.* **125** 49.
[5] Hu J J, Zhang G B, Xu H B and Chen Y F 2012 *Materials Science and Technology* **27 (4)** 300.
[6] Hao S Z, Qin Y, Mei X X, Gao B, Zuo J X, Guan Q F, Dong C and Zhang Q Y 2007 *Surf. and*

Coat. Technol. **201** 8588.

[7] Dong C, Wu A, Hao S [et. al.] 2003 *Surf. and Coat. Technol.* **163-164** 620.

[8] Zhang K M, Zou J X, Grosdidier T [et. al.] 2008 *The European Physical Journal: Applied Physics* **43 (3)** 343.

[9] Hao Y, Gao B, Tu G F, Cao H, Hao S Z and Dong C 2012 *Applied Surface Science* **258** 2052.

[10] Hao Y, Gao B, Tu G F [et. al.] 2011 *Materials Science Forum* **675-677** 693.

[11] Konovalov S V, Alsaraeva K V, Gromov V E, Ivanov Yu F and Semina O A 2016 *Key Engineering Materials* **675-676** 655.

[12] Ivanov Yu F, Alsaraeva K V, Gromov V E, Popova N A and Konovalov S V 2015 *J. of Surf. Investigation. X-ray, Synchrotron and Neutron Techniques* **9 (5)** 1056.

[13] Fellouz Dz. 1982 *Fractography and atlas of fracture patterns* (Moscow: Metallurgiya).

[14] Terentév V F 2003 *Fatigue of metallic materials* (Moscow: Nauka).

IOP Conf. Series: Materials Science and Engineering **142** (2016) 012081 doi:10.1088/1757-899X/142/1/012081

Research of Influence Electric Conditions Combined Electro-Diamond Processing by on Specific Consumption of Wheel*

D V Lobanov[1a], P V Arkhipov[1b], A S Yanyushkin[1c], V Yu Skeeba[2d]

[1]Bratsk State University, 40, Makarenko St., Irkutsk region, Bratsk, 665709, Russia
[2]Novosibirsk State Technical University, 20, Prospekt K. Marksa, Novosibirsk, 630073, Russia

e-mail: [a]mf_nauka@brstu.ru; [b]pavded@yandex.ru; [c]yanyushkin@brstu.ru; [d]skeeba_vadim@mail.ru

Abstract. The paper presents results of stability research of cutting properties of diamond abrasive tools in metal binder for grinding high-strength composite materials, as well as the need to reduce the specific consumption of wheel, as one of the most important economic indicators of processing. Shows a comparative analysis of the results of preliminary researches of various methods diamond processing, in which determined that the minimum specific consumption of wheel reached by combined electro-diamond grinding, combining electrochemical grinding with simultaneous continuous electrochemical correction surface wheel. Were conducted more research directed at identifying the specific consumption the diamond wheel on metallic binder depending from electric conditions combined electro-diamond processing. Researches have established the advantages of combined electro-diamond of hard alloys processing and define rational modes to ensure a satisfactory consumption of the diamond wheel, as well as significantly increasing quality and performance.

Introduction

Improving technology the processing of products from high-strength materials is an important task in conditions of modern machine-building production [1…10]. Methods final processing used in enterprises and are widely distributed abrasive tools in many cases do not allow reach the required performance characteristics of the finished products. Using the productive tool from super hard abrasive material, in this case limited, in view of the problems associated with the recovery of their cutting properties [1, 2, 7, 11]. As is known, wear resistance diamond tool determines performance and the cost of processing. Grinding of materials abrasive tool is accompanied by brining diamond layer of products processing, which leads to loss of cutting properties tool and affects the quality indicators of finished products. Achieve stability of cutting properties diamond grinding wheel probably due to additional corrections, but it increases the consumption of abrasive material and increases processing costs. Thus, one of the most important economic indicators of machining process is the specific consumption of diamond wheel. Using the combined technologies allowing combine in

*The reported study was funded by RFBR according to the research project No. 16-38-00123 мол_а.

a single process several methods in various combinations, leads an increase to processing efficiency [11, 12, 13, 14]. The authors noted that the use of electrochemical grinding with continuous electrochemical corrections wheel and combined electro-diamond grinding allows for a long time maintain the cutting ability of diamond wheels. Therefore considerable interest is the research the features of the influence of electric modes combined electro-diamond processing on the specific consumption of diamond grinding wheels.

Methods of research

Performed researches of the influence electrical modes of processing hard alloys on the specific consumption wheel at combined methods of diamond grinding. For an objective assessment are compared the following methods [13, 14]:

1. Combined method of electro-diamond processing, combining electrochemical diamond grinding with simultaneous continuous electrochemical corrections of wheel surface.

2. Grinding of with continuous electrochemical corrections diamond wheel. Herewith, diamond grinding wheel on metal bond is connected to the positive pole of a constant current source, a special cathode - to negative and processed material is electrically neutral.

3. Diamond electrochemical grinding in which the work piece is connected to the positive pole of a constant current source, and diamond grinding wheel on metal bond – to the negative. In this case, the electric circuit is closed through the electrolyte to be fed into the nip formed between the diamond grains and the work piece.

4. Grinding of diamond wheels on metal bond without the use of electro-physical and electrochemical processes, which is widely used in enterprises for processing products of hard alloys.

To determine the effect of diamond processing methods on consumption of wheel, selected widely used in engineering products and designs of the cutting tool hard alloy grades: WC3M, WC8, WC15 and TN20, which grinded using diamond wheels brand 12A2, abrasive AC6, metal binder M2-01, under similar conditions and processing modes.

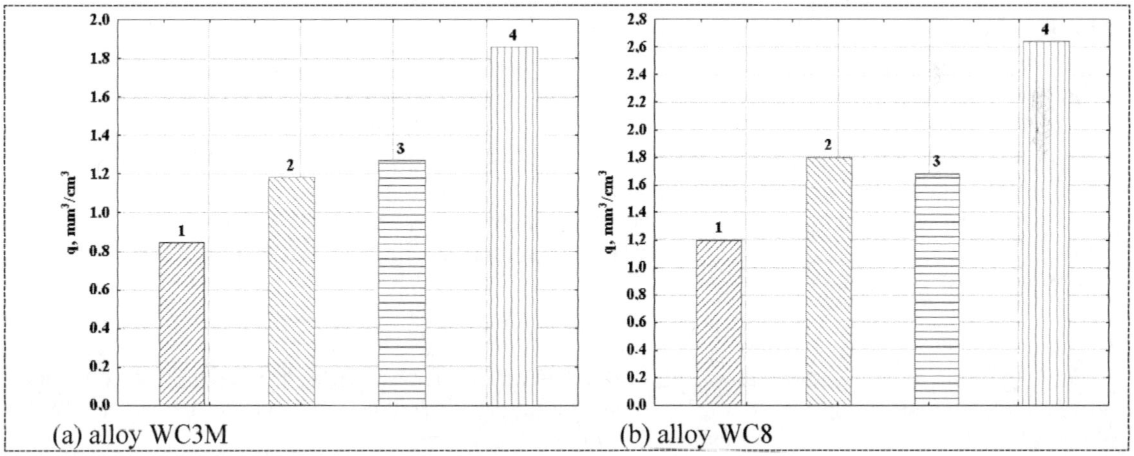

IOP Conf. Series: Materials Science and Engineering **142** (2016) 012081 doi:10.1088/1757-899X/142/1/012081

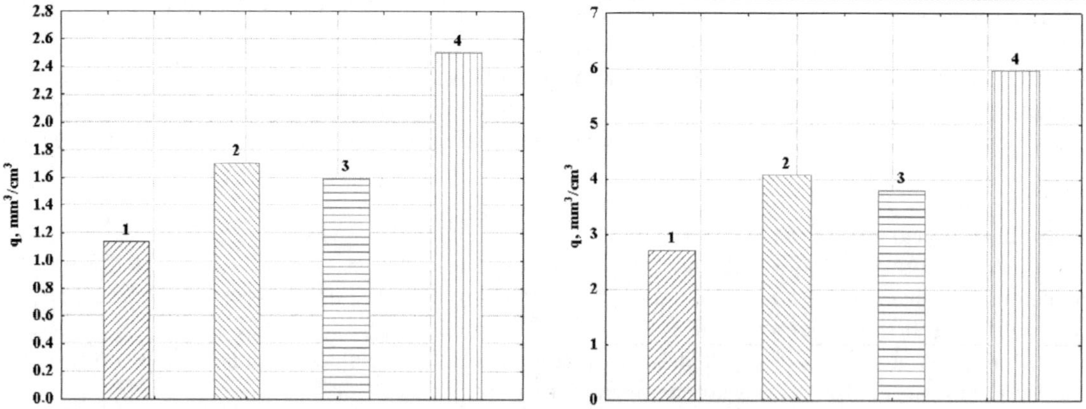

(c) alloy WC15 (d) alloy TN20

Figure 1. Consumption diamond wheel in processing of the hard alloys various methods diamond grinding
1 – combined electro-diamond grinding; 2 – diamond grinding with electrochemical wheel corrections; 3 – electrochemical diamond grinding; 4 – diamond grinding

Since at conventional diamond grinding (method 4) and diamond electrochemical grinding (method 3), intensively are salted down wheel, so from time to time was required periodical corrections surface wheel of multi-pass of rigid method – whetstone of carbide silicon green, to maintain the working capacity of the abrasive tool.

After each method of processing, wheel starred from the machine and using of the special adaptation by the linear method was measured by consumption, as the ratio of the volume consumed a diamond wheel (mm^3) to volume of removed material (cm^3).

The research results are presented in the form of comparative histograms for each brand of hard alloy (Figure 1).

According to results analysis of research results noted that the minimum consumption is observed in the method combined electro-diamond grinding (method 1).

Thus, for detection rational electrical conditions: current density of correction wheel i_{st} [A/cm^2] and the current density the etching i_{et} [A/cm^2], which mainly determined by the consumption of diamond wheel q [mm^3/cm^3], selected method of combined electro-diamond grinding.

As a rational technological modes of processing, are taken modes which defined by the results of previous researches [3, 4]: stride $S = 1.5$ [m/min], transversal feeding $t = 0.02$ [mm/d.s], rate of cutting $V = 35$ [m/s].

Results

According to the results of the research and follow statistical of processing of the data were obtained adequate mathematical dependences of the consumption of diamond wheel on the metal bond from electrical modes combined electro-diamond grinding (Figure 2).

IOP Conf. Series: Materials Science and Engineering **142** (2016) 012081 doi:10.1088/1757-899X/142/1/012081

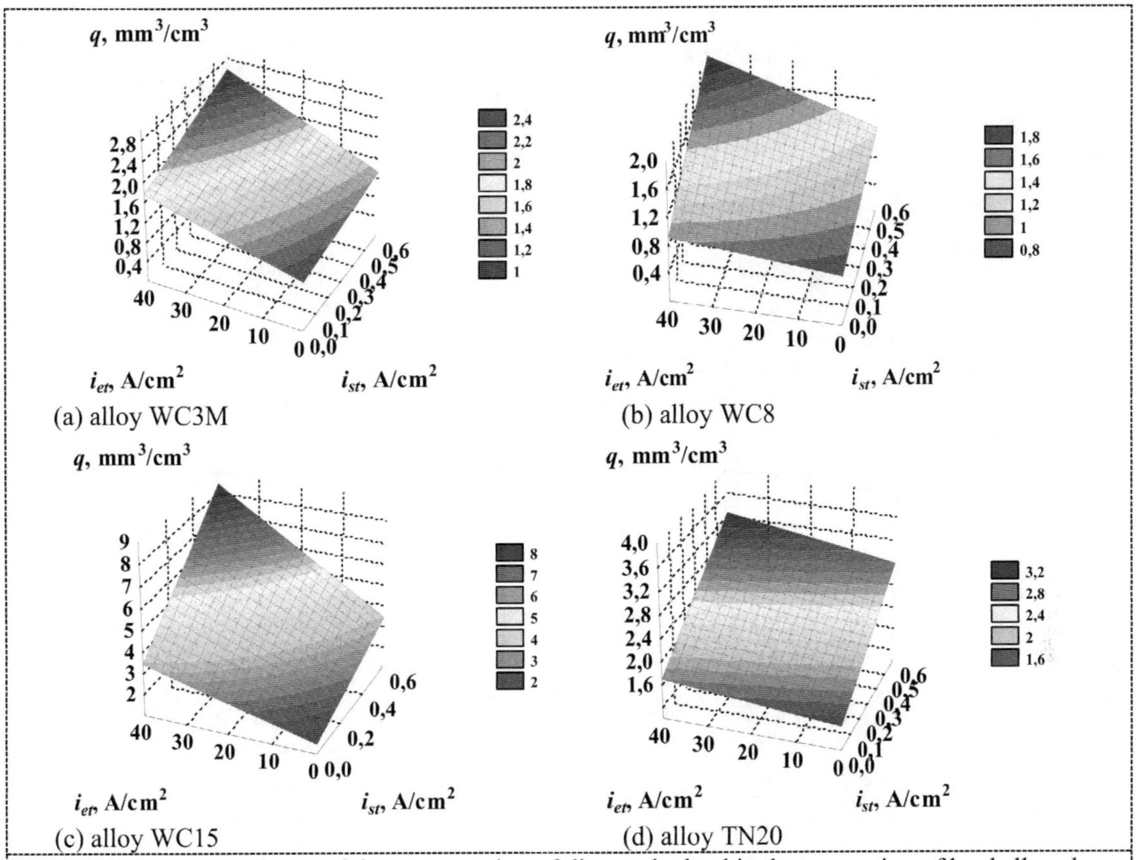

Figure 2. Dependence of the consumption of diamond wheel in the processing of hard alloys by electric modes combined electro-diamond grinding

Dependence the consumption of diamond wheel q [mm^3/cm^3] from electrical modes i_{st} [A/cm^2] и i_{et} [A/cm^2] for alloy WC3M has the form:

$$q = 0.9875 + 0.5611 i_{st} + 0.019 i_{et} + 0.0207 i_{st} i_{et}$$

The surface of a response in this case is presented in Fig. 2 (a).

Obtained dependence the consumption of diamond wheel q [mm^3/cm^3] from electrical modes i_{st} [A/cm^2] и i_{et} [A/cm^2] for alloy WC8 has the form:

$$q = 0.723 + 0.8325 i_{st} - 0.005 i_{et} + 0.0206 i_{st} i_{et}$$

The surface of a response of the examined parameter is presented in Fig. 2 (b).

Dependence the consumption of diamond wheel q [mm^3/cm^3] from electrical modes i_{st} [A/cm^2] и i_{et} [A/cm^2] for alloy WC15 has the form:

$$q = 1.4455 + 3.3383 i_{st} + 0.05 i_{et} + 0.125 i_{st} i_{et}$$

The surface of a response in this case is presented in Fig. 2 (c).

Obtained dependence the consumption of diamond wheel q [mm^3/cm^3] from electrical modes i_{st} [A/cm^2] и i_{et} [A/cm^2] for alloy TN20 has the form:

$$q = 1.417 + 2.2877 i_{st} + 0.007 i_{et} + 0.0078 i_{st} i_{et}$$

The surface of a response of the examined parameter is presented in Fig. 2 (d).

Discussion

Achieve high performance and quality of processing of high-strength materials is possible, on account of conditions that ensure the stability of the cutting properties of abrasive tools on metal bond. Such conditions is possible provide due to use of combined electro-diamond methods processing.

Research has revealed that the minimum consumption of diamond abrasive tools at the combined electro-diamond method of processing, that combines electrochemical grinding with simultaneous continuous electrochemical surface corrections wheel. During processing of this method occurs a continuous process of electrochemical corrections, which leads to the dissolution of a greasy layer formed on the surface of the wheel, diamond grains for a long time are operable because processed electrochemical softening surface. With the loss of cutting properties, diamond grains under the effect of increasing cutting forces leaving a binder.

Analysis of the results additional research allowed us to determine the rational electrical modes of combined processing that provide a satisfactory consumption diamond wheel: i_{st} = 0.2…0.3 A/cm^2 и i_{et} = 20…30 A/cm^2.

Processing of high-strength materials on recommended modes can significantly improve the quality and performance.

Reference

[1] Vasilyev E V Popov A Y Rechenko D S Diamond grinding of hard-alloy plates *Russian engineering research* 11-12 (2012) pp 730-732

[2] Nosenko V A Nosenko S V Deep grinding of titanium alloy with continuous wheel correction *Russian Engineering Research* 11 (2010) pp 1124-1128

[3] Jain V K Abrasive-based nano-finishing techniques: An overview. 12 (3) (2008) pp 257-294[4] Brinksmeier E Mutlugünes Y Klocke F et all Ultra-precision grinding *CIRP Annals – Manufacturing Technology* 59 (2) (2010) pp 652-671

[5] Gan W Chu H Xu H Xu B Research on NC electrochemical mechanical machining carbide alloy *16th International Symposium on Electro-machining* (2010) pp 379-383

[6] Nistoran Botis M Electrochemical discharge grinding of chrome alloyed steels in inorganic solutions 657 (2014) pp 201-205

[7] Nosenko V A Fedotov E V Nosenko S V Danilenko M V Probabilities of abrasive tool grain wearing during grinding *Journal of Machinery Manufacture and Reliability* 3 (2009) pp 270-276

[8] Soler Ya I Kazimirov D Yu Thermophysics of plane grinding *Journal of Machinery Manufacture and Reliability* 34 (5) (2005) pp 45-51

[9] Vasil'ev E V Popov A Y Renovation of hard-alloy end mills on numerically controlled grinding machines *Russian Engineering Research* 34 (7) (2014) pp 466-468

[10] Vasil'ev E V Popov A Y Bugai I A Nazarov P V Manufacture and design of special hard-alloy mills *Russian Engineering Research* 34 (8) (2014) pp 522-523

[11] Arkhipov P V Yanyushkin A S Lobanov D V Petrushin S I The effect of diamond tool performance capability on the quality of processed surface *Applied Mechanics and Materials* 379 (2013) pp 124-130

[12] Lobanov D V Yanyushkin A S Influence of sharpening on the quality of hard-alloy tools for the cutting of composites *Russian Engineering Research* 3 (2011) pp 236-239

[13] Yanyushkin A Lobanov D Arkhipov P Ivancivsky V Contact processes in grinding *Applied Mechanics and Materials* 788 (2015) pp 17-21

[14] Yanyushkin A S Lobanov D V Arkhipov P V Research of influence of electric conditions of the combined electro-diamond machining on quality of grinding of hard alloys *IOP Conf. Series: Materials Science and Engineering* 91 (2015) pp 1-6

Performance Potential of Grinding Tools on Flexible Backing Produced of Grains with the Controlled Form

D B Shatko[1, a], V S Lyukshin[1, 2, b], V N Bakumenko[2]

[1]Kuzbass State Technical University named after T.F. Gorbachev, 28 Vesennyaya St., Kemerovo, 650000, Russia
[2]Yurga Institute of Technology, TPU affiliate, 26 Leningradskaya St., Yurga, 652050, Russia

E-mail: [a]shdb.tm@mail.ru, [b] lwsfoa@rambler.ru

Abstract. The paper provides consideration to the approaches to designing new grinding tools on flexible backing – flap grinding wheels and grinding belts having abrasive grains with certain form and orientation in their structure. Methods to estimate the shape of abrasive grains have been analyzed. Experimental data has been presented how the form of a grain affects characteristics of tools on flexible backing. Recommendations on practical application of new tools have been given

1. Introduction

At present a great number of papers deal with the effect of grain form on the process of cutting. The results of these investigations emphasize the importance of «form factor» of an abrasive grain [7, 11, 20, 21, 23, 24]. In view of these results an assumption has been made designing new constructions of grinding tools on flexible backing using abrasive grains classified according to their form will be an urgent issue in the long-term outlook.

An integrated approach to investigation into the influence of abrasive grain form on operating characteristics of grinding tools on flexible backing made a detailed consideration of all phases in their life cycle possible. In particular, a wide range of problems focused on the analysis of single abrasive grains were solved, as well as issues of testing their cutting capacity and strength, manufacturing a preproduction batch of tools having a patent on invention, and investigating their properties were studied.

2. Methods of research

Let us follow main phases of our work.

As far as we know, abrasive grains get any shape provided that the production process of grinding grains involves fragmenting ingots of abrasive material with any machines proceeded by sieving. The form can vary greatly: disk-shaped (needle-shaped) to isometric (sphere-shaped) variants [6]. When it comes to abrasive materials the grain is shaped in the course of production in dependence on chemical composition and dimensions of the ingot to be molten, conditions of cooling and grinding etc. The influence of these parameters is important for form and geometry of a grain when grinding an ingot of abrasive material.

An up-dated oscillating separator [1, 7] and an electrostatic separator were applied to sort out equiform grain fractions from the mass of standard abrasive material. The principle of oscillating separator operation is shown below as an example (Figure 1). Precisely-sized abrasive material subject to separation is supplied evenly from the proportioning feeder 1 to the oscillating head 2. The grains of abrasive material 3 driven by directed oscillations from the vibration exciter and under own gravitational and frictional forces move on the oscillating head. Less coarse round particles, the so-called isometric grains, roll down on the surface into the corresponding collector-containers 4. Intermediate-formed grains move mainly in crosswise direction and are collected in the side containers. Disk-like and needle-shaped are raised on the inclined surface and get to the upper collector-containers.

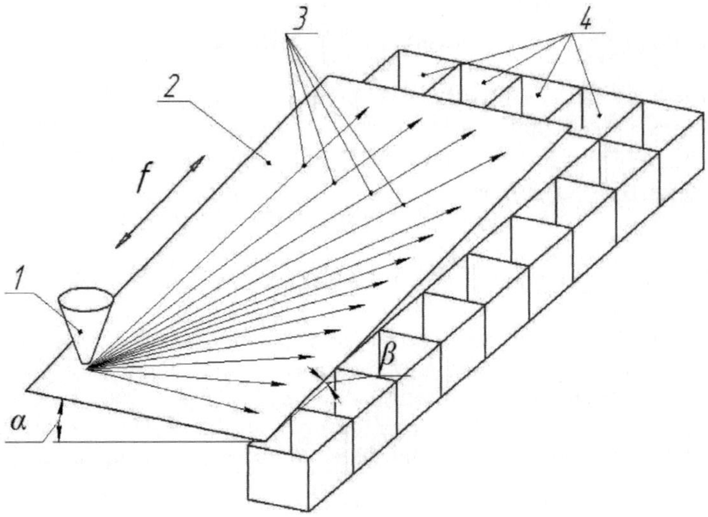

Figure 1. Form-based distribution of grinding grain on the oscillating head

After separation the form of each grain fraction was further evaluated on the base of quantitative and qualitative principles [23].

The qualitative evaluation procedure of a grain form involves grouping grains according to their forms. Three groups are divided: isometric, intermediate and needle (disk)-shaped. The qualitative evaluation procedure of a grain form has the advantage of promptness, although subjectivity of classification is its typical shortcoming as the form of a grain, which is a solid body, is evaluated according to the only projection. In view of this fact methods of qualitative evaluation can be recommended exclusively for the primary analysis of abrasive grains.

Methods of quantitave evaluation are more preferable to investigate how the form of a grinding grain affects the operational characteristics of grinding tools as they rely on specifying the parameter «form factor» K_f; its calculation necessitates a grain projection on the horizontal surface [15]. The most accepted approaches to specify the form factor are given below:

Form factor is calculated as a ratio of maximal length of a projection (l) to its breadth (b):

$$K_f = \frac{l}{b}. \quad (1)$$

Another approach to determine form factor is to calculate a ratio of a circle circumscribed around the grain projection (D) to the circle inscribed in it (d):

$$K_f = \frac{D}{d}. \quad (2)$$

Mentioned above methods to specify form factor take horizontal projection of a grain into consideration, that is why, inaccuracy of analysis is possible.

The evaluation procedure of a grain form involving analysis of horizontal and lateral projections provides exacter measuring results, eliminates inaccuracy as if compared with methods of grain form assessment based only on horizontal surface. The diagram of evaluation procedure of an abrasive grain form on the base of two projections is presented in Figure 2.

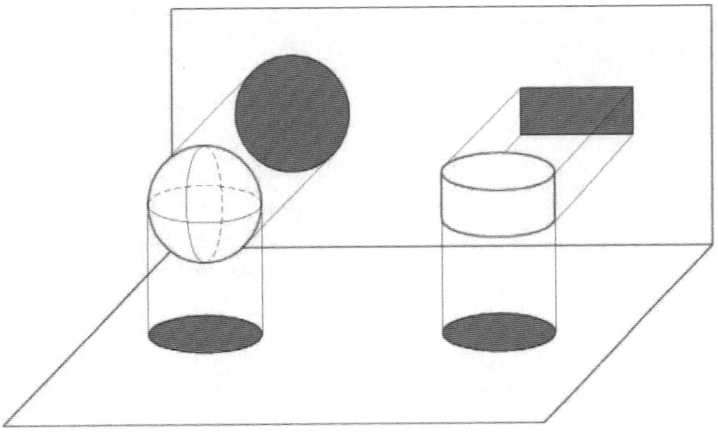

Figure 2. Evaluation of a grain form according to horizontal and lateral projections

Here, the quantitave characteristic of a grain form is described by the formula:

$$K_f^v = \frac{K_f^h}{K_f^l}. \quad (3)$$

where K_f^v - volume form factor;

K_f^h - form factor of a horizontal projection;

K_f^l - form factor of a lateral projection.

Cutting capacity, impact and static strength of single abrasive grains from each sorted out fraction were tested besides the evaluation of «form factor» parameter. As the result, regularity was revealed: both form and orientation angle of abrasive grains influence on their cutting capacity and impact strength [12, 22].

After required quantity of equiform abrasive grains had been sorted out, preproduction batches of grinding tools on flexible backing were made – flap grinding wheel and belts [5, 14]. Comprehensive studying the influence of a grain form on operational characteristics required three groups of tools produced in terms of State Standards. Precisely-sized grains were in the structure of the tools – isometric grains (mean form factor $K_f \approx 1.2$), intermediate ($K_f \approx 1.6$) and needle-shaped ones ($K_f \approx 2.2$).

Figure 3 demonstrates an operating section of a flap grinding wheel and a grinding belt. Structure numbers:

1 – flexible cloth backing,
2 – layer of coupling material,
3 – maker coat of adhesive material basic,
4 – size coat of adhesive material,
5 – abrasive grains with specified form.

With the purpose to improve cutting manufacture of preproduction batches of grinding tools involved electrostatic orientation of abrasive grains with the certain form on the flexible backing surface, performed by specially designed appliance and according to the original technology. The structure of this appliance makes possible full simulation of industrial technological process of abrasive cloth manufacture in laboratory conditions; moreover, its advantages are compactness, simplicity of construction, multifunctionality and flexibility. Furthermore, this appliance provides a possibility of coating grains by electrostatic filed with various inclination angels to the backing surface as opposed to the industrial production line.

Preproduction batches of tools on flexible backing were subject to abrasion and cutting capacity tests in order to identify the influence of grinding grain form on operational characteristics; roughness of the processed surface, temperature in the zone of cutting and forces of cutting when grinding were studied as well [8, 9, 10, 16, 17].

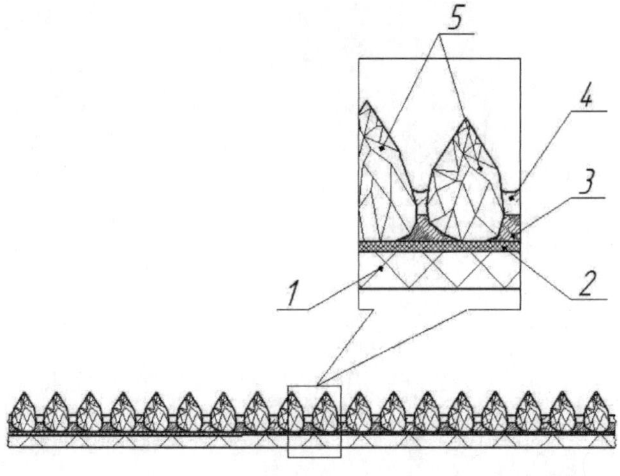

Figure 3. A section of a tool on a flexible backing produced of oriented grains with the controlled form

Grinding by a tool on a flexible backing is noted to be a complex and unstable phenomenon depending on numerous factors: properties of abrasive material and its granularity, characteristics and hardness of bonding material, features of a flexible backing, geometry of the tool, types and mode parameters of grinding etc. All the above mentioned is relevant both for flap grinding wheels and grinding belts as flexible backing amends grinding process and makes these tools different from dimensionally-stable grinding tools – cutoff and rough grinding wheels, bars, segments etc. [3, 4, 13, 18].

Operational characteristics of experimental grinding tools on flexible backing were tested by flat surface grinder 3G71.

The research was focused mainly on the effect abrasive grain form has on output parameters, models reliably for describing mechanisms of the processes were used for mathematical treatment. Form factor (K_f) of a grinding grain was used as an input variable parameter.

Mathematical treatment of obtained experimental data was performed with the software «STATISTICA 6.0» by means of stochastic simulation based on multiple correlation method. This software allowed processing experimental data set automatically and investigating into the dependence of input grinding parameters to the form factor (K_f) of a grinding grain [2, 7].

3. Results and discussion

A number of dependences experimentally obtained in the course of experimental flap grinding wheels are given below (Figure 4, 5).

IOP Conf. Series: Materials Science and Engineering **142** (2016) 012082   doi:10.1088/1757-899X/142/1/012082

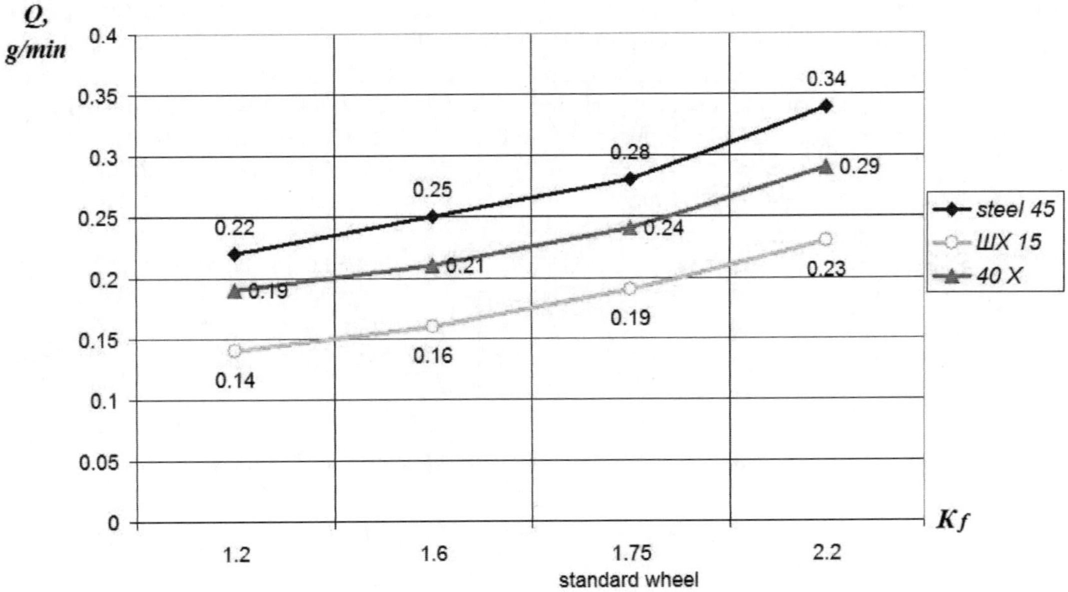

(*a*)

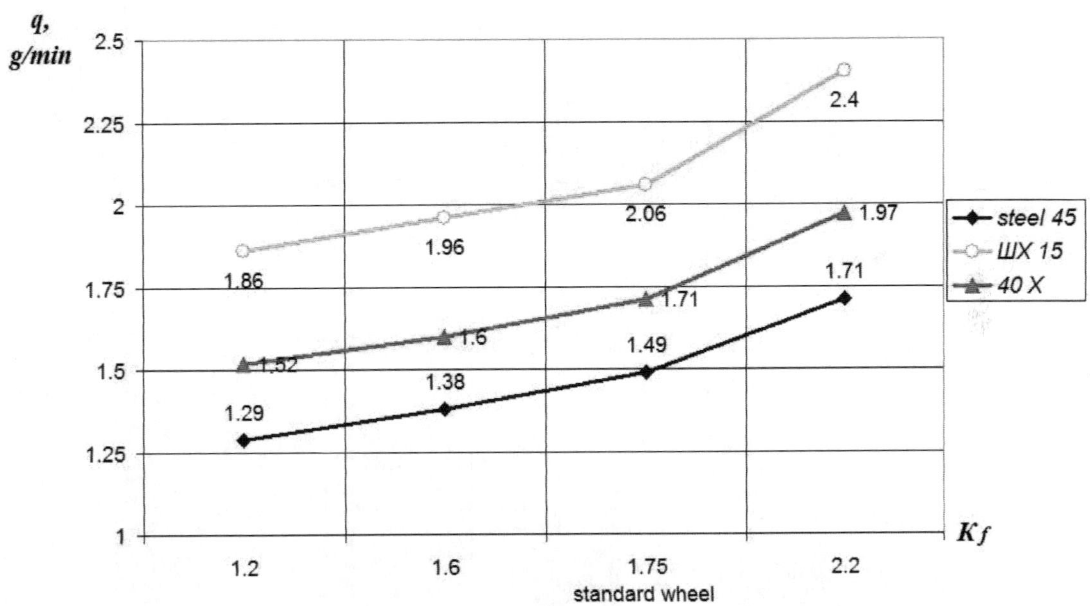

(*b*)

Figure 4. The dependence of cutting capacity *a)* and wear *b)* of a flap grinding wheel on form factor of a grinding grain when machining various steels

Given below results were obtained in the course of experiments [17, 19]:

1. When passing from the isometric ($K_f = 1.2$) to the needle-shaped ($K_f = 2.2$) form of grains in the structure of a flap grinding wheel, there is a monotonic increase by the factor of 1.56 in the curves of cutting capacity nearly for all steels. When passing from grains ($K_f = 1.2$) to the standard wheel ($K_f = 1.75$) cutting capacity increases to a very little degree, approximately by the factor of 1.29. In the range $K_f = (1.75 \div 2.2)$ cutting capacity tends to a slight increase.

With regard to the applied grain form cutting capacity goes down by the factor 1.51 on the average when increasing the hardness of machined half-product, that is, when passing from steel 45 (HB 187) in as-received state to the tempered ШХ 15 (HRC 65).

2. Wear of flap grinding tools when changing isometric grain form ($K_f \approx 1.2$) in needle-shaped grains ($K_f \approx 2.2$) goes up proportionally by the factor of 1.3. When passing to the grinding wheels produced according to the ordinary technology ($K_f \approx 1.75$) wear rises to a little degree – 1.12 times on the average, this trend is more or less relevant for machining all grades of steel.

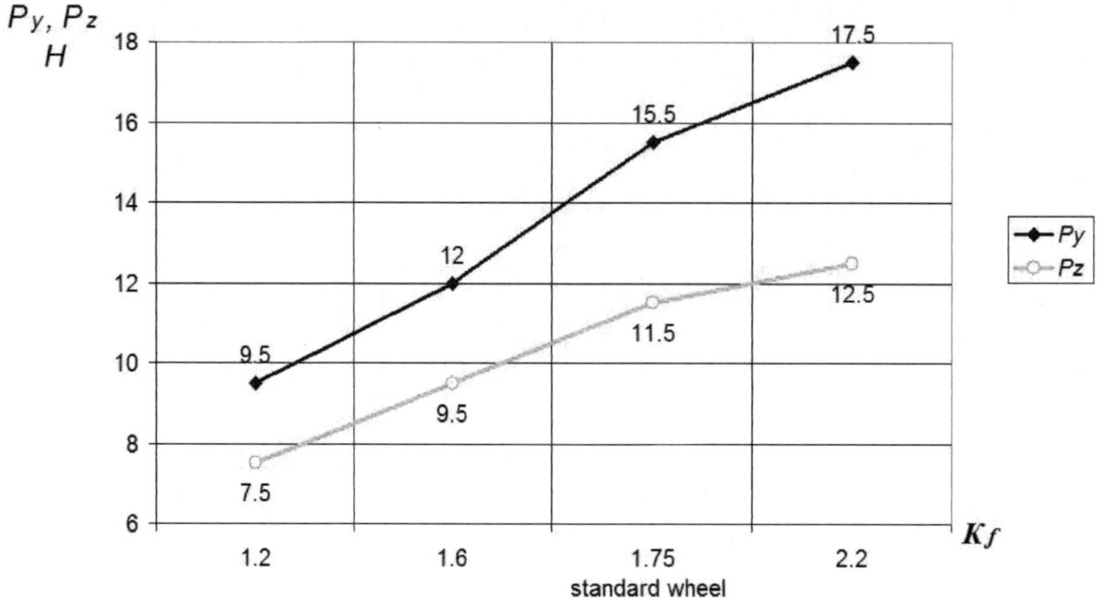

(*a*)

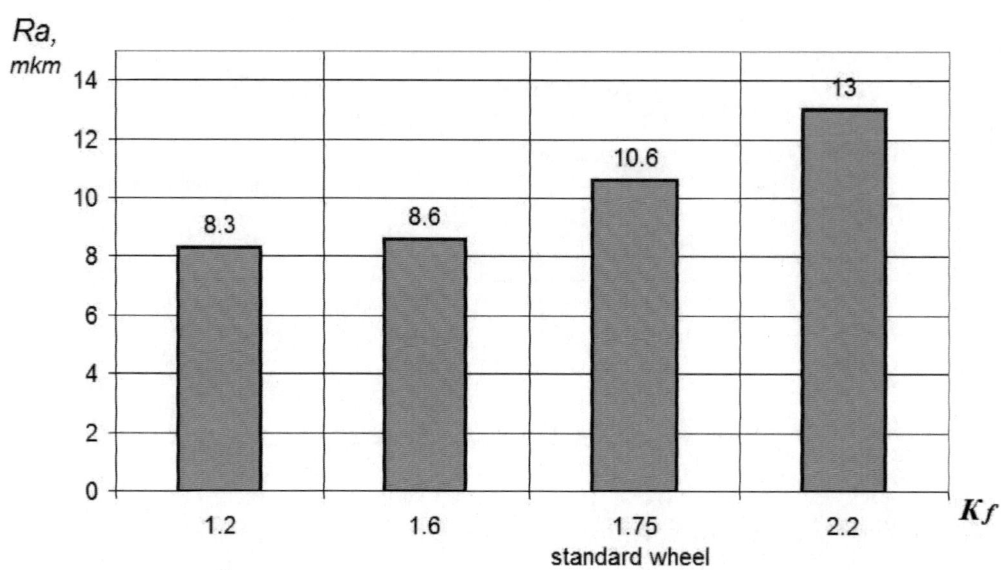

(*b*)

Figure 5. The influence of form factor of a grinding grain on:
a) cutting force components and *b)* roughness of the grinded surface

3. Evaluated force dependences of experimental flap wheel grinding demonstrate radial component of cutting force goes up by the approximate factor of 1.8, and its tangential component grows by the factor of 1.66 when passing from the wheels with isometric form of grains ($K_f = 1.2$) to wheels with needle-shaped grains ($K_f = 2.2$).

4. On the ground of obtained data we came to a conclusion flap wheel grinding is distinguished by low temperature in the zone of cutting due to the design features of these tools, low loads of grinding and effect of ventilation. The maximal registered difference in absolute values of temperature for wheels with isometric and needle-shaped grain forms is not significant, approximately 15 – 20 °C for applied modes of grinding.

The increase in hardness of the half-finished product to be grinded causes temperature rise in its surface layer by the factor of 1.2 on the average for all kinds of experimental wheels.

5. The quality of the surface is better if grinded with wheels having isometric grains in their structure as opposed to wheels with needle-shaped grains. When grinding samples of steel 45, the difference in roughness R_a is 36% for wheels with form factors $K_f = 1.2$ and $K_f = 2.2$, respectively. As opposed to the standard wheel roughness of needle-shaped grains amounts to 18.5 %, whereas isometric grains cause 17.5% decrease in roughness.

4. Conclusions

The results obtained in the course of experiments allow offering the following recommendations on more efficient use of newly designed grinding tools on flexible backing:

– tools made of grains with higher K_f (that is, needle-shaped and disk-shaped grains) are recommended to use for snagging and preliminary operations;

– tools with lower K_f (that is, isometric form of grains) can be applied for finishing operations requiring for high quality of the surface to be grinded [15, 17-20].

To sum up, operating characteristics of grinding tools on flexible backing can be improved significantly, as well as their potential can be used to the full provided that the most appropriate form and orientation of grinding grains are selected.

References

[1] Anakhin V D, Pliss D A, Monakov V N 1991 *Oscillating Separators* (Moscow: Nedra)
[2] Borovikov V P, Borovikov I P 1998 *«STATISTICA». Statistic Analysis and Data Processing in Windows* (Moscow: Informational publishing «Filin»)
[3] Gdalevich A I 1990 *Finishing Operation by Flap Wheels* (Moscow: Machinebuilding)
[4] Ivanov I Yu, Nosov N V 1985 *Efficiency and Quality of Machining by Tools on Flexible Backing* (Moscow: Machinebuilding)
[5] Korotkov A N, Shatko D B 2009 *Metal processing* **2** 34–35.
[6] Korotkov A N 1992 *Operational properties of abrasive materials* (Publishing Office of Krasnoyarsk University)
[7] Korotkov A N, Dubov G M, Shatko D B 2004 *Metal processing* **2** 43–44
[8] Korotkov A N, Lyukshin V S 2001 *Operational Characteristics of Grit Cloth with Grains of Various Forms* (Yurga: TPU Publishing) 74–76
[9] Korotkov A N, Shatko D B 2005 *Metal processing* **2** 37–39.
[10] Korotkov A N, Shatko D B 2008 *Assembling in machinebuilding, tool engineering* **10** 44–48
[11] Murdasov A V, Khshivo L N 1968 *Properties of Abrasive Grains Depending on Their Forms* (Ural Affiliate of Russian Research Institute of Abrasives and Grinding) 22–27
[12] Netrebko V P, Korotkov A N 1992 *Durability of Grinding Wheels* (Moscow: Agency of Russian Press)
[13] Pankov L A, Kostin N V 1988 *Machining with Tools Made of Grit Cloth* (Leningrad: Machinebuilding)
[14] Patent 2245240 C1 B 24 D 13/00. Flap Wheel / A N Korotkov, D B Shatko – applied 30.09.2003.; Publ. 27.01.2005, Bull. №3

[15] State Registration Certificate of Software №2004610227. Software to Calculate the Form of Grinding Grains (Programm) / V S Lyukshin, N A Alyekhin. - №2003612419; Applied 21.11.03; Published. 20.01.04
[16] Shatko D B 2008 *Metal processing* **1** 28–29
[17] Shatko D B 2005 *Improvement of Efficiency of Flap Grinding Wheels by Grains with the Controlled Form* (Thesis of PhD in Engineering: 05.03.01)
[18] Shchegolev V A, Ulanova M E 1977 *Elastic Abrasive and Diamond Tool* (Leningrad: Machinebuilding)
[19] Korotkov A N, Shatko D B 2015 *Applied Mechanics and Materials* **788** 313–317
[20] Korotkov A, Korotkova L, Gubaidulina R 2014 *Applied Mechanics and Materials* **682** 469–473
[21] Korotkov V, Minkin E 2015 *Materials Science and Engineering* 91 (2015) 012041. doi:10.1088/1757-899X/91/1/012041
[22] Lyukshin V S, Barsuk A V, Fazleev R R 2015 *Materials Science and Engineering* 91 (2015) 012047. doi:10.1088/1757-899X/91/1/012047
[23] Lyukshin V S 2014 *Applied Mechanics and Materials* **682** 148–153
[24] Korotkov V A, Petrushin S I 2014 *Applied Mechanics and Materials* **682** 224-230

Effects of Inert Nanoparticles of High-Melting-Point Compositions on Grain Structure and Strength of Ni3Al Intermetallic Compounds

V E Ovcharenko[1,2], E N Boyangin[1], V A Chudinov[1]

[1] Institute of Strength Physics and Materials Science (Siberian Branch of Russian Academy of Sciences), 2/4 Prospect Academicheskiy, Tomsk, 634021
[2] Yurga Institute of Technology, National Research Tomsk Polytechnic University, Leningradskaya street 26, Yurga, Russian Federation, 652055

e-mail: ove45@mail.ru

Abstract. The paper represents experimental findings both in the area of effects of nanoparticles of inert high-melting-point TiN compounds on a Ni3Al intermetallic grain structure creation in the conditions of high temperature synthesis under pressure, and in the area of impact of grain structure modification on intermetallic compounds' strength factor temperature dependence. It was demonstrated that appending a stoichiometric composition of nanosized particles of high-melting-point inert chemical compounds (TiN) initiates a many-fold loss of average size of grain of Ni3Al intermetallic compounds, synthesized under pressure, as well as a sufficient intermetallic compounds' strength rise within a wide range of temperatures (up to 1 000 degree C). Electron-microscopic evaluations of a synthesized intermetallic structure with TiN nanoparticles, showed that, during the process of intermetallic polycrystalline structure creation from high temperature synthesis products melts, TiN nanoparticles are mainly spread throughout the boundaries and joints of grain structure, acting as stoppers of grain boundaries migration.

Introduction

Physical and mechanical properties of nickel alloys, which are widely used in engineering, substantially depend on their grain structure dispersability and morphology [1, 2]. Grain structure dispersability growth causes a sufficient rise of flow and strength limits, and nickel alloys ductility advance. Grain refining is one of the most efficient ways to enhance durability in the conditions of cyclic stress [3, 4].

It is possible to purposefully increase nickel alloys grain structure dispersability via nucleation and grain growth processes control, when alloys are being crystallized by adding highly dispersive particles of infusible chemical compounds, which ay important parts of crystallization centers and efficient growth barriers of grains spread throughout the boundaries [5].

Above mentioned is of burning importance for heat-resisting alloys based on Ni3Al intermetallic compound, the amount of which can reach 89% in alloys. Extremely low ductility of intermetallic compounds causes their tendency for brittle intercrystalline failure and loss of alloys strength in the conditions of high temperatures [6, 7]. Intermetallic alloys mechanical properties enhancement technologies focus mainly at controlled alloying and crystallization of alloys; ways of producing monocrystal articles; and at reinforcement of alloys with refractory particles or fibers [8]. At that a possibility to increase intermetallic alloys' strength by means of a intermetallic compound (the main alloy ingredient) grain structure atomization has been left unattended. Currently the world practice does not possess any technologies that allow enhancing intermetallic compounds mechanical properties via their grain structure atomization.

The paper presents research findings that demonstrate the way nitride TiN refractory chemical compound's nanosized particles (as potential centers of intermetallic compound crystallization and definite grains growth stoppers within the process of a intermetallic compound grain structure development, when it is being crystallized from melt) influence a grain structure dimension, as well as Ni3Al intermetallic compounds physical and mechanical properties [9].

Study Materials and Methods of the Experimental Research

Ni3Al intermetallic compound patterns were obtained through the method of high temperature synthesis in a nickel (~2.0 μm) powder mixture with stoichiometric aluminum (~1.0 μm) in the conditions when steel mould was subjected to a continuous heating with high-frequency current up to the moment when a 30 % porosity powder compact autoignites with further pressure to the high temperature synthesis article [10, 11, 12]. Nanosized particles of a titanium nitride TiN (Figure 1) in a volume of 0.1 mas. % were added into an initial powder mixture of nickel and aluminum within the phase of mixing powders. To reach a distribution factor of TiN nanosized particles in a mixture of nickel and aluminum powders of a micron range of sizes, the mixture was carried out through two periods: by means of a vibromixer in a tank with a powder mass dissector, blending them in a multi knife homogenizer when they were in a dispersed dry-cleaning liquid (acetone) condition of a powder mixture. X-ray phase analysis of synthesized patterns of an intermetallic compound was carried out in the conditions of Cu Kα rays at 40 kV and 40 mA. Intermetallic alloys grain structure was investigated through the methods of optical metallography and scanning electron microscopy after metallographic specimen had been subjected to ion beam etching. Intermetallic compound patterns ware-resisting properties were tested with a BK -10 hard alloy counter body that moved along cutting tracks through the surface. The BK -10 hard alloy counterbody was a 3 mm ball that moved at 8N load with a linear speed equal to 2.5 cm per second with a 2.5 mm radius of a cutting ring, and with the total number of rotations that equaled to 3 500 revolutions. The intermetallic compound patterns stretch strength was evaluated at a 0.2 mm per minute rate of stretch.

Figure 1. TiN nanoparticles overview (transmission electron microscopy).

Results and Discussion

Figure 2 introduces difractograms obtained from the Ni3Al intermetallic compound reference specimen and from the intermetallic compound patterns synthesized under pressure in a powder mixture of nickel and aluminum with 0.1 % of nanosized particles of a TiN titanium nitride adding.

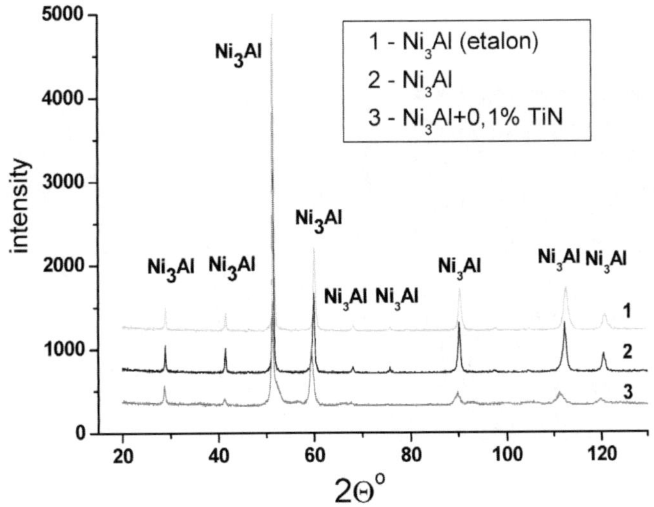

Figure 2. Difractograms obtained from the reference specimen (1), from the patterns synthesized under pressure (2), and from the patterns synthesized under pressure with a 0.1 % of nanosized particles of a titanium nitride adding (3) of Ni3Al intermetallic compound.

The analysis of difractograms introduced in Figure 2 allows making a conclusion that Ni3Al intermetallic compound patterns synthesized under pressure, including those which contain titanium nitride nano particles, have a one-phase composition.

Metallographic evaluation of intermetallic compound patterns' grain structures showed that adding 0.1% of nanosized particles of a titanium nitride into the initial powder mixture of nickel and aluminum, we sufficiently influence the size of an intermetallic compound grain structure, i.e. the average size of a grain reduces from 10 μm to 2 μm (Figure 3).

Figure 3. Metallography of a synthesized under pressure Ni3Al intermetallic compound grain structure (a) and a synthesized under pressure intermetallic compound which contains 0.1% of TiN nanoparticles (b), scanning surface electron microscopy of a synthesized under pressure Ni3Al intermetallic compound+0.1% of TiN nanoparticles (c).

IOP Conf. Series: Materials Science and Engineering **142** (2016) 012083 doi:10.1088/1757-899X/142/1/012083

Evaluations of intermetallic compound patterns (scanning electron microscopy) fracture relief thin structure showed that there are nanosized inclusions throughout the boundaries of intermetallic compound grains. An element compound energy-dispersive analysis technique helped to determine that nanosized inclusions consist of titanium and nitrogen, and are the nanoparticles of a TiN titanium nitride. Figure 4 represents micro photographs of a fracture surface of an intermetallic compound with TiN nanoparticles adding. Figure 4 b, d shows an energy spectrum of a grain body element compound and TiN particles at the grain boundary.

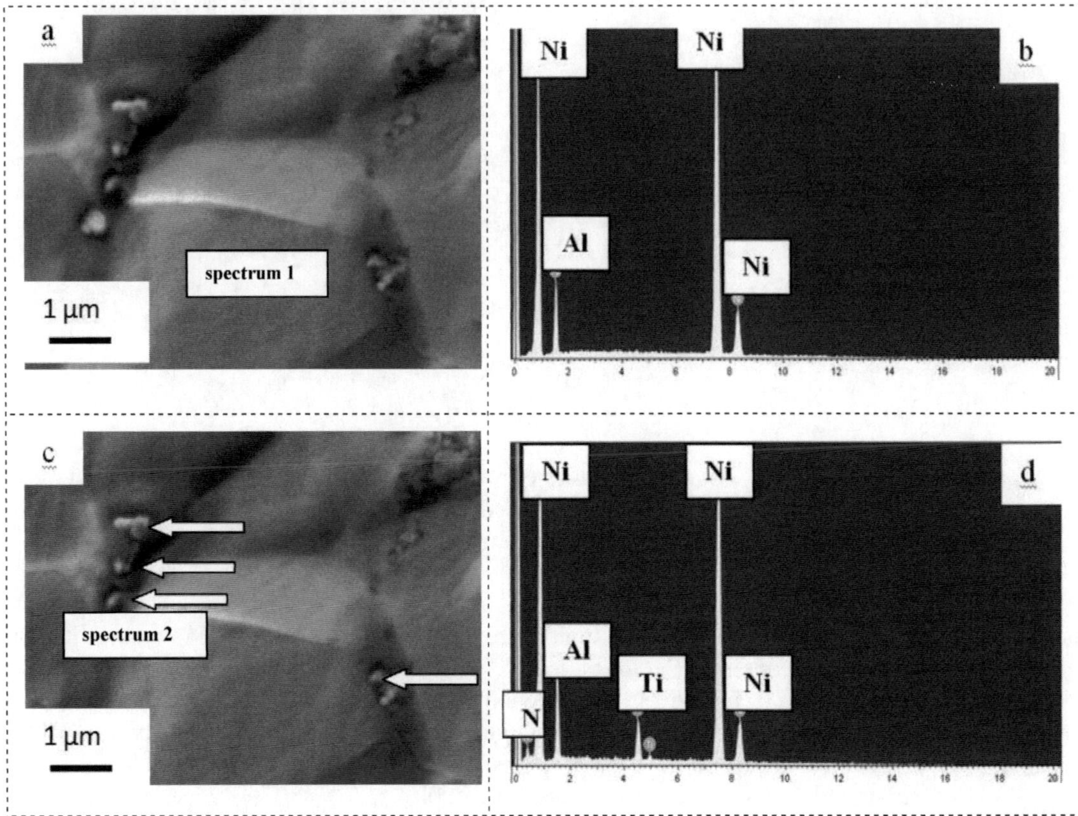

Figure 4. Fracture surface of a synthesized under pressure Ni3Al intermetallic compound with TiN nanoparticles (marked with arrows) (a, c) and energy spectra of a grain body element compound (b), and titanium nitride particles at the grain boundary (d).

TiN particles distribution throughout the boundaries and joints of grains in a synthesized Ni3Al intermetallic compound might be proved by the results of the investigations of intermetallic compound patterns' thin structure through the method of a transmission electron microscopy. Figure 5 presents both an electron-microscopic image of a TiN particle, situated within the joints of several grains of an intermetallic compound, and a particle element analyses findings.

Presented modification of a grain structure of a Ni3Al intermetallic compound, synthesized in a powder mixture of nickel, aluminum and TiN nanoparticles sufficiently effects intermetallic compound strength properties. Figure 6 represents temperature dependence of strength of a Ni3Al intermetallic compound, synthesized under pressure.

Strength temperature dependences of synthesized under pressure Ni3Al and Ni3Al+0.1%TiN intermetallic compounds show that an intermetallic compound grain structure modification with titanium nitride nanoparticles causes a sufficient (two times and more) increase of an intermetallic compound strength within wide range of test temperatures, which exceeds existing strength factor of nickel aluminide patterns, obtained through the latest technologies [12]. At that when the temperature comes at 1000 degrees C the effect of grain structure modification on an intermetallic compound strength stops completely, which reveals differences in fracture mechanisms of the investigated intermetallic compounds at temperatures below 1000 degrees C.

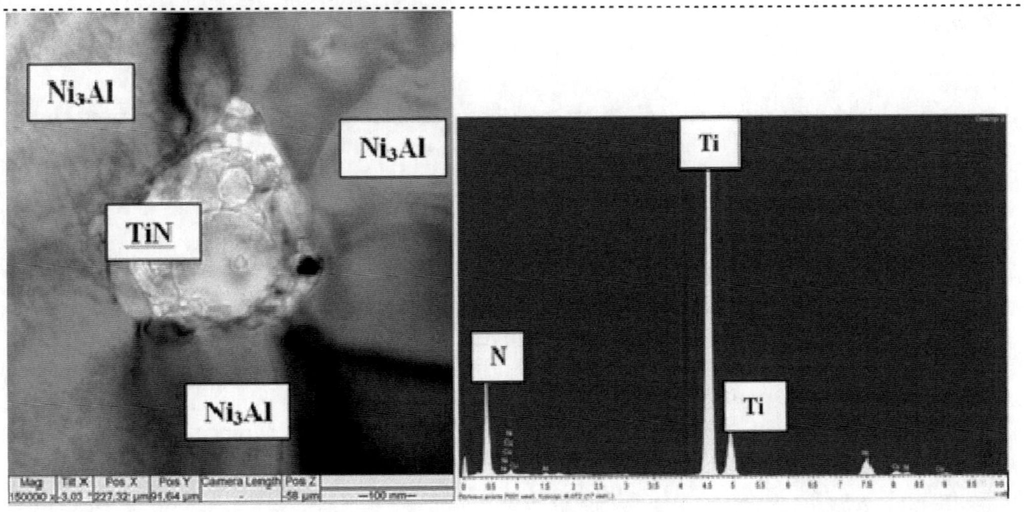

Figure 5. Electron-microscopic image of a TiN particle at the joint of several Ni3Al intermetallic compound grains; and an energy spectrum of a particle element compound.

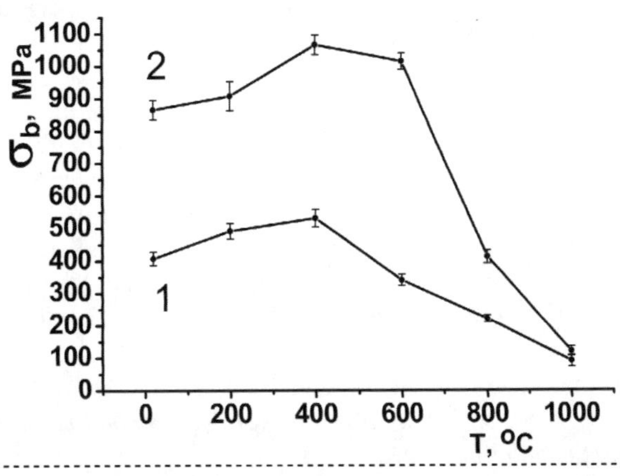

Fig.6. Temperature dependences of a Ni3Al intermetallic compound strength (1) and a Ni3Al intermetallic compound with 0.1% of TiN nanoparticles adding (2).

Conclusion

1. Appending a stoichiometric composition of nanosized particles of a titanium nitride into a powder mixture of nickel and aluminum causes a many-fold loss of an average size of a grain in Ni3Al intermetallic compound, synthesized under pressure.

2. A modificative effect of TiN nanoparticles on a Ni3Al intermetallic compound grain structure is determined by the character of their dissemination in a TiN particle, which appear during the process of intermetallic compound crystallization from melt. TiN nanoparticles are mainly placed along the

boundaries and in grain structures joints, preventing grain growth when a polycrystalline structure of an intermetallic compound is being developed.

3. A many-fold loss of an average size of a grain in Ni3Al intermetallic compound, synthesized under pressure, with TiN nanoparticles adding into the initial powder mixture of nickel and aluminum increases an intermetallic compound strength within a wide range of temperatures (up to 1 000 degree C) twofold and more.

References

[1] Logunov, A. V., Razumovskii, I. M., Stroganov, G. B., Ruban, A. V., Razumovskii, V. I., Larionov, V. N., ... & Poklad, V. A. (2008, August). Theoretical analysis of the alloying system and design of new nickel-base superalloys. In*Doklady Physics* (Vol. 53, No. 8, pp. 438-441). MAIK Nauka/Interperiodica.

[2] Stoloff, N. S. (1989). Physical and mechanical metallurgy of Ni3Al and its alloys.*International Materials Reviews*, *34*(1), 153-184.

[3] George, E. P., Liu, C. T., & Pope, D. P. (1992). Environmental embrittlement: The major cause of room-temperature brittleness in polycrystalline Ni3Al. *Scripta metallurgica et materialia*, *27*(3), 365-370.

[4] Ovcharenko, V. E., Ivanov, Y. F., Mohovikov, A. A., Baohai, Y., & Zhao, Y. (2014, November). Structural state scale-dependent physical characteristics and endurance of cermet composite for cutting metal. In *INTERNATIONAL CONFERENCE ON PHYSICAL MESOMECHANICS OF MULTILEVEL SYSTEMS 2014* (Vol. 1623, No. 1, pp. 455-458). AIP Publishing.

[5] Korostelev, A. B., Zherebtsov, S. N., Sokolov, I. P., & Chumak-Zhun, D. A. (2011). Modification of heat-resistant nickel alloy with a combined inoculator.*Metallurgist*, *54*(9-10), 711-713.

[6] Deevi, S. C., & Sikka, V. K. (1996). Nickel and iron aluminides: an overview on properties, processing, and applications. *Intermetallics*, *4*(5), 357-375.

[7] Stoloff, N. S., Liu, C. T., & Deevi, S. C. (2000). Emerging applications of intermetallics. *Intermetallics*, *8*(9), 1313-1320.

[8] Lapshin, O. V., Ovcharenko, V. E., & Boyangin, E. N. (2002). Thermokinetic and thermal-physics parameters of high-temperature synthesis of intermetallide Ni3Al by thermal shock of a powder mixture of pure elements. *Combustion, Explosion and Shock Waves*, *38*(4), 430-434.

[9] Komshukov, V. P., Foigt, D. B., Cherepanov, A. N., & Amelin, A. V. (2009). Modifying continuous-cast steel with refractory nanopowder. *Steel in Translation*,*39*(4), 363-365.

[10] Sohn, H. Y., & Wang, X. (1996). Mathematical and experimental investigation of the self-propagating high-temperature synthesis (SHS) of TiAl3 and Ni3Al intermetallic compounds. *Journal of materials science*, *31*(12), 3281-3288.

[11] Ovcharenko, V. E., Psakhye, S. G., & Boyangin, E. N. (2014, October). Bulk Nanostructured Ni3Al Intermetallic and Ni3Al-Base Alloy. In *Applied Mechanics and Materials* (Vol. 682, pp. 210-215). Trans Tech Publications.

[12] Jozwik, P., Polkowski, W., & Bojar, Z. (2015). Applications of Ni3Al Based Intermetallic Alloys—Current Stage and Potential Perceptivities. *Materials*, *8*(5), 2537-2568.

IOP Conf. Series: Materials Science and Engineering **142** (2016) 012084 doi:10.1088/1757-899X/142/1/012084

Development the Methods for Preventing Hot Cracking With Use Analysis of Temperature Fields

Evgeniy Ol'khovik

Associate professor,
Admiral Makarov State University of Maritime and Inland Shipping,
Dvinskaya str.5/7, Saint-Petersburg, Russia

E-mail: olhovick@gmail.com

Abstract. The reasons for formation of hot cracks in steel castings having difficult geometric shape of a revolution body are discussed in the paper. Mathematical model for the calculation of metal castings solidification and analysis of temperature conditions, which determines the conditions of hot cracking formation are proposed. The advantage of developed model is ability to use only results of thermal analysis without special calculation of stress-strain state. Criteria for evaluating of thermal conditions for directional solidification and heat dissipation in mold is developed, what allowing to correct technology of casting for defects prevention.

1. Introduction

Hot tearing in complex steel castings is well-known and urgent problem [1] for foundry production. Shrinkage strain in solidifying steel become very different from temperature strain for most metals, the initial stresses are developed in casting immediately upon its crystallization in the mold [2]. Due to complexity and difficulty of practical of experiments with study of process of curing shaped castings in the literature there are only a limited number of experimental studies [3] about temperature curing conditions, what complicates the development of physical and mathematical model of hot cracks formation.

Tasks of development analysis of thermal shrinkable strain of solidifying casting includes features which usually are absent at separate calculation of temperature field or development of strain, and are described by us in works [4-6]. Techniques for an assessment of the temperature conditions of solidifying influencing defects formation for the fittings preforms of the oil and gas industry are offered in these works. In article [7] we have considered in detail the problems of power interaction between the solidifying metal of casting and mold.

The corps details (fig. 1) which have big external diameter in comparison with thickness of the detail walls have been chosen as objects for research in this work. Such design is always a problem for technology development of her production as to ensuring of quality it is necessary to apply at once all complex of technological methods which includes the following tasks: completion of the detail design (use of shrink-resistant edges, technological stock, plate's ribs and so forth); development of technological means for creation of the volume or directed casting metal solidifying; development of special model and industrial equipment, methods for providing a necessary pliability of materials of the casting mold and cores.

IOP Conf. Series: Materials Science and Engineering **142** (2016) 012084 doi:10.1088/1757-899X/142/1/012084

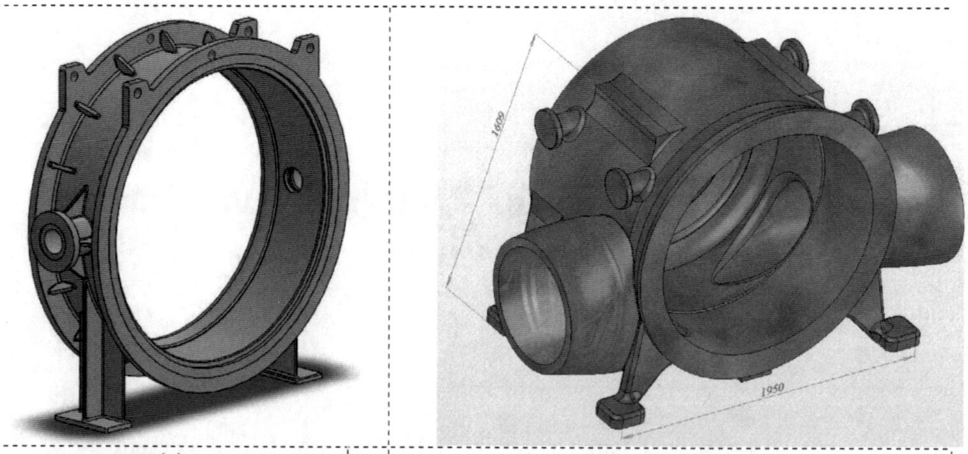

(a) (b)

Figure 1. Objects of research – corps steel castings: (a) corps of valve disk rotary lock with orifice of 800mm; (b) corps of the natural gas supercharger with orifice of 1200mm

Even application of vacuum-film mold technology which is the safest in terms of emergence of hot cracks not always solves a problem. Casting feed by liquid metal occurs through the feeders directly in the corps flange, and for high-quality improvement of feed the special thickenings are structurally made. Thus, walls of casting enough quickly solidify because of small cross section, and the flange exists long time in strongly warmed state under thermal influence of feeder, speeds of metal solidification can differ several times for wall (plate) and flanges.

2. Research methods

The strong temperature gradient in solidifying steel casting of corps on one hand is favorable to creation of volume or directed solidification, but on other hand existence of several centers of crystallization is a source of their competition, and at realization of the complicated metal shrinkage the intermediate zone which shrinkage is insufficiently compensated by liquid metal is formed. Such zone is presented in fig. 2. At decrease in metal temperature below the solidus point metal loses the plasticity, but doesn't gain sufficient durability yet, and due to uneven shrinkage between areas of the corps flange and wall there is a hot tearing.

Considering a real macrostructure (fig. 2.) of a hot crack in the casting wall (steel ISO 200), the existence of a crack between a zone of columnar dendrites and area of the equiaxed crystals was observed distinctly. The last one has been strongly warmed as a result of feed from feeders. The casting wall with columnar dendrites has smaller section and is cooled more intensively from mold and core. For forecasting of hot cracks formation was developed techniques of numerical modeling which reproduce all thermal process of metal solidification. Further the grid of final elements in the form of tetrahedrons is generated, for such operation any available generator FEM can be used.

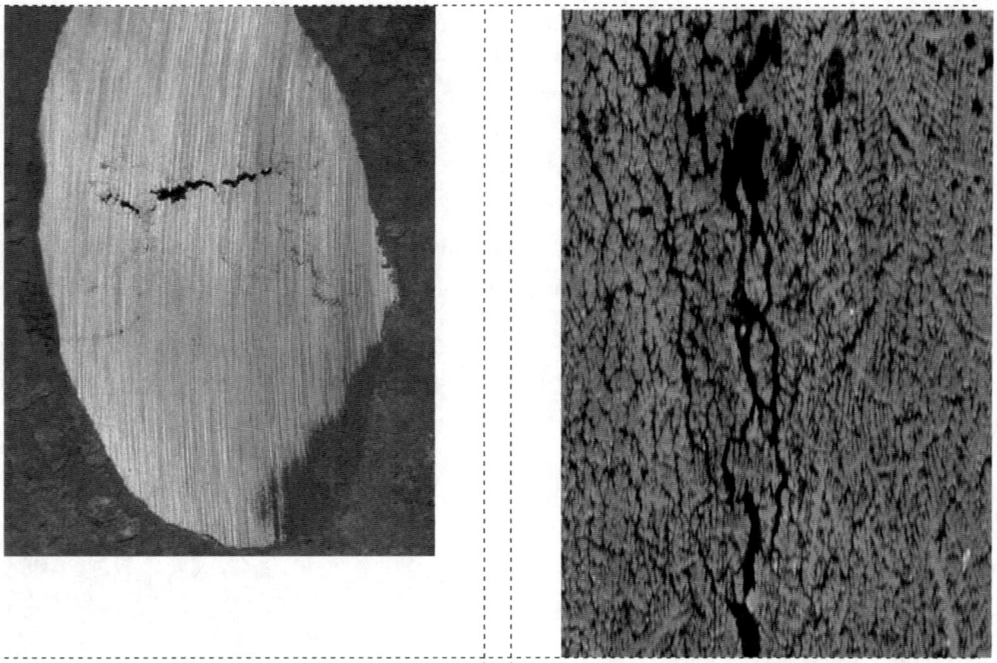

(a) (b)

Figure 2. A hot crack in casting of corps valve disk lock: (a) crack disposition on radial transition of the through passage section to a flange; (b) crack macrostructure, x10

Standard methods for calculating of heat conduction problems cannot be used for the simulation of solidification metal of casting in mold, because it does not take into account a number of physical processes, such as a phase transition or release of latent heat of crystallization.

For computer realization of calculation of cooling of casting in mold the special calculation procedure was developed where the equation of thermal balance for i-element of the grid splitting is as follows:

$$\sum_{j=1}^{m} k_{i,j}^{n}(T_i^n - T_j^n)F_{i,j}\Delta\tau = C_i G_i (T_i^n - T_j^n) \qquad (1)$$

where k – heat transfer coefficient between elements i and j; m – quantity of not-adiabatic sides of an element (m=4); F – the area contact between elements i and j; C_i и G_i – specific heat capacity of the element material and its weight; $\Delta\tau$ – time step; n – time index.

On contact of an i-element with an j-element the heat transfer coefficient is determined by a formula: $k_{i,j} = 1/(L_i/\lambda_i + L_j/\lambda_j)$, where L – the distance from the center of gravity of an "i" - element (with the heat conductivity coefficient λ) to the next element in the direction of the thermal stream movement. Preliminary all elements of FE grid are indexed (on belonging to materials and on boundary conditions of a heat transfer) that allows to change parameters of heat conductivity, a thermal capacity, etc. properties in process of calculations. For determination of heat and physical properties of materials which participating in a heat transfer their temperature dependences which are stored as piecewise linear dependences in the special database are used. As a result of calculations are temperature fields in metal, mold and core was formed on each time step. The Scheil-Gulliver model adjusted for the non-equilibrium of crystallization process was used as model of crystallization which allows to consider more precisely emission of latent heat of crystallization. Speed of solidification process was calculated as a ratio of a total time of element (i) solidification to the temperature range of solidification $V_i = \dfrac{T_{liq} - T_{sol}}{\tau_i}\left[\dfrac{^oC}{\sec}\right]$

As soon as temperature of metal has decreased lower than temperature of alloy liquids, there are conditions for firm phase formation according to the phase diagram. The process of crystallization begins and firm phase - g_S occupies some fraction $0 < g_S < 1$ from the metal volume which gradually increases with advance of the front of solidification. For the conditional "equilibrium" state of solidification comes to an end with achievement of solidus temperature and full solidifying of metal $g_S = 1$. For the case of equilibrium crystallization this range depends on the structure of an alloy and the inclination of the liquids line on phase diagram. Development of equilibrium crystallization process - $g_s(T)$ is described by the famous "lever rule". But in actual practice the temperature range of metal crystallization will also depend on the non-equilibrium factor – $g_s = 1 - \left(\frac{T_f - T_{liq}}{T_f - T}\right)^{\frac{1}{1-k}}$ first of all connected with local speed of solidification which for complex casting will have difficult distribution gradient of temperature (fig.3).

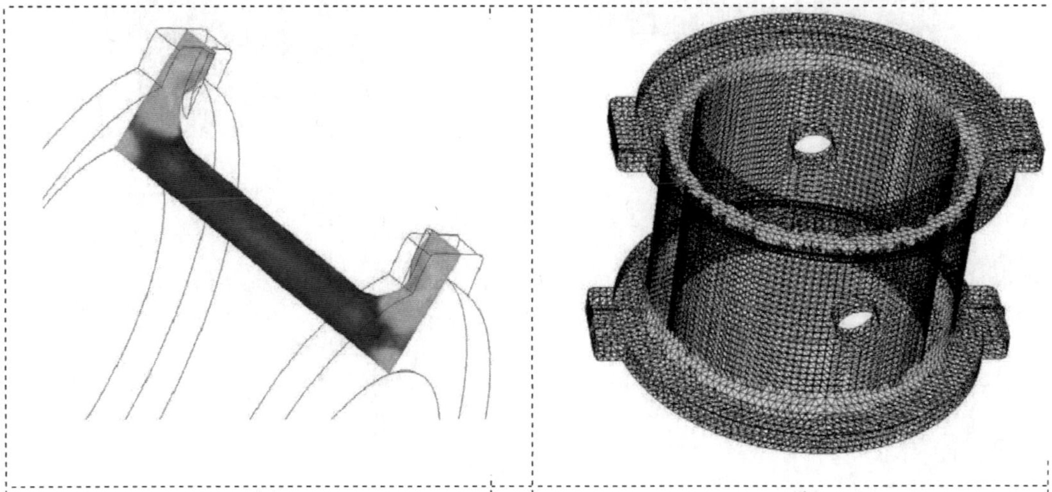

(a) (b)

Figure 3. Temperature gradient in the section of solidifying steel casting – (a) and computational area by criterion of danger of hot cracks formation – (b)

3. Results and Discussion

Existence of the powerful gradient of speeds of metal cooling leads to a situation when one local area of casting has completely solidified (passage section of the corps of the disk lock valve) at the same time it has managed to realize own shrinkage at the expense of still pliable liquid area. The casting zone with a low speed of cooling can't freely realize shrinkage without power consequences any more. Existence of the area which is in intermediate region between a zone of fast and slow solidifying is dangerous in aspect of the hot cracks formation. Distinction in cooling speeds, and hence in rate of solidification in turn seriously determines degree of feed by liquid metal for various local zones of casting. Such distinction in speeds of solidifying can be critical in terms of formation of thermal shrinkage of metal which at a strong gradient will lead to a hot tear in field of transition from one local zone to another. For casting of corps of rotary lock the dangerous zone on thermal gradient border (fig. 3(b), it is highlighted in the color) was revealed that corresponds to practice. Hot cracks usually hit such castings through on radial transition from cylinder to flange.

The analysis of thermal deformations calculated according to non-stationary temperature fields causes a number of complications for understanding of the mechanism of the complicated metal shrinkage. The incompressible nature of plastic deformation creates restriction in each element of FE grid. When the number of the restrictions arising because of incompressibility surpasses the number of degrees of freedom, blocking takes place as there is no possible decision for this case. The reduced order of calculated iterations for hydraulic components of tension (sphere tensor) can be the decision that can lead to inadequate results. The effect connected with it follows from an order of fields of thermal tension in the elements of FE grid. If nodal temperatures are interpolated for obtaining of temperature values in the element, they are used for determination of thermal tension, the field of thermal tension has the same order, as well as the field of movements in the element. The full deformation calculated from partial derivative movements, 10 times less. This incompatibility can also lead to problems at the decision. The need to avoid the fields of incompatible deformations is well-known, and in our case we always has within calculation a set of materials with various and changing in time elastic properties, and in case of the superficial accounting of such features the considerable mistakes can appear. The basis for the analysis of the stress-strained state of metal castings during hot cracks formation is the information on temperature fields [9] and microstructure [10] formation. The modeling of hot tearing formation during DC casting can be identified as a particular task [11,12].

However, the solution task about development of tension and deformations demands the special approach based on load mechanical model of interaction of solidifying metal casting with the sand mold offered by us in works [6, 7].

4. Conclusion

To prevent hot cracks formation it is possible at the stage of design process to carry out the modeling of solidification and to make the relevant amendments to technology of casting. The proposed criterion for assessing of the risk of hot cracking formation can be formulated as follows: the availability of areas in the casting containing several unrelated thermal centers as well as the presence of a strong gradient of cooling rates. Under such gradient it is possible to consider a difference in solidification speeds at distance not less than two minimum thickness of casting walls or existence of two or more thermal centers with a cooling speed difference in 30÷50 of °C/sec.

References

[1] Monroe C and Beckermann C 2005 Development of a hot tear indicator for steel castings *Materials Science and Engineering: A* **413** 30–36. (doi:10.1016/j.msea.2005.09.047)

[2] Chojecki A, Telejko I and Bogacz T 1997 Influence of chemical composition on the hot tearing formation of cast steel *Theoretical and applied fracture mechanics* **27.2** 99–105. (doi:10.1016/S0167-8442(97)89529-8)

[3] Liu B C, Kang J W and Xiong S M 2001 A study on the numerical simulation of thermal stress during the solidification of shaped castings *Science and Technology of Advanced Materials* **2.1** 157–164. (doi:10.1016/S1468-6996(01)00041-9)

[4] Desnitsky V V, Ol'khovik E O and Molchanyuk R A 2007 Investigations of deformative-strength properties of the molds materials during metal crystallization *Zagotovitelnye proizvodstva v mashinostroenii* **1** 16–18

[5] Desnitskaya L V, Ol'khovik Y O, Pirainen V Y, Glebov S M, Semyonova Y M and Matveyev Y A 2010 Problems of Casting Process Simulation *Foundry. Technologies and Equipment* **8** 25–28.

[6] Ol'khovik E O and Desnitsky V V 2007 Development of the mechanical concept of hot cracks formation in steel castings *Litejshhik Rossii* **11** 21–25.

[7] Ol'khovik E O, Desnitskii V V and Molchanyuk R A 2007 Interaction between casting and mold during solidification *Steel in Translation* **37.5** 422–424. (doi: 10.3103/S0967091207050051)

[8] Ol'khovik E O, Butsanets A A and Ageeva A A 2015 Use of the distributed computing at the castings solidification simulation *International Conference on Mechanical Engineering, Automation and Control Systems (MEACS)* pp. 1–4 (doi: 10.1109/meacs.2015.7414905)

[9] Liu J X and Liao R D 2010 Simulation of Thermal Stress and Hot Tearing in Engine Block Casting *Advanced Materials Research* **154-155** 1571–1574. (10.4028/www.scientific.net/amr.154-155.1571)

[10] Ridolfi M R 2014 Hot Tearing Modeling: A Microstructural Approach Applied to Steel Solidification *Metallurgical and Materials Transactions B* **45.4** 1425–1438. (doi: 10.1007/s11663-014-0068-1)

[11] Arth G, Ilie S, Pierer R and Bernhard C 2015 Experimental and Numerical Investigations on Hot Tearing During Continuous Casting of Steel *Berg Huettenmaenn Monatsh* **160.3** 103–108. (doi: 10.1007/s00501-015-0345-6)

[12] Alizadeh M 2015 Study on hot tearing tendency during continuous casting of steel by overall hot tearing susceptibility (OHTS) *International Journal of Cast Metals Research* **28.1** 20–27. (doi: 10.1179/1743133614Y.0000000124)

IOP Conf. Series: Materials Science and Engineering **142** (2016) 012085 doi:10.1088/1757-899X/142/1/012085

Ensuring Machine and Tractor Aggregates Operability

G V Redreev, PhD in Engineering Sciences

Omsk State Agrarian University named after P.A. Stolypin
644008, Omsk, Institutskaya Pl. st. 1, Tel. (8-3812) 650-172

E-mail: gv.redreev@omgau.org

Abstract: the operability of the machine and tractor aggregates is ensured by processes which occur in machine units and considered as technical systems. In order to develop theoretical understanding of the processes in technical systems as the basis and purpose of the repair-serving actions, the author's concept is presented which relies on on the basic concepts of "processes in technical systems", "maintenance and repair of performers", "maintenance and repair of technology" "objectives of the maintenance and repair". Analysis of the basic concepts of "processes in technical systems" made possible to distinguishing four types of relations: of order, stipulation, exactingness, and non-contradiction. It is shown that the implementation of maintenance and repair of technology should be conducted according to the assessment of the effectiveness of processes in technical systems, revealed in complex diagnosis. The perfection of the design of the machine in terms of its technical operation can be estimated according to the degree of consistency of processes in technical systems, purposes of maintenance and repair. In order to increase the efficiency of the lubrication system,the modernised design of the centrifugal oil filter with permanent control of its cleaning power is offered, which allows changing the technology of the maintenance of engine lubrication system by separating the operations of crankcase oil replacement and the rotor filter cleaning.

1. Introduction

Maintaining and restoring functionality of machine and tractor aggregates (MTA) is provided in the result of of employees work with certain qualification. This work is focused and goal oriented and done with taking into account the processes that go on in the units, systems and components of MTA. To form the concept of operations of a group of such people, we have defined the following, presented earlier, basic concepts taking into consideration works of Teslinov AG [1, 2]: maintenance and repair performers; technology of maintenance and repair; processes in technical systems (TS hereinafter); objectives that are achieved as a result of activities of maintenance and repair performers.

Since the processes occurring in TS are both the basis as well as the objective of maintenance and repair actions aimed to increase the efficiency, theirfurther profound and detailed analysis is needed.

When MTA are used as intended,gradually its technical condition worsens. These processes are related tothe deterioration of dynamic and kinematic modes of components, aggregates and separateunits of MTA, used under conditions of lubrication systems applications and temperature changes. They may be various for the same units included in the project of machine utilization [3] in the conditions of different climatic zones.

These processes are presented in picture 1.

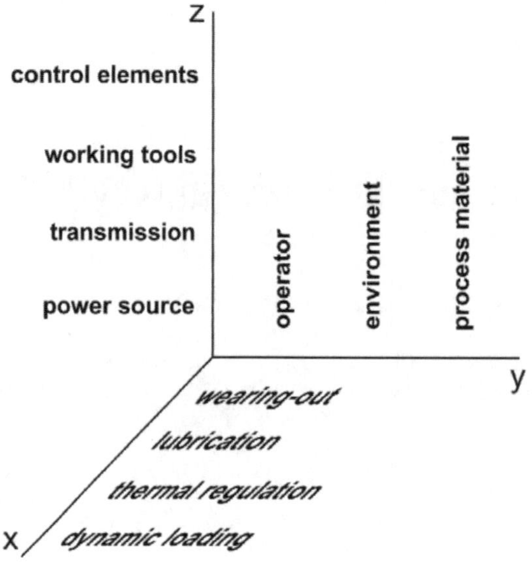

Picture 1. - Processes in TS

The diagram in Pic. 1: «x» axis represents aggregated groups of processes occurring in the TS units and aggregates; the «z» axis shows the necessary and sufficient components of autonomous TS [4]; axis «y» performs people interacting with machines and affecting them, the environment and the material being processed.

On the base concept of "processes in TS" the following relationships take place:

1. Relations of order

These are relations between the processes in the TS. Processes can follow both parallel and in series; may cross other processes such like, for example, the process of lubrication.

2. Relations of stipulation

These are relations between the concepts of "processes in TS" and “technology of maintenance and repair”. Each process that occurs in TS defines the demands for technology of maintenance and repair.

3. Relations of exactingness

These are relations between the concepts of "processes in TS" and "maintenance and repair performers". Activities of maintenance and repair performers depend on processes that take place in the TS.

4. Relations of non-contradiction

These are relations between the processes in the TS and objectives of the maintenance and repair. Processes in the TS (see Picture 1, the x-axis) worsen its technical condition; in this case we can say that they contradict the objectives of the M and R.

These relations on the basic concepts are represented in picture. 2.

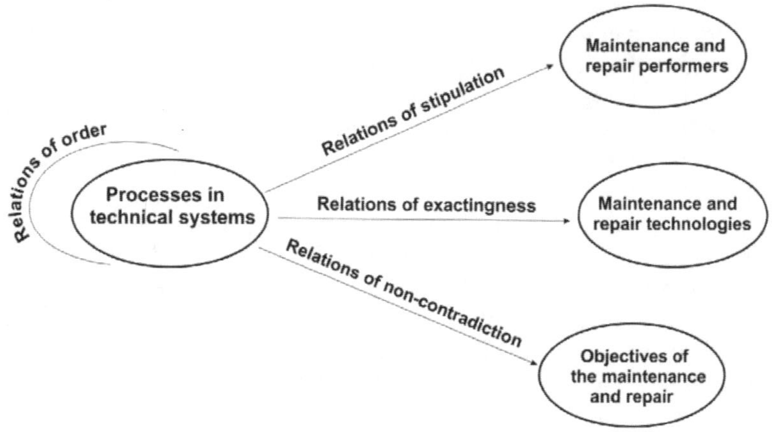

Picture 2. - Relationships to the basic concepts "in the processes of the TS"

2. Results of the research
2.1 Relations on the basic concepts
2.1.1 Relationship of order
The processes that take place in technical systems worsen their technical condition. We determined four types of processes (see Picture 1.): wear, lubrication, thermal regulation and dynamic loading. The process of wear, lubrication processes deterioration and thermoregulation lead to gradual failure. Change of processes of dynamic loading – vibration increase, the instability of the radial forces on shafts, torsional vibrations, - results insudden failures.

The processes we defined have certain temporal and spatial order. This order depends on the complexity of the units constituting the aggregate. Since the MTA is composed of individual units, these processes can be distinguished as for machines as well as for components of their units and aggregates [5]. At the same time, it is necessary to consider the influence that a person (operator of MTA) has on these processes, as well as the impact of the environment and the material. The complexity and uncertainty of this effect is amplified by the fact that the person is also exposed to the surrounding environment [6].

The set of processes of the TS (in accordance with the composition of TS see Pic. 1, the z axis) can also be structured as follows:

1) Main process that implements the main function of the TS.
2) Processes that facilitate the main process of the TS.
3) Processes that support providing processes.
4) Management processes.

This separation allows ranking processes based on their functions. The degree of approximation processes to the main options can be considered as an appropriate degree of correlation processes in the TS and objectives achievable with the work efficiency of the MTA. The degree of correlation is the basis for ranking of technology of maintenance and repair (maintenance and repair operations). It should be noted that the ranking is meaningful only when the target of maintenance and repairis defined.

2.1.2 Relations of stipulation
The processes which occur in the TS define the requirements to the technology of SP and P. When performing diagnostics of units and systems of the MTA, through which the main function is implemented, diagnostic information about the state of the components and assemblies implemented in supporting processes is necessary, and thensupported processes are needed. In this connection it is necessary to conduct a comprehensive diagnosis of machines that comprise the MTA, with the aim to determine the parameters for all structured processes. This will permit to determine required repair-

serving influences. The implementation of maintenance and repair technologies will be performed depending on the processes effectiveness evaluation in the TS defined via comprehensive diagnosing. Maintenance and repair technologies can be realized depending on the type of TS processes as well as on their effectiveness evaluation.

Processes in the TS can be divided in terms of the way they flow on a linear (continuous), cyclic and cyclic breaks. Linear processes require continuous monitoring the aim of which is to determine pre-failure condition and conducting maintenance and R timely. For cyclic processes the control of individual stages of the cycle can to be combined with the entire cycle control.

2.1.3 Relationship of exactingness
Processes in the TS determine the activities of maintenance and R performers. The design features of the machines that comprise the MTA, determine the specificity of processes in these machines. In this regard, diagnostic parameters with maximum informative value should be chosen. Obtained information will be dependent on the adequacy of diagnostic instrumentation of physical entities occurring during the processes in the machines, errors in measurement, and the correct interpretation of the set of measured diagnostic parameters.

2.1.4 Relationship non-contradiction
The degree of inconsistency is taken corresponding to the degree of deterioration of the technical condition. The longer the failure-free operation of a machine unit, the greater is the consistency (non-contradiction) of processes in this unit. According to the degree of consistency of processes in the TS, the maintenance and R., we can judge about the perfection of the construction of the technical system. Thus, the measure can be taken as a criterion of the perfection of machine construction in terms of its technical exploitation. This criterion has a qualitative difference from such indicator as maintainability. We will define it as the coefficient of the technical perfection of the structure. Numerically it can be determined through the degree of consistency of processes. On the basis of the criterion of the excellence of machine design it is possible to identify directions of perfection of the structure of such a machine. When using the machine for the purpose of it is simpler to find the solution of problems of optimization of process of technical exploitation.

2.2 Engine lubrication system
2.2.1 Lubrication system effectiveness
Tractor engines of Russian production are not sufficiently equipped with effective built-in diagnostic tools. This is of a current importance for the units of system of greasing of engines due the fact that it is known that about 90% of faults are the result of wear. To ensure trouble-free operation of the engines the constant monitoring of both the oil quality and condition of units lubrication systems is of a high importance [7, 8].

In practice of operation of automotive diesel engines during servicing of the lubrication system, the problem is observed, namely that the centrifugal oil filter, cleaning oil from abrasive particles, is an important component of engine lubrication system; however, the lack of filter state control means leads to delayed cleaning of the rotor, which may be a cause of the engine underutilization from 10 to 50% [9].

We hypothesized that the full use of automotive engines resource can be achieved through efficient operation of centrifugal oil filter via permanent operational control of its cleaning ability.

The study of the efficiency of the centrifugal filter can provide solution of two tasks that are of theoretical interest for us and that have indisputable practical application/

1) task of justification of a control device design in order to regulate the frequency of cleaning the rotor of the centrifuge deposits;

2) task of controlling oil replacement frequency.

IOP Conf. Series: Materials Science and Engineering**142** (2016) 012085 doi:10.1088/1757-899X/142/1/012085

2.2.2 Centrifugal filter control device

The basic parameter for the centrifugal oil filters is the rotor revolution frequency. Its limit value for different motors varies from 5000 to 5500 min^{-1}. The drop of rotor frequency can be explained by different factors:

- accumulation of significant amount of deposits in the rotor sleeve;
- decrease of inner dimensions of nozzle holes (in nozzle centrifuges);
- crankcase oil viscosity increase;
- oil pressure decrease in the rotor caused by oil leakages via worn out "rotor axis-rotor bushing" coupling;
- oil pressure decrease in the engine lubrication system.

According to the applied in diagnostics process division into two stages aimed to detect the rotor frequency drop lower than the limit value, the control of the frequency is necessary. From now forth, by using the algorithm developed by us, it is possible to discover causes of the rotor frequency drop.

Centrifuge rotor balancing is done on the manufacturing plant, and according to the recommendations it is forbidden to set any additional details on the rotor as it can put it out of balance and cause rotor axis break down as it becomes weakened by the port holes for oil feeding and removal.

We have developed centrifugal oil filters of two designs with variable reactance transducers which have utility model patents. Iron containing rotor details served as the source of impulses. In the first modification the steel nozzle of the rotor were used as the source [10]. The other modification employed the rotor nut [11]. The crankshaft position sensor of VAZ engine was used as the sensor.

To register the centrifuge rotor frequency drop lower than the limit value a transducer and an indicator were developed set together in one case (pic.3).

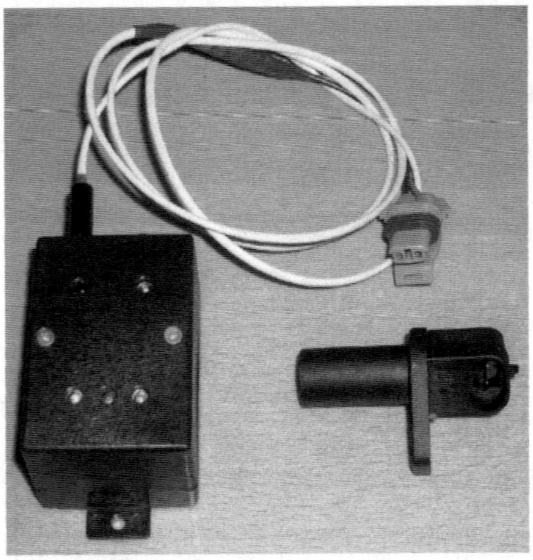

Picture 3. – Indicator with the transducer and the sensor (left)

The indicator is made using light emitting diodes, when the rotor frequency drops lower than the limit value the red diode lights up.

The transducer is made on a programmable microcontroller with specifically designed programming unit (see pic. 4) in order to make it usable on centrifuges of other manufacturers.

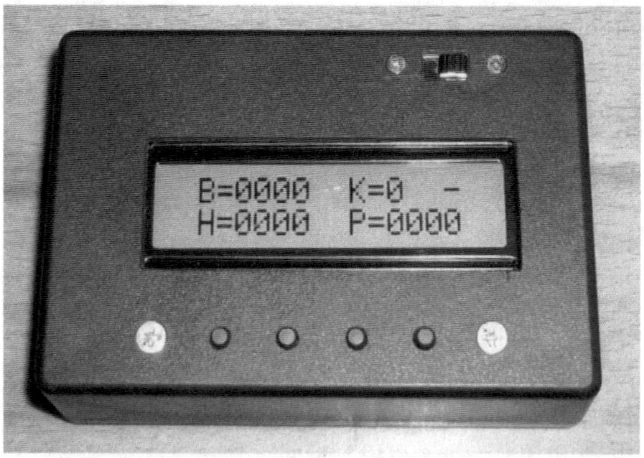

Picture 4. Programming unit

The oil replacement frequency is determined by its condition which is characterized by the interdependence between its numerical values and its qualitative characteristics and the motor operation time:

$$\Pi = F(t) \tag{1}$$

One of the basic characteristics of oil which can be the reason for its replacement is the change of oil viscosity. It is considered counterproductive to use oil if its viscosity increases by 35 %. In this paper we will not consider cases of crankcase oil viscosity decrease because of its contamination with fuel or coolant.

Oil viscosity influences centrifuge operation. One may state that viscosity increase can be allowed until it does not influence the centrifuge rotor frequency, that means until oil viscosity value does not decrease to the limit of the centrifuge's decontamination capacity (its rotor frequency). Thus the centrifuge rotor frequency may serve as an additional parameter for identifying the moment of oil replacement in the engine crankcase.

2.2.3 Oil replacement frequency control

There are two ways of oil replacement frequency control - by changing oil quantity and by using oil with different replacement frequency.

Let us consider oil replacement frequency control by changing the oil quantity. According to publications on the topic [12] the change of crankcase oil volume in internal combustion engines does not influence overall oil consumption. It means that the increase of oil feed leads to more frequent oil replacement in the same proportion. However, in the conditions of normal operation this method could be difficult to apply from the technical point of view because of equipment design features; oil leakages caused by oil feeding lines or sealing units failures can lead to significant oil losses.

In common situation the condition for carrying out oil replacement separately from continuous operation periods is:

$$C_{rec} + C_{diag} < \Delta C \times t_{rplc} + V \times C_i \times P_k(T_{\text{п}}) \tag{2}$$

where C_{rec} - cost of crankcase oil pan reconstruction in rubles;

C_{diag} - cost of a diagnosing device in rubles;

ΔC - specific value of losses because of tractor (vehicle) downtime, in rubles/h;

t_{rplc} - oil replacement time, h;

V - oil pan capacity, l;

C_i - cost of i-brand oil in rubles/l;

IOP Conf. Series: Materials Science and Engineering **142** (2016) 012085 doi:10.1088/1757-899X/142/1/012085

$P_k(T_n)$ - Probability of connecting oil lines and sealing parts failure during the period of oil usage.

Another parameter of oil that can cause its replacement is the content of mechanical impurities.

The speed of impurities separation in the centrifuge is calculated in the following way [13]:

$$\frac{dw}{dt} = 0.01x\gamma Q\varphi \qquad (3)$$

where x – is the impurities content on the centrifuge entry in % (of weight)

γ - suspension density, kg/l

Q – centrifuge flow rate, l/min

φ - oil decontamination index [13]

In general case the change of overall impurities content in oil can be represented as the difference between impurities content in the oil that enters the centrifuge and that leaves it.

As the speed of impurities separation is proportional to their concentration and, correspondingly, to the total content of impurities in oil, the following can be put forth:

$$\dot{w} = k - \alpha w \qquad (4)$$

where k - is the speed of impurities feed in the oil in g/h

a – is proportionality factor taken from the right part of equation (3)

The solution of the equation is as follows:

$$w(t) = \frac{k}{a}(1 - e^{-at}) \qquad (5)$$

Dividing both parts of the equation (5) on the weight of the crankcase oil we receive the dependence of mechanical impurities concentration in oil from the time of engine work.

Calculations for engine YMZ-238NB centrifuge were done. With crankcase capacity V=32 l, suspension density γ = 900 g/l, centrifuge flow rate Q = 720 l/h and value of oil decontamination index φ = 0,01 [10], the value of the proportionality factor a according to (3) comprises 0,225 hour^{-1}. The speed of impurities supply into oil comprises 50 mg/hp hour [13] or with engine YMZ-238NB power equal to 215 hp, k = 10,75 g/hour.

The solution of equation (4) with known values of a and k will be as follows:

$$w(t) = 47{,}48(1 - e^{-0{,}225t}) \qquad (6)$$

Likewise the values of numerical indices can be calculated with different increased values of YMZ engine crankcase capacity.

The graphs of the dependence (6) are presented on the picture 5.

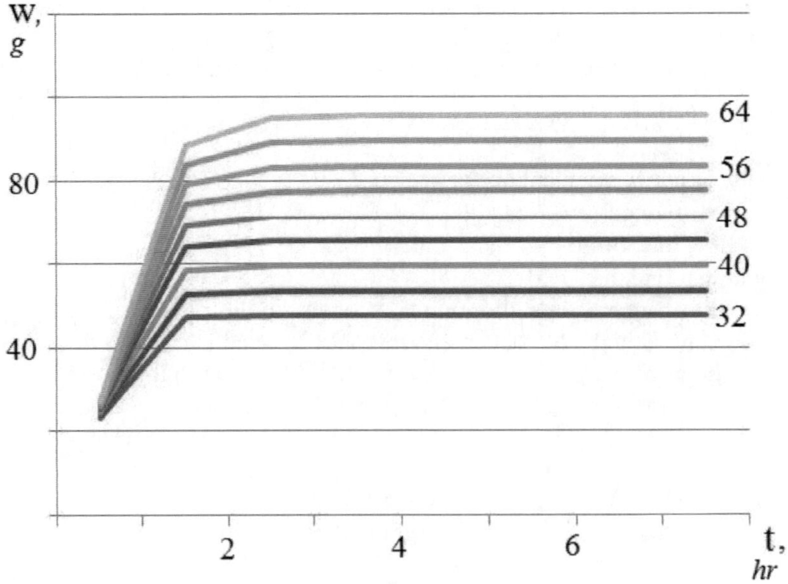

Picture 5. - Dependence of impurities content in the crankcase oil of the engine YMZ-238NB from the operation time with different crankcase capacity (from 32 to 64 l)

Certainly, the presented graphs illustrate somehow idealized dependences. In reality impurities content increases with time, that means *k=f(t)*. Dependence (4), with preliminary assumption of linear dependence *k* from *t*, will be:

$$\dot{w} = k(t + B) - \alpha w \quad (7)$$

where B is the proportionality factor.

The solution of the equation will be:

$$w(t) = C(1 + Dt - e^{-at}) \quad (8)$$

where *C,D* – are proportionality factors.

Experimental studies as well as the calculation of factor *B* numerical value in dependence (7) and applying the solution in equation (8) together with known limit values of impurities content in crankcase oil, makes it possible to calculate engine operation time until oil replacement with different crankcase capacities.

3. Conclusions

The concept, developed by the authors of the paper, relies on the basic concepts of "processes in technical systems", "performers of maintenance and repair, technology maintenance and repair, technical maintenance and repair", has permitted us to allocate four types of the arising relations: order, demand, dependence, and non-contradiction. The implementation of technologies and maintenance must be carried out depending on the assessment of the effectiveness of processes in technical systems, which is identified during the comprehensive diagnosis. Important improvements in the technical operation of the machines are the increase of efficiency of functioning of system of greasing. For this purpose, the improved design of centrifugal oil filter with continuous control of its cleansing abilities is proposed, which allows changing the technology of maintenance of the lubrication system of engines through the separation of actions of crankcase oil replacement and rotor filter cleaning.

IOP Conf. Series: Materials Science and Engineering **142** (2016) 012085 doi:10.1088/1757-899X/142/1/012085

Reference

[1] Conception thinking in solving complicated and knotted problems. / Teslinov A.G. – S. Petersburg: Piter, 2009. – 288 p.
[2] Conceptual design of complicated solutions. / Teslinov A.G. – S. Petersburg: Piter, 2008. – 288 p.
[3] Okunev G.A., Ruzhyev L.L. Technical reequipment of manufacturing enterprize, search for effective options // ChSAA Bulletin. – 2014. – Vol. 70. – Pp. 108-115.
[4] A chance for adventure. / Compl. by A.B. Selyutsky. – Petrozavodsk: Karelia. 1991. – 304 p. with pict. – (Technics – youth - creativity)
[5] Maintenance fundamentals, 2nd Edition / R. Keith Mobley - Elsevier Inc., 2004. – 418 p.
[6] Maintainability, Maintenance, and Reliability for Engineers / B.S. Dhillon - Taylor & Francis Group, LLC, 2006. – 214 p.
[7] Nakagawa T. Maintenance theory of reliability. - (Springer series in reliability engineering). Springer-Verlag London Limited. 2005, p.274
[8] Khairy A.H. Kobbacy, D.N. Prabhakar Murthy Complex System Maintenance Handbook. - (Springer series in reliability engineering). Springer-Verlag London Limited, 2008, p.648
[9] Alliluev V.A. Technical operation of machine and tractor fleet / V.A. Alliluev, A.D. Adanyin, V.M. Mikhlin. – Moscow: Agropromizdat, 1991. – 367 p.
[10] Utility model patent №100139 "Centrifugal oil filter" Published 10.12.2010 bull. №34 / Redreev G.V., Degtyarev A.A., Siryak A.S., Klyuev I.A.
[11] Utility model patent №134994 "Centrifugal oil filter" Published 27.11.2013 bull. №33 / Redreev G.V., Siryak A.S., Klyuev I.A.
[12] Under the general editorship of A.S. Orlin, M.G. Kruglov. Internal combustion engines: Systems of reciprocating and compound engines / S.I. Efimov, N.A. Ivashchenko, V.I. Ivin et al. – 3rd Edition, updated and revised. – Moscow, Mashinostroenie, 1985. – Pp. 243-244.
[13] Grigoryev M.A. Purification of oil and fuel in automobile and tractor engines. / M.A. Grigoryev. . – Moscow, Mashinostroenie, 1970. – 270 p.

IOP Conf. Series: Materials Science and Engineering **142** (2016) 012086 doi:10.1088/1757-899X/142/1/012086

Determination of Geometrical Parameters for Semi-Rolling Bevel Precessional Gears With Straight Teeth

V Syzrantsev[1], K Syzrantseva[2], M Milanovic[3] and A Pazyak[1,6]

[1] Department of Machines and Equipment for Oil and Gas Industry, Tyumen State Oil and Gas University, 38 Volodarskogo Street, Tyumen, 625000, Russian Federation
[2] Department of Cybernetic Systems, Tyumen State Oil and Gas University, 38 Volodarskogo Street, Tyumen, 625000, Russian Federation
[3] Faculty of Mechanical Engineering, University Of Belgrade, 16 Kraljice Marije Street, Belgrade, 11120, Serbia

E-mail: a.a.pazyak@gmail.com

Abstract. The aim of this paper is to analyse the semi-rolling bevel precessional gear with straight teeth. The given method of cutting pinion and gear teeth significantly simplifies its manufacture technology. Authors have given a system of equations to study the analysis of generating pinion tooth surfaces and geometry of meshing for semi-rolling bevel precessional gear with straight teeth. A computer program was developed to enable implement formulas provided in the paper. By using this program authors obtained contact lines on the pinion tooth surface and number of tooth pairs in gears mesh.

1. Introduction

The process of generation of a bevel gear [1, 2, 3] in comparison with cylindrical and worm (spiroid [4]) gears is more difficult. During the last decades, many papers [5, 6, 7] have been directed to the determination the optimal geometry of bevel gears which provide the required loading capacity for operating conditions. So we propose to use gearboxes based on bevel precessional gear (figure 1). Bevel precessional gears are applied in compact gearboxes for oil-and-gas equipment for transmitting rotation and torque [8, 9]. The most important characteristics for the quality of bevel precessional gearboxes are efficiency (about 0.9), wide range of gear ratios, low staring torque which is crucial for wide range of operating temperatures [8, 9]. Currently, the most fully researched bevel precessional gears are those with double-concave and double-convex teeth [8, 9]. The process of gear cutting in not only labor-intensive technology, but also is incredibly complex. It requires four re-settings of the cutting machine, as each tooth surface is cut separately. In this article we study semi-rolling bevel precessional gears with straight teeth. Despite the absence of localized contact in the studied gear, as a result of multiple teeth contact, it is rival product while manufacturing high-loaded gearboxes which operating under low angular velocity and short term duty.

---

[6] To whom any correspondence should be addressed.

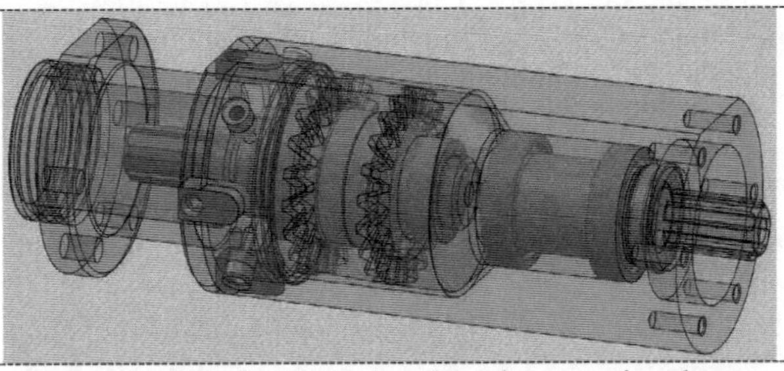

Figure 1. Gearbox based on bevel precessional gear

2. Analysis of gears mesh in the semi-rolling bevel precessional gears with straight teeth

Figure 2 describes design scheme of scheme of the semi-rolling bevel precessional gear with straight teeth. Cutting of the gear tooth space is done by a cutting tool with a straight-line cutting edge. Gear remains fixed during gear cutting.

Prior to gear cutting each of the following the gear tooth space, the gear rotates at an angle equal to tooth pitch angle, i.e. a method of «unit division» is realized. As cutting tools we can use cutters (cutting without generating on a gear-shaping machine), end-mill type cutters or side milling cutters (cutting on a milling machine with a turntable).

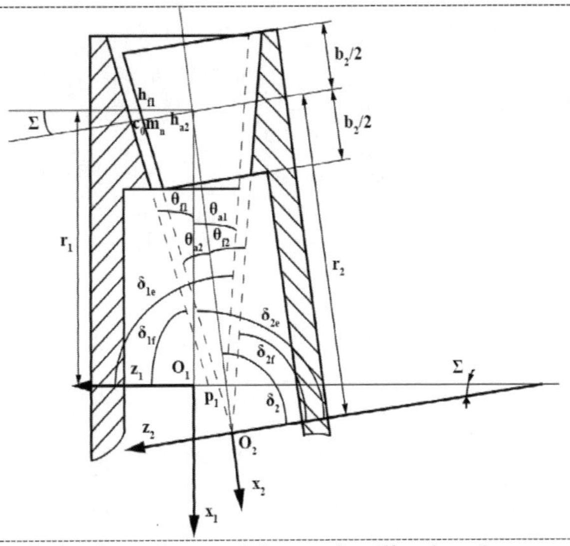

Figure 2. Design scheme of the semi-rolling bevel precessional gear with straight teeth

From the described gear cutting method it follows that the gear tooth surface is a plane. The reference surface of tooth is a plane that goes through point M (center of action) parallel to the plane $x_2o_2y_2$ (perpendicular to the axis z_2). The coordinate system S_2 (x_2, y_2, z_2) is rigidly connected to the gear. In the coordinate system $S_p(x_p, y_p, z_p)$ tooth surface, which axis y_p is directed along the normal towards the gear tooth surface, is a plane and calculated by equation (1):

$$x_p = u \;\; y_p = 0 \;\; z_p = h \qquad (1)$$

where u is the line coordinate along the tooth length; h is the line coordinate along the tooth profile.

Let us express the position vector $\overline{r_p}$ of the tooth surface («plane») in the coordinate system $S_p(x_p, y_p, z_p)$ as a row matrix $\tilde{r}_p = \left\| x_p; y_p; z_p; 1 \right\|$, elements of which are coordinates x_p, y_p, z_p, and the position vector of the tooth surface $\overline{r_2}$ in the system S_2 (x_2, y_2, z_2), - as a row matrix $\tilde{r}_2 = \left\| x_2; y_2; z_2; 1 \right\|$, elements of which are defined by coordinates (x_2, y_2, z_2). Then tooth surface in the system S_2 (x_2, y_2, z_2) is described as the formula:

$$\tilde{r}_2 = \tilde{A}_{2p} \cdot \tilde{r}_p \tag{2}$$

where $\tilde{A}_{2p}$ is the square matrix (4 x 4) [7, 8, 9], which is define coordinate transformation from coordinate system S_p to coordinate system S_2.

Solving equation (2), we obtain:

$$\left.\begin{aligned} x_2 &= u \cdot \cos\theta_{f2} - h \cdot \sin\theta_{f2} \cdot \cos\alpha_n - r_2 \\ y_2 &= -h \cdot \sin\alpha_n + t \\ z_2 &= u \cdot \sin\theta_{f2} + h \cdot \cos\theta_{f2} \cdot \cos\alpha_n \end{aligned}\right\} \tag{3}$$

where θ_{f2} is the gear dedendum angle; α_n is the pressure angle; r_2 is the mean cone distance of gear; t is a half of the tooth thickness.

In the studied semi-rolling bevel precessional gear with straight teeth pinion tooth surface is an envelope in a one-parameter motion. The imaginary generating surface is a gear tooth surface. Figure 3 describes the coordinate systems between the imaginary gear and work gears. The coordinate system S_2 (x_2, y_2, z_2) is rigidly connected to the gear and the coordinate system S_1 (x_1, y_1, z_1) is rigidly connected to the pinion during generation pinion surface.

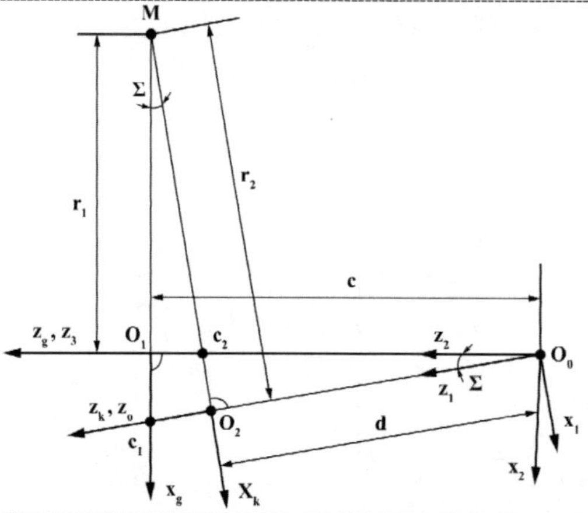

Figure 3. Scheme of coordinate transformation from coordinate system S_2 to coordinate system S_1

Due to the machine kinematics of gear cutting, when the gear tooth rotates around its axis at an angle φ_2 pinion rotates around its axis at an angle φ_1, related to the angle φ_2 by formulas:

$$\begin{gathered} \varphi_1 = \varphi_2 \cdot i \\ i = Z_2^* / Z_1^* \end{gathered} \tag{4}$$

where i is the gear ratio of the machine generating train, z_2^* is a number of gear teeth, z_1^* is a number of pinion teeth.

The coordinates of current point of the gear tooth surface (1) can be determined by two independent parameters: u and h, that is $\tilde{r}_2 = \tilde{r}_2(u, h)$.

Due to the rolling motion when generating the pinion tooth surface, the matrix of relative motion $\widetilde{A}_{12}$ is a function of parameter φ_1: $\widetilde{A}_{12} = \widetilde{A}_{12}(\varphi_1)$. A position vector $\widetilde{r}_1$ of the pinion tooth surface in the coordinate system S_1 (x_1, y_1, z_1) (figure 3), in matrix form is determined as:

$$\widetilde{r}_1 = \widetilde{A}_{12} \cdot \widetilde{r}_2 \tag{5}$$

where $\widetilde{r}_1 = \|x_1, y_1, z_1, 1\|$ is a row matrix, composed of coordinate projections of the position vector $\overline{r_1}$ of the pinion tooth surface.

According to figure 2 elements a_{ij}, $i = \overline{1,4}$, $j = \overline{1,4}$ of matrix $\widetilde{A}_{12}$ have the form:

$$\begin{gathered}
a_{11} = \cos\varphi_1 \cdot \cos\Sigma \cdot \cos\varphi_2 + \sin\varphi_1 \cdot \sin\varphi_2 \\
a_{12} = -\cos\varphi_1 \cdot \cos\Sigma \cdot \sin\varphi_2 + \sin\varphi_1 \cdot \cos\varphi_2 \\
a_{13} = \cos\varphi_1 \cdot \sin\Sigma \\
a_{14} = d \cdot \cos\varphi_1 \cdot \sin\Sigma \\
a_{21} = -\sin\varphi_1 \cdot \cos\Sigma \cdot \cos\varphi_2 + \cos\varphi_1 \cdot \sin\varphi_2 \\
a_{22} = \sin\varphi_1 \cdot \cos\Sigma \cdot \sin\varphi_2 + \cos\varphi_1 \cdot \cos\varphi_2 \\
a_{23} = -\sin\varphi_1 \cdot \sin\Sigma \\
a_{24} = -d \cdot \sin\varphi_1 \cdot \sin\Sigma \\
a_{31} = -\sin\Sigma \cdot \cos\varphi_2 \\
a_{33} = \cos\Sigma \\
a_{34} = d \cdot \cos\Sigma - c \\
a_{41} = a_{42} = a_{43} = 0 \\
a_{44} = 1
\end{gathered} \tag{6}$$

The values c and d included in the elements a_{ij} are calculated according to the formulas (1):

$$\begin{gathered}
c = r_1 \cdot (i - \cos\Sigma) \cdot (\sin\Sigma)^{-1}; \\
d = r_1 \cdot (i \cdot \cos\Sigma - 1) \cdot (\sin\Sigma)^{-1}
\end{gathered} \tag{7}$$

where r_1 is the mean cone distance of pinion, Σ is a shaft angle.

Solving formula (5), we obtain:

$$\widetilde{r}_1(u,h,\varphi_1) = \widetilde{A}_{12}(\varphi_1)\widetilde{r}_2(u,h) \tag{8}$$

As the tooth surface can have only two independent parameters, for mathematical description of the pinion tooth surface it is necessary to relate an additional relation among parameters φ_1, u and h. In the theory of gearing [10] such relation is referred to equation of meshing:

$$f(u,h,\varphi_1) = 0 \tag{9}$$

If the equation of meshing is known, the pinion tooth surface, as an envelope to the family of gear tooth surface, is described as follows [10, 11]:

$$\begin{gathered}
\widetilde{r}_1(u,h,\varphi_1) = \widetilde{A}_{12}(\varphi_1)\widetilde{r}_2(u,h) \\
f(u,h,\varphi_1) = 0
\end{gathered} \tag{10}$$

In this study, to determine the equation of meshing, we used the method suggested in the book [9]. Pinion tooth surface will be an envelope to the family of gear tooth surfaces in relative motion with parameter φ_1 only when relative velocity $\overline{V_\varphi}$ is perpendicular to the unit vector $\overline{m_2}$ of generating surface (gear tooth surface). This condition corresponds to the fact that the scalar product $\overline{V_\varphi}$ and $\overline{m_2}$ is equal to zero [10, 11]. Then, the equation of meshing may be derived as follows:

$$\begin{gathered}
\widetilde{r}_1(u,h,\varphi_1) = \widetilde{A}_{12}(\varphi_1)\widetilde{r}_2(u,h) \\
f(u,h,\varphi_1) = 0
\end{gathered} \tag{11}$$

Projections of m_{2x}, m_{2y}, m_{2z} on the unit vector $\overline{m_2}$ of gear tooth surface (3), determined as:

$$\begin{aligned} m_{2x} &= -\sin\theta_{f2}\cdot\sin\alpha_n; \\ m_{2y} &= \cos\alpha_n; \\ m_{2z} &= \cos\theta_{f2}\cdot\sin\alpha_n. \end{aligned} \qquad (12)$$

Where $\widetilde{A}_{21}$ is a square matrix (4 x 4), which is inverse to matrix $\widetilde{A}_{12}$ (6); $\frac{d\widetilde{A}_{12}}{d\varphi_1}\cdot$ is a square matrix (4 x 4), determined by differentiation of matrix $\widetilde{A}_{12}$ (6) with respect to φ_1, $\cdot\widetilde{r}_2$ is a column matrix, composed by coordinates of position vector gear tooth surface (3).

Relative velocity $\overline{V_\varphi}$ with parameter φ_1 is determined by equation, which based on the work [9]:

$$\widetilde{V}_\varphi = \widetilde{A}_{21}\cdot\frac{d\widetilde{A}_{12}}{d\varphi_1}\cdot\widetilde{r}_2 \qquad (13)$$

Equation (13) yield:

$$\begin{aligned} V_{\varphi x} &= -f_3\cdot\left(i^{-1}-\cos\Sigma\right)-\sin\Sigma\cdot\sin\varphi_2\cdot\left(f_2+d\right) \\ V_{\varphi y} &= f_1\cdot\left(i^{-1}-\cos\Sigma\right)-\sin\Sigma\cdot\cos\varphi_2\cdot\left(f_2+d\right) \\ V_{\varphi z} &= \sin\Sigma\cdot\left(f_1\cdot\sin\varphi_2+f_3\cdot\cos\varphi_2\right) \end{aligned}$$

Here

$$\begin{aligned} f_1 &= u\cdot\cos\theta_{f2}-h\cdot\sin\theta_{f2}\cdot\cos\alpha-r_2 \\ f_2 &= u\cdot\sin\theta_{f2}+h\cdot\cos\theta_{f2}\cdot\cos\alpha_n \\ f_3 &= t-h\cdot\sin\alpha_n \end{aligned}$$

As a result, formulas determining the pinion tooth surface described as:

$$\begin{aligned} x_1 &= A_1\cdot\cos\varphi_1+B_1\cdot\sin\varphi_1 \\ y_1 &= -A_1\cdot\sin\varphi_1+B_1\cdot\cos\varphi_1 \\ z_1 &= \sin\Sigma\cdot(f_3\cdot\sin\varphi_2-f_1\cdot\cos\varphi_2)+\cos\Sigma\cdot(f_2+d)-c \\ \varphi_2 &= \arcsin\left[-C_\varphi(\sqrt{A_\varphi^2+B_\varphi^2})^{-1}\right]-\xi \end{aligned} \qquad (14)$$

Here,

$$\begin{aligned} A_1 &= \cos\Sigma\cdot(f_1\cdot\cos\varphi_2-f_3\cdot\sin\varphi_2)+\sin\Sigma\cdot(f_2+d) \\ B_1 &= f_3\cdot\sin\varphi_2-f_1\cdot\cos\varphi_2 \\ A_\varphi &= \sin\alpha_n\cdot\sin\Sigma\cdot(u+d\cdot\sin\theta_{f2}-r_2\cdot\cos\theta_{f2}) \\ B_\varphi &= -\sin\Sigma\cdot(f_2\cdot\cos\alpha_n-f_3\cdot\sin\alpha_n\cdot\cos\theta_{f2}+d\cdot\cos\alpha_n) \\ C_\varphi &= (i^{-1}-\cos\Sigma)\cdot(f_1\cdot\cos\alpha+f_3\cdot\sin\theta_{f2}\cdot\sin\alpha) \end{aligned}$$

where angle ξ is determined based on its trigonometric functions:

$$\begin{aligned} \sin\xi &= B_\varphi\cdot(\sqrt{A_\varphi^2+B_\varphi^2})^{-1} \\ \cos\xi &= A_\varphi\cdot(\sqrt{A_\varphi^2+B_\varphi^2})^{-1} \end{aligned}$$

3. Conclusion

These formulas enable us, on the one hand, to perform the analysis of generating pinion tooth surfaces, and, on the other hand, to study geometry of meshing in the semi-rolling bevel precessional gearing with straight teeth, which is «matched» according to the way it generation of gear tooth surfaces.

By using formulas (14) and MathCAD software a computer program was developed to study the position and contact lines in gears mesh of the semi-rolling bevel precessional gear with straight teeth. The program showed its contact lines with the pinion tooth surface and the gear tooth surface of the semi-rolling bevel precessional gearing with straight teeth for a number of fixed values of the angle of action (φ_1=-0,172; -0,115; -0,057; 0,0; 0,088; 0,177; 0,265) (figure 4). Pinion and gear design parameters are given in table 1.

Table 1. Pinion and gear design parameters

	Number of teeth	Normal module (mm)	Pressure angle (degree, °)	Face width (mm)	Pitch angle (degree, °)	Addendum factor	Clearance coefficient
Pinion	64	5.0	20	0.593	88	1	0.5
Gear	65	5.0	20	0.796	90	1	0.5

In the studied semi-rolling bevel precessional gear with straight teeth a change in angle φ_1 from $\varphi_{1\min} = -0{,}172$ to $\varphi_{1\max} = 0{,}265$ is according to a maximum angle of action. Taking into account the value of the tooth pitch angle on the pinion ($t_1 = 2 \cdot \pi / z_1^* = 0{,}09817$), we determined that $(-\varphi_{1\min} + \varphi_{1\max})/t_1 = 4{,}45$ tooth pairs are simultaneously in gears mesh. A prototype of semi-rolling bevel precessional gear with straight teeth is shown in figure 5.

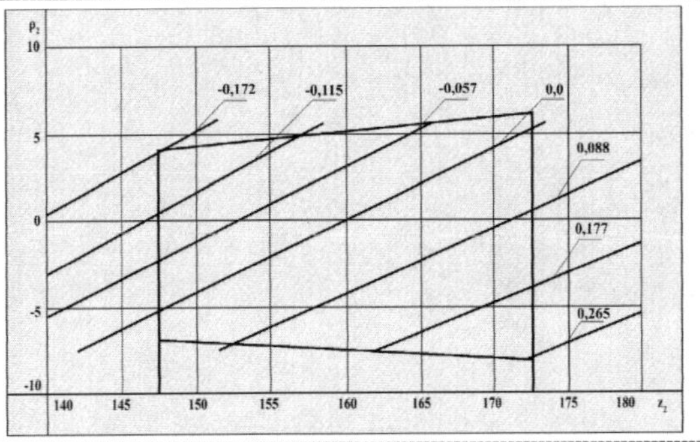

Figure 4. Contact lines on the pinion tooth surface

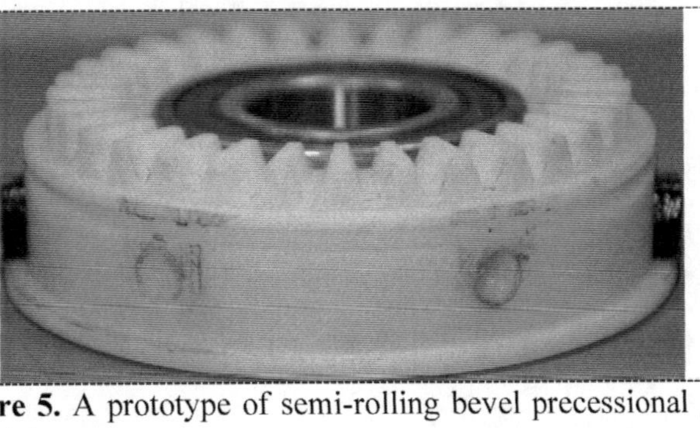

Figure 5. A prototype of semi-rolling bevel precessional gear with straight teeth

References

[1] Litvin F L, Petrov K M. and Ganshin V A 1974 The Effect of Geometrical Parameters of Hypoid and Spiroid Gears on Their Quality Characteristics *ASME Journal of Engineering for Industry* **96** 330-334

[2] Litvin F L, Fuentes A, Zanzi C, Pontiggia M and Handschuh R F 2002 Face Gear Drive with

Spur Involute Pinion: Geometry, Generation by a Worm, Stress Analysis *Computer Methods in Applied Mechanics and Engineering* **191** 2785-2813

[3] Fuentes A, Litvin F L, Mullins B R, Woods R and Handschuh R F 2002 Design and Stress Analysis of Low-Noise Adjusted Bearing Contact Spiral Bevel Gears *ASME Journal of Mechanical Design* **124** 524–532

[4] Goldfarb V I, Trubachev E S and Lunin S V 2008 System of hobs unification for gear-wheel cutting of worm type gears *Proc. of ASME International Design Engineering Technical Conferences and Computers and Information in Engineering Conference* (*Las Vegas, NV, USA, 4-7 September 2007* vol 7) pp 807-811

[5] Litvin F, Sheveleva G, Vecchiato D, Gonzalez-Perez I and Fuentes A 2005 Modified approach for tooth contact analysis of gear drives and automatic determination of guess values *Computer Methods in Applied Mechanics and Engineering* **194** 2927-2946

[6] Simon V 2013 Design of face-hobbed spiral bevel gears with reduced maximum tooth contact pressure and transmission errors *Chinese Journal of Aeronautics* **26** 777-790

[7] Kolivand M and Kahraman A 2010 An Ease-Off Based Method for Loaded Tooth Contact Analysis of Hypoid Gears Having Local and Global Surface Deviations *Journal of Mechanical Design* **132** 071004

[8] Syzrantsev V, Syzrantseva K and Pazyak A 2015 Method of Loading Capacity Calculation of Bevel Precessional Gear for Pipeline Valve Drives *Journal of Engineering and Sciences* **10** 243-246

[9] Syzrantsev V, Denisov J, Wiebe V and Pazyak A 2016 The Design and Production of Drives Based on Pan Precess Gear for Oil and Gas Machinery *Proc.of ASME 2015 International Design Engineering Technical Conferences and Computers and Information in Engineering Conference* (*Boston, Massachusetts, USA, 2-5 August 2015* vol 10) pp V010T11A057

[10] Litvin F L 1997 *Development of Gear Technology and Theory of Gearing* (NASA Reference Publication 1406)

[11] Yerikhov M L 1969 Synthesis of gearing on the condition of insensitivity to mounting errors *Automobile transport* **17** 2-36

IOP Conf. Series: Materials Science and Engineering **142** (2016) 012087 doi:10.1088/1757-899X/142/1/012087

Research of Adhesion Bonds Between Gas-Thermal Coating and Pre-Modified Base

Z Kovalevskaya[1,2], K Zaitsev[3], V Klimenov[1,4]

[1] National Research Tomsk Polytechnic University, 30, Lenin Av., Tomsk, 634050, Russia
[2] Institute of Strength Physics and Material Science SB of RAS, 2/4, Akademichesky Av., Tomsk, 634021, Russia
[3]Yurga Institute of Technology, TPU affiliate, 26, Leningradskaya Str., 652050, Yurga, Russia
[4]Tomsk State University of Architecture and Building, 2, Solyanaya Sq., Tomsk, 634003, Russia

E-mail: zaitsev81@rambler.ru

Abstract. Nature of adhesive bonds between gas-thermal nickel alloy coating and carbon steel base was examined using laser profilometry, optical metallography, transmission and scanning electron microscopy. The steel surface was plastically pre-deformed by an ultrasonic tool. Proved that ultrasound pre-treatment modifies the steel surface. Increase of dislocation density and formation of sub micro-structure are base elements of surface modification. While using high-speed gas-flame, plasma and detonation modes of coatings, surface activation occurs and durable adhesion is formed. Ultrasonic pre-treatment of base material is effective when sprayed particles and base material interact through physical-chemical bond formation. Before applying high-speed gas flame and plasma sprayed coatings, authors recommend ultrasonic pre-treatment, which creates periodic wavy topography with a stroke of 250 microns on the steel surface. Before applying detonation sprayed coatings, authors recommend ultrasound pre-treatment that create modified surface with a uniform micro-topography.

1. Introduction

In modern production, high-energy and high-speed gas-thermal spraying is widely used; before spraying surface preparation is performed by surface plastic deformation, designed for industrial production of machine parts. These methods are used for parts shape and surface finishing and compete with grinding and polishing [1]. Rolling and burnishing, diamond burnishing, regular surface microrelief formation, ultrasonic surface treatment - these methods simplify production of coated parts, since they do not require additional technological developments.

Before coating base layer surface pre-deformation is to be done with several objectives:

1. for gradient increase of base layer surface hardness so as to level material properties on the border between hard surface and soft base;
2. for base surface topography providing effective contact between a coating and a base material;

3. for mechanical activation of the surface by hardening and increasing the number of surface defects that increase adhesion bond areas and flow of physical and chemical processes as well as contribute to the bulk diffusion of materials;

4. for creation of compressive stress in subsurface layer and improving of fatigue resistance of coated parts [1-3].

Efficiency of surface treatment diminishes if temperatures, sufficient to remove effects of modifications, affect substrate while coating. However, in this case some beneficial effects may arise, for example, material structure improving, dislocation clusters redistribution and surface roughness optimization.

In the work a method of surface preparation for thermal coating is considered, namely, ultrasonic surface treatment with the indenter. The object of research is steel surface layer's structure after ultrasound treatment as well as interaction between coating and base material modified by surface deformation. Several methods of thermal coating were examined.

2. Materials and methods

Ultrasonic device UZGK-02 was used for preparation of surface basis for spraying. The ultrasonic device has a capacity of 200 W, indenter's vibration frequency 24 kHz, indenter's pressing force F_N = 75 N and vibration amplitude ξ = 7 microns. For surface treatment, a cylindrical steel sample was fixed in the lathe chuck and treated with ultrasonic instrument mounted in the lathe saddle. [4] Ultrasonic tool deformed and modified the surface layer as well as formed periodic topography. Microrelief controlled variables were set by rotary motion of the part at a rotation speed of 60 to 200 rev/min and ultrasonic instrument longitudinal feed at a rate of 0.05 mm/0.25 mm/rev. Samples were made from 20-carbon steel (0.2% of carbon) and 45-steel (0.45% of carbon).

Coating was performed by three methods: high-speed gas-flame, plasma and detonation [4-10]. Spraying modes were selected so as to achieve the most complete melting of the powder used and to provide the best coating quality. High-speed flame spraying was performed at the facility with the following characteristics: power 5 kW, fuel gas - propane; transporting gas - nitrogen, gas jet velocity at the nozzle exit - 800 m/s. Spraying was performed at a distance of 200mm; a linear velocity of torch movement 21 mm/s. The coating was formed by layers of 500-800mm [9].

Plasma spraying was performed with UPU-3D plasma torch with following specifications: power 30 kW; conveying plasma-forming gas - a mixture of argon and nitrogen; spraying distance of 100-150mm. Coating was performed in layers. Thickness of the sprayed coating was 1.000 micron [10].

A facility, developed at Institute of Hydrodynamics of SB RAS, was used for detonation spraying. The facility accelerates powder particles up to 1000 m/s. The coating was applied at a fixed (relative to the detonation gun jet) sample in the "spot" of 250 microns thickness. Sprayed surface was located at a distance of 100 mm from the nozzle [4]. A fuel mixture of acetylene and oxygen was used

For spraying powders of pure metals were used, such as nickel, chromium, molybdenum, and nickel-based alloy PR-N70X17CP4 (Fe – 5 mass.%; C – 1.5 mass.%; Cr – 26 mass.%; Si – 2.3 mass.%; B – mass.3%; Ni – remaining) with a particle size of 30-50 micron. Examinations of base's morphology and roughness and analysis of its features after the coatings' interfacial failure were performed with a profilometry device «MICRO MEASURE 3D station». With a graphical program calculation of cold-welded regions' area was performed, in accordance with a predicted adhesive strength. Base surface condition after failure of coatings was examined using focused-beam microscope SEM «Philips SEM 515». The micro-hardness of the coating and the base was measured with «Nano Hardness Tester». With optical metallographic microscope Olympus GX-71 images of the coating and the base micro-cross section were obtained. Fine structure was examined with electron microscope Tesla BS-540 by transmission microscopy.

3. Results and discussion

Ultrasound treatment performs plastic deformation of steel surface. Periodic topography, conditioned by deformation grooves' superposition, is formed along longitudinal feed of the ultrasonic tool.

Periodic topography, conditioned by consistent indenter strokes, is formed along rotational movement of the part.

By changing ultrasound tool feed rate and workpiece rotation speed, you can change periodic topography parameters in two directions: longitudinal feed of the ultrasonic tool and workpiece rotational movement. In our research two processing modes were selected, both with rotation speed of 150 rev/min. In the first mode, ultrasonic tool longitudinal feed was performed at a rate of 0.25 mm / rev. In the second mode ultrasonic tool traverse feed was performed at a rate of 0.05 mm / rev (fig. 1).

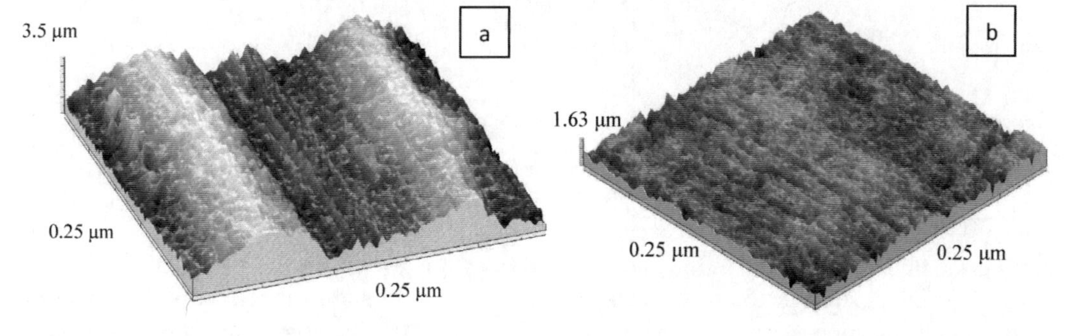

Figure 1. 3D- image of 45-steel surface after ultrasonic treatment with feed rate of 0.25 mm/rev (a) and 0.05 mm/rev (b).

In the first mode, after ultrasound treatment a topography is formed, which is a helical groove with micro-roughness interval of 0.25 mm. Roughness R_z 3.2 microns in direction of ultrasonic tool longitudinal feed; roughness R_z 0.7 microns in direction of rotational motion of parts (fig. 1a).

Topography with preset parameters provides high adhesion if coatings are formed from the highly heated liquids and particles sprayed at rates of 500 to 800 m/s while plasma and high-speed flame spraying. Flowing particles of sprayed material easily fill surface irregularities, crystallize or harden.

In the second mode, after ultrasonic treatment a uniform periodic topography is formed on the entire metal surface, without pronounced anisotropy, with roughness R_z 1.4 microns (fig. 1 b).

Topography with preset parameters provides high adhesion if coatings are formed from high- and low-heated particles deposited at rates of more than 800 m/s while detonation and gas-dynamic spraying. In this case, solid particles of sprayed material undergo plastic deformation on the base surface.

Adhesion of coatings, sprayed with high speed, is also determined by a surface activation level. While ultrasonic surface pre-treatment base surface structure modification occurs, resulting in surface activation. The process of sprayed particles adhesion to detail surface occurs as a first-order topochemical reaction. It occurs at interacting bodies' phase boundaries and is accompanied by substance's crystal lattice rearrangement [1, 11, 15]. The reaction rate and the completeness of its course depend on occurrence of defects in the reacting solid. These defects include dislocations, vacancies, grain boundaries, the surface itself and its sub micro-topography. Combinations of these defects, for example, surface dislocations, are the most favorable areas for topochemical reactions.

Let us see how ultrasonic treatment changes steel surface layer structure. Carbon steel structure is represented as a mixture structurally free ferrite and pearlite grains. Surface plastic deformation in the ultrasonic field causes grains' deformation reversal and shape changing in the surface layer to 30 microns. The upper parts of grains are extended by the forces of deformation displacement and adhesion between the workpiece and the tool, (fig. 2).

Nearer to the surface grains' deformation increases. On the surface grains become lamellae with length being ten times larger then thickness. By chemical etching, we identify light-colored ferrite and dark-colored pearlite lamellae, but formed sub-boundaries diminish image contrast, and quite often, the ferrite grains become dark (fig. 2). TEM shows what structural transformations occur with pearlite grains in the surface layer of steel (fig. 3). Initially pearlite structure is mainly lamellar consisting of alternating cementite and ferrite laminae (fig. 3 a-d). Chaotically distributed dislocations with a density of $3 \cdot 10^9$ cm^{-2} are observed in ferrite laminae.

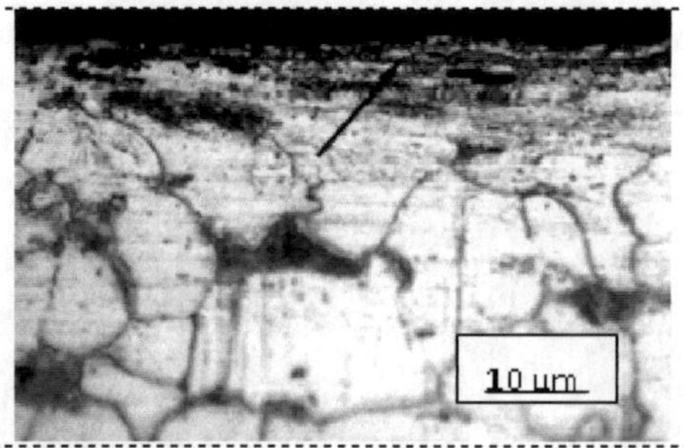

Figure 2. Metallographic image of 20-steel surface cross-section, after ultraviolet irradiation (region of plastic deformation is indicated by the arrow)

After ultrasonic treatment the following structural transformations (fig. 3 e-h) occur in the grains of pearlite. Fragmented substructure with fragments' size about 0.3 microns is formed in ferrite laminae. Cellular-mesh dislocation substructure with a scalar density $4 \cdot 10^{10}$ cm^2 (fig. 3 e) is observed inside the fragments

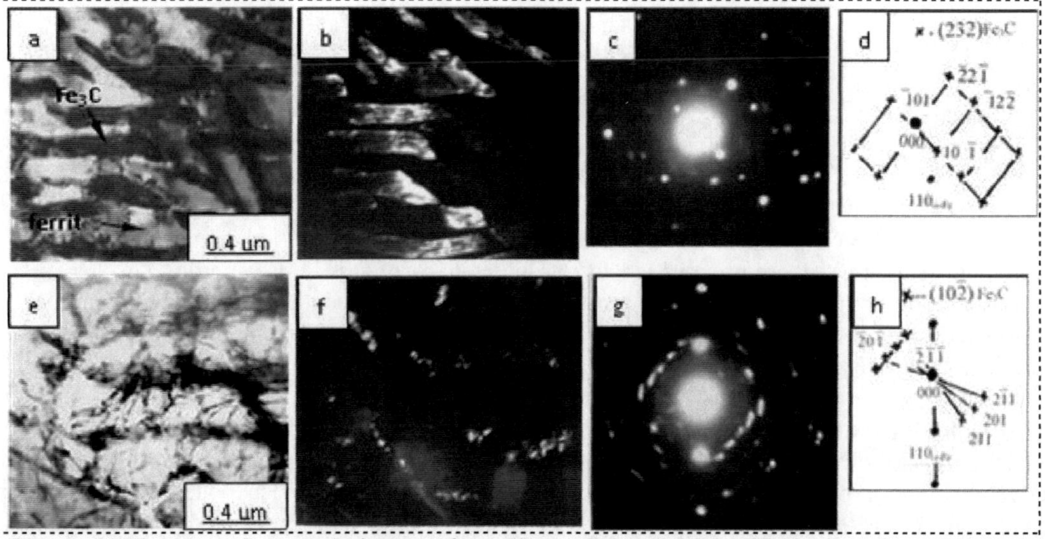

Figure 3. Electron microscopic image of pearlite grains in initial state (a-d) and on 20-steel surface after ultraviolet irradiation (e-h): a, e - bright-field image; b, f - dark-field image in one of Fe_3C reflexes; c, g - micro-diffraction picture; d, h - diagram of micro-diffraction picture

Cementite laminae are broken into individual nanoscale disordered particles (fig.3 f). Electron-diffraction examination of these particles shoes reflexes, which azimuthal deviation is about 10 degrees (fig. 3 g, h). Electron microscopy data confirm that plastic deformation with ultrasound vibrations modifies surface layer structure, namely, increases dislocations' density and forms sub microstructure. Described structural transformations provide activation of the steel surface [12].

Surface layer structure modification can be expressed in a gradient increase of micro-hardness values. On the surface micro-hardness can increase up by 100%. The hardened layer depth is as much as 100 microns.

Let us study how surface layer's structure and properties change after thermal sprayed coating. If any methods of coating, in the surface layer its structural features are sustained. As shown in fig. 4a, less contrast deformed area is sustained on the steel surface.

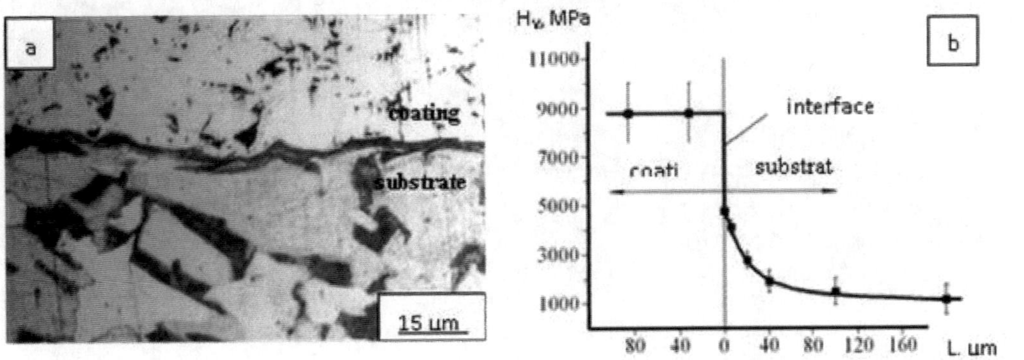

Figure 4. Metallographic image of sample cross-section of with gas-flame coating at interface between coating and 20-steel basis (a); and micro-hardness distribution in the same area (b).

Possibly, under applied shock and heating, additional deformation with returning and polygonization occurs in the steel surface layer. Available dislocation structure may transform into a perfect substructure with greater disorientation of sub grains and lower dislocation density. In any case, small size of structural objects, commensurate with sub microcrystalline size, is sustained. This creates a lower contrast while etching of surface layer (Figure 4 a). Also gradient increase of micro-hardness values is sustained in the steel surface layer (fig. 4b). As a result, after the gas-thermal sprayed coating, levelling of material properties on the boundary between hard surface and soft base occurs; that improves performance properties of the finished part. Sprayed coating adhesion was quantified by three-dimensional profilometry, which identified the total area of sprayed particles remaining in adhesion bonded areas after coating failure. Regions of adhesion between the surface and sprayed particles are characterized by cohesive coating failure and are interspersed with areas of adhesive coating failure. Proportion of these areas determines strength of adhesion of coating to base material.

In the first series of experiments, adhesion of nickel alloy coating, alloyed with chromium, silicon and boron, was tested. Before coating by gas-flame and plasma methods, ultrasonic pre-treatment was performed; this formed a periodic wavy topography, with strokes of 250 micron on the steel surface. Before coating by detonation method, pre-treatment was performed; that formed a modified surface with a uniform topography. The total area of adhesions between the coating and the base material was 38% after high-speed gas-flame spraying; and 40% after plasma spraying. This is comparable to adhesive strength of 40 MPa; it is considered to be a high-performance adhesion [9, 10]. After failure of the coating made by detonation method, the total area of seizures between coating and base material was 47%, which is comparable with adhesion strength of 47 MPa.

Greater adhesion between the detonation coating and the base is caused by specificity of particle deposition. High rate deposition of particles 1000 m/s provides additional surface activation due to shock plastic deformation [4].

A second series of tests was conducted to ascertain how particles' temperature affects interaction between the coating and the modified base; or how base heating with sprayed particles improves adhesion. For this task detonation coatings of refractory metals - nickel, chromium and molybdenum - were deposited on modified surfaces. Fig. 5 shows samples' surfaces after failure of the coatings; the image is obtained by scanning electron microscopy.

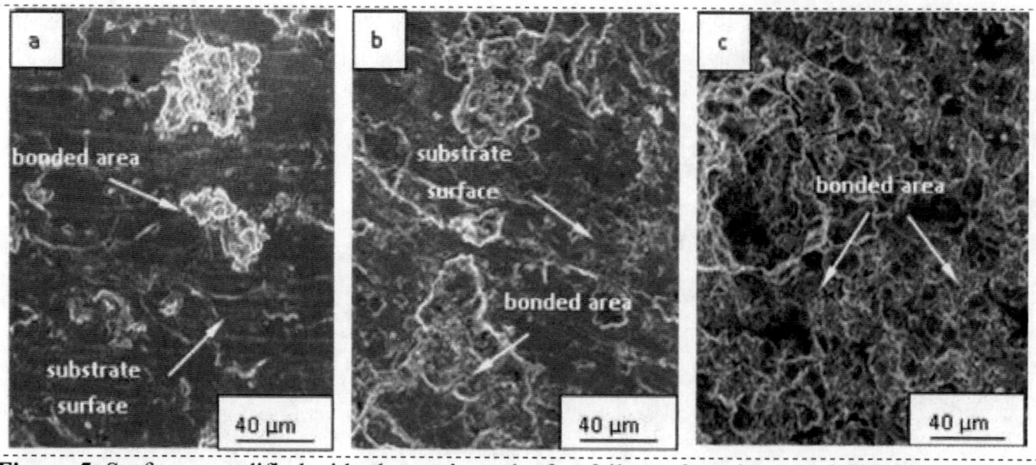

Figure 5. Surfaces, modified with ultrasonic tool, after failure of coatings a) nickel; b) chromium; c) molybdenum

There are a various number of bonded areas on the base surface after coating failure. After the nickel coating failure, some fragments, commensurate with sprayed particles (fig. 5a), are sustained on the base surface. In the chromium coating, particles are fixed with agglomerates (fig. 5 b), which are difficult to be differentiated from adherent particles. This is caused by higher activity of the areas adjacent to pre-fixed particles. In this case, a single fixed particle is a nucleus of the agglomerate [12-16]. After molybdenum coating failure, the whole base surface is covered with remaining particles (fig. 5 c), so while contact between sprayed particles and base, nuclei make progress in growing into agglomerates.

It has been found that, after molybdenum powder coating failure, the total area of adhesion bond was about 80%; after chromium powder coating failure - about 40%; after nickel powder coating failure - less than 20%. Such a difference is caused by different melting points of the metals. Molybdenum particles ($t_{пл} = 2620\ ^0C$) heat the base material to penetration, thus ensuring the creation of metallic bond. Chromium ($t_{пл} = 1860\ ^0C$) and nickel ($t_{пл} = 1455\ ^0C$) particles are not welded to the base material; they form a strong physical-chemical bond. With such a mechanism of interaction, additional surface activation is necessary. Hence, ultrasonic pre-treatment is a substantial stage of the coating process.

4. Conclusions

Ultrasonic surface modification contributes to formation of reliable adhesive bonds while high-speed gas flame, plasma and detonation coatings. Surface modification manifests itself as increase of dislocation density and formation of submicrostructure.

For high-speed flame and plasma coating ultrasonic pre-treatment is recommended, forming periodic waved topography with interval of 0.25 mm. For detonation coating ultrasonic pre-treatment is recommend, forming uniform periodic topography. Ultrasound pre-treatment is effective if sprayed particles and the base form physical-chemical bonding.

References

[1] Gissler W., Jehn H.A. Advanced Techniques for Surface Engineering. Springer Science & Business Media, 2013, 402

[2] Pawlowski L. Science and Engineering of Thermal Spray Coatings. Wiley-Blackwell, 2008, 656

[3] Hocking M. G., Vasantasree V., Sidky P. S. Metallic and ceramic coatings: production, high temperature properties and applications. Longman Scientific & Technical, 1989 – 670

IOP Conf. Series: Materials Science and Engineering **142** (2016) 012087 doi:10.1088/1757-899X/142/1/012087

[4] Klimenov V. A., Kovalevskaya Z. G., Ul'yanitskii V. Y.m Zaitsev K. V., Borozna V. Y. Effect of ultrasonic substrate treatment on formation of coating in detonation spraying. Welding International, 24 (2010) 803-807

[5] Hamidreza S. J. 2012, Advanced Plasma Spray Applications (InTech).

[6] J. Stokes, "Theory and Application of the High Velocity Oxy-Fuel (HVOF) Thermal Spray Process", Dublin City University, Dublin, 2008, 204 c.

[7] Gavrilenko T.P., Nikolaev Ju.A., Ulianitski V.Yu. D-gun "Ob" Detonation Spraying // Proceedings of International Thermal Spray Conference. Japan, Kobe, 1995.

[8] Gavrilenko T.P., Nikolaev Ju.A., Ulianitski V.Yu., Kim M.Ch., Hong J.W. Computational Code for Detonation Spraying Process // Proceedings of International Thermal Spray Conference. France, Nice, 1998.

[9] Klimenov V. A., Kovalevskaya Zh. G., Zaitzev K. V., Tolmachov A. I. Adhesion obtained by high velocity oxygen fuel thermal spray process. Bulletin of Tomsk Polytechnic University, 310 (2007) 52-55.

[10] Kovalevskaya Z. G., Klimenov V, A., Zaytsev K. V. Interfacial Adhesion between Thermal Spray Coating and Substrate Achieved by Ultrasonic Finishing. Applied Mechanics and Materials, 682 (2014) 459-463

[11] Matejicek J., Vilemova M., Musalek R., Sachr P., Hornik J. Influence of interface characteristics on adhesion/cohesion of plasma sprayed tungsten coatings. Coatings, 3 (2013) 108-125

[12] Kovalevskaya Zh. G., Ivanov Yu. F., Perevalova O. B., Klimenov V. A., Uvarkin P .V. Study of microstructure of surface layers of low carbon steel after rough cutting and ultrasonic finishing. Physics of Metals and Metallography, 114 (2013) 41–53

[13] Soboleva E. G., Igisheva A. L., Krit T. B. Reduction of longitudinal ultrasonic oscillations in NaClO3 crystals // IOP Conference Series: Materials Science and Engineering. - 2015 - Vol. 91, Article number 012032. - p. 1-7.

[14] Gawne D.T., B.J. Griffiths, and G. Dong, Splat Morphology and Adhesion of Thermally Sprayed Coatings, in Thermal Spraying: Current Status and Future Trends, Kobe, Japan (Pub.) High Temperature Society of Japan, (1995) pp. 779-784.

[15] A.A. Lasukov, P.A. Chazov, A.V. Barsuk Investigation on the Elemental Chip Formation Process in Hard-to-Machine Material Cutting, Applied Mechanics and Materials, Vol. 682 (2014), pp. 504-509

[16] V.A. Petrova, A.A. Bakanov, and A.V. Walter, Crack Resistance of Weld Seals, Hardening Methods and Composite Tools Quality, Applied Mechanics and Materials, vol. 682, pp. 431–437, 2014.

IOP Conf. Series: Materials Science and Engineering **142** (2016) 012088 doi:10.1088/1757-899X/142/1/012088

The Methodology of Calculation of Cutting Forces When Machining Composite Materials

D A Rychkov[1], **A S Yanyushkin**[1]

[1] Bratsk State University, 40 Makarenko street, Bratsk, Russia

e-mail: rychkovda@gmail.com

Abstract. Cutting of composite materials has specific features and is different from the processing of metals. When this characteristic intense wear of the cutting tool. An important criterion in the selection process parameters composite processing is the value of the cutting forces, which depends on many factors and is determined experimentally, it is not always appropriate. The study developed a method of determining the cutting forces when machining composite materials and the comparative evaluation of the calculated and actual values of cutting forces. The methodology for calculating cutting forces into account specific features of the cutting tool and the extent of wear, the strength properties of the processed material and cutting conditions. Experimental studies conducted with fiberglass milling cutter equipped with elements of hard metal VK3M. The discrepancy between the estimated and the actual values of the cutting force is not more than 10%.

Introduction

Widespread in modern industry are composite materials of polymer-based [1 – 3]. Their products in comparison with metals and alloys have a number of advantages, such as specific strength, resistance to corrosion, cost, etc. Machining composites has specific features in comparison with the processing of metals. These include increased wear of the cutting tool, the need to increase the front and rear corners of the cutting, the use of equipment with high-speed processing, the presence of local and general ventilation, etc. [4 – 10]. Modern production methods and technologies enable the manufacture of parts with high surface quality, performance, but do not always satisfy the requirements of form and dimensions of the product [11, 12]. For surfaces of different shapes is widespread milling. When milling of composite materials is recommended to use high-strength tool materials, such as cemented carbide and superhard materials [5, 13 – 15]. Difficulties in the use of these materials is the formation of a cutting edge with a small angle of sharpening, but this problem can be solved through the use of electro-technology diamond sharpening in a double etching [16 – 18].

The parameters include the milling process productivity, operability of the cutting tool, cutting power and surface finish. Urgent problem is to evaluate the cutting forces because they determine the processing capacity, the pressure on the cutting edge and chip character etc. Known research mainly aimed at finding the experimental methods for determining the cutting forces [19, 20], however, is not always advisable to conduct laboratory tests.

The aim of the study is to improve the technology for milling of composite materials, and the task is to develop methodology for calculating cutting forces, conducting an experiment for comparing the calculated and actual values of cutting forces.

1. Method for determining the cutting forces during milling

To determine the parameters influencing tool life in milling used the theory of strength of materials and cutting theory. We select the time when the forces acting on the "cutting tool - detail" system will have maximum values (Fig. 1). And refine them.

When cutting the material on the cutting edge of the tool by cutting layer the force of friction F_f^{ed}, which affects the degree of tool wear and the size of the chamfer wear. This force is related to the elastic resistance of the processed material and the appearance of the reaction N_{ed}. Frictional force F_f^{ed} direction is determined by the chamfer wear, which is taken tangent to the cutting surface, and the direction of reaction N_{ed} perpendicular to the direction of force F_f^{ed}.

When rotating milling cutter operates the centrifugal acceleration a_c^n and the presence of the tool imbalance can occur inertia force F^n directed in the opposite direction with respect to the centrifugal acceleration a_c^n. Since all rotating tools are subjected to balancing, the value of the imbalance in this case can be neglected, and take the position that the mass center of the cutter lies on the axis of rotation, $F^n = 0$.

Also at the cutting edge the force of friction F_f^{fr}, which is connected with the elastic resistance of the treated material and the appearance of the reaction N_{fr}. The force F_f^{fr} is directed along the front surface of the cutting wedge and the reaction N_{fr} is directed perpendicular to the friction force F_f^{fr}.

Removal of material from the surface is effected by the cutting force P_{cut} which can be decomposed into two components: P_z and P_x, directed along the axes OX and OZ respectively. The axes of the projections are directed as follows: OZ axis is parallel to the vector velocity V; and OX - perpendicular to the OZ axis.

Cutting force P_{cut} is determined by the formula, knowing its components:

$$P_{cut} = \sqrt{P_z^2 + P_x^2} \tag{1}$$

The cutting force P_{cut} affects flexural and compressive cutting edge of the tool axes $y_к$ and $x_к$ (Fig. 2). The $y_к$-axis is directed along the bisector of the angle wedge, on the $x_к$-axis perpendicular to the $y_к$-axis.

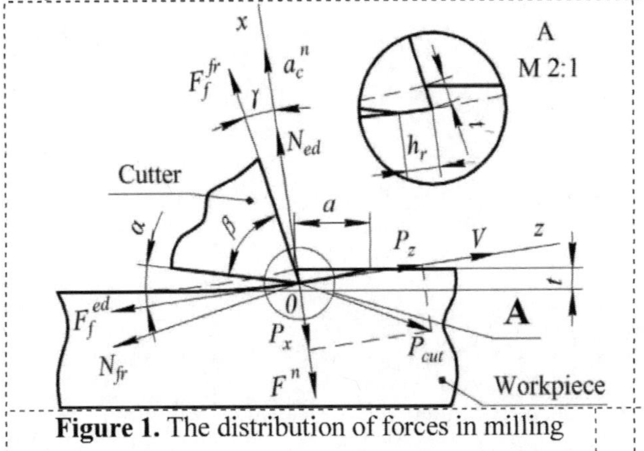

Figure 1. The distribution of forces in milling

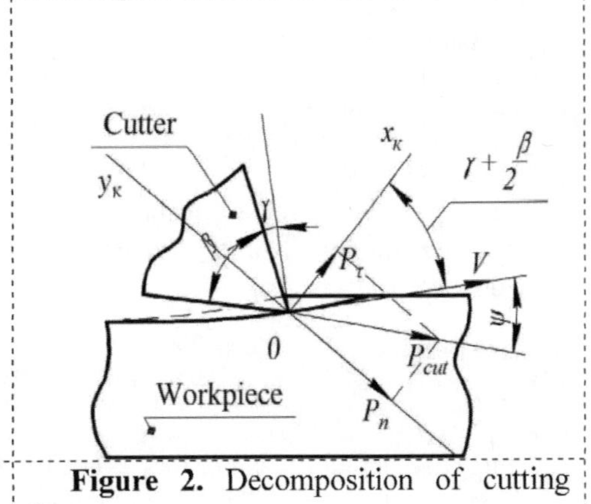

Figure 2. Decomposition of cutting force

Thus, the force P_n directed along the $y_к$ axis, will affect the value of the compressive stress in the cutting edge during milling, and the force P_τ , directed along the $x_к$ axis, - the value of the bending stress. Moreover, the direction of the cutting force P_{cut} is deflected from the direction of the velocity vector V on the angle ψ. According to the scheme in Fig. 2, this angle is $\psi = arctg\left(\frac{P_x}{P_z}\right)$.

The value of the stress caused by the influence of the cutting forces will be:

$$\sigma_n = \frac{P_n}{B \cdot h_r} = \frac{P_{cut} \cdot \sin(\psi + \gamma + \beta/2)}{B \cdot h_r} \tag{2}$$

$$\sigma_\tau = \frac{P_\tau}{B \cdot h_r \cdot \sin\gamma} = \frac{P_{cut} \cdot \cos(\psi + \gamma + \beta/2)}{B \cdot h_r \cdot \sin\gamma} \tag{3}$$

where h_r – value chamfer wear along the rear surface of the cutting tool, mm; B – the width of the shear layer, mm; γ – the rake angle of the cutting tool; β – sharpening angle of the cutting edge.

Under normal conditions, the cutting stress must not exceed the strength properties of the cutting blade.

Cutting forces are determined based on the fact that the sum of the projections of the forces on the axis OZ and OX is equal to zero:

$$\sum F_i(z) = 0\colon P_z = N_{fr} \cdot \cos\gamma + F_f^{fr} \cdot \sin\gamma + F_f^{ed} \tag{4}$$

$$\sum F_i(x) = 0\colon \ P_x = -N_{ed} - F_f^{ed} \cdot \cos\gamma + N_{fr} \cdot \sin\gamma \tag{5}$$

$$F_f^{ed} = N_{ed} \cdot f \tag{6}$$

$$F_f^{fr} = N_{fr} \cdot f \tag{7}$$

where f – the coefficient of friction between the tool and the workpiece.

Given that the reactions N_{ed} and N_{fr} are dependent on the elastic properties of the material being processed, we obtain:

$$N_{ed} = \sigma_{el}^{pr} \cdot S_{cut} \tag{8}$$

$$N_{fr} = \sigma_{com}^{pr} \cdot S_{pr} \tag{9}$$

where σ_{com}^{pr} - the compressive strength of the processed material, MPa; σ_{el}^{pr} - elastic stress arising from the cutting material, MPa; S_{pr} - sectional area of a cut-off stratum in the direction, perpendicular to the reaction N_{fr}, mm^2; S_{cut} - sectional area of a cutting element in the direction, perpendicular to the reaction N_{ed}, mm^2.

$$S_{pr} = B \cdot t' \tag{10}$$

where t' - the maximum thickness of cutting layer for milling, mm.

Maximum thickness of cutting layer depends on cutting conditions and cutter diameter and is defined by the formula:

$$t' = \frac{S_z \cdot t}{R \sin\left(\arccos\left(1 - \frac{t}{R}\right)\right)} \tag{11}$$

where S_z – feed to the cutter tooth, mm/tooth; R – cutter radius, mm; t – cutting depth, mm.

$$S_{cut} = B \cdot h_r \tag{12}$$

2. Method of the experiment

To establish the conformity of the calculated and the actual measurements of cutting forces set up an experiment to process composite glass fiber brand STEF-1 cutting tool equipped with inserts made of sintered carbide VK3M.

The study was conducted on the test bench, created on the base of the machine 3D642E model modernized under the processes of milling of composite materials with the possibility of establishing a high spindle speed and equipped with a number of local ventilation. The limits of variation of cutting conditions are set as follows: cutting depth $t = 0.5 - 1.5$ mm, feed to the cutter tooth $S_z = 0.16 - 0.33$ mm/tooth. Cutting speed V was set as much as possible that provide installation spindle speed $n = 6000$ min^{-1}. Given that the torque from the motor to the spindle of the machine is carried by a belt transmission, characterized by belt slippage, and the cutting diameter of 150 mm and the cutting speed $V = 45\pm2$ m/s.

The limits of variation, tool and processed materials are selected based on the recommendations on the treatment of polymeric materials reinforced with high-strength fibers and fabrics [5, 13 – 15].

When processing polymeric composite materials characterized by tool wear on the back surface, so the process stability criterion is the value chamfer wear along the rear surface of the cutting tool $h_r = 0.35$ mm. Its further increase would lead to poor quality of the machined surface [5]. The value of chamfer is optically controlled.

During the experiment were monitored using cutting power meter K506. Calculation of cutting forces when power measurements carried out according to the formula:

$$P_{cut} = \frac{N_{cut}}{V} \quad (13)$$

where V – cutting speed, m/s.

3. Results and Discussion

According to the experimental results obtained by mathematical relationship of the cutting force of the processing modes:

$$P_{cut} = 4.2 + 166S_z - 3t + 168S_z t \quad (14)$$

According to the above method calculated the cutting forces during milling fiberglass brand STEF-1 hard alloy VK3M with varying treatment regimens. The input parameters for the calculation were as follows: the rake angle $\gamma = 25°$; Compressive strength $\sigma_{com}^{pr} = 400$ MPa; Cutter radius $R = 62.5$ mm; The cutting width $B = 10$ mm; the coefficient of friction $f = 0.5$; chamfer wear along the rear surface $h_r = 0.35$ mm.

For comparison of the calculated and experimental data plotted as the cutting force of the feed and depth of cut (Fig. 3, 4).

With increasing feed and depth of cut, the cutting force increases due to the increased thickness of the shear layer of the processed material and the load on the cutting tool. The values of the calculated and experimental curves do not differ by more than 10%. This makes it possible to predict with sufficient certainty value cutting forces at varying grades of composite materials, process conditions and constructional features of the cutting tool.

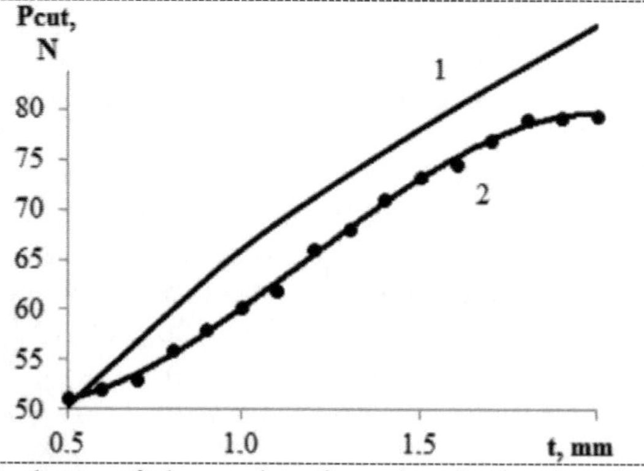

Figure 3. The dependence of the cutting force by the cutting depth: 1 – calculated dependence; 2 – experimental dependence

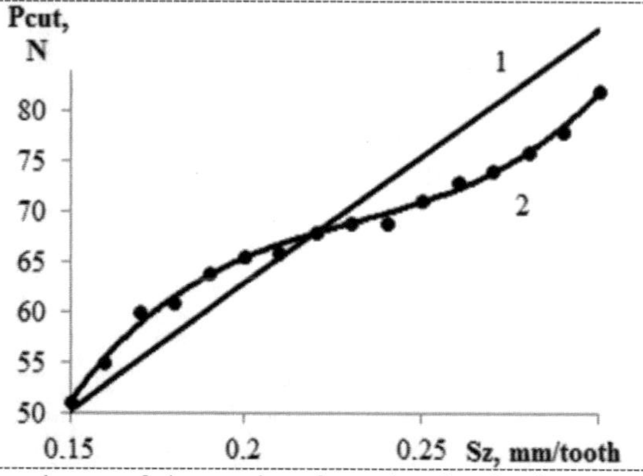

Figure 4. The dependence of the cutting force by the feed per tooth: 1 – calculated dependence; 2 – experimental dependence

Conclusions

Conclusions and recommendations were made on the results of research:

1. An analysis of the scientific literature shows that the composite cutting issues are not well understood, not fully presented studies of the characteristics of their processing.

2. A method for calculating cutting forces during milling of materials based on the provisions of the mechanics and the theory of cutting. The method allows to calculate the cutting force with high accuracy and is used to determine the optimum cutting performance composite materials.

3. The mathematical dependence of the cutting force on the treatment regimes in order to assess the difference between calculated and experimental values. It was found that the error rate of not more than 10%, which is acceptable to the theoretical studies of the process of milling of composite materials and designing manufacturing operations.

References

[1] Gallant F M, Bruck H A, Prickett S E and Cesarec M 2006 Graded polymer composites using twin-screw extrusion: a combinatorial approach to developing new energetic materials *Composites Part A: Applied Science and Manufacturing* **6** 957–969.

[2] Gorokhovsky A V, Escalante-Garcia J I, Gashnikova G Yu, Nikulina L P and Artemenko S E 2005 Composite materials based on wastes of flat glass processing *Waste Management* **7** 733–736.

[3] Dhand V, Mittal G, Rhee K Y, Park S J and Hui D 2015 A short review on basalt fiber reinforced polymer composites *Composites Part B: Engineering* **73** 166–180.

[4] Lobanov D V and Yanyushkin A S 2011 Analysis of structural precast solutions milling tools for machining composite materials *Engineering Technology* **5** 20–25.

[5] Lobanov D V, Yanyushkin A S, Rychkov D A and Petrov N P 2011 Optimal organization of tools for machining composites *Russian Engineering Research* **2** 156–157.

[6] Bailleul J L, Sobotka V, Delaunay D and Jarny Y 2003 Inverse algorithm for optimal processing of composite materials *Composites Part A: Applied Science and Manufacturing* **8** 695–708.

[7] Lemma E, Chen L, Siores E and Wang J 2002 Study of cutting fiber-reinforced composites by using abrasive water-jet with cutting head oscillation *Composite Structures* **1-4** 297–303.

[8] Fomin V N, Malyukova E B and Berlin A A 2004 Criteria for optimization of processing and fabrication of polymer composite materials *Doklady Chemistry* **4-6** 39–41.

[9] Matis I G 1991 Methods and means of inspecting the quality of composite materials *Russian Journal of Nondestructive Testing* **4** 277–285.

[10] Bakulin V N, Larin A A and Reznichenkod V I 2015 Improving the quality of manufactureof polymer-composite products using computed tomography as a nondestructive-testing method *Journal of Engineering Physics and Thermophysics* **2** 556–560.

[11] Skeeba V Yu, Ivancivsky V V and Pushnin V N 2015 Numerical modeling of steel surface hardening in the process of high energy heating by high frequency currents *Applied Mechanics and Materials* **698** 288–293.

[12] Plotnikova N V, Losinskaya A A, Skeeba V Yu and Nikitenko E 2015 Perspective of high energy heating implementation for steel surface saturation with carbon *Applied Mechanics and Materials* **698** 351–354.

[13] Yuanyushkin A S, Rychkov D A and Lobanov D V 2014 Surface quality of the fiberglass composite material after milling *Applied mechanics and materials* **682** 183–187.

[14] Davim J P, Reis P and António C C 2004 A study on milling of glass fiber reinforced plastics manufactured by hand-lay up using statistical analysis (anova) *Composite Structures* **3-4** 493–500.

[15] Yanyushkin A S, Lobanov D V and Rychkov D A 2015 Automation tool preparation in the conditions of production *Applied Mechanics and Materials* **770** 739–743.

[16] Arkhipov P V, Yanyushkin A S, Lobanov D V and Petrushin S I 2013 The effect of diamond tool performance capability on the quality of processed surface *Applied mechanics and materials* **379** 124–130.

[17] Yanyushkin A S, Saprykina N A and Medvedeva O I 2014 Mechanism of protective membrane formation on the surface of metal-bonded diamond disks *Applied Mechanics and Materials* **682** 327–331.

[18] Yanyushkin A S, Lobanov D V, Arkhipov P V and Ivancivsky V V 2015 Contact processes in grinding *Applied Mechanics and Materials* **788** 17–21.

[19] Kalla D, Twomey J and Sheikh-Ahmad J 2010 Prediction of cutting forces in helical end milling fiber reinforced polymers *International Journal of Machine Tools and Manufacture* **10** 882–891.

[20] Mckenzie W M, Ko P, Cvitkovic R and Ringler M 2001 Towards a model predicting cutting forces and surface quality in routing layered boards *Wood Science and Technology* **6** 563–569.

Improvement of the Sintered Surface and Bulk of the Product Via Differentiating Laser Sintering (Melting) Modes

N A Saprykina[1,a], A A Saprykin[1], D A Arkhipova[1], I F Borovikov[2,b]

[1]Yurga Institute of Technology, TPU Affiliate, Tomsk, Russia
[2]Bauman Moscow State Technical University, Moscow, Russia

E-mail: [a] saprikina@tpu.ru, [b] bif1986@mail.ru

Abstract. Selective laser sintering (melting) enables using metal powdered materials to manufacture products of any geometrical complexity, requiring no preliminary costs to prepare processing equipment. However, quality of the sintered surface is often inadequate as against the product manufactured traditionally. Manufacturing a high quality product requires solution of such vital task as prediction of the sintered surface roughness. The authors address to the effect of laser sintering modes on roughness of the surface, sintered of copper powdered material PMS-1 (ПМС-1). The dependence of roughness of the surface layer sintered of copper powder material PMS-1 upon sintering process conditions is expressed mathematically. The authors suggest differentiating sintering modes to improve the sintered surface and the bulk of the product and dividing them into rough, semi-finishing, and finishing ones.

Introduction

Rapid prototyping is widely applied in the present day economy to reduce costs of manufacturing competitive products. Products of any geometrical complexity can be constructed on the base of these technologies requiring no preliminary costs for preparation of processing equipment. A lot of rapid prototyping technologies, differing from each other by the applied material and methods of forming the product, are available these days [1-3]. They include stereolithography (SL), fused deposition modeling (FDM), laminated object manufacturing (LOM), selective laser sintering (SLS), used in foundry engineering, industrial designing, manufacturing products for medicine (implants) etc. Selective laser sintering (melting) is the most promising method to produce a physical copy of various objects using powdered materials on the base of a 3D CAD-model. So, this method enables manufacturing of functional products. This technology makes use of diverse raw materials, ranging from plastic to various metal alloys [4-6].

Alloys, which can be hardly produced relying on conventional technologies, are the products of mixing various powdered materials. The fields of SLS (SLM) application depend on physical and mechanical properties of the material and quality of the sintered surface. Synthesizing complex products of metal powdered materials with certain geometrical, physical and mathematical properties makes it possible to expand application fields of this technology [7]. Thermal effect of laser output on the powdered material is important for quality of the product and can cause rather complex physical phenomena of diverse nature, whereas high quality products can be manufactured in the certain conditions only [8, 9]. Post-processing is used in most cases to improve quality of the sintered products. However, the methods of post-processing do not support the control over geometrical and micro-geometrical characteristics of the synthesized surfaces, necessitating, therefore, application of technological methods to produce a high quality surface layer [10, 11]. Assessment of appropriate modes providing certain quality of the surface layer of the sintered metal powdered materials is a burning and difficult issue. In this paper the authors address to the effect of laser sintering conditions on roughness of the sintered surface of copper powdered material PMS-1. The dependence of roughness of the surface layer sintered of copper powder material PMS-1 upon sintering modes is expressed mathematically. The authors suggest differentiating modes of sintering to improve the sintered surface and bulk of the product and dividing them into rough, semi-finishing, and finishing ones [12-15].

Stabilized copper powder PMS-1 is widely applied in various branches of industry. Particles of the powder have a spherical form with the nominal dimension of 0.007 mm, bulk density is 1.25-1.9 g/cm^3. Thermal conductivity factor of copper powder is 3.6×10-3W/(m·°C), and melting temperature is 1030-1070 °C [16, 17].

Roughness of the sintered layer is assessed by the microscope LEXT OLS 4100 (OLYMPUS).

The experiments are carried out by the self-developed selective laser sintering facilities, which support adjustment of all processes of sintering. This machine is a technological laser jet to form surfaces of the products with complex spatial configuration, consisting of ytterbium fiber laser LK – 100 – V, three-dimensional desk, PC, system of the numerical control program and self-written software. Step motors are used to move the ball screw assembly along the coordinates X, Y, Z. The ytterbium fiber laser with the wavelength 1.07 μm makes it possible to adjust the power in the range 10-100 W. The constant output power and accuracy of fiber laser focusing ensure quality and precision of manufactured products. The laser beam is controlled by special software in the operation area of the size 100x100x100 mm, so scanning of any needed contour can be performed. As soon as a single layer is sintered the desk is lowered by the step motor as required (thickness of the layer) [18].

Results and Discussions

When manufacturing products by selective laser sintering, the quality of the surface differs from that of the bulk of the product. Synthesizing the bulk requires assignment of the sintering conditions, providing high roughness of the sintered surface; that is necessary to ensure stable cohesion between the layers and avoid delaminating. The outer surface of the product is to be synthesized in other conditions, providing low roughness. Roughness of the sintered surface depends on a number of factors, related to sintering conditions, properties of the powdered material, technical solutions. In the course of the experiments principal adjustable parameters are revealed, which influence on the formation of the sintered surface micro-section. They include laser output power (P), velocity of laser beam displacement (V), scanning step (s), preliminary pre-heating temperature of powdered material (t), impact of the shielding gas – argon and mechanical activation of the powder [19]. First, preliminary pilot experiments are conducted to determine the appropriate conditions of sintering, when a single layer does not go into pieces at a touch, i. d. it has some mechanical strength without considerable deformation. Insufficiency of laser power below 15 W and velocity of laser beam displacement faster than 3000 mm/min for sintering the powdered material has been revealed in the experiments aimed at detecting the appropriate conditions of sintering copper powdered material PMS-1. The power exceeding 30 W and velocity less than 200 mm/min cause intense oxidation and powder inflammation.

The increase in the scanning step more than 0.3 mm keeps single tracks from sintering, so a single layer fails to form. Increasing temperature of the powdered material pre-heating has a positive effect on the strength of a single layer. The samples of the powdered material PMS-1 are made in the following conditions: laser output power (P) is varied in the range 15 to 30 W, laser beam is moved with the velocity (V) 200 to 3000 mm/min, scanning step (S) is 0.1 to 0.3 mm, and temperature of the powdered material pre-heating (t) is 26 to 200 ^{0}C. Samples of 20 mm long and 10 mm wide sintered layer are produced for the purposes of experiments.

A four-factor experiment is conducted to identify a mathematical dependence of the sintered layer roughness upon the process conditions of sintering. Level of factors and variability intervals are assigned according to the results of preliminary pilot experiments. The resulting regression equation is in the form of an expression as follows:

$$R_Z = 356 + 15 \cdot P - 0.1 \cdot V - 0.057 \cdot t + 425 \cdot S \tag{1}$$

The dependencies of surface roughness on the laser sintering modes plotted using expression (1) are shown in figure 1.

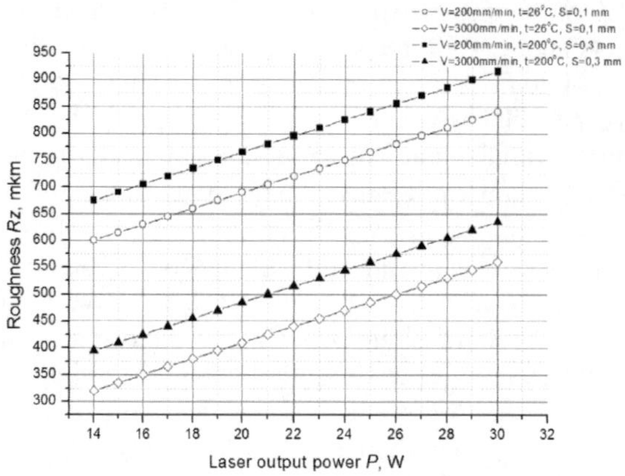

Figure 1. The roughness Rz vs. sintering conditions

Changing the beam power in the range 15 to 30 W resulted in increasing the surface layer roughness Rz from 600 to 830 μm, figure 2 and figure 3, and cracking appeared due to the higher thermal expansion stress. Thus, the beam power has a great effect on the surface roughness of laser deposited coatings. Changing the beam displacement velocity from 3000 to 200 mm/min results in increasing the roughness from Rz 320 to 600 μm, figure 4 and figure 5.

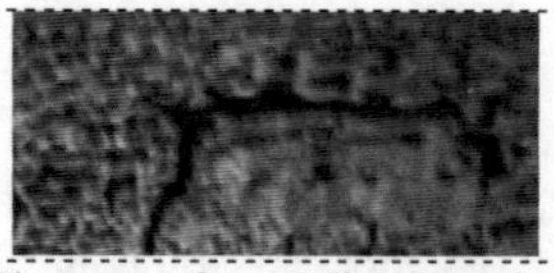

Figure 2. Micrographs, showing surfaces of the sintered copper samples (x2), process conditions of sintering: V=200 mm/min, t=26 °C, S=0.3 mm, P=30 W.

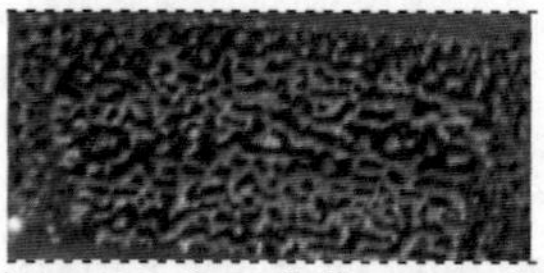

Figure 3. Micrographs, showing surfaces of the sintered copper samples (x2), process conditions of sintering: V=200 mm/min, t=26 °C, S=0.3 mm, P=15 W.

Figure 4. Micrographs, showing surfaces of the sintered copper samples (x2), process conditions of sintering: V=3000 mm/min, t=200 °C, S=0.3 mm, P=15 W.

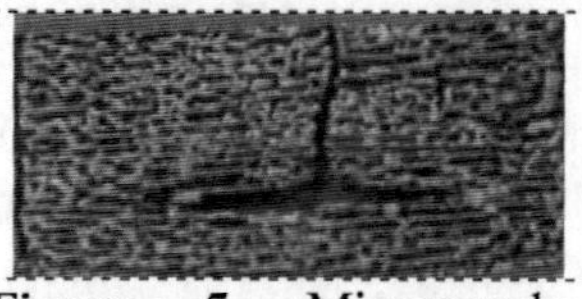

Figure 5. Micrographs, showing surfaces of the sintered copper samples (x2), process conditions of sintering: V=200 mm/min, t=200 °C, S=0.3 mm, P=15 W.

IOP Conf. Series: Materials Science and Engineering **142** (2016) 012089 doi:10.1088/1757-899X/142/1/012089

Increasing scanning step causes the augmentation of the sintered surface roughness *Rz* 600 to 700, figure 6 and figure 7.

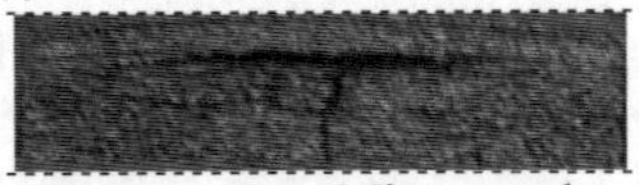

Figure 6. Micrographs, showing surfaces of the sintered copper samples (x2), process conditions of sintering: V=200 mm/min, t=200 °C, S=0.1 mm, P=30 W.

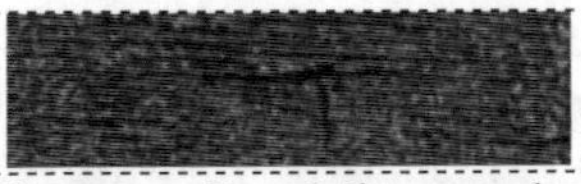

Figure 7. Micrographs, showing surfaces of the sintered copper samples (x2), process conditions of sintering: V=200 mm/min, t=200 °C, S=0.1 mm, P=30 W.

The temperature of pre-heating the powdered sample before sintering has only a slight effect on the roughness. Comparing the data of experiments and data calculated by formula 1, we have concluded that laser output power, velocity of laser beam displacement and scanning step are the main variables influencing on the sintered surface roughness, although the temperature of powder material pre-heating is to be taken into account.

The experiments are conducted to detect the dependence of the sintered surface layer on the shielding gas (argon), figures 8-11, and mechanical activation of the powder. Argon sintering results in improvement of the surface layer strength, without origination of cracks. Cross and length-wise cracks are not detected when sintering in argon, figure 9, sintering conditions: P=15 W, V=200 mm/min, S=0.3 mm, t=200 ^{0}C. The thickness of the sintered layer changes a little from 1.7 mm (air sintering) to 1.735 (argon sintering).

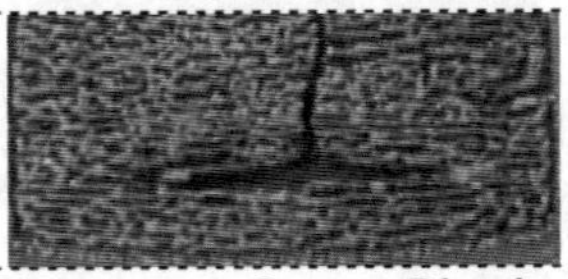

Figure 8. Physical configuration of the sintered surface PMS-1 (x2), sintering conditions: P=15 W, V=200 mm/min, S=0.3 mm, t=200 ^{0}C, air sintering

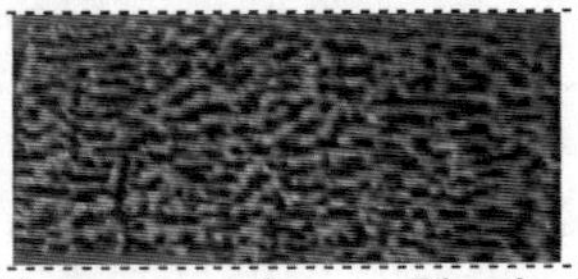

Figure 9. Physical configuration of the sintered surface PMS-1 (x2), sintering conditions: P=15 W, V=200 mm/min, S=0.3 mm, t=200 ^{0}C, argon sintering

A sharp decline of the surface layer quality and strength becomes apparent when comparing the samples sintered in argon and in air in conditions: P=30 W, V=200 mm/min, S=0.1 mm, t=200 ^{0}C, shown in figures 10-11. The roughness changes from 750 to 115 μm, the thickness of the sintered layer does not change much, varying from 1.0 to 0.915 mm. There are no cross and length-wise cracks detected in the sample, sintered in argon.

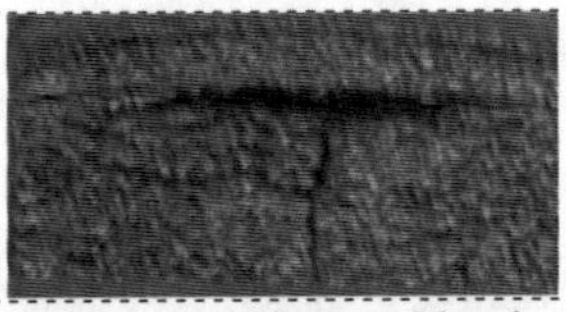

Figure 10. Physical configuration of the sintered surface PMS-1 (x2), sintering conditions: P=30 W, V=200 mm/min, S=0.1 mm, t=200 ^{0}C, air sintering

Figure 11. Physical configuration of the sintered surface PMS-1 (x2), sintering conditions: P=30 W, V=200 mm/min, S=0.1 mm, t=200 ^{0}C, argon sintering

IOP Conf. Series: Materials Science and Engineering **142** (2016) 012089 doi:10.1088/1757-899X/142/1/012089

Argon sintering has a significantly positive effect on the surface layer, wich becomes less rough and there are no defects detected.

Powders are subject to one-minute and three-minutes machining in a centrifugal planetary-type mill AGO-2 in order to detect the influence of mechanical activation of powder on the sintered layer. In figures 12-14 there are images of powder composite PMS-1 layers sintered in various conditions for non-activated powder and powder of different degrees of activation. In figures 12-14 one can see that surface roughness of sintered pre-activated samples varies from 700 µm (non-activated powder) to 426 µm (powder, activated for three minutes) in sintering conditions P=15 W, V=200 mm/min, S=0.3 mm, t=26^{0}C.

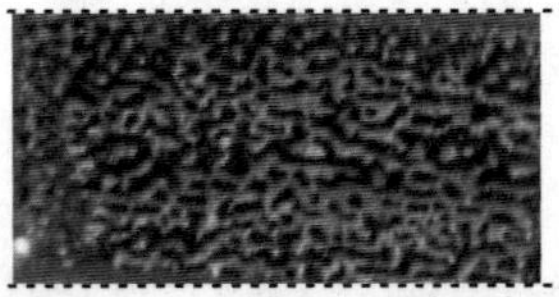

Figure 12. Physical configuration of the sintered surface PMS-1 (x2), sintering conditions: P=15 W, V=200 mm/min, S=0.3 mm, t=26 ^{0}C, non-activated

Figure 13. Physical configuration of the sintered surface PMS-1 (x2), sintering conditions: P=15 W, V=200 mm/min, S=0.3 mm, t=26 ^{0}C, – one-minute activation

Figure 14. Physical configuration of the sintered surface PMS-1 (x2), sintering conditions: P=15 W, V=200 mm/min, S=0.3 mm, t=26 ^{0}C, – three-minutes activation

Figure 15. Physical configuration of the sintered surface PMS-1 (x2), sintering conditions: P=30 W, V=200 mm/min, S=0.1 mm, t=26 ^{0}C, non-activated

Figure 16. Physical configuration of the sintered surface PMS-1 (x2), sintering conditions: P=30 W, V=200 mm/min, S=0.1 mm, t=26 ^{0}C, one-minute activation

Comparing the images of samples a conclusion can be drawn that mechanical activation has a significant influence on the sintered surface: coagulation and roughness decrease, the samples are more solid.

When analyzing the sintered layers of non-activated and activated powdered material, it is revealed that preliminary mechanical activation influences on the sintering process, improving the sintered surface, as the consequence, the diameter of coagulated particles gets smaller, roughness is decreased.

The comparison of graphs and observational results over the sintering process makes it obvious that increasing the power results in surface swelling, causing, therefore, shrinkage and melting. It is possible perhaps because of the rapidly growing thermal conductivity of the material at the moment of melting. Then the process is stabilized, and increasing power density has no effect on the thickness of the sintered layer.

Having processed a large amount of experimental data and detected an apparent change in the sintered surface of the powder material, figures 17-19 it is proposed to use the concepts of rough and finishing modes of laser sintering powder materials. This allows differentiating the modes of synthesizing the surface and the bulk of the prototype.

IOP Conf. Series: Materials Science and Engineering **142** (2016) 012089 doi:10.1088/1757-899X/142/1/012089

Figure 17. Images of the surface layer of sintered powder PMS-1 ×(150), high-quality surface

Figure 18. Images of the surface layer of sintered powder PMS-1 ×(150), swelling of the surface (increasing roughness),

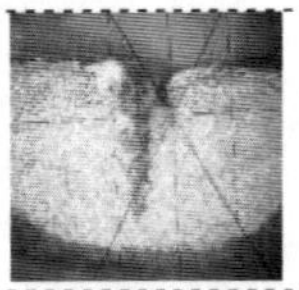

Figure 19. Images of the surface layer of sintered powder PMS-1 ×(150), melting

As it is revealed in the conducted experiments the thickness of the sintered surface layer and its roughness Rz can be varied in definite limits via changing the modes of laser treatment. To synthesize a sintered surface of different quality the authors suggest differentiating conditions of forming the prototype surface and its bulk into rough, semi-finishing and finishing ones, figure 20.

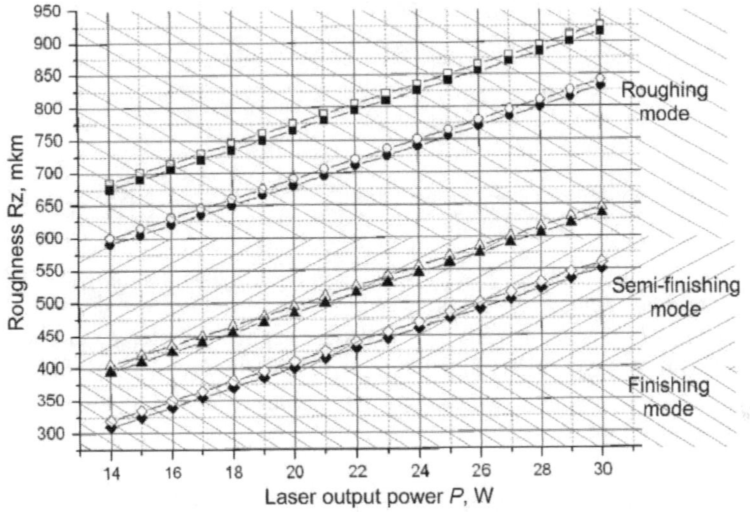

—□— V=200 mm/min, t=26 °C, S=0.3 mm; V=200 mm/min, t=200 °C, S=0.3 mm;
—○— V=200 mm/min, t=26 °C, S=0.1 mm; V=200 mm/min, t=200 °C, S=0.1 mm;
—△— V=300 mm/min, t=26 °C, S=0.3 mm; V=3000 mm/min, t=200 °C, S=0.3 mm;
—◇— V=3000 mm/min, t=26 °C, S=0.1 mm; V=3000 mm/min, t=200 °C, S=0.1 mm;

Figure 20 – Sintering modes for copper powder material

The rough sintering conditions are aimed at forming the bulk of the product. Increasing roughness of the surface (swelling) has a positive effect on the firm joint between the layers in this case.

The efficiency of layer-by-layer sintering is reduced considerably in semi-finishing conditions; however, no further processing of the prototype is required to make the surface of the specified quality and accuracy.

Finishing processing is performed in case of need. The finishing mode can be called accelerated one and it is based on the high density of laser output power in a short period of time. As the results, sublimation of micro-roughness peaks of the surface or plating the prototype surface are possible.

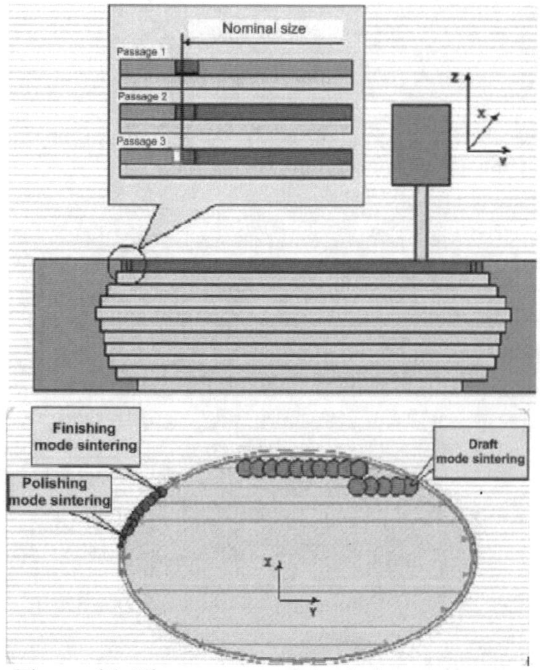

Figure 21. The strategy of single layer sintering

Conclusion

The paper describes the influence of laser sintering conditions and empirical dependence of the sintered surface layer on the sintering conditions, which allows controlling the manufacturing process of a high quality product. Controlled modifying the properties of the sintered surface layer is discussed via differentiating the process conditions into a semi-finishing mode for the surface and a rough mode for the bulk, that makes it possible to increase the efficiency of the sintering process without deteriorating quality of the product surface.

References

[1] C.R. Deckard, J.J. Beaman, "Recent advances in selective laser sintering" // Proceedings of the 14th Conference on Production Research and Technology, Michigan, 1987. P. 447-451.
[2] Polmear IJ. Light alloys: metallurgy of the light metals. 3rd ed. London: Arnold; 1995.
[3] Cam G, Kodak M. Progress in joining of advanced materials. Int Mater Rev 1998;43:1–44.
[4] Mondolfo LF. Aluminium alloys: structure and properties. London: Butterworths; 1976.
[5] Bias CT. In: Olsen FO, editor. Hybrid laser-arc welding. Woodhead Publishing Ltd: CRC Press; 2009. p. 216–69.
[6] Seyda V, Kaufmann N, Emmelmann C. Investigation of aging processes of Ti–6Al–4V powder material in laser melting. In: Physics procedia of the 7th international conference & exhibition on photonic technologies LANE, vol. 39, Fürth, Germany; 2012. p. 425–31
[7] Petrushin S. I. , Gubaydulina R. K. , Grubiy S. V. , Likholat A. V. On the Problem of Wear Resistant Coatings Separation From Tools and Machine Elements // IOP Conference Series: Materials Science and Engineering. - 2015 - Vol. 91, Article number 012048. - p. 1-7
[8] Niu HJ, Chang ITH. Selective laser sintering of gas atomized M2 high speed steel powder. J Mater Sci 2000;35:31–8.
[9] Asgharzadeh H, Simchi A. Effect of sintering atmosphere and carbon content on the densification and microstructure of laser-sintered M2 high-speed steel powder. Mater Sci Eng A 2005;403(1–2):290–8.
[10] Liu ZH, Zhang DQ, Sing SL, Chua CK, Loh LE. Interfacial characterization of SLM parts in

multi-material processing: metallurgical diffusion between 316L stainless steel and C18400 copper alloy. Mater Charact 2014;94:116–25.
[11] Saprykina N A, Saprykin A A, Matrunchik M S, Formation of Surface Layer of Cobalt Chrome Molybdenum Powder Products with Differentiation of Laser Sintering Modes, Applied Mechanics and Materials 682 2014 p 294-298
[13] Belomestnykh V N, Soboleva E G, Acoustic analogs of elasticity theory ratios for determining anisotropic Poisson ratios of cubic monocrystals 7th International Forum on Strategic Technology (IFOST - 2012) Tomsk: TPU Press 1 2012 p 499-502
[14] Galevsky G.V., Rudneva V.V., Garbuzova A.K., Valuev D.V., Titanium carbide: nanotechnology, properties, application, J. IOP Conference Series: Materials Science and Engineering. 91 (2015) p. 1-7
[15] Saprykina N A, Yanyushkin A S, Medvedeva O I, Mechanism of Protective Membrane Formation on the Surface of Metal-Bonded Diamond Disks, Applied Mechanics and Materials 682 2014 p. 327-332
[16] Babakova E V, Gradoboev A V, Saprykin A A, Ibragimov E A, Yakovlev V I, Sobachkin A V, Comparison of Activation Technologies Powder ECP-1 for the Synthesis of Products Using SLS Applied Mechanics and Materials 756 2015 p 220-224
[17] Saprykin A A, Ibragimov E A, Yakovlev V I, Influence of mechanical activation of powder on SLS process Applied Mechanics and Materials 682 2014 p 143-147
[18] Saprykin A A, Saprykina N A, Dudikhin D V, Emelyanenko S M, Influence of layer-by-layer laser sintering modes on the thickness of sintered layer of cobalt-chromium-molybdenum powder, Advanced materials research 1040 2014 p 805-808
[19] Saprykina N A, Saprykin A A, Borovikov I F, Sharkeev Y P, Influence of layer-by-layer laser sintering conditions on the quality of sintered surface layer of products, IOP Conf. Series: Materials Science and Engineering 91 2015 012031

Modeling the Digging Process of Tree Root System by the Mechanism with Hydropulse Drive

Michael V Drapalyuk[1], Petr I Popikov[1], Roman V Yudin[1], Anatoly A Fomin[2,a], Roman V Chernukhin[3]

[1] Voronezh State University of Forestry and Technologies named after G.F. Morozov, Timiryazeva street 8, Voronezh, Russian Federation, 394087
[2] Institute of engineering and automobile transport, Vladimir State University, Gorky street 87, Vladimir, Russian Federation, 600000
[3] Yurga Institute of Technology, National Research Tomsk Polytechnic University, Leningradskaya street 26, Yurga, Russian Federation, 652055

E-mail: [a] fomin1@mail.ru

Abstract. A mathematical model of the mechanism for digging large-sized seedlings or removing stumps on the undergrowth is developed, which is based on the idea of the root system and surrounding soil, as a combination of plurality of spherical elements, interacting with each other, and the knife mechanism, as a set of elementary surfaces that interact with elements of the stump and soil. The effect of blade vibration on the stump removal efficiency with the help of the model is studied.

1. Introduction

When digging or removing large-sized seedlings or stumps clearance on the undergrowth, roadside and shelter belts, machines of manipulator-type with replaceable technological equipment of discrete action have high efficiency. We proposed a mechanism with hydropulse drive as removable processing equipment for excavators and forestry cranes.

The purpose of this article is to develop a mathematical model of digging large-sized seedlings or stumps clearance in the undergrowth by mechanism with hydropulse drive, which will further assess the effectiveness of the proposed design and parameters of blade vibration in various modes. In the simulation of mechanism, shape and physical characteristics of the stump, forest soil, as well as the interactions of these environments with a knife must be properly represented in the model.

2. Justification and implementation of the process by the mechanism with hydropulse drive.

Timber and forest soil are extremely difficult to simulate by objects because of their typological diversity and large number of physical and mathematical parameters describing it, in particular, the type of soil or wood, humidity, anisotropy, friability, crispness, topography. Model of tree and soil is based on the method of discrete elements; environment is represented as the collection of large number of spherical elements (about 10.000) of small size, which are capable of reacting both between themselves and with the working surfaces of the mechanism.

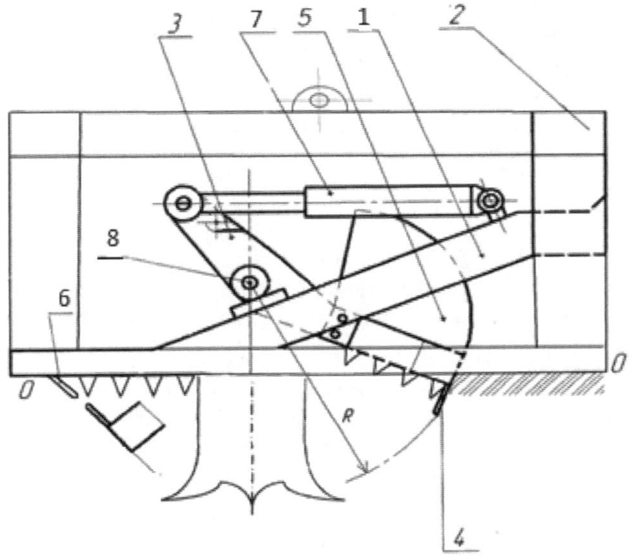

Figure 1. Scheme of mechanism for removing tree stumps: 1 – bearing beams; 2 – frame of mechanism; 3 – vertical rods; 4 – working body; 5 – semi-basket; 6 – spiked bumpers; 7 – hydraulic cylinder; 8 – hinges for mounting vertical rods

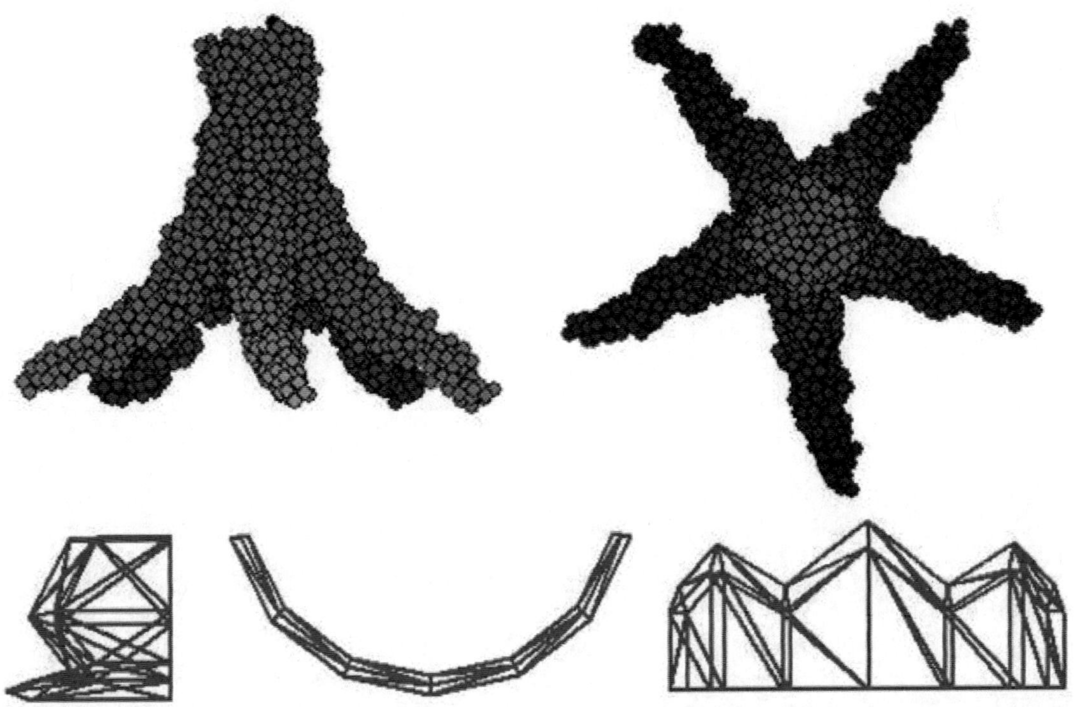

Figure 2. Representation of tree root systems (horizontal and vertical projection) and blade mechanism (three projections) in the model

3. Modeling the digging process of tree root system

The interaction of mechanism working surfaces with wood and soil, from the geometrical point of view, is the problem of finding the distance rв from a plane (the elementary plane of the working surfaces of the mechanism) to the surface of the any spherical element of environment. At this viscoelastic force $\vec{F}_{ij}^{BУ}$, acting on the *i*-th element of the soil from the *j*-th element is given by

$$\vec{F}_{ij}^{BУ} = c \cdot r_{в} \cdot \vec{n} - k\vec{v}, \tag{1}$$

where $\vec{n}$ and $\vec{v}$ – direction and speed of the interaction of a ball element and a given surface, calculated for each spherical element of the soil methods of analytic geometry; *c* and *k* - the coefficients of stiffness and viscosity of interaction.

The elastic component of interaction between the elements ensures both the repulsion of elements (distance between centers *rij* of *i*-th and *j*-th elements is less than the element diameter *dE*) and attraction of element over a narrow range of distances ($dE > rij > rк$) (Figure 2), where $rк = \alpha dE$ – critical distance to which elements interact with each other; α – coefficient of expression of critical distance in terms of the diameter the element (in most calculations it took the value of 1.02). When calculating the force $\vec{F}_{ij}$, exerted by the element *i* on the element *j*, it is considered, according to Newton's third law, that the force exerted by the element *j* on the element *i* is the same in magnitude and opposite in direction, i.e. $\vec{F}_{ij} = -\vec{F}_{ji}$. According to Newton's second law, the equations of motion of the *i*-th element are the following:

$$\begin{cases} m_i \dfrac{d^2 x_i}{dt^2} = \sum\limits_{\substack{j=1 \\ j\neq i}}^{N_E} \left(\begin{cases} c_{ij}(d_E - r_{ij}) \dfrac{(x_i - x_j)}{r_{ij}} + k_{ij}(r_{ij} - d_E)(v_{xi} - v_{xj}), & r_{ij} < d_E + \Delta d_l; \\ 0, & r_{ij} \geq d_E + \Delta d_1; \end{cases} \right) + \sum\limits_{k=1}^{N_s} F_{xik}^{S}; \\ m_i \dfrac{d^2 y_i}{dt^2} = \sum\limits_{\substack{j=1 \\ j\neq i}}^{N_E} \left(\begin{cases} c_{ij}(d_Э - r_{ij}) \dfrac{(y_i - y_j)}{r_{ij}} + k_{ij}(r_{ij} - d_E)(v_{yi} - v_{yj}), & r_{ij} < d_E + \Delta d_l; \\ 0, & r_{ij} \geq d_E + \Delta d_l; \end{cases} \right) + \sum\limits_{k=1}^{N_s} F_{yik}^{S}; \\ m_i \dfrac{d^2 z_i}{dt^2} = -m_i g + \sum\limits_{\substack{j=1 \\ j\neq i}}^{Ne} \left(\begin{cases} c_{ij}(d_E - r_{ij}) \dfrac{(z_i - z_j)}{r_{ij}} + k_{ij}(r_{ij} - d_E)(v_{zi} - v_{zj}), & r_{ij} < d_E + \Delta d_l; \\ 0, & r_{ij} \geq d_E + \Delta d_l; \end{cases} \right) + \sum\limits_{k=1}^{N_S} F_{zik}^{S}, \end{cases} \tag{2}$$

where *i* – element number; m_i – mass of the *i*-th element; x_i, y_i, z_i – Cartesian coordinates of the element; *t* – time; N_E – quantity of elements; *j* – element number, possibly contacting with the *i*-th element; c_{ij} and d_{ij} – coefficients of stiffness and viscosity of the interaction of elements *i* and *j*; r_{ij} – the distance between the centers of the elements *i* and *j*; v_{xi}, v_{zi} – the Cartesian components of the velocity of *i*-th element; d_l – distance of interaction limitations between the elements; *g* – gravitational acceleration; *k* – the number of elementary surface of the mechanism acting on the environment; N_S – the number of elementary surfaces; F^{S}_{xik}, F^{S}_{yik}, F^{S}_{zik} – Cartesian components of the force exerted by the *k*-th elementary surface on the *i*-th element.

The distance r_{ij} between the centers of the elements is calculated by center coordinates by

Pythagorean Theorem: $r_{ij} = \sqrt{(x_i - x_j)^2 + (y_i - y_j)^2 + (z_i - z_j)^2}$.

When simulating in the model the process of removing the stump pivotal movement of the working body and the vibration transmitted from the hydropulsator to the cylinder takes place. Turning the working body and its vibration are described in the following coordinate transformation of reference points of the elementary surfaces:

$$\begin{cases} x_i' = x_h + r_i \cos(\varphi_i + \omega_w t + \varphi_v \cos(2\pi f_v t)); \\ y_i' = y_i; \\ z_i' = z_h + r_i \sin(\varphi_i + \omega_w t + \varphi_v \cos(2\pi f_v t)); \\ r_i = \sqrt{(x_i - x_h)^2 + (z_i - z_h)^2}; \\ \varphi_i = \begin{cases} \arctan\dfrac{z_i - z_h}{x_i - x_h}, & x_i - x_h \ge 0; \\ \arctan\dfrac{z_i - z_h}{x_i - x_h} + 180^O, & x_i - xh < 0, \end{cases} \end{cases} \quad (3)$$

where i – index of reference point of elementary surface, which the working body consists of; (x_i, y_i, z_i) and (x_i', y_i', z_i') – coordinates of the i-th reference point before and after the coordinate transformation (in the initial and the current time); (x_h, z_h) – coordinates of axis of the working body passing through the hinges; r_i and φ_i – the polar coordinates of the i-th reference point in the coordinate system associated with the axis of the working body; ω_w – the angular velocity of rotation of the working body; φ_v – vibration amplitude (it is a maximum angle through which the operating element is deflected from the guideline in vibrations); f_v – the frequency of vibration of the working body.

In the proposed version of the model the vibration of the working body is considered only by addition of the term of sum $\varphi_v\cos(2\pi f_v t)$ into the expression for angular position of the working body, however, that already allows to study the influence of the amplitude and frequency of vibration on the stump clearance efficiency.

Blade and limit stop of mechanism are represented in the model as a set of a plurality of elementary triangles (figure 3). Before you start the computer experiment, location of reference points P1 ... P35 is performed in space according to the following formulas, which include the geometric parameters of the blade. After determining the current coordinates of the base points (for this step - time integration τ_i) the construction of the elementary triangles is made that define the working surfaces of the blade and the stop limit. So, for the blade we have the following 54 triangles:

$$\begin{array}{llll}
T_1 = P_1P_2P_{15}, & T_2 = P_2P_{15}P_{16}, & T_3 = P_2P_3P_{16}, & T_4 = P_3P_{16}P_{17}, \\
T_5 = P_3P_4P_{17}, & T_6 = P_4P_{17}P_{18}, & T_7 = P_4P_5P_{18}, & T_8 = P_5P_{18}P_{19}, \\
T_9 = P_5P_6P_{19}, & T_{10} = P_6P_{19}P_{20}, & T_{11} = P_6P_7P_{20}, & T_{12} = P_7P_{20}P_{21}, \\
T_{13} = P_8P_9P_{22}, & T_{14} = P_9P_{22}P_{23}, & T_{15} = P_9P_{10}P_{23}, & T_{16} = P_{10}P_{23}P_{24}, \\
T_{17} = P_{10}P_{11}P_{24}, & T_{18} = P_{11}P_{24}P_{25}, & T_{19} = P_{11}P_{12}P_{25}, & T_{20} = P_{12}P_{25}P_{26}, \\
T_{21} = P_{12}P_{13}P_{26}, & T_{22} = P_{13}P_{26}P_{27}, & T_{23} = P_{13}P_{14}P_{27}, & T_{24} = P_{14}P_{27}P_{28}, \\
T_{25} = P_{15}P_{16}P_{29}, & T_{26} = P_{16}P_{29}P_{30}, & T_{27} = P_{16}P_{17}P_{30}, & T_{28} = P_{17}P_{30}P_{31}, \\
T_{29} = P_{17}P_{18}P_{31}, & T_{30} = P_{18}P_{31}P_{32}, & T_{31} = P_{18}P_{19}P_{32}, & T_{32} = P_{19}P_{32}P_{33}, \\
T_{33} = P_{19}P_{20}P_{33}, & T_{34} = P_{20}P_{33}P_{34}, & T_{35} = P_{20}P_{21}P_{34}, & T_{36} = P_{21}P_{34}P_{35}, \\
T_{37} = P_{22}P_{23}P_{29}, & T_{38} = P_{23}P_{29}P_{30}, & T_{39} = P_{23}P_{24}P_{30}, & T_{40} = P_{24}P_{30}P_{31}, \\
T_{41} = P_{24}P_{25}P_{31}, & T_{42} = P_{25}P_{31}P_{32}, & T_{43} = P_{25}P_{26}P_{32}, & T_{44} = P_{26}P_{32}P_{33}, \\
T_{45} = P_{26}P_{27}P_{33}, & T_{46} = P_{27}P_{33}P_{34}, & T_{47} = P_{27}P_{28}P_{34}, & T_{48} = P_{28}P_{34}P_{35}, \\
T_{49} = P_{15}P_{22}P_{29}, & T_{50} = P_{21}P_{28}P_{35}, & T_{51} = P_1P_8P_{15}, & T_{52} = P_8P_{15}P_{22}, \\
T_{53} = P_7P_{14}P_{21}, & T_{54} = P_{14}P_{21}P_{28} & &
\end{array} \quad (4)$$

Stop limit in the model consists of four triangles:

$$T_{55} = P_{36}P_{38}P_{39},\ T_{56} = P_{36}P_{37}P_{39},\quad T_{57} = P_{36}P_{40}P_{41},\quad T_{58} = P_{36}P_{37}P_{41}.$$

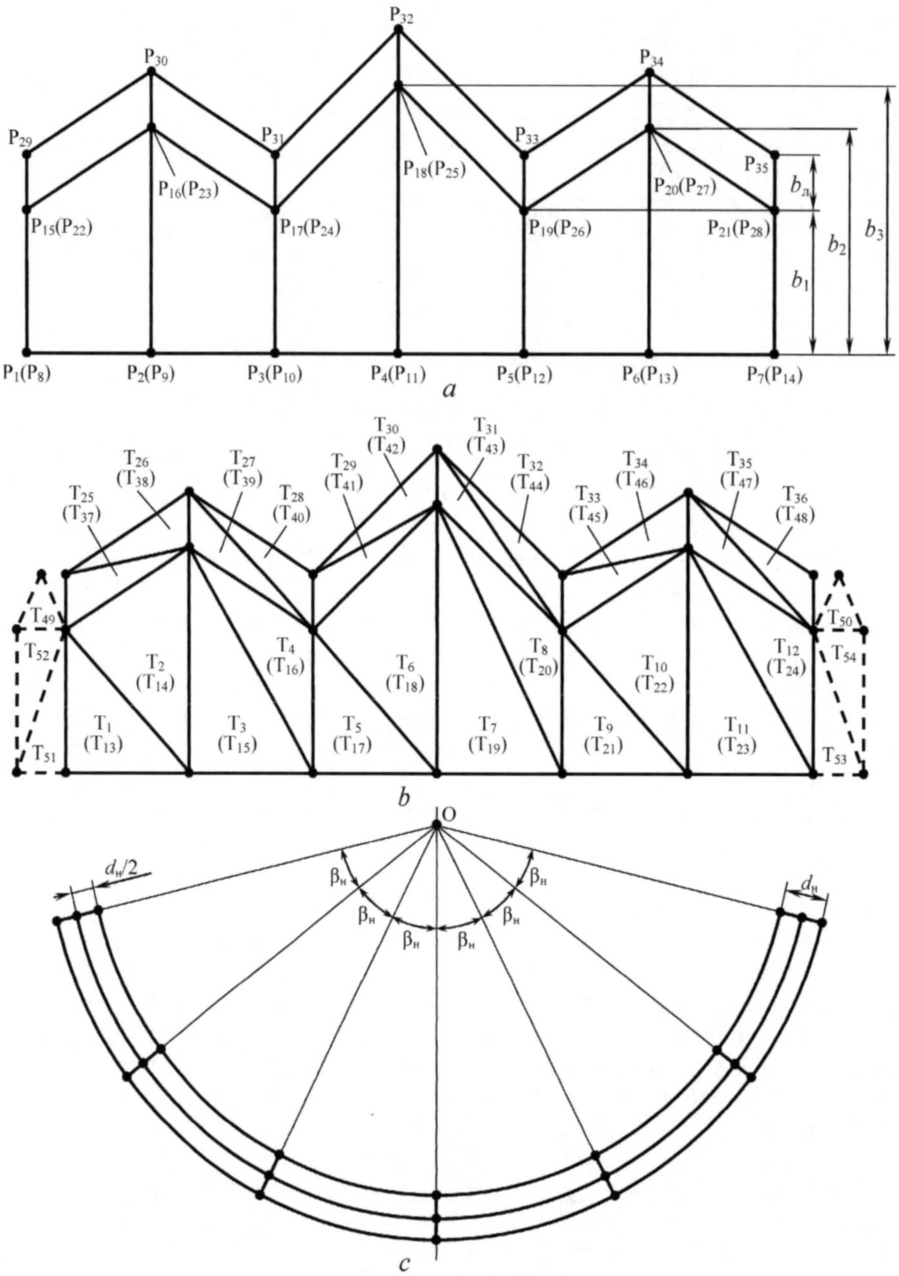

Figure 3. Presentation of blade of mechanism in the model: the basic point Pi (a) and elementary triangles Ti (b) to scan the blade, as well as a view of the front blade (c)

Modeling of hydraulic subsystem is based on analysis of changes in volume V_m of various cavities (m - means the index of the cavity) during the operation of the mechanism. The pressure P_m in the cavity m varies according to the relation:

$$\frac{dP_m}{dV_m} = -\frac{E}{V_m}, \tag{5}$$

where E – bulk modulus of the working fluid.

If pressures in two hydraulically interconnected cavities i and j are different, there is a flow of the working fluid in the model. The consumption Q_{ij} is determined by the known formula

$$Q_{ij} = k_{ij}\text{sign}(P_i - P_j)\sqrt{|P_i - P_j|}, \tag{6}$$

where i and j – indexes of cavities; k_{ij} – throttling coefficient; sign(x) – function that returns the sign of the variable x.

This formula is used both for throttles (throttling coefficient is sufficiently small), and piping (large throttling coefficient).

In the model it is assumed that all throttling elements have circular cross section, and then the throttling coefficient is determined through orifice diameter d_{ij} by the formula:

$$k_{ij} = \mu\frac{\pi d_{ij}^2}{4}\sqrt{\frac{2g}{\gamma}}, \tag{7}$$

where μ – discharge coefficient; g – gravitational acceleration; γ – relative density of the working fluid.

Ability of pipelines to expand elastically under the influence of pressure in the model is not directly considered, but the elasticity of working fluid is taken into account indirectly, i.e., through the ratio E.

The rotation of the spool and movement of the piston of the hydraulic cylinder causes a change in the volume of considered cavities, so at each step of integration recalculated of volume of cavities takes place.

To solve the system of differential and algebraic equations, which is laid in the basis for the model, a computer program "Program to simulate the operation of the mechanism with hydropulse drive for stumps clearance" is developed. The program is developed in Borland Delphi 7.0 environment, programming language Object Pascal. Before you start the computer experiment the required parameters of blade, geometric and physical and mechanical parameters of the stump, the parameters of the soil, parameters of hydropulsator and hydraulic system can be set. In the process, the program continuously displays a schematic representation of the mechanism of the stump in three dimensions, and performance indicators in numerical and graphical form (figure 4).

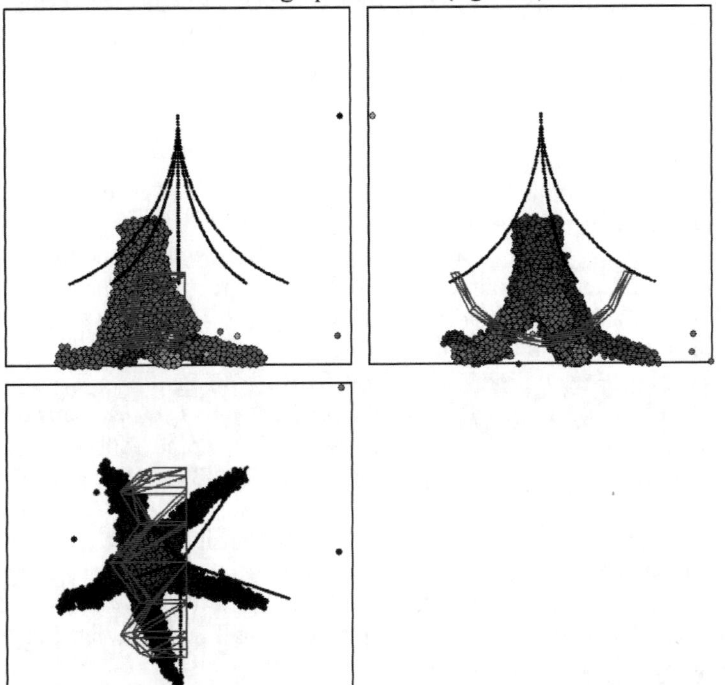

Figure 4. Displaying the simulation results in the developed program

The program is designed for repeated holding of computer experiments on the removal of stumps with mechanism of a new design and determining the optimal design and technological parameters on this basis.

- The main technical characteristics of the program:
- Number of elements in stump from 2000 to 5000;
- Estimated time of a computer experiment is about 5 minutes (at processor speed 3 GHz).

The first computer experiments have confirmed the effectiveness of the proposed design of the mechanism and efficiency of vibration to improve properties of cutting characteristics of the working body. The model provides a wide range of mechanical properties of the cutting process. In particular, the time dependence of modulus of resistance to the blade deepening was obtained (figure 5) which allows proving optimum shape and thickness of the blade at which the blade has sufficient strength and high cutting properties, but has small thickness and low metal content.

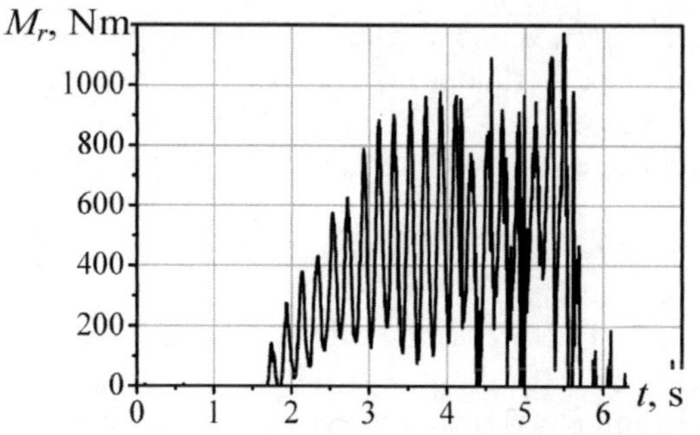

Figure 5. Dependence of the moment of resistance to blade deepening on time

Thus, the mathematical model of the mechanism with hydropulse drive for digging large-sized seedlings or removing stumps on the undergrowth, with its high spatial detail is developed, that allows studying the influence of parameters of blade, vibration, tree root systems and soil on the efficiency of work processes.

References

[1] Fomin, A. A. (2013). Vibrational motion of a complex mill under the action of the cutting force. Russian Engineering Research, 33(1), 57-60.

[2] Thies, W. G., Nelson, E. E., & Zabowski, D. (1994). Removal of stumps from a Phellinus weirii infested site and fertilization affect mortality and growth of planted Douglas-fir. Canadian Journal of Forest Research, 24(2), 234-239.

[3] Compton, J. E., & Boone, R. D. (2000). Long-term impacts of agriculture on soil carbon and nitrogen in New England forests. Ecology, 81(8), 2314-2330.

[4] Palviainen, M., Finér, L., Laiho, R., Shorohova, E., Kapitsa, E., & Vanha-Majamaa, I. (2010). Carbon and nitrogen release from decomposing Scots pine, Norway spruce and silver birch stumps. Forest Ecology and Management, 259(3), 390-398.

[5] Zabowski, D., Chambreau, D., Rotramel, N., & Thies, W. G. (2008). Long-term effects of stump removal to control root rot on forest soil bulk density, soil carbon and nitrogen content. Forest Ecology and Management, 255(3), 720-727.

[6] Morkovina, S. S., Santalova, M. S., Drapalyuk, M. V., & Panyavina, E. A. (2015). Concept of Creation of E-Platforms as Part of Support for Small Innovational Business. Mediterranean Journal of Social Sciences, 6(6), 57.

[7] Page-Dumroese, D. S., Harvey, A. E., Jurgensen, M. F., & Amaranthus, M. P. (1998). Impacts of soil compaction and tree stump removal on soil properties and outplanted seedlings in northern Idaho, USA. Canadian Journal of Soil Science, 78(1), 29-34.

IOP Conf. Series: Materials Science and Engineering **142** (2016) 012090 doi:10.1088/1757-899X/142/1/012090

[8] Fomin, A. A., & Gusev, V. G. (2013). Safe machining of blanks with nonuniform properties. Russian Engineering Research, 33(10), 602-606.

[9] Vasaitis, R., Stenlid, J., Thomsen, I. M., Barklund, P., & Dahlberg, A. (2008). Stump removal to control root rot in forest stands: a literature study. Silva Fennica, 42(3), 457-483.

[10] Teketay, D. (1997). The impact of clearing and conversion of dry Afromontane forests into arable land on the composition and density of soil seed banks. Acta Oecologica, 18(5), 557-573.

[11] Rabinowitsch-Jokinen, R., & Vanha-Majamaa, I. (2010). Immediate effects of logging, mounding and removal of logging residues and stumps on coarse woody debris in managed boreal Norway spruce stands. Silva Fennica, 44(1), 51-62.

[12] Prosvirnikov D.B., Ziatdinova D.F., Timerbaev N.F., Saldaev V.A. & Gilfanov K.H. 2016 Mathematical modelling of the steam explosion treatment process for pre-impregnated lignocellulosic material *IOP Conference Series: Materials Science and Engineering* **124**(1), 012087. IOP Publishing. DOI:10.1088/1757-899X/124/1/012087

[13] Iwald, J., Löfgren, S., Stendahl, J., & Karltun, E. (2013). Acidifying effect of removal of tree stumps and logging residues as compared to atmospheric deposition. Forest Ecology and Management, 290, 49-58.

[14] Staaf, H., & Olsson, B. A. (1994). Effects of slash removal and stump harvesting on soil water chemistry in a clearcutting in SW Sweden. Scandinavian Journal of Forest Research, 9(1-4), 305-310.

[15] Hyvönen, R., Olsson, B. A., & Ågren, G. I. (2012). Dynamics of soil C, N and Ca in four Swedish forests after removal of tops, branches and stumps as predicted by the Q model. Scandinavian Journal of Forest Research, 27(8), 774-786.

[16] Fomin, A. A., & Gusev, V. G. (2013). Spindle rigidity in milling blanks with nonuniform properties. Russian Engineering Research, 33(11), 646-648.

[17] Sadrtdinov A.R., Sattarova Z.G., Prosvirnikov D.B. and Tuntsev D.V. (2015, December) Modeling of thermal treatment of wood waste in the gasifiers. In *2015 Int. Conf. on Mechanical Engineering, Automation and Control Systems (MEACS)* pp 1-5. DOI:10.1109/MEACS.2015.7414914

[18] Aherne, J., Posch, M., Forsius, M., Lehtonen, A., & Härkönen, K. (2012). Impacts of forest biomass removal on soil nutrient status under climate change: a catchment-based modelling study for Finland. Biogeochemistry, 107(1-3), 471-488.

[19] Prosvirnikov D.B., Safin R.G., Ziatdinova D.F., Timerbaev N.F., and Lashkov V.A. 2016 Multifactorial modelling of high-temperature treatment of timber in the saturated water steam medium. *IOP Conference Series: Materials Science and Engineering* **124**(1), 012088. IOP Publishing. DOI:10.1088/1757-899X/124/1/012088

IOP Conf. Series: Materials Science and Engineering **142** (2016) 012091 doi:10.1088/1757-899X/142/1/012091

Mechanical Treatment of Raw Waste Lumber an Effective Way to Preserve the Ecology and Resources

Anatoly A Fomin[1,a], **Vladimir G Gusev**[1], **Roman V Yudin**[2], **Nail F Timerbaev**[3], **Oleg Yu Retyunskiy**[4]

[1] Institute of engineering and automobile transport, Vladimir State University, Gorky street 87, Vladimir, Russian Federation, 600000
[2] Voronezh State Academy of Forestry Engineering, Timiryazeva street 8, Voronezh, Russian Federation, 394087
[3] Institute of Petroleum and chemical Engineering, Kazan National Research Technological University, K. Marx street 68, Kazan, Republic of Tatarstan, Russian Federation, 420015
[4] Yurga Institute of Technology, National Research Tomsk Polytechnic University, Leningradskaya street 26, Yurga, Russian Federation, 652055

E-mail: [a] fomin1@mail.ru

Abstract. Alternative process flowsheet machining of the machining of raw waste lumber were analysed, and it was implemented in a real machine model based on the chosen scheme. The forming process of the treated surface of the stock material was examined, and consequently the mathematical models of the geometric errors in terms of independent factors of the profile milling process were defined. Based on these models is possible to construct a treatment process of the raw waste lumber with minimal errors on the surfaces which were treated. The manufacturing of products from raw waste lumber allows to reduce the volume of deforestation and helps to preserve the ecology and economize the material resources.

1. Introduction

A significant portion of the surface of the planet "Earth" is covered by forest, which plays a crucial role in the maintenance of the ecological balance, and, therefore, the conditions of existence of all life on the planet. At the same time, the wood is a unique material, which, thanks to its properties, is widely used in various sectors of the national economy. To satisfy the global market demands for wood products requires large volumes of wood-stock material, and this involves an intensive deforestation. Despite the application of modern technologies in the timber enterprises [1] significant amounts of wood waste are formed, which are not used for its original purpose.

Annually the wood-processing enterprises of the Russian Federation generate millions of cubic meters of barks, slabs, laths and other waste wood, of which they produce mainly wood pellets and industrial wood chip. However, the raw waste lumber consist of the highest quality peripheral wood grain from which it is appropriate not to make fuel, but the products of the original purpose: pieces of furniture, products for interior decoration, construction of modern houses, etc. The machining of the raw waste lumber and the production of quality wood products on this basis allow to reduce the irrational use of wood, save material and energy resources [2], stabilize and improve the environmental situation through the conservation of natural forestlands as a basis for the regeneration of oxygen and natural barrier for dust and sand storms, which turn oases into deserts.

The raw waste lumber are characterized by an unpredictable shape and sizes, heterogeneity of the properties of the threated material [3], the stochastic disposition of knots and the lack of developed technological bases, which provoke serious difficulties in their machining and to some extent explains the reason for the lack of high-performance of engineering processes for a deep wood recycling. In conditions of the mentioned instability to guarantee a high machining efficiency and obtain accurate finished products from raw waste lumber is problematic. Consequently, there is a necessity of the development of high-performance processes on scientific substantiation and equipment for machining raw waste lumber, which allows, in addition to what has been said above, to get a significant economic effect and increase the competitiveness of woodworking companies in the conditions of the today's market [4].

For machining peripheral segments remaining after the logs slitting, it is used a special equipment. The known equipment leaves considerable waste after cutting, including peripheral segments smaller and does not allow to optimize constantly the treatment regimens in the process changing of the cutting conditions.

2. Justification and implementation of the process flowsheet of machining raw waste lumber.

The authors have developed alternative process flowsheet of machining raw waste lumber, which have analysed the simplicities of a practical implementation of the flowsheet and resonance-free processing conditions from the perspective of a stable blank locating, on the basis of the selected flowsheet (figure 1), which more than any other satisfies the above criteria. The flowsheet is characterized by a content of machining steps: the milling of the unbarked curved surface and the subsequent cutting of the curvilinear lateral edges are performed simultaneously on one machine. As processing guide base, the machine for forming the unbarked working surface is selected, and further the one for milling the blank surface, which provides the greatest length of the processing guide base and eliminate defects of the blank during the treatement.

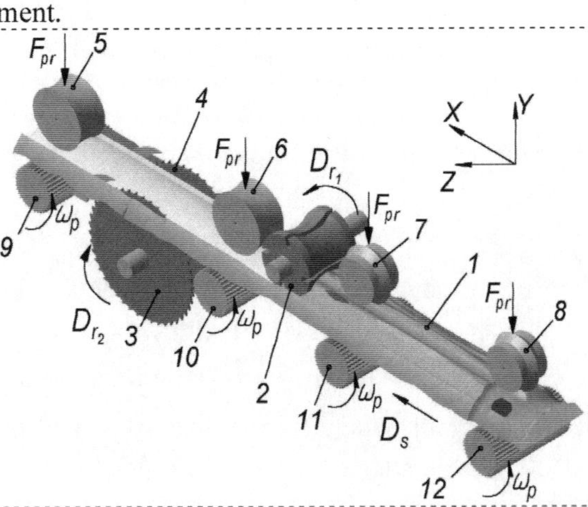

Figure 1. The process flowsheet selected for practical implementation: 1 – logging residues; 2 – form-milling cutter; 3-4 – saw wheels; 5-8 – pinch rollers; F_{pr} – pressing force; 9-12 – toothed rolling mill; ω_p – angular velocity of the rolling mill; D_{r1}, D_{r2}, D_s – working tool and stock material movement.

On the basis of the chosen scheme, the method and machine tool were developed for machining the raw lumber waste, equipped with a system of automatic speed control for supplying the working stock material and a device for cutting. The automatic speed control for supplying the working stock

IOP Conf. Series: Materials Science and Engineering **142** (2016) 012091 doi:10.1088/1757-899X/142/1/012091

material allows to stabilize the cutting power, avoid the machine tool overloading and the cutting tool damage at the unexpected appearance of knots or an overrate allowance on the cutting zone, and it stably operate at a jump cutting depth up to 15 mm, which satisfies the average change of the peripheral segments allowance.

The machine tool for machining the raw waste lumber (figure 2a) includes a base 1, a drive 2 of vertical motion of the slide mill, a control unit 3, claw-type protection 4 and exhauster system 5 for removal residues.

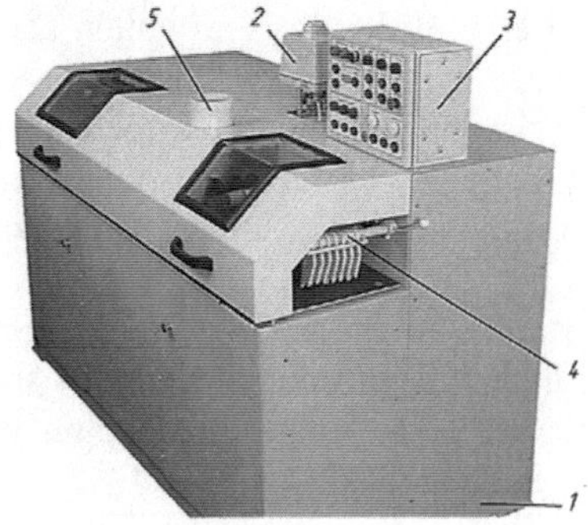

Figure 2. General view (a) and a working area (b) of the machine tool for machining raw waste lumber.

The rollers 1 are designed for clamping the stock material to the rolling mills 2 and 6 (figure 2b), the last move the stock material relatively to the rotating form-milling cutter 3, which removes the allowance from the unbarked surface stock material. The saws wheels 4 cut off the irregular side edges of the stock material, and the machined part of which is directed by the profiled rollers 7.

3. Analysis of the surfaces forming in the profile milling process of the raw waste lumber.

The developed equipment can be effectively used in the production, if they provide the high geometric accuracy to the finished products. Thereupon, we analyse the mechanism of the machined surfaces forming under the influence of kinematic and dynamic treatment process factors and determine the ways to reduce the geometrical errors, which arise during the profile milling of the unbarked surface of the raw waste lumber [5].

The material utilization coefficient will be maximized at removing it from the unbarked curved surface of the peripheral segment of the minimum allowance, which is achieved by a grinding cutting of the mill blade along a circumference KLM (figure 3) with the radius R_{pn}. The form-milling cutter is characterized by the current radius $R \in [R_{min}, R_{max}]$ The minimum radius R_{min} is located in the transversal plane of symmetry of the mill, and the maximum, R_{max} in the planes of both faces.

During the milling of the stock material with a cylindrical mill, the geometry error due to the kinematics of the milling process is described by the equation [6]:

$$\Delta = R - \sqrt{R^2 - \left(\frac{\pi S}{z\omega}\right)^2}, \quad (1)$$

where R is the radius of the cutting mill surface; S is the cutting feedrate of the raw material; $z\omega$ is the number of teeth and angular velocity of the mill respectively.

For the form-milling cutter the radius R of the cutting surface varies in height B, which leads to the formation of the cutting blade along different geometrical errors of the transversal stock material section. We determine the change of these errors, and for this we consider the contacting interaction of the form-milling cutter and the stock material (figure 3).

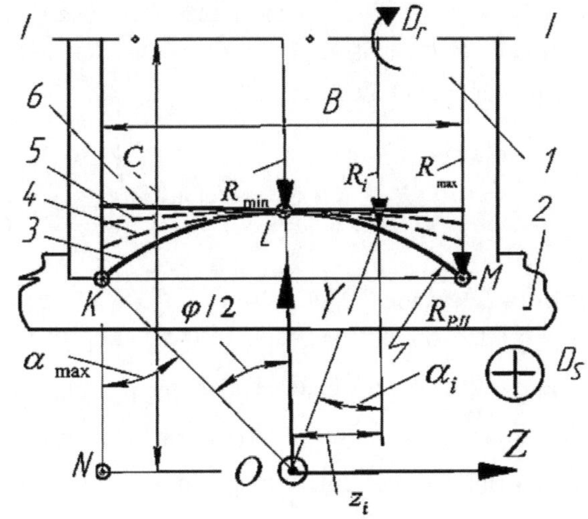

Figure 3. Scheme of contact interaction of the mill and the stock materail: 1 - form-milling cutter, 2 - stock material, 3 - curvilinear contour of the cutting blades 4-6 - contours of the cutting blade at a radius change of sharpening; R_{pn}, D_r, D_s - working motion of the tool and the stock material.

Distance from the center *O* of the circumference arc *KLM* to the axis I - I mill 1

$$C = R_{min} + O \cdot L = R_{min} + R_{pn} \tag{2}$$

In the current i-th transversal section of the mill, spaced at a distance z_i from the origin *O* of the coordinate system *YOZ*, the mill radius $R = R_i$, and the equation (2) of this section is given as:

$$C = R + R_{p\text{л}} \cdot \cos\left(\arcsin\left(\frac{z_i}{R_{p\text{л}}}\right)\right). \tag{3}$$

Equalling the right sides of (2), (3) and completing the conversion, we get:

$$R = R_{pn}\left(1 - \cos\left(\arcsin\left(\frac{z_i}{R_{p\text{л}}}\right)\right)\right) + R_{min} \tag{4}$$

The current angle α_i, is formed by the plane of the perpendicular tool axis and the radius R_{pn} of the cutting blade

$$\alpha_i = \arcsin\left(\frac{z_i}{R_{pn}}\right) \tag{5}$$

The numerical values of the angle $\alpha_i \in [0, \alpha_{max}]$, where $\alpha_{max} = 0{,}5\varphi$; φ – the central angle which corresponds to the arc of the circumference *KLM*. At the angle $\alpha_i = \arcsin\left(\frac{z_i}{R_{pn}}\right) = 0$ the coordinate

$z_i = 0$, and the plane perpendicular to the tool axis pass through the point L. In accordance with (4) at $z_i = 0$ the mill radius is equal to the minimum value R_{min}. At $\alpha_i = \alpha_{max} = \pm 0{,}5\varphi$ we have two transversal sections, one of which passes through the point *K*, and the other the point *M*. These sections are arranged symmetrically to the axis *Y*. The mill radius takes the maximum value R_{max} in these sections according to (4), which can be determined if the known minimum radius and height of the mill:

$$R_{\max} = R_{pn}\left(1-\cos\left(\arcsin\left(\frac{z_i}{2R_{pn}}\right)\right)\right) + R_{\min} \qquad (6)$$

The expression (4) allow us to determine the radius of the mill in any of its transversal section at a known minimum radius R_{min} and the sharpening radius of the cutting blade R_{pn}. If it is known the maximum radius R_{max}, the radius of the cutting blade sharpening R_{pn} and height *B*, so you can define R_{min}, i.e. to solve the inverse problem. The radius of the cutting blade is set by the working drawings products, manufactured from a peripheral segment. If are given R_{max}, R_{min} and α_{max}, so the radius of the cutting blade

$$R_{pn} = \frac{R_{max} - R_{min}}{1-\cos\alpha_{max}}. \qquad (7)$$

The central angle corresponding to the arc of the circumference *KLM* of the cutting blade is found from the equation:

$$R_{pn}\cos\alpha_{\max} = R_{pn} - (R_{\max} - R_{\min}) \qquad (8)$$

Solving the equality $\alpha_{max} = 0{,}5\varphi$ the central angle

$$\varphi = 2\cdot\arccos\left(1-\frac{R_{max}-R_{min}}{R_{pn}}\right) = 2\cdot\arccos\left(\frac{B}{2R_{pn}}\right) \qquad (9)$$

The contour of the cutting blade, delineate according to circumference *KLM*, is described by the equation:

$$Y^2 + Z^2 = \left(\frac{R_{max}-R_{min}}{1-\cos\left(\arcsin\left(\frac{B}{2R_{pn}}\right)\right)}\right)^2, \quad -\frac{B}{2} \le Z \le +\frac{B}{2} \qquad (10)$$

By (4) - (9) we can determine the radius of the form-milling cutter in any transversal section, as well as to establish the relation between its main structural elements. After substituting (4) into (1) and completing the conversion, we obtain a mathematical model which describes the geometric error of the peripheral segment, which pass the profile milling:

$$\Delta = -R_{pn}$$

$$\Delta = -R_{pn}\cos\left(\arcsin\left(\frac{z_i}{R_{pn}}\right)\right) - \sqrt{\left[-R_{pл}\cos\left(\arcsin\left(\frac{z_i}{R_{pn}}\right)\right) + R_{\min} + R_{pn}\right]^2 - \left(\frac{\pi S_x}{Z_\omega}\right)} + R_{\min} + R_{pn} \qquad (11)$$

The function (11) has not an extremum, so the graphics of the dependence of Δ on independent factors of the process was made: minimum radius of the mill $R_{min} = 20; 35; 50; 65; 80; 100; 120$ mm; radius of the cutting blade $R_{pn} = 100; 150; 200; 250; 300; 350; 400; 450; 500; 550; 600; 650$ mm; cutting feedrate of the stock material S = 300; 600; 900; 1200; 1500; 1800 mm/s; number of teeth of the mill Z = 4; 6; 8; 10; 12 pieces.; angular velocity of the mill ω = 300; 400; 500; 600; 700; 800 rad/s; current value of coordinates Z_i = 0; 20; 40; 60; 80; 100; 120 mm; height of the mill B = 100; 120; 140; 160; 180; 200 mm. The variation range of the values of these factors include the vast majority of the cutting conditions used in the processes of mechanical wood treatment.

Increasing the minimal radius R_{min} of irregularities formed during the milling process of the profiled surface of the stock material, the height is reduced (figure 4a), which is explained by the approaching of the intersection point of the adjacent cut to the treated surface.

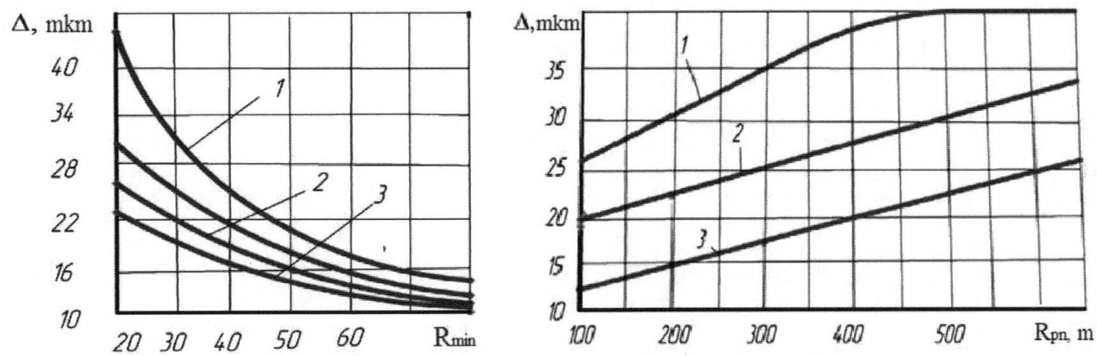

a b

Figure 4. Influence of the minimal radius R_{min} (a) and the radius of the cutting blade sharpening R_{pn} (b) mill on the machined surface accuracy.

The curves 1-3 are described by the equation (12), which at the automated calculation software environment ADVANCED GRAPHER the factor R_{min} is replaced by the argument x, and the error Δ on the value of the function *Y (x)*. The number of the curve in figure 4a corresponds to the number of the following equation:

$$1.\ Y(x) = -200\cos\left(\frac{50}{200}\right) - \left(\left(-200\cos\left(\frac{50}{200}\right) + X + 200\right)^2 - 2.465\right)^{\left(\frac{2}{4}\right)} + X + 200$$

at R_{pn}=200 mm; $B = 100$; $S_x = 100$ mm/s; $Z = 4$; $\omega = 500$ rad/s;

$$2.\ Y(x) = -100\cos\left(\frac{75}{100}\right) - \left(\left(-100\cos\left(\frac{75}{100}\right) + X + 100\right)^2 - 2.465\right)^{\left(\frac{2}{4}\right)} + X + 100 \qquad (12)$$

at R_{pn}=100 mm; B=150 mm; $S_x = 1000$ mm/s; $Z = 4$; $\omega = 500$ rad/s;

$$3.\ Y(x) = -150\cos\left(\frac{75}{150}\right) - \left(\left(-150\cos\left(\frac{75}{150}\right) + X + 150\right)^2 - 2.465\right)^{\left(\frac{2}{4}\right)} + X + 150$$

at R_{pn}=150 mm; $B = 100$; $S_x = 100$; $Z = 4$; $\omega = 500$ rad/s.

From the figure 4a and (12) it follows that to increase the accuracy of the surface after the profile milling of the stock material the radius R_{min} should be increased. The radius increase of the sharpening R_{pn} of the cutting blade leads to an increase in the geometric accuracy Δ of the treated surface. At the values R_{pn} = 100...350 mm the error Δ change linearly (figure 4b), increasing R_{pn} from 350 mm to 500 mm and Δ rise in accordance with the curve, but at $R_{pn} > 500\,\text{mm}$ Δ is stabilized at the maximum value (the curve 1 becomes parallel to the abscissa axis). The character of the curves 2 and 3 is similar to curve 1 with the difference that the sections are parallel to the abscissa and are arranged at a more remote distance from the origin of the coordinates distance.

As the increase of R_{pn} the current mill radius decreases, the contour of the cutting blade 3 is straightened (figure 2, curves 4, 5), reaching its limit position as a straight line 6, for which $R_{pn} = \infty$, that is the mill changes from profile into a cylindrical form with a radius. R_{min}. This explains the increase in error Δ during the rise of the sharpening radius R_{pn}.

The increase of the speed of the longitudinal feed S of the stock material leads to increment of the distance between the adjacent unit cuts [7, 8], the removal of the cut point of intersection from the treated surface and the rise of errors Δ (figure 5a). To improve the geometric accuracy of milled profiled surface, the feeding should be reduced [9, 10].

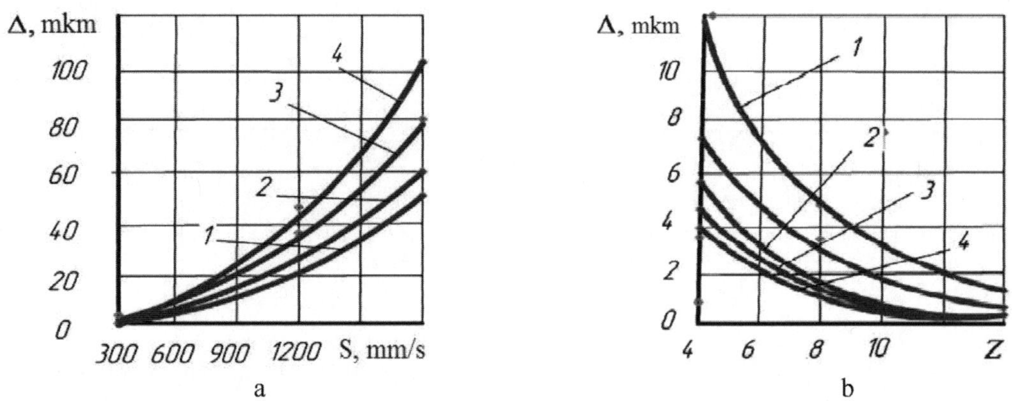

Figure 5. Influence of the cutting feedrate S of the stock material (a) and number of teeth z of the mill (b) to the error of the treated surface.

The increase of the number of cutting teeth of the profiled mill causes a reduction of the geometric errors of the treated surface (figure 5b), which is explained by a decrease in the time between the adjacent cuts, the reduction of the load on each tooth and a quieter work of the technological system. The number of teeth z shows a less influence on the kinematic error of the machined surface in comparison with other considered process factors.

During the increment of the angular velocity ω of the mill, the error Δ is reduced (figure 6a), which is associated with a decrease in the application time of the adjacent cuts, as well as the distance $\frac{2\pi \cdot S}{z\omega}$ between the points of intersection of the cuts [11]. The curves 1-4 characterize the change in error of the machined surface at different input kinematic factors of the process.

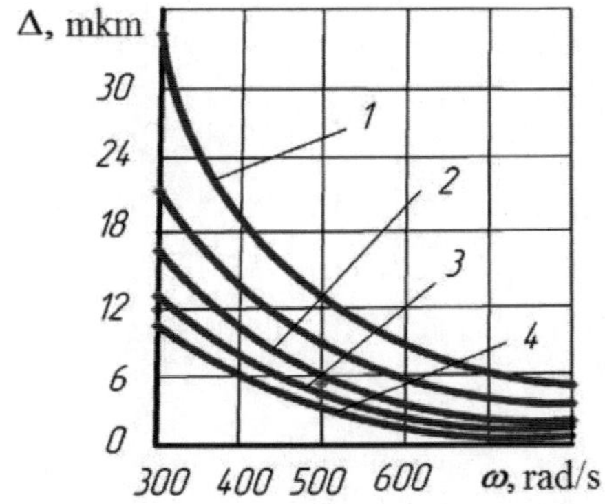

a

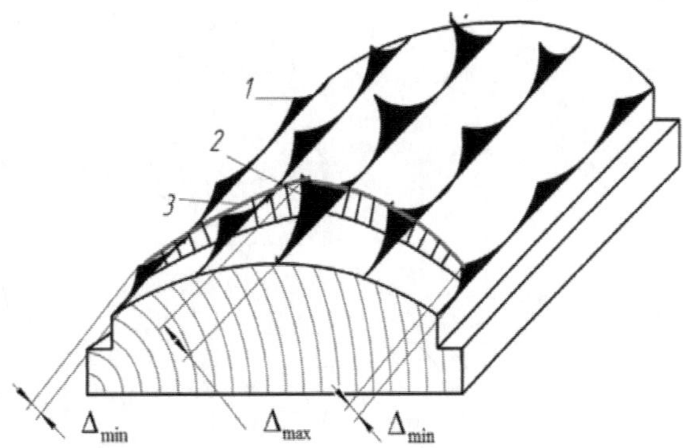

b

Figure 6. Influence of the mill angular velocity on the error (a) and disturbance of the errors on the treated of the profiled product surface (b): 1 - minimal error; 2 - maximal error; 3 - curve that defines the change in error in the transversal section.

However, it should be noted that the increase of ω, inevitably leads to the rise of the unbalanced centrifugal force $Q = m\omega^2\rho$, (m is the unbalanced mass instrumenta, ρ is the distance from the rotational axis to the unbalanced mass). At high angular velocities ω, even a small unbalanced mass m becomes a source of great value to the centrifugal force Q, causing fluctuations in the technological system [12, 13]. This creates a situation in which the reduction of the errors Δ by increasing ω leads to an rise of geometric errors due to the deterioration of the dynamics of the milling process. In addition, during the increment of ω the necessity for more accurate balancing of the tool increase, which leads to an additional labour costs. The increment of the height B of the mill causes a small reduction in errors [14].

Based on these equations and graphics, it was built a 3D-model of the geometric distribution errors on the machined profiled product surface (figure 6b). In the transversal section of the final product, the value of the error Δ is variable. The minimal error Δ_{min} occur at the edges of the product, and the maximal Δ_{max} in the middle, where the allowance is removed by the teeth of the minimal radius R_{min}. In the direction from the middle to the edges of the product, the errors are reduced according to the curve 3. The analysis of the geometric errors of the profiled product surface showed that at an incorrect construction of the milling process of the raw waste lumber on the treated surface generate significant irregularities in the height (4 mm or more).

Using the ADVANCED GRAPHER software analogous to the expression (12), the mathematical model errors was obtained Δ as a function of the radius of the cutting blade sharpening, R_{pn}, the cutting feedrate of the stock material S, the number of teeth z and the angular velocity ω of the mill, which allows to manage independent factors of the process to ensure the minimal error of the treatment. To minimize the geometric errors due to the kinematic profiled milling, it is necessary to increase R_{min}, ω and z, and also reduce R_{pn} and S.

At the profiled milling of the blank material in addition to the kinematic factors, also the dynamic ones act as principal vector and principal moment of the imbalances of the cutting tool and discrete milling force [15]. Under the action of the mill imbalances, the treated surface is formed as a sinusoidal surface, the error Δ_1 was formed by the end of the mill, which was removed from the leading foot of the of the console and Δ_2 by the close end to the front support (figure 7).

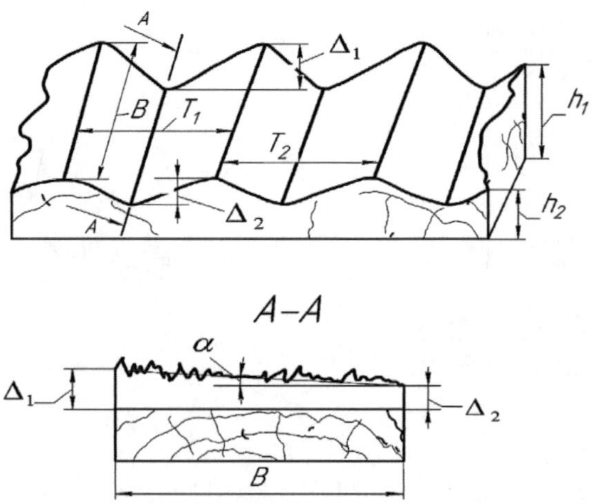

Figure 7. 3D - model of the surface treated by the unbalanced mill.

Error $\Delta_1 > \Delta_2$, treated profile inclined toward the horizontal plane at an angle, α , size $h_1 > h_2$ and pitch $T_1 - T_2$. The current sinusoidal error was made due to imbalances mills,

$$\Delta_1 = \frac{D_{CT}\omega^2}{J_B}(1 \pm tg\alpha), \tag{13}$$

where D_{CT} is the main vector of the imbalances mills; ω is the angular velocity of the mill; J_B is the stiffness of the spindle node with a perpendicular direction to the surface being treated. We receive a plus sign in (13) to the removed end from the front support of the mill, and negative one the to end, located close to the front support.

The geometrical error in an arbitrary *i*-th transversal section of the profiled mill was made due to the action of the milling force,

$$\Delta_{резi} = \frac{(J_B + J_3)P \cdot \cos\left[\frac{\pi}{2} - \text{arctg}\left(\frac{P_y}{P_z}\right) - 2\arcsin\left(\sqrt{\frac{t}{D\phi}}\right)\right]}{J_B J_3} \cdot (1 \pm \text{tg}\alpha), \tag{14}$$

where J_3 is the stiffness of the stock material in a perpendicular direction to the treated surface; P_y, P_z is the radial and the main component of the milling force $P = \sqrt{P_y^2 + P_z^2} \cdot t$ - is the milling depth; $D\phi$ is the current mill diameter.

The total geometric error of the treated surface includes the kinematic Δ_k and dynamic components Δ_{d1}, Δ_{d2}. The Δ_{d1} component is determined by the centrifugal force Q and the bending moment of inertia M, caused by imbalance mills, and the component Δ_{d2} by the discrete milling force P_{cut} (figure 8).

The sinusoid 1 formed by the vectors Q and M drive the burst 2 of the elastic displacements caused by the action of discrete milling forces P_{cut}, as well as tracks the cuts 3 left by the mill teeth. Pitch of the sinusoidal error Δ_{d1} is larger than the error Δ_{d2} in z times, z- is the number of the cutting teeth.

The thick curve line 4, which envelopes the error delineations Δ_k, Δ_{d1} and Δ_{d2}, constitutes the final profile of the longitudinal section of the machined surface formed by the tool imbalances, kinematic factors and shock pulses of the cutting tool teeth.

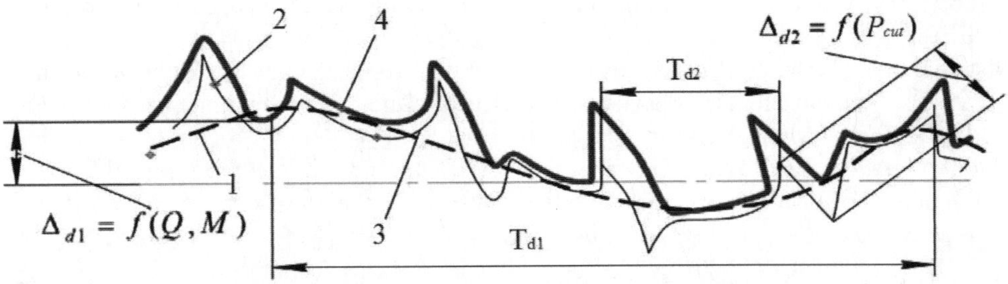

Figure 8. The profile of a longitudinal section of the treated surface is formed by the action of the kinematic and dynamic factors of the profile milling process of the raw waste lumber.

The analysis of the dynamics of the formation of geometrical errors in the longitudinal section under the action of the cutting tool imbalances showed that the value of the imbalances affect the height of the irregularities, and the angle of the principal vector and the principal moment of imbalance affects only the errors dislocation according to the machined surface. In order to reduce the errors caused by the imbalances, it is necessary to make the adjustments on the profiled mill masses, and for the errors caused by the cutting force to use the automatic power control of the milling.

The mathematical models of the geometrical errors of the treated surfaces were experimentally confirmed and are the basis for the development of an efficient machining process of the raw waste lumber. The designed and manufactured machine tool passed the validation in an accredited laboratory conditions, of which was received a document, which certifies its compliance with requirements according to the results. The machine tool and the treating raw waste lumber process were tested in production conditions with a positive result.

The widespread use of the developed equipment and technology in the timber enterprises will provide not only the saving in material and energy resources, but also improves the ecological environment by preserving forests in a permanently developing world of technological civilization.

4. Conclusion.

1. To ensure a high geometrical precision of the products from raw waste lumber, the profiled surface, the stock material and its geometrical errors were investigated, and the results showed that the disclosed mechanisms impact the kinematic and dynamic milling process factors on the final accuracy of the machined surfaces. On the basis of the obtained and experimentally validated mathematical models we can to predict the accuracy of the finished product in a wide range of the changes of the independent factors of the profile milling process, and to control the magnitude of the geometrical errors.
2. In order to minimize the geometric errors of the machined surface due to the kinematics of the profile milling and cutting tool geometry, it is necessary to increase the angular velocity ω, the number of cutting teeth z, the height B and the minimal radius R_{min} of the mill, and to reduce the radius of sharpening of the cutting blade R_{pn} and the cutting feed rate S of the stock material.
3. The reduction of the geometric error, caused by the discrete cutting force, is achieved by using a machine tool of the automatic power control of the milling and cutting feedrate control of the stock material. The widespread use of the developed equipment and machining process of the raw waste lumber will allow to obtain accurate products from raw waste lumber, saving material and energy resources, and it will help to preserve the forests on the planet.

References

[1] Mendes A., & Macqueen D. 2006. Raising forest revenues and employment: unlocking the

potential of small and medium forest enterprises in Guyana: Discussion Paper (No. 12). IIED.

[2] Hofstetter, K., Eberhardsteiner, J., Stürzenbecher, R., & Hackspiel, C. 2009 Wood and wood products-linking multiscale analysis and structural numerical simulations. In *7th European LS-DYNA Conference* Austria. (pp. 1-10).

[3] Sadrtdinov A.R., Sattarova Z.G., Prosvirnikov D.B. and Tuntsev D.V. (2015, December) Modeling of thermal treatment of wood waste in the gasifiers. In *2015 Int. Conf. on Mechanical Engineering, Automation and Control Systems (MEACS)* pp 1-5. DOI:10.1109/MEACS.2015.7414914

[4] Sadrtdinov A.R., Safin R.G., Gerasimov M.K., Petrov V.I., and Gilfanov K.K. 2016 The mathematical description of the gasification process of woody biomass in installations with a plasma heat source for producing synthesis gas. *IOP Conference Series: Materials Science and Engineering* **124**(1), 012092. IOP Publishing. DOI:10.1088/1757-899X/124/1/012092

[5] Fomin A. A., & Gusev V. G. 2013 Safe machining of blanks with nonuniform properties *Russian Engineering Research* **33** (10): 602-606.

[6] Fomin, A. A. 2013 Kinematics of surface formation in milling *Russian Engineering Research* **33** (11): 660-662.

[7] Azemović E., Horman I., & Busuladžić I. 2014 Impact of planing treatment regime on solid fir wood surface *Procedia Engineering* **69**: 1490-1498.

[8] Novák V., Rousek M., & Kopecký Z. 2011 Assessment of wood surface quality obtained during high speed milling by use of non-contact method *Drvna industrija* **62** (2): 105-113.

[9] Pinkowski G., & Szymanski W. 2010 Quality of profile milling on a CNC woodworking machine. *Annals of Warsaw University of Life Sciences-SGGW Forestry and Wood Technology* **72**.

[10] Pinkowski G., & Krauss W. 2010 Impact of the cutting edge condition on the results of oak wood milling *Annals of Warsaw University of Life Sciences-SGGW. Forestry and Wood Technology* **72**.

[11] Röbenack K., Ahmed D., Eckhardt S., & Gottlöber C. 2013 Peripheral milling of wooden materials without cutter-marks – A mechatronic approach *WSEAS Transactions on Applied and Theoretical Mechanics* **8** (2): 170-179.

[12] Fomin A.A., & Gusev V.G. 2013 Vibrational displacement of a spindle with static disequilibrium of the cutting tool *Russian Engineering Research* **33** (7): 412-415.

[13] Fomin A. A. 2013 Vibrational motion of a complex mill under the action of the cutting force *Russian Engineering Research* **33** (1): 57-60.

[14] Ramasamy G., & Ratnasingam J. 2010 A review of cemented tungsten carbide tool wear during wood cutting processes *Journal of Applied Sciences* **10** (22): 2799-2804.

[15] Chemborisov N.A., Sungatov I.Z., & Khisamutdinov R.M. 2013 Determining the contact zone in disk machining of a complex spherical mill *Russian Engineering Research* **33** (4): 243-243.

IOP Conf. Series: Materials Science and Engineering **142** (2016) 012092 doi:10.1088/1757-899X/142/1/012092

Modification of Structure and Strength Properties of Permanent Joints Under Laser Beam Welding with Application of Nanopowder Modifiers

A N Cherepanov [1], A M Orishich [1], A G Malikov [1], V E Ovcharenko [2,3]

[1]Institute of theoretical and applied mechanics named after S.A. Khristianovich SB RAS, Novosibirsk
[2]Institute of strength physics and materials science SB RAS, Tomsk
[3]Yurga Institute of Technology, National Research Tomsk Polytechnic University, Leningradskaya street 26, Yurga, Russian Federation, 652055

E-mail: ove45@mail.ru

Abstract. In the paper we present the results of experimental study of specially prepared nanosize metal-ceramic compositions impact upon structure, microhardness and mechanical properties of permanent joints produced by laser-beam welding of steel and titanium alloy plates.

1. Introduction

Laser beam welding of various materials has been widely studied in Russia and abroad since 1980s [1-6]. It has been established that heat efficiency of laser beam welding significantly exceeds that of arc welding and strength of the weld joint does not differ much from the strength of the metal. Ultrahigh rates of heating and cooling after laser implementation reduce the width of the heat affected area, reduce thermal deformation and gas saturation of titanium alloys especially under reliable shielding of fusion area with inert gases [7-8]. Production of permanent joints of iron-titanium alloys is an important technologic process when manufacturing parts and mechanisms which are widely applied in petrochemical engineering, power production and ship building. One of the main tasks here is improvement of strength, plasticity and resistance to cyclic loads in the process of their operation.

A well-known practice of weld strength improvement is increasing dispersion of structure and chemical homogeneity of the weld joint. Application of nanomodifying additives in the process of welding is one of the possible ways to solve this problem. Introduction of activated nanopowder particles of high-melting chemical compounds into the molten metal leads to atomization of matrix grains and proeutectoid constituents resulting from artificial increase of the number of crystallization centers in the molten metal. This leads to formation of highly dispersed globular structure in the molten metal during the cooling process which improves strength and plastic properties of the weld joint [9-14].

To study the peculiarities of structure, phase composition and mechanic properties formation in weld joints of “titanium-titanium”, “steel-steel” with application of nanopowder modifiers (NM) we completed experimental work on laser beam welding of titanium alloy and steel plates.

IOP Conf. Series: Materials Science and Engineering **142** (2016) 012092 doi:10.1088/1757-899X/142/1/012092

2. Results of the study and their discussion

Weld joint titanium – titanium. Samples produced by welding cold rolled 3 mm sheets made from VT 1-0 and VT 20 alloys were applied as the objects of study. Analysis of reference data shows that titanium alloys are highly prone to porosity under increased temperatures. Porosity is mainly caused by oxygen and hydrogen. Under laser treatment the molten metal absorbs gases due to convective motion. High crystallization rate does not allow gas bubbles come to the surface. That is why special attention was paid to protection of weld against oxidation during the welding process.

Helium was applied as shielding gas. It was supplied not only to the weld cup but also to its root. The power of the laser beam was 2.0 kW and the rate of welding was v=2 m/min. High-melting TiN, Y_2O_3 compounds and their mixture as well were used as modifying nanopowder materials. To improve the wettability the powders were previously treated in centrifugal planetary-type mill with addition of Cr powder with the ratio 1:3. Weight-part concentration of the modifying composition introduced into the weld pool was ~ 0.1% counting for the high-melting constituent, the size of particles was within 40-80 nanometer. The prepared composition was spread as suspension over the surface of the welded plates, its amount being ≤. 0.1% of the fused metal mass. It was established that application of nanopowder modifiers allows increasing the welding rate under the same power of the beam due to increasing the laser beam intensity absorption coefficient when metal is heated and begins to melt. The quality of the joint improves (morphology and structure of the weld), its mechanical properties significantly improve as well. Table 1 presents the results of mechanical tests of experimental and control samples modified with various powder compositions.

Table 1. Mechanical properties of the weld joint produced from VT 1-0 alloy

Composition of NM	Temporary strength σ_u, MPa	Yield strength σ_y, MPa	Percentage elongation, δ%
Without NM	288.1	134.6	0.1
TiN + Cr	365.8	277.1	0.197
TCN + Cr + Y_2O_3	4000	246.38	0.446

From table 1 we can see that all modifying compositions improved the properties of the weld joint. Thus, percentage elongation increased by 2 – 4.9 times, strength limit – by 1.23 – 1.35 times, yield strength – by 1.8 – 2.0 times.

In the similar manner we produced test samples of weld joints from VT 20 alloy. Figure 2 presents photographs of weld morphology and macrostructure made when welding VT 20 alloy with optical microscope NEOPHOT-21, without nanomodifying additive and with addition of TiN + Cr.

It can be seen that application of NM TiN + Cr notably changes the weld morphology and reduces the macrostructure of the hardened metal in the weld joint.

Fatigue tests of the welded samples produced from VT 20 alloy were carried out with universal servohydraulic test table "Instron 8810" equipped with software module for data acquisition and processing "Wavematrix". The testing was completed according to pulsating soft cycle of loading with the frequency of 5 Hz. The weld joints were being tested under the load varying from zero to 1500 N. Strain amplitude under such loading conditions was approximately 0.2 … 0.25 mm. During the test the time of complete destruction of the sample was recorded. Welding conditions: P = 3kW, v = 1 m/min. Average values of lifetime obtained from testing 3 samples are presented in Table 2.

As we can see from the table fatigue strength of weld joints NM modified significantly (by 1.5 – 2 times) exceeds that of the plates welded without nanomodifying additive.

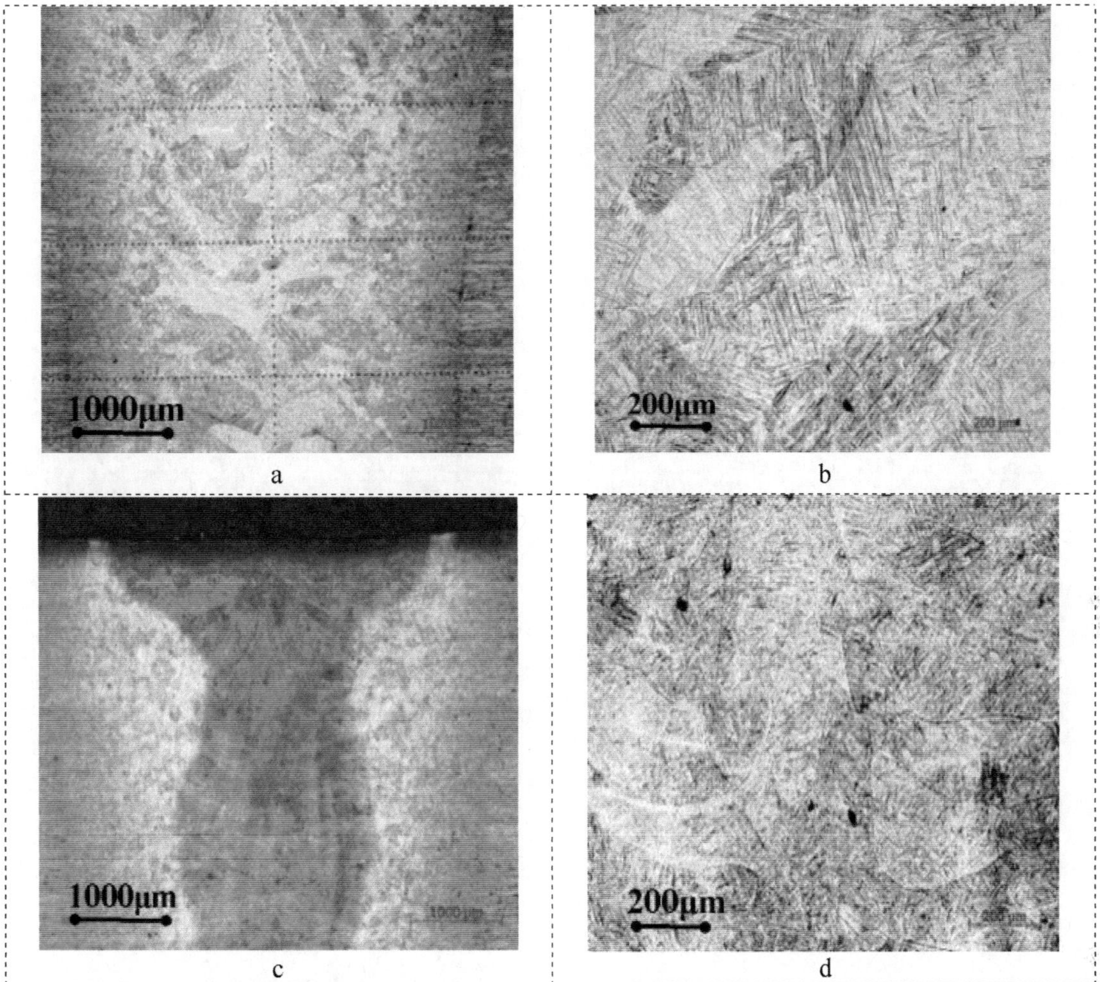

Figure 1. Weld morphology and structure without NM application (a), (b) and with NP application TiN + Cr (c), (d).

Table 2. Results of fatigue testing for VT 20

Sample №	Type of material	Lifetime τ before destruction, min
1	VT 20 alloy without the weld joint	τ = 220.5
2	Welded specimen without NM	τ =48
3	Welded specimen with additive TiN + Cr	τ = 130

Weld joint "steel – steel". We have studied the impact of nanostructured powder compositions $TiC_{0.5}N_{0.5}$ + Fe и $TiC_{0.5}N_{0.5}$+Cr upon the quality of weld joints produced from carbon steel 20. The thickness of the welded plates was 3 mm. Mechanically activated nanopowder $TiC_{0.5}N_{0.5}$, plated with Fe was spread as suspension on the top of the edges of welded plates. Welding was completed in helium atmosphere. The workpieces were butted. Power of the laser irradiation was 3.05 kW, welding rate – 1.7 m/min. Structure and microhardness of experimental and control samples were studied. In

Figure 2 we present photographs of morphology and structure of the weld produced without modifying additive (a, b) and with NM additive $TiC_{0.5}N_{0.5}$ + Fe (c, d) accordingly.

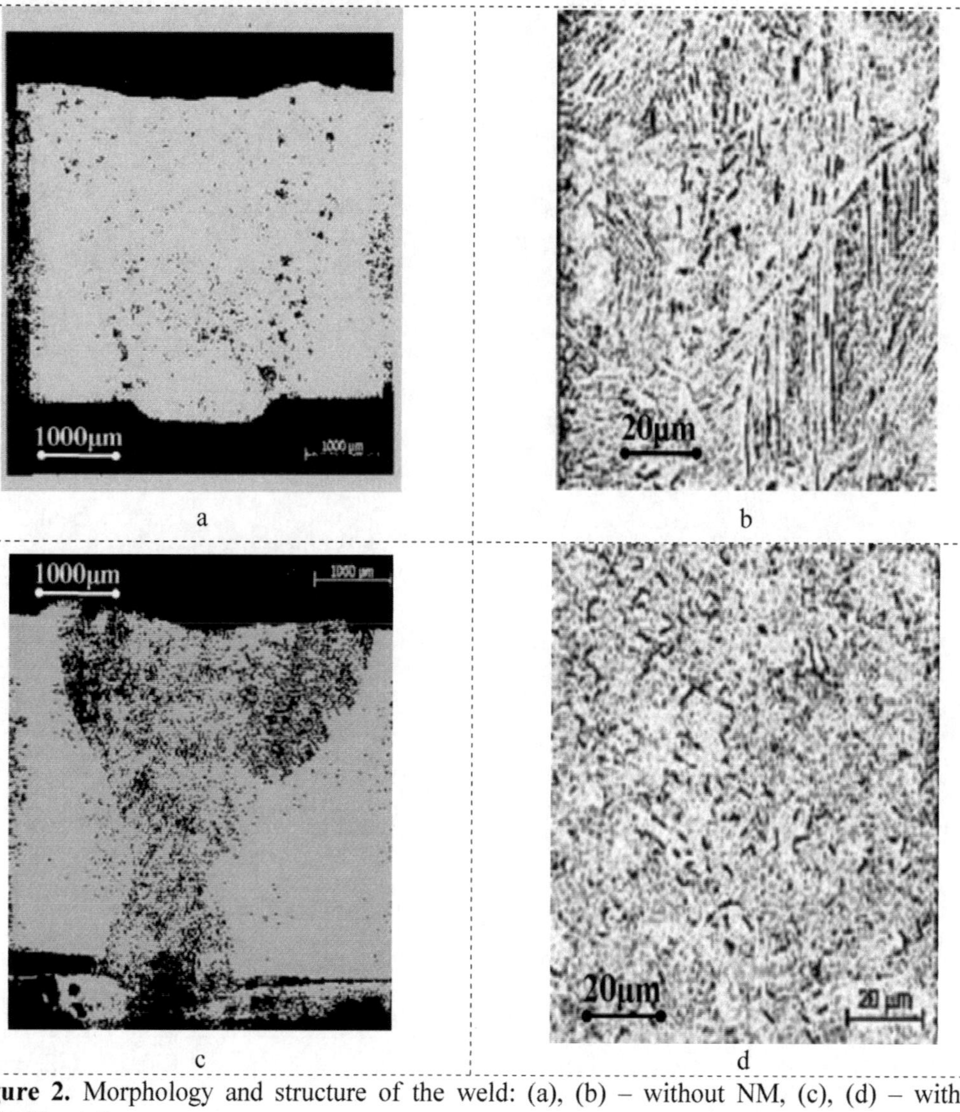

Figure 2. Morphology and structure of the weld: (a), (b) – without NM, (c), (d) – with NM $TiC_{0.5}N_{0.5}$ + Fe.

It can be obviously seen that application of nanomodifier changed both the weld morthology and the microstructure of the weld joint. The sizes of widemanstatten ferrite in the weld produced without the modifier reach 100÷150 micrometer lengthwise, while in the modified weld these grains do not exceed 30÷40 micrometer. This undoubtedly shows that application of NM additive leads to weld structure refining.

Figure 3 illustrates the grain structure of the weld when applying $TiC_{0.5}N_{0.5}$+Cr. Nanocomposition. As in the previous variant the structure of the weld produced without NM is coarse with large crystals extended in the direction of heat withdrawal. Application of NM additive leads to grain refinement, the length of crystallites reduces by 2-3 times.

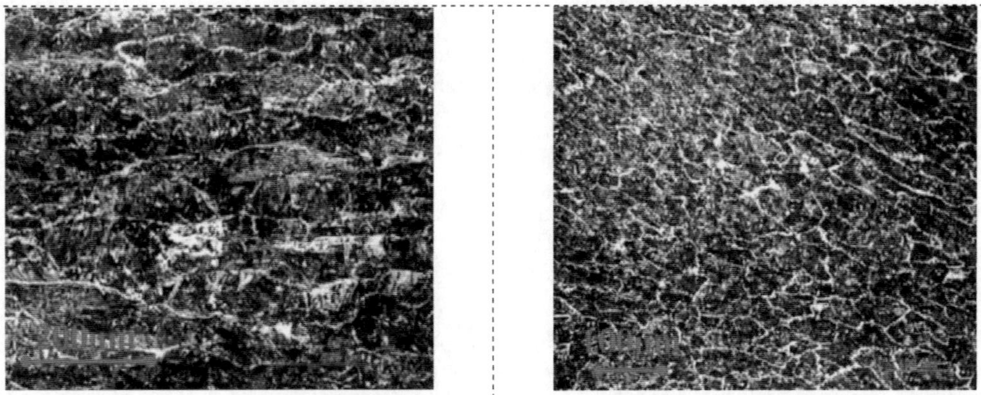

a b

Figure 3. Weld structure after laser-beam welding, a – without application of inoculator; b – with application of inoculator $TiC_{0.5}N_{0.5}$ + Cr.

Vickers microhardness distribution was studied under loading of 0.5N for the weld width and heat-affected zone (HAZ).

The results of the study are shown in Figure 4.

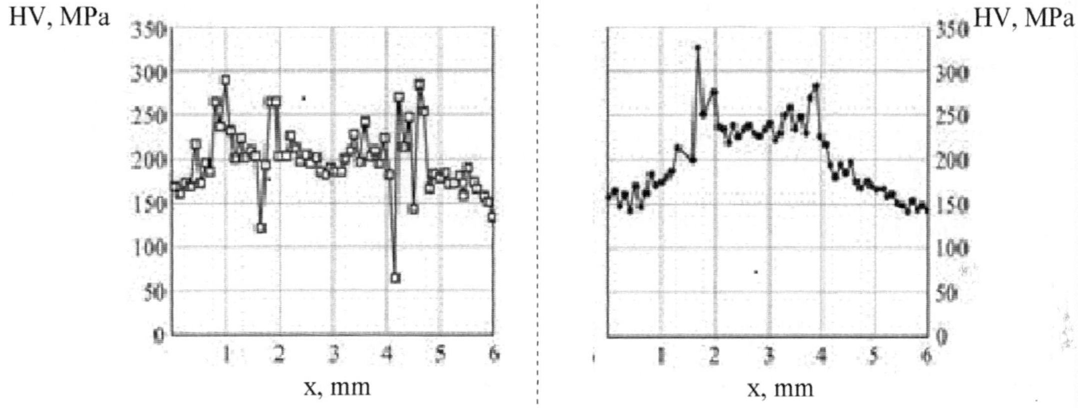

a b

Figure 4. Microhardness of welds without NM (a) and with addition of $TiC_{0.5}N_{0.5}$ +Cr powder (b).

Measurements of weld and HAZ microhardness showed notable smoothing of microhardness discontinuity observed under laser-beam welding without NM (Figure 4.b) which may be explained by improvement of structure homogeneity after introduction of the modifier into the weld pool.

3. Conclusions

Application of nanomodifying additive:

-improves welding rate under the same beam power due to increasing the laser beam intensity absorption coefficient when heating metal to achieve its fusion temperature;

-improves mechanical properties and quality of the joint due to changes in morphology and reducing the size of crystalline grains;

-changing of weld and HAZ microhardness showed significant smoothing of microharness discontinuity observed under laser-beam welding completed without NM application which is associated with structure dispersion and homogeneity increase resulting from application of modifier.

References

[1] Ready, John F. Industrial applications of lasers. Academic press, 1997.
[2] Duley, Walter. Laser processing and analysis of materials. Springer Science & Business Media, 2012.
[3] Ion, John. Laser processing of engineering materials: principles, procedure and industrial application. Butterworth-Heinemann, 2005.
[4] Grigor'iants, Aleksandr G., and Igor'N. Shiganov. "Laser welding of metals." Moscow Izdatel Vysshaia Shkola 5 (1988).
[5] Steen, William, and Jyotirmoy Mazumder. Laser material processing. Springer Science & Business Media, 2010.
[6] Atabaki, M. Mazar, et al. "High power laser welding of thick steel plates in a horizontal butt joint configuration." *Optics & Laser Technology* 83 (2016): 1-12.
[7] Torkamany, M. J., F. Malek Ghaini, and R. Poursalehi. "An insight to the mechanism of weld penetration in dissimilar pulsed laser welding of niobium and Ti–6Al–4V." *Optics & Laser Technology* 79 (2016): 100-107.
[8] Fang, Xiuyang, Hong Liu, and Jianxun Zhang. "Microstructure and mechanical properties of pulsed laser beam welded Ti-2Al-1.5 Mn titanium alloy joints." *Journal of Materials Engineering and Performance* 23.6 (2014): 1973-1980.
[9] Yunlian, Qi, et al. "Electron beam welding, laser beam welding and gas tungsten arc welding of titanium sheet." *Materials Science and Engineering*: A 280.1 (2000): 177-181.
[10] Caiazzo, Fabrizia, et al. "Investigation and optimization of laser welding of Ti-6Al-4 V titanium alloy plates." *Journal of Manufacturing Science and Engineering* 135.6 (2013): 061012.
[11] Katsnelson, S. S., G. A. Pozdnyakov, and A. N. Cherepanov. "Effect of nanoparticles on the structure and properties of an aluminum alloy poured into a mould with the use of a magnetohydrodynamic disk pump." *Journal of Applied Mechanics and Technical Physics* 56.5 (2015): 889-893.
[12] Cherepanov, A. N., et al. "Investigation of the structure and properties of titanium-stainless steel permanent joints obtained by laser welding with the use of intermediate inserts and nanopowders." *Thermophysics and Aeromechanics* 22.2 (2015): 135-142.
[13] Ovcharenko, V. E., A. A. Mokhovikov, and A. S. Ignatiev. "Improving durability of cermets for metal cutting by generation of subsurface multilevel structures." *Applied Mechanics and Materials*. Vol. 379. 2013.
[14] Orishich, A. M., A. G. Malikov, and A. N. Cherepanov. "Effect of Nanopowder Modifiers on Properties of Metal Laser-welded Joints." *Physics Procedia* 56 (2014): 507-514.

IOP Conf. Series: Materials Science and Engineering **142** (2016) 012093 doi:10.1088/1757-899X/142/1/012093

Nanostructured Hardening of Hard Alloys Surface Layers Through Electron Irradiation in Heavy Inert Gas Plasma Conditions

V E Ovcharenko [1,3], Yu F Ivanov [2], K V Ivanov [1], A A Mokhovikov [3], Yu Baohai[4], Xu Yun Hua[5]

[1]Institute of Strength Physics and Materials Science SB RAS, 634055, Tomsk, Academichesky pr., 2/4, Russia, tel. (3822)492771,
[2]Institute of High Current Electronics SB RAS, 634055, Tomsk, Academichesky pr., 4, Russia
[3]Yurga Institute of Technology, National Research Tomsk Polytechnic University, Leningradskaya street 26, Yurga, Russian Federation, 652055
[4]Institute of Metal Research,72 Wenhua Road, Shenyang, 110016, China, tel.+86-24-23843627
[5]Xi'an University of Technology, Xi'an, 5 Southern, Jinhua Road, Shaanxi, 710048, China,tel. +86-29-82312068

e-mail:[1]ove45@mail.ru, [3]maa28@rambler.ru, [4]bhyu@imr.ac.cn, [5]xuyunhua@vip.163.com

Abstract. The paper presents research and experimental findings which prove that metal ceramic composite surface layer contains micro constituents' hierarchies in the form of secondary nano sized inclusions inside ceramic phases. These inclusions have typical dimensions from several tens to several hundreds of nano meters. It has been shown that multi level structure-phase condition, developed in a nano sized area, effects physical and tribological properties of a metal ceramic composite surface layer.

1. Introduction

One of the burning issues of material engineering of today is to extend an operational life of metal working tools by means of nanostuctured hardening of industrial hard alloys surface layers, which contain up to 94 mass. % of a ceramic ingredient comprised out of very hard carbides. The method of nanostructured conditions inside a hard alloy surface layer is a promising one [1-9].

The research aimed at comparative evaluation of structure-phase conditions, which appear in a surface layer of a super ceramic hard alloy subjected to electron beam irradiation in the conditions of inert gases gas-discharged plasma with different atomic weight. Another goal of research was to study the way new structure-phase conditions influence a hard alloy cutting resistance.

2. Materials and research methodology

Two kinds of hard alloys were chosen to be evaluated, they are T40 (75%WC + 14% (Ti, Ta, Nb) C)+11%Co) and WC-15 (85%WC+15%Co). The hard alloys were subjected to pulsed electron beam irradiation with ES = (40…90) J/cm^2 electron beam energy density; τ = 150 μs electron beam influence pulse duration; N=15 number of irradiation impulses; and f=0.3 s^{-1} impulse frequency. The hard alloys were irradiated in the conditions of a residual pressure of 0.02 Pa plasma-supporting gas experimental chambers. Table 1 shows specifications of plasma-supporting gases used in the experiment.

Table 1 Specifications of plasma-supporting gases used in the experiment

Plasma-supporting gas	Ionization energy kJ/mol	Atomic weight g/mol
N	1401,5	14,00674
Ar	1519,6	39,948
Xe	1170,0	131,29

To evaluate structure-phase condition of hard alloys surface layers researchers applied methods of optical microscopy (Neofot 32), scanning (Zeiss LEO EVO 50) and transmission diffraction (JEOL JEM-2100) electron microscopy. Electron beam machined cutting plates were full-scale tested via turning a 40X steel up to its cutting edge wear rate came down to its 0.2 mm limit.

3. Results and Discussion

a) T40 hard alloy

Figure 1 shows the original condition of a hard alloy. Carbide particles are non equiaxial with dimensions changing from 0.5 μm to 3 μm.

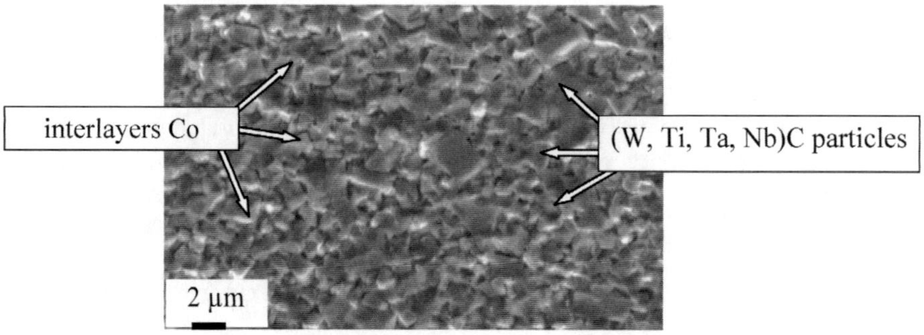

Figure 1. Original T40 hard alloy surface structure.

Impulse electron irradiation of a hard alloy surface layer caused microscopic pores in a surface layer as a result of a carbide phase decomposition and ariel carbon and oxygen compound development, which happen independent from gas-discharged plasma compound. This finding was proved by the evaluation of a hard alloy surface layer structure effected by impulse electron irradiation (Figure 2). It was observed that energy density in electron beam increases in parallel with microscopic pores dimension and number decrease.

Radiation causes melt development. Cristallyzation of the melt, in its turn, causes formation of a mesostructured level of a surface layer of electron irradiated hard alloy (Figure 3). It was found that mesolevel structure formation depends sufficiently on the kind of a plasma-supporting gas – in movingt from nitrogen to argon and then to xenon we observe a hard alloy ceramic ingredient particle size enlargement. Particle size grows in parallel with electron beam energy density increase. For example, if particle size changes within (4…10) μm when E_S = 40 J/cm^2, it changes within (2…5) μm when E_S = 70 J/cm^2.

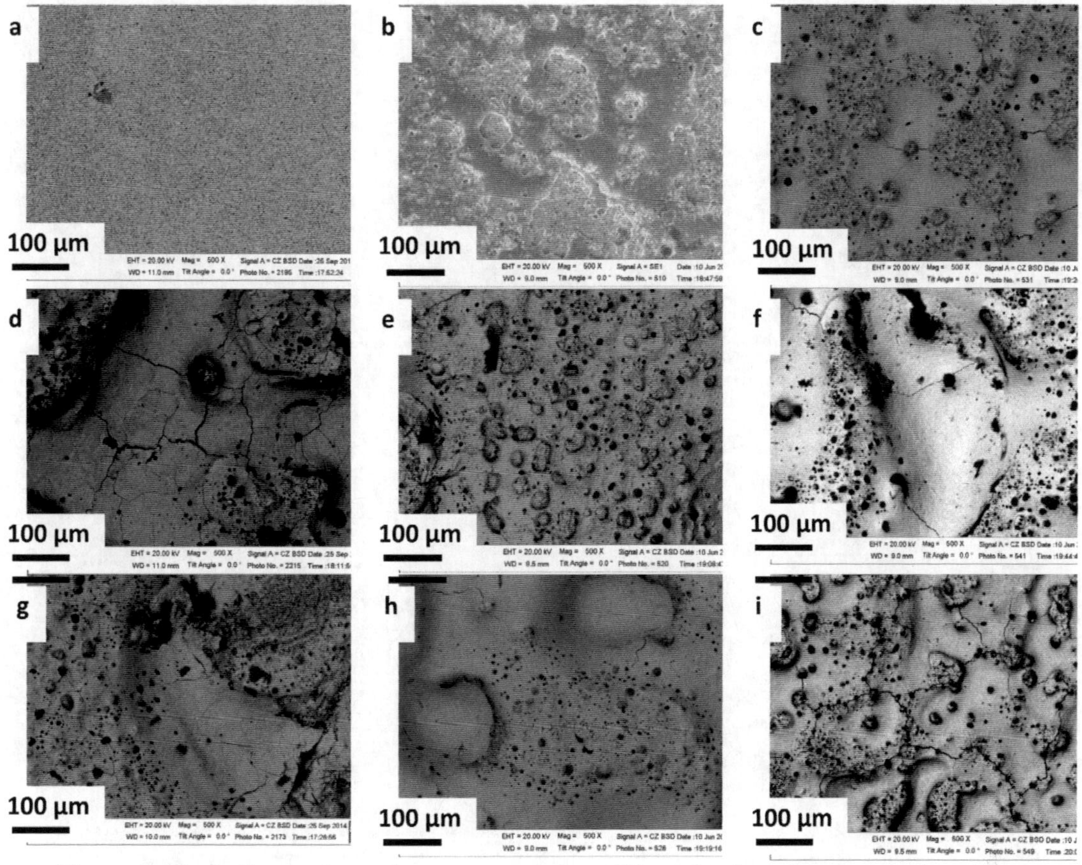

Figure 2. Hard alloy surface microstructure being subjected to impulse electron-ion-plasma irradiation (irradiation impulse duration – 150 ms, 15 impulses) in:
40 J/cm^2 in N (a), in Ar (b), and Xe (c) plasma;
60 J/cm^2 in N (d), in Ar (e), and in Xe (f) plasma;
70 J/cm^2 in N (g), in Ar (h), and in Xe (i) plasma.

Through the method of transmission diffraction electron microscopy a modified layer was evaluated to study the dependencies of a hard alloy surface layer micro level phase compound and defect substructure from impulse electron-ion-plasma irradiation modes.

Experience has shown that irradiation in nitrogen and argon gas-discharged plasma conditions results in ceramic particles fragmentation within a hard alloy surface layer. Sub grain structure with small angles of disorientation is formed inside sub micro sized (≈500 nm) ceramic particles (Figure 4 a, b). At that a xenon plasma irradiation causes appearance of a metal ceramic structure with nano- and sub micro particles with large angles of disorientation inside the surface layer (Figure 5 a); and gives rise to micro twins of deformation origin (Figure 5 b, plate like shape formation).

Nanostructured ceramic ingredients as well as twin structures in ceramic particles after impulse electron-ion-plasma irradiation are the attributes of ceramic materials crack resistance and their strength capacity enhancement in the conditions of shock loading [10-13]. Figure 6 presents experimental details of hard alloy plate's metal cutting ability examination held after impulse electron-ion-plasma irradiation in N, Ar and Xe plasmas in correlation with energy density in an electron beam. Experiments have allowed to establish that irradiation in N and Ar plasmas reduces a hard alloy surface layer resistance in the process of cutting metals, whilst irradiation in Xe plasma with electron beam energy density equal to 70 J/cm^2 increases a surface layer resistance capacity by 1.5 times [14].

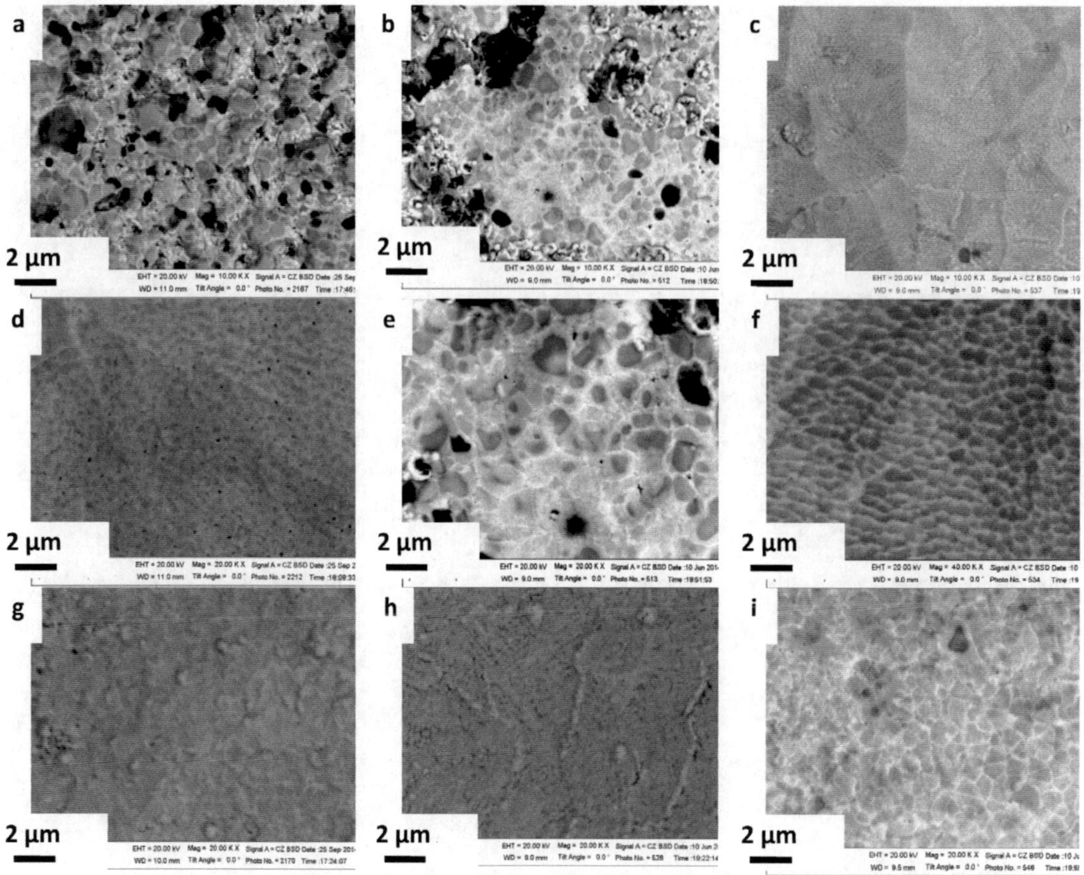

Figure 3. Mesostructured level of a hard alloy surface being subjected to impulse electron-ion-plasma irradiation (irradiation impulse duration – 150 ms, 15 impulses) in:
40 J/cm^2 in N (a), in Ar (b), and in Xe (c) plasma;
60 J/cm^2 in N (d), in Ar (e), and in Xe (f) plasma;
70 J/cm^2 in N (g), in Ar (h), and in Xe (i) plasma.

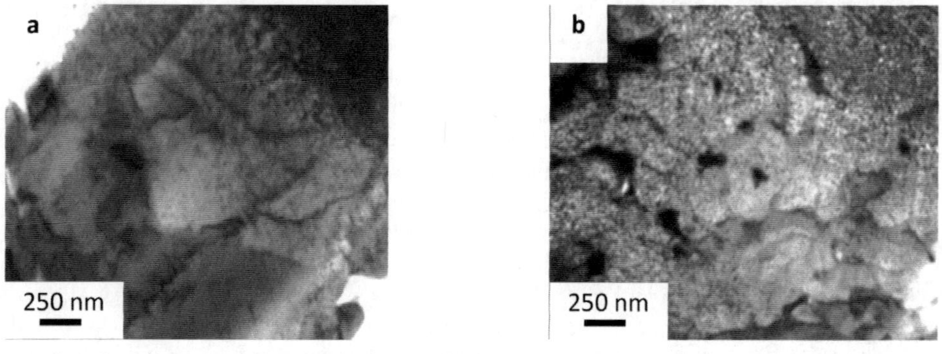

Figure 4. Microstructure of a surface layer of a hard alloy being subjected to impulse electron-ion-plasma irradiation in nitrogen (a), argon (b) and xenon (c) gas-discharged plasmas.

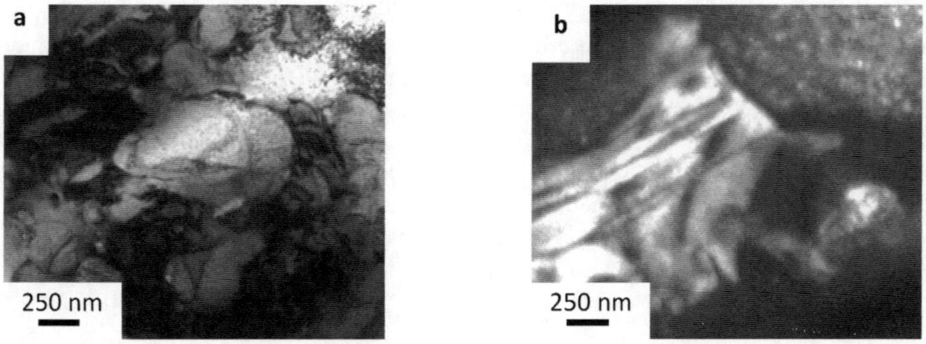

Figure 5. Microstructure of a surface layer of a hard alloy being subjected to impulse electron-ion-plasma irradiation in xenon gas-discharged plasma.

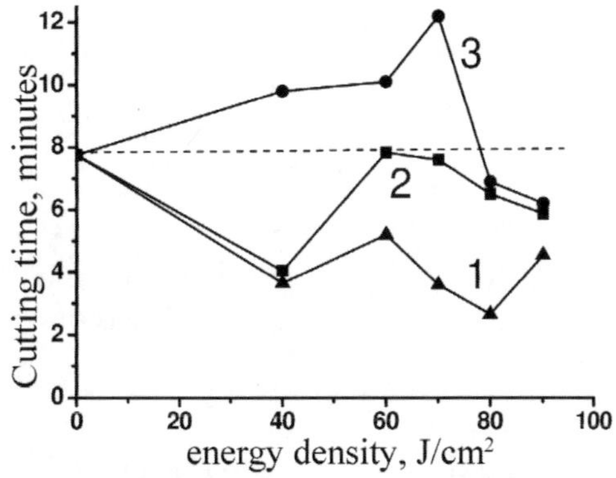

Figure 6. (75%WC+14%(Ti,Ta,Nb)C)-11%Co hard alloy surface layer resistance capacity during the process of metal (40X steel) cutting dependence from electron beam energy density during impulse electron-ion-plasma irradiation (150ms, 15 impulses) in N (1), Ar (2) и Xe (3) gas-discharged plasmas.

b) WC-15 hard alloy

Figure 7 presents an image of a hard alloy surface layer microstructure in its original condition and after it had been subjected to electron irradiation in Ar and Xe plasmas.

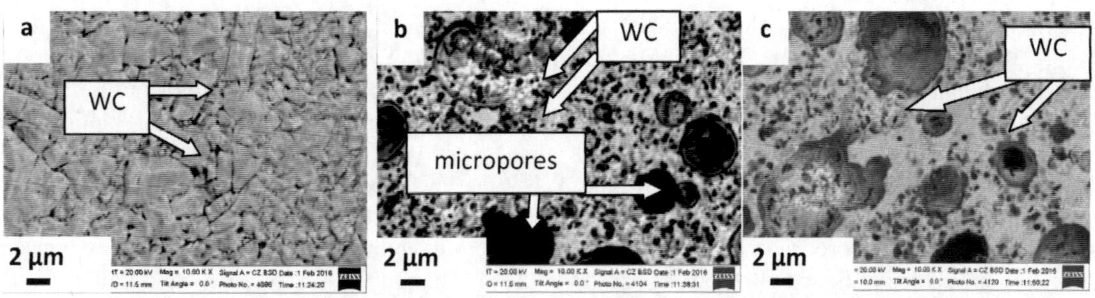

Figure 7. Microstructure of an original WC-15 hard alloy surface layer (a); and after electron irradiation in Ar (b) and Xe (c) gas-discharged plasmas
(with electron beam energy density equal to 40 J/cm^2).

It is important to note that a hard alloy surface layer microstructure does not contain micro pores after Xe plasma irradiation. That is the critical distinction of a hard alloy surface layer microstructure in moving from Ar irradiation to Xe gas-discharged plasma irradiation.

A hard alloy surface layer wear resistance was evaluated during the process of cutting metals by measuring the wear rate of the back cutting plate's edge, which was made of a hard alloy. The experiment has shown that electron-ion-plasma irradiation of a WC-15 hard alloy in Ar plasma decreases a surface layer wear resistance, whilst Xe plasma irradiation increases a surface layer wear resistance two times compared to that of the original hard alloy (Figure 8) [14].

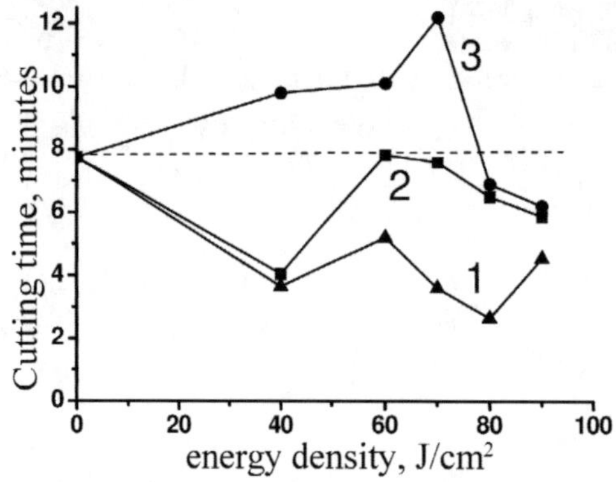

Figure 8. Dependence of the back hard alloy cutting plate's edge wear rate from energy density in an electron beam when irradiated in Ar (1) and Xe (2) gas-discharged plasmas.

3. Conclusions

When super ceramic hard alloys are being subjected to impulse electron-ion-plasma irradiation in Xe heavy inert gas plasma condition, ceramic particles in a hard alloy surface layer change their dimensions up to nano size level, and form nanostructured structure-phase conditions, which are specified by their high wear-resistance during the process of cutting metals. This modification of a structure-phase condition of a surface layer results in 1.5-2 time increase in hard alloy plates resistance capacity.

References

[1] Chawla, N., and K. K. Chawla. "Metal-matrix composites in ground transportation." JOM 58.11 (2006): 67-70.

[2] Ewsuk, Kevin, and Yury Gogotsi. Characterization, design, and processing of nanosize powders and nanostructured materials. *Wiley-American Ceramic Society*, 2006.

[3] Siegmann, Stephan, Marc Leparoux, and Lukas Rohr. "The role of nano-particles in the field of thermal spray coating technology." *OPTO-Ireland*. International Society for Optics and Photonics, 2005.

[4] Ovcharenko, V. E., A. A. Mokhovikov, and A. S. Ignat'ev. "Influence of surface nanostructure on the life of cermet in metal cutting." *Steel in translation* 43.6 (2013): 348-350.

[5] Klimenov, V., Borozna, V., Zaitsev, K. Influence of ultrasonic superficial processing on properties of titanic alloys METAL 2009 - 18th International Conference on Metallurgy and Materials, Conference Proceedings 2009, Pages 273-278 18th International Conference on Metallurgy and Materials, METAL 2009; Cerveny ZamekHradec Nad Moravici; Czech Republic; 19 May 2009 through 21 May 2009; Code 110530

[6] Baer, Donald R., Paul E. Burrows, and Anter A. El-Azab. "Enhancing coating functionality using nanoscience and nanotechnology." *Progress in Organic Coatings* 47.3 (2003): 342-356.

[7] Shon, In-Jin, et al. "Mechanical properties and consolidation of binderless nanostructured (Ti, Cr) C from mechanochemically-synthesized powder by high-frequency induction heating sintering." Ceramics International 39.8 (2013): 9721-9726.

[8] Ovcharenko, V. E., A. A. Mokhovikov, and A. S. Ignatiev. "Improving durability of cermets for metal cutting by generation of subsurface multilevel structures." *Applied Mechanics and Materials*. Vol. 379. 2013

[9] Ping, F. E. N. G., et al. "Influence of preparation process on sintering behavior and mechanical properties of ultrafine grained Ti (C, N)-based cermets." *Transactions of Nonferrous Metals Society of China* 17.3 (2007): 531-536.

[10] LLorca, Javier, et al. "Multiscale modeling of composite materials: a roadmap towards virtual testing." *Advanced Materials* 23.44 (2011): 5130-5147.

[11] Teng, Z. K., and X. M. Chen. "Edge cracking mechanism in two dual-phase advanced high strength steels." *Materials Science and Engineering*: A 618 (2014): 645-653.

[12] Mayrhofer, Erwin, et al. "Cracking resistance of Cr 3 C 2–NiCr and WC–Cr 3 C 2–Ni thermally sprayed coatings under tensile bending stress." *Surface and Coatings Technology* 281 (2015): 169-175.

[13] Shilko, Evgeny V., et al. "Overcoming the limitations of distinct element method for multiscale modeling of materials with multimodal internal structure." *Computational Materials Science* 102 (2015): 267-285.

[14] Ignatyev, A. S., A. A. Mokhovikov, and V. E. Ovcharenko. "Influence of Structural Phase State of the Surface Layer on Wear Resistance of Cutting Edge of Metal-ceramic Insert When Metalcutting." *Applied Mechanics & Materials* 682 (2014).

IOP Conf. Series: Materials Science and Engineering **142** (2016) 012094 doi:10.1088/1757-899X/142/1/012094

Mathematical Modeling for the Development of Equipment for Thermochemical Processing of Wood Waste in to Dimethyl Ether

Almaz R Sadrtdinov[1,a], Liliya M Esmagilova[1], Vladimir A Saldaev[1], Zulfiya G Sattarova[1], Alexey A Mokhovikov[2]

[1] Institute of Petroleum and chemical Engineering, Kazan National Research Technological University, K. Marx street 68, Kazan, Republic of Tatarstan, Russian Federation, 420015
[2]Yurga Institute of Technology, National Research Tomsk Polytechnic University, Leningradskaya street 26, Yurga, Russian Federation, 652055

E-mail: [a] almaz.sadrtdinov@gmail.com

Abstract. The paper describes the process of thermochemical wood waste processing in to dimethyl ether. The physical picture of the process of waste wood recycling was compiled and studied and the mathematical model in the form of differential and algebraic equations with initial and boundary conditions was developed on its basis. The mathematical model allows to determine the optimum operating parameters of synthesis gas producing process, suitable for the catalytic synthesis of dimethyl ether and to calculate the basic constructive parameters of the equipment flowsheet.

1. Introduction

During the development of new equipment there are always technological and engineering calculations to be done. Performing of these tasks in the modern world is unthinkable without the use of methods and means of mathematical modeling of developed processes. The implementation of the mathematical model allows: to solve the problem of choosing the optimal construction of the equipment and its main dimensions; to identify effective operational parameters of the process; to improve existing and develop new types of equipment [1]. The use of mathematical modeling is especially important in the development of new technological processes. The process of producing dimethyl ether at the thermochemical processing of wood waste can be referred to one of the new technologies. It should be noted that such a development direction of the wood raw material processing technologies is very important. This approach solves the environmental problem of the volume of waste wood and raw material reduction and a resource problem being as a result of dimethyl ether [2]. Dimethyl ether can be used as a motor fuel, fuel for power plants and as an intermediate for the synthesis of gasoline with improved environmental characteristics and minimum content of undesirable impurities [3].

During the development of new technologies due to their complexity and the danger it is often advisable to carry out their theoretical analysis and mathematical modeling. In such cases, the use of computer modelling based on the process of developed mathematical model is irreplaceable.

Based on the foregoing, the aim of the paper is forming a mathematical model of the process of thermochemical processing of wood waste in to dimethyl ether, to obtain data on the structural dimensions of the equipment and technological operation. The characteristics of wood waste, gasifying agent species, catalyst species, temperature and process pressure are taken as initial data.

2. Description and formalization of the process

In order to ease and simplify the development of a mathematical model, the installation scheme for processing wood waste in to dimethyl ether was charted (figure 1). The scheme reflects the physical processes that occur during the plant operation and basic operational parameters.

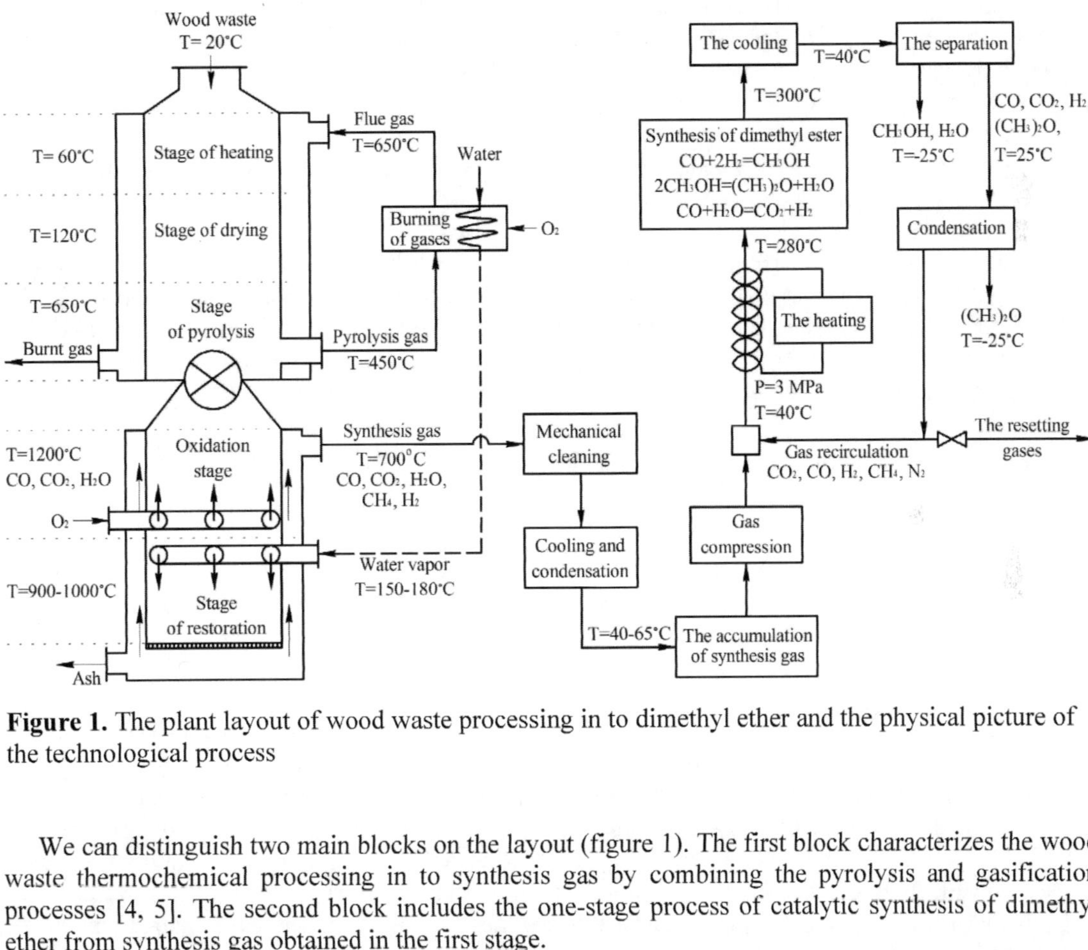

Figure 1. The plant layout of wood waste processing in to dimethyl ether and the physical picture of the technological process

We can distinguish two main blocks on the layout (figure 1). The first block characterizes the wood waste thermochemical processing in to synthesis gas by combining the pyrolysis and gasification processes [4, 5]. The second block includes the one-stage process of catalytic synthesis of dimethyl ether from synthesis gas obtained in the first stage.

We'll consider the process in more details. Wood waste successively passes through different phases after entering the installation: heating, drying, pyrolysis, oxidation and reduction step (synthesis gas stage) [6]. Further, to achieve the required parameters the formed synthesis gas passes the steps of purification from the mechanical impurities, is subjected to cooling and condensation of water vapour. As a result, the raw material - the synthesis gas is formed in the first block during thermal processing of waste wood which is then used for one-step dimethyl ether synthesis catalyst.

A catalyst mixture of a methanol synthesis (Katalco-58, composition: CuO / ZnO / Al_2O_3) and methanol dehydration (γ- Al_2O_3) is used as a catalyst for the synthesis of dimethyl ether [7]. The catalyst mixture is placed in three layers: the first layer - Katalco-58; second layer - a mixture of

Katalco-58 and γ-Al_2O_3; a third layer of γ- Al_2O_3. During the process of catalytic synthesis the reactions of joint formation as methanol so dimethyl ether are occurred. This process is exothermic and proceeds with decrease in volume, that is why a constant maintaining of the temperature and pressure within the required limits is needed [8, 9].

Next, the resulting product mixture leaving the synthesis reactor is cooled and separated into two phases: a liquid phase and a vapour one. The liquid phase consists of methanol and water, and the gas-vapour phase has dimethyl ether, which is released in the condensation stage. Not condensed gases are mixed with initial synthesis gas and are fed into the reactor again.

The description of the process shows that the processes of thermochemical processing of wood waste in to the dimethyl ether together are complicated for mathematical description. This is due to the complex chemical structure of wood waste and the influence of catalytic synthesis process of dimethyl ether as the component composition of the synthesis gas [10]. Therefore the receiving of a good quality synthesis gas has a very high value as a raw material for the synthesis of dimethyl ether. As a result, for the constructive mathematical description we must enter the following key assumptions:

- polydisperse wood particle system is considered as one particle of cylindrical shape, with the volume equal to the volume of the loaded material. Meanwhile the porosity of the layer is replaced with a porosity of the particles;
- when considering the heating step, drying and pyrolysis processes the imposition of time is not considered. Each stage is considered in sequence [6];
- wood pyrolysis process is described by the decomposition scheme on the main components of wood - coal and gas (figure 2). The mechanism of the thermal decomposition assumes, that pyrolysis of wood is held in two reactions. One of them is an endothermic reaction of gas formation, while the other - of the intermediate residue which is further decomposed into carbon and gas [11]. The reaction rate constant and the proportion of the substance components are reported in the literature [12]
-

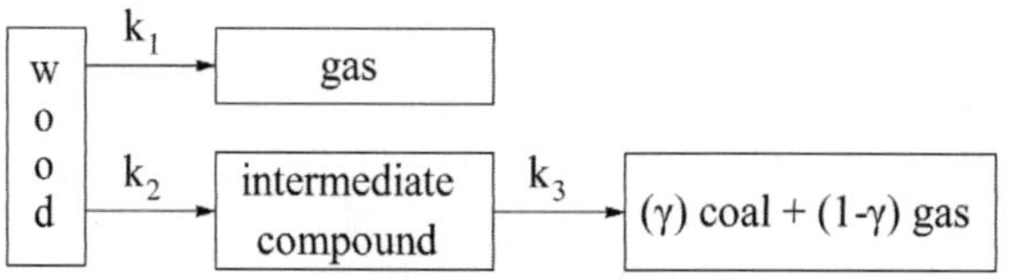

Figure 2. The mechanism of thermal decomposition of wood

- during the pyrolysis external volume of cylindrical particle remains constant, but the number of pores is increased, due to convective heat and mass transfer [13];
- we assume that there are only heterogeneous reactions of carbon and the oxidizing agent intercation occur in the oxidizing zone:

$$C + O_2 = CO_2, \qquad 2C + O_2 = 2CO;$$

- considering the high temperature on the oxidation stage we assume that the temperature of the coal is equal to gas temperature. This determines the initial and boundary conditions of the reduction stage;
- the molecular diffusion is neglected in the reduction zone [1];
- recovery zone is defined by the following set of reactions [14]:

$$C + CO_2 = 2CO; \qquad C + H_2O = CO + H_2;$$
$$CO + 2H_2O = CO_2 + 2H_2 \qquad CO + H_2O = CO_2 + H_2$$

- final component composition of the synthesis gas is taken in the form of CO, CO_2, H_2 and H_2O;
- steam and gas mixtures obey the ideal gas law;
- dimethyl ether synthesis on catalyst layers occurs only from carbon monoxide and hydrogen;

- changes do not occur in the chemical composition of the catalyst.

In view of the assumptions and hypothesis the mathematical model is described by the system of simplified differential equations of thermal and mass conductivity.

3. Description of mathematical model

Considering the description of the installation's scheme and physical picture of technological process (figure 1) at first the process unit that includes a complex of processes of warming, drying and pyrolysis of conditional cylindrical particle is considered. Change in the temperature of the particles at these processes is described by the differential equation of heat transfer [14]

$$c_{part} \cdot \rho_{part} \cdot \frac{\partial T_{part}}{\partial \tau} = \frac{\partial}{\partial l}\left(\lambda_{ef} \cdot \frac{\partial T_{part}}{\partial l}\right) + q_{part} \tag{1}$$

The initial and boundary conditions are formulated to solve the differential equation (1):

$$T_{part}(0,l) = T_{part.pr} \tag{2}$$

$$-\lambda_{part} \frac{\partial T_{part}}{\partial l}\bigg|_{l=0} = \alpha \cdot (T_g - T_{part}) \tag{3}$$

where T_g – gas temperature is determined from the following equation

$$T_g = \frac{q \cdot C_g}{\rho \cdot c} \tag{4}$$

When considering the drying step, the heat flux (q_{evap}) to the evaporation of moisture is given by

$$q_{evap} = r \cdot j = r \cdot \frac{m_{part}}{F} \cdot \frac{\partial W_{part}}{\partial \tau}\bigg|_{l=0} = r \cdot \frac{\rho_{part}}{f_{part}} \cdot \frac{\partial W_{part}}{\partial \tau}\bigg|_{l=0} \tag{5}$$

As a result of some transformations the drying step will be described by a system of differential equations

$$c_{part} \cdot \rho_{part} \cdot \frac{\partial T_{part}}{\partial \tau} = \frac{\partial}{\partial l}\left(\lambda_{ef} \cdot \frac{\partial T_{part}}{\partial l}\right) + r \cdot \varepsilon \cdot \frac{\partial W_{part}}{\partial \tau} \tag{6}$$

$$\frac{\partial W_{part}}{\partial \tau} = \frac{k_p}{\rho_0}\left(\frac{\partial^2 P_{part}}{\partial l^2}\right) \tag{7}$$

The pressure inside the particle is determined depending on the temperature of the particle, according to the equation [1]

$$P_{part} = \exp\left(A - \frac{B}{T_{part}}\right) \cdot \log_{10.6} \frac{W_{part}}{3.27 - 0.015 \cdot (T_{part} - 273)} \tag{8}$$

The initial and boundary conditions for solving the system of equations (6)÷(7) can be written as follows

$$W_{part}(0,l)=W_{part.pr} \quad (9)$$

$$P_{part}(0,l)=P_{atm} \quad (11)$$

$$T_{part}(0,l)=T_{part.pr} \quad (10)$$

$$P_{part}\Big|_{l=L}=P_{part} \quad (12)$$

and the boundary temperature conditions are written similar to the expression (3).

Description of the pyrolysis step is carried out based on the taken scheme of thermal wood decomposition (figure 2). As a result, the equation of mass changes of the substance can be written in the following form [4].

For the initial wood:
$$\frac{\partial m_w}{\partial \tau}=(-k_1+k_2)\cdot m_w \quad (13)$$

For intermediate substance:
$$\frac{\partial m_{i.c.}}{\partial \tau}=k_2\cdot m_w-k_3\cdot m_{i.c.} \quad (14)$$

For coal:
$$\frac{\partial m_{coal}}{\partial \tau}=\gamma\cdot k_3\cdot m_{i.c.} \quad (15)$$

From the assumptions the volume of the solid phase particles is determined by the equation

$$V_{sol}=V_{pr}-V_{voi} \quad (16)$$

Changes in the mass per volume unit of the gas phase can be expressed by convection and pyrolysis reactions

$$\frac{\partial(m_g\varepsilon)}{\partial \tau}+\frac{\partial(v_g m_g)}{\partial l}=(1-\gamma)\cdot k_3\cdot m_{i.c.} \quad (17)$$

where ε – is a porosity of material, that is determined from the expression [12]

$$\varepsilon=1-\left(1-\varepsilon_w^0\right)\cdot\frac{m_w+m_{coal}}{m_w^0} \quad (18)$$

Gas flow rate (v), is defined by Darcy's law [15]

$$v=-\frac{B_{part}}{\mu}\cdot\frac{\Delta P}{L} \quad (19)$$

In view of the equation (17) and the assumption that the gases behave as an ideal gas, we obtain the equation of gas pressure changes

$$\frac{\partial}{\partial \tau}\left(\frac{\varepsilon\cdot P_g}{T}\right)=\frac{\partial}{\partial l}\left(\frac{B\cdot P_g}{\mu T}\right)+\frac{R}{M_g}\cdot S_g \quad (20)$$

where mass formation of the gas (S_g) [4]:

$$S_g = (1-\gamma)\cdot k_3 \cdot m_{i.c.} \tag{21}$$

The equation of energy conservation for the pyrolysis step can be written as

$$\left(c_w^0 \cdot m_w + c_w^0 \cdot m_{i.c.} + c_{coal} \cdot m_{coal} + \varepsilon \cdot c_{Pg} \cdot m_g\right)\cdot \frac{\partial T}{\partial \tau} + c_{Pg} \cdot m_g \cdot v \frac{\partial T}{\partial l} =$$
$$= \frac{\partial T}{\partial l}\left(\lambda_{part} \frac{\partial T}{\partial l}\right) - \left(k_1 \cdot \Delta\alpha_1 + k_2 \cdot \Delta\alpha_2\right)\cdot m_w + k_3 \cdot m_{i.c.} \cdot \Delta\alpha_3 \tag{22}$$

where α – heat transfer coefficient, defined by Nusselt number [1]. The thermal conductivity coefficient (λ) is defined as the sum of the thermal conductivity of solid and volatile substances, considering the emission of heat transmitted through the pores [10]

$$\lambda_{part} = (1-\eta)\cdot \lambda_w^0 + \eta \cdot \lambda_{coal} + \varepsilon \cdot \lambda_g + \frac{13.5 \cdot \sigma \cdot T^3 \cdot d_{por}}{e} \tag{23}$$

To solve the system of equations (13)÷(15), (17), (20) and (22) the initial conditions are formulated (24)÷(29). The boundary conditions for the pressure (30) can be written on the basis of the ideal gas law [4], and the boundary temperature conditions are written similar to the expression (3).

$$m_w(0,l) = m_w^0 \tag{24}$$

$$m_{i.c.}(0,l) = 0 \tag{25}$$

$$m_{coal}(0,l) = 0 \tag{26}$$

$$m_g(0,l) = 0 \tag{27}$$

$$T_{part}(0,l) = T_{part.pr} \tag{28}$$

$$P_{part}(0,l) = P_{atm} \tag{29}$$

$$\left.\frac{\partial P_g}{\partial l}\right|_{l=L} = v \cdot \frac{\partial (v \cdot m_g)}{\partial l} \tag{30}$$

When considering the oxidation step, taking into account the assumptions, the equation changes of coal mass and concentrations of the reaction components can be written as

$$\frac{\partial C_i}{\partial \tau} = -k_i \cdot C_i \tag{31}$$

$$\frac{\partial m_{coal}}{\partial \tau} = -m_{coal} \cdot \sum_{m=1}^{z} k_i \tag{32}$$

wherein

$$\frac{\partial T_{coal}}{\partial \tau} = \frac{\partial T_g}{\partial \tau} \tag{33}$$

Then the equation of energy conservation for coal can be written as follows:

$$\rho_{coal} \cdot c_{coal} \cdot \frac{\partial T_{coal}}{\partial \tau} = \sum_{i=1}^{z} \left(q_i \cdot k_i (C_{i0} - C_i)\right) \tag{34}$$

To solve the system of equations (31)÷(34) the following points are taken:

- initial conditions:

$$T_g = T_{coal} = T_{part.pr} \tag{35}$$

$$m_{coal} = m_{coal0} \tag{36}$$

$$C_{(CO2)} = 0 \tag{37}$$

$$C_{(CO)} = 0 \tag{38}$$

$$C_{(O2)} = C_{(O2)pr} \tag{39}$$

- boundary conditions:

$$\left.\frac{\partial T_{coal}}{\partial y}\right|_{y=0} = T_{coal} \tag{40}$$

$$\left.\frac{\partial C_i}{\partial y}\right|_{y=0} = C_{i0} \tag{41}$$

$$\left.\frac{\partial m_{coal}}{\partial y}\right|_{y=0} = m_{coal0} \tag{42}$$

The reduction step based on the assumptions, is described by the following system of equations of thermal and mass conductivity [15]

$$w_{gs} \cdot \frac{\partial C_i}{\partial y} = -k_i \cdot C_i \tag{43}$$

$$w_{coal} \cdot \frac{\partial m_{coal}}{\partial y} = -m_{coal} \cdot \sum_{m=1}^{z} k_i \tag{44}$$

$$\rho_{gs} \cdot c_{gs} \cdot w_{gs} \cdot \frac{\partial T_{gs}}{\partial y} = -\alpha_{gs} \cdot \left(T_{coal} - T_{gs}\right) \cdot f + \sum_{i=1}^{z} \left(q_i \cdot k_i \cdot \left(C_{i0} - C_i\right)\right) \tag{45}$$

$$\rho_{coal} \cdot c_{coal} \cdot w_{coal} \cdot \frac{\partial T_{coal}}{\partial y} = \alpha_{gs} \cdot \left(T_{coal} - T_{gs}\right) \cdot f - \sum_{i=1}^{z} \left(q_i \cdot k_i \cdot \left(C_{i0} - C_i\right)\right) \tag{46}$$

To solve the system of equations (43)÷(46) following initial conditions are accepted

$$T_{coal} = T_{coal.pr} \tag{47}$$

$$T_g = T_{g.pr} \tag{48}$$

$$C_{(CO2)} = C_{(CO2)pr} \tag{49}$$

$$C_{(CO)} = C_{(CO)pr} \tag{50}$$

$$C_{(O2)} = C_{(O2)pr} \tag{51}$$

$$C_{(H2)} = 0 \tag{52}$$

$$C_{(H2O)} = 0 \tag{53}$$

$$m_{coal} = m_{coal.pr} \tag{54}$$

and the boundary conditions of temperature and mass are written in the form of

$$-\lambda_{coal} \frac{\partial T_{coal}}{\partial y} + v_{gs} \frac{\partial T_{coal}}{\partial y} + \sum_{i=1}^{z} q_i \cdot k_i \cdot \left(C_{i0} - C_i\right) = T_{coal.pr} \tag{55}$$

$$\left.\frac{\partial T_{coal}}{\partial y}\right|_{y=0} = \left.\frac{\partial T_g}{\partial y}\right|_{y=0} \quad (56)$$

$$v_{coal}\frac{\partial m_{coal}}{\partial y} = -\sum_{i=1}^{z} k_i \cdot C_i \quad (57)$$

The process of producing the dimethyl ether from synthesis gas, according to the process description proceeds on a mixture of various catalysts. Based on that the differential equation of heat and material balance, reflecting the change in flow key components: CO, CO_2, CH_3OH, C_2H_6O on various catalysts can be written as [9]:

$$\frac{dX_i}{d\tau} = \omega_{ik}(1-\varepsilon_\kappa) \quad (58)$$

$$\frac{dT}{d\tau} = \frac{-1}{\sum X_i c_{pi}\left(\sum \Delta H_{rk}\omega_k(1-\varepsilon_{MK}) + \sum \Delta H_{rk}\omega_k(1-\varepsilon_\kappa) - KS_x(T-T_x)\right)} \quad (59)$$

The initial conditions for solving the system of equations (58)÷(62) take the form

$$\tau = 0 \quad (60) \qquad C_i^{sur} = C_{i0} \quad (61) \qquad T = T_0 \quad (62)$$

and the condition of equality of the substance transfer rate to the chemical reaction rate on the catalyst can be written as [9]

$$\beta_i S_k\left(C_i^{sur} - C_i\right) = (1-\varepsilon_\kappa)\omega_{i\kappa} \quad (63)$$

Reaction rate is defined by the partial pressure of each component in the mixture.

4. Conclusions

The developed mathematical description of the process of thermochemical processing of wood waste in to the dimethyl ether is presented in the form of the main differential and algebraic equations of heat and mass transfer with the initial and boundary conditions. To simulate the process a numerical method of the decomposition of differential equations on finite-difference (mesh) schemes is used and a calculation algorithm is developed [6]. Modeling based on this mathematical description allows: to determine the optimum operating parameters of the process for producing synthesis gas suitable for catalyst synthesis of dimethyl ether; calculate the basic constructive parameters of technological scheme equipment. The following estimates can be attributed to the constructive parameters of equipment: adjusted volume of the pyrolysis chamber, which calculations are based on the productivity of equipment; the height of the pyrolysis chamber, which is determined depending on the time of the pyrolysis process; volume of the reaction chamber of the gas generator associated with the volume of the pyrolysis chamber; Further, knowing the volume of the reaction chamber we can calculate its diameter, height, number of nozzles, the size of the ash separation, etc.

It should be noted that the use of the assumptions during the development of the mathematical model of the process, imposes restrictions on it. Incorrect assumptions can lead to a divergence of the results with a real data or to gross errors, which are not acceptable in the development of equipment design.

The correct and perhaps the only solution of the problem of increasing the accuracy of the mathematical model is to check the estimates during the experimental research and the introduction of an empirical correction coefficients in to the model.

Notation

T – the temperature, K; τ – time, min; c – heat capacity, J/(kg·K); ρ – density, kg/m^3; λ – coefficient of thermal conductivity, W/(m·K); q – specific thermal energy, J/kg; α – the heat transfer coefficient, W/(m^2·K); r – latent heat of vaporization, J/kg; W – the moisture content of the material, %; f – a specific surface area in m^2/m^3; k_p – molar transfer coefficient, c; P – pressure, Pa; m – mass of a substance per unit volume, kg/m^3; k – rate constant of a chemical reaction, c^{-1}; γ – the proportion of the component, %; **V** – velocity, m/s; ε – porosity, m^3/m^3; η – the degree of pyrolysis, %; e – the coefficient of blackbody radiation, W/(m^2·K^4); C – concentration of substance, mol/m^3; i – the component (substance); j – water mass flow, kg/(m^2·s); F – surface area, m^2; A – calculated ratio; V – volume, m^3; S – mass formation, kg/(m^3·s); μ – the dynamic viscosity of the gas, Pa·c; B_g– the gas permeability of the particle, m^2; L – length, m; σ – Stefan-Boltzmann constant (5.67·10^{-8} W/m^{-2}·K^{-4}); d – diameter, m; M – molar mass, g/mol; R – universal gas constant, J/(mol·K); X – flow of matter, mol/s; ω – the velocity of chemical reaction, mol /(s·m^3); ΔH – the change in enthalpy of the system, J/mol; K – heat transfer coefficient W/(m^2·K); S_x – cooling surface area, m^2; β – mass transfer coefficient, kg/(m^2·s); S_k – the surface of the catalyst bed, m^2/m^3.

Subscripts

part – particle; ef – effective; g – gas; evap – evaporation; pr – primary; 0 – the initial; w – wood; i.c. – intermediate compound; coal – coal; sol – solid phase; voi – voids; por – pores; gs – the gas stream; z – the number of reactions; rz – reaction zone; ca – cooling agent; sur – surface; cat - catalyst; atm – atmosphere;

Acknowledgments

The presented materials are received during the realization of program for the grant of the President of the Russian Federation under state support of young Russian scientists. The cypher of the research topic is MK-3434.2015.8 "Development of theoretical foundations, technologies and equipment of integrated thermochemical processing of waste wood and plant biomass into raw materials for chemical synthesis and components of motor fuels".

References

[1] Timerbaev N.F., Safin R.R., Safin R.G. and Ziatdinova D.F. 2014 Modeling of the process of energy-technological treatment of wood waste by method of direct-flow gasification Journal of Engineering and Applied Sciences (EAAS) 9 (5): 141-146

[2] Saravanakumar A. and Haridasan T.M. 2013 A novel performance study of kiln using long stick wood pyrolytic conversion for charcoal production Energy Education Science and Technology A: Energy Science and Research 31 (2): 711-722

[3] Troy A. Semelsberger, Rodney L. Borup and Howard, L. Greene 2006 Dimethyl ether (DME) as an alternative fuel Journal of Power Sources 156 (2): 497–511. DOI:10.1016/j.jpowsour.2005.05.082

[4] Park W.C., Atreya A. and Baum H.R. 2010 Experimental and theoretical investigation of heat and mass transfer processes during wood pyrolysis Combustion and Flame, 157 (3): 481–

494. DOI:10.1016/j.combustflame.2009.10.006

[5] Saravanakumar A., Haridasan T.M. and Reed T.B. 2010 Flaming pyrolysis model of the fixed bed cross draft long-stick wood gasifier Fuel Processing Technology 91 (6): 669-675. DOI:10.1016/j.fuproc.2010.01.016

[6] Sadrtdinov A.R., Sattarova Z.G., Prosvirnikov D.B. and Tuntsev D.V. (2015, December) Modeling of thermal treatment of wood waste in the gasifiers. In 2015 Int. Conf. on Mechanical Engineering, Automation and Control Systems (MEACS) pp 1-5. DOI:10.1109/MEACS.2015.7414914

[7] Semelsberger T.A., Ott K.C., Borup R.L. and Greene,H.L. 2006 Generating hydrogen-rich fuel-cell feeds from dimethyl ether (DME) using physical mixtures of a commercial Cu/Zn/Al2O3 catalyst and several solid-acid catalysts Applied Catalysis B: Environmental 65 (3): 291-300. DOI:10.1016/j.apcatb.2006.02.015

[8] Gong-Xin Qi, Xiao-Ming Zheng, Jin-Hua Fei and Zhao-Yin Hou 2001 A novel catalyst for DME synthesis from CO hydrogenation: 1. Activity, structure and surface properties Journal of Molecular Catalysis A: Chemical 176: 195–203. DOI:10.1016/S1381-1169(01)00257-6

[9] Hadipour A. and Sohrabi M. 2008 Synthesis of some bifunctional catalysts and determination of kinetic parameters for direct conversion of syngas to dimethyl ether Chemical Engineering Journal 137 (2): 294–301. DOI:10.1016/j.cej.2007.04.039

[10] Di Blasi C. 2004 Modeling wood gasification in a countercurrent fixed-bed reactor AIChE Journal, 50 (9): 2306–2319. DOI:10.1002/aic.10189

[11] Tuntsev D.V., Filippova F.M., Khismatov R.G. and Timerbaev N.F. 2014 Pyrolyzates: Products of plant biomass fast pyrolysis Russian Journal of Applied Chemistry 87 (9): 1367-1370. DOI:10.1134/S1070427214090304

[12] Di Blasi C., Branca C., Lombardi V., Ciappa P. and Di Giacomo C. 2013 Effects of particle size and density on the packed-bed pyrolysis of wood Energy and Fuels 27 (11): 6781-6791. DOI:10.1021/ef401481j

[13] Safin R.R., Khasanshin R.R., Shaikhutdinova A.R. and Safina A.V. 2014 Research of heating rate while termo modification of wood World Applied Sciences Journal 30 (11): 1618-1621. DOI:10.5829/idosi.wasj.2014.30.11.14223

[14] Saravanakumar A., Hagge M.J., Haridasan T.M. and Bryden K.M. 2011 Numerical modelling of a fixed bed updraft long stick wood gasifier Biomass and Bioenergy 35 (10): 4248-4260. DOI:10.1016/j.biombioe.2011.07.012

[15] Saravanakumar A., Haridasan T.M., Reed B. Tomas and Bai R. Kasturi 2007 Experimental investigation and modelling study of long stick wood gasification in a top lit updraft fixed bed gasifier Fuel 86 (17): 2846–2856. DOI:10.1016/j.fuel.2007.03.028

IOP Conf. Series: Materials Science and Engineering **142** (2016) 012095 doi:10.1088/1757-899X/142/1/012095

The Development of Equipment for the Disposal of Solid Organic Waste and Optimization of Its Operation

Almaz R Sadrtdinov[1,a], Rushan G Safin[1], Nail F Timerbaev[1], Dilyara F Ziatdinova[1], Natalya A Saprykina[2]

[1] Institute of Petroleum and chemical Engineering, Kazan National Research Technological University, K. Marx street 68, Kazan, Republic of Tatarstan, Russian Federation
[2] Yurga Institute of Technology of National Research Tomsk Polytechnic University Affiliate, Tomsk, Russia

E-mail: [a] almaz.sadrtdinov@gmail.com

Abstract. The paper describes the developed system for the thermal utilization of solid organic waste, which can simultaneously process the paper, wood, rubber, plastic, etc. A method for improving the efficiency of the equipment, due to optimization of the gas extraction system was proposed. The influence of the characteristics of installed equipment and modes of their work on energy savings and efficiency of the gas extraction system was also determined. Optimization work which includes the introduction into the exhauster control system of the frequency converter, can save up to 70% of electricity and increases the life of the equipment.

1. Introduction

During the development of technological processes and equipment design there is always an issue of energy efficiency. This is especially important in the development of equipment for thermal disposal of various kinds of solid organic waste in the furnace and boiler units. In most cases, disposing of wastes has a small economic benefit, and sometimes is a very costly process [1].

During the designing of equipment for the disposal of solid organic waste, including solid wastes there are always lots of complicated problems. This is primarily due to the morphological composition of wastes that come to recycling. The waste may consist of the following organic materials: paper, wood (furniture), textiles, various plastics, rubber, etc. [2, 3, 4, 5]. Another important characteristic is the fractional composition of the waste, which can vary over a wide range [6, 7]. Such a diversity of species and sizes of raw materials greatly complicates the calculation and design of the equipment.

Secondly it's an environmental problem. Power plants that use solid waste as fuel are a continuous source of toxic emissions into the atmosphere. With the constant increase in the number of equipment disposing solid organic waste by thermal methods the problem of clearing of gases becomes very important [8, 9, 10, 11].

There are currently many samples of industrial installations for the thermal disposal of various solid waste using incineration technology, pyrolysis and gasification. However, not all technologies are energy-efficient and environmentally friendly when you need to dispose all types of waste. For example, some installations use only the process of burning. The disadvantage is the allocation of soot and toxic substances in the disposal of plastic waste, such as rubber, plastic resin, polyethylene, etc.

[12]. In other technologies after waste incineration, there is a neutralization of combustible components of waste gases only and non-flammable components (fly ash, heavy metals, etc.) are released into the atmosphere [13]. Some installations may not operate at high temperatures and as a result do not allow to dispose solid waste containing polymer materials such as rubber and hard combustible materials [14, 15].

All the above factors need to be considered in the development of new equipment for the processing and disposal of solid organic wastes of different origin. As a result, the aim is to develop an installation that will allow to process solid organic waste, comprising: wood, textiles, rubber, plastics, resins, polyethylene and other polymeric materials.

2. Description of the developed object

The developed equipment related to waste disposing technology combines two methods of thermal processing of raw materials: incineration and pyrolysis. The scheme of developed equipment is shown in figure 1.

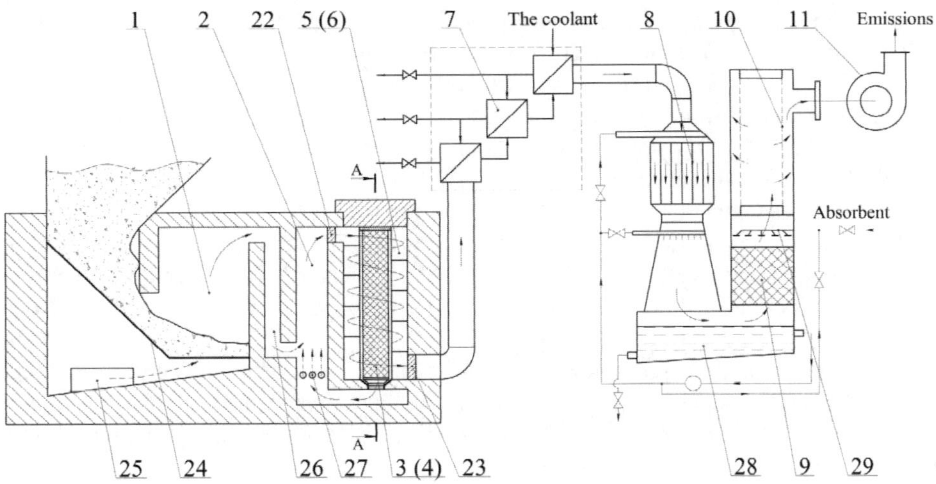

Figure 1. The installation for thermal disposal of solid wood wastes

Installation for the thermal disposal of solid organic waste consists of the incineration batch furnace 1; post-combustion chambers of exhaust and pyrolysis gases 2; removable pyrolysis chambers 3–4, installed in the cyclonic shaft 5–6 (figure 2); the heat exchanger 7 and cleaning unit of gases emitted into the atmosphere. The cleaning unit consists of series-connected scrubber as the Venturi tube 8, a packed absorber 9, the separator 10 and the exhauster 11.

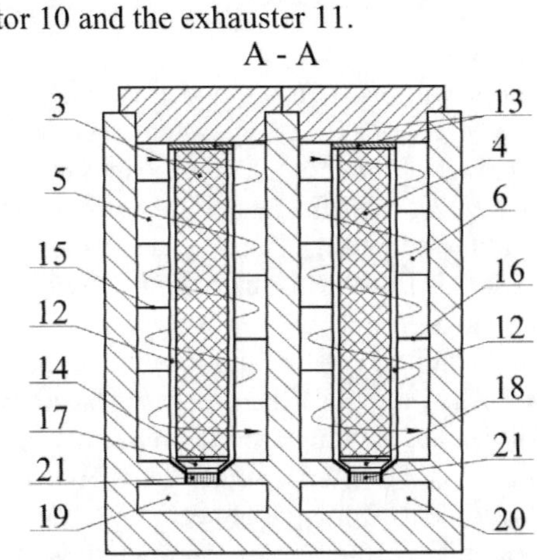

Figure 2. General view of the pyrolysis chambers

The device operates as follows: pre-sorting of waste occurs with the release of polymeric materials such as rubber, plastic, resin, polyethylene, etc. Further, wastes containing polymer materials are charged into the removable pyrolysis chamber 3–4, and the remaining waste is burned in the incineration batch furnace 1. Pyrolysis chambers 3–4 (figure 2) are the same and have the frame 12, hermetically closing lid 13 and support gas distribution grid 14 for extraction of the pyrolysis gases.

Removable pyrolysis chambers 3–4 are set in to the cyclonic shafts 5–6 which have the tangential entering of exhaust gases in the upper part and helical inserts 15–16 for even distribution of gas flow in the shaft. At the bottom of the shafts there are outlets (taps) 17–18, which are designed for sealing engagement of the pyrolysis chambers with pyrolysis gas outlet channels 19–20. The outlets (taps) 17–18 have a special partition 21 which is designed to prevent the leakage of exhaust gases from the shaft in the post-combustion gas chamber 2. After setting the pyrolysis chambers in to the cyclonic shafts they are sealed by lids. Cyclonic shafts operate alternately, and the process of switching on and off is done by valves 22–23. So when you turn on one cyclonic shaft the other is disconnected. An alternate operation of shafts is necessary for discharging waste pyrolysis chamber and its cleaning from carbon residue and loading of a new batch of solid wastes containing polymer materials.

Waste burning in the incineration batch furnace 1 is performed in the horizontal part of the grate 24. The supply of oxygen required for combustion and removal of ash is carried out through a hatch 25 at the bottom of the chamber. The exhaust gases formed during incineration and pyrolysis gases through outlets 26 and 19–20, respectively, enter the post-combustion chamber 2. There it is mixed with oxygen supplied via the channel 27. As a result of post-combustion of exhaust and pyrolysis gases in oxygen the gas temperature rises to 1400 °C. From the post-combustion chamber 2, the heated exhaust gases enter the cyclonic shaft 5–6 for warming of removable pyrolysis chambers 3–4. The exhaust gases from cyclonic shaft are fed into the heat exchanger 7 for cooling to a temperature of 40–55 ° C and production of thermal energy. For efficient operation the heat exchanger consists of three cascaded units operating in countercurrent mode.

Next, the cooled exhaust gases are cleaned in consecutively installed Venturi scrubber 8, the packed absorber 9 and the separator 10. The Venturi scrubber 8 is a frame with the tube unit placed in it, diffusers and nozzles placed at the top of the apparatus. The power fluid is sprayed evenly in the diffuser and interreacts with the exhaust gases. After that the power fluid flows into the receiving tub 28. The packed absorber 9 is located vertically above the receiving bath 28 and is provided with nozzles 29 for supplying the dispersed absorbent. Venturi scrubber 8 and a packed absorber 9 have a total absorbent circulation system. The packed absorber 9 at the top is connected to the separator 10 to remove moisture from the gas phase.

After separator 10 cleaned and cooled to 25–30 °C flue gases are removed from the exhauster 11 to the atmosphere.

3. Discussion of the developed results

Analysis of optimization methods of various fuels' combustion showed that most of the energy costs are associated with the operation of the gas extraction system. The costs depend on the gas flow rate and pressure loss on installed machinery [16]. Also, a significant impact on energy losses have modes of the equipment due to the creation of significant pressure oscillations before the exhauster [17]. Therefore, in most cases, the throttles and control valves are installed in the gas extraction system. These elements create an additional flow resistance and allow the pressure to ensure stable performance in the gas extraction system.

Not only throttle affects on the magnitude of the losses in the regulation but also the exhauster that has a certain power. For example, when the old exhauster is replaced by a new one its characteristics may be slightly different, that will lead to a change in the pressure range of the gas extraction system. All these factors can lead to the loss of up to 45% of the energy of the nominal capacity of the equipment, and as a result to reduction of the energy and economic efficiency [18].

To solve the problem of minimizing the losses associated with pressure control in the gas extraction system, it is necessary to eliminate hydraulic resistance by opening all control valves.

However, it is possible if the regulatory process will be carried out directly by exhauster. Theory of exhauster operation and experimental studies (figure 3) show that the change in the motor speed changes the characteristics of the equipment pressure [19, 20]. In figure 3 the curve 1 corresponds to the nominal pressure characteristic, while curves 2 and 3 – are to discharge characteristics at low speed.

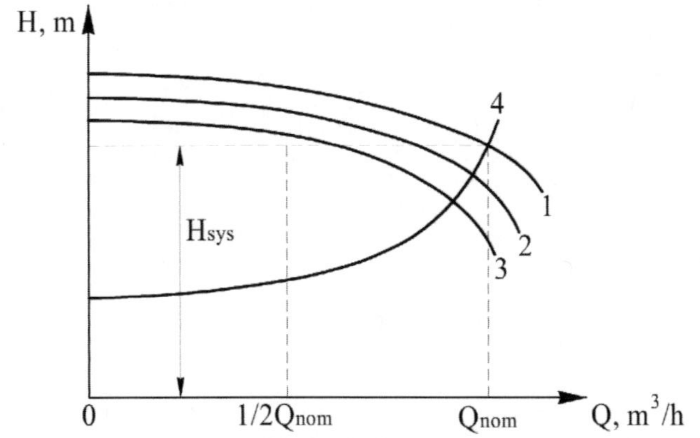

Figure 3. Changes in exhauster characteristics (1, 2, 3) due to the changes in rate of rotation: 1 – 1000 r/min 2 – 900 r/min; 3 – 750 r/min; 4 – system characteristic

If to set the operation of exhauster motor in a way that it would change its rotation rate during the changes of process parameters: gas flow rate and inlet pressure, it would be possible to adjust and stabilize the pressure in the gas outlet without energy loss. When using such regulation method the pressure loss is eliminated since there is no throttle elements.

Based on the analysis of the data the method of pressure regulation in the gas extraction system was developed. The method consists in introducing the frequency converter in power system of exhauster. Control scheme is shown in figure 4.

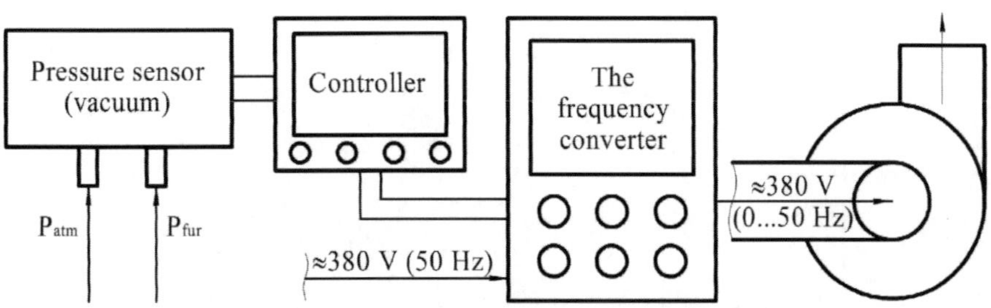

Figure 4. Pressure control scheme in the gas extraction system

The frequency converter is an electronic device which controls the rotation rate of motor due to variations in the amplitude and frequency of the electric voltage.

The proposed method of regulating pressure in the gas extraction system allows not only to reduce energy consumption but also to optimize the process [21]. For example, you can change the efficiency of the exhauster by changing the motor rate rotation, depending on the gas flow (figure 5).

From the graph (figure 5) it is seen that the maximum efficiency with a reduction in the rotation rate decreases and moves to the left. Analysis of changes in the exhauster rotation rate during the changes in flow rotation shows that a decrease in consumption may reduce the rotation rate. If we consider the exhauster operation for less than the nominal flow (vertical lines A and B), it is more efficient for these modes to operate at low speed. In this case, the efficiency of the exhauster is higher than when operating at the nominal speed. Thus, the decrease in rotation rate according to the load process does not only save energy consumption, but also improves the efficiency of exhauster.

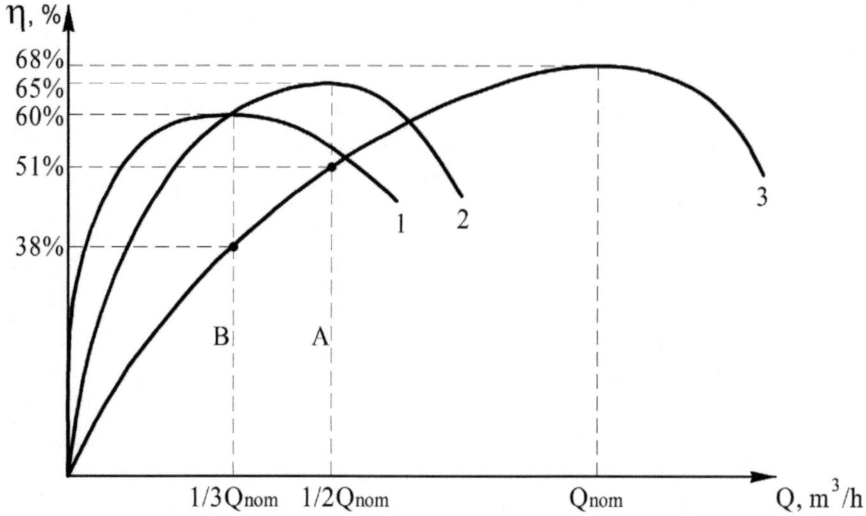

Figure 5. Changing in coefficient of exhauster efficiency depending on the performance when changing the rotation frequency: 1 – 750 r/ min; 2 – 900 r/ min; 3 – 1000 r/ min

Also, the dependence of power consumption on the relative gas flow at the throttle regulation and when using the frequency converter was analyzed (figure 6).

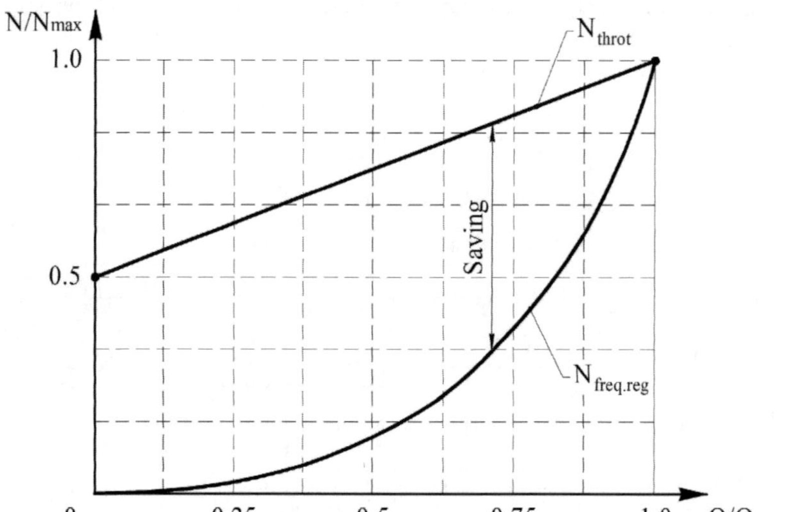

Figure 6. The dependence of the exhauster power consumption on the gas flow while using a frequency converter and throttle control

According to the received data the power consumption by using frequency control, as opposed to the throttle control is curvilinear and provides energy savings. To calculate the reduction value of the power consumption and obtain reliable calculations, it was proposed to use the scaling laws [1]:

$$\frac{N_1}{N_2} = \frac{n_1^3}{n_2^3} \tag{1}$$

$$\frac{H_1}{H_2} = \frac{n_1^2}{n_2^2} \quad (2)$$

$$\frac{Q_1}{Q_2} = \frac{n_1}{n_2} \quad (3)$$

That is, the reduction of power consumption is in proportion to the reduction in engine speed in the third degree. Data show that even in the systems with a constant flow rate it is possible to get an effect from the application of frequency converter. The total reduction in energy consumption by using frequency regulation can be up to 70%, even with perfectly matched equipment operating in variable flow systems.

4. Conclusions

The set of proposed technical solutions in the development of installation for thermal utilization of solid organic waste can simultaneously dispose different types of solid waste, including rubber, plastic, and others. Also, a joint post–combustion of the pyrolysis gases with the exhaust gases contributes not only to reduction of harmful emissions into the atmosphere, but also increasing of installation efficiency. In addition, the introduction of frequency control system of exhauster operation provides:

- reduction in valve wear;
- gradual inclusion of exhauster and access to the operating mode;
- full motor protection;
- increase in motor life;
- noise reduction, which is particularly important at the the exhauster location next to the residential or office space.

The practice of using frequency converters to control the exhausters, fans and pumps proves the feasibility of creating a special process control systems [22]. This approach can significantly reduce operating costs and improve working conditions. Modern frequency converters allow to monitor more than 20 parameters of electric motor status. Processing of these parameters allows for deep diagnostics of the equipment and the carried processes. There is an opportunity not only to respond to the evolving accident, but also to warn it which is more important for power assets [23].

Development of equipment for waste utilization using a frequency converter, in which the frequency control is carried out in conjunction with a variety of process parameters is very important. This not only reduces the consumption of electric energy, but also saves energy resources of the entire system. It is advisable to use frequency converters as components of technological process automation complexes. Such solutions make it possible to obtain an additional effect, which is obviously more than just saving electricity.

Notation

H – pressure, Pa; Q – gas expenditure, m^3/h; η - efficiency, %; N – power, Watt; n – rotation frequency, r/ min.

Subscripts

fur – furnace; atm – atmosphere; nom – nominal; max – maximum; sys – system; throt – throttle; freq.reg – frequency regulation.

Acknowledgments

The presented materials are received during the realization of program for the grant of the President of the Russian Federation under state support of young Russian scientists. The cypher of the research topic is MK–3434.2015.8 "Development of theoretical foundations, technologies and equipment of integrated thermochemical processing of waste wood and plant biomass into raw materials for chemical synthesis and components of motor fuels".

References

[1] Timerbaev N F, Safin R R, Safin R G and Ziatdinova D F, Modeling of the process of energy-technological treatment of wood waste by method of direct-flow gasification, *Journal of Engineering and Applied Sciences (EAAS)* 9 (5) 2014 p 141-146.

[2] Garcia A N, Marcilla A, and Font R, Thermogravimetric kinetic study of the pyrolysis of municipal solid waste, *Thermochimica Acta* 254 1995 p 277-304 DOI:10.1016/0040-6031(94)02002-6

[3] Lee C C, & Huffman G L, Medical waste management/incineration, *Journal of Hazardous Materials* 48 (1) 1996 p 1-30 DOI:10.1016/0304-3894(95)00153-0

[4] Wu C H, Chang C Y, Hor J L, Shih S M, Chen L W and Chang F W, On the thermal treatment of plastic mixtures of MSW: pyrolysis kinetics *Waste Management* 13 (3) 1993 p 221-235 DOI:10.1016/0956-053X(93)90046-Y

[5] Yufeng Z, Na D, Jihong L and Changzhong X, A new pyrolysis technology and equipment for treatment of municipal household garbage and hospital waste *Renewable Energy* 28 (15) 2003 2383-2393 DOI:10.1016/S0960-1481(03)00065-X

[6] Fomin A A, Gusev V G, Safin R G and Safin R R Dispersion of the margin removed in complex milling, *Russian Engineering Research* 35 2015 p 417-420

[7] Di Blasi C, Branca C, Lombardi V, Ciappa P and Di Giacomo C, Effects of particle size and density on the packed-bed pyrolysis of wood, *Energy and Fuels* 27 (11) 2013 6781-6791 DOI:10.1021/ef401481j

[8] Palacios-Orueta A, Chuvieco E, Parra A and Carmona-Moreno C, Biomass burning emissions: a review of models using remote-sensing data *Environmental monitoring and assessment* 104 (1-3) 2005 p 189-209.

[9] Austin C C, Wang D, Ecobichon D J and Dussault G, Characterization of volatile organic compounds in smoke at municipal structural fires *Journal of Toxicology and Environmental Health Part A* 63 (6) 2001 p 437-458 DOI:10.1080/152873901300343470

[10] Nolte C G, Schauer J J, Cass G R and Simoneit B R Highly polar organic compounds present in wood smoke and in the ambient atmosphere, *Environmental science & technology* 35 (10) 2001 p 1912-1919 DOI:10.1021/es001420r

[11] Michal J, Toxicity of pyrolysis and combustion products of poly - (vinyl chloride) *Fire and Materials* 1 (2) 1976 p 57-62 DOI:10.1002/fam.810010205

[12] Sharma V K, Mincarini M, Fortuna F, Cognini F and Cornacchia G, Disposal of waste tyres for energy recovery and safe environment, Review *Energy conversion and management* 39 (5) 1998 p 511-528. DOI:10.1016/S0196-8904(97)00044-7

[13] Suocheng D, Tong K W and Yuping W, Municipal solid waste management in China: using commercial management to solve a growing problem, *Utilities Policy* 10 (12001) p 7-11 DOI:10.1016/S0957-1787(02)00011-5

[14] de Marco Rodriguez I, Laresgoiti M F, Cabrero M A, Torres A, Chomon M J and Caballero B, Pyrolysis of scrap tyres *Fuel processing technology* 72 (1) 2001 p 9-22 DOI:10.1016/S0378-3820(01)00174-6

[15] Huang W C, Huang M S, Huang C F, Chen C C and Ou K L, Thermochemical conversion of polymer wastes into hydrocarbon fuels over various fluidizing cracking catalysts *Fuel* 89 (9) 2010 p 2305-2316. DOI:10.1016/j.fuel.2010.04.013

[16] Zhukov A V, Komarov A N, Safronov A N and Barsukov I V, Experience in connecting the power generating units of thermal power plants to automatic secondary frequency regulation within the united power system of Russia, *Power Technology and Engineering* 43 (3) 2009 p 185-193.

[17] Zhao P, Henze G P, Plamp S and Cushing V J, Evaluation of commercial building HVAC

systems as frequency regulation providers, *Energy and Buildings* 67 2013 p 225-235. DOI:10.1016/j.enbuild.2013.08.031

[18] Zorchenko N V, Rezinskikh V F, Suslov S Y, Zhukov A V, Safronov A N, Barsukov I V, Sobolev A N and Tikhobrazov V A Evaluating the effect of frequency regulation modes on the reliability and economic efficiency of thermal power generation units, *Power Technology and Engineering (formerly Hydrotechnical Construction)* 45 (2) 2011 p 132-136. DOI:10.1007/s10749-011-0236-2

[19] Sun G, Meng F, Guo X and Yang S, Numerical analysis on the characteristics of cavitation in the variable frequency regulation of centrifugal pump, In *Recent Advances in Computer Science and Information Engineering* 2012 p 281-286 Springer Berlin Heidelberg. DOI:10.1007/978-3-642-25766-7_37

[20] Bevrani H, Robust multi-objective control-based frequency regulation, In *Robust Power System Frequency Control* 2014 p 105-129 Springer International Publishing. DOI:10.1007/978-0-387-84878-5_6

[21] Monteiro L M and Igreja J M, On Performance of Distributed Model Predictive Control in Power System Frequency Regulation, In *CONTROLO'2014 – Proceedings of the 11th Portuguese Conference on Automatic Control* 2015 p 71-80 Springer International Publishing. DOI:10.1007/978-3-319-10380-8_8

[22] Zhao P, Henze G P, Brandemuehl M J, Cushing V J and Plamp S, Dynamic frequency regulation resources of commercial buildings through combined building system resources using a supervisory control methodology *Energy and Buildings* 86 2015 137-150 DOI:10.1016/j.enbuild.2014.09.078

[23] Dobganyuk I Y, Lokhmatov A P, Plotnikova T V, Sokur P V, Tuzov P Y and Shakaryan Y G Asynchronized machines as a means of counter-emergency control and frequency regulation in a power system *Power Technology and Engineering (formerly Hydrotechnical Construction)* 45 (6) 2012 466-470 DOI:10.1007/s10749-012-0296-y

IOP Conf. Series: Materials Science and Engineering **142** (2016) 012096 doi:10.1088/1757-899X/142/1/012096

The Development of Experimental Setups And Experimental Studies of The Process of Energy-Technological Processing of Wood

Nail F Timerbaev[1], Rushan G Safin[1], Dilyara F Ziatdinova[1], Anatoly A Fomin[2], Alexey A Mokhovikov[3]

[1] Institute of Petroleum and chemical Engineering, Kazan National Research Technological University, K. Marx street 68, Kazan, Republic of Tatarstan, Russian Federation, 420015
[2] Institute of engineering and automobile transport, Vladimir State University, Gorky street 87, Vladimir, Russian Federation, 600000
[3] Yurga Institute of Technology, National Research Tomsk Polytechnic University, Leningradskaya street 26, Yurga, Russian Federation, 652055

E-mail: tnail@rambler.ru

Abstract. The paper describes the experimental setups for the study of the various stages of the process of energy-technological processing of wood waste with the production of synthesis gas. The systems for the study of conjugated processes of drying, pyrolysis and gasification, that are an integral part of energy-technological processing of wood wastes were developed. Experimental studies of the processes have identified their basic properties and optimum operating parameters, allowing to obtain a synthesis gas suitable for the chemical synthesis of various olefins.

1. Introduction

Millions of tons of waste wood are produced at the enterprises of a timber industry complex of Russia every year. The easiest way to dispose the wood waste is their thermal processing by direct combustion to produce heat energy. A more complicated but more effective are the methods of conversion of wood waste in to liquid or gaseous state to obtain the products demanded by the chemical and other industries. One of such products, which can be obtained by energy-technological processing of wood waste - is synthesis gas which is widely used in the chemical industry [1].

For implementation and intensification of energy-technological processing of wood waste from the timber industry, with receiving of synthesis gas and chemical products of a certain quality, it is needed to study heat and mass transfer, which is complicated by chemical reactions at all stages of energy-technological processing of wood waste. It is necessary to develop analytical and experimental methods for prediction of parameters of the waste wood state and products of their thermal decomposition at different stages of their energy-technological processing [2]. To control the chemical processes and obtain the desired product of energy-technological processing, it is necessary to study the kinetics of thermal decomposition of wood waste at different temperatures and conditions, considering the basic properties of wood species, chemical transformation mechanisms, and wood state parameters during thermal exposure.

Thus, the aim of the research is the development of equipment for the study of the processes occurring during the energy-technological processing of wood waste with the use of direct-flow gasification in consideration with modern representations about the effect of temperature on the wood properties [3], about heat and mass transfer being complicated by chemical reactions, as well as with sorption kinetic, heat and mass conductivity and chemical properties of the wood [4].

2. Materials and methods

The variety of processes occurring during energy-technological processing of wood waste of the timber industry, was investigated in a number of established experimental setups that allow for physical modeling of individual processes of drying, pyrolysis and combustion processes in the reduction zone [5]. To study the gasification stage, an experimental setup was developed and is described in schematic diagram in figure 1. The installation consists of a series-connected loading feeder of wood waste 11, the gasifier 13, post-combustion of gas generator chamber 14, heat exchanger 20 and flue gas cleaning system in the form of absorber 26. To measure the parameters and control the process the experimental unit is equipped with temperature sensors, devices for determining the flow of gases and liquids, valves and piping. Registration data from all sensors and devices is carried out by monitoring system 40.

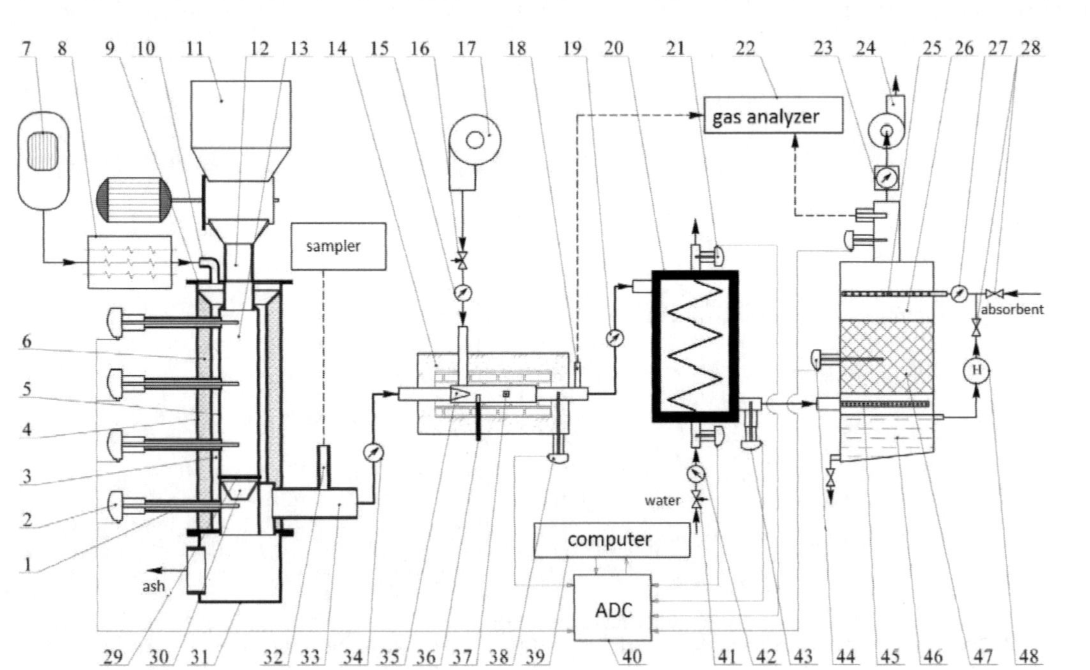

Figure 1. Experimental setup scheme for the research process of the direct flow gasification of wood waste

The gasifier 13 is a cylindrical reactor 4 on the inner surface lined with an inert, heat-resistant ceramic material. To ensure the thermal conditions of adiabatic approximation, the gasifier is thermally insulated from the outside with mineral insulator. The feeding in to gasifier is carried by air compressor 7 through a heater 8, which provides heating of the incoming air. The oxidant (air) is supplied to the afterburning chamber by blower 17. To control the generator gas flow the meter 34 is set up. The change of the temperature during the test is recorded at different levels of height of the gasification chamber by thermocouples, flow of the supply air (oxidant) to the combustion zone is controlled and is regulated by the sensors installed in the compressor 7. The consumption of produced generator gas is determined by using a flow meter 34. In order to determine the chemical composition of the product gas leaving the gasifier through the probe, there is a selection of gas samples and measuring of their temperature being done through the pipe 32. Determination of carbon monoxide gas, hydrogen, methane, carbon dioxide and water, is carried out to assess the quality and optimization of the gas generator mode parameters of the gasification process.

During the study of influence of wood waste moisture on gasification process the waste wood with the bulk density of 280 kg/m^3 and humidity of 10 % is used. Then for each subsequent experiment in this series the waste humidity is increased for 10 ± 1% to achieve the humidity values up to 70 % inclusive. When set humidity mass of wood waste is reduced to 0.4 kg. During the experiment the flow of gasified waste wood (GT) is 1.2 kg/h, with the oxidant flow in the combustion zone (Vvoz.) corresponding to the calculated value and equal to 1.5 m^3/h. The height of the reduction zone (HZ) is 100 mm.

To do a research of the influence of the oxidant flow in the combustion zone on the gasification process the flow of oxidant fed in to combustion zone of gasifier changes stepwise [6, 7]. The variation range of oxidant flow rate is from 1.25 to 0.5 m^3/h with a pitch variation of 0.25 m^3/h. Regulation and control of the oxidant supply to the combustion zone is carried out by the meter and valve located on the compressor 7.

An experimental study on the identification of the process parameters in the pyrolysis zone, such as the duration of the process, the kinetic temperature of the layer, and the mass loss was conducted. The experimental setup scheme for the study of the pyrolysis of wood waste is presented in figure 2.

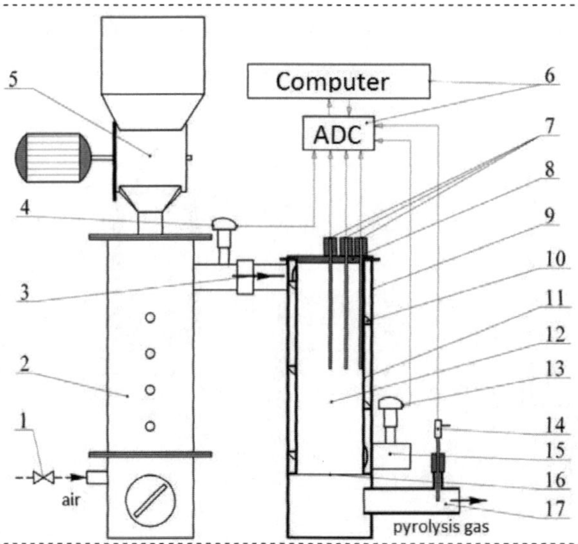

Figure 2. Experimental setup scheme for determining the process parameters of the pyrolysis of wood waste

The installation includes a furnace 2 with fuel feed hopper 5, waste conductive heating reactor 12, the gas sampler 14, thermocouples 4, 7 for temperature control, valve 1 for controlling the air supply to the furnace 2 and to the system of registration and data storage 6. The conductive heating reactor 12 has a cover 8 with thermocouples fixed on it 7, nozzles for supplying 3 and for removing 15 the heating flue gas generated during the combustion in the furnace 2, nozzle 17 for discharging the gas-vapour mixture from the heating chamber 11, which is a cylinder with the grate 16 on the bottom of it which prevents spillage of the material to the bottom of the reactor. Between the housing 9 and heating chamber 11 there is a helical guide 10 which serves to distribute the fuel gas and to wash the heating chamber's wall 11 by gas flow. The structural elements of the reactor are made of heat-resistant stainless steel. To prevent heat losses the unit has a double layer of insulation [8].

Experiments to determine the weight loss dependence on the temperature in the pyrolysis zone are carried out at temperatures ranging from 350 to 650 °C in increments of 50 °C on the samples having an initial moisture content of 10 %. During the pyrolysis process changing the temperature of the flue gases at the inlet and outlet of the reactor, respectively is controlled by thermocouples 4 and 7. The temperature of flue gases entering the reactor 12 is controlled by changing the air flow in the blast furnace 2. At the time of the parameters stabilization of the pyrolysis process by the probe 14, the pyrolysis gas being selected by samples, measured on the temperature and determined on chemical composition. Coal samples are subjected to chemical analysis to determine the content of the amount of volatile materials and pyrolysis gas concentrations in the coal.

An experimental installation was developed and built for the physical modeling and the study of heterogeneous chemical reactions occurring in the reduction zone of the gasification reactor which is shown in diagram of figure 3. The experimental installation consists of a loading hopper 26, the reactor 28, isolated from each other by vane feeder 27, gas supply system 11. Measurement of parameters and the management of the experimental installation are carried out via temperature sensors, devices for determining the gas flow, pressure reducers, closing and regulatory insert. Registration of data from all sensors and devices is done by the system of collection and storage of information, the gas analyzer 3, a calorimetric bomb 43.

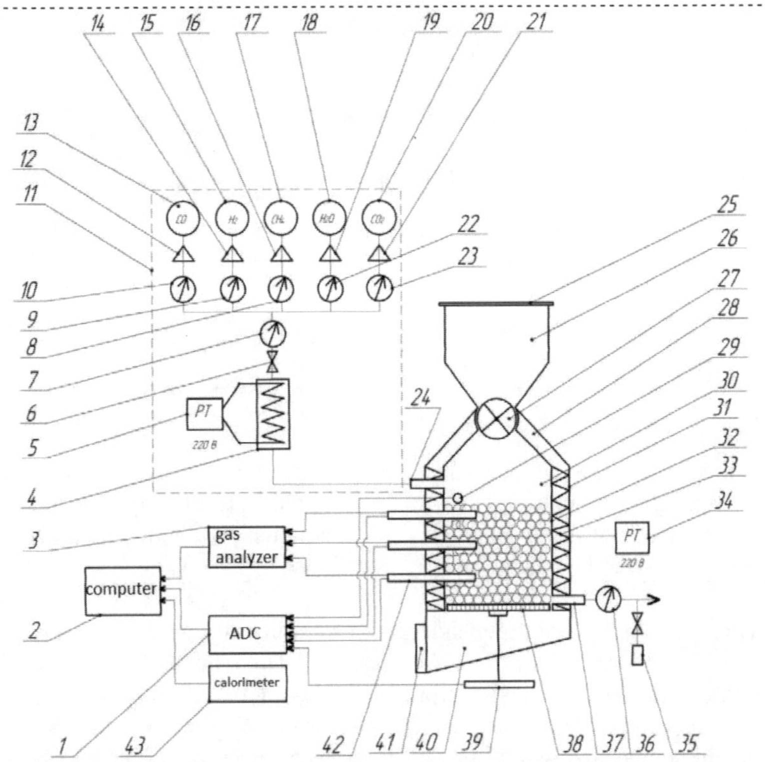

Figure 3. Installation of the study of reduction processes.

A gas supply system 11 is a system with a gas cylinder 13 (*SB*) 15 (H_2), 17 (CH_4) 18 (H_2O) 20 (CO_2), where the gas of predetermined pressure and composition is supplied to the heating element 4. The volume of the gas fed in to heating element is regulated by a flow meter 7 and a locking mechanism 6. The temperature of the heating element is regulated by a temperature controller 5. The heated gas with the temperature of 400-900 °C is sent into the reactor 28 and is determined by a pressure sensor 29. The gas selected from the sampler 35 for research of calorific value is sent in to the calorimetric bomb 43. The weight change of the material due to its response is fixed by means of strain gauge 39. Continuously recorded sample weight change during the experiment allows to determine the kinetics of thermal decomposition. The feed rate of gas into the chamber is regulated by the locking mechanism 44 and 48. The number of inlet and exhaust gas is fixed by flowmeters 43 and 47. During the study of dependance on the composition and quantity of gas from the particle diameter and the porosity of the layer the charcoal various fractions are prepared.

3. Results and Discussions

At a result of data obtained during the experiments, characteristic curves describing the process of energy-technological processing of wood waste at various regime parameters and at different stages of processing are constructed. The results are shown for wood waste of different fractional composition with different initial moisture content.

In the study of pyrolysis experimental data of layer's loss in weight, change in the layer's temperature, process time, the content of volatile substance in the coal residue.

The dependence of the specific gravity decrease on the process temperature in the pyrolysis zone, shown in figure 4 which shows that during the increase of process temperature the coal mass yield drops. This is due to higher degree of release of volatile substances contained in the coal mass [9]. The dependences on the mass loss of cellulose and hemicellulose in the pyrolyzed material on the temperature [10, 11], obtained by treatment of experimental data are also displayed in this figure.

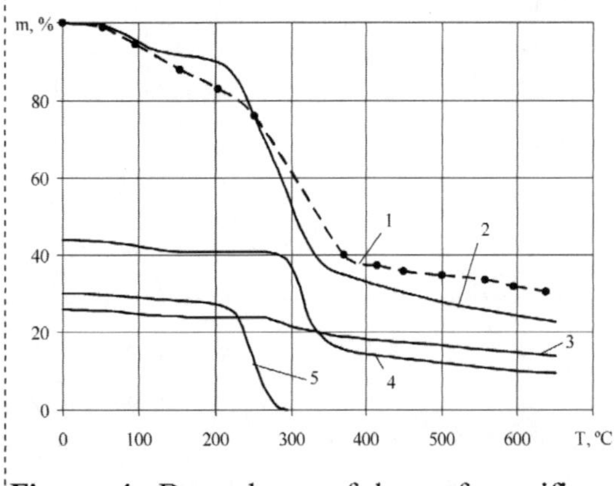

Figure 4. Dependence of loss of specific weight on the heating temperature: 1 - material (e); 2 - material (p), 3 - lignin (p) , 4 - cellulose (p), 5 - hemicellulose (p)

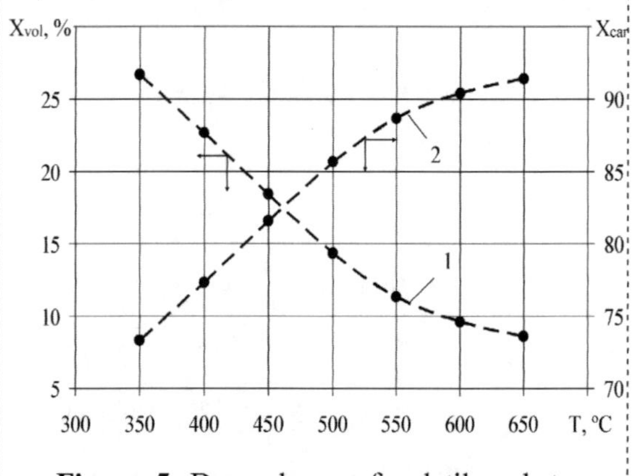

Figure 5. Dependence of volatile substances (curve 1) and carbon (curve 2) in the coke residue on the temperature in the pyrolysis zone

The experiments also revealed the dependence of volatile substances and carbon in coal on the temperature in the pyrolysis zone, as reflected in figure 5. The data show that with increasing of the processing temperature the mass of coal output decreases, and the proportion of carbon contained therein increases. This is explained by a greater degree of thermal degradation of lignocellulosic complex at high temperatures and, consequently, the release of more volatile substances contained in the pulp. During the analysis of pyrolysis gas sample the dependence of the effect of temperature on the pyrolysis process on the pyrolysis gas composition which is shown in figure 6 was also obtained. This relationship shows that within an increase in temperature a significant increase of methane formation occurs [12]. It is associated with a great increase in the reaction rate of methane formation in relation to the other components [13]. However, methane growth rate decreases at temperatures above 550 °C . The content of hydrogen also increases slightly with increasing of temperature, and its quantity increases intensively at temperatures above 550 °C. This happens because of formed char burning and mainly excreted hydrogen. Also, there is a decrease in carbon dioxide, which is explained by the relatively low temperature of pyrolysis process flow and a comparative increase in the content of the other components of the pyrolysis gas.

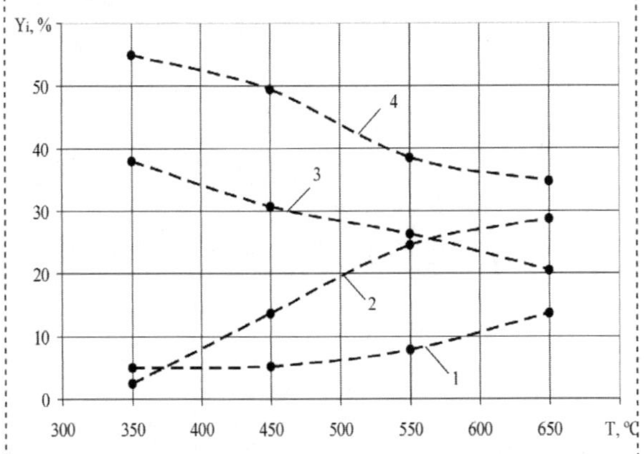

Figure 6. The composition dependence on the temperature of pyrolysis gas in the pyrolysis zone 1 - H_2; 2 - CH_4; 3 - CO; 4 - CO_2

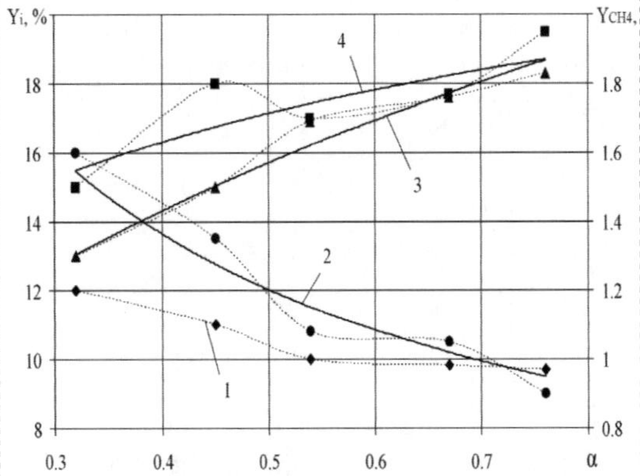

Figure 7. The dependence of the produced gas composition on oxidizer flow in the combustion zone of the gasifier 1 - CH_4; 2 - CO_2; 3 - SO; 4 - H_2

Experimental dependence of figure 7 shows that increasing the supply of oxidant into the combustion zone leads to a significant increase in temperature. This is because when the oxidizer flow rate and its outflow through the tuyere changes i.e. the gas flow rate in the combustion zone, rising the temperature on the surface of the waste layer in the incoming oxidant stream increases. In this case the reaction rates of interaction between the waste and the oxidant significantly increase. The intensity of the gasification process also has an impact on the composition and calorific value of the produced gas. The intensification of the process by increasing the oxidant flow results in an increase of the combustible components of carbon monoxide and hydrogen and the reduction of not recovered carbon dioxide, due to displacement of the equilibrium constants of the endothermic reduction reactions of carbon dioxide and the decomposition of water in the direction of increasing the content in the produced gas of carbon monoxide and hydrogen, connected, as it was noted above, with the increasing of the temperature and the rate of chemical reactions in combustion and reduction zones [14]. Furthermore, an increase in temperature in combustion and reduction zones leads to almost complete decomposition of the resulting resin, which improves the quality of the produced gas, but it increases the amount of ablative dust and the pressure resistance of the gas generator [15].

The study of the effect of reduction zone height on the parameters of the produced gas showed a significant value of this parameter during the gasification process, as it determines the contact time of gases and water vapour with coal made red-hot by carbon, which has a significant impact on the quality of the product gas [16].

The study results, shown in figure 8 indicate that with increasing the height of the reduction zone to 125-135 mm the content of combustible part of produced gas increases, and hence its calorific value. However, further increase in the height (over 135 mm) is followed by reduction of the quality of the produced gas, due to reducing the amount of carbon dioxide associated with the drop in temperature at the end of the reduction zone and the flow of reverse reactions. The quantity of hydrogen increases slightly. The reason is that the height of the reduction zone depends on the temperature at the end of the reduction zone. In addition, the gas flow rate substantively affects on the value of the height of the reduction zone. Thus, with increasing gas flow rate there is an increase in the height of the reduction zone. This is due to inadequate hold-up time of the gas stream in the reduction zone, which leads to its increase. It is also noted that the particle size influences at the increase of temperature and reaction ability of the waste. The smaller the particle size, the lower the height of the recovery zone, which provides the completion of reduction reactions.

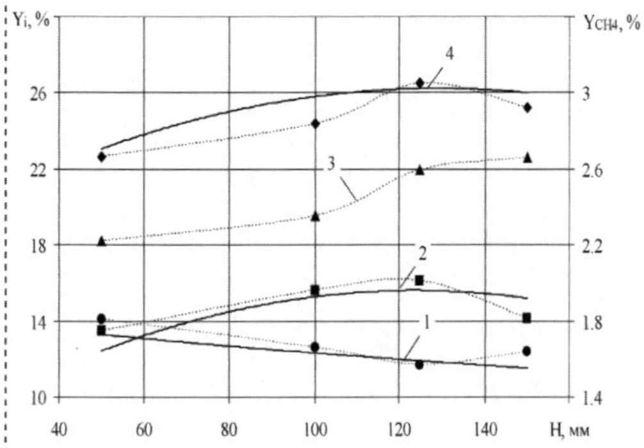

Figure 8. The dependence of composition of the produced gas from the height of the recovery zone 1 - CO_2; 2 - H_2; 3 - CH_4; 4 - CO

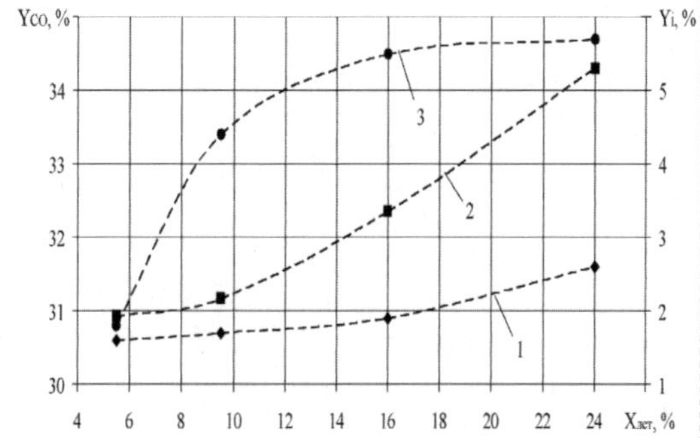

Figure 9. A produced gas composition dependence on the content of volatile substances in coal in the reduction zone 1 - CH_4; 2 - CO; 3 - H_2

The obtained results let us to suggest that the volatile substances contained in the coal in the reduction zone can influence on the quality of the produced gas. To confirm the hypothesis the study was conducted to determine the effect of volatile substances' content in coal in the reduction zone on the component composition and the quality of the produced gas. The obtained data are reflected in figures 9. The graphs show that the main combustibles content of the produced gas, with an increase of volatile substances in coal from 5 to 24 %, increases. As a result, the calorific value of the produced gas is also increased. However, the experience shows that a further increase in the content of volatile substances in the coal leads to the formation and growth of undecomposed tar content in the produced gas.

A generalized analysis of these dependences shows that even at a temperature of 400 °C in the pyrolysis zone, there is an achieved optimum value of the content of volatile substances produced in coal mass, but due to minor changes in the limits of the produced gas composition it was concluded that this option has no significant effect on the entire process of gasification. It is also noted that this option is difficult to manage and almost impossible to control.

4. Conclusions

Analysis of the results of physical modeling process of energy technological processing of wood waste showed that the volatile substances' content in the coal, the oxidant flow in the combustion zone and the height of the reduction zone are the main parameters that affect the work efficiency of the gasifier unit. The height of the reduction zone in co-current gasification depends on the flow rate, concentration and the oxidant temperature in the combustion zone and the amount of volatile substances in coal in the reduction zone. Increasing the height of the reduction zone, first, helps to increase in calorific value of produced gas, however, further increase reduces the calorific value of the produced gas. The duration of the thermochemical decomposition in the pyrolysis zone depends on the fractional composition of wood waste.

As a result of physical modeling are determined the following points: the optimum height of the drying zone and the pyrolysis depending on the initial humidity of wood waste, coal supply and pyrolysis gases generated during pyrolysis, the optimal parameters of the combustion zone, the effective height of the reduction zone and the parameters of the produced gas at the outlet of the installation.

An overall analysis of the data reveals the optimal modes of process of energy-technological processing of wood waste with the use of direct-flow gasification to develop the technology of thermochemical processing of wood waste using the gasification method and the project the gas generators' constructions, ensuring high quality of the received synthesis gas.

References

[1] Hu G., Li G., Zheng Y., Zhang Z., and Xu Y. 2015 Euler–Lagrange modeling of wood chip gasification in a small-scale gasifier. *Journal of the Energy Institute* **88**(3): 314-322.

[2] Sadrtdinov A.R., Safin R.G., Gerasimov M.K., Petrov V.I., and Gilfanov K.K. 2016 The mathematical description of the gasification process of woody biomass in installations with a plasma heat source for producing synthesis gas. *IOP Conference Series: Materials Science and Engineering* **124**(1), 012092. IOP Publishing. DOI:10.1088/1757-899X/124/1/012092

[3] Saravanakumar A., Haridasan T.M. and Reed T.B. 2010 Flaming pyrolysis model of the fixed bed cross draft long-stick wood gasifier *Fuel Processing Technology* **91**(6): 669-675. DOI:10.1016/j.fuproc.2010.01.016

[4] Janajreh I. and Al Shrah M. 2013 Numerical and experimental investigation of downdraft gasification of wood chips. *Energy Conversion and Managemen*, **65**: 783-792.

[5] Timerbaev N.F., Safin R.R., Safin R.G. and Ziatdinova D.F. 2014 Modeling of the process of energy-technological treatment of wood waste by method of direct-flow gasification *Journal of Engineering and Applied Sciences* **9**(5): 141-146

[6] Tuntsev D.V., Filippova F.M., Khismatov R.G., and Timerbaev N.F. 2014 Pyrolyzates: Products of plant biomass fast pyrolysis. *Russian Journal of Applied Chemistry*, **87**(9): 1367-1370.

[7] Pakdel H. and Roy C. 1991 Hydrocarbon content of liquid products and tar from pyrolysis and gasification of wood. *Energy & Fuels* **5**(3): 427-436.

[8] Saravanakumar A., Haridasan T.M., Reed T.B., and Bai R.K. 2007 Experimental investigation and modelling study of long stick wood gasification in a top lit updraft fixed bed gasifier. *Fuel* **86**(17): 2846-2856.

[9] Mahapatra S., Kumar S. and Dasappa S. 2016 Gasification of wood particles in a co-current packed bed: Experiments and model analysis *Fuel Processing Technology* **145**: 76-89.

[10] Bui T., Loof R. and Bhattacharya S.C. 1994 Multi-stage reactor for thermal gasification of wood *Energy* **19**(4): 397-404.

[11] Sadrtdinov A.R., Sattarova Z.G., Prosvirnikov D.B. and Tuntsev D.V. (2015, December) Modeling of thermal treatment of wood waste in the gasifiers. In *2015 Int. Conf. on Mechanical Engineering, Automation and Control Systems (MEACS)* pp 1-5. DOI:10.1109/MEACS.2015.7414914

[12] Dasappa S., Paul P.J., Mukunda H.S. and Shrinivasa U. 1994 The gasification of wood-char spheres in CO2 N2 mixtures: analysis and experiments *Chemical Engineering Science* **49**(2): 223-232.

[13] Bui T., Loof R., and Bhattacharya S.C. 1994 Multi-stage reactor for thermal gasification of wood. *Energy* **19**(4): 397-404.

[14] Lashkov V.A., Sattarova Z.G., Taymarov M.A., Gerasimov M.K. and Halitov R.A. 2016 Modeling of a reduction zone of the gasifier installation *IOP Conference Series: Materials Science and Engineering* **124**(1), 012111. IOP Publishing. DOI:10.1088/1757-899X/124/1/012111

[15] Palmer E.R. 1984 Gasification of wood for methanol production *Energy in agriculture* **3**: 363-375.

[16] Prosvirnikov D.B., Safin R.G., Ziatdinova D.F., Timerbaev N.F., and Lashkov V.A. 2016 Multifactorial modelling of high-temperature treatment of timber in the saturated water steam medium. *IOP Conference Series: Materials Science and Engineering* **124**(1), 012088. IOP Publishing. DOI:10.1088/1757-899X/124/1/012088

IOP Conf. Series: Materials Science and Engineering **142** (2016) 012097 doi:10.1088/1757-899X/142/1/012097

Equipment for the Production of Wood-Polymeric Thermal Insulation Materials

Vladimir A Saldaev, Dmitry B Prosvirnikov[1], Vladislav V Stepanov, Almaz R Sadrtdinov, Alexey N Kapustin[2,b]

[1] Institute of petroleum and chemical engineering, Kazan national research technological University, K. Marx Street 68, Kazan, Republic of Tatarstan, Russian Federation, 420015
[2] Yurga Institute of Technology, National Research Tomsk Polytechnic University, Leningradskaya street 26, Yurga, Russian Federation, 652055

E-mail: [a]c-vova@mail.ru; [b]lexcol@tpu.ru

Abstract. This article presents developed pilot-plant equipment for slabby patterns of wood-filled polyurethane foam insulation material and its specifications are presented. Based on the results of experimental studies of pilot models the allowable range of equipment's technological parameters was defined.

1. Introduction

Thermal insulation materials are the structural elements that reduce the rate and amount of heat transmitted through the layer of material, i.e. acting as main thermal resistance in the structure [1].

The need for thermal insulation materials is due, firstly, to the requirements of energy saving increase and a decrease of the energy consumption for space heating in winter and cooling in summer respectively [2]. Secondly, the outer insulation of the building provides ecqualized heat distribution throughout the whole volume of the facilities, thus there is no need for a progressive heating [3]. Third, the insulation improves indoor climate in winter time, preventing the formation of condensation and mold growth on the inner walls of the facilities [4]. Fourth, the insulation increases the service life of the building and its supporting elements, which significantly affects the cost of the building when sale or rent [5].

Currently when the thermal insulation of buildings and structures mainly mineral thermal insulation (fiberglass, rock wool, etc.) and gas-filled polymeric insulation materials (polystyrene, polyurethane foam, etc.) with high costs are used [2, 6, 7]. In recent years, there are insulation materials obtained by processing of unmerchantable wood – wood-fiber insulation, particleboard, cement particle boards that rapidly become very popular [8, 9, 10, 11]. However, from the point of modern requirements, they do not have sufficient heat-shielding properties and a level of hydrophobicity. In addition, nowadays there are no clear recommendations what should be the composition, conditions and technology of production of efficient thermal insulation materials, produced on the basis of shredded wood and wood waste [12, 13, 14].

Based on the above, the actual direction is the development and creation of new technologies for the production of thermal insulation materials in order to reduce their costs and improvement of the thermal insulation properties.

2. Case study

Given the high cost and unique properties of the polyurethane foam as insulation material, as well as a large amount of wood waste from wood processing complex, it was asked to fill polyurethane organic fillers, which are selected as shredded wood waste [15, 16]. The result was a method of producing wood-filled with insulating material. The method consists in mixing the filler and binder, followed by molding and curing. The wood is used as a filler pulpchips thickness of 5 ±2 mm, length 10–35 mm. The polyurethane foam used as the binder obtained from two components – a polyol and a polyisocyanate. Previously, the binder components are mixed, then binder is mixed with filler by layering (binder - filler - binder) to form at a ratio of mixture components (wt. %): polyol 22–24, isocyanate 33–36, pulp chips 40–45. After complete feed of components, the form is fixed by latch and held for 15–20 minutes. By this method there were prepared samples of easy and effective thermal insulation material based on wood waste with a high density, but the thermal conductivity properties were inferior to the regular polyurethane foam slabs, resulting in increased thickness of the plates [17, 18, 19].

In order to reduce the density of wood-filled polyurethane foam while maintaining its fractional composition the experimental research on the basis of the laboratory of the department of wood processing materials of FSEIHE "KNRTU" was carried out. The results showed that exposure to a vacuum of 0.3 bar effectively affects the foaming process the wood-filled polyurethane. Figure 1 shows the results obtained by determining the density of material depending on the weight ratio of components of the composition at atmospheric pressure (figure 1a) and a vacuum of 0.3 bar (figure 1b).

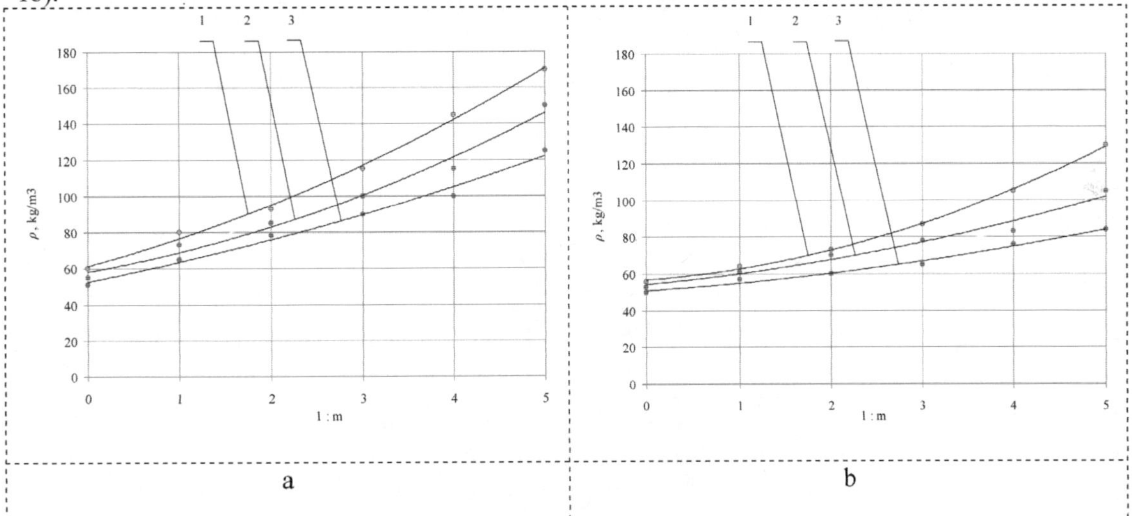

Figure 1. The density of the samples of wood-polymer thermal insulation material, depending on the mass ratio of the components "A – polyol:", "B – polyisocyanate" and the degree of dilution (ratio of "A" and "B" components: 1 – 1:1.1; 2 – 1:1.3; 3 – 1:1.5): a – 1 bar; b – 0.3 bar

The presented data show that the formation of the material at a vacuum of 0.3 bar with an increase in the proportion of wood filler densities increase (at a weight ratio of polyurethane to the wood filler 1:5 (ρ = 130 kg/m^3) – figure 1b, position 1), but one order less than the atmospheric pressure (at a weight ratio of polyurethane to the wood filler 1:5 (ρ = 160 kg/m^3) – figure 1a, position 1). From the data obtained it can be concluded that the introduction of wood filler in a polyurethane foam matrix, to achieve the desired density of the material, a necessary condition during its formation is the presence of vacuum.

Initially the density range of created wood-polymer insulation material is set as ρ = 60–80 kg/m^3. That means that in order to achieve the specified numbers, the following conditions must be observed during the formation of the composition:

- the weight ratio of polyurethane to the wood filler: from 1:2 to 1:4;
- the ratio of polyurethane foam components (polyol: polyisocyanate) from 1:1.3 to 1:1.5;
- the degree of vacuum: 0.3 bar.

Thus, foaming under vacuum enables cost reduction according to the standard slabs of polyurethane foam in 2.5 times.

Based on the research the pilot plant for wood-filled with polyurethane foam insulation material has been developed. The scientific novelty of this development is a patented method of layering components and the formation of plate products under vacuum (Pat. Russian RU2538004 (C1)).

3. Description of the developed object

The construction of pilot production unit is shown in figure 2.

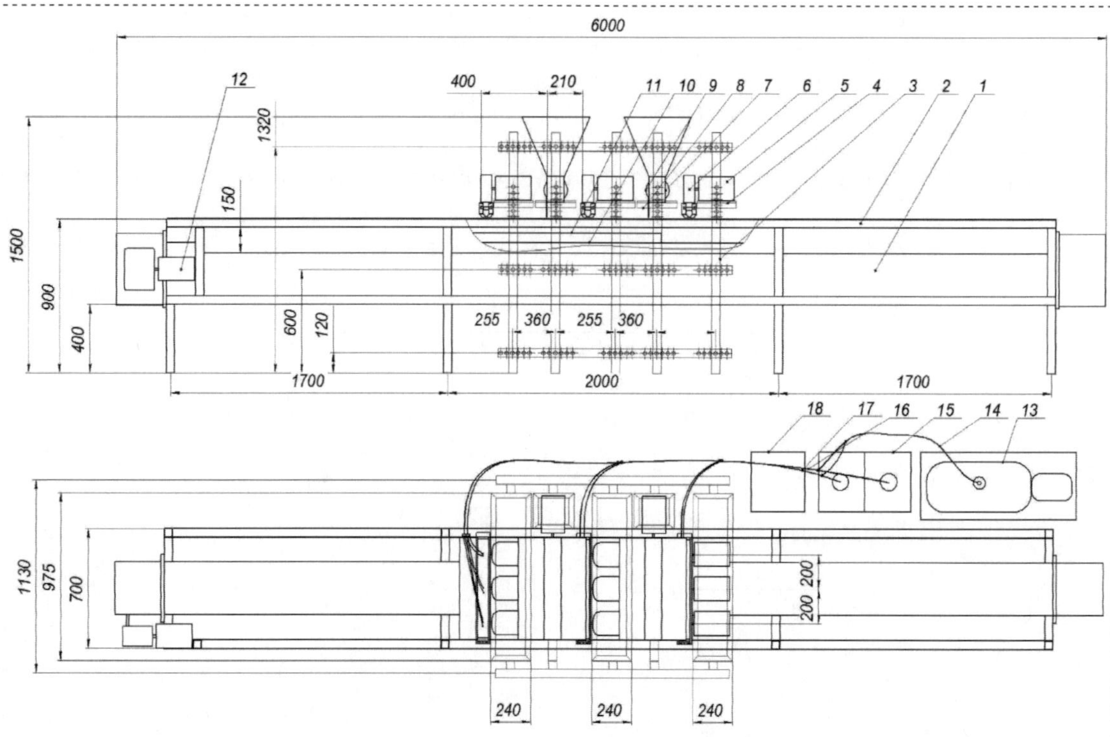

Figure 2. Sketch of the laboratory setup producing wood-filled thermal insulation material: 1 – conveyor, 2 – the conveyor frame, 3 – inlet unit frame, 4 – engine bearer, 5 – electric motor, 6 – filling head, 7 – chip batcher, 8 – batcher engine bearer, 9 – lock-gate hatch drive, 10 – conveyor belt, 11 – form, 12 – conveyor drive, 13 – compressor, 14 – compressed air hose, 15 – foamed polyurethane production machine, 16 – component "A" supply hose, 17 – component "B" supply hose, 18 – the vacuum pump.

The principle of operation of the unit is to organize the sequential process steps in the following order: a special form 11 for slabby samples of wood-filled polyurethane foam is placed on the belt 10 of conveyor 1, mounted on a frame 2 with boards to organize the movement of forms. The conveyor runs from the actuator 12. The form 11 is moving on the conveyor belt from left to right. At the beginning and end of the form the laser sensors are set for transmitting the response on signal sensors mounted on the filling heads 6 and lock feeders 7. The filling heads for feeding a mixture of polyurethane components operate from the actuator 9 and mounted on a support 4. The lock feeders are mounted under at the chip storage bins, placed on a support 8. The overshot assembled unit is placed on the frame 3. With the passage of the form under a first sealing head as a result of operation of sensors the compressor 13 supplying compressed air through a hose 14 to the infusion head is activated. At the same time with the compressor the pump of foam generator 15, the supplied component "A" through the hose 16, the component "B" through the hose 17 are turned on. The foaming in forms is carried out under an exhaustion of vacuum pump 18. The regulation of the flow of components and compressed air is carried out by automatic valves installed on the supply hose, and receiving signals from the sensors. Components fall in the filling head together with the compressed air, wherein are mixed by working from actuators mixers in the contact area. The mixer drives are switched simultaneously with the pump and the compressor. Since the mixing of the components in the filling head the countdown of polyurethane induction begins. The compressed air pressure pushes a liquid mixture of polyurethane components out through opening of the filling head in to the form of a continuous strip of a given thickness, which is given by the size of the flow cross section of the filling head. Thus, the first layer of polyurethane foam is filled in the form, the foaming start time is regulated by the equivalence ratio and its mixing speed in the contact zone of the head. When the form completely passes under the head, closing sensors are triggered, and the supply of component in the first filling head is stopped by closing the valve. The chip lock feeder works similarly, it is installed after the first filling head and performs the feeding of wood filler in to the form as a second layer. Likewise the second filling head is activated, the second chip lock feeder and the third filling head, which are switched off in the same order when passing beneath a laser off sensor, the latter off sensor doesn't only provide a signal to close the supply valves of the third filling head, but also disables pump of foam generator and compressor.

Thus there is a continuously layering of components of wood-filled polyurethane foam in a special form. The whole process of the technological cycle of layering is done during the induction of the polymer, that is, until the foaming of compositions starts spontaneously and a free volume remains in the form that will be filled during the time of foaming. After placing the components into the form, the form closing is performed with a special lid with the valve hole for supplying the vacuum pump hose, which performs the foaming and lifting of the composition till the predetermined volume is defined by the form after evacuating of air and creating of a vacuum in the form. After vacuuming of the form the cover valve hole is sealed and the form is sent to holding until test sample of wood-filled polyurethane foam thermal insulation material is fully cured.

The automatization of setup is presented by laser sensors mounted on the ends of the form, filling heads and chip lock-feeder. The optical position sensor with integrated microprocessor control system is made for use in monitoring of the alignment control and collocation of distant objects systems. It allows you to accurately measure the movement of the object under control without mechanical contact. Also the automation system includes the system of maintaining at a given temperature of components "A" and "B". It is possible to adjust the parameters that affect the final properties of the resulting thermal insulation material on the developed setup.

The experimental study on separate laboratory benches allowed to establish the boundaries of an experimental study of the parameters of the process and the final physical, mechanical and performance properties of wood-filled polyurethane foam, suitable for the manufacture of thermal insulation slabs [20].

The ranges of variation of process parameters on the setup are determined based on the allowable range of property variation of the resulting material. When you create a pilot plant to produce wood-filled insulation material it is necessary to set the allowable range of the basic physical, mechanical and performance properties of wood-filled polyurethane foam insulation material [21, 22, 23]:

- Thermal conductivity: 0.025–0.035 W/m·K;
- Density: 60–80 kg/m³;
- Flexural strength: 0.55–0.65 MPa;
- Absorption of water by volume: 1.4–1.6 %;
- Damaged length in combustion: 70–85 %;
- Flue gas temperature in combustion 150–200 °C;
- Damaged weight in combustion: 43–58 %;
- Time of independent combustion: 20–30 sec.

The ranges of variation of process parameters on the setup based on the results of the research were determined on a given range of properties of wood-filled polyurethane foam:

- the ratio of polyurethane foam components: chips – from 1:2 to 1:4 (governed engine RPM airlock feeder 9);
- the ratio of components a polyol : polyisocyanate – from 1:1.2 to 1:1.5 (adjusted in foam generator 15);
- vacuum - from 0.2 to 0.4 bar (regulated by the processing of vacuum pump 18);
- fire retarder flow – from 0.7 to 0.1 l/kg (regulated by engine speed of the fire retarder feeder);
- time of filling mixture to forms- from 22 to 37 sec (is regulated by capacity and the pressure of compressor 13).

4. Conclusions

Thus, the development of technology for wood-filled thermal insulation material is to determine the allowable range of process parameters that determine the rules of experimental studies on the experimental setup. These ranges allow to run the numbers in a right way, establish the limits of experimental variation and to carry a number of other research studies on the setup, purpose of which will be to determine the optimal regime parameters of the process of obtaining slabby wood-filled thermal insulation material.

Acknowledgments

The work was performed as part of the program "Start - 1" of Fund for Assistance to Small Innovative Enterprises in the scientific and technical sphere on a theme: "Development of technology and pilot production unit for wood-filled thermal insulation materials" (contract number 422GS1/9650).

References

[1] Yang C., Fischer L., Maranda S. & Worlitschek J. 2015 Rigid polyurethane foams incorporated with phase change materials: A state-of-the-art review and future research pathways *Energy and Buildings* **87**: 25-36. DOI:10.1016/j.cryogenics.2015.10.008

[2] Biswas K., Shrestha S.S., Bhandari M.S. & Desjarlais A.O. 2016 Insulation materials for commercial buildings in North America: An assessment of lifetime energy and environmental impacts *Energy and Buildings* **112**: 256-269. DOI:10.1016/j.enbuild.2015.12.013

[3] Moretti E., Belloni E. & Agosti F. 2016 Innovative mineral fiber insulation panels for buildings: Thermal and acoustic characterization *Applied Energy* **169**: 421-432. DOI:10.1016/j.apenergy.2016.02.048

[4] Sierra-Pérez J., Boschmonart-Rives J. & Gabarrell X. 2015 Environmental assessment of façade-building systems and thermal insulation materials for different climatic conditions *Journal of Cleaner Production* **113**: 102-113. DOI:10.1016/j.jclepro.2015.11.090

[5] Rehman H.U. 2016 Experimental performance evaluation of solid concrete and dry insulation

materials for passive buildings in hot and humid climatic conditions *Applied Energy* DOI:10.1016/j.apenergy.2016.01.026

[6] Roberts B.C., Webber M.E. & Ezekoye O.A. 2015 Development of a multi-objective optimization tool for selecting thermal insulation materials in sustainable designs *Energy and Buildings* **105:** 358-367. DOI:10.1016/j.enbuild.2015.07.063

[7] Sun Z., Shen Z., Ma S. & Zhang X. 2013 Novel application of glass fibers recovered from waste printed circuit boards as sound and thermal insulation material *Journal of materials engineering and performance* **22** (10): 3140-3146. DOI:10.1007/s11665-013-0587-y

[8] Matuana L.M., Stark N.M., Wacker J.P., Brashaw B.K., Jalinoos F., Bergman R., ... & Bergman R.D. 2015 The use of wood fibers as reinforcements in composites *Environmental Entomology* **44** (3): 890-897. DOI:10.1533/9781782421276.5.648

[9] Binici H. & Aksogan O. 2016 Eco-friendly insulation material production with waste olive seeds, ground PVC and wood chips *Journal of Building Engineering* **5**: 260-266. DOI:10.1016/j.jobe.2016.01.008

[10] Prosvirnikov D.B., Ziatdinova D.F., Timerbaev N.F., Saldaev V.A. & Gilfanov K.H. 2016 Mathematical modelling of the steam explosion treatment process for pre-impregnated lignocellulosic material *IOP Conference Series: Materials Science and Engineering* **124**(1), 012087. IOP Publishing. DOI:10.1088/1757-899X/124/1/012087

[11] Shaaban A., Se S.M., Ibrahim I.M. & Ahsan Q. 2015 Preparation of rubber wood sawdust-based activated carbon and its use as a filler of polyurethane matrix composites for microwave absorption *New Carbon Materials* **30** (2): 167-175 DOI:10.1016/S1872-5805(15)60182-2

[12] Binici H., Aksogan O. & Demirhan C. 2016 Mechanical, thermal and acoustical characterizations of an insulation composite made of bio-based materials *Sustainable Cities and Society* **20:** 17-26. DOI:10.1016/j.scs.2015.09.004

[13] Merle J., Birot M., Deleuze H., Mitterer C., Carré H., & Charrier-El Bouhtoury F. 2016 New biobased foams from wood byproducts *Materials & Design* **91:** 186-192. DOI:10.1016/j.matdes.2015.11.076

[14] Antoniadou P., Giama E., Boemi S.N., Karlessi T., Santamouris M. & Papadopoulos A.M. 2015 Integrated evaluation of the performance of composite cool thermal insulation materials *Energy Procedia* **78**: 1581-1586. DOI:10.1016/j.egypro.2015.11.214

[15] Fomin, A. A., Gusev, V. G., Safin, R. G., & Safin, R. R. 2015 Dispersion of the margin removed in complex milling *Russian Engineering Research* **35** (6): 417-420. DOI:10.3103/S1068798X15060040

[16] Safin R.R., Khasanshin R.R., Timerbaeva A.L., & Safina A.V. 2015 Study of the physical and energy properties of fuel granules based on a thermomodified wood raw material *Journal of Engineering Physics and Thermophysics* **88** (4): 958-961. DOI:10.1007/s10891-015-1270-y

[17] Matsagar V.A. 2016 Comparative performance of composite sandwich panels and non-composite panels under blast loading *Materials and Structures* **49** (1-2): 611-629. DOI:10.1617/s11527-015-0523-8

[18] Malinovskaya T.D., Suslyaev V.I., Melentyev S.V., & Dorozhkin K.V. 2014 Electrophysical and thermophysical characteristics of a multifunctional composite polyurethane-based material *Russian Physics Journal* **57** (8): 1094-1098. DOI:10.1007/s11182-014-0348-x

[19] Balo F. 2015 Feasibility study of "green" insulation materials including tall oil: Environmental, economical and thermal properties *Energy and Buildings* **86**: 161-175. DOI:10.1016/j.enbuild.2014.09.027

[20] Hamilton A.R., Thomsen O.T., Madaleno L.A., Jensen L.R., Rauhe J.C. M. & Pyrz R. 2013 Evaluation of the anisotropic mechanical properties of reinforced polyurethane foams *Composites Science and Technology* **87**: 210-217. DOI:10.1016/j.compscitech.2013.08.013

[21] Aranguren M.I., Marcovich N.E. & Mosiewicki M.A. Mechanical performance of polyurethane (PU)-based biocomposites *Biocomposites, A volume in Woodhead Publishing Series in Composites Science and Engineering* pp 465-485. DOI:10.1016/B978-1-78242-373-

7.00010-X

[22] Palumbo M., Avellaneda J., & Lacasta A.M. 2015 Availability of crop by-products in Spain: new raw materials for natural thermal insulation *Resources, Conservation and Recycling* **99**: 1-6. DOI:10.1016/j.proeng.2013.08.133

[23] Diascorn N., Calas S., Sallee H., Achard P. & Rigacci A. 2015 Polyurethane aerogels synthesis for thermal insulation-textural, thermal and mechanical properties *The Journal of Supercritical Fluids* **106**: 76-84. DOI:10.1016/j.supflu.2015.05.012

IOP Conf. Series: Materials Science and Engineering **142** (2016) 012098 doi:10.1088/1757-899X/142/1/012098

Experimental Research Into Generation of Acoustic Emission Signals in the Process of Friction of Hadfield Steel Single Crystals

D V Lychagin[1,2*], A V Filippov[2,3], O S Novitskaia[3], E A Kolubaev[2,3], and O V Sizova[3]

[1] National Research Tomsk State University, Tomsk, Lenin av. 36, Russia
[2] National Research Tomsk Polytechnic University, Tomsk, Lenin av. 30, Russia
[3] Institute of Strenght Physics and Materials Sciences SB RAS, Tomsk, per. Academicheskii 2/4, Russia

E-mail: dvl-tomsk@mail.ru

Abstract. The results of experimental research into dry sliding friction of Hadfield steel single crystals involving registration of acoustic emission are presented in the paper. The images of friction surfaces of Hadfield steel single crystals and wear grooves of the counterbody surface made after completion of three serial experiments conducted under similar conditions and friction regimes are given. The relation of the acoustic emission waveform envelope to the changing friction factor is revealed. Amplitude-frequency characteristics of acoustic emission signal frames are determined on the base of Fast Fourier Transform and Short Time Fourier Transform during the run-in stage of tribounits and in the process of stable friction.

1.Introduction

There are chaotic and unsteady phenomena registered between two objects in the process of friction. A relative slip of objects in a tribounit results in intense interaction between micro-roughnesses of surfaces, coming in contact. Seizure, separation and shear of micro-roughness elements are typical for this interaction. Small dimensions of actual contact areas are the cause of a considerable growth of stresses and temperature on certain roughnesses peaks; furthering generation of intense deformation processes in the surface layer, and elastic strain waves in particular. Elastic strains are amongst the principal sources of acoustic signal generation [1]. The development of elastic variations in the process of sliding friction is connected with plastic shears and twinning, intergranular sliding in a polycrystalline object, phase transformations and fractures, involving micro-cracks and separation. In conditions of friction the surface layer of a metal is subject to changes, caused by the accumulation of defects, formation and further fracture of a defect structure layer, this sequence recurs then. As these phenomena occur, the elastic deformations are formed with a varying intensity, changing, therefore, the acoustic emission intensity as well [2].

* Corresponding author: D V **Lychagin** - dvl-tomsk@mail.ru

IOP Conf. Series: Materials Science and Engineering **142** (2016) 012098 doi:10.1088/1757-899X/142/1/012098

The detailed study, based on experimental testing, of all formation and fracture phases in the surface layer seems to be impossible, since the contact area of objects in a tribounit is inaccessible for direct observation. Hence, additional research procedures are required, capable of direct or indirect providing information on phenomena, occurring in a tribounit. One of these procedures is to register acoustic emission (AE) signals in the process of friction. Registration and analysis of AE signals is widely used to study strain and fracture in materials that are loaded statically and dynamically [1-2]. At the same time one should take into account the diverse spectrum of acoustic variations typical for the dynamic processes, such as friction. These phenomena include both processes related to the surface evolution under friction and variations of tribounit elements. The AE spectrum includes sound and ultrasound range frequencies, distinguished by random and periodic character, dependent on loading conditions and friction regimes [3]. A number of diverse phenomena with various frequency and intensity occur on the contact areas over the period of registering a single signal (frame) of acoustic emission [3].

The authors of papers [4-8] studied in details AE signals, arising in the process of friction. AE signals versus type (and quantity) of generating wear particles [4]; wear mechanism [5-7]; and the state of friction surface correlation dependence was determined [8]. However, despite the works mentioned above, additional experiments, reproducing various aspects of friction and material states in condition of frictional interaction are necessary because of diverse phenomena, occurring in the process of friction. One of the aspects in focus of researchers is strain hardening of metals and alloys. It is interesting to study as a typical example Hadfield steel, distinguished by the high level of strain hardening under static compression, tension, torsion, and impact test [9-16].

The relation of AE signals to friction specifics of polycrystalline Hadfield steel has been already studied [3, 17-18]. The authors [17] have revealed the influence of AE signal parameters on friction behavior and wear. When forming the surface layer fiction factor is increased together with the growing median frequency of AE signal, which is conditional on origination of high-frequency components in the signal spectrum. The friction factor, median frequency and AE energy signals diminish as the surface layer fractures and wear particles are separated. On the base of frequency and time analysis [18] it is revealed that quasi-periodic bursts of AE intensity are interrelated with formation of wear particles. Since analysis and interpretation of AE signals generated in the process of friction are complex, there is a necessity to search for experimental procedures, reproducing certain material behavior in conditions of friction. One of these procedures is based on the use of single crystals, which makes it possible to control their strain behavior via selecting the crystal-lattice orientation and loading conditions.

The aim of the work is to study how generation of AE signals is related to deformation processes in Hadfield steel single crystals, arising in the surface and subsurface layers under dry sliding friction.

2. Methods of research

Single crystals are grown by Bridgman technique in helium atmosphere and homogenized in a noble gas for 24 hours at 1373 K. The specimens were cut by an electroerosion machine. As cutting out is completed, orientation of contraction axis in a single crystals is $[\overline{10}\ 7\ \bar{1}]$ and that of friction axis is $[\bar{3}\ 4\ 2]$. The specimen dimensions: height (L) – 10 mm; width (b) and thickness (a) – 1.3 mm. Cut out specimens are stored in helium atmosphere for an hour at 1050°C and water-tempered to fix the austenite structure. The defect surface layer is removed via grinding. The hardness of a tempered single crystal is 180 HV. Sliding friction of single crystals is carried out by a tribotester TRIBOtechnik according to the scheme "pin – on – disk" (see Figure 1) at ambient temperature 25°C. The normal load is 4 N (2.3 MPa, taking into account the friction surface area); sliding speed is 0.1 m/s. Each of three trials lasts 3 hours. A counterbody is a steel 40X disk with hardened surface, the hardness of which is 36 HRC. The friction surface topography of Hadfield steel single crystals and wear grooves on the counterbody are studied by confocal laser scanning microscope LEXT OLS4100. The parameters of AE signals, generated in the process of sliding friction are analyzed by a registering module EYA-2 (ЭЯ-2) developed at Togliatti State University.

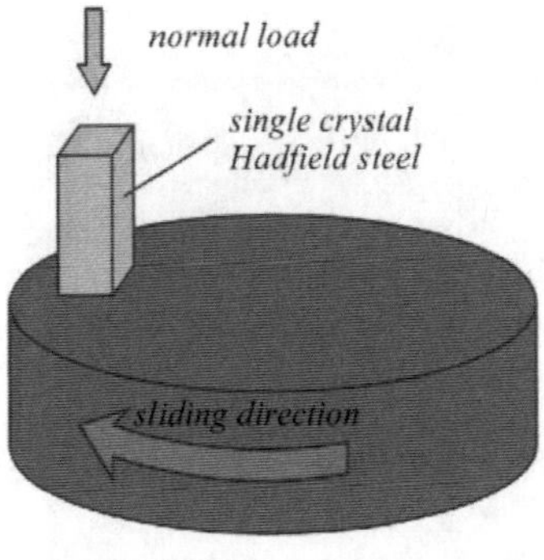

Figure 1. Scheme of «pin-on-disk» friction test

3. Results and discussions

The images of friction surfaces of Hadfield steel single crystal tested three times are shown in Figure 2.

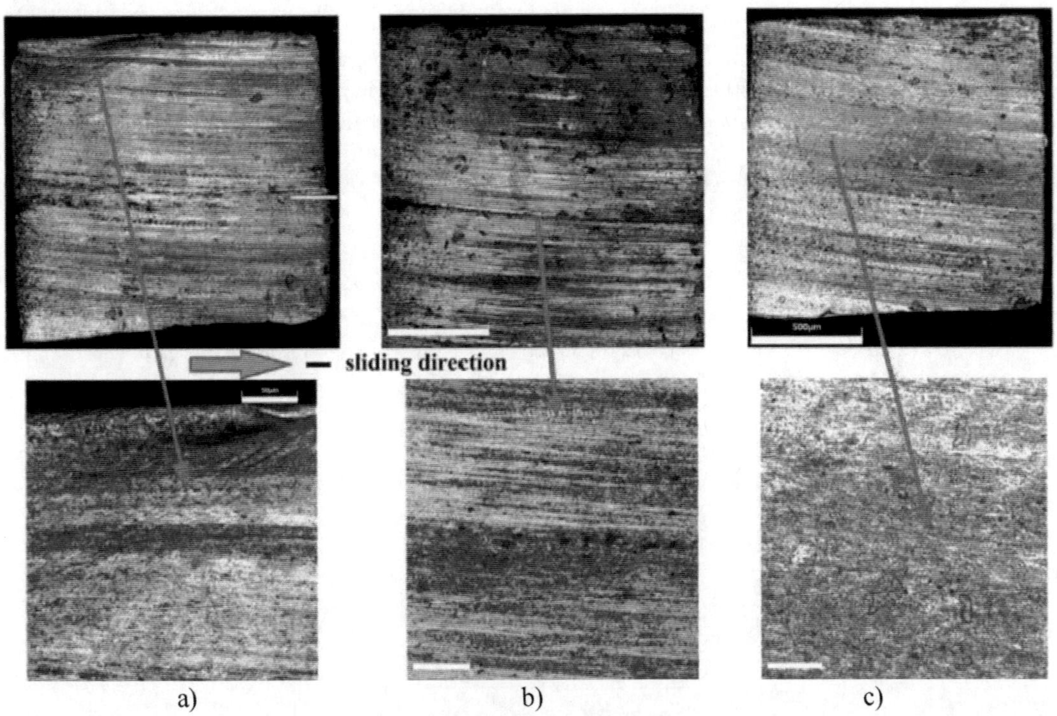

Figure 2. The friction surface of Hadfield steel single crystal: (a) - the first trial, (b) - the second trial, (c) the third trial

As the first trial is completed, there is a small area with strained relief on the friction surface. The surface roughness Sa=0.12 μm. After the second trial there are no strain lines (wrinkles) on the tested surface, the surface roughness rises up to Sa=0.35 μm. Moreover, there is some craterlike wear debris on the surface, which can be referred to oxidative fatigue wear. After the third trial abrasive and oxidative wear scars can be seen on the tested surface as well. The surface roughness Sa=0.5 μm.

The images of friction track surface on the counterbody after three trials are presented in Figure 3, the heights are distributed and typical profile of its cross section is given.

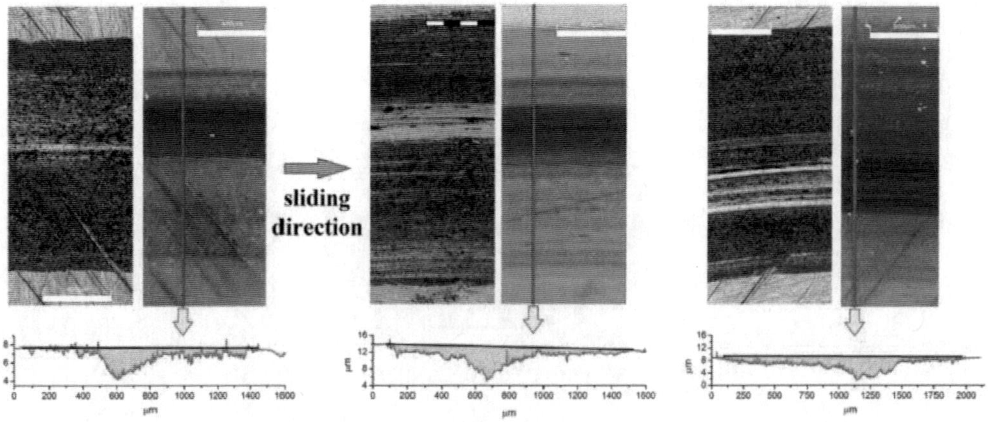

Figure 3. Images of the friction track, distribution of heights and cross section profile after the first (a), second (b) and third (c) trials.

The friction track profile on the counterbody is a non-uniform one in the cross section after the first trial, its area is 1100 μm^2, and the maximal depth is 3.6 μm. After the second trial the sectional area of the friction track rises up to 3500 μm^2, and the depth is 9.8 μm. The third trial results in the increase of the friction track sectional area up to 4400 μm ^{2}, although the profile depth diminishes to 7 μm. The sectional area of the friction track increases due to gradual but non-uniform wear of the counterbody surface. The decrease of the track depth is conditional, in its turn, on more uniform wear depth distribution in the track section as testing gets longer.

The graphs, displaying varying friction factor value and AE waveform envelope are given in Figure 4.

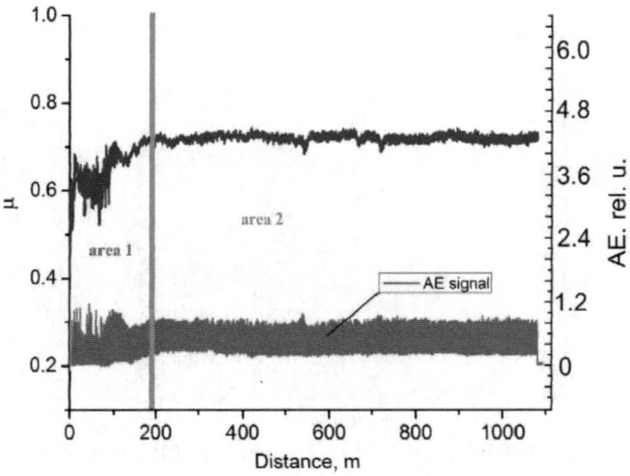

Figure 4. The change of friction factor over the three sequential trials and typical form of the waveform envelope of AE signal

Significant variations of friction factor values and AE signals are typical for the run-in area («area 1» see Figure 4). The value of AE signal waveform envelope rises because of the growing number of deformation events when running-in a tribounit. These strains are caused by intense formation and fracture of contact areas, separation and detachment of wear particles. In the area of stable friction («area 2» see Figure 4) slight variations of friction factor and AE signal are registered.

Periodic variations of friction factor are interconnected with the oxidative mechanism of the specimen wear. In the process of friction the surface layer of a specimen is deformed and a fine strain structure is formed. Accumulation of some strain results in failure of this layer and separation of tiny oxide and metal particles (see Figure 5).

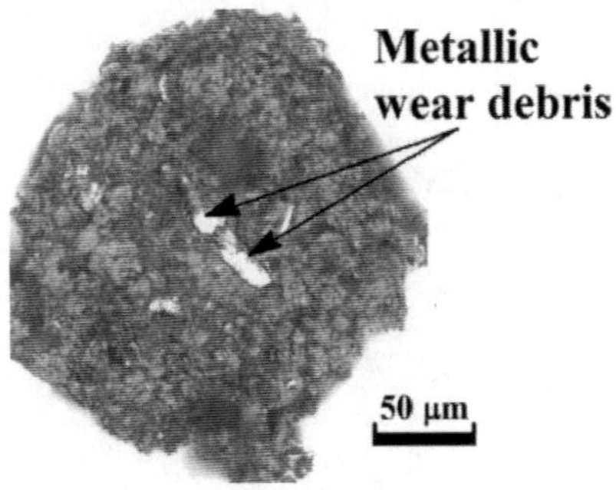

Figure 5. Optical image of wear particles

When analyzing frames of the registered AE signal in more details, it is revealed that type and amplitude-frequency characteristics of signals differ much in the areas of run-in and stable friction. A typical AE signal registered in the run-in area of a tribounit, intensity of the spectrum power and varying central frequency of signal are shown in Figure 6. Central frequency varies in the range 130...225 kHz. The highest intensity of the spectrum power (50 rel. u.) is in the frequency range ~175 kHz. In the area of stable friction the frame of AE signal consists of a sequence of AE peaks with various amplitudes (see Figure 7). The highest spectrum power intensity of this frame is twice as low as that of the frame in the run-in area of a tribounit. The central frequency of AE signal frame is in the range 125...305 kHz. The frame segment with a highlighted sequence of peaks agrees with the area of falling central frequency of signal (see signal area in Figure 7).

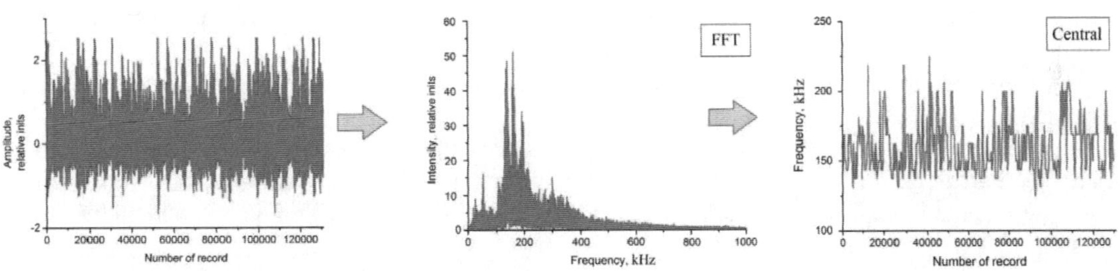

Figure 6. Typical AE signal in the run-in area in the process of Hadfield steel single crystals

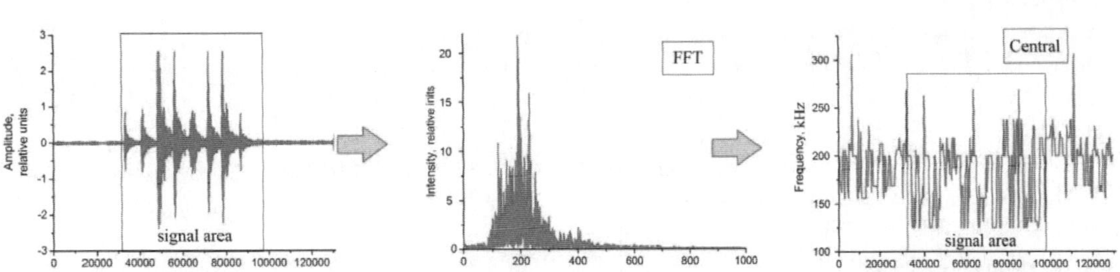

Figure 7. Typical AE signal in the stable friction area of Hadfield steel single crystals

The frequency-time analysis based on STFT (Short Time Fourier Transform) provides a more detailed visual image of varying frame frequency of AE signal over the period of its registration. Images of STFT AE signals in the areas of run-in and stable friction are given in Figure 8. As one can see in the presented graphs, frame frequencies of AE signal in the run-in area are distributed quite uniformly with well pronounced peaks over the whole period of time. At the same time only two pronounced AE signal frequency bursts are registered in the area of stable friction, they decay quickly in the frequency range 250…500 kHz, but drag on up to 1 MHz.

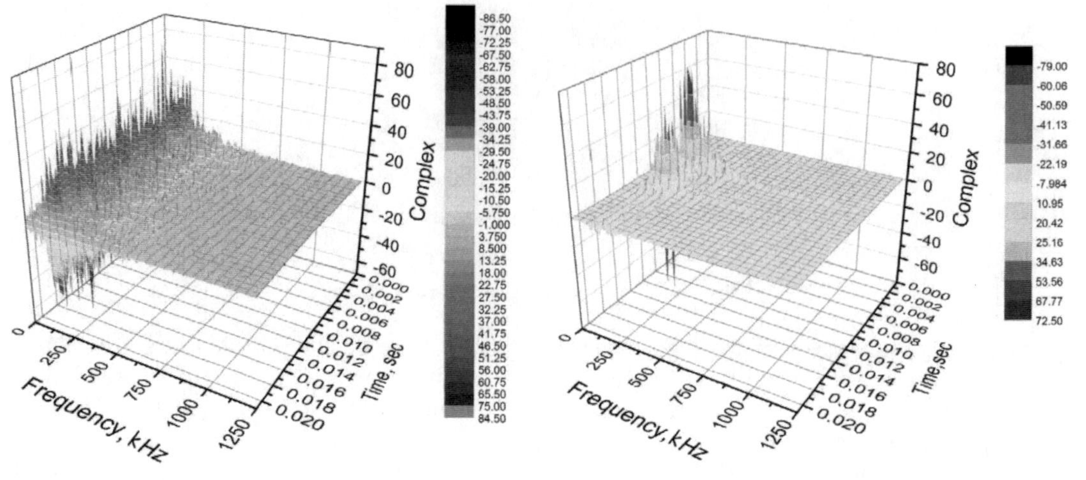

a) b)

Figure 8. Short time Fourier transform of AE signals in the run-in (a) and stable friction (b) areas

4. Conclusion

The experimental research into sliding friction of Hadfield single crystals based involving registration of AE signals allows of the following conclusions:

1. In the process of friction the fine surface layer of Hadfield steel is deformed and typical strain structures alongside with wear particles are formed.
2. A run-in to stable friction area transition moment of a tribounit can be detected quite precisely using registration of AE signals.
3. Amplitude-frequency characteristics of typical AE signals, generated in a tribounit at various wear phases are determined when analyzing AE signal frames relying on FFT and STFT.
4. The obtained results can be used for the development of a system to monitor the condition of a tribounit.

Acknowledgments

The reported study was partially supported by RFBR, research project No. 16-08-00377 a. The authors wish to thank Prof. Yu. I. Chumlyakov for providing test samples.

References

[1] Christian U and Ohtsu G M. 2008. *Acoustic Emission Testing.* (Springer)
[2] Baranov V, Kudryavtsev E, Sarychev G and Schavelin V 2007. *Acoustic emission in friction* (Elsevier)
[3] Rubtsov V E, Kolubaev E A, Kolubaev A V and Popov V L 2013 *Tech. Phys. Let.* **39 2** 223-225
[4] Haze A, Wada M and Mishina H 2008 *Wear* **265** 831-839
[5] Haze A, Mishina H and Wada M 2013 *Wear* **308** 142-147
[6] Haze A, Mishina H and Wada M 2012 *Wear* **292-293** 144-150
[7] Haze A, Wada M and Mishina H 2014 *Tribol. Int.* **72** 51-57
[8] Haze A, Mishina H and Wada M 2009 *Journ. of Adv. Mech. Design* **3 4** 333-344

[9] Karaman I, Sehitoglu H and Chumlyakov Yu I 1998 *Scr. Mater.* **38** **6** 1009-1015

[10] Chumlyakov Yu I, Kireeva I V, Litvinova E I, Zakharova E G, Luzginova N V, Karaman I and Sehitoglu H 2000 *The Phys. of Met. and Metallography* **90** **1** 1-17

[11] Karaman I, Sehitoglu H, Gall K and Chumlyakov Yu I 2000 *Acta Mater.* **48** 1345-1359

[12] Karaman I, Sehitoglu H, Beaudoin A J, Chumlyakov Yu I, Maier HJ and Tome C.N. 2000 *Acta Mater.* **48** 2013-2047

[13] Karaman I, Sehitoglu H, Chumlyakov Yu I, Maier H J and Kireeva I V 2001 *Metal. Mater. Trans. A.* **32A.** 695-706

[14] Karaman I, Sehitoglu H, Chumlyakov Yu I, Maier H J and Kireeva IV 2001 *Scripta Mater.* **44** 337-343

[15] Gumus B, Bal, B, Gerstein G, Canadic D, Maier H J, Guner F and Elmadagli M 2015 *Mat. Sci. Eng. A.* **648** 104-112

[16] Astafurova E G, Tukeeva M S, Zakharova G G, Melnikov E V and Maier H J 2011 *Mat. Charact.* **62** 588-592

[17] Kolubaev E A, Kolubaev A V and Sizova O V 2010 *Tech. Phys. Let.* **36** **8** 762-765

[18] Dobrynin S A, Kolubaev E A, Smolin A Y, Dmitriev A I and Psakhie S G *Tech. Phys. Let.* **36** **7** 606-609

Quality Estimation of Dry Grinding of Skiving Cutters With Organic Bonding Diamond Wheels

A V Filippov [1, 2,*], N N Shamarin [2], O A Podgornykh [2], V E Rubtsov [1, 2]

[1] Institute of Strenght Physics and Materials Sciencs SB RAS, Tomsk, per. Academicheskii 2/4, Russia
[2] Tomsk Polytechnic University, Tomsk, Lenina av. 30, Russia

E-mail: andrey.v.filippov@yandex.ru

Abstract. Engineering process preparation requires proper preparation of cutting tools. It influences not only the performance of the tools but also the quality of workpiece surface machining. One of the promising environmentally friendly trends of mechanical treatment is grinding without using lubricating cooling liquid. This method can considerably influence the quality of cutting tools grinding. Smoothing skiving turning is an effective treatment method providing high efficiency and workpiece quality. Proper preparation of cutting edges is especially important in this process. For that purpose we have carried out a research in grinding changeable carbide inserts for skiving turning by means of grinding wheels with different grain size. The influence of different combinations of wheels on roughness of the inserts front and rear surfaces and quality of cutting edge was studied with the help of laser confocal microscopy.

1. Introduction

Preparation of the tool surface and cutting edge is one of the key problems on the engineering process preparation [1-3]. It is generally known that providing proper preparation of the tools is an essential prerequisite of their effective application for machining operations [4, 5].

Machining without using lubricating cooling fluid called dry-cutting is one of ecologically-friendly machining methods [6, 7]. Alongside with dry cutting, dry grinding which is a process of grinding without lubricating cooling fluid attracts attention of researchers [8-12].

Skiving turning is a relatively new method of mechanical treatment which became spread due to a number of researches done in 70s-90s of the last century [13-16]. Based on these and some more recent works [17-20] we can determine a number of distinctive features of machining with skiving cutters which distinguish skiving turning from the conventional process (turning using a cutter with the tip):

1. Skiving cutters as they are called in [13-16] have no tip as a material body.
2. Bevel turning with a rectilinear skiving cutter produces one-sheeted hyperboloid of revolution while turning and boring with an elliptic skiving cutter provides a quartic surface [17-19].

* Corresponding author: A V Filippov - andrey.v.filippov@yandex.ru.

3. When in the process of bevel turning the edge of a skiving cutter is set at an angle (ω) to the workpiece rotation axis a wide range of cutting edge geometries along the working area of the cutting edge tip is produced [20].
4. Cross-cut section of the cut layer is characterized by cut layer thickness value much lower than cutting depth and feed while the same parameters of the cut layer width are much greater. Values of cross-cut section parameters depend not only on the feed, cutting depth and cutting edge shape but also on the workpiece diameter and the edge inclination angle [17-19].

The differences between skiving and conventional cutting identified above as well as the experimental research previously done [17-20] prove the necessity to carry out proper preparation of the cutting edge of a skiving cutter providing minimal possible radius of cutting edge tip round-off. The latest factor is greatly determined by the fact that when skiving cutters are used for finishing operations the cut-off layer has to be of a small width. The width of the cut-off layer is changes from zero at contact boundary of the tip with the workpiece to some maximum value which also does not exceed some tens of microns.

The aim of this research is to determine conditions for proper preparation of cutting edge tip of changeable carbide inserts used for skiving by means of grinding by diamond wheels with different grain size without treatment with lubricating cooling liquid.

2. Methods of research

Triangle inserts TNMG-160408 M2 of AP10AT alloy were used for the research. Grinding of changeable carbide inserts was carried out on universal tool grinder 3E642E using dish-shaped diamond wheels 5-0013 12A2-20 150-18-6-2-32 with organic bond B2-01 with 100% concentration of diamond powder AC4, AC6, ACH and grain size 100/80, 80/63 and 40/28 accordingly. Grinding powders AC4, AC6 are made of synthetic diamonds in which grains are separate crystals with developed surface, aggregates and crystal-jams. Micropowder ACH is made up of synthetic diamonds with high abrasive power. Organic bond B2-01 was chosen because it allows producing high quality grinding of carbide tools without treatment with lubricating cooling substance. Spindle revolutions per minute are 3,150. Table 1 shows feed values in rough finishing and smoothing grinding.

Grinding was carried out on the front and rear surfaces of changeable carbide inserts. Different combination of diamond wheels were applied in the process: the 1st wheel for rough grinding; the 1st and the 2^{d} wheels for rough grinding followed by finishing operation; the 1st, the 2^{d} wheels and the 3^{d} wheels for successive rough, finishing and smoothing grinding; the 1st and the 3^{d} wheels for rough and smoothing grinding. The latest combination is tested to analyze ways of make the grinding process more efficient.

Table 1. Feed values in grinding changeable carbide inserts

Machining operation	longitudinal feed, mm/double pass	longitudinal feed, m/min
Rough	0.03	1.0
Finishing	0.02	0.25
Smoothing	0.01	0.2

Figure 1 (a) shows a scheme of scanning of the grinded cutting edge of a changeable carbide insert for the skiving cutter performed with the help of 3D laser confocal microscope Olympus LEXT OLS 4100. Fig 1(b) shows an optical image of the insert.

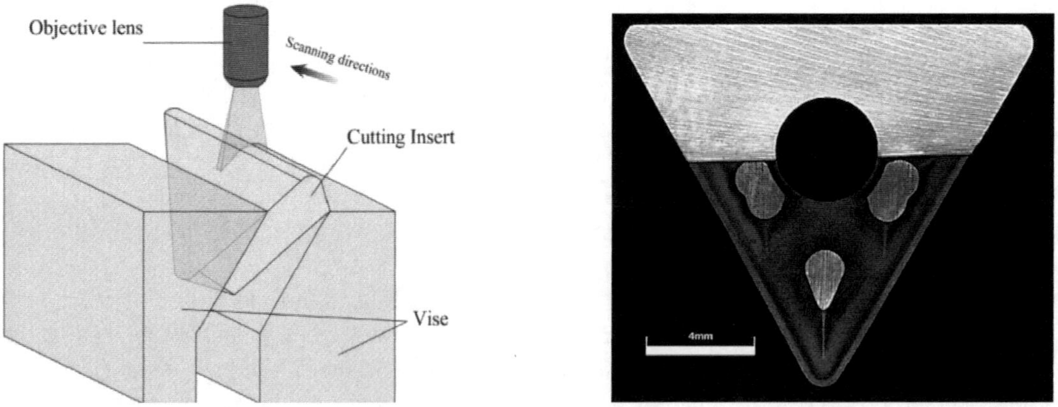

a) b)

Figure 1. Scanning of the grinded cutting edge of a changeable carbide insert for the skiving cutter performed with the help of 3D laser confocal microscope Olympus LEXT OLS 4100 (a) and optical image of an insert (b)

3. The main results

Figure 2 shows pictures illustrating the condition of the cutting edge of changeable carbide inserts grinded with different diamond wheels. Figures in the pictures show numbers of the wheels used: wheel 1 – grinding powder AC4 with grain size 100/80, wheel 2 – grinding powder AC6 with grain size 80/63, wheel 3 – grinding powder ACH with grain size 40/28.

a) b)

Figure 2. Cutting edge of changeable carbide inserts in the center (a), net to the tip (b) after grinding with diamond wheels: 1 – AC4 100/80, 2 –AC6 80/63, 3 –ACH 40/28

Figure 3 shows 3D profiles of the front surface of changeable carbide inserts after grinding with diamond wheels obtained by laser confocal microscope.

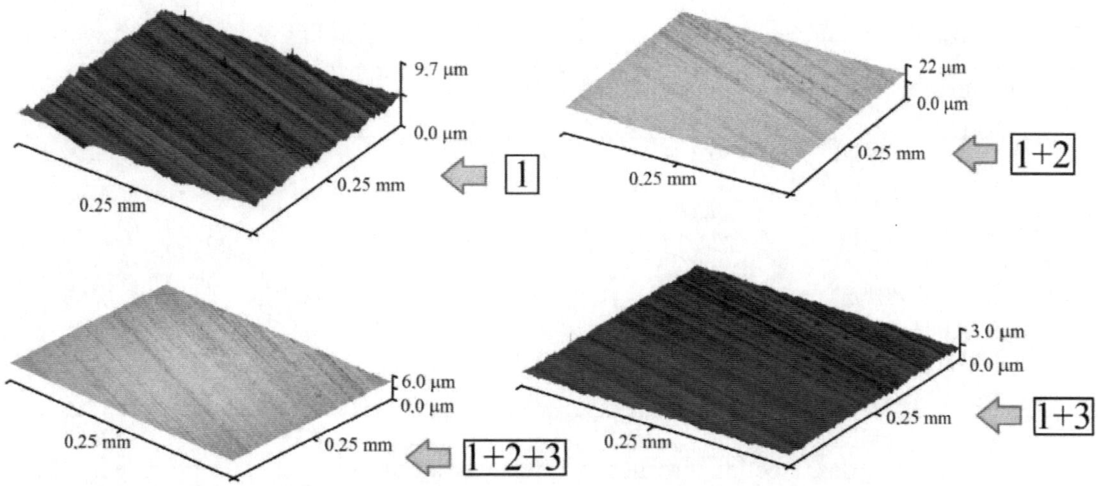

Fig. 3 Profile of the front surface of carbide inserts after ginding with diamond wheels: 1 – AC4 100/80, 2 –AC6 80/63, 3 –ACH 40/28

Figure 4 shows roughness graphs for the front surface of changeable carbide inserts for skiving cutters.

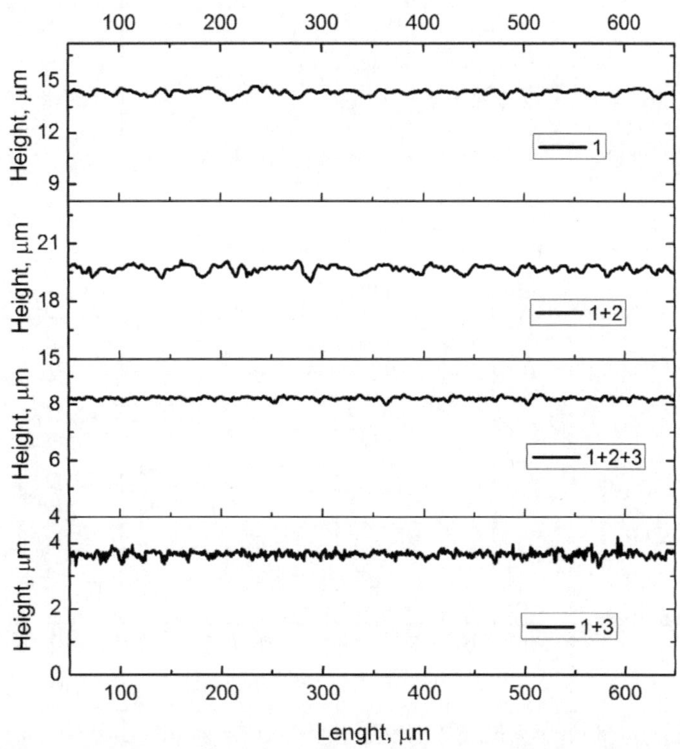

Figure 4. Roughness of the front surface of changeable carbide inserts after grinding with diamond wheels: 1 – AC4 100/80, 2 – AC6 80/63, 3 – ACH 40/28

Table 2 shows roughness values in longitudinal and cross directions (against grinding direction), obtained through an analysis of front surface topograms after grinding with diamond wheels 1, 2, 3 in different consequences.

Table 2. Roughness of changeable cutting inserts edge after grinding with diamond wheels

Diamond wheels	Longitudinal diretion		Cross direction	
	Ra, mkm	Rz, mkm	Ra, mkm	Rz, mkm
Wheel 1: AC4 100/80	0.13	0.9	0.17	1.2
Wheel 2: AC4 100/80 Wheel 2: AC6-80/63	0.14	1.1	0.18	1.3
Wheel 1: AC4 100/80 Wheel 2: AC6-80/63 Wheel 3: ACH-40/28	0.06	0.65	0.06	0.7
Wheel 1: AC4 100/80 Wheel 3: ACH-40/28	0.075	0.77	0.085	0.87

Figure 5 shows cross section profiles of changeable carbide inserts cutting edge after grinding with different diamond wheels obtained by laser confocal microscope.

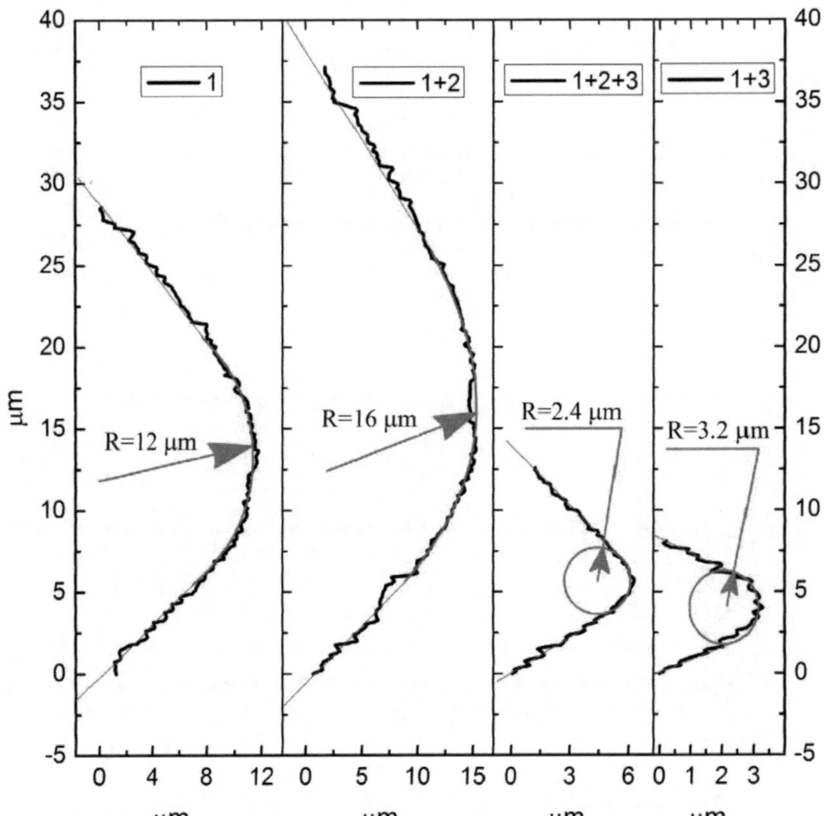

Figure 5. Front section of changeable carbide insert cutting edge after grinding with diamond wheels: 1 – AC4 100/80, 2 –AC6 80/63, 3 –ACH 40/28

4. Discussion

After grinding the front and rear surface of a skiving changeable carbide insert with wheel 1 for rough grinding (grinding powder AC4 with grain size 100/80) the cutting edge of the insert has chippings and the tip of the insert has a large chipping (over 40mkm in length and width). The cutting edge and the tip of the insert after successive grinding with wheels 1 and 2 (wheel 1 – grinding powder AC4 with grain size 100/80, wheel 2 – grinding powder AC6 with grain size 80/63) still has strongly marked large chippings. Extra grinding with wheel 3 (grinding powder ACH with grain size 40/28) after grinding with wheels 1 and 2 leads to considerable reduction of chipping on the cutting edge and on the tip of changeable carbide inserts. 3-D pictures of changeable carbide inserts prove that grinding with wheels 1, 2 and 3 successively provides the best quality of the product.

In case of successive grinding with wheels 1 and 2 the quality of the cutting edge and the tip is equitable with that obtained by successive grinding with the three wheels. Besides a good quality of the cutting edge, such combination allows reducing operational time due to skipping finishing grinding with wheel 2. However, some surface defects are found on the front and the rear surfaces of the changeable carbide insert (see Figure 2b marked as 1+3).

Together with providing a good quality of the cutting edge it is important to provide good quality of the front surface for skiving grinding. The front surface of the changeable carbide insert after grinding with wheel 1 has the greatest roughness. Roughness crosswise to grinding is 30% higher than that longwise. Successive grinding with wheels 1 and 2 did not lead to a better quality of the front surface. Surface roughness grew by 10% as compared to grinding with wheel 1. Successive grinding with wheels 1, 2 and 3 provides the best quality of the insert surface: traces of abrasive treatment are less seen, roughness is the lowest (more than twice less than that after grinding with wheel 1 and combination of wheels 1 and 2). Successive grinding with wheels 1 and 3 (without wheel 2) also gives good quality of changeable carbide insert surface. Surface roughness is only by 25% higher than that after grinding with combination of wheels 1, 2, 3 and by 50% less than that after grinding with wheel 1 or combination 1 and 2.

As it was mentioned above the radius of cutting edge tip round-off is one of important factors effecting deformation conditions and chip making in skiving grinding. In grinding with wheel 1 the radius of cutting edge tip round-off equaled R=12mkm. Fig.5 shows a typical profile of the cross section. Successive grinding with wheels 1 and 2 increased the round-off radius of changeable carbide insert cutting edge up to R=16mkm which is by 25% higher than after grinding with wheel 1 only. The smallest round-off radius R=2.4mkm (which is five times less than after grinding with wheel 1) was obtained as a result of successive grinding with wheels 1, 2, 3. Successive grinding with wheels 1 and 3 (without wheel 2) gave an average radius of cutting edge round-off equaling R=3.2mkm which is 25% more than after successive grinding with wheels 1, 2, 3, but 3.8 times less than that after grinding with wheel 1.

5. Conclusion

The results described above prove:

1. The best quality of grinding of the front and the rear surface of changeable carbide inserts for skiving cutters is achieved by applying successive grinding with wheels 1, 2 and 3 in which grain sizes decrease gradually.
2. Grinding with wheel 1 having maximum grain size (100/80) and grinding with wheels 1(grain size 100/80) and wheel 2 (grain size 80/63) successively gives comparably low quality of surfaces.
3. The best quality of surfaces was achieved by combination of wheels for rough and smoothing grinding (wheels 1 and 3). This combination provides surface roughness

Ra=0.075..0.085mkm and cutting edge round off radius R=3.2mkm which is only by 25% worse than that after grinding with wheels 1, 2 and 3. Besides this combination allows reducing grinding time due to eliminating wheel 2 from the grinding process.

4. Grinding of changeable carbide inserts without using lubricating cooling fluid allows obtaining high quality of working surfaces and cutting edges of skiving cutters.

6. References

[1] Lobanov D V and Yanyushkin A S 2011 *Rus. Eng. Res.* **31 (3)** 236-239

[2] Arkhipov P V, Yanyushkin A S, Lobanov D V and Petrushin S I 2013 *Appl. Mech. and Mat.* **379** 124-130

[3] Vasil'ev E V, Popov A Y and Rechenko D S 2012 *Rus. Eng. Res.* **32 (11-12)** 730-732

[4] B. Denkena, D. Biermann. Cutting edge geometries // CIRP Annals - Manufacturing Technology. Vol. 63, 2014, P. 631–653.

[5] Kovalevskaya, Z.G., Klimenov, V.A., Zaitsev, K.V. Interfacial adhesion between thermal spray coating and substrate achieved by ultrasonic finishing // Applied Mechanics and Materials.Vol. 682, 2014, P. 459-463.

[6] F. Klocke, G. Eisenblätter. Dry Cutting // CIRP Annals - Manufacturing Technology. Vol. 46, 1997, P. 519–526

[7] D.A. Dornfeld. Green Manufacturing. Fundamentals and Applications. Springer.2013. 289 p.

[8] J.F.G. Oliveira, E.J. Silva, R.T. Coelho, L. Brozek, A.C. Bottene, G.P. Marcos. Dry grinding process with workpiece precooling // CIRP Annals - Manufacturing Technology 64 (2015) 329–332.

[9] J.L. González–Santander. Analytic solution for maximum temperature during cut in and cut out in surface dry grinding // Applied Mathematical Modelling. Vol. 40 (2016) 2356-2367 - doi:10.1016/j.apm.2015.09.031.

[10] T. Tawakoli, B. Azarhoushang. Influence of ultrasonic vibrations on dry grinding of soft steel // International Journal of Machine Tools and Manufacture. Vol. 48, 2008, P. 1585–1591.

[11] S.Y. Luo, Y.C. Liu, C.C. Chou, T.C. Chen. Performance of powder filled resin-bonded diamond wheels in the vertical dry grinding of tungsten carbide // Journal of Materials Processing Technology. Vol. 118, 2001, P. 329–336.

[12] D. Anderson, A. Warkentin, R. Bauer. Experimental validation of numerical thermal models for dry grinding // Journal of Materials Processing Technology. Vol. 204, 2008, P. 269–278.

[13] Raphael G., Stone B.J. Boring with a Process Similar to Skiving // CIRP Annals - Manufacturing Technology. Vol. 39, Is. 1, 1990, P. 425–428

[14] Stone B.J., Bonikowski E.J., Chapple D.J., De Barr A.E. The Skiving of Ball-Bearing Tracks // CIRP Annals - Manufacturing Technology. Vol. 29, Is. 1, 1980, P. 275–280

[15] Nee A.Y.C., Venkatesh V.C. Form Accuracy of Tangentially Skived Workpieces // CIRP Annals - Manufacturing Technology. Vol. 34, Is. 1, 1985, P. 121–124

[16] Grzesik W. A real picture of plastic deformation concentrated in the chip produced by continuous straight-edged oblique cutting // Int. J. Mach. Tools Manuf. – 1991. – Vol. 31. – № 3. – P. 329-344.

[17] Filippov A.V. Cut-Layer Cross Section in Oblique Turning // Russian Engineering Research. 2014. Vol. 34. №11. P.718-721.

[18] Filippov A.V. Cut-Layer Cross Section in Oblique Turning by a Single-Edge Tool with a Curved Front Surface // Russian Engineering Research, 2015, Vol. 35, No. 5, pp. 381–384.

[19] Filippov A.V. Cut-Layer Cross Section in Oblique Turning by a Single-Edge Tool with a Curved Rear Surface // Russian Engineering Research, 2015, Vol. 35, No. 5, pp. 385–388.

[20] Filippov, A.V. Constructing a model of the equivalent wedge oblique cutting edge //Applied Mechanics and Materials. 2013. Vol. 379. P. 139-144.

Application of Artificial Neural Networks in the Heart Electrical Axis Position Conclusion Modeling

L N Bakanovskaya

Tyumen State Oil and Gas University, 38 Volodarskogo st., Tyumen, 625000, Russia

E-mail: mila27@tsogu.ru

Abstract. The article touches upon building of a heart electrical axis position conclusion model using an artificial neural network. The input signals of the neural network are the values of deflections Q, R and S; and the output signal is the value of the heart electrical axis position. Training of the network is carried out by the error propagation method. The test results allow concluding that the created neural network makes a conclusion with a high degree of accuracy.

1. Introduction

The Electric Axis of Heart (EAH) [1] – the term used in cardiology and functional diagnostics, reflecting the electric processes happening in heart. The electrocardiogram is a schedule of temporary dependence of a potential difference in the corresponding assignment, and temporary dependence of a projection of a complex electric vector of heart to the line of assignment. The direction of an electric axis of heart shows the total amount of the bioelectric changes proceeding in a cardiac muscle at each her contraction.

The conclusion "turn of an electric axis of heart around an axis" quite often can occur in descriptions of electrocardiograms. The provision of an electric axis of heart serves as an additional indicator at diagnosis of this or that disease. The provision of an electric axis in itself hearts isn't the diagnosis. However there is a number of diseases at which is observed shift of heart axis.

Currently, the development of medicine is based on the use of human-machine (automated) methods of information processing for the improvement of decision making in the disease diagnostics. Mathematical and algorithmic support of management and decision making, as well as obtaining expert information, analysis and processing methods are already known in different fields [2-9]. Now research works on application of artificial neural networks in the field of medicine and in particular in cardiology are conducted [10-12]. Enhancement of effective functioning of biotechnical systems for the scientific activity planning, data collection and analysis, using modern methods of information computer processing is becoming urgent, giving the most labor-consuming and important process in the diagnosis and prognosis of human life parameters to appropriate hardware and software means of simulating the medical experts activities.

2. Goal Setting

Research and study of artificial neural networks capabilities (ANN) is one of the promising areas of Medical Informatics. The objective of this work is modeling a conclusion about the heart electrical axis position using an artificial neural network. To achieve this goal it is necessary to accomplish the following tasks:

1. To form a training sample;
2. To select an ANN model;
3. To test the case study of the selected model.

The problem of diagnosing of deviations of an electrocardiogram from norm can be carried to problems of recognition of images. Image is the system of classification allocating a certain group of objects on some sign. Research in this work of an artificial neural network as the classifier is caused by her ability to processing of indistinct and difficult basic data for their classification [12]. When teeth of R are accurately visible, then it is easy for doctor or student to define the provision of an electric axis of heart. But happens that teeth R approximately identical amplitudes, data can be not full or inconsistent then the program can come to the rescue.

According to the available sources, there are no examples of making conclusion about the heart electrical axis position using an artificial neural network. Practical application of the MATLAB source environment is currently being actively developed and is the core technology for Biomedicine tasks simulations.

2. An artificial neural network design, modeling and training.

2.1 Training sample formation

To form a training sample the source data [1] were entered into MS Excel table (figure 1). The training sample is the values of deflections Q, R and S amplitudes in standard leads I, II and III calculated by the formula (1):

$$U = h \times S \qquad (1)$$

where U is the amplitude (mV),
h is the height of deflections (mm)
S is the height of the calibration signal (mm).

IOP Conf. Series: Materials Science and Engineering **142** (2016) 012100 doi:10.1088/1757-899X/142/1/012100

	A	B	C	D	E	F	G	H	I	J	L
1	№	IQ	IR	IS	IIQ	IIR	IIS	IIIQ	IIIR	IIIS	Conclusion
2	1	0,00	0,50	0,10	0,00	0,40	-0,10	0,00	0,10	-0,20	3
3	2	0,00	0,40	0,00	0,00	0,25	-0,40	0,00	-0,40	0,00	1
4	3	0,00	0,60	0,00	0,00	0,60	0,00	0,10	-0,10	0,10	3
5	4	-0,10	0,50	-0,10	-0,30	1,70	-0,10	-0,20	1,30	-0,10	5
6	5	-0,10	0,40	-0,10	0,00	0,50	-0,10	0,00	0,10	-0,10	3
7	6	-0,10	0,50	0,00	-0,50	0,60	-0,50	-0,10	0,40	-0,20	3
8	7	-0,10	0,20	-0,10	0,10	0,50	0,00	0,00	0,10	-0,10	3
9	8	-0,10	0,90	-0,20	-0,10	0,50	-0,30	0,10	-0,30	0,00	2
10	9	-0,10	1,10	-0,10	0,00	0,90	0,10	0,00	-0,20	0,00	3
11	10	-0,10	0,60	-0,20	-0,10	0,60	-0,20	-0,10	0,10	0,20	5
12	11	0,00	0,30	-0,10	0,00	1,50	0,00	-0,10	1,20	0,00	5
13	12	-0,20	0,60	-0,20	-0,10	0,60	0,00	-0,10	0,30	0,00	4
14	13	-0,10	0,80	-0,10	-0,20	0,40	-0,20	-0,10	-0,40	0,00	1
15	14	0,00	0,70	-0,30	-0,10	0,90	0,00	-0,20	0,60	0,00	4
16	15	0,00	0,60	0,00	-0,10	0,90	-0,10	-0,10	0,30	-0,10	3
17	16	0,00	0,20	-0,10	0,00	1,40	0,00	0,00	1,20	0,00	5
18	17	-2,00	-0,28	1,60	-1,36	-1,48	-0,12	1,92	1,92	-1,44	3
19	18	-0,80	-0,24	1,52	-0,52	1,80	-0,20	0,84	-1,60	-1,64	2
20	19	0,84	0,40	-0,56	0,24	0,72	-1,08	0,32	1,64	1,80	2
21	20	-0,92	0,52	-0,80	0,96	-1,16	0,08	1,24	1,44	-0,84	2
22	21	-1,72	1,08	-1,12	0,12	0,68	-0,12	1,32	-1,04	-1,76	2
23	22	-1,72	-1,40	-1,08	0,36	0,60	0,12	1,76	0,24	-0,08	1
24	23	-0,52	0,60	-1,92	1,76	-0,32	-1,04	0,12	1,40	1,68	1
25	24	0,00	0,04	-0,40	1,24	-0,36	1,88	-0,44	-1,60	0,04	3
26	25	-2,00	1,28	-1,60	-0,76	0,36	-1,76	1,28	1,72	0,08	2
27	26	-1,12	0,92	-0,76	0,68	0,24	-1,00	-1,32	-0,64	1,04	1
28	27	1,64	1,64	1,96	-1,40	1,04	2,00	1,68	-1,84	-0,72	3
29	28	1,64	1,52	1,48	-0,28	-0,72	0,96	-1,60	0,04	-1,00	1
30	29	-0,92	1,16	1,68	-1,76	0,68	1,60	1,36	-1,60	-1,52	2
31	30	0,92	-1,84	-1,12	-1,72	0,04	-1,24	-0,12	-1,20	0,96	3
32	31	-0,36	-0,76	-1,56	-0,84	0,84	-1,64	-1,24	-0,52	-1,68	1
33	32	-0,52	1,88	-0,24	-1,16	1,52	-1,84	1,00	-1,40	1,76	3
34	33	0,56	0,00	-0,32	0,44	-0,56	1,80	0,64	-0,24	1,92	1
35	34	-0,80	0,48	0,92	-1,84	-1,72	1,84	0,72	1,12	-0,40	3
36	35	1,92	1,56	1,92	0,16	-1,80	-0,56	0,16	-0,16	-1,32	4
37	36	1,88	0,24	-0,32	-1,24	-1,12	0,44	0,92	-1,52	0,20	3
38	37	0,08	-1,72	-1,92	1,92	1,12	0,96	1,28	1,96	0,16	5
39	38	0,64	-1,52	-1,60	-0,68	-1,04	-0,52	-1,36	1,20	-0,56	3
40	39	-0,72	-0,40	-2,00	-1,28	-1,16	0,12	0,20	1,00	-0,44	3
41	40	-1,92	-0,92	-1,80	0,16	-1,68	0,68	0,12	-0,40	0,16	3
42	41	1,84	1,76	-0,04	1,04	-0,84	1,96	-1,40	-0,84	-1,72	2
43	42	-1,92	1,68	0,20	-0,20	-0,20	-0,56	0,76	1,52	-1,44	3

Figure 1 - MS Excel sheet working area with source data

In the source data it is also necessary to calculate the algebraic sum of deflections Q, R and S for every standard lead and angle α (2), which connects the heart electrical axis with the first lead axis:

$$tg\alpha = \frac{1}{\sqrt{3}} \frac{\sum_{II} + \sum_{III}}{\sum_{II} - \sum_{III}} \quad (2)$$

where Σ_{II} is the algebraic sum of deflections Q, R and S in the second lead,
Σ_{III} is the algebraic sum of deflections Q, R and S in the third lead.

2.2 Model selection and artificial neural network training

Due to the fact, that in the source data there are 6 classes (table 1), i.e. values of the heart electric axis position angles (figures 2, 3), a multilayer neural network was selected for conclusion modeling. Such an ANN model will provide an effective solution to the problem of image classification.

MATLAB, which has several advantages [13] and has been widely used in intelligent and expert systems [14-16], was selected as an algorithm implementation source environment.

Table 1 – Heart electrical axis position by the value of angle α

Angle value	№	Heart electrical axis position
less (-30°)	1	Left anterior fascicular block
(-30°) - 0°	2	Sharp levocardiogram
0° - 50°	3	Levocardiogram
50° - 70°	4	Normocardiogram (normal axis position)
70° - 90°	5	Dextrogram
90° - 97°	6	Left posterior hemiblock

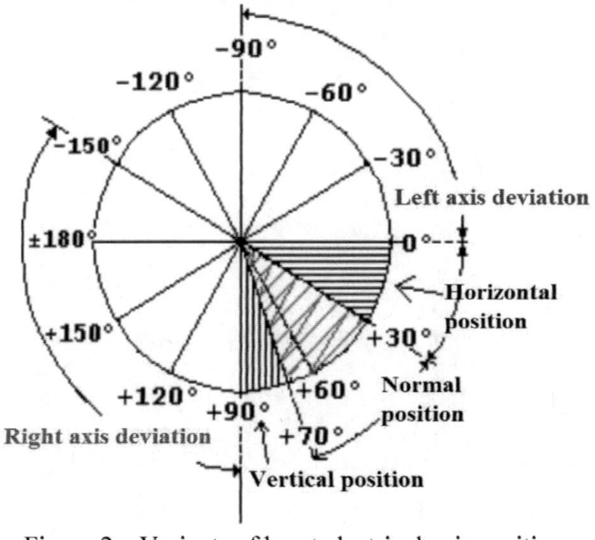

Figure 2 – Variants of heart electrical axis position

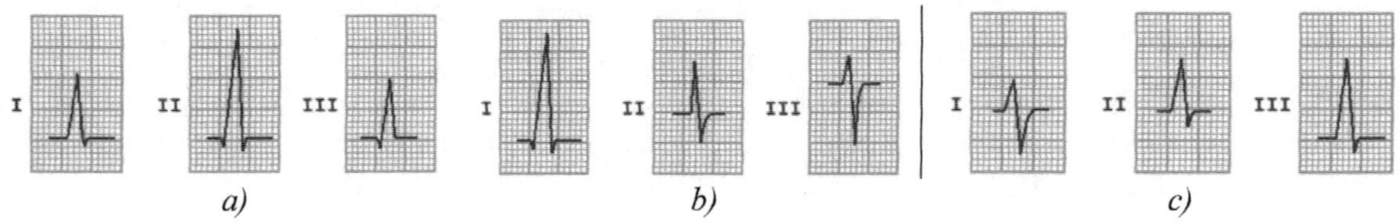

Figure 3 – *a)* Normocardiogram, *b)* Levocardiogram, *c)* Dextrogram

On the basic data is picked the architecture of an artificial neural network: multilayered perceptron, weight coefficients at this stage of work are accepted equal to zero, sigmoidalny activation function. The quantity of layers and amount of neurons in each layer retrieved experimentally.

The selected ANN model case study was accomplished using the built-in function *trainlm*, which implements the back propagation method with adjustment rate perturbation and adaptation, and the *learngdm* team realizing of algorithm of fine tuning of scales. In entrance, the first and second layers function of activation *tansig*, in the third – *logsig*, in the fourth – *tansig*, in an output layer – *purelin* is used. The assigned number of training cycles equaled to 1000 epochs (figure 4):

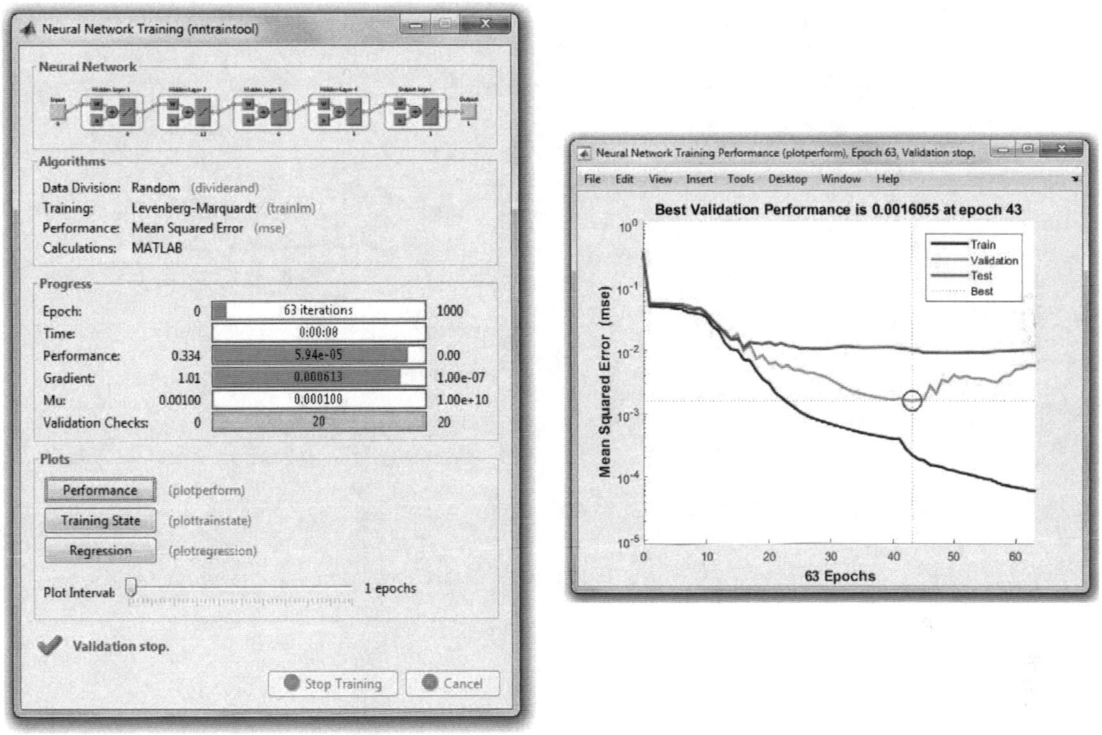

Figure 4 – Window of network training results

When the number of error count reaches the authorized number, the ANN training stops. In this case 63 epochs were processed; the best value was obtained in epoch 43 (which is shown on the MATLAB graph, figure 4). As soon as the training ended, the program opens a dialog window and an expert medic enters the electrocardiogram (ECG) initial data (figures 5, 6).

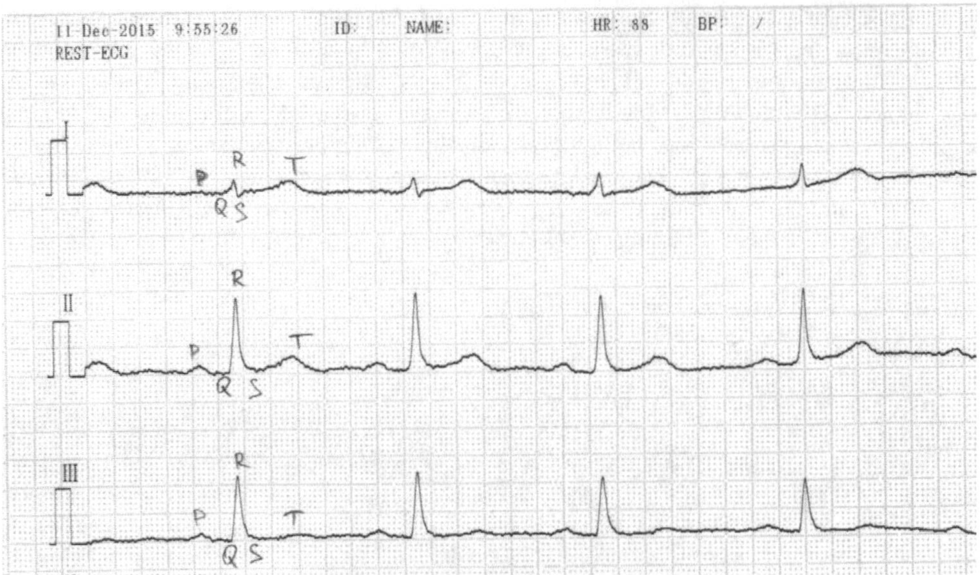

Figure 5 – Patient's ECG

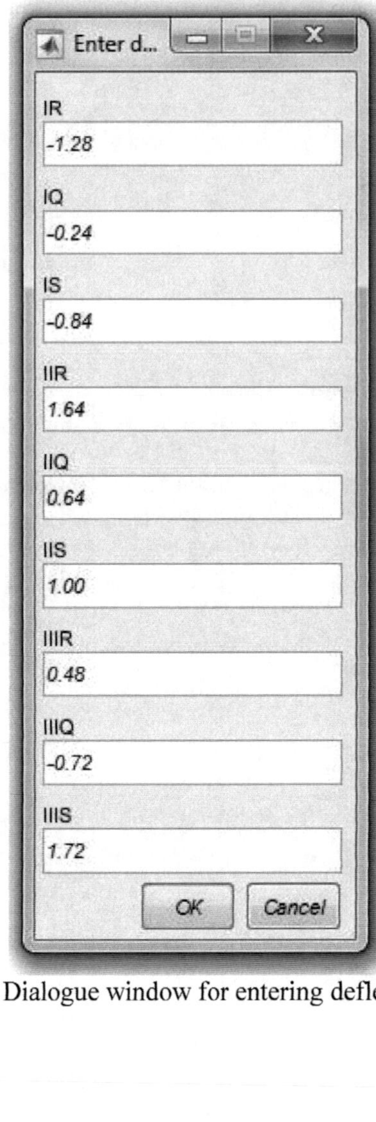

Figure 6 – Dialogue window for entering deflection values

After the patient's ECG values are entered the neural network outputs the result according to table 1 (figure 7).

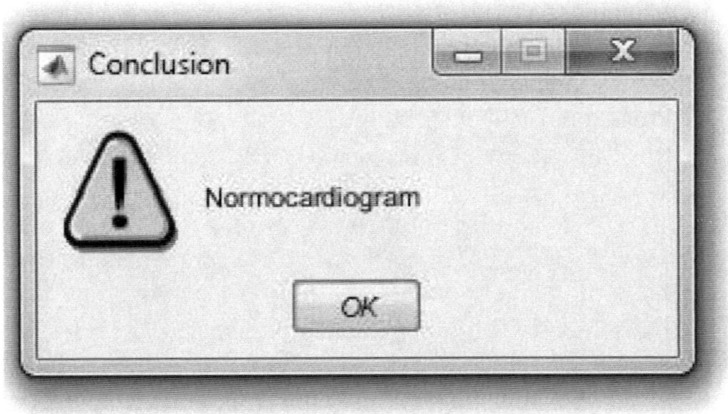

Figure 7 – Conclusion about a heart electrical axis position

3. Conclusion

The artificial neural network training with the use of the back-propagation method based on 2000 variants of deflections Q, R and S values showed the conclusions plausibility. The training results show that ANNs have high potential for the development of methods and algorithms of the heart electrical axis position calculation modeling and for further work to identify deviations in electrocardiograms.

The work is perspective for further development. So at the following stage is considered the possibility of automatic receiving basic data for performance of calculations from the digitized image of an electrocardiogram.

References

[1] Loskutov, E. D. (2002) *Methods of research of electric manifestations of life. Training manual.* Stavropol. CJSC Publishing house "The Press".

[2] Kitanina, K. Y., Khromushin, V. A., Dzasokhov A. S., Khromushin O. V. (2015) Features of construction of the expert system on the basis of algebraic models of constructive logic. *Bulletin of new medical technologies. Electronic edition,* 3 (Publication 1-1). http://www.medtsu.tula.ru/VNMT/Bulletin/E2015-3/5209.pdf doi: 10.12737/13073.

[3] Benderskaya E. N., Nikitin K. V. (2015) Using Spiking Recurrent Neural Networks for ECG Analysis. Information and control systems. 1(85-91). doi:10.15217/issn1684-8853.2015.1.85.

[4] Maass, W., Natschla, ger T., Markram, H. (2002) Real-Time Computing without Stable States: A New Framework for Neural Computations Based on Perturbations. *Neural Computation,* 11 (2531–2560). doi:10.1 162/089976602760407955.

[5] Izhikevich, E. M. (2004) Which Model to Use for Cortical Spiking Neurons? *IEEE Transactions on Neural Networks*, 15 N 5. (1063–1070). doi:10.110 9/TNN.2004.832719.

[6] Korshunov, S. A., Pavlov, A. I., Nikolaychuk, O. A. (2015) Web-oriented component of an expert system. *Software & Systems*, 2(110). (20-25). doi: 10.15827/0236-235X.109.020-025.

[7] Grif, M. G., Yumchmaa A. (2015) Application of expert systems by pulse diagnostics. *The application of expert systems of pulse diagnosis*, 3(81). (114-133). doi: 10.17212/2307-6879-2015-3-114-133.

[8] Akhmadulin, R. K. (2005) Design of quality indicators based on the expert presentations formalization. *Natural and technical Sciences*, 5 (19). (135-137).

[9] Bakanovska, L. N. (2011) Use of the "greedy" algorithm to optimize the order of launching of

apparel models into production. *Bulletin of computer and information technologies*. 2. (23-26).

[10] S. M. Kamruzzaman , Md. Monirul Islam, (2006) ”An Algorithm to Extract Rules from Artificial Neural Networks for Medical Diagnosis Problems”, International Journal of Information Technology, Vol. 12 No. 8.

[11] O. Er, N. Yumusak and F. Temurtas, (2010) "Chest disease diagnosis using artificial neural networks", Expert Systems with Applications, Vol.37, No.12, pp. 7648-7655.

[12] S. Moein, S. A. Monadjemi and P. Moallem, (2009) "A Novel Fuzzy-Neural Based Medical Diagnosis System", International Journal of Biological & Medical Sciences, Vol.4, No.3, pp. 146- 150.

[13] Bakanovskaya, L. N. (2015) On development of neural network expert systems. *Materials of XI International scientific-practical conference “Conduct of modern science”,* 3 (The economic science). (11-14).

[14] Glukhikh, I. N. (2006) Use of fuzzy logic in modeling of the plastic cards fraud detection information system. *Bulletin of the Tyumen State University*. 5. (204-210).

[15] Akhmadullin, R. K. (2016) Case history of the object-oriented programming principles in developing applications in Geology and Geophysics. *Automation, Telemechanization and Communication in Oil industry*. 1. (38-42).

[16] Babak Mohammadzadeh-Asl, Seyed Kamaledin Setarehdan.(2006) Neural network based arrhythmia classification using heart rate variability signal. *14th European Signal Processing Conference (EUSIPCO 2006),* Florence, Italy, September 4-8.

IOP Conf. Series: Materials Science and Engineering **142** (2016) 012101 doi:10.1088/1757-899X/142/1/012101

Measuring of Traction and Speed Characteristics as Well as of Fuel Economy of a Car in Road Conditions

Sergey N Krivtsov[1], Andrey P Syrbakov[2], Marina A Korchuganova[2]

[1]Irkutsk National Research Technical University
Irkutsk, Russia
[2]Urga Technological Institute Tomsk Polytechnic University
Urga, Russia

E-mail: [1]krivcov_sergei@mail.ru, [2]kormar@tpu.ru

Abstract. This article is devoted to the identification of traction and speed characteristics as well as of fuel economy of motor vehicles in road conditions. Among common variants of measuring of the above stated values, the preference was given to the immediate gaining of factors by means of a computer-aided measuring system. There is a theoretical justification given to the suggested approach as well as methods and results allowing to provide a practically sufficient solution accuracy of the problem.

Introduction

Vehicular transport is widely used in various areas of national economy. The utilization efficiency of vehicles affects production costs as well as delivery performance of passengers and cargos. It is also significantly defined by the type and the characteristics of internal combustion engines (ICE). The most widespread type of engines is diesel engine (over 90% in freight transport, and a constantly growing ratio among cars).

In practice, the lack of possibility to constantly control the technical conditions of vehicles results in the fact that they are often used with defects which are invisible to the driver but provoke the growing fuel consumption, raise transportation costs as well as increase the content of harmful substances in exhaust gases. Fuel economy as well as traction and speed characteristics of a motor vehicle are very important performance indicators that designate the exploitation efficiency of vehicles [1]. That is why a well-timed control, including diagnostics, is a vital task.

State-of-the-art.

Following measuring methods of motor vehicle parameters have been considered in road conditions:

1. Data acquisition by means of GSP/GLONASS technologies [2]
2. Recording of a data log by using a dealer scanner with subsequent data transfer onto an external medium [3, 4]
3. Immediate measuring of parameters by means of a computer-aided measuring system

IOP Conf. Series: Materials Science and Engineering **142** (2016) 012101 doi:10.1088/1757-899X/142/1/012101

The usage of satellite-assisted technologies and fuel consumption monitoring by fuel level in the tank are not suited for the given task because they deliver quite averaged data, while, in our case, during diagnostics, the process is considered to be very fast flowing. The second variant provides with quite exact data, however, these still need to be processed after being saved in a log. As for the fuel consumption registered by an on-board system, the calculation of fuel economy parameters can bring inaccurate results because of the difference between actual and expected injection rate. Taking into consideration all the above mentioned facts, it was decided to use the third variant with computer-aided technologies. The concept of this approach is depicted in the publication [5].

Mathematical model

Theoretic preconditions. We shall start to define the vehicle motion parameters by making up an analytical model of forces and moments that affect the motor vehicle (fig. 1) when accelerating uphill.

We shall define current net torque that is spent for the overcoming of motion resistance moment at every integration step on the basis of a force balance equation of a motor vehicle:

$$M_{ei}=M_{am}+(F_w+F_f+F_i\pm m_{eq}\cdot j_a)\cdot r_k/(i_t\cdot\eta_t) \quad (1)$$

where M_{am} is the moment spent for the gearing of auxiliary machinery, N·m

F_w is air resistance force, N

F_f is rolling resistance force, N

F_i is the force that affect the motor vehicle on the road inclination, N

m_{eq} is the equivalent mass of the motor vehicle, kg

j_a is the acceleration rate of the motor vehicle, m/s^2

r_k is the radius of rolling circle, m

i_t is the gear transmission ratio

η_t is the transmission efficiency factor

We shall admit that there is no crutch skidding and the crutch is rigidly closed, while the motor vehicle has elastic tires subject to force and speed deformation.

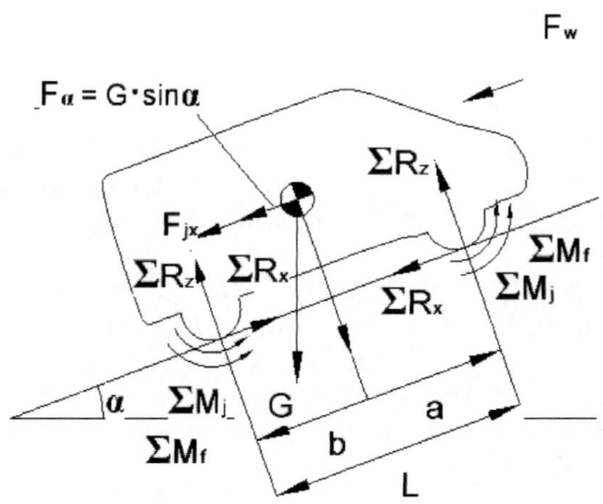

Fig. 1: Analytical model of forces and moments that affect a motor vehicle when accelerating uphill

The equivalent mass of the motor vehicle is defined taking into account the rotational inertia coefficient δ, by means of *a* familiar equation:

$$M_{eq}=m_a\cdot\delta=G\cdot\delta/g \quad (2)$$

where G is the weight of the motor vehicle, N

g is the gravitational acceleration, m/s^2.

In its turn, we shall define the rotational inertia coefficient the following way:

$$\delta=1+I_e\cdot i_t^2\cdot g\cdot\eta_t/(r_k^2\cdot G)+\Sigma I_k\cdot g/(r_k^2\cdot G)=$$
$$1+(I_e\cdot i_t^2\cdot\eta_t+\Sigma I_k)\cdot g/(r_k^2\cdot G) \quad (3)$$

I_e is the deduced inertia moment of the engine

I_k is the inertia moment of a car with a tire

Thus, taking into account equations (1-3), we shall write down the equation of motion of a motor vehicle:

$$j_a=(g/(\delta\cdot G))x$$

$$x(((M_{ei}-M_{am})\cdot\eta_t\cdot i_t/r_k)-G\cdot f\pm G\cdot sin\alpha-(c_x\cdot\rho\cdot S_x\cdot V_a^2)/(2\cdot 3.6^2)) \quad (4)$$

ρ is the air density

c_x is the aerodynamic drag factor

f is the coefficient of rolling resistance

α is the road inclination

S_x is the maximum mid-section of the motor vehicle

The speed of the motor vehicle at the ith step shall be calculated by means of numerical computation methods of differential equations. In our case, we used Euler's method:

$$V_i= V_{i-1}+j_a\cdot\Delta t \quad (5)$$

where V_{i-1} is the speed of the motor vehicle at the previous counting step, m/s

Δt is the counting step, c

The rotational rate of the crankshaft shall be defined on the basis of the current speed taking into account all made admittances:

$$n_{ei}= 2.65\cdot V_t\cdot i_t/r_k \quad (6)$$

Let us esteem the overall loss of capacity in the tires of leading wheels [6]. In a general case, the overall efficiency factor of a tire η_k is defined by the following equation:

$$\eta_k=\frac{N_m}{N_k}=\frac{R_x\cdot V}{M_k\cdot\omega_k}=\frac{M_k-M_f}{r_{ko}}\cdot\frac{V}{M_k\cdot\omega_k}=\frac{M_k-M_f}{M_k}\cdot\frac{r_k}{r_{ko}} \quad (7)$$

where N_k and N_m are the delivered and the withdrawn power of the wheel respectively; ω_k and V are the angular rate of the =resistance moment of the tire; R_x is the direct-axis reaction in the contact patch; r_k and r_{ko} are the rolling radius and the rolling radius in the conducted mode; M_k is the drive torque delivered to the wheel.

The efficiency factor of leading wheels depicts the loss of capacity and speed in the process of rolling. The efficiency factor of the tires of leading wheels η_k is the product of two multipliers:

$$\eta_k=\frac{M_k-M_f}{M_k}\cdot\frac{r_k}{r_{ko}} \quad (8)$$

The first multiplier reflects the friction force loss (hysteresis loss) of capacity in the tire while rolling. This component of capacity loss is calculated quite easily and exactly. There are usually no difficulties with its computation.

The second multiplier depicts speed losses of capacity connected with tangential elasticity of a tire, i. e. it takes into account the fact that the number of rotations made by a wheel as well as the covered distance do not always fit together exactly. The most of the hardships arise namely when determining the value of this multiplier. These complications are based on the definition of the kinematic radius r_k (the rolling radius of a wheel with an elastic tire):

$$r_k = \frac{V}{\omega_k} \tag{9}$$

The diagram below shows the dependence of the efficiency factor of tires (and in particular of every component of this) on the moment M_k delivered to the wheel.

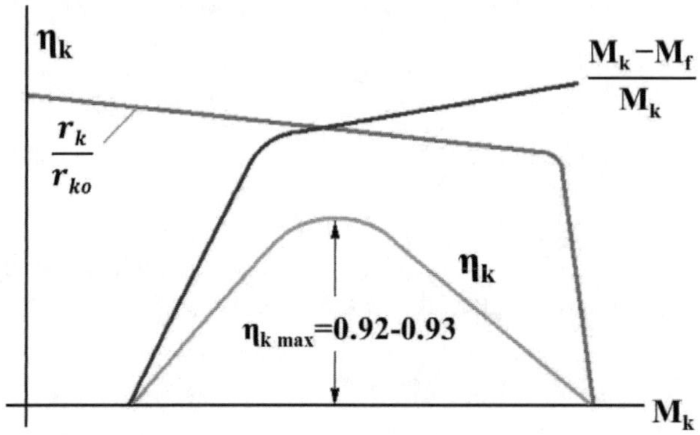

Fig. 2: The dependence of the efficiency factor of leading wheels of a motor vehicle on the value of the moment M_k delivered to the wheels (traction mode)

The maximum efficiency factor of tires of light vehicles, when the wheels are rolling in the conducted mode, tops $\eta_{k\ max}$≈0,92÷0,93. That means that at least 6…7% of engine capacity is lost in the tires of a light vehicle *for hysteresis and speed losses*. The maximum efficiency factor of tires of freight vehicles equals to $\eta_{k\ max}$≈ 0,90…0,87. It is obvious that the capacity loss in the tires of freight vehicles amounts to at least V 10…13% of engine capacity.

Thus, if we are given the parameters of the motor vehicle, the road type and its inclination, while measuring the speed of the motor vehicle, we find the drive torque that affects the axle tree and the capacity.

The given task can be solved by continuous measuring of the speed of the wheel(s) that work in a conducted mode.

Experimental procedure

The experiments were conducted on the motor vehicle "Sobol" (GAZ 2217) equipped with a diesel engine ISf 2.8 Cummins with Common Rail fuel injection system. The experiment was held on an even, dry road segment in a fine windless weather; the journey was made in two opposite directions. The method included the measuring of factors in the acceleration mode up to 100 km/h with a fully-pressed control pedal of fuel injection. The measuring system (fig. 3) consisted of an analog-digital converter to the input of which there were impulse signals delivered from the ABS wheel sensor of the front wheel, as well as of fuel flow meters mounted in the direct and in the reverse fuel-delivery lines. When calculating the fuel economy, the consumed fuel was determined as the difference between the direct and the reverse (returned into the tank) flow.

The data delivered to the input of the analog-digital converter were identically processed for fuel consumption as well as for the determination of the distance and the speed of the motor vehicle, because the signals are represented by rectangular pulses (fig. 4).

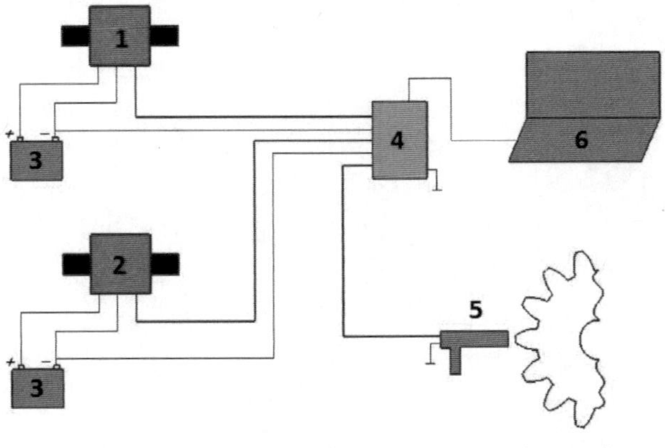

Fig. 3: Function chart of the instrumental complex
1 - fuel-flow sensor (delivery); 2 - fuel-flow sensor (return); 3 - accumulator batteries; 4 - analog-digital converter module; 5 - ABS sensor of the front right wheel; 6 - laptop

In order to facilitate data processing, there was a special script written, the essence of which is clarified in fig. 4. With a certain number of impulses given, the time of passing these intervals is registered, while the previous point is considered to be the initial point for the next interval.

Below, there is an example of equation for the calculation of line speed of a motor vehicle:

$$V_i = z_{ii}/(3600 \cdot k \cdot \Delta t_i) \tag{10}$$

where z_{ii} is the pre-defined number of impulses, imp

k is the calibration value, imp/meas. unit (for speed calculation of the motor vehicle GAZ 2217 there was the value assigned of 482 imp/km

Δt_i is the passing time of z_{ii}, s.

In order to measure the direct and the reverse flow, there were volume-flow rate sensors used with oval gears Microstream OF05ZAT by Aichi token (Japan) which are characterized by a high reproducibility of results and a practically sufficient measurement accuracy.

The metrological evaluation of the system allowed to conclude that the relative tolerance of line speed measurement of a motor vehicle and the angular rate of the wheel does not exceed 2.4%; the tolerance of fuel consumption does not exceed 3.8%.

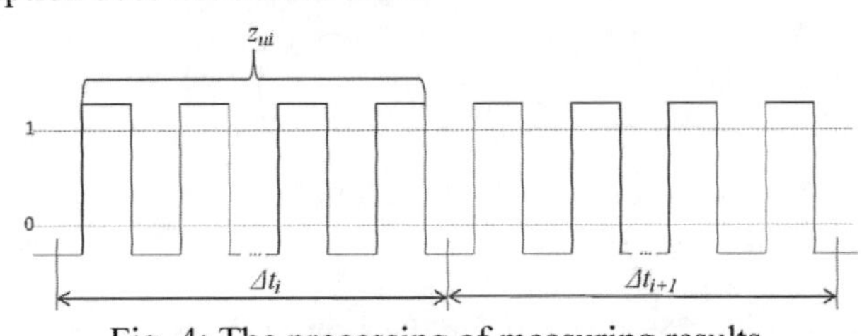

Fig. 4: The processing of measuring results

Results and discussion

The measurement and data processing results are represented in fig. 5, and 6. The dependence of speed of a motor vehicle on the time (fig. 5) was received when moving at each of the five gearings of the motor vehicle.

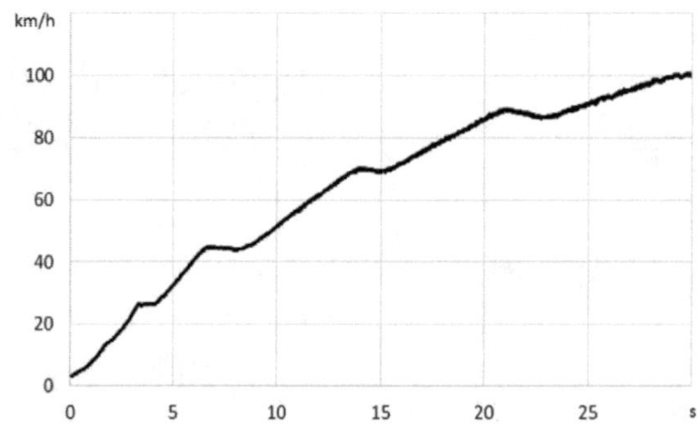

Fig. 5. The dependence of the speed change on the time when accelerating the motor vehicle GAZ 2217 up to 100 km/h

These data were used for the calculation of the effective drive torque and the effective capacity by means of an external velocity performance. There is a fact coming under notice that, on the basis of the given motion conditions (acceleration up to 100 km/h), there were points fallen out of the consideration which lie under 2000 rpm (fig. 6).

The calculated data available from experiments accord well with passport full-load curve of this diesel. In particular, the true values of an effective drive torque in the range of 2000-3500 rpm differ to 6.5% (in the initial specification, the maximum drive torque amounts to 310 Nm in the range of 2000-3500 rpm). The mentioned circumstance is not based on the calculation tolerance but on the difference in the versions as well as testing and exploitation conditions.

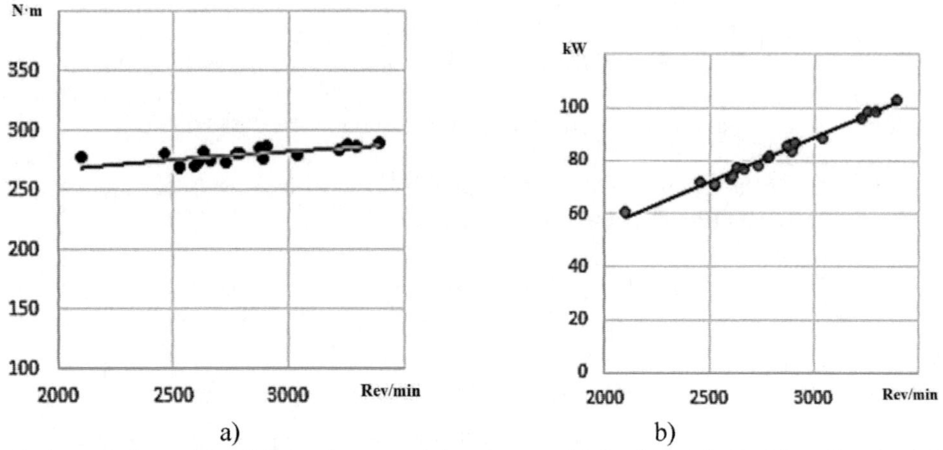

Fig. 6: Estimated data of the full-load curve of the motor vehicle GAZ 2217 with a diesel engine ISf 2.8 Cummins a – effective drive torque; b – effective capacity

The conducted research led to the following conclusions:

1. The elaborated computer-aided system for the measuring of traction and speed characteristics as well as of fuel economy of a motor vehicle in road conditions helps to estimate the acceleration, the

IOP Conf. Series: Materials Science and Engineering **142** (2016) 012101 doi:10.1088/1757-899X/142/1/012101

speed, and the distance covered by a motor vehicle with a practically sufficient accuracy (relative tolerance equals to 2.4%).

2. In order to collect the data of traction and speed characteristics of a motor vehicle, it is most viable to use a regular sensor of angular rate of the antilock braking system for a wheel working in a conducted mode. In the case of usage of a sensor of angular rate of the conducted wheel, it is necessary to take into account speed and force losses in the tire.

3. In common rail fuel injection systems, it is rational to use the engine fuel consumption method as the difference between the direct and the reverse flow, because they grow the number of diagnosis information about the technical state of fuel injection system by themselves.

References

[1] Head I M 1996 Reserves and methods of fuel saving in transport of goods by road. (Irkutsk: ISTU) 214 p

[2] Khudyakov D A 2015 Diagnostics of the technical state of vehicles on fuel consumption on the basis of information system (Irkutsk: Car for Siberia and the far North. Design, operation, Economics : Publishing house of ERNITA) pp 71-77

[3] Molev Y I, Moshkov P S, Sokolov D A, Tikhomirov A N, Shcherbakov V V 2015 Method of determining the fuel consumption of a vehicle using data diagnostic Protocol OBD II (Moscow: FUNDAMENTAL RESEARCH) pp 74-78

[4] Krivtsov S N, Zimin V G, Malashkin I V 2015 Using dealer diagnostic software to collect data about the operation of the vehicle with diesel engine (Irkutsk: Publishing house of ISTU, Tr. students and teachers of the Institute of Avialesookhrana and transport) pp 29-33

[5] Krivtsov S N 2009 The creation of a device for diagnosing diesel engines on the basis of computer technology (Novosibirsk: AGROINFO-2009) pp 172-176.

[6] Fedotov A I, Vlasov V G, Krivtsov S N 2012 Calculation of speed loss in the tires of the car in operation (Moscow: The automotive industry N. 9) pp 44-45

[7] Korchuganova M, Syrbakov A, Tkachev A, Zorina T 2015 Research of the pre-launch powered lubrication device of major parts of the engine D-240 (Published under licence by IOP Publishing Ltd: IOP Conference Series: Materials Science and Engineering,Volume 91, conference 1)

[8] Glasberger T, Peroutka Z, Janda M, Majorszký J 2012 Traction Drive with PMSM: Frequency Characteristics Measurement (Transactions on Electrical Engineering, Vol. 1) pp 13-18

[9] Xu Z. L., Yao Z. Y. 2013 Research on Control Methods of Speed Stability by AR Model Based on Linear Ultrasonic Motor" (Applied Mechanics and Materials, Vols. 385-386) pp 777-780

[10] Du X. 2012 The Kinematic Analysis and Simulation of the Large-Scale Linear Vibration Screen (Applied Mechanics and Materials, Vols. 215-216) pp. 726-729

IOP Conf. Series: Materials Science and Engineering **142** (2016) 012102 doi:10.1088/1757-899X/142/1/012102

Analysis of Organization of Production Process on the Basis of Value Stream Mapping

K T Sattarova, V V Kokareva, N D Pronichev

Department of the engine production technology, Samara State Aerospace University, 34, Moskovskoye shosse, Samara, Russia

E-mail: ks.mycity@gmail.com

Abstract. This article discusses the process of identifying the problem areas of the product cycle by the value stream. Mapping value stream mapping allowed the development of a number of management solutions to increase productivity, optimize the process and improve the competitiveness of products. For the study a product manufactured by one of the industrial enterprises of the city of Samara was selected. The production process as repeatedly optimized by services of the plant, but its cycle is still and unstable. To solve these problems the pull scheme of production was proposed. The proposed method for the improvement of the production process on the basis of value stream mapping allows optimizing the production process, reducing the production cycle, improving the quality and efficiency of production. The final results were expressed in value terms. The final result showed that the use of this method allows reducing the duration of the production cycle for 42.28%, the cost of products - by 57.71%.

1. Introduction

The modern industrial enterprise encountered the problem of the production process organization manufacturing products. Even a stable production process can often disrupt because of many factors such as machine idle time, lack of discipline in the workplace, bottlenecks, and many others. To hold the position in the competitive industrial market an enterprise needs to stabilize the production process. Lean manufacturing allows achieving not only the equilibrium state in the production, but also bringing it to a qualitatively higher level.

There are many tools and methods of lean production, which is being actively used by leading industrial enterprises. One of the methods of lean production, which enables us to analyze the production process and to identify its bottlenecks is the map of the value stream [1].

Value stream Mapping (VSM) - is a simple and clear presentation of material and information flows an order in the enterprise. Two maps are compiled: a map of the value steam currant state, which placing an order to delivery and available conditions and a map of the value steam future state, which presents an option for enhancing to achieve a higher level of performance. Today, the construction of value stream mapping is widely used in the world's leading industrial enterprises of different areas [2]. This method is used to change the logic of the technological process, yet what should we do has already been repeatedly optimized by specialists of the plant, but it still remains long and unstable? This issue was the focus of our research, which involved analyzing the feasibility of the method of value steam method within a planned technological process.

2. Materials and methods

Research an item, which is manufactured in an industrial plant in Samara was selected. The research was conducted throughout the entire production cycle, starting with the end of the stream (with the operations to determine the pace of work for other operations). It is important to note that to obtain reliable information the established norms for of the run-time operations were not used; the necessary data were obtained from personal observations, recorded directly in the operation run [3].

The concept value stream includes the production flow for transformation of raw materials into the final product, as were as design from a concept to launch in the production and provision of services from taking orders to its final execution. In our case, we considered the production flow within the enterprise, from getting raw material to shipment of the product to the customer.

To display the processes the cell of processes, which indicate to the process included in the material flow were used. In this case we use the cell's to determine a segment of the flow of materials, since their use for each process will lead to clutter on maps [4].

To simplify the following data were used:

C/T - cycle time. The time interval between the outputs of finished products from the process, or the time required for the operator all the work items before repeat they;

C/O - changeover time;

The state of operability is the required length of the machine operation in state of operability. Because during the entire processing cycle is used in operation and new equipment, the percentage of serviceable condition will be quite high.

120 pieces of acceptable, products are required per year to batches processing with once a month. Given the purported defect rate 188-192 pieces per are processed year. As a result, the cycle time is calculated to complete this order on condition of two-shift. Number of working days per year is 365 days (continuous production), and the duration of lunch breaks is 40 minutes per shift. Initial data are presented in Table 1.

Table 1. Calculation of the initial data for mapping

	parameter	value
(A)	Number of shifts	3
(B)	Number of minutes per shift	480
(C)	The duration of breaks, min.	40
(D)	The number of working shift per (D=B-C)	340
(E)	Number of working minutes per day (E=A×D)	3×340=1020
(F)	The volume of the monthly order, item	16
(Ty)	Cycle time Ty=E/F, min.	63,75

Item control does not add value to the final product, however, plays a significant role in the technological cycle. Thus, when a welded joint is discovered the item is returned back in the technological cycle for correcting and re-welding. Often item control is carried out directly in the workplace using scrap materials: caliper, gauge, micrometer, etc. Yet, the technology includes x-ray control, which is a bottleneck.

3. Results

The result was a map of the value stream in the current production, which is presented in Figure 1.

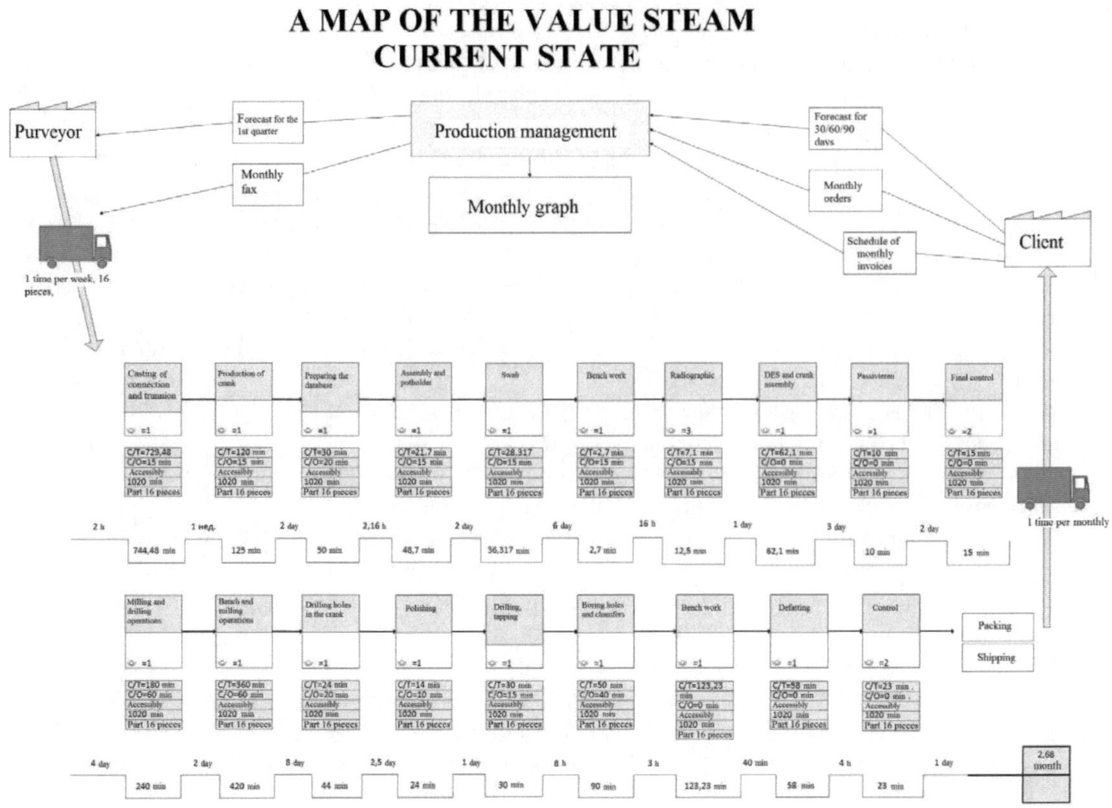

Figure 1. Map of the value stream in the current production

The study identified the bottlenecks, namely:

- X-ray control,
- Hydraulic test,
- Metallography,
- The operation of grinding,
- Other occasional problems.

Analysis of the current state showed that the cycle bath manufacture takes on average 2-3 months (in this case, the time was 2.68 months), but the time at which the cost of items is created, is 11.42 days. As we can see, about 87% of the time the value item is not created. Of course, part of this time is the time for transport, control of dimensions and geometry, etc., but even considering this, the time of "idleness" is not less than 70%. Therefore, measures to reduce the duration of queue product idleness should be taken. For this, we proposed to use the principles of smart production.

One of the main causes of downtime, in addition to mentioned above, is the problem of overproduction [5]. Before each subsequent process products transferred from the previous one are accumulated. The cause of overproduction in this case is that all manufacturing processes are isolated from each other due to the implementation of the ejector circuit. Each process produces at its own pace.

In the case of defects in any of the sections, the defective item will be located in the reserves until its transfer to the next process, and the cause of the defect which is not corrected, continues to exist, resulting in increasing the volume of defective items production. Such moment is present in the map of the value stream at the stage of the operation "weld on" when the assembly DES and the lever does not pass control, the item is sent to the re-assembly, welding, etc., at increases the manufacturing cycle. To resolve the problem of overproduction it is necessary to introduce a pull circuit (Figure 2), and this, in turn, will require the use of tools of smart production, which include:

1) The organization of production in accordance with the cycle time;
2) The organization of a continuous flow;
3) The use of supermarkets to control processes, fine adjustment by kanban;
4) Submission of the customer schedule only at specified process;
5) Alignment of production and regulation of product lines;
6) Regulation of the volume of production.

When analyzing the current state map was identified the following problems:

- Isolation of processes;
- long time fulfillment of an order compared to the time of processing;
- problems associated with machine idle time;
- shortage of workforce;
- lack of resources (steam and air under high pressure).

Introduction of the mentioned principles of intelligent production [6] resulted in creating a map of value stream future state (Figure 2).

A MAP OF THE VALUE STREAM
FUTURE STATE

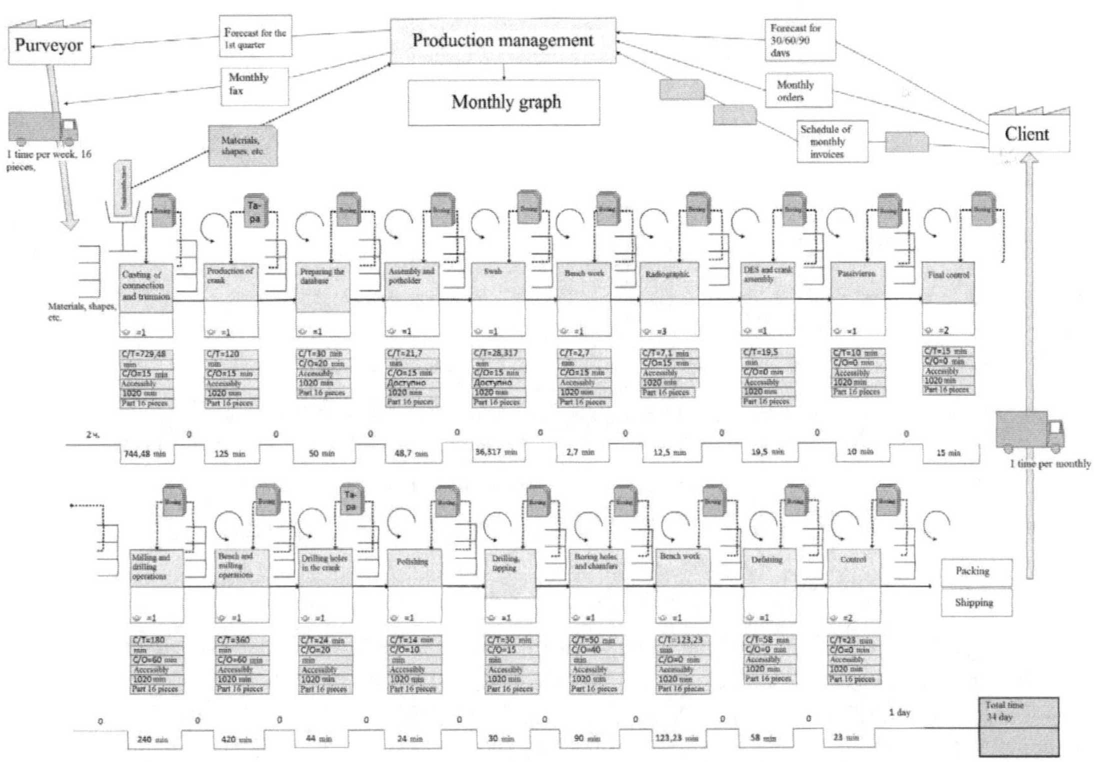

Figure 2. Map of the value stream future state

We note that the cycle of the item under investigation was repeatedly optimized. As a result, a change in the trajectory of the value stream by introducing a working cell and rational allocation of equipment is impractical. However, there are opportunities to reduce the item manufacture time by reviewing existing technologies [7]. The bottleneck, marked in the map of the current state as "casting nozzle and trunnion" could be eliminated by hiring qualified staff, and the total time stamping arm may be reduced by search intradepartmental reserves and reduce downtime.

The establishment of the future state map was found that the entire cycle of the kit of parts can be reduced to 34 days, i.e. at 42.28%. To evaluate the effectiveness of the work carried out, the results were evaluated in monetary terms. For this used cyclic schedule was, which is the most important document for inter departmental operational planning [8]. In our case, the cyclic graph of product assembly was built without considering payload on the basis of the cycle time of assembly and each assembly operation, taking into account the duration of the cycles of assembly and operation implementation.

4. Discussion

At the end of the study it was found that in a total manufacturing cycle, which stretches over 2.68 months, the manufacturing cost of the trunnion is 216 675,59 rubles. After the introduction of the principles of "smart" production it became possible was able to reduce the production cycle to 34 days. From the ratio, we can see that the cost of item manufacturing would be 95 407.85 rubles, i.e. it was to reduced the for 57.71%. Figure 3 shows the increase in total costs (cost of work in progress), and separately shows the material costs and wages, along with workshops and plant costs, determined by the cyclic schedule [9].

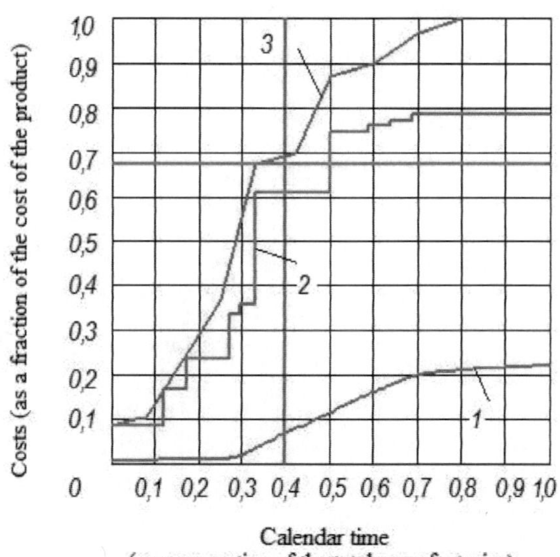

1 - manufacturing wages, along with workshops and overall plant costs; 2 - material costs; 3 - all costs.
Figure 3. The enlarged graph of increase of expenses on the trunnion

The value stream mapping is a proven tool that helps eliminate waste. This method can be used for the widest range of industries and processes, it is ideal for positive organizational change in companies, helps to develop optimal management solutions to reduce cost, improve the quality and flexibility of processes. Mapping helps to see and present a set of operations in a particular sequence in the future value stream. Research has shown that the application of the method of value stream mapping enables not only to significantly improve the existing system of item manufacturing, but also to reduce the final cost of the product. Further investigation in the present direction will help to optimize existing processes without violating the strict sequence of operations what is necessary in certain production cycles [10]. Even simple, at first glance, perfections can lead to radical improvements in the final result.

These studies were conducted on the equipment of CAM technology common use center (RFMEFI59314X0003).This work was supported by the Ministry of education and science of the Russian Federation in the framework of the implementation of the Program "Research and development on priority directions of scientific-technological complex of Russia for 2014-2020".

References

[1] S.M. Zahraee, A. Hashemi, A.A. Abdi, A. Shahpanah, J.M. Rohani, Lean manufacturing implementation through value stream mapping: A case study, Jurnal Teknologi (Sciences and Engineering) , vol. 68, Issue 3. (2014), p. 119-124.

[2] S.M. Zahraee, J.M. Rohani, Production line analysis via value stream mapping: a lean manufacturing process of color industry, 2nd International Materials, Industrial, and Manufacturing Engineering Conference, MIMEC2015, 4-6 February 2015, Bali Indonesia.

[3] J. Promngurn, N. Sirivongpaisal, S. Suthummanon, W. Kongkaew, P. Penchamrat, An application of lean supply chain management for cost reduction in block rubber industry, Advanced Materials Research, vol. 844. (2014), p. 450-453.

[4] T.T. Burton, Faster value stream maps, Industrial Engineer, vol. 46. (2014), p. 45-50.

[5] F Morlock, T Dorka, H Meier. Concept for a Performance Measurement Method for the Organization of the IPS^2 Delivery, Product Services Systems and Value Creation: Proceedings of the 6th CIRP Conference on Industrial Product-Service Systems. Procedia CIRP, Volume 16; 2014. P. 56 – 61.

[6] A.R. Rahani, M Al-Ashraf, Production flow analysis through Value Stream Mapping: A lean manufacturing process case study, 2nd International Symposium on Robotics and Intelligent Sensors 2012, Sarawak, Malaysia, vol. 41 (2012),p. 1727-1734.

[7] Wei Xia, Jiwen Sun, Simulation guided value stream mapping and lean improvement: A case study of a tubular machining facility, Journal of Industrial Engineering and Management, JIEM, 2013, vol. 6(2), p.456-476.

[8] V.G. Smelov, V.V. Kokareva, Lean organization of additive manufacturing of aircraft purpose products, International Journal of Engineering and Technology, 2014, vol. 6, Issue 5, p 2304-2309.

[9] V.G. Smelov, V.V. Kokareva, The organization of engineering shop “lean” management system, 5th International Scientific Practical Conference on Innovative Technologies and Economics in Engineering, 2014, Applied Mechanics and Materials, 2014, vol.682, p. 555-560.

[10] K. Venkataramana, B. Vijaya Ramnathb, V. Muthu Kumarc, C. Elanchezhiand, Application of Value Stream Mapping for Reduction of Cycle Time in a Machining Process, 3rd International Conference on Materials Processing and Characterisation (ICMPC 2014), Procedia Materials Science 6 (2014), p.1187 – 1196.

A System of Evaluation of Engineering Solutions Competitiveness of a Company Based on its Image

G O Tashchiyan[1], M A Suzdalova[1], S V Grichin[1], E S Goriaynova[1] and M S Kaz[2]

[1] Yurga Institute of Technology (affiliate) of National Research Tomsk Polytechnic University Leningradskaya 26, Yurga, 652055
[2] Tomsk State University Lenin Ave 36, Tomsk, 634050

E-mail: gtashiyan@mail.ru

Abstract. The paper presents a system of evaluation of engineering solutions competitiveness of a company based on its image using point rating criteria obtained through survey and questionnaire.

1. Introduction

The main task in achieving a stable position on the market is ensuring competitive engineering solutions.

Many Russian and western authors examined the problem of competitiveness of both a product and a company.

J.Robinson, for example, believed that competitiveness was not only the ability to withstand competition, but also to evade the fight, developing new markets of diversified products [1]. J.Schumpeter pointed out that a company's competitiveness depended on the ability to create new technologies, new markets and new ideas [2]. P. Drucker considered the human factor to be the basis of competitive advantage, that is a company can be more competitive if there are efficient managers in it and if the focus is made on knowledge [3]. M.Porter pointed out that competitive advantage was in a more efficient use of resources [4]. G.Hamel defined competitiveness as a notion based on skills, knowledge, experience, abilities, foresight, intellectual leadership [5].

Russian authors have also made their contribution in defining the concept of competitiveness. For example, E.A. Gorbashko believes that competitiveness is the company's ability to compete [6]. G.D. Antonov, O.P. Ivanova and V.M. Tumin defined competitiveness as the ability to stay ahead of the others, using one's own advantages to achieve the set goals [7]. I.A. Tarasova defined competitiveness as a complex of consumer and cost characteristics of goods, which determine its success in the market [8]. In an explanatory dictionary, competitiveness is defined as the ability to compete in the markets of goods and services [9].

In [10], the authors define the term competitiveness of a company as its advantage in relation to other companies in the industry at home and abroad. They also pointed out that competitiveness was not an intrinsic quality of a company. This means that a company's competitiveness can be applied to a group of companies operating in the same industry or producing similar goods (services). Thus, competitiveness is determined only by comparing the companies on the national and world scale.

2. Criteria of a company engineering solutions evaluation based on its image.

Evaluation of competitiveness of a company engineering solutions is rather a difficult task because there are certain economic, financial performance, and others indexes which characterize the performance of a company.

In this paper we consider the system of evaluation of competitiveness of a company engineering solutions based on its image, using the experience of assessing the competitiveness of a knowledge-based engineering products, as in [11].

Mechanical engineering in Russia is one of the most common industries in territorial aspect [12]. Currently, modern technological engineering systems [13] are required for the development and improvement of public production as a basis for achieving the economic power of the country.

Mechanical engineering is the main branch of industrial production, which affects the development of various spheres of economic activity of the country, it reflects the scientific and technical condition of the country [14]. The range of products is very diverse and includes shearers, shaft lining, Jib mobile cranes, etc.

At the beginning of 2011 the Russian mines used 134 shearers, 108 of which were imported, mainly from Ukraine, Poland, Germany and the United States [15]. The problem of low demand for Russian shearers is partly due to low competitiveness of domestic producers. This is not just the problem of mechanical engineering companies, it is characteristic of all the sectors of the Russian economy.

The image of a company in the eyes of its consumers plays an important role in the evaluation of its engineering solutions.

The image of a company is one of the important aspects of perception and evaluation, which produces a special impression on others. The image of a company is in people's minds. Formation of a positive image for a company is an advantageous and less laborious process than correcting a negative image, which is sometimes formed spontaneously.

The main condition to achieve sustainable business success by a company now is its positive image, because it gives a certain market weight, strength; it ensures the involvement of consumers and partners; it improves the company's competitiveness in the market; it protects the company from the substitute products of competing companies; it increases sales; it helps to maximize profit; it opens prospects for crediting; it provides a better access to various resources, etc.

The negative image not only arouses negative emotions of the public, but also leads to reducing the number of orders and sales volumes, which can lead to a shutdown.

Creating a positive image of a company includes a number of such activities as advertising, presentations, press conferences, symposia, meetings, patronage, sponsorship, public performance, publication of articles about the company, etc. Depending on the characteristics of a company, its scope of activities, location, customers, and other factors, specific actions for the formation of the image will be identified

Company image is formed of the following components:

- the image of the product, commodity or brand;
- the internal image of a company, which is the notion of employees on their company;
- the image of the company management team;
- the visual image, which is the opinions of others on the appearance of a company and personnel (interior, exterior, branding);
- consumers' image, when large customers may become the "face" of a company;
- service image, which is the notion of direct and indirect purchasers on the quality of service;
- social image, which is the notion of the general public about the social purposes and the role of a company in economic, social and cultural life;
- business image of a company, which is the notion of competent bodies and experts on the goodwill, good faith, performance of obligations, economic ties, patent and license protection, etc.

Survey and questionnaire have long been used for the assessment of different states of a controlled object.

Table 1. Criteria for the competitiveness of an engineering company solutions based on its image and evaluation score

Criteria of company image	Score
1. Image of the company head (I_{ch})	
1.1 Positive image of the company head	10
1.2 The indicator is above the average	7
1.3 Average	5
1.4 Below the average	3
1.5 Negative image	0
2. Image of the employees (I_e)	
2.1. Positive image of the employees	10
2.2. The indicator is above the average	9
2.3. Average	8
2.4. Below the average	5
2.5. Negative image of the employees	0
3. Social image of the company (S_{ic})	
3.1. A high level of social responsibility of the company	10
3.2. The level of social responsibility of the company above the average	8
3.3. Average level of social responsibility of the company	6
3.4. Low level of social responsibility of the company	3
3.5. The company does not care about the social level	0
4. Image of products or services (I_{ps})	
4.1. A high quality level of products and services is a major factor in the competitiveness of the company	10
4.2. The stability of production and services. It allows the company to be competitive.	7
4.3. Products and services partially affect the level of competitiveness of the company	3
4.4. Products and services do not affect the company's image	0
5. Business culture of the company and its style (B_c)	
5.1. Highly cultured style of the company	10
5.2. Average business culture of the company	7
5.3. Low business culture of the company	2
5.4. Lack of business culture in the company	0
6. Symbols and attributes (B_a)	
6.1. A highly professional and well considered symbols and attributes associated with the activities of the company	10
6.2. Bright and interesting symbols and attributes of the company	9
6.3. High-grade symbols and attributes of the company	8
6.4. Business style symbols and attributes	6
6.5. Simple and unvaried symbols and attributes	2
6.6. There are no symbols or attributes in the company	0
7. Business image (B_i)	
7.1. Highly professional business image	10
7.2. Middle level professional business image	8
7.3. Stable and reliable business image	5
7.4. Low level of stability, reliability and competitiveness of the company	1

For example, paper [16] presents a method of survey of engineering companies aimed at evaluating the performance of the staff from among the graduates from the Russian universities. In order to determine the competitiveness of the engineering company solutions based its image, the authors suggest using a questionnaire as a method of expert evaluation of the above indicators using score-rating methodology. As estimates, the 10-score evaluation scale can be used. The list of criteria for the competitiveness of an engineering company solutions based on its image and evaluation score (points) are presented in Table 1.

The score-rating technique has already been widely used. According to [17], the shortcomings of using expert techniques can be overcome by clearly articulating the task, by including universal performance indicators for expert evaluation, and by developing a dedicated decision-making software.

The authors of [18] also used the score-rating methodology. However, they faced a number of challenges that they offer to solve with the help of:

1. A clear formulation of a problem.
2. Inclusion of universal performance indicators into the procedure for expert evaluation.
3. Developing a dedicated decision-making software application.

In our case, the first task is defined. It is the evaluation of the competitiveness of the engineering company solutions.

The second task is the expert evaluation of competitiveness of a company engineering solutions based on its image using the proposed universal performance indicators.

Calculation of the competitiveness of the engineering company solutions based on its image (Kim) is carried out with the help of the following formula:

$$Kim = \frac{1}{10n}\sum_{i=1}^{n}\sum_{j=1}^{7} a_j Kio_{ij},$$

in which:

Kim is the index of competitiveness of the company's image;

Kio_{ij} is the assessment by *i*-expert of *j*-index of the image;

i is the number of experts;

j is the number of estimated criteria of the image;

a_j *is* the impact of *j*-criterion of the image;

10n is the maximum possible score that an evaluated criterion may get.

The third task is to use a special decision-making software "Automated monitoring of the competitiveness of a company image", registered in the Federal Service for Intellectual Property dated 5/14/2015, number 2015615283.

Studies to evaluate the competitiveness of engineering solutions of mechanical engineering companies based of their image were made using this method. The following companies were studied: «Caterpillar» (USA), «Joy» (USA), «Glinik» (Poland), «Ostroj» (Czech Republic) и «JSC PA Yurmash» (Russia, Yurga). The students of Yurga Institute of Technology took part in the survey (in total, 10 students were polled). The results are given in Table 2.

According to the results of the study, the highest level of competitiveness of engineering solutions of a company based on its image belongs to «Caterpillar» (USA) and amounts to 67.9 points. The runners up are «Joy» (USA) and «Glinik» (Poland) which scored 67.3 and 64.3 points, respectively. The next is «Ostroj» (Czech Republic) with the result of 62.3 points. The last in the list of five is «JSC PA Yurmash» with a score of 57.8 points.

The main reason for high results achieved by the leaders is the level of quality and low price for their products. The following reasons can be qualified as additional ones: highly qualified personnel;

compliance with the required standards and safety requirements; a well-developed marketing campaign.

Table 2. Evaluation of competitiveness of a company engineering solutions based on its image

Company name	Evaluation criteria							Total
	I_{ch}	I_e	S_{ic}	I_{ps}	B_c	B_a	B_i	
«Caterpillar» (USA)	9,4	9,5	9,6	9,8	9,9	9,9	9,8	67,9
«Glinik» (Poland)	9,1	9,2	9,1	9,3	9,1	9,4	9,1	64,3
«Joy» (USA)	9,5	9,5	9,4	9,6	9,8	9,7	9,8	67,3
«JSC PA Yurmash»	9,1	8,2	7,3	8,1	7,2	9,1	8,8	57,8
«Ostroj» (Czech Republic)	9,4	8,5	8,8	7,5	9,8	9,1	9,2	62,3

Consequently, «Caterpillar» can be described as dynamically growing company. The same can be said about «Joy» and «Glinik» companies. The position of «JSC PA Yurmash» in the market of mining equipment is rather weak.

Enterprise management in the modern market economy is an art that requires innovative solutions in the field of strategic management. Lack of such solutions will prevent a company from developing effectively in the long term.

Thus, «JSC PA Yurmash» should consider new engineering solutions for filling a new segment in the mining equipment market.

3. Conclusion

Any company should persistently pay attention to the creation of a loyal attitude to itself. Only in this case, it can count on the continued success and acceptance in the business community.

Creating a positive image of a company is a complex and multi-faceted process, which requires a serious attitude and tremendous effort.

Thus, the image of a company is one of the main components that determine the competitiveness of company engineering solutions

References

[1] Robinson G. 2006 The economic theory of imperfect competition. (M Economics) 390
[2] Schumpete J. 2005Capitalism, Socialism and Democracy. (M. Znaniye) 419
[3] Drucker P. 2002 Effective management. (M. FAIR PRESS) 288
[4] Porter M. 2003 Competition. (M. Williams) 496
[5] Hamel G. 2002 Competing for the Future. (M. Olymp-Business) 288
[6] Gorbashko E.A. 2014 Management of competitiveness. Theory and practice: the textbook for graduates / ed E.A, Gorbashko, I.A. Maksimtseva (M. Yurayt Publishers) 447
[7] Antonov G.D., Ivanova O.P., Tumin V.M. 2012 Management of competitiveness of the organization (M. INFRA-M) 300
[8] Tarasov I.A. 2006 The takes the choice. *Marketing Research in Russia and abroad.* **4** 15
[9] D. Blek 2000 Economics Dictionary ed Osadchaya I.M. (M. INFRA-M)
[10] Gruzinov V.P., Gribov V.D. 2000 Enterprise economy. (M.: Finance and Statistics) 208
[11] Tashchiyan G.O., Sushko A.V., Grichin S.V. 2015 Microsoft Business Solutions-Axapta as a basis for automated monitoring of high technology products competitiveness. *Paper presented at the IOP Conf. Series: Materials Science and Engineering* ***91***

[12] Suzdalova M.A., Politsinskaya E.V., Sushko A.V. 2015 About the problem of professional personnel shortage in mechanical engineering industry and ways of solving. *Procedia - Social and Behavioral Sciences.* **206** 394-398

[13] Osipov Y.M., Osipov O.Y., Trifonov V.A. 2015 Prospects for Establishment of Technological Complexes in Machine Building Industry on The Basis of Electromechatronic Propulsion Systems. *Paper presented at the IOP Conf. Series: Materials Science and Engineering* ***91***

[14] Loshchilova M.A., Lizunkov V.G., Zavjyalova A.V. 2015 Professional Training of Bachelors in Mechanical Engineering, Based on Networking Resources. *Procedia - Social and Behavioral Sciences.* **206** 399-405

[15] Prokopenko S., Sushko A., Kurzina I. 2015 New design of cutters for coal mining machines. *Paper presented at the IOP Conf. Series: Materials Science and Engineering* ***91***

[16] Lizunkov V. G. , Marchuk V. I. , Podzorova E. A. 2015 Identification of Criteria, Features and Levels of Economic and Managerial Competencies Development for Bachelors in Mechanical Engineering. *Procedia - Social and Behavioral Sciences.* **206** 388-393

[17] Trifonov V.A., Shabashev V.A., Verzhitsky D.G. 2014 The problems of developing the ecological market of the region. *Applied Mechanics and Materials.* **682** 591-595

[18] Nesteruk D.N., Momot M.V., Beglyakov V.Y., Politsinskaya E.V. 2015 Expert evaluation of the Quality of the Development of New Technics' Samples: Experience in the Development of a New Kind of Multipurpose Shield-Type Heading Machine. *Applied Mechanics and Materials.* **756** 709-713

IOP Conf. Series: Materials Science and Engineering **142** (2016) 012104 doi:10.1088/1757-899X/142/1/012104

Automatic Positioning System of Small Agricultural Robot

M V Momot*, A V Proskokov*, A S Natalchenko*, A S Biktimirov**

*Yurga Institute of Technology (Branch)
of National Research Tomsk Polytechnic University,
**Yurga-Technologies-Innovations, LLC
652055, Kemerovo Region, Yurga, 26 Leningradskaya St., tel. (38451) 7-77-62

E-mail: mikl_1239@mail.ru

Abstract. The present article discusses automatic positioning systems of agricultural robots used in field works. The existing solutions in this area have been analyzed. The article proposes an original solution, which is easy to implement and is characterized by high-accuracy positioning.

Introduction

Automatic motion control systems and parallel driving systems find wider application in production of agricultural machinery. These systems include electronic and hydraulic components to drive vehicles. Machinery can operate in fields using automatic pilot, however operator's work is still needed in this case. Operator helps machinery to turn, and levels out the route at significant deviations; in all other situations load on operator is reduced. Accurate field works completion enables to reduce time expenditure and overwork by 10%. It also allows saving fuel and spending mineral fertilizers reasonably. Processing of fields with herbicides becomes more effective. All the known automatic piloting systems imply mounting of direction indicator, controller and signal receiver of global satellite positioning GLONASS or GPS [1,2,5]. Scientific literature contains solutions based on using machinery vision systems for navigation [3,4], as well as integrated complex control systems including machinery of different use and controlling complex [6]. All the mentioned facilities are preferable when processing fields are large and boundless and minimum obstacles are observed. However accuracy of satellite navigation positioning without adjustment is known to be 10-15 meters. Thus it makes it hard to use them in agricultural machinery working on patches of land up to 1 hectare, particularly in rural areas where the territory is strictly divided between the land owners. Differential corrections obtained from geostationary satellites or from ground base stations are generally not provided for free and thus cannot be an option for agricultural machinery of small size.

Suggested solutions

When developing autonomous control system of agricultural robot one needs to define positioning and navigation concept. For that purpose the main checklist of algorithms of manual and automatic motion control should be developed. Robot system is to possess the following driving modes:

1. Remote manual control by operator through radio channel. Such control system should be able to give commands to onboard computer up to the distance of 500 m. The mentioned

IOP Conf. Series: Materials Science and Engineering **142** (2016) 012104 doi:10.1088/1757-899X/142/1/012104

commands are: start/stop engine, select direction "forward/backward", start/stop motion, wheel turn left/right. In this case operator is to monitor all the movements through video cameras mounted on the robot.

2. Automatic control enables the operator to set guiding points at the computer while the system calculates motion trajectory taking into consideration the complexity of land terrain. While soil processing it should also consider plowing modes (conventional, round, and flat).

Autonomous positioning system of agricultural robot can have a number of solutions. Simple and effective one is monitoring the robot position according to four tracking devices with infrared transmitters which are preliminary mounted on the angles of trapezoidal area. Each of the tracking devices generates original digital code. Infrared receiver with accurate rotating mechanism and narrow aspect angle in a horizontal plane is fixed on the robot body.

At the definite key moments the robot stops and coordinates its location in accordance with tracking devices. Information from indicators considering trajectory parameters and overall robot dimensions is transmitted to the electronic microcontroller and the system defines the current position of machinery relative to indicators. It is essential that the relative position (winding order) is not infringed.

Let us consider this system operation at the example. The first task for a robot before it starts operation is defining the initial area dimensions. Measuring of the area can be conducted by the following ways: 1) performing accurate motions of tractor and recording the alterations; 2) using several indicators on the tractor with the known distance between them.

The first way is marked by a significant drawback as calculating the length of the passed reference area is not an easy task due to rough and patchy land, and difference in soil resistance, also for different wheels. However in case we use internal system of distance measuring, based on e.g. accelerometer, than this drawback can be rectified. The measuring principles are similar in both cases, thus we first consider the fist way in detail and then turn to peculiarities of the second one.

Figure 1 illustrates schematically the robot displacement when defining parameters of the given area. I1-I4 – infrared tracking devices; A, B, C – tractor position; A2, B2, C2 – points located vertically lower (at the Figure) than the original points of robot position.

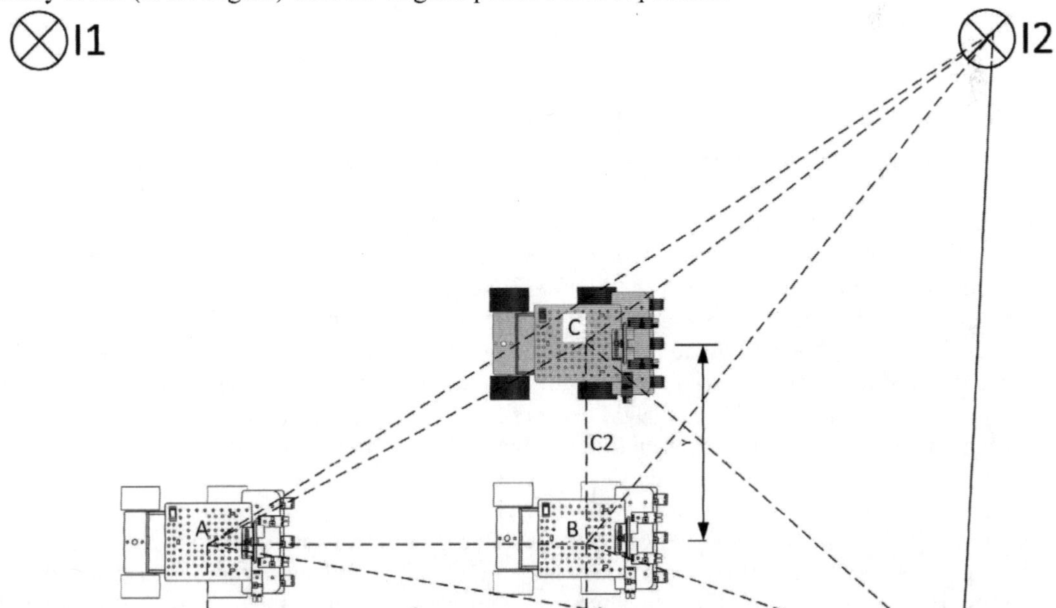

Figure 1: Robot displacement when defining parameters of a given area

Using infrared sensors robot scans 360° of space by means of servomotors (Fig. 2) to search for tracking devices. Upon finding the indicator the angle and number of the found tracking device is saved in the system memory.

At the point A we find angles A2-A-I1, A2-A-I2, A2-A-I3, A2-A-I4, than we move to the prescribed distance X to the point B, find angles B2-B-I1, B2-B-I2, B2-B-I3, B2-B-I4, at the right angle to the length A-B, after we move to point C, which is located from B at the distance Y, we find angles C2-C-I1, C2-C-I2, C2-C-I3, C2-C-I4. The obtained parameters are enough to define location of all four tracking devices regarding the tractor on the plane. Let us consider the example. We find distances C-I4 and B-I4.

Having the distance BC and two angles B_2BI_4 and C_2CI_4 , while $C_2BI_4 = 180^0 - B_2BI_4$,

We find the third angle of triangle:

$$CI_4B = 180^0 - C_2CI_4 - C_2BI_4. \qquad (1)$$

Now we have the necessary values to apply standard rule for solution of triangle:

$$CI_4 = BC\frac{\sin(C_2BI_4)}{\sin(CI_4B)}; \qquad (2)$$

$$BI_4 = BC\frac{\sin(C_2CI_4)}{\sin(CI_4B)}; \qquad (3)$$

All the distances and missing angles required for positioning can be defined the same way.

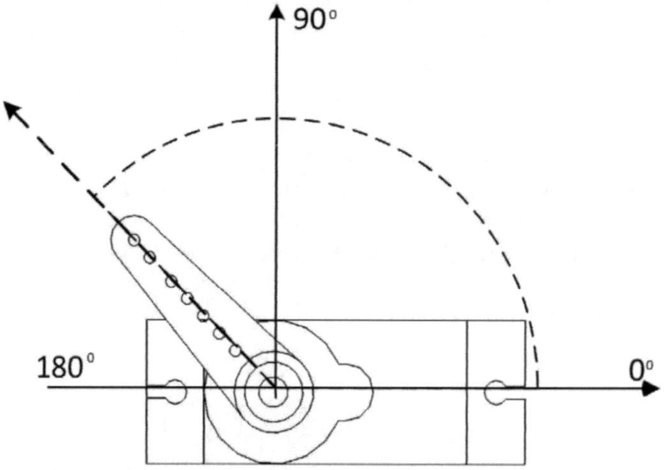

Figure 2: Operation principle of servomotor while finding the angle to tracking device

The second way can be implemented easier and faster by means of sensors placement along the angles of tractor roof. In this case the roof parameters are known, thus we only need to solve geometrical task and memorize parameters of the area (distances between tracking devices and angles), location and direction of the robot.

Two sensors mounted on servo drives with maximum turning angle 360° or four sensors with maximum turning angles 180° are enough. Sensors are fixed at the maximum possible distance within the robot body and the distance between them is considered for further calculations.

Generalized algorithm of measuring the angles between tracking devices and the robot is given in Fig. 3. Four infrared sensors with narrow field of vision in horizontal direction are used here, though other types of sensors are also possible to be used, e.g. radio position finding. Different types of sensors can be combined to control errors of random reflections. The algorithm suggests displacement of servo drive by ~0.01° each time, however it is experimental value, thus it can be changed within capacities of control circuit, servo drive resolution and area size. Each of the tracking devices generates its own original code which is known to our program. The unknown codes are rejected as invalid or interfering signals. After being launched the program should define coordinates of the robot in regards to the area and select the winding order suggesting that if some right/left side is missing than its processing has already been conducted before program restarting.

Considering that the angle of view of the sensor can be not that narrow, additional algorithm of center finding should be applied. Any other instrument can be used as a servo motor only in case its resolution capacity enables to position angles accurately on the area under study. How to control such instrument is described in details in [7].

Algorithm (Fig.3) implies accurate positioning of turning angles of servo motors with sensors that can lead to long-term waiting of process completion; however this process can be accelerated significantly by means of additional optimization.

Figure 4 illustrates a simplified algorithm of robot movement along the defined trajectory. Along with positioning system based on infrared sensors it is suggested to use electron gyroscope and accelerometer. Thus it makes possible to calculate the exact coordinates not constantly but in the definite distance intervals. The rest time, electron gyroscope is used to define turning angle while accelerometer is used to define the distance that has been passed. Magnetometer is also possible to be used. Low-cost instruments for not precise motion stabilization are also suggested for use. Systems of that kind have been considered in [8].

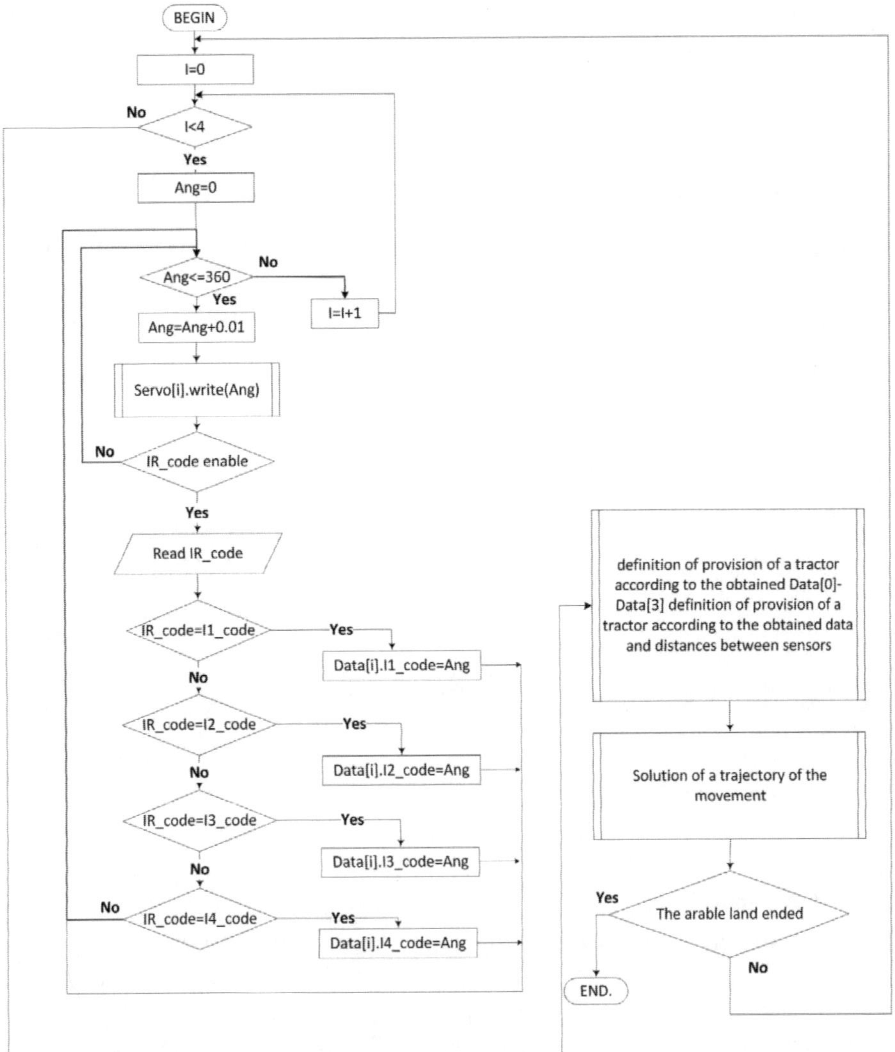

Figure 3: Generalized algorithm of angles finding for tracking devices

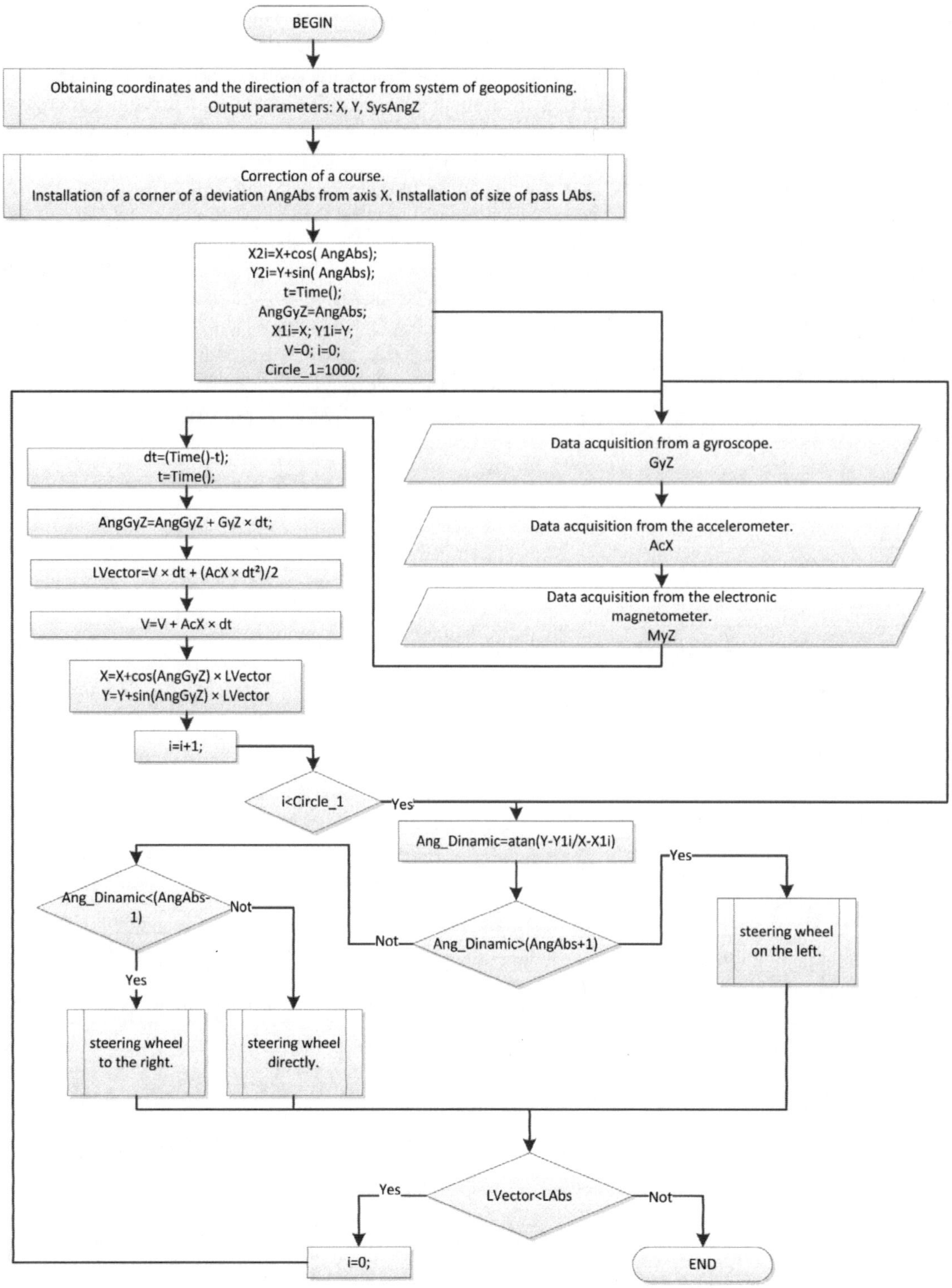

Figure 4: Simplified algorithm of robot movement along the defined trajectory

IOP Conf. Series: Materials Science and Engineering **142** (2016) 012104 doi:10.1088/1757-899X/142/1/012104

Conclusions

Resulting from use of the above mentioned solutions the system of robot automatic positioning has been developed. It allows conducting field works without using GPS and complex systems of machinery vision. The system is low-cost and has high quality indicators. In case the mentioned system is integrated into a complex system, it is possible to develop the maximally effective structure from the point of owning cost.

References

[1] Manuel Perez-Ruiz a,*, David C. Slaughter b, C. Gliever b, Shrini K. Upadhyaya b Tractor-based Real-time Kinematic-Global Positioning System (RTK-GPS) guidance system for geospatial mapping of row crop transplant. // biosystems engineering 111 (2012) p. 64 – 71

[2] Murilo Aparecido Voltarelli, Rouverson Pereira da Silva, Vicente Filho Alves Silva, Fábio Alexandre Cavichioli, Ariel Muncio Compagnon Performance Mechanized Set Tractor-Planter of Sugarcane Planting in Two Operation Shifts. //Journal of Agricultural Science; Vol. 5, No. 11; 2013 p.54-56

[3] K. E. Rumyantsev and S. V. Kravtsov, "Positioning autonomous mobile robot based on measurements onboard digital stereo vision system", International Journal of Robotics Applications and Technologies (IJRAT), vol. 2, No. 2, 37-77, (July-December 2014)

[4] Kümmerle, R., Ruhnke, M., Steder, B., Stachniss, C. and Burgard, W. (2015), Autonomous Robot Navigation in Highly Populated Pedestrian Zones. J. Field Robotics, 32: 565–589. doi: 10.1002/rob.21534

[5] T.Luettel, M. Himmelsbach, F.V. Hundelshausen, M.Manz, A. Mueller and H.-J. Wuensche. Autonomous Offroad Navigation Under Poor GPS Conditions, in Proc. of the IROS Workshop on Planning,Perception and Navigation for Intelligent Vehicles. St Louis, MO, USA. October 2009

[6] Luis Emmi, Mariano Gonzalez-de-Soto, Gonzalo Pajares, and Pablo Gonzalez-de-Santos, "New Trends in Robotics for Agriculture: Integration and Assessment of a Real Fleet of Robots," The Scientific World Journal, vol. 2014, Article ID 404059, 21 pages, 2014. doi:10.1155/2014/404059

[7] Momot, Mikhail Viktorovich. Algorithm of Controlling Servo Drives by Means of Atmega 8 Microcontroller [Electronic resource] / M. V. Momot, A. Biktimirov // Applied Mechanics and Materials : Scientific Journal. — 2014. — Vol. 682 : Innovation Technology and Economics in Engineering. — [P. 596-599]

[8] Kang Hou, Hanxu Sun, Qingxuan Jia, Yanheng Zhang, An Autonomous Positioning and Navigation System for Spherical Mobile Robot, Procedia Engineering, Volume 29, 2012, Pages 2556-2561, ISSN 1877-7058, http://dx.doi.org/10.1016/j.proeng.2012.01.350

IOP Conf. Series: Materials Science and Engineering **142** (2016) 012105 doi:10.1088/1757-899X/142/1/012105

Optical-Mechanical System for Stabilizing an Inverted Pendulum

O N Kuzyakov, M A Andreeva

Tyumen State Oil and Gas University
70 Melnikayte Str., 227, Tyumen 625027, Russia

e-mail: onkuzyakov@mail.ru

Abstract. Controlling open-loop unstable systems is a common benchmark for designing the algorithms to maintain the equilibrium state of anthropomorphous technical devices used within control theory. In this connection, considerable attention is currently being focused on the problem of stabilizing the inverted pendulum system. In this work, the execution of swinging-up the pendulum and, subsequently, maintaining its upward equilibrium state is presented with the help of the laboratory bench TP-802 by Festo Didactics and the movement control device. The configuration of dynamic system for stabilizing the inverted pendulum is offered. The algorithms to swing-up the pendulum and balance it around its upright position are offered as well.

1. Introduction

The problem of controlling the representatives of a pendulum type objects is fundamental for a range of scientific faculties. Its solution is reflected in the investigation of control techniques [1-3] used while modeling flying machines, rockets, anthropomorphous robots, segways etc [4].

A physical pendulum is one of the most popular physical models which consists of an oscillating massive object (pendulum bob) hung by an extensible weightless string or a thin vertical rod.

A special case of this type of systems is an inverted pendulum – an unstable physical object [5] with two equilibrium positions: in the highest and the lowest points, wherein an arbitrary small disturbance of any kind will accelerate it from the highest point down to the lowest point. Different kinds of feedback devices providing with pendulum movement control [6] may be added to the system so that the pendulum is stabilized at its upward equilibrium state.

The scientific works [7-11] are addressed to stabilizing the inverted pendulum.

A linear model of the system motion is described by two differential equations [12, 13], which represent the dependence of the vertical pendulum angle on force applied to the cart and the horizontal displacement of the cart.

$$\ddot{\theta} = \frac{1}{(I+ml^2)}\left(mglxin\theta - \ddot{x}mlcos\theta - c\dot{\theta}\right), \quad (1)$$

$$\ddot{x} = \frac{1}{M+m}\left(F - \ddot{\theta}mlcos\theta + \dot{\theta}^2mlxin\theta - k\dot{x}\right), \quad (2)$$

considering m is a mass of the pendulum, M is a mass of the cart, l is a length of the pendulum rod hinged on the cart, θ is a vertical pendulum angle, F is a horizontal force exerted on the cart, x is a displacement of the cart, I is a second moment of inertia, k is a viscous friction coefficient of the cart, c is a viscous friction coefficient of the pendulum.

2. Elaboration of balancing the inverted pendulum system

The task assigned in this work was to simulate and to implement the process of swinging-up the pendulum succeeded by stabilizing it in this position with the use of the laboratory bench TP-802 by Festo Didactics as a setup adjuster of the upward equilibrium state of the pendulum along with other stabilizing components.

Therefore, a physical control object is a pendulum inverted after a certain number of electromechanical drive carrier movements, operated by a step motor MTR-ST with PC-master by way of applying coordinate positioning controller SPC-200 [14].

The stabilization system startup follows swinging-up the pendulum. In order to complete this task, considering [15, 16], the algorithm represented in Fig.1 and the application program for positioning the cart were elaborated. The following assumptions have been made: (1) a cart moves along the X-axis between the points X = 0 mm and X = 300 mm; (2) the reference position of the cart is X = 150 mm; (3) N is a displacement value of the cart from the reference point; (4) K is a preset displacement value increment.

The code of the program for swinging-up the pendulum elaborated in software Festo WinPisa 4.41 [17] is presented below.

```
N000 #LR0 = 0
N001 G01 X150 FX10
N002 G00 G91 X-25
N003 G00 G91 X+50
N004 G00 G91 X-75
N005 G00 G91 X+100
N006 G00 G91 X-125
N007 G00 G91 X+150
N008 G00 G91 X-175
N009 G00 G91 X+200
N010 G00 G91 X-250
N011 G00 G91 X+30
N012 G00 G91 X-30
N013 M02
```

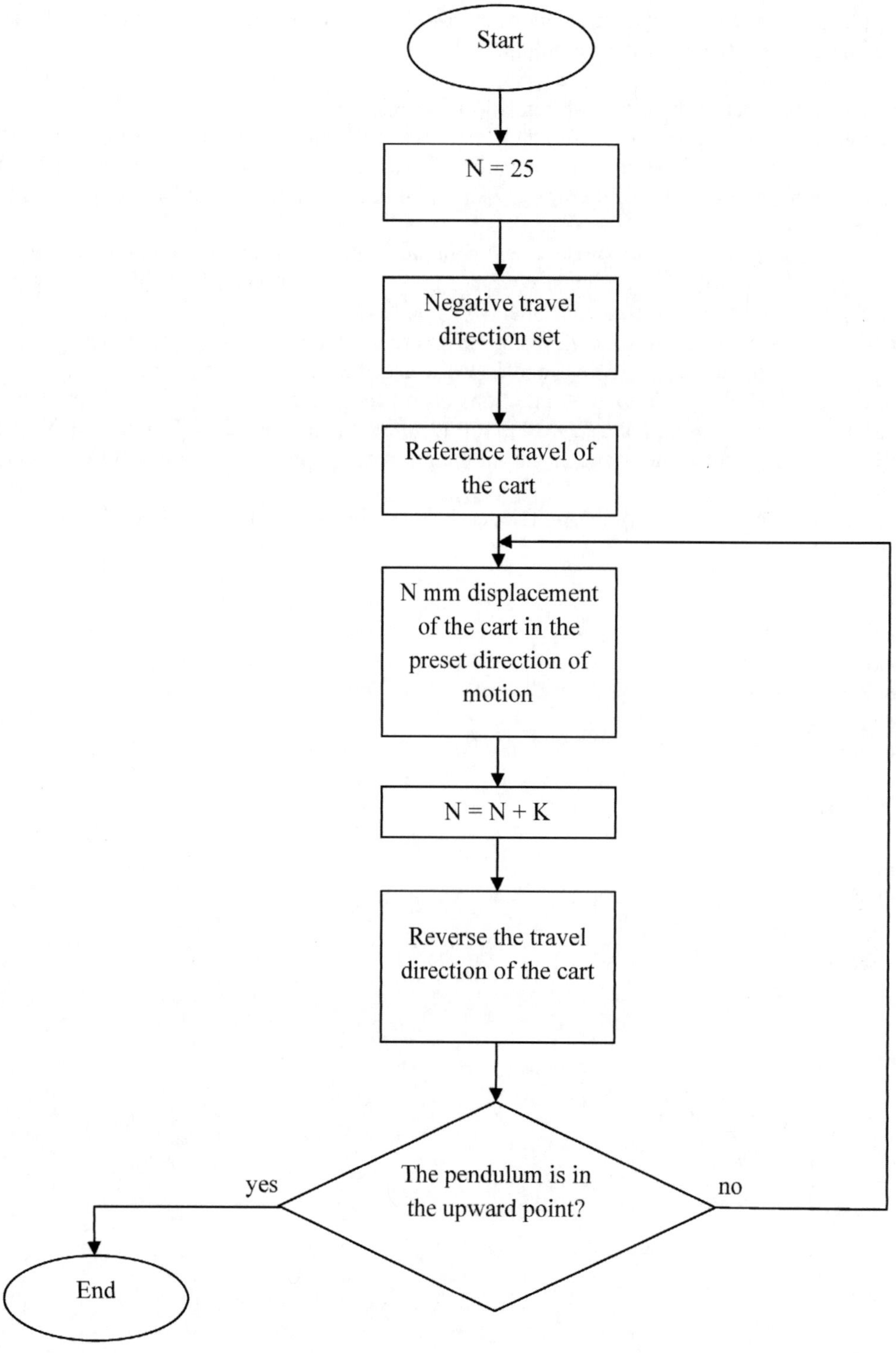

Figure 1. Algorithm of the swing-up procedure

After the program run the cart reference travel is executed. During running the strings 3-11 (in the code presented as N002...N010) of the program divergent oscillations appear, succeeded by two movements of the cart after which the pendulum is stabilized in the upright equilibrium position for a short period of time.

Immediately after the pendulum achieves the highest point the elaborated stabilization system starts controlling the cart with the pendulum hinged on the top of it. One of the key components of this system which structure is presented at the Fig.2 is an optical movement control device described in the paper [18].

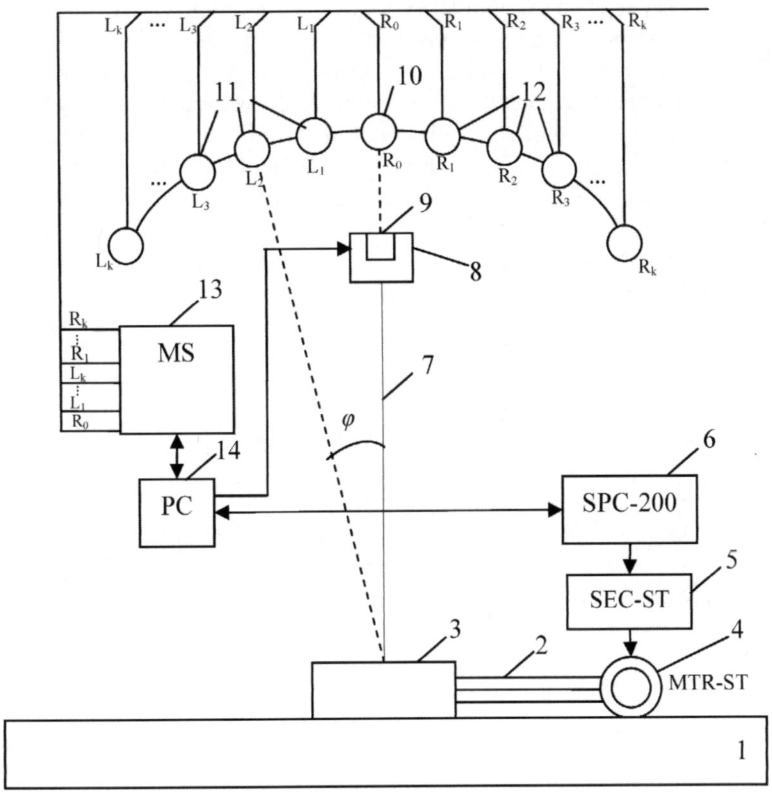

Figure 2. System structure

The cart 3 moves along the X-axis placed on a stable basement 1. A pendulum 7 consists of an oscillating massive object 8, hung by a thin vertical rod. At the top of the pendulum, there is a light emitter 9. A cart is rigidly bounded to the step motor 4 through the linear electromechanical drive 2. The step motor 4 is operated by positioning controller 6 with the help of a step motor controller 5. Computer 14 controls the light emitter 9 which include current translator mechanism [19]. The signals of the light sensors 10, 11, 12 are the input of the multiplexer switch 13 also controlled by the computer 14. The outputs of the light emitter and the multiplexer switch are connected to the computer 14. Moreover, the central light sensor 10 forms an output signal only in case the inverted pendulum is in vertical state.

The system works as follows: the pendulum 7 is swung-up into the upward unstable equilibrium state after a certain amount of moves of the cart operated by a step motor 5. Furthermore, the travel distance of the cart is 300 mm. The light emitter 9 attached to the pendulum bob 8 is switched on during the swinging-up. After a signal corresponding to the upward equilibrium state is produced (when the pendulum 7 is inverted) it flows through the multiplexer switch 13 to the computer 14 where it is finally captured.

There is no possibility the pendulum keeps the upward equilibrium state as there are physical forces applied to it. When the pendulum tilts with some angle, another light sensor captures the light emitter signal because of the displacement of the pendulum. The coordinates of the pendulum (the angular displacement of the vertical state) and the displacement direction are determined according to the index number of light sensor, which detected the light emitter signal the first. The number of light sensors and distance between each of them depends on the accuracy of measurements demanded. Measurement data flows to the computer 14 where it is processed according to the program being executed. Consequently, a control input for positioning controller 6 is formed, and a cart moves certain number of steps, which depends on the angular displacement of the pendulum.

Thus, an elaborated system stabilizes the upward equilibrium state of the inverted pendulum.

The algorithm of the system execution is presented at Fig.3. After the system startup an initialization of its components is carried out. The initialization includes microprocessor adjustment, positioning controller and step motor controller startup, forming switch-on emitter signal. Then the pendulum is inverted and the moment when the pendulum achieves the upward position is captured. The algorithm of this procedure is presented at Fig.1. After that, the direction and the amount of displacement from the vertical state are determined. Considering these, the control input is formed. The cart moves in the direction of the pendulum slope and the pendulum is eventually stabilized. The system is checked for the angular displacement presence again, and so on.

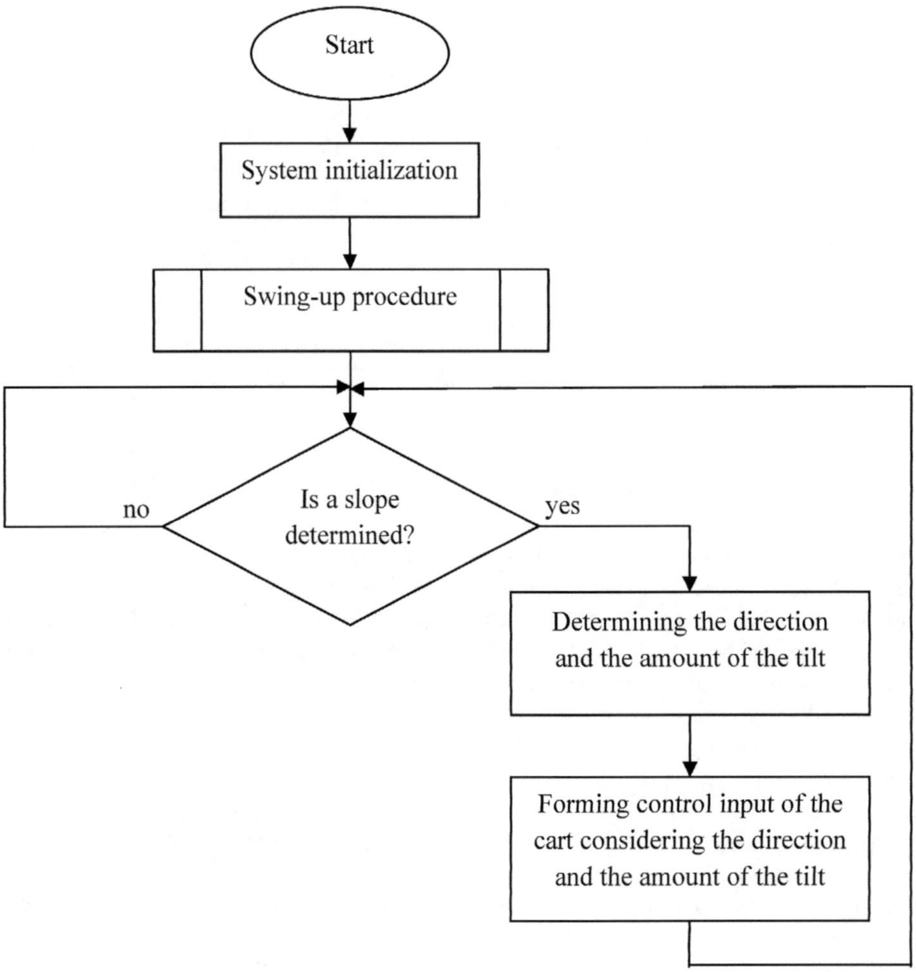

Figure 3. Algorithm of system execution

IOP Conf. Series: Materials Science and Engineering **142** (2016) 012105 doi:10.1088/1757-899X/142/1/012105

3. Conclusion

Thus, in this paper the application program of swinging-up the pendulum is elaborated, the algorithm of stabilization the pendulum in its upper equilibrium position is offered, the structure of optical-mechanical system for stabilizing the inverted pendulum consisting of computer with multiplexer switch, laboratory bench TP-802 by Festo Didactics and an optical movement control device is offered as well.

References

[1] Goldberg D E 1989 Genetic Algorithms in search, optimization, and machine learning, The University of Alabama
[2] Bugeja M 2003 Non-linear Swing-up and Stabilizing Control of an Inverted Pendulum System *IEEE Transactions on Industrial Electronics* (2003)
[3] Ngadengon R, Sam Y M, Osman J H S, Tomari R, Wan Zakaria W N 2015 Multirate Output Feedback with Discrete Time Sliding Mode Control for Inverted Pendulum System *Proc. Comp. Sci.* **76** 290-295
[4] Phuong N T, Loc H D, Thuan T Q 2013 Control of two wheeled inverted pendulum using sliding mode technique *IJERA* **3** 1276-1282
[5] Choi D, Oh J-H 2008 Human-friendly Motion Control of a Wheeled Inverted Pendulum by Reduced-order Disturbance Observer *2008 IEEE Robot. and Autom.* 2521-2526
[6] Ashish S Katariya 2010 Optimal state-feedback and output-feedback controllers for the wheeled inverted pendulum system *Georgia Institute of Technology* 72 p
[7] Kapitsa P L 1951 Dynamical stability of the pendulum with an oscillating pivot point *JETP* **21** (5) 588-597 (in Russian)
[8] Kapitsa P L 1951 The pendulum with an oscillating pivot point *Uspekhi Fizicheskikh Nauk* **44** (7) 7-20 (in Russian)
[9] Formalskiy A M 2006 Stabilization of the pendulum with an oscillating or non-oscillating pivot point *Doklady Akademii Nauk* **406** (2) 175-179 (in Russian)
[10] Breuer S 2005 Positioning system. Smart Positioning Controller SPC200. Manual, Festo AG & Co. KG, Dept. KI-TD
[11] Smith H T J, Blackburn J A 1992 Experimental study of an inverted pendulum *Am. J. Phys.* **60** (1992) 909-911
[12] Mahadi H, Chanchal S, Md Modtafizur R, Md Rabiual I S, Subrata K A 2012 Balancing of an inverted pendulum using PD controller *Dhaka Univ. J. Sci.* **60** (1) 115-120
[13] Landry M, Campbell S A, Morris K, Aguilar C O 2005 Dynamics of an inverted pendulum with delayed feedback control *SIAM J. Appl. Dynam. Sys.* **4** (4) 333-351
[14] Butikov E I 2001 On the dynamic stabilization of an inverted pendulum *Am. J. Phys.* **69** 755-768
[15] Bradshaw A, Shao J 1996 Swing-up control of the inverted pendulum systems *Robotica* **14** 397-405
[16] Astrom K J, Furuta K 2000 Swinging up a pendulum by energy control *Automatica* **36** 287-295
[17] Festo Ag&Co. KG 2009 Smart Positioning Controller SPC200. Software Package WinPISA. Manual
[18] Makarov A V, Kuzyakov O N 2000 Movement control device *Patent RF* 2150086
[19] Kuznetsov V P, Ivanov A A, Kudryashov B P 2013 Designing measuring facilities for technological objects based on fiber-optical converters, Kurgan State University (in Russian)

IOP Conf. Series: Materials Science and Engineering **142** (2016) 012106 doi:10.1088/1757-899X/142/1/012106

Optimization and Calculation of Probability Performances of Processes of Storage and Processing of Refrigerator Containerized Cargoes

A P Nyrkov, S S Sokolov, S G Chernyi, A A Shnurenko and L A Pavlova

Admiral Makarov State University of Maritime and Island Shipping, 5/7, Dvinskaya st., Saint-Petersburg, 198035, Russia

E-mail: sssokolov@mail.ru

Abstract. In the work the queueing system of the disconnected multi-channel type to which irregular, uniform or not uniform flows of requests with a unlimited latency period arrive is considered. The system is considered on an example of the container terminal having conditional-functional sections with a definite mark-to-space ratio on which the irregular inhomogeneous traffic flow with resultant intensity acts.

1. Introduction

Historically the mankind is indissolubly linked with navigation and for this reason the water transport is considered one of the most ancient aspects. Developing of a maritime transport in Russia is determined not only by geographical position, but also by its economic advantages. Marine cargo transportation remains the cheapest mode of transportation after pipeline, that leads to a demand and popularity of this mode of transport among others. As the cost price of a water transport remains the lowest it is competitive, despite a low velocity of delivery.

At present the maritime port is a large transport junction which links various modes of transports: marine, railway, automobile, pipeline, etc. Port activity is strategic aspect of developing of economy of the state and covers more than 50 % of all world cargo turnover [9]. The evidence of this fact is positive dynamics of growth of a cargo turnover of maritime ports of Russia According to statistical data during the period of January-May 2015 the cargo turnover of Russia has increased for 4,4 % in comparison with similar period of 2014, and in 2014 - for 5,7 % in comparison with the previous year [1].

A container terminals are being developed where large-capacity refrigerated containers with food stuffs are handled. Refrigerated containers are achievements modern technology. They are intended for transportation of goods demanding maintaining of special temperature and humidity. Basically such cargo containers are used for ocean carriages and storage of perishable cargoes. A temperature range of ref containers which they can maintain is from - 25°C up to + 25°C degrees of Celsius. According to the size refrigerated containers are divided into 20 and 40-foot. Refrigerated containers can be used for transportations practically for all modes of transports, and also for temporary storage of perishable cargoes as temporary when reloading on freight terminals, and long-term. Using of Refrigerated containers allows to deliver cargo «from door to door », not breaking a temperature condition of storage an route and during loading-unloading when changing a mode of transports.

For a diminution of costs part at a stage of transporting of cargoes up to a docks area and reducing the time of preliminary delivery container carriages are widely used, they have a very high degree of safety. Modern cargo transporters are designed developed and manufactured with observance of all existing norms and standards that warrants their durability and watertightness. It is one of the most convenient and popular aspects of transporting of cargoes. Popularity of usage of cargo transporters as universal package for a shipping goods all over the world grows [2-3].

Theoretical channel capacity of the terminal does not always coincide with actual. As the technical equipment of port in a direction of import and export, can be not always ready to storage or handling of containerized cargoes. In this connection, development of mathematical models which allow to optimize and moderate expenditures for the equipment and resources is necessary.

Process of storing goods in the container refrigerator terminal can be presented in the form of a queueing system. A conditional functional sections in port are considered as servicing devices in which the cargo-handling is carried out. The arriving boats which are a subject of handling are accepted as of a queueing system requests. If all moorings are occupied, the arrived ship will stop at the roadstead and wait its turn for mooring. Each vessel has the schedule of arrival, but because of unexpected circumstances it may be broken. Therefore it is considered, that arrival of boats is casual event that entails idle time of servicing devices or formation of turns. According to a response time of requirements in turn prior to the beginning of a upkeep of system in a classical theory of queues there are: systems of the "pure" and "mixed" type.

In system with "pure" (unlimited) expectation the figure of places in turns of boats and a latency period is not limited, i.e. each vessel during any time will be necessarily handled and a consignment will be located for storage. In limited expectation a refusal of the terminal to handle a vessel is possible. In real conditions the number of shipments in turn is always limited. To have analytical probabilities of separate condition statuses in many cases it is expedient to consider the figure infinite. Situations when in separate phases of time the part of ships can be transmitted for handling to other terminal because of excess of an admissible latency period are possible.

When using a of a queueing system of Markov model, the exponential (demonstrative) law of a upkeep is supposed. In calculations of probability performances of a queueing system, application of the exponential law, can lead to errors and an insufficient exactitude.

But, for today for exposition of a channel capacity of container terminals the determined methods are used. Given methods, actually, do not reflect all specificity of processing and storage of refrigerated cargoes: the interval of time between the moment of inflow of the requirement to the channel of maintenance and the moment of an exit of the requirement from this channel has no casual character, however, the moment of delivery consignments of containerized cargoes to the terminal presents a casual flow of events. In real conditions functioning of an infrastructure of the terminal from the point of view of processes of cargo handling are not adequate to above specified assumptions. Considering specificity of functioning of container terminals it is necessary to research stochastic processes of handling of ships, describing these processes on the basis of probability models, for a solution of research problems and optimization.

Let's consider other approach to calculation of performances of probability model of a queueing system.

2. Statement of a problem

The mathematical model of process of storage is formulated in view of irregular incoming of cargo consignments and their casual response time at depot in refrigerator container terminals. That leads in one case to idle time and overflow of places on storage depots, in others it leads to formation of turns.

To solve the research problem and to develop probability model it is necessary to research stochastic processes of storage of ships and to make their probability analysis. The given processes of a storing goods in casual instants transfer from one condition status to another. Passage of process happens during the moments when the new vessel approaches the terminal, or one of moorings is free. Thus the amount of vessel in turn for storing cargoes varies. The process maintains calculating amount

IOP Conf. Series: Materials Science and Engineering **142** (2016) 012106 doi:10.1088/1757-899X/142/1/012106

of condition statuses, occurrences of figure of requests: E_1, E_2 ... E_n, where n figure of ships (requests) which are in conditional-functional sector (the channel of a upkeep) in turn, and stored, during an interval of time. Probabilities of passages of system from condition status E_n to condition status E_{n-1}, i.e. servicing of one request depends on number of working channels of a servicing.

Condition statuses of process of a storing goods:

E_0 - all channels of a upkeep are not occupied;

E_1 - one channel of a upkeep is occupied;

E_i - it is occupied i channels;

E_f - all is occupied;

E_{f+d} - channels of a upkeep are occupied all and d shipments stand in a queue.

Process of a storing goods can be described through the block – the scheme which shows a principle of passage from one condition status to another (Figure 1).

Let's introduce labels of the refrigerator container terminal having F of conditional-functional sections with an identical duty cycle, on which the irregular traffic flow m (an amount of vessels with resulting intensity acts. Resulting intensity of incoming ships in the terminal is determined by intensity and number of the ships which are being in turn.

If all boats are identical from the point of view of their servicing then it is considered, that of a queueing system receives a uniform flow of requests or requirements (ships), i.e. all requests for a unloading are identical. In the previous works the queueing systems based on probability models with reference to a uniform flow of requests have been considered: all arriving boats possess identical characteristics, i.e. identical containers capacity and cargo consignment (number of containers on board a ship).

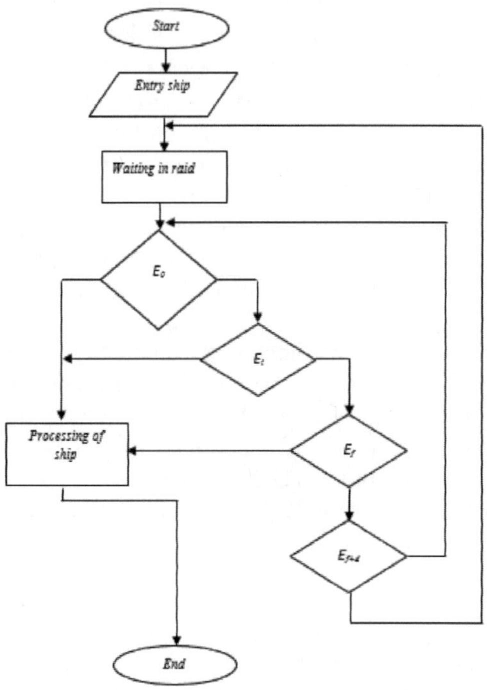

Figure 1. The chart of condition statuses of processes of a storing goods.

However, there may be specified heterogeneity of a traffic flow - boats can possess various performances: an amount of cargo containers on a vessel, capacity of cargo containers, etc.

Accepting, that the flow of consignments (ships) arriving to the port, submits to a Poisson distribution law, it is possible to consider it elementary (stationary Poisson), possessing three properties: absence of an aftereffect and a stationarity.

The system is characterized by following singularities:

- inflows of requirements (vessels) to the system for servicing happens one by one, that is the probability of inflow of two or more requirements (vessels) in one instant is very small, and it can be neglected - a flow of requirements is ordinary;
- the probability of inflow of the subsequent requirements (vessels) at any moment does not depend on probability of their inflow during the previous moments - a flow of requirements without aftereffects;
- the flow of requirements does not depend on a disposition of a considered interval of time for axis of time - a flow of requirements stationary.

Let's admit suppose, the multi-channel disconnected queueing system with a unlimited latency period and with the inhomogeneous elementary (stationary Poisson) a flow of requirements is set.

The inhomogeneous traffic flow possesses various performances, i.e. miscellaneous types of ships m_1 - boats of 1-st type, m_2 - ships of 2-nd type, m_3 - ships of 3-rd type, m_s - ships of s-th type; and intensity of incoming of a shipment: λ_1 - for ships of 1-st type, λ_2 - for ships of 2-nd type, λ_3 - for ships of 3-rd type, λ_s - for ships of s-th type.

The total number of ships in turn for storage is determined by expression:

$$m = \sum_{i=1}^{s} m_i \tag{1}$$

Intensity of a traffic flow can be discovered under the formula:

$$\lambda = \sum_{i=1}^{s} \lambda_i \cdot m_i \tag{2}$$

The process proceeding while handling and storage of container refrigerated cargoes in terminals, is a process of discrete type with final (or calculating in general) set of condition statuses. Thus the amount of the lots which are in turn for handling varies. At the casual instant, researched processes of consignments transfer from one condition status to another. Transfer of process happens during the moments, when a new consignments arrives, or one of conditional-functional sections is free.

Let's consider one of enumerated above processes, namely calculation of a latency period of batches of cargo containers in turn for storage.

Each vessel can appear in is conditional-functional sector (event A) with the same probability p. And movements of individual consignments may be considered as f of independent trials.

If the probability p is a determination of a shipment in the terminal q=1-p it is probability of that the given shipment to be in other condition status (in other phases), i.e. not occurrence of event A.

It is required to discover probability P_n of that in the terminal will treat n shipments, i.e. event A in these f experiences to appear exactly n time, as event.

Under the Bayes approach we shall consider the event B_n, consisting that event A to appear in f experiences equally n time. This event can be carried out by various modes. We shall spread out event B_n as the sum of product of the events consisting in occurrence or nonappearance of event A in separate experience. We shall designate A_i the event corresponding a determination i of lot in the terminal, and $\bar{A}_i$ - the event corresponding a determination i of lot outside of the terminal.

Really, each term of the sum B_n (occurrence of event) should consist of n events A_i and $(f$-$n)$ events $\bar{A}_j$ with various indexes. Then B_n it will be described by the following formula:

$$B_n = A_1 A_2 \dots A_n \overline{A}_{n+1} \dots + \overline{A}_1 \overline{A}_2 \dots \overline{A}_{f-n} A_{f-n+1} \dots A_f \tag{3}$$

In the given expression the number of possible modes by which it is possible from f experiences to choose n in which there was an event, figure such C^n_f is described.

Using the private theorem of a recurring of experiences, it is possible to discover probability $P_{n_1 n_2 \dots n_s}$

$$P_{n_1 n_2 \ldots n_s} = \prod_{i=1}^{s} C_{f_i}^{n_i} p^{n_i} q^{f_i - n_i} \tag{4}$$

Where $C_{f_i}^{n_i}$ - figure of modes with which from f_i experiences it is possible to choose n_i in which there was an event (the determination of a shipment of cargo containers in store).

The probability of a determination n_1, n_2, ..., n_s batches of cargo transporters stored corresponds to terms of expansion of a binominal formula in a degree, equal to figure of batches of cargo transporters.

Therefore the specified probability distribution calls binomial. The probability distribution is meant binomial allocating possible occurrences of events A (the determination of a batch of cargo transporters stored) at repeated independent trials, in each of some events A can be determined with the same probability.

For this purpose we shall present all possible condition statuses of a queueing system in the form of graduated column of condition statuses (Figure 2). Each rectangle column determines one of all possible condition statuses. P_n is a possibility of presence in system n requirements of vessels. Arrows on the column specify, in what condition status the system can transfer and with what intensity. Thus in multi-channel of queueing system it is necessary to distinguish two cases:

- the number of vessels n, arrived into system, is less than channels of upkeep N, that is all of them are on a upkeep ($0 \leq n < N$);
- the number of vessels n, arrived into system, is more or to equal to the number of channels of upkeep N ($N \leq n$), i.e. N vessels are serviced, and remaining r are expected to be in turn (r=1,2, ..., n-N).

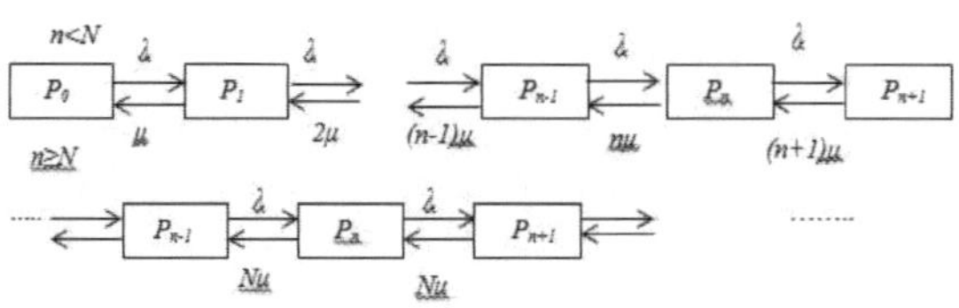

Figure 2. Labelled columns of condition statuses multi-channel broken of queueing system.

Columns with probability P_0 determines a condition status of system at which all channels of a upkeep stand because of lack of vessels. With intensity μ the system can transfer as in a condition status in P_1 when in it to appear one requirement, and also from condition status P_1 in condition status P_0 if the unique requirement was in system, has been serviced earlier, than has appeared new, etc.

Having specified probabilities of separate condition statuses of the terminal, it is possible to learn on the basis of the Bayes approach average value of a latency period of batches of cargo transporters in turn on the storage, corresponding defined values of number of ships m_s and figures is conditional-functional section F the researched specialized terminal.

Primal problem is determining average time expectations of lots in turn, i.e. a direct commitment - association of function on a duty cycle φ and is conditional-functional section F. And, the refrigerator terminal φ_1, φ_2, φ_3 … φ_i … φ_s, accordingly for each type of a vessel, and r_1, r_2, … r_i … r_s a size of a batch of containers (an amount of cargo containers) for vessels of s-th type. In work a duty cycle φ of directly proportional intensity of incoming of each lot λ_i'

The average reduced latency period of a location on storage of a batch of containers:

$$\bar{\tau}_{ож} = \frac{\bar{r}}{\sum_{i=1}^{s} \lambda_i' m_i r_i} \tag{5}$$

IOP Conf. Series: Materials Science and Engineering **142** (2016) 012106 doi:10.1088/1757-899X/142/1/012106

$$\overline{\tau}_{ож.ср} = \frac{\overline{r}}{\sum_{i=1}^{s} \lambda_i' m_i r_{ср}} \tag{6}$$

r_{sr} - the averaged value of a size of a batch of the containers, calculated under the formula:

$$r_{sr} = \frac{m_1 \cdot r_1 + m_2 \cdot r_2 + \ldots + m_s \cdot r_s}{\sum m_i} \tag{7}$$

The received probability model allows to make the analysis of processes of storage of containerized cargoes in refrigerator terminals, namely determining of an average latency period of lots in turn stored a refrigerator container terminal.

3. Outcomes

Using a mathematical model described above, the average latency period of a location has been calculated for storage of batches of cargoes.

In the subsequent, for fast and effective calculation of the specified indexes, the software product has been developed. By development of the program studio NetBeans IDE 7.2 maintaining gears of such languages, as java which was chosen [8]. NetBeans IDE - the free integrated medium of development of applications (IDE) in programming languages Java, JavaFX, Python, PHP, JavaScript, C, C ++, and of some others. For development of programs in medium NetBeans and for successful installation and work of medium NetBeans should be installed beforehand Sun JDK or J2EE SDK suitable version. The medium of NetBeans development by default supported development for platforms J2SE and J2EE.

Java - the object-oriented programming language, is high-power and now very popular. Has set of advantages: steady and reliable, multicontinuous, is architecturally-independent, highly effective, it is equipped by high-power library of classes and the methods, allocated, etc.

Java applications are usually compiled in special byte-code, therefore they can work by any virtual Java-machine (JVM) without dependence from computer architecture. That in turn, is very actual in the heterogeneous world of PO.

The choice of usage of the given toolkit, as development of software product, allows to solve all above tasks, and also provides simplicity and convenience of programming, and possibility of the future updating [4-7].

During development of the software, following objectives have been put and solved:

- implementing input of input datas: F - an amount of conditionally-functional section; m_s - boats of s-th type, where s - figure of "inhomogeneous" ships, φ_s - a duty cycle;
- calculation of an average number of container consignments in turn for storage;
- intensity of cargo containers arriving [10];
- average time of waiting of storage rooms for containers keeping.

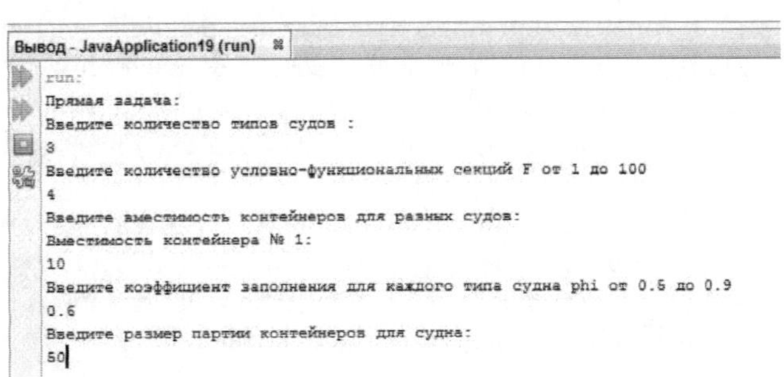

Figure 3. Input of input date's, for the first type of a vessel.

IOP Conf. Series: Materials Science and Engineering **142** (2016) 012106 doi:10.1088/1757-899X/142/1/012106

```
Вместимость контейнера № 2:
50
Введите коэффициент заполнения для каждого типа судна phi от 0.5 до 0.9
0.7
Введите размер партии контейнеров для судна:
70
```

Figure 4. Input of input date's, for the second type of a vessel.

```
Вместимость контейнера № 3:
100
Введите коэффициент заполнения для каждого типа судна phi от 0.5 до 0.9
0.8
Введите размер партии контейнеров для судна:
60
Среднее время ожидания на хранение, с использованием ri, tau: 0.1536698700441140З
Среднее время ожидания на хранение, с использованием rsr, tau: 0.15407426443896696
СБОРКА УСПЕШНО ЗАВЕРШЕНА (общее время: 3 минуты 23 секунды)
```

Figure 5. Input of input date's, for the third type of a vessel and a conclusion of outcome.

By results of calculations it is possible to tell, that the average time waiting significantly increases with magnification of a duty cycle and is moderated with magnification of number is conditional-functional section the refrigerator terminal. Calculations have shown, that at usage r_s and r_{sr} indexes slightly differ. That speaks that it is possible to use at the further calculations the averaged value r_{sr}.

4. Conclusion

With developing the terminal the new organizational structure of control was realized, both methods, and means of an estimation of job performance standards of the reloading terminal and other sites of multimodal transportations were simultaneously improved.

For simulation of work of the reloading terminal the means of a theory of queues is used. Thus in a basis of simulation of processes of cargo handling the broken queueing systems are supposed. Application of existing models of a queuing for determining performances of processes of processing of containerized cargoes not always is expedient as the specified models not correctly describe processes of functioning of work of port.

For today for describing a channel capacity of container terminals the determined methods are used. However these methods, actually, do not reflect all specificity of processing of container refrigerated cargoes. As the moments of delivery of cargo containers to the terminal are a casual flow of events.

Thus, the received probability model allows to manufacture the analysis of processes of storage of containerized cargoes in refrigerator terminals, considering specificity of functioning of processes of cargo handling. At calculation of a mathematical model, we can discover following parameters: an average number of cargo containers consignments in turn for storage; the average reduced latency period, intensity of incoming to the terminal of each cargo consignment and coefficient of filling of conditional-functional section. Outcomes are more precise, than when using of the "classical" determined method which does not completely reflects specificity of processing of refrigerator containerized cargoes. Because the usage of the given method some prominent aspects are not considered: an irregular flow of events, a casual response time of arriving of ships into the terminal (in is conditional-functional section) and casual period of cargo-handling time.

References

[1] Association of sea trade ports 2015 Association of marine commercial harbors. Available at: http://www.morport.com/rus/news/document1751.shtml

[2] Wiraj-perevozki 2011 Advantages of a container carriage. Available at: http://www.wiraj-perevozki.ru/text_7.html

[3] Tranzito 2014 Virtues of a container carriage and their basic minuses Available at: http://tranzito.com/osobennosti/1094-morskie-perevozki-v-chem-ix-osobennosti.html

[4] Nyrkov A, Sokolov S, Chernyi S and Mamunts D 2015 Using information technologies in dredging. *Russian Jour. Metallurgical and Mining Industry* **7(6)** 521–524

[5] Nyrkov A, Belousov A and Sokolov S 2015 Algorithmic support of optimization of multicast data transmission in networks with dynamic routing *Jour. Modern Appl. Science* **10(5)** 162

[6] Chernyi S, Zhilenkov A, Sokolov S and Titov L 2015 Self-contained drilling rig automatic control system efficiency improvement by means of assuring compatibility and integration methods development *Jour. Metallurgical and Mining* **7(3)** 66

[7] Korobeynikov A G, Gatchin Y A 2014 Fejer means application for determination of mail document spam level *Jour. Modern Appl. Science* **30(30)** 165

[8] Goloskokov D P 2014 Analyzing simply supported plates using maple system *Conf. International Conference on Computer Technologies in Physical and Engineering Applications, ICCTPEA*

[9] Pantina T A, Borodulina S A 2015 Methods for estimation of multiplier effect of investments in development of infrastructure of inland water transport in the russian federation in the frameworks of federal target programs *Jour. Review of European Studies*

[10] Pogodaev L I, Ezhov Y E 2014 Improving the durability of operating devices of technical fleet vessels using wearresistant facings *Juor. Journal of Machinery Manufacture and Reliability*

IOP Conf. Series: Materials Science and Engineering **142** (2016) 012107 doi:10.1088/1757-899X/142/1/012107

Methods Used to Support a Life Cycle of Complex Engineering Products

Alexandra A Zakharova[1,a], Olga A Kolegova[1,b], Maria E Nekrasova[1,a] and Andrey O Eremenko[1,a]

Yurga Institute of Technology National Research Tomsk Polytechnic University Affilate, ul. Leningradskaya 26, Yurga, Russia

E-mail: [a]aaz@tpu.ru, [b]Olga030188@mail.ru

Abstract. Management of companies involved in the design, development and operation of complex engineering products recognize the relevance of creating systems for product lifecycle management. A system of methods is proposed to support life cycles of complex engineering products, based on fuzzy set theory and hierarchical analysis. The system of methods serves to demonstrate the grounds for making strategic decisions in an environment of uncertainty, allows the use of expert knowledge, and provides interconnection of decisions at all phases of strategic management and all stages of a complex engineering product lifecycle.

1. Introduction

The process of product lifecycle management in complex engineering systems, associated with most challenging multistage production environments, is a prerequisite for good governance.

During each stage of the complex engineering product (CEP) life cycle (LC), a number of project-related conflicts occur, since project participants interact in an environment of incomplete information about the current and projected state of the related products. For solving this problem and eliminating contradictions between all project participants in the development of complex engineering products, Product Lifecycle Management systems are provided (PLMS) [1], [2].

Complex engineering products are developed in a variety of industries. For example, it is a nuclear power station in the electric power generating industry and the geohod [3] and other machinery in the mechanical engineering industry. Today in Russia, the issue relating to lifecycle management of complex engineering products becomes important and this importance is comprehensible. But experts have not arrived at consensus on means and tools for solving the problem. It is worth noting that not all Russian companies employ PLM systems in projecting and managing complex engineering products. Or they use similar tools only during certain stages of CEP LC. Thus, according to the research made by the Foundation Center for Strategic Research "North-West", less than 40% of the companies-respondents use software for product lifecycle management. The results, obtained during the market research among the companies that use specialized lifecycle management software [4], allow us to judge on a level of Product Lifecycle Management in Russian companies. Elements of the planning and developing processes are mostly often used (86%), as well as production or services (82%). Least of all companies are involved in implementing management tools for products' after-sales activities and products recycling (50% and 30% respectively). And only 18% of companies are associated with PLM fully involved [4].

Lifecycle management of complex engineering products is associated with certain difficulties. As a rule, any complex technical systems are characterized by long cycles in development, issues related to adaptability to manufacture, cost efficiency and technical risks [5]. When managing complex engineering products, it is necessary to solve a lot of ill-defined problems at different stages of the life cycle. Solving those problems requires the use of tools to support decision-making under conditions of high uncertainty, intelligent methods for information processing, and experts' involvement [6], [7]. There is a need for a set of tools to ensure the relationship of strategic decisions through all stages of CEP life cycles.

It should be noted that today the issues relating to LC strategic management is given insufficient attention, and existing PLM systems do not have the necessary set of tools to support the strategic level of complex engineering product lifecycle management.

There is no single platform developed with Russian specialists' involvement to address the issue of efficient systems to manage CEP LC. Commercial unavailability of specialized software designed to support the strategic level of complex engineering product lifecycle management may lead to the increased dependence of Russian software designers on solutions offered by foreign companies and producers of tools used to support individual LC processes while implementing complex systems.

Thus, an urgent need appears as to design and develop a system intended to support lifecycle management of complex engineering products in uncertain environments.

2. Survey of the existing PLM systems for complex engineering products

The results of researches show that Product Lifecycle Management (PLM) systems are most commonly used when dealing with related issues. For example, according to the market researches delivered by the Foundation Center for Strategic Research "North-West", the usage measured with regard to specialized software packages have been obtained as follows: PLM-systems - 40%; CAD / CAM / CAE systems - 17%, ERP-systems - 21%; 1S: Production management - 12%; in-house developments - 17% [2].

Today, there is a wide variety of PLM solutions in the IT market of the systems for lifecycle management of end products, ranging from simple tools that can aid you in product lifecycle managing within a small company to powerful systems used to effectively process amounts of engineering information to help manage complex technical products [8], [9].

Analysis of the researches delivered by TAdviser, a Russian analytical agency in the field of Product Lifecycle Management, shows that for last 3 years the leaders in terms of software used were the following PLM packages: 1S: Enterprise 8. PDM engineering data management, Teamcenter, Siemens NX, Windchill, Lotsman (Loodsman): PLM [10].

1S: Enterprise 8. PDM engineering data management is software that allows a company to reduce the production time by automating the process of selecting ready components and their adopting to designed products and providing tools to control the works flow on projects in design and technological preparation of manufacture.

Siemens NX is a PLM solution offering the broadest package of integrated, fully associative CAD/CAM/CAE applications in product design, manufacturing and simulation.

Teamcenter is an integrated PLM package that provides manufacturers with help in delivery complex products to market and streamlining global operations.

Windchill is PLM software that offers comprehensive capabilities to provide function ability required for effective global management of team works.

Lotsman: PLM is an effective solution for product lifecycle management, enabling centralized and structured storage of technical documentation on products.

All of above mentioned systems are the advanced solutions for product lifecycle management, and designed for a range of companies different in their size and tasks to be performed. The existing PLM systems offer the main processes automated with the purpose of full-scale professional management of the main stages of product life cycles: management of product structures, the processes of product developing, and the production process.

Despite the variety of tools for product lifecycle management, the ill-defined problems are not enough supported by available software relative to PLM in the environment of incomplete information about complex engineering products at different phases of their production.

Therefore, development of a system offering the methods for supporting lifecycle management of complex engineering products under conditions of uncertainty is relevant.

3. The conceptual approach to the development of a system of methods for supporting LC strategic management

The main stages of the process are distinguished in LC strategic management by a difference in a combination of methods applied, demands submitted to decision-makers (experts, analysts), types and sources of information, etc. (in total, it is a strategic analysis, choice, and control).

In strategic PLM, the typical stages of a life cycle are defined (marketing, concept, development, production, operation, maintenance, removal from operation).

It is proposed to develop two types of strategic decision-making support tools: universal and specialized.

The universal tools can be used in systems to support a life cycle of any complex engineering product, while the specialized tools are designed for accommodating peculiarities of each CEP in terms of different criteria to classify them (e.g., industry, type, level, etc.).

The software package for supporting CEP LC must meet the following functional requirements:

- to provide interconnection of decisions through all major phases of strategic management;
- to provide tools and interconnection of decisions made within typical stages of CEP life cycle;
- to facilitate the use of qualitative expert information along with quantitative information in evaluating alternatives and making a decision; to provide information on the relations between factors in the form of verbal expert statements, rank them and, on the basis of this information, determine the order of certain policies and measures to be implemented;
- to establish the relationship between the results of operating elements in the environment of the complex engineering product and the overall development of the organization;
- to enable the combination of management systems used to support CEP life cycles from libraries of universal and specialized tools.

The developed system of methods for supporting complex engineering product life cycles is intended for use in DSS-class systems at the strategic level [11], [12].

4. The concept of universal methods for supporting strategic management of CEP life cycles

Analysis of the main problems in strategic management relating to complex product life cycles allows us to point out the following typical decision-making problems, which are in the focus of universal models, methods and algorithms developed to support decisions based on experts' knowledge:

- evaluation of strategic factors with reference to the CEP life cycle driven by experts knowledge;
- formalization of expert knowledge on the relationship of strategic factors, internal and external environments;
- calculation of the importance of strategic factors and their combinations;
- assessment of strategic development projects for CEP according to objectives and effects due to acting external and internal forces in the environment;
- formalization of targets to be achieved while implementing the strategy for the CEP development and the results evaluation after implementing the strategy for CEP lifecycle management;
- effect assessment of strategic decisions made at various stages of CEP life cycle, and interrelation between decisions;
- organization and processing of the results of expert surveys.

The following approaches are proposed to solving these problems:

IOP Conf. Series: Materials Science and Engineering **142** (2016) 012107 doi:10.1088/1757-899X/142/1/012107

1. A SWOT-analysis on the basis of fuzzy sets. For this purpose, the following models and methods are used: the methods for constructing membership functions of linguistic variables; the systems of fuzzy production rules; models based on fuzzy logic. This approach allows us to calculate quantitative estimations according to the importance of opportunities, threats, strengths and weaknesses, as well as combinations thereof, on the ground of qualitative expert information.

2. A hierarchical model for evaluating development projects by analyzing alternatives (projects, programs) available for the CEP development, based on the influence exerted by the acting forces (the parties concerned).

3. A fuzzy integrated model for evaluating the CEP development strategy. It allows us to monitor the achievement of targets set in the strategy individually and to assess the overall success of implementing the selected strategy for CEP development.

4. A full cycle of evaluation made by experts, including the following: forming a group of experts, assessment of experts' competence, conducting expert surveys, consolidating evaluations made by a group of experts, assessing the degree of coordination among experts. The main feature of the proposed methods is the compiling of tables indicating the level of competences of experts that take into account the knowledge and experience of experts in different functional activities of socioeconomic systems.

The scheme in Figure shows how the decision-making models are applied and interact as to support strategic decisions on the basis of experts' knowledge, as well as the results of their application.

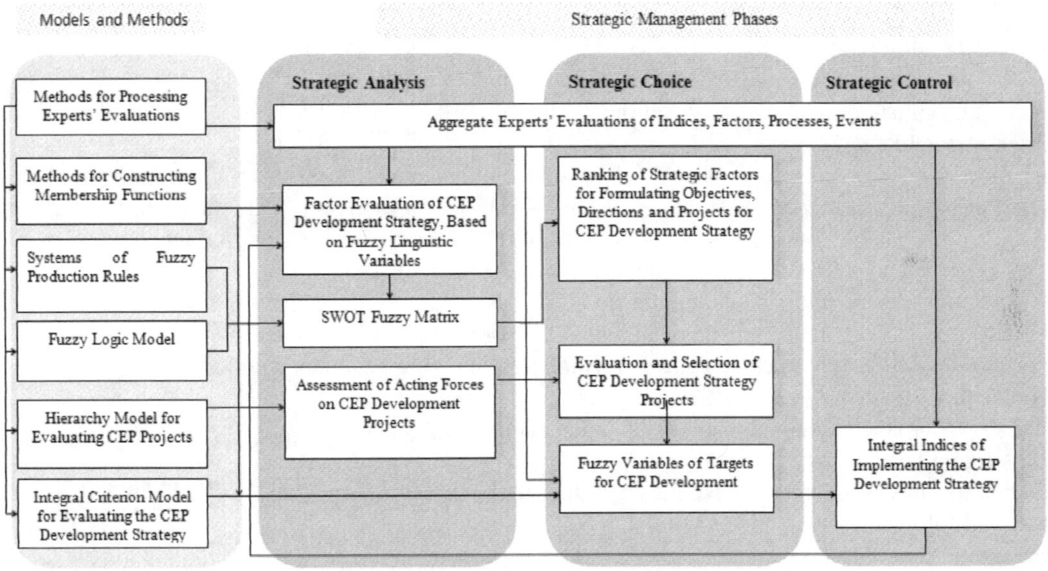

Figure 1. System of Universal Methods for Supporting Complex Engineering Product Life Cycle under Uncertainty

5. The main directions for the development of specialized methods for supporting strategic decision-making in lifecycle management of complex engineering products

According to the works, carried out earlier to develop the strategic Decision Support System, the following objectives were identified for strategic decisions based on experts' knowledge, requiring the development of specialized tools:

- Factors evaluating for strategic development of CEP (actual, planned, projected, etc.);
- Relations establishing between CEP elements and their external environment;
- Relations establishing between functional factors of SES elements and their environment;

- Influence evaluating of CEP elements on each other and on the whole development strategy for CEP;
- Forming a feedback loop in strategic management of CEP life cycle;
- Databases creating within the context of a particular CEP domain, etc.

The above list may be supplemented by the results of further research. For this purpose, the most frequently occurring problems will be analyzed, synthesized and formulated in relation with strategic decision-making that require the development of special tools to support decision-making on the basis of expert knowledge.

Conclusion:

This paper describes the significance of tools for use in supporting lifecycle management of complex engineering products in an environment of uncertainty at all stages of their lifecycle. The concept necessary to create the system of methods used to support life cycles of complex engineering products under uncertain conditions is proposed. The problems associated with the development of universal models for decision-making were analyzed and formulated. A system of methods is proposed for use in supporting life cycles of complex engineering products, based on fuzzy set theory and hierarchical analysis. The main directions were identified for the development of special methods for strategic decision-making in lifecycle management of complex engineering products, with specifying the scope of further research.

The main advantage of software is its capability to implement the principle of integrity and continuity in the process of strategic managing the CEP life cycle, which is achieved by:

- providing decisions interconnected through all major stages of strategic PLM;
- ensuring the relationship of strategic decisions over all stages of CEP LC;
- establishing relations between the results of CEP elements functioning to comply with the overall company strategy.

Implementing this principle allows us to expand the limits for possible decisions made by management while specifying the strategy of CEP lifecycle management, and thus to improve quality and validity of strategic decisions in PLM with regard to complex engineering products.

The proposed system of methods intended for the life cycle support of complex engineering products under the conditions of uncertainty is a basic part of the software package designed to support strategic lifecycle management of complex engineering products. Software for the system to support strategic lifecycle management of complex engineering products is an add-in component for PLM, PDM, CAD / CAM, PM, ERP packages.

Acknowledgments

The reported study was funded by RFBR according to the research project № 16-07-00299 a.

References:

[1] Essop I, Evans A R D, Wan S, Giddaluru M P, Gao J X, Baudry D, Mahdikhah S and Messaadia M 2016 Investigation into current industrial practices relating to product lifecycle management in a multi-national manufacturing company, *Computer-Aided Design and Applications* pp 1-5

[2] Bojcetic N, Salopek D and Marjanovic D 2015 PLM implementation: case study, *Design information and knowledge management* 15 (10)

[3] Efremenkov A B 2011 Forming the subterranean space by means of a new tool (geohod), *Proceedings of the 6th International Forum on Strategic Technology, IFOST 2011*, Code 86750.

[4] Center for Strategic Research North-West. Technical systems lifecycle management. Available at http://www.csr-nw.ru/publications/2011-2015/

[5] Tceplit A, Grigoreva A and Osipov Y 2014 Developing the Model for Assessing the Competitiveness of Innovative Engineering Products, *Applied Mechanics and Materials* 682 pp 623-630.
[6] Telipenko E V and Zakharova A A 2014 Bankruptcy Risk Management of a Machine Builder, *Applied Mechanics and Materials* 682 pp 617-622.
[7] Zakharova A A and Ostanin V V 2015 Formalization model of expert knowledge about a technical index level of engineering products. Paper presented at the IOP Conference Series: *Materials Science and Engineering,* 91(1) doi:10.1088/1757-899X/91/1/012070.
[8] Soto-Acosta P, Placer-Maruri E and Perez-Gonzalez D 2016 A case analysis of a product life cycle information management framework for SMEs, *International Journal of Information Management* 36 (2) pp 240-244.
[9] Messaadia M, Baudry D, Louis A, Mahdikhah S, Evans R, Gao J, Paquet T, Sahnoun M and Mazari B 2016 PLM adoption in SMEs context, *Computer-Aided Design and Applications* pp 1-10.
[10] Product Lifecycle Management. Available at http://www.tadviser.ru/index.php/PLM_
[11] Kalbar P P, Karmakar S and Asolekar S R 2016 Life cycle-based decision support tool for selection of wastewater treatment alternatives, *Journal of Cleaner Production* 117 pp 64-72.
[12] Balasubramaniam O A and Somu A 2012 IT-enabled supply chain management using decision support systems, *European Journal of Social Science* 32(4) pp 632-643.

IOP Conf. Series: Materials Science and Engineering **142** (2016) 012108 doi:10.1088/1757-899X/142/1/012108

Justification of Filter Selection for Robot Balancing in Conditions of Limited Computational Resources

M V Momot[1], E V Politsinskaia[1], A V Sushko[1], I A Semerenko[1]

[1]Yurga Institute of Technology (Branch) of National Research Tomsk Polytechnic University
652055, Kemerovo Region, Yurga, 26 Leningradskaya St., Tel. (38451) 7-77-62

e-mail:mikl_1239@mail.ru

Abstract. The paper considers the problem of mathematical filter selection, used for balancing of wheeled robot in conditions of limited computational resources. The solution based on complementary filter is proposed.

I. Introduction

Right selection of compensatory mathematical filters to control the motion of autonomous remotely piloted vehicles is a relevant objective, particularly in conditions of significant lack of computational resources. The right choice of such filters solves the energy efficiency problem, in terms of robot motion and with regard to expenses for analysis and calculation. Using low-quality filters will lead to excessive robot oscillations and thus additional energy consumption conditioned by that. Filters selection implying large number of calculations leads to significant energy consumption for calculations. It is essential to find the best option, considering all the important features and factors most effectively [1,2].

II. Subject and methods of researh

To conduct experiment a model of balance robot was developed, that was equipped with stabilization system based on electronic gyroscope and accelerometer.

Electronic gyroscope differs significantly from the classical one; it does not show the angle of deviation from the given direction. Electronic gyroscope is an instrument registering and giving back the current angular velocity. Knowing the instantaneous angular velocity one can calculate approximately the rotation angle. The formula in this case looks as follows:

$$Ang_2 = Ang_1 + \Delta t \times V_{ang}, \qquad (1)$$

where Ang_1 - initial angle, calculated before, Δt - short period of time, where one can consider that the velocity has not been changed, V_{ang} - value of angular velocity obtained from electronic gyroscope during Δt, Ang_2 - calculated angle, that object turned around the studied axis since the beginning of motion. For Fig.1 where axes X and Y of the gyroscope are directed parallel to the surface of motion, turning of robot is performed around the axis Z and therefore instrument readings of this axis should be studied. In practice, it all depends on how the instrument is installed; however, for MPU-6050, in case it is placed parallel to the robot motion, it will look exactly as follows.

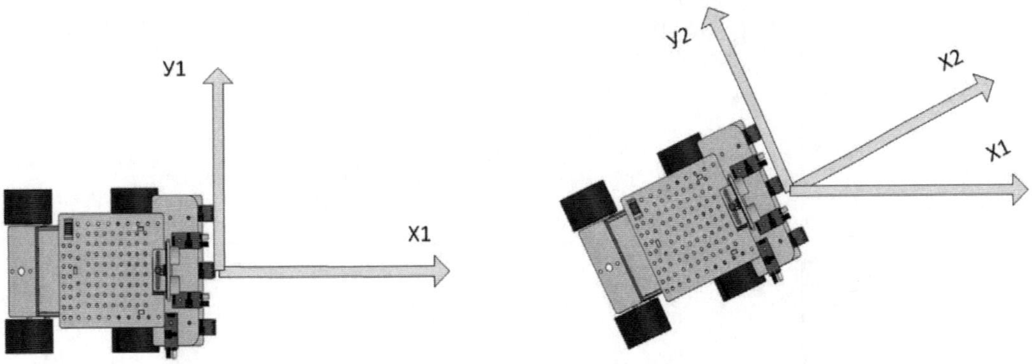

Figure 1: Demonstration of gyroscope axes and motion vector

Connection diagram of MPU-6050 to Arduino robot is given in Fig. 2.

Figure 3 demonstrates readings where the robot laid on its side, Fig.4 shows readings of the robot set parallel to the floor.

The gyroscope readings were not zero-point, even though the robot was motionless. However, they range around one value. This error can possibly be eliminated if the robot it is kept motionless for a while before it starts to move. The data is recorded at that time, the average value is calculated and then this value is subtracted from the readings obtained from the instrument. It could possibly be temperature drift, as these values are not always close to each other.

When the robot lay (Fig.3), the largest values of acceleration were observed on axis Y (>16000), while after robot turning into the normal position, the largest values were observed at the axis Z (Fig.4).

It should be taken to mind that electronic gyroscope-accelerometer MPU-6050 operates with relative values. It is produced by the company InvenSense. According to the documents, as standard, gyroscope has sensitivity of 250°/sec, to transfer gyroscope values to the scale °/sec, they are required to be divided by 131.

IOP Conf. Series: Materials Science and Engineering **142** (2016) 012108 doi:10.1088/1757-899X/142/1/012108

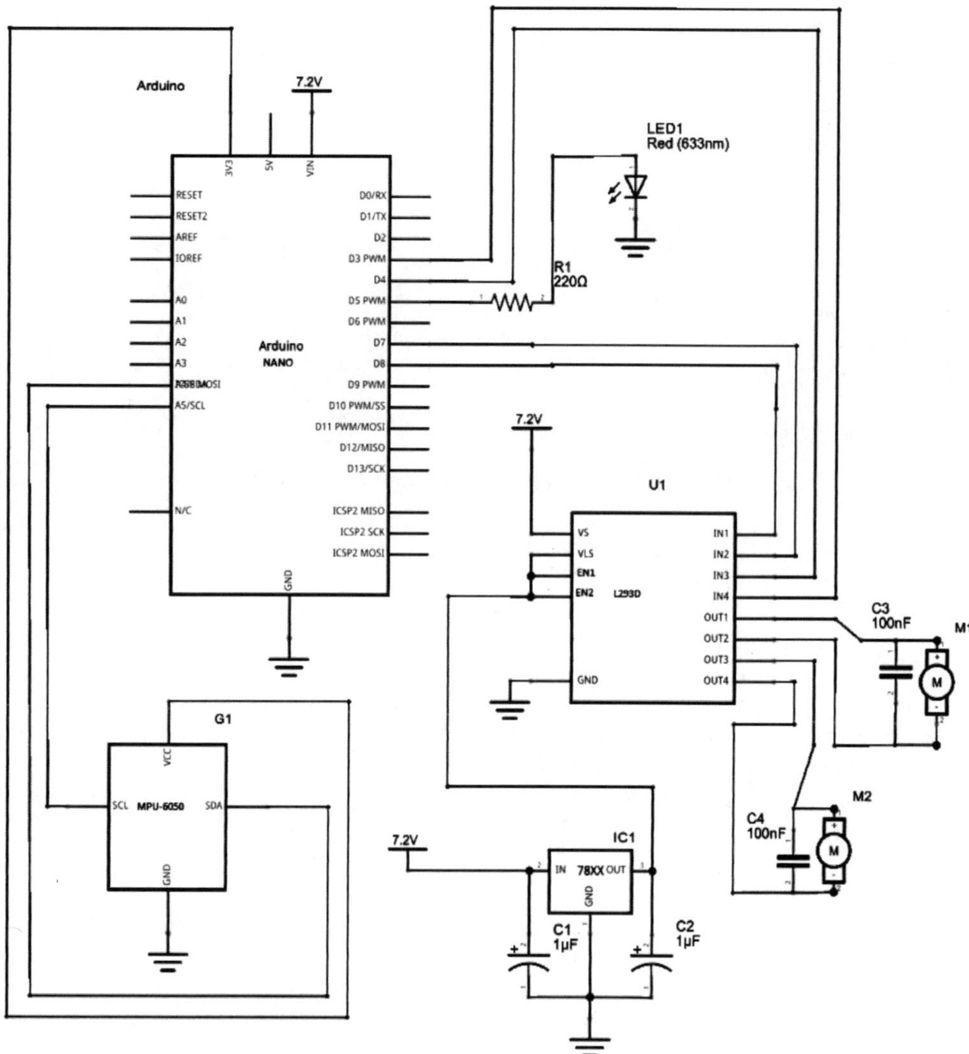

Figure 2: Connection diagram of MPU-6050 to Arduino robot

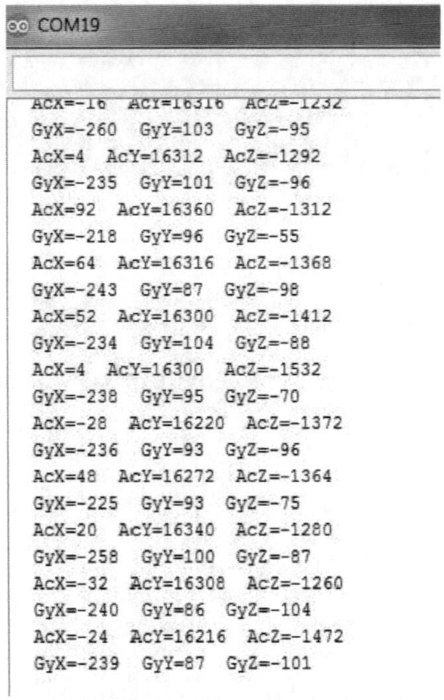

Figure 3: Readings of the robot instruments

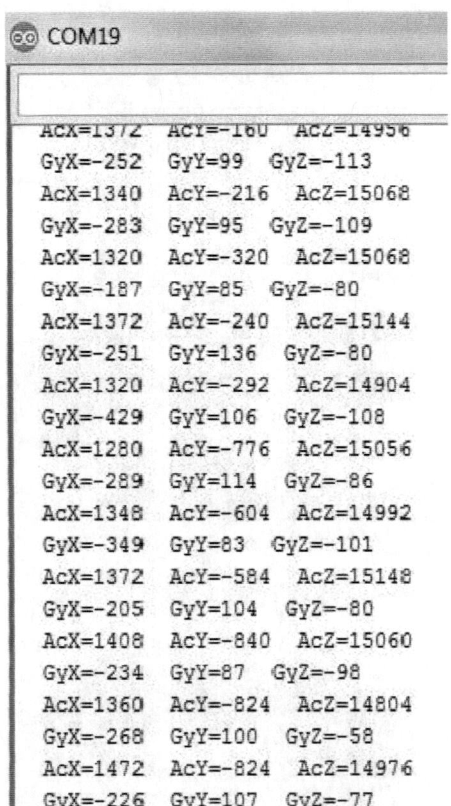

Figure 4: Readings of the robot instruments

Gyro Full Scale Range (°/sec)	Gyro Sensitivity (LSB/°/sec)	Gyro Rate Noise (dps/√Hz)	Accel Full Scale Range (g)	Magnetometer Full Scale Range	Accel Sensitivity (LSB/g)	Digital Output	Logic Supply Voltage (V)	Operating Voltage Supply (V +/-5%)
±250	131	0.005	±2	N/A	16384	I²C	1.8V±5% or VDD	2.375V-3.46V
±500	65.5	0.005	±4		8192			
±1000	32.8	0.005	±8		4096			
±2000	16.4	0.005	±16		2048			

Figure 5: Sensitivity of MPU-6050

Sensitivity of accelerometer normally is 2g (g – gravitational acceleration), 16384 is 1g=9.80665 m/s^2. This information is quite enough for work with the instrument [2].

For this robot 2 motors with reducing gear boxes from the previous projects were used, along with circuit board ArduinoNANO, as a motor driver microcircuit L293D was used. Microcircuit L293D was soldered on a tin base, using GND contacts for that purpose. The contacts at the same time performed the role of heat removal. L293D received logical power supply from a separate stabilizer on microcircuit L7805CV. Robot obtained information red light-emitting diode LED1 to inform about the availability. Ceramic capacitors were soldered on motors to connecting terminals. Without them the robot operation was unstable, and electromagnetic interfering signals took place. The robot body was modeled using engineering software and then assembled.

Figure 7 illustrates the robot participating in testings.

Figure 6: Robot, participating in the experiment

III. Survey results analysis

The first program (Fig.7) that was used in the present study had to analyze the angular velocity around the axis X and aimed to reduce it up to zero. In practice it looks as follows: when the robot starts to fall, it moves in that direction and compensates the bending reaching the upper part; when the robot stands still, its engines are turned off. Considering that the time intervals between polling of gyroscope are the same, time component has been reduced. The robot operates the data received directly from MPU-6050.

Zero drift. While zero angular velocity gyroscope showed non-zero velocity values around the axis X. It was compensated by introduction of the special compensator X aimed at reduction of zero drift. The robot should be set properly and then hold fix, bringing its angular velocity to zero. It is quite hard to realize in practice, however the robot response in general was correct. At the beginning of falling it was trying to get back to the initial state. However, full trim balance was not achieved.

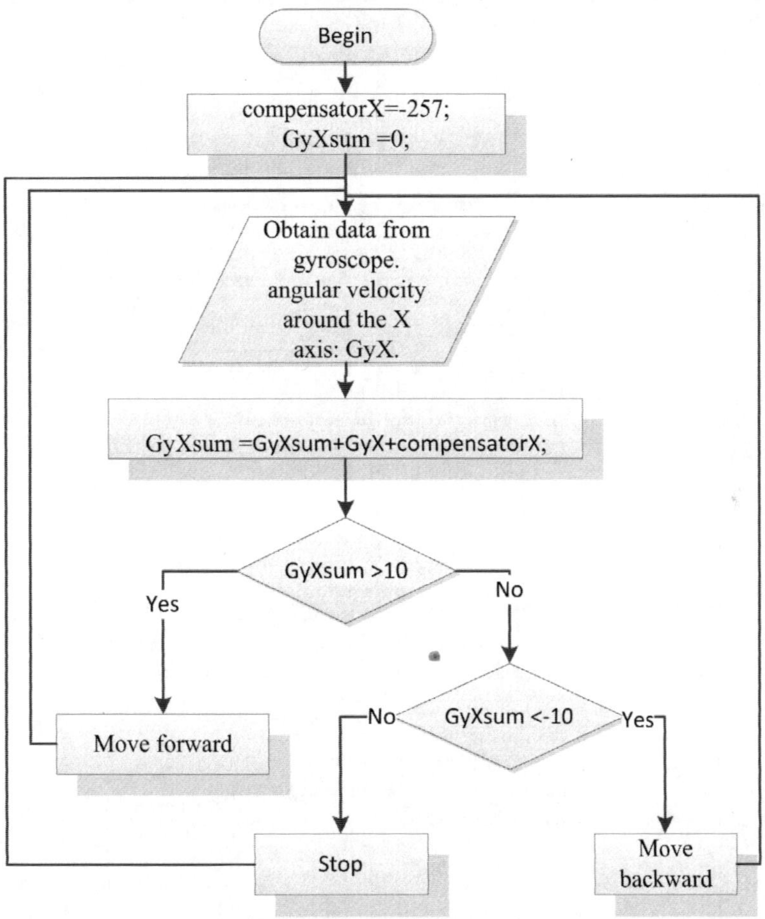

Figure 7: Simplified algorithm

We had to turn to literature and apply mathematics. Kalman filter is a recursive filter evaluating the current state of dynamic system using the data with interfering signals. Kalman filter levels jump signals successfully, transforming the values of system condition to the predictable values. The filter developer is American scientist Rudolf Kalman. While calculating the current values filter considers the previous ones and generates the current value taking them into account.

The algorithm is divided into two stages. The first stage is forecasting, the filter extrapolates the values of state variables and their indeterminancy. During the second stage the result is confirmed based on the obtained measurements.

Developer Kristian Lauszus posted on the website his own library implementations of Kalman filter [3]. In order to trim balance the robot using Kalman filter (Fig.8) one needs to record three values from MPU-6050, in this case the Y axis looks ahead while the X axis looks to the right in regards to the robot motion. These values are: acceleration values along the axes Y and Z and angular velocity along the X axis. Considering the acceleration values along the axes Y and Z we find angle of depression according to accelerometer readings (Fig.9). After that the angular velocity along the X axis is recalculated from relative units to degrees per second [4,5,6]. Then Kalman filter is called with the input parameters: angle of depression according to accelerometer readings, angular velocity along the X axis, time increment on the given area [7]. Research has shown that the filter works well, the readings are gradual but they are behind the real values thus leading to engines response being incorrect (Fig.10).

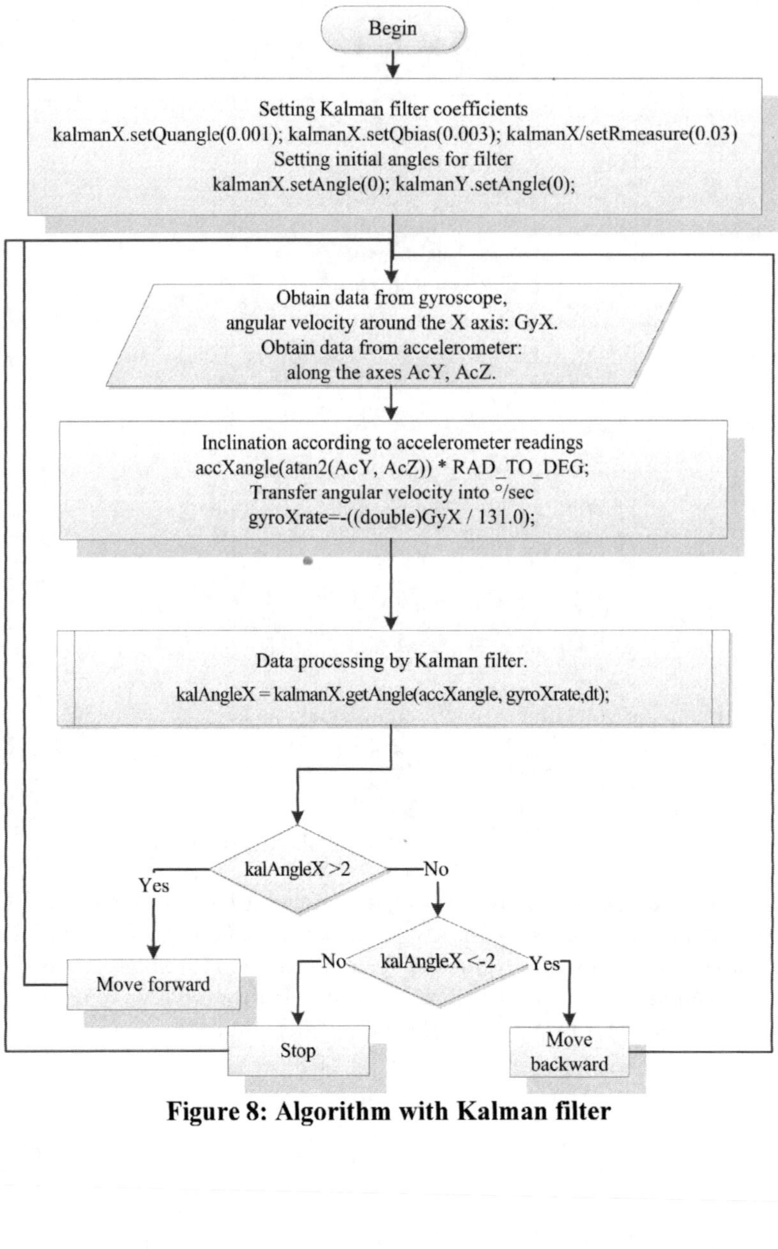

Figure 8: Algorithm with Kalman filter

The angle calculated by the program coincides visually with the preset angle (inclination angle of the robot). Further experiments demonstrated that the velocity is not enough to the filter. Filter is quite energy-consuming thus leading to problems connected with large expenses for mathematics.

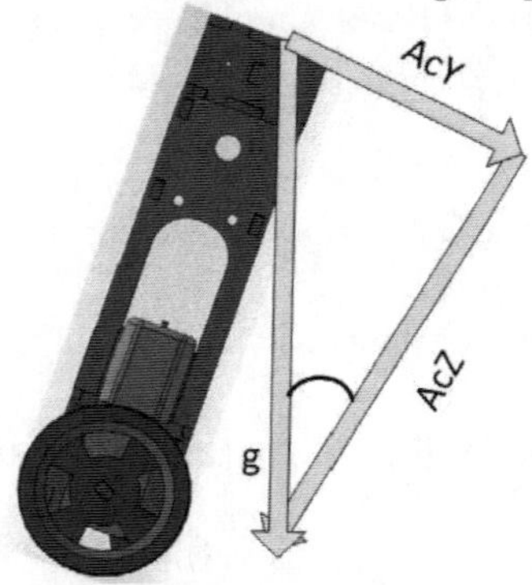

Figure 9: Demonstration of accelerometer work

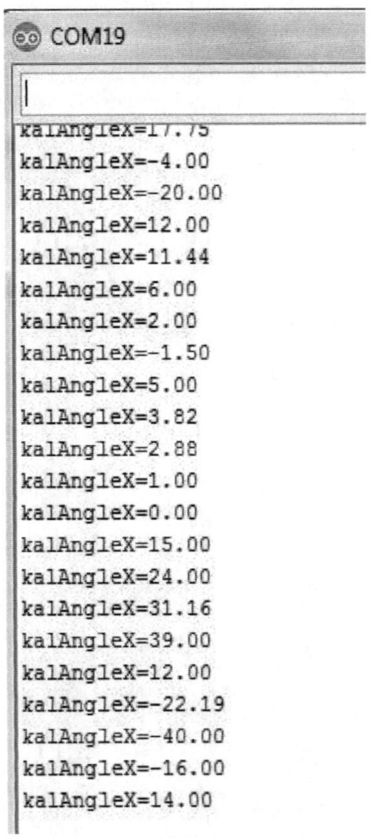

Figure 10: Instrument readings according to Kalman filter

There are other filters which are less inertial and comparatively fast, enabling to get the normal output values combining the readings from different instruments. Complementary filter refers to such filters [8,9,10].

When combining the accelerometer and gyroscope readings based on complementary filter the following formula is obtained:

$$GyXsum_1 = (1 - alfa) \times (GyXsum_0 + GyX \times \Delta t) + alfa \times AcYsum, \qquad (2)$$

IOP Conf. Series: Materials Science and Engineering **142** (2016) 012108 doi:10.1088/1757-899X/142/1/012108

where $GyXsum_1$ - the calculated value of inclination angle, $GyXsum_0$ - the value of inclination angle from the previous calculation, $alfa$ - complementary filter coefficient (is selected experimentally from 0.001 to 0.05), GyX - angular velocity, Δt - iteration time, $AcYsum$- angle calculated according to accelerometer readings.

It is observed that the influence of gyroscope values on the final result is higher (0.99-0.95), however thanks to accelerometer the errors of calculations are compensated, and errors occur anyway if one uses only gyroscope. The final value is participatory sum of integrated value of gyroscope and instantaneous value of accelerometer.

Coefficient $alfa$ is accepted as 0.001; that was enough to compensate the errors of gyroscope integration and zero drift.

Thanks to using the complementary filter we have managed to stabilize the robot and it handles the objective of trim balance successfully.

During testing one peculiar feature was observed – the robot constantly moved one way. The instrument was fixed a little bit not steady in regards to the center of gravity of the robot (it was tilted). It was possible to compensate that feature by constant component added to the value of angle obtained from accelerometer. After that the robot stood still. Thus we solved the problem of motions of balancing robot, to move forward or backward one needs to change this coefficient.

The algorithm of complementary filter operation is given below (Fig.11); it is simpler for understanding and from the point of computational power consumed.

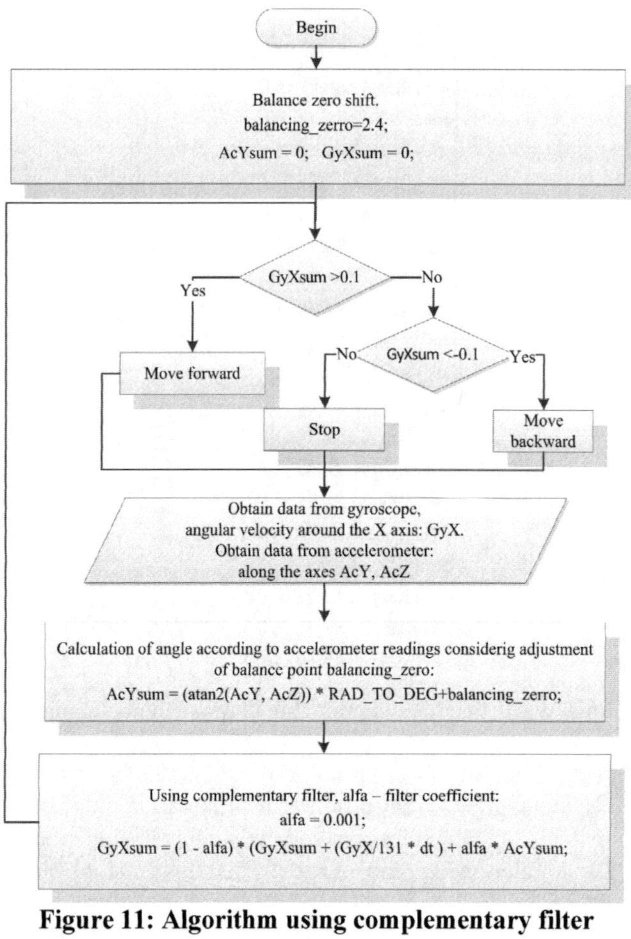

Figure 11: Algorithm using complementary filter

IOP Conf. Series: Materials Science and Engineering **142** (2016) 012108 doi:10.1088/1757-899X/142/1/012108

IV. Conclusions

Resulting from the conducted experiments it was established that taking into account the conditions of limitation of computational resources, namely when using controllers of ATMega, simple complementary mathematical filter should be used. It will allow reaching the optimum parameters of robot balancing. Filters under study behave the same way in other situations as well, e.g. while robot positioning in 3D space, defining its motion vector. Application of complementary filter enables solving the tasks of machinery positioning without using high-powered computation processors.

References

[1] Momot, Mikhail Viktorovich. Algorithm of Controlling Servo Drives by Means of Atmega 8 Microcontroller [Electronic resource] / M. V. Momot, A. Biktimirov // Applied Mechanics and Materials : Scientific Journal. — 2014. — Vol. 682 : Innovation Technology and Economics in Engineering. — [P. 596-599]

[2] Ooi, Rieh Chi (2003). Balancing a Two-Wheeled Autonomous Robot. University of Western Australia: Thesis B. Mechatronics Engineering.

[3] https://github.com/TKJElectronics/KalmanFilter

[4] Bose, I. (2000). How to Make Good Homemade PCBs? Electronics for You Magazine.Pg 37-40.

[5] R. Fierro, F. Lewis, and A. Lowe, "Hybrid control for a class of underactuated mechanical systems", IEEE Transactions on Systems, Man and Cybernetics, Part A: Systems and Humans, vol. 29, no. 6, pp. 649-4, nov 1999.

[6] N. M. A. Ghani, F. Naim, and P. Y. Tan, "Two wheels balancing robot with line following capability", World Academy of Science, Engineering and Technology, vol. 55, pp. 634-8, 2011.

[7] X. Ruan, J. Liu, H. Di, and X. Li, "Design and LQ control of a two-wheeled self-balancing robot", in Control Conference, 2008. CCC 2008. 27th Chinese, july 2008, pp. 275-279.

[8] T. Nomura, Y. Kitsuka, H. Suemistu, and T. Matsuo, "Adaptive backstepping control for a two-wheeled autonomous robot", in ICROS-SICE International Joint Conference, Aug. 2009, pp. 4687-92.

[9] K.-H. Su and Y.-Y.Chen, "Balance control for two-wheeled robot via neural-fuzzy technique", in SICE Annual Conference 2010, Proceedings of, aug. 2010, pp. 2838-2842.

[10] V. Georgitzikis and I. Akribopoulos, and O. Chatzigiannakis, "Controlling physical objects via the internet using the Arduino platform over 802.15.4 networks", IEEE Latin America Transactions, vol. 10, no. 3, pp. 1686-9, 2012.

IOP Conf. Series: Materials Science and Engineering **142** (2016) 012109 doi:10.1088/1757-899X/142/1/012109

Influence of Ambient Temperature on the CO_2 Emitted With Exhaust Gases of Gasoline Vehicles

D Chainikov, E Chikishev, , I Anisimov, A Gavaev

Motor transport operation, Tyumen State oil and Gas University, 625039, Tyumen, Russian Federation

E-.ail: 4ainikov@mail.ru

Abstract. This article focuses on the regulation of CO_2 emitted in the exhaust gases of gasoline vehicles. Based on comparing the world practices of restrictive measures on greenhouse gas emissions with Russian legislation, we conclude that there is a need to adjust the limits of CO_2 emission taking into account the negative impact of ambient temperature on CO_2 emission. The climatic conditions of many countries stipulate the use of vehicles in temperatures below zero. At the same time, the existing regulations fully take into account the temperature features of the various countries, which casts doubt on the existence of uniform emission standards for all countries. Here, we conduct an experiment on one of the most popular cars in Russia: the Mitsubishi Lancer 9. We establish that lower temperatures are correlated with larger concentrations of CO_2 in the exhaust gases. We draw a conclusion about the need to account for the actual operating conditions when establishing limit values on CO_2 emissions of vehicles.

1. Introduction

Significant attention worldwide is paid to environmental issues such as the negative impact of emissions from anthropogenic stationary and mobile sources; these emissions have a negative impact on the Earth's climate. As the 20th century began, active international activity aimed to fight climate change. The most well-known conference on the subject took place in December 1997 in Kyoto (Japan), and there has been a considerable extension of the United Nations Framework Convention on Climate Change (UN FCCC) in 1994. The UNFCCC defined legal obligations for emission reductions. The Kyoto Protocol outlined the basic rules, but it did not provide details about how such rules would be applied. The Kyoto Protocol also required a separate formal process of signature and ratification before taking effect. The obligatory purposes and terms of reducing greenhouse gas (GHG) emissions in the majority of the countries have been modelled on the Kyoto Protocol.

The mandate involved maintaining GHG emissions at a level of 8–10% compared with 1990 levels.

The Kyoto Protocol was the first global agreement about environmental protection.

The primary GHG include water vapour (H_2O), carbon dioxide (CO_2), methane (CH_4) and ozone (O_3).

Of all of these gases, we focus on CO_2. The sources of CO_2 in the atmosphere of Earth include volcanic activity and the activity of living organisms.

Anthropogenic sources of CO_2 include the burning of fossil fuels in both stationary and mobile sources. A number of countries have reduced the negative impact on the nature from the operation of motorised transportation; exhaust gases (EG) consist of a significant amount of CO_2.

The European Union (EU) imposes some of the strictest restrictions on CO_2 emissions in the world. By 2020, new cars in Europe will be required to emit only 95 g/km of CO_2 [1]. In the United States and Japan, there are similar programs to decrease GHG emissions [2]. This standard is the same for countries with different climatic conditions (i.e., the EU, United States, Japan and others). CO_2 emissions are monitored along with carbon monoxide (CO), hydrocarbons (HC), nitrogen oxides (NO_x) and soot; but in Russia, an assessment of the component structure of EG the CO_2 emissions is not considered at all [3,4].

Moreover, according to a letter of the Ministry of Natural Resources and Environmental Protection of the Russian Federation from October 3, 2015 (number 12-47/5413) entitled "About a payment for negative impact from mobile sources," the following is reported:

Enacted since January 1, 2015 of the Federal Law (FL) of July 21, 2014 (number 219-FL) "On Amendments to the Federal Law On Environmental Protection and separate acts of the Russian Federation" the 28th article of the FL of May 4, 1999 (number 96-FL) "On air protection" is presented in a new version, according to which legal entities and individual entrepreneurs will be charged for emissions of harmful (polluting) substances (HS) into the atmosphere from stationary sources.

Therefore, since January 1, 2015 the charge for emissions of HS into the atmosphere from mobile sources from legal entities and individual entrepreneurs is not provided by the legislation of the Russian Federation.

The Ministry of Natural Resources and Environmental Protection of the Russian Federation is currently preparing regulations specifying the rules for calculating, collecting and adjusting charges for negative impacts on the environment.

However, in 2014 the Russian Federation became one of the top five emitters of CO_2 emissions [5]. Annually, the country emits 1,766,000kt of CO_2 due to human activities. At the same time, a substantial fraction of these emissions derives from motor transport.

Considering the absence of Russian CO_2 emissions restrictions and the total annual level of emissions, there is a need for CO_2 emission reduction programs that applied to mobile sources. At the same time, it is necessary to take into account the peculiarities of the conditions in which Russian motor transport is operated. Therefore, according to Köppen V.P. climate classification [6], more than 70% of country consists of areas with average annual ambient temperatures below 0°C (figure 1).

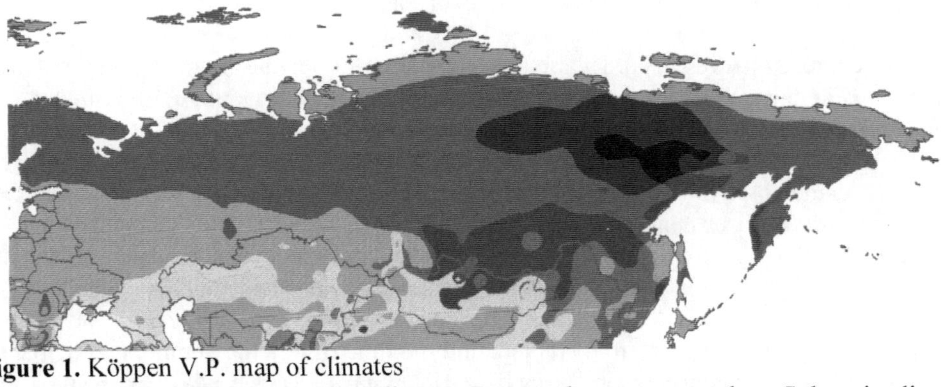

Figure 1. Köppen V.P. map of climates
Gray colour corresponds to Tundra climate, Green colour corresponds to Subarctic climate, Blue colour corresponds to Humid continental climate.

Research about the influence of low ambient temperatures on the emissions of HS in the EG of vehicles has been conducted [7-9]. However, research about the influence of sub-zero ambient temperatures (SZAT) on the CO_2 emissions has not yet been undertaken. At the same time as the

IOP Conf. Series: Materials Science and Engineering **142** (2016) 012109 doi:10.1088/1757-899X/142/1/012109

introduction of a program dedicated to decreasing and controlling CO_2 emissions in Russia, other countries with similar climatic conditions may also enact such programs., Emissions must be normalised differently given different conditions.

Therefore, the establishment of the level of SZAT influence on CO_2 emissions from the EG of vehicles is an urgent task.

The aim of this research is to investigate changes in CO_2 emissions in the EG of vehicles under the influence of SZAT.

2. Methodology

Experimental studies were carried out in Tyumen city on a gasoline-powered car (Mitsubishi Lancer 9). This car is one of the most popular models of foreign cars in Russia. There are currently approximately 4,000 such cars in Russia. The Mitsubishi Lancer 9 is equipped with a catalytic converter. The technical characteristics of this vehicle are presented in table 1.

Table 1: Technical characteristics of Mitsubishi Lancer

Index	Value
Volume of engine	1584 cm^3
Fuel System	Multi-point injection
Position of cylinders	Inline
Number of cylinders	4
Number of Gears (automatic transmission)	4
Max. weight	1750 kg
Engine power, kW (h.p.)	(98)

The car was technically sound. Prior to carrying out the experiment, we ensured that the vehicle passed a routine maintenance test.

We used an AVG-4 gas analyser (Garo Company) to measure CO_2 emissions, engine speed and oil temperature.

An AVG-4 gas analyser was used to measure the volume fraction of carbon monoxide, hydrocarbons, CO_2 and oxygen in the EG of vehicles with gasoline engines. There are channels in the gas analyser for measuring the engine speed and the oil temperature. The maximum measurement error for CO_2 is ±0.5%. The operating principle of the CO_2 volume fraction sensor is optical absorption.

The instrument readings were recorded after steady-state values in the range of 15–30 seconds.

We measured ambient temperatures using a Checktemp1 (Hanna, Germany) thermometer with a measurement accuracy of ±0.5°C.

By choosing the engine operation mode, we can analyse and determine that vehicles operating in cities spend a considerable amount of time in a stationary condition when the engine of the vehicle is ticking over for various reasons, including:

1. Warming up the engine prior to driving. This warm-up time period depends on the ambient temperature.

2. Downtime in congestion. At the same time, the downtime depends on the total number of vehicles in the settlement, transport infrastructure and the time of day (downtime in anticipation of a traffic light signal, yielding to pedestrians, peak times, interpeak times, weekdays, weekends, holidays, etc.).

3. Frequent technological stops connected with downtime and the loading/unloading of passengers and (or) freight.

At the same time, the HS emitted with EG during the movement of the vehicle dissipate evenly as the car moves. However, in the case in which a car is in a steady state (e.g., congestion), HS concentrate in much larger quantity in one area (i.e., in the space of the congestion). Therefore, the real question pertains to rationing the amount of HS emitted with EG, in particular CO_2, while the engine of a vehicle is ticking over.

We focused on a vehicle in tick over mode with an engine speed of 800–850 rpm. The temperature of the engine (oil) was 75–80°C. The studied ambient temperatures ranged from 20–0°C. We used 92 octane gasoline.

3. Results and analysis

Figure 2 shows the results of the actual CO_2 emissions depending on the ambient temperature. Within each temperature interval, we carried out approximately 5 measurements.

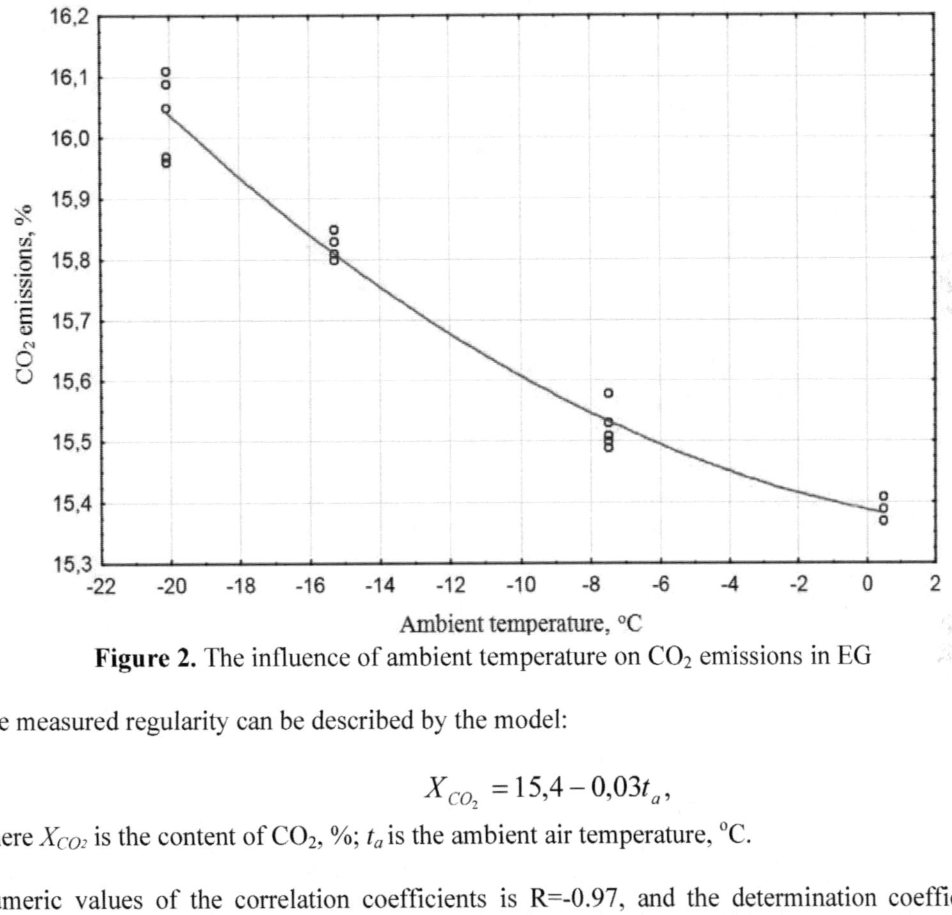

Figure 2. The influence of ambient temperature on CO_2 emissions in EG

The measured regularity can be described by the model:

$$X_{CO_2} = 15{,}4 - 0{,}03t_a, \quad (1)$$

where X_{CO_2} is the content of CO_2, %; t_a is the ambient air temperature, °C.

Numeric values of the correlation coefficients is R=-0.97, and the determination coefficient is R^2=0.94.

We are able to draw the conclusion that a decrease in air temperature from 0 to -20°C results in a regular increase in CO_2 emissions. This finding can be explained by the fact that a decrease in temperature corresponds to a small increase in fuel consumption, which entails a certain increase in the harmful components in EG, including CO_2.

4 Conclusions

On the basis of the obtained data, it is possible to note that within the considerable country of Russia vehicles are operated in SZAT conditions, which leads to an increase in CO_2 emissions with EG. The lower the temperature, the higher the emission.

The introduction of a program to decrease and control CO_2 emissions is necessary must take into account the climatic features of the place in which the vehicles are operated. The regular increase in

CO_2 emissions with decreasing ambient temperatures demonstrates the need for the introduction of correction coefficients to existing standards for vehicles that are operated in SZAT. Depending on the climatic region and the average daily temperature, it may be necessary to differentiate the level of payment for environmental pollution via CO_2 emission.

To reduce the negative impact of gasoline vehicles on the environment, one can use alternative motor fuels (e.g., natural gas). In some studies, it has been noted that the operation of transport on natural gas is comparable to gasoline and achieves better environmental performance by separate components of EG, including CO_2 [10-20]. One should also take into account that CO_2 emissions associated with gaseous fuels will increase at lower temperatures.

Therefore, expanding use of natural gas use to transport in various regions of Russia will reduce the negative environmental impact of EG on the environment.

References

[1] Regulation (EU) No 333/2014 of the European Parliament and of the Council of 11 March 2014 amending Regulation (EC) № 443/2009 to define the modalities for reaching the 2020 target to reduce CO_2 emissions from new passenger cars http://eur-lex.europa.eu/legal-content/EN/TXT/?uri=uriserv:OJ.L_.2014.103.01.0015.01.ENG (accessed March 1, 2016).

[2] Model Year 2012-2016 Light-Duty Vehicle Greenhouse Gas Emissions Standards and Corporate Average Fuel Economy Standards https://www.gpo.gov/fdsys/pkg/FR-2010-05-07/pdf/2010-8159.pdf (accessed March 1, 2016).

[3] GOST R 41.24-2003 (Regulation № 24) *Uniform provisions concerning: I. The approval of compression ignition (C.I.) engines with regard to the emission of visible pollutants; II. The approval of motor vehicles with regard to the installation of C.I. engines of an approved type; III. The approval of motor vehicles equipped with C.I. engines with regard tothe emission of visible pollutants by the engine; IV. The measurement of power of engines* – M.: Standartinform, 36 p., 2007.

[4] GOST R 52033-2003. *Vehicles with gasoline engines. Emissions of pollutants with the exhaust gases. Norms and methods of control when assessing the technical condition* - M .: Standartinform, 10 p., 2007.

[5] http://edgar.jrc.ec.europa.eu/overview.php?v=CO2ts1990-2014&sort=des9 (accessed March 4, 2016).

[6] https://www.climate.gov/file/worldkoppenmappng (accessed March 4, 2016).

[7] Anisimov, I., Ivanov, A., Chikishev, E., Chainikov, D., Reznik, L. Assessment of gas cylinder vehicles adaptability for operation at low ambient temperature conditions // *WIT Transactions on Ecology and the Environment*, **190** volume 1, pp. 685-695, 2014, doi: 10.2495/EQ140651.

[8] Karnaukhov N.N., Reznik L.G., The differentiated correction of charge for air pollution // *Advances in transport*, Volume **16**, pp. 701-705, 2004, doi: 10.2495/UT040681.

[9] Ertman J.A., Ertman S.A., Anisimov I.A., Chainikov D.A., Chikishev Y.M. Complex characteristic estimation of vehicle adaptability for low temperature operation conditions // *Research Journal of Pharmaceutical, Biological and Chemical Sciences*. Volume **6**, Issue 1, 2015, Pages 1761-1770.

[10] Aslam, M.U., Masjuki, H.H., Kalam, M.A., Abdesselam, H., Mahlia, T.M.I., Amalina, M.A. An experimental investigation of CNG as an alternative fuel for a retrofitted gasoline vehicle // *Fuel* - Volume **85**, Issue 5-6, pp. 717-724, 2006, doi: 10.1016/j.fuel.2005.09.004.

[11] Bielaczyc, P., Woodburn, J., Szczotka, A. An assessment of regulated emissions and CO2 emissions from a European light-duty CNG-fueled vehicle in the context of Euro 6 emissions regulations // *Applied Energy* - Volume **117**, pp. 134-141, 2014, doi: 10.1016/j.apenergy.2013.12.003.

[12] Dronniou, N., Kashdan, J., Lecointe, B., Sauve, K., Soleri, D. Optical Investigation of Dual-fuel CNG/Diesel Combustion Strategies to Reduce CO_2 Emissions // *SAE International Journal*

of Engines - Volume **7**, Issue 2, pp. 873-887, 2014, doi: 10.4271/2014-01-1313.

[13] Jacqueline L. Kirk, Abigail L. Bristow, Alberto M. Zanni Exploring the market for Compressed Natural Gas light commercial vehicles in the United Kingdom // *Transportation Research Part D: Transport and Environment* - Volume *29*, pp. 22-31. 2014, doi: 10.1016/j.trd.2014.03.004.

[14] Jahirul, M.I., Masjuki, H.H., Saidur, R., Kalam, M.A., Jayed, M.H., Wazed, M.A. Comparative engine performance and emission analysis of CNG and gasoline in a retrofitted car engine // *Applied Thermal Engineering* - Volume **30**, Issue 14-15, pp. 2219-2226, 2010, doi: 10.1016/j.applthermaleng.2010.05.037.

[15] Khan, M.I., Yasmin, T., Shakoor, A. International experience with compressed natural gas (CNG) as environmental friendly fuel // *Energy Systems* – Volume **6**, Issue 4, pp. 507-531, 2015, doi: 10.1007/s12667-015-0152-x.

[16] Lim, C., Kim, D., Song, C., Kim, J., Han, J., Cha, J.-S. Performance and emission characteristics of a vehicle fueled with enriched biogas and natural gases // *Applied Energy* - Volume **139**, pp. 17-29, 2015, doi: 10.1016/j.apenergy.2014.10.084.

[17] Oprešnik, S.R., Seljak, T., Bizjan, F., Katrašnik, T. Exhaust emissions and fuel consumption of a triple-fuel spark-ignition engine powered passenger car // *Transportation Research Part D: Transport and Environment* - Volume **17**, Issue 3, 2012, pp. 221-227, doi: 10.1016/j.trd.2011.08.002.

[18] Rose L., Hussain M., Ahmed S., Malek K., Costanzo R., Kjeang E. A comparative life cycle assessment of diesel and compressed natural gas powered refuse collection vehicles in a Canadian city // *Energy Policy* - Volume **52**, pp. 453-461, 2013, doi: 10.1016/j.enpol.2012.09.064.

[19] Wang, R., Wu, Y., Ke, W., Zhang, S., Zhou, B., Hao, J. Can propulsion and fuel diversity for the bus fleet achieve the win-win strategy of energy conservation and environmental protection? // *Applied Energy* - Volume **147**, pp. 92-103, 2015, doi: 10.1016/j.apenergy.2015.01.107.

[20] Yao, Z., Cao, X., Shen, X., Zhang, Y., Wang, X., He, K. On-road emission characteristics of CNG-fueled bi-fuel taxis // *Atmospheric Environment* - Volume **94**, pp. 198-204, 2014, doi: 10.1016/j.atmosenv.2014.05.027.

IOP Conf. Series: Materials Science and Engineering **142** (2016) 012110 doi:10.1088/1757-899X/142/1/012110

Prospects of and Problems in Using Natural Gas for Motor Transport in RUSSIA

E Chikishev[1], A Ivanov[2], I Anisimov[1], D Chainikov[1]

[1] Motor transport operation, Tyumen State oil and Gas University, 625039, Tyumen, Russian Federation
[2] Technical systems in the agricultural sector, Northern Trans-Ural State Agricaltural University, Tyumen, Russian Federation

E-mail: 4ainikov@mail.ru

Abstract. This article is devoted to increasing the use of natural gas in Russia as a measure to decrease the negative influence of motor transport on the environment. A brief analysis of the global fleet of natural gas vehicles is provided above. The documents accepted in Russia to promote public awareness of compressed natural gas in transport are submitted. The basic reasons keeping the growth of natural gas vehicle fleets in Russia consist of weak branching of refuelling stations; difficulty in determining the actual amount of compressed natural gas required; and control methods of the consumption of gas fuel. The offers promoting the growth of the fleet of natural gas vehicles are given.

1. Introduction

The rapid growth of global motor transport is having an increasingly negative impact on the environment. Against this background, addressing the problems of replacing traditional motor fuels, such as gasoline and diesel, with alternatives are all the more real. There is a set of options to do so, beginning from electric cars and finishing with hydrogen fuel. But for Russia, it is more appropriate and logical to use natural gas (NG), as the country possesses 30% of the world's NG deposits.

There is a world-wide successful transportation industry that operates on NG, which shows the best ecological indicators in comparison with both gasoline and diesel fuel [1-4].

According to the NGV journal [5] global leaders in terms of the number of natural gas vehicles (NGV) and natural gas refuelling stations (NGV RS) are Argentina, Brazil, Germany, India, Iran, Italy, China, Colombia, Pakistan, the United States, Thailand, and others. In total, in the world, there are about 22 million methane NGVs (2% of the total global vehicle fleet) (figure 1). According to NGV journal [5] the global vehicle fleet operating on gaseous fuel annually increases by 30%.

Today, Russia holds a fairly modest position in terms of number of NGVs (90050) and NGV RSs (250). In comparison, Iran has about 4 million NGVs, and China, 6500 NGV RSs (table 1).

However recently at the legislative level, several attempts of widespread use of NG on the motor transport have been made that includes the:

- The order of the Deputy Prime Minister of the Russian Federation I.I. Sechin № IS P9-5878 of 25 August 2010 on the establishment of a permanent working group to develop proposals for an increase in the demand for liquefied petroleum and NG as a motor fuel.

- The order of the President of the Russian Federation № 1923 from 27.06.2011 on the organization of stage-by-stage replacement of public motor transport by NGVs and on the expansion of a network of NGV RSs.

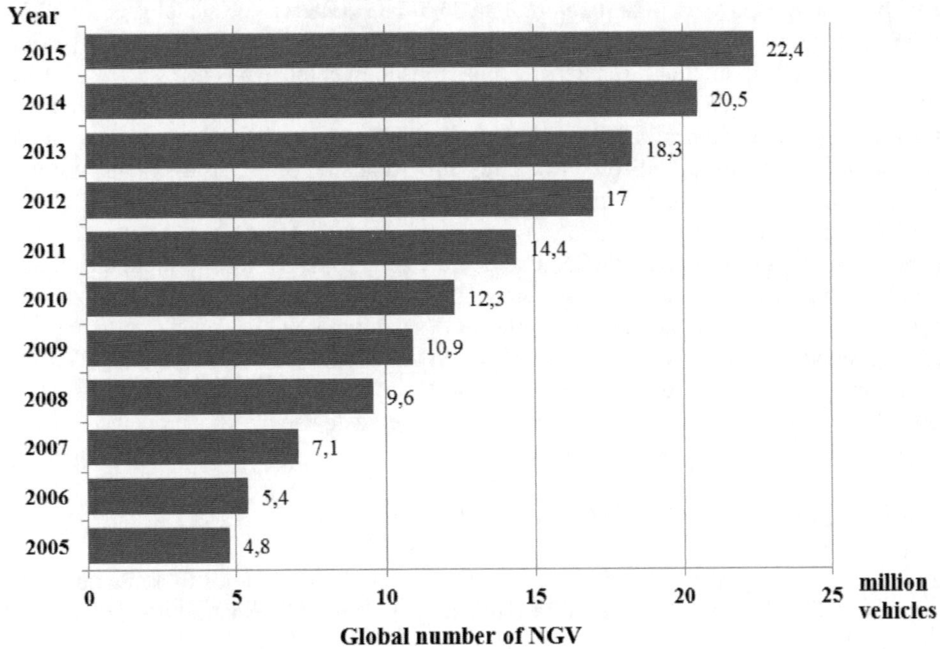

Figure 1. Dynamic of global NGV fleet change

Table 1. Global lieders in number of NGV, fuelled with compressed natural gas and the amount of NGV RS according to NGV journal

Country	N/m³ monthly sales average	Number of NGV, units	Amount of NGV RS, units
Iran	630000000	4068632	2268
China	no data	3994350	6502
Pakistan	245750000	3700000	2997
Thailand	184200000	462454	497
Argentina	239815000	2487349	1939
India	163210000	1800000	936
Brazil	144535636	1781102	1805
Bangladesh	91550000	220000	585
USA	77520000	150000	1615
Italy	75000000	885300	1060
Colombia	45000000	500000	800
Germany	23000000	98172	921
Uzbekistan	no data	450000	213
Russia	33750000	90050	253

- The order the Prime Minister of the Russian Federation D.A. Medvedev № 767-r "On regulation of the relations in the sphere of the use of gas motor fuel" (from 13.05.2013).
- The comprehensive plan of actions for expansion of use of NG as motor fuel from November 14, 2013 № 6819p-P9 (it is approved as the Deputy Prime Minister of the Russian Federation A.V. Dvorkovich).

On the basis of the submitted documents of PJSC "Gazprom" and LLC "Gazprom gas fuel" program for the construction of compressed natural gas refuelling stations CNG RSs and liquefied natural gas refuelling stations (LNG RS)s was developed. According to a related document submitted for Russia with a plan of action until 2030, more than 2600 NGV RS will be established. It is also noted that the servicing of NGV at each NGV RS must be planned post maintenance and warranty period. This action plan also includes transferring 524 thousand vehicles to NG.

In the JSC "Oil Company "Rosneft," one of three main marketing strategy priorities of gas business is active participation in federal programs of development of the market of gas motor fuel, including creation of a corporate network of the NGV RS. Thus, the expansion of NG usage in the motor transport is a real prospect of the upcoming future of Russia. And the questions connected with operation of NGV are actual.

NG can be transported in compressed (CNG; at compression to 25 MPa) or liquefied (LNG; when cooled to -162 °C) states. This requires installing the corresponding gas equipment at the same time. When using CNG, vehicles refuel at CNG RSs. The NGVs using a LNG gas need a refueling procedure based on cryogenic gas stations (LNG RSs). In this article we will focus on NGV running on CNG and, respectively about the CNG RSs.

Objective of this research is studying of prospects of large-scale use of NG on the motor transport in Russia and identification of limiting factors.

2. Factors affecting the growth of the NGV fleet in Russia

2.1 Locations of CNG RSs

The primary issue interfering with the full-scale transfer of vehicles to gaseous fuel in a number of the countries, including Russia, is the necessity of establishing appropriate infrastructure and in particular CNG RSs [6-7].

The network of CNGRSs is distributed in the following federal districts of Russia (the real number of CNGRSs may slightly differ from the submitted data): 1. Volga Federal District (FD) – 62; 2. Central FD – 54; 3. Ural FD – 42; 4. Southern FD – 35; 5. North Caucasus FD – 24.

From a technical point of view, the small number of CNG RSs in other FDs can be attributed to several factors, most importantly:

1. Weak branching (or total absence) of gas pipelines in Eastern Siberia and the Far East. It is known that the main part of Russia's gas transmission system is located in the western, central and southern parts, as well as in the Urals. Supplying fuelling stations with NG requires a gas pipeline.

2. The significant extent of the specified FDs territory and the resulting problematic nature of the creation of so-called "blue corridors" (i.e., vehicles moving in different settlements and using NG only). As noted by professor Ya. S. Mkrtychan, the distance between CNG RSs should be in the range of 70–250 km. Between Tyumen (Urals FD) and Novosibirsk (Siberian FD) (i.e., a distance of 1278 km; federal highway R402–R254), there are no CNG RSs. Furthermore, in the Siberian and Far Eastern FD one cannot count more than 20 gas stations.

3. The problematic nature of NGV operation on NG given low ambient temperatures. The majority of Russian is in a cold climate zone (about 70% - northern parts of Russia, Western Siberia, Eastern Siberia and the Far East) and is accordingly subject to low ambient temperatures. Such conditions can lead to unstable operation of transport. However, research [8] has noted that such temperature changes can affect NGV output parameters; even so, the operation of vehicles, with appropriate preparation, is possible.

2.2 Determination of the actual quantity of NG

There are a number of organisations in Russian currently involved in the design, construction, and refurbishment of old CNG RSs using the latest technologies. However, there is a complexity of determination of actually released gas to vehicle owners.

The quantity and consumption of NG is typically determined by reference tables presented in RD 3112199-1095-03 [9]. Therefore, the owner of a vehicle, knowing the cylinder volume, the gas pressure in it (measured according to indications of the manometer installed with the gas equipment in kgf/cm^2 or MPa) and the ambient temperature can determine the quantity of CNG in m^3.

However, determining exactly how much gas is in the cylinders after refuelling is difficult. Volumetric accounting of methane fuel in this case may not be sufficiently reliable because the volume of a gaseous product depends on both the gas pressure and temperature. In RD 3112199-1095-03 [9], it is indicated ambient temperature. However, the gas temperature is not always equal to the ambient temperature.

It is likely for this reason that CNG RSs are sometimes faced with the problem of selling gas to vehicle owners that differs slightly from the volume of gas received at the station.

To determine a solution to this problem, organisations engaged in construction and reconstruction of CNG RSs have developed and certified a measuring and control system for refuelling NGVs. This system, which is based on a mass meter for measuring the mass flow of liquids and gases, is constructed on the principle of a Coriolis flow sensor. This method yields the most objective data for determining the air volumes of gaseous fuel. Nevertheless, at the present time, not all CNG RSs possess this equipment.

For the majority of CNG RSs in Russia, the information board resembles what is shown in figure 2.

However, the only constant parameter of a gaseous substance is its weight. Transitioning to directly measuring the mass of the fuelled NG makes it possible to not only avoid the influence of various factors affecting the measurement accuracy but also simplifies the organisation of the account when refuelling at the compressor station.

Considering this fact, some CNG RSs offer refuelling with methane not in volume sizes (m^3) but in mass (kg). This refuelling scheme is shown in figure 3. Some CNG RSs offer refills in m^3 and some offer refills in kg, depending on the choice of the vehicle owner.

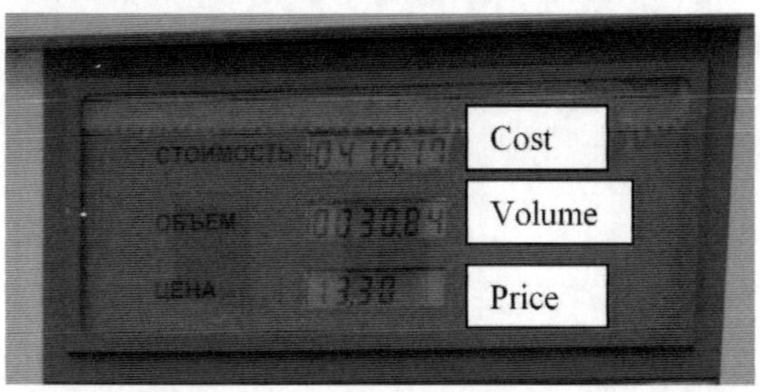

Figure 2. An example of an information board at a CNG RS with actual volume indications of the fueled NG (Tyumen)

Figure 3. These two figures have been placed side-by-side to save space. Justify the caption.

In standard climatic conditions, converting from the volume of methane to mass is rather simple: 1 m^3 corresponds to 0.717 kg.

Therefore, the actual amount of NG in the cylinders of a vehicle can be defined in three ways:

1. According to the manometer indications in kgf/cm^2 or MPa during vehicle refuelling;
2. According to indications of an information display on the CNG RS in m^3 (when refuelling the vehicle);
3. According to indications of an information display on the CNG RS in kg (when refuelling the vehicle).

2.3 Design features of NGV and methods of gaseous fuel consumption control

Internal combustion engines using CNG as a motor fuel can be divided into two categories:

a) dual-fuel engines with a universal power supply system and a spark ignition system that includes two equivalent power supply systems running on gas and liquid oil fuel (gasoline);

b) gas-liquid – engines with power supply system, in which a portion of liquid engine (diesel) fuel during the engine operation on CNG is used as an ignition dose for the ignition of an air-gas mixture in the engine (gas diesel engine);

c) gas – engines, convertible only for work on NG via ignition of an air-gas mixture in the cylinders from an electric spark or a glow plug.

The majority of Russian NGVs are currently dual-fuel, and they possess two equivalent power supply systems running on NG and liquid oil fuel (gasoline or diesel fuel). Furthermore, the majority these vehicles which do not possess regular gas equipment are petrol (diesel) and are only equipped additionally with such equipment for methane use. However, recent trends both in Russia and worldwide have dictated that more and more vehicles have regular gas equipment installed by the manufacturer. In this case, traditional oil fuel is used as supplemental fuel.

Audi, Chevrolet, Citroen, Daimler-Benz, Fiat, Ford, Iveco, Honda, MAN, Opel, Peugeot, Scania, Toyota, Volkswagen, Volvo, etc. today offer factory vehicles with engines with regular gas equipment that is compatible with CNG. Moreover, the design features of these engines are adapted to work mainly on methane fuel.

In Russia, the most popular producers of NGV on methane are are KAMAZ, Liaz, KAvZ, GAZ, VAZ and others. However, compared with foreign automakers, domestic automakers produce very few modifications of such vehicles. Moreover, almost all Russian plants install imported equipment, from the gas equipment to the gas engines.

In Europe, Volkswagen cars are very popular among NGV cars. This company considers constructive decisions that allow it to estimate how useful such vehicles will be in operation.

A summary of the technical characteristics and the dashboards of the considered brands are presented in table 2.

It should additionally be noted that all of the NGV brands presented in table 2 are started and warmed at low air temperatures (up to the temperature of the cooling liquid plus 10–15 °C) on gasoline and then automatically switched to power with NG.

Table 2. Characteristics of methane NGVs Volkswagen

Characteristics	Touran EcoFuel	Caddy EcoFuel	Golf Variant Blue Motion	Passat EcoFuel
	Engine Type dual fuel working on CNG and gasoline			
Engine capacity	1.4	2.0	1.4	1.4
Power, kW / hp	80/109	80/109	81/110	110/150 2 boost
Number of cylinders, pcs.	4	4	2	3
The total volume of cylinders, l	115	160	94	132
The mass of gas at 200 bar, kg	18	26 (37 m^3)	15	21 (31 m^3)
Volume of gasoline tank, l	13	13	50	31
Consumption kg / 100 km	5.9	6.0	3.4 H[a]	4.4 H[a]

[a]NG of group H (high) contains more methane than gas of group L (low). The higher the content of the methane, the better the quality of the gas and the greater the distance that can be driven on the fuel.

When discussing NGV converted from traditional motor fuels to natural gas-powered vehicles, it is possible to note that the determination of the fuel amount in the cylinders can be made via indications of the manometer (in kgf/cm^2 or MPa). It causes inconveniences for two reasons:

1. The absence of the information dashboard of the manometer (fuel consumption sensor). The dashboard is, as a rule, located in a location of cylinders.

2. Reading the manometer indications, it is difficult to determine the actual amount of gas in the cylinders. For this purpose, it is necessary to use reference tables [9].

For the Volkswagen vehicles that we consider, the relevant design decisions for this purpose are provided.

Touran EcoFuel and Caddy EcoFuel have the following indicators of a gas operating mode, and these makes also show the mode of emergency work on gasoline on the dashboard:

- the control indicator of the emergency work mode on gasoline;
- a segment indicator of the gasoline level;
- an analogue indicator of the gas level.

There are two various options for a dashboard:

1. Midline cars (figure 3). If the arrow flashes near or over the segment indicator, it means that the car works with gasoline. When the arrow dies out, the car switches to gas. Operation on gas is possible at temperatures exceeded the cooling liquid by +15°C. The engine control unit automatically switches from a gas-operating mode to the emergency work mode on gasoline.
2. Highline cars (figure 4).

Figure 5 shows an optional dashboard for a Volkswagen Passat EcoFuel.

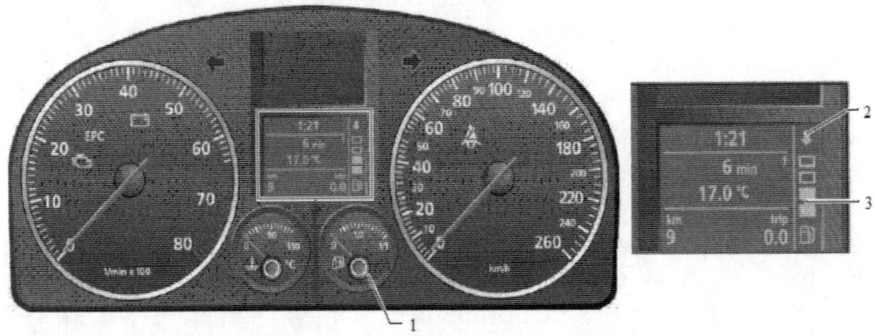

Figure 3. A dashboard in the Volkswagen Touran EcoFuel and Caddy EcoFuel (Midline), where 1 corresponds to the gas level indicator, 2 corresponds to the control indicator of the emergency work mode on gasoline, and 3 corresponds to the fuel level indicator

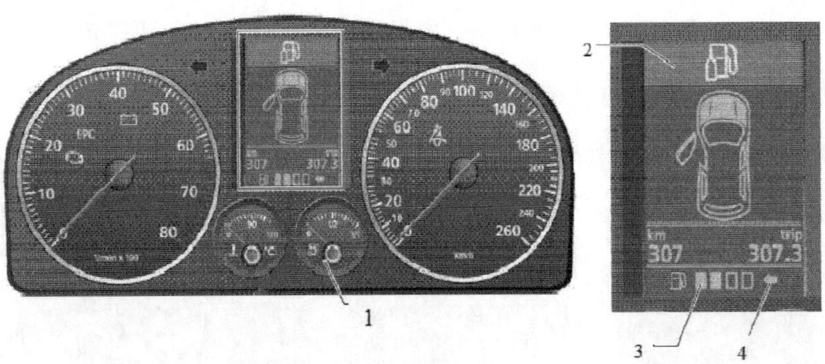

Figure 4. A dashboard in the Volkswagen Touran EcoFuel and Caddy EcoFuel (Highline)), where 1 corresponds to the gas level indicator, 2 corresponds to the gas reserve indicator, 3 corresponds to the fuel level indicator, and 4 corresopnds to the control indicator of the emergency work mode on gasoline

Information about the actual NG consumption (in kg) or gasoline (in litres) per 10 km and a total fuel distance possible when using a particular type of fuel is shown on the multifunction display. In addition, the information about the ambient temperature is located on the display. The green indicator lamp of running on gaseous fuel means that the engine is running on NG.

IOP Conf. Series: Materials Science and Engineering **142** (2016) 012110 doi:10.1088/1757-899X/142/1/012110

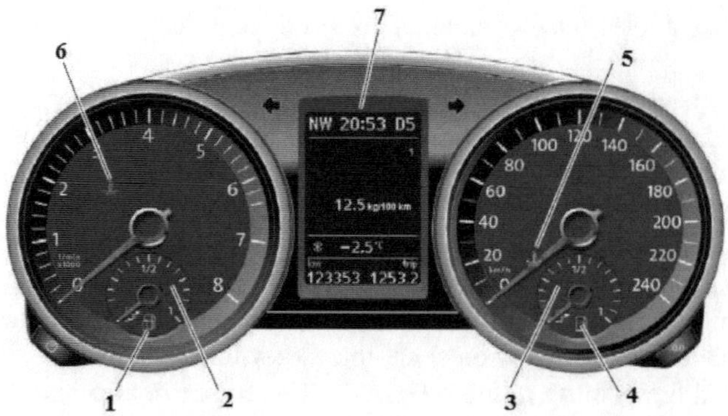

Figure 5. A dashboard in Volkswagen Passat EcoFuel, where 1 corresponds to a control lamp of work on NG, 2 corresponds to the gas quantity indicator, 3 corresponds to the fuel level indicator, 4 corresponds to a control lamp of gasoline reserve, 5 and 6 are control lamps of the coolant temperature, and 7 is a multifunction display

One of the most modern NGV is the Volkswagen Golf Variant Blue Motion. The dashboard on this brand of vehicles looks practically identical to that shown in figure 5. However there is one feature, the multifunction display. Information pertaining to the quality of NG (figure 6) is shown on this feature.

The quality of gaseous fuel is displayed as a percentage, with a step size of 10% between 70% and 100%. After each refuelling, an adjustment to the quantity of NG is carried out. During refuelling, the adaptation of the content of the gaseous fuel of methane, which characterises the quality of NG (heat of combustion), is defined. An indication of 100% means that the content of methane in the gas is higher than 90%, and an indication of 90% means that the content of methane in the gas is higher than 80%. Since the content of methane is not measured directly but rather calculated, an indication of NG quality can differ from the actual content of the methane.

Furthermore, for convenience, all Volkswagen NGVs indicate NG consumption in kg. With regard to domestic producers, it is worth noting that they are also attempting to improve their dashboards with the addition of information about the quantity of NG in the cylinders. Figure 7 shows a bus KAvZ-4238-72 dashboard that indicates the residual amount of NG as a percentage ratio (level sensor).

Figure 6. The indicator showing the NG quality in Volkswagen Golf Variant Blue Motion vehicles

Figure 7. Bus KAvZ-4238-72 dashboard

However, after analysing figures 3–5, one can note that the dashboard on Volkswagen cars is more convenient and informative than the dashboard shown in figure 7.

3. Conclusions

To promote NGVs in Russia operating on CNG, it is necessary to promote the widespread expansion of the CNG RS network. Additionally, it is necessary to:

1. Develop an easy-to-understand system of defining the actual amount of gas in a cylinder. We propose to adopt foreign conventions and measure methane in mass units (i.e., kg).
2. Increase the Russian production of domestic vehicles with regular gas-cylinder equipment and engines. Particular emphasis, in accordance with Government Order № 767 [10], should be placed on buses, transport of road utilities, trucks and the most in-demand brands of domestic cars. At the same time, it is worth paying attention to the dashboard and placing on it the quantity of gas in the cylinder.
3. In view of the intensive gasification of Eastern Siberia and the Far East on the national level, it is necessary to work out the issue of creating a "blue corridor" in these regions.
4. Improve the legislative base pertaining to stimulation of converting transportation to NG via various benefit programs (e.g., providing instalment payments for the installation of gas equipment, lowering interest rates on loans to buy new NGVs, cost reductions on toll roads and parking areas, etc.).
5. Increase the awareness of vehicle owners about the advantages of NG by running television ads, distributing booklets, etc.

Replacing traditional fossil fuels with NG will enable reductions in the general emissions of harmful substances emitted with exhaust gases from vehicles, which will make it possible to improve the ecological situation, particularly in large cities.

References

[1] Hassan, M.H., Kalam, M.A., Mahlia, T.M.I., Aris, I., Nizam, M.K., Abdullah, S., Ali, Y. Experimental test of a new compressed natural gas direct injection engine (2009) *Energy and Fuels* - Volume **23**, Issue 10, pp. 4981-4987, doi: 10.1021/ef8011382.

[2] Lim, C., Kim, D., Song, C., Kim, J., Han, J., Cha, J.-S. Performance and emission characteristics of a vehicle fueled with enriched biogas and natural gases (2015) *Applied Energy* - Volume **139**, pp. 17-29, doi: 10.1016/j.apenergy.2014.10.084.

[3] Tong, F., Jaramillo, P., Azevedo, I.M.L. Comparison of Life Cycle Greenhouse Gases from Natural Gas Pathways for Light-Duty Vehicles (2015) *Energy and Fuels* – Volume **29**, Issue **9**, pp. 6008-6018, doi: 10.1021/acs.energyfuels.5b01063.

[4] Wang, R., Wu, Y., Ke, W., Zhang, S., Zhou, B., Hao, J. Can propulsion and fuel diversity for the bus fleet achieve the win-win strategy of energy conservation and environmental protection? (2015) *Applied Energy* - Volume **147**, pp. 92-103, doi: 10.1016/j.apenergy.2015.01.107.

[5] NGV journal [http://www.ngvjournal.com/worldwide-ngv-statistics/] [accessed 20.02.2016].

[6] Engerer, H., Horn, M. Natural gas vehicles: An option for Europe (2010) *Energy Policy* - Volume **38**, Issue 2, pp. 1017-1029, doi: 10.1016/j.enpol.2009.10.054.

[7] Ma, L., Geng, J., Li, W., Liu, P., Li, Z. The development of natural gas as an automotive fuel in China (2013) *Energy Policy* - Volume **62**, pp. 531-539, doi: 10.1016/j.enpol.2013.06.066.

[8] Anisimov, I., Ivanov, A., Chikishev, E., Chainikov, D., Reznik, L. Assessment of gas cylinder vehicles adaptability for operation at low ambient temperature conditions (2014) *WIT Transactions on Ecology and the Environment*, **190** volume 1, pp. 685-695, doi: 10.2495/EQ140651.

[9] Guidance document 3112199-1095-03. *Guidelines on the organization and operation of gas-cylinder vehicles that run on compressed natural gas.* – Ministry of Transport, Road Department, JSC SaRIoMT - 2002. 59 p.

[10] Order of the Russian Federation Government dated 13/05/2013 № 767-r *On regulation of relations in the sphere of gas motor fuel use.* – Moscow, 2013. 3 p.

A Technique for Decreasing Reactivity of Coal Material to Suppress the Oxygen Absorption Process

S S Timofeeva [1a], N Yu Lugovtsova [2b], A R Gubanova [2c]

[1a]Doctor of Technical Sciences, Prof. National Research Technical University. Russia, Irkutsk, Lermontov str., 83
[2b]Assistant lecturer. Yurga Institute of Technology (affiliate) of National Research Tomsk Polytechnic University. Russia, Yurga, Leningradskaya str., 26.
[2c]Student. Yurga Institute of Technology (affiliate) of National Research Tomsk Polytechnic University. Russia, Yurga, Leningradskaya str., 26.

E-mail: [1a]timofeeva@istu.edu, [2b]lnyu-70583@bk.ru

Abstract. The paper describes the mechanisms of self-ignition formation in coal liable to spontaneous combustion, on the basis of experimental works performed to analyze heat and mass transfer in the coal-air system. A new approach was developed to the coal self-heating suppression and thermodynamic control of the oxidation process. The influence of coal moisture content and thermal behaviour of air in the cooling process was studied during moisture evaporation.

1. Introduction

For establishing the determining factors that favour oxidation and self-heating of coal, extensive studies were conducted to analyze the influence of combined parameters (rate of flow, temperature, humidity, oxygen concentration) of air on reactivity of coal material, which has various natural moisture contents, in a laboratory environment and underground [1-5].

It has been experimentally proved that self-heating of coal (up to 80 °C) occurs at a relatively low rate of air filtration $(3\text{-}7)\cdot 10^{-3}$ m/s, and the determining factor is its moisture exchange in the heterogeneous coal-air system, which is dependent on aerodynamic and thermo physical properties of air, and moisture content of coal. At relative humidity being constant, the oxidative process and self-heating of coal rapidly slow down, accompanied with a simultaneous decrease in rate, temperature and concentration of oxygen in the intake ventilation air flow. However, these parameters, used to quantify any ventilation loss underground in order to prevent spontaneous combustion of coal in a goaf at working areas, are difficult to control. Therefore, decelerating the oxidation process by increasing a relative humidity up to 100 % at the constant rate and oxygen concentration has a practical importance. The physical nature of the effect that results in inactivating coal compaction is associated with changing the direction of moister exchange in the coal-air system, thereby reducing the rate of drying and chemical activity of coal.

According to the studies conducted for this purpose, an increase in coal moisture decreases its ability to absorb oxygen, with the slow down effect on the coal oxidation taking place for a long time, what is important for underground operations. Earlier researches established that, at the initial phase of oxidation, the moist coal absorbs less oxygen than the dry coal (Figure 1). However, the amount of oxygen absorbed by the moist and dry coals becomes equal after 144 hours.

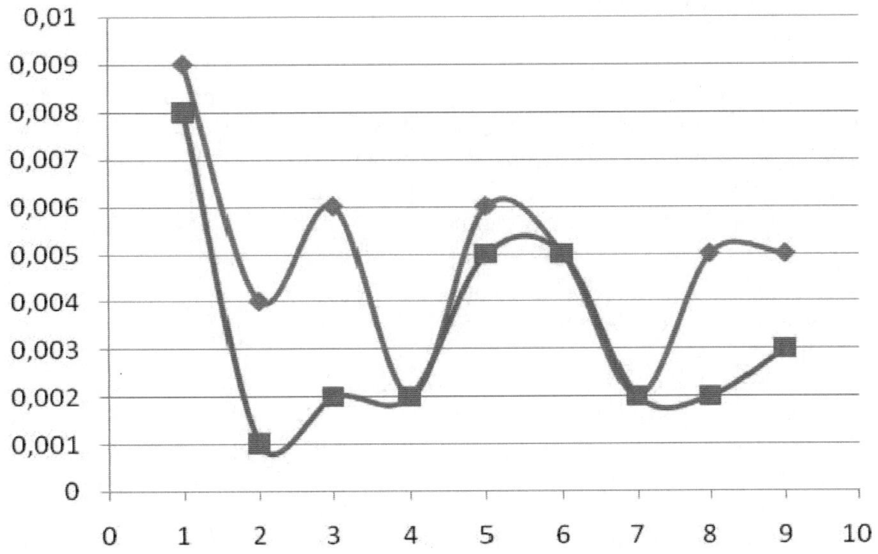

Figure 1. The time dependence of the amount of oxygen absorbed by coal
—— Prokopyevsk mine dry coal at +25
—— Prokopyevsk mine moist coal at +25

The previously watered and then dried coal becomes more active to oxygen. The activating effect of water that is removed from coal by drying can be explained by the increased availability for coal-derived substances to reach both methane and oxygen. It is known that, in all cases, the residual content of methane in the preliminarily watered coal is 30-50 % higher than that of the un-watered. The absorbable by coal methane behaves as if it screens a highly developed surface, prevents oxygen molecules to connect with coal particles. Besides, the higher moisture of coal results in increasing its thermal parameters that accelerates a dissipation of heat released due to oxidation and oxygen adsorption, and inhibits the development of an autocatalytic process of oxidation and self-heating of coal.

2. Results and Discussion

On the basis of theoretical and large-scale laboratory studies relating to heat and mass transfer in the coal-air system [6], it has been established that the mechanisms of self-ignition formed in the coal susceptible to spontaneous combustion, allowing for real-time factors, are as follows:

- Density of the resulting moisture flow (from coal to air) and heat released during coal oxidation have an identical progressive pattern over time in reference to the rates of air filtration through the compact coal in the range of $(2.5\text{-}4.6)\cdot10^{-3}$, with corresponding to both laminar and transitional modes;
- At a coal temperature of 32-45 °C and a rate of air filtration $4.7\text{-}6.0)\cdot10^{-3}$ m/s, the decrease in moisture exchange takes place that results in initiating thermodynamic stabilization of the heterogeneous oxidation reaction in the coal compaction due to the deepening of evaporation area inside lumps of coal and the increasing of a moisture diffusion pathway;

- Lowering a temperature of the intake air by 3-5 °C and its relative humidity down to 70 % at the second phase of intensive moisture evaporation from coal leads to an extremum in the coal-air system shifting to the zone with a higher risk of occurring spontaneous coal combustion processes;
- The self-heating process in the coal compaction changes into the self-ignition process when the coal moisture content is decreased by on average 60%, if compared to the natural bed moisture, under the turbulent air flow conditions.

The mechanisms established in forming places of spontaneous coal combustion allowed the investigators to come to a scientifically based conclusion that, during the self-heating phase, the moisture of coal is the most important parameter in determining this phase duration. In this context, a new approach was developed to the spontaneous combustion inhibition to improve thermodynamic stability of the oxidation process by controlling this parameter.

This approach relates to maintaining a thermodynamic state of the heterogeneous coal-air system within a rigidly fixed range, and shifting its extremum to the area with a low propensity for spontaneous coal combustion by means of changing the orientation of moisture exchange processes between the phases in this system with a purpose of regulating thermal parameters of air by using artificial or natural aerosols.

It is known that any rank of coal has the so-called critical moisture content of coal, and when this content is reached, the rate of coal oxidation sharply increases by 1.5-2 times. It is established for coals to have a critical moisture content varying from 2 or 3 % to 17-18 % for brown coals. Thus, the results of theoretical and laboratory studies, aimed at identifying a relationship between critical moisture of coal and distribution of moisture in the micro- and super microporous structure of coal, can be used to describe mechanisms of oxidation reactions in coal as well as calculate temperatures of coal self-heating.

To date, there are a considerable number of mathematical models and software developed and employed to simulate coal self-heating processes [7]. Some of them allow for either the time dependence of absorption rate or an increase in the rate of oxygen absorption by coal at critical moisture contents. At the same time, experimental studies prove the combined influence of these factors on the process of coal self-heating. In this regard, it is relevant to design a procedure for the calculation of temperatures in coal compactions to improve the existing models and take into account both the processes of deactivating coal and impacts associated with the critical moisture content of coal.

A number of methods and techniques are known that used for quantification of thermal behaviour of coal while coal self-heating and various natural and man-induced factors, having an influence on this behaviour. Those methods served as the basis for the development of a technique to be used for preventing spontaneous coal combustion.

However, there is currently no procedure that used to analyze the combined impact of coal deactivation depending on time and coal activation at critical moisture content on the thermal behaviour of the self-heating coal. The work in [8] describes the procedure for quantifying the thermal behaviour of the self-heating coal, allowing for the coal deactivation over time and higher rates of coal oxygen absorption, with a moisture loss through super micropores.

According to the research of V.M. Maevskaya, there are three phases of the spontaneous coal combustion process: heating, moisture evaporation, active oxidation. In general, the thermal behaviour modeling verifies the characteristics peculiar to each phase of spontaneous combustion. However, it should be noted that during the moisture evaporation phase the temperature decreases due to coal deactivation over time, not intensive moisture evaporation.

The phase of active oxidation starts with the moisture loss through super micropores.

The tendency towards a thermodynamic equilibrium at constant environmental conditions is characteristic of thermodynamic systems, including the heterogeneous coal-air system. With regard to the issues related to the spontaneous combustion prevention in coal, the problem of regulating these processes reduces to the explanation of the mechanism and establishment of parameters ensuring the equilibrium state with decreasing a temperature in self-ignition sites.

Coal Moisture is the first and most important parameter that influence the duration of coal self-heating to a stabilized temperature in the coal compaction. Therefore, to improve thermal stability in the places propitious for coal self-heating and self-ignition, it is recommended to influence the concentration of oxygen and the rate of air filtration in mined-out areas [9].

The significant slowing down of self-ignition in the oxidizing material can be also achieved by reducing the initial temperature of interacting components [10]. In this case, the duration of low-temperature oxidation increases. Simultaneously, according to the Arrhenius equation, a decrease in temperature causes the reduction of reaction rate in oxidation. Therefore, coal cooling also reduces the amount of heat released that will raise potential to prevent self-ignition from occurring and spontaneous coal fires.

Coal temperature in the worked-out area can be lowered by supplying low-temperature medium such as cooling gas. However, gas compositions have a low density and specific heat content, so large volumes of gas are required for cooling the coal. Cooling medium consumption can be significantly reduced by adding particles of a frozen fluid to the gas [11,12]. The intensity of heat removal from coal and rocks increases dramatically due to the absorption of heat during phase transition, as well as higher density and specific heat of fluid particles. It is recommended to use nitrogen as cooling gas, as it does not interact with carbon.

For an effective formulation designed to suppress spontaneous coal combustion. It is important to investigate how the coal moisture and air thermal parameters influence the cooling process and its intensity, when moisture evaporates from the surface, and determine how wetting fluids penetrate the coal mass. The following experiment was conducted for the purpose.

The crushed coal in an amount of 1 kg was poured into a container in the form of a funnel (Figure 2), the air that has various intensity rates and humidity was supplied from the bottom. The outlet air temperature, flow rate and humidity were then measured. Samples of coal with natural bed moisture and samples with a 10 % and 20 % increase in moisture content were used in the course of experiment.

For the first series of experiments, the samples of coal with natural bed moisture were used under the following conditions: the rate of air flow was 65 l/min; the temperature of supply air was 25 °C; the air humidity was 20 %. The temperature was measured every 5 minutes.

According to the results obtained in the course of experiments, the temperature of the coal watered by 10 % decreased by 3 °C during the period of 15 minutes, and watered by 20 % decreased by 4 °C (Figure 3). Consequently, the mechanism can be assumed to be evident: when the coal compacted around worked-out areas is watered to have a higher content of moisture, then blown with dry air, the decrease of coal temperature can be achieved down to a safe value. This can be implemented by periodic water spraying into the air flow at the worked-out area. As a result, coal and rocks become wet. After injecting the fluid into the mined-out area, it is necessary to supply a dry gas, such as nitrogen.

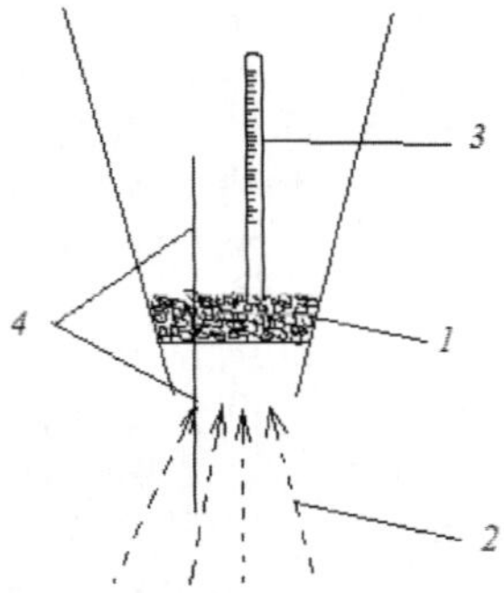

Figure 2 – An installation used for determining the rate of evaporation in the coal compaction: 1 – Compacted and crushed coal; 2 – Air supplied for blowing (of various humidity); 3 – A temperature and humidity meter; 4 – An air flow meter

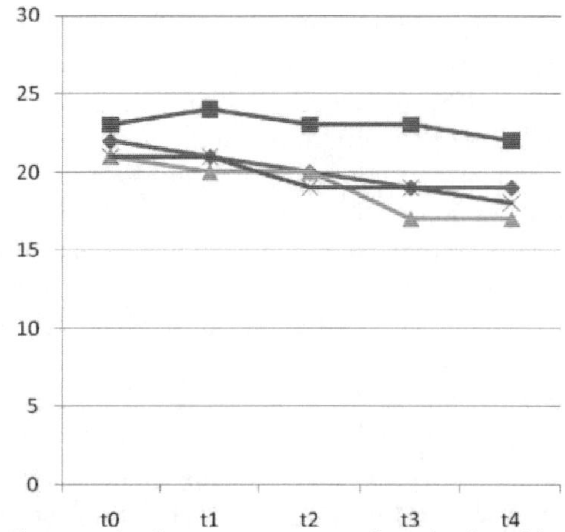

Figure 3 Changes in the temperature of coal when blown with air

Temperature of the coal compaction watered by 10 %, 1 point
Temperature of the coal compaction watered by 10 %, 2 point
Temperature of the coal compaction watered by 20 %, 1 point
Temperature of the coal compaction watered by 20 %, 2 point

The blown dry gas makes particles of previously injected fluid evaporate more actively that result in a decrease in the coal temperature.

3. Conclusions

Thus, it can be concluded that the deactivation of coal over time is an important parameter in modeling self-heating of coal which makes the simulation process more accurate.

The preliminarily watered and then dried coal becomes more active to oxygen. A higher moisture content of coal promotes the increase of its thermal parameters that contribute to more rapid dissipation of heat released in the oxidation process due to oxygen absorption, and suppress the progress of autocatalytic oxidation process and coal self-heating.

It is established in a laboratory environment, when the coal compaction is blown with absolutely dry gas and the humidity of air is high, a decrease in coal temperature is not observed for both cases with using dry and wetted air.

References

[1] Chernov O I, Belaventsev L P, Skritskiy V A and Einer F F Vlijanie vlagi na razvitie samovozgoranija uglja // Bezopasnost' truda v promyshlennosti [Influence of moisture on spontaneous combustion of coal // Safety in the industry]. – 1982. –№ 5. – pp. 34-36.

[2] Belaventsev L P Vlijanie teplofizicheskih parametrov rudnichnogo vozduha na process okislenija uglja [Influence of thermal parameters of air in a mine on coal oxidation] / L P Belaventsev, V A Boldin and A G Stepanov // Bor'ba s gazom, vnezapnymi vybrosami i pozharami v ugol'nyh shahtah. [Fighting against sudden gas emissions and fires in coal mines]. – Kemerovo, 1975. – pp. 210-214.

[3] Belaventsev L P Vlijanie gidravlicheskoj tehnologii ugledobychi na jendogennuju pozharoopasnost' [Influence of hydraulic coal mining technology on the endogenous fire risk] / L P Belaventsev, V A Skritskiy and Yu I Donskov // Journal Ugol Ukrainy [Coal of Ukraine] – 1983. № 3. – pp. 28-30.

[4] Belaventsev L P Vozniknovenie ochagov samovozgoranija uglja v shahtah // Bezopastnost Truda v Promyshlennosti [Emergence of places of spontaneous combustion of coal in mines // Safety in the industry]. – 1986. № 2. – pp. 43-45.

[5 ortola V A and Galsanov N L Inert compositions for underground fire fighting in mines / Taishan Academic Forum – Project on Mine Disaster Prevention and Control. – 2014, pp. 356-360.

[6] Igishev V G, Belaventsev L P, Portola V A, Fedorovich A P and Chubarov B V Razrabotka i vnedrenie novyh sposobov profilaktiki, lokacii i lokalizacii ochagov samovozgoranija uglja v dejstvujushhih vyemochnyh poljah shaht Kuzbassa [Developing and introducing new methods for preventing, locating and localizing places of spontaneous combustion of coal in active working areas of Kuzbass mines] // Kemerovo: Publ.: Kuzbass State Technical University, 2006. – 98 p.

[7] Voroshilov A S Issledovanie zakonomernostej vlijanija vlagi na process samonagrevanija uglja [Investigating the mechanisms of influence of moisture on the process of coal self-heating: Doctoral Thesis in Tech. Sc.] – Kemerovo, 2012.

[8] Portola V A, Galsanov N L and Lugovtsova N Yu Razvitie processa samovozgoraniya v scoplenii predvaritel'no ohlagdennogo uglya // Vestnik Kuzbasskogo gosudarstvennogo tehnicheskogo universiteta. – 2013 - № 1(95). – p.49-52.

[9] Portola V A Detection of sites of self-ignition and evaluation of their temperature according to moisture emission / Bezopastnost Truda v Promyshlennosti. – Issue 1. – 2001. P. 51-53.

[10] Portola V A Preduprezhdenie samovozgoranija uglja snizheniem ego nachal'noj temperatury [Prevention of spontaneous combustion of coal by decreasing its initial temperature] // Bezopastnost Truda v Promyshlennosti [Safety in the industry]. – 1995. – № 6. – pp. 29-30.

[11] Portola V A and Galsanov N L Povyshenie jeffektivnosti primenenija azota dlja podavlenija samovozgoranija uglja [Enhancing the efficiency of nitrogen in the use for suppressing spontaneous combustion of coal] // Kemerovo: Vestnik KuzGTU. – 2011. – № 5. – pp. 59-63.

[12] Lugovtsova N Yu and Portola V A Vliyanie predvaritel'no ohlagdennogo uglya na razvitie processa samovozgoraniya [electronnii resurs] // Sovremennie problemi nauki I obrazovaniya. – 2013 - № 6. – p.1. – Regim dostupa: http://www.science-education.ru/113-10985

IOP Conf. Series: Materials Science and Engineering **142** (2016) 012112 doi:10.1088/1757-899X/142/1/012112

Determination of the Absorption Coefficient and Cloudiness Multiplicity Attenuation During the Gamma-Radiation Passage

K N Orlova [1, a], I F Borovikov [2], M A Gaidamak[1]

[1]Yurga Technology institute, Tomsk Polytechnic University, Leningradskaya 26, 652055, Yurga, Russia
[2]Bauman Moscow State Technical University, ul. Baumanskay 2-ya, 5, 105005, Moscow, Russia

e-mail: kemsur@rambler.ru

Abstract. The paper presents background value equivalent dose of gamma-radiation investigation in different weather: clear cloudy and overcast. The change of the dose rate of gamma radiation, depending on the weather and the ability cloudiness to shield gamma rays is shown. A new method for eliminating the consequences of accidents at nuclear power plants or plants using radioactive elements is proposed. A calculation method of cloudiness coefficient absorption and cloudiness gamma-radiation multiplicity attenuation is developed. The gamma-radiation multiplicity attenuation and the absorption coefficient of gamma radiation were calculated.

1. Introduction

In today's world there are many unresolved problems [1-3], one of which is to find engineering solutions for protection against to exposure to ionizing radiation, for example, in the case of a nuclear explosion or leakage on the radiation-hazardous facility. It is known that an important spread factor of radiation is the atmosphere, in particular, cloudiness [4-6].

Basic methods of radiation exposure protection can be divided into three types (Fig.1).

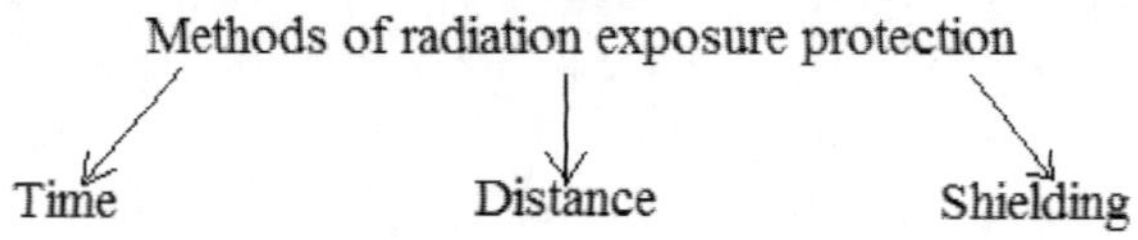

Figure 1. Basic methods of protection against exposure to ionizing radiation

Protection time is to residence time reducing the in areas with high background radiation. The distance protecting is increasing distance from the radiation contamination source to the spacing at which radiation level will reach background values. The ionizing radiation intensity dependence with distance from the source is represented by the formula:

$$Radiation\ intensity \infty \frac{1}{(Distance^2)} \quad (1)$$

In some situations, just need to carry out any activities in the zone with high background radiation. An example would be the elimination of the nuclear power plants accident consequences or industrial jobs, where radiation sources take place. In this case is the use of shielding, such as sarcophagus, radiation-resistant enclosures, personal protective equipment.

According to the International Commission on Radiological Protection, in most countries, the ionizing radiation sources of natural origin provide about 50% of the average human dose irradiation [7](Table 1).

Table 1. Contribution of different sources to the total human exposure

Radiation sources	Contribution to the annual exposure of human,%
Natural background	23
Irradiation of human by decay products of radon and thoron indoors	42
Nuclear testing products fallout	0,77
Nuclear power	0,03
Motor vehicles using	0,1
Consumption of radio luminescent products	0,1
Usage of ionizing radiation in medicine	34

At the same time according to the research about 30% of the natural background make it cosmic particles [7].

Cosmic radiation bumps into Earth's upper atmosphere, which provides effective protection for all living creatures without missing much of the radioactive particles. Cosmic rays consist of «galactic» particles which occur outside the solar system and the "sun" particles emitted by the sun. The solar radiation is energetic charged particles, electrons, protons and nuclei injected Sun into interplanetary space Galactic radiation consist of the various chemical elements nucleus having a kinetic energy of more than a few tens of MeV/nucleon, and the electrons and positrons with E> 10 MeV.

The cosmic rays comprised of high-energy atomic particles, the percentage contribution of which is as follows [8]:

- 87% protons.
- 11% α-particles,
- ≈1% heavy atoms,
- ≈1% electrons.

In the atmosphere "solar" particles produce cascades of nuclear interactions, which give a lot of secondary particles that play an important role in the production of space radionuclides (Fig. 2).

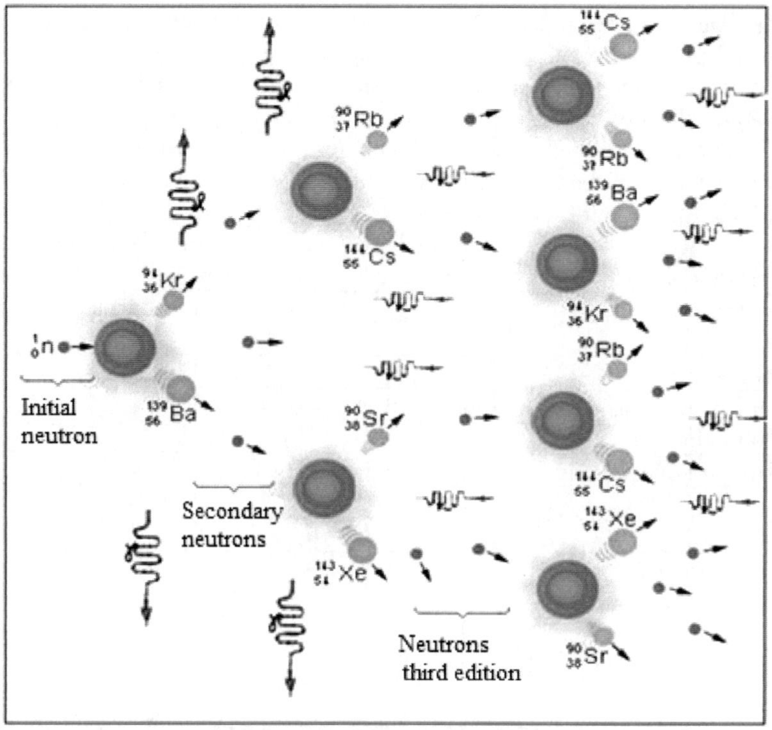

Figure 2. Basic Cascades interaction of nuclear particles

Radioactive transformation is a chaotic process and often leaves the transforming nucleus in an excited state in which the protons and neutrons in the shells of the nucleus are not in the most tightly bound state possible. This excitation energy will be emitted as electromagnetic radiation as the protons and neutrons in the nucleus rearrange themselves to the desired lowest energy state. The shell model of the nucleus suggests discrete energy states for neutrons and protons; it is this difference between energy states that is emitted as a gamma photon when rearrangement takes place. Thus, the emitted gamma ray is characteristic of that particular nucleus [7].

At the passage of radiation through the atmosphere, there are altitude, barometric and geomagnetic effects:

1. The altitude effect - dependence of the radiation intensity I and the height of the observation point above sea level H. The higher point of observation, the greater will be the contribution of cosmic radiation. The observations were made at a height of 200-300 meters above sea level.

In absolute terms the absorbed dose increasing with the height above sea level:

Table 2. The correlation of the absorbed dose rate and height above sea level

H(km)	0	4	8	20
D (μGy/h)	0,032	0,14	0,84	8,72

2. The barometric effect - dependence of the cosmic radiation intensity I from the atmospheric pressure p at the observation point.

Change *p* means that a substance amount varies over the observation point. This leads to a change in cosmic radiation intensity *I*. The barometric effect is expressed barometric coefficient quantitatively:

$$b = \frac{1}{I} * \frac{\Delta I}{\Delta p} \quad (2)$$

The coefficient b is uniquely associated with an average of cosmic radiation absorption coefficient in the air $b = -\mu$. The b and μ values are different for the various components of cosmic radiation. The value $b \geq 1{,}5\%$ by 10 mm Hg. Maximum variation with a period of several days more than 10% for the charged particles and up to 15% of neutrons.

3. The geomagnetic effects (latitude and longitude) - dependence of the intensity and cosmic radiation energy spectrum from the geomagnetic coordinates of the observation point. The coordinates of the observation point *55 °44'00"* N *84 °54'00"* E:

At this development society stage is quite technically feasible is the creation or deletion of cloud clusters. Therefore the aim of this research is: ability determining of cloudiness to shield of gamma-radiation and can therefore be used for gamma-radiation protection.

Tasks

- To determine the gamma radiation dose rate in clear weather
- To determine the gamma radiation dose rate in cloudy weather
- To determine the dose rate in overcast weather
- To calculate the cloud absorption coefficient and the multiplicity attenuation when gamma radiation passes.

2. Results and Discussion

The gamma radiation dose rate is investigated during the research. It has the highest penetration ability. And it is a byproduct of the radioactive particles decay majority, including galactic particles. So gamma radiation dose rate can be used to evaluate the cosmic radiation influence on the radiation situation.

Detector dosimeter is used to convert the phenomena caused by ionizing radiation into an electrical or other easily accessible to measurements signal. The detector of the dosimeter is the ionization counter, by type of Geiger counter with a sensitivity of 20,000 pulses/μSv. The gamma radiation dosimeter has two measurement channels: to measure ambient dose equivalent rate of gamma radiation and to measure the dose with independent reset these values. The measurement carried out continuously with the constant refinement of the result, at what the statistical error displayed on the screen. So we can get a result with the necessary statistical accuracy. The measurements were carried out with not more than 5% statistical error.

Carrying dosimetric characteristics investigation performed during the spring and autumn. The measurements were made every day at 14 o'clock in the afternoon, as it is the time of greatest solar activity [9-11].

The dosimetric characteristics were obtained at atmospheric pressure 730-760 mm Hg. The indications are taken at the same distance of 1 meter from the ground is strictly horizontal relative. Measurements at a height of 1 meter correspond to the peculiar standards related to the gamma and beta radiation ratio and the location of the most sensitive body parts standing person.

The experimental data were processed by the ORIGIN program. And approximation of the resulting curve is conducted. The results are shown in Fig. 3 and Fig. 4.

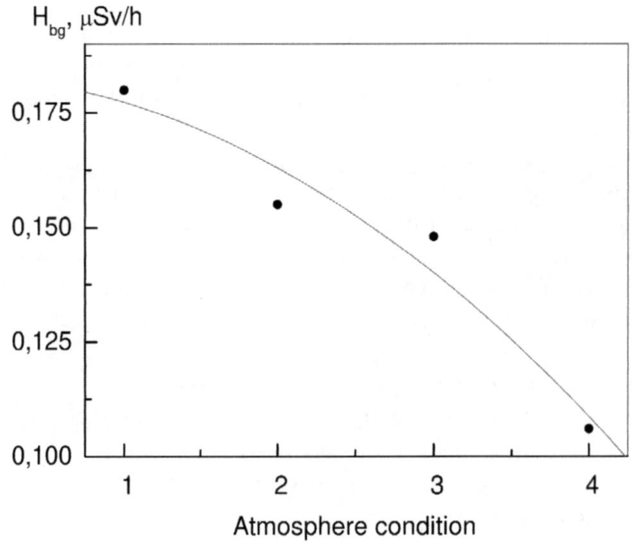

Figure 3. Background radiation value dependence from the weather type, spring: 1 - clear weather, 2 - cloudy weather, 3 - overcast weather; H_{bg} – background dose rate of gamma-radiation.

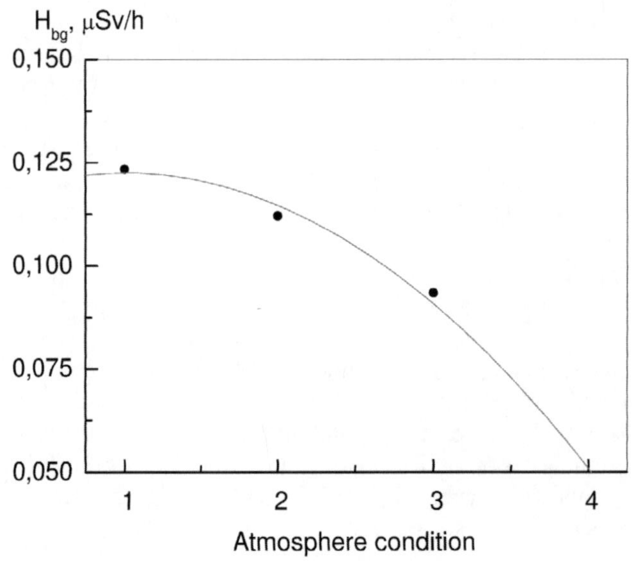

Figure 4. Background radiation value dependence from the weather type, autumn: 1 - clear weather, 2 - cloudy weather, 3 - overcast weather; H_{bg} – background dose rate of gamma-radiation.

In order to determine cloudiness absorption coefficient formula for attenuation intensity of narrow gamma radiation monoenergetic beam with a layer thickness in the matter is used.

$$I = I_0 * (\exp(-\mu * \chi)) \quad (3)$$

where:

- I - the intensity,
- I_0 - initial intensity of the gamma radiation,
- μ - linear (mass) absorption coefficient,
- χ- thickness.

And for a parallel gamma radiation stream:

$$I = I_0 * (\exp(1 - \mu * \chi)) \tag{4}$$

The gamma radiation dose is proportional to the intensity and dose rate of gamma-radiation, then the following expressions are valid:

$$H = H_0 * (\exp(1 - \mu * \chi)) \tag{5}$$

where:

- H - the dose rate of gamma-radiation after passing the thickness of cloud clusters,
- H_0 - initial dose rate of gamma-radiation before passing the thickness of cloud clusters,

Whence by means of mathematical transformations, we obtain an expression for the absorption coefficient:

$$\mu = \frac{1 - ln\frac{H}{H_0}}{\chi} \tag{6}$$

Substituting the values of the dose rate of gamma radiation and the average height of the middle layer cloud clusters (2000 - 3000 meters), we obtain the value of the cloud absorption coefficient:

$$\mu_{cloudiness} = 6.3*10^{-4} m^{-1}$$

For comparison, the absorption coefficient for air and lead:

$$\mu_{air} = 4*10^{-4} m^{-1}, \ \mu_{lead} = 1{,}18 m^{-1}$$

Thus, the absorption coefficient of lead, of course, more, but the cloudiness about 3000 meters can attenuate gamma radiation in two or more times. Thus, it is possible to use cloudiness to restrain the radioactive contamination spread and this will solve the global issues of environmental safety [12]. The use of special decontamination additives for spraying will allow to precipitate the radioactive substances and to delay the radiation spread effectively.

Similarly, the gamma radiation multiplicity attenuation was calculated - as one of the main shielding characteristics from gamma - radiation, showing how many times the need to reduce the exposure dose rate to get set (limits) the gamma radiation dose rate H_{lim}:

$$K = \frac{H}{H_{lim}}$$

$$K_{cloudy} = 0{,}90$$
$$K_{overcast} = 0{,}80$$

Conclusions.

The following conclusions can be made on the basis of the study:

1. A result of research revealed that average equivalent dose rate of gamma radiation higher in clear weather than in cloudy weather. Thus cloud clusters may be used to eliminate the consequences of accidents at nuclear power plants or plants using radioactive elements.
2. There is no doubt, we have to be shielding and absorption of gamma radiation in the thickness of the cloud clusters.
3. The absorption coefficient of gamma radiation and gamma radiation multiplicity attenuation for cloudy and cloudy weather are calculated.

References

[1] Portola, V.A. Indirect Negative Influence of Coal Mine Motor Vehicles on the Environment [Electronic resource] / V. A. Portola, E. S. Torosyan, A. S. Kuznetsova // Applied Mechanics and Materials : Scientific Journal. — 2015. — Vol. 770 : Urgent Problems of Up-to-Date Mechanical Engineering. — [P. 690-694].

[2] Torosyan, V.F. Updating of sewage - purification facilities of electroplating enterprises with counterflow ion-exchange filters [Electronic resource] / V. F. Torosyan [et al.] // IOP Conference Series: Materials Science and Engineering. — 2015. — Vol. 91: VI International Scientific Practical Conference on Innovative Technologies and Economics in Engineering, Yurga, Russia, 21-23 May 2015. — [012077, 8 p.].

[3] Orlova, K.N. Analysis of air pollution from industrial plants by lichen indication on example of small town [Electronic resource] / K. N. Orlova, I. R. Pietkova, I. F. Borovikov // IOP Conference Series: Materials Science and Engineering. — 2015. — Vol. 91: VI International Scientific Practical Conference on Innovative Technologies and Economics in Engineering, Yurga, Russia, 21-23 May 2015. — [012072, 7 p.].

[4] Andersen, P.A. Environmental variables associated with vacationers' sun protection at warm weather resorts in North America/ P.A. Andersen [et.al]//Environmental Research. — 2016. — V. 146. — [P. 200-206].

[5] Tesfaye, M. Simulation of bulk aerosol direct radiative effects and its climatic feedbacks in South Africa using RegCM4 / M. Tesfaye [et.al] //Journal of Atmospheric and Solar-Terrestrial Physics. — 2016. — V. 142. — [P. 1-19].

[6] Allen, R.J., Sherwood, S.C. Aerosol-cloud semi-direct effect and land-sea temperature contrast in a GCM// Geophysical Research Letters. —2010. —V. 37 (7), art. no. L07702.

[7] James E.Martin. Physics for radiation protection, WILEY-VCH Verlag GmbH & Co. KGaA, Weinheim. — 2006. — 844 p.

[8] Gradoboev, A.V. Radiation Model of Light Emitting Diode Based on AlGaInP Heterostructures with Multiple Quantum Wells [Electronic resource] / A. V. Gradoboev, K. N. Orlova // Advanced Materials Research : Scientific Journal. — 2014. — Vol. 880 : Prospects of Fundamental Sciences Development (PFSD-2013) . — [P. 237-241].

[9] Kim, D.A, Ramanathan, V.B. Improved estimates and understanding of global albedo and atmospheric solar absorption / D. A. Kim, V.B. Ramanathan // Geophysical Research Letters. — V. 39 (24). — Article number L24704.

[10] Marshak, A. Biases in shortwave column absorption in the presence of fractal clouds/ A. Marshak [et.al] //Journal of Climate. — 1998. — V. 11 (3), — [P. 431-446].

[11] Kinne, S. Clear-sky atmospheric solar transmission: An analysis based on FIRE 1991 field experiment data / S. Kinne [et.al] //*Journal of Geophysical Research Atmospheres.* — 1998. — V.103 (D16), art. no. 98JD01540. — [P. 19709-19720].

[12] Malchik, A. G. Investigations of physicochemical properties of bottom-ash materials for use them as secondary raw materials [Electronic resource] / A. G. Malchik, S. V. Litovkin, P. V. Rodionov // IOP Conference Series: Materials Science and Engineering. — 2015. — Vol. 91: VI International Scientific Practical Conference on Innovative Technologies and Economics in Engineering, Yurga, Russia, 21-23 May 2015. — [012081, 7 p.].

Propagation of Magnetic Fields from Electrical Domestic Appliances

K N Orlova [1, a], M A Gaidamak [2], and I F Borovikov [3]

[1,2]Yurga Technology institute, Tomsk Polytechnic University, Leningradskaya 26, 652055, Yurga, Russia
[3] Bauman Moscow State Technical University, ul. Baumanskay 2-ya, 5, 105005, Moscow, Russia

e-mail: kemsur@rambler.ru

Abstract. The article presents a research into propagation of magnetic fields from electrical domestic devices. A safe distance at which magnetic induction does not exceed the background level is determined for each type of devices. It is proved that there are two stages of increasing magnetic induction as the distance from the source increases. At the first stage magnetic induction rises and electromagnetic field is formed. At the second stage exponential decrease of magnetic field induction takes place. Mathematical regularities of propagation of magnetic field from electrical domestic devices are experimentally educed.

1. Introduction

It is generally known that all electricalal household appliances generate electromagnetic fields. Nowadays the influence of electromagnetic rays on humans is an open subject for research [1-5]. It would not be true to claim that electromagnetic radiation has only a bad effect on humans because people live their lives in electricalal and magnetic fields formed by natural sources (for example the Earth and the Sun) [1,2,6].

Natural electromagnetic radiation is an essential condition for existence of humans and other biological species. It accelerates life processes, cell regeneration, metabolism and so on. Nowadays electromagnetic radiation is widely used in medicine for treating different diseases [3].

However it is equally thoughtless to believe that electromagnetic radiation is absolutely safe and can't be escaped. Though the latest statement can be recognized true natural electromagnetic influences cannot be given the same status as the impact of a range of modern electricalal appliances. Firstly, electromagnetic radiation has a great wavelength range. Secondly, it can have different energies. And thirdly, there is a great number of various types of electromagnetic rays [4,5]. Each type can possess some special impact and peculiar electromagnetic fields. In this view the problem of understanding the influence of different electromagnetic fields on people is urgent and requires extensive theoretical research and practical experiment in this area.

Table 1. The basic types of electromagnetic radiation.

Length	**Radiation**	**Frequency**
More than 100km	Low-frequency oscillations	0-3 kHz
100 km – 1mm 1 m – 1 mm	Radiowaves including microwaves	3kHz -3 THz 300Mhz – 300GHz
2 mm - 760 nm	Infrared radiation	150GHz - 400 THz
760 - 380 nm	Visible radiation (optical spectrum)	400 - 800 THz
380 - 3 nm	Ultraviolet radiation	800 - 100 PHz
10 nm - 1 pm	X-rays	30 PHz - 300 EHz
More than 10 pm	Gamma-rays	More than 30 EHz

According to world researches the following effects of electromagnetic fields on humans have been revealed:

- Changing DNA [6]
- Increasing the number of children suffering from asthma by 15% [7]
- Reducing melatonin (antioxidant and antitumoral inhibitor) and some other hormones [8].
- Increasing cancer risk by 40% [9].
- Histolytic changes causing worsening of reproductive function [10]
- Reducing sensibility of hair cells leading to hearing impairment
- Sleep problems
- Changes in protein structure

Two basic impacts revealed by modern science lead to such consequences of exposition to electromagnetic radiation:

- Heat impact
- Non-heat impact (reorientation of charged particles).

When studying the influence of electromagnetic fields on humans special attention should be paid to superhigh frequency waves (SHF). SHF (microwave) radiation is applied in microwave ovens, radio navigation, satellite television, cell phones. The Sun is a natural source of microwaves. Thus devices emitting superhigh frequency waves are widely used nowadays. There are data which prove especially strong effect of SHF on body organs with low heat circulation (brain tissues and eyes) [11-16].

The aim of the research is to estimate the impact of electromagnetic radiation of electricalal domestic appliances on humans.

To achieve this purpose the following tasks have been performed:

- Measuring magnetic induction at different distances from electricalal domestic appliances
- Determining the distance at which magnetic induction is on the background leel which is relatively safe for people
- Determining physical laws of magnetic field generation for different electricalal domestic appliances
- Determining mathematical dependences of magnetic induction and the distance from a radiation source.

2. Results and Discussion

Measurements were carried out by means of magnetic induction sensor LabQuest Vernier with relative error about 3%. Measurement range was 6.3 mT which is enough to study the generation of magnetic fields by different electricalal domestic appliances and determine their safety levels. The measurement pitch was about 0.1m which enabled finding the necessary graphic dependences.

The measurements of magnetic field induction were summarized in a table. The measurement unit is *mT*.

Table 1. Measurements of magnetic field induction, *mT*.

Electricala l domestic appliances	Capa-city	Distance										
		0,01m	0,1m	0,2m	0,3m	0,4m	0,5m	0,6m	00,7m	00,8m	0,9m	1,0m
Mobile phone in waiting mode	2100 W	0,0569	0,0561	0,0566	0,0571	0,0522	0,0521	0,0545	0,0459	0,0423	0,0358	0,0329
Mobile phone in conversati on mode		0,0659	0,0627	0,0626	0,0574	0,0633	0,0644	0,0628	0,0644	0,0633	0,0612	0,06
Monitor	34W	0,0497	0,0484	0,0052	0,0027	0,0067	0,0053	0,0055	0,0043	0,0057	0,0022	0,0021
System unit		0,0425	0,0508	0,0574	0,0633	0,0644	0,0628	0,0576	0,0423	0,0374	0,0081	0,0075
Electrical pot	2200 W	0,0623	0,0244	0,0236	0,0226	0,0220	0,0211	0,0178	0,0089	0,0021	0,0057	0,0032
Outlet		0,1307	0,1670	0,0529	0,0274	0,0623	0,0216	0,0197	0,0097	0,0052	0,0028	0,0026
Desk lamp	100W	0,0269	0,0211	0,0189	0,0135	0,0121	0,0134	0,0075	0,0053	0,0075	0,0043	0,0038
Microwav e oven	2300 W	0,0727	0,0646	0,0657	0,0453	0,0478	0,0433	0,0421	0,0450	0,0417	0,0210	0,0109
Electrical cooker	1500 W	0,2081	0,210	0,1307	0,0890	0,0463	0,0129	0,0044	0,0053	0,0089	0,0031	0,0029
Refrigerat or	1000 W	0,0609	0,0452	0,0339	0,0258	0,0148	0,0078	0,0073	0,0032	0,0057	0,0021	0,0075
Washing machine	2100 W	0,0612	0,0556	0,0331	0,0238	0,0149	0,0063	0,0085	0,0014	0,0083	0,0048	0,0038
Electrical iron	1800 W	0,0261	0,0278	0,0131	0,0039	0,0079	0,0053	0,0047	0,0073	0,0026	0,0028	0,0025

We have studied the basic types of electrical domestic appliances such as an electrical iron, a washing machine, a computer, a microwave oven, an electrical cooker, a refrigerator and others. All the devices were placed into the same room at the same time. For each measurement only one device was switched on, so the influence of the natural electromagnetic radiation was the same in all measurements and the relative error caused by a natural background and experiment conditions was minimized. As far as the aim of the research was to find physical regularities of changing magnetic induction as the distance from an induction source increases and to determine an average safe distance

IOP Conf. Series: Materials Science and Engineering **142** (2016) 012113 doi:10.1088/1757-899X/142/1/012113

we have not compared different brands of domestic appliance and only one sample of each device has been tested.

An analysis of the experimental data revealed the following regularities:

1. Induction exceeding limit value has not been registered. (Limit value for population equals 0.8mT for 8 hours working day according to state standard specifications).
2. It has been found out that electromagnetic field decreases as the distance from the source increase. For the most devices a safe distance i.e. the distance at which electromagnetic field reaches the natural background value is about 50cm in average. For low capacity devices (less than 100W) it is 20cm. Fig. 1 and 2 show graphic dependences of electromagnetic induction and the distance from the source.
3. For superhigh frequency (microwave) devices such as mobile phone and microwave oven there are oscillations of magnetic field induction but at the distance of 1m this decrease is insignificant. Fig. 3 and 4 show graphic dependences of electromagnetic induction form SHF devices and the distance from the source.

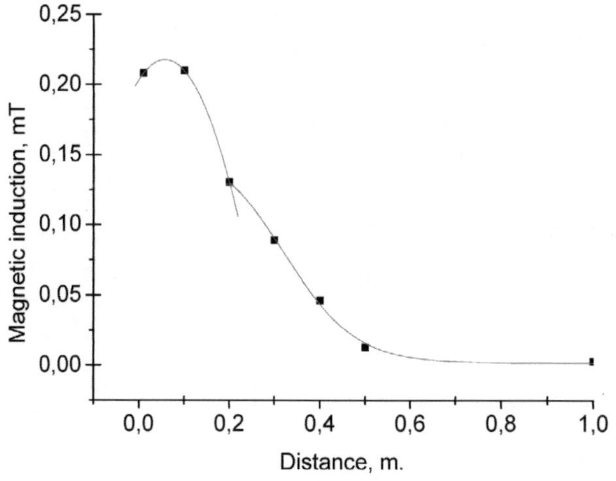

Figure 1. Dependence of electromagnetic induction and the distance from an electrical oven

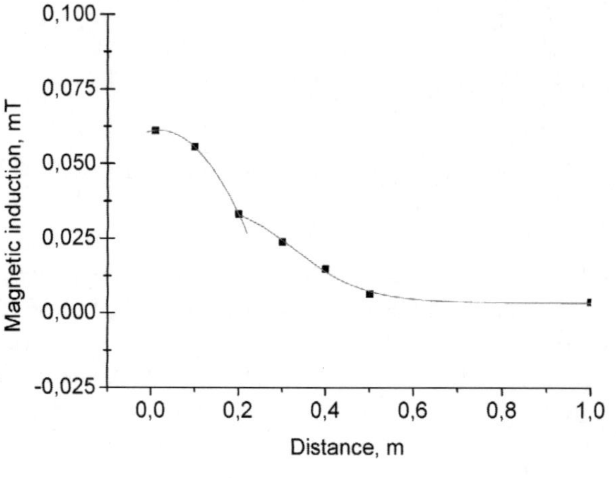

Figure 2. Dependence of electromagnetic induction and the distance from a washing machine.

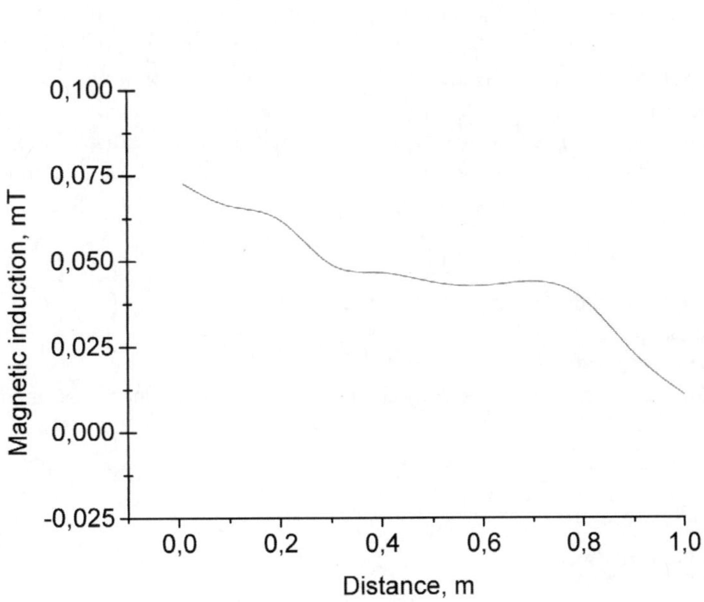

Figure 3. Dependence of electromagnetic induction and the distance from a microwave oven

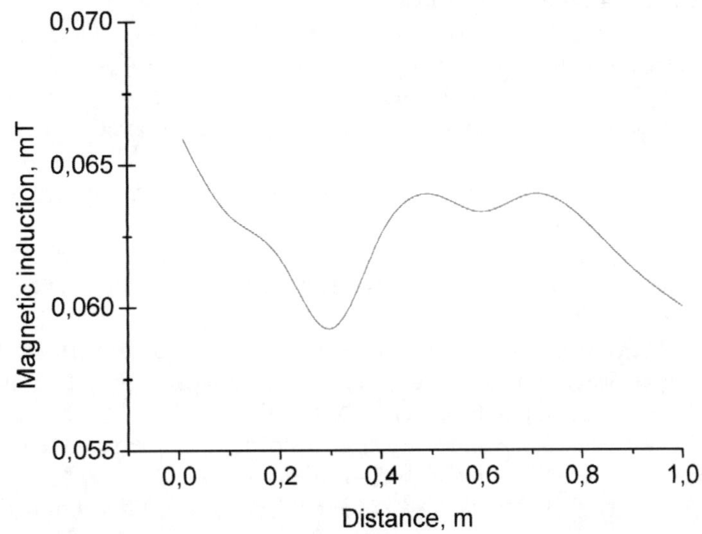

Figure 4. Dependence of electromagnetic induction and the distance from a mobile phone

Let us consider the special features which have been found out and peculiarities of radiation of magnetic fields by these devices. First of all in Fig.1 and 2 alongside with an evident decrease of magnetic induction as the distance increases we notice an initial increase of induction (at the distance of 20cm) which cannot be caused by a measurement error. This increase can be explained by the induction zone in which electrical and magnetic fields act separately, electromagnetic field being generated at the same time. The induction zone is generated by all electrical domestic appliances which are not based on superhigh frequency waves. So, we can conclude that there are two stages of propagation of electromagnetic induction from electrical domestic appliances: at the first stage there is an induction zone, at the second stage there is an electromagnetic field propagation zone.

Devices based on superhigh frequency waves have a small induction zone and a large propagation zone (Fig.3 and 4).

Mathematical dependences of magnetic field induction and the distance from a device which is not based on SHF were determined by the methods of interpolation and approximation.

At the first stage we have determined polynominal dependence which is described by the formula:

$$B=A_1r^{-2}+A_2r+A_3 \quad (1)$$

where:

A_1, A_2, A_3- coefficients for the device, rel.units,
r – the distance, *m,*
B – magnetic induction, *T.*

At the second stage we have determined a hyperbolic dependence described by the formula:

$$B=X_1r^{-1}+X_2 \quad (2)$$

where:

X_1, X_2 - coefficients for the device, rel.units.,
r – the distance, *m*,
B – magnetic induction, *T.*

Conclusions

The following conclusions can be made on the ground of the research:

1. Differences in generation and propagation of electromagnetic waves from different electrical devices have been found.
2. Devices based on SHF proved to have two stages of changing magnetic induction as the distance from the source increases.

   At the first stage, a slight increase of magnetic induction at an increasing distance takes place and polynominal analytical dependence can be observed.

   At the second stage, a hyperbolic dependence of induction decrease at an increasing distance takes place. And this research helps to solve the global issues of environmental safety [17-20].
3. A safe distance at which magnetic induction does not exceed the background level has been determined for each type of devices. It makes up ≈0.5m in average.

References

[1] Lagroye, I. Measurements of alkali-labile DNA damage and protein-DNA cross links after 2450 MHz microwave and low-dose gamma irradiation in vitro / I. Lagroye, G. J. Hook, Wettring [et al]. B. A.//Radiation Research. – 2004. – V. 161, №2. – P. 201–214.
[2] . Lee, G.M, A nested case-control study of residential and personal magnetic field measures and miscarriages/ G.M. Lee [et al] // Epidemiology. – 2002. – V. 13(1). – P.21-31.
[3] Li D.K, Checkoway H, Mueller BA. Electrical blanket use during pregnancy in relation to the risk of congenital urinary tract anomalies among women with a history of subfertility // Epidemiology. – 1995. – V. 6(5). – P.485-489.
[4] Burch, J. B. Reduced excretion of a melatonin metabolite in workers exposed to 60 Hz magnetic fields/ J. B. Burch, J. S. Reif, M. G. Yost, T. J. Keefe// American Journal of Epidemiology. - 1999. -V. 150, №1. – P. 27–36.
[5] Schüz, J. Cellular telephone use and cancer risk: update of a nationwide Danish cohort/ J. Schüz, R. Jacobsen, J. H. Olsen., Jr. J. D. Boice,// Journal of the National Cancer Institute. -2006. - V. 98, №23. – P. 1707–1713.
[6] Hamada, A. J. Cell phones and their impact on male fertility: fact or fiction/ A. J. Hamada, A. Singh, A. Agarwa// The Open Reproductive Science Journal. -2011. -V. 5, -P. 125–137.
[7] Wood, A. W. Changes in human plasma melatonin profiles in response to 50 Hz magnetic field exposure / A.W. Wood [et al] // Journal of Pineal Research. – 1998. – V. 25, N. 2. – P. 116–

127.

[8] Kaprana, A. E. Studying the effects of mobile phone use on the auditory system and the central nervous system: a review of the literature and future directions/ A. E. Kaprana, A. D. Karatzanis, E. P. Prokopakis [et al]// European Archives of Oto-Rhino-Laryngology. – 2008. –V. 265, N. 9. – P. 1011–1019.

[9] Röösli, M. Cardiovascular mortality and exposure to extremely low frequency magnetic fields: a cohort study of Swiss railway workers/ M. Röösli, Egger, D. Pfluger, and C. Minder Environmental Health. – 2008. – V. 7, article 35.

[10] .Zeni, O. Formation of reactive oxygen species in L929 cells after exposure to 900 MHz RF radiation with and without co-exposure to 3-chloro-4-(dichloromethyl)-5- hydroxy-2(5H)-furanone/ O. Zeni, R. di Pietro, G. d'Ambrosio [et al.]// Radiation Research. – 2007. – V. 167, N. 3. – P. 306–311.

[11] Schüz, J., R. Cellular telephone use and cancer risk: update of a nationwide Danish cohort/ J.R. Schüz [et al] // Journal of the National Cancer Institute. – 2006. – V. 98, N. 23. – P. 1707–1713.

[12] Hook,G. J. Evaluation of parameters of oxidative stress after in vitro exposure to FMCW- and CDMA-modulated radiofrequency radiation fields/ G. J. Hook, D. R. Spitz, J. E. Sim [et al.] //Radiation Research. –2004. –V. 162, N. 5. – P. 497–504.

[13] Singh S.. Health Implications of Electromagnetic Fields, Mechanisms of Action, and Research Needs/ Sarika Singh, Neeru Kapoor // Advances in Biology. – 2014. – V 2014. – Article ID 198609.

[14] Sarkar,S. Effect of radiofrequency radiation on biological system/ S. Sarcar [et al.] // In Topics in Electromagnetic Waves: Devices, Effects and Applications, J. Behari, Ed. – 2005. – P. 42–73.

[15] Garaj-Vrhovac, V. Assessment of ˇ cytogenetic damage and oxidative stress in personnel occupationally exposed to the pulsed microwave radiation of marine radar equipment/ V. Garaj-Vrhovac// International Journal of Hygiene and Environmental Health. –2011. – V. 214, N. 1. – P. 59–65.

[16] Regel, S. J. Cognitive performance measures in bioelectromagnetic research critical evaluation and recommendations/ S. J. Regel and P. Achermann// Environmental Health. – 2011. – V. 10(10).

[17] Portola, V.A. Indirect Negative Influence of Coal Mine Motor Vehicles on the Environment [Electronic resource] / V. A. Portola, E. S. Torosyan, A. S. Kuznetsova // Applied Mechanics and Materials : Scientific Journal. — 2015. — Vol. 770 : Urgent Problems of Up-to-Date Mechanical Engineering. — [P. 690-694].

[18] Torosyan, V.F. Updating of sewage - purification facilities of electroplating enterprises with counterflow ion-exchange filters [Electronic resource] / V. F. Torosyan [et al.] // IOP Conference Series: Materials Science and Engineering. — 2015. — Vol. 91: VI International Scientific Practical Conference on Innovative Technologies and Economics in Engineering, Yurga, Russia, 21-23 May 2015. — [012077, 8 p.].

[19] Orlova, K.N. Analysis of air pollution from industrial plants by lichen indication on example of small town [Electronic resource] / K. N. Orlova, I. R. Pietkova, I. F. Borovikov // IOP Conference Series: Materials Science and Engineering. — 2015. — Vol. 91: VI International Scientific Practical Conference on Innovative Technologies and Economics in Engineering, Yurga, Russia, 21-23 May 2015. — [012072, 7 p.].

[20] Gradoboev, A.V. Radiation Model of Light Emitting Diode Based on AlGaInP Heterostructures with Multiple Quantum Wells [Electronic resource] / A. V. Gradoboev, K. N. Orlova // Advanced Materials Research : Scientific Journal. — 2014. — Vol. 880 : Prospects of Fundamental Sciences Development (PFSD-2013) . — [P. 237-241].

Solution of the Applied Task to Determine Total Exposure for Mechanical Engineering Workers

O V Medvedeva[1], K N Orlova[1], I F Borovikov[2]

[1]Yurga Technology institute, Tomsk Polytechnic University, Leningradskaya Street, 26, 652055, Yurga, Russia
[2]Bauman Moscow State Technical University, Baumanskay 2-ya Street, 5, 105005, Moscow, Russia

e-mail: alisasten@mail.ru

Abstract. In everyday life, people repeatedly exposed to ionizing radiation. Radiation sources is as natural background radiation due to the natural radioactivity of the earth, cosmic and solar radiation and artificial emitters in enterprise engineering. Therefore, the determination of the total human exposure in engineering industry will consist of a variety of factors will be strictly defined for a specific person, and undoubtedly true. Neural network technology can completely resolve this problem, follow neyroalgoritm presented in this paper.

1. Introduction

In today's world there are many unresolved problems [1-2], one of which is to find engineering solutions for protection against to exposure to ionizing radiation, for example, in the case of a nuclear explosion or leakage on the radiation-hazardous facility. It is known that an important spread factor of radiation is the atmosphere, in particular, cloudiness [3-5].

One of the important applications is to determine the total exposure of Machinists by mathematical modeling. Indeed, modern means of calculating possible to obtain the value of the radiation dose due to the action of one factor (finding and manipulation of sources of ionizing radiation while working at the enterprises). However, in the course of business, the employee is subjected to multivariate exposure.

To assess the exposure of workers and algorithmization engineering industry, we apply neural network.

Neural network - electrical impulses between neurons of the human brain, which a man tries to structure. The brain is the biggest "computer" for the storage and processing of information. Nervous system and the human brain consists of neurons interconnected by nerve fibers. Nerve fibers themselves are capable of transmitting electrical impulses between neurons. All transfer processes stimuli from the skin, eyes and ears to the brain, thought processes and management actions - all this is done in a living organism as the transmission of electrical impulses between neurons [6].

After the appearance of the term «neural network» is an allocation of a separate field of science - Simulation. Currently, simulation of neural activity is carried out successfully in many areas of science, such as economic processes and phenomena, algorithmization, organize data, etc. But it is not every possible application of neural networks.

Currently neural network - a kind of algorithm; mathematical model is a system of connected and interacting neurons.

Artificial neural networks are built on the principles of organization and functioning of their biological counterparts. They are able to solve a wide range problems of pattern recognition, identification, prediction, optimization and control of complex objects. Further increase the performance of computers is increasingly associated with artificial neural network, in particular, with Neurocomputers which are based on artificial neural network.

The aim of this study is to solve the problem of determining the application of the total exposure of Machinists.

In this connection were as follows:

1 Analysis of the current state of the methods of neural network technology in the activities of Machinists;

2 Analysis of the processes and phenomena that contribute to the total annual dose of gamma rays engineering industry workers;

3 Construction of an algorithm to determine the result of radiation exposure of Machinists.

According to the International Commission on Radiological Protection, in most countries, the ionizing radiation sources of natural origin provide about 50% of the average human dose irradiation [7] (Table 1).

Table 1. Contribution of different sources to the total human exposure

Radiation sources	Natural background	Irradiation of human by decay products of radon and thoron indoors	Nuclear testing products fallout	Nuclear power	Motor vehicles using	Consumption of radio luminescent products	Usage of ionizing radiation in medicine
Contribution to the annual exposure of human,%	23.0	42.0	0.77	0.03	0.1	0.1	34.0

At the same time according to the research about 30% of the natural background make it cosmic particles [7].

Cosmic radiation bumps into Earth's upper atmosphere, which provides effective protection for all living creatures without missing much of the radioactive particles. Cosmic rays consist of «galactic» particles which occur outside the solar system and the "sun" particles emitted by the sun. The solar radiation is energetic charged particles, electrons, protons and nuclei injected Sun into interplanetary space Galactic radiation consist of the various chemical elements nucleus having a kinetic energy of more than a few tens of MeV/nucleon, and the electrons and positrons with E> 10 MeV.

The cosmic rays comprised of high-energy atomic particles, the percentage contribution of which is as follows [9]:

- 87% protons.
- 11% α-particles,
- ≈1% heavy atoms,
- ≈1% electrons.

In the atmosphere "solar" particles produce cascades of nuclear interactions, which give a lot of secondary particles that play an important role in the production of space radionuclides (Fig. 2).

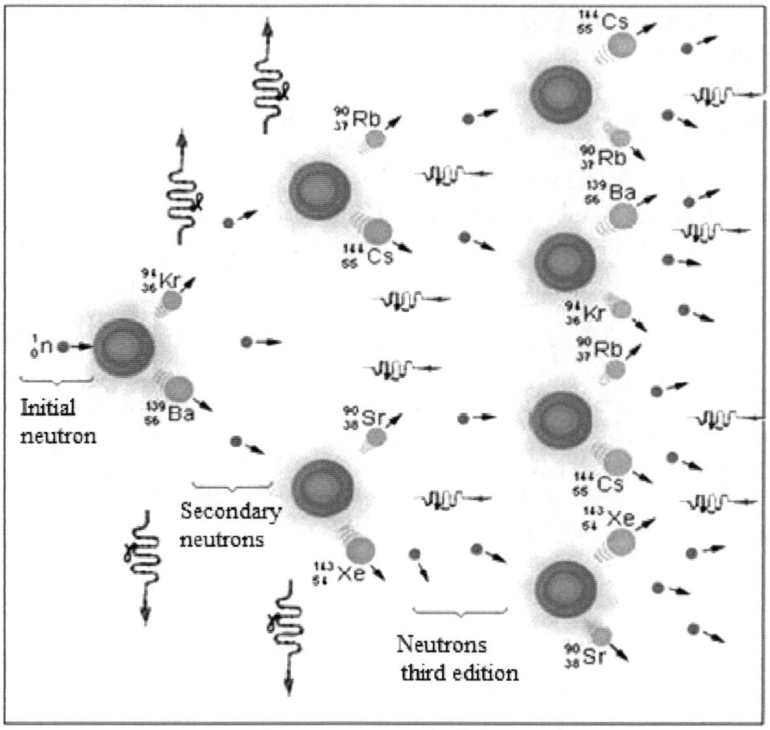

Figure 2. Basic Cascades interaction of nuclear particles

Radioactive transformation is a chaotic process and often leaves the transforming nucleus in an excited state in which the protons and neutrons in the shells of the nucleus are not in the most tightly bound state possible. This excitation energy will be emitted as electromagnetic radiation as the protons and neutrons in the nucleus rearrange themselves to the desired lowest energy state. The shell model of the nucleus suggests discrete energy states for neutrons and protons; it is this difference between energy states that is emitted as a gamma photon when rearrangement takes place. Thus, the emitted gamma ray is characteristic of that particular nucleus [7].

Person is exposed to ionizing radiation from various sources throughout their lives. Irradiation may be both of synthetic and natural sources of radiation. With an increase in the technical development of the planet to the overall human exposure to artificial radiation sources is constantly growing.

Natural sources of ionizing radiation is primarily the natural background of the Earth, that is, the soil, the earth's crust and other layers of the Earth have a sufficient amount of a radionuclide that decompose emit ionizing radiation.

Dose that gets the population will vary depending on the geographic location of residence. Radiation levels in some areas of the world where lie especially radioactive rocks (the Caucasus and Altai) is considerably above the average, and in other places - is correspondingly lower.

In everyday life, people repeatedly exposed to ionizing radiation. Radiation sources is as natural background radiation due to the natural radioactivity of the earth, cosmic and solar radiation and artificial emitters (household appliances, building materials, fuel, medical equipment, power plants and nuclear power plants). Therefore, the determination of the total human exposure in everyday life will consist of a variety of factors will be strictly defined for a specific person, and undoubtedly true. Neural network technology can completely resolve this problem, follow neyroalgoritm presented in this paper.

One of the important applications is to determine the total exposure of Machinists by mathematical modeling. Indeed, modern means of calculating possible to obtain the value of the radiation dose due to the action of one factor (finding and manipulation of sources of ionizing radiation while working at the enterprises). However, in the course of business, the employee is subjected to multivariate exposure.

To assess the exposure of workers and algorithmization engineering industry, we apply neural network.

Also, throughout all history of existence of Earth different types of ionizing radiation fall on Earth surface from the Sun from space and a galaxy, despite the protective layers of Earth formed by the atmosphere. The size of galactic and solar radiation depends on existence of atmospheric overcast (the high overcast, moderate weather or clear).

It is worth carrying a problem of radonic pollution to natural sources of radiation also. More exact than migration and accumulation of radionuclide of radium which as a result of disintegration forms radon, and it in turn toron and other affiliated products of disintegration including gripping gamma radiation. And poor such radionuclides as potassium, calcium, sodium, phosphorus, size of migration of radionuclides considerably grows in soils. In a life this problem is aggravated with existence of plastic windows, insignificant frequency of ventilation of rooms, sealing of walls, existence of full-fledged outlets at basements, etc. If the person works in machine-building branch, his body is exposed to additional serious radiation [8]. There is a number of separate types of radiation of workers of mechanical engineering on a workplace, it is possible to carry to them: radiation by the working laser, radioactive materials, air pollution by harmful volatiles, etc. Besides, in mechanical engineering sources of intensive radiation by electromagnetic waves of an infrared range are: heated surfaces of walls, furnaces and their open apertures, and also trenches, the final trays, rolled metal of a stream of the melted metal, heated processed details and preparations, different types of welding and plasma processing, etc. All this significantly influences efficiency of the person, his health and health. For definition of result of radiation of workers of mechanical engineering it is necessary to process the following factors (Fig. 3):

1 Space radiation: 1.1 Galactic radiation (1.1.1 Cloudy weather, 1.1.2 Sunny weather, 1.1.3 Moderate weather); 1.2 Solar radiation (1.2.1 Cloudy weather, 1.2.2 Sunny weather, 1.2.3 Moderate weather);

2 Using air transport: 2.1 Yes, 1-2 times a year; 2.2 Yes, 6 and more times a year; 2.3 Isn't present;

3 Ionizing radiation use in medicine: 3.1 didn't pass in a year; 3.2 passed 1 time in a year; 3.3 passed several times in a year;

4 Radio engineering: 4.1 TV; 4.2 Personal computer (laptop, tablet);

5 Global loss of products of nuclear tests;

6 Radiation by disintegration products toron and radon: 6.1 Frequency of airing of the room (6.1.1 2 times a day, 6.1.2 Once a day, 6.1.3 Not daily are more often); 6.2 Sealing of walls (6.2.1 Existence of repair with isolation of walls, 6.2.2 Isn't present repair with isolation of walls); 6.3 Accommodation on first floors of buildings (6.3.1 Ventilation of basements, 6.3.2 Lack of fragrances); 6.4 Top floors;

7 Natural radiation background: 7.1 Accommodation in places of the raised radiation background; 7.2 Accommodation in places of a moderate radiation background; 7.3 Accommodation in places of a usual radiation background;

8 Construction material of the house: 8.1 Tree; 8.2 Brick; 8.3 Slag stone (including bulk); 8.4 Panel;

9 Radiation on production: 9.1 Laser; 9.2 Heated surfaces of walls, furnaces and their open apertures; 9.3 Welding and plasma processing; 9.4 Rolled metal of a stream of the melted metal, heated processed details and preparations; 9.5 Volatiles in air; 9.6 the radioactive materials participating in production;

10 Number of the people living together in one room;

11 Wireless communication indoors: 11.1 Cell phones; 11.2 Wi-Fi zone.

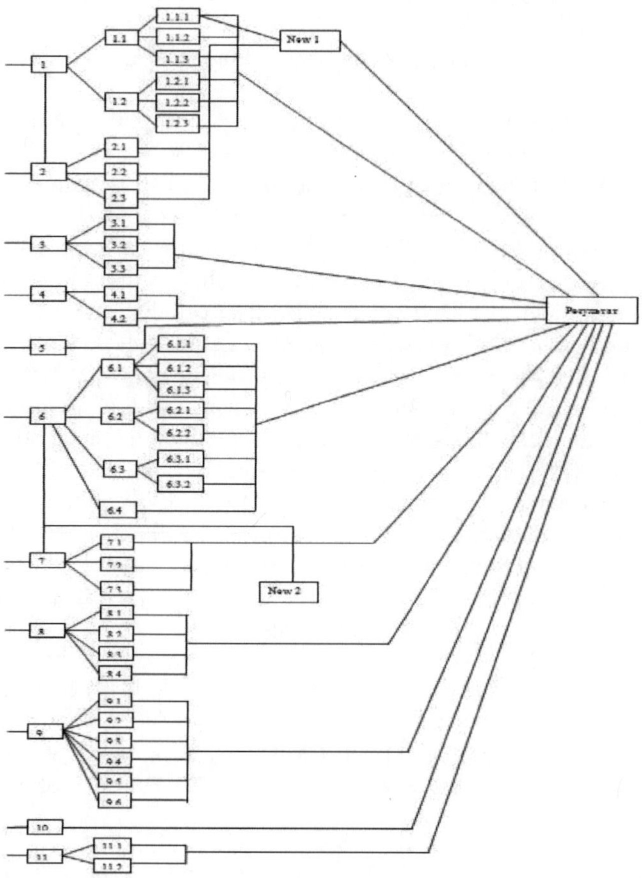

Figure 3. Diagram of a neural network of natural processes related to everyday human life

Thus, the radiation exposure is subject to any citizen of the world, especially if he carries out his professional activities with sources of ionizing radiation.

Important factors of radiation exposure in engineering is the time taken to complete the work in the enterprise, while overtime hours.

The value of annual dose also depends on the lifestyle of people. So if a person often uses air (especially more than 4-6 times in a year), the total dose received will be much higher.

Studying the interaction data in the neural network of natural processes, it was found that there is an important influence on the outcome of the process, based on the interacting factors in the direction of strengthening, weakening or compensation exposure. Was built by the algorithm of the neural network to analyze the data (Fig.4).

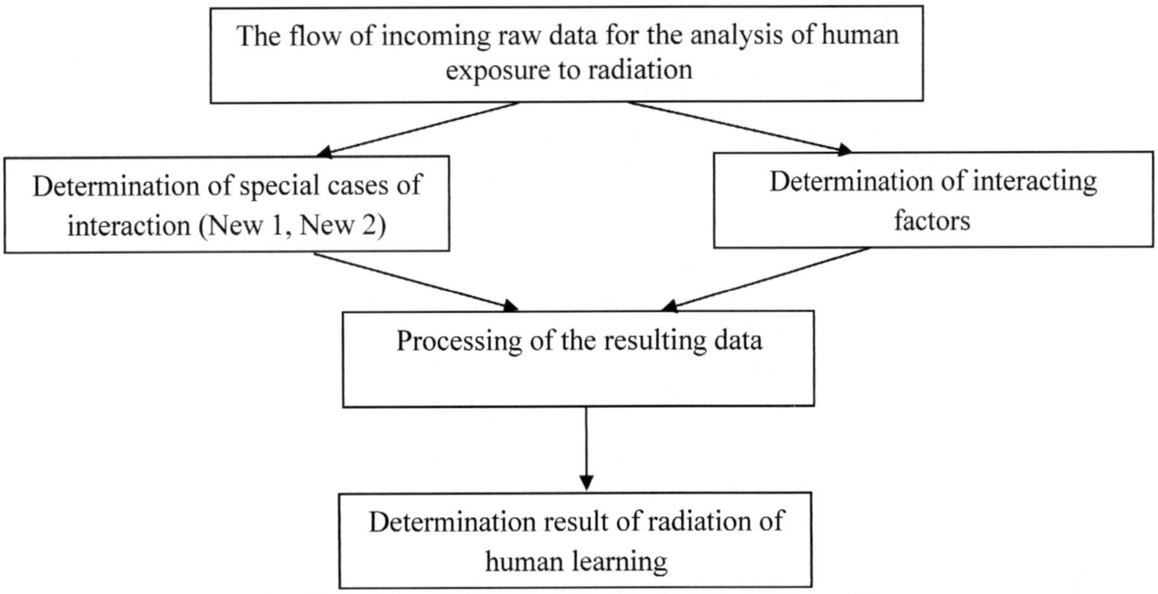

Figure 4. Algorithm to determine the result of radiation exposure of human

Based on this, the first paragraph of the algorithm is to find the particular cases and to determine the relationship factors. After the data is determined by the result - the annual exposure dose of gamma rays.

Mathematical modeling was solved important problems of the process of total human exposure in the company, allowing to calculate the radiation dose caused by multiple factors.

2. Results and Discussion

Detector dosimeter is used to convert the phenomena caused by ionizing radiation into an electrical or other easily accessible to measurements signal. The detector of the dosimeter is the ionization counter, by type of Geiger counter with a sensitivity of 20,000 pulses/μSv. The gamma radiation dosimeter has two measurement channels: to measure ambient dose equivalent rate of gamma radiation and to measure the dose with independent reset these values. The measurement carried out continuously with the constant refinement of the result, at what the statistical error displayed on the screen. So we can get a result with the necessary statistical accuracy. The measurements were carried out with not more than 5% statistical error.

Conclusions

The following conclusions can be made on the basis of the study:

1) based on neural network technology algorithmization allow to resolve the issue of private to determine the total exposure of the worker in the field of engineering activities;

2) radiation exposure in engineering is going from the set of interdependent natural phenomena and factors in the classification and algorithmization which revealed the dependence of such a neural;

3) based on neural network technology built algorithm resolves the issue of search and compute the total radiation dose of Machinists in the course of their professional activities.

References

[1] Portola, V.A. Indirect Negative Influence of Coal Mine Motor Vehicles on the Environment [Electronic resource] / V. A. Portola, E. S. Torosyan, A. S. Kuznetsova // Applied Mechanics and Materials : Scientific Journal. – 2015. – Vol. 770 : Urgent Problems of Up-to-Date Mechanical Engineering. – [P. 690-694].

[2] Torosyan, V.F. Updating of sewage - purification facilities of electroplating enterprises with counterflow

ion-exchange filters [Electronic resource] / V. F. Torosyan [et al.] // IOP Conference Series: Materials Science and Engineering. – 2015. – Vol. 91: VI International Scientific Practical Conference on Innovative Technologies and Economics in Engineering, Yurga, Russia, 21-23 May 2015. – [012077, 8 p.].

[3] Andersen, P.A. Environmental variables associated with vacationers' sun protection at warm weather resorts in North America/ P.A. Andersen [et.al] // Environmental Research. – 2016. – V. 146. – [P. 200-206].

[4] Tesfaye, M. Simulation of bulk aerosol direct radiative effects and its climatic feedbacks in South Africa using RegCM4 / M. Tesfaye [et.al] //Journal of Atmospheric and Solar-Terrestrial Physics. – 2016. – V. 142. – [P. 1-19].

[5] Allen, R.J., Sherwood, S.C. Aerosol-cloud semi-direct effect and land-sea temperature contrast in a GCM// Geophysical Research Letters. – 2010. – V. 37 (7), art. no. L07702.

[6] Kashirina I.L. Neural network technology. Teaching aid for higher education. - Publishing House of the Voronezh State University, 2008 – [72p.]

[7] Kostenko O.V., Orlova K.N. Building neyroalgoritm to determine the total human exposure // Scientific and Technical Gazette of the Volga region. - 2013. № 2. – [pp 142-145]

[8] James E.Martin. Physics for radiation protection, WILEY-VCH Verlag GmbH & Co. KGaA, Weinheim. – 2006. – [844 p.]

IOP Conf. Series: Materials Science and Engineering **142** (2016) 012115 doi:10.1088/1757-899X/142/1/012115

Improved Operating Performance of Mining Machine Picks

S Prokopenko [1,2,4] , A Li[2], I Kurzina[3] and A Sushko[1]

[1] Yurginsky Technological Institute (Branch), Tomsk Polytechnic University 652055, Kemerovo Region, Yurga, Leningradskaya str., 26, Russia;
[2] JSC "NC VostNII", 650003, Kemerovo, Institutskaya str., 3, Russia;
[3] Tomsk State University, 634050, Tomsk, Lenin Ave. 36, Russia;

[4]E-mail: sibgp@mail.ru

Abstract. The reasons of low performance of mining machine picks are stated herein. In order to improve the wear resistance and the cutting ability of picks a new design of a cutting carbide tip insert to be fixed on a removable and rotating pick head is developed. Owing to the new design, the tool ensures a twofold increase in the cutting force maintained longer, a twofold reduction in the specific power consumption of the breaking process, and extended service life of picks and the possibility of their multiple use.

1. Introduction

The main cutting tools used for breaking rock and minerals in coal, salt, ore mines and etc. by mining machines are conical rotary picks. Having a pointed body and being reinforced with a hard-alloy tip, such picks ensure considerable cutting force and a relatively long service life. Easy replacement, operational reliability, self-sharpening while rotating promote a wide use of this type of picks. Thus, for the last two decades shearers and road headers in the Russian Federation are fitted mainly with these picks. Annually about 80 mln. tons of coal in the Russian mines are broken by conical rotary picks. Their consumption rate is up to 250-300 thousand pcs per year. Milling drums of surface miners widely used in quarries of the United States, Australia, South Africa and other countries for layer-by-layer breaking of rock are also fitted with conical picks. Over 200 picks can be mounted on a milling drum at a time. The consumption rate of wear resisting picks produced by *BETEK* is up to 1.6 pcs per every 1,000 tons of broken rock mass [1]. Due to a poor quality of picks, the consumption rate may increase up to 14-27 pcs per 1,000 ton of broken rock mass [2].

The pick is designed as a steel cylinder shaped body with a tapered head. A conical reinforcing hard-alloy insert tip for rock breaking is soldered in the top of the head. The practice of using picks in coal mines of Russia, India, and Australia shows that one of the disadvantages significantly reducing the service life of picks mounted on mining machines is early wear of the pick body [3,4] resulting in reduced efficiency of the joint and the reinforcing insert falling out which leads to premature pick failure. Thus, the consumption of cutting tools increases and mines incur significant purchase costs.

2. Results and Discussion

The wear of picks mounted on roadheaders and shearers was tested in Kuzbass mines during 2008-2015. Road headings and coal faces of twenty mines were examined. The wear degree of picks mounted on a mining machine is presented in Fig. 1.

Picks are used on a mining machine until a hard-alloy tip wears out; then picks are removed and replaced with new ones. The picks in Fig. 1 are worn evenly after long time operation. However, in terms of research different wear patterns were identified which determine their operation periods. Table 1 shows the reasons of conical rotary picks failure [5].

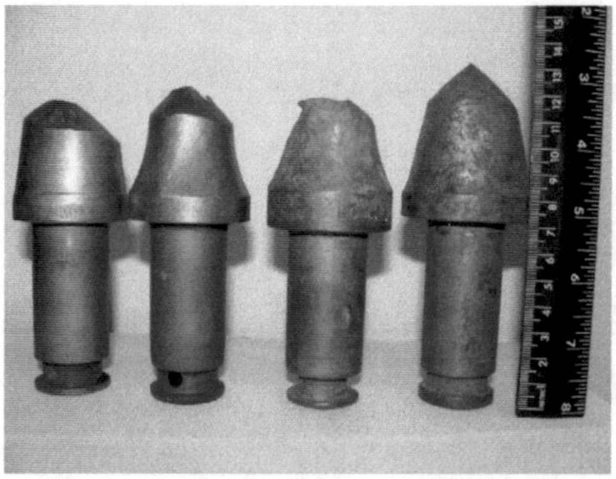

Figure 1. Picks wear over time

Table 1. Percentage of *RSH* picks failure by types

№	Failure type	Failure percentage, %
1	One-sided wear of pick body with further breaking-out of cutting insert	45
2	Even wear of pick body	27
3	Loss of picks	25
4	Break of pick body	3
	TOTAL	**100**

According to these experimental results, only 27% of RSH picks are worn out during rock breaking operations. Almost 50% of picks break down prematurely and do not pay back; and 25% are lost, and mines bear direct losses.

One of the reasons of early breaking out of a cutting insert is a high wear of a steel body around it. Premature thinning of the body walls results in the inability of the metal to resist the side force affecting the tip, therefore, the insert is broken out of the holder (Fig. 2) [2].

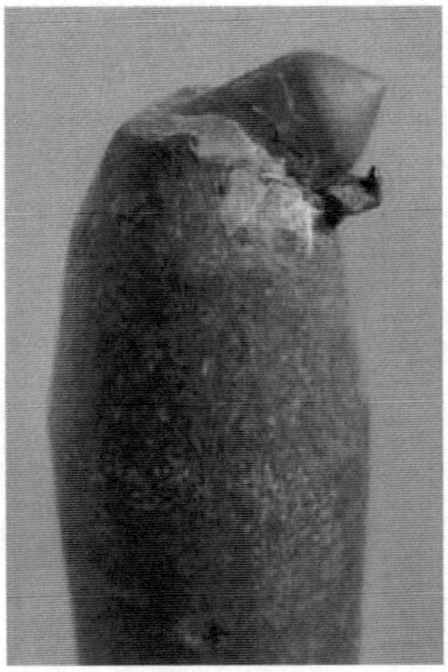

Figure 2. RKS-1 pick with a hard-alloy insert broken out

Similar deformation of picks in the Indian and Australian mines are identified and described by foreign researches (Fig. 3) [3, 4]. The examination of failures shows an early wear of a steel body around the hard-alloy tip. At the same time, the tip itself has an insignificant wear rate, retains its shape, and can be used again. However, the strength of steel is considerably lower than the strength and wear resistance of ceramic alloys. Another uncovered reason is uneven hardening of the pick head during heat-treatment [6].

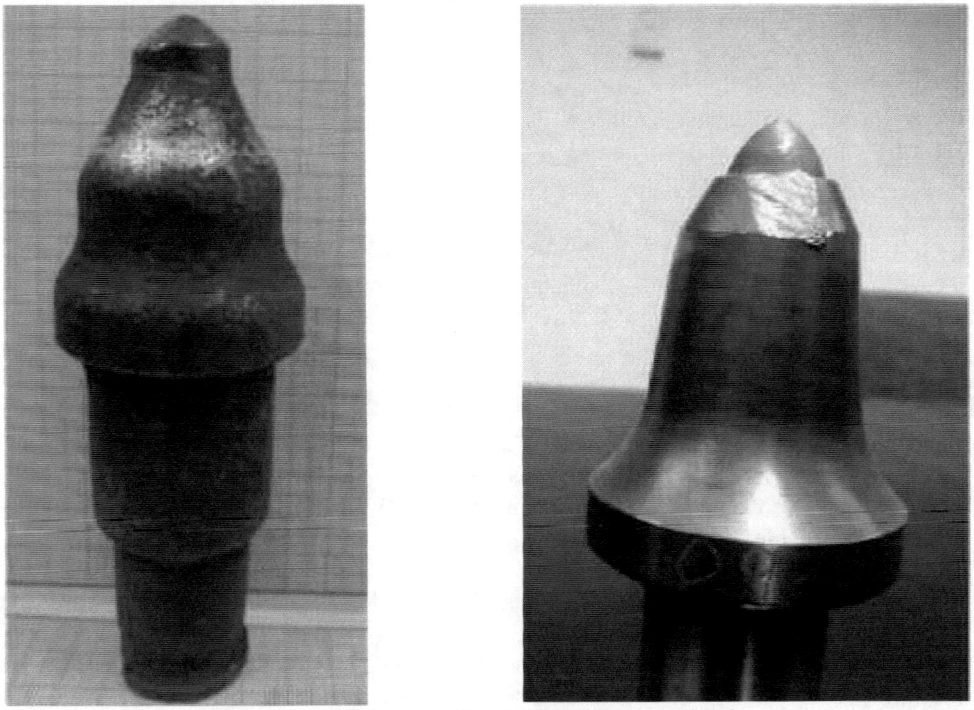

Figure 3. Early wear of a pick body around the tip

Many research papers of scientists around the world deals with examination and elimination of such types of picks deformation [7-10]. The solutions aim at ensuring uniformity of heat treatment and increasing strength of the steel body, balancing the wear resistance of the inserts and the holder, providing additional head protection around the tip, etc. One of the recent practical solutions is that of *Sandvik* offering a carbide ring *TriSpecTM* placed on the pick body and protecting an insert joint area (Fig. 4) [11]. The ring covers the front part of the pick head in the place of soldering the tip, protects the steel body from premature wear, and extends the service life of the hard-alloy insert.

A

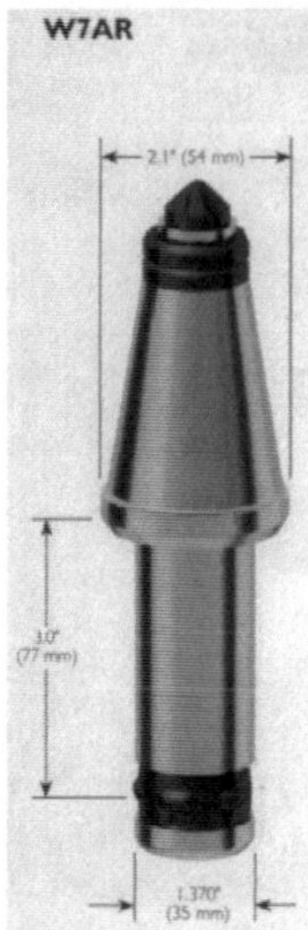

B

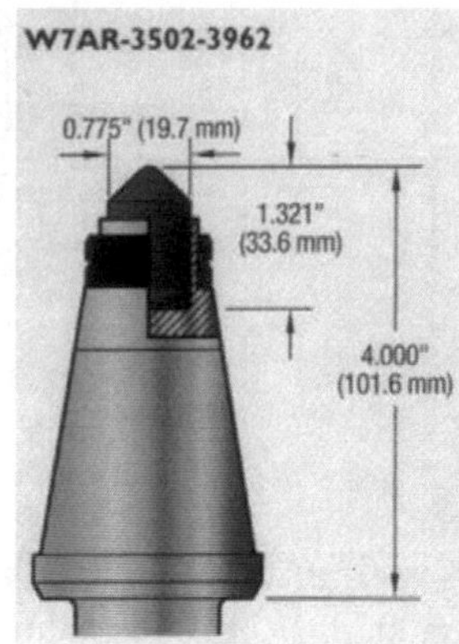

Figure 4. Pick with *TriSpecTM* protecting ring:
A – general view; B – pick head in section

A similar solution was developed by *Kennametal* offering *U119HF 80 16* and *AM521HF.74* picks with a protecting ring on the head nose. Despite the innovative nature of the solution, it has significant disadvantages. The first disadvantage is that this solution increases the pick cost and does not allow its further utilization after the hard-alloy insert wears out. Besides that, a conical insert is not suitable for cutting rock mass. Its breaking mechanism is deformation and chipping out that requires high power consumption and mechanical efforts that result in overgrinding and lower-grade coal being produced [4, 12]. High dust generation contaminates the atmosphere and poses a risk of explosion in mines. During such contact with the rock mass the pick is exposed to considerable friction resulting in sparkling, thus high water consumption is required to reduce sparkles and dust.

The new design of the pick with a head and a tip of a special shape protected by patent in Russia contributes to elimination of the disadvantages above (Fig. 5). The basis of the tool is an advanced design of the pick with a removable head [13,14]. Mine tests showed that such picks can be used up to 10 times, and only the replaceable head need to be changed.

The proposed pick consists of high-strength steel rotary body; the main conical head has a replaceable part changed when worn out and enabling multiple use of the pick holder. The replaceable head is fitted to the main one by an axial lug secured in the main head. A distinguishing feature of the new tool is a ceramic tip which is not soldered in the replaceable head (as it used to), but put on it enabling rotation. A tip is secured in the replaceable head by a small metal pin. Thus, for the picks mounted on roadheaders *KP-21* and *KSP-35* diameter and height of the tip is 35mm, and a securing pin is 40-45 mm long and has a diameter of 5 mm.

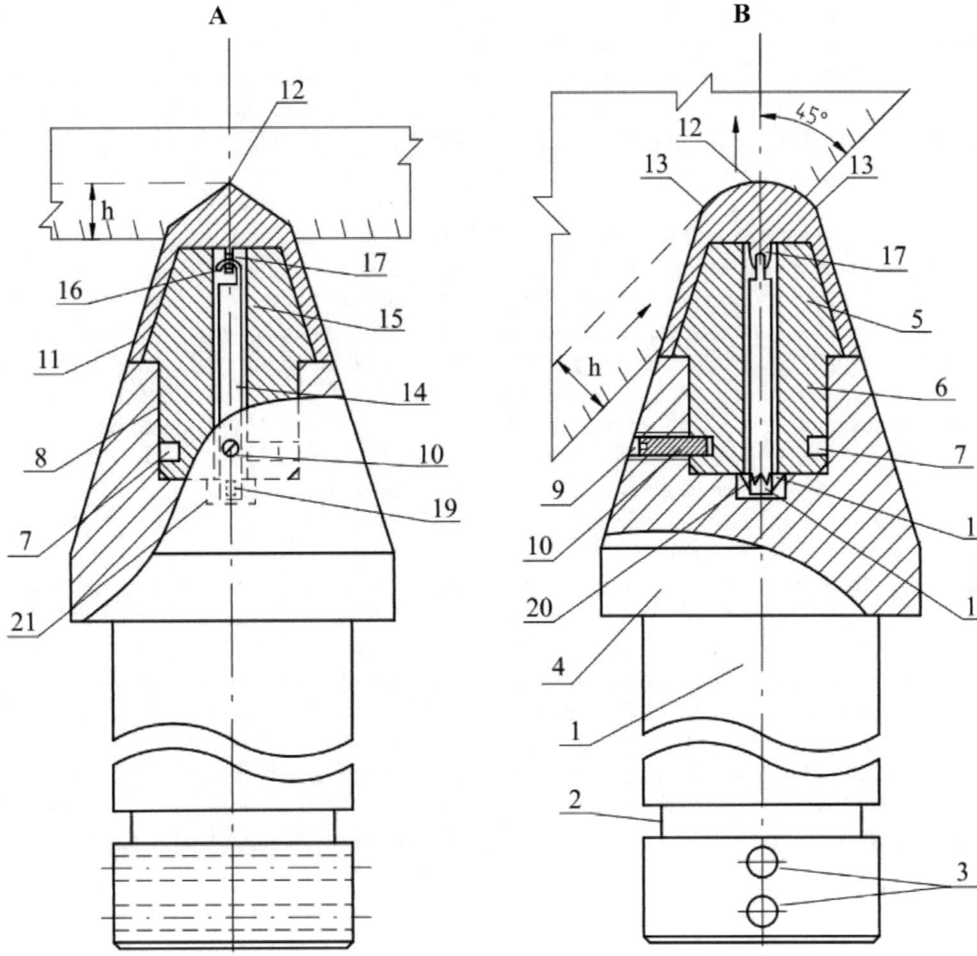

Figure 5. Pick with hard-alloy cutting tip:
A – front view; B – side view;
1 – cylinder shaped holder; 2 – circular groove; 3 – split pin holes; 4 – main head; 5 – replaceable head; 6 – axial lug of replaceable head; 7 – circular groove on axial lug; 8 – socket; 9 – screw hole; 10 – retaining screw; 11 – hard-alloy tip; 12 – cutting edge of tip; 13 – rounded face; 14 – securing pin; 15 – axial bore; 16 –hook; 17 – tip eye ring; 18 – lock; 19 – locking petals; 20 – spring; 21 – pin receiver

It is proposed that a new tip should be shaped as a semi-circle with a cutting edge, rather than as a cone. A cutting edge is sharpened at 50-80 degree angle with due account to hardness of rock to be broken.

The solution follows the modern line of improving rock breaking methods by using disk tools acknowledged recently as the most promising [15]. In addition to that, the service life of the proposed cutting tip made of tungsten-cobalt alloy will be longer than that of a steel disk tool. Currently its production of hard alloys is deemed cost-inefficient due to large size (200-250 mm).

A pick produced in the way above is utilized as follows. A pick mounted on the cutting body of a mining machine or a surface miner is brought to the coal or rock surface and a breaking process begins, wherefore the pick is forced into the rock to a depth of *h* and is moved there. The angle of inclination of the pick mounted on modern machines to the cutting surface (angle of attack) is 45-55 degree. Once the pick contacts the rock mass, owing to friction and three rotary groups (pick in block; replaceable head in the main head; tip in replacement head), the tip 11 rotates along the line of least resistance. Its cutting edge 12 is aligned with the line of pick movement over the rock surface, penetrates into it and cuts. Rock is broken with less effort in a light duty mode as the penetrating power and the cutting ability of the cutting edge 12 of the tip 11 is higher than those of the conical tip. Mining machine mechanisms operate without dynamic overloads and consume less energy. Beveled cutting edge 13 of the tip encourages easier cutting and movement of the pick in the rock mass. A special pattern of mounting picks on the cutting tools of a mining machine ensures the required frequency of cutting lines to separate parts of rock.

The tip 11, owing to its conical shape, covers the replaceable steel head 5 and protects it from the contact with the rock and wear. After a long-lasting operation of the pick and inevitable wear of its tip 11 a retaining screw 10 is unscrewed, and the replaceable head 5 is taken out of the main head. Then by clasping petals 19 of the lock 18 the securing pin 14 is released and taken out of the axial bore 15 of the replaceable head. The worn out tip 11 is replaced with a new one, the tip is attached in the reverse order to the replaceable head secured to the main head 4 by the retaining screw 10, and another operating cycles begins. It ensures extended service life of the replaceable head. In case of unforeseen excessive wear of the tip 11 and damage to the replaceable head, the latter is replaced with a new one. However, the holder 1 and the main head 4 are utilized multiple times. The service life of the pick increases significantly as compared to the picks used at this moment.

Three rotating parts of the proposed pick (holder, replaceable head, and tip) prevents jamming of the tool in one position and significantly reduces its accelerated one-sided wear. Two parts (tip and replaceable head) replaced in terms of gradual wear ensure extended service life of the pick.

A cutting edge of the tip encourages cutting and chipping of rock without its crushing, and no compacted core is formed under the pick. As the length of the cutting edge 12 of the tip greatly exceeds that of the cutting edge of the conical carbide insert, the cutting ability of the new tip is maintained longer. Destruction of the rock mass with a blunt pick results in increased operational load of mining machines and high energy consumption. According to the research, in order to penetrate into the rock mass at the depth of 1mm the force of 4.7 kN is required for a sharp tool, the double force is required for a blunt tip (Fig. 6) [5].

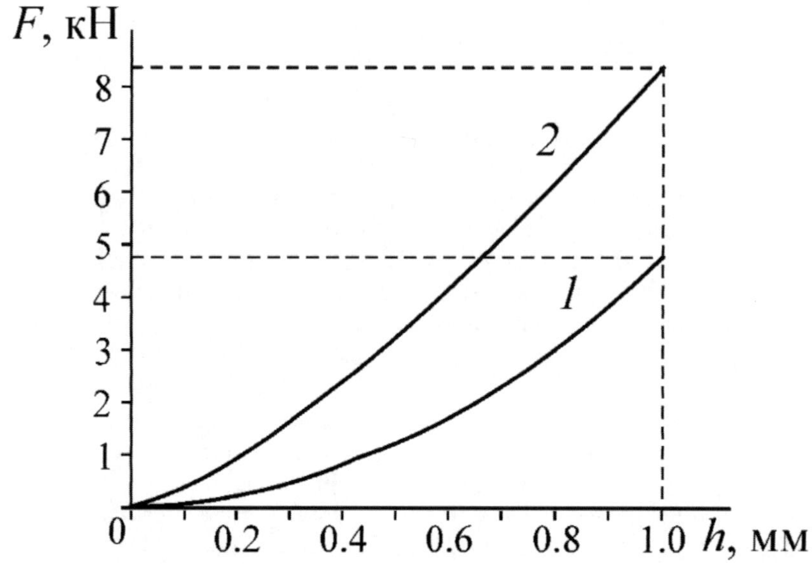

Figure 6. Force required for penetration of a sharp (1) and blunt (2) tools into the core

Other tests showed impact of the cutting angle of the tip on the cutting force with ultimate uniaxial compression strength of 52.3 MPa and sandstone density of 2.5. t/m^3. It was revealed that rock breaking with *Sandvik* conical pick *P9QA-2560-3562* having a tip of 12 mm ∅ and cutting angle of 70 degree at the depth of 15-20 mm (normal depth for mining conditions) requires the cutting force of 40-70 kN, and for a pick with the cutting angle of 110 degree (blunt) the cutting force should be 80-120 kN (Fig. 7). It was also revealed that the specific energy consumption of rock breaking increased two- or threefold with a pick blunting from 70 to 110 degree [16].

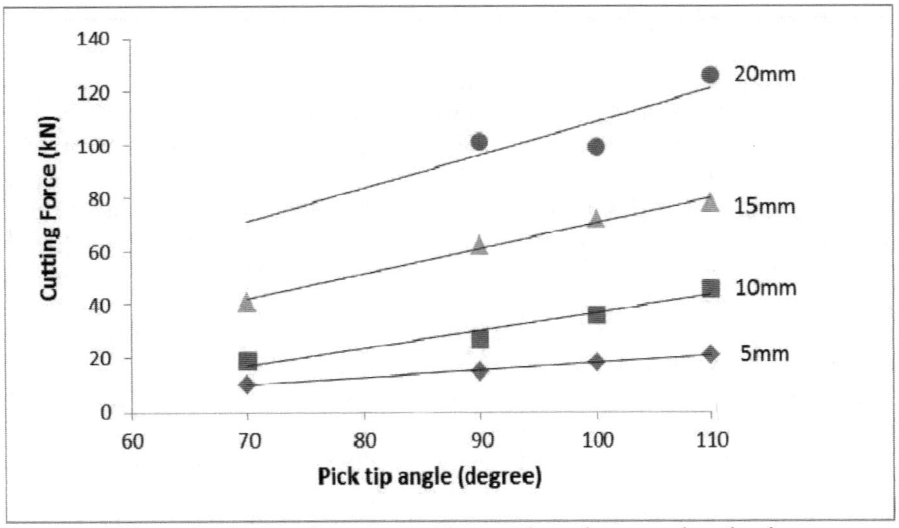

Figure 7. Cutting force vs. cutting angle and penetration depth

The tests conducted in mines showed that the specific energy consumption of KPD mining machines equipped with conical picks increases at least by 20% during the first operation as picks become blunt [17].

Great and continuous cutting ability of the proposed pick considerably reduces the energy consumption of rock breaking operations, frictional sparkling of the pick, enhances security of underground mining. Switching from destruction and chipping out to cutting reduces dust and fines, and increases the coal grade. The given advantages of the designed tool determine the prospects of its utilization in coal, sole, and gypsum mines, etc.

3. Conclusion

Wide use of conical rotary picks on mining machines and surface miners all over the world is accompanied by high consumption of cutting tools. One of the reason of frequent replacement of picks is early wear of the steel body around the carbide pick that results in chipping out of the tip from the holder. The recent innovative solutions are not perfect yet. Besides, the conical shape of the pick tip determines the destruction and chipping nature of the rock breaking process that results in high energy consumption.

A design of the pick with a cap tip and a cutting half-disk edge was developed and patented. The carbide tip is not soldered, but mounted on the replaceable head of the pick and secured by a small pin through an axial bore which enables rotation of the tip. The replaceable rotary head is secured by an axial lug in the socket of the main head and retained by a screw. Once the pick contacts the rock mass, owing to the attack angle and three rotary groups (pick in block; replaceable head in the main one; tip on replacement head), the tip rotates along the line of the least resistance; its cutting edge is aligned with the line of pick movement over the rock surface, penetrates into it and cuts. Rock is broken with less efforts in a light duty mode as the penetrating power and the cutting ability of the pick cutting edge are maintained longer with the wear rate twice as less as of the blunt conical insert.

Mining machine mechanisms equipped with the proposed picks will operate with less dynamic loads and energy consumption. The specific energy consumption of the rock breaking process can be reduced two- or threefold. Besides, cutting of coal and rock, instead of crushing, results in reduced friction and sparkling in coal faces, less production of fines and dust, reduced risk of explosions in the dust-methane-air atmosphere in coal faces. The possibility to replace worn tips and changeable heads ensures operation of a pick holder during 5-10 cycles and reduces purchase costs.

4. References

[1] Pichler M et al 2013 Commissioning of combine Wirtgen 2200SV at JSC "Kovrov Kar'eroupravlenie" *Mining industry* vol 2 pp 110-116

[2] Romanovich A S and Kalukiewich 2015 The requirement for energy efficient tangential cutter used for the development of potassium salt *Mining industry* vol 1 pp 2-4

[3] Saurabh Dewangad, Somnath Chattopadhyaya and Sergey Hloch 2015 Critical damage analysis of WC-Co tip of conical pick due to coal excavation in mines *Advances in Material Science and Engineerin* 292046

[4] Sun Y and Li X S 2014 Ineffective rock breaking and its impacts on pick failures *The 31st International Symposium on Automation and Robotics in Construction and Mining* (ISARC 2014)

[5] Khoreshok A A et al 2013 *Production and operation of destructive mining machines tool* (Tomsk) p 296

[6] Bolobov V I, Bobrov V L, Talerov M P and Bochkov V S 2012 The reason of quick wearing of tangential cutters *Notes of Mining Institute* vol 195 pp 238-240

[7] Sarwary E and Hagan P 2015 The effect of tool geometry on cutting performance *Mining education Australia – Journal of research projects revie* vol 4 number 1 pp 43-48

[8] Xiaohui Liu, Songyong Liu, Lie Li and Xinxia Cui 2015 Experiment on Conical Pick Cutting Rock *Material Assisted with Front and Rear Water Jet Materials Science and Engineering* 506579

[9] David R. Hall 2014 Patent US 8,777,326 B2 Pick with hardened core assembly. Data of Patent Jul.15

[10] Prokopenko S, Sushko A and Kurzina I 2015 New design of cutters for coal mining machines *IOP Conference Series: Materials Science and Engineering* 91 012058

[11] Cutting tool for rocks 2014 Mining RC 100 *Sandvik product catalogue* p 52

[12] Kolesnichenko E A, Kolesnichenko I E, Lyubomischenko E I and Demura V N 2012 Techniques for Face Ventilation According to Dust-making Ability of Roadheader *Coal* 6 pp 39 -42

[13] Prokopenko S A, Ludzish V S, Kurzina I A and Sushko A V 2015 Results of industry testing of multiple use rock-cutting picks *Gornyi zhurnal/ Mining Journal* 5 pp 67-71

[14] Prokopenko S A and Ludzish V S 2014 Problems of innovative development of the mining Enterprises of Russia *Gornyi zhurnal/ Mining Journal* **1** pp 47-49

[15] Khoreshok A A, Mametev L E, Borisov A Yu and Vorobev A V 2015 Stress state of disk tool attachment points on tetrahedral prisms between axial bits *Applied mechanics and materials* vol 770 pp 434-438

[16] Esmat Sarwary and Paul Hagan 2015 Changes in Cutter Performance with Tool Wear *15th Coal Operators' Conference (*University of Wollongong, The Australasian Institute of Mining and Metallurgy and Mine Managers Association of Australia) pp 283-290.

[17] Shabaev O E, Hitsenko N V and Bridun I I The formation of the cutting forces on the incisors of the Executive body of roadheaders based blunt Available at: http://ea.dgtu.donetsk.ua:-8080/jspui/bitstream/123456789/26246/1/shhicbri.pdf (Last accessed date 20.03.2015)

IOP Conf. Series: Materials Science and Engineering **142** (2016) 012116 doi:10.1088/1757-899X/142/1/012116

Assessment of Spatial Unevenness of Road Accidents Severity as Instrument of Preventive Protection from Emergency Situations in Road Complex

A Petrov[1], D Petrova[2]

[1]Associate professor, *Tyumen State Oil and Gas University,*
Melnikaite street, 72-221, Tyumen, Russian Federation
[2]Undergraduate, *Tyumen State Oil and Gas University,*
Melnikaite street, 72-221, Tyumen, Russian Federation

E-mail: [1]ArtIgPetrov@yandex.ru, [2]DaArtPetrova@mail.ru

Abstract. Emergency situations in road complex are road traffic accidents (RA) with severe consequences. These are incidents connected with the death and injury of large number of people. The most common reasons for this are the collision of three or more cars, the collision of buses with trains at railroad crossings, the fall of the buses in the mountain gorge, and other similar cases. Is it possible to predict such events? How to build a preventive protection against such emergencies? We have to understand that emergencies in a road complex are qualitative expression of the quantitative processes that characterize the general state of road safety in the region. In this regard, at the level of state monitoring of emergency situations it is important to understand in general - in which region the situation is more complicated and in which is more favorable. This knowledge helps to more efficiently reallocate resources intended to solve the problems of road safety provision. The consequence of this is improvement of the quality of preventive protection from the emergencies in the road complex. The article presents quantitative values of severity of accidents in the Russian Federation regions and the Pareto chart distribution of cumulates of the accident severity for the Russian Federation. On the basis of the complex assessment of the spatial non-uniformity of the accident severity results it offers two important recommendations, implementation of which will alleviate the issue of formation of emergency situations in the road of the Russian Federation on the basis of the complex assessment of the spatial non-uniformity of the accident severity results.

1. Introduction.

According to the State Report «On the state of protection of population and territories of Russian Federation from natural and technogenic emergency situations in 2014» [12] 262 emergency situations of diverse nature happened on the territory of Russia in 2014. 567 people killed our country as a result of disaster in 2014 [12, 13]. Structure of quantitative indicators by type of disaster in the Russian Federation (2014) is shown in Fig. 1.

A large proportion of emergency situations - 186 cases (72 % of total) - were the incidents of technogenic nature. These include road traffic accidents (RA) with serious consequences [12]. The term «accident with serious consequences» is used in relation to the accident, which killed 5 or more people, or 10 or more people were injured.

Table 1 shows the main characteristics of such accidents that occurs in the territory of Russia in 2014.

Data of Table 1 shows that technogenic accidents with severe consequences occupy a priority position among other emergencies on all indicators except material damage. At the same time, the number of traffic accidents with severe consequences is only a small fraction of the total number of accidents with victims.

In 2014 in Russia, there were 199720 accidents with victims, which killed 26963 people and 251785 people were injured [14].

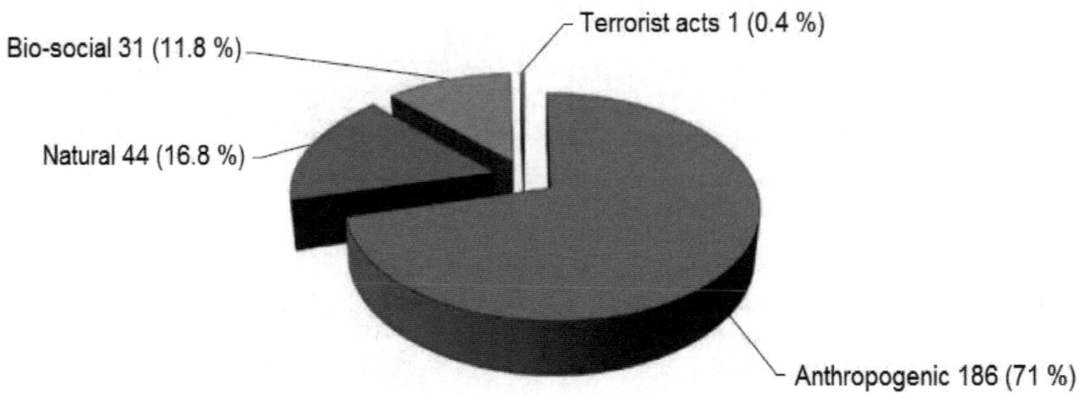

Figure 1. The structure of the quantitative indicators by type of emergency situations, units (%) [12]

Table 1.The main characteristics of the accident with severe consequences in Russia (2014) [12]

Estimated characteristic	Indicator of transport accidents in the Russian Federation (2014)				
	The number of traffic accidents with grave consequences, units / year	The number of deaths in road accidents with grave consequences, pers. / year	The number of victims in road accidents with serious consequences, pers. / year	The number of survivors in an accident with serious consequences, pers. / year	Material damage, mln. rub / year
Absolute value of RA with severe consequences, units.	87	381	968	586	31,64
Absolute value of emergencies of technogenic nature, units.	186	567	1620	1063	16321,53
Relative value, %	47	67	60	53	0,2

Analysis shows that the number of traffic accidents with severe consequences doesn't exceed 0.045 % of the total number of accidents with victims. The death toll of such road accidents is 1.4 % of the total number of people killed on the roads of the country. In this regard, for an understanding of the degree of danger of the situation with traffic accidents in different regions of Russia we need to use other indicators [1, 2, 3, 5].

The best choice [9, 10, 11] is recognized among road safety experts Factor accident severity $K_{severity\ RA}$ (1):

$$K_{severity\ RA} = [\frac{Killed_{RA}}{Victims_{RA}}] \times 100 = [\frac{Killed_{RA}}{(Killed + Injured)_{RA}}] \times 100 , \tag{1}$$

wher $K_{severity\ RA}$ factor of severity RA in units from 100;
$Victims_{RA}$ amount of killed and injured in RA in a year, pers./year;
$Killed_{RA}$ amount of killed in RA in a year, pers./year;
$Injured_{RA}$ amount of injured in RA in a year, pers./year.

The hypothesis that is proved in this article, is formulated as this: ***The severity of the accident, or the relative proportion of road deaths among the victims (killed + injured) in an accident on the territory of Russia is formed unevenly.*** The thesis, which follows from the proof of this hypothesis, allows to develop the tools of probabilistic forecasting of emergencies [4, 6, 7, 8, 9] connected with accidents in certain regions of Russia.

2. Basic data.

The data on the spatial (in the regions of Russia) distribution of accidents with victims and all related information can be found in the archives of the Russian Interior Ministry traffic police [13].

Table 2 shows the Russian regions rating of largest severity of traffic accidents in 2013 [9, 14].

Table 2. The Russian regions rating of largest severity of traffic accidents in 2013 [9, 14]

The subject of Russian Federation (2013)	Rating place in the Russian Federation (from the most dangerous to safest)	The numerical values of indicators	
		Severity of RA, killed/victims, $K_{severity\ RA}$	% from the national average of index $K_{severity\ RA}$
Republic of Dagestan	1	20.42	209.0
Chechen Republic	2	18.39	188.2
Tuva Republic	3	17.27	176.8
Kabardino-Balkar Republic	4	17.20	176.0
Republic of Ingushetia	5	16.24	166.2
Republic of Adygea	6	16.18	165.6
Belgorod Oblast	7	14.50	148.4
Krasnodar Krai	8	14.11	144.4
....	....	....	....
....	....	....	....
Moscow	76	6.10	62.4
Komi Republic	77	5.67	58.0

Ivanovo Oblast	78	5.55	56.8
Jewish AO	79	5.15	52.7
Murmansk Oblast	80	4.60	47.1
Saint-Petersburg	81	4.24	43.4
Nenets AO	82	3.41	34.9
Chukotka AO	83	3.13	32.0

Even a simple visual examination of the data in Table 2 makes it clear that the severity of an accident varies in a very wide range of values in different regions. The ratio between the value of the severity of the accident in the most accident-prone region of the Russian Federation - the Republic of Dagestan (20.4) and the value of the severity of the accident in the accident-safe region of the country - Chukotka (3.13) is up to 6.5-fold, which is a very large value.

3. Methods of assessing the severity of the spatial unevenness of the accident.

Analysis [9] shows that the spatial distribution of the severity of the accidents in Russian regions is convenient to estimate by constructing a Pareto chart. Features of Pareto diagrams are determined by the bending of the Lorenz curve, which describes the distribution of cumulates accident severity of the Russian Federation space. Fig. 2 shows a Pareto diagram of distribution of severity of the accident characteristics in the space of the Russian Federation regions [9].

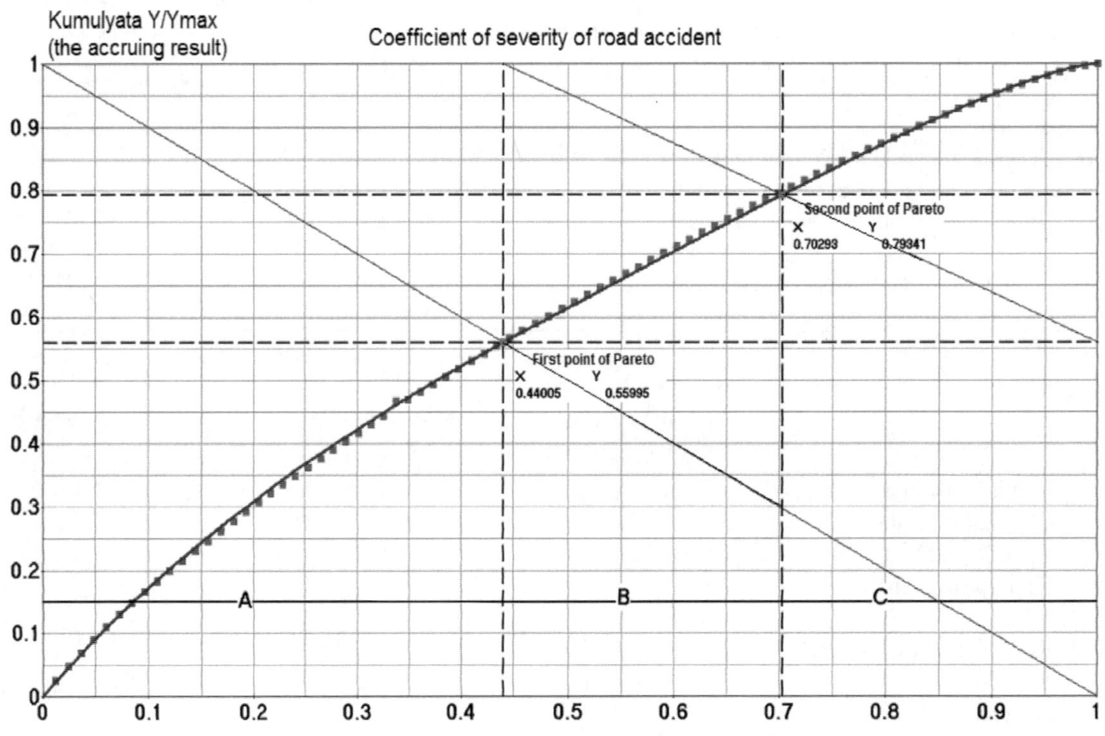

Figure 2. Pareto diagram of distribution of severity of the accident characteristics in the space of the Russian Federation regions [9]

4. Results and Discussion

Visual examination of this diagram leads to the conclusion that the bending of the Lorenz curve explicitly is typical only for the end portions of the curve, that is for the first eight (1 ... 8) and the last eight (76 ... 83) regions of Russia in the ranking index of the accident severity. The list of regions where the accident severity is the highest includes five Caucasus republics, Tuva Republic, the Krasnodar region and Belgorod region (Table 1).

These eight regions (10 % of the composition of the Russian Federation) form about 18 % cumulative result for the severity of the accident in the Russian Federation [9].

The eight regions where the severity of the accident is minimal, are federal cities - Moscow and St. Petersburg, the four northern regions and sparsely populated Jewish autonomous region and Ivanovo region. These eight regions form about 5 % of the cumulates severity of road accidents in Russia [9].

In fact, the severity of the accident, or the relative proportion of road deaths among the victims (killed and injured) in an accident on the territory of Russia is formed unevenly. Thus, we can conclude that the irregularity of the formation of an accident severity in the territory of the Russian space is determined by geographical indication [6, 7, 8, 9]. The geography forms quality of people's life, that reflects on the circumstances related to the features of the trucking accident [6, 7, 8, 9].

5. Conclusion

Preventive protection from emergency situations in the road-transport complex can be built only on the basis of forecast assessment based on probability statistics [3, 5, 6, 7, 8, 9]. Presented in this article, Pareto chart of accidents severity in the Russian space [9] allows us to offer two important recommendations, implementation of which will allow to multiple alleviate the issue of formation of emergency situations in the road complex.

- In order to analyze the seriousness of the situation in the sphere of road complex from the point of the possibility of forming an emergency situation it is necessary to adjust some approaches of the Ministry of Emergency Situations of the Russian Federation to the classification of emergency situations. Particularly, the factor of «an accident with serious consequences», that is used by Emergency Department as the most important assessment indicator, should be supplemented with an indicator «The severity of the accident».
- In the space of Russia should be paid special attention to traffic safety in the road complex to the Caucasus republics and the Tuva Republic. There are where the severity of accidents relative to other Russian regions is maximal. There are the most likely emergencies related to the accident.

Reference

[1] Adams J G D 1985 *Risk and Freedom: the record of road safety regulation* (Cardiff: Transportation Publishing Projects) p 202
[2] Cerwenka P et al 2012 *Traffic Engineering and Control* **53(4)** 149
[3] Chiou Y and Fu C 2013 *Accident Analysis & Prevention* **vol 50** 73
[4] Cole T B 2004 *The Journal of American Medical Association* **vol 291** 2531
[5] Elvic R 2015 *Accident Analysis & Prevention* **vol 82** 36
[6] Haight F et al 1988 *Reducing Traffic Injury - A Global Challenge* (Melbourne Australia: Volvo Traffic Safety Award) p 156
[7] Jin T G and Saito M and Eggett D L 2008 *Accident Analysis & Prevention* **vol 40 iss 6** 2015
[8] Kerner B S 2013 *Traffic Engineering and Control* **54 (4)** 47
[9] Petrov A I 2015 *Features of formation of motor transportation accident rate in space and time* (Tyumen: TSOGU) p 254
[10] Smeed R J 1949 *The Journal of Royal Statistical Society. Series A (Statistics in Society)* **1** 1
[11] Smeed R J 1968 *Traffic Engineering and Control* **10** p 364
[12] State Report 2015 On the state of protection of population and territories of Russian Federation from natural and technogenic emergency situations in 2014 (Moscow: Emercom of Russia. Federal State Budgetary Institution) p 352
[13] The Report 2014 Traffic safety in Russia: current state and urgent measures for improvement of a situation (Moscow: High School of Economics) p 25

[14] Statistical information // Traffic police of the Ministry of Internal Affairs of Russia – Traffic security status indicators – archive http://www.gibdd.ru/stat/archive/

IOP Conf. Series: Materials Science and Engineering **142** (2016) 012118 doi:10.1088/1757-899X/142/1/012118

Computer Analysis of Durability and Leakproofness of Multilateral Junction of Wells

K Syzrantseva[1] and M Dvoynikov[2]

[1]Tyumen State Oil and Gas University, 38 Volodarskogo st., Tyumen, 625000, Russia
[2]National Mineral Resources University (Mining University), 2, 21st Line, St Petersburg, 199106, Russia

E-mail: kv.syzr@gmail.com

Abstract. The paper considers method of computer simulation of well multilateral junctions corresponding to TAML Level-4. The base stages of algorithm of stress-strain condition calculation of cased main and side pipes are described. Durability of junction depending on internal pressure is established in a kind of functional dependences. The influence of side pipe thickness and whipstock angle on junction deformation is researched.

Introduction

Now in Russia construction of multilateral wells is promptly developing segment of oil and gas service. Technologies of multilateral drilling allow to increase considerably efficiency of development of oil and gas fields that has special importance for operation of oil-pool with hardly recoverable resources.

Today there are a lot of technical and technological decisions for a construction of multilateral wells which can be related to six levels in accordance with the standard international classification TAML (Technology Advancement for Multi-Laterals). Each TAML level is characterized by specific type of a junction design of side and main boreholes [1]. The most universal and economically justified decision for construction of multilateral wells on the land is the type of junction design corresponding to TAML Level-4 [2], which is characterized by the fact that the main and side boreholes are cased and cemented. This type of junction provides mechanical fixing of connection, but has not hydraulic isolation, therefore is characterized by insufficiently reliable bracing.

For the analysis of TAML-4 junction functioning in order to determine the directions of this technology development and identify the technical and technological restrictions of junction design authors developed the method of carrying out computing experiment for stress-strain condition estimation of multilateral junction in the conditions of a well.

1. Computer simulation

As the Lame's formula [3] used for calculation of stresses in casing columns is applicable only to the infinite hollow cylinder and doesn't allow to consider assembly module, for the solution of this task it is reasonable to use a numerical method.

The Finite Element Method [4, 5] realized in ANSYS has the powerful capabilities for the solution of 3-dimensional tasks [6, 7]. This method is successfully applied for durability and reliability estimation of drilling [8, 9] and oil equipment [10].

As object of computer simulation the place of junction of the main and side pipes of a multilateral well was chosen because this element of module is the most difficult at construction and the least reliable at the subsequent operation. In the analysis process the five stages of modeling and calculation are realized.

Stage 1. <u>The geometrical model creation.</u>
To have an opportunity to research the influence of side pipe parameters on strength characteristics of module in whole, it is expedient to create its parametrical model in the ANSYS Parametric Design Language [11]. Therefore geometrical model is created with the help of Boolean operations [12].

The main case pipe was accepted steel (Strength grade E, yield strength is 552 MPas, tensile strength is 689 MPas) that corresponds to widespread conditions at multilateral wells drilling. The upper pipes of side pipe shank are made of an aluminum alloy (V95t1 aluminum, Russian State Standard 18482-79, yield strength is 400 MPas, tensile strength is 510 MPas). This feature of module is caused by a possibility of further drilling of cement and shank in the main column. Properties of materials were described by multilinear laws of deformation (Fig. 1).

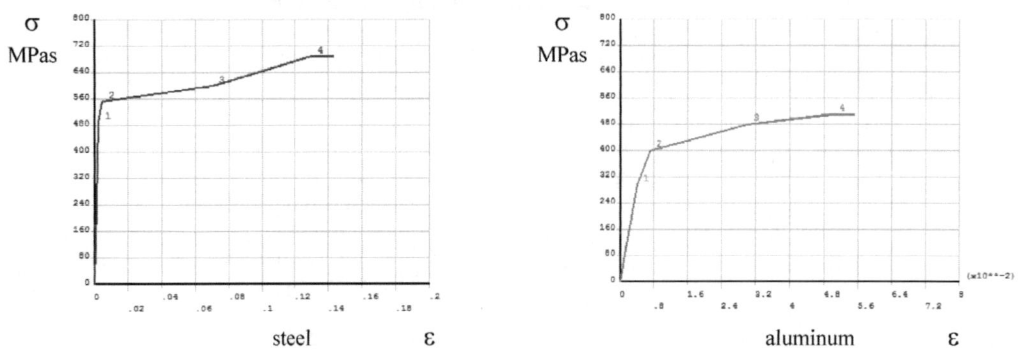

Figure 1. Deformation curves of pipes materials

Stage 2. <u>Finite element meshing.</u>
For model meshing finite elements SOLID95 describing elastic and plastic strains by quadratic functions were used. In order to decrease the random access memory size and to reduce the calculation time the sizes of elements in junction volumes were set small, and in constrained areas of model – larger [13].

Stage 3. <u>The boundary conditions definition.</u>
Because of symmetry of a model, only ½ its part was subjected to the analysis. On the section planes of model the symmetry boundary conditions were defined, upper and lower areas are constrained (Fig.2). As external loads the rock pressure equal 22 MPas was accepted, at junction position of the main and side pipes of well at depth 2000 m. As it was specified earlier, the TAML Level-4 junction has no hydraulic isolation and the cavity of case pipe and rock in junction place are connected. Therefore the internal pressure depending on degree of non-leakproofness of a junction, on rock properties and on the works performed in well, was varied during simulation.

Stage 4. <u>Solution of the task.</u>
For the solution of this task the Sparse Solver was chosen. Because of considering of elastic-plastic strains the task is characterized by physical nonlinearity, therefore for acceleration of nonlinear decision convergence the 4 substeps were set.

Stage 5. <u>The results analysis.</u>
During computer simulation eight calculations at different internal pressure values were carried out. For check of calculation model adequacy it is necessary to estimate the finite element grid quality [14]. The finite element grid quality estimation was realized similar to paper [8], in accordance with formula:

$$\Delta_1 = \frac{ESOL - NSOL}{ESOL} \cdot 100\% , \tag{1}$$

where ESOL – element solutions; NSOL – nodal solutions. The estimation has shown quite satisfactory results presented in Table 1.

Table 1. The results of computer simulation

Internal pressure, MPas	Maximum radial displa-cements, mm	Von Mises stress in steel, MPas		Grid quantity in steel, %	Von Mises stress in aluminum, MPas		Grid quantity in aluminum,%
		NSOL	ESOL		NSOL	ESOL	
10.2	-1.777	519.228	521.679	0.47	368.967	403.732	8.611
11	-1.651	484.523	486.803	0.468	358.578	397.412	9.772
12	-1.4963	441.549	443.618	0.466	343.456	372.118	7.702
13.6	-1.2537	373.868	375.606	0.463	318.646	341.421	6.671
15.3	-1.003	303.086	304.478	0.457	271.454	275.312	1.401
17	-0.877	232.992	234.025	0.441	203.995	206.759	1.337
18	-0.856	192.074	192.911	0.434	164.482	166.643	1.297
19	-0.8361	151.295	151.937	0.423	125.103	126.711	1.269

The results of stress-strain condition estimation of multilateral junction at internal pressure 12 MPas are presented on Fig. 3. Deformations are magnified by a factor of 10. Contours of von Mises stress [15] the most illustrative for this task are shown on Fig.4.

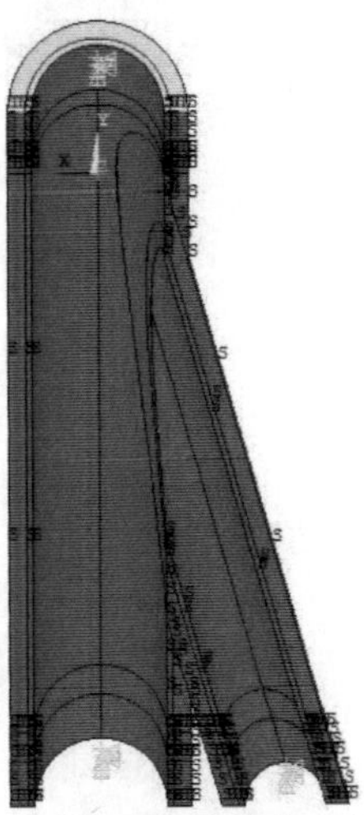

Figure 2. Model boundary conditions

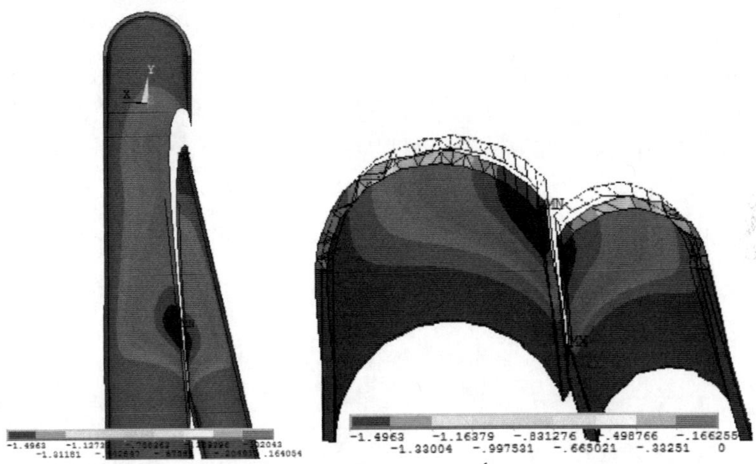

a b

Figure 3. Deformation of model at internal pressure 12 MPas
a- contours of radial displacements, b- model deformation

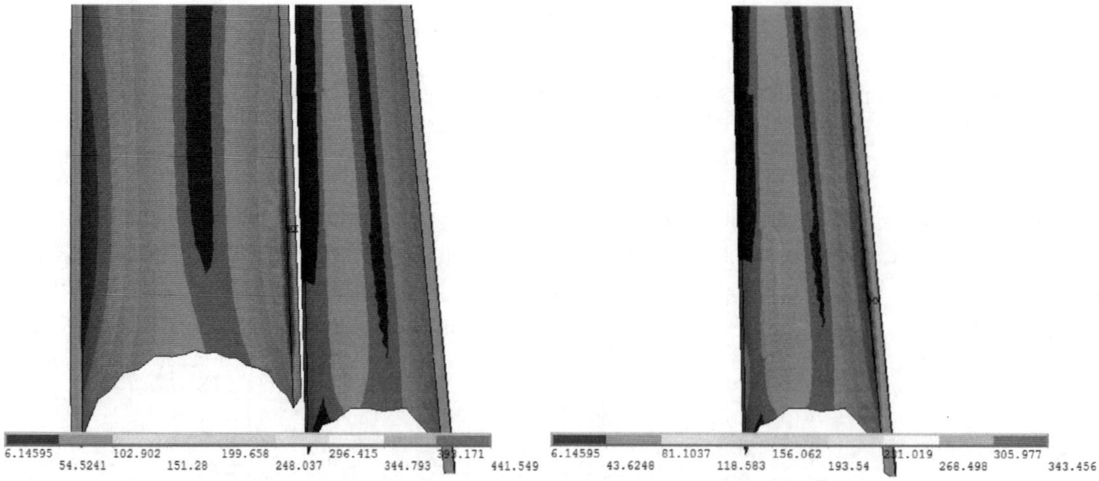

a B

Figure 4. Contours of von Mises stress at internal pressure 12 MPas
a – main and side pipes, b – side pipe only

By results of computer simulation with the help of regression analysis following functions of displacements and stresses in pipes materials of pipes depending on internal pressure value were established. The obtained dependences are presented as polynomials.

Radial displacements:

$$Ur(p) = -5.060313 + 2.230299 \cdot p - 0.244619 \cdot p^2 + 0.010638 \cdot p^3 - 0.0001621682 \cdot p^4 \quad (2)$$

Von Mises stress in steel pipe:

$$\sigma_{EQV}(p) = 976.707 - 46.539 \cdot p + 0.164 \cdot p^2 \quad (3)$$

Von Mises stress in aluminum pipe:

$$\sigma_{EQV}(p) = 208.641 + 38.991 \cdot p - 2.293 \cdot p^2 \quad (4)$$

Fig. 5 illustrates the computer simulation data showed by labels and plotted functions showed by solid curves.

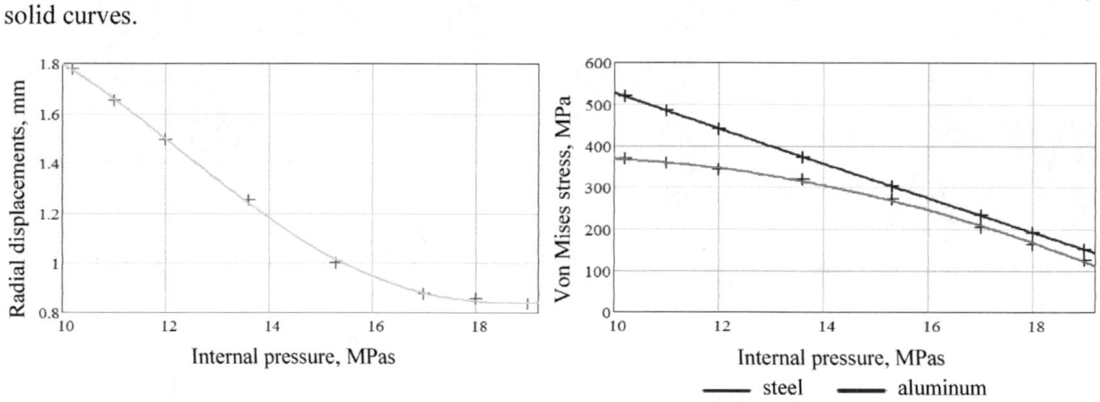

Figure 5. Influence of internal pressure on model deformation and material stress

2. Research of influence of multilateral junction geometrical parameters on its durability and leakproofness

The developed by authors algorithm of stress-strain condition estimation of multilateral junction realized in APDL language allows to research the influence of side pipe parameters (sidetracking angle and side pipe thickness) on distribution of displacement and stresses in module.

Now whipstocks of angles less than 3 degrees are generally presented at the market of oilservices [9]. However in the carried by authors researches also larger angles were considered to estimate possible options of junction optimization. The results of calculations are presented in Table 2.

On Fig.6 contours of von Mises stress till yield strength for junction of angles 1.92 degrees and 3 degrees correspondingly are shown. Zones of plastic strain are marked by gray color. It is well visible that even small increase of sidetracking angle provides considerable shortening of plastic zone length.

Authors of paper [16] recommend to choose the tools for sidetracking of angle 5 degrees to avoid big zones of the plastic strains capable to lead to column buckling. At the same time, researches of this paper showed that dependence of plastic strains zone length on whipstock angle has cubic character:

$$L(\alpha) = 2573.914 - 1319.991 \cdot \alpha + 276.523383 \cdot \alpha^2 - 20.503049 \cdot \alpha^3 . \quad (5)$$

Table 2. The results of junction stress-strain condition estimation depending on whipstock angle

Whipstock angle, degree	Maximum total displacements, mm	Length of plastic zone along side pipe, mm
1.92	3.64765	917
2.5	3.37134	670
3	3.17468	561
4	2.85901	402
5	2.609	325

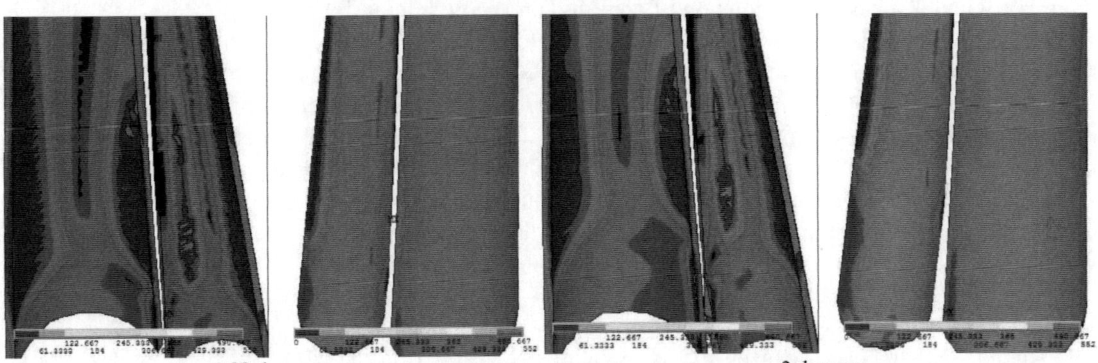

1.92 degrees 3 degrees

Figure 6. Contours of von Mises stress for junctions of different angles

The curve of established by polynomial regression dependence (5) is submitted on Fig. 7. Its well illustrates that increase of whipstock angle more than 3 degrees is ineffective. Moreover, use of whipstock of big angle can cause complications at well drilling and at subsequent shank descent in a side borehole.

Other way to increase the junction working capacity is magnification of side pipe thickness. Results of computer simulation have shown that side pipe thickness also has influence on working capacity of multilateral junction in whole. The established function of plastic strains zone length depending on side pipe thickness is quadratic, but is very close to linear: The curve of the function (6) is presented on Fig.7.

$$L(t) = 2639.086 - 222.142857 \cdot t - 1.428571 \cdot t^2 . \quad (6)$$

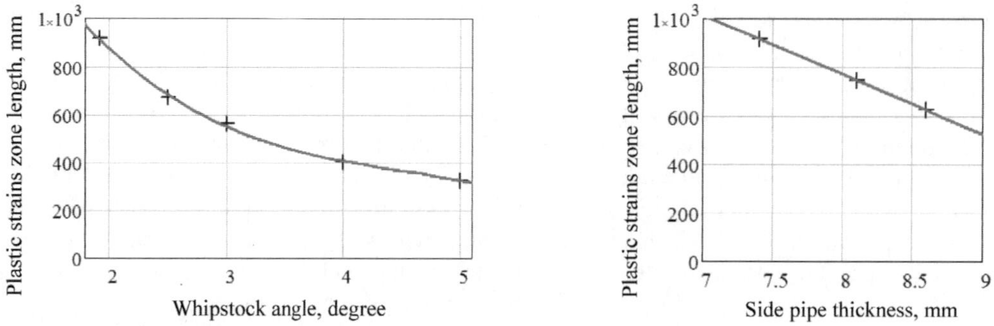

Figure 7. The influence of whipstock angle and side pipe thickness on plastic strains zone length

Results and discussion

Carried out computer experiment allowed to establish that the TAML Level-4 junction is subjected to extreme loads depending on difference between rock pressure and internal well pressure. At the same time high values of differential pressure apply stresses in pipes metal, close to material yield strength.

It follows that candidate wells for multilateral drilling must have the casing column of corresponding strength grade, and it is necessary to consider that casing column durability of old stock wells can decrease as a result of corrosion.

The obtained results testify that even at small pressure difference the junction is subjected to deformations. This factor will inevitably cause integrity violation of pipes and cement contact in the considered model. During operating in case of pressure decrease in well lower than limit pressure there is column displacement, violation of adhesion and cement integrity.

As a result at operation the prolapse of cement stone fragments and rock is observed that leads to frequent jammings and wear of pumps. Also at such operation the considered junction provides risk of intercolumn cross-flows and gas inrushes. This factor needs to be considered at design of construction of a multilateral well and to choose the place of sidetracking taking into account that the cement outside of main column and side borehole will be damaged on extended region.

Important design problem of multilateral well is quest for of reasonable compromise at the choice of whipstock angle and side pipe thickness, and also their combination with the purpose of multilateral well design optimization and increase of its reliability.

References

[1] Liland, G., Prebeau-Menezes, L., & Mjolhus, S. J. de C. (2013, October 29). World's First TAML Level 5 Multi-lateral Well with Individual Remote Inflow Control of Three Branches on Troll Oil Field. Offshore Technology Conference. DOI:10.4043/24427-MS.

[2] Zhang Yanping, Ren Rongquan, Wang Hui, Wang Jun. Multilateral drilling & completion technology based on Solid Expandable Tubular fixing system. PETROLEUM EXPLORATION AND DEVELOPMENT 36(6): 768-775 · DEC, 2009. DOI:10.1016/S1876-3804(10)60008-0.

[3] A Vakhrushev , M Wu, A Ludwig, Y Tang , G Hackl, G Nitzl. Modeling of the flow-solidification interaction in thin slab casting MCWASP XIII. IOP Conf. Series: Materials Science and Engineering 33 (2012) 012014. DOI:10.1088/1757-899X/33/1/012014.

[4] Ivo Babuska, Uday Banerjee, and John E. Osborn. Generalized Finite Element Methods — main ideas, results and perspective. International Journal of Computational Methods 01, 67 (2004). DOI: 10.1142/S0219876204000083.

[5] N. A. Makhutov, M. M. Gadenin, O. F. Chernyavskii, A. O. Chernyavskii, S. V. Evropin. Low-cycle deformation of the structures in a nuclear power plant and methods for calculating them. Atomic Energy, Sept. 2009, Volume 107, Issue 3, pp 173-179. DOI: 10.1007/s10512-010-9212-4.

[6] K.V. Syzrantseva. Computer aided engineering of load ability and deformability of oil and gas equipment units. Monograph. – Tyumen, TSOGU, 2009. – 124p.

[7] K.V. Syzrantseva, V.A. Arishin. The stress-strain condition estimation of the motive section of screw downhole motor in program complex ANSYS. Bulletin of the Tomsk Polytechnic University. 2009. Vol. 315, № 5. Pp.57-61. http://earchive.tpu.ru/handle/11683/2886.

[8] Oshibkov, A.V., Vodorezov, D.D., Syzrantseva, K.V., Frolov, S.A., Dvoynikov, M.V., Svetashov, V.N., Bakirov, D.L. (2015) The stress-strain condition estimation of construction of multilateral well junction. Oil Industry, 2015, no.6, p. 83-85.

[9] Oshibkov A.V., Vodorezov D.D., Syzrantseva K.V., Frolov S.A., Dvoynikov M.V., Bakirov D.L. Analysis of sidetracking characteristics influence on multilateral well construction reliability. Oil Industry, 2015, no.11, p. 121-123.

[10] Syzrantsev V., Syzrantseva K. Nonparametric approach to the durability estimation task of details with complicated geometry. Monograph "MACHINE DESIGN 2008", Novy Sad, Republic of Serbia, – 2008. P.139-144.

[11] Y. J. Wang, F. T. He, Y. Zhao, "The General APDL Program for Calculation of Structures under Seismic Response", Advanced Materials Research, Vols. 482-484, pp. 2396-2399, 2012. DOI: 10.4028/www.scientific.net/AMR.482-484.2396.

[12] Aldo Iannetti, Matthew T. Stickland, and William M. Dempster. An advanced CFD model to study the effect of non-condensable gas on cavitation in positive displacement pumps. 11th International symposium on compressor & turbine flow systems theory & application areas SYMKOM 2014 IMP, Lodz, 20 – 23 October, 2014. DOI 10.1515/eng-2015-0027.

[13] Liu, Y. and Glass, G. Effects of Mesh Density on Finite Element Analysis. SAE Technical Paper 2013-01-1375, 2013, DOI:10.4271/2013-01-1375.

[14] Timothy A. Burkhart, David M. Andrews, Cynthia E. Dunning. Finite element modeling mesh quality, energy balance and validation methods: A review with recommendations associated with the modeling of bone tissue. Journal of Biomechanics. 2013. Volume 46, Issue 9, Pages 1477–1488. DOI: http://dx.doi.org/10.1016/j.jbiomech.2013.03.022.

[15] P.V. Burkov, S.P. Burkova , S.A. Knaub Stress and Strain State Analysis of Defective Pipeline Portion. YIT-ITEE. IOP Conf. Series: Materials Science and Engineering 91 (2015) 012055 DOI:10.1088/1757-899X/91/1/012055.

[16] Bayfield, M., Fisher, S.L., and Ring, L.M. 2000. Burst and Collapse of a Sealed Multilateral Junction: Numerical Simulations. SPE Drilling and Completion 15 (2): 75-83. SPE-63813-PA. DOI: 10.2118/63813-PA.

Raising of Operating a Motor Vehicle Effects on Environment in Winter

S A Ertman, J A Ertman, D A Zakharov

Tyumen State Oil and Gas University, 38, Volodarskogo street, Tyumen, 625000, Tyumen, Russia

E-mail: ertmans@list.ru

Abstract. Severe low-temperature conditions, in which considerable part of Russian Motor Park is operated, affect vehicles negatively. Cold weather causes higher fuel consumption and CO2 emissions always. It is because of temperature profile changing of automobile motors, other systems and materials. For enhancement of car operation efficiency in severe winter environment the dependency of engine warm-up and cooling time on ambient air temperature and wind speed described by multifactorial mathematical models is established. -On the basis of experimental research it was proved that the coolant temperature constitutes the engine representative temperature and may be used as representative temperature of engine at large. The model of generation of integrated index for vehicle adaptability to winter operating conditions by temperature profile of engines was developed. the method for evaluation of vehicle adaptability to winter operating conditions by temperature profile of engines allows to decrease higher fuel consumption in cold climate.

1. Introduction

Motor vehicles play an important role in shaping our natural environment. Motor vehicles collectively contribute to a number of important and pressing ecology problems. Vehicles are a major source of air pollution, climate change, and urban sprawl.

Cold weather causes higher fuel consumption and CO2 emissions from the nature of the things [1, 2, 3, 4, 5, 6, 7, 8, 9, 10, 11]. It is because of temperature profile changing of automobile motors, other systems and materials [1, 3, 6, 7, 8, 9, 10].

The major part of Russia territory resides in the boreal, rigorous and cruel cold climate zones, accordingly the severe winter environment affecting the negative influence on performance indicators of rolling stock are typical for exploitation of motor vehicle fleet of the Russian Federation. The winter conditions primarily characterized by low temperature of ambient air which impact is compounded by the simultaneous wind exposure [1, 6].

Under the influence of winter environment the temperature profile of automobile motors varies essentially that can be attributed to rise of unit surface conductance to environment. In this regard, the warm-up time after start rises sharply and the shutdown cooling period thereof declines [6].

It has been established that the engines of vehicles of various brands and models under identically severe service conditions have different rate of heat exchange, in other words marked by different values of rates of heating and cooling. This can be attributed to different automobile suitability to winter environment that is poorly considered in service thereof and leads to decline in efficiency of motor transport under these conditions [1, 3, 6].

The objective appraisal of automobiles suitability by temperature profile of motors is necessary for decreasing of negative influence on environment by motor-vehicle operation in wintertime.

Researchers associate the higher fuel consumption in winter conditions with deterioration of motors temperature profile [1, 3, 6, 7, 8, 9, 10]. The majority of researchers consider the ambient temperature and wind speed as critical parameters of ambient medium having an impact on thermal state of motors. The basic parameters characterizing the motor temperature profile are steady-state temperature as well as heating and cooling time. The impact of winter environment to period of heating and cooling of engine of different vehicles has a differential nature that is conditioned by various levels of adaptability of these vehicles to severe conditions. The presence of quantitative index of adaptability thereof by temperature profile of engines is necessary for obtaining of possibility for adequate evaluation of vehicle adaptability level.

2. Methodology

2.1. The models of change of engine cooling and heating time

The air speed blowing the car, degree of heat insulation and thermophysical properties of engine determine the rate of engine heat exchange and consequently affect the period of its heating and cooling [6].

Engine cooling period $\tau_{охл}$ after stopping down is determined by thermophysical properties of unit and conditions o1f cooling process behavior on its surface. However the theoretical determination of engine cooling time is hindered by determination of dependence for irregularity ration of temperature distribution in engine from the cooling conditions on its surface. The differential equation of thermal conductivity for ball may be applied for determination of type of dependence of irregularity ratio of temperature distribution ψ. Solution of this equation allows determining the dependence of irregularity ratio of temperature distribution on *Bi* number and provide this dependence in the form of adequate model (1):

$$Bi \cdot \psi = b'_1 \cdot \exp\left(\frac{b'_2}{Bi}\right), \tag{1}$$

where Bi – is a *Biot* number,

b'_1, b'_2 – is a model parameters depending on engine (ball) properties.

Taking into account the formula (1) and that for each engine the heating capacity *C*, mass *M*, thermal conductivity λ, determining the amount of l and surface area *F* are constant – the rate of cooling of individual engine depends only on heat-transfer coefficient α. The model of change of engine cooling time τ_{cool} from initial $t_н$ to final temperature $t_к$ under the influence of ambient temperature $t_в$, wind speed w and thermal properties of engine is elaborated (2):

$$\tau_{охл} = \frac{1}{b_1}\exp\left(-\frac{b_2}{\alpha}\right) \cdot \ln\frac{t_H - t_в}{t_K - t_в}, \tag{2}$$

where b_1 – is a model parameter, c^{-1};

b_2 – is a model parameter, W/m^2·°C.

Based on analysis of heat-balance equation the problem of theoretical determination of engine warm-up time $\tau_{пр}$, depending on capacity of internal heat source *P*, ambient temperature, rate of cooling m_{cool} and thermal properties of engine was solved and takes the form (3):

$$\tau_{пр} = \frac{730 + 0.5 \cdot (t_K - t_H)}{d_1 \cdot P - d_2 \cdot m_{охл}} \cdot \ln\left(\frac{t_y - t_K}{t_y - t_H}\right), \tag{3}$$

where d_1 – is a constant of model depending on engine thermal properties, °C/J;

d_2 – is a constant of model, °C.

The change of engine temperature at a time in the process of warm-up and cooling is described by co-exponential models of adaptability (4) and (5):

$$t_K = t_y - (t_y - t_H) \cdot e^{-f(M, m_{охл})}, \quad (4)$$

$$t_K = t_в - (t_в - t_H) \cdot e^{-f(a, b, w, \tau_{охл})}. \quad (5)$$

2.2. The characteristics of car adaptability to severe operation conditions

The adaptability coefficient reflecting the change of time for warm-up and cooling under these operating conditions towards its values under standard conditions is applied for characteristics of car adaptability to severe operation conditions. It has been established that the car adaptability coefficients to winter working conditions by engine warm-up time from initial temperature $t_н$ to final $t_к$ is determined from the formula (6):

$$K_{пр} = \frac{\tau_{пр}^{ст}}{\tau_{пр}}, \quad (6)$$

And by the cooling time from the formula (7):

$$K_{охл} = \frac{\tau_{охл}}{\tau_{охл}^{ст}}, \quad (7)$$

where $\tau_{пр}^{ст}, \tau_{охл}^{ст}$ – is a time of engine warm-up and cooling from temperature $t_н$ to temperature $t_к$ under standard conditions, *c*;

$\tau_{пр}, \tau_{охл}$ – is an actual time of engine warm-up and cooling from temperature $t_н$ to temperature $t_к$, *c*.

Herewith the conditions under which the technically sound vehicle realizes the rating values of own performance index are considered as *standard.*

The applicability of two criterial model of aggregated factor of vehicle adaptability to winter service conditions by the temperature profile of engines which may be called as "adaptability parameter" *a* is substantiated.

It is proposed to determine the parameter of vehicle adaptability using the following formula:

$$a = \frac{2 \cdot K_{пр} \cdot K_{охл}}{K_{пр} + K_{охл}}. \quad (8)$$

The adaptability parameter varies from 0 to 1. The case $a = 1$ takes place under complete vehicle adaptability to winter service conditions, namely when the warm-up at idle speed and cooling with engine out is performed during the same period that under standard conditions $\tau_{пр} \rightarrow \tau_{пр}^{ст}$, а $\tau_{охл} \rightarrow \tau_{охл}^{ст}$. The car is absolutely poorly adapted for service to actual operating conditions under $a = 0$, if the car engine is not warmed up at all or its warm-up is performed during long time $\tau_{пр} \rightarrow \infty$, and cooling time is precious little $\tau_{охл} \rightarrow 0$. The value of *a* parameter is also tends to zero if one of the adaptability coefficients included thereto tends toward zero.

2.3. The assessment of cars operating conditions severity

For the purposes of assessment of cars operating conditions severity, for ensuring the comparability of various severe factors of operating conditions the universal severity scale *R* was applied. The value of *H* severity index varies from *0* to the largest value $H_{max} = 12R$, corresponding to absence of severity and maximum possible deviation of operating conditions factor from the standard value.

For ease of application in practice the full range of ambient air temperature values is divided into six equal intervals and the range of wind speed values in four intervals: temperate, moderately-severe, sever and highly severe.

For convenience of application in practice the range of car adaptability indicators values may be split into three levels – with low, medium and high adaptability. The specified breakdown may be performed using the standard deviation value σ subject to conformity with allocation of the car adaptability coefficients values to normal law. $\overline{a} \pm 2\sigma$ is accepted as limits for indicators values range, corresponding to confidence coefficient 0.95 (Table 1).

Table 1. Characteristics of car adaptability levels to winter operating conditions by temperature profile of engine

Adaptability level	Range of adaptability coefficient values	Mean value of adaptability coefficient
Low	$(\overline{a} - 2\sigma; \overline{a} - 0.5\sigma]$	$\overline{a} - 1.25\sigma$
Medium	$(\overline{a} - 0.5\sigma; \overline{a} + 0.5\sigma]$	$\overline{a}$
High	$(\overline{a} + 0.5\sigma; \overline{a} + 2\sigma]$	$\overline{a} + 1.25\sigma$

For conducting of experimental study the approved methods and up-to-date measuring units have been applied, including the multimeter MAS-838 and α-calorimeter with differential thermocouple.

The calibration of engine coolant temperature sensor and calibration of thermocouple have been performed in response to carrying out of preliminary experiments, as well as overall heat retention of thermocouple and mutimeter were determined.

3. The results of applied research

3.1. Methodology of applied research

The Russian-manufactured vehicles Ural 4320, Ural 5557 (YEP 236), Ural 4420 (YEP 238), КрАЗ 6444, КрАЗ 255 Б1, ZiL 131, ZiL 130, GAZ 53, GAZСАЗ 3507, GAZ 2705, GAZ 31029, UAZ 31512, VAZ 2106, VAZ 21213, KAvZ 3976 with various degree of engine winterizing have served as a subject of experimental study. The range of ambient temperature study is from 20 above zero to 42 °C, below zero and wind speed –0 to 13 m/s.

For determination of engine temperature measurability by temperature gage value of cooling fluid the experimental data (as exemplified by engine of ZMZ-402) were obtained, which allowed to draw deduction on availability of functional (non-stochastic) relation between non-random variables – value of mean engine surface temperature and its cooling fluid. Therefore the engine coolant temperature is deterministically predicated upon engine medium temperature and may act as a representative temperature for engine generally.

The multimeter was connected to engine temperature sensor for determination of period for engine warm-up after care parking in the open air and whereat the engine starting was performed. The stop watch was switched upon reaching the coolant temperature of 20 °C. The warm-up was performed up to the temperature of 60 °C.

During conducting of experiment for determination of engine cooling time the engine was shut down after continuous operation providing the equal warmup of all minor components. During temperature depression up to 80 – 90 °C the stop watch was stopped. The cooling was performed up to the temperature of 40 °C. The statistical analysis of experimental values of integrated index has allowed to find that the probability of obtained distribution conformance to normal low equals 0.95.

The normal law of distribution was confirmed as a result of statistical analysis of empirical values of adaptability parameters of the vehicles under investigation (Figure 1).

3.2. The allocation of vehicles by adaptability levels

For ease of application in practice the whole range of the vehicles adaptability values parameters was split to three levels – with low, medium and high adaptability, at that $\overline{a} = 0.479$, $\sigma = 0.109$. The magnitude of adaptability levels difference was established based on statistical check. The allocation of vehicles by adaptability levels is presented in the 'figure 1' and Table 2.

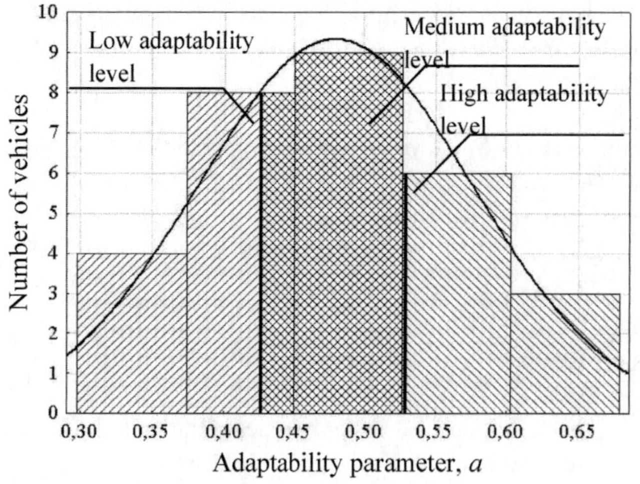

Figure 1. Histogram, distribution curve and vehicles adaptability levels to winter conditions by temperature profile of engines

Table 2. Vehicles allocation by levels of adaptability to winter operating conditions by temperature profile of engine

Level of vehicle adaptability		
Low	Medium	High
GAZ 2705	VAZ 2106	VAZ 2106 [c]
GAZ 53	VAZ 21213	VAZ 21213 [c]
ZiL 130	GAZ 31029	GAZ 2705 [c]
ZiL 131	GAZ 3507	GAZ 31029 [c]
KrAZ 255 Б1 [a]	GAZ 53 [c]	GAZ 3507 [c]
KrAZ 6444 [a]	ZiL 130 [c]	KAvZ 3976 [c]
UAZ 31512	ZiL 131 [c]	KrAZ 6444 [ac]
Ural 4320	KrAZ 255Б1 [ac]	UAZ 31512 [c]
Ural 4420 [a]	Ural 4320 [ac]	Ural 4420 [ac]
Ural 5557 [b]	Ural 5557 [bc]	

a – installed with YEP 238 engine
b – installed with YEP 236 engine
c – engine insulant applied

Numerical values of integrated index of the vehicle adaptability with high, medium and low adaptability are presented in the Table 3.

Table 3. Numerical values of integrated adaptability index

Level of vehicle adaptability	Severity interval H_V by wind speed, R	Value of integrated index in the severity interval H_t by ambient temperature, R					
		[0 ... 2)	[2 ... 4)	[4 ... 6)	[6 ... 8)	[8 ... 10)	[10 ... 12]
High	[0 ... 3)	0.79	0.71	0.64	0.56	0.48	0.40
	[3 ... 6)	0.75	0.67	0.59	0.51	0.43	0.35
	[6 ... 9)	0.70	0.62	0.54	0.47	0.39	0.31
	[9 ... 12]	0.66	0.58	0.50	0.42	0.34	0.26
Medium	[0 ... 3)	0.64	0.58	0.52	0.46	0.40	0.34
	[3 ... 6)	0.61	0.55	0.49	0.42	0.36	0.30

Level of vehicle adaptability	Severity interval H_V by wind speed, R	Value of integrated index in the severity interval H_t by ambient temperature, R					
		[0 ... 2)	[2 ... 4)	[4 ... 6)	[6 ... 8)	[8 ... 10)	[10 ... 12]
	[6 ... 9)	0.57	0.51	0.45	0.39	0.33	0.27
	[9 ... 12]	0.54	0.47	0.41	0.35	0.29	0.23
Low	[0 ... 3)	0.55	0.49	0.43	0.37	0.31	0.25
	[3 ... 6)	0.48	0.42	0.36	0.30	0.24	0.18
	[6 ... 9)	0.41	0.35	0.29	0.23	0.17	0.11
	[9 ... 12]	0.34	0.28	0.22	0.16	0.10	0.04

The analysis of values of integrated index for vehicles adaptability with different adaptability levels represented in the Table 3, suggests that the higher the adaptability level and lower operating conditions are then the value of adaptability integrated index is closer to one.

3.3. Definition of numerical values of models parameters

The models of cooling time change after engine stoppage and its warm-up at idle speed from the ambient air temperature, wind speed and availability of engine winterization for various car brands and models have been confirmed at the second stage of experimental research. The numerical values of (2) and (3) models parameters are presented in the Table 4.

Table 4. Numerical values of (2) and (3) models parameters

Level of vehicle adaptability	Range of models parameters value variation			
	$b_1 \cdot 10^5$, s^{-1}	b_2, $m^2 \cdot °C/W$	$d_1 \cdot 10^7$, $(°C)/J$	$d_2 \cdot 10^{-2}$
High	21.7 … 36.7	0.8 … 4.2	5.8 … 93.0	1.19 … 3.74
Medium	21.9 … 56.4	1.6 … 5.3	5.9 … 111.5	1.49 … 3.57
Low	31.4 … 48.0	1.5 … 3.8	6.0 … 57.5	1.29 … 2.65

For determination of the engine warm-up time at idle speed and engine cooling time after stoppage dependence from ambient air temperature, wind speed and availability of engine winterization for various car brands and models the correlation-regression analysis was conducted. The model verification was performed by Fischer ratio test F and average approximation error $\overline{\varepsilon}$. The calculations of specified statistical characteristics have testified that the values of Fischer variance ratio are bigger than tabulated values and the average approximation error fall within the range of 1.4…11.2%, that is evidenced by the model adequacy. The experimental results in graphic form by the example of GAZ 2705 are presented in the Figures 2 – 5.

3.4. The model of generation of adaptability integrated index

Two-factor model of generation of adaptability integrated index was obtained at the subsequent stage:

$$a = 1 - a_1 \cdot h_t - a_2 \cdot h_t \cdot h_w^{0,73} + a_3 \cdot h_t^2 \cdot h_w^{0,73}, \qquad (9)$$

where h_t – is severity of operating conditions by ambient air temperature, R;
h_w – is severity of operating conditions by wind speed, R;
a_1 – is a parameter reflecting the degree of ambient air temperature impact, °C-1;
a_2, a_3 – are the parameters reflecting the degree of wind speed and ambient air temperature impact.

Numerical values of model parameters are presented in the Table 5.

Table 5. Numerical values of model (9) parameters

Level of vehicle adaptability	Range of models parameters value variation		
	$\eta_1 \cdot 10^{-2}$	$\eta_2 \cdot 10^{-2}$	$a_3 \cdot 10^{-3}$
High	3.63 ..... 8.78	1.65 ... 5.28	2.63 ..... 8.02
Medium	4.03 ... 13.33	2.93 ... 6.05	3.60 ... 11.38
Low	7.06 ... 15.02	3.60 ... 7.36	5.82 ... 12.14

According to the findings of investigation the higher severity of operating conditions and lower the level of vehicle adaptability, the more time it needs for engine warm-up at idle speed and the less duration of cooling thereof after shutdown.

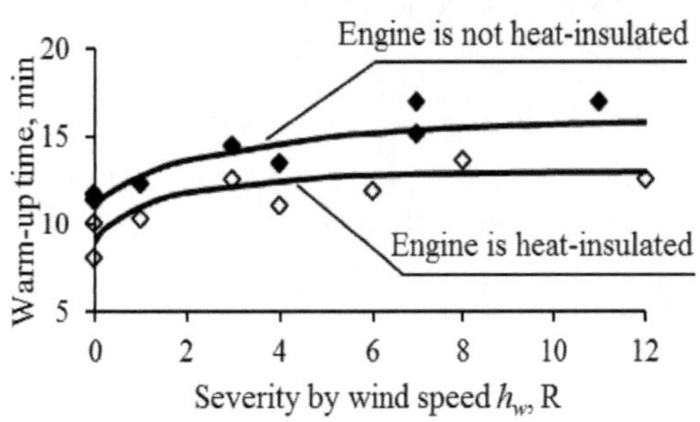

Figure 2. Dependence of time for GAZ 32705 engine warm-up on wind speed under ambient air temperature of minus 30 °C

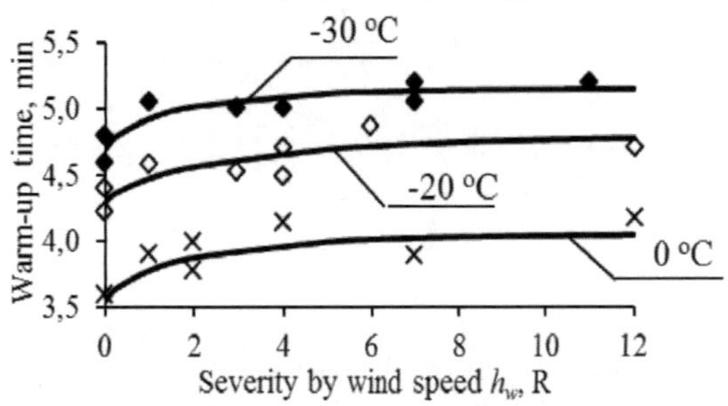

Figure 3. Dependence of time for GAZ 32705 engine warm-up (20 to 60 °C) on air temperature under various wind speed (without heat-insulation of engine)

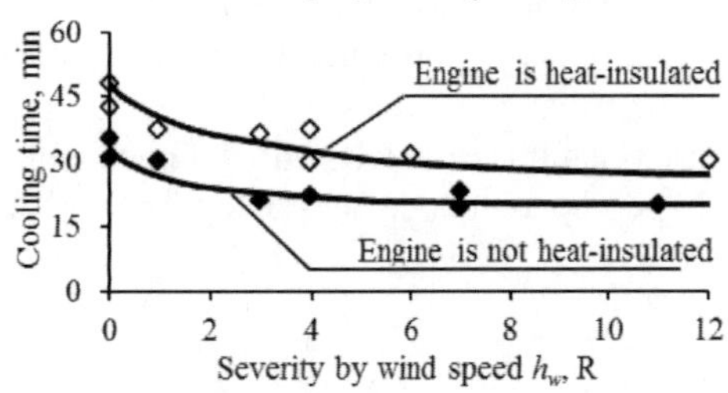

Figure 4. Dependence of time for GAZ 32705 engine cooing (90 to 40 °C) on wind speed at ambient temperature 30 °C below zero

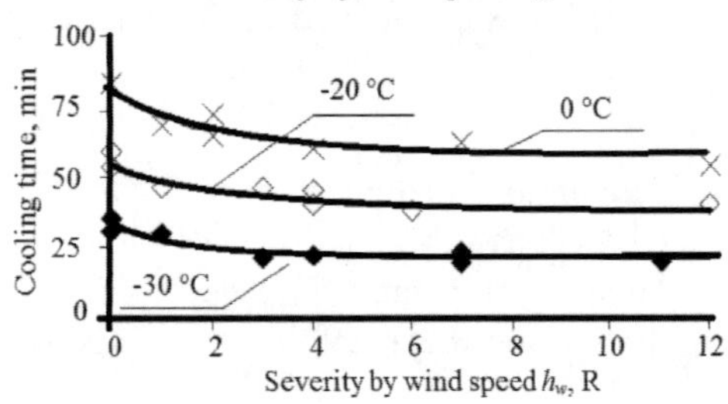

Figure 5. Dependence of time for GAZ 32705 engine cooing from 90 to 40 °C on ambient air temperature under various wind speeds (without heat-insulation of engine)

4. The directions of practical implementation

Numerical values of models parameters established as a result of researches enable developing the following principal directions of practical implementation:

1. Determination of engine cooling under designed conditions of operating conditions severity and level of car adaptability.
2. Determination of rational warm-up period within specified severity conditions under known level of vehicle adaptability.

It is proposed to implement the developed methods for engine temperature maintenance in the range of 40 to 80 °C using the vehicle as standby and in the range of 10 to 60 °C in case of intershift storage of vehicle in isolation from production plant.

The recommendations to operators and industry for improvement of fuel efficiency of vehicles in winter operating conditions are enunciated.

5. Conclusions

In such a way as a result of conducted researches

- The dependency of engine warm-up and cooling time on ambient air temperature and wind speed described by multifactorial mathematical models is established. Numerical values of parameters of established models for vehicles under investigation are defined and adequacy thereof was proved on the basis of experimental research;
- The model of generation of integrated index for vehicle adaptability to winter operating conditions by temperature profile of engines was developed;
- On the basis of experimental data it is found that distribution of integrated index values for adaptability to winter environment by temperature profile of different cars engines is governed by normal law of distribution;
- It is determined that the conformity of the change in integrated index of car adaptability to winter operating conditions by temperature profile of engines is described by the multifactor model, the numerical values of parameters included thereto are also established;
- The method for evaluation of vehicle adaptability to winter operating conditions by temperature profile of engines based on mathematical model of change in engine warm-up and cooling time and implemented on designed program-methodical provision is developed;
- Ecological effect is achieved by means of decrease in fuel consumption that leads to reduction in environmental pollution with toxic agents contained in exhausted gas.

References

[1] Anisimov I, Ivanov A, Chikishev E, Chainikov D and Reznik L 2014 Assessment of gas cylinder vehicles adaptability for operation at low ambient temperature conditions. *Energy production and management in the 21st Century.* 685-96

[2] Alvarez R and Weilenmann M 2012 Effect of low ambient temperature on fuel consumption and pollutant and CO2 emissions of hybrid electric vehicles in real-world conditions *Fuel* **97** 119-24.

[3] Clairotte M, Adam T, Zardini A, Manfredi U, Martini G, Krasenbrink A and Astorga C 2013 Effects of low temperature on the cold start gaseous emissions from light duty vehicles fuelled by ethanol-blended gasoline *Applied Energy* **102** 44-54.

[4] Duarte G, Gonçalves G and Farias T 2016 Analysis of fuel consumption and pollutant emissions of regulated and alternative driving cycles based on real-world measurements. *Transportation Research* D *Transport and Environment* **44** 43-54.

[5] Durbin T, Johnson K, Miller J, Maldonado H and Chernich D 2008 Emissions from heavy-duty vehicles under actual on-road driving conditions *Atmospheric Environment* **42(20)** 4812-21.

[6] Ertman J, Ertman S, Anisimov I, Chainikov D and Chikishev Y 2015 Complex characteristic estimation of vehicle adaptability for low temperature operation conditions *Research Journal of Pharmaceutical, Biological and Chemical Sciences* **6(1)** 1761-70

[7] Irimescu, A. 2010 Study of cold start air–fuel mixture parameters for spark ignition engines fueled with gasoline–isobutanol blends *International Communications in Heat and Mass Transfer* **37(9)** 1203-07.

[8] Martins A, Rocha R and Sodré J 2014 Cold start and full cycle emissions from a flexible fuel vehicle operating with natural gas, ethanol and gasoline *Journal of Natural Gas Science and Engineering* **17** 94-8.
[9] Reiter M and Kockelman K 2016 The problem of cold starts: A closer look at mobile source emissions levels. *Transportation Research D Transport and Environment* **43** 123-32.
[10] Sales L and Sodré J 2012 Cold start emissions of an ethanol-fuelled engine with heated intake air and fuel *Fuel* **95** 122-5.
[11] Shancita I, Masjuki H, Kalam M, Fattah I, Rashed M and Rashedul H 2014 A review on idling reduction strategies to improve fuel economy and reduce exhaust emissions of transport vehicles *Energy Conversion and Management* **88** 794-807.

IOP Conf. Series: Materials Science and Engineering **142** (2016) 012120 doi:10.1088/1757-899X/142/1/012120

Study of Falling Roof Vibrations in a Production Face at Roof Support Resistance in the Form of Concentrated Force

G D Buyalich[1,2,3,a], **K G Buyalich**[1,b], **V Yu Umrikhina**[1,c]

[1] T.F. Gorbachev Kuzbass State Technical University, Kemerovo, Russia
[2] Yurga Institute of Technology of Tomsk Polytechnic University, Yurga, Russia
[3] Institute of Coal of SB RAS, Kemerovo, Russia

E-mail: [a]gdb@kuzstu.ru, [b]konstantin42@mail.ru, [c]Umrevgen @yandex.ru

Abstract. One of the main reasons of roof support failures in production faces is mismatch of their parameters and parameters of dynamic impact on the metal structure from the falling roof during its secondary convergences. To assess the parameters of vibrational interaction of roof support with the roof, it was suggested to use computational models of forces application and a partial differential equation of fourth order describing this process, its numerical solution allowed to assess frequency, amplitude and speed of roof strata movement depending on physical and mechanical properties of the roof strata as well as on load bearing and geometry parameters of the roof support. To simplify solving of the differential equation, roof support response was taken as the concentrated force.

1. Introduction

Operation of powered roof supports of longwall sets of equipment in conditions of hard-to-cave and hard-to-manage roof strata, presented especially by thick fine-grain sandstones and siltstones, is accompanied by increased dynamic loads occurring during regular secondary roof falls. At the same time, the magnitude and the rate of main roof blocks movement depend on physical and mechanical properties of the strata, their thickness, load-bearing features of the roof support and its location in the face.

Such impact of the roof on support elements causes their failure due to improper manufacturing technology [1–3], swelling of cylinders on hydraulic props both due to their design features [4–6] and high rate of the movement process development [7–9], leading inevitably to improper operation of seals and loss of their tightness [10, 11]. The magnitude and intensity of secondary roof convergences depend on roof support resistance and its distribution over the working area width [12].

Also, vibration frequency of a falling rock block may coincide with the frequency of own vibrations of roof support steel structure, and as a result resonance evolves which may lead to its destruction with the external loads not exceeding the rated operating resistance.

Therefore, it is important to know the frequency of external load on the roof support during secondary roof convergence as well as the maximal amplitude and rate of rock strata movement during its impact on the support with the aim to choose the required parameters of support design at the stage of its designing.

2. Methods

Solution of a partial differential equation describing the beam oscillating process was obtained by a numerical computation at given initial and boundary conditions corresponding to the agreed computational models before and after brittle rock failure.

This method is widely applied when using advanced highly efficient computing systems.

3. Work Description

To study the oscillating process of a roof block at a sudden (brittle) rock collapse, the following hypothesis was assumed as the basis for the schemes of roof block interaction with external environment.

In the process of coal seam cutting a roof block is getting deformed under its own weight, under uniformly distributed load from overlying strata (additional load) and from the concentrated force which is numerically equal to the rated operating resistance of the roof support. This concentrated force is located at a distance from the face equal to the distance from the face to the resultant force from the roof support.

One end of the beam above the face is restrained, the other end on the goaf side is hanging free. This scheme is shown in fig. 1. Maximal beam movements for the given scheme take place at the free end, while maximal stresses are in the restraint.

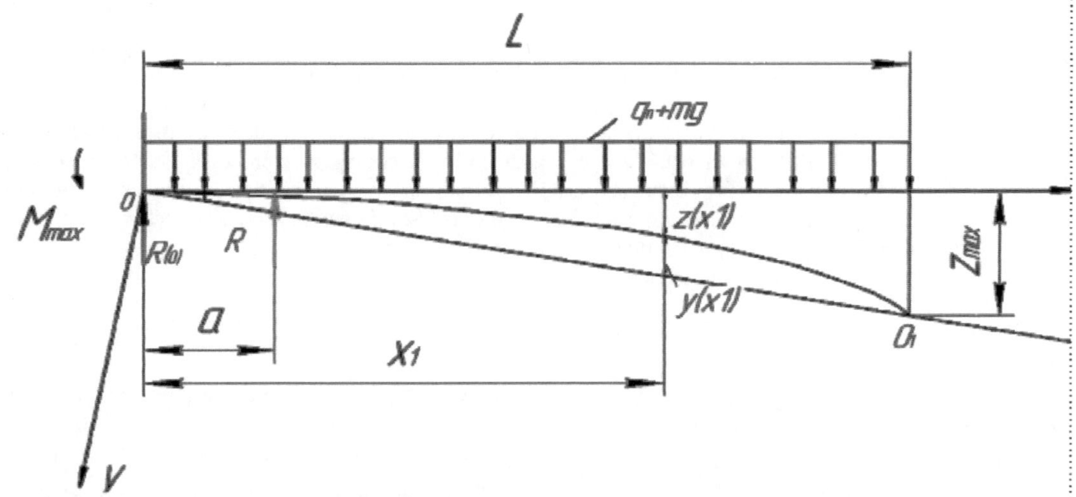

Figure 1. Computational module for loading of (roof) block prior to its brittle destruction.

When admissible stresses are reached in the beam restraint, brittle destruction of roof strata takes place (a secondary roof convergence occurs), and the restraint breaks at once. As a result, the beam tends to straighten and the stored potential energy of the deformed beam converts in the kinetic energy of the oscillating process.

The kinetic energy magnitude is pro rata to geometrical dimensions of the block, the mass of its strata, the size of deformations prior to destruction, and the magnitude of applied forces.

The model for computation after failure of the restraint is shown in fig. 2 and it becomes two-point, since upon straightening of the bent beam it will rest on the overlying roof strata.

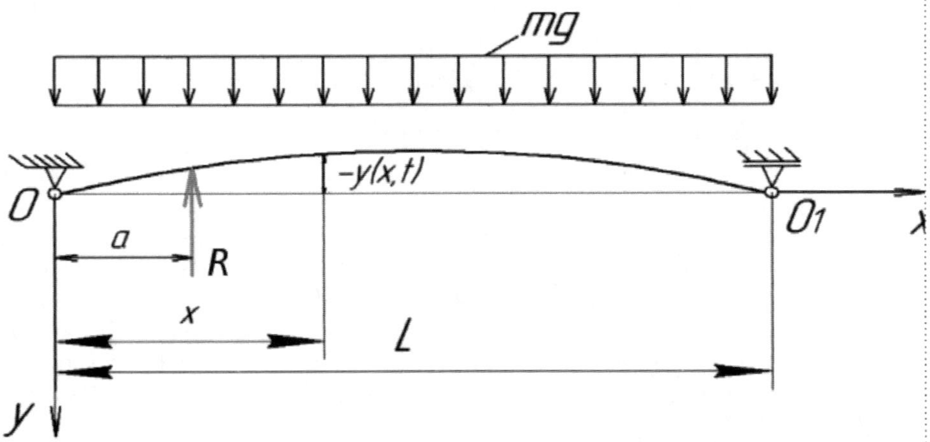

Figure 2. Computational module of the roof block after its brittle destruction.

The schemes show uniformly distributed additional load from the overlying strata side and deflections y corresponding to the deformed block at the moment preceding brittle destruction. The concentrated force corresponding to the rated resistance of the roof support acts on the block from the roof support side.

Legend (fig. 1 and 2):

R – roof support force in the form of a concentrated force equal to the rated operating resistance; a – distance from the face to the total force from forces acting from the support side (concentrated force); qп – magnitude of additional load from the overlying strata side; mg – magnitude of additional load from the blocks's own weight; g - free fall acceleration; m - distributed mass of the roof block; Mmax – maximal clamped end bending moment at the moment preceding brittle destruction of roof strata; R(0) – response in the clamped end at the moment preceding brittle destruction of roof strata; L – length of roof block; zmax – maximal bending of the roof block at the moment preceding brittle destruction of roof strata in z-x1 coordinates; x1 – axis direction for determination of roof block sagging at the moment preceding brittle destruction of roof strata; y and x – direction of axes for determination of parameters of roof block vibrations at the moment after brittle destruction of roof strata.

Vibrations of the roof block after brittle destruction of strata can be described by the non-uniform differential 4th order partial equation

$$\frac{\partial^2 y}{\partial t^2} + \frac{E_0 J}{m} \cdot \frac{\partial^4 y}{\partial x^4} = g,$$

where $\partial^2 y/\partial t^2$ – the second derivative of the roof block deflection in time; J – moment of inertia of the block cross-section; E_0 – E- modulus of roof strata; $\partial^4 y/\partial x^4$ – the fourth derivative of the roof block deflection along its length.

The equation of beam deflection at the moment preceding brittle destruction of roof strata (fig. 1) reduced to x – y coordinate system and used in review of oscillating process after brittle destruction is the initial condition for numerical solution of the given differential equation

$$y(x,0) = \begin{cases} \left(\frac{x}{24E_0 J}\right) \cdot [(q_п + mg)(6L^2x - 4Lx^2 + x^3 - 3L^3) + 2Rx(x - 3a)], \text{ при } x < a, \\ \left(\frac{x}{24E_0 J}\right) \cdot [(q_п + mg)(6L^2x - 4Lx^2 + x^3 - 3L^3)], \text{ при } x \geq a. \end{cases}$$

In accordance with fig. 2, the equal-zero deflection and deflection moments in rocker bearings of the beam (roof block) will be boundary conditions in solving of the induced differential equation

$$y(0,t) = 0,$$

$$y(L,t) = 0,$$
$$\frac{\partial^2 y(0,t)}{\partial x^2} = 0,$$
$$\frac{\partial^2 y(L,t)}{\partial x^2} = 0$$

The roof support response taken in the form of a concentrated force facilitates solving of the given differential equation describing the oscillating process of the roof falling and allows to assess at a first approximation the parameters of block vibrations (frequency, amplitude and rate of displacement) during secondary convergences.

As a result of numerical solution of the given differential equation, there were received dependencies of amplitudes and forms of the roof block vibrations after its brittle destruction in relation to physical and mechanical properties of rock, rock parameters and roof support parameters (fig. 3–5) in which one can see that maximal value of the block vibrations amplitude is located in the point of application of roof support force R.

Also, an increase in the rated operating resistance R of roof support and in the distance from the face to the place of location of resultant response a of the support causes an increase in the amplitude and rate of roof displacement. While an increase in the specific weight of roof strata decreases the amplitude and the rate of roof displacement due to increase in inertial forces.

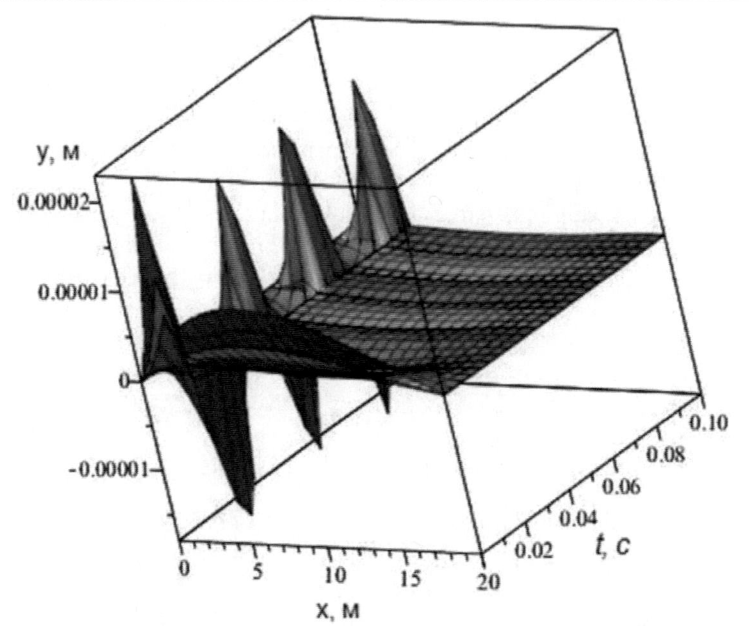

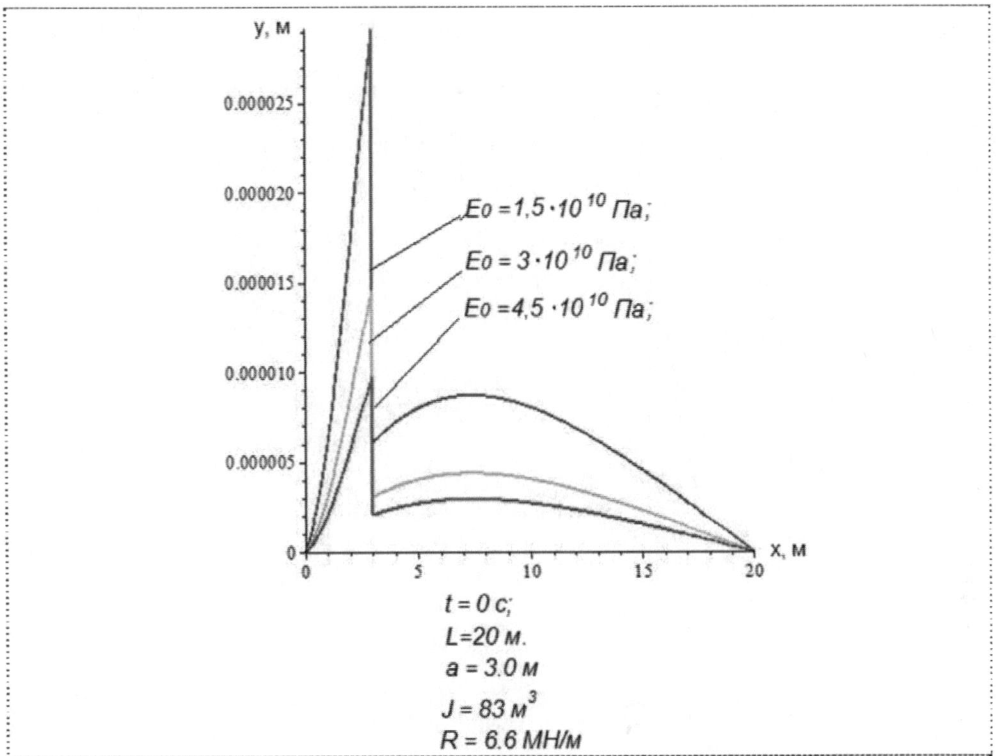

Figure 3. Change in roof vibration amplitude (y) in time (t) and in length of the block (x) depending on elasticity modulus of roof strata (E_0) at the rated operating resistance of roof support R = 6.6 MN/m, at the moment of inertia moment of the block cross-section J = 83 m^3, length of the block L = 20 m and distance from the face to roof support response a = 3.0 m.

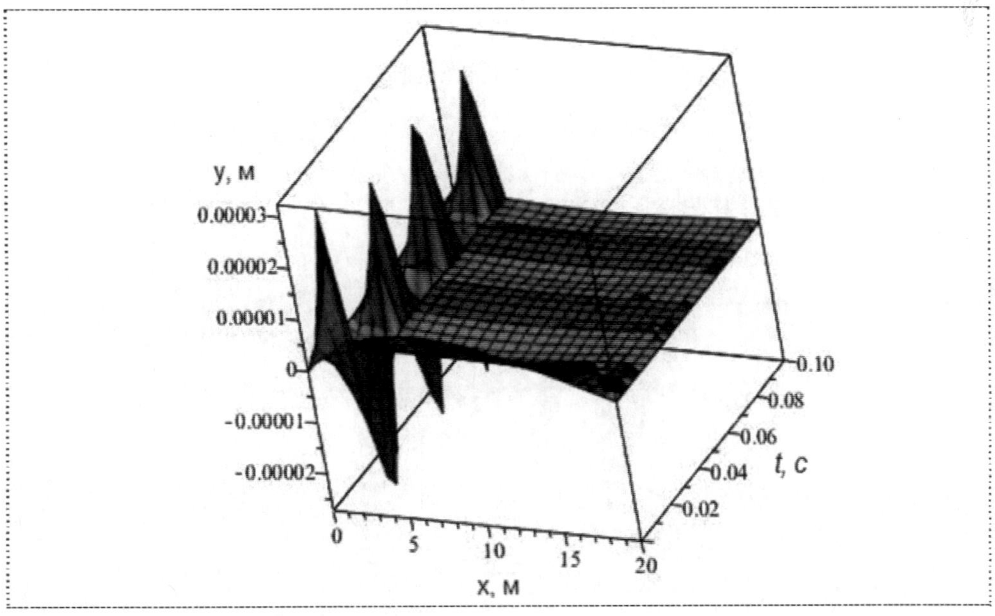

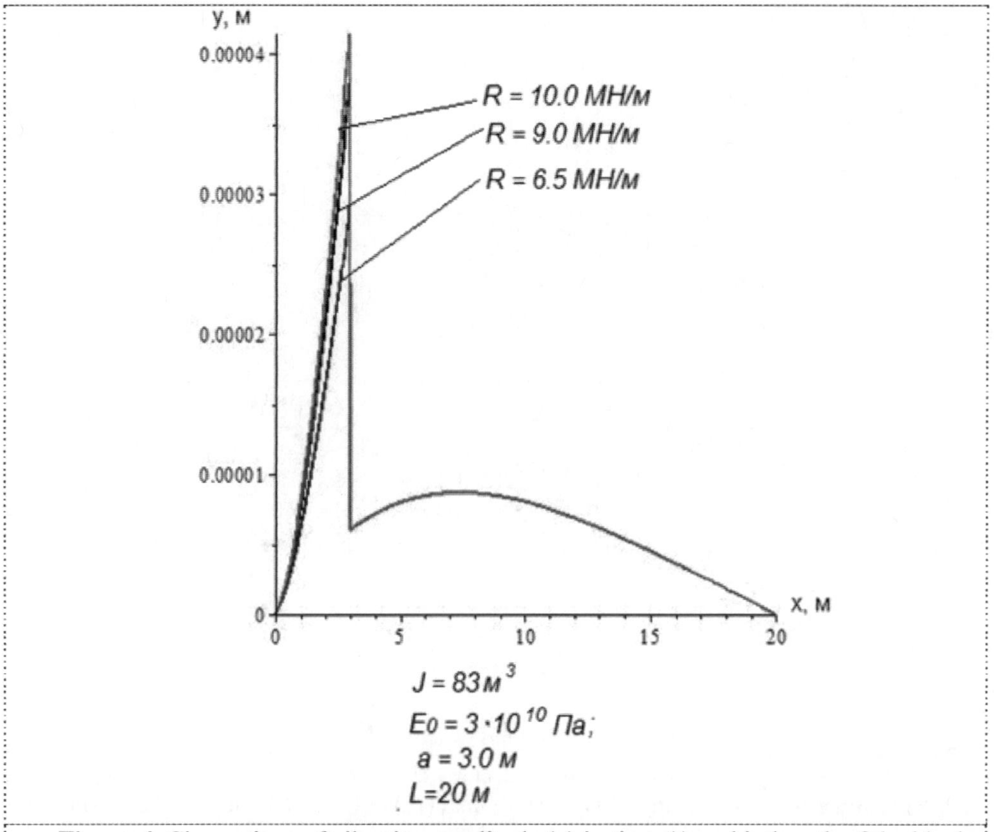

Figure 4. Change in roof vibration amplitude (y) in time (t) and in length of the block (x) depending on the rated operating resistance of roof support R at elasticity modulus of roof strata $E_0 = 3 \cdot 10^{10}$ Pa, inertia moment of the panel cross-section J = 83 m^3, length of the panel L = 20 m and distance from the face to roof support response a = 3.0 m.

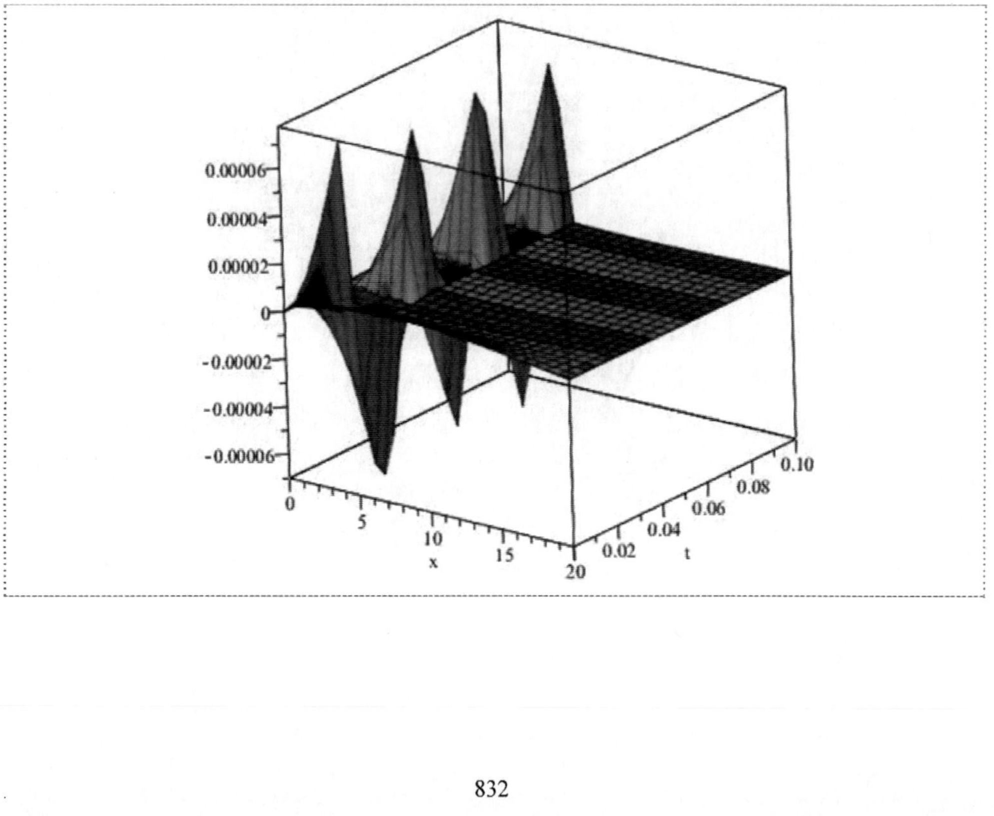

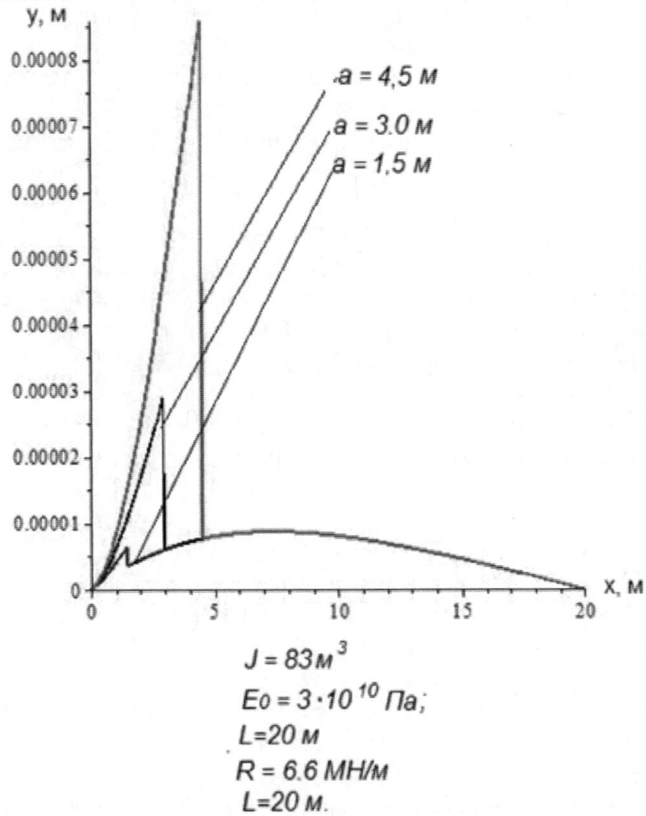

Figure 5. Change in roof vibration amplitude (y) in time (t) and in length of the block (x) depending on the distance from the face to the roof support response a at the rated operating resistance of the roof support R = 6.6 MN/m, elasticity modulus of roof strata $E_0 = 3{\cdot}10^{10}$ Pa, inertia moment of the block cross-section J = 83 m^3, length of the block L = 20 m.

4. Conclusions

The parameters of roof block vibrations during secondary roof convergences are defined by physical and mechanical properties of strata (specific weight m, elasticity modulus E_0), roof block dimensions (length L, thickness h) as well as by force (rated operating resistance R) and geometrical parameters of the roof support (distance from the face to the point of application of the resulting response from support a).

The values of roof vibration parameters for the roof support response in the form of the concentrated force vary within the following range:

- Maximal amplitude in the point of application of roof support response A = 0.00001–0.00008 m;
- Frequency of vibrations f = 25–100 Hz;
- Rate of roof displacement in the place of location of roof support response V = 0.002–0.008 m/sec.

References

[1] Chinakhov D A 2011 Study of Thermal Cycle and Cooling Rate of Steel 30ХГСА Single-Pass Weld Joints J.*Applied Mechanics and Materials* **Vol 52–54** pp 442–447
[2] Chinakhov D A, Vorobyov A V, Davydov A A and Tomchik A A 2012 Simulation of active shielding gas impact on heat distribution in the weld zone of consumable electrode welding

In the Proceedings of the 7th International Forum on Strategic Technology IFOST2012, Tomsk Polytechnic University, **Vol II**, pp 136–138

[3] Chinakhov D A 2013 Simulation of Active Shielding Gas Impact on Heat Distribution in the Weld Zone J. *Applied Mechanics and Materials* **Vol 762** pp 717–721

[4] Buyalich G.D., Anuchin A.V., Dronov A.A., 2015. The Numerical Analysis of Accuracy of Hydraulic Leg Cylinder in Modeling Using Solid Works Simulation, *Applied Mechanics and Materials Vol 770* (2015) pp 456–460pp. 167–170. DOI:10.4028/www.scientific.net/AMM.770.456.

[5] Burkov, P.V., A.V. Vorobiev and A.V. Anuchin, 2011. Analysis of Stress Concentrators and Improvement in Designs of Hydraulic Legs, *Mining Informational and analytical Bulletin (scientific and technical journal).* **Separate issue # 2**, pp: 172–183.

[6] Buyalich, G.D., Buyalich K.G., Voyevodin V.V., 2015. Radial deformations of working cylinder of hydraulic Legs depending on their extension, *IOP Conf. Series: Materials Science and Engineering*. **Vol. 91**. 012087, DOI:10.1088/1757-899X/91/1/012087.

[7] Klishin V I and Tarasik T M 2001 Stand tests of hygraulic supports with respect to dynamic loads *Journal of Mining Science* **Vol 37 # 1** pp: 77–84

[8] Klishin V I 1995 Inertial means of protecting hydraulic props from dynamic loads *Journal of Mining Science* **Vol 30 # 4** pp: 390–394

[9] Klishin S V, Klishin V I and Opruk GYu 2013 Modeling coal discharge in mechanized steep and thick coal mining *Journal of Mining Science* **Vol 49 # 6** pp: 932–940

[10] Buyalich G D and Buyalich K G 2014 Modeling of Hydraulic Power Cylinder Seal *Assembly Operation, Mining 2014 : Taishan Academic Forum – Project on Mine Disaster Prevention and Control: Chinese Coal in the Century: Mining, Green and Safety*, China, Qingdao, 17–20 October 2014. Amsterdam, Paris, Beijing. Atlantis Press, pp. 167–170

[11] Buyalich G.D., Buyalich K.G., 2015. Comparative Analysis of the Lip Seal in Hydraulic Power Cylinder, *Applied Mechanics and Materials.* **Vol 770**, pp: 402-406. DOI:10.4028/www.scientific.net/AMM.770.402.

[12] Buyalich G D, Aleksandrov B A, Antonov Yu A and Voyevodin V V 2000 Increasing the Resistance of Powered Support Brackets J. *Journal of Mining Science* **Vol 36** # 5 pp 487–492

The Algorithm of Development the World Ocean Mining of the Industry During the Global Crisis

Anatoliy Nyrkov, Vladislav Budnik, Sergei Sokolov and Sergei Chernyi

Admiral Makarov State University of Maritime and Inland Shipping. Saint-Petersburg, Russia

E-mail: sergiiblack@gmail.com

Abstract. In the article reviewed extraction effect of hydrocarbons on the general country's developing, under the impact of economical, demographical and technological factors, as well as it's future role in the world energy balance. Also adduced facts which designate offshore and deep water production of unconventional and conventional hydrocarbons including mining of marine mineral resources as perspective area of development in the future, despite all the difficulties of this sector. In the article considered the state and prospects of the Russian continental shelf, in consideration of its geographical location and its all existing problems.

1. Introduction

Every country needs the natural raw materials and, in particular, the well-being of humans is of associated with the possibility to dispose of these or those natural resources. The economic development of certain countries (Russia, Norway, the United Arab Emirates, Venezuela, Bahrain, etc.) are also directly connected with the possession of natural resources, inter alia, hydrocarbons. To this end, in these countries, a special attention is paid to the production and exploration of new deposits of natural resources. Other countries, using imported resources, are engaged in forging connections for supply of raw materials from the countries possessing them. Among all the available prospects for the development of the industry of extraction of natural resources in many countries at the moment, the most viable options are the modernization of means and methods of extraction of natural resources, as well as more intensive research work to find new fields of energy and mineral resources. The main role is still given to the production of hydrocarbon fuels, as no nation can escape dependence on natural energy resources in return, the renewable energy that regenerated naturally (sunlight, water flows, wind, tidal and tidal phenomena and geothermal heat) though global energy is moving in that direction.

2. The state of the industry

For the reason that at this stage there is no oil and gas energy analog, which could be their alternative, further extraction of hydrocarbons will continue and production volumes will grow. Taking into account the fact that most of the land hydrocarbon fields loses its relevance due to their exhaustion, the offset of their production and research on the seas and oceans is very urgent. The following ones are the main factors influencing the need for offshore and deepwater fields (Demographic, Technological, Economic).

The main problem of development of the sea shelf deposits is its knowledge-based and technological complexity of the exploration process. We should not forget the extreme conditions of the work done on the surface of the sea, which can cause very dangerous consequences for the oil and gas fleet. Such conditions include tides, severe storms, ice conditions, hydrogen sulfide, and sea salt.

Despite all the problems, expediency of the sea shelf exploration is approved by numerous studies, according to which, the availability of hydrocarbon seabed deposits exceed all expectations, being about half of the global total deposits. Today only about 2% of the waters of the world's continental shelf has been explored. Most of the reserves remains still hidden, being located in an offshore zone (exclusive economic zone). It is assumed that more than 60% of the total area of the sea continental shelf have coal, oil and gas in the interior. Over 100 thousand boreholes has already been drilled by 2013. Their number increases annually. Over the past 20 years, the number of new, large offshore fields significantly exceeded the number of large, discovered hydrocarbon sources on land. According to certain estimates, 20-30% of the world hydrocarbons are produced from offshore fields even now [3-8].

Based on the data [2-5] for 2010, provided by IFP Energiesnouvelles (IFPEN) and IHS Energy, reserves of the world ocean continental shelf hydrocarbons may be assessed and distribution of the world hydrocarbon extraction may be evaluated (Table 1).

Table 1. Global distribution of oil and gas extraction on the continental shelf

Region	Gas,%	Oil,%
Russia and CIS countries	2	4
Asia and Pacific Area	24	11
Oceania	5	2
The Middle East (Western Asia)	15	22
Africa	11	20
South America	11	16
North America	8	8

Europea	24	17
Grand total	100	100

More than 150 countries have an access to the sea, which gives them the opportunity to develop their offshore zone. Moreover, the main prospect of the future research of hydrocarbon reserves, of course, lies with the deep ocean. It is expected to be truly enormous.

3. Research of problems in development

Mining, including its environmental and technological issues, as well as a geological survey of the seabed is described in details in papers by Yu.V. Bruyakin, Yu.V. Bubis, L.N. Molochnikov, K.V. Yablokov (1979), Yu.V. Kafidov, N.G. Putya (2005), V.B. Dobretsov, V.A. Rogalev, D.S. Opryshko (2007), I.S. Kalinin, V.P. Drobadenko, N.G. Malukhin (2010). Review and analysis of these sources determines the trends of further development, popular new techniques and technical solutions [15-21]. They are described in detail in [20-26]. Resources are distributed at present in the following way. Most phosphorites concentrate offshore the coasts of North and South Americas. With the current rate of their consumption, these reserves will hold for more than 1,000 years, although marine phosphorites of more poor quality are mined on land. Barite nodules. The main deposits are currently on the shelves of Sri Lanka, Singapore, in the Sea of Japan, in the waters of Alaska, as well as in other areas of the ocean.

Large deposits of sulfur were found in the Mexico and the Persian Gulfs, the Caspian Sea and the Red Sea. Its underwater extraction is carried out by means of specialized drilling platforms. Underwater iron ore deposits are actively mined offshore Canada, Finland, in the Hudson Gulf and the Gulf of Finland. Shelves of Africa, Australia, South America, the Indian subcontinent are also promising in this respect. Sea sources of lead, mercury ore, zinc, gold and silver have been also discovered. In the deeper parts of the ocean with a depth of 2000 m, glauconite, deposits of rare nickel, cobalt, copper, manganese ores, as well as some other minerals are found. Today, technology is not enough to develop deep-sea mining, and it requires further study.

Russian shelf zone is also rich in minerals. The bottom of the central Kara Sea is also rich in iron-manganese nodules [13-15]. In the Russian part of the Baltic Sea iron-manganese nodules have been found, as well as deposits of amber [12-15]. In the northwestern Black Sea continental shelf, deposits of titanomagnetite sands were found (the Taman Peninsula). All of this suggests that the extraction on the bottom of the seas and oceans, in the near future, will have an even greater share in the global mining. The most urgent task at the moment is the development of technologies that will provide cost-effective and environmentally friendly development of bottom deposits of various natural resources, as the experience of such work is still quite small. The available equipment is inefficient for marine mining, being a major obstacle to their beginning. To this end, for example in Russia, the "Concept of development of deep forces and means of the Russian Federation for the period up to 2021" has been developed and approved [21-27]. Under this program, various morphostructural parts of the ocean are explored to identify and evaluate in the geological respect large concentrations of non-traditional types of minerals.

IOP Conf. Series: Materials Science and Engineering **142** (2016) 012121 doi:10.1088/1757-899X/142/1/012121

There is an alternative trend in the hydrocarbon production from unconventional sources, which created significant structural changes and changes in the oil and gas sector of the world economy.

4. Indicators. The prospects for the industry

Every trend in production of unconventional hydrocarbons has a number of advantages and disadvantages. The main disadvantage is the cost of production at the stage of the development and production from such deposits. The advantage is that because the development takes place directly in the vicinity of a consumer, it allows to save on transportation. Most of their production is concentrated in continental deposits, the development of the offshore area is still not profitable and is not justified, but has options in the future. In the next 10-15 years, non-conventional hydrocarbons will remain the local fuel, including in Russia [21-26]. Moreover, the investment in accordance with [14, 24-28] will be spent on the traditional sources of hydrocarbons mainly over the next 20 years.

It should be remembered that the rapid decline in production costs associated with the production of unconventional oil and gas, as well as the emergence of breakthrough technologies could change the picture of oil and gas production dramatically.

While analyzing Table 1, it is clear that Russia is at the initial stage of industrial development of its marine continental shelf, which is 20-22% of the world's sea shelf and at the same time is the biggest shelf of the world with an area of 6.2 million km2. About 4.2 million km2 of this shelf line is within the Russian exclusive economic zone. According to [2-7] resources of the Russian continental shelf can be estimated at the level about 100 billion tons of fuel taking into account the resources of the Crimean sector of the Black Sea and Azov Sea, of which oil is 16.7 billion tons, and gas is approximately 79 trillion m3.

Offshore development of hydrocarbons is able to compensate for the production decline in depleted oil and gas fields in the traditional areas of Russia, namely Volga-Urals and Western Siberia. Already in 2014, according to [10-14] oil production on the continental shelf of Russia amounted to 16.1 million tons, which is 3.1% of the total volume of oil produced in Russia, and gas is 31.5 billion m3, i.e. 4.9% of Russia's total amount. These volumes fall on the Caspian Sea and the Sea of Okhotsk mostly.

Based on the data presented in the journal "Oil and Gas Vertical", "Oil of Russia", the survey results from "Shelf of Russia» and «Oil and Gas Journal», it is evident that the main projects of the Russian Federation in the field of energy, actively discussed today, is the Arctic shelf water area (the Barents and Kara Seas), which accounts for approximately 83-85% of already explored hydrocarbon reserves of the total available volume. At the moment, more than 20 provinces and basins are discovered on the Arctic shelf hydrocarbon, with 50% of deposits are proved and justified. The bulk of the resources in the Arctic is in the western part, the eastern part remains still little explored. The Far Eastern continental shelf comprises 12-14% of hydrocarbon reserves, the remaining 5-3% fall on the shelves of other seas (the Caspian, the Black Sea, the Baltic Sea, the Sea of Azov) [2-7]. With the view of the allocation of resources in the Russian waters, it is clear the future direction for development of Russian offshore hydrocarbon production. It will be aimed at the waters of the Arctic shelf in the long run, where most of oil and gas reserves are concentrated. The Far Eastern shelf will be paid attention to as well, where at the moment one of the most important strategic oil fields (shelf of Sakhalin) are located, and the Caspian, the Black and Azov Seas, marine and terrestrial deposits (mostly gas ones).

If we take into account the assessment by O.M.Prischepa, General Director of the Federal State Unitary Enterprise "All-Russia Petroleum Research Exploration Institute" (VNIGRI), it is still too early to speak about the guaranteed prospects for the Russian shelf, because of the complexity of production and the lack of the required technology and expertise to develop the Arctic shelf, as well as all the environmental hazards which can be associated with its development. Severe operational conditions (storm, movement and the formation of ice, the ice cover of the sea, the critical temperatures, etc.) and the borders of areas of the Arctic waters under dispute,cause problems, due to which the permanent suspension of the plans of works, carrying out for the development of the Arctic shelf and legal restrictions of the development activities will last for more than one year.

According to the General Director of VNIGRI and results of survey to 108 respondents [9-12], among which oil and gas giant Russian market research companies and organizations engaged in services and equipment, it becomes clear that at the moment the most promising hydrocarbon deposits are located on the Caspian shelf (Korchagin field V. Filanovsky (12-15 million tones/year), as well as Sarmatian, and cross-border Khvalynskoe Rakushechnoye). Rights for the development of these fields belong to the JSC "Lukoil". The second promising area is the northern Sakhalin shelf, namely the projects "Sakhalin-1" and "Sakhalin-2" (oil extraction is estimated as 20 million tons / year and gas as 30 billion cubic m / year). Operators are Royal Dutch Shell and JSC "Gazprom. Further, the Arctic shelf and the shelf of the Black and Azov seas are next in the ranking.

Analyzing the assumption by O.M.Prishchepa and the survey results, we can understand that the main focus is aimed at the development of Russian oil and gas fields already available by their natural (available depth, the impact of weather conditions), technical (available to the technical equipment) and economic (profitability of production) factors. Now, the technical equipment, required for the development of the shallow shelves in the Caspian, Black, Azov Seas and the Sea of Okhotsk, as well as natural and weather conditions, are conducive to the normal development of fields located in these waters almost without any problems. These factors are the primary cause of slow development of the Arctic shelf, though the bulk of offshore hydrocarbons are found there.

Implementation of innovative techniques and equipment capability for the development of the Arctic shelf is still poor. The reason for this is the insufficient knowledge about areas and lack of world practical experience gained over the years of the offshore production in the Arctic with its harsh conditions. However, this does not in any way will be the cause of refusal in favor of Arctic hydrocarbon production in the long term. For example, Prirazlomnoye field in the Pechora Sea is still quite promising. It is located 55 km offshore. This field was put into service in 2013 and currently is the only Russian Arctic offshore field. For the first time production of hydrocarbons is carried out in the Arctic shelf the in a similar way, with the use of a fixed platform "Prirazlomnaya".

The Ministry of Natural Resources and Environment also provides the data (Table 2) on the resource potential of the Russian Arctic waters in terms of billion tons of oil equivalent, according to which you can see the actual distribution of offshore resources in 2013 [35].

Table 2. The Resource level of certain artic water bodies in the Russian Federation in the billion tons of the oil equivalent.

Arctic waters	Resource level	Booked reserves
The Barents and Pechora Seas	30.3	4.95
The Bering Sea	1.08	0
The East Siberian Sea	5.58	0
The Kara Sea	41.2	4
The Laptev Sea	3.26	0
The Sea of Okhotsk	8.7	1.7
The Chukchee Sea	3.3	0
Totally by water bodies	93.42	10.65

Thus, with the further development of marine energy production, the priority will be sea hydrocarbon production in the Russian continental shelf, which share in the Russian total production will grow from year to year. Over the period from 2010 to 2014, there were drilled 18 prospecting and 17 exploration wells on the Russian continental shelf with the 26 discovered fields, mainly in the Far East region and almost 122 licenses were issued for development in the period from 2015 to 2020. Most (60%) of the licenses shall apply to the waters of the Arctic seas. Main companies, that will develop these fields, were Rosneft, Gazprom and Lukoil. At the beginning of 2015, twenty-two deposits were discovered in the Russian Western Arctic seas and eighty-eight exploratory wells were drilled. Total discovered reserves amounted to more than 10 trillion m3 of gas and more than 500 million tons of oil. M. [17-25]

Taking into account the geographical position of Russia, where a large part of the shelf is located in the Arctic and northern regions, one should expect the implementation and development of breakthrough technologies there. This is one of the important aspects in solving the problem how to extract the challenged Arctic hydrocarbons. At the same time if you really look at the situation in the near future, Russia will not be able to switch to the Arctic hydrocarbon reserves as an alternative energy supply. It is necessary to arrange the deep-water exploration and production of oil and gas reserves at this stage, as most of the shelf and deep waters remain unexplored (Black Sea). According to [23-27], 60% of experts believe that to develop Russian offshore projects only on their own, without the use of foreign technologies is not possible. This fact should be taken into account as well. These technologies include: technologies and equipment providing subsea extraction, drilling equipment; support fleet with available heavy lift capacity; high-frequency and wideband 3D seismic survey; information systems; technology of controlled deep-sea drilling; the equipment capable of being operated at extreme temperatures and icing conditions, for year-round drilling in the Arctic and Far Eastern regions; drilling platforms.

Ecological safety of oil and gas fields and the transport through the Arctic Sea should not go unmentioned, as any mistake in the development of mineral resources of the Arctic shelf may cause catastrophic consequences and a further ban on future work for the extraction of Arctic hydrocarbons. This can result in significant blow to the economy of the Russian Federation, as resources of the Arctic shelf are a prerequisite for the economic power of our country. For this reason, navigational and environmental safety should be paid particular attention.

5. Conclusions

In 2015, the focus is preparatory drilling operations and processing of earlier accumulated data concerning the Russian continental shelf. Drilling is scheduled to begin in 2016 in the areas of the Okhotsk and Barents Seas and the Crimean sector of the Black Sea. There is also a long-term plan for exploration and development of the Arctic seas, based on recently acquired experience in the field Prirazlomnoe and information obtained by the results of the search and reconnaissance operations. Investment and cooperation with foreign companies is top-mind concern, but due to the sanctions, there may be some challenges. The problem of domestic production of platforms by Russian companies is very crucial as well, but a final decision has not yet been made.

According to the governmental program of the Russian shelf development until 2030, it is planned to increase oil and gas production fourfold and fivefold respectively as compared with the current level.

Based on trends for the last decades, we can conclude about the unprecedented growth of the world's costs for exploration and production. It is governed by an ever-growing demand for natural resources, which is caused by a number of economic, demographic and technological factors. Maintenance of production at a safe level for the staff and the global environment as a whole is one of the priorities. The future belongs to these trends in the natural resources production.

References

[1] S Mastepenov Fracture of energy philosophy 2004 Russian Oil. **11-12** 18–26
[2] Maritime zone, http://www.ncbi.nlm.nih.gov
[3] Energy land info, http://energyland.info/news-show-neft_gaz-technology-105001
[4] Circum-Arctic Resource Appraisal: Estimates of Undiscovered Oil and Gas North of the Arctic Circle. U.S. Geological Survey. Fact Sheet 2008–3049
[5] Gautier Donald L., Bird K,J., Charpentier R.R., et al: 2009 Assessment of Undiscovered Oil and Gas in the Arctic. **324(5931)** 1175–1179
[6] IHS, http://press.ihs.com/press-release/energy-power/ihs-cera-role-deepwater-production-global-oil-supply
[7] V Bogoyavlenskii, I. Bogoyavlenskii, 2015 On the threshold of the Arctic era. Russian Oil. **4** 25-30
[8] Rig zone, www.rigzone.com
[9] Y. Grama. 2012 The analysis of Russian Oil and Gas Reserves. International Journal of Energy Economics and policy. **2(2)** 82–91

[10] Krukov, V., et al: 2013 Familiarization with Arctic- only together. Shelf resources. **4** 40-46
[11] Offshore, http://offshore-mag.ru/news-detail.php?ID=518
[12] A.M. Ashihmin. The results of the survey. The forecast of development of Russian shelf 2015-2020. Technical report, http://www.russianshelf.com/ru/page/shelf-rossii-skachat-resultati-oprosa
[13] Great Soviet Encyclopedia, http://bse.sci-lib.com/article093165.html
[14] Y.G. Nikifipov, A.O. Shpaiher. 1978 Kara sea. Great Soviet Encyclopedia. **11**
[15] Ju.P. Doronin. 2007 Shelf zone oceanography. St. Petersburg. **128**
[16] M. Pilipchuk. 2001 Environmental impacts of nodule mining in the deep areas of the World Ocean. Moskow. **8** 32-36
[17] Chernyi S G, Budnik V Yu 2015 Elements of the introspective analysis to evaluate software in navigation. 22nd Saint Petersburg International Conference on Integrated Navigation Systems, ICINS 2015. 147-150
[18] I O Korolev. 2013 Substantiation of parameters of the sea bed unit, taking into account the added mass of water in the extraction of iron-manganese nodules of offshore zones. 143
[19] Ju V Bruyakin. 1989 Scientific basis for the development of underwater mineral field and improve jetting rocks. Moskow. **197** 137
[20] Ju V Bruyakin. 1984 Experience of development of sea field of non-metallic building materials. Moskow. 48
[21] Ju V Bubis, A V Grishin, D V Semenuk. 1999 The prospect of the mining technical complex. **1** 88-92
[22] Ju V Bubis, N G Kafidov. 2005 Ways to improve the efficiency of development of underwater fields. Construction Materials. **4** 12-13
[23] Zhilenkov A, Chernyi S 2015Investigation performance of marine equipment with specialized information technology. Procedia Engineering. **100** 1247–1252
[24] World Energy Investment Outlook. 2014 Special Report OECD/IEA. 19-45
[25] Dmitrievskii A N, Visockii V I 2010 Shale gas -a new vector of the global oil market development. **2**
[26] IHS ,http://press.ihs.com/press-release/energy-power/ihs-cera-role-deepwater-production-global-oil-supply
[27] Chernyi S, Zhilenkov A 2015 Analysis of complex structures of marine systems with attraction methods of neural systems. Metallurgical and Mining Industry, **1** 37–44
[28] Chernyi S, Zhilenkov A 2015 Modeling of complex structures for the ship's power complex using XILINX system. Transport and Telecommunication. **16 (1)** 73–82
[29] Chernyi S 2015 The implementation of technology of multi-user client-server applications for systems of decision making support. Metallurgical and Mining Industry, **3** 60- 65

IOP Conf. Series: Materials Science and Engineering **142** (2016) 012122 doi:10.1088/1757-899X/142/1/012122

Innovative Technics of Managing Engineers' Global Competencies

A A Khoreshok[1], S A Zhironkin[2], M A Tyulenev[3], G A Barysheva[4], V Yu Blumenstein[5], M C Hellmer[6], S V Potyagailov [7]

[1]Dr.Sc. Professor of Mining Equipment department, National Research Tomsk Polytechnic University. Russia, 634050, Lenina av., 30
[2Dr].Sc. Professor of Economy department, National Research Tomsk Polytechnic University. Russia, 634050, Lenina av., 30
[3]PhD. Assistant professor of Open pit mining department, Kuzbass State Technical University, Russia, 650000, Kemerovo, Vesennya st., 28
[4]Dr.Sc. Chairman of Economy department, National Research Tomsk Polytechnic University. Russia, 634050, Lenina av., 30
[5]T.F. Gorbachev Kuzbass State Technical University, 28 Vesennyaya str., Kemerovo, 650000, Russia
[6]MSc in Hydrology and Water Resources Management., Department of Civil and Environmental Engineering, Imperial College London. United Kingdom, South Kensington Campus, London SW7 2AZ, UK
[7]Post-graduate student, National Research Tomsk Polytechnic University, National Research Tomsk Polytechnic University. Russia, 634050, Lenina av., 30

E-mail: [3]tma.geolog@kuzstu.ru

Abstract. Higher education modernization in the CIS countries takes place under the conditions of dynamic changes in economy and society. These changes are determined by the social and economic development of the country and the world globalization processes - cross-border intercultural communication, knowledge transparency, and the establishment of information society. Educational globalization is a continuous process of creating a unified global educational system, in which the distinctions between its member educational systems are being blended.

1. Introduction

Globalization of education is not possible without the development of regional and interregional cooperation in the field of educational, information and communication technologies, establishment of virtual universities, equality and access to education [1]. Globalization of education reflects the strengthening of the links between international business and nations, by means of multinational corporations to become the main customer of educational services, the development of the Internet, the growth of cities, and the standardization of industrial products reporting system [2]. In the process of creating a unified cross-cultural business environment the value of foreign language, mostly English, increases

IOP Conf. Series: Materials Science and Engineering **142** (2016) 012122 doi:10.1088/1757-899X/142/1/012122

many times as it becomes a means of transmission of the global economic, social and cultural information.

Thanks to the universality and integrity - the basic principles of educational globalization - modern specialist, who speaks professional foreign language, attains global competitiveness in the labor market. This was an important element of Bologna Process, which Russia joined in 2003 [3].

Today the engineering universities of the former Soviet Union countries go through the following set of integration processes:

1. At the educational and methodical level of a particular university implementation of international experience of practical engineering activities in the educational process is expanding. It is realized in the course of the project and case-study training, interaction of teachers and students.

2. At the university level, the degree course scheme and curriculum are being adapted to the needs of specific businesses - educational customers; target training of the engineering staff is being developed. International manufacturers' entering Russian markets makes higher educational institutions world requirements for quality of engineers' training, encourages universities to cooperate with scientific organizations, initiates the development of original copyright courses, gives impetus to the development of in-house training directly in the workplace.

3. At the regional level the conditions for the development of inter-university, scientific-and-educational cooperation are being created. At the same time, increased competition in the domestic and international markets of educational services raises the question of improving the competitiveness of technical higher schools. Its primary factors are innovative educational activities and the ability of universities to participate in the development of the national innovation system. But it is impossible without having foreign language skills by teachers and students.

4. At the national level the system of universities is being adapted to international standards of quality training of engineers, to the level of graduates' competence, to the technical equipment and teaching materials provision of educational process. It's no doubt that the leading role in this process belongs to the technical universities, which should preserve and increase the scientific and industrial potential of the national economy.

5. At the international level, globalization of the education system has some sort of response to the integration processes in the world economy, the formation of a global information space, international labor migration. For the technical universities of CIS countries educational globalization opens up the possibility to act within the framework of the space-knowledge paradigm of social-and-economic development. This means that higher technical education, due to its applied nature, becomes an important factor of production and brings its owner a guaranteed income regardless the country in which the engineer works.

In this regard as significant directions for further Russian higher education modernization European benchmarks of educational systems consistent with the objectives of internationalization and creation of a common European Higher Education Area became of the first priority. During higher education reforms in the CIS countries, there was a definite rethinking of the previous system of training of professionally significant qualities of future specialists. Today they are called "competencies", which are the ability to apply knowledge, skills and personal qualities to be a successful specialist in a certain professional area [4].

2. Material and Method

The homeland of competence approach can be considered the UK, where this term had been proposed and studied in detail by J. Raven, a professor at the University of Edinburgh [5]. The competence approach to the training at higher school, according to the first researchers, is based on such a system which provides the quality of training consistent with the requirements of modern global labor market. In connection with scientific and technological progress and, as a consequence, the emergence of new professions higher educational systems in a number of countries have turned their attention to the need to align the requirements of the labor market, training and education.

For example, T. Brinker defines competence as the key abilities of the future specialist and gives them global character. As key competencies she considers the following abilities: taking the initiative (orientation to the result, ability to change, ability to make decisions, etc.), attracting partners (the ability to implement, communication skills, customer orientation, etc.), developing strategy for success (learning ability, strategic thinking, etc.), team managing (the ability to motivation, the ability to appreciate the position and resolve conflicts, etc.) [6]. As we can see, the key skills of specialist are unchanged for any kind and place of professional activity and, therefore, they are global.

Former Principal of Cumbria Institute of Arts, D. Vaughan emphasizes the global role of key competencies by defining them as a personal "portable" skills - time management, teamwork, communication, presentational skills, information technology, business skills, independence and autonomy, self-criticism, the ability to present themselves and others [7].

In today's global economy, these competencies can be realized by specialist working in any country. This is especially important for Mining Engineers, because mining industry is the basic one for most countries. And the important global competence, integrating all others, is a professional foreign language skill.

Russian federal state educational standards define the professional and common cultural competences. General professional and specialized skills which characterize a person as a professional are referred to the professional competencies. But in modern conditions it is not enough to train only professional competences in order to meet the demand of the market, so there is a need to form a common cultural competence at students.

General cultural competences include the following: social and personal, organizational, administrative and general scientific skills necessary for specialists to carry out their professional duties and to be successful and in demand in the labor market.

For example, Russian state educational standard for mining engineers training of "Open pit mining" specialty represented 22 common cultural and 28 professional competences, 10 of which we consider global.

To the latter we refer such common cultural competence as "to speak foreign language in order to study foreign experience in professional activities, as well as to contact on professional level." It is tightly connected with other common cultural competences demanded in today's global world:

- to be able logically, convincingly and clearly express thoughts, to speak orally and write properly;
- readiness to cooperate with colleagues, work in a team;
- the ability to negotiate, establish contacts, to resolve conflicts of interest ;
- to strive to self-development, upgrade qualifications and skills;
- the ability to analyze the philosophical, social and personal problems, independently form and defend personal ideological positions;
- readiness for social interaction in various spheres of social life, for cooperation and tolerance.

Future engineer will be able to implement the global competences while working in multinational teams abroad or in multinational corporations. The following professional competences, which have global character, become the condition for successful employment in the world labor market:

- provide technical management of mining activities in the exploration and extraction of minerals, manage the processes at production facilities;
- to develop and bring to the performers of mining operations the tasks; to carry out the control of work quality and to ensure the correct implementation of it by the performers; to prepare work schedules and long-term plans, regulations, estimates, requests for materials and equipment, to fill in the necessary reporting documents;
- to study the scientific and technical information in the field of operational exploration, extraction, processing of solid minerals.

But the "bridge" to the formation of global competence, in our opinion, is the use of innovative training technics in foreign language teaching of future mining engineers. The key to success here is the integrated formation of language, common cultural and professional skills while teaching them foreign languages, and using pedagogical innovations.

Structuring criteria of innovation process in higher education are diverse enough. Thus, based on the spread scale global, national and university innovations are distinguished. Such changes in the educational systems as the use of the Internet, the establishment of virtual universities, online conferences, and webinars are referred to global innovations for CIS countries. The transition to two-stage system of training, the use of the ECTS (European Credit Transfer System), the introduction of rating system of knowledge and students' independent work assessment should be named among educational innovation.

3. Results and Discussion

The use of interactive teaching technics, computer technologies, the creation of student community, the work of the tutors, etc. become integral elements of innovation at the university level. Based on functional criterion, innovations can be divided into *intra-subject* (identified by the peculiarities of particular subject and activity of the teacher), *methodological* (they are connected with the organization and methodical support of the educational process), *administrative* (decisions taken by the managers of different levels in order to improve the efficiency and quality of the educational process) [8].

In the context of the competence approach conscious rethinking of the content, technics and teaching techniques of foreign language at technical higher schools is taking place. Teachers are increasingly applying to innovative technologies and technics such as interactive teaching technics and the use of computer technology. Variability, which consists in an adequate choice of a particular technic aimed at the formation of a certain professional or common cultural competence, performs an essential element in increasing the effectiveness of the educational process. Among the variety of interactive training technologies of foreign languages special attention should be paid to: role-plays, case-study, brainstorming, "653" method, as well as the project technic, which enhances students' independent and extracurricular work.

In the scientific and methodological literature there is a number of studies devoted to theoretical and practical understanding of innovation activity in the development of competence approach to the educational process [9-15]. Therefore, we will omit the theoretical issue, but merely note that, within the higher school training of mining engineers special attention deserves role-plays, brainstorming, case-study and project technic.

"The point of application" of innovative teaching in the formation of global competencies should become foreign language learning by the future mining engineers. We believe role-play to be a means of simulation the professional activity, and its implementation in training provides conditions for the formation of both professional and common cultural competences.

Starting with the first lessons students prepare for the play gathering factual, relevant information relating to the topic for its further using in the scenario. Language (ability to build sentences grammatically and lexically correct in a foreign language) and communication (ability to interact with a partner, choosing the appropriate(verbal and non-verbal) means of communication) skills are being trained in the classroom activity during the entire study period. This process implies the use of special tasks which help students to prepare to spontaneous communication on professional topics.

These tasks include the preparation and presentation of the role dialogue, listening comprehension, etc. After the role-play teacher, together with students conducts its detailed analysis, identifies the achievement of educational and play's goals, comments on the students' progress in language and communicative skills, corrects mistakes and makes recommendations and suggestions for the future. Using role-play in the classroom enhances students' interest to the language learning, removes psychological barrier in foreign language communication and contributes to the formation of common cultural and professional competences.

Recently, such technic as case-study has been widely spread in teaching foreign language at technical higher schools, some researchers and textbooks' authors call it problem solving or dilemma and decision. Analyzed the situation, offered to students must have a definite structure consisting of a number of components. The first component is a problem background or brief history, which describes the main actors of the conflict, their economic and financial situation, relationships with each other, etc.

This is followed by the information detailing with the problem itself and containing a specific task for students.

The tasks may be the following: working in small groups (group), analyze the problem on behalf of the participants, present your solution, prepare the report (presentation), conduct negotiations to resolve the existing conflicts between companies, etc. The final component may be a written task (write it up), which includes the preparation of a report, memo or minute of negotiations, depending on the content of a problem situation.

Using case-study in learning process helps students not only to raise their language level, but also shows them how to build the paradigm of problem-solving, participate in language and professional interaction. That is why the particular importance here acquires the possibility to combine individual and group work of students on solving professional problems in educational environment.

It is important to provide these problems by mining science, and with their help, students can learn about promising lines of mining development. To realize that we used the results of scientific research of higher schools -members of Consortium of Russian raw material universities and Imperial College London in the preparation of cases for study.

At Kuzbass State Technical University and Tomsk Polytechnic University we conducted a number of training seminars for teachers of foreign languages and specialized engineering disciplines for the Mining specialization. Their goal was the development of problems for role-plays, brainstorming, and case-studies. They were based on the results of scientific research in the field of geo-ecology, technology and complex mechanization of mining operations, business management.

In particular, based on the results of geo-ecological studies conducted by the Department of Geology of Kuzbass Technical and Tomsk Polytechnic University, as well as the Department of Civil and Environmental Engineering of Imperial College London, we have developed a number of problem situations: "The use of overburden rocks for the treatment of mine and quarry water", "Formation of artificial filter arrays", "Low-cost technologies of cleaning quarry and mine waters" [16-19].

Based on the research of Mining Machines and Complexes Department of Kuzbass State Technical University, we have formulated the following topics for problem discussion: "Underground Geo-Walker", "Active cutting tool" [20-23] and "Multiple use of rock cutters" [24].

Scientific research conducted by Open Cast Mining Department of Kuzbass State Technical University, allowed offering the following topics: "Land-saving open geo-technologies", "Transportless development of coal deposits by open pit mining", "Coal loss reduction during the work of excavator" [25].

To expand the horizon of knowledge and get acquainted with the works of foreign scientists, we have prepared topic "Neo-industrial technologies" [26-32].

Research in the field of Engineering Economics, conducted by the Department of Economy of Tomsk Polytechnic University, allowed offering the following ideas for problem discussion: "The development of state leasing and factoring in coal industry", "The establishment of Coal Regional Bank", "Attracting foreign investment in coal industry" [33-36].

Brainstorming, being very interesting and productive technic in terms of efficiency of professional and common cultural competence formation, is also frequently used in teaching process of subjects in a foreign language at Kuzbass State Technical University and Tomsk Polytechnic University. Without dwelling on the details of the organization and conducting brainstorming session, we would like to describe, as an example, the task offered to students for the brainstorming discussion.

The problem is the following: a small coal mine with spillway pit water into the stream is forced to carry out their treatment, obeying environmental requirements. However, the construction of sewage treatment plants is expensive enough project that can endanger other investment projects. The construction of artificial filter arrays can cost much cheaper, but requires the machinery and equipment used for another purposes. What should the technical management of the coal mine do?

The advantage of the brainstorming technic is an ability to "think loud" in a foreign language provided to students during the brainstorming session. Since the rules of brainstorming provide finding a significant number of solutions (including absurd ones) to professional problem even students with low language level could take part in the discussion.

We would like to touch briefly method "653", which is mainly based on the principles and rules of brainstorming. It's better to use it when there is a need to develop collective solution of the problem and the group demonstrates high language skills [37].

Method "653" is carried out in a group of six people, which generate three solutions of a given problem within five minutes. These ideas are recorded on paper and then transferred in a clockwise direction to each other. Each participant studies the received information next five minutes and details it. The process continues as long as each member of the group has not studied and corrected all the proposed ideas. Finally the group has 18 proposals of detailed solutions to the problem. Further the students discuss, specify and select the best decision.

The advantage of using method "653" is to obtain in a short time a large number of ideas in a foreign language. The disadvantage is frequent duplication of ideas, as at the first stage obvious ideas are expressed by many students. You must also bear in mind that not all students will be able to offer 3 ideas in 5 minutes written in a foreign language. So it should be emphasized once more that this technic should be used in group with higher language skills which allows achieving the best results.

Along with those mentioned above, one of the changes in the system of higher education should be growing mobility of students, teachers and staff, and an international open policy for universities. This increases the role and importance of foreign language skills. However, we observe a significant reduction of in-class learning in new curriculum for the future mining engineers, and increase of their individual work. Consequently, there is a serious problem of competence formation at the foreign language classes. In connection with that this process moves towards an independent and extracurricular work. At the lesson the teacher can use interactive technics such as role-play, brainstorming, case-study, and one of the forms of extracurricular work is the project technic.

The result of students' individual work in "Foreign Language" discipline is the development of projects for technical re-equipment of enterprises, promotion of coal products in the market, and others, as well as the presentation of these projects. Practice shows that individual self-study provides an opportunity not only to realize oneself as a researcher, but also to train future professional competence during presentation and discussion of projects, reports, abstracts.

Each of the above-described technics for the formation of future mining engineers' competencies at technical higher school has its advantages and disadvantages that should be considered by a teacher while organizing the educational process. Therefore, sharing experiences and teaching materials in this area can greatly help teachers to make learning process more interesting, and mostly focused on the formation of particular competences.

4. Conclusion

Thus, the use of innovative technics for the formation of common cultural and professional future mining engineers' competencies is intended not only to accelerate their entry into the global educational and cultural space, but also improve the quality of mining higher education.

5. Acknowledgement

The research was made in National Research Tomsk Polytechnic University with financial support of the Ministry of Education and Science of the Russian Federation in the framework of scientific research "The evaluation and improvement of the social, economic and emotional prosperity of senior citizens" contract № 14.Z50.31.0029. Some parts of research were made in Kuzbass State Technical University with financial support of the Ministry of Education and Science of the Russian Federation in the framework of scientific research "Exploring the parameters of technological and technical solutions for selection and development of improving cutting and extracting mining machines operation efficiency in Kuzbass", contract №632. The authors would like to thank to prof. Yury Lesin and prof. Magerram Gasanov for their discussions during the study. We thank anonymous referees for constructive and useful comments on the papers sent via e-mail.

References

[1] Corban E and Lowson D 2009 Globalization and modern education *Journal of ECUP* **3** pp 56–62

[2] Irvine E 2007 New approach to high education modernization: a world competition *Journal of ECUP* **6** pp 91–102
[3] Hirvonen M 2006 Bologna challenges: a new outlook *Journal of ECUP* **4** pp 76–85
[4] Alatartseva E and Barysheva G 2015 Well-being: Subjective and Objective Aspects *Procedia - Social and Behavioral Sciences* **166** pp.36-42
[5] Raven J 1984 *Competence in Modern Society: Its Identification Development and Release* (Oxford England: Oxford Psychologists Press) 354 p
[6] Brinker T 2009 Key competences professional ability and Bologna process *Journal of ECUP* **6** pp 106–115
[7] Vaughan D *Learning outcomes and competencies. The results of teaching and competence approach* (London: Collman Pub.) 288 p
[8] Christensen C M and Eyring HJ 2011 *The Innovative University: Changing the DNA of Higher Education from inside out* (Denver: Jossy-Bass) 513 p
[9] Bloom B S 1971 *Handbook on formative and summative evaluation of student learning* (New York Mc Grow-Hill Book Co) 211 p
[10] Bötteher W 2011 *Wissen Kompetenz Bildung Erziehungoder was? ZurDiskussion um Standardisierung in der allgemeinbilden Schule* (Opladen: Leske&Budrich) 320 p
[11] Herzog R 1998 *Redezur Eröffnung des Paderborner Podiums* (Heinz Nixdorf Museums Forum am 9 Juni 1998 Hannover) 548 p
[12] Hutmacher W 2009 *Key competencies for Europe* (Report of the Symposium Berne Switzerland 06-07 May 2009 Council for Cultural Cooperation CDCC a Secondary Education for Europe Strasburg) 103 p
[13] Kohler J 2004 *Universität Grecfswald Schlüsselkompetenzen und "employability" in Bologna Prozess Konferenz* Schlüsselkompetenzen (Schlüsselzumehr Aus Bildungsqualität und Beschäftigungsfähigkeit Heidelberg 22 und 23 Januar 2004) pp 241-248
[14] Romiszowski A J 2011 *Producing instructional systems Lesson planning for individualized and group learning activities* (Liverpool: Kogan Page Nichols Pub) 322 p
[15] Schwarz-Hahn S and Rehburg M 2003 *Bachelor und Master in DeutschlandEmpirische Befundezur Studienstrukturreform Wissenschaftliche Zentrumfür Berufs – und Hochschulforschung* – (Kassel: Universität Kassel) 356 p
[16] Lesin Y V Luk'yanova S and Tyulenev M 2010 Mass transfer of dispersed particles in water filtration in macro-grained media *Journal of Mining Science* **46 (1)** pp 78-81
[17] LesinY V, Luk'yanova S Y and Tyulenev M A 2015 Formation of the composition and properties of dumps on the open-pit mines of Kuzbass *IOP Conference Series: Materials Science and Engineering* **91 (1)** 012093
[18] Tyulenev M A and Lesin Y V 2014 *Justification complex purification technology open-pit mines wastewater* (Symposium of the Taishan academic forum – Project on mine disaster prevention and control) pp 441-444
[19] Tyulenev M, Zhironkin S and Litvin O 2015 The low-cost technology of quarry water purifying using the artificial filters of overburden rock *Pollution Research* **34 (4)** pp 825-830
[20] Aksenov V V, Khoreshok AA and Beglyakov VY 2013 Justification of Creation of an External Propulsor for Multipurpose Shield-Type Heading Machine – GEO-WALKER *Applied Mechanics and Materials* **379** pp 20-23
[21] Khoreshok AA 2002 On side cutting bit when operating at sheerer drums *Ugol'* **7** pp 10-12
[22] Lekontsev Yu M, Sazhin P V Temiryaeva O A and Ushakov S Yu 2013 Two-side sealer operation *Journal of Mining Science* **49(5)** pp 757-762
[23] Kovalev V A, Gerike B L and Gerike P B 2014 *Preventive maintenance of mining equipment based on identification of its actual technical state* (Symposium of the Taishan academic forum – Project on mine disaster prevention and control) pp 184-189
[24] Prokopenko S A 2014 Multiple service life extension of mining and road machines' cutters *Applied Mechanics and Materials* **682** pp 319-323

[25] Zhironkin S A, Tyulenev M A, Zhironkina O V, Hellmer M 2016 The global determinants of mining higher education development *Metallurgical and Mining Industry* **2** pp 62-67
[26] Tyulenev M, Lesin Y, Zhironkin S and Garina E. 2016 Coal producers waste water purification M*etallurgical and Mining Industry* **2** pp 52-56
[27] Gasanov M, Gasanov E, Egorova M The Technologies of the Network Prosperity in Russia *Procedia - Social and Behavioral Sciences* **166** pp 103-106
[28] Gasanov M A, Gasanova N V, Zhavoronok A V 2016 Information Technologies in Ensuring Continuous Wellbeing of the Person *The European Proceedings of Social & Behavioural Sciences* **VII** *pp* 323-336
[29] Klemasheva E, Gasanov M and Zeremskaya Y 2016 Neo-Industrialization of Russian Economy as a Source of Social Well-Being Growth *The European Proceedings of Social & Behavioural Sciences* **VII** *pp* 150-154
[30] Alatartseva and Barysheva G A What is Well-Being In The Modern Society: Objective View *The European Proceedings of Social & Behavioural Sciences* **VII** *pp* 375-384
[31] Baryshev A A and Barysheva G A Social Entrepreneurship: Metaphysics of Entrepreneurship in Practice *The European Proceedings of Social & Behavioural Sciences* **VII** *pp* 352-365
[32] Barysheva G A and Novoselova E G 2014 Methodology of Application of the Structural Shift Mechanism for Regulation of the National Economic Management System *Applied Mechanics and Materials* **682** pp 550-554
[33] Zhironkin S A 2002 Prospects and new possibilities investment attracting to Kuzbass coal mining industry *Ugol'* **6** pp 31-36
[34] Zhironkin S A 2001 Factoring and leasing development at coal mining industry of Kuzbass as an important element of its financial part *Ugol'* **4** pp 29-30
[35] Zhironkin S A 2001 Governmental factoring development of TEK Kuzbass *Ugol'* **6** p 62
[36] Zhironkin S A 2002 About measures of vexel circulation development and vexelability definition of fuel-and-power complex' enterprises *Ugol'* **4** pp 47-48
[37] Kohn N and Smith S M 2011 Collaborative fixation: Effects of others' ideas on brainstorming *Applied Cognitive Psychology* **25** **3** pp 359-371

IOP Conf. Series: Materials Science and Engineering **142** (2016) 012123 doi:10.1088/1757-899X/142/1/012123

The Experience of Implementation of Innovative Technology of Quarry Waste Water Purifying in Kuzbass Open Pit

Yu V Lesin[1], M C Hellmer[2]

[1]Dr.Sc. Professor of Economy department, National Research Tomsk Polytechnic University. Russia, 634050, Lenina av., 30
[2]MSc in Hydrology and Water Resources Management, Department of Civil and Environmental Engineering, Imperial College London, South Kensington Campus London SW7 2AZ, UK

E-mail: [1]tma.geolog@kuzstu.ru

Abstract. Among all industries in Kuzbass (Western Siberia, Russia) the coal industry provides the most environmental threat. However, the construction of new and maintenance of existing open pit mines do not often correspond to the tasks of improving the environmental safety of surface mining. So the article describes the use of innovative quarry waste water purifying technology implemented in Kuzbass open pit mine «Shestaki». This technology is based on using artificial filter arrays made of overburden rock.

1. Introduction
In the twenty-first century environmental protection and rational use of natural resources is one of the major problems of society. The most actual problem of environmental protection is in major industrial regions such as Kemerovo region [1, 2]. The strategy of social-and-economic development of the region provides to increase the volume of coal production in the region up to 250-300 million tons by 2025 [3, 4].

One of the constraints for the increase of coal produced volume is an intense pollution that leads to a change in the landscape, disruption of air quality. However, the greatest harm open pit mining does to hydrosphere. This is due to the complex composition of the open pit waste water, which is contaminated with suspended solids, oils, heavy metals, oil and explosion products simultaneously. Every year the water bodies of Kemerovo region receive 2.033.800.000 m^3 of quarry water [5], more than 500 million m3 of them coal enterprises discharge without any treatment [6]. Aggregate weight of pollutants reaches 606.12 thousand tons [7]. The main polluting substances in pit waste water are suspended solids, represented by finely-dispersed rock and coal particles. The concentration of suspended solids in pit water is 50-250 mg / l, reaching in some cases 2000-3000 mg / l [8].

2. Material and Method
For purifying quarry waste water in Kuzbass various methods and techniques of treatment, including the settling ponds and filtering arrays built of overburden rock are currently used [9, 10].

The advantages of such artificial filter in comparison with settling ponds are small land capacity, stationarity, simplicity of construction, low cost of waste water treatment, as evidenced by the conducted researches [11, 12].

IOP Conf. Series: Materials Science and Engineering **142** (2016) 012123 doi:10.1088/1757-899X/142/1/012123

Water purification process is based on the colmatation phenomenon [11] (Fig. 1).

In rock masses the sedimentation of suspension in the pores and lightening of filtering quarry waste water take place, suspended particles are trapped in a porous medium as a result of getting stuck in narrow pores, and sticking to the wide pore channels.

The design of the filter depends on the terrain and the properties of the upper layers of the soil of underlying surface, and the parameters are determined by the quantity of water inflows, its contamination and filtration characteristics of the filter material.

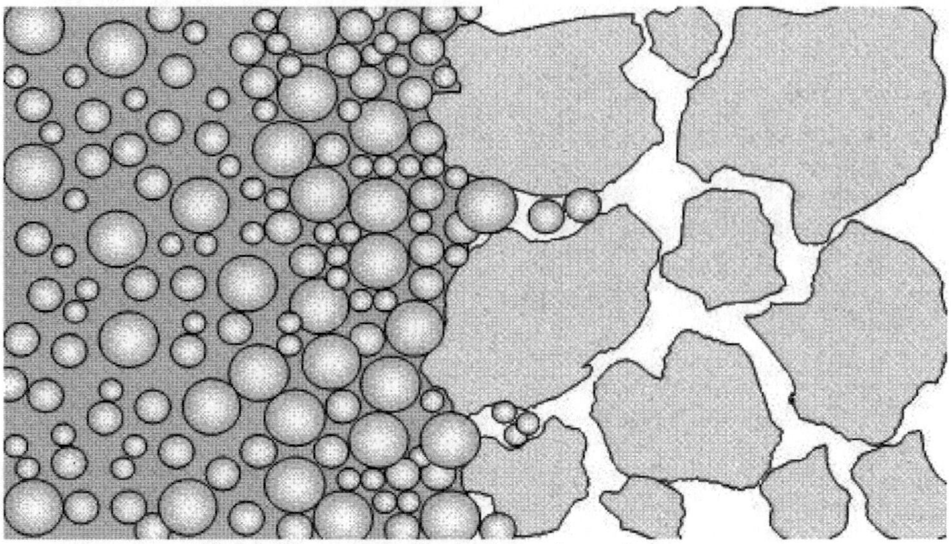
Fig. 1. The scheme illustrating colmatation process

The main filter elements are the unit for supplying polluted water, filtering array, filter body, a device for collecting and draining purified water (Fig. 2).

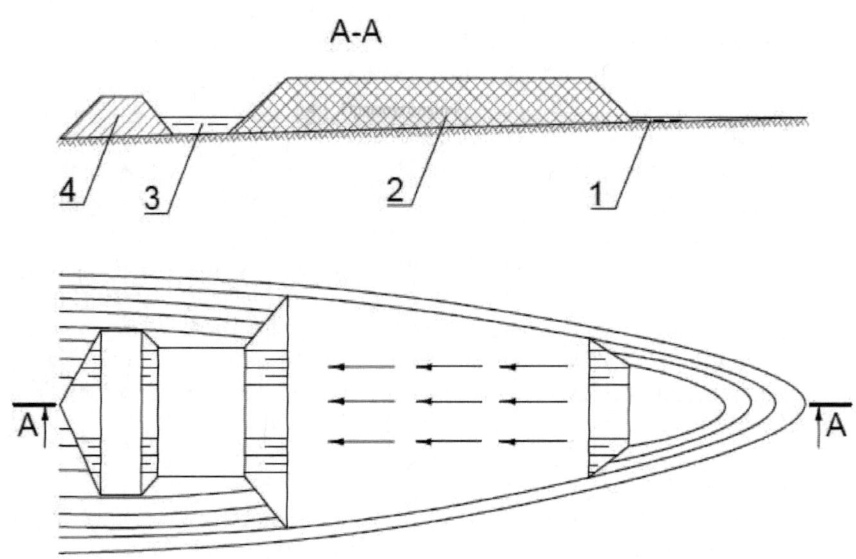

Fig. 2. Common scheme of filtering array: 1 – the pond in the place where waste water enters the filter; 2 – filtering array; 3 – collector for purified water; 4 – water collecting dam

The filtering array consists of crushed rock and semi rock with grain size of 20 to 200 mm or more, the process of its creation doesn't differ from the peripheral dumping technology. The rock laying into the filter array must have a softening ratio not less 0.8 and not dissolve in water. When dumping bottom of such filter semi rocks are used, when they are softened form a dense layer with low permeability.

The filter itself works in the following way: contaminated water (suspended mixture) flows by gravity through a pipeline in to the water intake, from where it drains through the filter array. After mudding of lower layers of the filtering array water level in the water intake increases, and new fresh layers of rock are being involved in the process of filtration. The content of suspended solids in the influent quarry waste water is not limited, but in the purified water is 0-10 mg / l. Filter size is determined by the volume of water flow and the required degree of purification.

3. Results and Discussion

We considered quarry waste water purification problem by the example of JSC "Open pit Shestaki". JSC "Open pit Shestaki" has been a part of the companies group JSC "Strojservis" since 2000. It is located on Bachatski coal deposit (Guryev municipal district, Kemerovo region (Kuzbass), Western Siberia, Russia). The company produces high-quality coal of coking and energy brands. Production capacity of this open pit is 1 million tons of coal per year.

On the 1st of April, 2015, the sewage treatment plant of "Open pit Shestaki" represent a settling pond, which is organized by dumping earth dam of open pit overburden rock, with impervious screen of loam. Purification of quarry waste water takes place in the settling pond by settling (mechanical precipitation).

Qualitative characteristic of the water, discharged after treatment into the river Small Bachat, is given in Table 1.

Table 1. Quantitative characteristics of quarry waste water and the water disposed after purification using artificial filter array on JSC "Open pit Shestaki"

	Defining Ingredients (Pollutants)	*Units*	*Maximum Permeated Concentration (MPC)*	*Concentration of Pollutants Before Purifying*	*Concentration of Pollutants After Purifying (in Disposed Water)*
1	pH		6.5-8.5	7.5	7.21
2	Chlorides	mg / l	300	86.85	19.1
3	Ammonium ion	mg / l	0.5	1.18	0.44
4	Nitrate ion	mg / l	40	321.84	8.5
5	Nitrite ion	mg / l	0.08	6.42	0.19
6	Ferrum	mg / l	0.1	0.62	0.17
7	Sulphates	mg / l	100	389.42	76.1
8	Petroleum products	mg / l	0.05	3.45	0.13
9	Suspended solids	mg / l	Exceeding of concentration in the water body not more than by 75 mg / l	154.24	6.0
10	Dissolved oxygen	mg / l	More than 4	4.5	8.5
11	Cuprum	mg / l	0.001	0.006	0.0012
12	Manganese	mg / l	0.01	0.029	0.001

Taking into account the values given in Tab.1 we defined the efficiency of the existing on "Open pit Shestaki" water purifying facilities for major pollutants, which is:

- for suspended solids - 70.0% ;
- for petroleum products - 99.5% ;
- for sulphates - 50.0%;

- for chlorides - 50.0%.

Actual and permissible concentration of polluting substances in discharged waste water is accepted on the basis of the materials used in MPC developed in Russia.

The volume of water wasted in open pit mining of coal deposits in Segment number 2 of "Open pit Shestaki" constitutes 6,118.56 m^3 per year. Using treated waste water for technological purpose (moistening of exploded block, hydrodedusting of dumps' surface, roads irrigation) will reduce the amount of waste water discharged. Water consumption for technological purposes amounts 452.09 thousand m3 per year. Consequently, the volume of waste water discharged into the river Small Bachat after purification will be 5666.47 thousand m^3 per year.

Based on the purpose of the artificial filter array, its possible location on the JSC "Open pit Shestaki" must meet the following criteria:

1. Positive slope (water entry point to an array must be above the exit point of it).
2. Favorable location to existing water pipes to minimize costs.

Since the places with the natural groove are preferable to form the filtering array but not easy to find, for "Open pit Shestaki" we choose a site 150 meters long with a mellow slope (46.6 ‰) and relatively flat terrain.

A general view of artificial filter array implemented on JSC "Open pit Shestaki" is shown in Fig. 3.

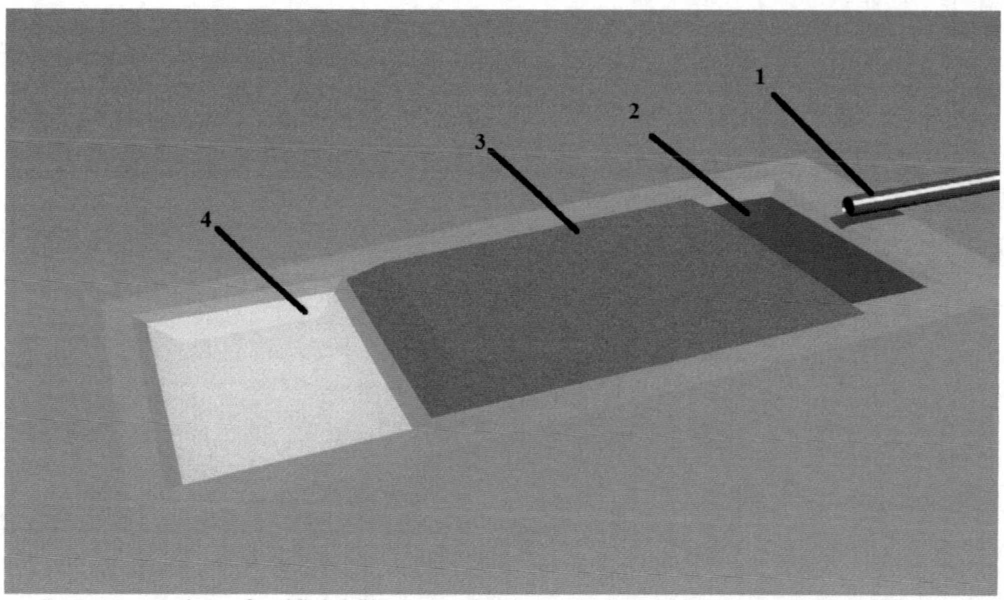

Figure 3. Common view of artificial filter array implemented on JSC "Open pit Shestaki": 1 – quarry waste water feeding pipe; 2 - quarry waste water collector; 3 – filtering array of overburden rock; 4 – purified water reservoir

Parameter determination of artificial filter arrays was performed by the method described in [12, 13] as:

- determination the length of the filter array (required filter length L);
- defining the width and height of the filter array;
- determination of the filter life-cycle;

The determination of filter array length is based on the quarry waste water purification to the Maximum Permitted Concentration (MPC) and can be calculated as:

$$L = \frac{1}{\eta}\ln\frac{C_{in}}{C_{mpc}} = \frac{1}{0.035}\ln\frac{154.24}{6} = 92.7\ m,$$

wherein: C_{in} – the initial concentration of suspended solids in quarry waste water (for 09/30/2015 C_0 = 154.24 mg / l);

C_{mpc} – MPC for suspended solids (C_{mpc} = 6 mg / l);

η – rate of filtration (for Kuzbass is in average 0.035).

We took the length of the top of filtering array as L = 95 m.

Determination of the width and height of the filter array is based on the conditions of purifying all incoming for treatment water without overflow over the top of the array or over the dam edge (in case of the construction of the filter on a horizontal surface). This sets the width of the filter array B.

Taking into consideration the topography, we assume the width of the filtering array B = 50 meters.

We used previously collected data of the filtering arrays on many open pits of Kuzbass observed in our later articles [8, 14] and defined the exact depth of filtration as 4.2 m.

We defined the average life-time (T) of filtering array accordingly its given height as:

$$T = \frac{1000 \times K_{use} \times L \times B \times h_{d.f} \times \rho_d}{Q_{av.an}(C_{in} - C_{mpc})} = \frac{1000 \cdot 0.15 \cdot 95 \cdot 50 \cdot 4.2 \cdot 1500}{646.85 \cdot (154.24 - 6)} = 46812\ hours = 5{,}34\ years.$$

wherein K_{use} – coefficient of using the volume of filtering array; $h_{d.f}$ – depth of filtration, m; ρ_d – density of sludge, $Q_{av.an}$ – average annual volume of waste water.

So, it can be set as 5 years and 5 months.

Conclusion

With this artificial filter array at the JSC "Open pit Shestaki" we can significantly improve the quality of water discharged into the river Small Bachat. Together with the existing waste water treatment plants such filter will allow to reduce the concentration of suspended substances, bringing them to the level significantly below the MPC. Finally, technological innovation in the open coal mining make it possible to increase the investment attractiveness of mining enterprises, to accelerate the modernization of Kuzbass coal cluster and to bring the environmental component of mining operations on the modern international level [15-18].

Acknowledgement

The research was made in National Research Tomsk Polytechnic University with financial support of the Ministry of Education and Science of the Russian Federation in the framework of scientific research "The evaluation and improvement of the social, economic and emotional prosperity of senior citizens" contract № 14.Z50.31.0029. Some parts of research were made in Kuzbass State Technical University with financial support of the Ministry of Education and Science of the Russian Federation in the framework of scientific research "Exploring the parameters of technological and technical solutions for selection and development of improving cutting and extracting mining machines operation efficiency in Kuzbass", contract №632. The authors would like to thank to prof. Alexey Khoreshok, prof. assistant Maxim Tyulenev and prof. Sergey Zhironkin for their discussions during the study. We thank anonymous referees for constructive and useful comments on the papers sent via e-mail.

References

[1] Zhironkin S A, Tyulenev M A, Zhironkina O V, Hellmer M 2016 The global determinants of mining higher education development *Metallurgical and Mining Industry* **2** pp 62-67

[2] Barysheva G A and Novoselova E G 2014 Methodology of Application of the Structural Shift Mechanism for Regulation of the National Economic Management System *Applied Mechanics and Materials* **682** pp 550-554

[3] Zhironkin S A 2002 Prospects and new possibilities investment attracting to Kuzbass coal mining industry *Ugol'* **6** pp 31-36

[4] Zhironkin S A 2001 Factoring and leasing development at coal mining industry of Kuzbass as an important element of its financial part *Ugol'* **4** pp 29-30

[5] Maksim Tyulenev, Yuriy Lesin, Sergey Zhironkin, Ekaterina Garina 2016 Coal producers waste

water purification *Metallurgical and Mining Industry* **2** pp 52-56

[6] Tyulenev M, Zhironkin S and Litvin O 2015 The low-cost technology of quarry water purifying using the artificial filters of overburden rock *Pollution Research* **34 (4)** pp 825-830

[7] Tyulenev M A and Lesin Y V 2014 *Justification complex purification technology open-pit mines wastewater* (Taishan: Academic Forum – Project on Mine Disaster Prevention and Control) pp 441-444.

[8] Lesin Y V, Lukyanova S Y and Tyulenev M A 2010 Mass transfer of dispersed particles in water filtration in macro-grained media *Journal of Mining Science* **46-1** pp 78-81

[9] Zhironkin S A 2002 About measures of vexel circulation development and vexelability definition of fuel-and-power complex' enterprises *Ugol'* **4** pp 47-48

[10] Zhironkin S A 2001 Governmental factoring development of TEK Kuzbass *Ugol'* **6** p 62

[11] Khoreshok A 2002 On side cutting bit when operating at sheerer drums *Ugol'* **7** pp 10-11.

[12] Karwan D L and Saiers J E 2012 Hyporheic exchange and streambed filtration of suspended particles *Water Resourse Research* **48** pp15-19

[13] Karwan D L, Gravelle J A and Hubbart J A 2007 Effects of timber harvest on suspended sediment loads in Mica Creek, Idaho. *For. Sci.* **53** pp181-188

[14] LesinY V, Luk'yanova S Y and Tyulenev M A 2015 Formation of the composition and properties of dumps on the open-pit mines of Kuzbass *IOP Conference Series: Materials Science and Engineering* **91 (1)** pp 1044-1049

[15] Gasanov M, Gasanov E and Egorova M 2016 The Technologies of the Network Prosperity in Russia *Procedia - Social and Behavioral Sciences* **166** pp 103-106

[16] Kovalev V, Gerike B, Khoreshok A, Gerike P 2014 *Preventive maintenance of mining equipment based on identification of its actual technical state* (Taishan: Academic Forum – Project on Mine Disaster Prevention and Control) pp 184-189.

[17] Aksenov V V, Khoreshok A A, Beglyakov V Yu 2013 Justification of creation of an external propulsor for multipurpose shield-type heading machine – GEO-WALKER, *Applied Mechanics and Materials*, **379**, pp. 20-23.

[18] Lekontsev Yu M, Sazhin P V, Temiryaeva O A Khoreshok A A (2013) Two-Side Sealer Operation, *Journal of Mining Science*, **49(5)**, pp. 757–762.

IOP Conf. Series: Materials Science and Engineering **142** (2016) 012124 doi:10.1088/1757-899X/142/1/012124

Air Processes Resulting in a Surface Layer Pollution in Industrial Regions

Yu V Lesin[1], T V Leshukov[2]

[1]Dr.Sc. Professor of Mining equipment department, Yurga Technological institute (branch) of National Research Tomsk Polytechnic University. Russia, Yurga, 652055, Leningradskaya Str., 26
[2]Post-graduate student, Kemerovo State University, Russia, Kemerovo, 650099, Krasnaya street, 6.

E-mail: [1]lyuv.geo@kuzstu.ru

Abstract. The article describes the air quality research in Western Siberia (Kemerovo region). The problem of air quality preservation in the conditions of mining industry intensive development is caused by the weather conditions which lead either to the concentration of pollutants in the surface layer, or to their migration to other geosphere, or to dissipation as a result of convective mixing or advection of air masses. Zoning of the territory in view of the research results provides insight into areas where the greatest risk to human health and life is formed.

1. Introduction

The problem of environmental changes as a result of the mining activity is very important for the Kemerovo Region. It is know that the success of the conservation of local ecosystems depends on the success of the transition of the coal industry from extensive to more efficient methods of mining taking into account the various constraints (technology, seismic, transportation, environmental, energy, demographic, etc.). In the works of V.Oparin and A.Ordin [1] the problem of considering all factors of limiting the volume of coal production in the Kemerovo region was observed. One of the limiting factors of mined coal volumes sums ecological and demographic problems. Also the problem of providing mining industry with human resources [2] has a major influence on limits of mined coal.

One of the geospheres that changes as a result of mining activities is the atmosphere, in particular, its surface layer, which leads to adverse consequences for the population: increased mortality, morbidity and economic losses of the individuals. In highly urbanized regions, most of the population is located in cities, where industries are concentrated.

2. Material and Method

The air quality depends not only on the issue of contaminants (pollutants) from industry, road transport and other sources, but also on weather events that can neutralize the emissions of pollutants or cause their concentration near the source. The frequency and duration of periods of high air pollution depends on the mode of emissions (single, emergency, etc.), and on the nature and duration of the meteorological conditions that increase the impurity concentration in the surface layer.

The main pollutants of the air basin in the Kemerovo region are mining companies, energy and fuel manufacturers [3-5]. The share of enterprises extracting energy minerals in total pollution reach up to 61.1% and of manufacturing enterprises - 20.4% (including the steel industry – 17.9%), power generation and gas and water distribution - 16.3% [6]. The increase in emissions is observed for power generation and distribution, gas and water distribution (more than 10% of 2010-2014).

When a temperature inversion decreases the speed of air floods convection, the accumulation of pollutants in the surface layer of the atmosphere (SLA) increases. The long temperature inversion increases the concentration of pollutants in the atmosphere to levels that are dangerous to the people.

Meteorological conditions (wind, precipitation, humidity, etc.) are important for the dispersion of pollutants [7-9]. Kemerovo region has severely continental climate so anticyclones are often occur in the summer and winter, forming calm weather with low rainfall and low wind speed [10, 11]. That reduces the self-cleaning ability of the surface layer of the atmosphere. Also in the winter over the territory the temperature inversion phenomenon in the atmosphere is formed due to the special orography and climate of the region.

In order to avoid increasing air pollution level in unfavorable for dispersion of pollutants meteorological phenomena it is necessary to predict and define conditions of severe pollution. In this research zoning of the Kemerovo region territory was conducted to determine the most disadvantaged areas of human activity. Also, these areas were correlated with the areas with the highest pollutant emissions [12] and the number of people living in this territory [13]. All data have been reduced to municipalities' borders of in order to create proposals to improve the situation.

The method of assessment the scattering ability of the surface atmosphere layer was proposed by Selegey T.S. [14]. The calculation was made according to the formula (1):

$$MPAD = \frac{Rc + Rf}{Rp + Rw}, \quad (1)$$

where: MPAD - meteorological potential atmospheric dispersion; Rc - repeatability of calm days (wind speed is less than 1 m/sec); Rf - repeatability of days with fog (humidity above 80%); Rp - repeatability of days with precipitation at least 0.5 mm; Rw - repeatability of days with wind speeds at least 6 m/sec.

From the values of the potential of air pollutants scattering by the atmosphere the following grading was proposed [14]:

- for MPAD > 2, it means the extremely adverse conditions (EUC);
- for MPAD > 1, it means adverse conditions (AC);
- for MPAD ≤ 1, it means favorable conditions (FC);
- for MPAD ≤ 0.5, it means an extremely favorable conditions for dispersion of pollutants in the atmospheric boundary layer.

The indicator can be used for calculating the self-cleaning capacity of the atmosphere at any time.

3. Results and Discussion

For MPAD index calculation the data from 21 meteorological stations interpolated in the Mapinfo software was used.

For the spatial analysis of inversion phenomena occurrence in the atmosphere, we used in this study a 6-hourly data reanalysis by NCEP/NCAR, presented at National Oceanic & Atmospheric Administration website (the NOAA) [15], converted into the local time (+7 hours). NCEP/NCAR data have a horizontal extension of 2,5°×2,5°. Geographical grid of research was 50.0° N - 57.5° N, 82.50° E - 90.00° E.

To study the inversion phenomena in the atmosphere there were selected three baric surfaces:

- 1000 hPa - isobaric surface of the Earth;
- 925 hPa - isobaric surface at an altitude of 750 m;
- 850 hPa - isobaric surface at an altitude of 1500 m.

Accordingly A.N. Akhmetshina methodology of layers selection [16] which is usually used for searching the presence of inversions, we use the following set of formulae:

$$\Delta T1 = T1000 - T925$$

$$\Delta T2 = T925 - T850$$

$$\Delta T = T1000 - T850$$

It is considered that the negative values of ΔT, ΔT1, ΔT2 indicate the presence of a temperature inversion.

The database was created using the MATLAB software, ncBrowser, and Microsoft Office Excel. The day with the inversion is the day in which there was either a high-altitude, or surface inversion [17].

Nevertheless this level of detailing is not sufficient to analyze the thermal structure of the different baric levels for the complex orographic [17]. Since the coal industry enterprises located within the Kuznetsk depression, having small elevation differences, we can roughly estimate the time and the frequency of inversions, their duration, and their impact on the air quality [18].

Emissions of pollutants into the atmosphere and population were transferred to municipal boundaries.

For the processing and presentation of data we used the following software: MATLAB, ncBrowser, Microsoft Office Excel, Mapinfo.

Results and discussion

The maps of comprehensive study are shown in Fig. 1.

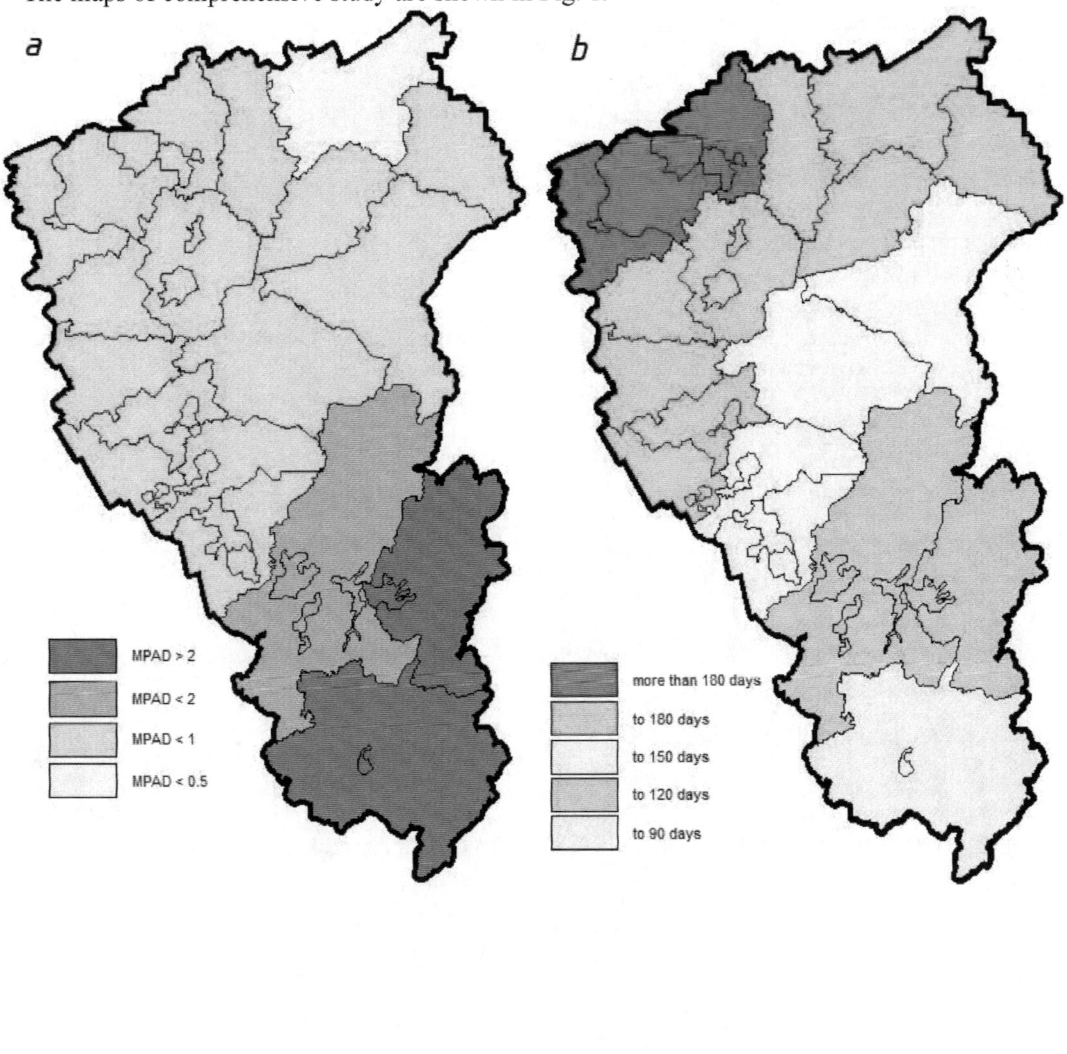

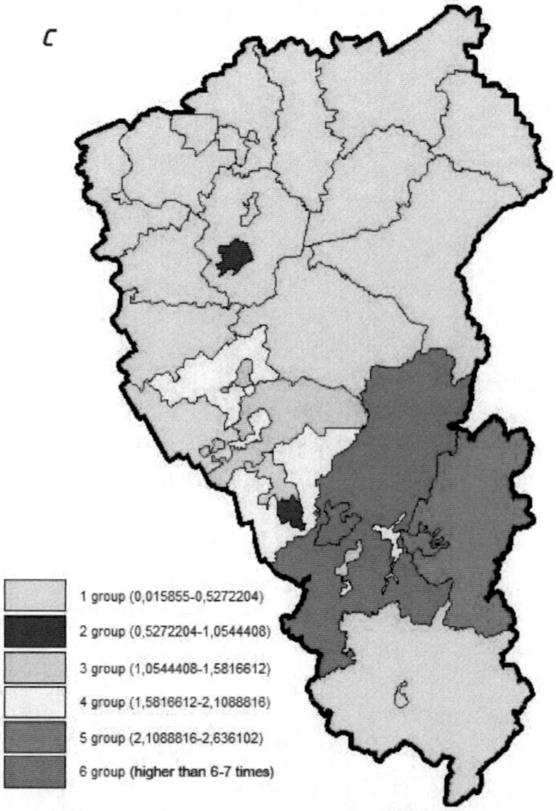

Figure 1. Schematic maps of Kemerovo region on various parameters: a – the map of meteorological potential of atmosphere pollutant scattering (MPAD); b – schematic map of number of days with temperature inversions in the atmosphere; c – schematic map of changes of standardized pollutants emitted to the atmosphere by stationary sources.

Over the territory of Kemerovo region in different times of the year various conditions for the accumulation of pollutants in the atmosphere is being formed. Fig. 2 shows the levels of MPAD in Kemerovo and Novokuznetsk – the main cities of the region.

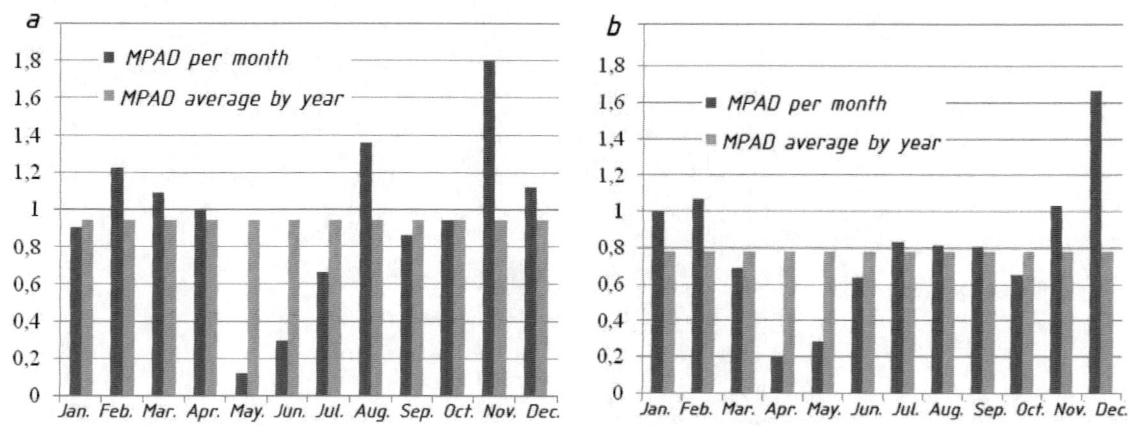

Figure 2. Change index MPAD (in points) in Kemerovo (A) and Novokuznetsk (B) during the year

As a result of meteorological potential research we concluded that the worst conditions for pollutants scattering for used parameters are formed during the winter, which worsens the condition of the surface atmosphere layer in the cities. The most favorable conditions for pollutants scattering are formed over the north area (Yaya, Izhmorsky, Mariinsky, Yurga areas). Deterioration of MPAD in the south-east of Kemerovo region threatens the whole environment, as well as they are the main areas of various pollutants' emission into the atmosphere [19-21].

Research of formation of inversions in Kemerovo region has shown that an increase in inversions from south to north, to north-west and to east-west. Temperature inversions were mostly close to surface in winter. There was also a simultaneous formation of surface and elevated inversions. Most often their joint forming took place in the winter months, and due to the air compression process in powerful anticyclones. Formation of prolonged inversions prevents convective mixing of air masses, which leads to pollutants concentrations increase in the surface layer and above the maximum permissible concentration.

The main areas of atmosphere pollutants' emission are located with mining specialized districts [22] (Novokuznetsk, Prokopevsk, Kiselevsk districts). The location of these areas in the MPAD interval of 1 - 2 prevents the scattering of pollutants in the atmosphere.

4. Conclusion

During the town-planning work not only the wind rose but the frequency of incidence and annual variation of temperature inversions over the planning area should be taken into account. This is especially important for Kemerovo region, as over its territory anticyclones' weather is often formed, which is associated with surface and elevated temperature inversions. If this phenomenon is not taken into account there is a risk for population and the local ecosystem. The problem of the formation of photochemical smog is available in Moscow, Tokyo, Sydney and others.

The use of MPAD indicators, the number of days with temperature inversions and the amount of pollutants emitted over the territory gives an overview of the changes in air quality, and allows determining the areas with the most favorable environment for people' living. In the transition to sustainable regional development these atmospheric processes must be considered.

5. Acknowledgement

The authors would like to thank to prof. Alexey Khoreshok, prof. assistant Maxim Tyulenev and prof. Sergey Zhironkin for their discussions during the study. We thank anonymous referees for constructive and useful comments on the papers sent via e-mail.

References

[1] Oparin VN, Ordin AA 2011 Strategic Mine Planning Under Uncertainty *Journal of Mining Science* **47(2)** pp 254-266.
[2] Leshukov TV 2013 Demograficheskaya bezopasnost gornopromyishlennogo regiona (na primere Kemerovskoy oblasti). *Vestnik Kemerovskogo gosudarstvennogo universiteta* **4(2)** pp 14-18.
[3] Zhironkin SA 2002 Prospects and new possibilities investment attracting to Kuzbass coal mining industry *Ugol'* **6** pp 31-36
[4] Zhironkin SA 2001 Factoring and leasing development at coal mining industry of Kuzbass as an important element of its financial part *Ugol'* **4** pp 29-30
[5] Tyulenev M, Lesin Y, Zhironkin S, Garina E 2016 Coal producers waste water purification *Metallurgical and Mining Industry* **2** pp 52-56
[6] Tyulenev M, Zhironkin S and Litvin O 2015 The low-cost technology of quarry water purifying using the artificial filters of overburden rock *Pollution Research* **34 (4)** pp 825-830
[7] Tyulenev MA and Lesin YV 2014 Justification complex purification technology open-pit mines wastewater *Taishan Academic Forum – Project on Mine Disaster Prevention and Control* pp 441-444.
[8] Lesin YV, Lukyanova SY and Tyulenev MA 2010 Mass transfer of dispersed particles in water

filtration in macro-grained media *Journal of Mining Science* **46-1** pp 78-81

[9] Zhironkin SA 2002 About measures of vexel circulation development and vexelability definition of fuel-and-power complex' enterprises *Ugol'* **4** pp 47-48

[10] Zhironkin SA 2001 Governmental factoring development of TEK Kuzbass *Ugol'* **6** p 62

[11] Khoreshok AA 2002 On side cutting bit when operating at sheerer drums *Ugol'* **7** pp 10-11.

[12] Lesin YV, Luk'yanova SY and Tyulenev MA 2015 Formation of the composition and properties of dumps on the open-pit mines of Kuzbass *IOP Conference Series: Materials Science and Engineering* **91(1)** 012091

[13] Aksenov VV, Khoreshok AA and Beglyakov VY 2013 Justification of Creation of an External Propulsor for Multipurpose Shield-Type Heading Machine – GEO-WALKER *Applied Mechanics and Materials* **379** pp 20-23

[14] Selegey TS, Zinchenko GS, Bezuglova NN 2005 Uchet meteorologicheskogo potentsiala samoochischeniya atmosferyi pri reshenii zadach promyishlennogo osvoeniya territoriy *Polzunovskiy vestnik* **4** p.23.

[15] NOAA Earth System Research Laboratory.http://www.esrl.noaa.gov/psd/data/reanalysis/reanalysis.html.

[16] Akhmetshina AS 2013 Assessment of thermal structure of boundary layer atmosphere of western Siberia B*ioclimland (biota,climate, landscapes,* **1b** pp.5-8.

[17] Stiperski I, Rotach MW 2016 On the Measurement of Turbulence Over Complex Mountainous Terrain *Journal of labor research* **159** pp.97–121.

[18] Antony Z. et al. 2005 Effects of changing surface heat flux on atmospheric boundary-layer flow over flat terrain *Boundary-Layer Meteorology* **116** pp.331–361.

[19] Khoreshok AA 2002 On side cutting bit when operating at sheerer drums *Ugol'* **7** pp 10-12

[20] Lekontsev Yu M, Sazhin P V Temiryaeva O A, Ushakov S Yu, Khoreshok AA 2013 Two-side sealer operation *Journal of Mining Science* **49(5)** pp 757-762

[21] Prokopenko SA 2014 Multiple service life extension of mining and road machines' cutters *Applied Mechanics and Materials* **682** pp 319-323

[22] Kovalev VA, Gerike BL, Khoreshok AA. and Gerike PB 2014 Preventive maintenance of mining equipment based on identification of its actual technical state *Symposium of the Taishan academic forum – Project on mine disaster prevention and control* pp 184-189.

IOP Conf. Series: Materials Science and Engineering **142** (2016) 012125 doi:10.1088/1757-899X/142/1/012125

Determination of Dynamic Characteristics of the Frame Bearing Structures of the Vibrating Separating Machines

V V Piven and O L Umanskaya

Tyumen State Oil and Gas University 38, volodarskogo St., Tyumen, 625000 Russia.

E-mail: pivenvv@yandex.ru

Abstract. Within the vibrating separating machines the vibration displacement of the members is transferred to a frame bearing structure, and over it the movement is transferred again to the suspension brackets of the sieve separating surfaces and to the foundation, on which the machine is fixed. The forced oscillations of the sieve separating surfaces ensure the separation process, and the vibration, transferred from the frame structure, disturbs this process. It is necessary to ensure the vibration displacement of the separating surfaces within the fixed limitations by means of optimal design of the frame bearing surfaces. The aim of the work is to decrease adverse vibrations towards the technological separation process. The calculated and graphical relations, acquired according to the presented methods, enable to estimate the influence of various structure solutions on vibration displacements of the structure elements at the stage of design.

1. Introduction

The vibrating machines are widely used in the industry for physical and chemical processes intensification, transportation, mixing, compaction, and dozing of bulk materials, as well as during their division into fractions [1, 2]. Numerous researches have established a distinctive relation of the quality of performance of vibration technological process from the rational operating modes and parameters of the vibration equipment [3, 4, 5]. The most challenging and demanding task is separation of bulk granular materials in postharvest grain. This is because the active material is a foodstuff and requires a high quality of separation [3, 6]. The main working members of vibrating separating machines are flat sieves, making vibration displacements. The sieves are connected to the frame bearing structure through elastic elements (Fig. 1).

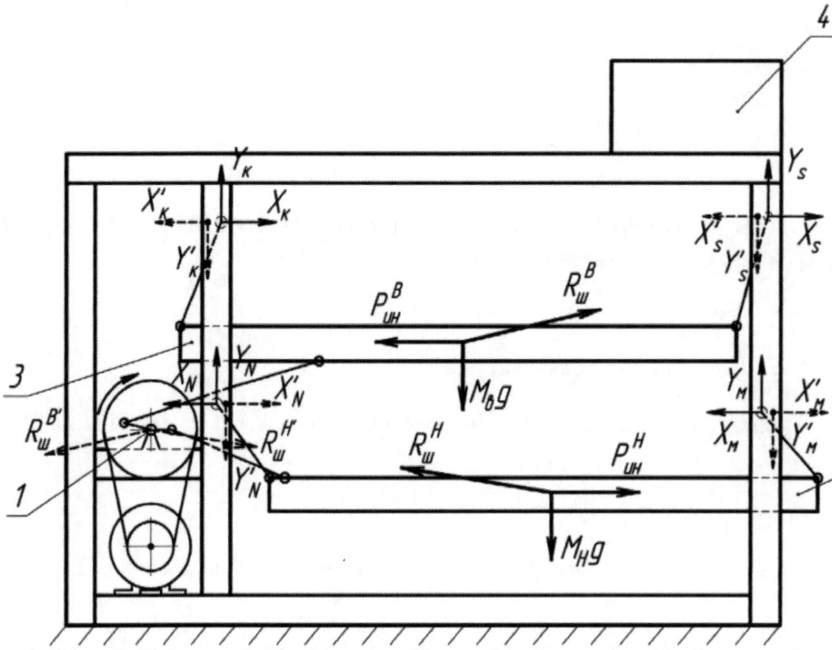

Figure 1. Structural scheme of vibrating separating machine's structure (the forces, acting towards the bearing construction from the drive mechanism parts, are shown with dotted lines): 1. Drive mechanism 2. Lower sieve case 3. Upper sieve case 4. Aspiration system.

The vibration displacement of the working members is transferred to the framework bearing construction. Then, over the bearing structure vibration is transferred to a closing of the structure building, where the machines of this type are fixed [7]. Vibration enables the technological process implementation, but its maximum values are limited by ergonomic requirements. Besides, the vibration displacement in the surface, perpendicular to the separating surface, may lead to a decrease of technological process quality.

In order to choose the optimal structure parameters for vibrating machines design it is necessary to investigate all the vibration characteristics of these machines' elements [7, 8, 9, 10]. One of the most important dynamic characteristics of the frame bearing structure is eigenfrequencies of oscillations of the mechanical system.

2. Materials and Method.

To assess the impact of the design features on the vibratory displacement of the frame structure is a necessity to determine the natural eigenfrequencies of the structure. We perform these calculations by the example of an elementary design with two weights of m1 and m2 (Fig. 2). The frame structure taken into consideration may be presented as an asymmetric frame with sealed poles, loaded with two concentrated masses.

Disregarding the proper mass of the system, we apply the inertia forces to the points of the frame structure taken into consideration may be presented as an asymmetric frame with sealed poles, loaded with two concentrated masses location of the masses m_1 and m_2. For the system with two degrees of freedom the formulae for calculating the eigenfrequencies will be as follows [11]:

$$\begin{aligned} &\lambda_1 \cdot (m_1 \delta_{11} \omega^2 - 1) + \lambda_2 m_2 \delta_{12} \omega^2 = 0; \\ &\lambda_1 m_1 \delta_{21} \omega^2 + \lambda_2 \cdot (m_2 \delta_{22} \omega^2 - 1) = 0, \end{aligned} \tag{1}$$

where λ - amplitude of oscillations, m; δ_{ik} – shift towards *i*, caused by the single force, acting towards *k*, m; ω - oscillations' eigenfrequencies, rad/sec.

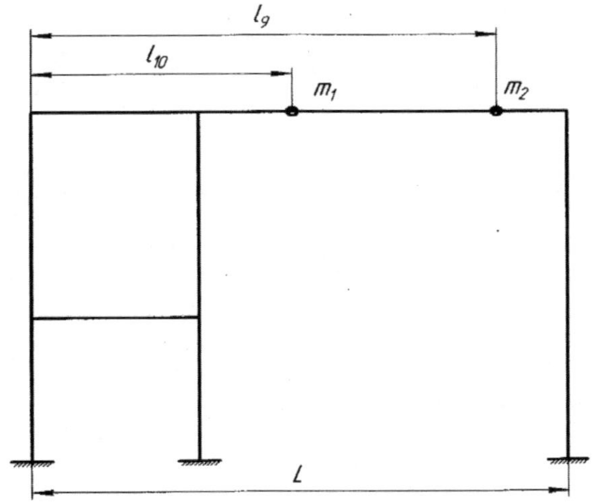

Figure 2. Diagram for calculating eigenfrequencies of oscillations of a framework construction in a vertical direction.

Taking into consideration that the amplitude, in case of oscillation doesn't vanish (does not turn to zero), the determiner, made up from coefficients of the system of equations, is equal to zero. The expanded determiner may be presented with the following equation:

$$\omega^4 \cdot (\delta_{11}\delta_{22} - \delta_{12}^2) \cdot m_1 m_2 - \omega^2 \cdot (\delta_{11}m_1 + \delta_{22}m_2) + 1 = 0 . \tag{2}$$

In order to determine the shifts from the single loads it is necessary to draw distribution diagrams of bending moments under the single forces $P_1 = 1$ and $P_2 = 1$, applied at the points of location of the loads with the masses m_1 and m_2 the values of the shifts δ_{ik} are determined with the help of the Vereschagin's rule and the shifts reciprocity theorem (Maxwell's theorem) [11].

3. Results and Discussion.

The solution of the equation (2) was carried out in the "MATHCAD" program. There were the first and second eigenfrequencies of the frame bearing structure's oscillations obtained, which were equal to 110 rad/sec and 185 rad/sec correspondingly.

In order to pick up the optimal construction of the frame it is necessary to have a look at the influence of various factors on the vibration displacements of the frame. The value of the vibration displacements of the bearing structure depends on the value of eigenfrequencies of the structure. According to the equation (2), the influence on the eigenfrequencies value is made by the value of concentrated masses and the shift of the beam under the influence of the loads with the masses m_1 and m_2 (Fig. 2), as well as the deviation of direction of the gravitation force from the m_1 mass under the influence of the m_2 mass. The values of the given shifts depend on these masses location points. If the masses are shifted towards the center, the deflections increase and the non-dimensional dynamic factor, i.e. the relation of dynamic shift to the static, determined by the formula [12], is growing:

$$K_{\partial} = 1 / |(1 - p^2/\omega^2)| , \tag{3}$$

where p – frequency of the exciting force, rad/sec.

The relation between the dynamic factor and the point of application of the load with the m_1 mass, in case the value of l_9 is constant, is presented in the figure 3 (curve 1). The relation between the dynamic factor from the point of

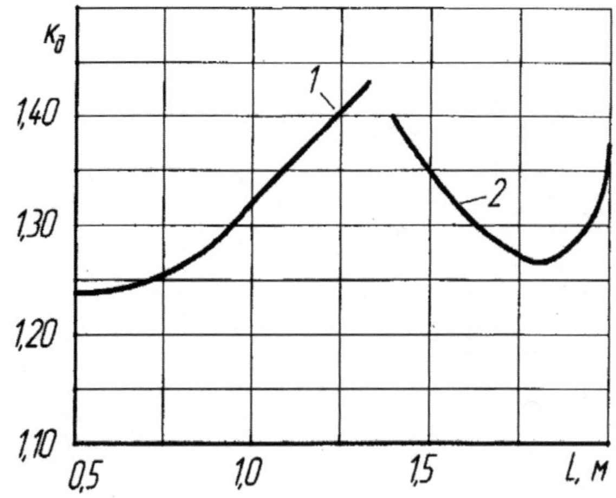

Figure 3. Relation of the dynamic factor *($k_д$)* from the masses location points: 1.For the load with the m_1 mass 2. For the load with the m_2 mass.

location of the load with the m_2 mass, in case the value of l_{10} is constant, is described with the curve 2. The value of dynamic factor is maximal, if the masses are simultaneously applied to the center of the construction. In this case the dynamic factor takes the value of 1.78, in case of the initial diagram it is equal to 1.26 and its increase in this case is 40%. In case of the maximal remoteness of the given loads from one another, the dynamic factor decreases for 10%.

Let us investigate the relation of the dynamic factor from the values of the masses, applied to the given structure. As either of the masses grows, all the eigenfrequencies decrease, which leads to the grow of the dynamic factor. The relations of the dynamic factor from the values of the loads' masses are presented in the figure 4. The increase of the m_1 mass for 40% causes the growth of the dynamic factor for 8% (curve 1).

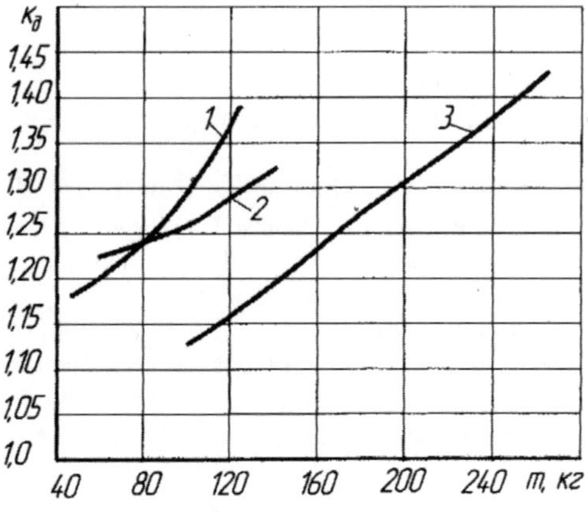

Figure 4. Relation of the dynamic factor *($k_д$)* from the value of masses *(m_1, m_2)*: 1. For the load with the m_1 mass
2. For the load with the m_2 mass 3. For the both loads with the sum of masses $m_1 + m_2$.

As the m_2 mass grows for 47%, the value of the dynamic factor grows for 9.5% (curve 2). The simultaneous growth of the masses causes the maximal growth of the dynamic factor. In case of increasing the sum of the masses for 43% the dynamic factor grows for 13.2% (curve 3).

The data presented indicate that the value and position of concentrated masses make a considerable influence on vibration displacements of the separation machine's bearing structure. The most significant factor is location of the masses. If the masses are positioned in the middle of the upper horizontal beam, the value of its vibration displacement in a vertical direction increases for 40%. If the mass of the system is increased due to the increase of concentrated masses, the values of eigenfrequencies decrease, and the vibration displacements increase (for the case when the forced frequencies of oscillations are less then the eigenfrequencies).

In order to determine the influence of additional stiffeners on the vibration displacements of the bearing structure there were calculations made for the maximal values of the vibration displacements of changed structures (Fig. 5 – 6). The additional stiffeners are marked with dotted lines. The values of axial inertia torques of additional stiffeners sections are equal.

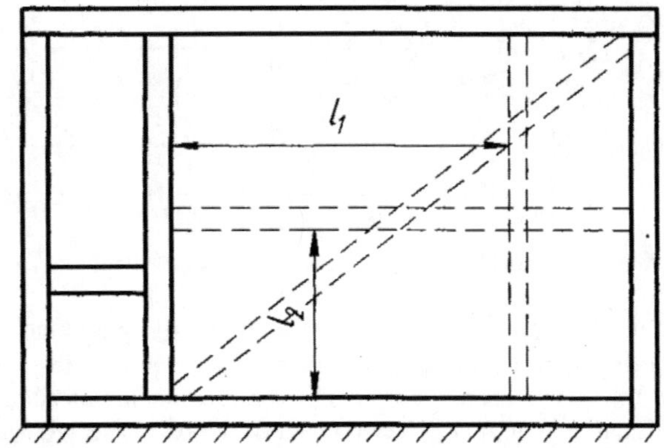

Figure 5. The framework structure diagram with the elements, implementing its stiffness (the length of the construction L= 2.27 m; height H= 1.59 m).

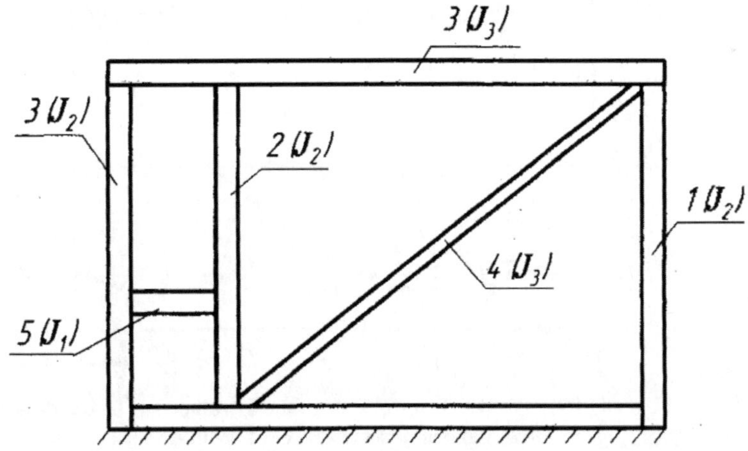

Figure 6. Example of the bearing structure diagram with an indication of axial inertia torques of the section (J) and numbers of beams constituents: J_1= 88.35·10^{-8} m^4 (length of the beam with inertia moment of the section J_1 – 0.4 m),

J_2= 87.88·10^{-8} m^4, J_3= 29.96·10^{-8} m^4.

The calculations of maximal vibration displacements were carried out with the help of the "MATCAD" program, adding the introduced internal force factors to the initial diagram. The statistical indeterminateness of the changed structure was found. The determination of shifts in the bars was carried out by the initial parameters method, taking into consideration, that the bars have the same section. The obtained values approximation was carried out with the help of the "MAPLE" program.

The values obtained were decomposed into the first, second, and third degree polynomials, and the function with the least error was taken. Secondary members of rigidity of the frame structure have different effects on the vibration displacement of the frame structure. Examine the impact of changes in the value of l2 vibration characteristics for horizontal stiffener (see Fig. 5).

We will investigate the influence of the l_2 value change on the vibration characteristics. The dependence of the dynamic fac tor from the l_2 value is presented in the figure 7.

If a horizontal element is added, the dynamic factor decreases with the minimal value in the middle of the vertical pole's height. The dynamic factor's decrease, compared to the initial diagram, is 15%.

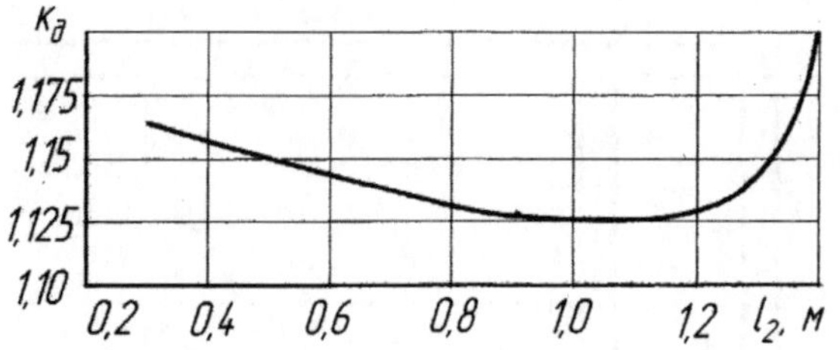

Figure 7. Relation of the (k_∂) dynamic factor from the l_2 value.

If a horizontal element is added, the dynamic factor decreases with the minimal value in the middle of the vertical pole's height. The dynamic factor's decrease, compared to the initial diagram, is 15%.

At change of the size l2 there is a change of magnitude of the maximum vibration displacement of the top horizontal beam in the vertical direction. It is expressed by the relation, presented in the Fig. 8 (curve 2).

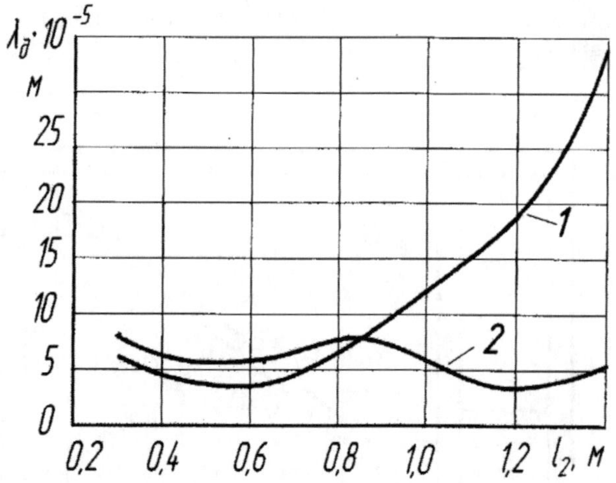

Figure 8. Relation of the vibration displacement (λ_∂) from the l_2 value: 1. For the right vertical pole in the horizontal direction 2. For the upper horizontal beam in the vertical direction.

The vibration displacement's decrease of the horizontal beam, compared to the initial diagram is 30%. For the operation of the machine with the presented structure diagram the most important issue is the vibration displacement of the vertical pole, on which the sieve frame's suspension brackets are fixed. This vibration displacement makes a considerable influence on the technological process. The influence of the l_2 value change on the vibration displacement of the right vertical pole in the horizontal direction is shown in the Fig. 8 (curve 1).

The additional horizontal stiffener increases the vertical shift of the vertical pole for 70%. Thus, installation of the given additional stiffening element is not efficient for the presented structure.

Let us change the initial diagram, placing an additional vertical stiffening element into it (see Fig. 5). Let us have a look at the influence of the given element on the vibration displacements of the same vertical and horizontal beams. The space location of the additional bar is determined by the l_l variable quantity. In the Fig. 9 the relation of the dynamic factor from the l_l value is presented. Placement of the additional vertical bar increases the construction's dynamic factor within a certain range. The maximal increase of the dynamic factor, compared to the initial diagram is 70%.

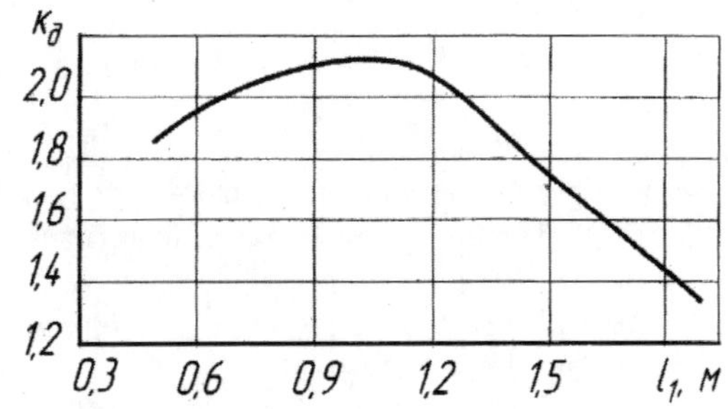

Figure 9. Relation of the dynamic factor (k_d) from the l_l value.

The dependence of the maximal vibration displacement of the upper horizontal beam from the l_l value is shown in the Fig. 10 (curve 2). The dependence of the vibration displacement of the vertical pole from the l_l value is presented in the Fig 10 (curve 1).

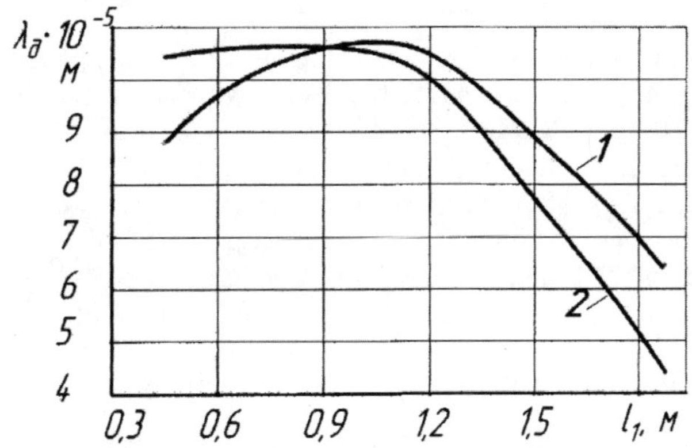

Figure 10. Relation of the vibration displacement (λ_d) from the l_l value: 1. For the right vertical pole in the horizontal direction 2. For the upper horizontal beam in the vertical direction.

The maximal value of the horizontal beam's vibration displacement, if the location of the stiffening element is within the range of l_l= 0.6 – 1.2 m is 80% and the vertical beam – 44%.

The efficiency of application of any stiffening elements in statically undetermined frame structures should be identified by means of improving the structure in terms of a number of necessary parameters.

The introduction of additional inclined stiffener (see Fig. 5) causes a reduction of vibration displacement of horizontal beam by 11% and the vertical one by 89%. Thus, the given diagram proves to be the most efficient.

By the similar method we analyze the vibration displacements of the frame structure members with various methods of connecting the elements between each other (for example, in case of a joined connection). In this case, disclosing the static indeterminateness of the construction, it is necessary to make the corresponding internal force factors to be equal to zero.

4. Conclusion.

As a result of the conducted research, there is established the influence of capacity of the concentrated masses and a place of their location on the coefficient of dynamism of a frame bearing structure of the vibrating separating machines. There are obtained dependences describing influence on vibration displacements of a frame structure of various elements of rigidity. The results of the research are recommended to be used at design of bearing structures of vibrating separating machines for decrease of adverse vibrations.

Further researches in the field should be directed on studying of the influence of vibration displacement of the basis on which the separating machine is installed. Accountability of all the presented factors enables one to provide a more qualitative description of a process of moving of a separated component on a perforated separating surface and to estimate a possible technological separation effect more precisely.

References

[1] Piven' V V and Umanskaya O L 2013 Classification of vibration separating machines *Modern scientific researches and innovations* **3** [Electronic journal] URL: http://web.snauka.ru/en/issues/2013/03/22592
[2] Chumakov V G 2011 Division of grain heap into fractions by a pneumatic screen separator *Tractors and Agricultural Machinery* **12** 31 http://www.tismash.ru/archives/5582
[3] Piven V V and Umanskaya O L 2014 Problems of decrease in harmful vibration in vibration separating machines *Modern scientific researches and innovations* **3** [Electronic journal]. URL: http://web.snauka.ru/en/issues/2014/03/31171
[4] Lapshin N P 2016 Self-cleaning a sieve at vertical vibrations sieve *International Research Journal* **2-2(44)** 28 http://dx.doi.org/10.18454/IRJ.2016.44.107
[5] Ivanov N M and Lapshin N P 2015 Stability of motion of the sieve cleaner MVR-8 (RP-50) *International Research Journal* **10-2(41)** 73 DOI: 10.18454/IRJ.2015.41.091
[6] Tarasenko A P, Orobinsky V I and Giyevsky A M 2008 The results of the state certification and periodic testing of devices of the family OSF *Vestnik of Voronezh State Agrarian University* **3-4** 32 http://elibrary.ru/item.asp?id=11741667
[7] Piven' V V end Umanskaya O L 2007 Determining the dynamic parameters of the structures of the vibration machines on an elastic foundation *Russian Engineering Research* **27** **Issue 5,** 241 DOI: 10.3103/S1068798X07050012
[8] Lapshin I P 2004 Modernization of grain screening machines as one of the ways to maintain the industrial technology of postharvest grain processing *Izvestia Orenburg State Agrarian University* **2** **(№ 2-1)** 241 http://elibrary.ru/item.asp?id=15263947
[9] Lapshin I P, Lapshin N P and Babaev I H 2014 Reducing sieve dirt collection by circular vibrations in a longitudinal vertical plane *Vestnik Bashkir State Agrarian University* **2 (30)** 84 http://elibrary.ru/item.asp?id=22299233
[10] Vozmilov A G and Yarullin R B 2011 The definition of the law and parameters of regulation inertial vibro drive of a vibro grain cleaner *Bulletin of South Ural State Universiti. Series: Power Engineering* 15 (232) 46
[11] Feodosyev V 1973 Strength of materials (Moscow: Mir Publishers).
[12] Targ S 1968 Theoretical Mechanics: A Short Course (Moscow: Mir Publishers)

IOP Conf. Series: Materials Science and Engineering **142** (2016) 012126 doi:10.1088/1757-899X/142/1/012126

The Mutual Effect of Reciprocally Moving Geokhod and Geological Environment Studied by the Discrete Element Method in Software PFC3D 5.00

V.Yu. Timofeev[1], T.S. Kust[1], A.A. Dronov[1], I.I. Beloglazov[2], D.A. Ikonnikov[2]

[1] Yurga Institute of Technology, TPU Affiliate, 26, Leningradskaya Str., 652050, Yurga, Russian Federation
[2] National Mineral Resources University (Mining University), 21 line, Vasilevsky Island, 199106, Saint Petersburg, Russian Federation

E-mail: tv-ytitpu@mail.ru

Abstract. A numerical experiment procedure of geokhod traverse in the geological environment, based on software PFC3D 5.00 is presented in the paper; the interpretation of numerical experiment results is provided.

Introduction

At present a new generation of geokhod is being developed by researchers of Yurga Institute of Technology, TPU affiliate [1]. The principles of operation, regularities of its traverse in the geological environment are studied [2], parameters of its transmission [3-6], working units [7-8], pump station [9], as well as those of casing of geokhod sections [10] are being determined.

A pilot sample of geokhod is to be tested soon when tunneling mines with a diameter of 3.2 m. Results of tests will be used for making a decision regarding its further modernization or serial production organization of the developed sample, as well as outlooks of prospective investigations should be done. However, it is obvious at this moment geokhod, having a new operation principle based on the mutual effect with the geological environment requires further study in order to improve its constructions.

One of the promising ways of research into the traverse of geokhod in the geological environment is a quite new method of discrete elements (MDE). MDE is a set of numerical methods intended for calculating a lot of discrete particles. Having been used in 1971 first to solve problems of rock mechanics, this method is widely applied now to solve geo-mechanical problems. A number of software is available currently for MDE-simulation of processes and solution of engineering, applied and fundamental scientific problems. The most accepted and widely used software is listed below: YADE; LIGGGHTS; Chute Maven; EDEM; PFC3D etc.

A wide range of software is available as they support solution of special tasks. The software programs differ mainly in functional capabilities, tasks to be solved, process, used operation system, a possibility of splitting up calculation, which is necessary for the complete use of a multi-process equipment (efficient computing machines).

One of leading companies, manufacturing software for MDE implementation, is Itasca (Itasca Consulting Group, Inc., USA). This company is oriented at the development of following software

IOP Conf. Series: Materials Science and Engineering **142** (2016) 012126 doi:10.1088/1757-899X/142/1/012126

programs: FLAC/FLAC3D, PFC2D/PFC3D, KUBRIX, 3DEC, UDEC. A wide range of offered programs support solution of almost all MED-solved geo-mechanical problems. We used software PFC3D, version 5.00 to carry out numerical experiment of geokhod traverse in the geological environment. Simulation of various physical processes for solution of scientific and engineering problems is based on this software, used in diverse fields of science and production: process industry, mining industry, and engineering and geological prospecting sciences etc. This software is selected due to its functional capabilities in the field of modelling the mutual effect of solids and discrete particles. Operation system Windows is used to run this program, making it widely applied all over the world, moreover, it supports splitting up the calculations. An advantage worth emphasizing is compatibility of CAD-system file with accepted formats (format STL).

Methodology of numerical experiments

When carrying out a numerical experiment by software PFC3D 5.00 assumptions listed below are made:

- particles have six degrees of freedom;
- particles are rigid and resist to deformation when coming into contact;
- the environment to be modeled is isotropic and homogenous;
- particles of any geometrical form can be applied in principle, but spherical particles are most efficient for the case under consideration;
- particles interact according to the pre-set contact law when contacting;
- external forces can be applied to particles, e.g. gravity force;
- contacts between particles can be broken or re-formed when carrying out experiments.

The experiment comprises six phases:

1. Generation of particles (development of the geological environment model) and assignment of boundary and starting conditions.
2. Determination of a contact type of particle to particle and particle to boundary.
3. Calculation of contact forces and torsion moments of all interacting particles.
4. Calculation of accelerations, speeds, values of displacement and rotation frequencies of all particles.
5. Calculation of new positions for all particles.
6. Phases 2 to 5 are repeated at time interval till approaching the boundary conditions.

The development of a design model

A design model of geokhod was developed in order to conduct MDE-experiments of geokhod traverse in the geological environment by software PFC3D 5.00. A design model of geokhod is made in software SolidWorks and exported in format STL (Figure 1).

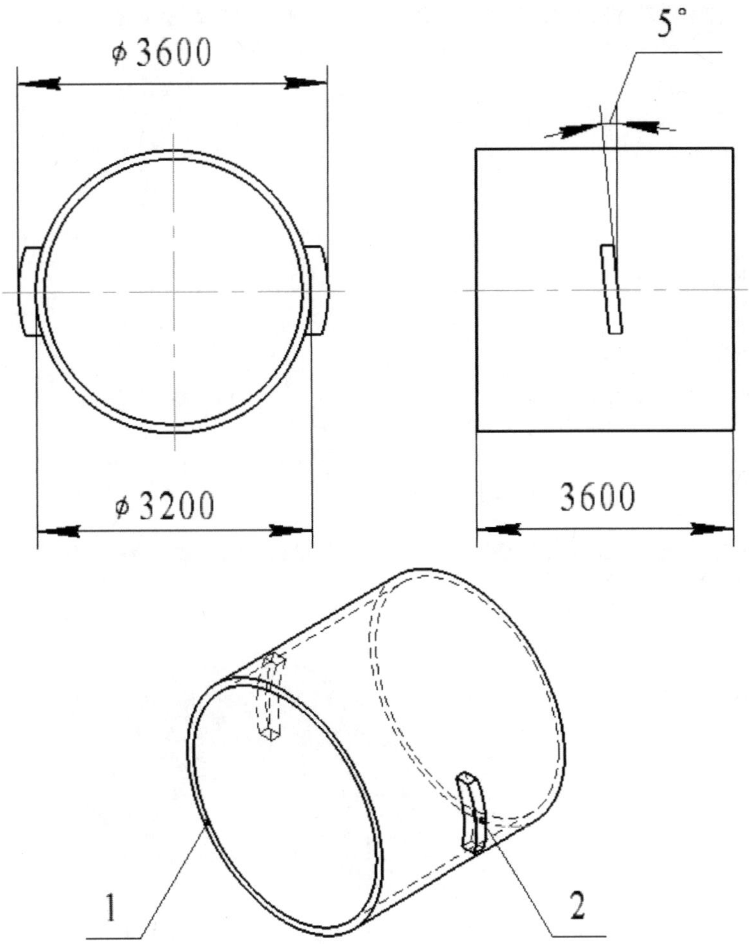

1 – body; 2 – propulsive agent
Figure 1. A design model of geokhod

A design model is a simplified structure of geokhod head section with two propulsive agents. The dimensions of the design model taken in absolute magnitudes are in line with those of geokhod. Propulsive agents are sections of a helical surface; an inclination angle to the perpendicular of traverse axis equals the same angle of geokhod. The model exported into format STL is loaded as an imported geometry in software PFC3D 5.00.

Assignment of modelling boundary conditions
A distinctive feature of system PFC3D 5.00 is a capability of function setting by program using pre-set algorithms and commands in special text files instead of visual function setting by the interface. The development of modelling boundary conditions is based on the same principle. Figure 2 provides a command sequence, which imports the model of geokhod and modelling boundary conditions are developed in form of a domain (command "domain") and "walls" (command "wall").

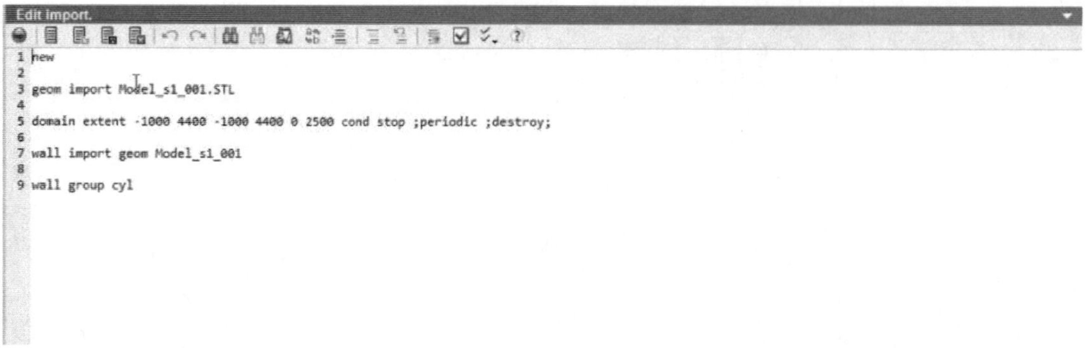

Figure 2. A sequence of commands for import of a design model and development of boundary conditions

The result of implementing commands is given in Figure 3.

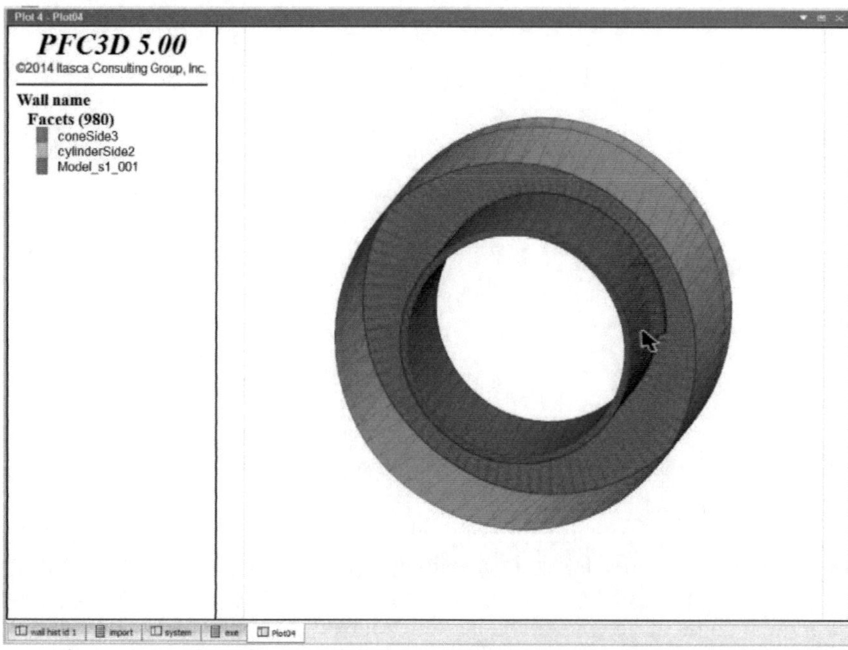

Figure 3. Import of the model and development of boundary conditions

An imported design model of geokhod is given blue in Figure 3, green is boundaries of a “domain” – field, where particles will be developed (area of geological environment modelling), a “wall” is given red – a boundary of discrete elements traverse when modelling the geological environment. Properties of a “domain” differ from those of a “wall” because the “wall” can be permeable or it tends to destroy discrete elements out of its boundaries. In this case particles out of the “wall” are cancelled due to its properties. Dimensions of a “domain” are determined taking into consideration a probable area, where the mutual effect of propulsive agent and geological environment influences displacement and deformation of discrete elements. The dimensions of the domain: diameter (4400 mm), length (2500 mm).

Modelling the geological environment

Modelling is one of the most important experimental phases, since parameters, which are as alike those of real rocks as possible, are to be built into the geological environment to be simulated. Moreover, a number of discrete elements have a significant effect on computer capacity and its operation time when calculating experimental results. It is impossible to take into account all the parameters because of deficient data on geokhod operation at this stage; so a simplified model of the geological environment is accepted. This model includes a minimal set of parameters of the geological environment to be modeled, although accurate data can be obtained on their base; the geological environment is anisotropic and homogenous as well. In this case a geological environment with 75 000 discrete elements is simulated in the "domain". The number of particles is selected empirically while carrying out similar experiments by computer with CPU clock 3.7 GHz, RAM memory space 16 GB; and the experiment lasted approximately 6-8 hours. An element size is varied from 5 to 17 mm, on assumption of complete filling "domain" – "wall" – and design model space.

In Figure 4 there is a sequence of commands presented to create discrete elements, which make a geological environment to be modeled, as well as an algorithm of using these discrete elements to fill the space of geokhod design model and "walls" of "a domain".

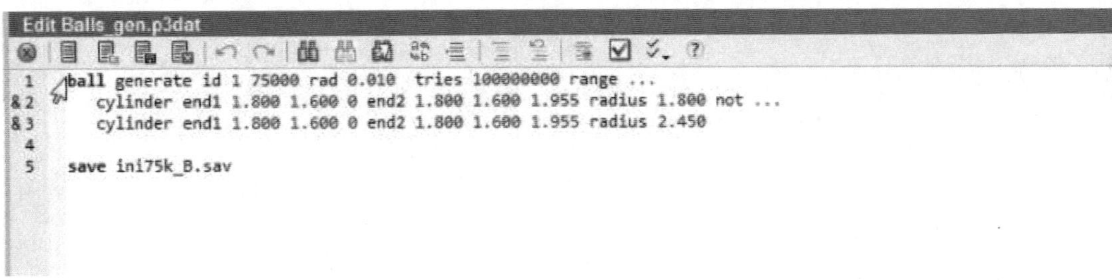

Figure 4. Development of discrete elements

In Figure 5 there is a fragment of a file presented, which includes commands assigning properties of experiment elements and ways of their mutual effect:

- a rule of discrete element mutual effect («linear bond» [1]);
- a rule, describing the mutual effect of discrete elements and the model edge («linear»);
- properties of contacting elements (variables «kn», «ks», «dp_ratio» [1]);
- properties of contacting discrete elements (variables «pb_kn», «pb_ks», «pb_ten», «pb_coh» [1]);
- properties of model "walls" (variables «kn», «ks», «froc» [1]).

An algorithm is also presented, which increases cyclically the diameter of a ball-shaped discrete element 5 to 17 mm with 1 mm interval. This algorithm supports gaining the sizes of discrete elements as soon as they fill the required space, although the contact force between them does not exceed pre-set values.

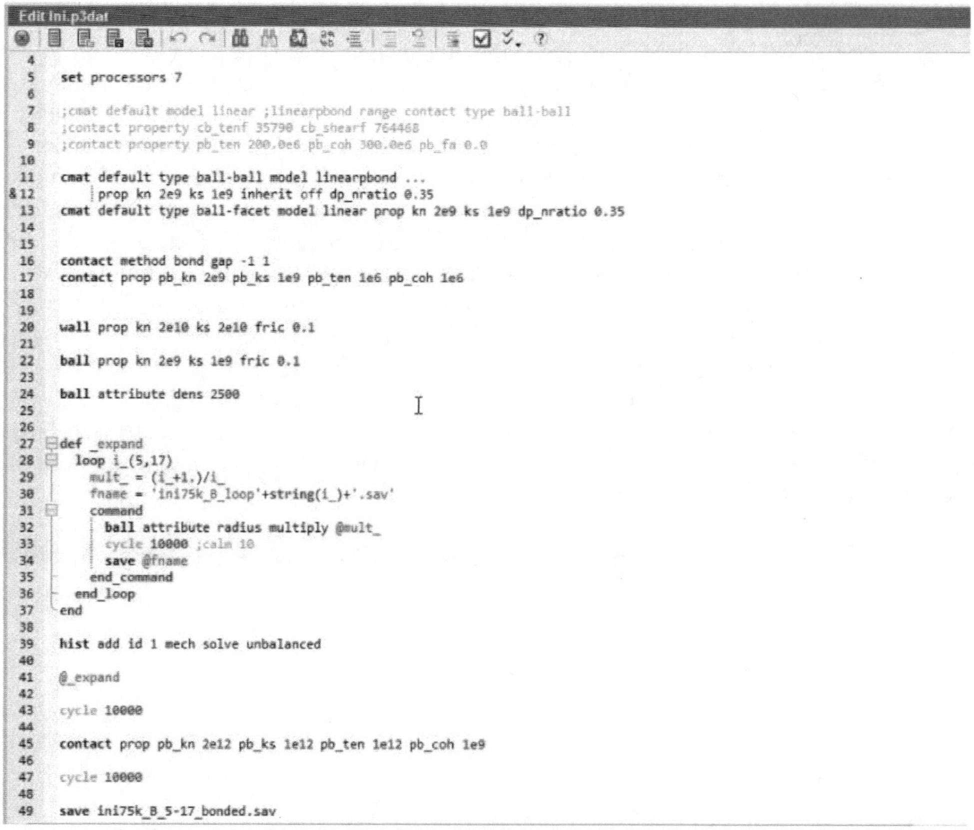

```
set processors 7

;cmat default model linear ;linearpbond range contact type ball-ball
;contact property cb_tenf 35790 cb_shearf 764468
;contact property pb_ten 200.0e6 pb_coh 300.0e6 pb_fa 0.0

cmat default type ball-ball model linearpbond ...
    prop kn 2e9 ks 1e9 inherit off dp_nratio 0.35
cmat default type ball-facet model linear prop kn 2e9 ks 1e9 dp_nratio 0.35


contact method bond gap -1 1
contact prop pb_kn 2e9 pb_ks 1e9 pb_ten 1e6 pb_coh 1e6


wall prop kn 2e10 ks 2e10 fric 0.1

ball prop kn 2e9 ks 1e9 fric 0.1

ball attribute dens 2500


def _expand
  loop i_(5,17)
    mult_ = (i_+1.)/i_
    fname = 'ini75k_B_loop'+string(i_)+'.sav'
    command
      ball attribute radius multiply @mult_
      cycle 10000 ;calm 10
      save @fname
    end_command
  end_loop
end

hist add id 1 mech solve unbalanced

@_expand

cycle 10000

contact prop pb_kn 2e12 pb_ks 1e12 pb_ten 1e12 pb_coh 1e9

cycle 10000

save ini75k_B_5-17_bonded.sav
```

Figure 5. Assigning properties of numerical experiment constituents and algorithm of filling the model space by discrete elements

As soon as an algorithm is developed a model of the geological environment is created. Intermediate calculations of the geological environmental properties are given in Figure 6.

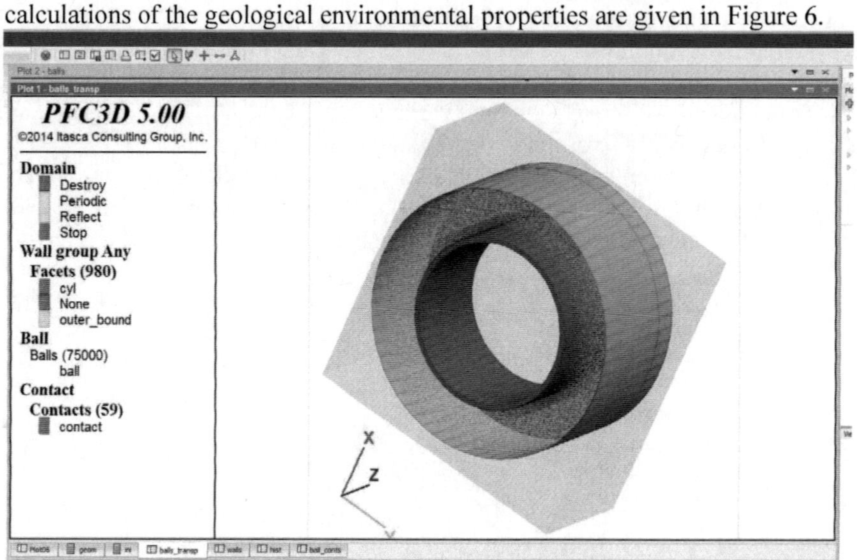

Figure 6. Intermediate calculations of geological environmental properties

Minimal discrete elements are shown white in Figure 6 at the start of modelling. There is a result of modelling the geological environment shown in Figure 7.

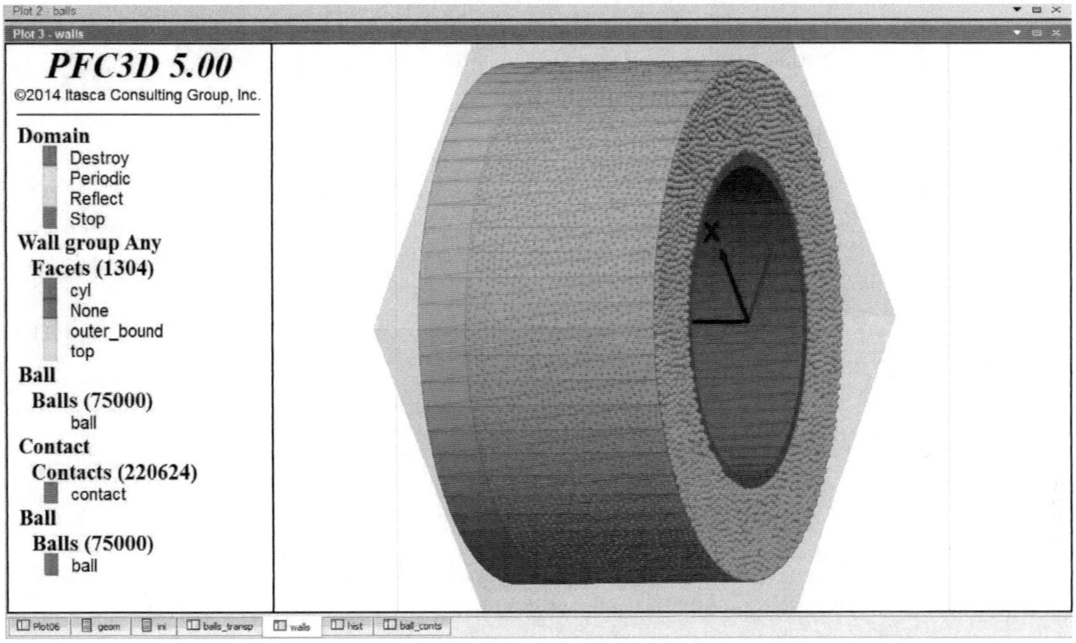

Figure 7. Computational results of geological environmental properties

It is seen in Figure 7 ball-shaped discrete elements fill the space of design model and modeling edges.

Numerical experiment

A numerical experiment is required to simulate the mutual effect of moving geokhod and geological environment. The software supports visualization of geokhod traverse in the modeled geological environment, as well as real-time tracking parameters of phenomena, occurring in the process of traverse such as diagrams of breaking ties between discrete elements, graphs of changing force parameters both on geokhod surface and in the geological environment etc. A simulating algorithm of the geokhod head section rotation with a real frequency (Figure 8) is developed for translational displacement and rotation. A real rotation frequency of geokhod is set 6 revolutions per hour.

Edit Rotation.p3dat

```
res ini75k_8_5-17_bonded.sav

wall generate group inner_bound cylinder base 1.800 1.600 0 ...
    axis 0 0 1 height 2.500 radius 1.600 cap false
wall prop kn 2e10 ks 2e10 fric 0.1 range group inner_bound

wall attr centrotation 1.800 1.600 0 range group cyl

wall history xcontactf id 1

wall history contactforce id 1

wall history velocity id 1

wall history spin id 1

def inirot
  loop i(1,10)
    rotvel_ = i*0.01309/10.
    forwardvel_ =  i*0.00167/10.
    command
      wall attr spin 0 0 @rotvel_ range group cyl
      wall attr vel 0 0 @forwardvel_ range group cyl
    end_command
  end_loop
end
@inirot

set timestep scale

solve time 480
save rotation_end.sav
```

Figure 8. Algorithm of rotation and translational displacement of a geokhod model

As soon as an algorithm of rotation is developed a numerical experiment is conducted. Carrying out an experiment the algorithm involves one full revolution of a geokhod model round its axis.

Results of the numerical experiment

As a result of numerical experiment a pattern is developed by software PFC3D 5.00, which represents the mutual effect of a design model with discrete elements, modelling the geological environment, and supports displaying various parameters of this interaction in form of diagrams.

A diagram of contact forces is given in Figure 9, which arise in the geological environmental model (in discrete elements), when geokhod rotating.

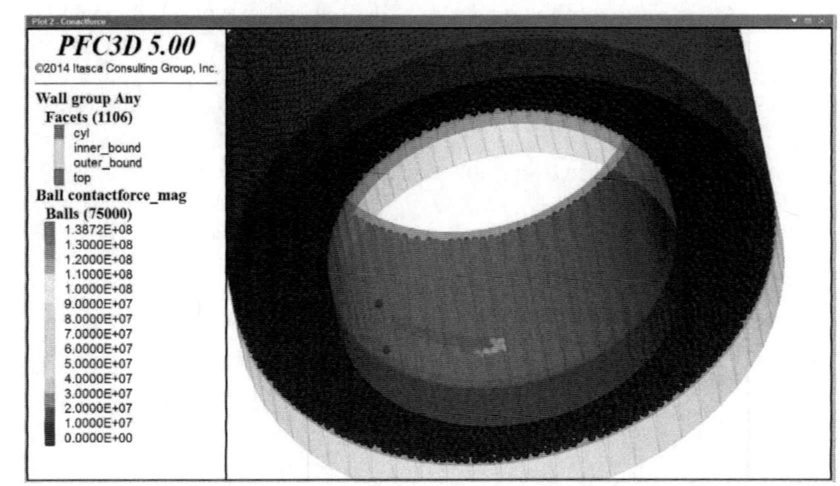

Figure 9. A diagram of contact forces in the geological environment (in discrete elements) when propulsive agent moving

The design model of geokhod is presented transparent (Figure 9) for better visualization on the diagram. The diagram shows a propulsive agent of geokhod model, moving among discrete elements of the rock mass. Software PFC3D 5.00 allows of following a changeable numerical value of variable to be measured. Varying values of forces on the contact surface of geological environment and propulsive agent, as well as in the geological environment are given in various colors. Minimal contact forces are given blue in this diagram, they equal 0.0000E+00 (Figure 9), and gradient colors of red present rising contact forces. Maximal contact forces of 1.3872E+08 or 1.3872·108 N are given in red. The maximal contact forces shown on the diagram are there in font of the propulsive agent, where geological environment is broken.

Software PFC3D 5.00 supports detecting forces arising when breaking the ties of discrete elements (internal stresses in geological environment). Other possibilities for representation of geological environmental model are provided. A diagram of contact forces, representing the geological environment by the ties between discrete elements not by elements themselves is given in Figure 10.

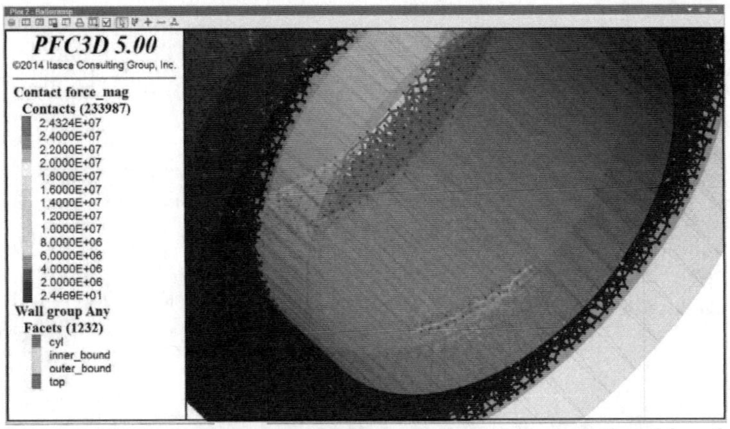

Figure 10. A diagram of contact forces in the geological environment (breaking the ties) when a propulsive agent moving

The maximal contact forces of 2.4324·107 N arise in the zone of moving propulsive agent when breaking the ties between discrete elements (Figure 10).

The software makes it also possible to determine other variables, e.g. a velocity diagram of geological environmental particles (discrete elements) is obtained when moving a geokhod model (Figure 11).

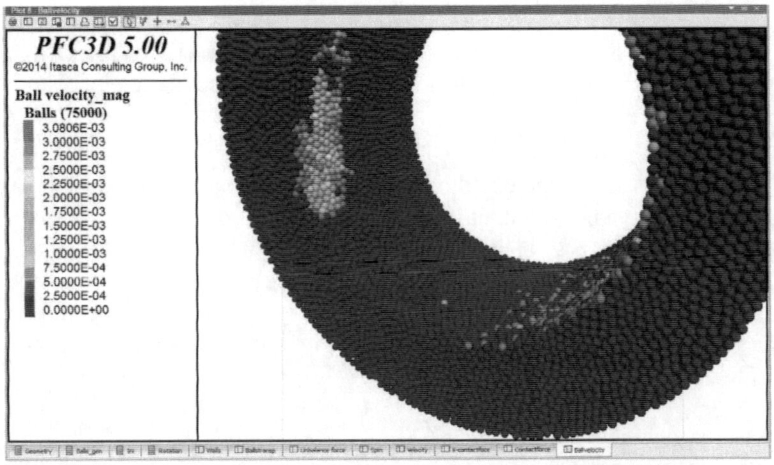

Figure 11. A velocity diagram of discrete elements when geokhod traverse

The maximal velocity of discrete elements hardly exceeds 3.0806·10-3 m/s.

The obtained diagrams and values represent only some capabilities of the software. It can be used to determine such parameters as stresses in the geological environment along various axes, stresses on the model surfaces, and value and velocity vector of discrete elements etc.

Processing the results of numerical experiment

Software PFC3D 5.00 supports plotting the results of numerical experiment. Here, the results are given as plotted functions of calculation steps (variable "Step"). The value of steps of calculation (variable "Step") is directly dependent on computer capacity and can be seen as analogue of experiment period.

Values to be measured vary according to the rotation angle of the propulsive agent in this experiment; however, angular velocity of rotation is uniform and period or the quantity of computation steps (variable "Step") are accepted as argument of function.

A plot of changing contact forces on the cylindrical surface of a design model is given in Figure 12 (on the body surface of the geokhod head section) according to the calculation steps ("Step"), in this case the length of X-axis is in line with 360° revolution of a model.

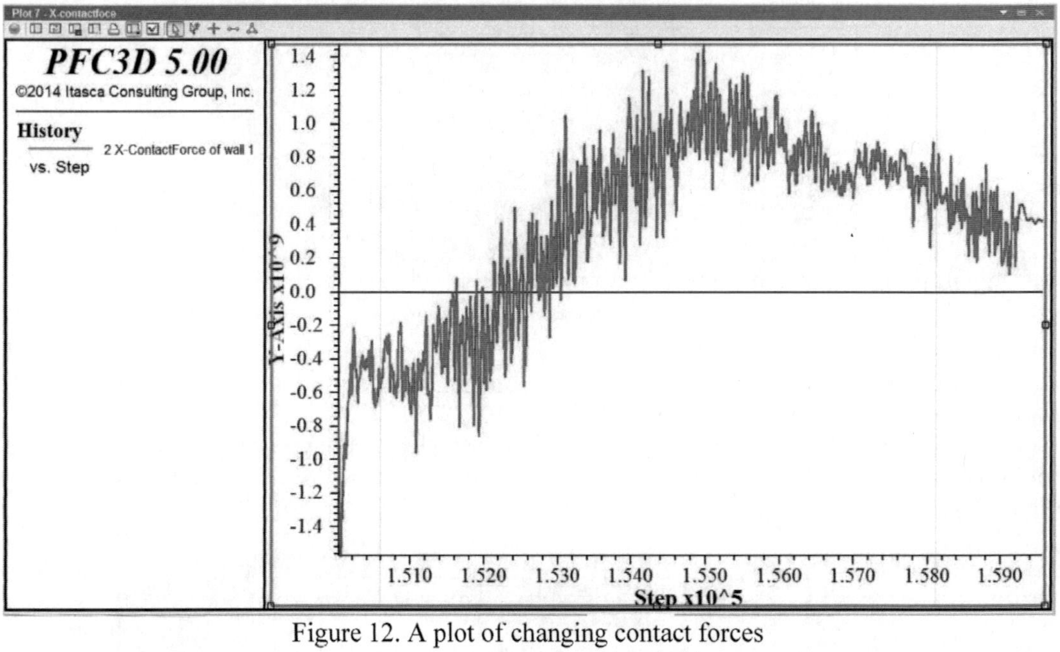

Figure 12. A plot of changing contact forces

The value of contact forces on the geokhod surface is changeable, reaching its maximum 1.4·109 N when rotating. A negative value of a contact force indicates its direction, which is reverse to traverse of a design model.

A force, arising on the inner surface of a propulsive agent is also an important force parameter. This force is important when forming forces required for traverse. A plot showing the change of contact forces on the surface of the propulsive agent is given in Figure 13.

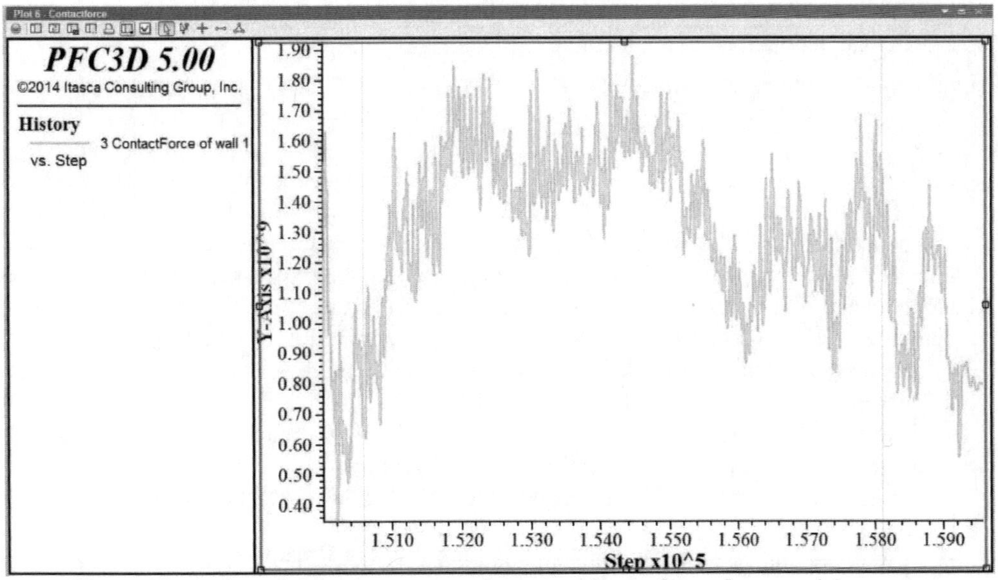

Figure 13. A plot of changing contact forces on the surface of a propulsive agent

Contact forces vary in a wide range when rotating. An average value of contact forces is 1.2·109 N with its maximum of 1.9·109 N.

Conclusions

A possibility in principle to use software PFC3D 5.00 for modelling geokhod traverse in the geological environment was revealed in the work. The principles of numerical experiment based on discrete element method are outlined to simulate the mutual action of moving geokhod and geological environment. As a result the following shortcomings of software PFC3D 5.00 were detected:

- the software is strongly dependent on computer capacity, changing quantity of discrete elements to be processed requires longer experiments (the experiment under consideration involved 75 000 discrete elements, taking 10-12 hours for one rotation of geokhod), improvement of experimental accuracy or input data variation necessitates the increase in discrete elements, resulting in significant growth of computation time;
- the experiments detected incomplete compliance of the software with real physical processes, for instance, an axial force arises and pushes geokhod ahead in real conditions when inclination angles of propulsive agents are 5 degrees (Figure 1) and the model of geokhod rotates, but specifics of software PFC3D 5.00 allows no axial force.

The following measures were proposed to eliminate these shortcomings:

- parallel computations are available in software PFC3D 5.00 when carrying out experiments, making possible its use by powerful supercomputers like "cluster" and reducing the time of computations;
- software PFC3D 5.00 supports control over obtained data and development of an algorithm of progressive traverse of geokhod according to forces arising on the propulsive agent.

The advantages of software PFC3D 5.00 when modelling geokhod traverse in the geological environment are as follows:

- substantial reduction of costs as against field simulation;
- a wide range of obtained experimental data;
- a possibility of varying parameters of the geological environment to be modeled (anisotropy of the geological environment can be pre-set);

IOP Conf. Series: Materials Science and Engineering **142** (2016) 012126 doi:10.1088/1757-899X/142/1/012126

- values of experimental data both in the geological environment and on the contacting surfaces of geokhod can be obtained;
- simplicity of obtaining experimental data.

References

[1] Aksenov, V. V., Walter, A. V., Gordeyev, A. A., & Kosovets, A. V., Classification of geokhod units and systems based on product cost analysis and estimation for a prototype model production, In IOP Conference Series: Materials Science and Engineering. 379 (2015).

[2] A.B. Efremenkov, V.Yu. Timofeev. Determination of Necessary Forces for Geokhod Movement, 7th International Forum on Strategic Technology IFOST2012, (2012) 211-214.

[3] Timofeev V. Y., Aksenov V. V., Galjamova J. I. Determination of parameters of roller of wave transmission with intermediate rolling bodies with hollow shaft for geokhod // Applied Mechanics and Materials. – Trans Tech Publications, 2014. – T. 682. – C. 246-250.

[4] Begljakov V. Y., Timofeev V. Y., Dokhnenko M. V. Parameters of Force Interaction of Elements the Wave Transmission with Intermediate Rolling Bodies in Geokhd's Transmission //Applied Mechanics and Materials. – Trans Tech Publications, 2014. – T. 682. – C. 282-287.

[5] Aksenov, Vladimir V., Mikhail Yu Blaschuk, and Mikhail V. Dubrovskii. "Estimation of torque variation of geohod transmission with hydraulic drive." Applied Mechanics and Materials. Vol. 379. 2013.

[6] Efremenkov, Andrey Borisovich, V. V. Aksenov, and M. Yu Yurga. "Force parameters of geohod transmission with hydraulic drive in various movement phases." Strategic Technology (IFOST), 2012 7th International Forum on. IEEE, 2012.

[7] Sadovets, V. Yu, V. Yu Beglyakov, and V. V. Aksenov. "Development of math model of geokhod bladed working body interaction with geo-environment." IOP Conference Series: Materials Science and Engineering. Vol. 91. No. 1. IOP Publishing, 2015.

[8] Aksenov, Vladimir V., Andrey Borisovich Efremenkov, and V. Yu Begljakov. "The influence of relative distance between ledges on the stress-strain state of the rock at a face." Applied mechanics and materials. Vol. 379. 2013.

[9] Chernukhin, R. V., A. A. Dronov, and M. Y. Blashchuk. "The application of the analytic hierarchy process when choosing layout schemes for a geokhod pumping station." IOP Conference Series: Materials Science and Engineering. Vol. 91. No. 1. IOP Publishing, 2015.

[10] Walter, A. V., and V. V. Aksenov. "Determining deviations in geometry of the geokhod shells." Applied Mechanics and Materials. Vol. 770. 2015.

IOP Conf. Series: Materials Science and Engineering **142** (2016) 012127 doi:10.1088/1757-899X/142/1/012127

Economic and Technological Role of Kuzbass Industry in the Implementation of National Energy Strategy of Russian Federation

S A Zhironkin[1], A A Khoreshok[2], M A Tyulenev [3], G A Barysheva [4], M C Hellmer[5]

[1]Dr.Sc. Professor of Economy department, National Research Tomsk Polytechnic University. Russia, 634050, Lenina av., 30
[2]Dr.Sc. Professor of Mining Equipment department, National Research Tomsk Polytechnic University. Russia, 634050, Lenina av., 30
[3]PhD. Assistant professor of Open pit mining department, Kuzbass State Technical University, Russia, 650000, Kemerovo, Vesennya st., 28
[4]Dr.Sc. Chairman of Economy department, National Research Tomsk Polytechnic University. Russia, 634050, Lenina av., 30
[5]MSc in Hydrology and Water Resources Management., Department of Civil and Environmental Engineering, Imperial College London. United Kingdom, South Kensington Campus, London SW7 2AZ, UK

E-mail: [3]tma.geolog@kuzstu.ru

Abstract. This article describes the problems and prospects of development of coal mining in Kuzbass - the center of coal production in Siberia and Russia, in the framework of the major initiatives of the National Energy Strategy for the period until 2035. The structural character of the regional coal industry problems, caused by decline in investment activity, high level of fixed assets depreciation, slow development of deep coal processing and technological reduction of coal mining is shown.

1. Introduction

Prospects of fuel and energy complex development in Russia, as described in the Energy Strategy for the period until 2035, are based on the main direction of the global energy market transformation – self-providing and import substitution for countries – the largest importers of energy producing materials [1]. Therefore, the strategic initiatives of fuel and energy complex of Russia technological upgrade include the development of energy saving technologies, export diversification and the development of deep processing of coal and oil [2-5]. These initiatives are specified in the key tasks of National Energy Strategy of Russian Federation as the following:

1. Upgrading the power industry means development of oil refining, "smart" power grids, decentralized power generation, upgrading the heating systems.
2. The development of domestic energy infrastructure means eliminating the steady investment imbalance between export, domestic-oriented projects and infrastructure.

IOP Conf. Series: Materials Science and Engineering **142** (2016) 012127 doi:10.1088/1757-899X/142/1/012127

3. The growth of the domestic energy market means reducing the level of monopolization, establishment of more efficient power market regulation by government and regional authorities, supporting the competition and trade on raw material exchanges.

4. Improving accessibility and quality of energy producing materials, providing modern customer services through the embedding of international technology standards.

5. Increasing the flexibility and diversification of energy producing materials exports.

6. Implementation of the principles of socially and environmentally responsible management of mineral enterprises and holding companies.

2. Material and Method

The territories of Russia where energy sources and power producing are concentrated must become the "locomotives" of the National Energy Strategy implementation. Kuzbass (West Siberia, Russia) is one of the famous clusters of Russian heavy industries. Mining industry output of Kuzbass includes coal, iron stone and polymetallic ores (35% of gross regional product in average for 2005-2015). Processing industrial complex of Kuzbass is represented mainly by ferrous and nonferrous metallurgy, primary chemical products and constructing materials (21% of gross regional product).

Significant part of Russian coal production is concentrated in Kuzbass. According to the forecasts, regional coal mining will continue to ramp up the volume of output and it is expected to reach 270 million tons by 2025 [6]. The reason of it is the increasing share of coal in power production in Russia by 2020, at least twice. Today its share is only 12%, while in Western Europe - 40-60%, in USA - 56%, in China - up to 70%. At the same time, Russia has at least one third of proven coal deposits in the world, most of which (more than 690 billion tons) are located in Kuzbass [7]. High quality coal from Kuzbass is consumed in more than 40 countries [8].

In 2012 the total volume of coal mined in Kuzbass was about 200 million tons, including 116 million tons (60.3%) produced in open way. In 2013, it was produced 203 million tons of coal (in 2014 – 211, 2015 - 215 million tons). In addition, there is no reason to believe that in the near future coal consumption will decline. There are 45 coal mines and 56 open pits, 27 coal-processing plants and installations involved in Kuzbass coal industry, 133 thousand people are employed in it [9].

At the same time, technologic and economic development of Kuzbass coal industry is constrained by the problems caused by structural misbalances of regional and national economy.

The first problem is the high level of the active part of fixed assets depreciation in coal mines and open cuts (more than 60% averagely) [10]. Therefore, the achievement of strategic targets for coal production requires complete re-engineering of coal enterprises and massive capital investments in the industry. But there are some issues for the capital inflow to Kuzbass industry such as unstable coal prices on the global market, low labor productivity in regional coal industry, a high accident rate, poor investment infrastructure and undercapitalized segment of investment fund market in coal industry. So despite steady growth of coal mining in Kuzbass, total revenue of its sales remains unstable [11]. Operating income of coal companies tends to shortening and investment problems are strengthening consequently.

The second problem is insufficient quality of the exported coal from Kuzbass. This is mainly due to the dominance of primarily coal preparation, while whole process of coal preparation up to the most expensive sorts of carbon fuel is developed insufficiently. This suppresses the competitiveness of regional coal industry and its investment potential. So the following "vicious circle" of investment problems of coal industry continues to maintain: "The lack of competitiveness of the coal enterprises - difficulties in fund rising - the preservation of low competitiveness." It leads to shortening the part of unaffiliated investors in coal companies' funds – from 42% in 2010 to 26% in 2014 and then – to keeping Kuzbass economy underinvested [12-13].

The third problem is the existence of a significant number of unincorporated enterprises (limited companies) engaged in coal mining in Kuzbass. In 2012, they mined 29 million tons of coal, or 14.3% of total coal production in the region. At the same time they raised their funds by less than 45 million dollars, or 4% of the total investment in Kuzbass coal industry. Along with the unincorporated coal enterprises, which cannot enter the stock market for capital rising, there is the high proportion of closed corporations, which shares cannot be easily sold to the interested investors. Their part of the total coal output in Kuzbass in 2012 amounted to 15.5%, or 31 million tons. Among them we can name such giants of Russian coal industry as "Raspadskaya Coal Company" with the annual output in 2006 to 10.6 million tons of coal (6.1% of the total volume of Kuzbass), "Holding company "Siberian Business Union" - 9 million tons (53%). Consequently, the limited coal companies and closed corporations summed almost 30% of the total coal output in Kuzbass, and they cannot raise their funds using the stock market in the near future [14].

The fourth problem is retarding of new technologies of coal mining and processing implementation, and lag in using of new high-performance equipment. As a result the productivity of Kuzbass coal open pits is in three times less than in the United States (for coal mines is in five times less) [15].

Despite it, coal companies' funds have been used for construction of 25 technological inter-mine rail roads and supporting stations. Special attention is paid to the constructing of coal sea port terminals. Thus coal companies and enterprises of Kuzbass continue to create the processing chain from coal mines and open pits to the ship charter. This allows exporting of 33% of the coal mined in Kuzbass in 2012-2015 [6].

Nevertheless the remaining structural problems of Kuzbass economy lead to the gap between investments and productivity growth. It means that new investments do not sufficiently improve the productivity. The reason of it is using fund rising for out-of-date equipment and restoring the existing machinery by coal companies and enterprises in other industries of Kuzbass economy.

Securing Kuzbass economy specialization on coal mining is supported by dominance of minerals extraction and its primary processing in the regional export. In 2005-2014 up to 90% of the Kuzbass industrial export was represented by three commodity groups [1]:

- fuel and energy products - hard coal, coke and semi-coke. The export of them grew in 2005-2014 from 48.3 to 83 million tons, or more than 3.8 billion dollars per year;
- metals and products from them - semi-finished products, bars, rods, special profiles, flat-rolled products of iron or non-alloy steel, pig iron, aluminum, ferroalloys. During 2005-2014 these articles of Kuzbass export doubled up to 2.9 billion dollars;
- chemical products (mainly heterocyclic compounds and nitrogen fertilizers). Their exports of decreased in 2005-2014 by 26% and amounted to 0.3 billion dollars.

To achieve the goals set by National Energy Strategy of Russian Federation it is necessary to initiate a deep structural transformation of the industry, which cannot be carried out with reasonable-priced raw materials, new productive technologies of processing minerals and investment institutions.

3. Results and Discussion

It is impossible to enhance the role of Kuzbass coal industry in the implementation of National Energy Strategy of Russian Federation without structural changes in the power and mining industries, especially in the segment of production energy producing materials. The following parity: "High technology in processing sector of coal industry - new technologies of fuels production - new ways of fund rising - new investment institutions" is in demand today more than ever. The most important role in the formation of a "bridge" between restructuring of national economy as a whole and its financial sector as a part should be given to the industrial clusters where output of energy producing materials is concentrated.

Such neo-industrialization of Kuzbass economy suggests implies some directions of governing the restructuring of its dominating industries – coal, power, chemistry and metallurgy.

The main direction is providing a transition from the "participation in governing" model of relations between regional business companies and authorities to a model of "participation in innovation". This means a comprehensive encouragement of innovation activity of raw-materials extracting and processing companies and power producers, technology transfer and cross-sectorial reallocation of investment [16-18].

IOP Conf. Series: Materials Science and Engineering **142** (2016) 012127 doi:10.1088/1757-899X/142/1/012127

The second (institutional) direction means purposeful change of the most inertial segments of Kuzbass economy structure, mainly formal and informal institutions, mechanisms of reproduction of fixed assets.

The third (resource) direction means a transition from reliance on natural resources as the main source of regional budget revenues to the deep processing of coal, metals, organic compounds based on technological upgrading of machine building, instrument-making, synthetic materials producing plants. It requires, above all, setting the tax and finance incentives to increase productivity of existing processing enterprises along with technological upgrade of coal mines and open pits. The expected results of these measures are reducing of the costs and improving competitiveness of regional industrial companies what was proved many times in different countries [19-23].

The fourth (social) strategic direction is enhancing employment of staff with post-industrial competences. It requires state guarantees in the labor market for training specialists for the companies upgrading their technologies to modern level .

The fifth (integration) direction suggests conglomeration of raw materials and processing enterprises, financial companies, R&D organizations in the high-tech processing holdings. Their competitiveness should be provided by cooperation with national technological platforms and investing breakthrough innovations by commercial banks by long-term project financing [24-26].

Setting goals of structural transformation of Kuzbass economy should be closely linked to implementation of National Energy Strategy of Russian Federation. Therefore, among the main objectives of regional industry complex structural transformation, we emphasize the following.

The primary goals can be reached all together and depend on the solution of main structural problems of the Kuzbass economy:

- modernization of Kuzbass economy industrial sector by deepening of coal processing , increasing productivity of mining, renovation of machine-building plants;
- increase the share of the 5th and 6th technological layers in Gross Regional Product, in tax revenues, in the structure of employment; changes in the sector composition of exports in favor of processing and high-tech enterprises;
- forming competitive on domestic and foreign markets raw-material processing and mechanical engineering clusters;
- developing R&D and innovation activity in basic industries of Kuzbass economy to expand technology transfer from science to industry and business. It requires diffusion of innovations between small, medium and large companies.

Secondary goals, which are also aimed to help implementing structural changes of Kuzbass economy, are achievable only within primary goals:

- developing of market competition in the industry, the financial sector, and infrastructure sectors;
- overcoming the dependence of the state budget revenues of raw materials extracting;
- reduction of the shadow and inefficient non-market segments of regional economy;
- decrease in the proportion of poor people by increasing the role of labor as a factor of producing in high-tech industries.

Determination of the conditions for achieving goals of structural transformation of Kuzbass economy suggests the following.

Firstly, the existence of an institutional framework of structural reforms: regulatory framework, public and private market agents, mechanisms of effective investment redistribution between raw-material and processing sectors of regional economy.

Secondly, the presence of innovation investment infrastructure which helps to promote regional enterprises in domestic and foreign fund markets, small business support, training of the staff of processing enterprises. It should be noted that currently investment infrastructure in Russia are being made by state and occur in the transport and energy systems. Their effectiveness is recognized extremely low by both Russian and foreign analysts [27-29].

Third, the availability of necessary funds and forms of inter-sector capital flow.

Fourth, the formation of the system of scientific and methodological support of Kuzbass economy structural transformation based on ambitioned targets of National Energy Strategy of Russian Federation like development of green power, increasing of energy efficiency in industry, extending alternative power sources.

Fifthly, establishing of state-and-private partnerships in R&D and applied scientific research in coal mining and processing, the power producing.

Considering the fact that Kuzbass coal industry is not only producer of the important energy source, but also the driver of the other sectors' of economy development, it is necessary to preserve and strengthen existing trend of growth in coal production and investment activity. In current structure of Kuzbass economy with raw-material industries domination, it is necessary to create conditions for expansion of brand new segments related to the deep processing of minerals. The leading role in the breading the new segments of Kuzbass economy should play convergent technologies.

We define the development of convergent technologies as an important condition for the implementation of National Energy Strategy of Russian Federation (along with raw-material production technological development and expanding high-tech industries). The content of the convergence of technologies consists of interpenetration and combination of different technological innovations, united by common forms of design, implementing and diffusion [30-31]. Also, technological convergence means the development of industrial clusters like Kuzbass, aggregated development of innovations created in different technological platforms but bringing common product. We have identified the following fist of convergent technologies perspective in Kuzbass:

- technologies uniting biochemistry and power producing (bioenergy), allowing going to the non-waste technologies of environmentally clean fuel producing to replace non-renewable hydrocarbon resources in the foreseeable future [32-33]. Prospects for the development of these technologies we see in the interactions between such technological platform as "Medical and Bio-Technologies", "Energy", "Extraction of natural resources and oil and gas", approved by Russian Government;
- technologies joining producing computer hardware and software (for distributed and "cloud" computing). They will be able to create the integrated telecommunication networks, significantly advancing diagnostic medicine and accelerating development of new machinery by rapidly performing design and multivariate simulation modeling. Their cross-platform character should be achieved with the interaction of such technological platforms as "Information and Communication Technologies", "Electronics and Mechanical Engineering";
- laser informatics technologies that can create a data repository, surpassing existing ones by thousands times. This will speed up the digitization of technological basis and automation of such traditional Kuzbass industries such as coal mining, chemical industry, power producing. Cross-platform interactions can be implemented here in the platforms like "Photonics", "Information and Communication Technologies", "Electronics and Mechanical Engineering."

Productions which can "grow up" from industries dominating in existing structure of Kuzbass economy should become investment "donors" for new processing industries. Therefore, to the promising directions of development of coal mining, deep seated in Kuzbass economy, should be referred the following:

- implementation of non-waste plasma-based technologies of conversion coal organic components to synthetic gas and recovering microelements from it;
- creating of power systems for combined producing of methanol and electricity based on coal gasification and organic synthesis;
- start-up and development of extracting methane from coal seams. Reducing mining safety problems, this gas is used around the world as valuable source of energy. In the USA 40 billion m3 of methane is produced annually [34];
- development of coal and natural gas processing technologies for manufacturing of high-quality engineering plastics, synthetic materials for electronics and machinery;

- set up of high-tech industry which is rather new for Kuzbass economy – producing of equipment for reducing environmental damage caused by power companies, chemical plants, coal production and foundry.

On the level of the single industry and inter-industries management the coherent and flexible institutional policy of mineral and power sectors technological development is required. Its main elements are the following:

1. The setting up effective national energy market with a high level of competition and market pricing mechanism.
2. Forming a stable tax system for mineral and power sectors, creating balanced systems of custom duties to encourage foreign investment in the deep coal processing.
3. Creating a legal framework that protects the investors' rights and supports the development of competition, removes the excessing administrative barriers.
4. Promote the implementation of advanced coal extracting and processing, energy-saving technologies in Kuzbass mining companies, and encourage the development of technological innovation activity in power sector.
5. Improvement of business relations with traditional European consumers of Russian energy sources producers and establishment of new relations on South Asia market, mainly with China and Thailand partners.
6. Active participation of Kuzbass mining companies in Eurasian power and energy resource market observing general principles of market regulation, providing free distribution of power, energy producing materials, technologies and funds.

4. Conclusion

Thus, Kuzbass coal industry is to become a driver of structural reforms of Siberian and Russian economy. In accordance with Energy Strategy of Russian Federation, huge technological transformation of coal industry is required, based on transition to deep coal processing, producing new kinds of mineral fuel, improving industrial management and market institutions.

5. Acknowledgement

The research was made in National Research Tomsk Polytechnic University with financial support of the Ministry of Education and Science of the Russian Federation in the framework of scientific research "The evaluation and improvement of the social, economic and emotional prosperity of senior citizens" contract № 14.Z50.31.0029. Some parts of research were made in Kuzbass State Technical University with financial support of the Ministry of Education and Science of the Russian Federation in the framework of scientific research "Exploring the parameters of technological and technical solutions for selection and development of improving cutting and extracting mining machines operation efficiency in Kuzbass", contract №632. The authors would like to thank to prof. Yury Lesin and prof. Magerram Gasanov for their discussions during the study. We thank anonymous referees for constructive and useful comments on the papers sent via e-mail.

References

[1] Zhironkin S A, Tyulenev M A, Zhironkina O V, Hellmer M 2016 The global determinants of mining higher education development *Metallurgical and Mining Industry* **2** pp 62-67
[2] Zhironkin S A 2001 Factoring and leasing development at coal mining industry of Kuzbass as an important element of its financial part *Ugol'* **4** pp 29-30
[3] Tyulenev M, Lesin Y, Zhironkin S and Garina E. 2016 Coal producers waste water purification *Metallurgical and Mining Industry* **2** pp 52-56
[4] Gasanov M A, Gasanova N V, Zhavoronok A V 2016 Information Technologies in Ensuring Continuous Wellbeing of the Person *The European Proceedings of Social & Behavioural Sciences* **VII** pp 323-336
[5] Zhironkin S A 2002 Prospects and new possibilities investment attracting to Kuzbass coal mining

industry *Ugol'* **6** pp 31-36

[6] Tyulenev M, Zhironkin S and Litvin O 2015 The low-cost technology of quarry water purifying using the artificial filters of overburden rock *Pollution Research* **34 (4)** pp 825-830

[7] Tyulenev M A and Lesin Y V 2014 *Justification complex purification technology open-pit mines wastewater* (Taishan: Academic Forum – Project on Mine Disaster Prevention and Control) pp 441-444.

[8] Lesin Y V, Lukyanova S Y and Tyulenev M A 2010 Mass transfer of dispersed particles in water filtration in macro-grained media *Journal of Mining Science* **46 (1)** pp 78-81

[9] Zhironkin S A 2002 About measures of vexel circulation development and vexelability definition of fuel-and-power complex' enterprises *Ugol'* **4** pp 47-48

[10] Khoreshok A 2002 On side cutting bit when operating at sheerer drums *Ugol'* **7** pp 10-11.

[11] Zhironkin S A 2001 Governmental factoring development of TEK Kuzbass *Ugol'* **6** p 62

[12] Barysheva G A and Novoselova E G 2014 Methodology of Application of the Structural Shift Mechanism for Regulation of the National Economic Management System *Applied Mechanics and Materials* **682** pp 550-554

[13] Klemasheva E, Gasanov M and Zeremskaya Y 2016 Neo-Industrialization of Russian Economy as a Source of Social Well-Being Growth *The European Proceedings of Social & Behavioural Sciences* **VII** pp 150-154

[14] Lesin Y V, Luk'yanova S Y and Tyulenev M A 2015 Formation of the composition and properties of dumps on the open-pit mines of Kuzbass *IOP Conference Series: Materials Science and Engineering* **91 (1)**, 012093

[15] Wei G L 2011 Statistical analysis of Sino–US coal mining industry accidents *International Journal of Business Administration* **2 (2)** pp. 82–86

[16] Cushtar C, Matti M and Veron S. 2011 Industrial coal demand in China: a provincial analysis Resource and Energy Economics **33(1)** pp12–35

[17] Newton A 2011 The impact of ICT and human capital on achieving knowledge-based economy: applications in Malaysia's economy *Int. J. of Green Economics* **5 (3)** pp 231 - 247

[18] Wang J L, Feng L Y and Tverberg G E 2013 An analysis of China's coal supply and its impact on China's future economic growth *Energy Policy* **57** pp 542–551

[19] Danuta Kopycinska and Bruno S Sergi 2008 Economic development and prospects in Poland: an introduction *Int. J. of Economic Policy in Emerging Economies* **1 (2/3)** pp 127-136

[20] Xinjie L and Hebing L 2015 A brief analysis of the circumstances of the current coal industry and its trends in China *Int. J. Mining and Mineral Engineering* **6(1)** pp 87-96

[21] Wang Shicheng 2016 Investment Decision-Making on Renewable Energy Based on Improved Real Option Model *Metallurgical and Mining Industry* **1** pp 36-41

[22] Yu Wang and Chengqun Yu 2016 Research on the Database Marketing in the Big Data Environment Based on Ensemble Learning *Metallurgical and Mining Industry* **1** 41-49

[23] Letucha O, Samoilenko I and Tkachenko N 2016 State regulation of ecological and economic process as a factor of activation of metallurgical production in Ukraine *Metallurgical and Mining Industry* **2** pp 58-61

[24] Gasanov M, Gasanov E, Egorova M The Technologies of the Network Prosperity in Russia *Procedia - Social and Behavioral Sciences* **166** pp 103-106

[25] Plotnikova I, Red'ko L and, Yanushevskaya V Social Responsibility of Business *The European Proceedings of Social & Behavioural Sciences* **VII** pp 173-180

[26] Engelbrecht-Zenkina Z Analysis of Wellbeing of the Population in the Metropolitan Region Rhein-Neckar (Germany) *The European Proceedings of Social & Behavioural Sciences* **VII** pp 188-192

[27] Elsadig Musa Ahmed 2006 The green economy and the knowledge economy: exploring the interface *World Review of Science, Technology and Sustainable Development* **3 (3)** pp 270-283

[28] Fernandes B, Cunha H and Ferreira P 2011 The use of real options approach in energy sector

investments *Renewable and Sustainable Energy Reviews* **19 (12)** pp 71-80
[29] Kang P, Cho S and Mac Lachlan D L 2012 Improved response modeling based on clustering, undersampling, and ensemble *Expert Systems with Applications* **39(8)** pp 6738-6753
[30] Deng Xiaoyi, Jin Chun and Higuchi Yoshiyuki 2011 KSP: A Hybrid Clustering Algorithm for Customer Segmentation in Mobile E-commerce *Journal of Management Science* **24(4)** pp 54-61
[31] Maclin R and Opitz D. 2011 Popular ensemble methods: An empirical study *Journal Of Artificial Intelligence Research* **11** pp 169-198
[32] Bose I and Chen X 2009 Quantitative models for direct marketing: A review from systems perspective. *European Journal of Operational Research* **195(1)** pp 1-16
[33] Antti Haapala and Janne Härkönen 2015 Bioeconomy potential - focus on Northern Finland *Int. J. of Sustainable Economy* **7(1)** pp 66- 90
[34] BP 2011 *BP Statistical Review of World Energy 2011* (British Petroleum, London)

IOP Conf. Series: Materials Science and Engineering **142** (2016) 012128 doi:10.1088/1757-899X/142/1/012128

Calculation of Geometrical Parameters of Geokhod Transmission With Hydraulic Cylinders

M Yu Blaschuk, A A Dronov, S S Ganovichev

Yurga Institute of Technology, TPU Affiliate, 26, Leningradskaya Str., 652050, Yurga, Russia

E-mail: mby.tpu@gmail.com

Abstract. Developed analytical expressions for determining parameters of transmission hydraulic cylinders' arrangement are considered, as well as the conditions for internal arrangement of a required number of hydraulic cylinders.

Introduction

Currently, creation of geokhods' prototypes has been actively conducted. The geokhod is a new kind of mining heading machines embodying innovative technical solutions and working modes [1,2]. A transmission (power drive) with hydraulic drive on the basis of hydraulic cylinders, placed chordwise inside the sections, is the most promising design for the geokhods'. In order to ensure the requirements for a new generation of geokhods' power drives [3], design solutions should be made on the basis of layout diagrams, representing hydraulic cylinders' operation in different phases [4,5].

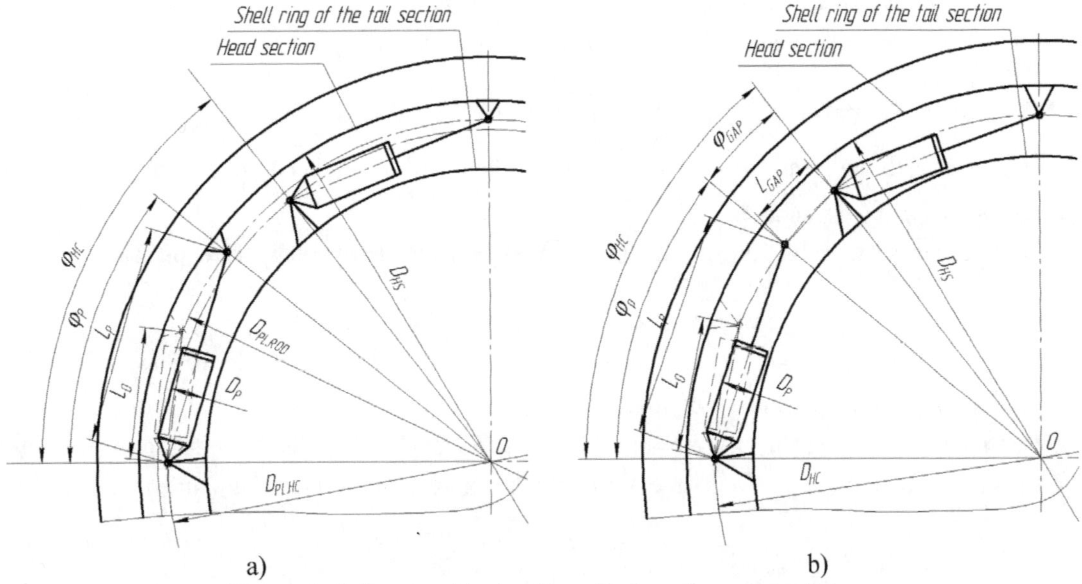

a) b)

Figure 1. Schemes of hydraulic cylinders characterization

Results and Discussion

The hydraulic cylinders are arranged in the space between two cylindrical shells (Figure 1). Hydraulic cylinder' body supports are fixed on the outer surface of the tail section shell, and rod supporting members periodically engage into mesh with the ratchet rack on the inner surface of the head section. While rod cylinders are advancing, sliding members engage the ratchet rack and drive it into revolving. Body and rod's spatial position changes while advance and return motions of the hydraulic cylinder rod; at that, head and tail shell should not impede changes in spatial positions of hydraulic cylinders' elements. Power drive layout schemes have shown that the most rational parameters of hydraulic cylinders arranged between the shells are obtained with the following proportions

$$D_{PL.HC} = D_{HS} - (3...4)D_P, \qquad (1)$$

$$D_{PL.ROD} = D_{HS} - (2...3)D_P. \qquad (2)$$

When arranging hydraulic cylinder supports at one circumference (Figure 1b), that is when $D_{PL.ROD} = D_{PL.HC} = D_{HC}$ it can be taken:

$$D_{HC} = D_{HS} - (3...4)D_P, \qquad (3)$$

Most commercially available cylinders [6,7] has the following ratio of a piston stroke length L_x to a size along axes of mounting brackets L_0

$$L_X = (0.3 \div 0.8)L_0, \qquad (4)$$

The maximum distance between the hydraulic cylinder supports L_p can be represented as the sum of

$$L_p = L_0 + L_X, \qquad (5)$$

Stroke length value L_X can be obtained from the formulas (4) and (5)

$$L_X = \frac{L_p}{1 + \frac{1}{(0.3 \div 0.8)}}, \qquad (6)$$

Approximate value of the maximum possible distance L_p between the hydraulic cylinder supports can be obtained from the scheme, taking into account the total number of cylinders – n_{HC} [8] (Figure 1b). To simplify the calculation, we assume that the hydraulic cylinder support will be arranged on one circumference [9], that is $D_{PL.ROD} = D_{PL.HC} = D_{HC}$. The angle φ_{HC} between bodies (rods) supports of adjacent cylinders will be determined by the number of cylinders n_{HC}

$$\varphi_{HC} = \frac{360^\circ}{n_{HC}}, \text{ degrees} \qquad (7)$$

Cylinder with maximally advanced rod will cut the chord of L_p length on the circumference D_{HC} based upon the central angle φ_P.

Between these geometric parameters there is a relationship, described by the expression:

$$L_P = D_{HC} \cdot \sin\frac{\varphi_P}{2}, \qquad (8)$$

The central angle φ_P will be taken for reasons of leaving a required angular gap φ_{3A3} between rod supports and bodies of adjacent cylinders. The angle φ_{GAP} will be based on a chord L_{GAP} on the circumference D_{HC}. For this scheme (Figure 1b) a gap magnitude can be set equal to the diameter of the piston D_P, that is $L_{GAP} = D_P$, while, similar to the expression (8), it can be written:

$$L_{GAP} = D_P = D_{HC} \cdot \sin\frac{\varphi_{GAP}}{2}, \qquad (9)$$

Let us express φ_{GAP} from (9)

$$\frac{D_P}{D_{HC}} = \sin\frac{\varphi_{GAP}}{2}, \text{ from which } \arcsin\frac{D_P}{D_{HC}} = \arcsin(\sin\frac{\varphi_{GAP}}{2});$$

$$\varphi_{GAP} = 2\cdot\arcsin\frac{D_P}{D_{HC}}, \text{ degrees} \tag{10}$$

The central angle φ_P will be equal to

$$\varphi_P = \varphi_{HC} - \varphi_{GAP}, \text{ degrees} \tag{11}$$

Substituting in (11) the expressions (7) and (10) we find

$$\varphi_P = \frac{360^\circ}{n_{HC}} - 2\cdot\arcsin\frac{D_P}{D_{HC}}, \text{ degrees} \tag{12}$$

Now, substituting the expression (12) in (8) we find the expression for L_p

$$L_P = D_{HC}\cdot\sin\left[0{,}5\left(\frac{360^\circ}{n_{HC}} - 2\cdot\arcsin\frac{D_P}{D_{HC}}\right)\right] = D_{HC}\cdot\sin\left(\frac{180^\circ}{n_{HC}} - \arcsin\frac{D_P}{D_{HC}}\right), \tag{13}$$

In view of the expression (3), the expression (13) takes the form

$$L_P = [D_{HC} - (3...4)D_P]\cdot\sin\left(\frac{180^\circ}{n_{HC}} - \arcsin\frac{D_P}{(D_{HC} - (3...4)D_P)}\right), \tag{14}$$

Another distance limitation L_p can be longitudinal stability of compressible rod. Maximum distance between the supports can be determined by the formula

$$L_{P\max} = \frac{356.8\cdot d_{ROD}^2}{D_P\cdot k_H\sqrt{p\cdot[n]}}, \tag{15}$$

where d_{ROD} – is cylinder rod diameter, m;

k_H – fixing coefficient for hinge mounts on both ends of the cylinder k_H=1;

$[n]$ – safety margin (usually $[n]$ = 3.5…4);

p – working pressure in the hydraulic cylinder, Pa.

The rod diameter is usually taken to be based on the ratio [10]

$$d_{ROD} = (0.3 \div 0.7)\cdot D_P, \tag{16}$$

For a compressive load it is recommended to take larger values of the expression (15).

While the hydraulic cylinders operate in different stroke phases, it is necessary to ensure the reverse of the piston of one cylinder or group of cylinders while advancing of other pistons at a distance of amount of difference between adjacent phases' strokes. This occurs when the reverse speed υ_{REV} is above the stroke rate $\upsilon_{P.X}$ by n_{PH}, that is

$$\upsilon_{REV} \geq n_{REV}\cdot\upsilon_{P.X}, \tag{17}$$

Based on this condition, rod diameter d_{ROD} at a constant working fluid's flow rate Q_{Pi} will be determined by the formula

$$d_{ROD} = D_P\sqrt{1 - \frac{1}{n_{PH}}}, \tag{18}$$

Substituting (18) into (15) we obtain

$$L_{P\max} = \frac{356.8\cdot\left(D_P\sqrt{1-\frac{1}{n_{PH}}}\right)^2}{D_P\cdot k_H\sqrt{p\cdot[n]}} = \frac{356.8\cdot D_P\cdot\left(1-\frac{1}{n_{PH}}\right)}{k_H\sqrt{p\cdot[n]}}, \tag{19}$$

When the number of reversing cylinders is n_{REV} = 1, number of intermediate positions (phases) will be

$$n_{PH} = n_{HC}, \tag{20}$$

Then the expression (19) takes the form:

$$L_{P\max} = \frac{356.8 \cdot D_P \cdot \left(1 - \dfrac{1}{n_{HC}}\right)}{k_H \sqrt{p \cdot [n]}}, \tag{21}$$

Figure 2 shows diagrams the maximum possible (safe) distances between hydraulic cylinder supports – L_p against the number of cylinders n_{HC} and the piston diameter D_P, plotted for a section with diameter D_{HC} = 3.7 m by design arrangement condition (Fig. 2) and stability condition (Fig. 3).

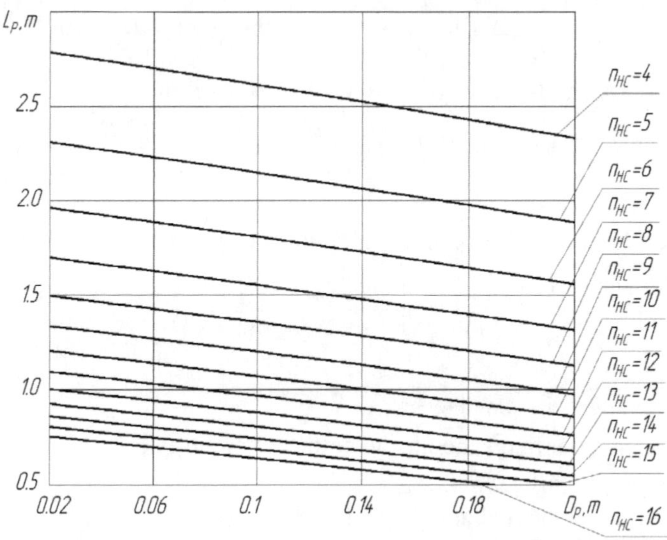

Figure 2. Dependence of the safe distance by stability condition

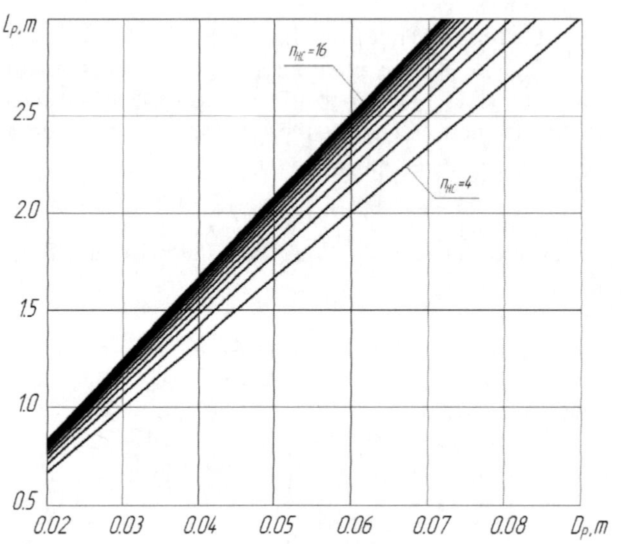

Figure 3. Dependence of the safe distance by design condition

Figure 4 shows a diagram plotted on both conditions for various numbers of cylinders.

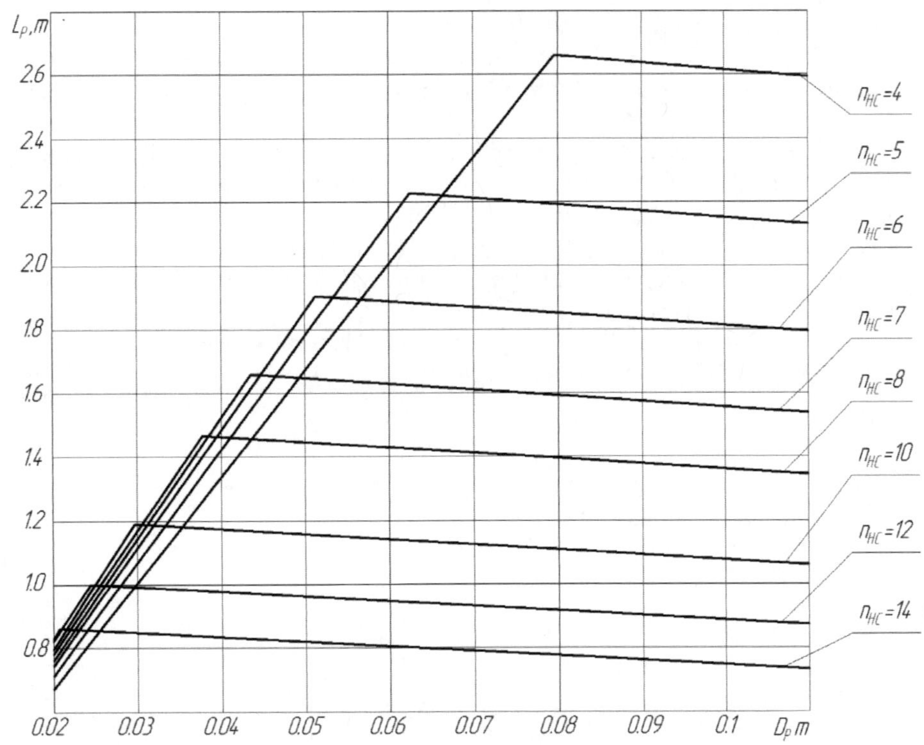

Figure 4. Dependence of safe distance by design arrangement condition and stability condition

Conclusion

Diagrams of dependences of the safe inscribable distance between the hydraulic cylinder supports by stability condition show that for small piston diameter safe distances between the hydraulic cylinder supports by stability condition are significantly less than the maximum safe distance by design arrangement condition; thus, as a number of cylinders increases, the stability condition is predominant.

In addition to the length limitations by design conditions and stability condition, the required inside structure gage may also limit geokhod size. For the purpose of increasing its size, it is necessary to diminish the distance between supports and, correspondingly, the stroke length.

References

[1] Efremenkov A.B., Aksenov V.V., Blashchuk M.Y. Force parameters of geokhod transmission with hydraulic drive in various movement phases // 7th International Forum on Strategic Technology (IFOST - 2012): Proceedings: in 2 vol. IEEE, 2012. Vol. 2. P. 159–163.

[2] Efremenkov A.B. Forming the subterranean space by means of a new tool (geokhod) // 6th International Forum on Strategic Technology (IFOST - 2011): Proceedings: in 2 vol. IEEE, 2011. Vol. 1. P. 348–350.

[3] Efremenkov A.B., Timofeev V.Y. Determination of necessary forces for geokhod movement. IEEE, 2012. P. 1–4.

[4] Aksenov V.V., Blaschuk M.Y., Dubrovskii M.V. Estimation of Torque Variation of Geokhod Transmission with Hydraulic Drive // Applied Mechanics and Materials. 2013. Vol. 379. P. 11–15.

[5] Blashchuk M.Y., Kazantsev A.A., Chernukhin R.V. Capacity Calculation of Hydraulic Motors in Geokhod Systems for Justification of Energy-Power Block Parameters // Applied Mechanics and Materials. 2014. Vol. 682. P. 418–425.

[6] Buyalich G.D., Buyalich K.G., Voyevodin V.V. Radial deformations of working cylinder of hydraulic Legs depending on their extension // IOP Conference Series: Materials Science

and Engineering. 2015. Vol. 91. P. 012087.
[7] Buyalich G.D., Anuchin A.V., Dronov A.A. Numerical Analysis of Accuracy of Hydraulic Leg Cylinder in Modeling Using Solid Works Simulation // Applied Mechanics and Materials. 2015. Vol. 770. P. 456–460.
[8] Chernukhin R.V., Dronov A.A., Blashchuk M.Y. Application of analytic hierarchy process when choosing layout schemes for geokhod pumping station // IOP Conference Series: Materials Science and Engineering. 2015. Vol. 91. P. 012086.
[9] Sadovets V.Y., Beglyakov V.Y., Aksenov V.V. Development of math model of geokhod bladed working body's interaction with geo-environment // IOP Conference Series: Materials Science and Engineering. 2015. Vol. 91. P. 012085.
[10] Buyalich G.D., Buyalich K.G. Modeling of hydraulic power cylinder seal assembly operation // Taishan Academic Forum - Project on Mine Disaster Prevention and Control. 2014. P. 167–170.

IOP Conf. Series: Materials Science and Engineering **142** (2016) 012129 doi:10.1088/1757-899X/142/1/012129

Research on Geometrical Errors of Geokhod Prototype Shell Based on Coordinate Control Data

A V Walter[1], I F Borovikov[2] and M F Savin[1]

[1] Yurga Institute of Technology, TPU, 26, Leningradskaya Str., 652050, Yurga, Russia
[2] Bauman Moscow State Technical University, 5/1, ul. Baumanskaya 2-ya, 105005, Moscow

E-mail: avwalter@rambler.ru

Abstract. The article contains results of a research on geometric accuracy of a geokhod prototype shell. The article outlines the general structural features of geokhod bodies, and the main principles of manufacturing in test production. An overview of approaches to modeling of shell error occurrence is given. The researches were conducted on the basis of data obtained by coordinate control over the stabilizing section shell. The data were studied by statistical methods and analyzed in terms of their compliance with previously proposed mathematical models of formation of geokhod shell inaccuracies. It is shown that available mathematical models can not completely explain the origin of all the errors. The authors attribute unexplained geokhod shell errors as deformations caused by welding.

1. Introduction

Currently, preparations are underway for commercializing of a new class of mining machines namely geokhods [1]. Geokhod is an apparatus intended for underground works of various purposes and locations in the rock space, and progressing in the rock mass using geo-environment. Design novelty of the machine requires new technologies for production of specific high-precision components and systems. A critical element of the geokhod is a shell (outer surface of the machine body), which interacts with the geological environment. Geometric precision of the shell affects the efficiency and service life of the machine as a whole, since it affects the resistance of rock mass to geohod progressing.

Three of the machine bodies are shelled: the head section, the adapter module outer body and the stabilizing section (figure. 1). The sections are structurally similar, have a cylindrical shape and are composed of shells, stiffening elements and flanges (figure. 2). Each shell is divided into four sectors. The sectors are lock-jointed and fastened with pins and bolts. The sections are large-sized and impact significantly on the production costs of the machine as a whole [2]. The nominal diameter of the shell is 3200 mm.

In order to assure assemblability, an assembly procedure, common in heavy engineering, is used, namely, pre-assembled locks are installed on the assembly plate in the specific mutual position. The locks form a frame to which the remaining structural elements are welded sequentially. This procedure reduces the amount of machining [3]; geometrical errors of the shell are mainly formed during assembly of the body. Inaccuracies of the shell are also affected by manufacturing errors. [4] describes how errors of mutual arrangement of sectors in the shell assembly and errors of sector shells affect the geometry errors of the shell shape. In [4] - [6], models for determining of ring shell errors are given, based on correction for geometric tolerances of mating surfaces. In [7] it is stated that the sequence of assembling of ring elements affects the assembling accuracy.

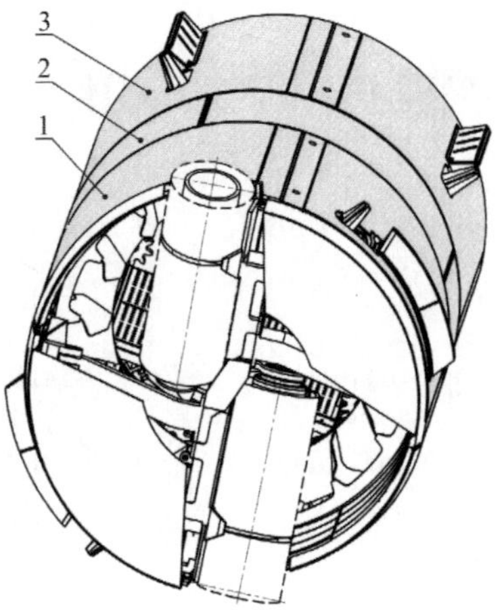

Figure 1. General view of geokhod prototype (shell is highlighted in gray): 1 - head section; 2 - adapter module outer body; 3 - stabilizing section.

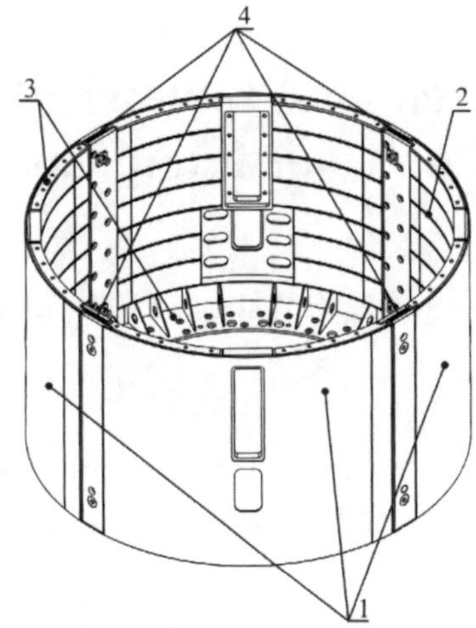

Figure 2. General view of stabilizing section: 1 - shell; 2 - stiffeners; 3 - flanges; 4 - locks.

The study is based on a geometrical approach only and do not include some error-factors, in particular, the deformation of shell parts caused by welding and pressure treatment. The objective is to study the actual shell accuracy of the geokhod prototype and identify factors, significantly affecting its accuracy.

2. Research Methods

For objective purposes the shell of the stabilizing section was studied. The body was subjected to coordinate control [8] using a mobile CMM "FARO Arm Edge 9". For controlling the body frame was placed on the control plate by a particular flange. Next to the body a coordinate measuring machine (CMM) was installed. To expand the examination zone, control was performed from four directions. To obtain the coordinates of recorded points in a single coordinate system, affixment of CMM by calibrating cones was used for each position. Monitoring was carried out by determining of coordinates of point aggregates, evenly distributed over the shell surface.

To analyze and determine values of sector arrangement errors, measurement data was imported into a specially designed program. The program is based on development and analysis of regression models of cylindrical surfaces. Models of cylindrical surfaces were developed by approximation of point aggregates obtained by the measurements; they are expressed by the following system of equations:

$$\begin{aligned}
&\sqrt{A^2+B^2+C^2}-r+\varepsilon_i=0\\
&A=-a_y z_i-\sqrt{1-a_x^2-a_y^2}(y_0-y_i)\\
&B=\sqrt{1-a_x^2-a_y^2}(x_0-x_i)+a_x z_i\\
&C=a_x(y_0-y_i)-a_y(x_0-x_i)\\
&a_z=\sqrt{1-a_x^2-a_y^2}
\end{aligned} \quad (1)$$

x_i, y_i, z_i – coordinates of approximate points; a_x, a_y, a_z, x_0, y_0, r – unknown regression coefficients, having the following geometric meaning: a_x, a_y, a_z – coordinates of direction vector of approximating cylinder axis; x_0, y_0 – coordinates of the point of approximating cylinder axis path; r – radius of the approximating cylinder; ε_i – residues of the regression model.

The regression coefficients in equations (1) were determined by Gauss method [9], that is, from the condition

$$\sum_{i=1}^{n} \varepsilon_i^2 \rightarrow \min \tag{2}$$

For determination of total geometrical deviations of the active surface points from the nominal shell geometry, a cylindrical surface model was developed based on approximation of the control dataset (figure. 3).

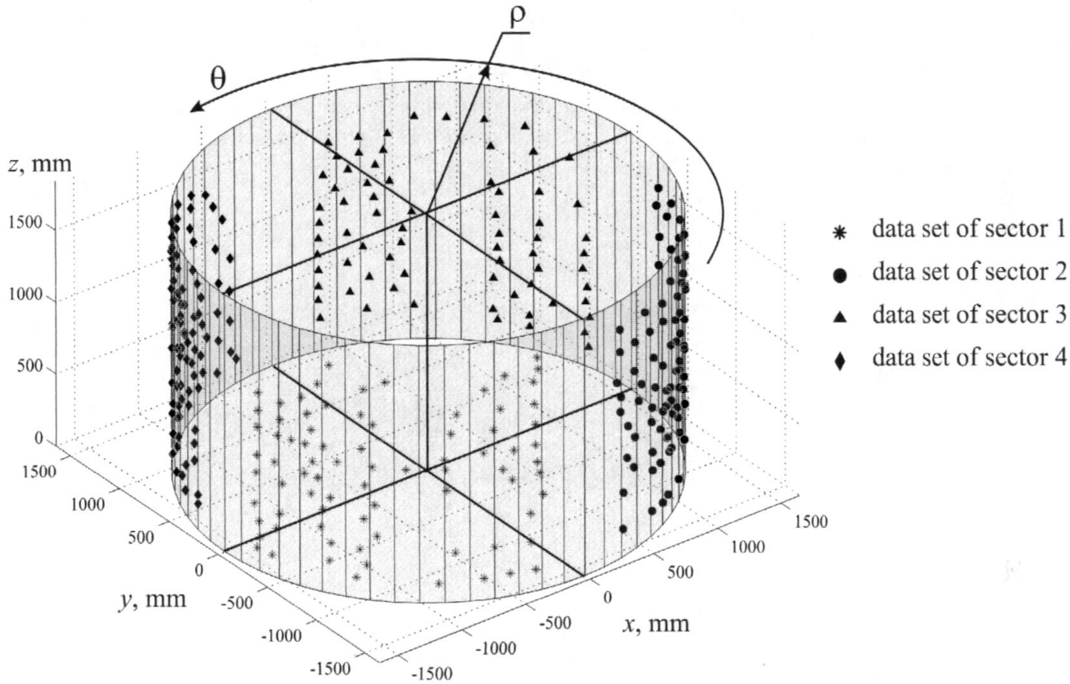

Figure 3. Data sets of measurements and a model of a cylindrical surface.

Model residuals were studied by correlation and regression analysis to identify systematic components in shell geometric errors in a sequence similar to that described in [10]. For this, point coordinates of each sector were transferred into a cylindrical coordinate system, associated with the sector axis $\theta z\rho$, and transformed into the first octant.

To determine values of deviations of separate sectors and deviations of sector radii, models of cylindrical surfaces were developed for each sector separately. Model residuals were also studied by correlation and regression analysis to identify systematic components in shell geometric errors [11].

3. Results and discussions

Table 1 shows model parameters of shell surfaces both completely and of separate sectors. Analysis shows that the body sectors are significantly different in their actual size and shape errors. However, errors are within the limits specified by the design documentation, but close to the critical values. The latter means that in mass production by this manufacturing method, there may be problems with assuring the accuracy of the geokhod shell assembly.

IOP Conf. Series: Materials Science and Engineering **142** (2016) 012129 doi:10.1088/1757-899X/142/1/012129

Table 1 Parameters of surface models

Parameter name	Shell in the whole	Sector 1	Sector 2	Sector 3	Sector 4
Radius of approximating cylinder, mm	1595.25	1595.01	1591.89	1604.45	1610.18
Standard deviation, mm	2.10	1.42	1.11	1.32	0.99
Maximum deviation, mm	5.25	2.76	2.85	3.23	1.83
Minimum deviation, mm	-4.50	-4.04	-3.10	-3.33	-2.82
Radius error, mm	9.75	6.80	5.95	6.56	4.65

We also note that errors in manufacturing and positioning of sectors are only a part of production errors. That is, in manufacturing process accuracy is conditioned by other factors; character and importance of these factors should be identified.Figure 4 shows the series of residuals of regression models of sector surfaces along angular coordinate θ. Specific changes of values in the set of residuals are noticeable for each sector. In each sector there are two "waves"; wave **bottoms** fall on sector edges and their middles. Similar regularities are observed in sectors 1, 2 and 4 particularly clearly. In sector 3 it is less pronounced.

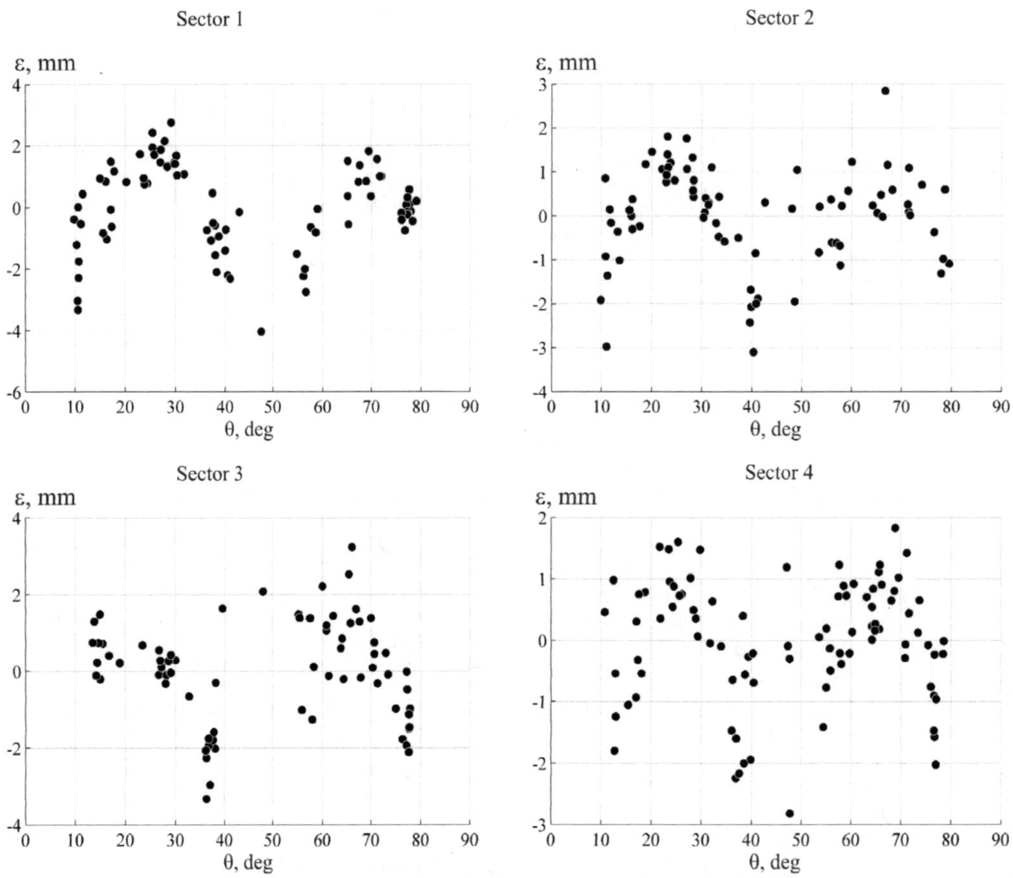

Figure 4. Sets of residuals along the angular θ-coordinate.

Figure 5 shows sets of residuals of sector surface regression models along z-coordinate. In the sets a dependence of residual values is less visible, however, reducing of residual magnitudes in the middle of each sector, can be observed. The results of the statistical analysis (see Table 2) indicate that mode

IOP Conf. Series: Materials Science and Engineering **142** (2016) 012129 doi:10.1088/1757-899X/142/1/012129

of residuals is non-random. Critical values of statistics in this table are given for level of significance $\alpha = 0.05$.

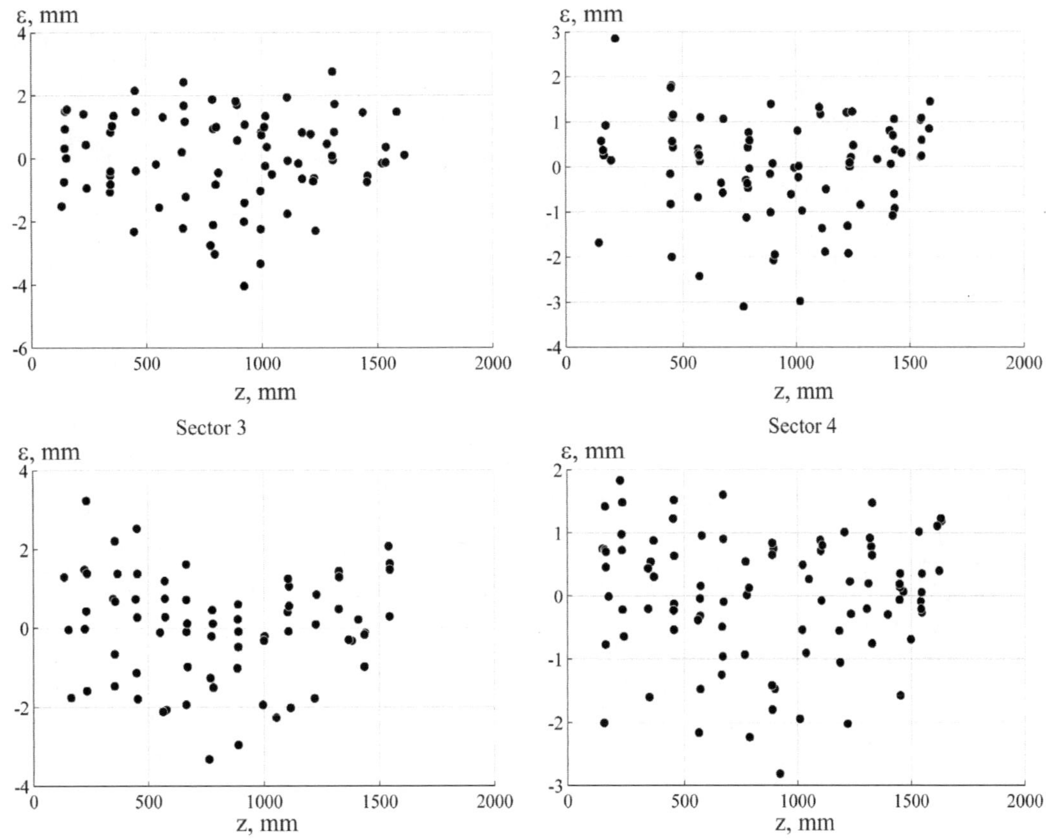

Figure 5. Sets of residuals along *z*-coordinate.

Statistic Epps-Pally criterion has shown that the residual sets in sector 1 and sector 2 do not correspond to the normal distribution law. Criterion of turning points has not determined the existence of a trend in residual sets $\varepsilon(\theta)$ and $\varepsilon(z)$ for any sector. Durbin-Watson statistic [12] has revealed that in residual sets $\varepsilon(\theta)$ autocorrelation is observed in each sector, in the same time there is no autocorrelation in residual sets $\varepsilon(z)$. Consequently, there are some regularities in shell point deviations from sector models; they depend on shell radius errors and assembling inaccuracy. In order to identify causes of regularity deviations, sets of control points were placed on sector scanning (figure. 6). In the scans structural elements (ribs, locks and massive elements) are traced. Magnitude of deviation points is designated with markers, corresponding to a specific range of deviations from the mean radius.

Image analysis (Figure 6) shows that negative-valued deviation points are concentrated around massive and rigid structural elements of sectors. Thus, "wave" bottoms in the middle of the sector (see Figure 4) are caused by a massive element namely a counter-rotation support member; and bottoms at the sector edges are caused by rigid elements namely locks. In this way, regularity in deviation changes is based on the regularity of the actual sector structure.

Table 2. Results of statistical analysis of model residuals.

Parameter name	Sector 1	Sector 2	Sector 3	Sector 4
Epps-Pally statistic	0.3363	0.6347	0.1949	0.7006
Epps-Pally statistic critical value	0.3754	0.3754	0.3751	0.3757
Number of turning points in $\varepsilon(\theta)$	51	51	44	57
Number of turning points in $\varepsilon(z)$	57	52	42	62
Critical number of turning points	44.7	44.7	39.1	51.5
Durbin-Watson statistic for $\varepsilon(\theta)$	0.7655	0.9767	0.7242	1.1818
Durbin-Watson statistic for $\varepsilon(z)$	1.9726	1.9843	2.0224	1.9245
	1.4650	1.4650	1.4330	1.4989
	1.5140	1.5140	1.4884	1.5433
	2.4860	2.4860	2.5116	2.4567
Durbin-Watson statistical significance bands	2.5350	2.5350	2.5670	2.5011

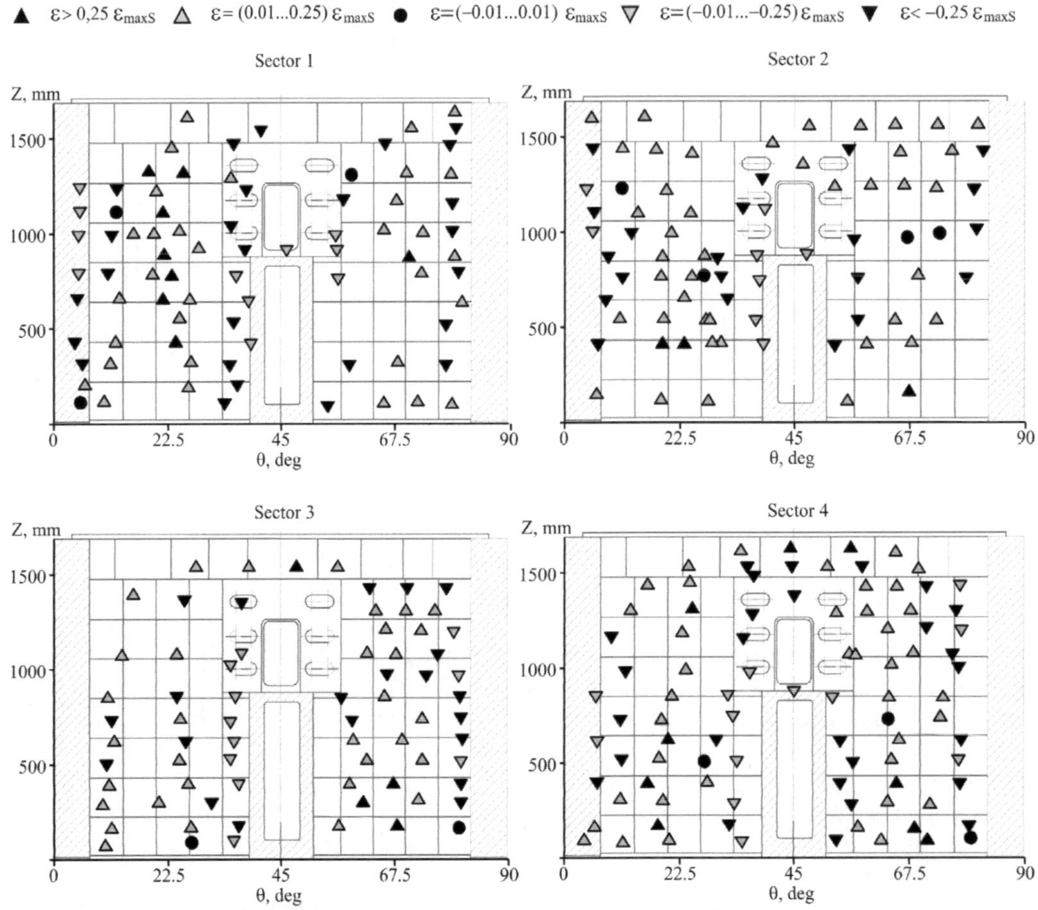

Figure 6 Deviation of measuring points from model mean radius on sector scans (ε_{maxS} – maximum difference between residual set terms for current sectors)

The only objective reason for such distribution of deviations in sectors may be deformations arising during welding [13]. Uneven rigidity leads to the fact that massive structural elements, without changing their geometry, change their position towards other elements and cause considerable deformations of less rigid elements. The results generally correlate with the results presented in [14], and require further development of shell error modeling by methods similar to those in [15], [16].

4. Conclusion

- Analysis of coordinate measuring of the stabilizing section shell shows that deviations from the geometric precision, close to critical values, are foreseeable at manufacturing of geokhod bodies.
- The causes of deviations are shell manufacturing errors, assembling inaccuracy of sector positioning and other causes, that have not been determined until now.
- It has been established that mode of shell surface deviations is caused by sector structure which is composed of elements of different rigidity. This gives reason to believe that shell deviations that are not caused by shell manufacturing errors and assembling inaccuracies, are deformations caused by welding.

References

[1] Timofeev V Y, Aksenov V V and Galjamova J I 2014 Determination of Parameters of Wave Transmission Roller with Intermediate Rolling Bodies with Hollow Shaft for Geokhod *Applied Mechanics and Materials* **682** 246–250

[2] Aksenov V V, Walter A V, Gordeyev A A and Kosovets A V 2015 Classification of geokhod units and systems based on product cost analysis and estimation for a prototype model production *IOP Conference Series: Materials Science and Engineering* **91** 012088

[3] Putilova U S, Nekrasov Y I and Lasukov A A 2014 Loading of the Manufacturing Systems Elements in the Process of Unsteady Mode Cutting and the Models of their Arrangement Deviations *Applied Mechanics and Materials*. **682** 192–195

[4] Walter A V and Aksenov V V 2015 Determining deviations in geometry of the geokhod shells *Applied Mechanics and Materials* **770** 439–444

[5] Lowth S and D. A. Axinte 2014 An assessment of 'variation conscious' precision fixturing methodologies for the control of circularity within large multi-segment annular assemblies *Precision Engineering* **38** 379–390

[6] Hussain T, Yang Z, Popov A A and McWilliam S 2011 Straight-build assembly optimization: A method to minimize stage-by-stage eccentricity error in the assembly of axisymmetric rigid components (two-dimensional case study) *Journal of Manufacturing Science and Engineering, Transactions of the ASME* **133-3** 031014

[7] Memon M, Hussain T and Memon Z A L I 2012 Minimizing Assembly Errors by Selecting Optimum Assembly Sequence in the Assembly of a Rigid Circular Structure *Mehran University Research Journal of Engineering & Technology* **31-4** 743–754

[8] Saunders P, Wilson A, Orchard N, Tatman N and Maropoulos P 2014 An Exploration into Measurement Consistency on Coordinate Measuring Machines *Procedia CIRP* **25** 19–26

[9] Lemeshko B Y, Lemeshko S B, Nikulin M S and Saaidia N 2010 Modeling statistic distributions for nonparametric criteria for testing complex hypotheses with respect to the inverse Gaussian law *Automation and Remote Control* **71-7** 1358–1373

[10] Walter A V 2012 Modeling of Track Overlapping Effect on Layer Integrity in the Process of Laser Sintering of Polymer Suspension *7th International Forum on Strategic Technology (IFOST - 2012)* **2** 125–128

[11] Kendall M G, Stuart A and Ord J K 1968 *Design and analysis, and time-series*

[12] Durbin J and Watson G S 1971 Testing for Serial Correlation in Least Squares Regression *Biometrika* **58-1** 1

[13] Petrova V A, Bakanov A A and Walter A V 2014 Crack Resistance of Weld Seals, Hardening Methods and Composite Tools Quality *Applied Mechanics and Materials* **682** 431–437

[14] Camelio J A, Hu S J and Ceglarek D **2004** Impact of Fixture Design on Sheet Metal Assembly Variation *Journal of Manufacturing Systems* **23-3** 182–193

[15] Chen Z, Chen Z and Shenoi R A 2015 Influence of welding sequence on welding deformation and residual stress of a stiffened plate structure *Ocean Engineering* **106** 271–280

[16] Deng D, Murakawa H and Liang W 2007 Numerical simulation of welding distortion in large structures *Computer Methods in Applied Mechanics and Engineering* **196-45-48** 4613–4627

IOP Conf. Series: Materials Science and Engineering **142** (2016) 012130 doi:10.1088/1757-899X/142/1/012130

Mathematical Description of THE Traction Characteristics of the Driving Devices at Spatial Stiffness Compensators of the Vibration Isolation Installations

E G Gurova

Mechatronics and automation department,
Novosibirsk State Technical University,
Av. K. Marx, 20, Novosibirsk, 630073, Russia

e-mail: lena319@mail.ru

Abstract. During the researches the mathematical description of the traction characteristics of the stiffness compensators of the vibration isolation devices, relatively of the each axis, has been done. Representation of the compensators properties considers the variable load, thereby provide the wide enough spectrum of the action of the suggested vibration isolators. The derived expressions are valid for all three axes of space at the different stiffnesses, i.e. basic basic and two compensating. The research was supported by the scholarships of Russian Federation President for young scientists №184 from 10th of March 2015.

Introduction

The use of the power machine not only in industry, but and in any type of the vehicles (auto, water-craft, metro, trams, trolleybuses) inevitably leads to vibration. The vibration negatively impacts on reliability, lifetime of the machines, buildings, installations, where it were installed, and also on the system of automatic control. It is not rarely, that vibration is one of the reasons of the accidents. But the particularly harmful impacts on human body. The human, unwittingly falls under the negative influence of the vibration, during the driving of the all types of the vehicles, and which subsequently causes a variety of diseases. Therefore the issue of the decreasing of the transmitted vibrations stands very acute.

During the last time many types of the vibroprotection devices have been made [3, 7-11], but many of it does not decrease the level of the vibration down to normal level, satisfying to State Standards. Most of perspective method of decreasing of the vibration levels, created by power installations is using of the vibration isolation suspensions with floating area of zero stiffness [1, 2, 5]. The principle of the operation of the vibroprotection mechanisms is shown on figure 1. Under the limited values of the vibration isolation movement of the suspension H and under the certain range of the traction changing from P_{max} to P_{min}, transmitted from protected object to vibroprotection device, the traction characteristics of the vibroisolating mechanisms, provides the ideal vibration isolation of the system, are infinite number of the segments. These segments have a length $2A$ (magnitude of the oscillation), which in parallel of abscise axis and placed by its middles on AB line, which is ramped to abscises axis under the certain angle. The tangent of this angle is equal of the stiffness of the suspension.

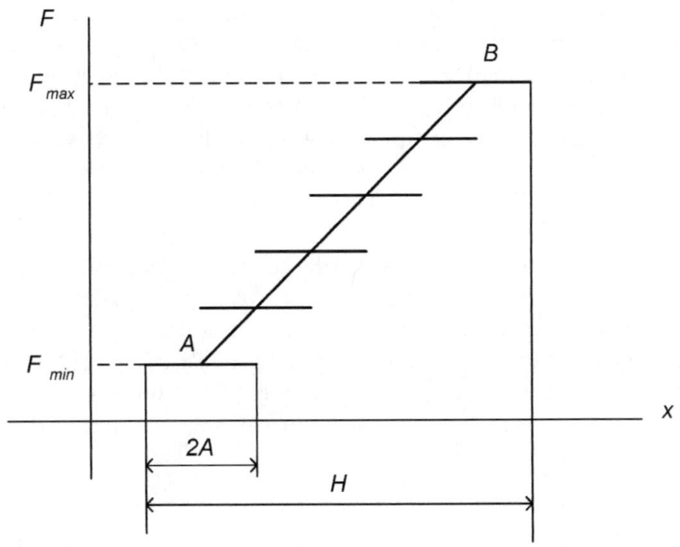

Figure 1. The traction characteristic of the tunable vibration isolation device

In such suspensions there are two basic components: resilient element and parallel compensator of the stiffness, which is device, had falling traction characteristic, with negative coefficients of the stiffness. The total stiffness of the vibration isolator is the sum of the stiffnesses of the resilient element and compensator (figure 2). The stiffness of the suspension can be decrease to zero, it can allow provide the ideal vibration isolation [3].

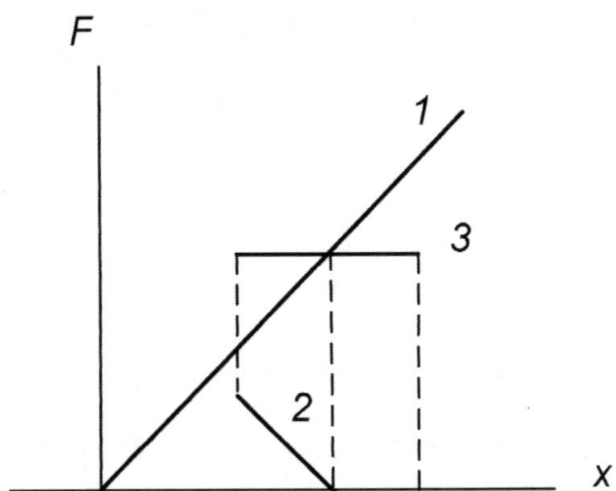

Figure 2. Characteristic of the vibration isolator with stiffness compensator, of the:
1- resilient element; 2 - stiffness compensator; 3 - vibration isolator

Calculation and simulations

In researches [4, 6] the vibration isolation device with electromagnet compensator of the stiffness was suggested, which best meets the requirements of ideal vibration isolation and has the some advantages above the early suggested of the mechanic compensators of the stiffness. Such compensator is characterized by absence of the friction forces and inertia, easy to automatization, and also has tuning system, redistributed the voltage on the electromagnets coils during the changing of the load. It allows the exclude the resonance and provides the zero of the stiffness. However, under the all its achievements the application of the vibration isolation with corrector, based on constant current electromagnets, is limited by big enough sizes. In [4, 6] for vibroprotection devices with correctors of the stiffness the use of the supermagnets was suggested. These magnets have the less size at the same traction forces; it sufficiently expands the application range of such devices to decrease the vibration: vehicles, space equipment, aircraft and etc. However, mentioned above devices, can exclude the vibration oscillations only respectively of one axis, and as is known, the vibration is chaotic spatial oscillations. To exclude the spatial vibrations the vibration isolation devices with electromagnetic and supermagnet compensator of the stiffness were suggested. The principle of such devices has been built on effect of the zero stiffness [4, 5], as is shown on figure 3, under the changing of the load the traction characteristic of the stiffness compensator is moving in parallel itself with safe constant stiffness. For automatic tuning of the vibration isolation device to changing load, the compensator of the stiffness has tuning system, which taking to account the magnitude of the load [6]. At the development of the tuning system of the spatial stiffness compensator we need to complete the description of its forces characteristic with accounting the nonlinearity. Consider the description of the corrector characteristic on example of one of the axis, as is shown on figure 3. For another axis the similar description of the characteristic, but the value of the stiffness will have name compensation.

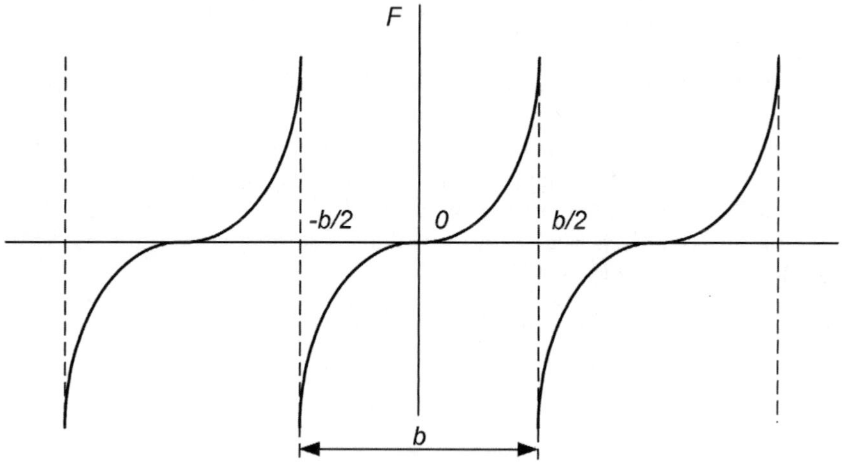

Figure 3. The traction characteristics of the stiffness compensator

In research [6] the equation, describing the traction characteristic of the compensator in the space, was obtained. The mathematical description which was completed, allows us to make representation of the characteristic and consider the stiffness respectively to each spatial axis, under the constant load. The traction characteristic can be described by odd polynomial of *n-th* degree:

$$F(x) = c_1 \cdot x + c_3 \cdot x^3 + \ldots + c_{2n-1} \cdot x^{2n-1} + c_n \cdot x^n, \tag{1}$$

where c_1, c_3, c_{2n-1}, c_n - are constant coefficients.

Under the changing of the load of the vibration isolation device, the tuning system should tune the system, so that the traction characteristic of the compensator moves in parallel itself with constant stiffness. At the description accept the coefficients c_1, c_3, c_{n-1}, c_n are equal *1*.

For the accounting of the traction characteristic in developing of the tuning system we should make series Fourier expanding in range (*-b/2; b/2*), which is equal interpolar distance of the stiffness compensator (*b*):

$$F(x) = a_0 + \sum_{n=1}^{\infty}\left(a_n \cdot cos\frac{n\pi x}{b/2} + b_n \cdot sin\frac{n\pi x}{b/2}\right) \tag{2}$$

For representation of the function (1) in Furrier series (2) derive the coefficient:

$$a_0 = \frac{1}{b} \cdot \int_{-b/2}^{b/2} \left(x + x^3 + \ldots + x^{n-1} + x^n\right) \cdot dx = \frac{2}{b} \cdot \left[\frac{x^2}{2} + \frac{x^4}{4} + \ldots + \frac{x^n}{n} + \frac{x^{n+1}}{n+1}\right]_0^{b/2} = \\ = \frac{2}{b} \cdot \left[\frac{b^2}{8} + \frac{b^4}{64} + \ldots + \frac{b^n}{2^n \cdot n} + \frac{b^{n+1}}{2^{n+1} \cdot (n+1)}\right] \tag{3}$$

$$b_n = \frac{1}{l} \int_{-l}^{l} (x + x^3 + \ldots x^{n-1} + x^n) \cdot \sin\frac{k\pi x}{l} dx = \begin{vmatrix} \frac{k\pi x}{l} = b \\ x = \frac{bl}{k\pi} \end{vmatrix} = 2\frac{1}{l} \int_0^l \left[\frac{lb^1}{k\pi} + (\frac{l}{k\pi})^3 b^3 + (\frac{l}{k\pi})^5 b^5 + (\frac{l}{k\pi})^n b^n\right] \sin b db \cdot \frac{l}{\pi k} = \\ = \frac{2}{\pi k} \int_0^b \left[(\frac{l}{k\pi})^1 b^1 + (\frac{l}{k\pi})^3 b^3 + (\frac{l}{k\pi})^5 b^5\right] \sin b db = \frac{-2}{\pi k} b^n \cos b + \frac{2n}{\pi k} \int_0^l b^{n-1} \cos db = \sum_{n=0}^{n-1} -\frac{2}{\pi k} \frac{(k\pi)^n (x)^n}{l^n} \cos(\frac{k\pi x}{l}) + \\ +2\frac{n}{\pi k} \int_0^l \frac{(\pi k)^{n-1} x^{n-1}}{l^{n-1}} \cos\frac{k\pi x}{l} d\frac{k\pi x}{l} = \sum_{n=0}^{n-1} -\frac{2}{l^n} \cdot (k\pi)^{n-1} x^n \cos\frac{k\pi x}{l} \Big|_0^l + 2n\frac{(\pi k)^{n-1}}{l^n} \int_0^l x^{n-1} \cos\frac{k\pi x}{l} dx; \tag{4}$$

Finally, the coefficients for series expansion of furrier have a view:

$$a_0 = \frac{2}{b} \cdot \left[\frac{b^2}{8} + \frac{b^4}{64} + \ldots + \frac{b^n}{2^n \cdot n} + \frac{b^{n+1}}{2^{n+1} \cdot (n+1)}\right] \tag{5}$$

$$a_n = \sum_{n=0}^{n-1} -\frac{2}{l^n} (k\pi)^{n-1} x^n \sin\frac{k\pi x}{l} \Big|_0^l + 2n\frac{(\pi k)^{n-1}}{l^n} \int_0^l x^{n-1} \sin\frac{k\pi x}{l} dx; \tag{6}$$

$$b_n = \sum_{n=0}^{n-1} -\frac{2}{l^n} (k\pi)^{n-1} x^n \cos\frac{k\pi x}{l} \Big|_0^l + 2n\frac{(\pi k)^{n-1}}{l^n} \int_0^l x^{n-1} \cos\frac{k\pi x}{l} dx; \tag{7}$$

On the basis of the obtained coefficients (4)-(6) the function (1)-(2) after the series expansion of furies, has a view:

$$F(x) = \frac{1}{b} \cdot \left[\frac{b^2}{8} + \frac{b^4}{64} + \ldots + \frac{b^n}{2^n \cdot n} + \frac{b^{n+1}}{2^{n+1} \cdot (n+1)}\right] + \sum_{k=1}^{\infty} \left[a_n \cdot cos\frac{\pi nx}{b/2} + b_n \cdot sin\frac{\pi nx}{b/2}\right] \tag{8}$$

The represented characteristic, described by function (8), can be used during the description of the characteristic of the stiffness compensator under the changing of the load on the vibration isolator with taking to account of the movement:

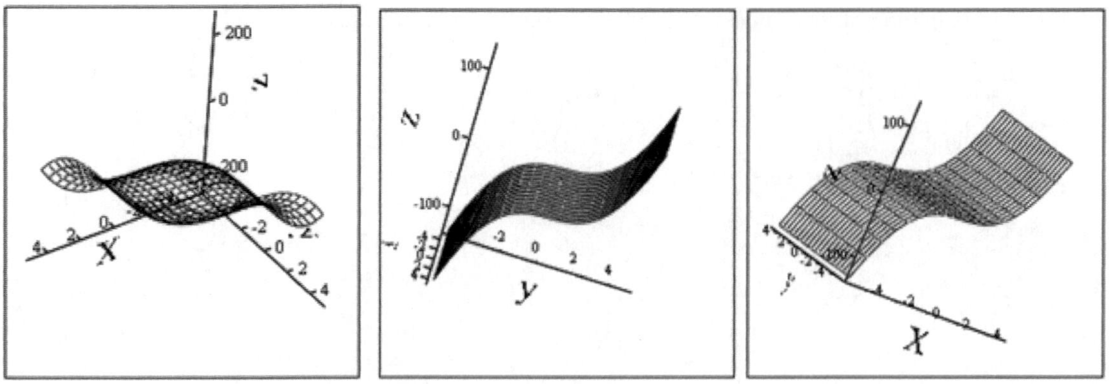

Figure 4. The moving characteristic of the stiffness compensator with respect to each space axis under the changing of the load

The equation (8) graphically shows the traction characteristics of the compensator in the plane during the variation of the load, which will be taking to account for development of the *3D* stiffness compensator tuning system. The derived equations further become the basis for designing of the regulators and the tuning system of the *3D* electromagnetic and supermagnets stiffness compensators of the vibration devices.

Conclusion

In this paper the equation, described the moving in parallel to each other traction characteristics of the stiffness compensators, connected in parallel to resilient element were displayed. The stiffness in such characteristics is constant on account of control devices of the tuning correctors as under the constant loads, so and under the variable loads. The final stiffness of the vibration isolators, including the resilient element and the stiffness compensator is zero, thereby theoretically ideal vibration isolation of the protected device is provided. This derived law will be used in the development of the control device of the stiffness compensator of the vibration isolator.

The spatial vibration isolator can be used in any area of the machinery and equipment, and also can be very efficient solution for the human protection from vibration, generated by the power installation of the vehicles.

The research was supported by the scholarships of Russian Federation President for young scientists №184 from 10th of March 2015.

References

[1] Alabuzhev P M, Zuev A K, Negodaev M N, Yarunov A M, Kargin V A, Kopeikin G F and Khan V B 1965 Use of high-speed cinematography to study the action of an electromechanical detent hammer *Journal of mining science* **2 № 1** pp 139-141

[2] Alabuzhev P M, Demin V N, Demin O N, Kimarckii M S, Nagayev R F, Sokolov V S and Yushin V V 1988 Vibratory tightening *Journal of mining science* **1 № 2** pp 67-74

[3] Bartel T, Gaisbauer S, Stohr P, Herold S and Melz T 2014 Development of functionally integrated mounts for three- and six-axial vibration isolation of sensitive equipment *Proceedings of international conference on noise and vibration engineering (ISMA2014) and international conference on uncertainty in structural dynamics (USD2014)* pp 119-130

[4] Gurova E G and Gurov M G 2014 Vibro isolator with neodymium magnets compensator of the stiffness *Applied Mechanics and Materials* **682** pp 118-121

[5] Gurova E G, Gross V Y, Kurbatov V S, Makarov S V, Sergeev A A and Shchurov N I 2013 Development of spatial vibration protection devices *World Applied Sciences Journal* **22** pp

44-48

[6] Gurova E G 2015 Improving of the operation efficiency of the vehicle due to using of the neodymium magnets inside the vibration isolation devices *Conf. Series- IOP Materials Science and Engineering* **91** 012091

[7] Liu C C, Jing X J and Chen Z B 2015 Band stop vibration suppression using a passive X-shape structured lever-type isolationsystem *Mechanical systems and signal processing* **68-69** pp 342-353

[8] Spiewak S, Karpachevskyy T and Zaiss C 2014 High performance isolation and excitation of vibration for enhanced identification of inertial sensors *Proceedings of the asme international mechanical engineering congress and exposition* **4B** V04BT04A061

[9] Prisekin V L, Pustovoi N V and Rastorguev G I 2014 Algorithms for controlling fatigue tests of airplanes *Journal of Applied Mechanics and Technical Physics* **55 № 1** pp 164-171

[10] Xu Z D, Xu F X and Chen H 2015 Vibration suppression on a platform by using vibration isolation and mitigation devices *Nonlinear dynamics* **83** pp 1341-1353

[11] Wei H, Jian X, Da-Yong Z, Ying-Lei W, Jian-Wei L, Kun-Lin L 2016 Ulti-objective optimization of parameters and location of passive vibration isolation system excited by clamped thin plate foundation *Engineering review* **36 № 1** pp 19-27

IOP Conf. Series: Materials Science and Engineering **142** (2016) 012131 doi:10.1088/1757-899X/142/1/012131

Simulation of the Magnetic Characteristics and Properties of the Neodymium Compensator of the Stiffness

E G Gurova[1], M G Gurov[2], Y V Panchenko[3]

Mechatronics and automation department,
Novosibirsk State Technical University,
Av. K. Marx, 20, Novosibirsk, 630073, Russia

e-mail: lena319@mail.ru

Abstract. This research is devoted to consideration of the possibility to use the software ELCUT for development of the magnetic compensator of the stiffness based on neodymium magnets. The software represents precision enough apparatus to solve the issues of the magnetostatic. The solution of these issues is the most important phase at the stage of the designing and calculation of the magnetic compensator of the stiffness, so as at the beginning we need to find the traction force of the interaction between magnet and magnetic materials to provide necessary falling traction characteristic of the compensator. In this paper the simulated models of the neodymium magnets are shown; the view of the field are calculated, the plots of the distribution and directions of the magnetic field strength and induction vectors are presented. Results, which were obtained during of the simulation, further will be used for designing and creation of the magnetic compensator of the stiffness based on supermagnets. Research & Development is under the scholarship of the President of Russian Federation, order №184 from 10th of March 2015.

Introduction

Nowadays the one of the actual issue is vibrations, created by power installations of transport vehicles. Frequently enough the vibration leads to premature damages of the machinery and equipment, emergency situation. At that moment the different research groups are developing big enough amount of the vibroprotectors and vibration isolation devices [1, 5-10]. However, the vibration isolators, based on zero stiffness effect, are most perspective. Accordingly the theory of the vibration isolation, the devices to reduce of the mechanical oscillations, must operate in conformity with effect of the zero stiffness, when ideal vibration isolation from random spatial oscillations of the absolute hard body will be provide, if at any time the sum of the projections of all forces applied to it on arbitrary coordinate points and the amount of the axis with relatively to these axes are equal to zero [3, 4]:

$$\sum P_x = 0, \quad (1)$$

$$\sum P_y = 0, \quad (2)$$

$$\sum P_z = 0, \quad (3)$$

$$\sum M_x(P) = 0, \quad (4)$$

$$\sum M_y(P) = 0, \quad (5)$$

$$\sum M_z(P) = 0, \tag{6}$$

To provide the area of the zero stiffness within amplitude of the oscillations, the vibration isolators must consist of resilient element and compensator of the stiffness connected in parallel with the same stiffness falling traction characteristic so as resilient element, but with negative sign. So, the linear increasing characteristic is compensated by falling characteristic under the same angle relatively abscise axis. It will leads to comply of the condition (1)-(6) at any spatial movements of the vibrating object relatively of the protected object.

Nowadays, there are many different designs of the correctors of the stiffness, which are presented in different versions, based on principles of mechanics, hydraulics, and pneumatics [7]. However, the most efficiency device should be considered the electromagnetic compensator of the stiffness, because it has many benefits over the previously proposed: automatic tuning to variable load, absence of the friction forces and additional inertia.

With all of these achievements the application of this compensator of the stiffness is a bit difficult due to significant sizes and additional supply. In [2] the change of the basic components of the stiffness compensator (electromagnet coils with cores on neodymium magnets, which are rare-earth element magnets, consists of alloy neodymium-boron) was suggested. On the figure 1 the schematic of the supermagnets position is shown. Here the distance between two magnets is calculated from parameters of the supposed vibrations, and metallic plate is thin enough (the thickness of the plate is negligible in contrast of thickness of the magnets).

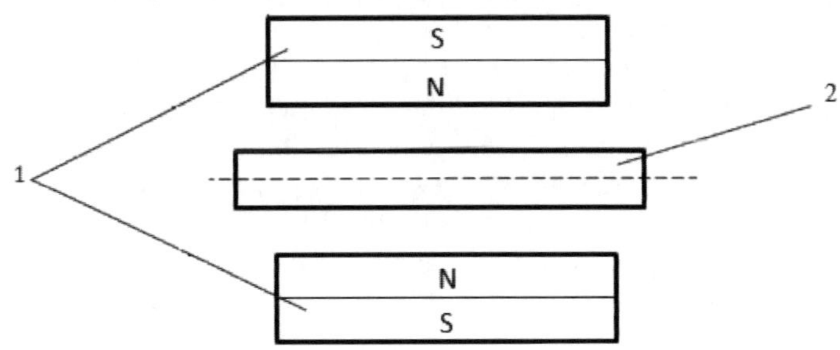

Figure 1. Interaction of the two magnets with flat anchor:
1 – is the pair of the oppositely directed magnets; 2 – is the flat iron anchor

The application of these components, which have magnetic properties, leads to necessity of the solving of the issues of the magnetostatics. For the solution of the similar tasks we can use the software «ELCUT».

Calculation and simulations

To make precise traction characteristic of the compensator of the stiffness based on permanent magnets, there is neediness to make some simulation with software «ELCUT». The creation of the compensator model allows determining of the strength of fields magnets interaction. Further simulation for different values of the air gap between the bodies, allows us to determine the force acting on the anchor at the different movements relative neodymium magnets. These data will be the basis for building, the necessary falling traction stiffness compensator characteristics relatively from the position of the anchor on the axis.

This software allows making the engineer analysis and simulation by the finite element method (FEM), so as 2D and 3D models of the different magnets, and compensators of the stiffness based on it. The capability of «ELCUT» allows determining necessary properties and parameters of the magnetic fields as electromagnets and permanent magnets [1, 2]. This software consider neodymium magnet as element, which has permeability, coercive force and remanence. In software, it is possible to find the strength of the magnetic field at any point of the space around simulated object, its induction, strength and direction by integration method along the contour of the element.

Simultaneously, with mentioned above, there are possibility to plot the deviation of the induction, strength, permeability and energy axially.

Let us to use the software «ELCUT» for simulation of the neodymium magnet cylinder of the type N42 and analysis of its magnetic properties.

The parameters is used as original data:

- radius of the magnet (steel cylinder):2,25 cm;
- height of the magnet (steel cylinder):2 cm;
- remanence: 1.36 T;
- the coercive force of the magnet: 955 kA/m;
- relative magnetic permeability of the medium: 1;
- relative permeability of the steel: 1000;
- the distance between the magnet and the magnet (steel):1 cm.

The first step in simulation of the magnetic system of the stiffness compensator elements is creation of the magnetic field of the elements. On the figure 2, 4 and 5 the shape and values of the magnet fields, created by neodymium magnet (interaction between magnet, steel cylinder and two magnets) is presented.

On the figures the models of the cylinder bodies are presented. These models is presented with help of the areas with permanence, coercive force and residual induction, and the lower bound is the axis of rotation of the body that allows us to simulate portion of the magnetic body, and the result is a parameter that is valid for the entire body.

From figure 2, 4 and 5 we can see, that the plot of the magnetic field is changing from parametric data: single magnet, interaction with steel cylinder or another similar magnet. It is caused by difference of the magnetic conduction, and also by distribution of the induction vectors and strength, created by components of the devices.

Let us consider these issues. The modelling system «ELCUT» allows showing the vector of the magnetic induction for single magnet, and also the plot of the distribution of the magnetic induction axially. The results are shown on figures 1 and 2.

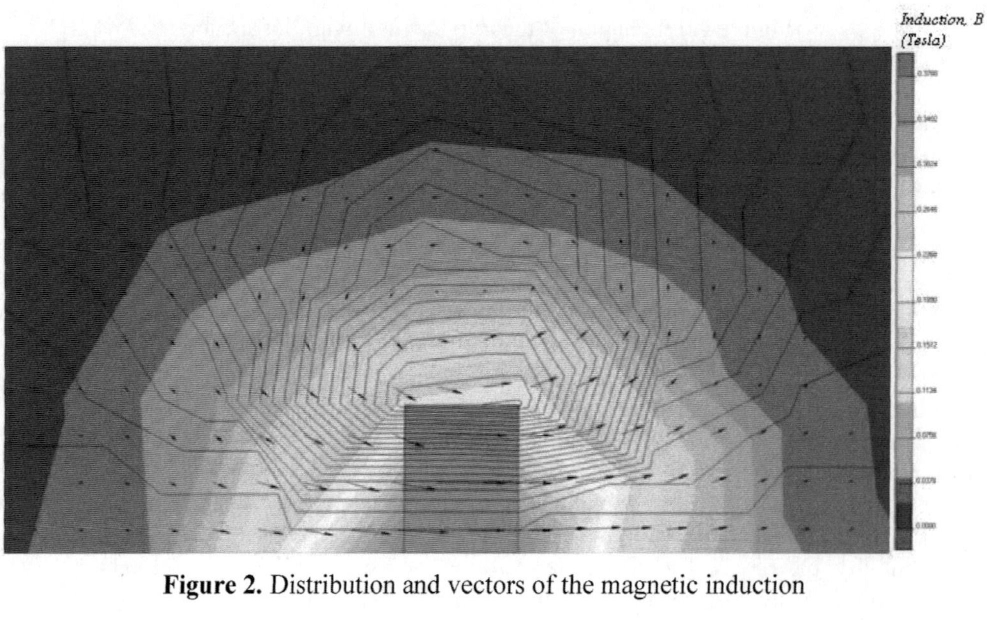

Figure 2. Distribution and vectors of the magnetic induction

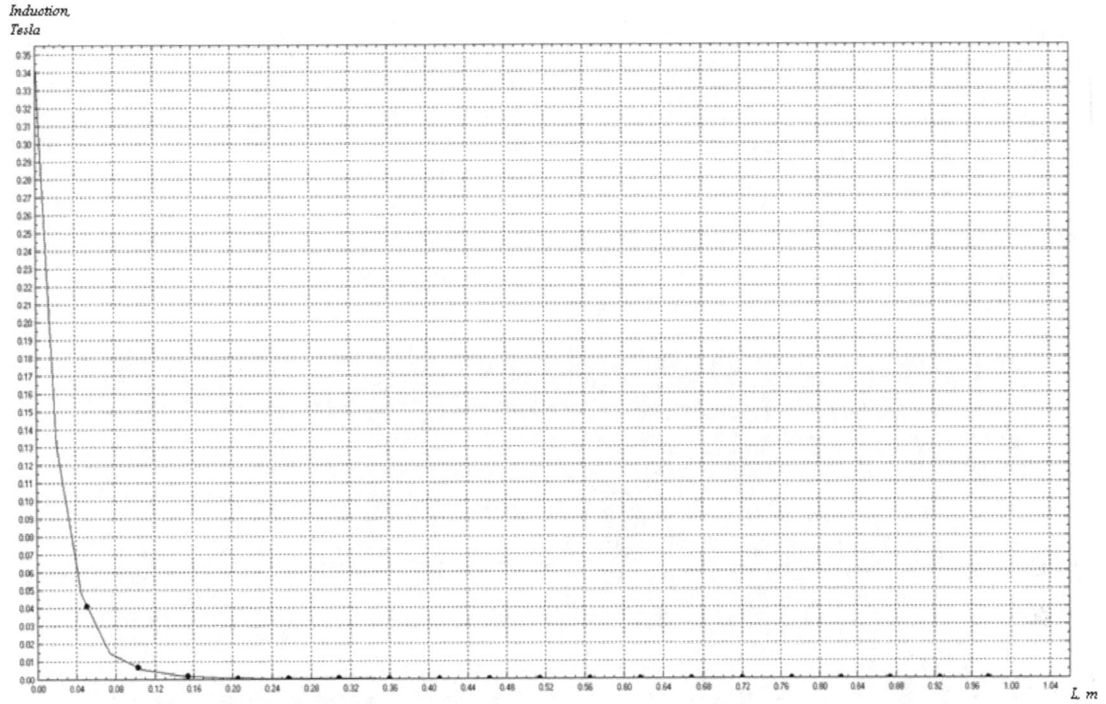

Figure 3. Plot of the distribution of the magnetic induction along cylinder magnet axis

The plot of the dependency, which is presented on figure 3, allows preliminarily calculate the value of the induction of the magnetic field along axis or contour. Also we can plot the model of the field, created by two similar magnets with counter directed vectors of the induction and magnetic field strength. The results are shown on figure 3. On the basis of this model we can obtain the direction and magnitude of the induction vectors and strength, and also the force, acted on these magnets. To obtain the results, the simulation of the two magnets with previously specified parameters and changed direction of the coercive force of the one magnet has been done.

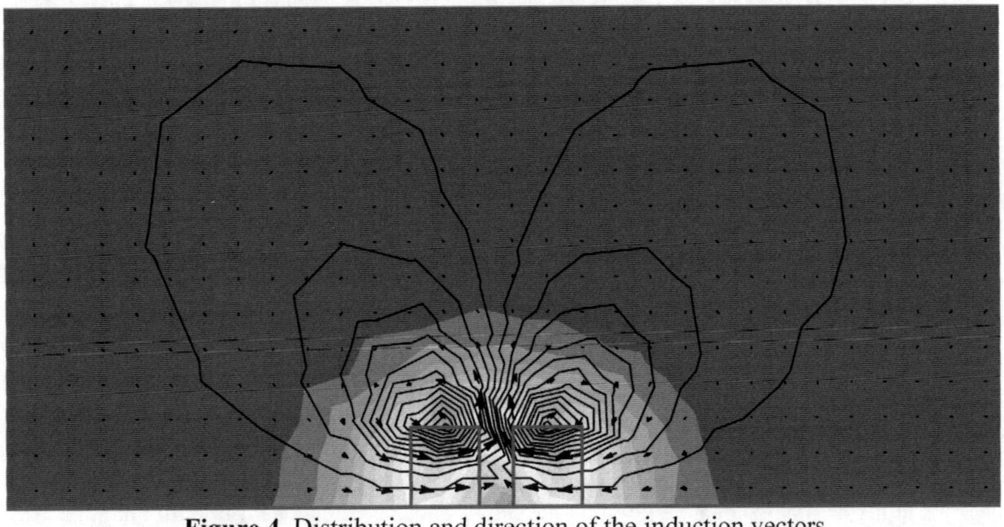

Figure 4. Distribution and direction of the induction vectors

Parameters, obtained from software are precise data of the interaction of the two magnets, which are precisely enough to give the required number of magnets to create the desired strength of their interaction (attraction/repulsion), that allow to have resilient elements of the necessary stiffness to create the compensator of the stiffness with necessary characteristics.

At the modelling of the magnets this software allows investigation of the ferromagnetic materials, such as iron and steel. Results of the modelling of the interaction of the steel cylinder and neodymium magnet are shown on figure 5 and 6. The lines of the magnetic field, which is created from interaction of the magnet and steel, are shown on these figures. The vectors of the inductions and magnetic field are shown also. This model, as previous one, allows to determine the direction and value of the force vector of the interaction of the steel and magnet.

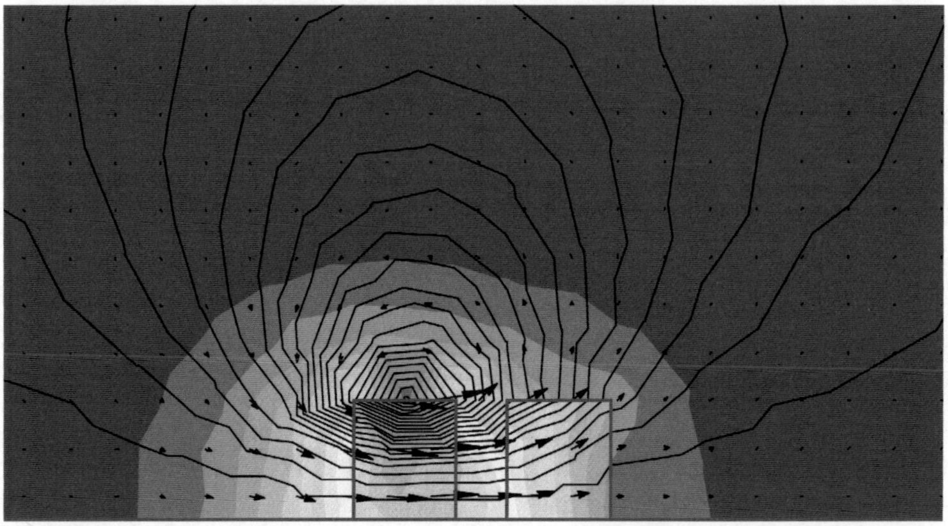

Figure 5. Distribution and vectors of the magnetic field induction of the magnet and steel cylinder

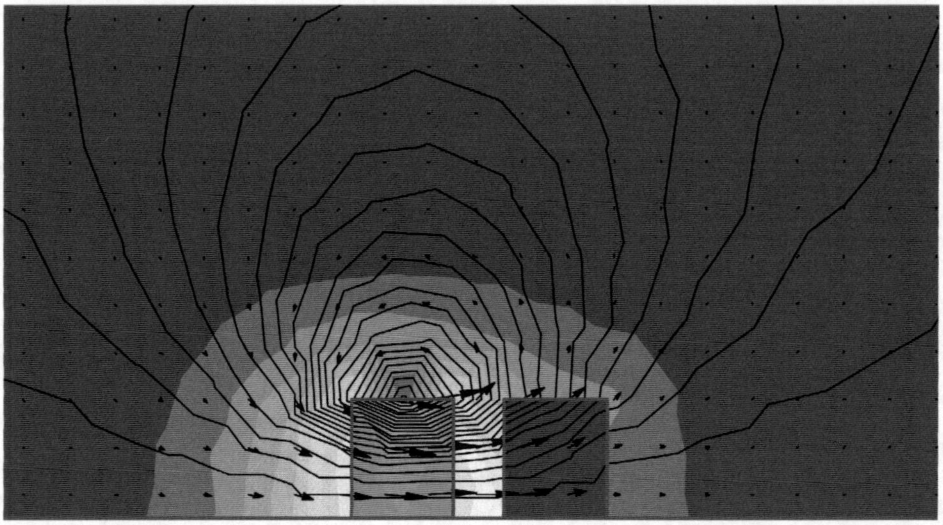

Figure 6. Distribution and vectors of the magnetic field induction of the magnet and steel cylinder

The lines of the elements net are shown on the plots. These elements net are built automatically and it is a basis for solving of the magnetostatics task by finite elements method, and gives the possibility to estimate the heat transfer processes inside of the device. The core of the method of the finite elements is separation of the finite amount of the sub areas (elements) and calculation of the differential equations for each sub area. In simple case is the first-degree polynomial. Outside of the element the approximation function is zero. The values of these functions on the boards (at the nods) are solutions.

Conclusion

Based on our results we can make the conclusion, that the software «ELCUT» is efficient tool to find the solution of magnetostatic tasks, one of this is development of the neodymium compensator of the stiffness based on neodymium magnets. The date, which was obtained during the simulation of the interaction of the different magnetic elements, will be the one of the bases for designing of the compensators of the stiffness based on the neodymium magnets.

The research was performed under the scholarship of the President of Russian Federation for young scientists, order №184 from 10th of March 2015.

References

[1] Bartel T, Gaisbauer S, Stohr P, Herold S and Melz T 2014 Development of functionally integrated mounts for three- and six-axial vibration isolation of sensitive equipment *Proceedings of international conference on noise and vibration engineering (ISMA2014) and international conference on uncertainty in structural dynamics (USD2014)* pp 119-130

[2] Gurova E G and Gurov M G 2014 Vibro isolator with neodymium magnets compensator of the stiffness *Applied Mechanics and Materials* **682** pp 118-121

[3] Gurova E G, Gross V Y, Kurbatov V S, Makarov S V, Sergeev A A and Shchurov N I 2013 Development of spatial vibration protection devices *World Applied Sciences Journal* **22** pp 44-48

[4] Gurova E G 2015 Improving of the operation efficiency of the vehicle due to using of the neodymium magnets inside the vibration isolation devices Conf. Series-Materials Science and Engineering **91** 012091

[5] Lin S Y, Xu Jand Cao H 2015 Analysis on the ring-type piezoelectric ceramic transformer in radial vibration IEEE Transactions on power electronics **31 № 7** pp 5079-5088

[6] Liu C C, Jing X J and Chen Z B 2015 Band stop vibration suppression using a passive X-shape structured lever-type isolationsystem *Mechanical systems and signal processing* **68-69** pp 342-353

[7] Spiewak S, Karpachevskyy T and Zaiss C 2014 High performance isolation and excitation of vibration for enhanced identification of inertial sensors *Proceedings of the asme international mechanical engineering congress and exposition* **4B** V04BT04A061

[8] Prisekin V L, Pustovoi N V and Rastorguev G I 2014 Algorithms for controlling fatigue tests of airplanes Journal of Applied Mechanics and Technical Physics **55 № 1** pp 164-171

[9] Xu Z D, Xu F X and Chen H 2015 Vibration suppression on a platform by using vibration isolation and mitigation devices *Nonlinear dynamics* **83** pp 1341-1353

[10] Wei H, Jian X, Da-Yong Z, Ying-Lei W, Jian-Wei L, Kun-Lin L 2016 Ulti-objective optimization of parameters and location of passive vibration isolation system excited by clamped thin plate foundation *Engineering review* **36 № 1** pp 19-27

Rolling Friction in Loose Media and its Role in Mechanics Problems

S V Klishin[1], A F Revuzhenko[1], A A Kazantsev[2]

[1] - Institute of Mining, SB of RAS, Russia, Novosibirsk, Krasnii av, 54
[2] - Yurga Institute ofTechnology, TPU affiliate, Russia, Kemerovo region, Yurga, Leningradskaya str. 26

E-mail: sv.klishin@gmail.com, revuzhenko@yandex.ru, kazantsev@tpu.ru

Abstract: Rolling friction between particles is to be set in problems of granular material mechanics alongside with sliding friction. A classical problem of material passive lateral pressure on the retaining wall is submitted as a case in point. 3D method of discrete elements was employed for numerical analysis. Material is a universe of spherical particles with specified size distribution. Viscose-elastic properties of the material and surface friction are included, when choosing contact forces. Particles' resistance to rolling relative to other particles and to the boundary is set into the model. Kinetic patterns of medium deformations are given. It has been proved that rolling friction can significantly affect magnitude and nature of passive lateral pressure on the retaining wall.

1. Introduction

It is known, that the mechanics of granular material is based on Coulomb's law:

$$|\tau_n| = -tg\varphi \cdot \sigma_n + k, \tag{1}$$

где σ_n, τ_n – respectively are normal and tangential stresses at limit state area, k – adherence, tgφ – internal friction coefficient (in fact, sliding friction only). While solving continuum mechanics problems, proceeding to invariants is to be done in equation (1), and on this basis various options of closed mathematical models are built [1-4]. Since the original equation (1) involves sliding friction only, corresponding models and numerous solutions are also based on sliding friction. It is known that while deformation of soils and granular materials (including some rocks), freedom degrees, associated with particles rotation are significant. Therefore, along with sliding friction, rolling friction should be involved. Within the framework of continuum mechanics, it significantly complicates models because of additional internal variables [5-9].

The purpose of this paper is to demonstrate the importance of rolling friction by means of the numerical method of discrete elements (DEM). As an example, we choose a retaining wall classical problem, which is used in studies of natural phenomena (landslides), and in design of engineering structures (road and rail embankments, quarries sides). Consider the following problem: let a rigid vertical wall to hold heavy loose ground. Evaluate the required pressure P, acting on the wall by the material if its displacement towards the ground (passive pressure case).

In practice, the limit equilibrium theory is widely used for calculations of slope and retaining walls stability; according to this theory, wall's displacement forms sliding surfaces in the material, wherein displacement towards the material forms a sliding wedge. This is supported by numerous laboratory experiments and field observations. Depending on a model, the sliding wedge is limited either by a plane or by a curved surface. Coulomb, Rankine, Sokolovsky's models are conventional for calculations of retaining walls [1, 10 -12]. For example, according to Rankine, in absence of friction between ground and the retaining wall, horizontal pressure P is expressed by:

$$P = \frac{1}{2}\rho g H^2 tg^2\left(\frac{\pi}{4} \pm \frac{\varphi}{2}\right), \qquad (2)$$

where ρ is material density ; φ - internal friction angle; g - gravity acceleration; H - height of the filling; "+" corresponds to passive pressure, and "-" to active pressure.

A limitation of these approaches is that they do not consider a material stress history, preceding a limit state. On the other hand, need for limit state hypotheses is eliminated in continuum statements, based on the theories of elasticity and plasticity. However, there is a need for formulation of closed systems of equations, which have not been set up yet. Numerous experiments on various stress modes show formation of localized currents in discrete medium [13-15].

We suggest employing, as a research tool, of the method of discrete elements, which allows determining evolution of stress-strain state of loose medium both at a pre-limiting stress stage and at the stage of localized displacements formation. The advantage of this method is that it does not require either limiting state hypotheses or continuum equations.

At present, method of discrete elements is widely used to solve problems of geo-mechanics [16–19]. Its essence is that a real medium is replaced with a pack of discrete particles, among which particular interaction laws are postulated. Particles have free multi-dimensional parameters which are to be chosen. Under this method, engineering problems of large deformations and rotations are solved without further difficulties. In addition, localization displacements and physically nonlinear effects can be described without major complications, while data on the constitutive equation are not required. Thus, this method is a fundamental alternative to the classic methods based on traditional concepts of the continuum mechanics.

2. Research method

DEM represents discrete material as a universe of N spherical particles (discrete elements). Each i-th element has a radius r_i and a set of physical properties: density, elastic and viscous moduli, friction, adherence, etc. $(i = 1, \ldots, N)$. Single discrete element motion is both translation and rotation and is described by the following equations:

$$m_i \ddot{\boldsymbol{x}}_i = \sum_j \boldsymbol{f}_{ij} + m_i \boldsymbol{g}, \qquad (3)$$

$$I_i \ddot{\boldsymbol{\theta}}_i = \sum_j (\boldsymbol{r}_{ic} \times \boldsymbol{f}_{ij} + \boldsymbol{M}_{r,ij}). \qquad (4)$$

Here the dots denote differentiation of time t; $\boldsymbol{x}_i$ - radius-vector of i-th particle's gravity center; $\boldsymbol{\theta}_i$ - its rotation relative to coordinate axes; m_i - mass; I_i - moment of inertia; $\boldsymbol{g}$ - gravity acceleration; $\boldsymbol{r}_{ic}$ - vector directed from the i-th particle center to the point of contact; contact force $\boldsymbol{f}_i$ affects to i-particle with another j-particle (or border), (it depends on their overlapping as well as on elastic and viscous moduli), its normal component is calculated according to Hertz law [20]; $\boldsymbol{M}_{r,ij}$ - resisting moment of particles (while particles interaction or along the wall).

In (3) and (4) all elements and borders, being in contact with a current i-particle, are summed up. Since forms of discrete elements are supposed to be constant during the contact, deformation degree is described by magnitude of contacting particles overlapping. A special feature is a spherical shape of elements, which does not change during the entire numerical experiment. This imposes certain limitations on the scope of this method, since, in an actual situation, rolling resistance of particles, when they are in reciprocal contact or contact with a border, is conditioned by their shapes and deformations. To remove this limitation, you can either use particles of any form or create clusters, which are considered as separete elements. These methods require larger computational resources.

In this paper we use the method presented in [21-24]; we propose the model of rolling resistance as an adjunct to the contact interaction model; rolling resistance hinders relative particles' mutual motion or their motion along the border.

From the right side of the equation (4) the contact moment is represented as two summands. The first summand is the moment of direct contact or relative mutual sliding of particles. The second summand is the moment, hindering relative rotation of discrete elements:

$$\boldsymbol{M}_{r,ij} = -\frac{\boldsymbol{\omega}_i - \boldsymbol{\omega}_j}{\left|\boldsymbol{\omega}_i - \boldsymbol{\omega}_j\right|} \mu_{ij} r_{ij} \left|\boldsymbol{f}_{ij} \cdot \boldsymbol{n}_{ij}\right|, \qquad (5)$$

where ω_i and ω_j - angular velocities of contacting particles; $\boldsymbol{n}_{ij}$ - unit vector lying on the straight line joining their centers; r_{ij} - reduced radius: $1/r_{ij} = 1/r_i + 1/r_j$.
Dimensionless parameter μ_{ij} is rolling resistance coefficient ($\mu_{ij} \cdot r_{ij}$ - rolling friction coefficient of length dimension). If to draw an analogy with dry friction angle between two bodies, just depict that this ratio can be represented as $\mu_{ij} = \mathrm{tg}\,\psi_{ij}$.
Here ψ_{ij} is the maximum angle of plane inclination at which particles will be in equilibrium due to rolling resistance (if no slipping).

To examine the effect of a set parameter on kinematics of particle motion, let us perform the following numerical experiment. Suppose that radius sector of 0.01m rests on a horizontal plane in gravity field. Contact friction angle between the particle and the boundary is equal to 30.0°. Particle mass center's velocity v_0 = 1.0 m/s and zero angular velocity are set at zero time parallel to the plane. If there are no rolling ($\psi_{ij} = 0$) and viscous component of contact force, particle mass center's velocity will be equal to the initial ($v(t) \equiv v_0$) at any time t; and the traveled distance will be a linear function of time: $s(t) = v_0 \cdot t$.

If $\psi_{ij} > 0$ solution will differ. Fig. 1 shows results of numerical experiments for different values of ψ_{ij}.

Higher magnitudes for ψ_{ij} were chosen to demonstrate the prominent effect of rolling friction. Fig. 1 shows that the distance, traveled by the particle, is finite and diminishes with increasing of rolling resistance angle ψ_{ij}.

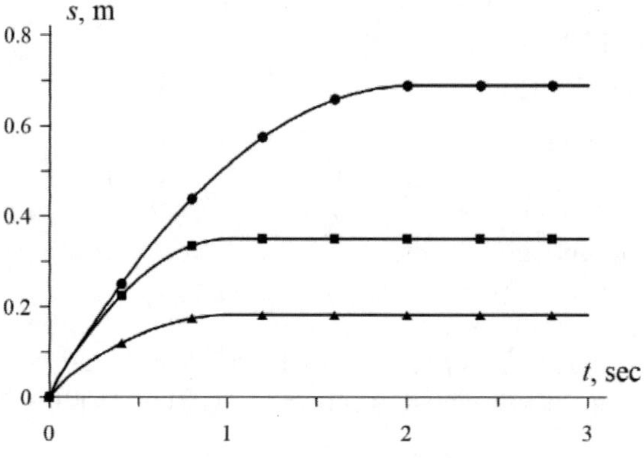

Fig. 1 Distance-time relationship (s is passed by a spherical particle with 0.01m radius) if rolling resistance angles ψ_{ij} are: 5.0° (●); 15.0° (■); 25.0° (▲)

Thus, if there is data on laws of motion and particles contact interaction (equation (3) - (5)) as well as initial and boundary conditions, you can set up and solve equations and determine environment's stress-strain state evolution.

3. Numerical DEM simulation

Based on the model considered above, a software has been developed for numerical analysis of various modes of motion of discrete material consisting of separate spherical particles [25, 26].

Let us consider the following three-dimensional problem. Suppose that Ω volume is set in *Oxyz* space (Ω is a parallelepiped container for loose material, constrained with planes (walls) oriented along coordinate axes); the upper boundary is non-strained (Fig. 2). Length l is equal to 0.2 m, width = 0.01 m and filling height $h = 0.1$ m.

Discrete material is a universe of discrete elements, its density $\rho_i = 2500$ kg/m^3, elastic moduli E_i =3 GPa, Poisson's ratio $v_i = 0.25$; diameters are chosen in uniform distribution over the range 0.0015m to 0.0025m.

Restorability ratio of particles' interaction and that between particles and a retaining wall $e_r = 0.75$. Gravity vector $\mathbf{g} = (0, 0, -9.81)$ is directed downwards *Oz*. Front and rear walls are rigid and smooth, i.e. the condition is close to that of planar deformation along *Oy*.

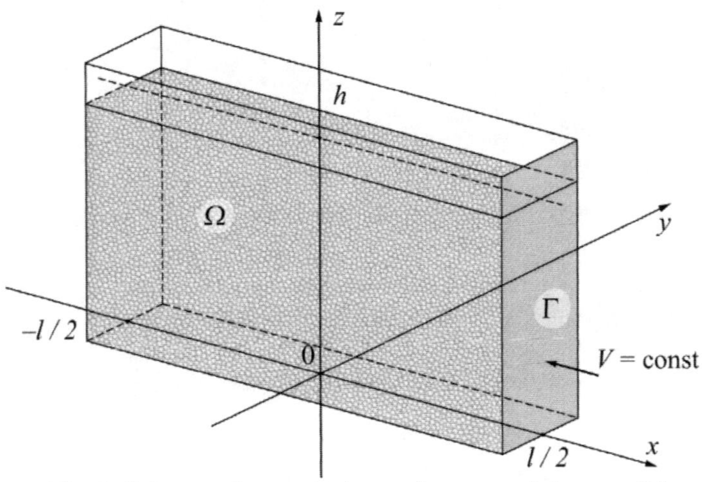

Fig. 2 Scheme of numerical experiment, straining condition

Creating of equilibrium pack of solid spherical particles is the initial step in solving problems with use of DEM and is to be investigated as a separate independent task. In our study, filling of the volume (container) was performed under dynamic algorithm of spherical particles packing. The initial layer of non-segregated discrete elements was created at a certain height from the fixed lower boundary $z = 0$. Further, under gravity, particles deposited in the container interacting together and with boundaries. On reaching equilibrium state, elements were removed if vertical coordinates of their gravity centers did not correspond to $z_i \leq h$. We suggested frictionlessness between the particles and between the particles and the walls; that gives maximum density magnitude.

When equilibrium pack was completed, the material constituted a universe of discrete elements with total weight of 0.3kg. Angle φ_{ij} of contact friction between the particles was set (in all subsequent numerical experiments $\varphi_{ij} = 30°$). For evaluation of rolling resistance coefficient's effect on localized flow pattern of discrete material and on evolution of lateral pressure P, numerical calculations were performed for the case $\psi_{ij} = 0°$ and $\psi_{ij} = 30°$. Here, greater ψ_{ij} magnitude was chosen to denote a prominent effect of rolling friction. Pressing displaced the wall Γ towards the discrete material at a constant low speed V; that is the case of passive pressure. The distance of wall displacement was 0.02m; total deformation of the sample $\varepsilon_x = 0.1$.

4. Results and Discussion

Authors examined effect of rolling resistance angle on material deformation. Fig. 3 shows the final state of discrete material in non-rolling conditions and while high rolling resistance. Kinematics deformation is colored for visualization. We see that if $\psi_{ij} = 0°$ (Fig. 3a) free surface configuration is close to a straight line, i.e. without rolling resistance particles, under action of gravity, can roll down to the fixed left edge, forming a smooth slope.

If $\psi_{ij} > 0°$, strain state of the material differs. In this case, the sliding wedge may form a horizontal area of free surface (Fig. 3b).

a *b*

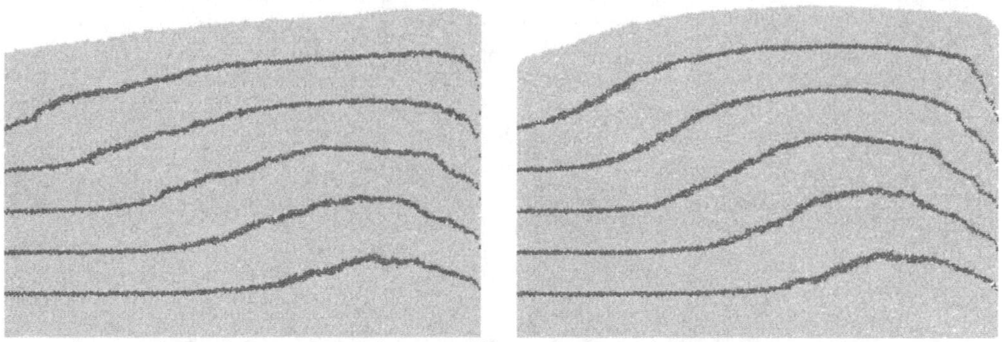

Fig. 3 Deformed state of sample; no rolling resistance (a) high rolling resistance (b)

Sliding wedges are formed by distribution of vertical components of displacement vectors, as shown in Fig. 4. Low magnitudes of vertical displacements are colored dark; great magnitudes are colored light. In the material two or more sliding lines are formed; that differs significantly from the theories based on hypotheses for granular medium ultimate state.

a *b*

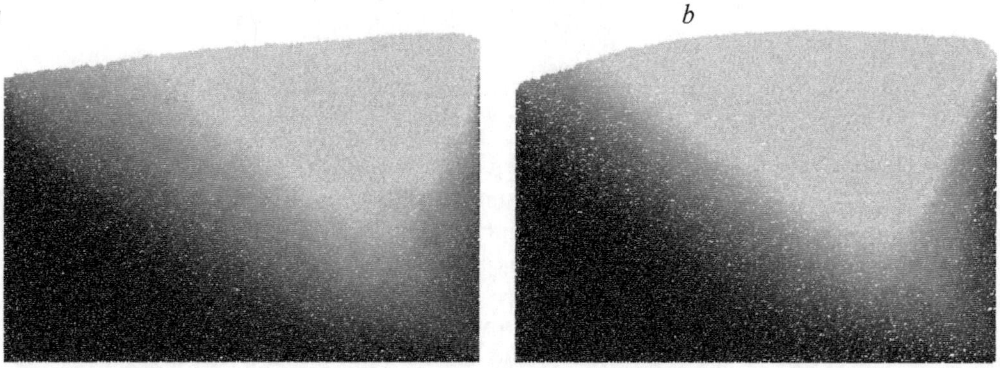

Fig. 4 Vertical component of particles' displacement vector; no rolling resistance (a) high rolling resistance (b)

Let us define material lateral pressure $P(t)$, acting on the retaining wall Γ during deformation. At a fixed time t we suggest:

$$P(t) = \frac{1}{h}\sum_{i=1}^{N_p} F_{x,i}, \qquad (6)$$

where $F_{x,i}$ - a horizontal component of i-th particle's contact force vector acting on Γ (summing is performed only over particles being in contact with a boundary (wall)); N_p - number of inter-contacts between particles and boundaries; h - maximum vertical contact coordinate.

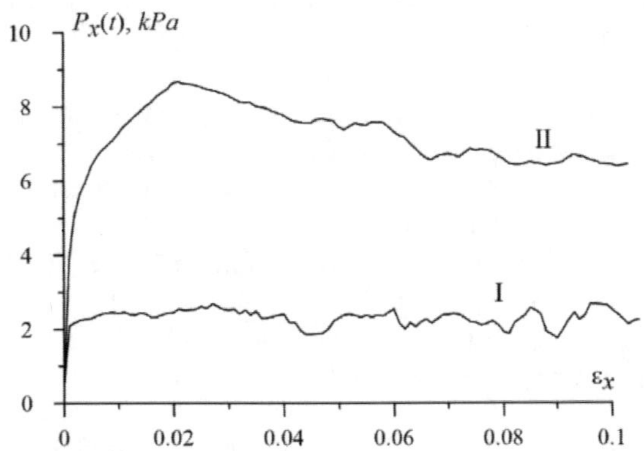

Fig. 5 Change of horizontal pressure *P(t)* on retaining wall during straining; rolling resistance angle $\psi_{ij} = 0°$ (I); $\psi_{ij} = 30°$; (II)

Fig. 5 shows variables of *P*(*t*) during all the experiment. Diagram I corresponds to rolling resistance angle $\psi_{ij} = 0°$, diagram II corresponds to rolling resistance angle $\psi_{ij} = 30°$.

It is evident that if specifying external friction between particles (without rolling resistance), material passes to limiting state immediately (Figure I). If rolling friction, material state can be considered as pre-limiting with deformation magnitude 0.02. If further wall's displacement, granular medium passes to limit state; resulting pressure varies significantly.

5. Conclusions
In problems of loose material mechanics rolling friction should be considered as well as sliding friction. In the 3D-model of discrete material deformation caused by displacement of the retaining wall (passive pressure), effect of rolling resistance coefficient on emergence of localized flow patterns has been confirmed as well as change of horizontal lateral pressure towards the retaining wall while further straining.

References
[1] Sokolovski V V 1965 Statics of Granular Media. Pergamon Press, London.
[2] Ivlev D D 2001 Mechanics of Plastic Media. Theory of Perfect Plasticity [in Russian] (Moscow Fizmatlit, Vol. 1)
[3] Ivlev D D 2002 Mechanics of Plastic Media. General Problems. Rigid-Plastic and Elastoplastic State of Bodies. Reinforcement. Strain Theories. Compound Media [in Russian] (Moscow Fizmatlit, Vol. 2)
[4] Revuzhenko A F 2006 *Mechanics of Granular Media* (Springer).
[5] Verveyko N D 1977 Micropolar theory of granular flow medium [in Russian] (Voronezh University)
[6] Valyuhov S G, Verveyko N D, Smotrova O A 1999 *Micropolar model of cohesive bulk materials [in Russian]* (Voronezh University)
[7] Varygina M P, Sadovskaya O V, Sadovskii V M 2010 Resonant properties of moment Cosserat continuum. *J. Journal of Applied Mechanics and Technical Physics*, vol 51, **3**, pp 405–413.
[8] Tejchman J 2008 Shear Localization in Granular Bodies with Micro-Polar Hypoplasticity (Springer).

[9] Dyszlewicz J 2004 Micropolar theory of elasticity (Springer)
[10] Mazindrani, Z H, Ganjali M H 1997 Lateral earth pressure problem of cohesive backfill with inclined surface. *J. Journal of Geotechnical and Geo-environmental Engineering* ASCE, **123** **(2)**, pp 110–112.
[11] Terzaghi K 1943 Theoretical Soil Mechanics (New York: John Wiley & Sons)
[12] Terzaghi K, Peck R B 1967 Soil Mechanics in Engineering Practice, Second Edition (New York: John Wiley & Sons)
[13] Bobryakov A P, Kosykh V P, Revuzhenko A F 1990 Temporary structures during the deformation of granular media. *J. Journal of Mining Science* vol. 26, Issue 2, p. 128–137
[14] Kosykh V P 2012 Experimental investigation into variable density of granular material under deformation cycling. *J. Journal of Mining Science* vol. 48, Issue 6, pp 990–995.
[15] Bobryakov A P 2012 Modeling trigger effects in faulting zones in rocks. *J: Journal of Mining Science* vol. 49, Issue 6, pp. 873–880.
[16] Kramadzhyan A A, Rusin E P, Stazhevsky S B, Khan G N 2015 Enhancement of load-bearing capacity of ground anchors with flexible tendon. *J: Journal of Mining Science* vol.51, Issue 2, pp. 314–322.
[17] Lavrikov S V, Revuzhenko A F 2015 DEM code-based modeling of energy accumulation and release in structurally heterogeneous rock masses. *AIP Conference Proceedings* 1683, 020121
[18] Shilko E V, Psakhie S G, Schmauder S, Popov V L, Astafurov S V, Smolin A Yu 2015 Overcoming the limitations of distinct element method for multiscale modeling of materials with multimodal internal structure. *J: Computational Materials Science* vol.102, pp. 267–285
[19] Olukayode I. Imole, Nishant Kumar, Vanessa Magnanimo, Stefan Luding 2013 Hydrostatic and Shear Behavior of Frictionless Granular Assemblies under Different Deformation Conditions. *J: KONA Powder and Particle Journal* vol. 30, pp. 84–108.
[20] Johnson K L 1985 Contact mechanics (Cambridge University Press)
[21] Ai J, Chen J-F, Rotter J M, Ooi J Y 2011 Assessment of rolling resistance models in discrete element simulations *J: Powder Technology* vol. 206, Issue 3, pp. 269–282
[22] Fukumoto Y, Sakaguchi H, Murakami A 2013 The role of rolling friction in granular packing *J: Granular Matter* vol. 15, Issue 2, pp. 175–182
[23] Bo Zhou, Runqiu Huang, Huabin Wang, Jianfeng Wang 2013 DEM investigation of particle anti-rotation effects on the micromechanical response of granular materials. *J: Granular Matter* vol. 15, Issue 3, pp. 315–326
[24] Jiang M J, Yu H-S, Harris D 2005 A novel discrete model for granulates incorporating rolling resistance. *J: Computers and Geotechnics* vol. 32, Issue 5, pp. 340–357
[25] Klishin S V, Klishin V I, Opruk G Yu 2013 Modeling coal discharge in mechanized steep and thick coal mining. *J: Journal of Mining Science* vol. 49, Issue 6, pp. 932–940
[26] Klishin S V, Revuzhenko A F 2014 3D discrete element approach to Janssen's problem. *J: Journal of Mining Science* vol. 50, Issue 3, pp. 417–422

Can Cross-Listing Mitigate the Impact of an Information Security Breach Announcement on a Firm's Values?

Yong Chen[1,a], Feng Dong[1,b], Hong Chen[1,c] and Li Xu

Old Dominion University, USA

E-mail: [a]y7chen@odu.edu, [b]fdong@odu.edu, [c]hchen001@odu.edu, lxu@odu.edu

Abstract. The increase in globalization in the markets has driven firms to adopt online technologies and to cross-list their stocks. Recent studies have consistently found that the announcements of information security breaches (ISBs) are negatively associated with the market values of the announcing firms during the days surrounding the breach announcements. Given the improvement in firms' information environments and the better protection for investors generated by cross-listing, does cross-listing help firms to reduce the negative impacts caused by their announcements of ISBs? This paper conducts an event study of 120 publicly traded firms (among which 25 cross-list and 95 do not), in order to explore the answer. The results indicate that the impact of ISB announcements on a firm's stock prices shows no difference between cross-listing firms and non-cross-listing firms. Cross-listing does not mitigate the impact of ISBs announcement on a firm's market value.

1. Introduction

In the past two decades, globalization has driven the use of online technology and the spread of Internet connectivity around the world (Garg, Curtis, & Halper, 2003).The rapid growth of e-business, built on the Internet, makes information security a great concern for firms (Garg, Curtis, & Halper, 2003; Hovav & D'Arcy, 2003). The Internet offers convenience and business for firms, but security issues can be serious. In a 2006 survey that examined privacy policies of 47 U.S. and E.U. multinational companies, 94 percent of the E.U. companies reported that they had experienced an information security breach (ISB) in the prior three years. That number can be compared with 86 percent of their American counterparts who reported experiencing a similar breach (Cline, 2006).

Firms are usually reluctant to release information about ISBs for fear of embarrassment, increased expense, information provided to competitors and to hackers, the possible dropping of their stock prices and values as well as their profits and dividends, and the diminishment of their customers' confidence in them (Cavusoglu, Mishra, & Raghunathan, 2004; Garg, Curtis, & Halper, 2003; Goel & Shawky, 2009; Gordon & Loeb, 2002; Hoffer & Straub, 1989; Power, 2001; Power, 2003). Recent studies have consistently found that the announcements of ISBs are negatively associated with the market values of the announcing firms during the days surrounding the announcements (e.g., Acquisti, Friedman, & Telang, 2006; Campbell, Gordon, Loeb, & Zhou, 2003; Cavusoglu, Mishra, & Raghunathan, 2004; Garg, Curtis, & Halper, 2003; Goel & Shawky, 2009; Hovav, & D'Arcy, 2003; Kannan, Rees, & Sridhar, 2007; Telang & Wattal, 2007). The decline in market value of breached firms reflects investors' concerns about possible financial damage caused by the announcements of ISBs.

The globalization in the financial markets drives firms to choose to cross-list their stocks on exchanges outside pf their domestic markets (Baker, Nofsinger, & Weaver, 2002). Cross-listed firms aim to have higher valuations, a lower cost of capital, and increased liquidity (King & Segal, 2009). By cross-listing, firms can increase their liquidity, since their shares become more accessible to investors (Baker, 1992; Euromoney, 1986; Fanto & Karmel, 1997; Mittoo, 1992; Sarkissian & Schill, 2009). Furthermore, cross-listing can increase investor recognition by widening cross-listed firms' shareholder bases and improving their information environments (Baker, 1992; Euromoney, 1986; Fanto & Karmel, 1997; Merton, 1987; Mittoo, 1992). In addition, cross-listing provides better investor protection (Coffee, 1999; Coffee, 2002; Stulz, 1999). As Doidge, Karolyi, and Stulz (2004) point out, cross-listing helps controlling shareholders commit to limiting their expropriation from minority shareholders and increases the ability of firms to take advantage of growth opportunities. Given the improvement in firms' information environments and the better protection for investors generated by cross-listing, does cross-listing help firms to reduce the negative impact caused by announcements of ISBs? The answer remains unknown. In order to explore the impact of ISBs on the market value of publicly traded firms, this paper conducts an event study of 120 firms which suffered ISBs between 2005 and 2014. Among them, 25 cross-list and 95 do not. The stock prices of these firms are compared between a 120-day pre-event period and a six-day event window. The results show that the impact of an ISB announcement on firm's stock price shows no difference between those firms that have chosen to cross-list those firms that do not cross-list. Cross-listing does not mitigate the impact of the announcement of an ISB on a firm's market value.

The remainder of this paper is organized as follows. Section 2 reviews the relationship between ISBs and firms' value and discusses the benefits of cross-listing. Based on the review, a hypothesis is developed. Section 3 describes the data and methodology, and presents the empirical results of the event study. Section 4 discusses the findings based on the results and concludes the paper.

2. Background and hypothesis development

ISBs and firm value

An ISB is defined as "a malicious attempt to interfere with a company's business and its information" (Cavusoglu, Mishra, & Raghunathan, 2004, p72). ISBs often include theft or modification of computer programs, embezzlement or modification of data, unauthorized use of computer services, purposeful interruption of computer services, unauthorized access to passwords, and destruction of data by computer viruses (Ettredge & Richardson, 2001; Garg, Curtis, & Halper, 2003; Straub, 1990).

Even though firms have gradually realized the importance of information security, making an assessment of the value of information technology security is challenging because of the difficulty of measuring the cost of ISBs (Cavusoglu, Mishra, & Raghunathan, 2004; Garg, Curtis, & Halper, 2003). The costs associated with ISBs include transitory costs, permanent costs, tangible costs, and intangible costs (Cavusoglu, Mishra, & Raghunathan, 2004; Goel & Shawky, 2009). Transitory costs include lost business and decreased productivity due to the unavailability of the breached resources; labor and material costs required to detect, contain, repair, and reconstitute breached resources; costs associated with evidence collection and prosecution of attackers; government sanctions; and the cost (in media coverage) of providing information to customers and to the public (D'Amico, 2000; Garg, Curtis, & Halper, 2003; Goel & Shawky, 2009). In contrast, permanent costs, which can have far-reaching effects on a breached firm's future cash flow, include the loss of customers who switch to competitors, the loss of ability to attract new customers due to perceived poor security, the loss of trust of a firm's customers and business partners, the loss of competitive edge, legal liabilities arising from the breach, the cost of attackers' access to confidential or proprietary information, possible increased insurance costs, and higher capital costs in debt and equity markets (Cavusoglu, Mishra, & Raghunathan, 2004; Goel & Shawky, 2009). Tangible costs include the cost of lost sales, material and labor, and insurance, whereas intangible costs include costs related to trust. ISBs damage customers' trust and confidence about the breached firm. They are harmful for long-term relationship building. Once an ISB occurs, investors will doubt if a firm is not concerned enough about its customers' privacy and may assume that the firm's internal security practices ae poor. They tend to question firms' long-term performance (Cavusoglu, Mishra, & Raghunathan, 2004; Fama, Fisher, Jensen, & Roll, 1969). ISBs also negatively impact investors' expectation of the value of a firm. For example, immediately following the February 2000 denia; of service (DOS) attack, Yahoo, eBay, and buy.com lost 15 percent, 24 percent, and 44 percent, respectively, of their market values (Atomic Tangerine).

Recent studies have consistently found that the announcements of ISBs are negatively associated with the market value of the announcing firms during the days surrounding the announcements (e.g., Acquisti, Friedman, & Telang, 2006; Campbell, Gordon, Loeb, & Zhou, 2003; Cavusoglu, Mishra, & Raghunathan, 2004; Garg, Curtis, & Halper, 2003; Goel & Shawky, 2009; Hovav, & D'Arcy, 2003; Kannan, Rees, & Sridhar, 2007; Telang & Wattal, 2007). For example, Campbell, Gordon, Loeb, and Zhou (2003) document that firms which suffer a breach of confidential information have a five percent drop in their market values over a two-day window. Cavusoglu, Mishra, and Raghunathan (2004) find that the breached firms lose 2.1 percent of their market value within two days of the announcement -- an average loss in market capitalization of $1.65 billion per breach. Garg, Curtis, and Halper (2003) document that firms' market values suffer a decline of nine to fifteen percent if their customers' credit card information is stolen. Telang and Wattal (2007) find that firms lose about 0.6 percent of their market values when the vulnerabilities of their software vendors are reported. Goel and Shawky (2009) find that the announcements of ISBs cause firms to lose one percent of their market values during the days surrounding the announcement. These studies prove that the decline in market value of breached firms reflects investors' concern about the possible financial damage caused by the announcements of ISBs.

Cross-listing

Cross-listing can help firms to increase their liquidity, can increase investor recognition (Merton, 1987), and can improve investor protection (Coffee, 1999; Coffee, 2002; Stulz, 1999). As Sarkissian and Schill (2009) point out, cross-listing helps firms to overcome capital and informational barriers; to gain market characteristics, such as liquidity, tax treatment, disclosure; and to contain the risk of shareholder expropriation.

More specifically, cross-listed firms can get greater media visibility and analysis following (Baker, Nofsinger, & Weaver, 2002; Foerster & Karolyi, 1999; Lang, Lins, & Miller, 2003). In addition, cross-listing makes shares more accessible to investors. As a result, more investors are attracted, and cross-listed firms' shareholder bases become larger (Bancel & Mittoo, 2001; Fanto & Karmel, 1997; Mittoo, 1992). A broader shareholder base brings cross-listed firms higher volume of liquidity. However, the increased liquidity caused by cross-listing is not the focus of this paper. Instead, this paper concerns itself more with the improvement in investor recognition and investor protection generated by cross-listing.

Other than widening a firm's shareholder base, cross-listing enhances a firm's value by improving its information environment (Lang, Lins, & Miller, 2003). As such, investor recognition is increased. As Lang, Lins, and Miller (2003) point out, cross-listed firms are required to provide more disclosure due to greater regulatory and investor scrutiny, disclosure requirements, and potential legal exposure. Pagano, Roell, and Zechner (2002) also argue that the greater information transparency generated by cross-listing increases the willingness of international and local investors to commit capital. In addition, King and Segal (2009) point out that cross-listing increases the quantity and/or quality of information about firms, because it reduces the shadow cost of incomplete information as well as the information asymmetry between controlling and minority shareholders.

Cross-listing provides better investor protection as well. Coffee (1999), Coffee (2002), and Stulz (1999) argue that when firms from a jurisdiction with weaker investor protection of minority shareholders cross-list in a jurisdiction with reputational intermediaries, tougher regulation, and better enforcement, their valuation will increase. Meanwhile, cross-listing is likely to improve firm-level corporate governance because the greater transparency of a firm required by cross-listing might reduce the potential diversion of that firm's cash flow to managers and to controlling shareholders (Coffee,

1998; Stulz, 1999). In addition, Doidge, Karolyi, and Stulz (2004) point out that cross-listing increases the quality and/or the quantity of information available to minority shareholders and limits the expropriation of controlling shareholders from minority shareholders. They further note that cross-listing increases the ability of firms to take advantage of growth opportunities.

These aforementioned studies indicate that increasing the quality and/or the quantity of information available to minority shareholders is a key requirement of cross-listing. The greater information transparency generated by cross-listing is vital to improving investor recognition and investor protection. If this is the case, the market value of cross-listed firms will not be affected by the announcement of ISBs to the same degree as that of non-cross-listed firms. There are two facets to the reason for this: on one hand, information transparency requires cross-listed firms to routinely release reports about their ISBs. Therefore, it is not easy for their investors to be surprised by reports about ISBs in cross-listed firms. In contrast, non-cross-listed firms might not be required to release reports about their ISBs routinely, so their investors may tend to be surprised when receiving reports about their ISBs. On the other hand, the routinely released ISB reports may force cross-listed firms to improve their information system management. As a result, cross-listed firms may eventually suffer fewer and fewer ISBs. Among non-cross-listed firms, however, this trend might not exist. Investors tend to be more concerned about the future information security of these firms. This concern is easy to be magnified by release of ISB reports.

According to the efficient market hypothesis, which asserts that financial markets are informationally efficient, that markets recall to all publicly available information, and that stock prices reflect all publicly available information, all of the present and future effects of a publicly reported ISBs are captured in a firm's stock price (Garg, Curtis, & Halper, 2003). In other words, when a report of an ISB is publicly released, it will be immediately absorbed by investors and incorporated into the firm's stock price (Garg, Curtis, & Halper, 2003). This indicates that in an efficient capital market, the true impact of an event will be quickly and completely reflected in the value of the firm. As such, Goel and Shawky (2009) argue that the impact of ISBs can be measured by observing changes in the market values of firms in response to announcements of ISBs. This leads to the hypothesis of this paper:

H1: *The announcement of ISBs causes a smaller drop in the stock prices of cross-listed firms than in those of non-cross-listed firms.*

3. Data and Methodology

Data on ISBs were collected by searching in the LexisNexis database with the term "information security breaches." The search consisted of all public announcements of ISBs between January 1, 2005 and December 31, 2014. The list of firms that suffered ISBs was compared with the records of data breaches provided by Privacy Rights Clearinghouse (https://www.privacyrights.org) for triangulation. The initial list included 515 ISB announcements. Some firms were reported by more than one announcement during this time period. For these repeat-reported firms, only the report of the largest total breached customer record was kept. Next, those ISBs in which fewer than 1000 total customer records were breached were eliminated, given the small impact they would generate. Then, the firms that are not publicly traded were eliminated, and 126 firms remained on the list. After checking the listing status of these 126 firms in LexisNexis, the six without any record were removed. The final sample listed 120 firms (25 cross-listing ones and 95 non-cross-listing ones). Each of the 120 firms is publicly traded either on the New York Stock Exchange (NYSE) or on the NASDAQ stock exchange. Twenty-five of them cross-list on other exchanges, such as London Stock Exchange (LSE), the Toronto Stock Exchange (TSE), and the Italian Stock Exchange (ISE).

Each of the 120 firms was considered as one event. The date of the public announcements of the ISB was the event day. Daily firm information data, which includes date, firm name, ticker, and daily last stock price information, was collected from the Bloomberg database. Summary of the firm-level variable statistics, including market capitalization, market-to-book ratio (M/B), return on assets (ROA), price per earning (P/E), and debt to equity ratio (D/E), are listed in table 1. The estimation period started 120 days before the ISB announcement date and ended one day before the

IOP Conf. Series: Materials Science and Engineering **142** (2016) 012133 doi:10.1088/1757-899X/142/1/012133

announcement date. The event window started one day before the announcement date and ended five day after the announcement date. Altogether, this study comprises 120 firm-event observations, each of which contains 125 trading days' information.

Table 1. Firm-level variables summary statistics.

Variable	Mean	Std. Dev.	Min	Max
Market Cap (M)	59173.95	87457.24	168.85	647361.02
M/B	6.29	17.16	-13.99	129.35
ROA	6.63	6.34	0.75	28.3
P/E	40.18	112.46	5.54	1014.67
D/E	219.7	1477.26	-357.61	14397.26

An event study was adopted in order to explore the market reaction surrounding the announcements of ISBs. The market model was applied to estimate normal returns. The event date was set as day 0. The parameters for calculating firms' normal returns were calculated based on firm data from day -121 to day -1 (120 non-holiday trading days before the event window). The equation for normal return estimation is:s

$$R^*_{ijt} = \alpha_{ij} + \beta_{ij} R^*_{mjt} + \varepsilon_{it} \quad (1)$$

where $E(\varepsilon_{it}) = 0$ and $Var(\varepsilon_{it}) = \sigma^2_{\varepsilon i}$. R^*_{ijt} is the return of stock i in time t listed in market j. R^*_{mjt} is the market j's return in time t. ε_{it} is the error term. α_{ij} and β_{ij} are parameters to be estimated by ordinary least squares regression.

An abnormal return was calculated as the difference between the real return of each stock and the expected return of this stock on the same day, based on the parameters gotten from the estimation period, $\hat{\alpha}_{ij}$ and $\hat{\beta}_{ij}$. The abnormal returns for the event window were calculated with:

$$AR_{ijt} = R_{ijt} - \hat{\alpha}_{ij} - \hat{\beta}_{ij} R_{mt} \quad (2)$$

where AR_{ijt} denotes the abnormal return of stock i listed in market j on time t. R_{ijt} and R_{mt} are real returns of stock and market on time t. The cumulative abnormal returns (CARs) were calculated with:

$$CAR_{ij} = \sum_{T_1}^{T_2} AR_{ijt} \quad (3)$$

where T_1 to T_2 is the event window.

Table 2. Cumulative abnormal return (CAR) of cross-listing firms and non-cross-listing firms.

Days	Mean CAR (%)	t	Positive/Negative
Cross-Listing Firms			
(-1,-1)	0.0024	0.256	15/10
(0,0)	-0.0044	-0.3146	12/13
(1,5)	0.0029	0.0369	12/13
(-1,5)	0.0009	0.0091	12/13
Non-Cross-Listing Firms			
(-1,-1)	-0.0008	-0.0986	35/42
(0,0)	-0.0019	-0.1718	31/46
(1,5)	-0.0091	-0.0169	36/41
(-1,5)	-0.0118	-0.0174	33/44

All Firms			
(-1,-1)	-2.90E-05	-0.0117	50/52
(0,0)	-0.0025	-0.2068	43/59
(1,5)	-0.0062	-0.0116	48/54
(-1,5)	-0.0087	-0.1967	45/57

Abnormal returns were computed by using an event study approach. Not all of the 120 selected firms completed data for running event study. Table 2 presents the CAR analysis around the ISB announce date for 25 cross-listing firms and 77 non-cross-listing firms. Results are presented for the windows (−1, −1), (0, 0), (+1, +5) and (−1, +5), where day 0 represents ISB announcement day. The third column is the results of t tests on the cumulative mean abnormal return. The positive/negative column lists the number of firms having positive CARs versus the number of firms having negative CARs in the event window.

The mean of a CAR did not change much throughout the event window across the two groups of firms. In addition, all of the t values were not significant. For cross-listing firms, the number having positive CARs was larger than that of those having negative CARs before the ISB announcement date. However, the ratio reversed on the announce date and continued unchanged for the duration of the event window. In contrast, for non-cross-listing firms, the ratio between firms having positive CARs and those having negative CARs remained the same throughout the event window. And the latterwas always bigger than the former.

4. Discussion and conclusion

The small changes of the mean of CAR throughout the event window and a cross the two groups of firms indicate that the announcement of ISB impacted the market value of firms in the same way. Although the means of CAR for cross-listing firms dropped on the ISB announcement day and rose after that day, and the means of CAR for non-cross-listing firms continued to drop throughout the event window; all of the t values are not significant. There is no difference in the impact that the announcement of ISBs had on cross-listing firms and non-cross-listing firms. Therefore, H1 is rejected. Cross-listing does not mitigate the impact of ISBs announcement on firm market value.

However, even though all of the t values were not significant, the numbers of firms having positive CARs was larger than or close to those having negative CARs for cross-listing firms throughout the event window. In contrast, for non-cross-listing firms, the number of firms having negative CARs was always larger than those having positive CARs. What does this difference mean? The answer(s) are not provided in this study. Future research is need to explore the question.

In addition, the ratio between firms having positive CARs and those having negative CARs for cross-listing firms on the day before the ISB announcement day (15/16) was found to be quite different from those of the other days in the event window (12/13). As previous study (e.g. Goel & Shawky, 2009) pointed out, this difference might be caused by insiders leaking information. This introduces a new question: does cross-listing cause more insider leaking information?

References:

[1] Acquisti A, Friedman A and Telang R 2006 Is there a cost to privacy breaches? An event study. *ICIS 2006 Proceedings* p 94.

[2] Atomic Tangerine (2000). NPV: Information Security. Atomic Tangerine, Pittsburgh, available at http://www.ttivanguard.com/risk/netpresentvalue.pdf.

[3] Baker H K 1992 Why U.S. Companies List on the London, Frankfurt and Tokyo Stock Exchanges, *Journal of International Securities Markets* 6 pp 219–227.

[4] Baker H K, Nofsinger J R and Weaver D G 2002 International cross-listing and visibility, *Journal of Financial and Quantitative Analysis* 37(03) pp 495-521.

[5] Bancel F and Mittoo C 2001 European managerial perceptions of the net benefits of foreign stock listings, *European Financial Management* 7(2) pp 213-236.
[6] Campbell K, Gordon L A, Loeb M P and Zhou L 2003 The economic cost of publicly announced information security breaches: empirical evidence from the stock market, *Journal of Computer Security* 11(3) pp 431-448.
[7] Cavusoglu H, Mishra B and Raghunathan S 2004 The effect of internet security breach announcements on market value: Capital market reactions for breached firms and internet security developers, *International Journal of Electronic Commerce* 9(1) pp 70-104.
[8] Cline J 2006 Why isn't Europe suffering a wave of security breaches? Computerworld Security. Available at http://www.computerworld.com/action/article.do?command=viewArticleBasic&articleId=9001176&pageNumber=2. Accessed 23 September 2013.
[9] Coffee Jr J C 1998 Future as history: The prospects for global convergence in corporate governance and its implications, Nw. UL Rev. 93 p 641.
[10] D'Amico A D 2000 What does a computer security breach really cost? Secure Decisions, Applied Visions Inc.
[11] Denis D J, Denis D K and Yost K 2002 Global diversification, industrial diversification, and firm value, *The Journal of Finance* 57(5) pp 1951-1979.
[12] Doidge C, Karolyi G A and Stulz R M 2004 Why are foreign firms listed in the US worth more?, *Journal of financial economics* 71(2) pp 205-238.
[13] Doukas J A and Kan O B 2006 Does global diversification destroy firm value? *Journal of International Business Studies* 37(3) pp 352-371.
[14] Euromoney 1986 To List or Not to List, *Euromoney* pp 43–45.
[15] Fama E F, Fisher L, Jensen M C and Roll R 1969 The adjustment of stock prices to new information, *International economic review* 10(1) pp 1-21.
[16] Fanto J A and Karmel R S 1997 Report on the Attitudes of Foreign Companies Regarding a U.S. Listing, *Stanford Journal of Law, Business and Finance* 51 pp 37–58.
[17] Foerster S R and Karolyi G A 1999 The effects of market segmentation and investor recognition on asset prices: Evidence from foreign stocks listing in the United States, *The Journal of Finance* 54(3) pp 981-1013.
[18] Garg A, Curtis J and Halper H 2003 Quantifying the financial impact of IT security breaches *Information Management & Computer Security* 11(2) pp 74-83.
[19] Goel Sand Shawky H A 2009 Estimating the market impact of security breach announcements on firm values, *Information & Management* 46(7) pp 404-410.
[20] Gordon L A and Loeb M P 2002 The economics of information security investment, *ACM Transactions on Information and System Security* (TISSEC) 5(4) pp 438-457.
[21] Hoffer J A and Straub D W 1989 The 9-to-5 underground-Are you policing computer crimes? *Sloan Management Review* 30(4) pp 35-43.
[22] Hovav A and D'Arcy J 2003 The impact of denial of service attack announcements on the market value of firms, *Risk Management and Insurance Review* 6(2) pp 97-121.
[23] Kannan K, Rees J and Sridhar S 2007 Market reactions to information security breach announcements: An empirical analysis, *International Journal of Electronic Commerce* 12(1) pp 69-91.
[24] King M R and Segal D 2009 The long-term effects of cross-listing, investor recognition, and ownership structure on valuation, *Review of Financial Studies* 22(6) pp 2393-2421.
[25] Kuppuswamy V and Villalonga B 2007 Does Diversification Create Value in the Presence of External Financing Constraints? Evidence from the 2007–2009 Financial Crisis (November 29, 2010). Harvard Business School Finance Working Paper No. 10-101. Available at SSRN: http://ssrn.com/abstract=1571255

[26] Lang M H, Lins K V and Miller D P 2003 ADRs, analysts, and accuracy: Does cross-listing in the United States improve a firm's information environment and increase market value? *Journal of Accounting Research* 41(2) pp 317-345.

[27] Mittoo U R 1992 Managerial perceptions of the net benefits of foreign listing: Canadian evidence, *Journal of International Financial Management & Accounting* 4(1) pp 40-62.

[28] NIST (National Institute of Technical Standards) (1995). An introduction to computer security: the NIST handbook, Special Publication 800-812.

[29] Pagano M, Röell A A and Zechner J 2002 The geography of equity listing: why do companies list abroad? *The Journal of Finance* 57(6) pp 2651-2694.

[30] Power R 2001 CSI/FBI computer crime and security survey, *Computer Security Journal* 17(2) pp 20-51.

[31] Power R 2003 2002 CSI/FBI computer crime and securuty survey, *Computer Security Issues and Trends* 8(1) pp -21.

[32] Reuters 2011 Sony PlayStation suffers massive data breach. http://www.reuters.com/article/2011/04/26/us-sony-stoldendata-idUSTRE73P6WB20110426. Accessed 06 November 2013.

[33] Sarkissian S and Schill M J 2009 Are there permanent valuation gains to overseas listing? *Review of Financial Studies* 22(1) pp 371-412.

[34] Stulz R M 1999 Globalization of Equity Markets and the Cost of Capital, *Journal of Applied Corporate Finance* 12 pp 8–25.

[35] Telang R and Wattal S 2007 An empirical analysis of the impact of software vulnerability announcements on firm stock price, *Software Engineering, IEEE Transactions on 33*(8) pp 544-557.

Social Media and eBusiness: Cultural Impacts on the Influence Process in Consumer Communities

Yong Chen[1,a], Hong Chen[1,b], Li Xu

Old Dominion University, USA

E-mail:[a]y7chen@odu.edu, [b]hchen001@odu.edu, lxu@odu.edu

Abstract. Social media has been used as an important tool by firms to influence consumers' attitude and behavior. Influence occurs in consumer communities in social media because community members have the control of discovering, producing, sharing, and distributing information and because the spread out of their experiences and opinions in the format of electronic word-of-mouth forms emerging conformance. Prior research has explored how the influence occurring in online social media communities impacts consumers' attitude and behavior (e.g., product attitude and purchase decision, effectual thinking and behavior, brand trust and brand loyalty). But because social media has the ability of global reach, cross-border factors should not be neglected in studying the influence process. As such, this paper adopts national cultural dimensions identified by Hofstede (1984), individualism/collectivism and power distance particularly, the index of cultural distance, and the social influence theory to explore how culture impacts the influence occurring in consumer communities in social media.

1. Introduction

The use of social media among Fortune 500 companies has surged in recent years. A study conducted by the University of Massachusetts Dartmouth in 2012 show that as many as 73 percent of these companies have an official corporate account in Twitter, and that 66 percent have a corporate Facebook page (Okazaki & Taylor, 2013). Meanwhile, the expenditures firms spend on social media are increasing as well. According to VanBoskirk (2011), social media marketing in the U.S. is expected to grow at an annual rate of 26 percent from 2011 to 2016, reaching an estimated worth of USD 5 billion. Social media has been used as an important tool by firms to influence consumers' opinions and purchase intentions (Algesheimer, Dholakia, & Herrmann, 2005), increase marketing communication effectiveness (Dholakia & Durham, 2010; Kozinets, de Valck, Wojnicki, & Wilner, 2010; Trusov, Bucklin, & Pauwels, 2009), interact with customers and manage customer relationship (Boyd & Ellison, 2008), monitor a marketplace (Berinato, 2010), and gain competitive advantages (Cook, 2008).

Social media's unique characters make it important for marketing. On one hand, it facilitates online communities where consumers can share experiences and opinions with progressively larger audiences in the format of electronic word-of-mouth (Thackeray, Neiger, Hanson, & McKenzie, 2008; Okazaki & Taylor, 2013). On the other hand, consumers have the control of discovering, producing, sharing, and distributing information in the online communities (Solis & Carroll, 2008; Thackeray, Neiger, Hanson, & McKenzie, 2008). In social media, community members influence each other during the process of interaction and conversation. Recent studies show that the influence occurring in online social media communities has impacts on consumers' product attitude and purchase decision (Algesheimer, Dholakia, & Herrmann, 2005; Wang, Yu, & Wei, 2012), effectual thinking and behavior (Fischer & Reuber, 2011), adoption and use of products and services (de Valck, van Bruggen, & Wierenga, 2009; Subramani & Rajagopalan, 2003), and brand trust and brand loyalty (Laroche, Habibi, Richard, & Sankaranarayanan, 2012; Laroche, Habibi, & Richard, 2013).

However, prior studies explore the influence from the perspective of domestic marketing only. Because social media has the ability of global reach (Okazaki & Taylor, 2013), consumers across the world can access the online communities in it. For firms operating globally, understanding influence happening between members with cross-border background in social media is crucial for their marketing strategies. As such, this paper adopts national cultural dimensions identified by Hofstede (1984), individualism/collectivism and power distance particularly, the index of cultural distance, and the social influence theory to explore how culture impacts the influence occurring in consumer communities in social media. The main contributions of this paper are these: (1) theoretically, it originally combines the theory of social influence and national cultural dimensions to analysis the influence occurring in consumer communities in social media; and (2) practically, it helps international marketing practitioners better understand the mechanism of influence occurring in consumer communities in social media from a new perspective, so that they make better use of social media in their cross-border marketing operations.

2. Literature Review

Social Media

Social media and Web 2.0 are often used as synonyms, but actually they are not (Kaplan & Haenlein, 2010; Weinberg & Pehlivan, 2011). Social media is "a group of Internet-based applications that build on the ideological and technological foundations of Web 2.0, and that allow the creation and exchange of User Generated Content" (Kaplan & Haenlein, 2010, p61), whereas Web 2.0 is comprised of computer network-based platforms, on which social media run or function (Weinberg & Pehlivan, 2011). Although there is no definitive typology of social media (Kaplan and Haenlein, 2010), it is common to classify social media as social networking (e.g. Facebook), social bookmarking (e.g. Digg), video-sharing (e.g. Youtube), picture-sharing (e.g. Flickr), professional networking (e.g. LinkedIn), user forums, weblogs (or blogs), and microblogging (e.g. Twitter) (Fischer & Reuber, 2011).

Social media is built on the idea of sharing information openly and fostering a sense of community. It brings people together for interaction and conversation (Peypoch, 1998; Rheingold, 2000). In social media, people meet others with common interests, share ideas, and interact with each other on a variety of topics (Winer, 2009). User-generated content and discussions are used by community members to educate each other about products, brands, services, personalities, and issues (Blackshaw & Nazzaro, 2006), and to generate some type of influence (Hanna, Rohm, & Crittenden, 2011). As a result, powerful communities are created and they further facilitate the interactions among community members (Winer, 2009). Because emerging conformance exist in social media (De Bruyn & Lilien 2008; Iyengar, Han, & Gupta 2009; Trusov, Bodapati, & Bucklin 2010), social media has been used for influencing on community members' opinions and purchase intentions (Algesheimer, Dholakia, & Herrmann, 2005; Wang, Yu, & Wei, 2012), guiding consumers' adoption and use of products and services (de Valck, van Bruggen, & Wierenga, 2009; Subramani & Rajagopalan, 2003), managing customer relationship, facilitating the establishment and maintenance of relationships between consumers and brands/products (Weinberg & Pehlivan, 2011), and gaining competitive advantage (Cook, 2008) in the forms of collaboration-based productivity (Soriano, Lizcano, Cañas, Reyes, & Hierro, 2007) and customer-driven innovation (Tapscott & Willams, 2008).

In social media communities, huge number of members is engaged collectively in a conversation that leads to the generation of online content (Thackeray, Neiger, Hanson, & McKenzie, 2008). For example, Twitter, one of the fastest growing social networks in the Internet, has experienced a burst of popularity consisting of several tens of millions of users who actively participate in the creation and propagation of content (Asur & Huberman, 2010). As such, social media makes it possible to reach large amount of people with common interests globally (Hanna, Rohm, & Crittenden, 2011; Okazaki & Taylor, 2013). Furthermore, the wide range of online and word-of-mouth forums encompassed by social media provides an unparalleled platform for community members to publicize and share their product experiences and opinions (Chen, Fay, & Wang, 2011). Electronic word-of-mouth, which has a large impact on consumer decisions (Feick & Price 1987; Leonard-Barton, 1985), increases the speed of the spreading of community members' experiences and opinions (Thackeray, Neiger, Hanson, & McKenzie, 2008). More specifically, the electronic word-of-mouth in social media can create a group's influence faction by facilitating information exchange, providing recreational pastime, and bringing users together (Hennig-Thurau, Walsh, & Walsh, 2003; Lee & Youn, 2009; Okazaki & Taylor, 2013). Therefore, social media has fundamentally altered marketing's ecosystem of influence (Walmsley, 2010).

The main characters of social media are participation, interaction, collaboration, sharing, and relationship (Falls, 2008; Kaplan & Haenlein, 2010). In social media, community members are not passive recipients any longer. Instead, they become active participants (Thackeray, Neiger, Hanson, & McKenzie, 2008). Two-way discussions bring them together and they have the control of discovering, producing, sharing, and distributing information (Solis 2008, Thackeray, Neiger, Hanson, & McKenzie, 2008). In Twitter, for example, users are both consumers of tweets (followers) and producers of tweets (followed)(Fischer& Reuber, 2011). This change of information flow indicates a user's influence on others (Cha, Haddadi, Benevenuto, and Gummadi, 2010). Because relationships are the foundation for social media (Waters, Burnett, Lamm, & Lucas, 2009), the two-way information flow mode in social media influences the relationship between community members (Weinberg & Pehlivan, 2011).

3. Social Influence

Scholars from diverse fields, such as sociology, communication, marketing, and political science, have studied influence for a long time (Katz & Lazarsfeld, 1955; Rogers, 2010). Social influence refers to "a pervasive and familiar phenomenon in our economic and social life: namely, that individuals tend to conform their conduct to that of other individuals" (Kahan, 1997, p352). Social influence occurs when an individual's behavior is influenced by surrounding people. According to Bearden, Netemeyer, and Teel (1989), there are two forms of peer influence: normative and informational. Normative influences push group members to conform to group norms and modify their attitudes and behaviors according to peers' expectations, whereas informational influences drive group members to learn about something by searching for information from knowledgeable peers or observing peers' behavior (Bearden, Netemeyer, &Teel, 1989).

According to Kelman (1958), an individual's resulting overt behavior may appear the same after he adopts induced behavior under the impact of social influence. But the underlying processes in which he engages may be different. Kelman (1958) points out that there are three reasons why individuals adopt social influence: compliance, identification, and internalization. Compliance happens when an individual adopts the induced behavior because he expects to gain rewards or to avoid punishments, not because he believes in its content. Identification refers that an individual accepts influence for establishing or for maintaining a satisfying self-defining relationship to another person or group. Internalization means that an individual accepts influence because it is congruent with his value system (Malhotra & Galletta, 1999).

Regarding the source where social influence comes from, there are two conflicting theories. The traditional influential theory argues that it is a minority of members (influential) who possess the qualities that make them persuasive in spreading ideas to others in a society (Rogers 1962). These individuals, who are labeled as opinion leaders, social hubs, boundary spanners, lead users, heavy users, or experts in literature (Nejad, Sherrell, & Babakus, 2014), disproportionately impact the spread of information or some related behavior or interest (Gladwell, 2000; Goldenberg, Han, Lehmann, & Hong, 2009; Keller & Berry, 2003; Weimann, 1994). Thus, one can target these influential to achieve a large-scale chain-reaction of influence driven by word-of-mouth with small cost (Katz & Lazarsfeld, 1955). This theory spread well beyond and has been adopted widely by practitioners (Gladwell, 2006; Keller & Berry, 2003).

A newer view, in contrast, argues that influential do not have that much power and that interpersonal relationship among ordinary users and the readiness of society to adopt an innovation determine influence (Domingos & Richardson, 2001; Watts & Dodds, 2007). In new information age, people make choices based on the opinions of their peers and friends rather than by influential (Dominogos & Richardson, 2001). According to this view, anyone can be influential and rumors spread only if there is societal willingness to accept them (Quercia, Ellis, Capra, & Crowcroft, 2011). As a result, supposed influential do not work for word-of-mouth marketing any longer.

So far, these two theories have not been validated yet because quantification for influence study is not available and its essential components, such as human choices and the ways in which our societies function, cannot be reproduced in labs (Cha, Haddadi, Benevenuto, & Gummadi, 2010).

4. Social Influence in Social Media

Social influence, in the context of social media, refers to "the concept of people modifying their behavior to bring them closer to the behavior of their friends" (Yu, Asur, and Huberman, 2011, p2). Social media provides a perfect platform for people to check the opinions of their peers and friends. In social media, a small percentage of users have a large and active audience who consumes and multiplies the content published by them. As such, these contents are distributed in a number of micro-networks and receive attention from large amount users (Anger & Kittl, 2011). This character of social media makes it not only a typical place where social influence occurs, but also a platform on which quantification for social influence study can be conducted. Recent studies have explored how social influence works in social media. For example, by analyzing 355 brand posts from 11 international brands on Facebook, de Vries, Gensler, and Leeflang (2012) find that positioning brand posts on top of brand fan pages on Facebook enhances brand post popularity, and that the share of positive comments is positively related to the number of likes and the number of comments. Agarwal, Liu, Tang, and Yu (2008) conclude that the most influential bloggers are not necessarily the most active ones. In an empirical study of dynamics of users influence across topics and time in Twitter, Cha, Haddadi, Benevenuto, and Gummadi (2010) reach the conclusion that popular users who have big numbers of followers are not necessary influential, that most influential users can hold significant influence over a variety of topics, and that influence is gained through concerted effort.

In a similar study, Kwak, Lee, Park, and Moon (2010) find that the ranking of the most influential users differed depending on the measure of influence-number of followers, page-rank, and number of retweets. Weng, Lim, Jiang, and He (2010) reached the same conclusion although they use different measures of influence. Romero, Galuba, Asur, and Huberman (2010) argue that users must obtain attention and overcome their passivity to become influentials. These empirical evidences show that influential in social media, specifically Twitter, are not accidental or spontaneously, but rather individuals who exhibit specific behaviors (Cha, Haddadi, Benevenuto, and Gummadi, 2010; Romero, Galuba, Asur, and Huberman, 2010). Influence on Twitter is strongly depends on audience engagement and partly on linguistic qualities that reflect one's personality and mood (Quercia, Ellis, Capra, & Crowcroft, 2011).

Although some scholars argue that people are more affected by opinions of their peers and friends than influential, most recent online social networks studies still support the concept that influential exert disproportionate amount of influence (Cha, Haddadi, Benevenuto, and Gummadi, 2010; Kwak, Lee, Park, & Moon, 2010; Weng, Lim, Jiang, & He, 2010; Romero, Galuba, Asur, & Huberman, 2010). Therefore, identifying influential and then targeting them to achieve a large-scale chain-reaction of influence is still an important approach for applying social media in marketing (Gruhl & Liben-nowell, 2004; Kempe, Kleinberg, & Tardos, 2003).

5. Theoretical Development

National Culture

The national cultural dimensions identified by Hofstede (1984) provide scholars and practitioners a valuable insight into the dynamics of cross-cultural relationships. The four cultural dimensions (uncertainty avoidance, individualism, masculinity, and power distance) have been widely adopted as the most dominant framework for theory development and validation in cross-cultural studies (Myers & Tan, 2003). Because of its unique character of community building, social media has been used as an important tool for firms to influence community members' opinions and purchase intentions (Algesheimer, Dholakia, & Herrmann, 2005; Wang, Yu, & Wei, 2012), guide consumers' adoption and use of products and services (de Valck, van Bruggen, & Wierenga, 2009; Subramani & Rajagopalan, 2003), manage customer relationship, facilitate the establishment and maintenance of relationships between consumers and brands/products (Weinberg & Pehlivan, 2011).

In these communities, firms' current and potential customers from different counties interact with each other and build relationships. Influence takes place in this process, either from influential or from their peers. Since consumers in social media come from diverse national cultures, the influence on them might be different. In other words, they react differently when facing conformity pressures caused by individualism/collectivism and power distance specifically. Thus, this paper predicts that national cultures surrounding consumers impact the influence process in consumer communities in social media. Particularly, this paper predicts that individualism cultures negatively impact the influence process, whereas collectivism cultures positively impact the influence process.

Individualism/Collectivism

The individualism dimension of culture refers to the degree to which individuals in a society are integrated into groups (Hofstede, 1984). Individualism cultures value individual freedom and achievements, and the relationship between individuals is tied loosely in such society (Rothaermel, Kotha, & Steensma, 2006). Collectivism cultures are less individualistic. They value the collective good instead of the individual, and members are strongly tied to one another via some kinds of relationship, such as birth and family (Rothaermel, Kotha, & Steensma, 2006). In collectivism cultures, interpersonal relationships play an important role, so trust is required to develop on an individual-to-individual basis (Rothaermel, Kotha, & Steensma, 2006). In addition, deviations from group norms are much less tolerated by members in collectivism cultures than by members in individualism cultures (Triandis & Bhawuk, 1997). As such, when members in collectivism cultures are facing the influence coming from influential or peers, they weigh the pros and cons based on group's reaction and tend to be the same as other group members. Members in individualism cultures, however, consider less about other people's reactions and might not accept influence from others. Therefore, this paper proposes the following propositions:

Proposition 1a. Individualism cultures negatively impact the influence process in consumer communities in social media.

Proposition 1b. Collectivism cultures positively impact the influence process in consumer communities in social media.

Power Distance

The power distance dimension of culture refers to the degree to which equality is distributed among members in a society (Chaula, 2006). Power distance can be determined by the level of hierarchy in societies (Jones & Alony, 2007). High power distance cultures allow inequalities in opportunity, power, status, and wealth among members. In contrast, low power distance cultures intervene to create an equal distribution among members (Hofstede, 1984; Rothaermel, Kotha, & Steensma, 2006). As for the influence process in consumer communities in social media, members in high power distance cultures tend to follow the influence, especially from influential, namely opinion leaders, social hubs, boundary spanners, lead users, heavy users, or experts. Community members tend to think that they are in a subordinate position and that it is acceptable to follow the influence coming from influential.

In contrast, community members in low power distance cultures tend to think that each member has the right to make his own decision, regardless the opinions from influential. Thus, the above arguments lead to the following proposition:

Proposition 2. Power distance positively impacts the influence process in consumer communities in social media.

Cultural Distance

Culture distance was first proposed by Kogut and Singh (1988) to explore the effect of national culture on multinational enterprises' entry mode choices. They built a composite index to calculate culture distance between countries based on the deviation along each of the four cultural dimensions in Hofstede's study. According to Kogut & Singh (1988), the larger the cultural differences between countries are, the larger the difference in their firms' organizational and managerial practices exists. In consumer communities in social media, the cultural differences between countries that consumer come from play a role in consumers' interaction process. The larger the cultural difference is, the larger the difference in consumers' attitude and reaction for conformity pressures. Thus, this paper proposes the following proposition:

Proposition 3. The cultural differences between countries that consumers come from negatively impact the influence process in consumer communities in social media.

Research Model

In summary, this paper expects that the adoption of national cultural dimensions and cultural distance will yield a better perspective on the influence process in consumer communities in social media for international marketing. The relationships among constructs include individualism/ collectivism, power distance, cultural distance and influence. Figure 1 shows the research model.

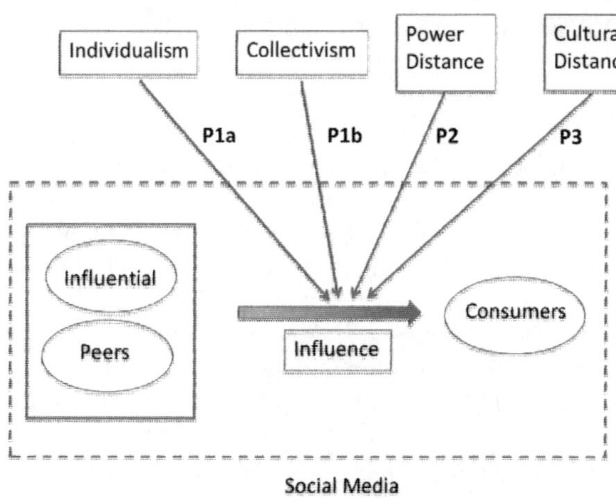

Figure 1. Cultural impacts on the influence process in consumer communities in social media

6. Discussion and Conclusion

Social media facilitates communities in which members interact and collaborate with others who share common interests. The communities can engage huge number of members and have the ability of global reach. In the communities, members are not passive recipients. Instead, they have the control of discovering, producing, sharing, and distributing information. Their experiences and opinions spread out in the format of electronic word-of-mouth and form some sort of emerging conformance in social media. The influence generated by communities in social media is proved to have impacts on consumers' product attitude and purchase decision (Algesheimer, Dholakia, & Herrmann, 2005; Wang, Yu, & Wei, 2012), effectual thinking and behavior (Fischer & Reuber, 2011), adoption and use of products and services (de Valck, van Bruggen, & Wierenga, 2009; Subramani & Rajagopalan, 2003), and brand trust and brand loyalty (Laroche, Habibi, Richard, & Sankaranarayanan, 2012; Laroche, Habibi, & Richard, 2013).

However, prior research neglects the role of cross-border factors in the influence process. Because community members in social media may come from several countries, their cultural background might impact the influence occurring in the communities. This is crucial for firms operating globally. Therefore, this paper adopts national cultural dimensions identified by Hofstede (1984), individualism/collectivism and power distance particularly, the index of cultural distance, and the social influence theory to explore how culture impacts the influence occurring in consumer communities in social media. More specifically, this paper proposes that individualism cultures negatively impact the influence process, whereas collectivism cultures positively impact the influence process. This paper also proposes that the cultural differences between countries that consumers come from negatively impact the influence process in consumer communities in social media.

Among the four national cultural dimensions (uncertainty avoidance, individualism, masculinity, and power distance) identified by Hofstede (1984), this paper does not cover uncertainty avoidance and masculinity. Although these two dimensions do not appear in the propositions, it does not mean that they do not play a role in the influence process. Instead, this paper thinks that their roles are smaller than the other two: individualism and power distance. Future studies need to explore how they impact the influence process in consumer communities in social media. Furthermore, besides cultural factors, what other cross-border factors impact the process? Future studies are needed to clarify this question.

Prior research uses the theory of social influence to analyze social media (Algesheimer, Dholakia, & Herrmann, 2005; de Valck, van Bruggen, & Wierenga, 2009; Fischer & Reuber, 2011; Laroche, Habibi, Richard, & Sankaranarayanan, 2012; Laroche, Habibi, & Richard, 2013; Subramani & Rajagopalan, 2003; Wang, Yu, & Wei, 2012). This paper makes contributions to analyzing the influence process in consumer communities in social media by combining the theory of social influence and national cultural dimensions. The primary practical implication of this study is that firms operating globally need to take wise use of social media for their cross-border marketing operations. For managers, this paper allows them to better understand the role that consumers' cultural backgrounds play in social media. When they make and implement marketing strategies via social media, they can take into consideration of the cultural factors.

References:

[1] Agarwal N, Liu, H, Tang L & Yu P S 2008 Identifying the influential bloggers in a community. *In Proceedings of the 2008 international conference on web search and data mining* pp 207-218). ACM.

[2] Algesheimer R, Dholakia U & Herrmann A 2005 The social influence of brand community: Evidence from European car clubs *Journal of Marketing* 69(3) pp 19–34.

[3] Anger I, & Kittl C 2011 Measuring influence on Twitter. In Proceedings of the 11th *International Conference on Knowledge Management and Knowledge Technologies* (p 31). ACM.

[4] Asur S & Huberman B A 2010 Predicting the future with social media. In Web Intelligence and Intelligent Agent Technology (WI-IAT), 2010 IEEE/WIC/ACM *International Conference* on (Vol. 1 pp 492-499). IEEE.

[5] Bearden W O, Netemeyer R G.& Teel J E 1989 Measurement of consumer susceptibility to interpersonal influence *Journal of consumer research* pp 473-481.

[6] Berinato S 2010 Six ways to find value in Twitter's noise *Harvard Business Review* **88** (6) pp 34–35.

[7] Blackshaw P & Nazzaro M 2006 Consumer-generated media (CGM) 101: Word-of-mouth in the age of the web-fortified consumer. A Nielsen BuzzMetrics White Paper, Second Edition, Spring.

[8] Boyd D M & Ellison N B 2008 Social Network Sites: Definition, History, and Scholarship *Journal of Computer-Mediated Communication* (13:1) pp 210-230.
[9] Cha M, Haddadi H, Benevenuto F & Gummadi P K 2010 Measuring User Influence in Twitter: The Million Follower Fallacy. ICWSM, 10 pp 10-17.
[10] Chaula J A 2006 A socio-technical analysis of information systems security assurance: A case study for effective assurance (Unpublished doctoral dissertation, Stockholm University, Sweden).
[11] Chen Y, Fay S & Wang Q 2011 The role of marketing in social media: how online consumer reviews evolve *Journal of Interactive Marketing* 25(2) pp 85-94.
[12] Cook N 2008 Enterprise 2.0: How Social Software Will Change the Future of Work, *Gower Publishing* Ltd., pp 164.
[13] De Bruyn A & Lilien G L 2008 A multi-stage model of word-of-mouth influence through viral marketing. *International Journal of Research in Marketing* 25(3) pp 151-163.
[14] de Valck, Kristine, van Bruggen, Gerrit H & Wierenga Berend 2009 Virtual communities: A marketing perspective. *Decision Support Systems* 47(3) pp 185–203.
[15] De Vries L, Gensler S & Leeflang P S 2012 Popularity of brand posts on brand fan pages: an investigation of the effects of social media marketing. *Journal of Interactive Marketing* 26(2) pp 83-91.
[16] Dholakia U M & Durham E 2010 One café chain's Facebook experiment. *Harvard Business Review* 88 (3) pp 26.
[17] Domingos P & Richardson M 2001 Mining the network value of customers. In Proceedings of the seventh ACM SIGKDD international conference on Knowledge discovery and data mining (pp 57-66). ACM.
[18] Falls J 2008 Social Media in Two Words or Less. Retrieved April 13, 2014, from http://www.socialmediaexplorer.com/social-media-marketing/social-media-in-two-words-or-less/
[19] Feick L F & Price L L 1987 The market maven: A diffuser of marketplace information. *The Journal of Marketing* pp 83-97.
[20] Fischer E & Reuber A R 2011 Social interaction via new social media:(How) can interactions on Twitter affect effectual thinking and behavior? *Journal of business venturing* 26(1) pp 1-18.
[21] Gladwell M 2006 The tipping point: How little things can make a big difference. Hachette Digital, Inc
[22] Goldenberg J, Han S, Lehmann D R & Hong J W 2009 The role of hubs in the adoption process. *Journal of Marketing* 73(2) pp 1-13.
[23] Gruhl D, Guha R, Liben-Nowell D & Tomkins A 2004 Information diffusion through blogspace. In Proceedings of the 13th international conference on World Wide Web (pp 491-501). ACM.
[24] Hanna R, Rohm A & Crittenden V L 2011 We're all connected: The power of the social media ecosystem. *Business Horizons* 54(3) pp 265-273.
[25] Hennig-Thurau T, Walsh G and Walsh G 2003 Electronic word-of-mouth: motives for and consequences of reading customer articulations on the Internet, *International Journal of Electronic Commerce* Vol. 8 No. 2 pp 51-74.
[26] Hofstede G 1984 Culture's consequences. International differences in work-related values. Beverly Hills, CA: Sage.
[27] Iyengar R, Han S & Gupta S 2009 Do friends influence purchases in a social network. Harvard Business School.
[28] Jones M, & Alony I 2007 The Cultural Impact of Information Systems--Through the Eyes of Hofstede--A Critical Journey. Issues in Informing Science & Information Technology, 4.
[29] Kahan D M 1997 Social influence, social meaning, and deterrence. *Virginia Law Review* pp 349-395.

[30] Kaplan A M & Haenlein M 2010 Users of the world, unite! The challenges and opportunities of Social Media. Business horizons 53(1) pp 59-68.

[31] Katz E & Lazarsfeld P F 1970 Personal Influence, The part played by people in the flow of mass communications. Transaction Publishers.

[32] Keller E, & Berry J 2003 The influentials: One American in ten tells the other nine how to vote, where to eat, and what to buy. Simon and Schuster.

[33] Kelman H C 1958 Compliance, identification, and internalization: Three processes of attitude change. *Journal of conflict resolution* pp 51-60.

[34] Kempe D, Kleinberg J & Tardos É 2003 Maximizing the spread of influence through a social network. In Proceedings of the ninth ACM SIGKDD international conference on Knowledge discovery and data mining (pp 137-146). ACM.

[35] Kogut B & Singh H 1988 The effect of national culture on the choice of entry mode. *Journal of international business studies* 19(3) pp 411-432.

[36] Kozinets R V, De Valck K, Wojnicki A C & Wilner S J 2010 Networked Narratives: Understanding Word-of-Mouth Marketing in Online Communities. *Journal of marketing* 74(2) pp 71–89.

[37] Kwak H, Lee C, Park H & Moon S 2010 What is Twitter, a social network or a news media?. In Proceedings of the 19th international conference on World wide web (pp 591-600). ACM.

[38] Laroche M, Habibi M R & Richard M O 2013 To be or not to be in social media: How brand loyalty is affected by social media? *International Journal of Information Management* 33(1) pp 76-82.

[39] Laroche M, Habibi M R, Richard M O, & Sankaranarayanan R 2012 The effects of social media based brand communities on brand community markers, value creation practices, brand trust and brand loyalty. *Computers in Human Behavior* 28(5) pp 1755-1767.

[40] Lee M and Youn S 2009 Electronic word of mouth (eWOM) – how eWOM platforms influence consumer product judgement. *International Journal of Advertising* Vol. 28 No. 3 pp 473-499.

[41] Leonard-Barton D 1985 Expert as Negative Opinion Leaders in the Diffusion of a Technological Innovation. *Journal of Consumer Research* 11(4) pp 914–26.

[42] Malhotra Y & Galletta D F 1999 Extending the technology acceptance model to account for social influence: theoretical bases and empirical validation. In Systems Sciences, 1999. HICSS-32. Proceedings of the 32nd Annual Hawaii International Conference on (pp 14-pp). IEEE.

[43] Myers M D & Tan F B 2003 Beyond models of national culture in information systems research. *Advanced topics in global information management* 2 pp 14-29.

[44] Nejad, M. G., Sherrell, D., & Babakus, E. (2014). Influentials and Influence Mechanisms in New Product Diffusion: An Integrative Review. Journal of Marketing Theory and Practice, 22(2), 185-208.

[45] Okazaki S & Taylor C R 2013 Social media and international advertising: theoretical challenges and future directions. *International Marketing Review* 30(1) pp 56-71.

[46] Peypoch R 1998 The case for electronic business communities. *Business Horizons* 41(5) pp 17—20.

[47] Quercia D, Ellis J, Capra L & Crowcroft J 2011 In the mood for being influential on twitter. In Privacy, security, risk and trust (passat), 2011 ieee third international conference on and 2011 ieee third international conference on social computing (socialcom) (pp 307-314). IEEE.

[48] Rheingold H 2000 The virtual community: Homesteading on the electronic frontier (No. 28). MIT press.

[49] Rogers E M 2010 Diffusion of innovations. Simon and Schuster.

[50] Romero D M, Galuba W, Asur S & Huberman B A 2011 Influence and passivity in social media. In Machine learning and knowledge discovery in databases (pp 18-33). Springer Berlin Heidelberg.

[51] Rothaermel F T, Kotha S & Steensma H K 2006 International market entry by US internet firms: an empirical analysis of country risk, national culture, and market size. *Journal of Management* 32(1) pp 56-82.

[52] Solis B & Carroll B 2008 Customer service: the art of listening and engagement through social media. Business, 2.

[53] Soriano J, Lizcano D, Cañas M A, Reyes M and Hierro J J 2007 Fostering Innovation in a Mashup-oriented Enterprise 2.0 Collaboration Environment, *System and Information Sciences Notes* (1:1) pp 62-68.

[54] Subramani M R & Rajagopalan B 2003 Knowledge-sharing and influence in online social networks via viral marketing. *Communications of the ACM* 46(12) pp 300-307.

[55] Tapscott D & Williams A D 2008 Wikinomics: How mass collaboration changes everything. Penguin.

[56] Thackeray R, Neiger B L, Hanson C L & McKenzie J F 2008) Enhancing promotional strategies within social marketing programs: use of Web 2.0 social media. *Health promotion practice* 9(4) pp 338-343.

[57] Triandis H C & Bhawuk D P S 1997 Culture theory and the meaning of relatedness, in Earley, P. C. & Erez, M. (eds.), New Perspectives on International Industrial/organizational Psychology (The New Lexington Press, San Francisco, CA) pp 13–52.

[58] Trusov M, Bucklin R E & Pauwels K 2009 Effects of word-of-mouth versus traditional marketing: findings from an internet social networking site. *Journal of marketing* 73(5) pp 90-102.

[59] Trusov M, Bodapati A V & Bucklin R E 2010 Determining influential users in Internet social networks. *Journal of Marketing Research* 47(4) pp 643-658.

[60] VanBoskirk S 2011 US interactive marketing forecast, 2011 to 2016. Retrieved April 20 2014 from http://metrogroupagency.com/media/whitepapers/Interactive_Marketing.pdf

[61] Walmsley A 2010 New media needs new PR. Retrieved April 28, 2014, from http://www.marketingmagazine.co.uk

[62] Wang X, Yu C & Wei Y 2012 Social media peer communication and impacts on purchase intentions: A consumer socialization framework. *Journal of Interactive Marketing* 26(4) pp 198-208.

[63] Waters R D, Burnett E, Lamm A & Lucas J 2009 Engaging stakeholders through social networking: How nonprofit organizations are using Facebook. *Public Relations Review* 35(2) pp 102-106.

[64] Watts D J & Dodds P S 2007 Influentials, networks, and public opinion formation. Journal of consumer research 34(4) pp 441-458.

[65] Weimann J 1994 Individual behaviour in a free riding experiment. *Journal of Public Economics* 54(2) pp 185-200.

[66] Weinberg B D & Pehlivan, E 2011 Social spending: Managing the social media mix. *Business Horizons* 54(3) pp 275-282.

[67] Weng J, Lim E P, Jiang J & He Q 2010 February Twitterrank: finding topic-sensitive influential twitterers. In Proceedings of the third ACM international conference on Web search and data mining (pp 261-270). ACM.

[68] Winer R S 2009 New communications approaches in marketing: Issues and research directions. *Journal of Interactive Marketing* 23(2) pp 108-117.

[69] Yu L, Asur S & Huberman B A 2011 What trends in Chinese social media. arXiv preprint arXiv:1107.3522.

Modeling Business Processes of the Social Insurance Fund in Information System Runa WFE

M Yu Kataev[1,2], L A Bulysheva[3], Li D Xu[3], N V Loseva [4]

[1] Yurga Institute of Technology of National research Tomsk Polytechnic University, Russia
[2] Tomsk state university of control systems and radioelectronics, Russia
[3] Old Dominion University, USA
[4] Social insurance fund of the Russian Federation, Tomsk, Russia

Abstract. Introduction – Business processes are gradually becoming a tool that allows you at a new level to put employees or to make more efficient document management system. In these directions the main work, and presents the largest possible number of publications. However, business processes are still poorly implemented in public institutions, where it is very difficult to formalize the main existing processes. Us attempts to build a system of business processes for such state agencies as the Russian social insurance Fund (SIF), where virtually all of the processes, when different inputs have the same output: public service. The parameters of the state services (as a rule, time limits) are set by state laws and regulations.
The article provides a brief overview of the FSS, the formulation of requirements to business processes, the justification of the choice of software for modeling business processes and create models of work in the system Runa WFE and optimization models one of the main business processes of the FSS. The result of the work of Runa WFE is an optimized model of the business process of FSS.

1. Introduction

Application of modern methods and information technologies required for the effective solution of problems arising in connection with the description, design, modeling, analysis of business processes (BP) for the effective development of the enterprise. For this you can use existing tools and technologies to save resources of the enterprise while solving the tasks.

One of the most common tools is the methodology of the SADT, which includes a number of IDEF models. Yet another way organizations manage business processes in organizations is the use of BPMS (Business Process Management System) is a system for managing business processes.

The objectives of the use of such systems are improving the quality of drafting of business processes, reduction of time costs, the possibility of monitoring the activities to improve quality management and continuous improvement of internal business processes with their quick changes.

Information technology implementation is RunaWFE – information system management business processes, allowing you to build effective interaction of employees of the organization and to monitor their activities with the aim of improving the quality of the whole organization with the resources of the local computer network of the organization.

The RunaWFE system implements the concept of BPM (Business Process Management) that allows you to build flexible adaptive information systems, able to quickly change with the change in BP organization.

The object of the research institution of the state social insurance Fund, and the research object, the model of work organization in the performance of one of the main business processes.

Building such models will allow you to:

- effectively manage current activities of the organization;
- to ensure fulfilment of services provided to policyholders within specified timeframes;
- to calculate the possible solutions to improve and optimize;
- to consider the necessary number of employees and the ability to reorganize parts of the organization.

2. General characteristics of the organization of the social insurance fund

The Fund is a specialized structure that ensures the functioning of the whole multi-level system of state social insurance. The Fund finances the payment of benefits for temporary disability, maternity, child birth and monthly allowances until the child reaches the age of eighteen months, as well as benefits for burial of the dead.

Also, the Fund is financing SANATORNO-resort service workers and members of their families, children's summer recreation campaign [1]. In 2000, the social insurance Fund began to pay for the damage to victims in the workplace in the framework of a new form of insurance, the Federal law of the Russian Federation [2]. Department of the Fund established in the territories of all subjects of the Russian Federation [1].

In Fund of social insurance of the Russian Federation consists of the management bodies, Executive bodies and controlling (figure 1).

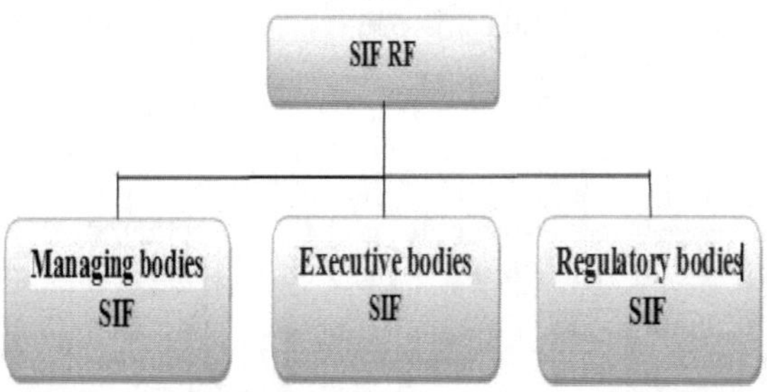

Figure 1. Organizational structure of the FSS

The social insurance Fund supervises the implementation of the budget of state social insurance adopted on a yearly basis the Federal law that controls the use of funds of social insurance. In necessary cases, the Foundation redistributes funds social insurance between regions and industries, supporting financial system stability. In addition, the Foundation develops and implements state program for improvements in social insurance, protection of workers ' health.

3. Business processes

At present, there are many definitions of a business process:

- a set of different activities, in which the input uses one or more types of resources, and as a result of this activity "output" is the product that represents value for the customer [3].
- a set of interrelated or interacting activities that transforms inputs into outputs [4].
- a set of interrelated activities or tasks aimed at the creation of a particular product or service to consumers. As a graphic description of the activities used in the flowcharts of business processes [5].

The analysis of the concepts and related situations, when describing business processes of the FSS, we stopped at a definition [5].

In modeling the business processes of the organization an important part of the information management and resources, the output we receive the necessary information. (Figure 2)

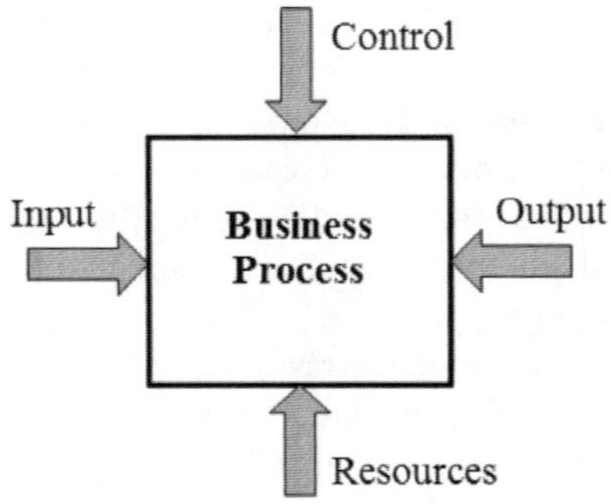

Figure 2. General structure of a business process

The input process receives control information. A resource is a material or informational object that is required to complete the process. Under control action refers to the service resource required for performing the process, but not converted in the course of a process.

The output is converted in the process, the resource. Work with the business processes can be divided into graphical models (schedules) and cost-benefit analysis, in which the modeling of business processes and reengineering [6].

4. Statement of the problem

Statement of the problem is the development model of the business process of FSS. The development is based on applying methods of process management. The manufacturing process consists of many stages and phases, which involved different employees and departments of businesses that require efficient organization, coordination and high speed transmission of information

During planning you need to assess the current condition of the organization through modeling the business processes of the FSS. Take into account the time required to execute the business process. To analyze the model, identify challenges, to improve and accelerate service time policyholders.

The model allows the FSS:

- to define the role of an employee;
- to determine the time required to perform the work;
- to identify problems in the work structure of the enterprise;
- to analyze the enterprise and identify issues;
- formulate objectives to address the identified problems.

When choosing a suitable environment for modeling, you need to compile a list of essential parameters to be met by the system:

1) graphical development environment to accelerate the creation of models;
2) intuitive graphical interface;
3) the possibility of exporting and importing data - communication with other programs;
4) the display of information in the Internet;
5) speed of work;
6) the presence of a constant support system developers.

Also important to the user are ease to understand, learn and use in practice.

5. Simulation of business processes

The main purpose of modelling and simulation tools BP — ensuring understanding at all levels of the organization, bridging the gap between the strategic vision of the organization and practical implementation.

To this end, in modern modeling tools use special languages, understandable and easy to learn and senior managers, including financial Directors, analysts, and managers of IT-departments, each with their vision of solving business problems.

With the help of these languages are constructed graphical models, diagrams, demonstrating, level by level, step by step, as built in the company's business processes, how the interaction between people and what needs to be changed to optimize the architecture of the organization as a whole [10]. RunaWFE-free, end user oriented management system, business processes of the enterprise [http://sourceforge.net/projects/runawfe].

The main objective of the system: distribute tasks to performers and control performance. The job sequence is determined by the graph of the business process that a programmer or business analyst can quickly modify using a "development Environment".

The system is open source, which reduces the cost of acquisition and ownership, enables the organization who installed the system, independently to develop and improve. Using variables in a business process is the transfer of information between the executors of the tasks. If variables in the business process to store documents, the system can be used for automating the workflow of the enterprise. [11]

One server runs RunaWFE – server. On client computers, run the customer-notification about incoming jobs, or the browser, which opens the web interface of the system. On client computers can be running the development environment, and simulation of business processes.

RunaWFE contains definitions loaded in it business processes and running instances of business processes. Using web interface system, the user can:

- –Receive, filter, perform the tasks generated by the instances of business processes;
- –To launch new instances of business processes;
- –View the status of running instances of business processes;
- –Load the archive files containing the definitions of the business processes in the system.

Using web interface system administrator can:

- –Create-delete users and user groups;
- –Include (exclude) users in the group;
- –To give rights to system objects to users and groups of users;
- –To forcibly stop copies of business processes;
- –Forcibly stop instances of business processes.

Using a "development Environment" analysts will be able to develop business processes and export them to archive files in file system. With the help of the client-the siren of the job came user can receive alerts about incoming jobs. With the help of simulation of business processes allows to test the developed business processes for conditional configuration on the client computer analyst, not loading them into the industrial system.

The project RunaWFE placed on the website developers, open source software sourceforge at the address. On this site you can download system both in source code and in compiled form, view and send messages in the forum, get acquainted with the latest news on the project. [11]

Modeling business processes in the organization can be directed to the solution of a large number of different tasks:

– Accurately determine the outcome of the business process and evaluate its importance to the organization.

To determine the set of actions constituting a business process. Clear identification of a set of tasks and activities that must be performed, it is extremely important for a detailed understanding of the process.

– To determine the order of steps. Actions within a single business process can be executed either sequentially or in parallel. Obviously, parallel execution, if it is valid, reduces the overall run time of the process and therefore increase its effectiveness.

– Perform segregation of duties: to determine, and then monitor, any employee or division of the company is responsible for the execution of an action or process as a whole.

– Identify resources consumed by a business process. Knowing who uses which resources and for what operations, you can improve the efficiency of resource use through planning and optimization.

– To understand the interactions between participating in the process by employees and departments of the organization and evaluate and then enhance the effectiveness of communication between them.

– To see the movement of documents during the process. Business processes produce and consume various documents (in paper or electronic form). It is important to understand, whence the documents or information flows, and to determine optimally whether their movement and whether they are all necessary.

– Identify potential bottlenecks and opportunities for process improvement, which will be used later to optimize it.

– More effectively implement quality standards, such as ISO 9000 and successfully pass certification.

– Use models of business processes as a guide for new employees.

– Effectively manufacture automation of business processes in General or their individual steps, including the automation of interaction with external environment — customers, suppliers, partners.

– Understand complex business processes, to understand and describe the enterprise as a whole.

In turn, the main challenge when modeling business processes is to describe existing processes for the construction of their models "as is".

For this purpose it is necessary to collect all available information on the process necessary, as a rule, possess only employees directly involved in the implementation process. Thus, we come to the modeling business processes in the FSS environment RUNA WFE e need for detailed survey (interviews) of all involved in the business process employees.

It should be emphasized that only information about the process provided by the head of Department and managers. Usually only the conversation with the employee directly involved in actions in the framework described in the business process, gives a good idea of how the process in reality [10].

6. Modeling business processes in the fss environment runa wfe

In the social insurance Fund studies have been conducted of the business processes. The research was carried out directly observing the work of professionals in the workplace during the working day. In the observation were over-fixed the actions of employees, when taking the insurers, who appealed to the organization to produce public services. Fixed the execution time of each action.

The main objectives of the study:

– The definition of real time required to perform actions necessary to obtain the insurer of the state service;

– Obtaining of initial data for time for rest and personal need and standards of service;

– The study of personal experience of employees at work;

– Determination of the degree of interaction between units FSS

In the FSS organization there are many processes that are associated with only one output - the provision of public services or assistance in obtaining this service. The main business process associated with the admission of the report on form 4FSS and 4AFCC.

The first workflow (figure 3) shows the process of treatment of the insurant in the social insurance Fund for public service.

The insurant must send the report on form 4FSS and 4AFCC executed in accordance with the requirements established by the regulations and orders of the government. Transmission is possible in three ways:

1) Personally, at the reception of the specialist in the FSS. On average, delivery of the report will take 10-15 minutes;
2) By e-mail, transfer time is 1 minute;
3) By mail, transfer time 2 days.

Consider a situation in which the policyholder himself personally goes to the social insurance Fund. If the report finished on time, the technician checks the reports against the requirements, verification takes about 10 minutes. If reports are specialist reports for signature of supervisor, time-7-10 minutes.

After obtaining signatures, the specialist starts processing the report programmatically. Treatment lasts 8-10 minutes. When you put in time and correctly decorated report the time of receiving the state service is 35-45 minutes. If the report is framed is not true, the expert is the necessary requirements (5 minutes) to modify the report.

To change the report, the insurer takes 7-10 minutes. Thus the time the service is 47-60 minutes. In the situation if the policyholder was late or did not pass the penalty, he pays a fine.

Overall, we have 6 possible cases:

1) Transfer personally in hands right feature records of 35-45 minutes;
2) the Transfer by hand is not correctly drawn up the report 47-60 minutes;
3) Transmission by e-mail right a report 26-32 minutes;
4) Transmission by electronic mail is not correctly drawn up the report 41-47 minutes
5) Transfer by mail correctly drawn up a report from 4 days;
6) e-mail is not correctly drawn up the report from 1 week.

From the obtained models it is possible to identify the problems and challenges identified in the first business process:

1) Long wait when using email, the minimum time to report is 4 days.
2) it is possible to automate the inspection and to expedite the signing of the report. Thus, to reduce the waiting time.
3) the ability to change the report if not the correct design directly with the specialist electrician change request.

Options for solving problems and tasks. In the first case, the ability to send report via e-mail, regulated by government decree. Cancel the transmission by mail of the FSS is not entitled. However, due to the long wait by mail, to offer the FSS to inform the government of the Russian Federation on the abolition of reports by mail.

Automate report verification is possible by checking with a computer that will speed up checking in 10 times. To expedite the signing of the report may, if you sign the report directly by the specialist. The opportunity to sign the report due to a specialist can improve his skills. Will reduce the wait time to 10 times. We find that by automating the verification and signing of the report by the specialist waiting time is reduced to 18-19 minutes.

If the report is framed is not true. The policyholder in the presence of a specialist, immediately changes the report. You can use this method in online mode. For example, via Skype. The modification time will be 5 minutes. Imagine a modified model of the business process with all the decisions (figure 4).

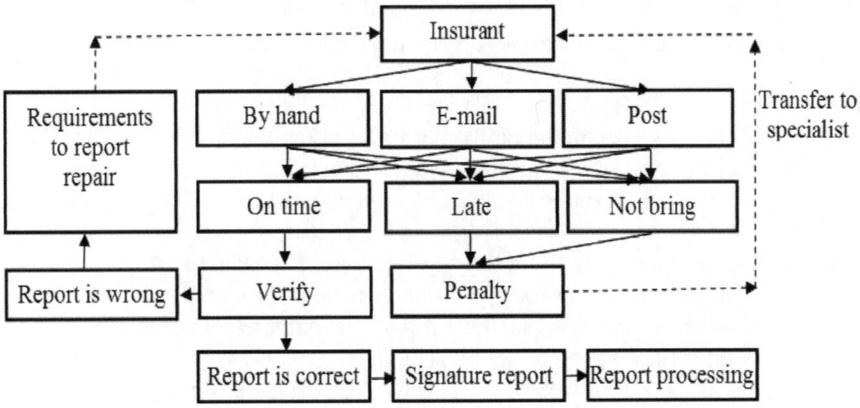

Figure 3. Receive report on form 4FSS and 4AFS

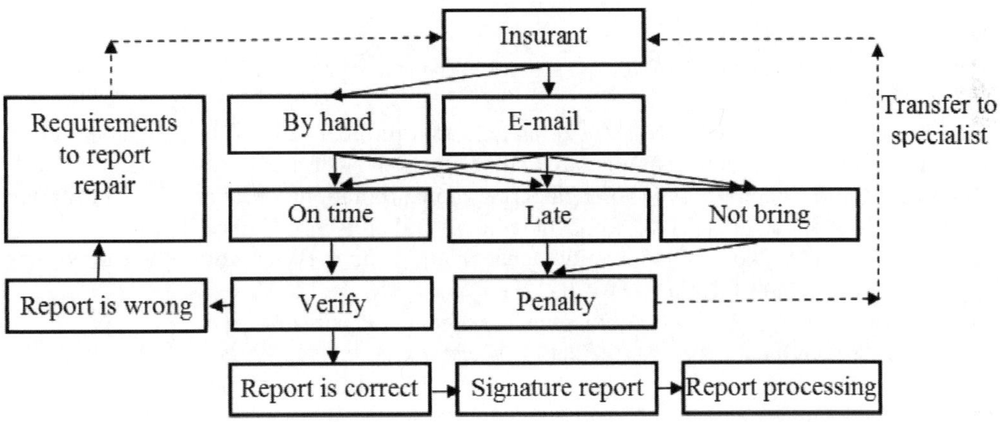

Figure 4. Optimized receipt of report on form 4FSS and 4AFSS

Based on the applied changes, the speed of obtaining service is 25-28 minutes when visiting the insurant FSS and true structured reports. In percentage terms the rate of service receipt increased by 28-37%.

The change in time is presented in figure 5.

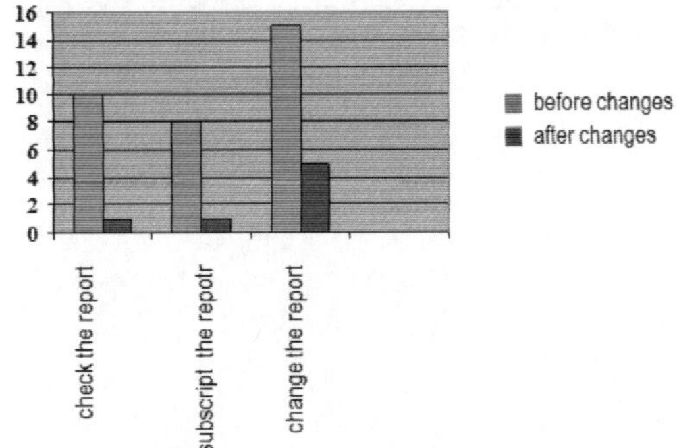

Figure 5. is Graph comparing changes time after optimization

Increase speed to adopt the report will increase the number of insurants; reduce the waiting time to the policyholders in the queue, in the case of a personal appeal in the FSS.

Conclusion

Modeling business processes — a method allowing to build the models describing processes how they are actually using real options, and would take place, using the optimization settings. The model is composed of the problem and objectives of the FSS in the speed of service for policyholders and their number.

In the result of the work the following results were obtained:

1. the analysis of existing business processes social insurance Fund;
2. modeled business processes on the basis of real parameters of the FSS;
3. produced the optimization of the built models.
4. the model proposed for the consideration of the decision of problems and challenges in the social insurance Fund;
5. speed of service increased by 28-37%.

References

[1] The social insurance Fund of the Russian Federation [Electronic resource]. – Access mode: http://r70.fss.ru/.

[2] Federal law of 24 July 1998 N 125-FZ (as amended on 21.07.207) "On obligatory social insurance from accidents on manufacture and occupational diseases".

[3] Hammer M and Champy J 1997 Reengineering the Corporation: a Manifesto for business revolution. TRANS. angl. — SPb.: Publishing house S.-Petersburg University p – 332 .

[4] GOST R ISO 9000-2001 "of the quality management System. Basic provisions and vocabulary" – M.: Gosstandart of Russia, 2001, p – 25.

[5] The free encyclopedia Wikipedia [Electronic resource]. – Access mode: https://ru.wikipedia.org/wiki/Бизнес-процесс.

[6] Kataev M Y, Zholobov O A 2013 Program analysis Business Studio 3.6 *Scientific session TUSUR (Tomsk)* Part 3 pp 270-272.

[7] Vendrov A M 1998 *CASE-technologies. Modern methods and tools for information systems design, Finance and statistics* p –176

[8] G. N. The Kalyanov Consulting during enterprises automation. [Electronic resource]. - M.: SINTEG, 1997 – Access mode: http://www.interface.ru/fset.asp?Url=/case/defs0.htm

[9] The tools of the ARIS 2000 *News – Metatechnology* p - 227.

[10] Repin V V 2013 *Business processes. Modeling, implementation, management* - Moscow: Mann, Ivanov and Ferber p 512.

[11] General description RunaWFE [Electronic resource]. – Access mode: http://www.runawfe.org/olddoc/RunaWFE._General_description.